DIAGNOSTIC HISTOCHEMISTRY

PATHOLOGY is an inherently subjective discipline, and therefore is often referred to as both an art and a science. Over the years, laboratory physicians have implemented special tissue stains and molecular techniques to limit subjectivity in the discipline. Beginning in the late 19th century, histochemical stains were developed to assess diagnostic biochemical reactions in tissue. Histochemistry has recently seen a resurgence in popularity because of the higher costs of other newer methods. Today, this technique is used by almost every pathology laboratory across the world. This book comprehensively covers all diseases for which that technique plays a central role in diagnosis. Every anatomic region is covered in detail with examples of appropriate staining techniques, and the book is heavily illustrated with more than 850 color photomicrographs. This is the first monograph to be published on histochemistry in 15 years, and it is the only one that is diagnostically oriented.

Mark R. Wick (M.D. University of Virginia, Charlottesville) is a board-certified anatomic pathologist and the Associate Director of Surgical Pathology at the University of Virginia Health Sciences Center, as well as a professor of pathology at the University of Virginia School of Medicine. Dr. Wick is a renowned pathologist with expertise in several sub-disciplines of pathology. He is the Editor-in-Chief of the *American Journal of Clinical Pathology* and has edited or co-edited several other books in surgical pathology.

DIAGNOSTIC HISTOCHEMISTRY

Edited by

Mark R. Wick, M.D.

University of Virginia, Charlottesville

CAMBRIDGE
UNIVERSITY PRESS

CAMBRIDGE UNIVERSITY PRESS
Cambridge, New York, Melbourne, Madrid, Cape Town, Singapore, São Paulo, Delhi

Cambridge University Press
32 Avenue of the Americas, New York, NY 10013-2473, USA

www.cambridge.org
Information on this title: www.cambridge.org/9780521874106

First published 2008

Printed in Hong Kong by Golden Cup

A catalog record for this publication is available from the British Library.

Library of Congress Cataloging in Publication Data

Diagnostic histochemistry / edited by Mark R. Wick.
p. ; cm.
Includes bibliographical references and index.
ISBN 978-0-521-87410-6 (hardback)
1. Histochemistry. 2. Molecular diagnosis. I. Wick, Mark R., 1952–II. Title.
[DNLM: 1. Histocytochemistry–methods. 2. Diagnosis. 3. Histological Techniques–methods. 4. Pathology. QS 531 D536 2008]
RB43.7.D524 2008 572–dc22 2007052025

ISBN 978-0-521-87410-6 hardback

To

JANE, ROBERT, & MORGAN,

who allowed me to take time from them

to spend on completing this volume.

And, to all of my teachers in Pathology,

who long ago gave me a sound foundation in laboratory science

and stimulated my interest in Histochemistry.

Contents

Contributors

WILLIAM K. BRIX, MD
Department of Pathology
University of Virginia Health System
Charlottesville, VA

GIANNI BUSSOLATI, MD, F.R.C. PATH
Department of Biomedical Sciences and Human Oncology
University of Turin
Italy

HELEN P. CATHRO, MB, CHB, MPH
Department of Pathology
Division of Surgical Pathology and Cytopathology
University of Virginia Medical Center
Charlottesville, VA

JOANNE M. DAVIS, MT (ASCP), SH
Department of Pathology
University of Virginia Health System
Charlottesville, VA

PATRICK L. FITZGIBBONS, MD
Department of Pathology
St. Jude Medical Center
Fullerton, CA

ROBERT FOLBERG, MD
Department of Pathology
University of Illinois–Chicago
Chicago, IL

JULIA C. IEZZONI, MD
Department of Pathology
Division of Surgical Pathology and Cytopathology
University of Virginia Health System
Charlottesville, VA

AMY LIN, MD
Department of Pathology
University of Illinois–Chicago
Chicago, IL

M. BEATRIZ S. LOPES, MD
Department of Pathology
Division of Neuropathology
University of Virginia School of Medicine
Charlottesville, VA

NANCY C. MILLS, HT, QIHC (ASCP)
Department of Pathology
University of Virginia Health System
Charlottesville, VA

STACEY E. MILLS, MD
Division of Surgical Pathology and Cytopathology
University of Virginia Medical Center
Charlottesville, VA

ELIZABETH MONTGOMERY, MD
Department of Pathology
Johns Hopkins University School of Medicine
Baltimore, MD

AZORIDES R. MORALES, MD
Department of Pathology
University of Miami Miller School of Medicine and
Jackson Memorial Hospital
Miami, FL

JAY M. PACKER, MD
Department of Pathology
Good Samaritan Hospital
Los Angeles, CA

MAURO PAPOTTI, MD
Department of Clinical and Biological Sciences
University of Turin at San Luigi Hospital
Orbassano (Torino), Italy

JAMES W. PATTERSON, MD
Division of Surgical Pathology and Cytopathology
University of Virginia Medical Center
Charlottesville, VA

RICHARD A. PRAYSON, MD
Department of Anatomic Pathology
Cleveland Clinic Foundation
Cleveland, OH

XIAOFEI QIU, MD, PHD
Department of Pathology
Johns Hopkins University School of Medicine
Baltimore, MD

JOSEPH T. RABBAN, MD, MPH
Division of Surgical Pathology
Department of Pathology
University of California–San Francisco Medical Center
San Francisco, CA

PHILIP G. ROBINSON, MD
Department of Pathology
University of Miami Miller School of Medicine
Miami, FL

STEVEN S. SHEN, MD, PHD
Department of Pathology and Laboratory Medicine
The Methodist Hospital
Houston, TX
Department of Pathology
Weill Medical College of Cornell University
New York, NY

NATHAN R. SHUMAKER, MD
Department of Pathology
University of Virginia Health System
Charlottesville, VA

JAMIE SHUTTER, MD
Department of Pathology and Cell Biology
University of South Florida
Tampa, FL

LUAN D. TRUONG, MD
Departments of Pathology and Medicine
Baylor College of Medicine
Houston, TX
Methodist Hospital
Houston, TX

MARCO VOLANTE, MD
Department of Biomedical Sciences and Human Oncology
University of Turin
Italy

MARK A. WEISS, MD
The Urology Center
Cincinnati, OH

BRUCE M. WENIG, MD
Department of Pathology and Laboratory Medicine
Beth Israel Medical Center
New York, NY

MARK R. WICK, MD
Department of Pathology
University of Virginia Health System
Charlottesville, VA

CHARLES J. ZALOUDEK, MD
Division of Surgical Pathology
Department of Pathology
University of California–San Francisco Medical Center
San Francisco, CA

Preface

READERS (OR BROWSERS) OF THIS TEXT may have opened it with the expectation that the title was not completely accurate, and that *immuno*-histochemistry would be at least one of the subjects it considered. That assumption is incorrect. *The book you have in your hands is devoted to histochemistry, and that discipline alone, as it applies to anatomic pathology.*

Why histochemistry, in the twenty-first century? will be the next question. Surely, with the current availability of immunohistology, molecular diagnostics, cytogenetics, proteomics, etc., one can obtain needed diagnostic information through a variety of other technological avenues. That statement is true enough, but it does not mean that one should discard older disciplines simply because they are no longer novel.

The application of histochemical methods to diagnostic pathology has a long and rich history. It began in the late 1800s, linked to such names as Ehrlich, Mallory, Koch, Heidenhain, Masson, Biebrich, Warthin, Bielschowsky, Rio de Hortega, and other greats in laboratory science. Advances in this area of study accrued steadily, and, up until roughly 1970, histochemistry was "state of the art" for anatomic pathologists. Even the growth of electron microscopy at that point did not significantly blunt the use of histochemical stains in pathologic diagnosis.

In the decade after that, however, the burgeoning availability of diagnostic immunohistochemistry did begin to eclipse histochemical applications in laboratory medicine. Slowly but surely, trainees in pathology—and their teachers—eschewed the daily use of the latter methods in favor of newer procedures. Knowledge concerning the mechanisms of histochemical stains—and the principles of tissue processing and tissue chemistry—began to evaporate in the collective consciousness of pathologists. Thus, today, practitioners may know the *names* of a few stains that are not immunologically based, but they usually have very little detailed understanding of how the stains are done, what their biochemical bases are, or what their case-based applications can be.

Hence, we come to the reason for this book. The editor firmly believes that *no* area of scientific endeavor should be forgotten or abandoned if it still has utility and can add to diagnostic certainty. Furthermore, if effective older methods are quicker and cheaper than new ones, there is all the more reason to maintain their viability and implementation. Given the fact that automated commercial platforms are now available to produce histochemical preparations—as is also true of immunohistochemical ones—such stains can be done rapidly indeed. They are less expensive than most other evaluations in pathology, they are still reimbursed by the healthcare system, and, most importantly, they still add to diagnostic conclusions when used in the correct manner.

The format of this volume is quite different from that of preceding texts on histochemistry. They have typically focused only on the *technical* aspects of the stains in question, often adopting a "procedure manual"–like approach, with little or no discussion of practical applications. Instead, only a general discussion of the actual methodology of histochemistry is presented here; the main emphasis is on how the stains can be used in practice and what information they can provide. Authors of chapters in this book have been asked to provide several case-based presentations to further those goals.

Concerted efforts have been made to populate this text with good-quality color illustrations of various pathologic entities and corresponding stains that can be applied to them. However, it should be said that not all histochemical preparations readily yield themselves to the taking of perfect photographs, because of technical attributes of the stains themselves. Therefore, some forebearance of that fact is asked of readers.

For those who wish to have more technical details than those presented here—in line with those contained in past books on the topic—several of the latter are currently in print and they represent excellent procedural resources (1–3). In addition, a variety of Internet sites are devoted to histotechnological procedures and are additional valuable references (4–7).

The main goals of this presentation are to educate young pathologists on a subject with which they likely have only superficial familiarity, and to refresh the histochemical knowledge of more experienced practitioners. The staining techniques presented here are still worthy of clinical use on a regular basis, and they undeniably augment diagnostic information that can be obtained using other analytical evaluations.

References

1. Filipe MI, Lake BD (Eds): *Histochemistry in Pathology* (2nd Ed.), Churchill Livingstone, New York, 1990.
2. Carson FL: *Histotechnology: A Self-Instructional Text* (2nd ed.), ASCP Press, Chicago, 1997.
3. Bancroft JD, Gamble M: *Theory & Practice of Histological Techniques*, Churchill-Livingstone, New York, 2002.
4. *Cytochemistry & Histochemistry?* http://focosi.immunesig.org/histochemistry.html (Accessed 5/4/07).
5. Histological Staining Techniques http://www.bris.ac.uk/vetpath/cpl/histmeth.htm#Top (Accessed 5/4/07).
6. Histonet Archives http://www.histosearch.com/histonet/ (Accessed 5/4/07).
7. The Stains File http://home.primus.com.au/royellis/stains.html (Accessed 5/4/07).

Acknowledgments

The editor is grateful to all of the chapter authors of this text, who enthusiastically and skillfully contributed their expertise. In addition, the faculty members, housestaff, and histotechnologists in the Department of Pathology at the University of Virginia were uniformly supportive of, and helpful with, this project. Finally, much gratitude is due to Mr. Marc Strauss, medical acquisitions editor of Cambridge University Press, for his help and patience during the writing of this volume.

Mark R. Wick, M.D.
Divisions of Surgical Pathology-Cytopathology & Autopsy Pathology
University of Virginia Health System
Charlottesville, Virginia

May 2007

CHAPTER 1

Tissue Procurement, Processing, and Staining Techniques

Mark R. Wick, M.D.,
Nancy C. Mills, H.T., QIHC (ASCP), and
William K. Brix, M.D.

It is an unfortunate reality that many pathologists have only a rudimentary knowledge of the effects of surgical technique and tissue processing on the final results that will be obtained in stained microscopic sections. All too often, one is faced with a sample that has been obtained crudely, fixed badly, or mishandled in the histology laboratory, making morphologic interpretation of it needlessly complex. These faults typically occur not through willful neglect of proper methodology but rather because of ignorance of the sequence of steps that constitute the science of histotechnology. Most trainees in pathology are not well versed in the details of this laboratory discipline, making them totally dependent on the expertise of their technicians, or the lack of it.

Accordingly, this chapter will provide an outline of recommendations for the procurement and subsequent handling of histopathologic specimens. Potential reasons for poor results are also included.

BIOPSY TECHNIQUES

The specific procedures that are used in performing biopsies of clinical lesions are usually left to the discretion of the attending clinician. This provision is not a problem if the operator has been adequately educated on the specific advantages and disadvantages of various techniques, as applied to specific diseases. However, it may prove to be a disaster if the surgeon is inexperienced in such matters. Conventional or enzyme histochemistry, immunohistology, or electron microscopy—all of which are greatly affected by nuances in tissue preservation—may be necessary in some instances to obtain a firm diagnosis. Because the clinician may not be able to anticipate these possibilities before obtaining the tissue sample, a predetermined routine should be followed in doing so (1).

There are basically four generic categories of procedures that may be used in any given case. These include punch biopsies, using circular cutting devices of several sizes; partial or complete excisions done with a scalpel; electrosurgical excisions; and laser-mediated biopsies. In choosing one of these options, the operator should be cognizant of the two opposing "forces" that affect his or her final decision on this matter. On the one hand, the patient is often preoccupied with the cosmetic effects of a biopsy, and this typically induces the surgeon to limit the size of the sample as much as possible. The opposing consideration is represented by the degree of difficulty with which the microscopic diagnosis is made by the pathologist—a factor that is often predictable by the amount of material that will be required to study the disease process adequately.

The cardinal rule to be remembered on this topic is that a properly done biopsy is virtually never cosmetically deforming, if it can be accomplished in an outpatient setting by a competent operator. In contrast, specimen inadequacy and artifactual changes in tissue are problems that relate to faulty procedure, and these account for the great majority of diagnostic obstacles that pathologists encounter. There is nothing quite so aggravating for the clinician (and the patient) as to be informed that a second biopsy will be necessary because of these deficiencies, causing additional expense and anxiety.

As an example of these contentions, it is well known that malignant hematolymphoid proliferations and certain metastatic carcinomas are composed of extremely fragile cells that are exquisitely susceptible to the compressive or shearing effects of some biopsies (Figure 1.1) (2). Moreover, it is probable that several cubic millimeters of tissue will be necessary for the complete pathologic characterization of such lesions. Hence, a small biopsy specimen would be predictably unsuitable in these circumstances. When in doubt, the clinician should contact the pathologist *before* the procedure is done and inquire about recommended handling of the tissue sample and its minimally acceptable size based on the likely diagnostic possibilities.

Other procedures causing reproducibly detrimental physical effects on tissue specimens are represented by

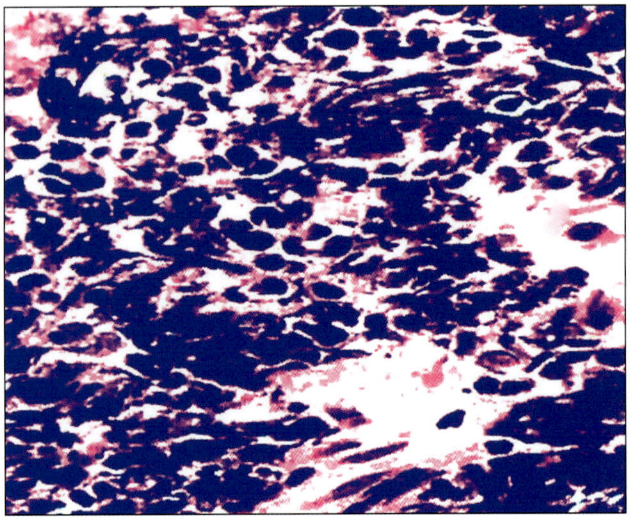

FIGURE 1.1: Smudging or "crush" artifact results from the mishandling of tissue from fragile neoplasms such as lymphoma or small cell carcinoma (shown here).

electrocautery and laser excision. These methods enjoy wide clinical usage at present because of their ease of performance and the limitation of surrounding tissue damage that they afford. Nonetheless, lesional cells in the specimen are often rendered unrecognizable because of widespread thermal coagulation, precluding histologic interpretation altogether. It should therefore be obvious that cauterizing techniques must be avoided for diagnostic purposes. Several adjunctive pathologic studies require the availability of specimens that have been handled in a special manner (Table 1.1). Again, these can be obtained prospectively following preprocedural consultation with the laboratory.

IDENTIFICATION AND ORIENTATION OF THE BIOPSY SPECIMEN

There is nothing quite so exasperating for the pathologist as to receive a specimen that is unoriented and for which no anatomic location is given on the request form for pathologic examination. A lack of meaningful clinical history or a failure to list potential clinical diagnoses often compounds such omissions. These problems usually cannot be solved by the pathologist and typically require a laboratory visit by, or a telephone conversation with, the responsible physician. In many instances, it would be medicolegally dangerous to attempt a morphologic interpretation in the absence of such information. On occasion, a specimen may be received that is so poorly labeled that the identity of the patient is even in question. Such a submission should never be accepted by the laboratory unless the clinician is willing to provide written documentation verifying its origin and accepting exclusive legal responsibility for its interpretation.

If a lesion is a suspected malignancy for which a diagnostic biopsy is also intended to be a complete excision, the clinician should provide some means of identifying the superior, inferior, medial, and lateral borders of the tissue sample. This can be accomplished by attaching sutures of differing lengths or types to the specimen, and sending a corresponding "map" of the tissue to the laboratory along with the pathology request form (3). Alternatively, indelible (e.g., tattoo) ink of various colors can be affixed to the borders of the specimen and identified accordingly (Figure 1.2). As a minimal requirement—for example, in very small excisional biopsies—at least one pole of an elliptical or circular tissue fragment should be labeled by such means.

TABLE 1.1: Specimen Processing

Pathologic Technique	Recommended Fixative	Processing Time	Comments
Conventional histology	NBF or FA*	1 day	Tissue should be sectioned at 2–3 mm for good fixation
Immunohistology	NBF or FA**	2–3 days	Technique can be applied to frozen or fixed sections
Electron microscopy	2% phosphate-buffered glutaraldehyde	3–4 days	Tissue must be minced into 1- or 2-mm cubes
Immunofluorescence	None, if tissue is flash frozen; 95% ethanol or acetone for touch preparations; Michel's medium for transportation	1–2 days	Tissue can be held in Michel's medium for up to 48 hours. Frozen tissue must be kept at −70°C until use
In situ hybridization	NBF or FA for DNA studies; frozen tissue preferred for RNA studies	1 week	DNA studies can be done on frozen or fixed tissue

NBF = neutral-buffered 10% formalin, FA = NBF-ethanol (50%:50%).

* Tissue for routine histology can be fixed in B5 or Bouin's solutions to improve nuclear morphology, but these preservatives require special processing and compromise immunohistology.

** Certain tissue antigens (e.g., light chain immunoglobulins) are detectable only by frozen section immunohistochemistry.

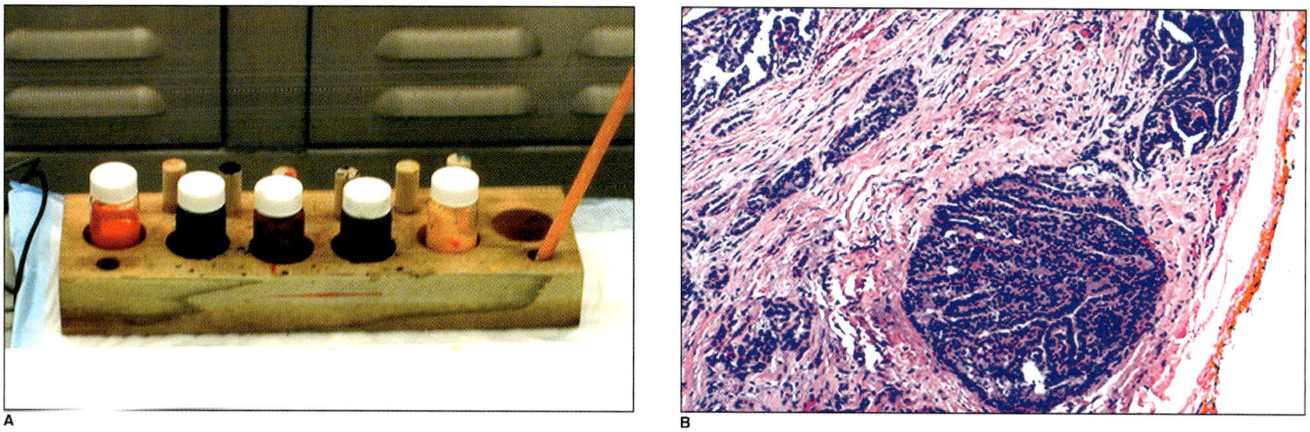

A B

FIGURE 1.2: ***Gross room inking stations (A) contain indelible inks of several colors, which can be applied to specimens with cotton-tipped swabs and fixed in place with Bouin's solution. The ink can then be seen in an FS (B) or permanent section.***

The clinician should be discouraged from attempting to prosect the specimen further before it is examined by the pathologist, except in very well-defined settings. When they are improperly performed, transections of small biopsy samples often confound subsequent orientation steps and may mechanically damage the lesion that is intended for study. The only acceptable reason for undertaking further clinical manipulation of the tissue sample is that of preparing cellular "touch" preparations in examples of suspected hematolymphoid disease. The latter can be obtained if the operator bisects a lesion at its bulkiest point and touches the cut surface of the tissue gently to glass slides in a serial fashion (Figure 1.3) (2). When this is done, special care should be taken subsequently to orient both halves of the resulting two-part specimen for the pathologist. Moreover, all air-dried or fixed touch preparations must be labeled with the patient's name, his or her date of birth or medical record number, and the date on which the procedure was performed.

PREPARATION OF FROZEN SECTIONS

Intraoperative consultations, generically termed "frozen sections" (FSs) by many surgeons, are often requested in treating presumed or proven malignancies (3,4). Procedural aspects of the FS method are familiar to all anatomic pathologists, but these will be reviewed briefly in this section.

The purposes of obtaining FS examination are twofold; it may be used to secure a rapid diagnosis for a lesion with unknown histologic attributes, or the technique may be employed to confirm that margins of excision are uninvolved by the pathologic process in question. Because of the potential distortion of morphologic detail that this procedure may induce, the first of the cited applications is not one that should be used frequently. With respect to the analysis of excisional margins, the operator must be certain to supervise the orientation and labeling of all specimens,

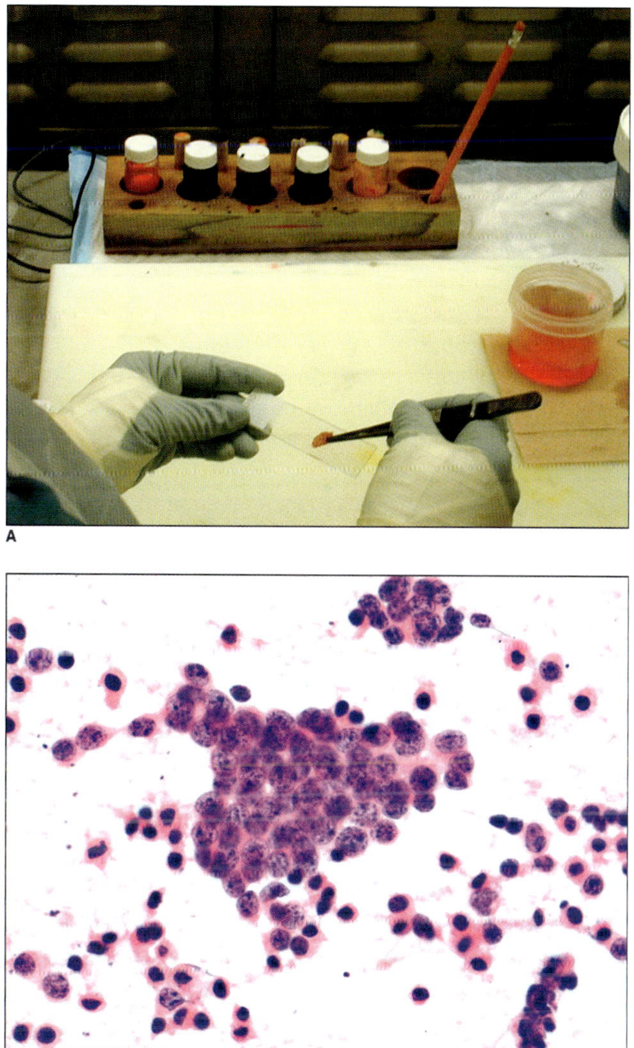

A

B

FIGURE 1.3: ***Touch preparations can be made from fresh tissue by serially touching a cut surface from it to adhesive-coated glass slides (A). The touch imprint sections can then be air dried and stained with Romanowsky dyes, or briefly fixed in alcohol and stained with H&E (B).***

as outlined above. This makes the availability of indelible ink an absolute requirement.

Following such steps, one must be certain that the tissue sample is small enough to assure rapid and uniform freezing, and ease of sectioning with the cryomicrotome (cryostat). The specimen is usually placed in a small pool of gelatinous, water-soluble mounting medium (e.g., "optimum cutting temperature" medium; Cryogel[R]) that has been applied to a precooled Teflon or metal "chuck." After making sure that the tissue is properly oriented on the flat surface of this implement, it is then totally covered with additional mounting medium, fashioned into a circular pledget. Immediately thereafter, best results are obtained if the chuck is immersed in a bath of isopentane suspended in an outer container of liquid nitrogen. These devices are available commercially, and they allow for virtually instantaneous freezing of the mounting medium with minimalization of ice crystal formation. The latter eventuality is undesirable because entrapment of ice in the specimen (caused by slowly decreasing temperature) will cause significant distortion (Figure 1.4) and may interfere with microscopic interpretation. For this reason, the utilization of metal cooling "plates," which are incorporated into many cryostats by their manufacturers, is not recommended as a means whereby initial freezing is accomplished. However, these plates are acceptable for maintaining the chucks in a frozen state while sections are being cut.

The microtome in any cryostat must be set in such a manner that uniform sections of reproducible thickness (approximately 5 μm) can be prepared. Regular maintenance regarding the sharpness and integrity of microtomy blades is essential to this process. After "facing" the frozen block with the blade—to obtain a smooth, flat tissue surface—the operator cuts a "ribbon" of several individual sections that can be kept flat by manipulation with a camel-hair brush or with a Teflon-coated panel. These are then apposed to acid-cleaned glass slides that have been kept at ambient temperature, causing the tissue to adhere to them quickly. To eliminate concerns about the subsequent loosening of this bond, slides that have been precoated with albumin, poly-L-lysine, or a chrome-alum gel may be utilized (5).

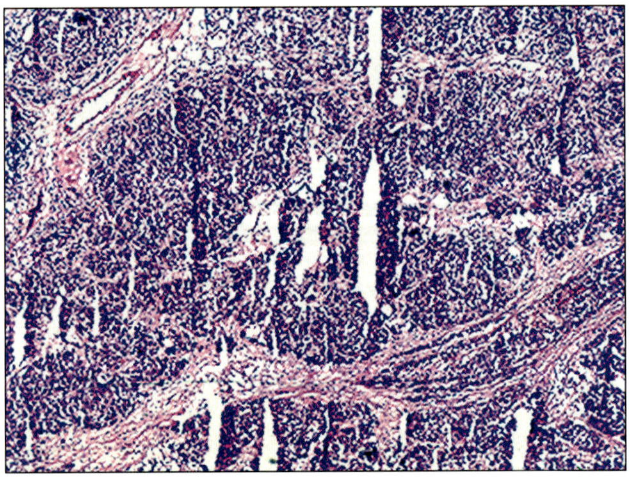

FIGURE 1.4: ***If ice crystals are allowed to form in tissue during the freezing process in a cryostat, linear defects will appear in the final FS.***

Most FS laboratories employ a brief (30–60 second) fixation step immediately after mounted sections are prepared, in Copland jars containing absolute acetone or 95% ethanol. The slides may then be stained with hematoxylin and eosin (H&E), a "polychromatic" or metachromatic reagent such as methylene blue, or other reagents. Following dehydration in graded alcohols and xylene, a synthetic mounting medium is placed over the tissue, and a glass coverslip is applied. Addition of a few drops of xylene to the mounting medium will slightly lessen its viscosity and help to prevent the entrapment of air bubbles under the coverslip.

Alternatively, one may wish to keep some unstained FSs for future studies. This aim is best served by removing slides from the acetone or alcohol fixative and placing them promptly in a freezer at −20 or −70°C. These can be kept in such devices indefinitely for further analysis at a later date.

Specific problems connected with poor microtomy technique will be considered subsequently in this discussion. However, the most common difficulty that is seen in the FS area can be ascribed to improper calibration of the cutting interval between successive sections. Overly thick sections may result in consumption of the tissue before a suitable slide is obtained for microscopic examination; in contrast, it is extremely hard to obtain very thin sections without causing them to fold on themselves or shred. Thus, it is essential for the cryostat to be checked frequently to make certain that it is set up properly from a technical viewpoint. Also, there is no substitute for practice and experience on the part of the operator, in regard to preparation of optimal FS slides. The labeling of specimens used for FS examination should be no different than that used for other samples. The remnant tissue should be placed in a plastic cassette that is suitably inscribed with the accession number of the case (preferably using a CasMark[R]-type labeler) and kept together with corresponding paperwork for transmittal to the histology laboratory. Under no circumstances should unlabeled frozen tissue be allowed to accrue in the FS laboratory, lest disastrous errors in identification occur.

FIXATION OF SPECIMENS

Questions that are often asked of the pathologist concern the choice of one fixative solution over another for the preservation of various cutaneous specimens. There is no "universal" fixative in pathology because tissue samples may be used for an ever-growing number of investigative analyses, many of which demand that special processing measures be applied in order to procure optimal results. Selected immunohistologic studies, electron microscopy, and genotypic assessment represent three advanced modalities of pathologic evaluation that are associated with specific fixation requirements. Laboratory specialists are continuing to develop procedural modifications to lessen the need for such provisions, but they still do exist.

In the most optimistic of scenarios, it would be best to submit all biopsies in their fresh state in physiologic saline solution, and for the pathologist to subdivide these specimens into several parts for future diagnostic eventualities. Nevertheless, this is often not practical for two main reasons. First, outpatient specimens are commonly submitted over long distances from the pathology laboratory, increasing the likelihood that unfixed tissue will undergo autolysis before it is received. Second, many biopsies are limited in size, making judiciousness in the selection of special studies an important point. The latter issue again emphasizes the wisdom of preprocedural consultation with the pathologist, if unconventional evaluations are desired.

Fixatives Used for "Routine" Histopathologic Examination

In the great majority of cases, the clinician requesting histologic examination of a biopsy is interested in a "traditional" interpretation based on microscopic findings as seen with the H&E stains. With this stipulation in mind, most laboratories have advocated the use of formalin as the fixative of choice. Nonetheless, the following sections will briefly review the chemical characteristics of preservative solutions in a broader sense, so that exceptions to the above-cited situation may be addressed.

General Considerations

The preservative effects of certain chemicals have been recognized for thousands of years, dating back to the ancient Egyptians. On an empiric basis, therefore, various fixatives have been employed to preclude bacterially mediated putrefaction of human tissues since the inception of pathology as a discipline.

In the past century, detailed studies of these agents have elucidated the probable mechanisms responsible for these beneficial effects (6–10). In addition to antibacterial effects, fixatives also enhance the differences in refractive indices between dissimilar tissue constituents, allowing for greater resolution upon light microscopy. Moreover, they augment the affinity that chemical dyes have for particular cellular elements. It is now known that chemical fixatives may be divided into two broad categories—coagulating and noncoagulating—with respect to their effects on proteins, which form the framework of virtually all cells. Further subdivision into aqueous and nonaqueous agents, as well as additive or nonadditive preservatives, is also possible (11).

Noncoagulative fixatives are the most widely used, and these include formaldehyde (called *formalin* when prepared in aqueous dilution and *paraformaldehyde* when employed in polymeric form), glutaraldehyde, acetic acid, potassium dichromate, and osmium tetroxide. In contrast, acetone, alcohols, chromium trioxide, mercuric chloride, and picric acid exemplify the *coagulative* preservatives. In the process of denaturation and coagulation, a network of altered protein is formed in tissue; in contrast, noncoagulative agents act to produce a stable intracellular "gel." Acetone and alcohol are the major nonaqueous reagents, with most others being soluble in water. *Additive* fixatives react with tissues by combining with them chemically, whereas nonadditive reagents (primarily alcohols and acetone) do not.

Various combinations of these chemicals [e.g., formalin-alcohol or Carnoy's solution (a mixture of ethanol, chloroform, and acetic acid)] are sometimes utilized as fixatives that are intended to augment the stainability of predefined tissue components. Moreover, metal salts—such as those containing zinc and mercury—may be added to aqueous solutions, as in zinc-formalin or "B5" fixative (a mixture of mercuric chloride, sodium acetate, and formalin). The apparent effect of the latter agents is to stabilize complexes that are formed by nucleic acid and protein, yielding improved preservation of nuclear detail. By convention, many fixatives are named for the laboratory investigators who devised them. Thus, one may encounter such designations as Bouin's, Hollande's, Zenker's, Helly's, Zamboni's, Orth's, and Carnoy's solutions. Most of these are mixtures of chemicals in different classes or showing differing effects on proteins, with or without metal salts. Selected reagents in this list will be alluded to later in this discussion.

One important concept to be borne in mind is that *all* fixatives induce chemical artifacts in tissue sections. This effect has two potential ramifications for pathologists. First, we all become inured to the artifacts that we are accustomed to seeing with routine use of a particular preservative solution; indeed, we may even rely on such changes as diagnostic features. Changing the fixative one uses will also alter the tissue artifacts, often leading to interpretative confusion with any given staining method. Second, one artifact produced by a preservative may be desirable, whereas others are detrimental. For example, B5 solution yields excellent nuclear detail on H&E stains, but it virtually destroys the integrity of some cellular proteins that may be the targets of immunohistochemical studies (12). Lastly, the optimal period of fixation varies greatly from one solution to another; tissue placed in formalin may be allowed to remain in it for days with no compromise of morphologic features, whereas specimens in B5, Zenker's, and Bouin's fixatives *must* be transferred to other chemical solutions after predefined periods of time to avoid a serious loss of cellular definition (7,8). Thus, the ultimate choice of a preservative solution is not one to be made indiscriminately.

Specific Fixatives

Formalin: Formalin represents a 37–40% aqueous solution of formaldehyde, the latter of which is marketed commercially in the United States. Because the former reagent is characteristically used at a 10% dilution, the final formaldehyde concentration is 3.7–4%. Various other chemicals have been added to formalin to alter its stability and preservative capabilities, including calcium chloride, calcium carbonate, ammonium bromide, sodium chloride, sodium phosphate, sodium hydroxide, and absolute ethyl alcohol. Among these mixtures, that consisting of formalin, distilled water, and monobasic/dibasic sodium phosphate is the most widely employed and is known as "10% neutral-buffered formalin" (NBF). *Paraformaldehyde* is a polymerized form of formaldehyde admixed with methanol; it is generally employed as a fixative for specialized immunohistologic procedures, particularly when combined with periodate and lysine ("PLP" solution) (12,13).

Although it is a general-purpose fixative and yields good morphologic detail when prepared properly, NBF does have some disadvantages in tissue pathology. First, any solution containing formaldehyde is potentially carcinogenic, and levels of formalin vapor in the ambient air of the laboratory must be measured regularly by governmental mandate. The maximum permissible exposure limit for any individual employee is 1 part per million over an 8-hour period, as established by the Occupational Safety and Health Administration (14). Second, poorly prepared NBF, which has been buffered erroneously and has a pH outside of the physiologic range, may cause unwanted precipitates of "black acid hematin pigment" in tissue sections. The latter has a dark particulate appearance, and may simulate microorganisms on a histologic slide. These two possibilities can be distinguished through the use of polarization microscopy because hematin pigment is birefringent, whereas microbes are not (11). Third, NBF that is allowed exposure to ambient air for prolonged periods of time (as with large "batches" that are diluted for use in the gross laboratory) will develop high levels of formic acid. The latter is detrimental to protein substructure and may accentuate the formation of methylol bonds between polypeptides. This effect can "mask" proteinaceous epitopes that correspond to the targets of immunohistologic antibody reagents (15). Lastly, formalin has a limited capacity for penetration of bulky pieces of tissue, and specimens fixed in it must be no thicker than 4–5 mm.

Despite these drawbacks, formalin is inexpensive and widely available, and is therefore ubiquitously employed as the fixative of choice for clinical specimens. The above-cited failings of this preservative can be prevented by careful technique in its preparation, adherence to environmental monitoring requirements, and application of proper prosection and fixation techniques for the submission of tissue sections. Some laboratories prefer to use NBF ethanol (mixed in equal volumes) because it affords a greater degree of tissue penetration than formalin alone.

B5/Zenker's/Helly's Solutions: B5, Zenker's, and Helly's solutions were introduced because of their superiority over NBF in the preservation of nuclear detail (7,8). They are fixatives based on the inclusion of mercuric chloride, with or without sodium acetate, potassium dichromate, sodium sulfate, acetic acid, and formaldehyde as additional constituents. Because of the excellent morphologic detail that is achievable with these solutions, many laboratories prefer them for the routine preparation of H&E-stained sections. Nevertheless, there are three distinct disadvantages of B5, Zenker's, or Helly's reagent, as compared with NBF. Tissue sections must be removed from the former three fixatives after no more than 8 hours and placed into 70% ethanol; if this is not done, specimens will become extremely brittle and virtually impossible to section (11). Also, the presence of mercuric chloride will cause deposition of pigment in microscopic preparations, which must be removed with iodine before final staining procedures are done. Lastly, mercury-based solutions are powerful coagulating agents and therefore damage many cytoplasmic proteins. This effect commonly renders tissue sections unsuitable for a variety of immunohistochemical studies (12).

Bouin's Solution: Bouin's fixative is again based on formaldehyde as a major component, together with picric and acetic acids in aqueous solution. Like B5, this reagent affords excellent preservation of nuclear morphology but suffers from failings pertaining to brittleness of tissue, pigment deposition, and adverse effects on cytoplasmic polypeptides. In addition, Bouin's-fixed specimens acquire a yellow color (because of the effects of picric acid) that must be removed by postfixation washing in alcohol and lithium carbonate. Bouin's fixative is preferred for visualization of delicate mesenchymal tissues because of its superior differentiating abilities in regard to these elements (11). Accordingly, some "stromal" special stains (such as the Masson trichrome method) are best performed on specimens preserved in this solution.

Acetone and Alcohols: Acetone and alcohols are rapidly acting fixatives with good penetration of tissue. They also afford better preservation of some cytoplasmic enzymes than formaldehyde-based solutions do, in paraffin sections. However, two major disadvantages attend the use of these organic reagents. They cause striking shrinkage of tissue because of their dehydrating effects, thereby altering morphologic details appreciably. Also, acetone and methyl or ethyl alcohol are relatively expensive, and they require special storage and inventory procedures because of possible use by laboratory workers as inebriants. In current practice, these agents are usually applied only in the fixation of touch preparations and are not commonly utilized in the processing of biopsy specimens. Similar comments apply to Carnoy's solution, which is constituted by ethyl alcohol, chloroform, and acetic acid.

Decalcifying Solutions: Some biopsy samples may contain obvious foci of calcium salts, as suggested by anatomic location, clinical findings, or difficulty in performing the biopsy procedure. In these circumstances, two main methods exist for the removal of such minerals from the specimen. One employs simple acids (hydrochloric or nitric), which rapidly solubilize calcium deposits. The other technique is based on the ability of certain chelating agents—such as ethylenediaminetetraacetic acid (EDTA)—to accomplish this task. The second of these methods is much gentler and does not cause the loss of microscopic detail that acid decalcification may incur. Fixation is allowed to progress in concert with decalcification with both acidic and EDTA reagents because they are commercially marketed as mixed solutions containing formaldehyde.

Glutaraldehyde: Glutaraldehyde is similar in chemical activity to formaldehyde; both cause cross-linkage of proteins in tissue (7). However, glutaraldehyde penetrates specimens very slowly, making the size of the tissue sample a critical determinant of fixation with this reagent. Moreover, 2–4% glutaraldehyde (representing the usual working concentration) has a propensity to cause brittleness of specimens that are immersed in it for more than 2–3 hours; transfer to a buffer solution is absolutely necessary after this point. For these reasons, among others, glutaraldehyde is not used often for the preservation of biopsy samples that are intended for light microscopy. However, it is the preferred fixative for electron microscopy, wherein specimens are very small and limited "hardening" of tissue may actually be morphologically advantageous.

Other Factors Influencing Fixation

As outlined by Carson (11), there are several other considerations in the fixation of tissue besides one's choice of preservative solution. These include temperature, size of the sample, the volume ratio of tissue to fixative solution, the duration of fixation, and the pH of the solution.

Recently, the rapid but controlled elevation of temperature with microwave ovens has been utilized as an independent means of fixation, by coagulation of tissue proteins (16). Surprisingly, this process appears to have little if any adverse effect on staining characteristics, even with immunohistologic methods. However, it must be emphasized that careful control is the key to thermal fixation; overheating may completely destroy the specimen if it is allowed to reach an extreme level (e.g., over 65°C). In a more conventional context, there are really no compelling reasons to employ fixative solutions at one temperature versus another.

Specimen size is, in contrast, a potentially crucial factor affecting quality of fixation, and this determinant goes hand in hand with the volumetric relationship between a tissue sample and the solution in which it is immersed. Large, extremely thick specimens will be inadequately penetrated by most fixatives, allowing autolysis to proceed unchecked in their central areas. This problem results in eventual loss of the unfixed foci during microtomy, yielding microscopic sections that resemble doughnuts (Figure 1.5). Because penetration is facilitated by minor thermal or mechanical currents in the fixative solution, large specimens that are covered with an inadequate volume of preservative will predictably be underfixed. An experienced histotechnologist typically detects this problem upon attempting microtomy of the tissue and will "run the specimen back" for more prolonged fixation and reprocessing. However, this consumes additional time and should be unnecessary.

As noted at several points in the foregoing discussion, there is a maximum recommended period of fixation with several preservatives, over which unwanted changes reproducibly occur in tissue biochemistry. Overfixed specimens are difficult to cut and often demonstrate alterations in morphologic definition or antigenic integrity. In contrast, underfixation allows bacterial putrefaction to proceed, similarly damaging the tissue sample. Specimens that are immersed in the most commonly used preservative—NBF—should ideally be processed further within 8–12 hours.

The pH of fixatives is not critical for light microscopy, except that certain unwanted pigmentary deposits may be seen with unduly acidic preservatives. Nonetheless, hyperacidity is extremely detrimental to cellular ultrastructure and also to the maintenance of tissue antigenicity

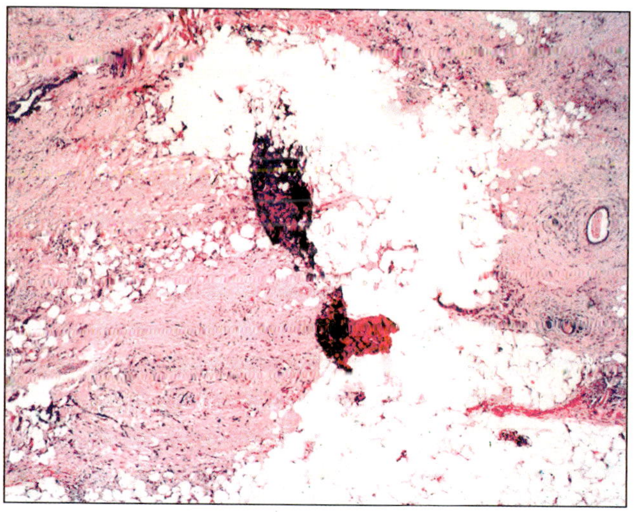

FIGURE 1.5: Inadequate fixation of tissues—especially fatty ones—will often result in loss of the central areas of tissue blocks in final microscopic sections because they "fall out" during processing.

(11,12). For these reasons, it would be wise to control pH within the physiologic range during fixation, in the event that electron microscopy or immunohistology is necessary diagnostically.

Tissue Processing and Preparation of Microscopic Sections

Because most commonly employed fixatives are aqueous in nature, the next step in tissue processing is usually that of dehydration and "clearing" (removal of all water from the specimen). Graded solutions of ethanol are used for this purpose, and these must be changed frequently to maintain their desiccating properties. A variety of clearing agents are available, but the most common are xylene and limonene derivatives. In likeness to the alcohols, such reagents may be contaminated by water with repeated use and should be monitored closely for this problem.

Xylene is inexpensive and does not leave a residue on glassware or other instrument parts in the histology laboratory. In light of these virtues, it is the most popular clearing agent. However, xylene fumes are potentially toxic to technologists, making careful storage, controlled disposal, and environmental monitoring mandatory. In addition, we have found that xylene may damage the protein substructure of certain fragile tissue antigens (12). Limonene-type clearing agents are derived from plants and are therefore biodegradable. They have a strong odor—like that of lemons or oranges—which is alternatively perceived as pleasant or noxious by various people. Other disadvantages of limonenes are that they leave a residue on mechanical tissue processors and may sometimes interfere with the adherence of tissue sections to glass slides. The microtomy of specimens cleared in limonenes has been said to be easier than that encountered with xylene (11).

In the relatively early days of histotechnology, all dehydration and clearing steps were done by hand. Over the past 45 years, however, a variety of automatic tissue processors have been engineered and marketed. These are used widely at present and may be divided into two main groups—"open" and "closed." Open processors mechanically transfer baskets containing tissue cassettes from one "station" (chemical bath) to another, on a computer-driven schedule. The latter may be altered by the operator to change the time of dehydration, clearing, or other steps. Closed instruments vary the solutions to which each specimen basket is exposed by pumping chemicals in and out of fixed chambers, again according to a programmed schedule. In other words, open processors move specimens, whereas closed processors move chemical solutions.

Each of these two types of instruments has advantages and disadvantages. Open processors show a low incidence of reagent contamination from one station to another, but they are subject to the mechanical "hang-up" of specimen baskets in transit. Closed processors do not suffer from the latter drawback, but they are subject to chemical carryover from one reagent pumping step to another. This potentially compromises the dehydration-clearing sequence. On balance, individual experience on the part of technologists and pathologists ultimately determines which type of processor will be chosen.

EMBEDDING AND SECTIONING OF BIOPSY SPECIMENS

The final stations in any tissue processor infiltrate all specimens with paraffin or another wax-based embedding medium. Thereafter, the technologist removes each biopsy (one at a time) from its metal or plastic cassette and proceeds to embed it in a rectangle of additional liquid wax, with attention to the proper orientation of the tissue sample. The pathologist may direct this process by notching or inking one or several surfaces of the specimen (Figure 1.6) and providing a "map" in accompanying paperwork that indicates whether these should be placed facedown, faceup, or in parallel with the lateral aspects of the cassette. Such provisions are usually necessary only with large pieces of tissue. For example, technologists accustomed to handling skin biopsies will, as a matter of routine, orient the epidermis perpendicularly to the bottom of the cassette mold and facing one of its long sides. If several pieces of tissue are included in the same block, these are best arranged diagonally.

The embedding step is a potential source of great irritation (and medicolegal liability) for the pathologist if it is done by an inexperienced or careless laboratory worker. With few exceptions, small biopsy specimens that are oriented improperly cannot be interpreted microscopically (Figure 1.7), necessitating that the block be remelted and

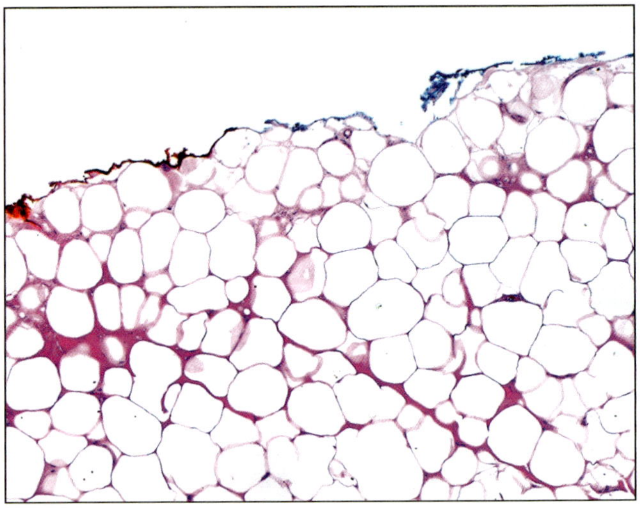

***FIGURE 1.6:** This microscopic section of a breast biopsy shows the intersection of two anatomic planes that were labeled by the submitting surgeon, and inked with two different colors by the prosecting pathologist.*

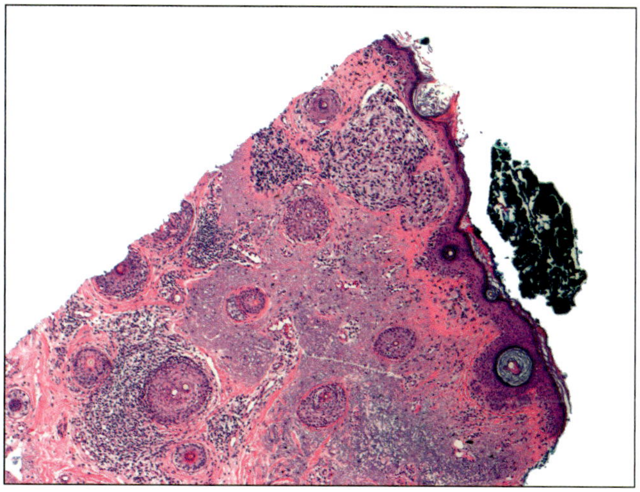

FIGURE 1.7: ***Malorientation of biopsy "tips" or bisected biopsy specimens at the embedding station, as shown here, will compromise the pathologists' ability to evaluate the true specimen margins for tumor involvement. Reprocessing may or may not solve this problem, and every effort should be made to avoid it in the first place.***

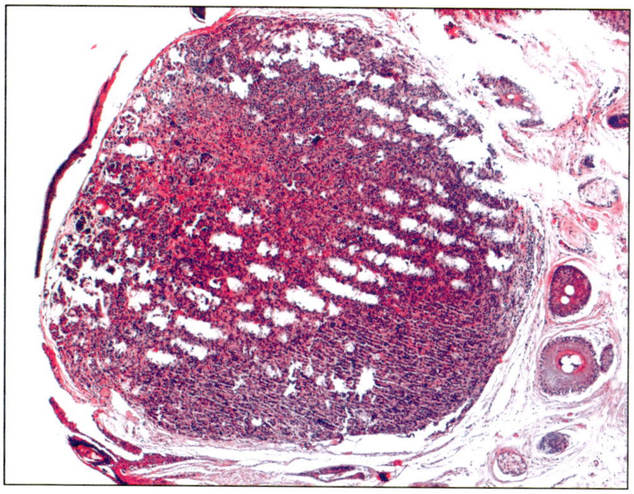

FIGURE 1.8: ***"Chatter" artifact in tissue sections is the result of loose microtome blades or poor microtomy technique.***

re-embedded. This takes time, and in the process of facing the poorly oriented specimen for preparation of initial sections, valuable tissue may be lost (17).

In order to circumvent embedding difficulties, some pathologists have taken to *pre*-embedding small biopsies in agar before they are put in cassettes for fixation. This does assure proper orientation, but agar will not "fix" in the same manner that tissue does, nor will it respond similarly to dehydration, clearing, and infiltration by wax. All these factors may cause the tissue to "pop" free of the surrounding agar after embedding and during tissue sectioning, defeating the purpose of the agar impregnation step altogether. Therefore, we do not advocate this procedure, rather preferring to educate technologists on the details of orientation during wax embedding. Even a very small biopsy can be appropriately configured in the wax block, with the use of a magnifying lens or dissecting microscope.

Paraffin is still the most widely utilized embedding medium, but some laboratories have opted to employ "Carbowax" as a substitute. The latter compound is a water-soluble wax, making dehydration and clearing of the tissue unnecessary and allowing for direct infiltration of formalin-fixed tissue with embedding medium in the tissue processor (11). This element of simplicity is attractive, but Carbowax has its drawbacks. One concerns the dissolution of the embedding medium when microtomized tissue ribbons are placed in a water bath prior to mounting them on glass slides. This unwanted eventuality makes it difficult for the technologist to keep the tissue section flat, resulting in undesirable folds in the final stained slide. Second, we have noted irregularities in antigen preservation when Carbowax-embedded tissues are studied immunohistologically. The temperature of paraffin or Carbowax stations in the tissue processor, and at the embedding center, must be monitored closely. Overheating the wax will cause unwanted thermal artifacts in the tissue and compromise its cellular detail. Excessively cool wax fails to infiltrate the specimens adequately.

Another class of embedding compounds that is presently in vogue in some centers is represented by polymeric plastic resins such as glycol methacrylate or epoxy. Disadvantages of these compounds include the necessity of cutting corresponding tissue sections with a glass or diamond knife microtome, and the requirement for a transitional fluid, such as propylene oxide, to embed the tissue after dehydration and clearing (11). Moreover, plastic sections are difficult to stain with the same intensity as that seen in paraffin-embedded preparations. The main advantage of plastic media is that extremely thin, flat sections may be prepared by experienced microtomists, providing exquisite cellular detail. In addition, some enzyme-histochemical staining methods that otherwise require the use of FSs are possible with specimens embedded in epoxy or glycol methacrylate.

Histomicrotomy is a seemingly straightforward process, representing the cutting of serial paraffin-embedded sections with a tissue microtome. Nevertheless, this technique has many hidden traps that relate to the proper maintenance, calibration, and orientation of cutting blades; preparation of paraffin blocks; and dexterity of the technologist. Microtome blades that are dull loose or nicked will produce "chatter" or "venetian blind" artifacts in tissue sections (Figure 1.8). In addition, the "clearance angle" (between the tissue block and the microtome knife) is crucial to good technique. It should be approximately 3–8°. If the angle is too narrow, alternately thick and thin sections are cut, or they are folded on themselves (18–20). An excessive clearance angle causes chattered or otherwise hideous sections and may preclude the ability of the technologist to obtain a tissue ribbon. Even worse are the effects of loose microtome blades or tissue blocks in the microtome chuck. These deficiencies may shatter the paraffin block entirely or deeply groove the tissue specimen. A block that is mounted crookedly in the microtome chuck will produce irregular ribbons or cause individual sections in the ribbon to break free from one another.

Regardless of whether one uses paraffin or Carbowax as an embedding medium, there is still a need to refrigerate tissue blocks before microtomy is attempted. This step hardens the wax slightly and allows for crisp sections to be cut. Warm blocks will yield wrinkled ribbons or cause successive sections to anneal to one another. In addition, failure to moisten the surface of blocked tissue suitably before cutting it yields an excessive number of knife marks or fragmented sections. The technologist can simply rub a wet finger over the block several times prior to microtomy, if the specimen is small. If it is large, and particularly if the tissue is heavily cornified, a wet piece of cloth or cotton soaked in 5% ammonium hydroxide may be applied for 2 or 3 minutes to rehydrate the tissue face (18).

Another problem that is sometime seen at this step is the tendency for ribbons to "fly" onto the knife blade. This is the result of static electricity between the wax or tissue and the metal blade, and also may be avoided by slightly moistening the knife and the block surface before each ribbon is prepared.

MOUNTING OF TISSUE SECTIONS

The wax ribbon of serial tissue sections can be removed from the microtome knife as it is cut, by using a wooden tongue depressor blade. In this process, the operator exerts slight traction on the end of the ribbon, stretching it gradually over the wooden blade, and subsequently depositing it on the surface of a warm water bath at the cutting station. The temperature of such flotation devices should be kept at 5–10°C below the melting point of the embedding wax. If it is too hot, desiccated-looking sections will result; in contrast, cool flotation baths produce excessive wrinkling of the tissue.

To facilitate the process of obtaining a smooth, unwrinkled, paraffinized ribbon of tissue, it can be stretched by slight traction on its ends while floating in the warm water bath. Also, we have found that adding a few milliliters of ethyl alcohol to the water is beneficial in this regard. The ribbon must not be left in the bath for more than 1 or 2 minutes, or spurious overhydration of the tissue will be produced. This effect simulates the appearance of edema fluid microscopically (17). Because tissue sections do not adhere well to untreated glass slides, a bonding agent also must be a component of the water bath. Elmer's-Glue^{R}, albumin, and poly-L-lysine are all suitable additives of this type.

One of the most dangerous of all mistakes in the histology laboratory can take place when mounting sections from flotation baths. Friable tissue may "shed" small fragments that float free on the surface of the water, and these may be inadvertently picked up when mounting slides from subsequently processed unrelated cases. Derisively known as "floaters," these rogue pieces of tissue commonly cause agonizing interpretative problems for the pathologist (Figure 1.9). For example, it is not difficult to envision a small piece of a prostatic carcinoma that may find its way onto slides of another prostate biopsy, a ribbon of which is mounted subsequently in the same water bath. Technologists must be impressed with the tremendous medicolegal liabilities that such a mistake incurs, and they must routinely skim, or otherwise clear, the surface of the water bath between cases. An alternative source of floater-type artifacts is the "tongue blade metastasis," wherein tissue adheres to a wooden applicator stick that is used to float successively prepared ribbons from two different cases (11). Needless to say, this practice is highly inadvisable.

With respect to optimizing the cost of slide preparation, we recommend that as many individual sections as possible should be mounted on one slide, from the same ribbon. It is not difficult for adept technologists to include three to five cuts of a specimen on each slide, arranged in a serial fashion. Also, in light of the limited size of many skin, bronchoscopic, and gut biopsies, it is advisable to save any unmounted paraffin ribbons (with appropriate identification) from such cases for 1 week after they are accessioned. Remounts can be prepared from these directly, without the need for further microtomy of the tissue block.

Finally, the identification of tissue sections must be scrupulously maintained throughout the remainder of their sojourn in the histology laboratory. Such a necessity is assured by having the technologist scratch the case and

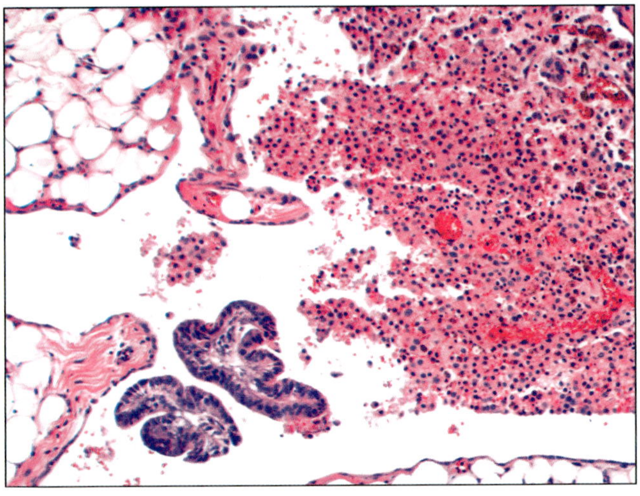

***FIGURE 1.9:** "Floaters" (lower left) in microscopic sections are unwanted pieces of tissue from other cases. They may cause serious diagnostic mistakes and again should be avoided at all costs.*

block numbers onto one end of all mounted glass slides with a diamond knife.

GENERAL STAINING AND COVERSLIPPING PROCEDURES

Once the tissue has been affixed to glass slides, paraffin is removed prior to the staining procedure. This is accomplished by placing the sections in a carrier, heating them to 56°C for at least 1 hour (preferably longer—to evaporate water from the glass slides and further anchor the tissue), and immersing them for 3–5 minutes in each of three or four successive containers of a clearing agent (xylene or a limonene derivate). Most histologic dyes penetrate tissue best if it has been rehydrated; thus, passage of the slides through containers of graded ethanol solutions (absolute, 95%, 70%) and distilled water is necessary before staining can be undertaken. The importance of proper paraffin clearance cannot be overemphasized, with regard to final results. If a sizable quantity of wax remains in the sections, dyes will not be able to penetrate them and impregnate constituent tissues.

The majority of diagnostic pathology laboratories employ H&E as the stains of choice for standard microscopic interpretation. There are two principal procedures for applying these dyes—the "progressive" method and the "regressive" technique. The former is usually preferred, and it is characterized by sequential staining with hematoxylin (for roughly 15 minutes) and eosin (for 1–3 minutes). These steps are separated by application of ammonia water or lithium carbonate as "bluing" agents for hematoxylin, which is actually a chemical "lake" of hematein and a mordant such as ammonium aluminum sulfate (21). In addition, copious rinsing with water must be assured after exposure of the tissue to hematoxylin and ammonia or lithium. The progressive H&E method is so named because the technologist can monitor the depth of staining as it develops and terminate each step appropriately. This procedure is desirable because it can compensate for the effects of various fixatives that enhance the uptake of either hematoxylin (formaldehyde, osmium, heavy metals) or eosin (picric acid). "Retrogressive" H&E staining refers to a technique wherein tissue is purposefully overimpregnated with hematoxylin and then modified to suit the operator by decolorization with dilute hydrochloric acid. Final results are much more difficult to control in that procedure, and it requires a highly experienced technologist to achieve good nuclear detail in the ultimate product. In regard to eosin as a counterstain for hematoxylin, two points merit further mention. One is that eosin is a "differentiating" stain that dissimilarly impregnates tissues of varying chemical structure and molecular density. As much, it requires the same amount of technical attention as hematoxylin in a progressive H&E method. Different

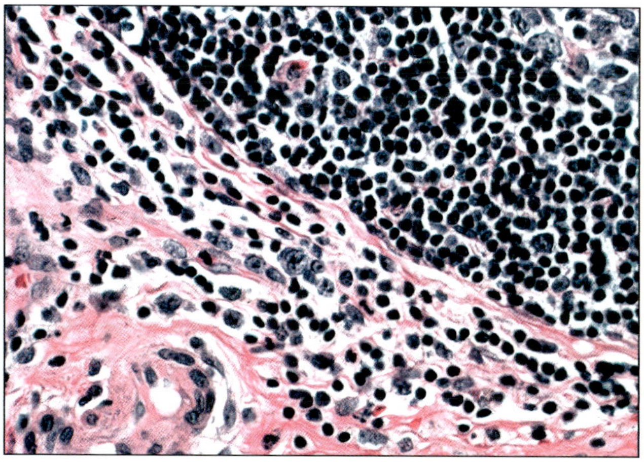

***FIGURE 1.10:** This optimally sectioned and stained slide shows good differentiation of acidic and basic components of the H&E method.*

connective tissues should be stained to variable degrees in a properly performed procedure of this type (Figure 1.10); however, overly rapid passage of sections through ethanol solutions, *after* exposure to eosin, interferes with this effect and should be discouraged. Second, some workers prefer to use a *combined* eosin reagent (incorporating such dyes as phloxine or safranin) to enhance its differentiating properties (11).

In addition to the H&E technique, there are many other specialized staining methods that are pertinent to pathologic diagnosis. These will be discussed specifically later in this chapter.

Inasmuch as most laboratories utilize Harris' hematoxylin—which is insoluble in alcohol—the final preparation of H&E-stained microscopic sections involves dehydration in graded alcohols and several changes of a clearing agent. Subsequently, a drop of synthetic mounting medium (such as Permount^R) is placed over the tissue after blotting away excess clearing solution, and a glass coverslip is attached.

The latter steps sound simple enough, but they actually require some technical finesse. Experienced technologists often dilute the mounting medium slightly with a few drops of clearing agent and also leave a very small amount of clearing solution on the glass surface surrounding the tissue. These provisions assure that the mounting medium will disperse evenly under the glass coverslip, to the exclusion of entrapped air bubbles. Excess mounting medium must be carefully blotted from the area around the attached coverslip because it may otherwise diffuse over the surface of the slide. This unwanted phenomenon results in "blobs" of Permount on the coverslips (interfering with microscopy) or on the sides of the glass slides (causing them to stick to other surfaces or to one another). Under no circumstances should newly mounted sections ever be stacked on top of one another, or immediately placed in vertical slide boxes or small plastic carriers. These maneuvers make it virtually certain that sections will anneal to one another, to plastic surfaces, or to struts in slide boxes, as the mounting medium hardens. The latter process can be accelerated by brief and gentle warming in a drying oven into which forced air is pumped.

In recent years, automatic coverslipping machines have been marketed by several firms. These are attractive because they potentially save technologists expenditure of time in a rote procedure. However, our experience with such devices has been less than satisfactory. Quite often, the amount of mounting medium dispensed is inadequate, and air bubbles rapidly appear under the coverslips after the slides are delivered to the pathologist (Figure 1.11).

With regard to labeling, some laboratories attach gum-backed labels to coverslipped sections, on which corresponding case and block numbers have been written or typed. Others prefer to use slides with a "frosted" area on one end, where these numbers can be written in indelible ink. Whichever one of these procedures is followed, laboratory workers must be certain to check all labels against block designations and diamond knife etchings on the slides. The crucial nature of this step cannot be overstated in pathology laboratories that process many similar biopsies because the gross characteristics of small specimens are of limited help in distinguishing one from the other. Nearly all "shave" skin biopsies, prostatic needle biopsies, and endoscopic gut biopsies look alike macroscopically. If a mistake occurs that compromises the identity of a given case, the pathologist and responsible clinician should be informed immediately. Such errors may result in the need to rebiopsy the patients who are concerned, in order to avoid potential diagnostic disasters.

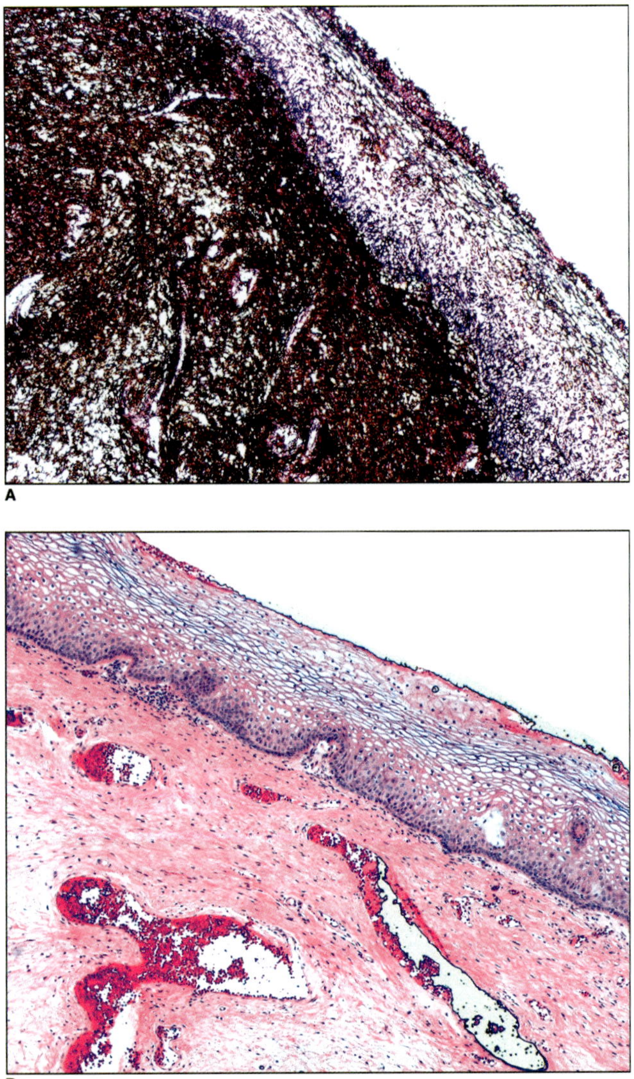

A, **B**

***FIGURE 1.11:** Air bubbles under section coverslips cause sufficient refractility to interfere with interpretation of the slide (A). Compare with a properly mounted version (B) of the same slide.*

SPECIFIC SPECIAL HISTOCHEMICAL STAINS

For purposes of differential diagnosis, it is often desirable to evaluate biopsy specimens for their content of several tissue constituents, including glycogen, glycoproteins, mucosubstances, iron, melanin, calcium, lipofuscin, elastin, collagen, amyloid, muscular proteins, fat, nucleic acids, nucleoproteins, neural elements, specific enzymes, microorganisms, and others. These moieties comprise the targets of predefined histochemical stains, in which the chemical affinities of various biologic components for several dyes have been exploited. A comprehensive discussion of the specific chemical reactions that are involved, the detailed methods behind such stains, and all of the circumstances in which they might be applied is beyond the scope of our consideration here. These topics have been well covered in other textbooks, Internet sites, and review articles to which the interested reader is directed (11,21–29). However, general principles and pertinent brief summaries of selected histochemical methods are provided in the following sections.

General Principles

As summarized by Stoward (30), the conceptual bases of histochemical techniques can be divided into four groups. These include ionic reactions involving the association of cations and anions, the action of aldehydes with Schiff's reagent (sulfuric acid + pararosaniline hydrochloride) to yield a colored product, the association of aromatic diazonium salts with aromatic constituents of proteins, and the mediation of reactions wherein dyes are precipitated in colored form by target tissue enzymes. Biochemical moieties that are not directly evaluable by these methods can be altered to make them so. For example, glycogen can be oxidized by periodic acid to allow it to combine with Schiff's reagent [in the *p*eriodic *a*cid-*S*chiff (PAS) procedure], whereas it would not otherwise be reactive with the latter compound.

Specific Histochemical Methods

Glycogen, Glycoproteins, and Mucosubstances: As just stated, the PAS procedure is used to visualize deposits of glycogen in tissue, as well as to label glycoproteins or mucinous cell products that contain vicinal glycol residues. In order to distinguish between simple glycogen and the latter chemical substances, one usually exposes two serial tissue sections to the PAS reaction. One of them is, however, modified by prior incubation with diastase or amylase; this step dissolves glycogen (Figures 1.12 and 1.13) but does not interfere with the ability of sugars in glycoproteins or vicinal glycols in mucin to combine with Schiff's reagent and yield a bright red reaction product (Figure 1.14). Thus, pathologists commonly request that PAS stains be done "with and without" diastase or amylase.

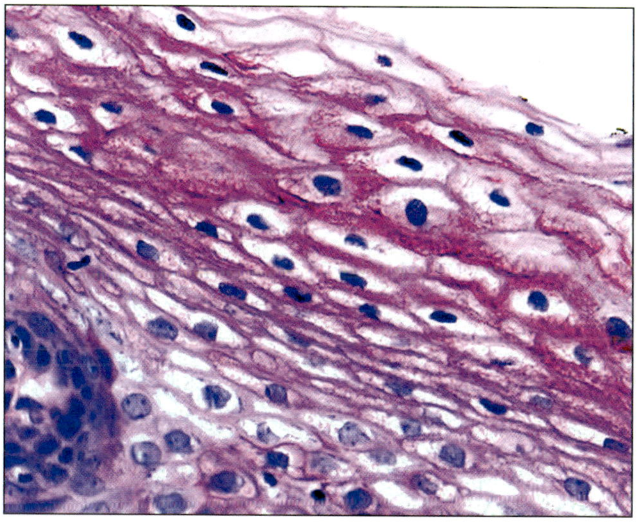

FIGURE 1.12: ***The PAS stain, done without prior diastase digestion of the mounted tissue, is capable of labeling intracellular glycogen. Vaginal mucosa is seen here.***

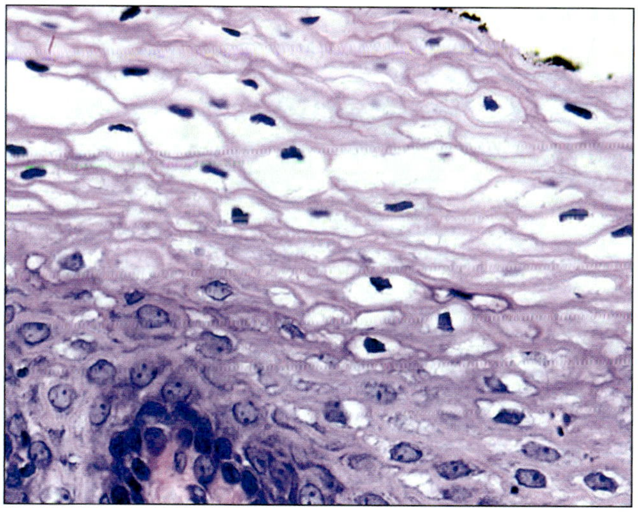

FIGURE 1.13: ***The same tissue as seen in Figure 1.12, but digested with diastase before performance of the PAS method, no longer demonstrates any glycogen content.***

Glycogen is seen in many normal structures and in diverse neoplasms (23). It plays a particular role in the distinction between poorly differentiated epithelial neoplasms (which are potentially PAS positive) and malignant hematopoietic tumors (usually PAS negative) with which they may be confused.

The glycoproteins of greatest interest are those that are components of basement membranes, together with collagen type IV and laminin. These are well seen with the PAS technique after prior diastase digestion of tissue sections (PAS-D). Diseases showing prominent alterations in basement membranes include, but are not limited to, lupus erythematosus, diabetes mellitus, and porphyria (Figure 1.15).

Mucosubstances that react with PAS-D are commonly called "epithelial mucins" (more appropriately, *glandular* mucins), and they tend to contain sialic acid residues (31). As one would expect, connective tissue mucins are PAS-D

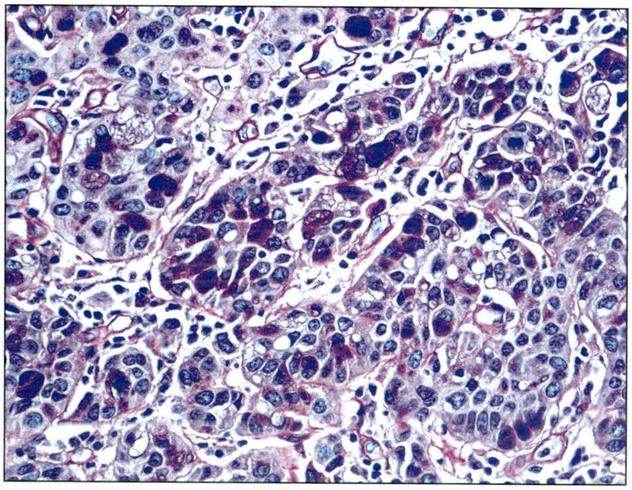

FIGURE 1.14: ***Diastase-digested tissue stained with the PAS procedure can still be labeled for epithelial mucin, as seen in this adenocarcinoma.***

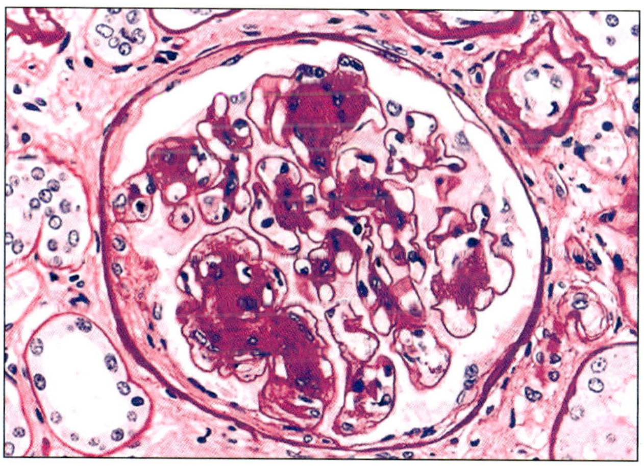

FIGURE 1.15: ***Glycoprotein, as shown in this example of nodular diabetic glomerulosclerosis, is also reactive with the PAS technique after prior diastase digestion.***

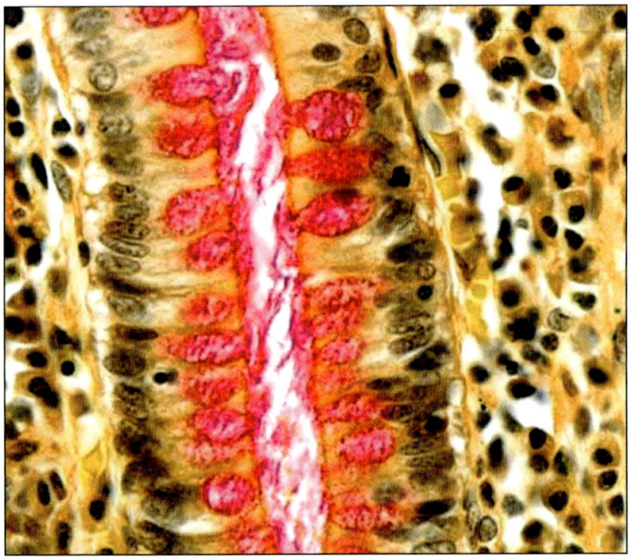

FIGURE 1.16: Best's mucicarmine method is another stain for epithelial mucin, as seen in this section of colon.

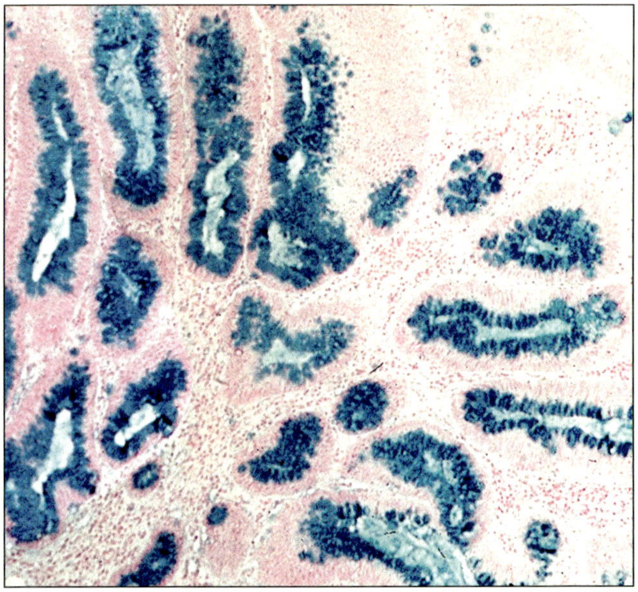

FIGURE 1.17: Alcian blue binds to sulfomucins at pH 2.5, as seen in this section of colon.

nonreactive. Stains in this category may distinguish cellular proliferations that produce glandular mucin (e.g., adenocarcinomas) from others that synthesize "stromal" (mucopolysaccharide- or sulfated acid-rich) mucin. One more procedure that is sometimes used to recognize epithelial mucin—the *mucicarmine* or *Best* technique (Figure 1.16)—is not as well characterized as PAS-D in regard to its biochemical *modus operandi*. Nevertheless, staining results with these two methods are largely equivalent (including the color of the reaction product), with the possible exception that mucicarmine is less sensitive.

If a distinction between the nonglandular mucins is desirable, the *aldehyde-fuchsin* stain (AFS) can be performed along with the PAS-D, *colloidal iron*, *alcian blue*, *azure-A*, and *toluidine blue* procedures. If it is done at a pH below 1.0, the AFS yields green-blue positivity in tissues that contain primarily sulfated acid mucosubstances (11). In contrast, acid mucopolysaccharides such as hyaluronic acid, heparin, and chondroitin/dermatan/keratin sulfates constitute the major components of other stromal mucins (24). All these moieties are potentially visualized with colloidal iron or alcian blue, both of which yield light blue positivity (Figure 1.17); azure-A, showing red reactivity; and toluidine blue, producing lavender or purple positivity. A side benefit of toluidine blue is that it can be used to label mast cells in tissue, owing to the presence of heparin sulfate in mastocytic granules.

By manipulating either the tissue or the reaction conditions, further information may be obtained. Prior digestion of sections with hyaluronidase greatly decreases staining of ground substances with both the alcian blue or colloidal iron procedures. Pretreatment with neuraminidase will abrogate the labeling of sialomucins, but they are intensely stained with the alcian blue method when it is done in a milieu containing high concentrations of magnesium chloride at pH 2.5 (25). Similarly, decreasing the pH to <1 results in preferential alcian blue staining of sulfated glycosaminoglycans such as heparin or dermatan sulfate, but this same step abolishes the recognition of acid mucopolysaccharides by toluidine blue (24).

The stromal mucins are important to the diagnosis of several diseases. In the skin, for example, follicular and focal mucinoses feature the deposition of colloidal iron/alcian blue–positive mucosubstance in the hair follicles or dermis, respectively. Myxedema, myxoid cysts, and lichen myxedematosus show regional or diffuse increases in stromal mucin content in various tissues. Necrobiotic processes may be separated from one another with the methods that label stromal mucin (32). Likewise, lupus erythematosus, dermatomyositis, and malignant papulosis all may demonstrate matrical mucin deposition (33).

Another exciting category of histochemical reagents is represented by the lectins; these proteinaceous products of various plants show preferential and differential affinity for particular carbohydrate residues (34) (Table 1.2). It is now known that some—such as peanuts, soybeans, and wheat germ agglutinins—bind to sulfated or sialated mucins (Figure 1.18), whereas others (concanavalin-A and *Lens culinaris* agglutinin) do not (35). In addition, *Ulex europaeus* I agglutinin recognizes poly-L-fucose residues that appear to be integral components of a pan-endothelial determinant, and they are also seen in epithelium in patients of blood group O. Many lectins are commonly marketed in biotinylated form, lending themselves to the use of the avidin-biotin-peroxidase complex method to visualize their tissue binding sites (34,36–38). This technique has many similarities to immunohistochemical procedures (39).

TABLE 1.2: Lectins Used in Pathologic Diagnoses*

Common Abbreviations	Source	Nominal Saccharide Specificity
Glucose/mannose group		
Con A	Jack bean (*Canavalia ensiformis*)	α-Man > α Glc ≥ GlcNAc
LCA**	Lentil (*Lens Culinaris*)	α-Man > α Glc > GlNAc
PSA	Pea (*Pisum sativum*)	α-Man > α Glc = GlcNAc
N-Acetylglucosamine group		
BSA II	*Bandeirea simplicifolia* seed	β- and α-GlcNAc
GSA-II	Griffonia seed (*Griffonia simplicifolia*)	α- and β-GlcNAc
DSA	Jimson weed (*Datura stramonium*)	GlcNAc (β1, 4-GlcNAc)$_{1\text{-}3}$ = Galβ1, 4-GlcNAc
PWM	Pokeweed (*Phytolacca americana*)	GlcNAc (β1, 4-GlcNAc)$_{1\text{-}5}$ = Galβ1, 4-GlcNAc$_{2\text{-}5}$
STA	Potato (*Solanum tuberosum*)	GlcNAc (β1, rGlcNAc)$_{1\text{-}4}$
UEA-II	Gorse seed (*Ulex europaeus II*)	L-Fucα1, 1-Galβ1, 4-GlcNAc> GlcNAc (β1,4-GlcNAc)$_{1\text{-}2}$
WGA	Wheat germ (*Triticum vulgaris*)	GlcNAc(β1, 4-GlcNAc)$_{1\text{-}2}$ ≥ GlcNAc> Neu5-Ac
N-Acetylgalactosamine/galactose group		
BPA	*Bauhinia purpurea* seed	α- and β-GalNAc > α- and β-Gal
BSA 1-B4	*Bandeirea simplicifolia* seed	α-D-Gal > α-D-GalNAc
DBA	Horse gram (*Dolichos biflorus*)	α-D-Gal-NAc>> α-D-Gal
GSA-1	Griffonia seed (*Griffonia simplicifolia*)	α GalNAc > α-D Gall-A4
HPA	Edible snail (*Helix pomatia*)	GalNAc α 1, 3GalNAc > α GalNAc
LBA	Lima bean (*Phaseolus lunatus limensis*)	GalNAc α 1, 3[L-Fuc α 1, 2] Gal β GalNAc
MPA	Osage orange seed (*Maclura pomifera*)	αGalNAc > α Gal
PNA	Peanut (*Arachis hypogaea*)	Gal β 1, 3 GalNAc > α and β Gal
RCA 1	Castor bean (*Ricinus communis*)	β Gal > α Gal ≥ GalNAc
SBA	Soybean (*Glycine max*)	α and β GalNAc> α and β Gal
SJA	Japanese pagoda tree (*Sophora japonica*)	α and β GalNAc > α and β Gal
VVA	Hairy vetch (*Vicia villosa*)	GalNA α 1, 3Gal = α GalNAc α GalNAc
WFA	Wisteria seed (*Wisteria floribunda*)	GalNAc α 1, 6Gal > αGalNAc> βGalNAc
L-Fucose group		
AAA	Orange peel fungus (*Aleuria aurantia*)	α-L-Fuc
LTA	Asparagus pea (*Lotus tetragonolobus*)	α-L-Fuc > L-Fuc α 1, 2 Gal β 1, 4 GlcNAc ≥ L-Fuc α 1, 2 Gal β-1, 3GlcNAc
UEA-1	Gorse seed (*Ulex europaeus*)	α-L-Fuc
Sialic acid group		
LFA	Slug (*Limax flavus*)	α-Neu 5Ac > α-Neu 5Gc
LPA	Horseshoe crab (*Limulus polyphemus*)	Neu 5Ac (or Gc) α-2, 6GalNAc > Neu 5Ac

Ac = acetyl, Fuc = fucose Gal = galactose GalNAc = *N*-acetylgalactosamine Gc = glucuronic acid Glc = glucose GlcNAc = *N*-acetylglucosamine Man = mannose Neu = sialic acid (neuraminic acid)

* Reprinted with permission from: Stoward PJ: Histochemical methods for routine diagnostic histopathology. In: *Histochemistry in Pathology* (Filipe MI, Lake BD, Eds.), Churchill-Livingstone, New York, 1990; p. 21.

** Not to be confused with leukocyte common antigen.

Amyloid: Amyloid is a peculiar tissue substance that is unique to pathologic conditions. It is principally a fibrillar protein that forms rods that are of variable length and 7.5–10 nm in diameter and that tend to aggregate extracellularly in a "tangled" configuration. A secondary, minor component of amyloid is sulfated proteoglycan, accounting for approximately 10% by weight (40). It accounts for the fact that amyloid has some chemical similarities to polysaccharides and certain mucosubstances, such as an affinity for iodine and metachromatic labeling with toluidine blue (11).

Detailed studies on linear amyloid fibrils have shown that they aggregate in beta-pleated sheets as seen by X-ray crystallography (40). This property has been exploited histopathologically, together with dyes that preferentially impregnate amyloid protein and proteoglycan. The latter stains, such as *Congo red*, *sulfated alcian blue*, *Sirius red*, *Lieb's crystal violet*, and *thioflavine-T*, are used to impart a red, magenta, or fluorescent yellow-green color to amyloid deposits, respectively. Polarization microscopy of sections labeled with Congo red shows a characteristic "apple-green" dichroism in amyloid deposits because of its physicochemical attributes (Figures 1.19 and 1.20).

There are advantages and disadvantages to some of the above-cited methods. Sections stained with Congo red must be meticulously prepared at a thickness of 8–10 μm; if they are too thin or thick, the amyloid deposits will be a faint red or yellow upon polarization, rather than apple-green. We also have noticed that the use of unmatched polarizing microscopic lenses may cause similar problems. Lieb's stain is soluble in organic compounds, necessitating that sections cannot be coverslipped in a conventional fashion. Glycerol has been employed as a mounting medium, but it allows for the diffusion of crystal violet from the amyloid with the passage of time. Modified Apathy's medium (a mixture of acacia, cane sugar, water, sodium chloride, and thymol) is a suitable alternative, as are polymeric compounds such as Crystal Mount^{R} (11). As stated above, the thioflavine-T procedure must be interpreted by fluorescence microscopy (Figure 1.21) and is therefore not widely used.

Several biochemical classes of amyloids exist. These include one type derived from fragments of immunoglobulins in plasma cell dyscrasias (AL amyloid), another produced by the apparent modification of plasma proteins by macrophages in chronic inflammatory diseases (AA amyloid), a familial form (AF amyloid), a fourth that represents altered neurohormonal cell products and is confined

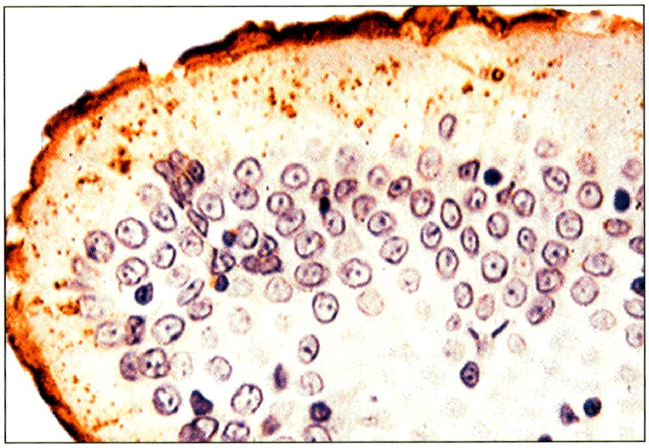

FIGURE 1.18:** Wheat germ agglutinin* **(Triticum vulgaris** *lectin)** ***binds to sulfated mucins as well, as shown in this section of metaplastic esophageal epithelium.***

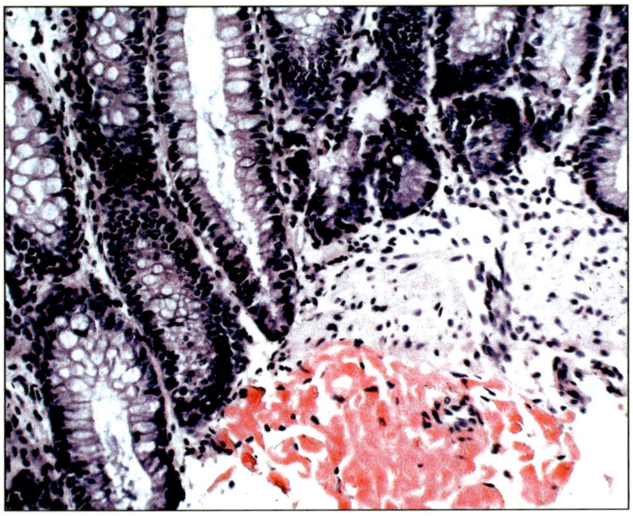

***FIGURE 1.19:** Amyloid has an affinity for Congo red, as seen in this section of rectum in a patient with amyloidosis.*

***FIGURE 1.20:** Polarization of Congo red–stained amyloid deposits yields "apple-green" birefringence.*

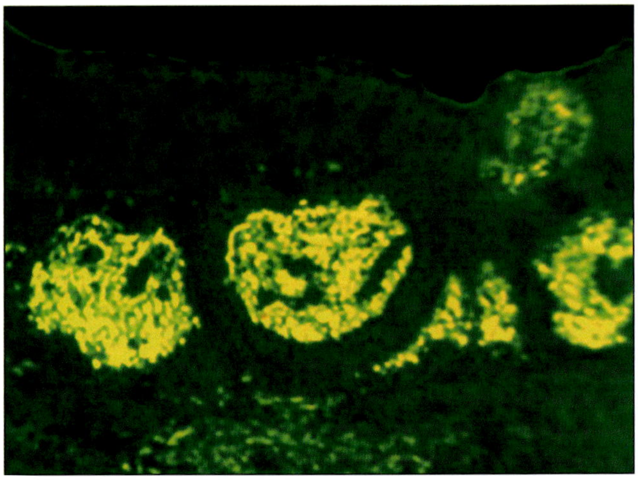

***FIGURE 1.21:** Thioflavine-T is another stain with affinity for amyloid, but detection of positive labeling requires fluorescence microscopy, as shown here in a section of skin.*

to the interstices of endocrine neoplasms, and another that is related to apoptotic transmogrification of keratin. Regardless of which specific type of amyloid one encounters in an individual case, all of them show essentially identical histochemical staining properties (40), with one exception. Labeling for amyloid AA, which is implicated when chronic inflammation is the etiology of amyloidosis, is greatly diminished by pretreatment of tissue sections with 0.5% potassium permanganate solution (41,42). Nevertheless, staining for other forms of amyloid protein—including AL and tumor-related types—may rarely be abrogated by permanganate as well.

Minerals and Pigments: There are a number of minerals and pigments in tissue that have pathologic relevance. These can be localized and identified with histochemical methods.

Iron is detected with *Perl's* reaction, also known as the *Prussian blue* procedure. When loosely complexed with protein, ferric ions will combine with acidified potassium ferrocyanide to yield a deep blue product that is insoluble (21). Because of potential confusion with structures that hematoxylin may impregnate, Prussian blue tissue sections are counterstained with nuclear fast red.

Copper deposition is of interest in several tissues in the context of a hereditary condition, Wilson's disease, in which the metabolism of that element is abnormal. In addition, liver diseases that feature cholestasis may also be associated with the retention of copper. Three main histochemical methods are available to label this substance; the *rhodanine* technique, *Shikata's orcein* stain, and the *rubeanic acid* procedure (21,25). Rhodanine is a chelating agent that demonstrates strong binding to copper-protein tissue complexes; the resulting precipitates are red-orange

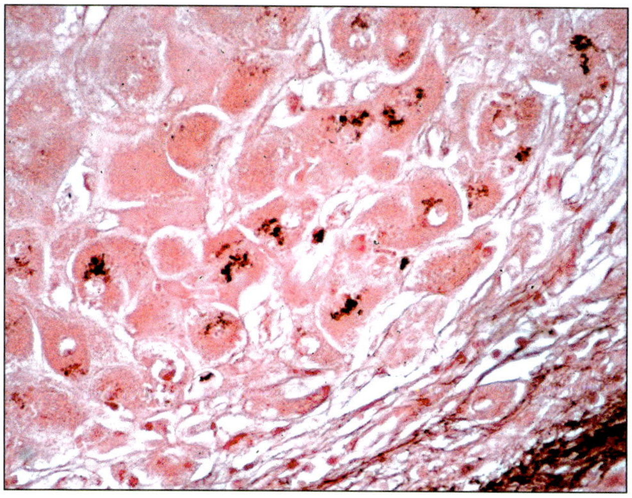

FIGURE 1.23: ***The orcein (Shikata) technique is another stain for tissue copper; this section of liver is from a patient with Wilson's disease.***

(Figure 1.22), against a background that is typically lightly stained with hematoxylin. In contrast, the Shikata and rubeanic acid procedures yield a black-brown precipitate in a reddish background (Figure 1.23). There is some debate as to whether these methods recognize elemental copper or copper-protein conjugates (43). Incidentally, the orcein method will also label hepatocytes that are infected with hepatitis-B virus, and it stains sulfated mucins as well (44).

Melanin is derived from tyrosine-containing compounds and it is typically bound to protein in tissue. This pigment is capable of reducing silver nitrate solutions to metallic silver, which is deposited in tissue sections as a black precipitate in the *Fontana-Masson* technique (Figure 1.24) (25). Alternatively, solutions containing ferrous ions can be used to form complexes with melanin,

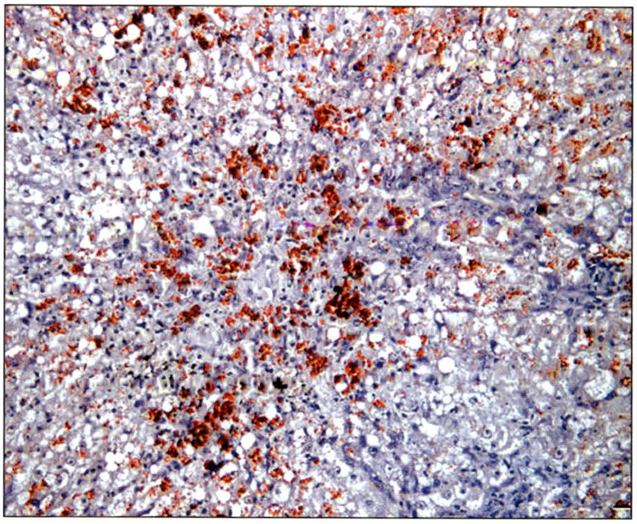

FIGURE 1.22: ***Copper deposits in tissue are detectable with the rhodanine method, as seen here in a liver biopsy from a patient with primary biliary cirrhosis.***

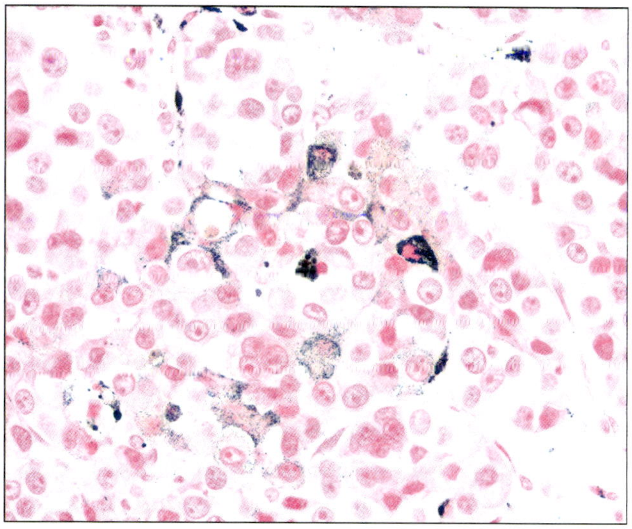

FIGURE 1.24: ***The Fontana-Masson procedure is often used to label melanin, as seen in this section of a metastatic melanoma that was essentially amelanotic in H&E stains.***

and these are subsequently detectable as a blue product with *Turnbull's* or *Schmorl's* reactions (11). In the assessment of heavily melanized lesions, it is sometimes necessary to bleach the pigment to allow for evaluation of cellular detail. This can be accomplished by treating tissue sections with 10% hydrogen peroxide for 1–2 days, or 0.25% potassium permanganate for 1–4 hours, followed by 1% oxalic acid (25). One also can capitalize on knowledge of the biochemical pathway whereby melanin is formed by melanocytes, in order to visualize nonpigmented cells of this type. The *DOPA* reaction involves the exposure of tissue sections to a dilute solution of 3,4-dihydroxyphenylalanine. Melanocytes that contain DOPA oxidase will convert this compound to melanin, which may then be detected by one of the above-cited methods. Unfortunately, the DOPA reaction mandates the use of FSs, inasmuch as melanocytic enzymes are largely inactivated by routine tissue processing (45).

In selected circumstances, one may wish to label melanin with a metachromatic dye, to avoid its confusion with other brownish pigments in tissue sections. Azure-B is capable of accomplishing this task, in that it produces a green product complexed with melanin (46,47).

Calcium typically is hematoxylinophilic on H&E stains. In order to confirm its presence definitively in such conditions as calcinosis cutis and pseudoxanthoma elasticum, one of two procedures should be utilized—the *von Kossa* technique and the *alizarin red* method (22). The first of these employs a solution of silver nitrate; calcium carbonate will combine with the latter compound to form silver carbonate, which, upon exposure to ultraviolet light or sunlight, is reduced to elemental silver in a black precipitate (Figure 1.25). Unfortunately, formalin pigment will produce an identical result. This problem makes the alizarin red technique, wherein the dye chelates directly with calcium ions to form an orange-red complex, preferred. Alizarin red–stained sections are not counterstained, and the resulting calcium complexes show birefringence on polarization microscopy (11).

Uric acid deposits are seen in several organs in gouty tophi. These are difficult to visualize in sections that are fixed with aqueous solutions because urates are dissolved in the course of routine processing. Alcohol-fixed tissues are therefore best used in this context. Uric acid reduces the silver ions in a 5% silver nitrate solution to their metallic form, giving a black precipitate over the gouty deposits (27). When employed as a part of the *Gomori methenamine silver* procedure, this reaction is useful diagnostically. Sections are counterstained with light green solution.

Bile pigments are potentially seen in several tissue sites in patients with hepatic failure. These include the basal ganglia of the brain, the renal tubules, and the liver itself. The demonstration of bile in hepatic neoplasms can also be instrumental in recognizing variants of hepatocellular carcinoma. Three histochemical methods are used to recognize bile: the Van Gieson procedure, the Sirius red technique, and Fouchet's (Hall's) stain. These methods are predicated on the oxidation of bile pigments to yield colored moieties or derivatives that complex with a chromogen. Bile is green-blue with Fouchet's reaction (Figure 1.26), and bright green with the Van Gieson and Sirius red procedures (21,25).

Lipids: Lipids in tissue may be of diagnostic importance in several settings. For example, it may be desirable to confirm the biochemical nature of apparent xanthomas, and some granulomas characteristically contain a central area of lipid deposition (32). Cholesterol emboli may cause microinfarcts in several organs, and sebaceous tumors are

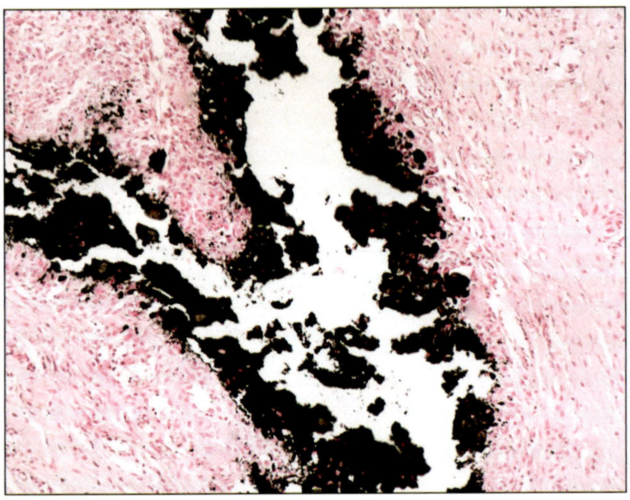

FIGURE 1.25: ***Calcium deposition in tissue is demonstrable with the von Kossa procedure, as seen in this section of an artery showing Monckeberg's vascular sclerosis. Another procedure for calcium labeling is the alizarin red stain (see text).***

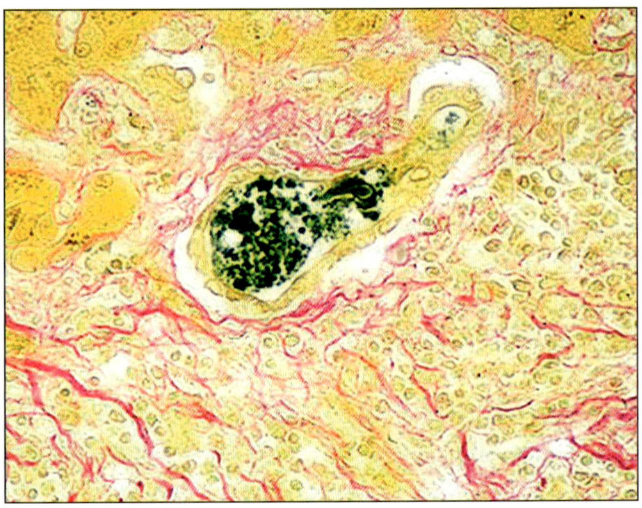

FIGURE 1.26: ***Fouchet's procedure or Hall's method is used to label bile, as seen in this section of liver.***

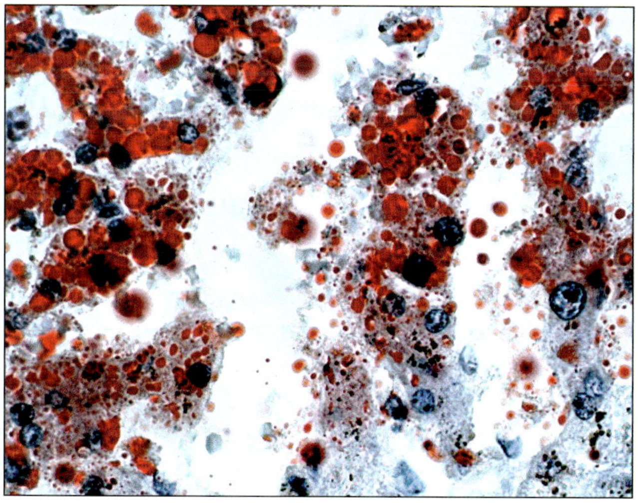

***FIGURE 1.27:* Assessment of fat content in tissue necessitates the use of frozen tissue because lipid will dissolve during routine processing. The oil red O method (shown here in a section of steatotic liver), the Sudan IV procedure, or the Sudan black-B technique can be used for this purpose.**

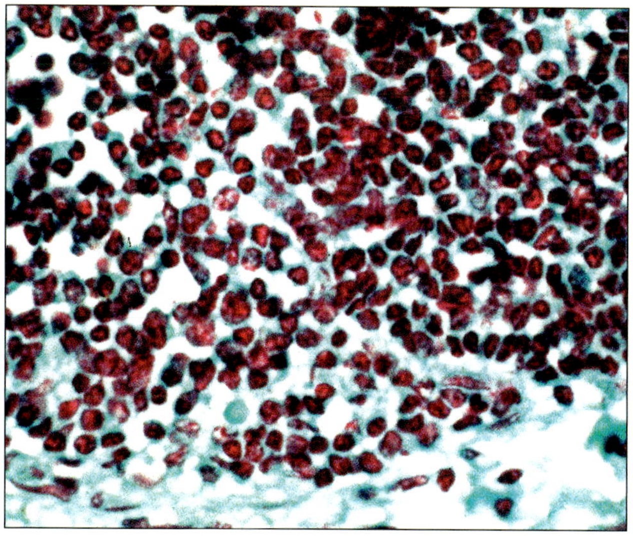

***FIGURE 1.28:* DNA is preferentially labeled with the Feulgen stain, as shown here in a section of lymph node.**

definitively identified by the demonstration of cytoplasmic fat in the neoplastic cells.

An inviolate requirement for visualization of lipids in clinical specimens is that frozen tissue or, at the very least, unembedded formalin-fixed tissue must be available for study. This is because the alcoholic and clearing agent steps in tissue processors otherwise dissolve all fat before sections are prepared for microscopic analysis.

Three major staining techniques are applicable to this problem. The *oil red O* and *Sudan IV* methods utilize dyes that are miscible in fat and therefore impregnate it (Figure 1.27) (25). These yield red and black products, respectively, as their names suggest. The third procedure is probably the best from a technical perspective because it provides microscopic preparations of higher quality. This is the *osmium tetroxide* stain with periodic acid as a differentiating agent, wherein osmium fixes lipids and stains them black. Because this chemical is insoluble in organic compounds, frozen tissues labeled with 1% osmium can be postfixed in formalin and processed routinely (11). The result is a histologic section with good cytologic detail, as opposed to the suboptimal architecture that one sees in FSs. Nevertheless, procedures employing osmium tetroxide do have a significant disadvantage. It is a very caustic and potentially toxic compound, and technologists must work in a fume hood to avoid such dangers. Disposal of the reagent is accordingly difficult as well.

Nucleic Acids and Nucleoproteins: Before the advent of flow cytometry, in situ hybridization, and the polymerase chain reaction, histochemical methods were developed to aid in assessing the amount and nature of nucleic acid that malignant cells contain. The latter procedures are still useful in selected instances, particularly when coupled with morphometric (densitometric) techniques. There are two stains that are appropriately categorized as semispecific for nucleic acids—the *Feulgen* method and the *methyl green–pyronine* (MGP) procedure.

The chemical basis for the Feulgen stain is predicated on the fact that controlled acid hydrolysis can be used to digest purine bases in deoxyribonucleic acid (DNA), producing aldehyde groups that are capable of combining with the Schiff reagent (11). Therefore, the method is an indirect one requiring modification of the target compound. The reader will recall that Schiff conjugates yield a red color, representing the desired end product of the Feulgen stain (Figure 1.28). The hydrolytic step in this procedure is fixative dependent; formalin yields that most consistent results and requires only short exposure of tissue sections to acid, whereas specimens preserved in heavy metal fixatives must be digested for a longer period of time. Bouin's solution is itself hydrolytic and therefore renders tissue unsuitable for studies with the Feulgen reaction. Ribonucleic acid (RNA) is not sensitive to hydrochloric acid and is not labeled in this procedure.

In contrast, the MGP technique depends on differential affinities of DNA and RNA for the two dyes used in this reagent. These affinities are presumably based on the tendency for methyl green to associate with large polymers (DNA), whereas pyronine intercalates with smaller polymeric compounds (RNA). As one would expect, nuclei are stained green with MGP, and the cytoplasm—where the majority of intracellular RNA resides—is red (Figure 1.29) (48). It should be remembered that pyronine is not specific for RNA; this dye also labels keratin and the granules of mast cells or eosinophils. Therefore, neoplasms showing mastocytic, granulocytic, or keratinocytic differentiation are not reliably assessable with MGP. It is most often used in helping to support a plasmacytic or immunoblastic

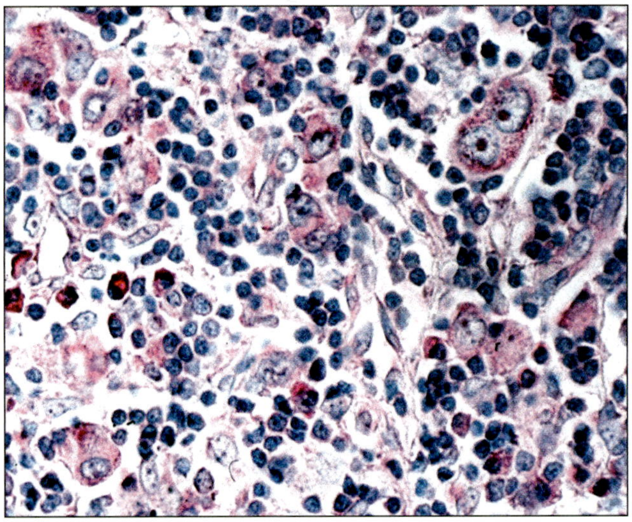

***FIGURE 1.29:* The MGP procedure differentially labels DNA and RNA. DNA is stained green-blue, whereas RNA is stained red. This procedure can be used to identify rapidly replicating cell populations, especially as seen in hematopoietic neoplasms.**

lymphoid phenotype among hematopoietic tumors, where high levels of both DNA and RNA are expected.

Exclusive of immunohistochemical evaluation for cell phase–dependent nuclear antigens, or flow cytometric measurement of S-phase populations, there is only one method for the estimation of actively dividing cells. It is an indirect one and is known as the "Ag-NOR" procedure. This acronym stands for silver (Ag)–stained *n*ucleolar *o*rganizer *r*egions, which are complexes of nuclear proteins and ribosomal RNA genes. These have been localized to five chromosomes in man, and tend to increase in number, size, and density in proliferative cell populations (49). Published studies have suggested, for example, that cells in benign melanocytic lesions have few, compact Ag-NORs, whereas those in malignant melanomas tend to show multiple coarse intranuclear bodies of this type (50).

The Ag-NOR reagent is somewhat complicated to prepare and utilize. It consists of freshly dissolved gelatin and formic acid, which are mixed with silver nitrate solution prepared with deionized distilled water. The final product is a colloidal suspension; it is applied to deparaffinized-rehydrated or frozen sections of tissue and incubated in the dark for 30 minutes. Ag-NORs are visible in the sections as distinct, dark, intranuclear bodies when the procedure is performed properly (Figure 1.30) (51). However, slight variations in incubation time or reagent preparation can result in abundantly but nonspecifically deposited silver, producing "dirty" slides that are difficult to interpret.

Mesenchymal Structures: Several connective tissues and derivatives of embryonal mesenchyme figure into the assessment of selected inflammatory, degenerative, and neoplastic diseases. These include collagen types, elastic tissue, muscle, and nerve, as well as fat and ground substances, which were discussed earlier.

One of the most widely used histochemical methods for the evaluation of these tissue components is the *Masson trichrome* technique. It is a mixture of three dyes—Biebrich's scarlet, hematoxylin, and aniline blue—as well as phosphotungstic or phosphomolybdic acid. The latter compounds alter the affinity of various mesenchymal tissues for the dyes, causing differential staining (11). Collagen and elastin are stained blue and muscle is red with the trichrome procedure, whereas nuclei are black and nonmuscular cytoplasm is also red (Figure 1.31). Attention to fibrillation of the cytoplasm is necessary on high-power microscopy to distinguish myogenous from nonmuscular cells.

Elastic tissue is best seen with *Verhoeff's method* (and a modification known as the Verhoeff–Van Gieson

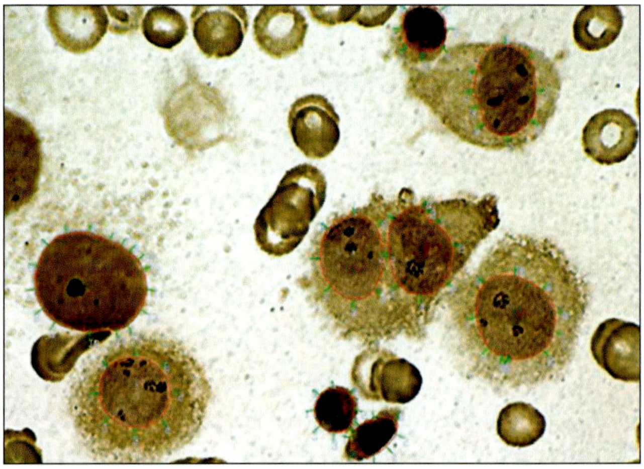

***FIGURE 1.30:* The Ag-NOR method impregnates nucleolar areas in nuclei that again reflect the presence of rapid cell replication.**

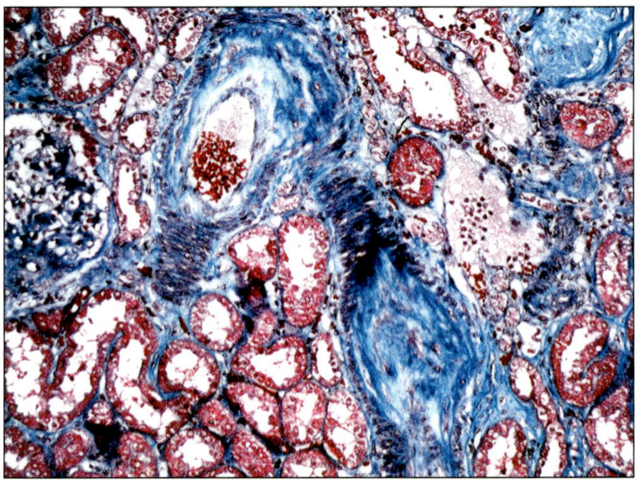

***FIGURE 1.31:* Masson's trichrome method can be used to differentially label muscle or epithelium (red), collagen or elastin (blue), and nuclei (dark blue or black), as shown in this section of renal cortex.**

method) or the *aldehyde fuchsin* elastic stain. In the first of these, a chemical "lake" of hematoxylin and ferric chloride is the principal staining reagent. Elastic tissue has a strong affinity for the latter complex and therefore resists decolorization better than other tissue components in this regressive staining technique. Elastic fibers are blue-black (Figure 1.32), collagen is red, nuclei are blue, and other tissues are yellow in the version of Verhoeff's method, whereas Van Gieson's reagent is used as a counterstain (25). This procedure is an excellent one for the demonstration of elastic tissue abnormalities, such as those seen in elastofibroma dorsi, large-vessel vasculitis, pseudoxanthoma elasticum, and elastosis of various tissues.

In the aldehyde fuchsin elastic stain (52), basic fuchsin is combined with hydrochloric acid and paraldehyde to yield aldehyde fuchsin. This reagent has the ability to form Schiff bases and shows a strong affinity for elastic fibers. The latter acquire a dark blue or purple color, and light green is usually used as a counterstain.

Nerve fibers may be visualized either through staining their neuronal processes or the myelin surrounding them. Silver methods such as the *Bodian* stain are utilized to accomplish the former task, whereas the *luxol fast blue* technique is employed to delineate myelin in tissue (Figure 1.33). The Bodian method is a complicated procedure using silver proteinate, gold chloride, hydroquinone, sodium thisulfate, hydrochloric acid, and aniline blue. It was designed as a regressive staining technique to selectively impregnate neuronal processes with elemental silver, yielding a black precipitate on these elements (53). Nuclei are black, and surrounding tissues are blue. In the luxol fast blue procedure, sulfonated copper phthalocyanine is used in an acid–base reaction to replace the bases of lipoproteins in myelin. As its name suggests, this dye has a blue color in the final staining product (27).

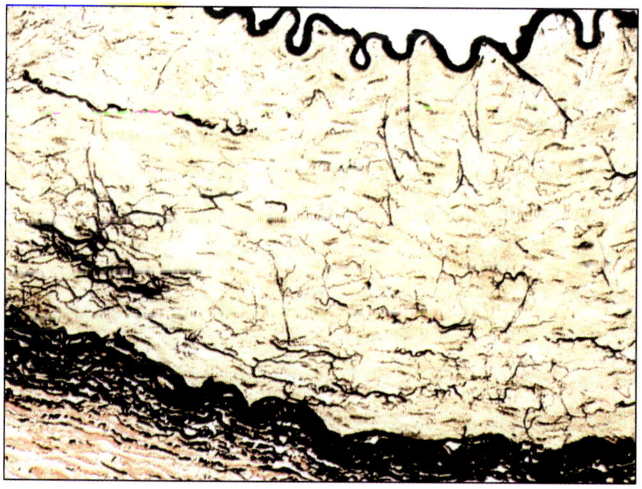

FIGURE 1.32: *The Verhoeff–Van Gieson technique selectively labels elastic tissue black; this is a section of the aorta.*

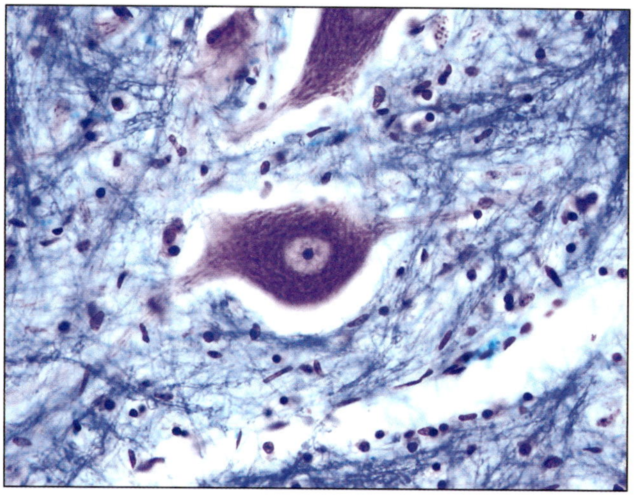

FIGURE 1.33: *Luxol fast blue is a method that preferentially decorates myelin in neural tissue. This is a section of spinal cord.*

Neuroendocrine Proliferations

Another cellular group of interest in diagnostic pathology is that called the "diffuse neuroendocrine system." Silver impregnation techniques also are employed to delineate neurosecretory granules in neuroendocrine cells and neoplasms. The *Grimelius* and *Churukian–Schenk* procedures are useful in this context (54,55). Both of them are predicated on the knowledge that neurosecretory granules have the ability to bind silver ions (i.e., they are *argyrophilic*); a reducing substance is then applied to tissue sections to yield a precipitate of black elemental silver (Figure 1.34). Argyrophilic techniques are influenced by fixation. Formaldehyde-based solutions produce inconsistent positivity, whereas this is one situation where Bouin's fixative gives

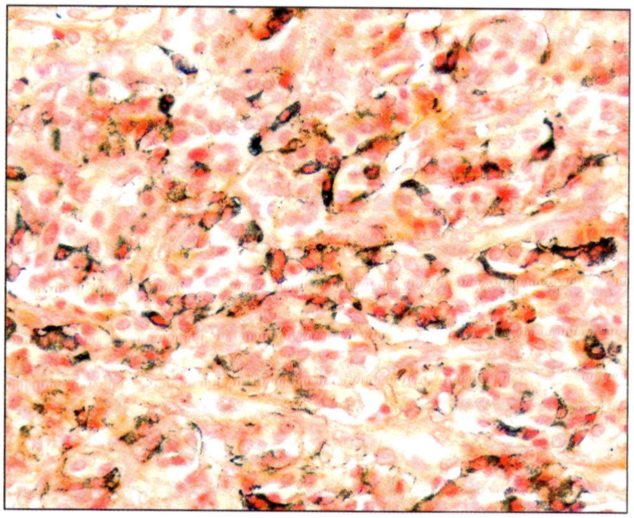

FIGURE 1.34: *Several argyrophilic procedures (Grimelius, Sevier-Munger, Churukian-Schenk) can be utilized to label neurosecretory granules black-brown, in neuroendocrine tumors. This is a section of grade 2 neuroendocrine carcinoma in the lung.*

superior results. Also, melanosomes are potentially reactive with the Grimelius and Churukian–Schenk methods, necessitating concomitant performance of the Fontana–Masson stain before interpreting an argyrophilic cellular proliferation as neuroendocrine in nature.

Microorganisms

The range of infectious agents that are evaluated in human diseases is broad. It includes conventional and atypical bacteria, spirochetes, mycobacteria, fungi, viruses, protozoan organisms, and nematodes (11). An exhaustive discussion of the histochemical methods that are used to identify these microbes is not within the scope of this chapter presentation; hence, only selected techniques in this group will be considered here. A summary of stains for commonly sought organisms is given in Table 1.3, and organ-specific infections are also considered throughout the remainder of this monograph.

Bacteria: DNA is common to all bacteria, which also have cell walls that contain carbohydrates and complex lipids. Indeed, the density of lipid in the cell wall and the particular composition of the fatty substances therein govern the success with which histochemical stains are able to label the organisms in question (56). In particular, mycobacterial cell walls contain mycolic acids of variable lengths, which likely govern the degrees of acid fastness seen in this group of microbes. The most common acid-fast method used in clinical practice is the Ziehl–Neelsen procedure (57,58), which employs basic fuchsin as the dye used to impregnate the mycobacteria. Fuchsin is

TABLE 1.3: Histochemical Methods for Identification of Microbes

Class of Organism	Stains	Biochemical "Targets"	Comments
"Conventional" bacteria	Brown-Hopps modification of Gram stain	Peptidoglycans and lipopolysaccharide in bacterial walls	Superior to Brown-Brenn method
Fungi	Hotchkiss-McManus version of PAS technique	Polysaccharides in fungal walls	Diastase digestion suggested
	Grocott's methenamine silver nitrate procedure	Polysaccharides in fungal walls	Also will stain *Nocardia* and *Actinomyces*
	Mucicarmine method	Mucinous material in capsules of cryptococci	Only used for *Cryptococcus neoformans*
	Biotinylated lectins (soybean and succinylated wheat germ)	Polysaccharides in fungal walls	Requires detection by avidin-biotin-peroxidase complex technique
Mycobacteria	Kinyoun's method	Complex lipids in bacterial walls	Does not label "atypical" organisms; will stain *Nocardia*
	Fite's technique	Complex lipids in bacterial walls	Preferred method for *Mycobacterium leprae* and other atypical organisms
	Auramine-rhodamine procedure	Unknown	Requires fluorescent microscopy
Spirochetes	Warthin-Starry technique	Argyrophilic proteins in bacterial walls	
	Dieterle method	Argyrophilic proteins in bacterial walls	
Nematodes	Grocott's methenamine silver nitrate technique	Polysaccharides in chitinous coat of organisms	Organisms usually seen well with H&E
	PAS-D method	Glycoproteins in chitinous coat of organisms	Organisms usually seen well with H&E
Rickettsiae	Giemsa or methylene blue technique	Unknown	Specific identification requires immunohistology
Leishmania	Giemsa or methylene blue technique	Unknown	Specific identification requires immunohistology
Viruses (*Herpes simplex*, cytomegalovirus, Orf, Varicella-zoster virus)	MGP Y method	Viral DNA or RNA	Used for confirming generic nature of viral inclusions; specific identification requires immunohistology

a rosanilin, a cationic moiety that binds to nuclear chromatin and other acidic groups. Depending on which dewaxing solvents are used and how long sections are exposed to alcohols during staining, several species of mycobacteria demonstrate different degrees of labeling with variations of the Ziehl-Neelsen procedure (such as the Fite method) (Figure 1.35) (59).

A fact that is often forgotten is that mycobacteria are commonly stained with the tissue Gram method and the Giemsa technique as well (60,61). Gram stain variations used in tissue sections—known as the Brown and Brenn or Brown and Hopps methods—are predicated on aqueous crystal violet as the principal dye, which is mordanted using iodine solution and then differentiated with alcohol or acetone. Gram-positive organisms retain their affinity for crystal violet after differentiation (Figure 1.36), whereas Gram-negative bacteria do not, and can then be counterstained with another dye such as neutral red or safranin. However, Gram-negative microbes are particularly difficult to visualize optimally in tissue sections. The Giemsa technique produces blue-violet labeling of bacteria.

In accord with their aforementioned content of various carbohydrates and DNA, other histochemical procedures can be used to recognize bacteria as well. The PAS technique, the Feulgen or MGP methods, and argyrophilic silver impregnation stains [e.g., the Warthin-Starry, Steiner, Dieterle, and Levaditi procedures, which recognize cell wall glycoproteins (62)] are all potentially effective in this regard. In fact, some uncommon bacterial pathogens—including *Helicobacter*, *Calymmatobacteria*, *Treponema*, and *Leptospira*—can *only* be recognized using these alternatives to Gram or acid-fast techniques. Among the mycobacteria, *Mycobacterium leprae* is particularly prone to demonstrate reactivity with the alcian blue or colloidal iron stains (Figure 1.37) because it uses hyaluronic acid for active growth.

Some filamentous nonmycobacterial Gram-positive bacteria—especially *Nocardia*—and *Rhodococci* (63) may be partially acid fast as well. Tissue colonies of filamentous organisms also can show a peculiar in vivo encrustation by host proteins, known as the Splendore-Hoeppli phenomenon (64), resulting in their relative unavailability to dyes in the Gram or Ziehl-Neelsen procedures (Figure 1.38).

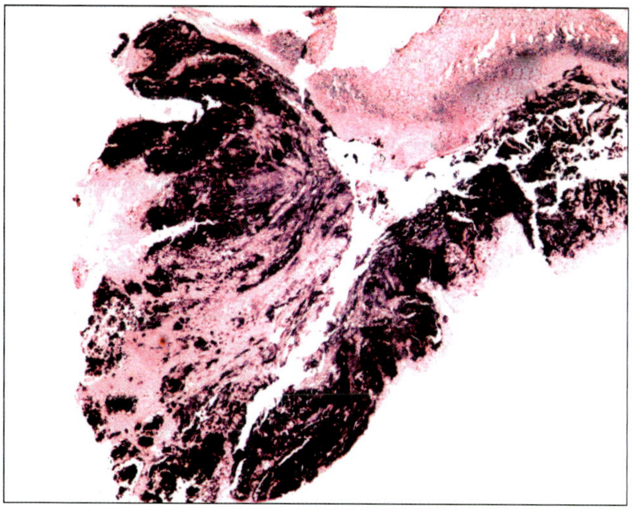

FIGURE 1.36: ***Tissue Gram stains—known as the Brown-Hopps and Brown-Brenn methods—can label many bacteria in microscopic sections. Gram-positive organisms are detected with greater sensitivity than gram-negative ones. This is a section of mitral valve in bacterial endocarditis, showing innumerable gram-positive cocci with the Brown-Hopps technique.***

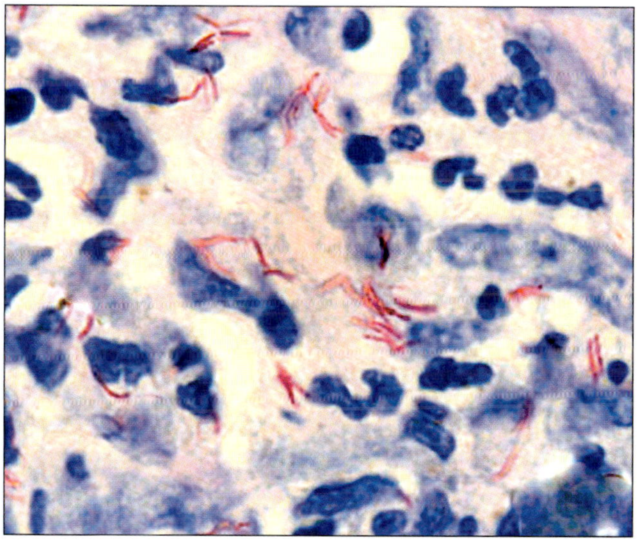

FIGURE 1.35: ***Acid-fast bacteria can be recognized with the Ziehl-Neelsen procedure (shown here), the Kinyoun technique, or the Fite procedure.***

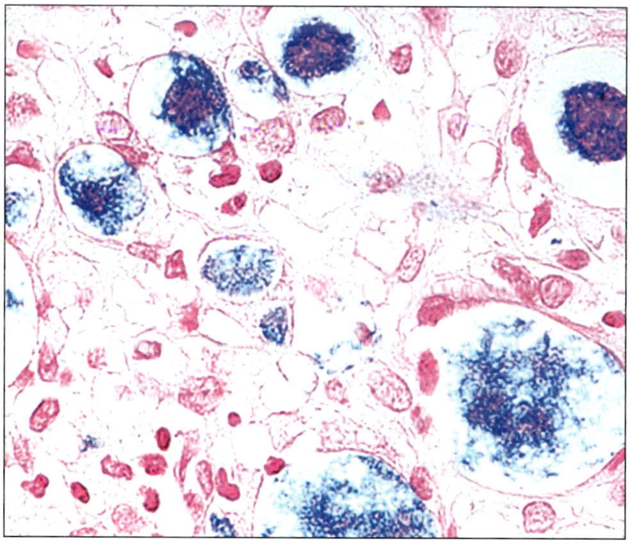

FIGURE 1.37: ***Some microorganisms can be visualized indirectly through the biochemical products associated with them. In this section from a patient with lepromatous leprosy, colonies of* Mycobacterium leprae *are recognizable with the colloidal iron method because they are surrounded by stromal mucin.***

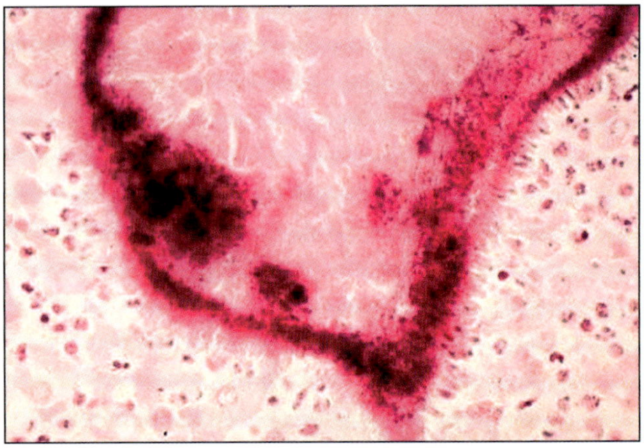

FIGURE 1.38: The Splendore-Hoeppli phenomenon is a peculiar host response to filamentous bacteria, in which the organisms are encrusted by protein. This process may interfere with the affinity of histochemical methods for the organisms. A section of uterine cervix is shown here, in a patient with actinomycosis.

Small atypical intracellular bacteria such as the rickettsiae and chlamydiae can be recognized by nucleic acid stains. They also show an affinity for the Machiavello stain, which uses basic fuchsin as its principal active dye (65). Cells that are infected with the organisms in question are stained red.

Fungi: Fungi, including *Pneumocystis carinii*, also contain abundant glycoproteins in their cell walls, and, as such, are consistently labeled by the PAS or silver-methenamine techniques (Figure 1.39). One may wish to perform both of those stains, however, because some fungal organisms are demonstrated better—but unpredictably—by one or the other of them.

Three particular fungal diseases have special histochemical properties. Chromoblastomycosis and phaeohyphomycosis are "deep" mycoses that are caused by dematiaceous (pigmented) fungi. The pigment in these organisms is closely related to melanin, and clustered fungi in tissue (Medlar bodies) or individual microbes can therefore be labeled with the Fontana-Masson or Schmorl procedures (66). In contrast, *Cryptococcus neoformans* contains abundant complex glycoproteins in its capsule; these are uniquely stained with Best's mucicarmine technique (Figure 1.40) and the colloidal iron or alcian blue methods (67).

Protozoan Parasites: Protozoan parasites, including *Entamoeba*, *Acanthamoeba*, *Giardia*, *Trichomonas*, *Plasmodia*, *Leishmania*, *Trypanosoma*, and *Toxoplasma*, often contain glycogen and can be labeled with the PAS method. In addition, the Giemsa or Feulgen stains are effective at visualizing such organisms (Figure 1.41).

Viruses: Depending on whether they are DNA or RNA based, the Feulgen or MGP stain is useful in accentuating inclusions of several viruses in tissue. Examples include *Herpes*, *Varicella*, *Cytomegalovirus*, *Molluscum*, *Adenovirus*, and *Rubella*. Moreover, some viral inclusions, for example, as in measles, can be labeled with the Machiavello stain, as discussed earlier.

Nematodes and Trematodes: Nematode and trematode parasites can typically be seen adequately in tissue with simple H&E stains. Deposits of their eggs are also

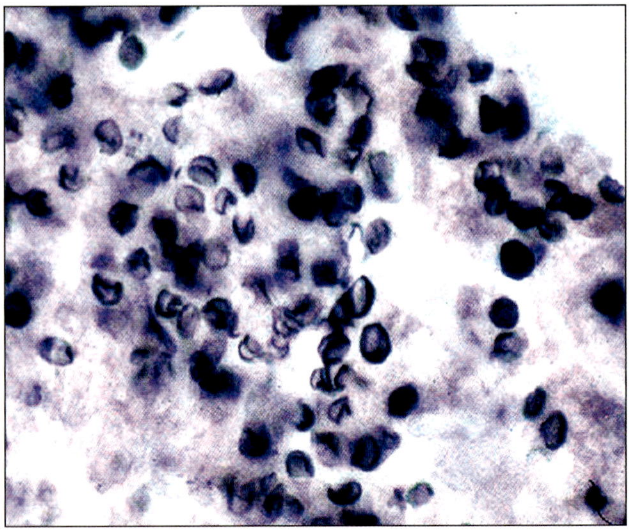

FIGURE 1.39: Silver impregnation methods such as the Gomori methenamine silver or Grocott procedure can label fungal organisms in tissue. A section of lung is shown here, from a patient with pneumocystosis.

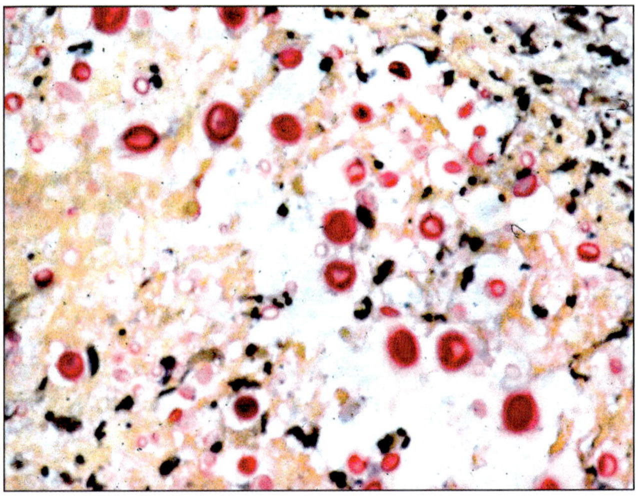

FIGURE 1.40: Selected microorganisms have distinctive histochemical profiles. In this section of lung from a patient with cryptococcosis, the pathogenic yeast-form fungi were stained by the Gomori methenamine silver and the mucicarmine procedures. The latter is shown here.

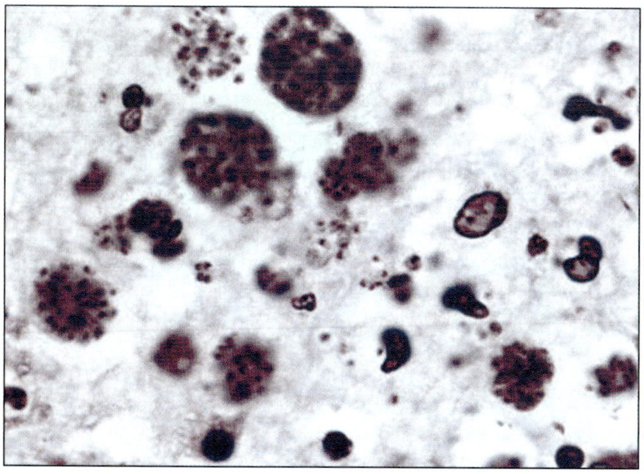

***FIGURE 1.41:** Protozoans in tissue are often best visualized with Romanowsky dyes. Here, a section of brain stained with the Giemsa procedure shows numerous* **Toxoplasma gondii** *organisms.*

similarly visible, but can, if desired, be shown to better advantage with the PAS stain.

Cytoplasmic Enzymes

A number of cytoplasmic enzymes have been assessed in many tumor types using histochemical procedures. These include phosphorylase, succinic dehydrogenase, leucine aminopeptidase, indoxyl esterase, cytochrome oxidase, acid and alkaline phosphatases, ATPase, beta-glucuronidase, acetate esterase, and monoamine oxidase, among others. The methodology for each of these techniques has been outlined in other sources (43,68) and will only be referenced here. Most of them are based on the ability of any given enzyme to hydrolyze or oxidize a colorless insoluble compound to yield a colored precipitate. Appropriate counterstains are then applied to yield the final histological product.

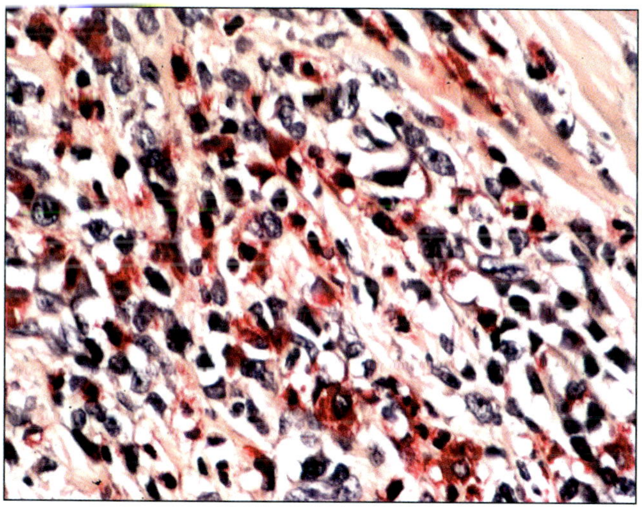

***FIGURE 1.42:** Some enzymes are relatively restricted among different cells and tissues, and are preserved in routinely processed specimens. Here, a stain for chloroacetate esterase (Leder's method) is helpful in characterizing a hematopoietic malignancy as granulocytic sarcoma (extramedullary myeloid tumor).*

Generally, enzyme histochemistry is used routinely today only in the evaluation of muscle biopsies for medical neuromuscular diseases. However, one stain in this category that was not cited above is the naphthol AS-D chloroacetate method, also known as the *Leder* stain. It is very useful in identifying immature granulocytes in suspected cases of extramedullary myeloid tumor or in recognizing tissue mast cells (Figure 1.42). The red product in this stain derives from the inclusion of Fast Garnet as the active dye (69,70). Contrary to metachromatic techniques such as the toluidine blue stain—which is also employed for the labeling of mast cells—the Leder procedure is not dependent on the integrity of mastocytic granules to produce a positive result.

REFERENCES

1. DeLellis RA, Resnick M, Frable WJ: General and special techniques in surgical pathology and cytopathology. In: Silverberg's Principles & Practice of Surgical Pathology and Cytopathology (4th ed., Silverberg SS, Ed.), Churchill-Livingstone, New York, 2006; pp. 15–54.
2. Banks PM: Technical aspects of specimen preparation and introduction to special studies. In: Surgical Pathology of the Lymph Nodes & Related Organs (2nd ed., Jaffe ES, Ed.), W.B. Saunders, Philadelphia, 1995; pp. 1–21.
3. Zarbo RJ, Hoffman GG, Howanitz PJ: Interinstitutional comparison of frozen section consultation: a College of American Pathologists Q-probe study of 79,647 consultations in 297 North American institutions. *Arch Pathol Lab Med* 1991; 115: 1187–1194.
4. Mair S, Lash RH, Suskin D, Mendelsohn G: Intraoperative surgical specimen evaluation: frozen section analysis, cytologic examination, or both? A comparative study of 206 cases. *Am J Clin Pathol* 1991; 96: 8–14.
5. Huang WM, Gibson SJ, Facer P: Improved section adherence for immunocytochemistry using high molecular weight polymers of L-lysine as a slide coating. *Histochemistry* 1983; 77: 275–279.
6. Hopwood D: Fixatives and fixation: a review. *Histochem J* 1969; 1: 323–360.
7. Hopwood D: Cell and tissue fixation. *Histochem J* 1985; 17: 389–442.
8. Leatham A, Atkins NJ: Fixation and immunohistochemistry of lymphoid tissue. *J Clin Pathol* 1980; 33: 1010–1012.
9. Hopwood D, Yeaman G, Milne G: Differentiating the effects of microwave and heat on tissue proteins and their crosslinking by formaldehyde. *Histochem J* 1988; 20: 341–347.
10. Hixson DC, Yep JM, Glenney JR Jr., Hayes T, Walborg EF: Evaluation of periodate/lysine/paraformaldehyde fixation as a method for cross-linking plasma membrane glycoproteins. *J Histochem Cytochem* 1981; 29: 561–566.
11. Carson FL: Histotechnology: A Self-instructional Text (2nd ed.), ASCP Press, Chicago, 1997.

12. Reisner HM, Wick MR: Theoretical and technical considerations for the use of monoclonal antibodies in diagnostic immunohistochemistry. In: Monoclonal Antibodies in Diagnostic Immunohistochemistry (Wick MR, Siegal GP, Eds.), Marcel Dekker, New York, 1988; pp. 1–50.
13. McLean IW, Nakane PK: Periodate-lysine-paraformaldehyde fixative: a new fixative for immunoelectron microscopy. *J Histochem Cytochem* 1974; 22: 1077–1083.
14. NIOSH 85–114: NIOSH Pocket Guide to Chemical Hazards, U.S. Department of Health and Human Services, Washington, D.C., 1985.
15. Puchtler H, Meloan SN: On the chemistry of formaldehyde fixation and its effects on immunohistochemical reactions. *Histochemistry* 1985; 82: 201–204.
16. Morales AR, Nassiri M, Kanhoush R, Vincek V, Nadji M: Experience with an automated microwave-assisted rapid tissue processing method: validation of histologic quality and impact on the timeliness of diagnostic surgical pathology. *Am J Clin Pathol* 2004; 121: 528–536.
17. Luna LG: Specimen preparation. In: Pathology of the Skin (Farmer ER, Hood AF, Eds.), Appleton & Lange, Norwalk, CT, 1990; pp. 30–41.
18. Luna LG: Amorphous bevel surface and moisture: important factors in microtomy. *Microviews (Surgipath Medical Industries Newsletter)* 1988; 4: 5–9.
19. Molner LM: Eliminate wrinkles on paraffin section slides. *Histology* 1975; 5: 59–62.
20. Thompson SW, Luna LG: An Atlas of Artifacts, CC Thomas, Springfield, IL, 1978.
21. Bancroft JD, Stevens A: Theory and Practice of Histological Techniques, Churchill-Livingstone, New York, 1982.
22. McGee-Russell SM: Histochemical methods for calcium. *J Histochem Cytochem* 1958; 6: 22–42.
23. Johnson WC: Histochemistry of the cutaneous adnexa and selected adnexal neoplasms. *J Cutan Pathol* 1984; 11: 352–356.
24. John WC, Helwig EB: Histochemistry of the cutaneous interfibrillar ground substance. *J Invest Dermatol* 1964; 42: 81–85.
25. Luna LG: Manual of Histologic Staining Methods (3rd ed.), Armed Forces Institute of Pathology, Washington, D.C., 1968.
26. Vacca LL: Laboratory Manual of Histochemistry, Raven Press, New York, 1985.
27. Sheehan DC, Hrapchak BB: Theory and Practice of Histotechnology (2nd ed.), Battelle Press, Columbus, OH, 1980.
28. University of Bristol Histological Staining Techniques (http://www.bris.ac.uk/vetpath/cpl/histmeth.htm#Top) accessed 2/27/2007.
29. Protocol Online (http://www.protocol-online.org/prot/Histology/Staining/) Accessed 2/27/2007.
30. Steward PJ: Histochemical methods for routine diagnostic histopathology. In: Histochemistry in Pathology (Filipe MI, Lake BD, Eds.), Churchill-Livingstone, New York, 1990; pp. 13–30.
31. Johnson WC, Helwig ED: Histochemistry of primary and metastatic mucin-secreting tumors. *Ann NY Acad Sci* 1963; 106: 794–803.
32. Wood MG, Beerman H: Necrobiosis lipoidica, granuloma annulare, and rheumatoid nodule. *J Invest Dermatol* 1960; 34: 139–147.
33. Panet-Raymond G, Johnson WC: Lupus erythematosus and polymorphous light eruption: differentiation by histochemical procedures. *Arch Dermatol* 1973; 108: 785–787.
34. Alroy J, Ucci AA, Pereira MEA: Lectin histochemistry: an update. In: Advances in Immunohistochemistry (DeLellis RA, Ed.), Raven Press, New York, 1988; pp. 93–131.
35. McNeal JE, Alroy J, Villers A, Redwine EA, Freiha FS, Stamey TA: Mucinous differentiation in prostatic adenocarcinoma. *Hum Pathol* 1991; 22: 979–988.
36. Ordonez NG, Batsakis JG: Comparison of *Ulex europaeus* I lectin and factor VIII-related antigen in vascular lesions. *Arch Pathol Lab Med* 1984; 108: 129–132.
37. Wick MR, Manivel JC: Vascular neoplasms of the skin: a current perspective. *Adv Dermatol* 1989; 4: 185–254.
38. Ulrich W, Horvat R, Krisch K: Lectin histochemistry of kidney tumours and its pathomorphological relevance. *Histopathology* 1985; 9: 1037–1050.
39. Cattoretti G, Berti E, Schiro R, et al.: Improved avidin-biotin-peroxidase complex (ABC) staining. *Histochem J* 1988; 20: 75–80.
40. Robbins SL, Cotran RS, Kumar V (Eds.): Pathologic Basis of Disease (3rd ed.), W.B. Saunders, Philadelphia, 1984; pp. 195–205.
41. Wright JR, Calkins E, Humphrey RL: Potassium permanganate reaction in amyloidosis: a histologic method to assist in differentiating forms of this disease. *Lab Invest* 1977; 36: 274–281.
42. Janssen S, Elema D, van Rijswijk MH, et al.: Classification of amyloidosis: immunohistochemistry vs. the potassium permanganate method in differentiating AA from AL amyloidosis. *Appl Pathol* 1985; 3: 29–38.
43. Sethi S, Khodaskar MB: Diagnostic potential of histochemical demonstration for copper-orcein stain. *J Indian Med Assoc* 2000; 98: 434–438.
44. Shousha S, Boxer GM: Orcein as a mucin stain for gastrointestinal tissue sections. *J Clin Pathol* 1981; 34: 1063–1064.
45. Filipe MI, Lake BD: Appendices. In: Histochemistry in Pathology (Filipe MI, Lake BD, Eds.), Churchill-Livingstone, New York, 1990; pp. 433–488.
46. Kamino H, Tam ST: Immunoperoxidase technique modified by counterstain with azure B as a diagnostic aid in evaluating heavily pigmented melanocytic neoplasms. *J Cutan Pathol* 1991; 18: 436–439.
47. Kligora CJ, Fair KP, Clem MS, Patterson JW: A comparison of melanin bleaching and azure blue counterstaining in the immunohistochemical diagnosis of malignant melanoma. *Mod Pathol* 1999; 12: 1143–1147.
48. Potvin C: A simple, modified methyl green–pyronine Y stain for DNA and RNA in formalin-fixed tissues. *Lab Med* 1979; 10: 772–776.
49. Egan MJ, Crocker J: Nucleolar organizing regions in cutaneous tumours. *J Pathol* 1988; 154: 247–254.
50. Crocker J, Skilbeck N: Nucleolar organizing region-associated proteins in cutaneous melanotic lesions: a quantitative study. *J Clin Pathol* 1987; 40: 885–889.

51. Smith R, Crocker J: Evaluation of nucleolar organizing region-associated proteins in breast malignancy. *Histopathology* 1988; 12: 113–123.
52. Gomori G: Aldehyde fuchsin: a new stain for elastic tissue. *Am J Clin Pathol* 1950; 20: 665–672.
53. Holmes W: Silver staining of nerve axons in paraffin sections. *Anat Rec* 1943; 86: 157–169.
54. Lack EE, Mercer L: A modified Grimelius argyrophil technique for neurosecretory granules. *Am J Surg Pathol* 1977; 1: 275–280.
55. Churukian CJ, Schenk EA: A modification of Pascual's argyrophil method. *J Histochem Cytochem* 1979; 2: 102–111.
56. Lyon H: Theory & Strategy in Histochemistry, Springer-Verlag, Berlin, 1991.
57. Ziehl F: Zur Farbung des Tuberkelbacillus. *Dtsch Med Wochenschr* 1882; 8: 451–462.
58. Neelsen F: Ein Casuistischer Beitrag zur Lehre von der Tuberkulose. *Zentralbl Med Wiss* 1883; 21: 497–501.
59. Nair SP, Moorthy KP, Suprakasan S, Jayapalan S, Mini G: Histoid leprosy: unusual presentation. *Int J Dermatol* 2006; 45: 433–434.
60. Giemsa G: Uber die Farbung von Feuchtpraparaten mit meiner Azur-Eosin methode. *Dtsch Med Wochenschr* 1909; 35: 1751–1756.
61. Gram C: Uber die isolirte Farbung der Schizomyceten in Schnitt und Trockenpraparaten. *Fortschr Med* 1884; 2: 185–189.
62. Levaditi C: Sur la coloration du sprochaete pallida Schaudinn dans les coupes. *Compt Rend Soc Biol* 1905; 59: 326–335.
63. Echeverri C, Matherne J, Jorgensen JH, Fowler LJ: Fite stain positivity in *Rhodococcus equi*: yet another acid-fast organism in respiratory cytology—a case report. *Diagn Cytopathol* 2001; 24: 244–246.
64. Rodig SJ, Dorfman DM: Splendore-Hoeppli phenomenon. *Arch Pathol Lab Med* 2001; 125: 1515–1516.
65. Weiss E, Pietryk HC: Growth of *Coxiella burnettii* in monolayer cultures of chick embryo entodermal cells. *J Bacteriol* 1956; 72: 235–241.
66. Gomez BL, Nosanchuk JD: Melanin and fungi. *Curr Opin Infect Dis* 2003; 16: 91–96.
67. Lazcano O, Speights VO Jr., Strickler JG, et al.: Combined histochemical stains in the differential diagnosis of *Cryptococcus neoformans*. *Mod Pathol* 1993; 6: 80–84.
68. Wick MR: Appendix I. In: Pathology of Unusual Malignant Cutaneous Tumors (Wick MR, Ed.), Marcel Dekker, New York, 1985; pp. 399–404.
69. Nelson DA: Leukocytic disorders. In: Clinical Diagnosis and Management by Laboratory Methods (16th ed., Henry JB, Ed.), W.B. Saunders, Philadelphia, 1979; pp. 1036—1100.
70. Leder LD: The chloroacetate esterase reaction: a useful means of histological diagnosis of hematological disorders from paraffin sections of skin. *Am J Dermatopathol* 1979; 1: 39–42.

CHAPTER 2

Diagnostic Histochemistry in Diseases of the Head and Neck

Bruce M. Wenig, M.D.

INFECTIOUS DISEASES OF THE HEAD AND NECK

Undoubtedly, the identification of an infectious agent is the single most common reason for utilizing histochemical stains in diseases of the head and neck. The following discussion is based on categories of pathogens affecting this anatomic site.

Fungal Disorders

Sinonasal Aspergillosis: Sinonasal diseases associated with *Aspergillus* sp. are divided into three distinct entities—allergic fungal sinusitis (AFS), aspergillus mycetoma (fungus ball), and invasive fungal sinusitis. *Aspergillus* organisms are characterized by relatively thin (2–5 μm) hyphae with acute (45°) branches and septation. They are members of the Moniliaceae family, Hyphomycetes class, Deuteromycota phylum. *Aspergillus* is abundant in soil and in decaying matter; the mode of transmission is via inhalation. Sinonasal mycosis was originally thought to be associated exclusively with *Aspergillus* infection (1), but other fungi may be causative as well. They include dematiaceous organisms such as *Bipolaris*, *Exserohilum*, *Curvularia*, *Drechslera*, and *Alternaria* (2–7).

Allergic Fungal Sinusitis: AFS represents the noninvasive collection of impacted mucus and cellular debris (8–13). It is similar to allergic bronchopulmonary aspergillosis in that pathologic findings result from an allergic response to topical fungal colonization of a sinus cavity (1). AFS occurs in both immunocompetent and immunosuppressed patients, as well as children. The density of fungal elements is variable and often is scanty (14). Nevertheless, even a few organisms may elicit an allergic response (1). The identification of so-called allergic mucin, characterized by amorphous chondroid-like material and a prominent inflammatory cell infiltrate, is considered diagnostic for AFS even in the absence of identifiable microorganisms (14). Fungal forms are notoriously difficult to visualize by light microscopy but can be seen by Gomori methenamine silver (GMS) staining (Figure 2.1) or with the periodic acid–Schiff (PAS) method. Correlation should be made with results of microbiological cultures (15).

Mycetoma: In contrast to AFS, mycetoma (fungus ball) represents the indolent mass-forming growth of fungal hyphae in a sinus cavity (16). Histologically, the hyphae are so densely packed that they are distorted and may not be recognized immediately. GMS and PAS methods delineate them satisfactorily (Figure 2.2).

Invasive Sinonasal Aspergillosis: Invasive sinonasal aspergillosis is an acute fungal infestation of the sinonasal tract that often results in destruction of the involved sinuses in days (17–20). Unlike AFS and mycetomas, invasive fungal sinusitis features infiltration of the organisms into the mucosa, submucosa, and blood vessels (angioinvasion). Tissue necrosis is evident but an inflammatory response often is limited. GMS and PAS stains are positive (Figure 2.3).

Sinonasal Mucormycosis: Sinonasal mucormycosis is a rapidly evolving infection caused by the Zygomycetes class of fungi (8). The human pathogens in that group include the orders Mucorales and Entomophthorales. Three genera, designated *Rhizopus*, *Mucor*, *and Absidia*, are included in the order Mucorales, family Mucoraceae. Infections with those agents are also labeled for the higher level class grouping of Zygomycetes, formerly Phycomycetes. Some species in Entomophthorales can cause mild infections, but this is not the case in the usual form of mucormycosis. *Mucorales* sp. have relatively broad (10–20 μm) nonseptate hyphae that branch at haphazard angles. The mode of transmission is thought to be via inhalation. The usual clinical scenario includes patients with diabetes mellitus, particularly those with ketoacidosis, or immunocompromised individuals (21–23). *Rhizopus* is the most common isolate. The diagnostic finding in mucormycosis is the presence of broad, glassy fungal hyphae, usually in necrotic tissue; they are usually identifiable with hematoxylin and eosin (H&E) stains (24) and are angioinvasive as well as neutrotropic (24,25). Luminal involvement of blood vessels often causes infarctive (gritty) necrosis with a geographic pattern. Similar histologic findings are seen in invasive aspergillosis, Wegener's granulomatosis (WG), and nasal-type NK/T-cell non-Hodgkin lymphomas.

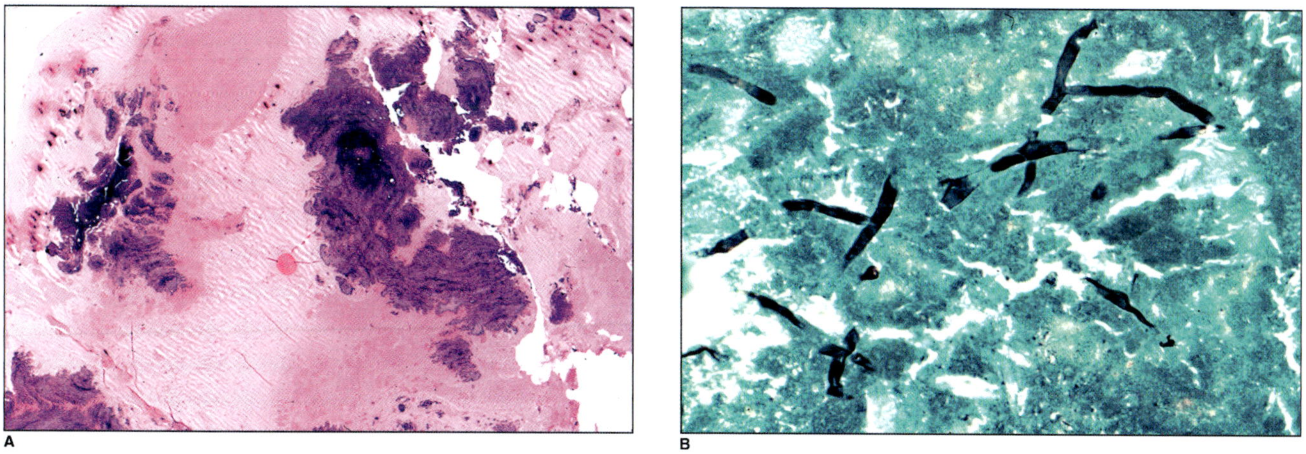

FIGURE 2.1: Allergic fungal sinusitis. (A) Inspissated mucus composed of amorphous collection of eosinophilic material associated with nests of inflammatory cells (H&E); (B) GMS stain assists in the identification of the scattered fungal forms showing septated hyphae and acute angle branching.

GMS and PAS stains are helpful in recognizing mucormycosis (Figure 2.4); however, one must take care not to confuse degenerate forms of *Aspergillus*, in which hyphal septations tend to disappear, with Zygomycetes.

Rhinosporidiosis: Rhinosporidiosis is a chronic infectious disease of the nasal cavity and nasopharynx, with the formation of polypoid masses. It is caused by a sporulating organism, *Rhinosporidium seeberi*, which is thought to be a zoonotic agent. Rhinosporidiosis is also seen in cattle, horses, and mules (8,26–29). The mode of transmission is thought to be respiratory, via water or dust. The endospores penetrate the nasal mucosa, mature into sporangia in the submucosa, and then burst into the surrounding tissue. The most common sites of involvement are the inferior turbinate along the lateral nasal wall and the nasopharynx. Less commonly, infection involves the larynx, tracheobronchial tree, esophagus, oropharynx, palpebral conjunctiva, and ears. *R. seeberi* is a fungus that does not grow on synthetic media, although it has been propagated in cell culture. Mucosal and submucosal sporangia range in size from 10 to 300 μm in diameter, and contain innumerable sporangiospores (endospores) in H&E sections. Smaller cysts are also seen. The sporangia and trophocysts have thick eosinophilic walls. The rupture of cysts induces an acute inflammatory response but granulomas are lacking. These organisms stain with the PAS, GMS, and mucicarmine methods (Figure 2.5). Given the polypoid configuration of rhinosporidiosis, diagnostic confusion may occur with the cylindrical cell

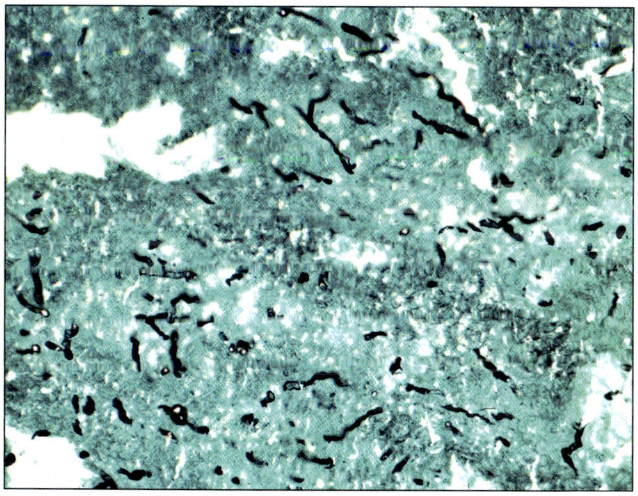

FIGURE 2.2: Mycetoma (aspergilloma or fungus ball). Amorphous debris within which a GMS stain delineates the fungal spores and hyphae.

FIGURE 2.3: Invasive fungal sinusitis. GMS stain showing fungal (aspergillus) forms in sinonasal mucosa; residual necrotic respiratory epithelium is seen at the left of the image.

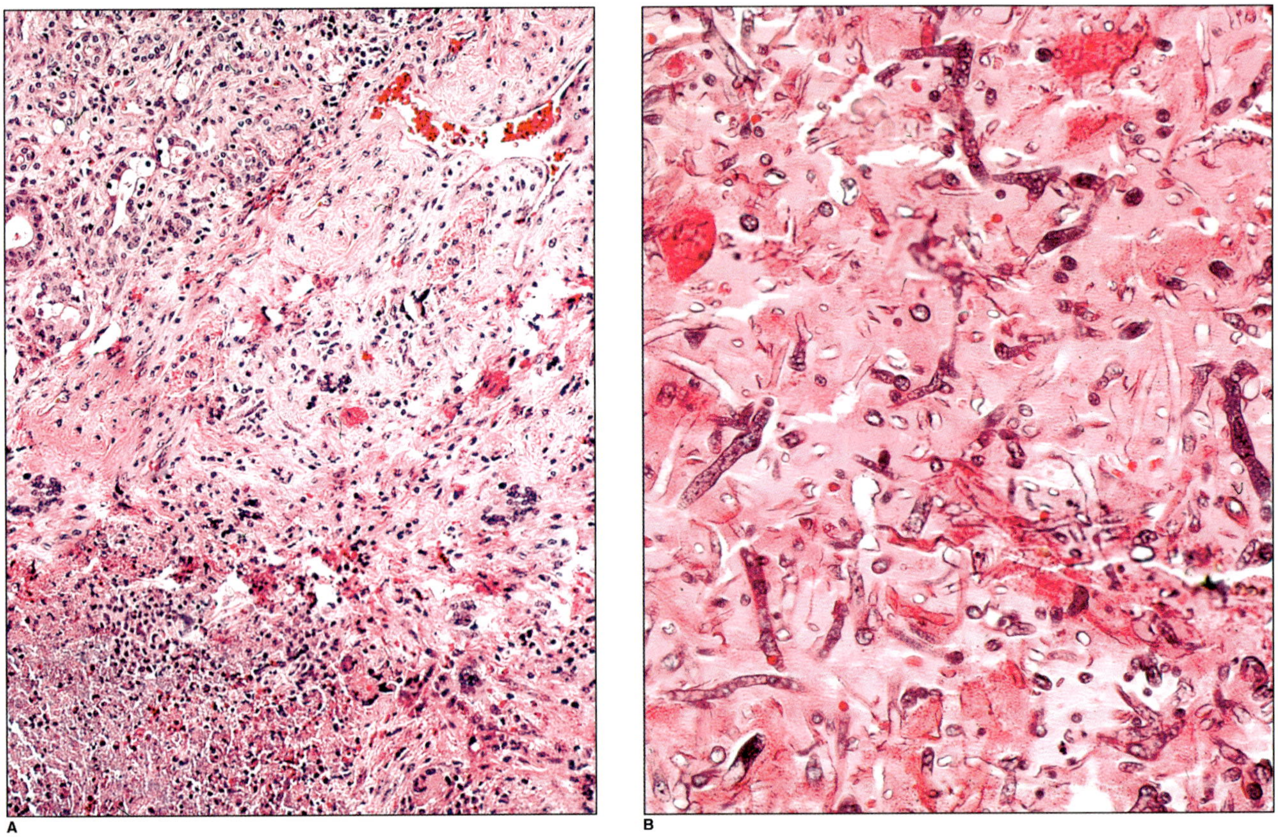

FIGURE 2.4: *Sinonasal mucormycosis. (A) At low magnification, the appearance is that of "ischemic-type" geographic necrosis (lower left) with effacement of the normal mucosal architecture although residual mucoserous glands are present; (B) PAS stain showing the presence of fungi (*Mucorales sp.*) characterized by hyphae that are broad in diameter and nonseptated with branching at haphazard angles.*

variant of Schneiderian papilloma. However, the latter tumor manifests only intraepithelial cysts, whereas those seen in rhinosporidiosis are both intraepithelial and submucosal.

Candidiasis: Candidiasis is one of the most common fungal infections in humans, especially in the mouth (30–33). Its clinical forms include pseudomembranous candidiasis (also known as "thrush"), erythematous candidiasis, mucocutaneous candidiasis, and chronic hyperplastic candidiasis (candidal leukoplakia) (31). Oral candidiasis typically affects immunocompromised or debilitated patients, but occasionally may be seen in otherwise healthy individuals. Pseudomembranous candidiasis is characterized by adherent white mucosal plaques (130) Chronic erythematous candidiasis has also been called *median rhomboid glossitis* or "central papillary lingual atrophy" (34–37).

There are several members of the *Candida* genus, but the most common cause of upper aerodigestive disease is *Candida albicans*. This organism is a dimorphic fungus with both yeast and hyphal forms; the latter are responsible for tissue invasion. Hyphae are variable in length and branched; the PAS and GMS methods are effective in delineating them (Figure 2.6). In oral candidiasis, the microorganisms typically are found in a parakeratotic layer of the squamous epithelium and in its superficial cellular strata. One usually sees an associated neutrophilic infiltrate, and when acute inflammation is seen in a parakeratotic stratum corneum, fungal infection should always be considered. Additional features of mucosal candidiasis include hyperkeratosis, irregular epithelial hyperplasia, and a nonspecific submucosal chronic inflammatory infiltrate. Fungi are occasionally present in deeper epithelial layers as well, particularly in immunocompromised patients.

Other Fungal Diseases: Other rarer fungal infections of the sinonasal tract and mouth include sporotrichosis, blastomycosis, coccidioidomycosis, cryptococcosis, and histoplasmosis (8,38,39). In these conditions, epithelial pseudoepitheliomatous hyperplasia (PEH) is relatively common and may suggest the presence of a squamous neoplasm. GMS and PAS stains are important in avoiding this error; cryptococci are also reactive with the mucicarmine procedure.

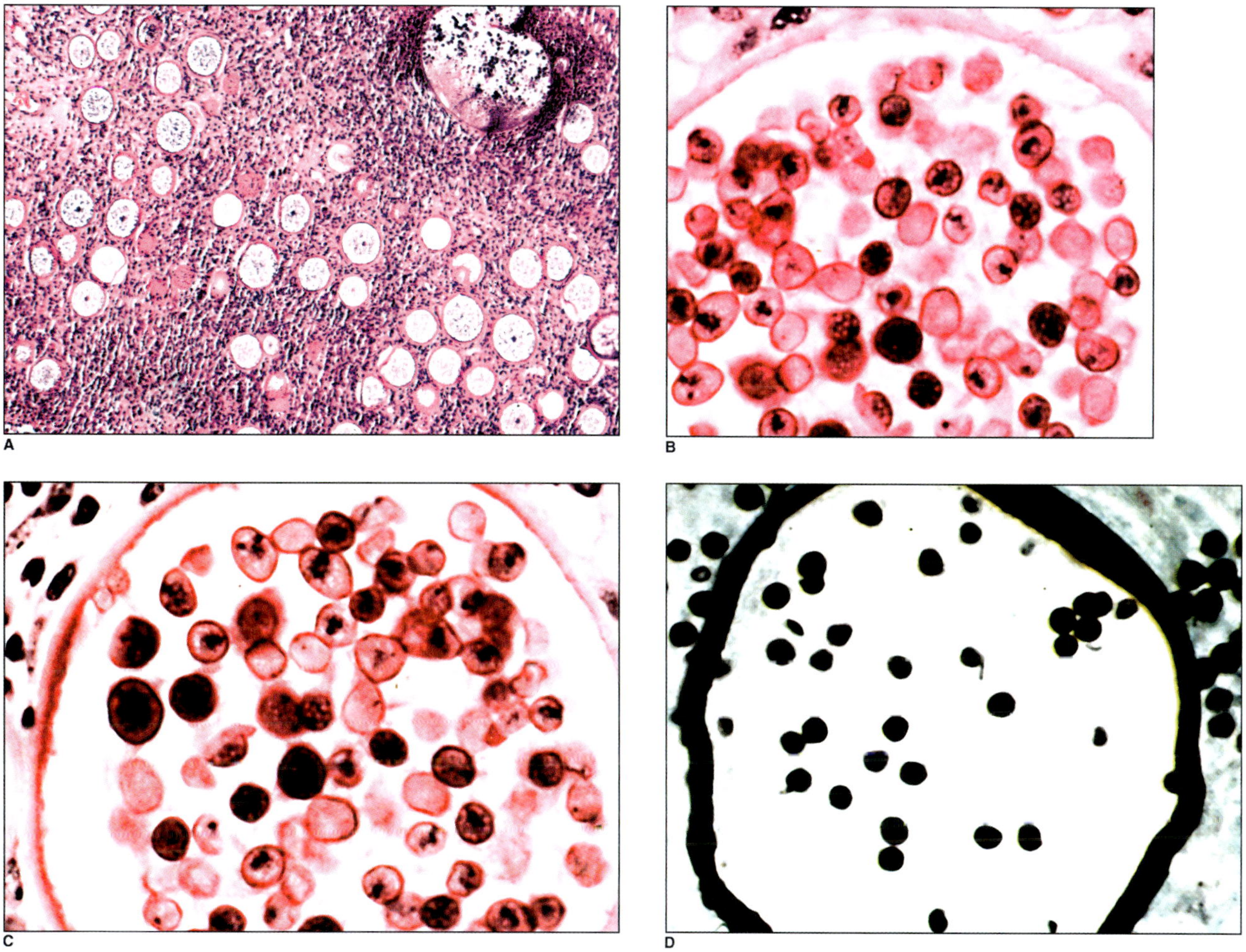

FIGURE 2.5: Rhinosporidiosis. (A) Sinonasal rhinosporidiosis characterized by the presence of multiple submucosal cysts (sporangia); the sporangia contain endospores, while the microorganisms (Rhinosporidium seeberi) are readily identifiable by H&E staining, additional stains may be utilized to detail these fungi, including (B) PAS, (C) mucicarmine, and (D) GMS.

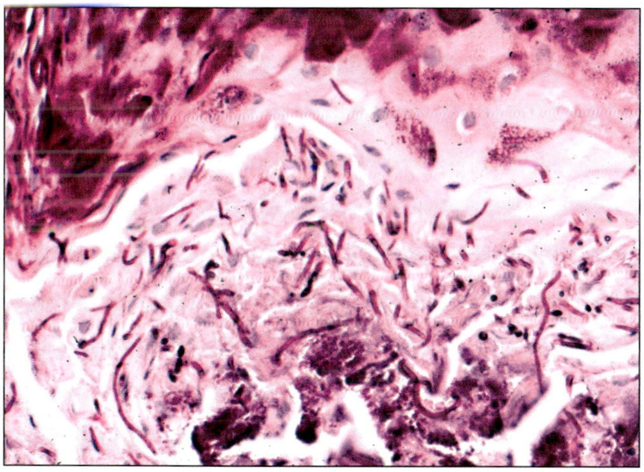

FIGURE 2.6: Oral candidiasis (thrush). PAS stain showing the Candida budding yeasts and pseudohyphae within the squamous epithelium.

Bacterial Disorders

Bacterial diseases of the head and neck are potentially caused by group A β-hemolytic streptococci, *Hemophilus influenzae*, *Stapyloccocus aureus*, *Corynebacterium diphtheriae*, *Bordetella pertussis*, *Streptococcus pneumoniae*, and α-hemolytic streptococci. Necrotizing external otitis is another possibility, caused by *Pseudomonas aeruginosa*. Gram-negative bacteria are not easily demonstrated with tissue Gram stains, in contrast to most gram positive organisms. In general, however, all these diseases are recognized clinically and principally identified using microbiologic cultures. Hence, one rarely receives surgical pathology specimens in such cases, with selected exceptions as considered below.

Rhinoscleroma: Rhinoscleroma is a chronic granulomatous disease of the nasal cavity and nasopharynx. It is caused by *Klebsiella rhinoscleromatis*, a gram-negative

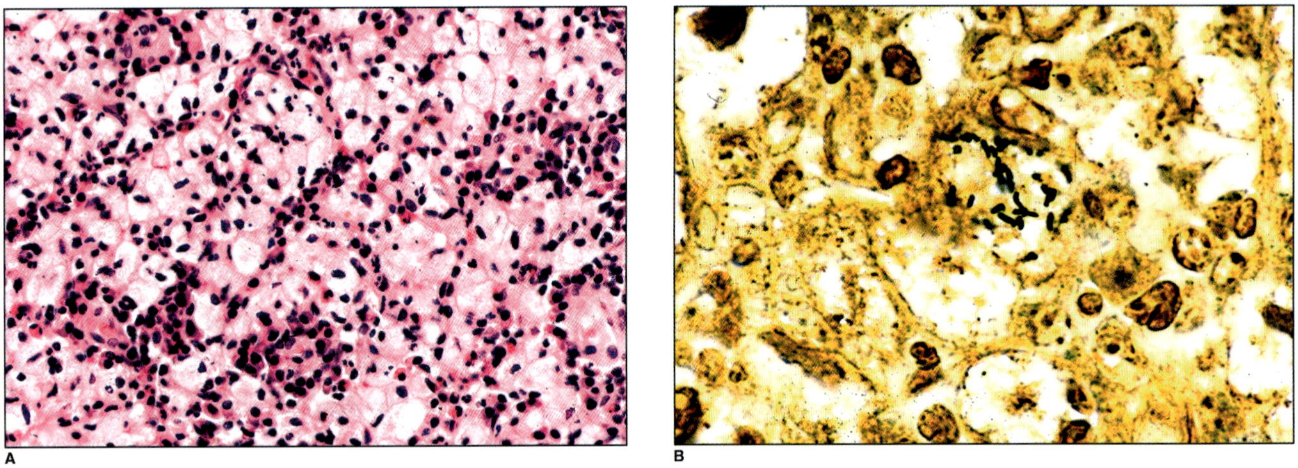

FIGURE 2.7: Sinonasal rhinoscleroma. (A) Submucosal infiltrate showing characteristic macrophages with clear to foamy cytoplasm (so-called Mikulicz cells); (B) Warthin-Starry silver stain demonstrates the presence of the Klebsiella rhinoscleromatis, a gram-negative bacteria, within the macrophages.

diplobacillus in the family *Enterobacteriaceae* (28). Rhinoscleroma is endemic in Egypt, Central and South America, North and Central Africa, and Eastern Europe, but it is very uncommon in the United States. This infection may be the cause of upper airway obstruction (40,41), and also can affect the paranasal sinuses, orbits, larynx, tracheobronchial tree, and middle ear (28,42). Histologically, rhinoscleroma is typified by a submucosal granulomatous infiltrate with clear to foamy histiocytes (Mikulicz cells). An admixture of lymphocytes and plasma cells is also seen. The Mikulicz cells are "parasitized" by the causative bacteria, which are best visualized with the Warthin-Starry or Steiner procedures (Figure 2.7); the Giemsa stain may also label the organisms, but with lesser sensitivity.

Gonorrhea: Gonorrhea is caused by *Neisseria gonorrheae*, a pyogenic gram-negative diplococcus. Patients with acquired immunodeficiency syndrome (AIDS) have an increased incidence of orolaryngeal gonorrhea, as well as syphilis. Upper airway manifestations include pharyngitis, tonsillar enlargement, and cervical lymphadenopathy. Gram-stained smears from the pharynx are unreliable diagnostically because of the interfering effect of other organisms on interpretation. Diagnosis is predicated on the results of formal cultures.

Syphilis: The pathogenic organism in syphilis (lues) is *Treponema pallidum*, a member of the family *Spirochaetaceae*. Related agents include *T. pertenue* (causing yaws) and *Treponema carateum* (the etiologic bacterium in pinta). Syphilis of the head and neck may produce pharyngotonsillitis or a painless solitary chancre, which may simulate neoplasia (43). The lips, tongue, palate, gingiva, and tonsils may all be affected. Luetic mucositis features a relatively nonspecific inflammatory infiltrate, composed of plasma cells, scattered histiocytes, lymphocytes, and polymorphonuclear leukocytes. The inflammation has a tendency to involve small blood vessels, with endovasculitis. The microorganisms can be demonstrated with the Warthin-Starry or Steiner method, as elongated, thin, rod-like structures with a "corkscrew" curvilinear configuration (Figure 2.8).

Bacillary Angiomatosis: Bacillary angiomatosis (BA) is a pseudoneoplastic vasoproliferative lesion that usually represents a complication of human immunodeficiency virus (HIV) infection. It can involve the skin, mucosae, or viscera, and is caused by an opportunistic bacterial infection with *Rochalimaea* species (especially *Rochalimaea henselae*) or *Bartonella quintana* (44–47). BA also may produce systemic symptoms, including fever, chills, weight loss, and night sweats (48,49). Clinically, lesions of BA are similar to those of lobular capillary hemangioma or Kaposi's sarcoma (KS) (50–53). Indeed, KS may coexist as an independent condition (54,55).

The histologic features of BA include a well-circumscribed lobular capillary proliferation (48,56). Constituent vessels have prominent endothelium, in which granular bluish aggregates may be seen. Cytologic atypia, mitotic activity, and necrosis are usually absent. Another important histologic feature is the presence of intralesional neutrophilia, especially associated with the granular cytoplasmic inclusions in endothelial cells. Warthin-Starry or Steiner stain shows that the granular aggregates are groups of bacterial organisms (Figure 2.9).

Cervicofacial Actinomycosis: Cervicofacial actinomycosis is a chronic granulomatous and suppurative infection caused by gram-positive, microaerophilic, and anaerobic bacteria. The most common isolate in man is *Actinomyces*

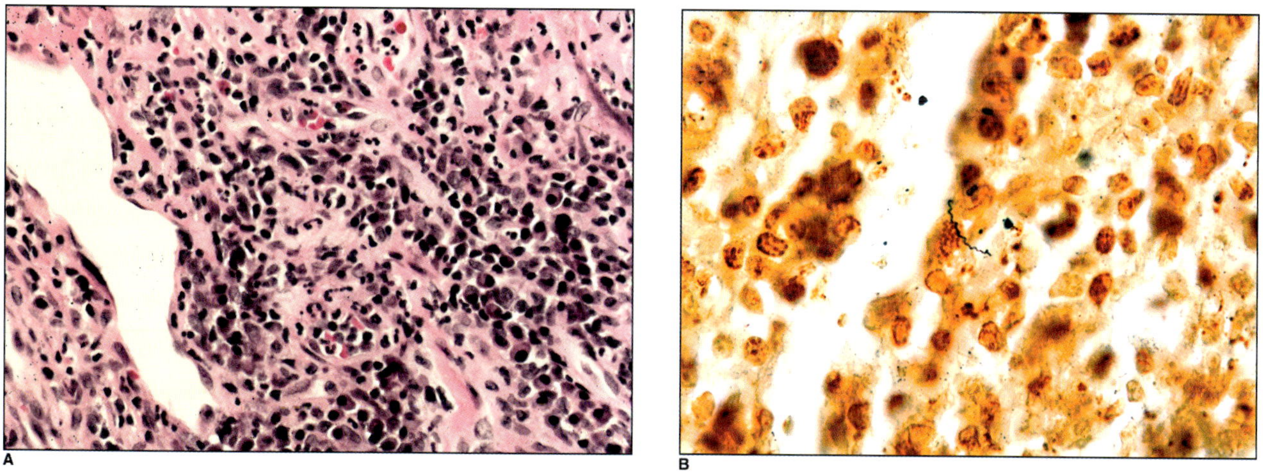

FIGURE 2.8: Oral syphilis. (A) Mixed chronic inflammatory infiltrate including mature plasma cells with a tendency to cluster in proximity to blood vessels (left); (B) Spirochete is seen on Warthin-Starry stain appearing as an elongated, thin corkscrew-shaped structure.

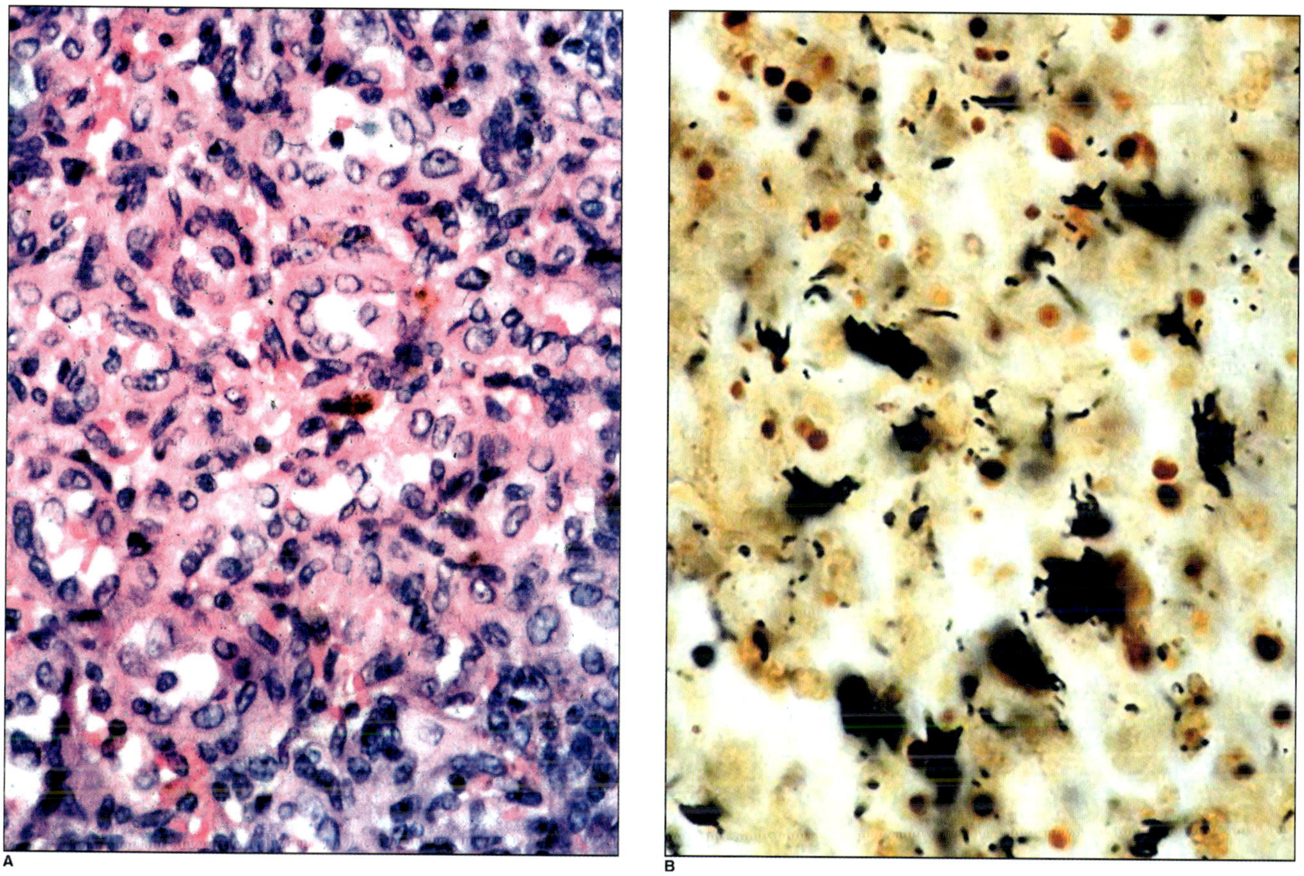

FIGURE 2.9: BA. (A) Lymph node replacement by well-circumscribed lobular vascular proliferation; (B) Warthin-Starry stain showing the granular material to contain bacteria; bacteria are interstitially located.

israelii. *Actinomyces* are often seen in the tonsillar crypts as saprophytes, unaccompanied by an inflammatory response. In contrast, cervicofacial actinomycosis is an invasive necrotizing process, usually caused by dental manipulation or trauma to the head and neck (57–60). The area around the angle of the mandible is the most commonly involved, but any site in the head and neck is a possible focus. Osteomyelitis of the jaw is a common complication and involvement of the skull, meninges, and brain is another possibility. Histologically, one observes a necrotizing granulomatous reaction with central aggregates of neutrophils, forming microabscesses.

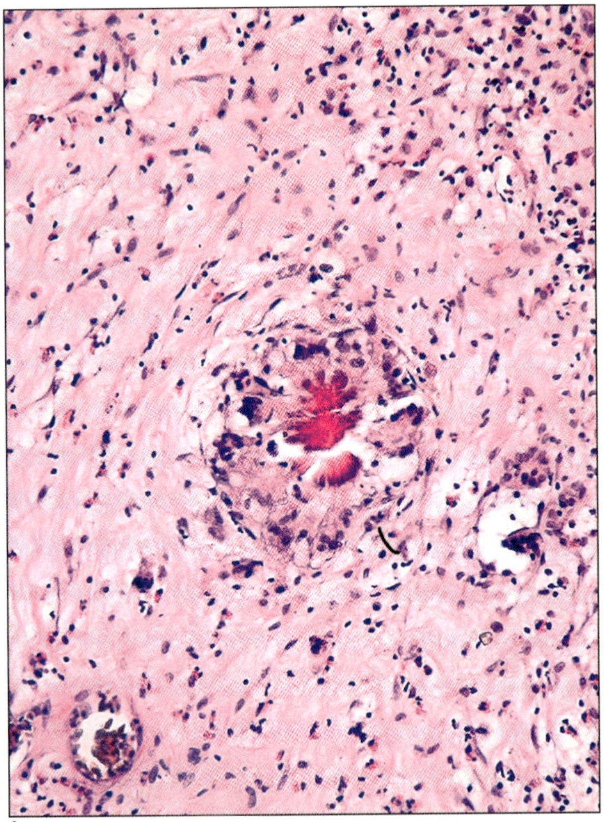

A

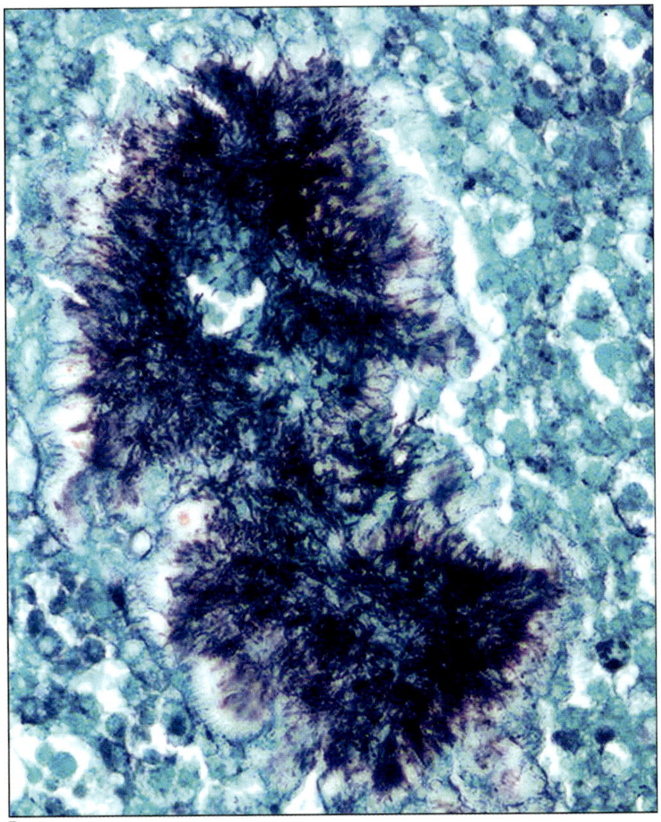

B

***FIGURE 2.10:** (A and B) Cervicofacial actinomycosis. Gram stain showing the presence of actinomycotic colony with a characteristic appearance referred to as "sulfur granules" within an acute suppurative inflammatory infiltrate.*

Bacterial colonies are typically situated inside of the neutrophilic collections, and they form characteristic structures that have been called "sulfur granules." These are rounded or elongated, deep purple aggregates composed of filamentous organisms. The sulfur granules often have eosinophilic club-shaped ends and are often encrusted with protein in the "Hoeppli-Splendore" phenomenon. The internal bacteria stain best with the Gram and GMS methods (Figure 2.10); the PAS procedure may occasionally label them as well.

Cat-Scratch Disease: Cat-scratch disease is caused by infection with a pleomorphic, gram-negative bacterium—*Bartonella henselae*—resulting in regional lymphadenopathy (61–64). The mode of transmission is direct transdermal inoculation from a scratch, bite, or lick by the animal. In the majority of cases, clinical disease develops 7–12 days after contact with the pet. The enlarged nodes are often tender; those in the submental, submandibular, cervical, occipital, and supraclavicular chains can all be affected (65).

In early nodal lesions, one sees follicular lymphoid hyperplasia and a histiocytic proliferation. Granulomatous inflammation follows, and late lesions also exhibit microabscess formation with a "stellate" pattern. Nodal sinuses are often filled with monocytoid B-lymphocytes. Cat-scratch bacilli can be identified with the Warthin-Starry or Steiner procedure, as extracellular pleomorphic coccobacilli (Figure 2.11). Tissue Gram, acid-fast, GMS, and PAS stains are all negative.

Botryomycosis: Botryo- [Greek, *botryos* (bunch of grapes)] mycosis is a chronic bacterial infection that is analogous to mycetoma and is sometimes called "bacterial ball." A mass of bacteria, usually *Pseudomonas* sp., is seen in the skin, subcutis, or mucosa of the head and neck (66) in this condition. Patients with sinonasal disease may have persistent headache, exophthalmos, and radiographic evidence of an expansile lesion in the paranasal sinuses with bony erosion (66). The histopathologic findings include amorphous, acellular deposits of proteinaceous material; rounded eosinophilic grains or granules; and a neutrophilic infiltrate. The granules contain the causative microorganisms. Club-like or radial projections are seen along the periphery of the granules, again representing the Splendore-Hoeppli phenomenon. Tissue Gram or GMS stains can delineate the bacteria (Figure 2.12), but microbiologic cultures are necessary in most instances.

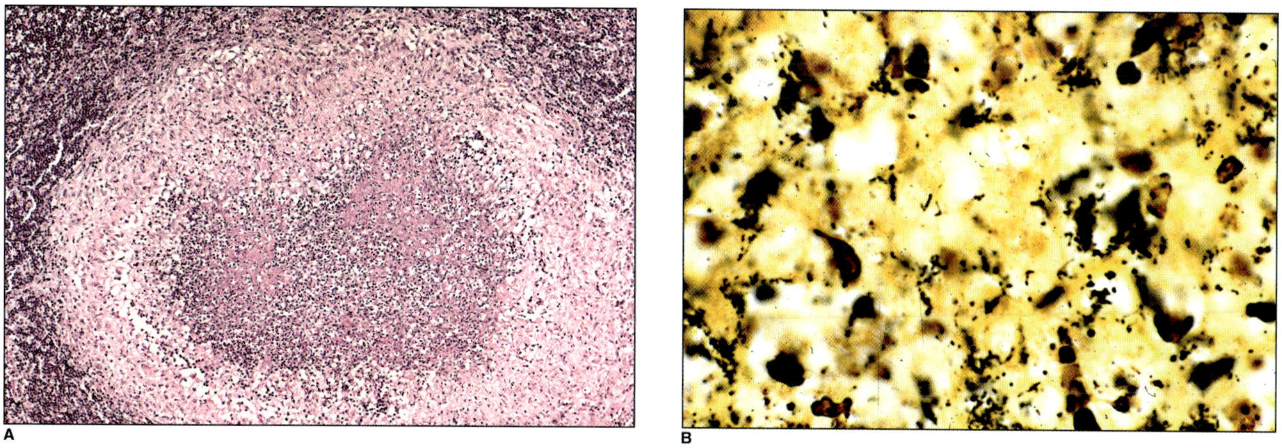

FIGURE 2.11: ***Cat-scratch disease. (A) Lymph node showing characteristic but not pathognomonic appearance of the abscess composed of a central area of necrosis with a stellate pattern admixed with polymorphonuclear leukocytes surrounded by palisading histiocytes; (B) Warthin-Starry stain showing the presence of pleomorphic coccobacilli (*****Bartonella henselae*****).***

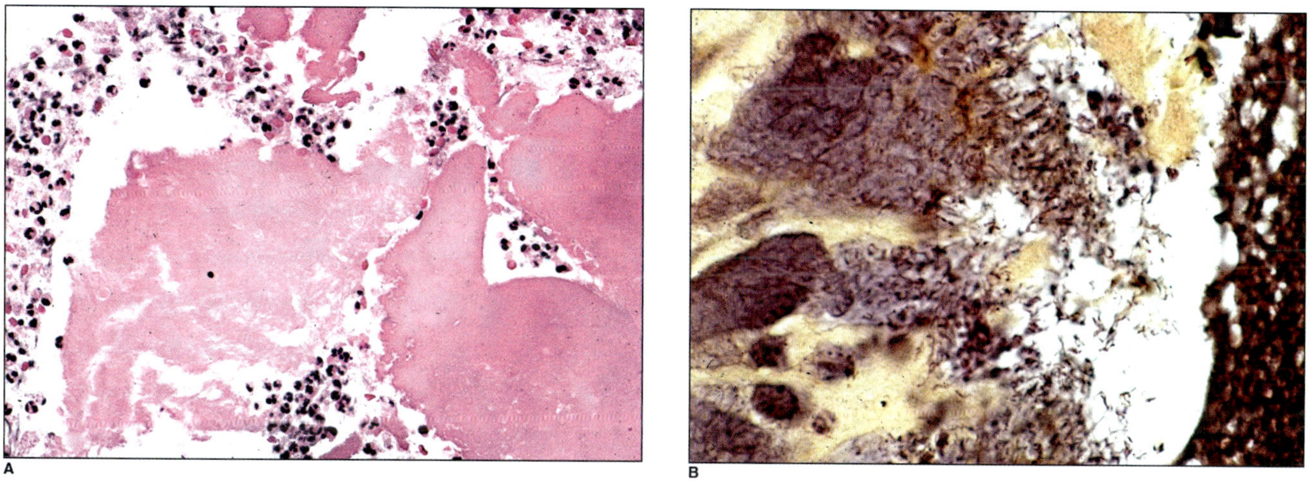

FIGURE 2.12: ***Sinonasal botryomycosis (bacterial ball). (A) Amorphous, acellular material with deposition of proteinaceous material and eosinophilic grains or granules associated with a neutrophilic infiltrate (H&E stain); the grains or granules contain the causative bacterial organisms; (B) filamentous gram-negative bacilli morphologically compatible with*** **Pseudomonas aeruginosa** ***are present (Gram stain).***

Mycobacterial Disorders

Mycobacteria that cause infections of the head and neck include *Mycobacterium tuberculosis* and nontuberculous (atypical) organisms. Representatives of the latter group are *Mycobacterium avium-intracellulare*, *Mycobacterium scrofulaceum*, *Mycobacterium kansasii*, and *Mycobacterium ulcerans*.

Overall, the incidence of infections with *M. tuberculosis* has decreased over the past five decades. Among reported cases of extrapulmonary disease, 12% are situated in head and neck sites (67,68). The AIDS pandemic and other causes of immunocompromise have augmented the frequency of "atypical" mycobacterial infections (69). In relatively immunocompetent individuals, such nontuberculous organisms do not cause pulmonary disease but often are the etiologic agents for lymphadenitis (e.g., scrofula) or subcutaneous infection. More rarely, the tonsils, pharynx, oral cavity, sinonasal structures, larynx, salivary glands, middle ear, and temporal bone are affected (68,69). Standard culture techniques are still employed for identification of the pathogenic agents, but nucleic acid amplification tests on surgical specimens are also now available for more rapid diagnosis (70–76).

The histologic profile of mycobacterial infections is comparable regardless of the specific causative organisms. It is characterized by caseating (necrotizing) granulomatous inflammation, featuring histiocytic arrays, multinucleated giant cells, other chronic inflammatory cells, and geographic areas of necrosis. The microbes are often difficult to find histologically, despite extensive efforts. In any case, their identification is strictly dependent on the use of aryl methane dyes (Ziehl-Neelsen, Kinyoun, Fite) that confer "acid fastness." Mycobacteria are beaded with a red or purple color in all these preparations (Figure 2.13).

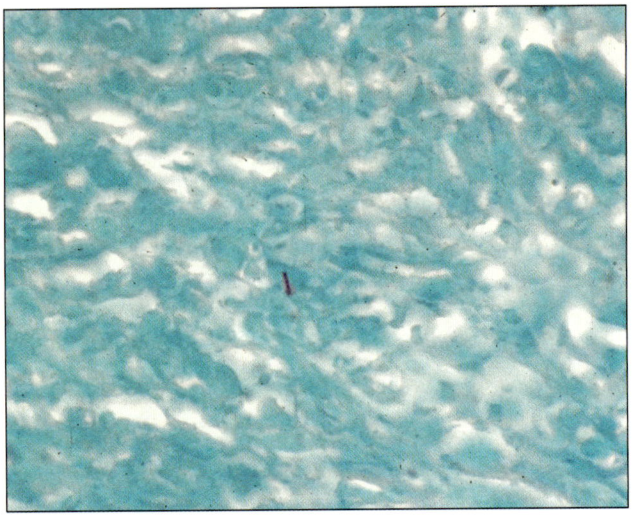

FIGURE 2.13: Mycobacterial infection in which there was caseating necrosis. Mycobacteriae are extremely difficult to identify and require the aid of special histochemical stains (AFB stain), with the microorganism appearing as a slender, beaded, red/purple rod seen here centrally in the necrotic material.

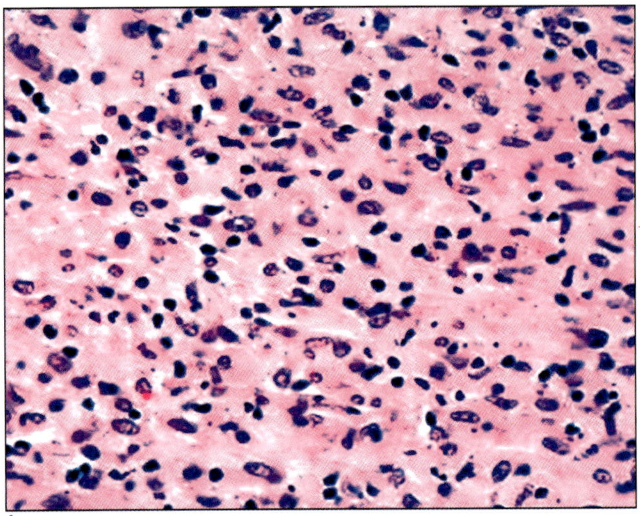

A

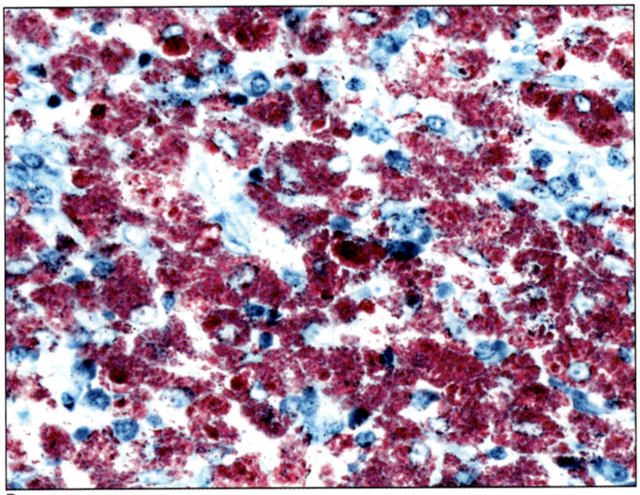

B

FIGURE 2.14: A. Mycobacterial infection of cervical lymph nodes (scrofula) in an immunocompromised patient. There were diffuse sheets of foamy histiocytes within which appearing spindle-shaped cells in a storiform pattern with multinucleated giant cells and foamy histiocytes are not present; B. AFB stains show the presence of numerous AFB-positive microorganisms within the spindle cell infiltrate.

Scrofula: Cervical lymph nodal mycobacterial infection is referred to as "scrofula," most commonly involving the high cervical nodes in the region of the submandibular gland. Periparotid, periauricular, and submental nodes are less often affected (77,78). Scrofula usually is unilateral when caused by nontuberculous mycobacteria and bilateral if related to systemic infection with *M. tuberculosis*. In up to 25% of atypical mycobacterial infections, a caseating granulomatous response is absent histologically, especially in immunocompromised patients. Instead, one sees sheets of foamy histiocytes in which numerous acid-fast bacteria (AFB) are present (Figure 2.14).

Mycobacterial Spindle Cell Tumor: Mycobacterial spindle cell tumor (MSCT) is a pseudoneoplastic spindle cell proliferation that is seen almost exclusively in HIV-infected patients (79–82). The causative microorganism is *M. avium-intracellulare*. Lymph nodes of the head and neck, mucosal surfaces, skin, spleen, brain, and bone marrow may all be affected by MSCT (83). Histologically, this lesion demonstrates a proliferation of bland spindle cells with a storiform growth pattern. Multinucleated giant cells and foamy histiocytes are typically absent, and there is partial or complete effacement of the underlying tissue architecture. Histochemical stains for AFB show the presence of innumerable organisms in the spindle cells (Figure 2.15).

Leprosy: Leprosy (Hansen's disease) is caused by *Mycobacterium leprae* (84), with involvement of the skin, mucosae, and peripheral nerves. Sinonasal disease is fairly common in leprosy patients (85); oral and laryngeal lesions are also possible. In "tuberculoid" leprosy (TL), which is seen in patients with relatively intact immunity, the disease tends to be localized and microorganisms are typically absent in biopsy specimens (86). In contrast, "lepromatous" leprosy (LL), as seen in individuals with immunocompromise, manifests many lesional microorganisms (86).

Histologically, TL is characterized by well-formed granulomas with an admixture of histiocytes, multinucleated giant cells, and lymphocytes, often centering on peripheral nerves. In contrast, LL shows sheets of lymphocytes and vacuolated histiocytes (lepra cells) and an absence of granulomas. The Ziehl-Neelsen or Kinyoun method for AFB is often falsely negative in leprosy, making the Fite procedure preferred (Figure 2.16).

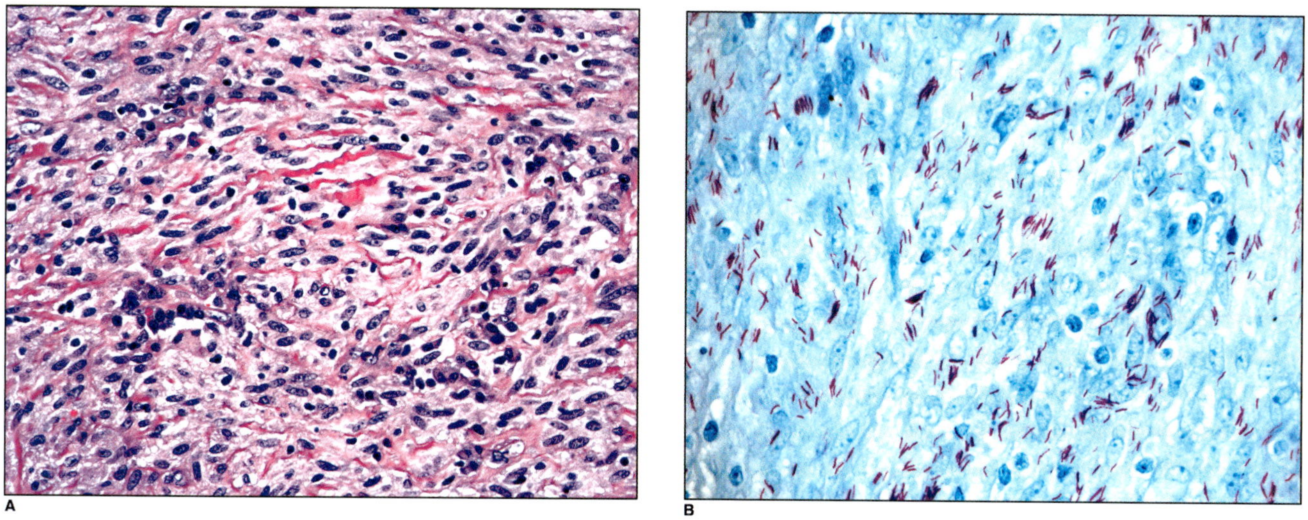

FIGURE 2.15: Mycobacterial spindle cell pseudotumor. (A) Cellular proliferation composed of bland spindle cells with a storiform configuration. (B) Ziehl-Neelsen stain shows numerous mycobacteria in the lesional cells.

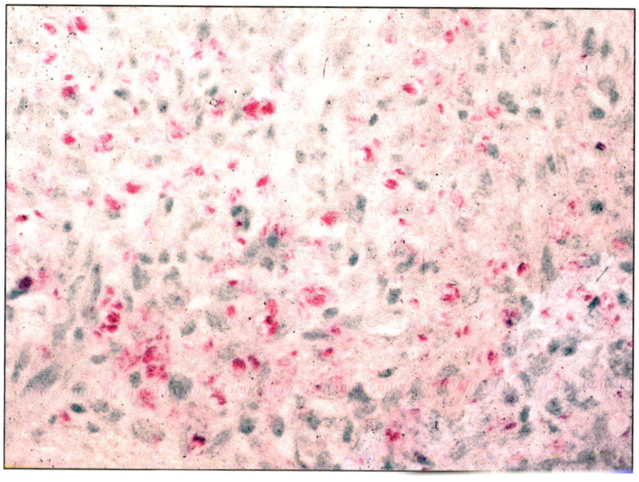

FIGURE 2.16: LL. Fite stain (modified AFB stain) showing the presence of abundant microorganisms.

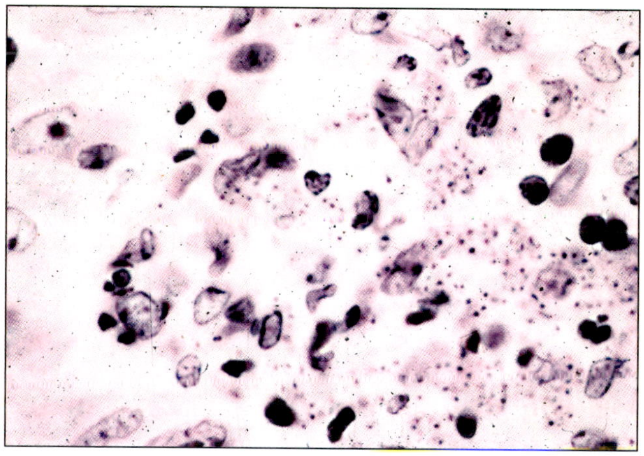

FIGURE 2.17: Mucocutaneous leishmaniasis caused by Leishmania braziliensis. *The organisms of* L. braziliensis, *referred to as amastigotes, are found within histiocytes often localizing along the periphery of the cells (so-called marque sign), the organisms are small, ovoid to round with a thin cell membrane, large nucleus, and capsule, which is a differentiating feature from* Histoplasma capsulatum *(H&E stain).*

Viral Disorders

Viral infections of the head and neck are common causes of rhinosinusitis and other conditions. Potentially causative agents include *rhinoviruses*, *influenza* and *parainfluenza viruses*, *adenoviruses*, *paramyxovirus*, and *respiratory syncytial virus*. Viruses producing tonsillitis are represented by *adenoviruses* and *Epstein-Barr virus* (EBV). Other diseases of the head and neck are related to infection with human papillomavirus (HPV) and the *Herpes* virus group [*Herpes simplex*, EBV, Varicella-zoster, cytomegalovirus (CMV)] or the HIV-1. Most of these disorders are diagnosed on clinical grounds, but occasionally may require biopsies. Unlike the infectious diseases discussed thus far, the histologic diagnosis of viropathic processes is usually based on morphologic features and supported by immunohistochemistry or molecular diagnostic techniques. Histochemical stains are used only uncommonly. Lendrum's phloxine tartrazine method, the Machiavello procedure, the Feulgen method, and the methyl green–pyronine stain can be used to label nucleic acids in intracellular organisms. Cytoplasmic CMV inclusions may also sometimes label with silver impregnation techniques (e.g., GMS) and the PAS stain.

Protozoan Infestations

Leishmaniasis: Leishmaniasis is a protozoan infestation (90); mucocutaneous disease is caused by *Leishmania braziliensis*, which is endemic to Central and South America. Patients may develop destructive ulcerative lesions

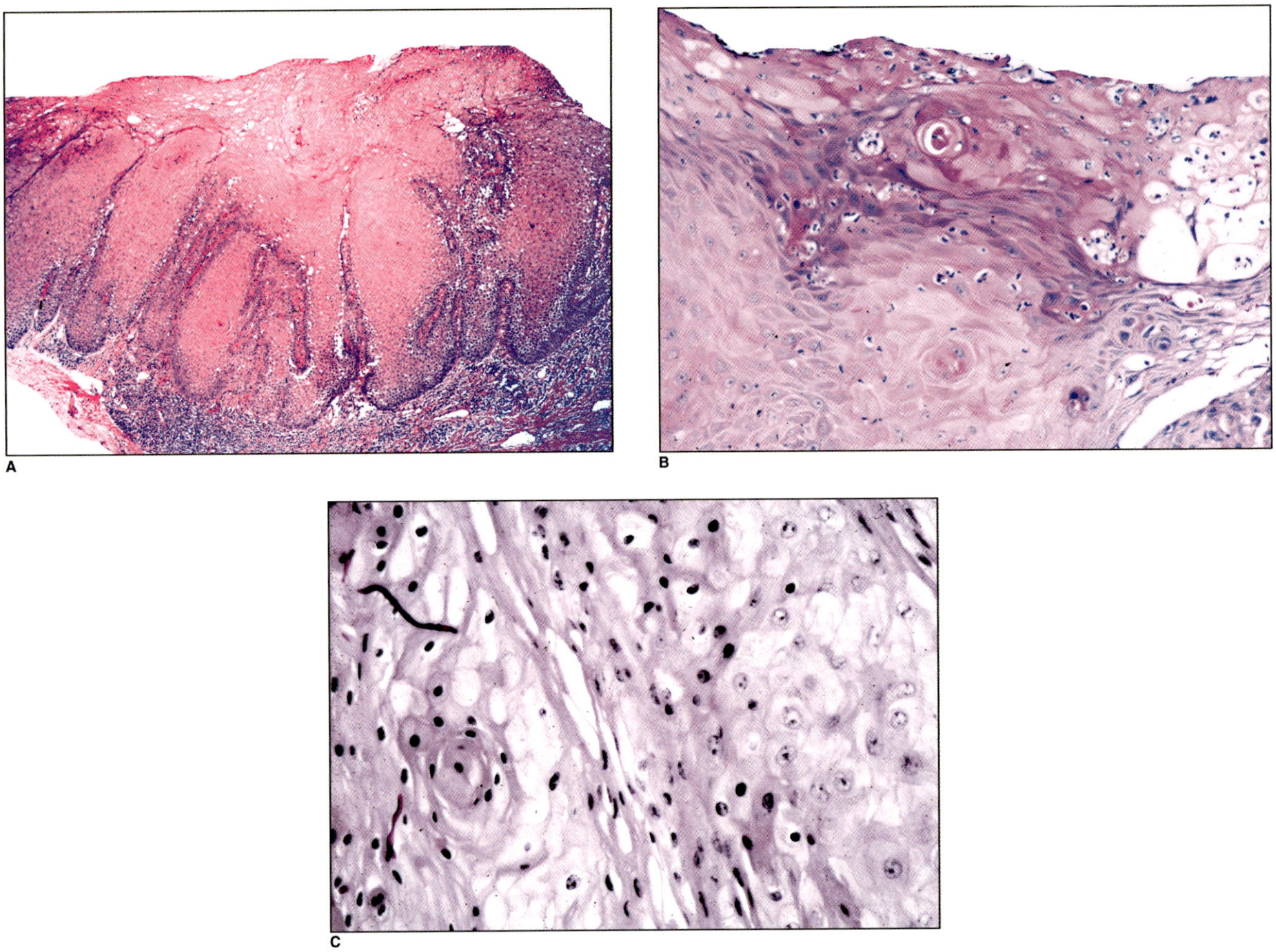

FIGURE 2.18: Oral PEH. (A) At low magnification, the epithelium is hyperplastic and hyperkeratosis that in conjunction with its clinical appearance of a verrucoid lesion raises concern for a possible diagnosis of verrucous carcinoma (H&E stain); (B) at higher magnification a neutrophilic infiltrate is identified (H&E stain); (C) PAS stain shows the presence of a fungal hyphae within the depth of the epithelium (left). The histology suggests Candida species that was confirmed by microbiologic evaluation.

of mucous membranes that are termed "espundia." Aerodigestive sites of involvement also include the sinonasal tract, nasopharynx, and tongue (91).

The histologic image of mucocutaneous infection is that of ulceration with a marked lymphoplasmacytic cell infiltrate and subjacent granulation tissue. PEH of the adjacent surface epithelium also may be present. In immunocompetent patients, the inflammatory response is vaguely granulomatous with epithelioid histiocytes and scattered giant cells. Necrotizing foci may be present as well. In diffuse anergic leishmaniasis, granulomatous inflammation is absent and many large histiocytes are filled with microorganisms. They can be seen with H&E staining (Figure 2.17) and tend to be localized at the periphery of parasitized macrophages as ovoid to round structures measuring 1.5–3 μm in diameter. Nuclei and kinetoplasts of the organisms are accentuated by Giemsa staining but often require oil immersion microscopy for identification. The absence of a cellular capsule aids in distinguishing *Leishmania* from *Histoplasma*. Molecular diagnostic techniques are now available to confirm the diagnosis of leishmaniasis (92).

Pneumocystis carinii: *Pneumocystis carinii* (PC) is an opportunistic microorganism that typically causes pneumonia in the context of immunodeficiency. In the head and neck, it may rarely infect the skin, the external auditory canal, and the middle ear (43), via vascular dissemination from a pulmonary source (93). Histologic findings are similar to those seen in the lung; they feature submucosal or dermal "foamy" exudates of proteinaceous material, in which the organism can be identified with the GMS stain.

ACQUIRED LESIONS OF THE HEAD AND NECK

Nonneoplastic Lesions

Pseudoepitheliomatous Hyperplasia: PEH is the exuberant reactive or reparative overgrowth of squamous epithelium with no cytologic evidence of malignancy. However, it may be mistaken clinically and histologically for invasive squamous carcinoma. PEH is a morphologic pattern, not a distinct clinicopathologic entity. It may be isolated and idiopathic, or be induced by infections (e.g., candidiasis, leprosy, and blastomycosis) and tumors (e.g., granular cell tumor) (94). Hence, GMS and PAS stains are recommended before making a diagnosis of unqualified PEH (Figure 2.18).

Necrotizing Sialometaplasia: Necrotizing sialometaplasia (NS) is a benign, self-healing, reactive process involving minor salivary glands. It may clinically and histologically simulate a malignant neoplasm (95,96). NS most commonly involves the palatal salivary gland tissue but other intraoral sites can be affected and major salivary gland involvement has been reported (97–99). The histologic appearance is that of lobular necrosis of the salivary glands with preservation of their architecture. The microscopic hallmark of NS is squamous metaplasia of residual acini and ducts. Pools of mucin also may extend into adjacent tissue and elicit a granulation tissue reaction. Mucicarminophilic material is seen with Best's stain in glandular lumina and the cytoplasm of residual mucocytes (Figure 2.19). This finding helps in the differential diagnosis between NS and squamous cell carcinoma.

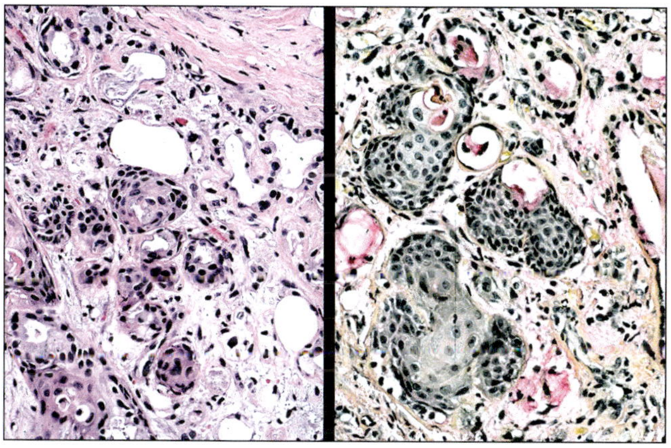

FIGURE 2.19: ***NS. Left: Rounded lobules of the minor salivary glands being replaced by squamous epithelium (H&E stain); Right: mucin stain shows the presence of intracytoplasmic mucin–positive material indicative of residual mucocytes although the majority of the lobular units are replaced by squamous epithelium.***

Amyloidosis: Amyloidosis of the head and neck is most often "primary" (immunoglobulin related), and, in that location, it preferentially affects the tongue (100–104). The larynx may be affected in "secondary" (AA protein related) amyloidosis as well (104–106). Histologically, one observes extracellular, eosinophilic, amorphous material that is deposited randomly throughout the submucosa or around blood vessels. Associated plasma cells, lymphocytes, and histiocytes may be seen, together with a foreign body giant cell reaction. Stains for amyloid include Congo red, crystal violet, thioflavine-T, pagoda red, and sulfated alcian blue. Characteristic apple-green birefringence is seen in polarization microscopy after Congo red staining (Figure 2.20).

Amyloid deposition is also seen in "amyloid goiter" and medullary carcinoma of the thyroid gland (107,108). These conditions are discussed in the chapter in this monograph on endocrine diseases.

Mucus Retention Cyst, Ranula, and Mucus Extravasation Phenomenon: Mucus retention cyst (MRC), also referred to as "mucus escape phenomemon," is an obstructive disorder of salivary glands, usually affecting those in the

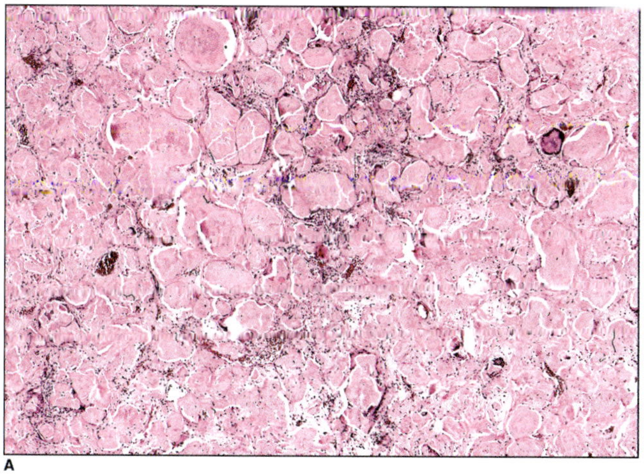

A

B

FIGURE 2.20: ***Laryngeal amyloidosis. (A) Amyloid appearing as eosinophilic, acellular, amorphous, somewhat nodular material effacing the submucosal architecture; (B) Congo red staining for amyloid showing the characteristic apple-green birefringence under polarized light.***

mouth, buccal mucosa, and lips. It results in ductal dilatation, producing unilocular cysts lined by cuboidal, columnar, or nonkeratinizing squamous epithelium. Occasional mucus-secreting cells also can be found. MRCs are filled with eosinophilic material that stains with Best's mucicarmine method.

"Ranula" is related to MRC and specifically occurs in the floor of the mouth. "Simple" ranulas are true cysts with an epithelial lining; "plunging" ranulas are pseudocysts because they lack epithelial components and consist only of mucus pools and a fibroinflammatory reaction. Mucicarmine or PAS-D stain is helpful in identifying the mucus in MRCs and ranulas. A closely related lesion is the "mucus extravasation phenomenon" as usually seen in the lower lip. It is likewise characterized by the presence of mucus in soft tissue, intermixed with granulation tissue and inflammation.

Histochemical Exclusion of Infection in Other Diseases: Selected idiopathic inflammatory diseases have a histologic appearance that suggests a possible infectious etiology. They therefore require histochemical studies to exclude the presence of microorganisms. Examples of such lesions in the head and neck include sarcoidosis, WG, and myospeherulosis.

Sarcoidosis

Sarcoidosis is a multisystem chronic granulomatous disease of unknown etiology. Any organ site may be involved, but the most common ones are the lungs, skin, and lymph nodes. Isolated extranodal sarcoidal lesions of the head and neck can also be seen in the pharynx, tonsils, ear, temporal bone, sinonasal cavity, salivary glands, and larynx (109,110).

Histologic features include noncaseating granulomas and a mixed, usually scanty, chronic inflammatory infiltrate. Necrosis is typically absent or focal, and all histochemical stains for microorganisms (Ziehl-Neelsen, Fite, GMS, Warthin-Starry) are negative. Therefore, a diagnosis of sarcoidosis is generally one of exclusion.

Wegener's Granulomatosis

WG is an idiopathic nonneoplastic inflammatory disease with a predilection for the upper and lower respiratory tract and the genitourinary system. It is characterized by the presence of granulomatous inflammation, vasculitis, and parenchymal destruction. The initial biopsy in cases of WG may be taken from lesions of the upper aerodigestive system (111).

Histologic features include vasculitis, granulomatous inflammation, and necrosis, which is usually "ischemic" or "geographic" in nature. Histochemical stains for microorganisms (Ziehl-Neelsen, Fite, GMS, Warthin-Starry) must be negative to render a diagnosis of WG. In addition, trichrome stains and others for elastic tissue (e.g., the Verhoeff–Van Gieson method) are useful in defining the presence of vascular disruption in this disease (Figure 2.21).

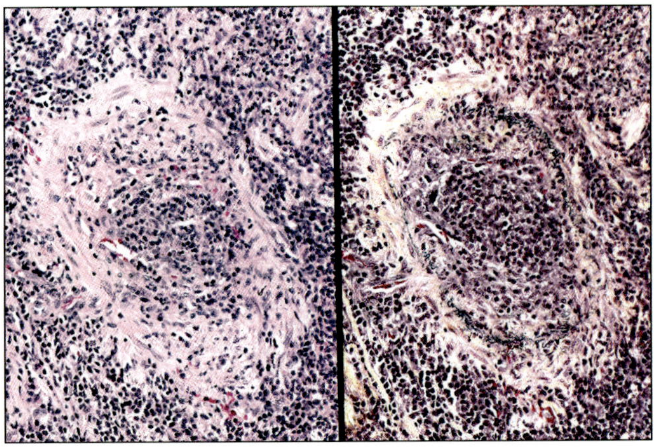

FIGURE 2.21: WG of the sinonasal tract. Left: One of the hallmarks of this condition is vasculitis of small blood vessels with a mixed inflammatory infiltrate surrounding and infiltrating the involved blood vessel; Right: the identification of blood vessels can be difficult and is assisted by elastic stain showing disruption of the elastic membrane (appearing black and discontinuous) by the inflammatory infiltrate that has penetrated the vessel wall and completely fills the lumen resulting in the ischemic-type necrosis seen in this disease.

Myospherulosis

Myospherulosis is an iatrogenic pseudomycotic disease that results from the interaction of red blood cells with ointments in cloth packs that are placed in the airway or ear after surgical procedures (112–115). Histologically, pseudocysts or microcysts are embedded in fibrotic tissue, creating a "Swiss-cheese" appearance (Figure 2.22). The "spherules" are dark brown in H&E stains, and, in keeping with their erythrocytic derivation, they are reactive with the peroxidase method. Stains for fungi are negative.

NEOPLASMS OF THE HEAD AND NECK

The most common neoplasms of the head and neck are epithelial in nature; specifically, they are represented by squamous papillomas and squamous cell carcinomas. Histochemistry has limited application in the diagnosis of these particular lesions. However, other neoplasms do lend themselves to histochemical characterization.

Sinonasal (Schneiderian) Papillomas

Schneiderian papillomas are benign neoplasms that demonstrate differentiation toward the Schneiderian sinonasal

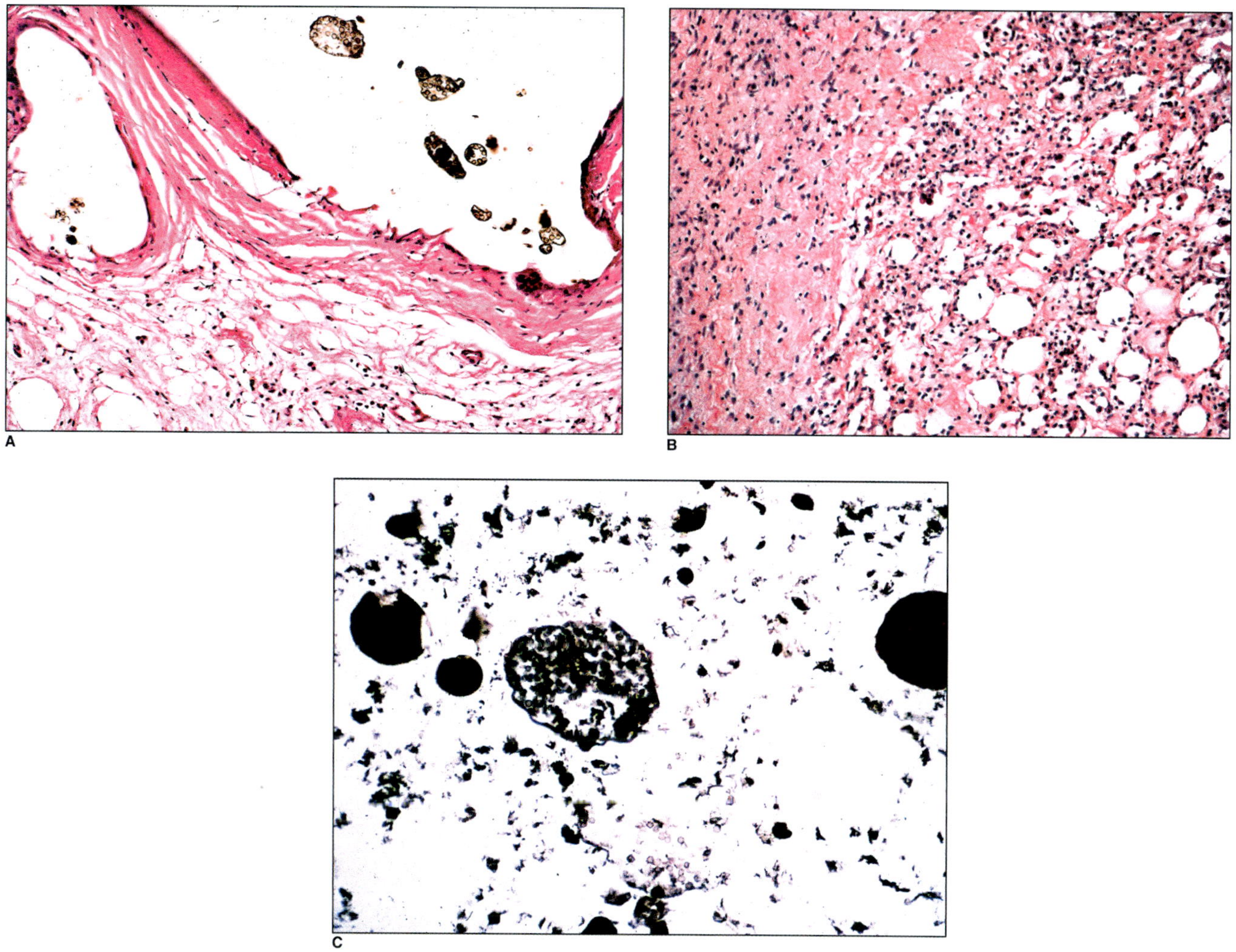

FIGURE 2.22: Myospherulosis of the sinonasal tract. Pseudocystic spaces contain round, saclike structures (parent bodies); the "parent bodies" are brown staining and contain numerous spherules or endobodies (H&E stain) (A and B). Stains for microorganisms are negative, but the peroxidase method is positive (C).

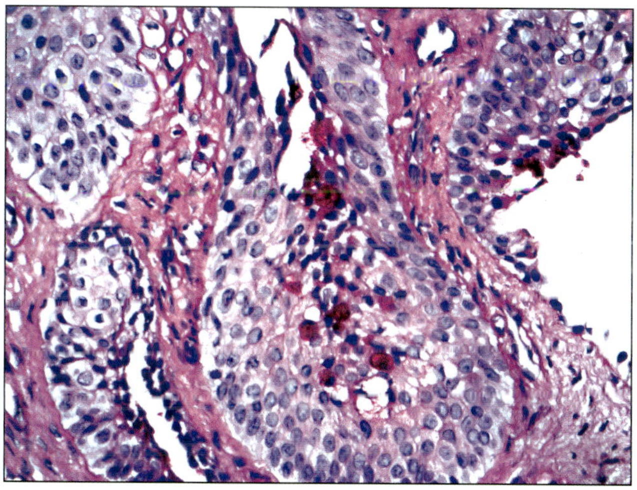

FIGURE 2.23: Schneiderian papilloma. This lesion was located near the nasal vestibule but the presence of intraepithelial mucocytes showing intracytoplasmic diastase-resistant, PAS-positive material assists in identifying this lesion as being of the sinonasal mucosa rather than a cutaneous-based lesion.

mucosa. As such, they are composed of squamoid, transitional, or columnar epithelial elements, often with associated mucous cell differentiation (116). The tumors in this category include Schneiderian papillomas of the inverted, cylindrical, and exophytic types. Histochemical stains can be used to identify intraepithelial mucocytes, which demonstrate the presence of cytoplasmic mucin with the mucicarmine and PAS-D methods (Figure 2.23).

Adenosquamous Carcinoma

Adenosquamous carcinoma (ASC) in the head and neck is a high-grade surface epithelial malignancy with histologic features of both squamous cell carcinoma and adenocarcinoma (117). The squamous cell carcinoma component varies from well to poorly differentiated and may have an in situ surface component. In contrast, adenocarcinomatous elements are identified only in the submucosa. Overt mucous cell differentiation usually is absent, but cytoplasmic and intraluminal mucin can be labeled with the mucicarmine and PAS-D procedures in the glandular tumor component. ASC is a poorly differentiated surface tumor, in contrast to high-grade salivary gland–based mucoepidermoid carcinoma (MEC). The latter is another "hybrid" neoplasm that will be discussed subsequently.

Salivary Gland Neoplasms

The use of histochemistry may be useful selectively in the diagnosis of salivary gland neoplasms.

Basal Cell Adenoma: BCA is a histologic variant of monomorphic adenoma comprising basaloid cells. No mesenchyme-like stromal component is present, as seen in pleomorphic adenomas. BCAs are subdivided into four subtypes: solid, trabecular, tubular, and membranous. Membranous BCA may be multilobular, multifocal, and unencapsulated. This variant is characterized by thick eosinophilic hyaline membranes that surround and separating cellular islands and create a "jigsaw puzzle" appearance. The hyaline membranes are reactive with the PAS-D stain (Figure 2.24).

Oncocytoma: Oncocytoma is a benign salivary gland neoplasm exclusively composed of oncocytes. These cells are granular and eosinophilic because of the presence of innumerable mitochondria. Accordingly, the phosphotungstic acid–hematoxylin (PTAH) stain is positive in oncocytoma, represented by blue-black granular labeling of the cytoplasm. Intracellular glycogen may also be present in oncocytomas, as demonstrated with the undigested PAS procedure.

Mucoepidermoid Carcinoma: MEC is a low-grade salivary glandular malignancy showing an admixture of epidermoid, mucous, and intermediate cells. Other elements may be presented as well, such as columnar, oncocytic, and clear cells. MEC is the most common malignant salivary gland tumor, principally being encountered in the parotid gland, submandibular gland, and sublingual gland in order of frequency (118). Minor salivary glands may be affected as well. Mucocytes are readily identifiable in low- and intermediate-grade MECs but are not obvious in poorly differentiated lesions. In these tumors, they often require application of PAS-D or mucicarmine stains for delineation (Figure 2.25).

Acinic Cell Adenocarcinoma: Acinic cell adenocarcinoma (ACA) manifests a variety of histologic growth patterns. However, all of them show at least partial recapitulation of serous acinar-type salivary gland differentiation, as characterized by the presence of cytoplasmic zymogen-type secretory granules (119). The majority of ACAs arise in the parotid gland. Histochemistry shows that acinar cells and intercalated duct cells in ACA contain diastase-resistant, PAS-positive cytoplasmic granules. Mucicarmine and alcian blue stains are typically negative or, at most, weakly positive. In addition, ACAs may demonstrate cytoplasmic hemosiderin granules, as identified with Perl's stain.

Adenoid Cystic Carcinoma: Adenoid cystic carcinoma is a basaloid, monomorphous tumor with a tendency to invade nerves and a protracted but relentless clinical course. Histochemical stains have only limited utility in the characterization of this lesion and its differential diagnosis from other neoplasms. The mucoid cylinders and pseudocysts in adenoid cystic carcinoma contain diastase-resistant, PAS-positive material that also labels with the mucicarmine, alcian blue, and colloidal iron methods. Unfortunately, this staining profile is not specific.

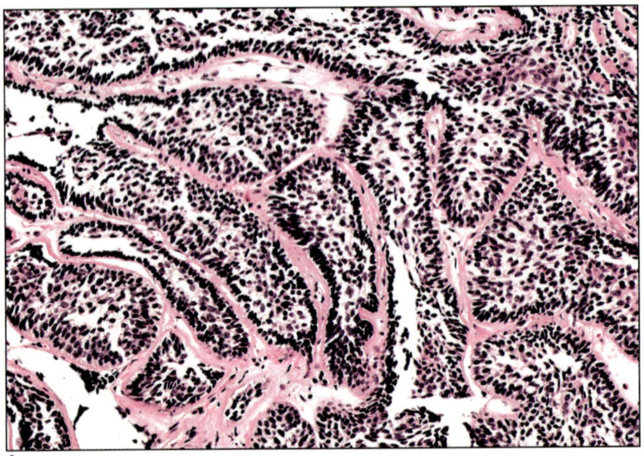
A

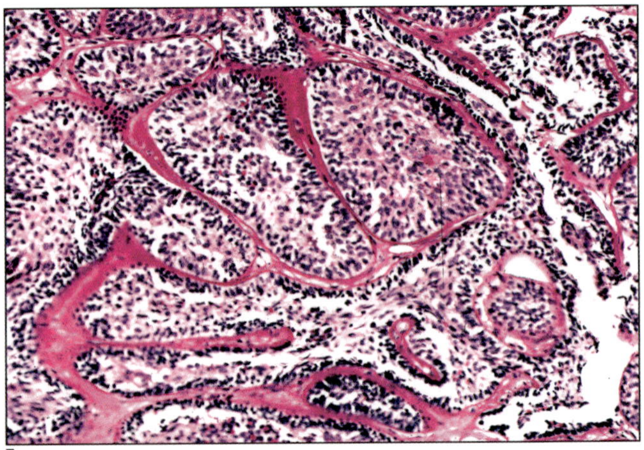
B

FIGURE 2.24: BCA, membranous type, of the parotid gland. (A) This tumor is characterized by the presence of thick eosinophilic hyalin membranes surrounding and separating cell islands creating a jigsaw puzzle appearance representing reduplicated basal lamina (H&E stain); (B) the eosinophilic hyaline material is diastase resistant, PAS positive.

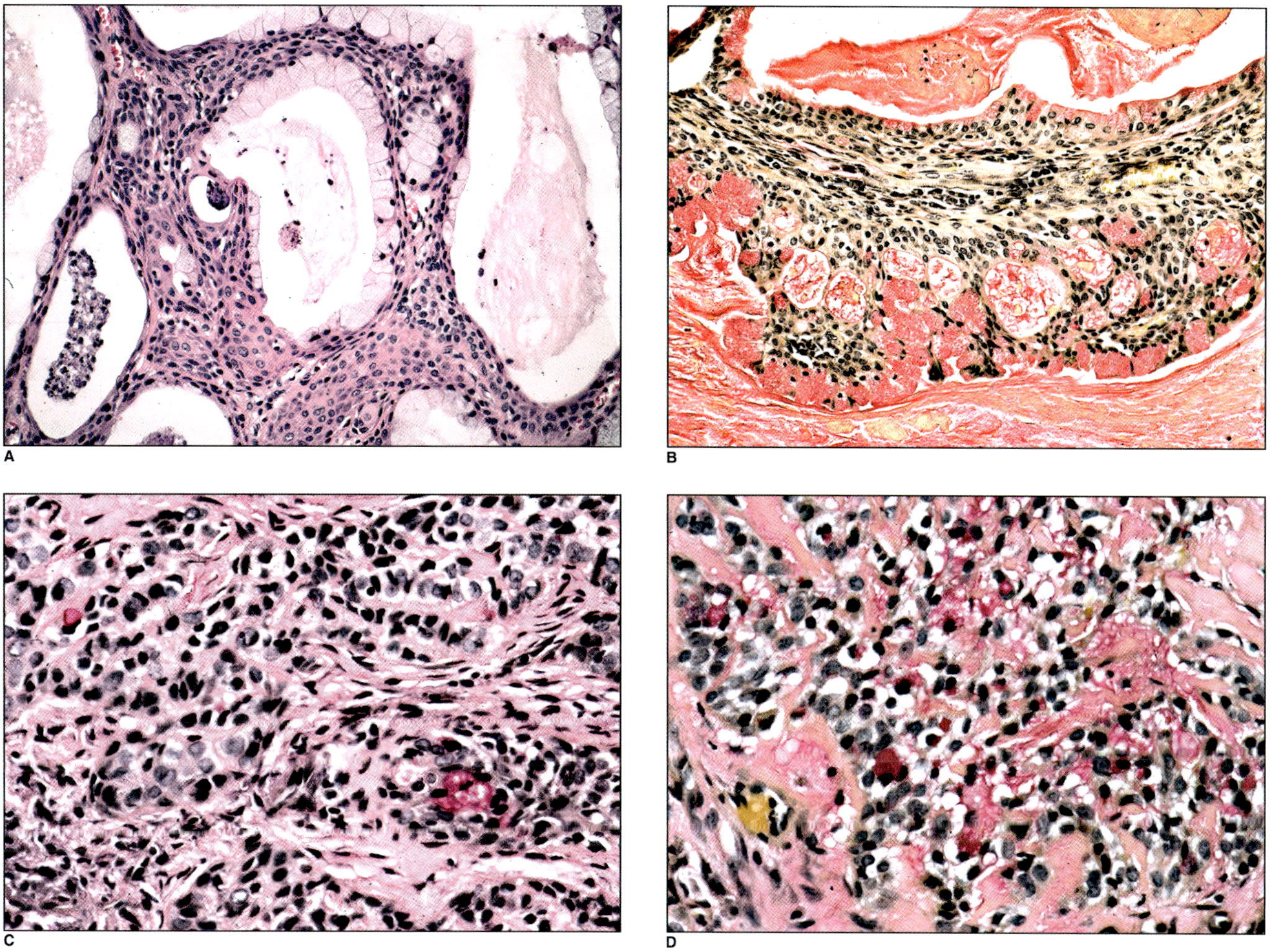

FIGURE 2.25: MEC of salivary glands. (A) The combination of cell types including epidermoid cells, intermediate cells, and mucocytes, the latter readily identifiable in this illustration is diagnostic for a MEC, low grade (H&E stain); (B) if needed, mucicarmine stain will delineate the presence of the mucocytes containing intracytoplasmic mucin appearing as bright red material within the cells; (C) high-grade MEC may be difficult to diagnose within identifying isolated mucin-positive cells within an otherwise poorly differentiated malignant neoplasm; (D) this MEC was predominantly composed of clear cells raising the differential diagnosis with other salivary gland neoplasms comprised of clear cells; the presence of intracytoplasmic mucin–positive material assisted in the diagnosis of clear cell MEC.

Other Malignant Salivary Gland Neoplasms: Histochemical evaluations also contribute to the recognition of other malignant salivary gland neoplasms, including adenocarcinoma not otherwise specified (ACANOS), clear cell adenocarcinoma, salivary duct carcinoma, cystadenocarcinoma, and mucinous adenocarcinoma. In ACANOS, cytoplasmic mucin can be seen with mucicarmine and PAS-D stains. Clear cell adenocarcinoma manifests glycogen content with the undigested PAS procedure (Figure 2.26); the mucicarmine stain is usually negative. In salivary duct carcinomas, cytoplasmic PAS-D–positive material can be present; the mucicarmine stain is typically negative or positive only in intraluminal material. All mucin stains in cystadenocarcinoma are characteristically negative. The copious extracellular mucinous material in mucinous adenocarcinomas is mucicarminophilic, PAS-D–positive, and reactive with the alcian blue and colloidal iron methods.

Non–Salivary Glandular Adenocarcinomas of the Head and Neck

Other than salivary glandular tumors, other forms of adenocarcinoma are uncommon in the head and neck. Of these, the principal location for non–salivary gland adenocarcinomas is the sinonasal tract, where two main categories are encountered—intestinal-type adenocarcinoma and nonintestinal adenocarcinomas (120). Regardless of subclassification, stains for epithelial mucin are positive in such lesions with the mucicarmine method, the PAS-D technique, and the alcian blue and colloidal iron procedures.

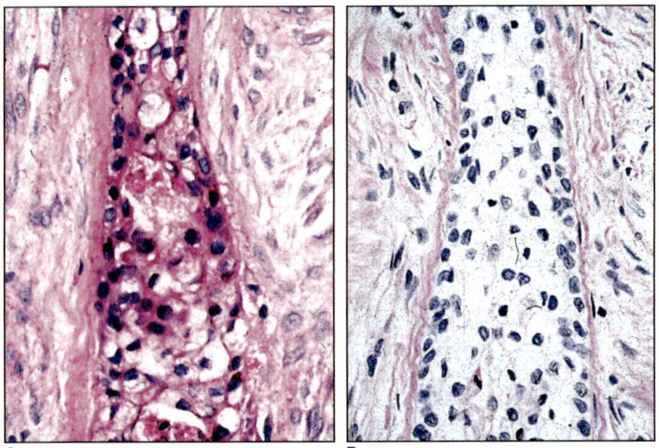

A B

FIGURE 2.26: Clear cell carcinoma of intraoral minor salivary glands. This infiltrative neoplasm was exclusively composed of cells with clear-appearing cytoplasm (i.e., clear cells). The cells showed (A) the presence of intracytoplasmic PAS-positive material that (B) was lost following diastase digestion. The presence of diastase-sensitive, PAS-positive intracytopalsmic material is indicative of glycogen and assists in the diagnosis of this clear cell carcinoma.

Melanocytic Neoplasms and Other Melanin-Containing Lesions

Mucosal malignant melanoma (MMM) demonstrates melanocytic differentiation; approximately 15–25% of these tumors arise in the head and neck (121). The cytomorphologic features of MMM include small, round, epithelioid, fusiform, and pleomorphic cells, or mixtures thereof. One may occasionally see dense melanin deposition in MMM, but approximately one third of cases contain very scanty pigmentation or are amelanotic altogether in H&E stains. Histochemical studies can help to identify "latent" melanin in melanomas, using argentaffinic methods such as the Fontana-Masson procedure (Figure 2.27), or Schmorl's technique. Melanin-containing cells also can be seen in nonmelanomatous tumors including melanotic schwannoma, pigmented neurofibroma, glandular malignant peripheral nerve sheath tumor, and melanotic neuroectodermal tumor of infancy (progonoma) (122,123).

Neuroendocrine and Neuroectodermal Neoplasms

Neuroendocrine-neuroectodermal neoplasms of the head and neck include paragangliomas (PGs), neuroendocrine carcinomas (NECs), olfactory neuroblastomas (ONBs) (esthesioneuroblastomas), and primitive neuroectodermal tumors (PNETs). In addition, other tumor types may show "occult" neuroendocrine differentiation, as seen in middle ear adenomas.

Paraganglioma: PGs are nonepithelial lesions that arise from the paraganglia of the autonomic nervous system.

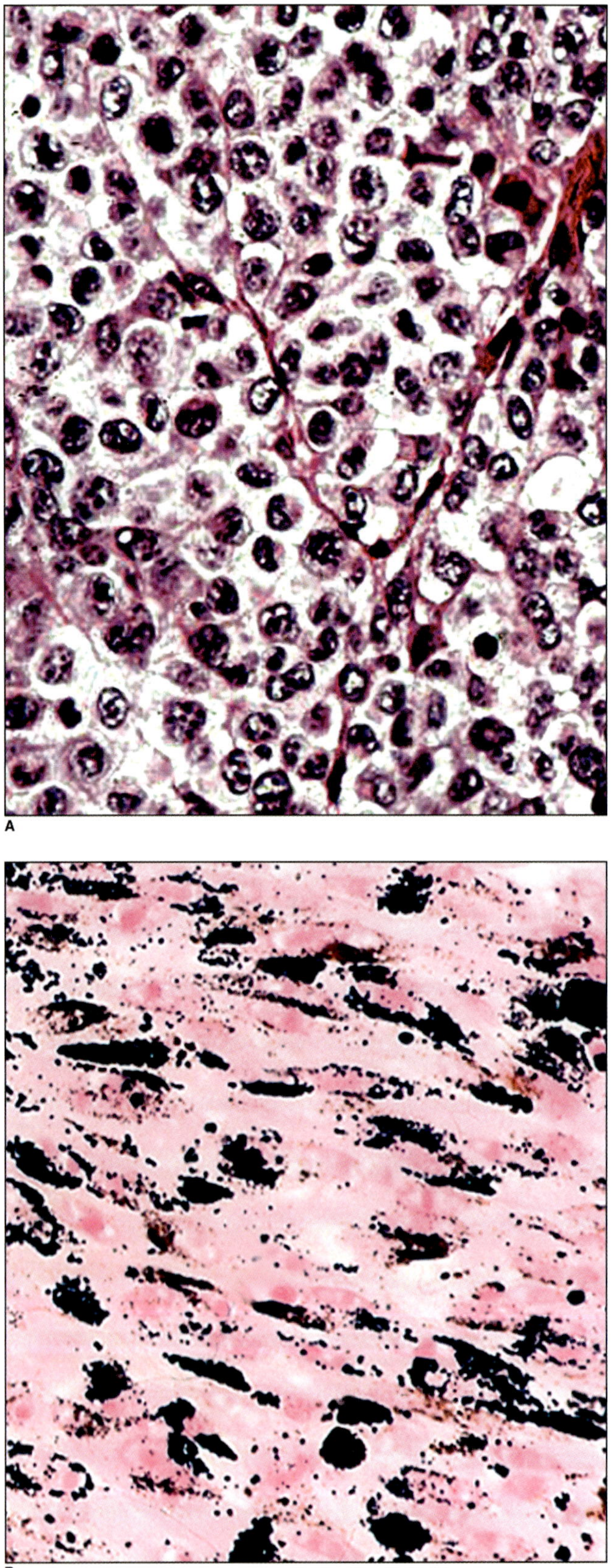

A

B

FIGURE 2.27: MMM. The melanin within the neoplastic cells (A) stains with the Fontana-Masson method (B).

In the head and neck, the most common representatives of this tumor type are localized along branches of the vagus nerve (carotid body PG) and the glossopharyngeal nerve (jugulotympanic PG) (124,125). Regardless of site of origin, their histologic appearances are comparable, featuring prominent cell nesting in an organoid (zellballen) pattern; a delicate fibrovascular stroma; and the presence of sustentacular elements at the periphery of cell nests. The "chief" cells in PGs are argyrophilic (Figure 2.28); reticulin (Snook's, Wilder's) stains delineate the organoid architecture and may be useful when biopsies of PG are artifactually distorted (Figure 2.29). Argentaffin and mucin stains are nonreactive with PGs of the head and neck.

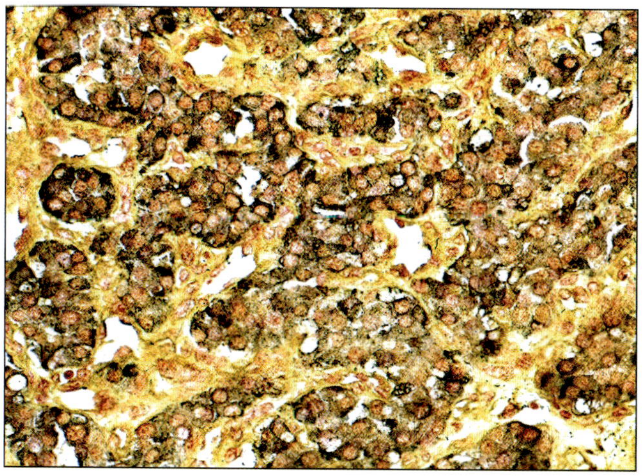

FIGURE 2.28: *Carotid body tumor (PG). The neoplastic (chief) cells show intracytoplasmic argyrophilic (Churukian-Schenk)–positive material.*

Middle Ear Adenomas with "Occult" Neuroendocrine Differentiation: In some middle ear adenomas, the tumor cells show dispersed nuclear chromatin with a "salt-and-pepper" pattern, as seen in other neoplasms with neuroendocrine differentiation. These lesions tend to demonstrate reactivity with argyrophil stains, such as the Sevier-Munger, Grimelius, and Churukian-Schenk procedures.

Neuroendocrine Carcinomas: NECs of the head and neck are a heterogenous group of epithelial malignancies that manifest "pure" neuroendocrine differentiation. Three tumor grades are generally recognized. These lesions may be encountered in the larynx, sinonasal cavity, salivary glands, mouth, and middle ear (126,127). The diagnosis of NEC is

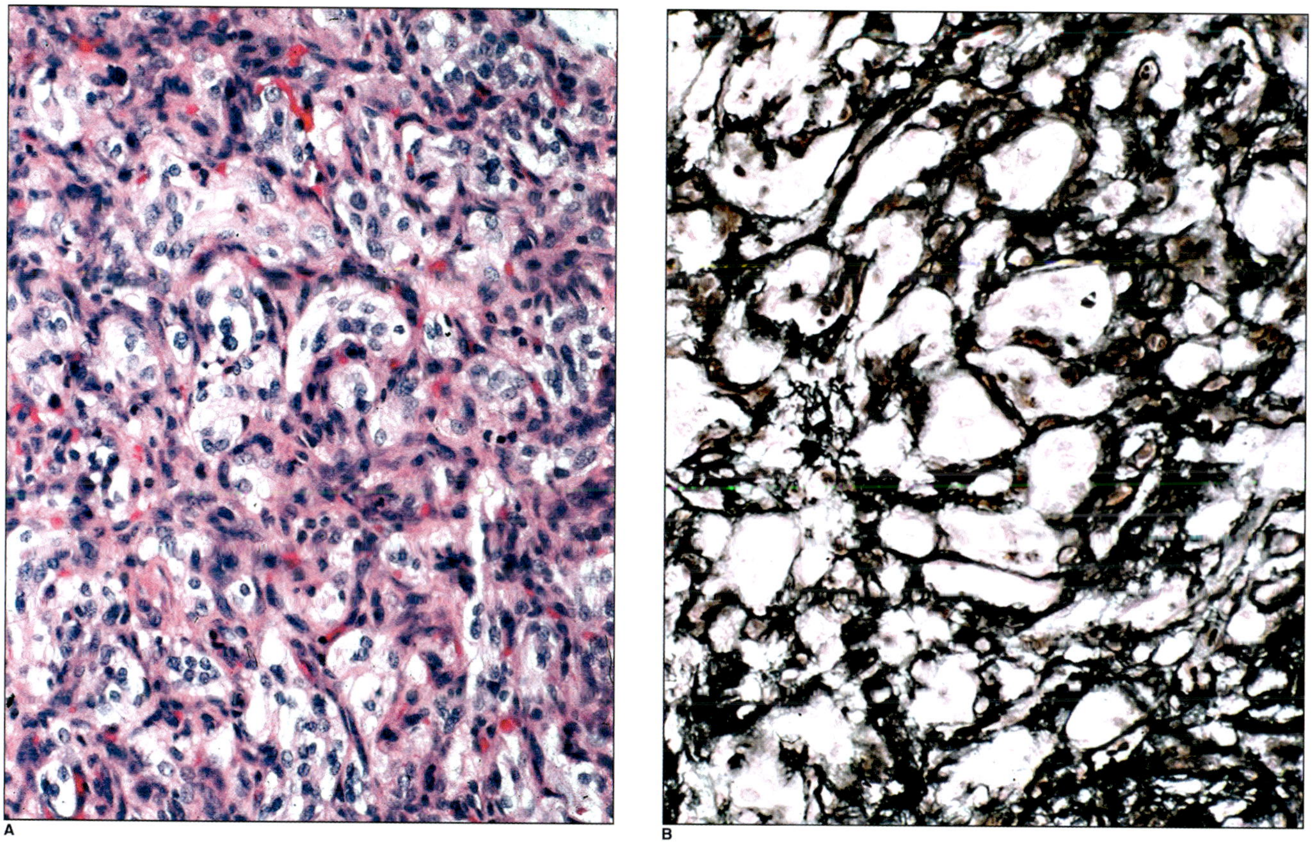

A B

FIGURE 2.29: *Jugulotympanic paraganglioma. (A) Not uncommonly, the classic cell nest pattern of this neoplasm may be somewhat obscured; (B) reticulin stain nicely delineates the cell nest pattern as seen by the presence of black-appearing fibers enveloping individual tumor nests.*

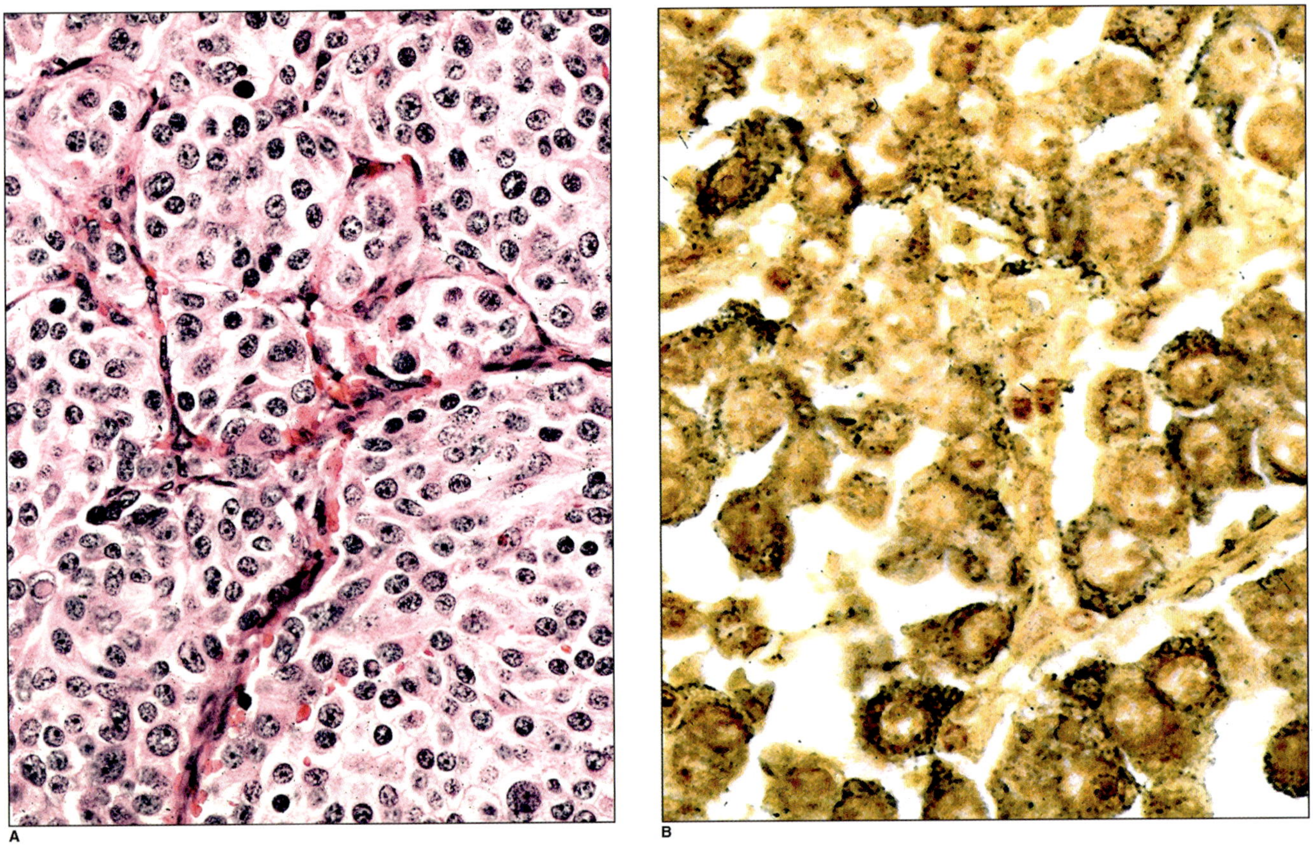

FIGURE 2.30: Atypical carcinoid tumor. (A) Characteristic neoplastic infiltrate (H&E stain); (B) argyrophilic (Churukian-Schenk)–positive intracytoplasmic material.

predicated on light microscopic features and histochemical or immunohistochemical properties. The argyrophilic procedures cited above label NECs with varying degrees of intensity, generally inversely paralleling tumor grade (Figures 2.30 and 2.31); uncommonly, argentaffin methods (e.g., the Fontana-Masson stain) are also positive. Because some NEC may demonstrate divergent glandular differentiation, focal luminal reactivity with the mucicarmine, alcian blue, colloidal iron, and PAS-D techniques can be present.

Olfactory Neuroblastoma: ONB is a malignant, small cell neuroectodermal neoplasm that is related to the olfactory membrane of the sinonasal tract. Immunostains have generally supplanted histochemistry in the diagnosis of this tumor. However, positive staining with the Bodian procedure—which labels nerve fibers and nerve endings—and the Grimelius method can be observed in ONBs. In addition, cytoplasmic PAS-positive material may be present, and reticulin stains can be used to outline the organoid growth pattern of these lesions.

Primitive Neuroectodermal Tumors: PNETs represent a group of high-grade primitive small round cell neoplasms with neuroectodermal differentiation. They are largely defined by the presence of a balanced chromosomal translocation between the Ewing's sarcoma (*EWS*) gene on chromosome 22 and the *FLI-1* gene on chromosome 11 (128). PNETs are rare in the head and neck (129) but can potentially arise in the skull and jaws, the sinonasal tract, orbits, and various mucosal sites. Histochemical studies often show abundant cytoplasmic glycogen in the tumor cells with the undigested PAS procedure.

Mesenchymal and Lymphoreticular Tumors

Myogenic Neoplasms: Myogenous tumors of the head and neck include neoplasms with skeletal muscle differentiation (rhabdomyoma and rhabdomyosarcoma) and smooth muscle lesions (leiomyoma and leiomyosarcoma). In all such proliferations, the PTAH stain may assist in the identification of cytoplasmic cross-striations and myofibrils (Figure 2.32); the Masson trichrome stain is similarly helpful (Figure 2.33). Moreover, intracellular glycogen can be identified with the PAS method, particularly in striated muscle tumors.

Synovial Sarcoma: Synovial sarcoma (SS) is a malignant mesenchymal neoplasm of uncertain histogenesis; it

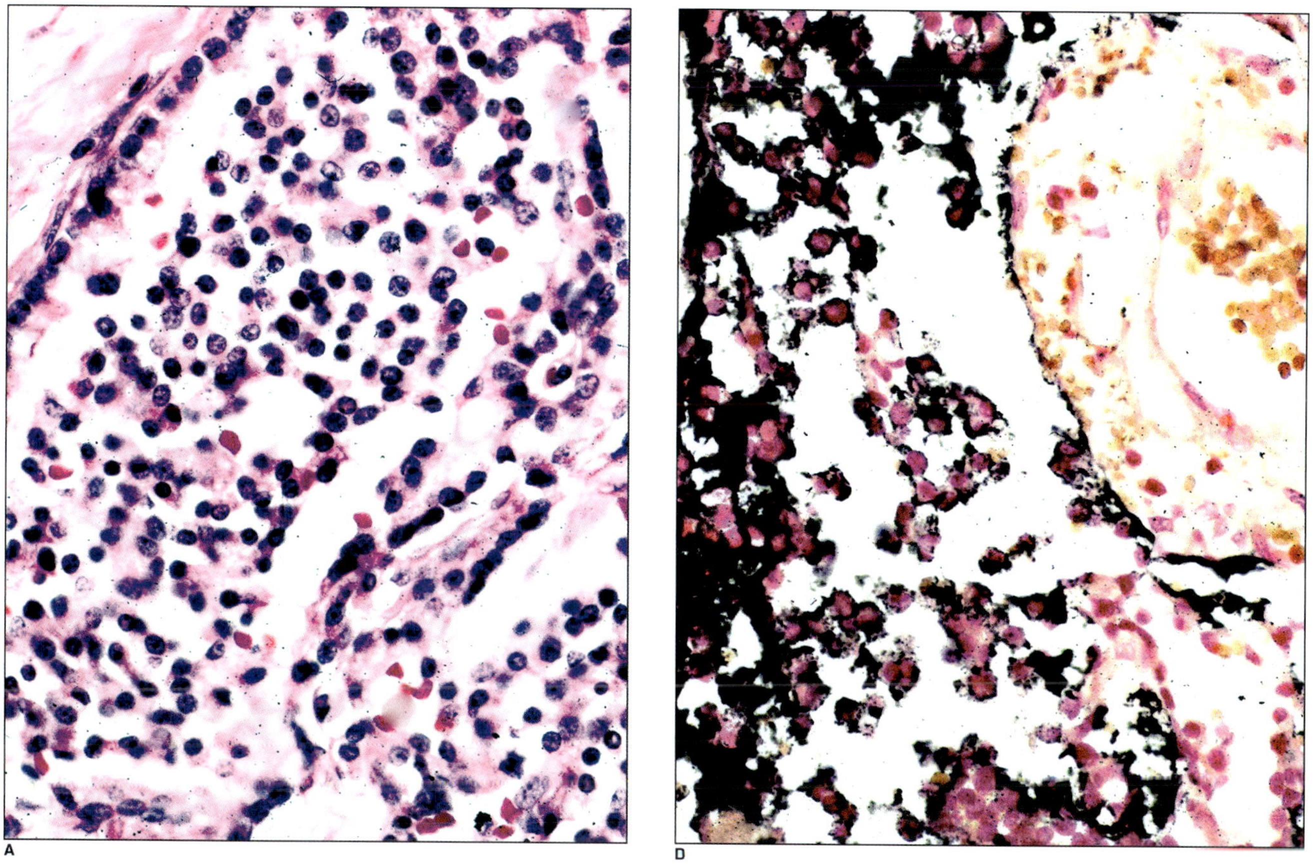

FIGURE 2.31: Carcinoid tumor. (A) Characteristic neoplastic infiltrate (H&E stain); (B) argentaffin (Fontana-Masson)–positive intracytoplasmic material.

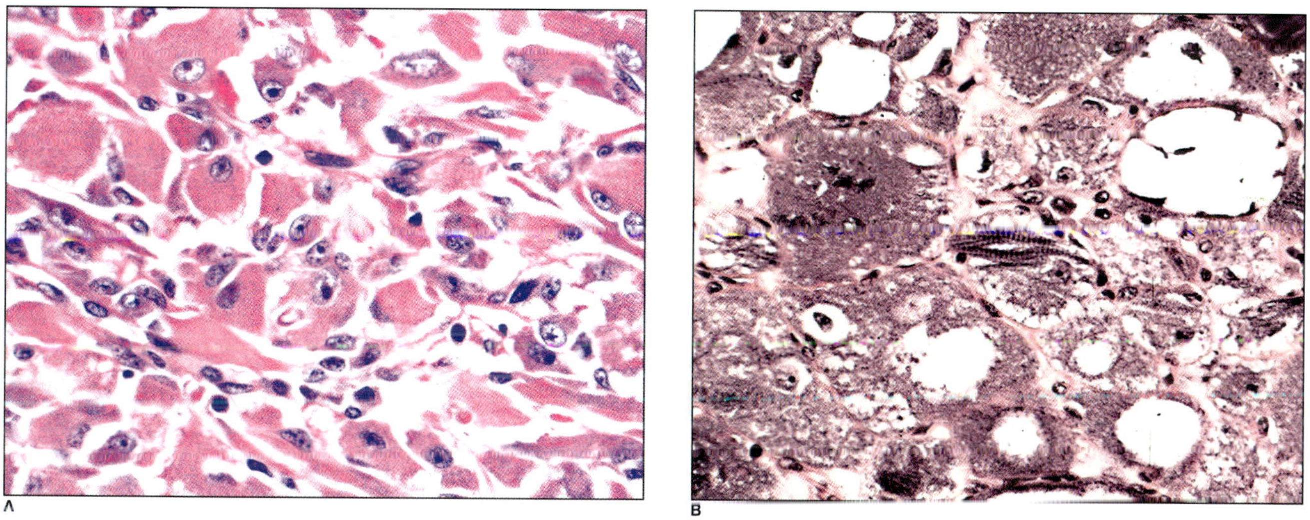

FIGURE 2.32: (A and B) Rhabdomyoma. PTAH stain assists in the identification of intracytoplasmic cross striations.

primarily arises in the soft tissues in close association with joints, but is not confined to such locations. The most common sites of occurrence for SS in the head and neck are the hypopharynx and retropharynx (130). This tumor comprises two cell types, which are epithelioid and fusiform; depending on the relative prominence of each, SS is classified as biphasic or monophasic. A small cell variant is also recognized. Histochemical stains for mucin, including PAS-D, mucicarmine, alcian blue, and colloidal iron, show positivity in tumoral pseudoglands and intercellular spaces,

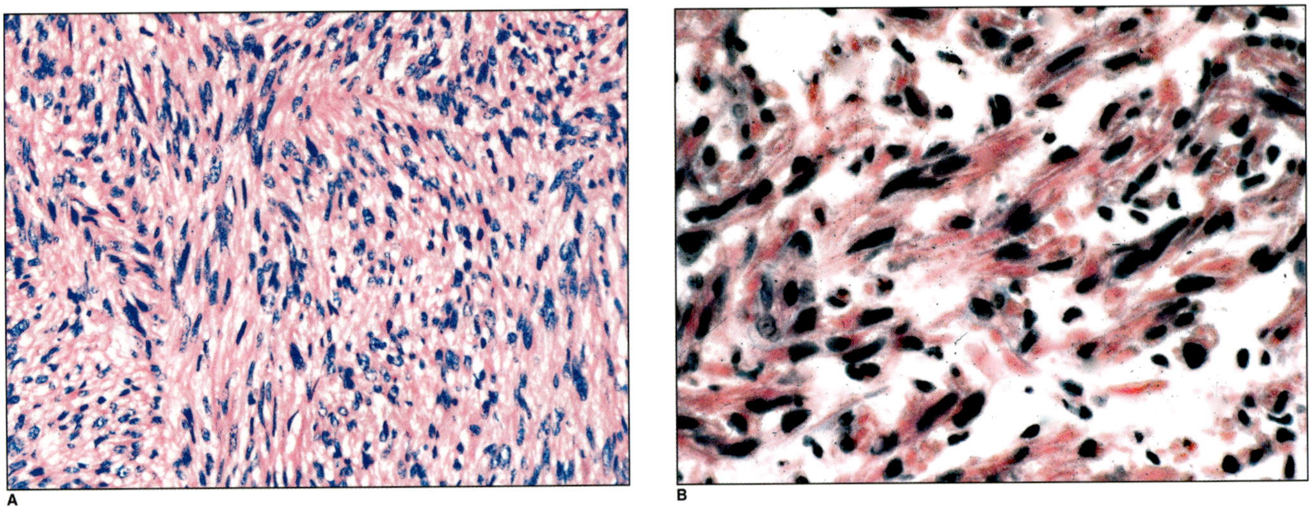

FIGURE 2.33: (A and B) Leiomyosarcoma. Masson trichrome stain showing the presence of intracytoplasmic myofibrils.

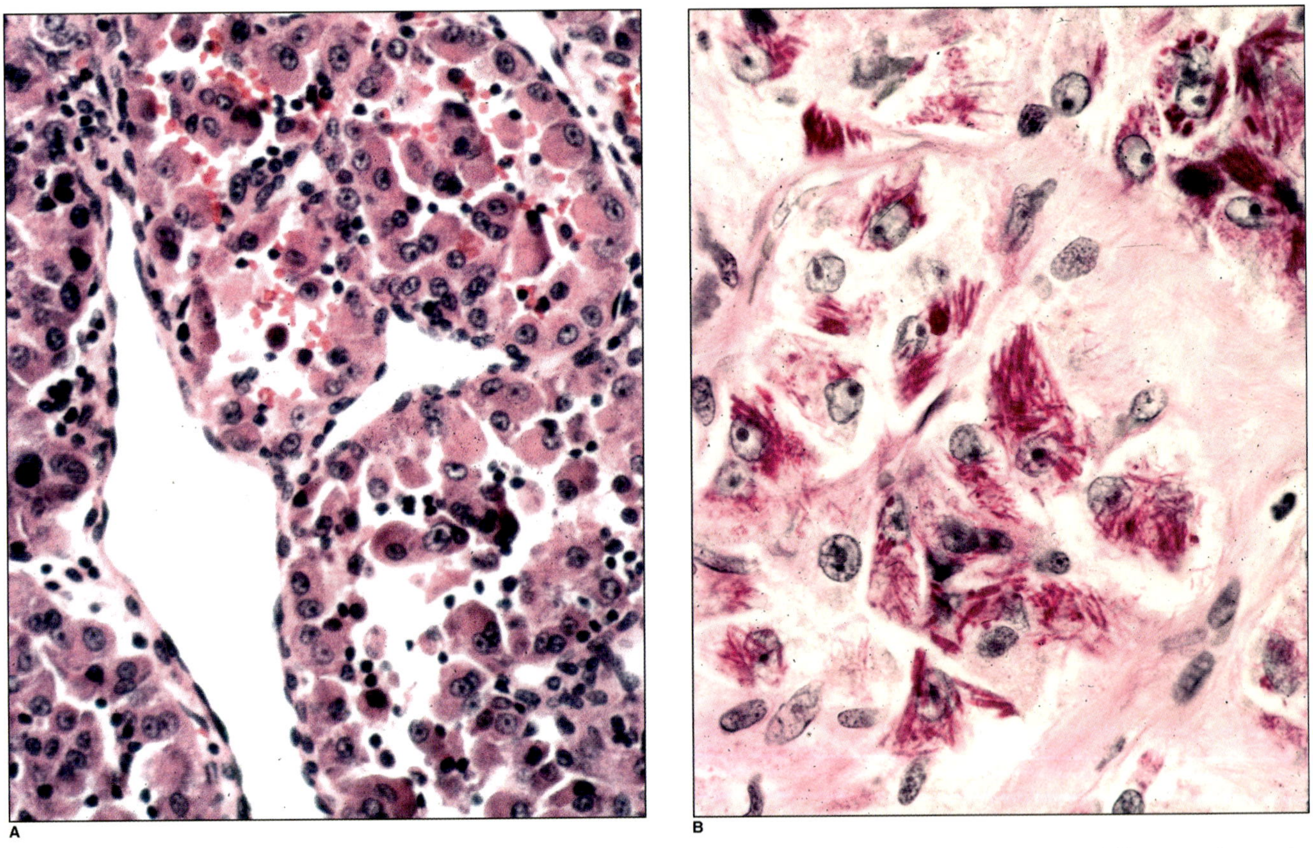

FIGURE 2.34: Lingual ASPS. (A) Alveolar, organoid, or nest-like growth with nests separated by thin-walled fibrovascular septa; (B) intracytoplasmic PAS-positive, diastase-resistant rhomboid or rod-shaped crystalline material is seen, which is considered to be a diagnostic feature for this tumor.

even in monophasic lesions. This reactivity remains despite pretreatment of sections with diastase and hyluronidase. Numerous mast cells can also be seen in the stroma of SS; they can be labeled with the toluidine blue method or the Leder (chloroacetate esterase) stain.

Alveolar Soft Part Sarcoma: Alveolar soft part sarcoma (ASPS) represents a rare and slowly growing but lethal malignancy of the soft tissue (131). In infants and children, ASPS most often arises in the head and neck, especially in the tongue and orbits. Histologically, this

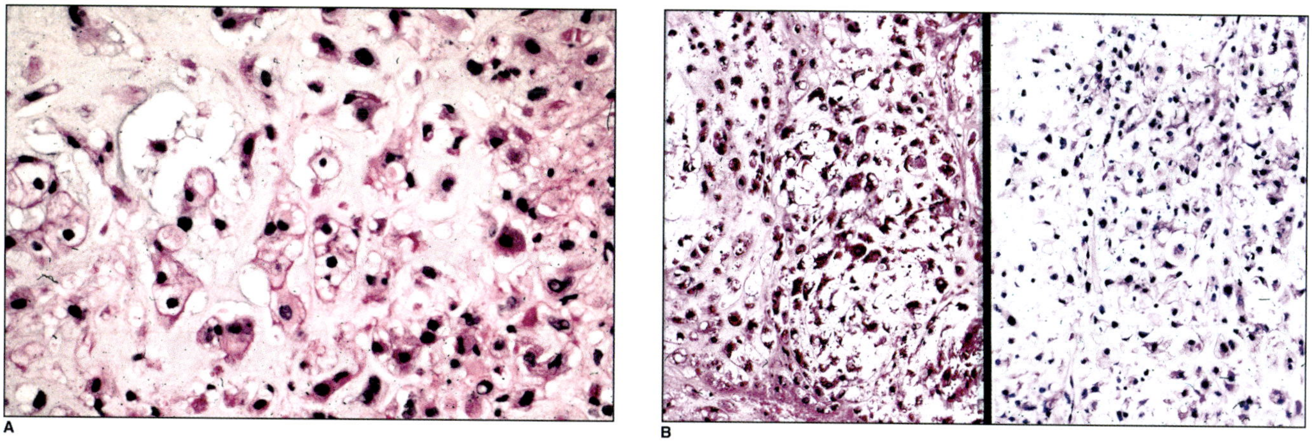

FIGURE 2.35: Chordoma. (A) Characteristic cell infiltrate including bubbly (physaliferous) cells; (B) the neoplastic cells show the presence of (left) PAS-positive intracytoplasmic material that is (right) sensitive to diastase digestion.

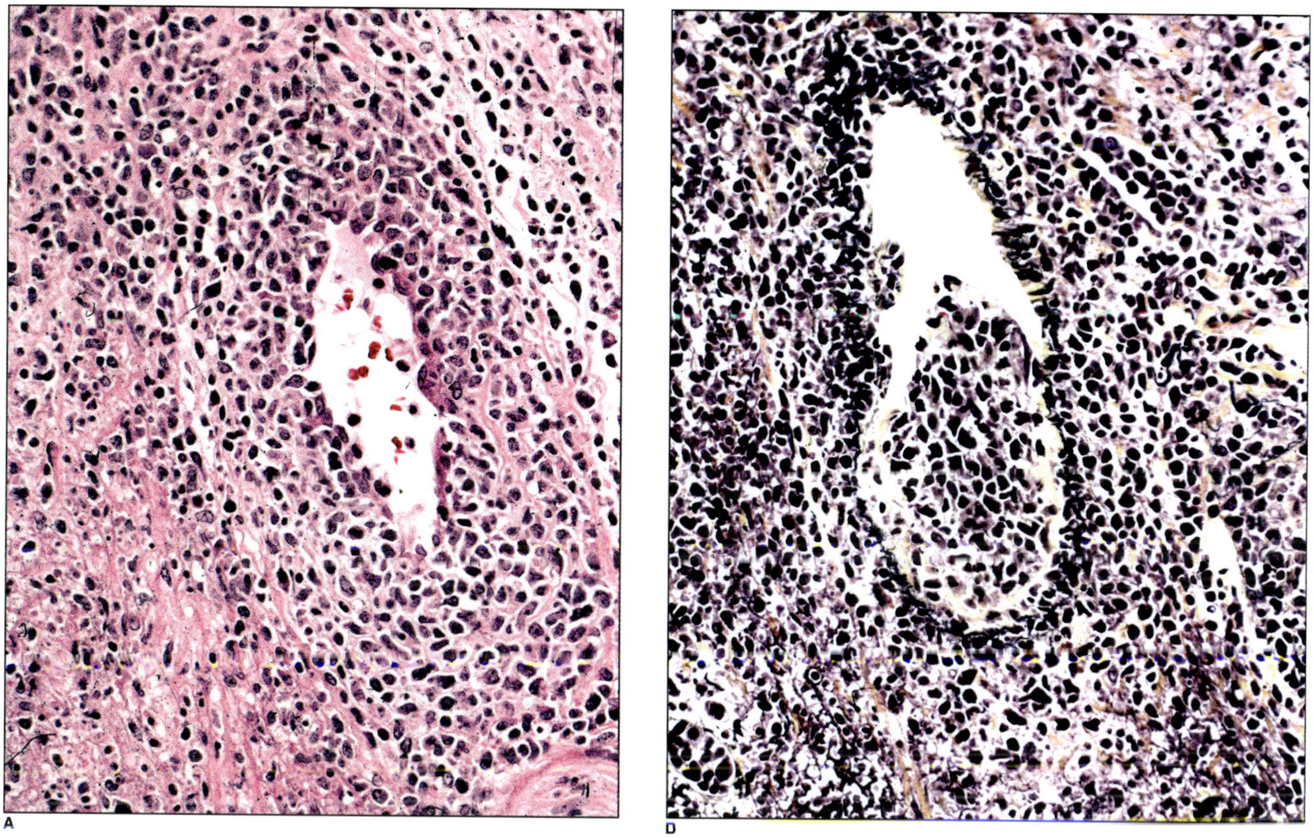

FIGURE 2.36: Nasal-type NK/T-cell lymphoma. Left: the malignant cells are angiocentric and angioinvasive; Right: elastic stain assists in identifying vascular structures with disruption of the elastic membrane (appearing black and discontinuous) by the malignant cells penetrating the vessel wall and filling the lumen resulting in the ischemic-type necrosis seen in this disease.

tumor is characterized by an alveolar, organoid growth pattern wherein thin fibrovascular septa separate groups of plump epithelioid cells with eosinophilic cytoplasm. In children, one sees a greater tendency toward solid growth in ASPS and the nesting pattern is often inconspicuous (132). The tumor cells contain vesicular nuclei and one or more nucleoli. Histochemical studies show cytoplasmic PAS-D positivity in ASPS, in the form of rhomboidal or rod-shaped crystalline material. This feature is considered to be diagnostic for this tumor (Figure 2.34). The crystals can be identified in both primary and metastatic lesions. In addition, ASPS cells contain glycogen. Reticulin stains or the trichrome method can be applied to outline the alveolar architecture of ASPS;

they are especially helpful in the diagnosis of its "solid" variant.

Kaposi's Sarcoma: KS is a locally aggressive endothelial tumor that is associated with infection by human herpesvirus 8 infection. The presence of diastase-resistant, PAS-positive hyaline globules in this tumor may assist in its diagnosis. These inclusions also label with the peroxidase method.

Chordoma of the Craniocervical Region: Chordomas recapitulate the formation of embryonic remnants of the notochord (133,134). Histochemical stains of these tumors reveal cytoplasmic PAS-D positivity in most cases, and their myxoid stroma often labels with alcian blue and colloidal iron (Figure 2.35).

Extranodal NK/T-Cell Lymphoma of Nasal Type: Extranodal NK/T-cell lymphoma (ENKTCL) of the "nasal" type usually presents as a destructive process of the midfacial region (135). Histologically, it may show a spectrum of appearances but cytologic atypicality is a constant. In addition, tissue necrosis is consistently seen, often with a geographic pattern. The neoplastic cells demonstrate angioinvasion and angiodestruction; elastic stains such as the Verhoeff–Van Gieson method can be used to good advantage to demonstrate this feature (Figure 2.36). The differential diagnosis may also include mycobacterial or fungal infections, and histochemical studies for microorganisms are therefore prudent before making a diagnosis of ENKTCL.

Secondary Tumors of the Head and Neck

Secondary malignant tumors of the head and neck originate in other anatomic locations. Renal cell carcinoma and malignant melanoma are among the most common of these. Histochemical studies have relatively limited use in the localization of the primary sites for such lesions. Nevertheless, renal cell carcinoma is a possible exception because it is singular in its concurrent content of abundant glycogen and lipid (demonstrable with PAS and oil red O or Sudan IV stains) in the tumor cells.

ILLUSTRATIVE CASE STUDIES

CASE 1— *Extramedullary Plasmacytoma:* A 59-year-old man developed nasal stuffiness that was unresponsive

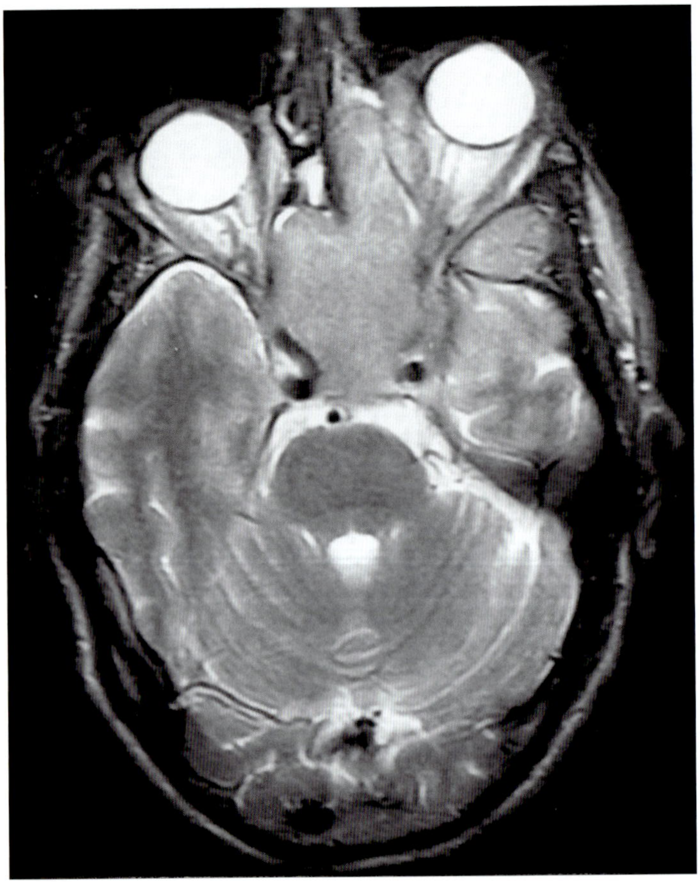

FIGURE 2.37: ***This magnetic resonance image of the head and neck shows a mass in the nose, left orbit, and paranasal sinuses.***

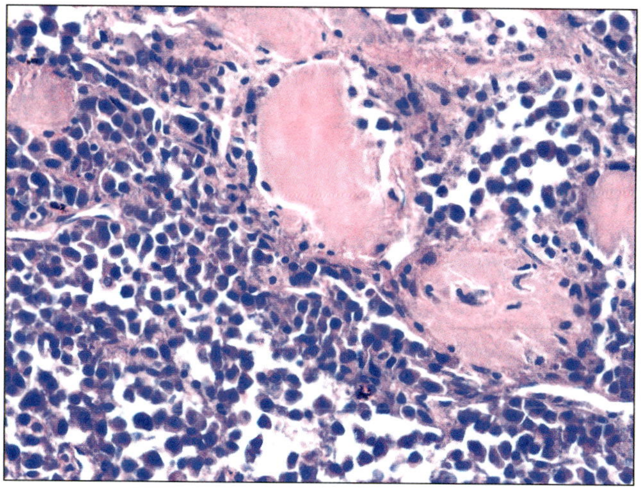

FIGURE 2.38: ***Sheets of ovoid cells with high nucleocytoplasmic ratios, eccentric nuclei, and "blocky" chromatin are interspersed with amorphous eosinophilic extracellular deposits in this nasal biopsy.***

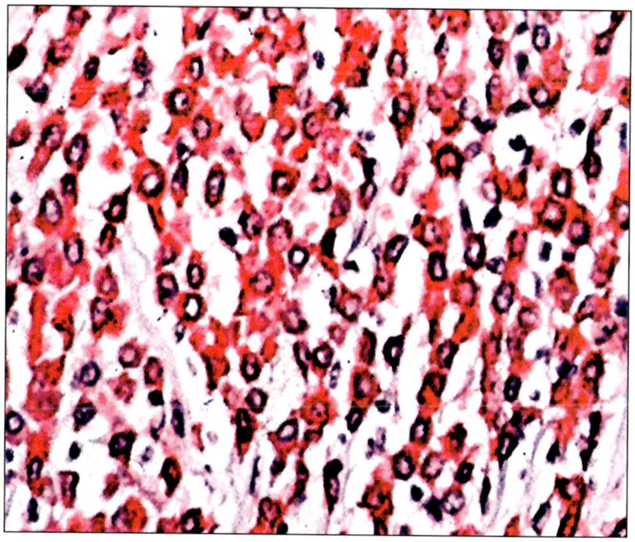

FIGURE 2.40: ***Intense reactivity is present in the neoplastic cells of plasmacytoma with the methyl green-pyronine stain.***

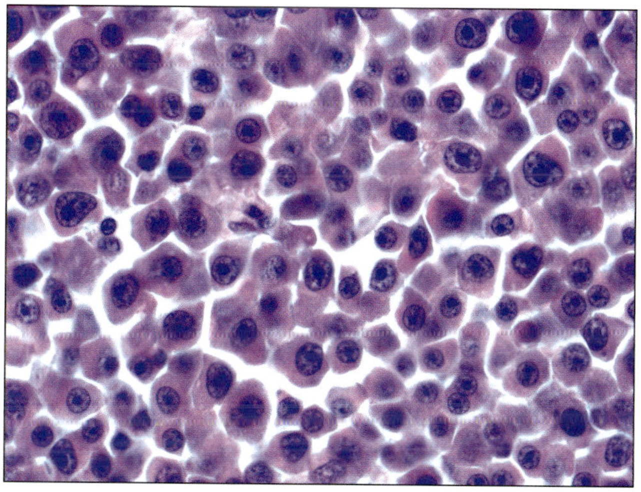

FIGURE 2.39: ***The tumor cells have eccentric nuclei, amphophilic cytoplasm, and coarse chromatin.***

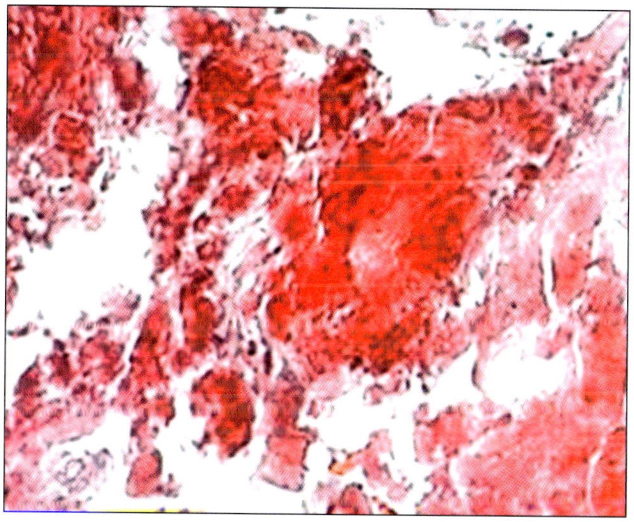

FIGURE 2.41: ***A Congo red stain of the amorphous intralesional material shows intense positivity, confirming its identity as amyloid.***

to medication, along with frontal headaches. Radiographic studies of the nose and paranasal sinuses showed a mass in the left nose and orbit and both ethmoid sinuses, with involvement of the surrounding bones (Figure 2.37). A biopsy was obtained, demonstrating sheets of rounded epithelioid cells with high nucleocytoplasmic ratios, eccentric nuclei, "blocky" chromatin, and amphophilic cytoplasm. Amorphous eosinophilic extracellular material was also present in the lesion multifocally (Figures 2.38 and 2.39). The histologic differential diagnosis included undifferentiated carcinoma, melanoma, and a malignant lymphoreticular neoplasm.

Histochemical stains showed intense positivity of the tumor cells with the methyl green-pyronine method (Figure 2.40); they were also weakly and diffusely reactive with the PAS-D technique. The mucicarmine, colloidal iron, and Fontana-Masson methods showed no positivity in the neoplasm. The extracellular eosinophilic deposits were labeled by the Congo red stain (Figure 2.41), and they demonstrated apple-green birefringence on polarization microscopy. A diagnosis of extramedullary plasmacytoma was made. Subsequent immunofixation studies of the serum and urine showed the presence of monotypic IgG-kappa immunoglobulin. ■

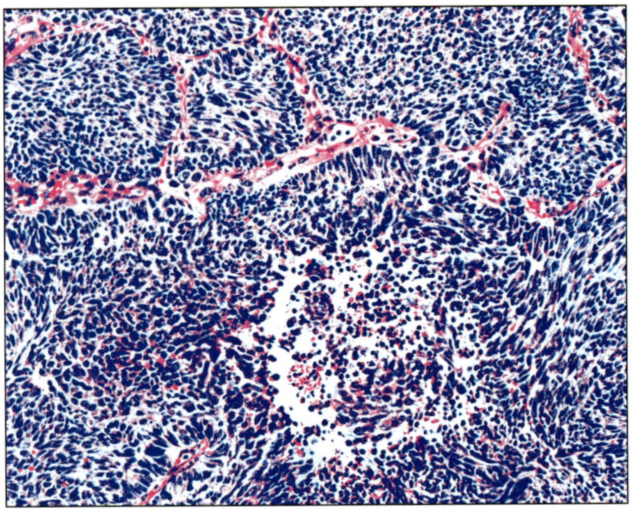

FIGURE 2.42: ***This laryngeal tumor comprises basaloid cells that are arranged in a nested pattern, with foci of necrosis.***

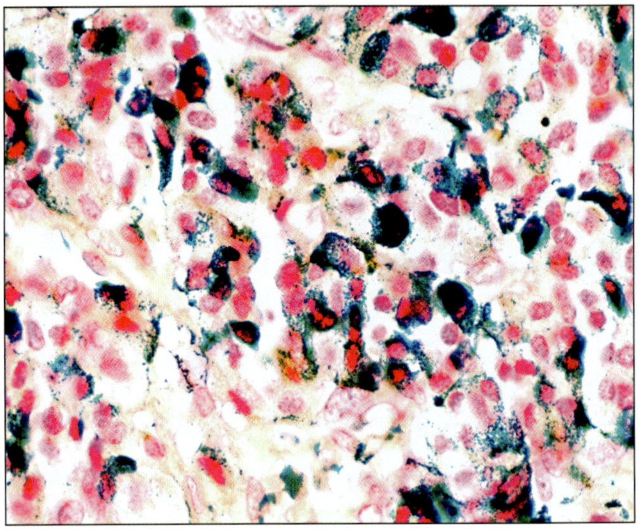

FIGURE 2.44: ***Multifocal granular positivity is present in grade 2 NEC of the larynx, in this Churukian-Schenk argyrophil stain.***

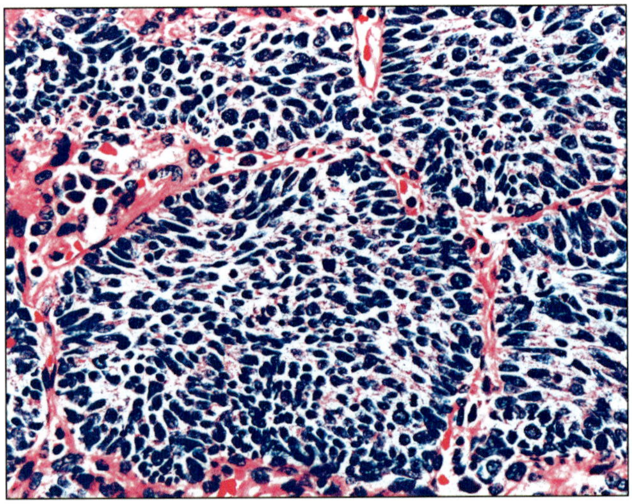

FIGURE 2.43: ***The neoplastic cells have dispersed chromatin, and numerous mitoses are seen.***

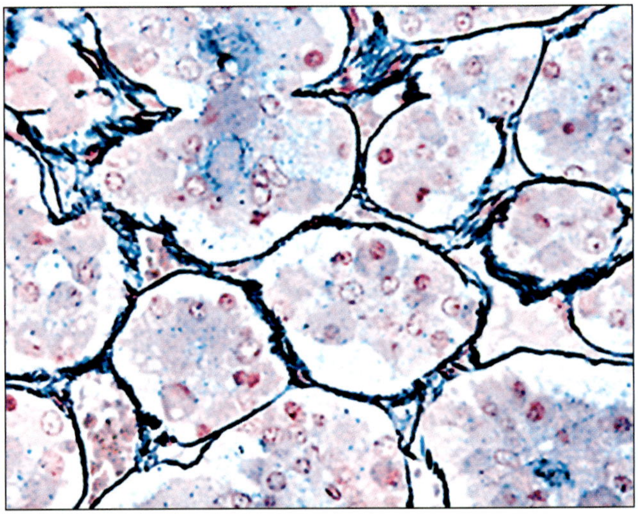

FIGURE 2.45: ***Wilder's stain for reticulin accentuates the organoid growth pattern of NEC.***

CASE 2— ***Laryngeal Neuroendocrine Carcinoma:*** A 59-year-old man presented with progressive hoarseness and odynophagia. He had a long history of cigarette smoking and alcohol abuse. Laryngoscopy showed a mass in the left supraglottis that replaced the entire vestibular fold. A biopsy of the lesion demonstrated a basaloid malignant neoplasm with an organoid growth pattern, brisk mitotic and apoptotic activity, and foci of necrosis. Nuclei were ovoid with dispersed chromatin and inconspicuous nucleoli (Figures 2.42 and 2.43). A Churukian-Schenk stain showed granular positivity in many of the tumor cells (Figure 2.44) and Wilder's method showed distinctly organoid architecture (Figure 2.45); PAS, mucicarmine, and colloidal iron stains were negative. A diagnosis of grade 2 NEC was made, and the patient was advised to undergo laryngectomy and cervical lymphadenectomy. However, he declined any active treatment whatsoever, preferring to pursue "alternative" medical care. Widespread metastases to the lungs, liver, and bones developed subsequently, followed by death 9 months after diagnosis. At autopsy, a 4-cm mass was found that effaced the left larynx (Figure 2.46). ■

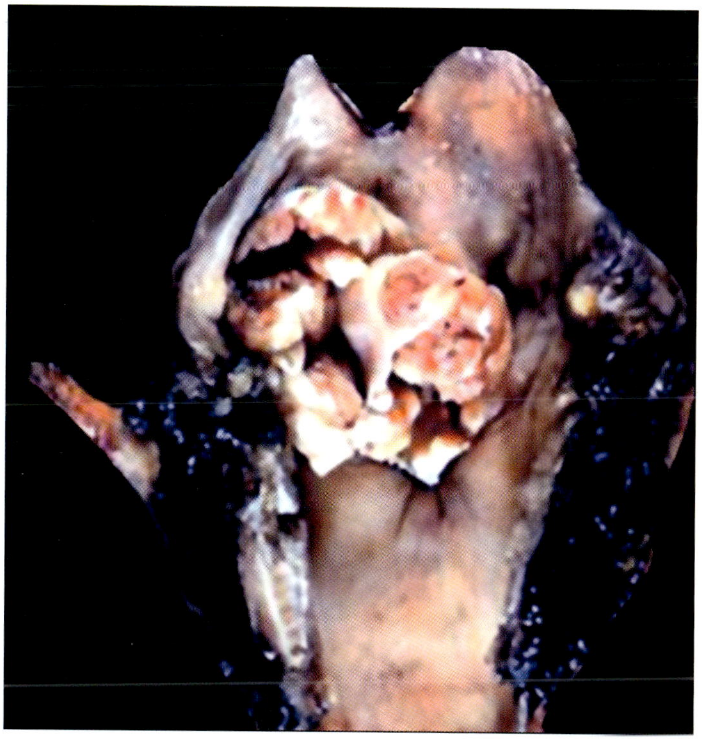

FIGURE 2.46: A large left laryngeal mass is present at autopsy, representing grade 2 NEC.

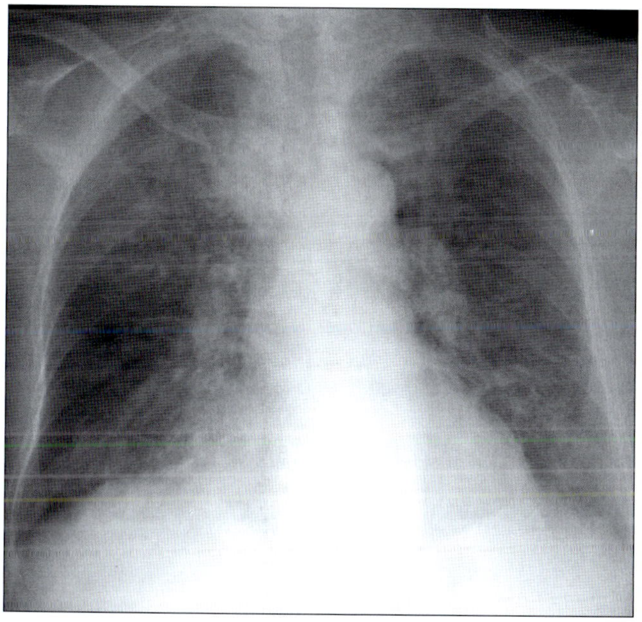

FIGURE 2.47: This chest radiograph shows diffuse bilateral interstitial and alveolar pneumonic infiltrates.

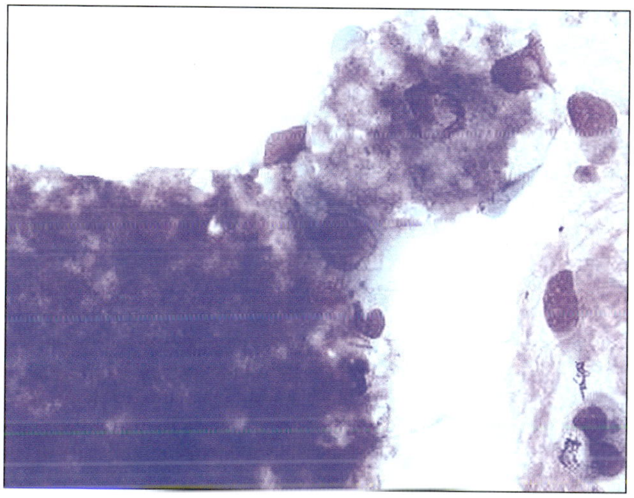

FIGURE 2.48: A bronchoalveolar lavage specimen demonstrates the presence of granular alveolar casts, as seen in this Romanowsky stain.

CASE 3— *Extrapulmonary Pneumocystosis:* A 41-year-old man had been infected with the HIV for 15 years but had only become refractory to antiretroviral chemotherapy in the past 12 months. During that time, he developed an episode of acute pneumonia (Figure 2.47) and a bronchoalveolar lavage specimen showed the presence of PC (Figures 2.48 and 2.49). Treatment with intravenous trimethoprim and sulfamethoxazole was successful, and the patient was subsequently given daily aerosolized pentamadine for PC prophylaxis. Six months thereafter, a painless red plaque developed on the bridge of his nose, which

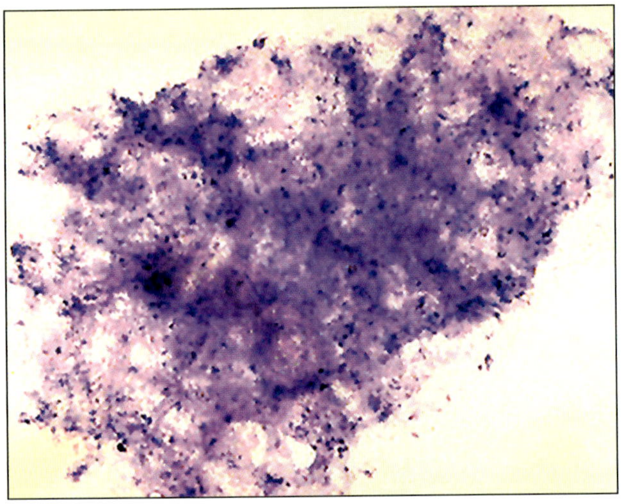

FIGURE 2.49: *Giemsa staining of the bronchoalveolar lavage demonstrates the presence of numerous darkly labeling ovoid organisms, representing PC.*

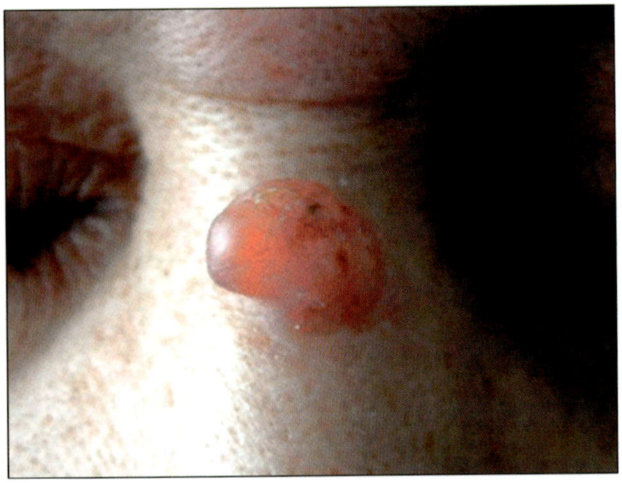

FIGURE 2.50: *A 1-cm reddish nodule is present in the skin of the nasal bridge.*

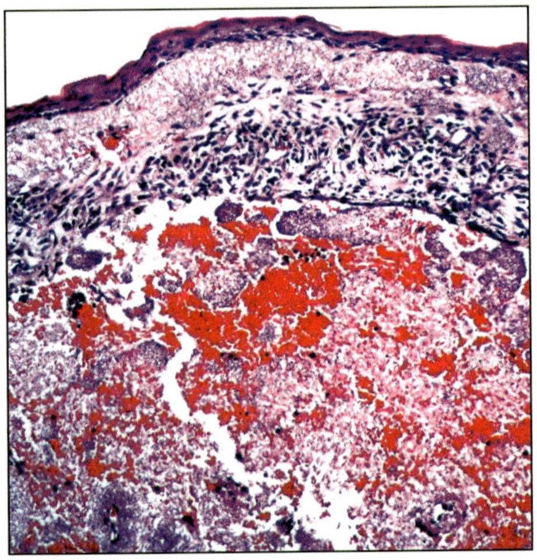

FIGURE 2.51: *A coarsely granular, hypocellular dermal lesion is present in this skin biopsy from the nasal bridge.*

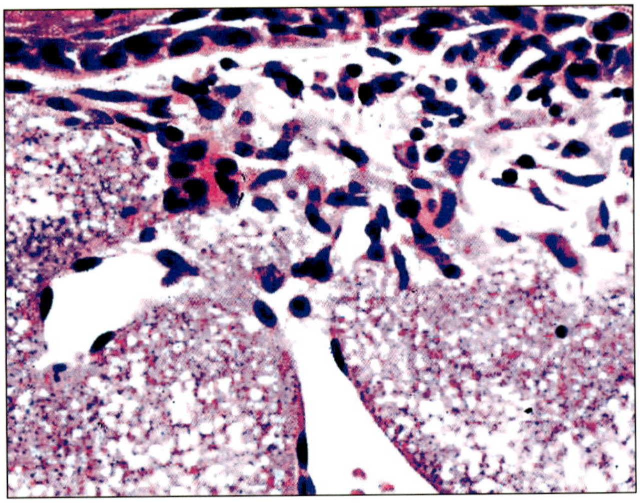

FIGURE 2.52: *A higher magnification view of the previous figure shows a flocculent acellular zone in the corium. There is no vascular proliferation and inflammation is virtually nonexistent.*

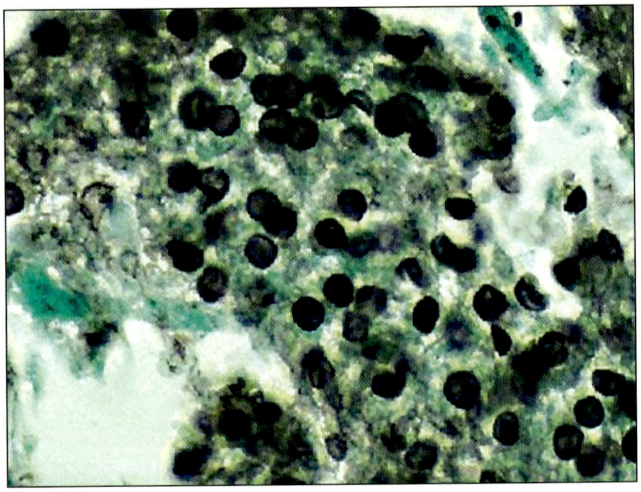

FIGURE 2.53: *Positive silver methenamine staining of the skin biopsy establishes the diagnosis of extrapulmonary pneumocystosis.*

enlarged to become a 1-cm nodule (Figure 2.50). The diagnosis of KS was suspected clinically, and a biopsy of the skin lesion was done. It showed a relatively circumscribed, hypocellular, coarsely granular lesion of the deep dermis, with only scant chronic inflammation. No vascular proliferation was present (Figures 2.51 and 2.52). A GMS stain demonstrated numerous intralesional ovoid organisms with central densities, having the appearance of PC (Figure 2.53). The final diagnosis was that of extrapulmonary pneumocystosis.

REFERENCES

1. Katzenstein AL, Sale SR, Greenberger PA. Pathologic findings in allergic aspergillus sinusitis. A newly recognized form of sinusitis. *Am J Surg Pathol* 1983;7:439–43.
2. Adam RD, Paquin ML, Petersen EA, et al. Phaeohyphomycosis caused by the fungal genera Bipolaris and Exserohilum. A report of 9 cases and review of the literature. *Medicine (Baltimore)* 1986;65:203–17.
3. Macmillan RH, Cooper PH, Body BA, Mills AS. Allergic fungal sinusitis due to curvularia lunata. *Hum Pathol* 1987;18:960–4.
4. Rinaldi MG, Phillips P, Schwartz JG, et al. Human curvularia infections: report of five cases and review of the literature. *Diagn Microbiol Infec Dis* 1987;6:27–39.
5. Friedman GC, Hartwick WJ, Ro JY, et al. Allergic fungal sinusitis: report of three cases associated with dematiaceous fungi. *Am J Clin Pathol* 1991;96:368–72.
6. Zieske LA, Kopke RD, Hamill R. Dematiaceous fungal sinusitis. *Otolaryngol Head Neck Surg* 1991;105:567–77.
7. Brandt ME, Warnock DW. Epidemiology, clinical manifestations, and therapy of infections caused by dematiaceous fungi. *J Chemother* 2003;2:36–47.
8. Blitzer A, Lawson W. Fungal infections of the nose and paranasal sinuses. *Part I. Otolaryngol Clin North Am* 1993;26:1007–35.
9. Cody DT 2nd, Neel HB 3rd, Ferreiro JA, Roberts GD. Allergic fungal sinusitis: the Mayo Clinic experience. *Laryngoscope* 1994;104:1074–9.
10. deShazo RD, Swain RE. Diagnostic criteria for allergic fungal sinusitis. *J Allergy Clin Immunol* 1995;96:24–35.
11. Saeed SR, Brooks GB. Aspergillosis of the paranasal sinuses. *Rhinology* 1995;33: 46–51.
11a. Schubert MS. Allergic fungal sinusitis. *Otolaryngol Clin North Am* 2001;37.301–26.
12. Torres C, Ro JY, el-Naggar AK, Sim SJ, Weber RS, Ayala AG. Allergic fungal sinusitis: a clinicopathologic study of 16 cases. *Hum Pathol* 1996;27:793–9.
13. Uri N, Cohen-Kerem R, Elmalah I, Doweck I, Greenberg E. Classification of fungal sinusitis in immunocompetent patients. *Otolaryngol Head Neck Surg* 2003;129:372–8.
14. Heffner DK. Allergic fungal sinusitis is a histopathologic diagnosis; paranasal mucocele is not. *Ann Diagn Pathol* 2004;8:316–23.
15. Granville L, Chirala M, Cernoch P, Ostrowski M, Truong LD. Fungal sinusitis: histologic spectrum and correlation with culture. *Hum Pathol* 2004;35:474–81.
16. Ferreiro JA, Carlson BA, Cody DT 3rd. Paranasal sinus fungus balls. *Head Neck* 1997;19:481–6.
17. deShazo RD, O'Brien M, Chapin K, Soto-Aguilar M, Gardner L, Swain R. A new classification and diagnostic criteria for invasive fungal sinusitis. *Arch Otolaryngol Head Neck Surg* 1997;123:1181–8.
18. Anselmo-Lima WT, Lopes RP, Valera FC, Demarco RC. Invasive fungal rhinosinusitis in immunocompromised patients. *Rhinology* 2004;42:141–4.
19. Dufour X, Kauffmann-Lacroix C, Roblot F, et al. Chronic invasive fungal rhinosinusitis: two new cases and review of the literature. *Am J Rhinol* 2004;18:221–6.
20. Morgan MA, Wilson WR, Neel HB, Roberts GD: Fungal sinusitis in healthy and immunocompromised individuals. *Am J Clin Pathol* 1984;82:597–601.
21. Morrison VA, McGlave PB. Mucormycosis in the BMT population. *Bone Marrow Transplant* 1993;11:383–8.
22. Gaziev D, Baronciani D, Galimberti M, et al. Mucormycosis after bone marrow transplantation: report of four cases in thalassemia and review of the literature. *Bone Marrow Transplant* 1996;17:409–14.
23. Ferguson BJ. Mucormycosis of the nose and paranasal sinuses. *Otolaryngol Clin North Am* 2000;33:349–65.
24. Frater JL, Hall GS, Procop GW. Histologic features of zygomycosis: emphasis on perineural invasion and fungal morphology. *Arch Pathol Lab Med* 2001;125:375–8.
25. Rangel-Guerra R, Martinez HR, Saenz C. Mucormycosis. Report of 11 cases. *Arch Neurol* 1985;42:578–81.
26. Satyanarayana C. Rhinosporidiosis with a record of 225 cases. *Acta Otolaryngol* 1960;51:348–56.
27. Lasser A, Smith HW. Rhinosporidiosis. *Arch Otolaryngol* 1976;102:308–10.
28. Batsakis JG, el-Naggar AK. Rhinoscleroma and rhinosporidiosis. *Ann Otol Rhinol Laryngol* 1992;101:879–82.
29. Makannavar JH, Chavan SS. Rhinosporidiosis—a clinicopathological study of 34 cases. *Indian J Pathol Microbiol* 2001;44:17–21.
30. Lehner T. Oral thrush, or acute pseudomembranous candidiasis: a clinicopathologic study of forty-four cases. *Oral Surg Oral Med Oral Pathol* 1964;18:27–37.
31. Holmstrup P, Axéll T. Classification and clinical manifestations of oral yeast infections. *Acta Odontol Scand* 1990;48:57–9.
32. Fotos PG, Vincent SD, Hellstein JW. Oral candidosis: clinical, historical and therapeutic features of 100 cases. *Oral Surg Oral Med Oral Pathol* 1992;74:41–9.
33. Epstein JB, Polsky B. Oropharyngeal candidiasis: a review of its clinical spectrum and current therapies. *Clin Ther* 1998;20:30–57.
34. Wright BA. Median rhomboid glossitis: not a misnomer. Review of the literature and histologic study of twenty-eight cases. *Oral Surg Oral Med Oral Pathol* 1978;46:806–14.
35. van der Waal I, Beemster G, van der Kwast WA. Median rhomboid glossitis caused by Candida? *Oral Surg Oral Med Oral Pathol* 1979;47:31–5.
36. Carter LC. Median rhomboid glossitis: review of a puzzling entity. *Compendium* 1990;11:446–51.
37. Kolokotronis A, Kioses V, Antoniades D, et al. Median rhomboid glossitis. An oral manifestation in patients infected with HIV. *Oral Surg Oral Med Oral Pathol* 1994;78:36–40.
38. Morgan M, Reves R. Invasive sinusitis due to Sporothrix schenckii in a patient with AIDS. *Clin Infect Dis* 1996; 23:1319–20.
39. Kumar R, Kaushal V, Chopra H, Gupta R, Chhina DK, Kaushal RK, Mohan U. Pansinusitis due to Sporothrix schenckii. *Mycoses* 2005;48:85–8.
40. Goldman SN, Canalis RF. Rhinoscleroma as a cause of airway obstruction. *Ear Nose Throat J* 1980;59:6–14.
41. Verma G, Kanawaty D, Hyland R. Rhinoscleroma causing upper airway obstruction. *Can Respir J* 2005;12:43–5.

42. Hyams VJ, Batsakis JG, Michaels L. Rhinoscleroma. In: Tumors of the Upper Respiratory Tract and Ear. Series 2, Fascicle 25. Armed Forces Institute of Pathology, Washington, D.C., 1988:24–6.
43. Frankel SS, Wenig BM. Human immunodeficiency virus and acquired immunodeficiency syndrome. In: Fu Y-S, Wenig BM, Abemayor E, Wenig BL, eds. Head and Neck Pathology with Clinical Correlations. Churchill Livingstone, New York, 2001:65–80.
44. Koehler JE, Quinn FD, Berger TG, LeBoit PE, Tappero JW. Isolation of *Rochalimaea* species from cutaneous and osseous lesions of bacillary angiomatosis. *N Engl J Med* 1992;327:1625–31.
45. Koehler JE, Sanchez MA, Garrido CS, Whitfeld MJ, Chen FM, Berger TG, Rodriguez-Barradas MC, LeBoit PE, Tappero JW. Molecular epidemiology of *bartonella* infections in patients with bacillary angiomatosis-peliosis. *N Engl J Med* 1997;337:1876–83.
46. Spach DH, Koehler JE. *Bartonella*-associated infections. *Infect Dis Clin North Am* 1998;12:137–55.
47. Maguina C, Gotuzzo E. Bartonellosis. New and old. *Infect Dis Clin North Am* 2000;14:1–22.
48. LeBoit PE, Berger TG, Egbert BM, Beckstead JH, Yen TS, Stoler MH. Bacillary angiomatosis. The histopathology and differential diagnosis of a pseudoneoplastic infection in patients with human immunodeficiency virus disease. *Am J Surg Pathol* 1989;13:909–20.
49. LeBoit PE. Bacillary angiomatosis: a systemic opportunistic infection with prominent cutaneous manifestations. *Semin Dermatol* 1991;10:194–8.
50. Batsakis JG, Ro JY, Frauenhoffer EE. Bacillary angiomatosis. *Ann Otol Rhinol Laryngol* 1995;104:668–72.
51. Greenspan D, Greenspan JS. Oral manifestations of HIV infection. *AIDS Clin Care* 1997;9:29–33.
52. Lopez de Blanc S, Sambuelli R, Femopase F, et al. Bacillary angiomatosis affecting the oral cavity. Report of two cases and review. *J Oral Pathol Med* 2000;29:91–6.
53. Matar GM, Koehler JE, Malcolm G, Lambert-Fair MA, Tappero J, Hunter SB, Swaminathan B. Identification of *bartonella* species directly in clinical specimens by PCR-restriction fragment length polymorphism analysis of a 16S rRNA gene fragment. *J Clin Microbiol* 1999;37:4045–7.
54. Steeper TA, Rosenstein H, Weiser J, Inampudi S, Snover DC. Bacillary epithelioid angiomatosis involving the liver, spleen, and skin in an AIDS patient with concurrent Kaposi's sarcoma. *Am J Clin Pathol* 1992;97:713–8.
55. Rosales CM, McLaughlin MD, Sata T, Katano H, Veno PA, de Las Casas LE, Miranda RN. AIDS presenting with cutaneous Kaposi's sarcoma and bacillary angiomatosis in the bone marrow mimicking Kaposi's sarcoma. *AIDS Patient Care STDS* 2002;16:573–7.
56. Tsang WY, Chan JK. Bacillary angiomatosis. A "new" disease with a broadening clinicopathologic spectrum. *Histol Histopathol* 1992;7:143–52.
57. Bennhoff DF. Actinomycosis: diagnostic and therapeutic considerations and a review of 32 cases. *Laryngoscope* 1984;94:1198–217.
58. Nielsen PM, Novak A. Acute cervico-facial actinomycosis. *Int J Oral Maxillofac Surg* 1987;16:440–4.
59. Nagler R, Peled M, Laufer D. Cervicofacial actinomycosis: a diagnostic challenge. *Oral Surg Oral Med Oral Pathol Oral Radiol Endod* 1997;83:652–6.
60. Oostman O, Smego RA. Cervicofacial actinomycosis: diagnosis and management. *Curr Infect Dis Rep* 2005;7:170–4.
61. Wear DJ, Margileth AW, Hadfield TL, Fisher GW, Schlagel CJ, King FM. Cat scratch disease: a bacterial infection. *Science* 1983;221:1403–5.
62. Carithers HA. Cat-scratch disease: an overview based on a study of 1200 patients. *Am J Dis Child* 1985;139:1124–33.
63. Miller-Catchpole R, Variakojis D, Vardiman JW, Loew JM, Carter J. Cat scratch disease: identification of bacteria in seven cases of lymphadenitis. *Am J Surg Pathol* 1986;10:276–81.
64. English CK, Wear DJ, Margileth AW, Lissner CR, Walsh GP. Cat-scratch disease: isolation and culture of the bacterial agent. *JAMA* 1988;259:1347–52.
65. Lamps LW, Scott MA. Cat-scratch disease: historic, clinical, and pathologic perspectives. *Am J Clin Pathol* 2004;121:S71–80.
66. Wenig BM, Smirniotopolous J, Heffner DK. Botryomycosis of the sinonasal tract: a report of two cases. *Arch Pathol Lab Med* 1996;120:1123–8.
67. Alvarez S, McCabe E. Extrapulmonary tuberculosis revisited: a review of experience at Boston City and other hospitals. *Medicine* 1984;63:25–55.
68. Cleary KR, Batsakis JG. Mycobacterial disease of the head and neck: current perspective. *Ann Otol Rhinol Laryngol* 1995;104:830–3.
69. Srirompotong S, Yimtae K, Srirompotong S. Tuberculosis in the upper aerodigestive tract and human immunodeficiency virus coinfections. *J Otolaryngol* 2003;32:230–3.
70. Pao CC, Yen TS, You JB, Maa JS, Fiss EH, Chang CH. Detection and identification of *Mycobacterium tuberculosis* by DNA amplification. *J Clin Microbiol* 1990;28:1877–80.
71. Liu YP, Behr MA, Small PM, Kurn N. Genotypic determination of *Mycobacterium tuberculosis* antibiotic resistance using a novel mutation detection method, the branch migration inhibition *M. tuberculosis* antibiotic resistance test. *J Clin Microbiol* 2000;38:3656–62.
72. Fluit AC, Visser MR, Schmitz FJ. Molecular detection of antimicrobial resistance. *Clin Microbiol Rev* 2001;14:836–71.
73. Glennon M, Cormican M. Detection and diagnosis of mycobacterial pathogens using PCR. *Expert Rev Mol Diagn* 2001;1:163–74.
74. Shrestha NK, Tuohy MJ, Hall GS, Reischl U, Gordon SM, Procop GW. Detection and differentiation of *Mycobacterium tuberculosis* and nontuberculous mycobacterial isolates by real-time PCR. *J Clin Microbiol* 2003;41:5121–6.
75. Noordhoek GT, Mulder S, Wallace P, van Loon AM. Multicentre quality control study for detection of *Mycobacterium tuberculosis* in clinical samples by nucleic amplification methods. *Clin Microbiol Infect* 2004;10:295–301.

76. Vernet G, Jay C, Rodrigue M, Troesch A. Species differentiation and antibiotic susceptibility testing with DNA microarrays. *J Appl Microbiol* 2004;96:59–68.
77. Albright JT, Pransky SM. Nontuberculous mycobacterial infections of the head and neck. *Pediatr Clin North Am* 2003;50:503–14.
78. Munck K, Mandpe AH. Mycobacterial infections of the head and neck. *Otolaryngol Clin North Am* 2003;36:569–76.
79. Brandwein M, Choi HS, Strauchen J. Spindled nontuberculous mycobacteriosis in AIDS mimicking neoplasia. Evidence of dual histiocytic and fibroblastic-like characteristics of spindle cells. *Virchows Arch A* 1990;416:281–6.
80. Umlas J, Federman M, Crawford C, O'Hara CJ, Fitzgibbon JS, Modeste A. Spindle cell pseudotumor due to *Mycobacterium avium-intracellulare* in patients with acquired immunodeficiency syndrome (AIDS). Positive staining of mycobacteria for cytoskeleton filaments. *Am J Surg Pathol* 1991;15:1181–7.
81. Chen KT. Mycobacterial spindle cell pseudotumor of lymph nodes. *Am J Surg Pathol* 1992;16:276–81.
82. Logani S, Lucas DR, Cheng JD, Ioachim HL, Adsay NV. Spindle cell tumors associated with mycobacteria in lymph nodes of HIV-positive patients: 'Kaposi sarcoma with mycobacteria' and 'mycobacterial pseudotumor'. *Am J Surg Pathol* 1999;23:656–61.
83. Gunia S, Behrens MH, Stosiek P. Mycobacterial spindle cell pseudotumor (MSP) of the nasal septum clinically mimicking Kaposi's sarcoma: case report. *Rhinology* 2005;43: 70–1.
84. Monot M, Honore N, Garnier T, et al. On the origin of leprosy. *Science* 2005;308:1040–2.
85. Barton RP, McDougall AC. The paranasal sinuses in lepromatous leprosy. *Lepr India* 87 1979;51:481–5.
86. Binford CH, Meyers WM. Leprosy. In: Binford CH, Connor DH, eds. Pathology of Tropical and Extraordinary Disease: An Atlas. Armed Forces Institute of Pathology, Washington, D.C., 1976: 205–25.
87. Torres P, Camarena JJ, Gomez JR, et al. Comparison of PCR mediated amplification of DNA and the classical methods for detection of *Mycobacterium leprae* in different types of clinical samples in leprosy patients and contacts. *Lepr Rev* 2003;74:18–30.
88. Almeida EC, Martinez AN, Maniero VC, et al. Detection of *Mycobacterium leprae* DNA by polymerase chain reaction in the blood and nasal secretion of Brazilian household contacts. *Mem Inst Oswaldo Cruz* 2004;99:509–11.
89. Williams DL, Torrero M, Wheeler PR, et al. Biological implications of *Mycobacterium leprae* gene expression during infection. *J Mol Microbiol Biotechnol* 2004;8:58–72.
90. Connor DH, Neafie RC. Cutaneous leishmaniasis. In: Binford CH, Connor DH, eds. Pathology of Tropical and Extraordinary Disease: An Atlas. Armed Forces Institute of Pathology, Washington, D.C., 1976:258–64.
91. Carbonell Casasus J, Canovas Llorens DL, Soler Vilarrasa R, Mas Mercant S. Leishmaniasis and ENT. *An Otorrinolaringol Ibero Am* 1995;22:139–49.
92. Singh S, Sivakumar R. Challenges and new discoveries in the treatment of leishmaniasis. *J Infect Chemother* 2004;10:307–15.
93. Sandler ED, Leboit PE, Wenig BM, et al. *Pneumocystis carinii* otitis media in AIDS: a case report and review of the literature regarding extrapulmonary pneumocystosis. *Otolaryngol Head Neck Surg* 1990;103:817–21.
94. Wenig BM. Atlas of Head and Neck Pathology. Elsevier, Edinburgh, 2007.
95. Abrams AM, Melrose RJ, Howell FV. Necrotizing sialometaplasia: a disease simulating malignancy. *Cancer* 1973;32: 130–5.
96. Fechner RE. Necrotizing sialometaplasia: a source of confusion with carcinoma of the palate. *Am J Clin Pathol* 1977;67:315–7.
97. Brannon RB, Fowler CB, Hartman KS. Necrotizing sialometaplasia: a clinicopathologic study of sixty-nine cases and review of the literature. *Oral Surg Oral Med Oral Pathol* 1991;71:317–25.
98. Batsakis JG, Manning JT. Necrotizing sialometaplasia of major salivary glands. *J Laryngol Otol* 1987; 101:962–6.
99. Jensen JL. Necrotizing sialometaplasia. In: Ellis GL, Auclair PL, Gnepp DR, eds. Surgical Pathology of Salivary Glands. W.B. Saunders Co., Philadelphia, 1991:60–6.
100. Kyle RA, Bayrd ED. Amyloidosis: review of 236 cases. *Medicine* 1975;54:271–99.
101. Kyle RA, Gertz MA. Primary systemic amyloidosis: clinical and laboratory features of 474 cases. *Semin Hematol* 1995;32:45–59.
102. Falk RH, Comenzo RL, Skiner M. The systemic amyloidoses. *N Engl J Med* 1997;337:898–909.
103. Gertz MA, Lacy MQ, Dispenzieri A. Amyloidosis: recognition, confirmation, prognosis and therapy. *Mayo Clin Proc* 1999;74:490–4.
104. Lewis JE, Olsen KD, Kurtin PJ, Kyle RA. Laryngeal amyloidosis: a clinicopathologic and immunohistochemical review. *Otolaryngol Head Neck Surg* 1992;106:372–7.
105. Kerner MM, Wang MB, Angier G, Calcaterra TC, Ward PH. Amyloidosis of the head and neck. A clinicopathologic study of the UCLA experience, 1955–1991. *Arch Otolaryngol Head Neck Surg* 1995;121:778–82.
106. Zundel RS, Pyle GM, Voytovich M. Head and neck manifestations of amyloidosis. *Otolaryngol Head Neck Surg* 1999;120:553–7.
107. Hamed G, Heffess CS, Shmookler BM, Wenig BM. Amyloid goiter. A clinicopathologic study of 14 cases and review of the literature. *Am J Clin Pathol* 1995;104: 306–12.
108. Ozdemir BH, Akman B, Ozdemir FN. Amyloid goiter in Familial Mediterranean Fever (FMF): a clinicopathologic study of 10 cases. *Ren Fail* 2001;23:659–67.
109. Popper HH, Klemen H, Hoefler G, Winter E. Presence of mycobacterial DNA is sarcoidosis. *Hum Pathol* 1997, 28:796–800.
110. Li N, Bajoghi A, Kubba A, Bhawan J. Identification of mycobacterial DNA in cutaneous lesions of sarcoidosis. *J Cutan Pathol* 1999;26:271–8.
111. Barnes L. Midfacial destructive diseases. In: Barnes L, ed. Surgical Pathology of the Head and Neck. Second edition revised and expanded. Marcel-Dekker, New York, 2001: 759–86.

112. Rosai J. The nature of myospherulosis of the upper respiratory tract. *Am J Clin Pathol* 1978;69:475–81.
113. Kyriakos M. Myospherulosis of the paranasal sinuses, nose and middle ear: a possible iatrogenic disease. *Am J Clin Pathol* 1977;67:118–30.
114. Manni JJ, Schaafsma HE, Vennix PP. Myospherulosis in the operated temporal bone. *Eur Arch Otorhinolaryngol* 1992;249:231–2.
115. Mace AT, Marshall JN. Foreign body reaction to an oil-based ointment: a cause of persistent otorrhoea following mastoid surgery. *J Laryngol Otol* 2003;117:496–7.
116. Hyams VJ. Papillomas of the nasal cavity and paranasal sinuses: a clinicopathologic study of 315 cases. *Ann Otol Rhinol Laryngol* 1971;80:192–206.
117. Cardesa A, Zidar N, Alos L. Adenosquamous carcinoma. In: Barnes L, Eveson JW, Reichart P, Sidransky D, eds. World Health Organization Classification of Tumours. Pathology and Genetics of Head and Neck Tumours. IARC Press, Lyon, France, 2005:130–1.
118. Ellis GL, Auclair PL. Mucoepidermoid carcinoma. In: Rosai J, Sobin LH, eds. Tumors of the Salivary Glands. Atlas of Tumor Pathology. Third series, Fascicle 17. Armed Forces Institute of Pathology, Washington, D.C., 1996:155–75.
119. Ellis GL, Auclair PL. Acinic cell adenocarcinoma. In: Rosai J, Sobin LH, eds. Tumors of the Salivary Glands. Atlas of Tumor Pathology. Third series, Fascicle 17. Armed Forces Institute of Pathology, Washington, D.C., 1996:183–203.
120. Franchi A, Santucci M, Wenig BM. Adenocarcinoma. In: Barnes L, Eveson J, Reichart P, Sidransky D, eds. World Health Organization Classification of Tumours. Pathology and Genetics of Head and Neck Tumours. IARC Press, Lyon, France, 2005:20–3.
121. Barnes L. Malignant melanoma of the nasal cavity and paranasal sinuses In: Barnes L, ed. Surgical Pathology of the Head and Neck. Second edition, revised and expanded. Marcel Dekker, New York, 2001:523–7.
122. Pettinato G, Manivel JC, d'Amore ESG, Jaszcz W, Gorlin RJ. Melanotic neuroectodermal tumor of infancy: a reexamination of a histogenetic problem based on immunohistochemical, flow cytometric, and ultrastructural study of 10 cases. *Am J Surg Pathol* 1991;15:233–45.
123. Kapadia SB, Frisman DM, Hitchcock CL, Ellis GL, Popek EJ. Melanotic neuroectodermal tumor of infancy. Clinicopathological, immunohistochemical, and flow cytometric study. *Am J Surg Pathol* 1993;17:566–73.
124. Barnes L, Tse LLY, Hunt JL, Michaels L. Tumours of the paraganglionic system: introduction. In: Barnes L, Eveson JW, Reichart P, Sidransky D, eds. World Health Organization Classification of Tumours. Pathology and Genetics of Head and Neck Tumours. IARC Press, Lyon, France, 2005:362–3.
125. Barnes L, Tse LLY, Hunt JL. Carotid body paraganglioma. Barnes L, Eveson JW, Reichart P, Sidransky D, eds. World Health Organization Classification of Tumours. Pathology and Genetics of Head and Neck Tumours. IARC Press, Lyon, France, 2005:364–5.
126. Wenig BM, Hyams VJ, Heffner DK: Moderately differentiated neuroendocrine carcinoma of the larynx: a clinicopathologic study of 54 cases. *Cancer* 1988;62:2658–76.
127. Wenig BM, Gnepp DR. The spectrum of neuroendocrine carcinomas of the larynx. *Semin Diagn Pathol* 1989;6:329–50.
128. Ladanyi M. The emerging molecular genetics of sarcoma translocations. *Diagn Mol Pathol* 1995;4:162–73.
129. Mills SE. Neuroectodermal neoplasms of the head and neck with emphasis on neuroendocrine carcinomas. *Mod Pathol* 2002;15:264–78.
130. Schmookler BM, Enzinger FM, Brannon RB. Orofacial synovial sarcoma. *Cancer* 1982;50:269–76.
131. Ordonez NG. Alveolar soft part sarcoma: a review and update. *Adv Anat Pathol* 1999;6:125–39.
132. Fanburg-Smith JC, Miettinen M, Folpe AL, Weiss SW, Childers EL. Lingual alveolar soft part sarcoma; 14 cases: novel clinical and morphological observations. *Histopathology* 2004;45:526–37.
133. Hyams VJ, Batsakis JG, Michaels L. Chordoma. In: Hyams VJ, Batsakis JG, Michaels L, eds. Tumors of the Upper Respiratory Tract and Ear. Fascicle 25, second series. Armed Forces Institute of Pathology, Washington, D.C., 1988:192–6.
134. Mirra JM, Nelson SD, Della Rocca C, Mertens F. Chordoma. In: Fletcher CDM, Unni KK, Mertens F, eds. World Health Organization Classification of Tumours. Pathology and Genetics Tumours of Soft Tissue and Bone. IARC Press, Lyon, France, 2002: 316–7.
135. Jaffe E S, Chan J K C, Su I-H, et al. Report of the workshop on nasal and related extranodal angiocentric T/Natural killer cell lymphomas. Definitions, differential diagnosis, and epidemiology. *Am J Surg Pathol* 1996;20:103–11.

CHAPTER 3

Diagnostic Histochemistry in Pulmonary Pathology

Mark R. Wick, M.D.

The lungs are the seat of diverse disease processes probably because of their constant interface with the ambient environment and consequent exposure to many inhaled substances. In addition, the richness of the pulmonary vasculature often involves the lungs in diseases of an infectious or immunologic nature, and the peculiar balance between the pulmonary and systemic circulations makes primary or secondary vasculopathic disorders another important consideration in this location. The lungs are now routinely transplanted at large medical centers, and singular diseases relating to opportunistic infections or allograft rejection may eventuate thereafter. Finally, neoplasms of the lungs and pleura assume a variety of histologic appearances and cellular lineages, many of which are difficult to separate from one another on conventional microscopy. In this broad context, histochemical analysis has much to offer in regard to differential diagnosis and specific microscopic interpretations.

PULMONARY INFECTIOUS DISEASES

The most common application of histochemical stains in lung disease is in the detection and identification of infectious disorders. These are discussed subsequently as broadly categorized by the generic nature of the causative organisms.

Bacterial Disorders

Bacterial pneumonias are still extremely common as community-acquired and nosocomial diseases. There are many causative organisms in this group, and, therefore, several histologic methods are necessary to visualize all of them.

Tissue Gram Stains: The Brown and Brenn (BB) and Brown and Hopps (BH) procedures represent modifications of the Gram stain that are applicable to formalin-fixed tissues (Figures 3.1 and 3.2) (1). However, in comparison with direct preparations from infected specimens, the BB and BH methods are markedly less sensitive in the detection of causative organisms. In particular, gram-negative bacteria are rarely seen in tissue sections unless the density of those microbes is extremely high. It should also be remembered that positivity with the BB or BH technique does not equate with a bacterial pathogenesis; some fungi—in particular, *Candida* and *Torulopsis*—also appear as gram-positive organisms (Figure 3.3). Lastly, selected bacteria are notoriously resistant to labeling with the BB

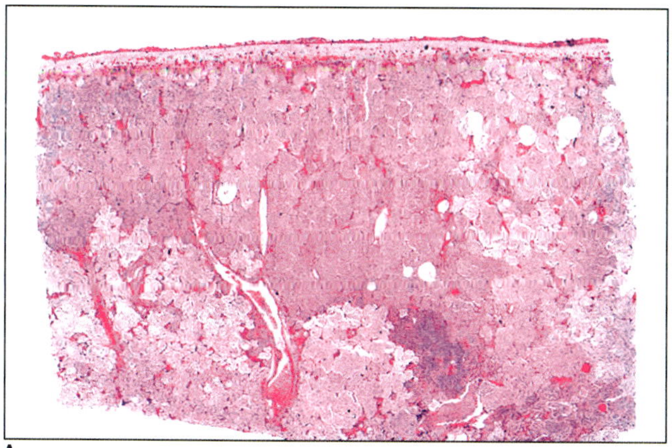

A

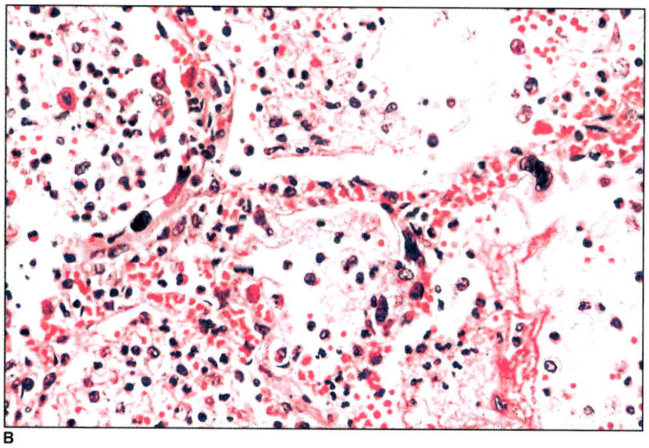

B

FIGURE 3.1: (A) Acute pneumonia, showing diffuse neutrophilic consolidation of alveolar spaces (H&E, 60×). (B) Neutrophilic exudates are seen in the alveoli in acute pneumonia (H&E, 200×).

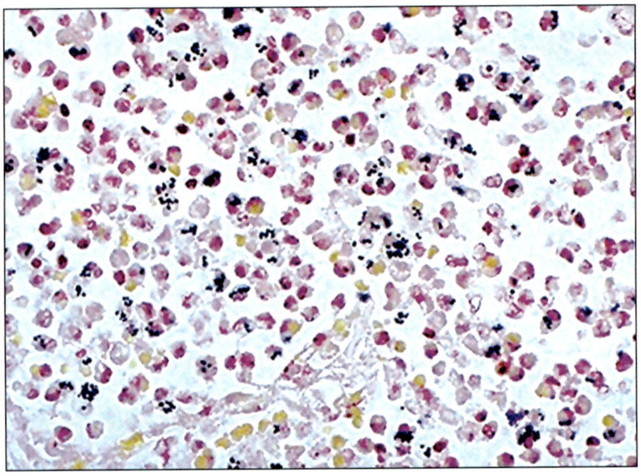

FIGURE 3.2: ***Gram-positive cocciform bacteria are evident among the intraalveolar neutrophils in this example of acute pneumonia (BB, 400×).***

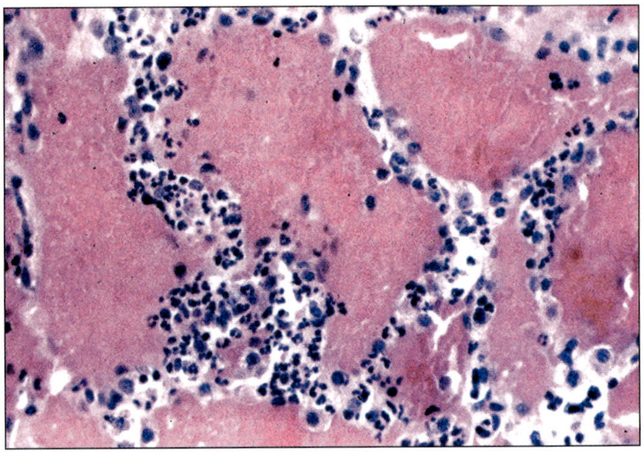

FIGURE 3.4: ***Diffuse necrotizing and hemorrhagic pneumonia is characteristic of causation by gram-negative and gram-resistant bacteria (H&E, 200×).***

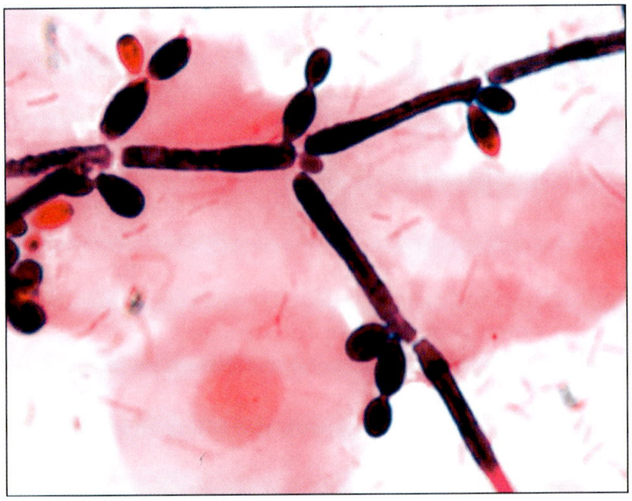

FIGURE 3.3: ***Gram-positive*** **Candida** ***species are seen in this sputum cytology specimen (BB, 600×).***

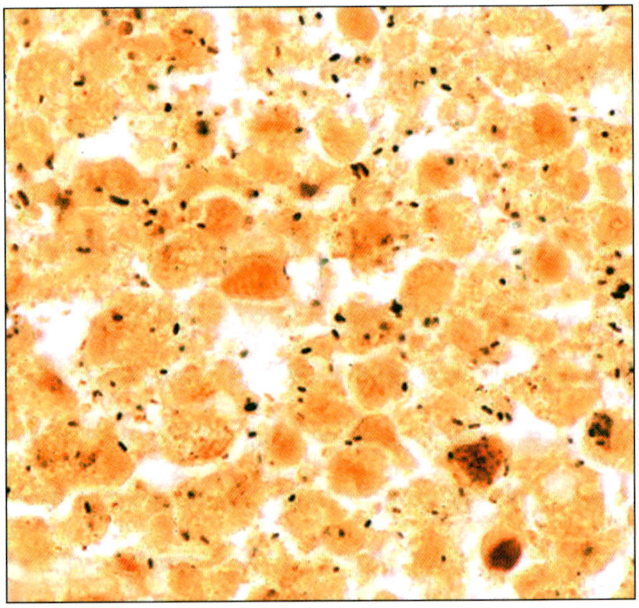

FIGURE 3.5: ***Stubby rod–shaped bacilliform organisms are present in this Steiner stain, typical of*** **Legionella** ***species. The Gram stain was negative (Steiner stain, 600×).***

or BH procedure. *Treponema* and *Borrelia* species as well as *Bartonella*, most *Yersinia*, and *Legionella* organisms are virtually universally unstained by tissue Gram techniques (2). Overall, the BB and BH techniques have a reasonable degree of specificity for bacteria and selected fungi (which can generally be distinguished from one another by their relative sizes and morphologic features), but the sensitivity of those procedures is probably not much greater than 50%.

Other Specialized Stains for Non–Acid-Fast Bacteria: As mentioned above, there are several genera of bacteria in pulmonary infections that are poorly recognized by the BB and BH methods, if at all. In general, many of those microorganisms *are* satisfactorily labeled by the Steiner or Warthin-Starry silver impregnation techniques. They include spirochetes, *Bartonella*, *Yersinia*, *Helicobacter*, *Legionella*, and *Yersinia* (Figures 3.4 and 3.5) (3). Filamentous bacteria such as Actinomyces and Nocardia are also well seen with silver stains (Figure 3.6) (4). Although these methods are technically challenging and are associated with high "background" labeling of cellular detritus, it could be argued that they should be included routinely in the evaluation of any potentially infectious lesion in the lung.

Mycobacterial Diseases

Pulmonary mycobacterial infections continue to be a major source of morbidity and mortality worldwide, particularly in the context of the human immunodeficiency virus

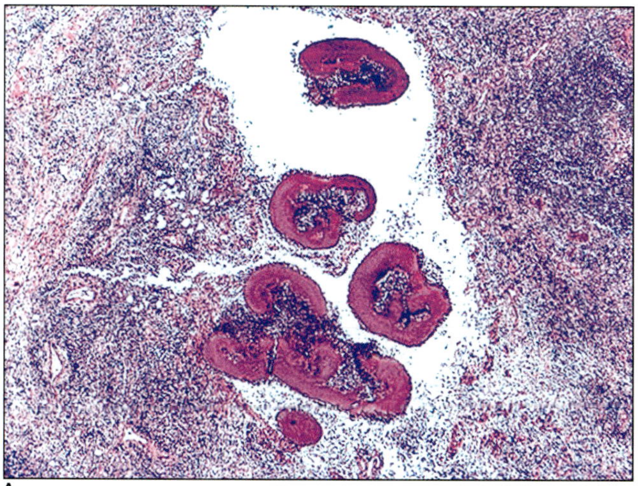

A

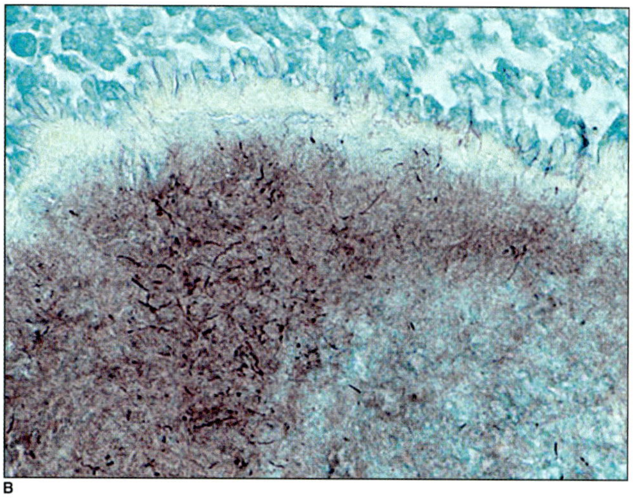

B

***FIGURE 3.6:** (A) "Sulfur granules" are apparent within a pulmonary abscess cavity in this photograph (H&E, 100×). (B) Faintly silver-positive filamentous bacteria are present inside the sulfur granules, representing* Actinomyces *(MS stain, 400×).*

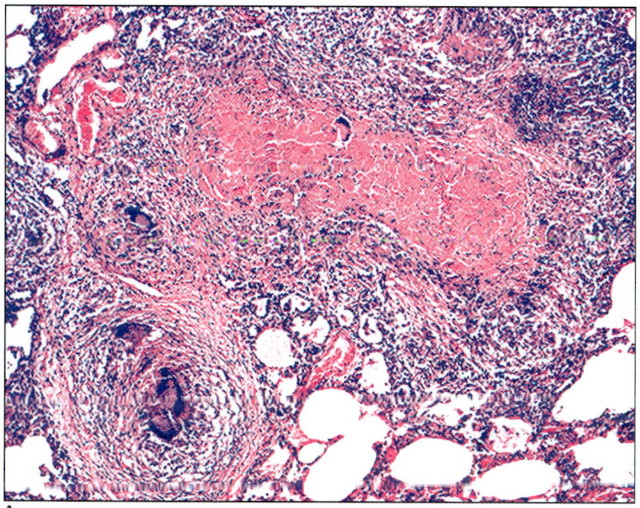

A

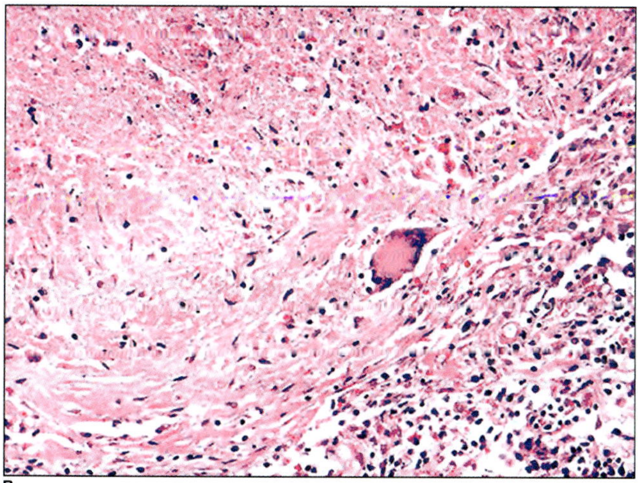

B

***FIGURE 3.7:** (A and B) Necrotizing granulomatous inflammation in the lung may represent either mycobacterial or fungal infection (H&E, A: 150×; B: 250×).*

pandemic. Although they had decreased in the United States and other developed countries up to 1990, a steady increase in mycobacterial illnesses has been observed since then (5). *Mycobacterium tuberculosis* (MTB) is still the principal agent in this group of disorders. Nevertheless, other mycobacteria—notably, *M. avium-intracellulare* and *M. kansasii* [so-called atypical mycobacteria (AMB)]—are significant pathogens as well, particularly in immunocompromised individuals and those with cystic fibrosis (6–9).

There are three standard "acid-fast" methods for visualization of mycobacteria in tissue sections; these are the Ziehl-Neelsen/Kinyoun (ZNK) procedure, the Fite stain, and the auramine-rhodamine technique (Figures 3.7 and 3.8) (10). The last of those is the most sensitive, but unfortunately, it is suboptimally applicable to paraffin sections and requires the use of fluorescence microscopy for visualization (11). The ZNK method is best at detecting MTB and related mycobacterial species (*M. africanum*, *M. bovis*, *M. microti*), and the Fite stain is intended to recognize AMB preferentially. These techniques differ procedurally in the degree and strength of acidification (12,13).

The efficiency of acid-fast staining depends on the causative organism and its density in tissue. Many mycobacteria are often seen in immunosuppressed patients, whereas their identification commonly requires painstaking evaluation in other cases. Overall, our experience has shown no better than 50% sensitivity for ZNK and Fite stains, as compared with the results of cultures on the same tissue specimens. Both methods may be used concurrently in tissue sections depending on the clinical details of the case at hand, but the Fite stain should not be used preferentially because of its inferior sensitivity as a screening procedure.

Nonmycobacterial organisms also may demonstrate acid-fast reactivity. In particular, *Cryptosporidium* (a parasite),

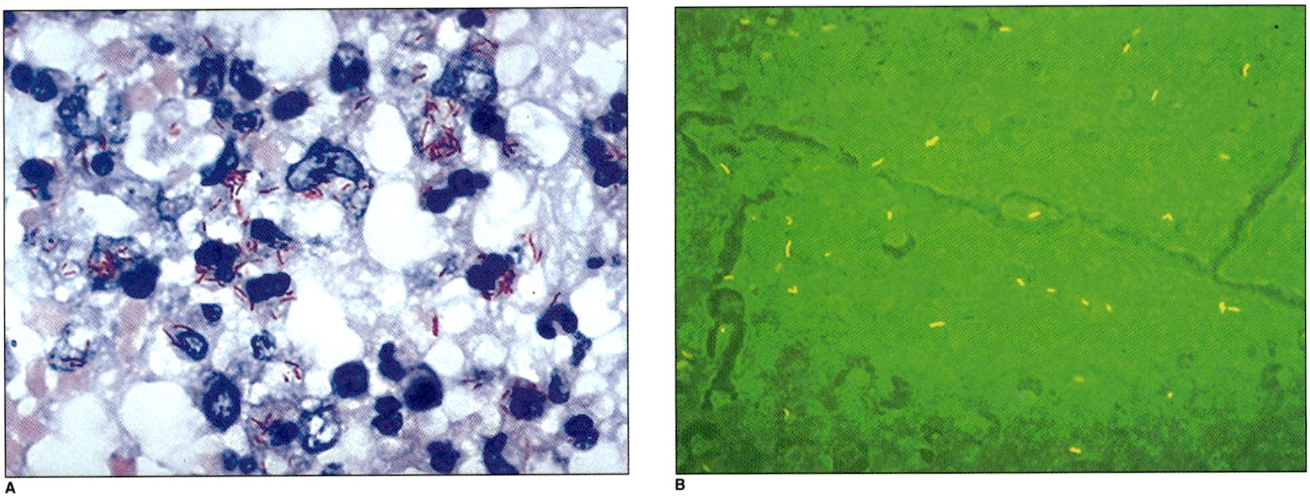

A B

FIGURE 3.8: ***(A) Acid-fast mycobacteria are present in a focus of granulomatous inflammation. (B) Similar mycobacteria are visible using fluorescence microscopy (A: Ziehl-Neelsen stain, 1000×; B: auramine-rhodamine stain, 1000×).***

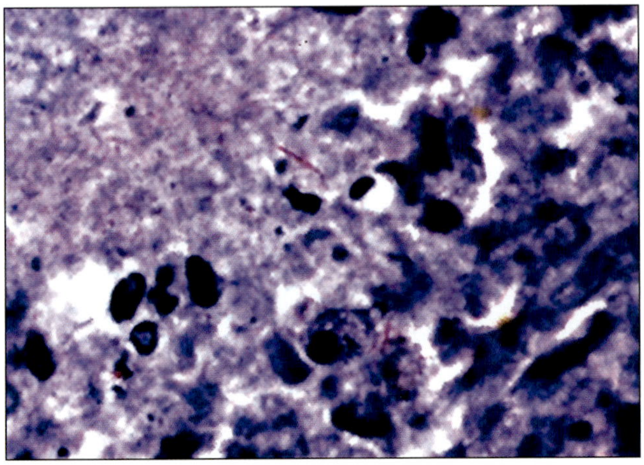

FIGURE 3.9: **Nocardia** ***species may also demonstrate weak acid-fast positivity, as shown here (Ziehl-Neelsen stain, 1000×).***

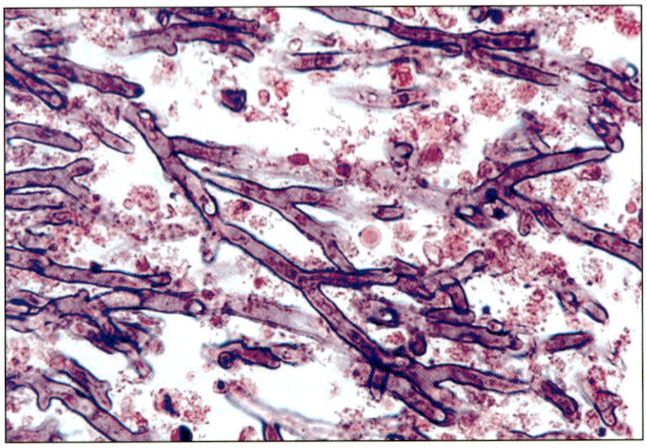

FIGURE 3.10: **Aspergillus** ***species are often visible with H&E stains, as branching and septated hyphal organisms (H&E, 600×).***

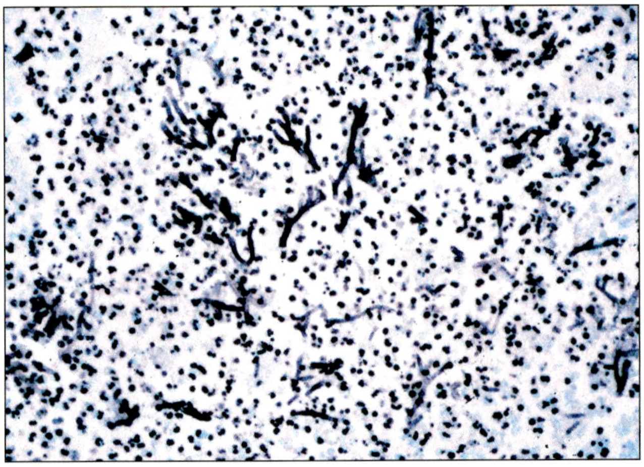

FIGURE 3.11: **Aspergillus** ***is seen well in this silver stain (MS stain, 300×).***

Nocardia, *Rhodococcus*, and *Legionella micdadeii* (Pittsburgh agent) do so (Figure 3.9) (2,14–18).

Fungal Infections

Several fungal organisms are capable of causing pulmonary disease—which is typically granulomatous in character—and they vary considerably in their morphologic characteristics and specialized biochemical properties. Some are intracellular (usually intrahistiocytic), such as *Histoplasma*, whereas others preferentially grow in exudates in the airspaces.

Routine hematoxylin and eosin (H&E) stains are often sufficient to visualize *Aspergillus* and *Zygomyces* species (Figure 3.10), but most others require additional histochemical methods for detection. The periodic acid-Schiff (PAS) technique is often effective, as are methenamine

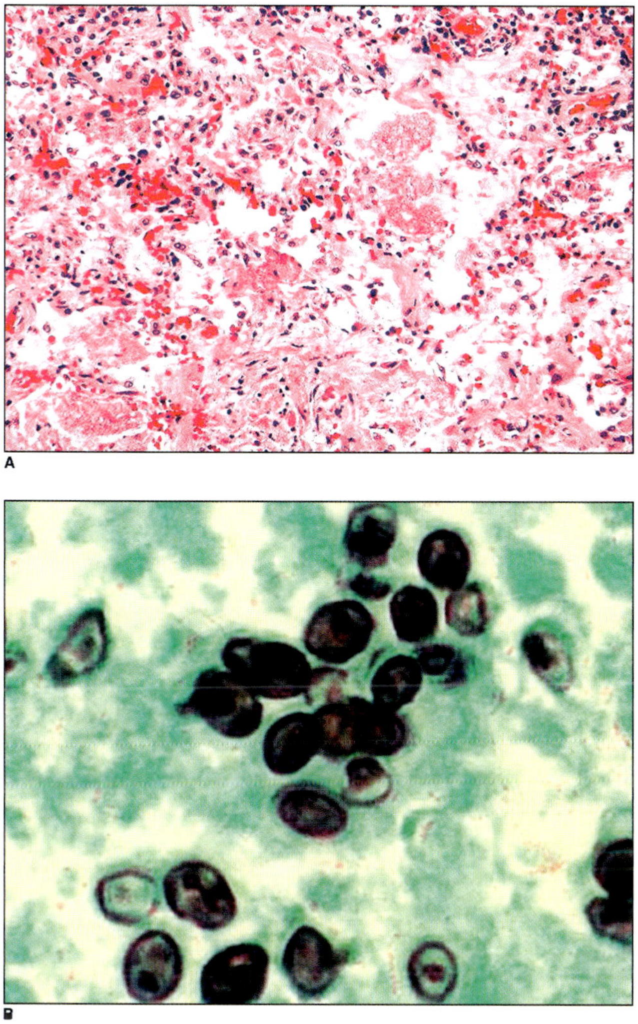

FIGURE 3.12: ***(A)*** **Pneumocystis carinii** ***pneumonia is characterized by "fluffy" exudates in the alveolar spaces, forming "casts" of those structures (H&E, 200×). (B) Silver stains demonstrate the*** **Pneumocystis** ***organisms as disciform structures with central densities (MS stain, 600×).***

silver [MS (Gomori's, Grocott's)] stains or the Gridley procedure (19–24). The first of these stains produces red staining of fungal organisms, whereas they are black-brown with the MS procedures (Figure 3.11) and red-purple with Gridley's technique. MS methods are probably the most sensitive among this group, but they have the disadvantage of also labeling erythrocytes, cellular debris, and "naked" nuclei in tissue sections. The overall efficiency of fungal stains is higher than that of acid-fast techniques relative to culture results, but it again parallels the density of organisms and the particular genus in question (25). Moreover, positive results are distinctly unusual in reference to organized-hyalinized fungal granulomas in the lung.

Dematiaceous (pigmented) fungi contain melanin or melanin precursors in their cell walls, accounting for the fact that they are potentially labeled with the Fontana-

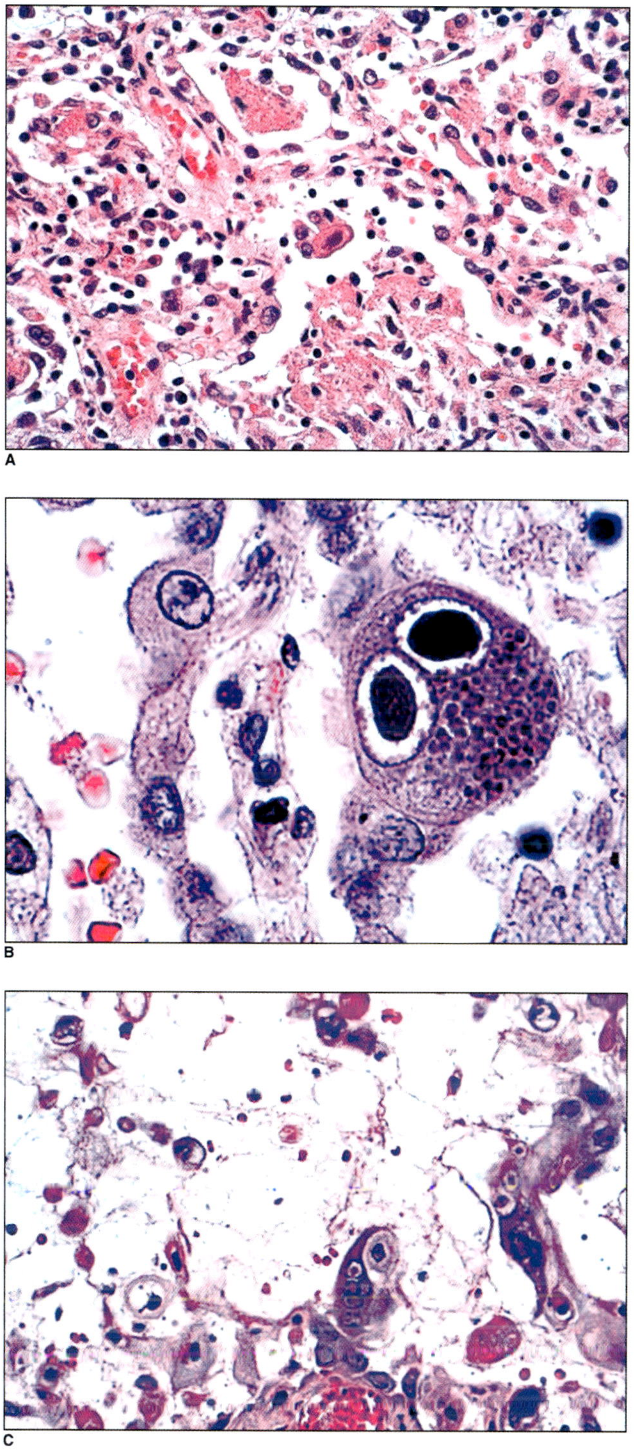

FIGURE 3.13: **Cytomegalovirus** ***(A and B) and*** **Herpes** ***or*** **Zoster** ***viruses (C) demonstrate typical intranuclear inclusions in infected epithelial or endothelial cells in the lungs (H&E, A: 150×; B: 1000×; C: 300×).***

Masson (FM) silver stain (26). Also, because mucin is argentaffinic as well, the FM method will recognize the capsules that surround individual organisms of *Cryptococcus*. Alcian blue, colloidal iron (CI), FM, and mucicarmine stains

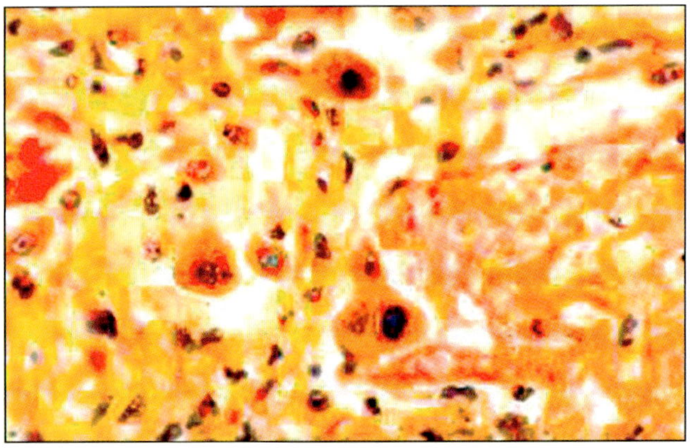

FIGURE 3.14: Lendrum's phloxine-tartrazine method can be used to highlight viral nuclear inclusions and distinguish them from histologically similar degenerative or reactive nuclear alterations in the lung (Lendrum stain, 300×).

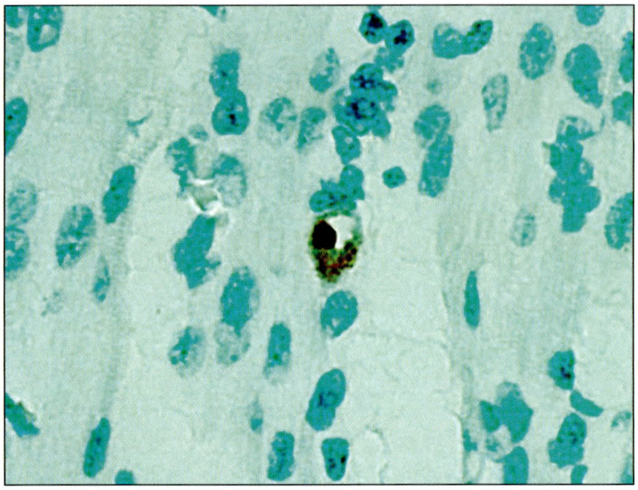

FIGURE 3.16: The Steiner stain also may label Cytomegalovirus *inclusions in some cases (Steiner method, 400×).*

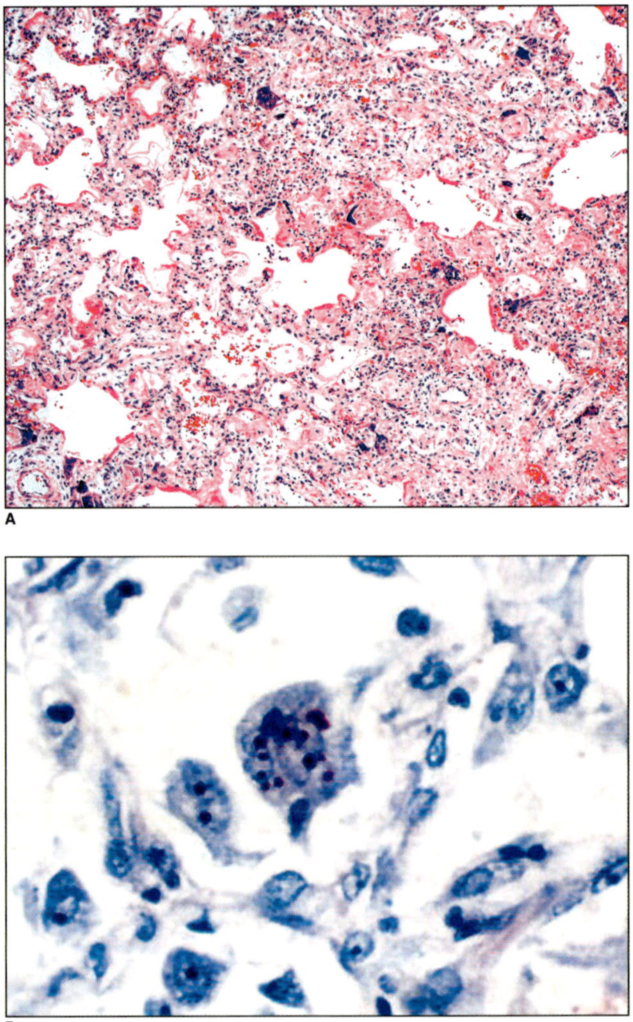

A

B

FIGURE 3.15: Measles pneumonia features the presence of multinucleated giant cells that contain viral inclusions (A), and those can be labeled with the Macchiavello technique (B) (A: H&E, 150×; B: Macchiavello method, 400×).

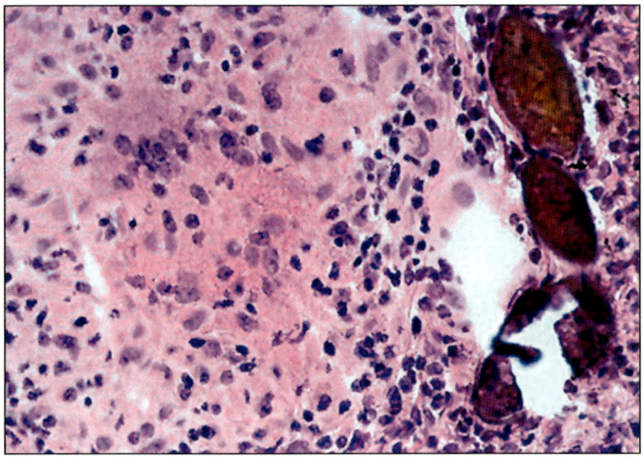

FIGURE 3.17: Infestation with Paragonimus westermani *may produce granulomatous "coin" lesions in the peripheral lung, as shown here (H&E, 400×).*

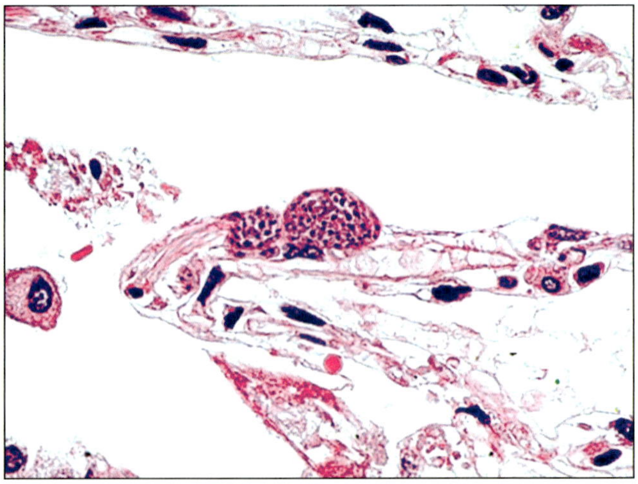

FIGURE 3.18: In immunocompromised patients, infestation with Toxoplasma gondii *may affect many organs, including the lung, in which parasitic microcysts may be seen by conventional microscopy (H&E, 400×).*

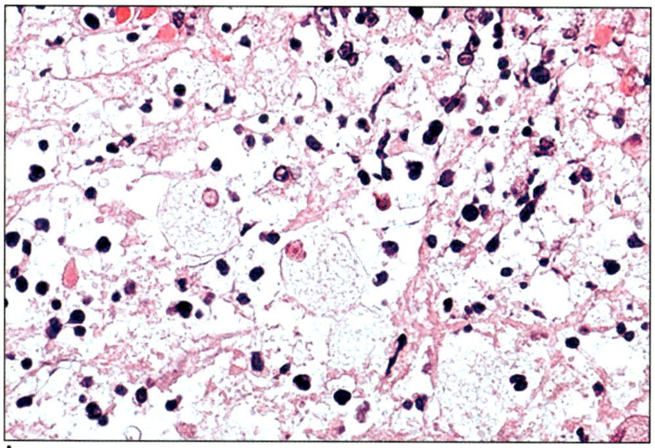

A

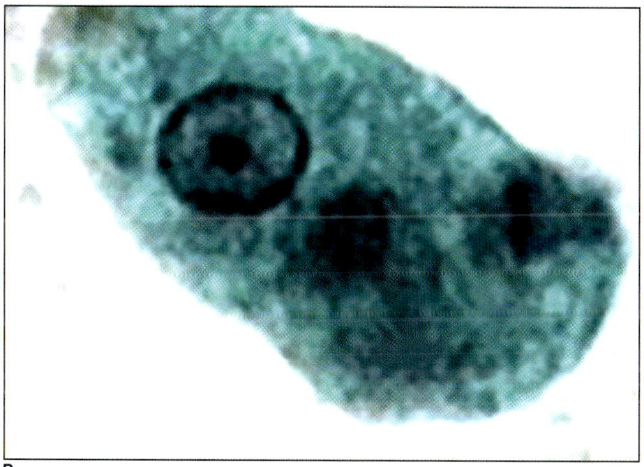

B

FIGURE 3.19: ***Multiorgan infestation by*** **Entamoeba histolytica** ***may produce abscesses in the lung, which contain amebic trophozoites (A). These can be labeled with the Masson trichrome method (B) (A: H&E, 400×; B: Masson trichrome, 1000×).***

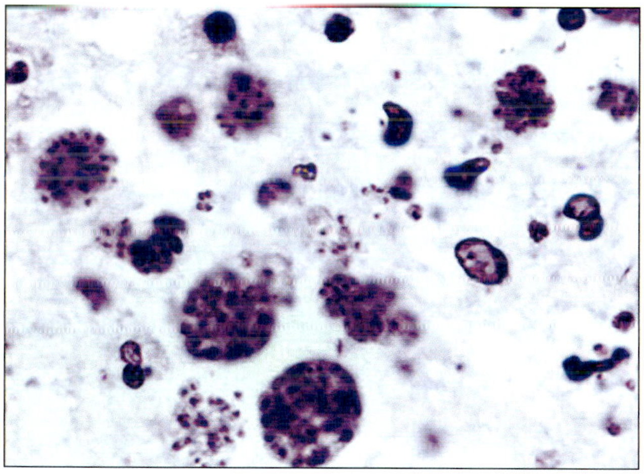

FIGURE 3.20: ***Protozoan parasitic organisms can often be stained with the Giemsa method, as shown in this case of toxoplasmosis (Giemsa technique, 1000×).***

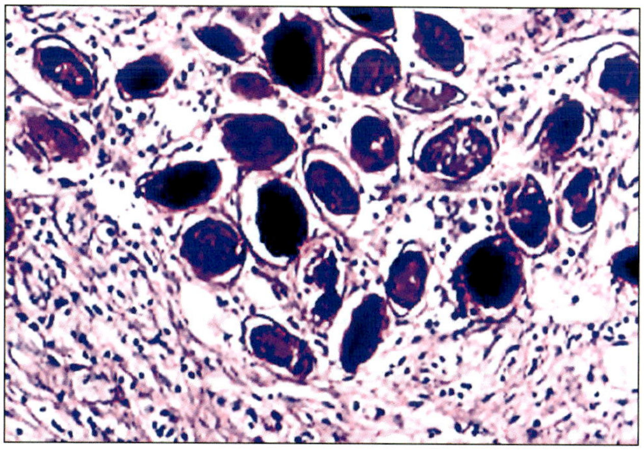

FIGURE 3.21: ***The eggs of trematodes and nematodes can often be decorated by the PAS method, as shown in this case of paragonimiasis (undigested PAS technique, 400×).***

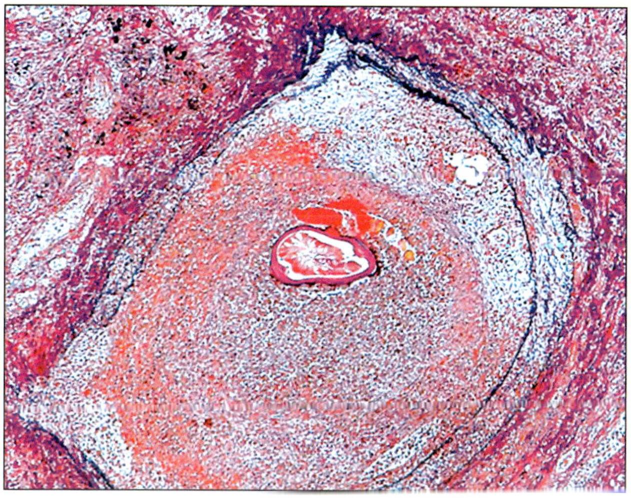

FIGURE 3.22: ***Elastic stains may be used to better show the relationship between parasites and surrounding structures. Here, an adult*** **Dirofilaria immitis** ***organism is present in the lumen of a pulmonary artery (Van Gieson method, 200×).***

similarly label cryptococci and provide a means for definitive identification of those agents in tissue sections (27,28).

Pneumocystis organisms are also a potential cause of pneumonia in immunosuppressed patients. This genus was originally considered protozoan taxonomically but has now been reclassified with the fungi. It may be labeled with MS methods or the Giemsa technique, with the former usually being employed in paraffin sections (Figure 3.12) (19,20,29).

Viral and Rickettsial Diseases

Viral pneumonias are relatively common, and most are diagnosed on clinical and radiologic grounds. However, when unusual viral pathogens are involved or a particular

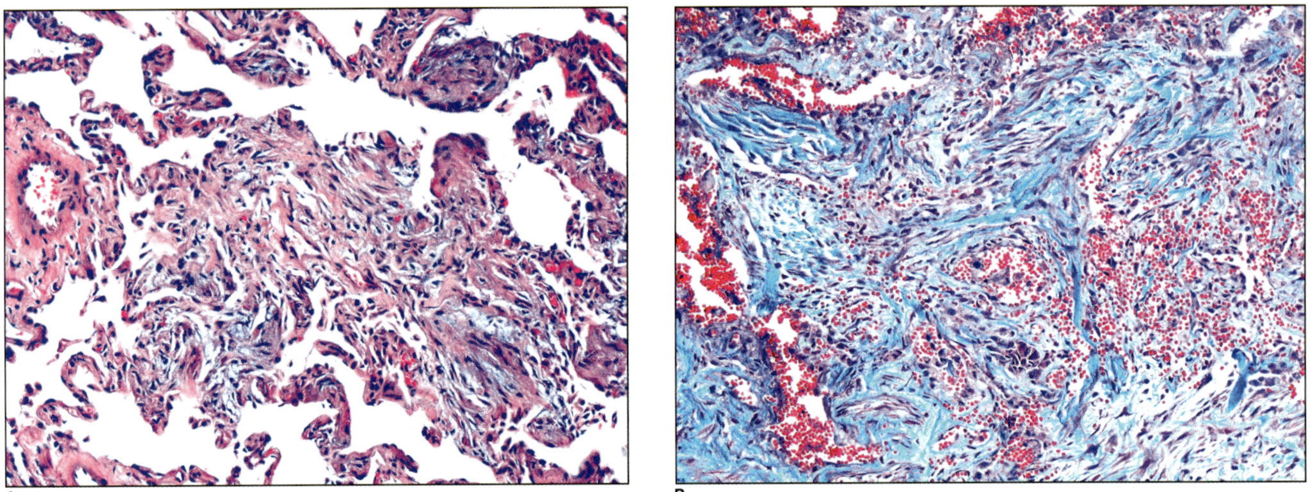

A B

FIGURE 3.23: Organizing pneumonia may produce proliferations of fibrovascular tissue in alveolar ducts and respiratory bronchioles, as shown here (A). This pattern was formerly known as "bronchiolitis obliterans organizing pneumonia." The fibroblastic component of the proliferation is better seen in a Masson trichrome stain (B) (A: H&E, 350×; B: Masson trichrome, 250×).

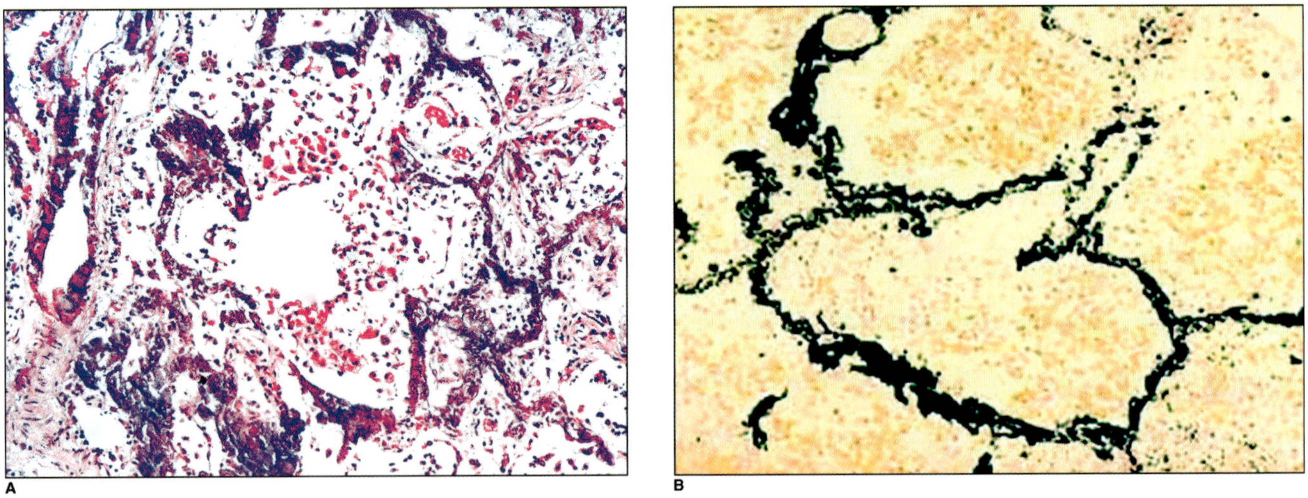

A B

FIGURE 3.24: (A) "Metastatic" calcification in the lung is manifest as deposition of calcium phosphate salts in the alveolar walls. The latter process can be visualized specifically with the von Kossa method (B) (A: H&E, 350×; B: von Kossa stain, 250×).

infectious agent may respond to a specific therapy, pathologic specimens are often obtained for morphologic evaluation (30–32). Similarly, patients with rickettsial pneumonias (e.g., Q-fever, typhus, Rocky Mountain spotted fever) also commonly undergo lung biopsies (33,34).

Some viral agents produce characteristic cytopathic changes that are recognizable with a high degree of diagnostic certainty using H&E stains. These include the *Herpesvirus* group (*Herpes simplex*, *Varicella-zoster*, *Cytomegalovirus*) (Figure 3.13), as well as *Rubeola* and *Adenovirus*. Nonetheless, the intracellular changes—including viral inclusions—that are associated with such viruses may be simulated by noninfectious alterations in bronchial epithelial cells and pneumocytes, particularly owing to inflammatory atypia or degenerative processes. Under these circumstances, it may be desirable to attempt to confirm the presence of viral inclusions histochemically. Rickettsiae are typically invisible in H&E sections, and special techniques also must be employed to demonstrate them.

Several of the histochemical methods used to label viruses and rickettsiae target the proteinaceous components of those microbes. Such techniques include Mann's methyl blue–eosin stain, the phosphotungstic acid–eosin procedure, and Lendrum's phloxine-tartrazine method (Figure 3.14) (35,36). In contrast, the Macchiavello procedure, the Feulgen method, and the methyl green–pyronine (MGP) stain can be used to label nucleic acids in intracellular organisms (Figure 3.15) (37,38). In the protein-target procedures, the final stains most commonly yield magenta, red, or purple inclusions; the Feulgen reagent labels DNA red, and the MGP procedure stains DNA blue green and RNA pink red.

Unexpectedly, cytoplasmic cytomegaloviral inclusions may also label with MS techniques and the PAS stain (Figure 3.16) (39). This observation probably reflects the envelopment process of such inclusions, which likely involves enwrapment of virions by layers of cellular plasmalemma or endoplasmic reticulum (40). The latter structures contain glycoprotein, explaining their affinity for silver impregnation or PAS stains.

Parasitic Infestations

The lungs are potentially involved in parasitic infestations caused by protozoans, trematodes, and nematodes of various types. Either parasite eggs or the mature organisms may be the agents of pulmonary disease. In particular, *Entamoeba histolytica*, *Cryptosporidium parvum*, *Toxoplasma gondii*, *Paragonimus westermani*, *Strongyloides stercoralis*, and *Dirofilaria immitis* can cause human infestations in the lungs, especially in immunocompromised patients (Figures 3.17 and 3.18). Usually, these organisms can be satisfactorily visualized by H&E stains alone; however, the trichrome and PAS methods are helpful in further defining amoebae (Figure 3.19) (41), and the Giemsa stain aids in demonstrating *Toxoplasma* or *Cryptosporidium* (Figure 3.20) (42–44). The last of those organisms is also potentially acid fast, as mentioned in an earlier section. Parasite eggs can often be highlighted with the PAS method (Figure 3.21) (45), and the elastic Van Gieson (EVG) technique assists in better showing relationships between adult organisms and large blood vessels or conducting airways (Figure 3.22).

NONNEOPLASTIC STRUCTURAL DISORDERS OF THE PULMONARY PARENCHYMA

The pulmonary parenchyma comprises a diversity of tissues, which are epithelial, vascular, fibroblastic, myogenous, and neural in nature. Each of those elements may participate in various nonneoplastic disorders, as outlined below, and histochemical methods are often helpful in further defining such processes.

Emphysema and Chronic Obstructive Airway Disease

Emphysema is an extremely common disorder, and one which is virtually always definable with H&E stains. Nonetheless, in selected cases where questions center on the *type* of emphysema (centrilobular versus panacinar), EVG, reticulin, and trichrome stains may help in demonstrating relative retention of interstitial structure in the subpleural parenchyma in centrilobular emphysema (46). In addition, trichrome stains aid in demonstrating smooth muscle hyperplasia in bronchi affected by chronic obstructive disease.

Idiopathic Interstitial Pneumonias

Several forms of idiopathic interstitial pneumonia are now recognized nosologically. In general, however, these have the common thread of producing interstitial fibrosis as they progress toward end-stage lung disease. This process can again be better delineated with trichrome and reticulin stains (47–49). In particular, the disorder now known as cryptogenic organizing pneumonia (formerly called "bronchiolitis obliterans organizing pneumonia") features a combination of interstitial fibrosis and fibrovascular proliferation in alveolar ducts and respiratory bronchioles. The trichrome stain provides a clear image of this process (Figure 3.23).

Diseases Featuring Heterotopic or Abnormal Parenchymal Deposits

Three principal conditions are characterized by the abnormal deposition of minerals or other materials in the lung parenchyma—"metastatic" calcification (pulmonary calciphylaxis) (50,51), mineralizing elastosis of chronic heart failure (52), and pulmonary amyloidosis (53–55). The first of these disorders features linear deposits of calcium salts in the interalveolar septa, which are well demonstrated by the von Kossa or alizarin-red (AR) method (Figure 3.24). Mineralizing elastosis likewise shows deposits of calcium, together with iron, on abnormal elastic fibers in the interstitium. This constellation can be defined by conjoint application of the von Kossa or AR method with Perl's stain and the EVG procedure. Finally, several stains are available for amyloid, the most familiar of which are the Congo red, crystal violet, and thioflavine-T techniques (Figure 3.25). The last of these requires visualization by fluorescence microscopy.

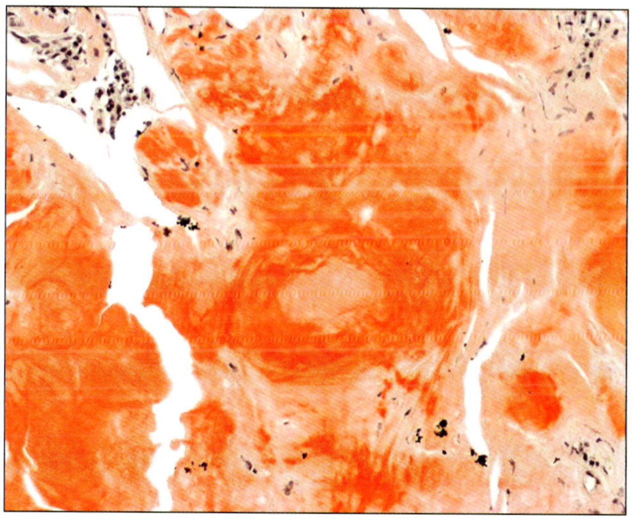

***FIGURE 3.25:** Amyloid deposits in the lung can be labeled by the Congo red method, among others (Congo red technique, 300×).*

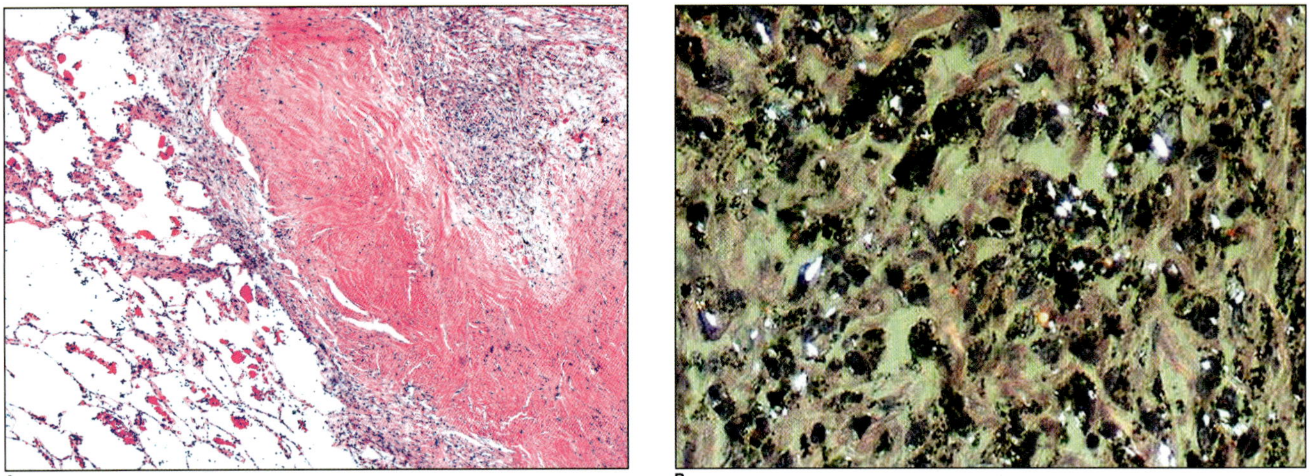

A B

FIGURE 3.26: Silicosis is represented by nodular fibrosis, usually along with anthracosis (A). The silica deposits are inert to histochemical methods but can be demonstrated by polarization microscopy of H&E-stained sections (B) (H&E, A: 150×; B: 400 × with polarization).

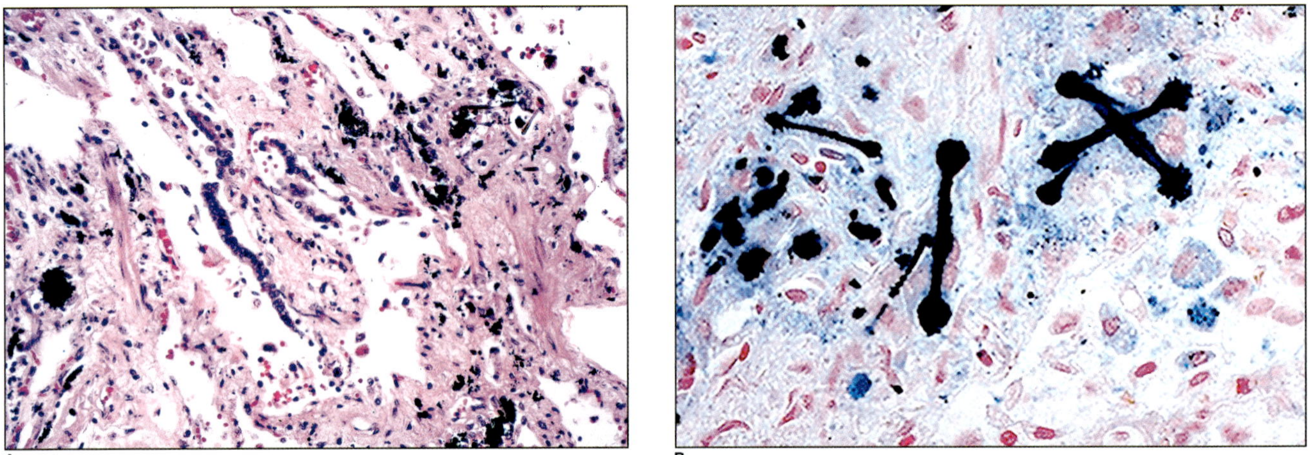

A B

FIGURE 3.27: Asbestosis demonstrates interstitial fibroelastosis on H&E stains, with golden-brown asbestos bodies enmeshed in the sclerotic foci (A). Perl's method for iron can be used as a means of labeling the asbestos bodies for purposes of rapid microscopic screening (B) (A: H&E, 200×; B: Perl's iron stain, 400×).

Pneumoconioses

The pneumoconioses are pulmonary diseases caused by occupational inhalational exposure to exogenous materials that incite parenchymal fibrosis (56). Although several of them are recognized, the two most common disorders in this group are silicosis and asbestosis. Silica is inert biochemically and cannot be emphasized with histochemical stains, but it is visible on polarization microscopy (Figure 3.26) (57). In addition, carbon pigment and iron salts are often co-deposited in areas of the lung where silica is present, and the second of those substances can be documented with Perl's method. Asbestos bodies represent mineral fibers that are variably coated with protein and iron by the host; hence, Perl's stain can be used as a tool for rapid screening of tissue sections for their presence (Figure 3.27). Asbestosis is distinctive among the pneumoconioses—and, for that matter, most other interstitial sclerosing diseases of the lung—in that it is a fibroelastosis (58). Therefore, the EVG stain is expected to demonstrate an increased density of interstitial elastic tissue in that disorder (Figure 3.28).

Vascular Abnormalities

The most common vascular abnormality in the lungs is that of chronic pulmonary venous hypertension, in the setting of congestive heart failure. With time, secondary *arterial* hypertension will supervene as well in this disease. These alterations can be well seen using the trichrome and EVG methods, showing both subintimal fibrosis and medial muscular hyperplasia in the affected vessels. *Primary* pulmonary hypertension or idiopathic veno-occlusive disease has dissimilar pathogeneses and clinical profiles, but they also show comparable vascular abnormalities to those just described (Figure 3.29) (59–63).

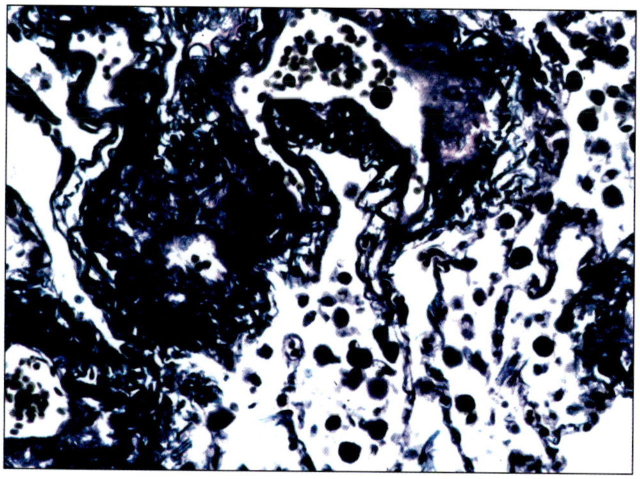

FIGURE 3.28: The interstitial elastotic component of asbestosis can be demonstrated well with an elastic stain (Van Gieson method, 250×).

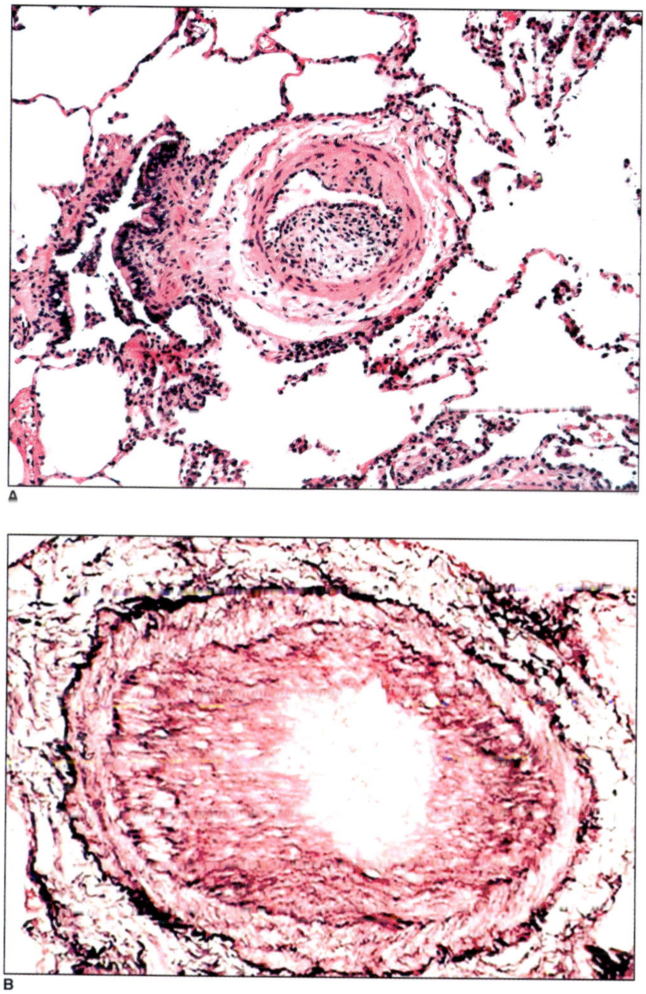

A

B

FIGURE 3.29: Pulmonary hypertension—whether primary or secondary in nature—is accompanied by eccentric or concentric subintimal fibrosis in pulmonary arterial branches (A). This process can be seen well with elastic stains (B) (A: H&E, 250×; B: Van Gieson method, 350×).

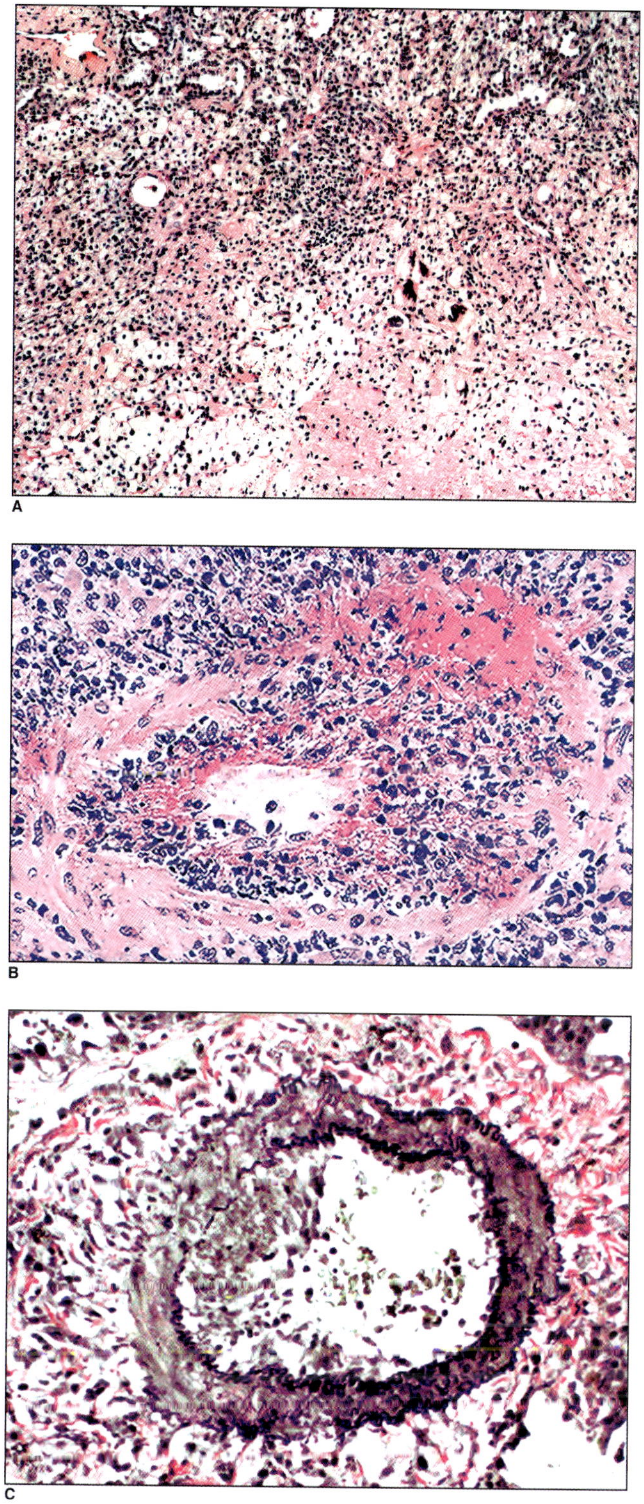

A

B

C

FIGURE 3.30: Arteritides in the lung are represented by this example of Wegener's granulomatosis, showing granulomas and confluent necrosis (A), along with granulomatous and lymphocytic vasculitis (B). The latter process can be confirmed with an elastic stain, demonstrating defects in the vessel walls (C) (A: H&E, 250×; B: H, 350×; C: Van Gieson method, 300×).

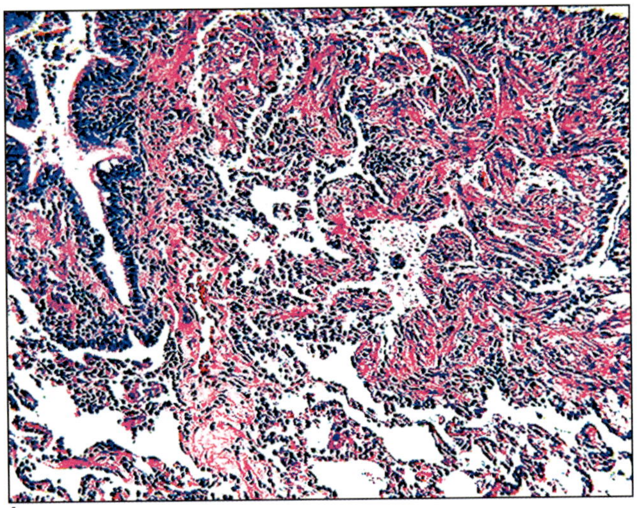
A

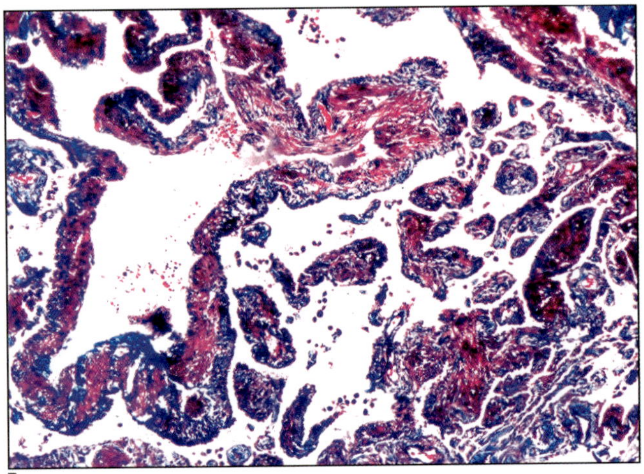
B

FIGURE 3.31: LAM shows a proliferation of smooth muscle cells around small airways and blood vessels in the lungs (A). The trichrome stain can be used to further demonstrate this process (B) (A: H&E, 350×; B: Masson trichrome method, 300×).

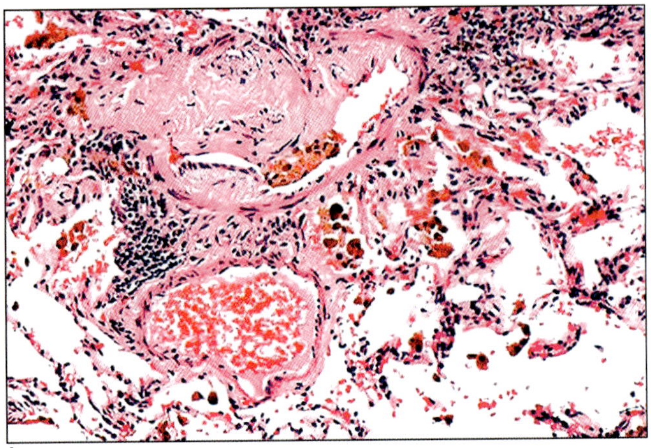
A

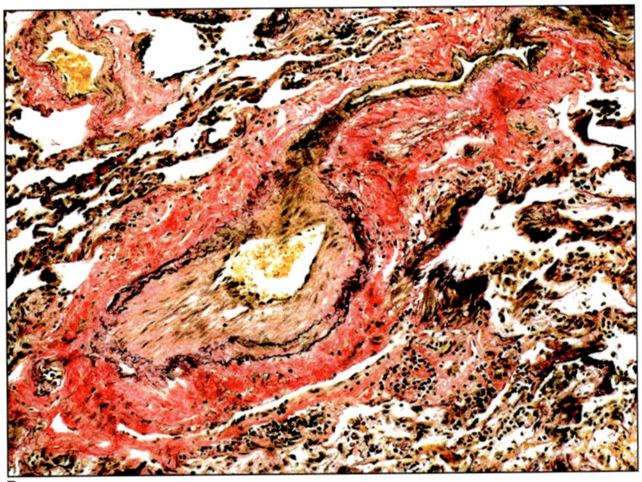
B

FIGURE 3.32: Chronic rejection of lung allografts is typified by sclerotic obliteration of small airways (A) or subintimal sclerosis of pulmonary arterial branches (B), or both (A: H&E, 200×; B: Van Gieson method, 350×).

Primary vasculitis affecting the pulmonary circulation is rare. Wegener's granulomatosis, "lymphomatoid granulomatosis" (angiocentric immunoproliferative lesions), and Churg-Strauss disease share a tendency to affect medium- and large-sized pulmonary arteries, with interruption of the vascular elastica (Figure 3.30) and evolution of luminal fibrosis (64–67). These changes are demonstrable with the EVG and trichrome procedures. Polyarteritis nodosa virtually never involves the lungs. Importantly, however, infectious vasculitis may closely simulate the image of Wegener's granulomatosis morphologically (68), making routine application of stains for acid-fast and fungal organisms mandatory in this setting.

Lymphangioleiomyomatosis

Lymphangioleiomyomatosis (LAM) is a distinctive disorder that—at least in some cases—may represent a *forme fruste* of tuberous sclerosis. It produces cystic alteration in the lung parenchyma, accompanied by distortion of pulmonary lymphatics, blood vessels, interstitium, and airways by a fibromusuclar cellular proliferation (69–71). LAM shows reactivity with the trichrome method (Figure 3.31).

Chronic Allograft Rejection

Chronic rejection of pulmonary allografts is typified by subepithelial fibrosis of small airways, leading to eventual luminal obliteration. This evolution can be demonstrated well using a combination of H&E and trichrome or EVG stains (72–74). Chronic *vascular* rejection in the transplanted lung is less well characterized, but similar sclerotic alterations may occur in small arteries of the graft over time (Figure 3.32).

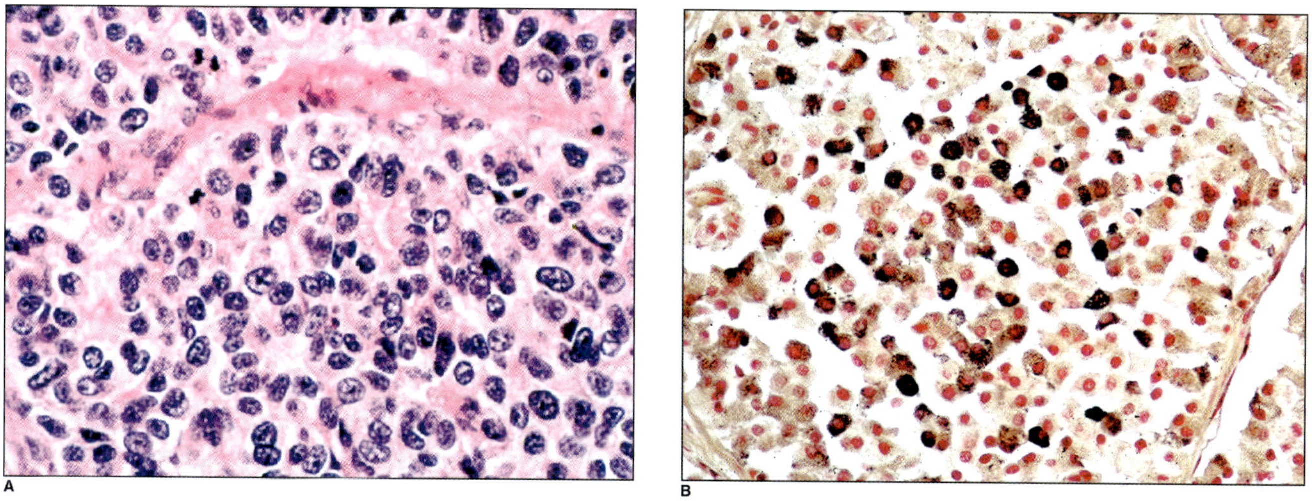

FIGURE 3.33: Neuroendocrine tumors of the lung show a composition by monomorphous tumor cells with dispersed nuclear chromatin and granular cytoplasm (A). The peculiar "neurosecretory" granules that they contain can be labeled with any of several argyrophil methods (B) (A: H&E, 400×; B: Churukian-Schenk stain, 400×).

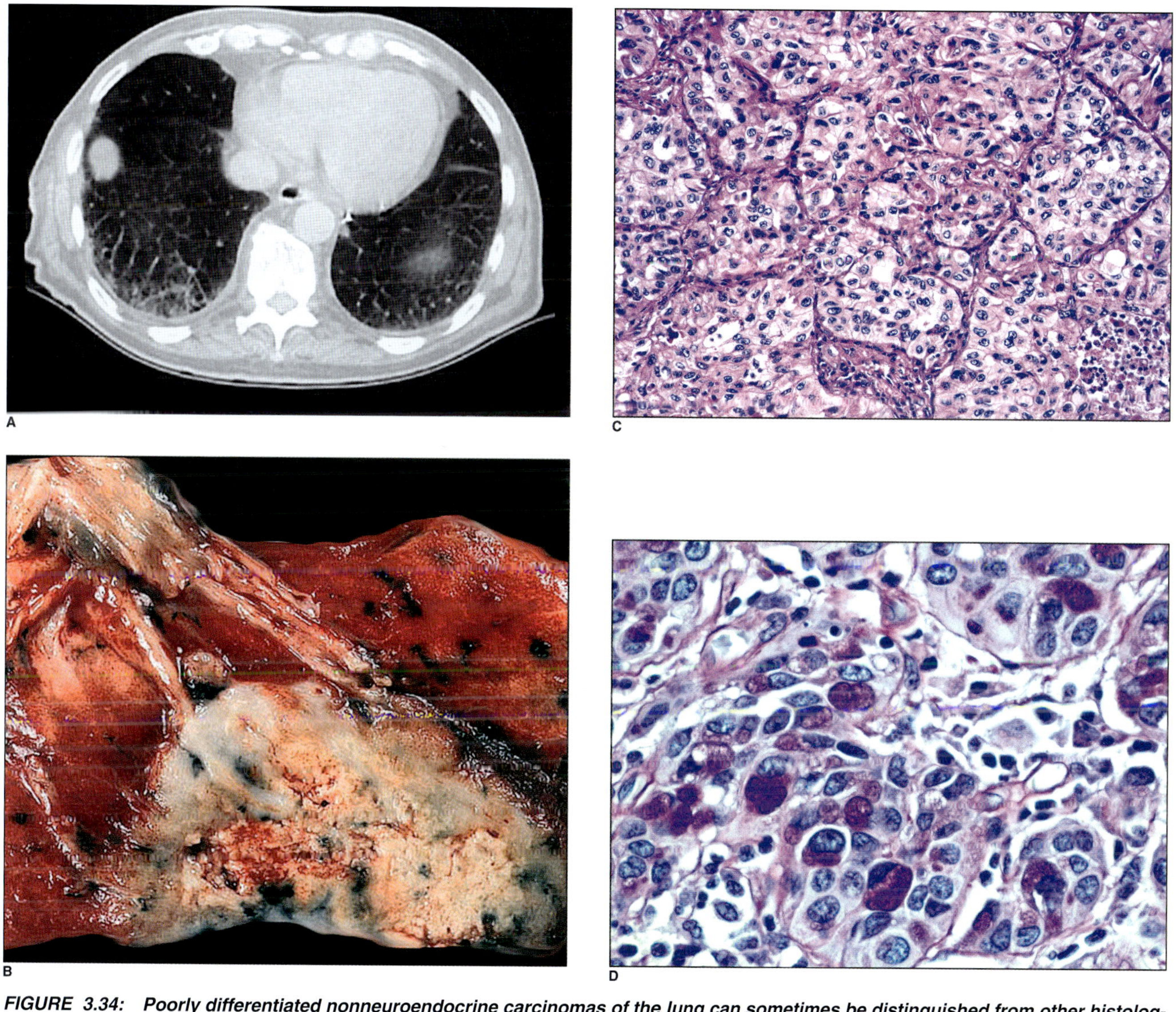

FIGURE 3.34: Poorly differentiated nonneuroendocrine carcinomas of the lung can sometimes be distinguished from other histologically similar nonepithelial tumors with mucin stains. A high-grade adenocarcinoma of the right lung is shown here (A–D) (A: thoracic computed tomogram; B: gross photograph of resected tumor; C: H&E, 300×; D: digested PAS stain, 400×).

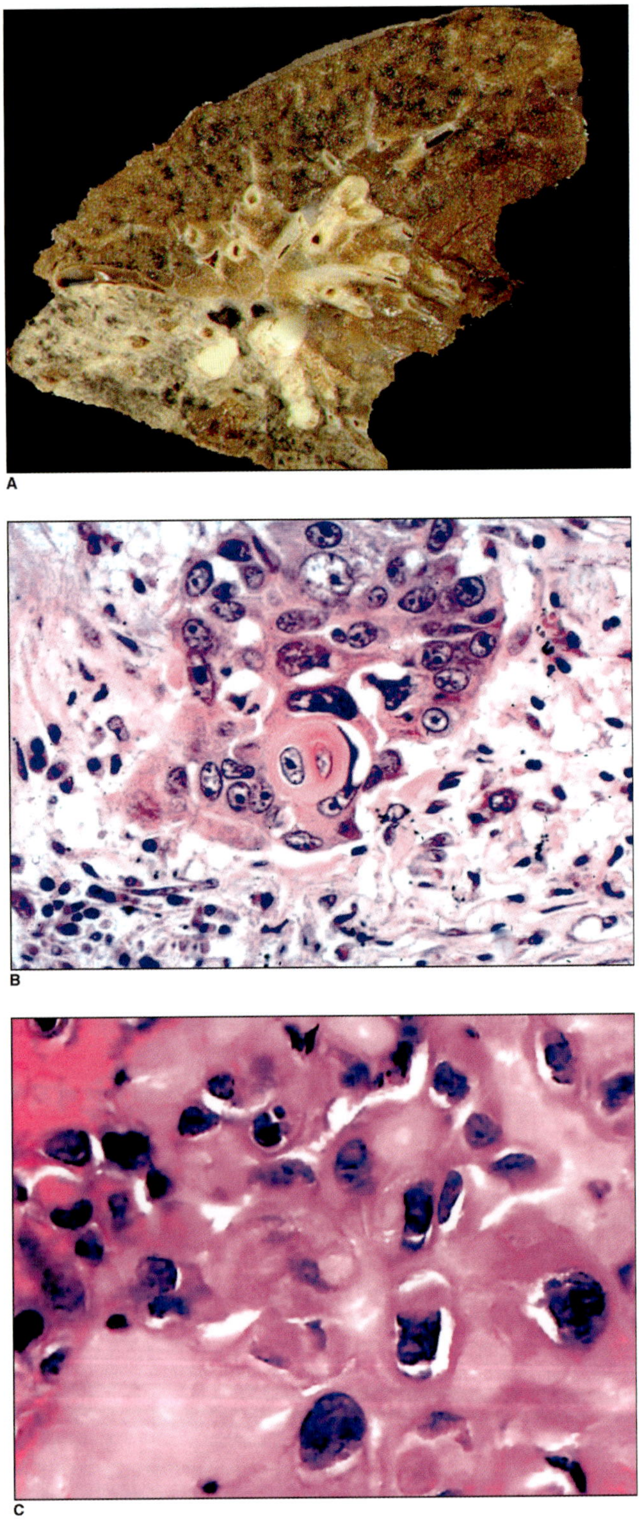

FIGURE 3.35: *Histochemical methods used for the general recognition of keratin are less satisfactory, as shown in this case of pulmonary squamous carcinoma (A and B). The histochemical techniques that label keratin (C) also are capable of decorating nonepithelial intermediate filament proteins (A: gross photograph of resected left lower lobe tumor; B: H&E, 300×; C: Levafix Red-Violet E-2BL stain, 600×).*

NEOPLASMS OF THE LUNG

Virtually all categories of neoplasia are represented in the lung, either as primary or secondary processes. Histochemical methods may aid in a determination of lineage for such lesions, or in the subcategorization of tumors within a particular group.

Neuroendocrine Tumors

Neuroendocrine carcinomas of the lung (including so-called carcinoid, atypical carcinoid, small cell carcinoma, and large cell neuroendocrine carcinoma) are distinctive because of their potential endocrine functionality and their content of cytoplasmic neurosecretory granules (75,76). Those inclusions can be selectively labeled by specialized silver impregnation methods, including the Grimelius, Sevier-Munger, and Churukian-Schenk techniques, which yield stippled black pigmentation in positive preparations (Figure 3.33). The latter three procedures are known as "argyrophil" stains, whereas the FM method is an "argentaffin" technique (as well as a melanophilic method). The significance of this distinction is that *primary* neuroendocrine tumors of the lung may be argyrophilic but are rarely if ever argentaffinic (77). The likelihood of silver positivity in a neuroendocrine carcinoma of the lung is inversely proportional to tumor grade (75).

Nonepithelial neuroendocrine tumors—for example, paraganglioma—may be seen as metastases in the lung or, very uncommonly, as primary neoplasms. They likewise share the potential for argyrophilia but are almost always argentaffin negative (78).

Primary Nonneuroendocrine Tumors

The greatest utility of histochemistry in the analysis of primary pulmonary neoplasms is in the identification of glandular differentiation. This feature can additionally be used in high-grade tumors as a generic indicator of epithelial lineage (79,80). In our experience, the most sensitive histochemical marker of epithelial mucin production is the PAS stain after diastase digestion of the tissue (Figure 3.34). Best's mucicarmine method is also potentially helpful in the same regard but has a much lower degree of sensitivity (81). Incidentally, the undigested PAS stain also has value in evaluating tumors for which both carcinoma and lymphoma are differential diagnostic possibilities. PAS positivity tends to exclude the latter of those choices. Histochemical methods otherwise have very limited application in the definitive delineation of epithelial differentiation. Some reports have linked the presence of keratin production to reactivity with Levafix Red-Violet E-2BL, fluorescent rhodamine-B, and Pauly's reagent (Figure 3.35). However, these stains generically recognize intermediate filament proteins of *all* types and are not restricted to epithelial lesions (82,83).

Stromal mucin—as recognized by the CI or alcian blue methods—has a much wider distribution among tumors of diverse lineages. However, one particular diagnostic setting lends itself to application of these procedures, namely, the separation of "pseudomesotheliomatous" adenocarcinoma in the pleura (84–86) from epithelioid mesothelioma. Lesions in both groups may in fact show positivity with CI or alcian blue stains, but only mesotheliomas are associated with abrogation of this reactivity by prior digestion with hyaluronidase (86–88). In addition, mesothelial tumors typically are not stained with the PAS-diastase or mucicarmine methods, although exceptions do exist (89).

Borczuk et al. (90) described an example of primary sebaceous carcinoma of the lung, a lesion that is much more often observed in the skin or salivary glands. The oil red O (ORO) stain was utilized in documenting the lipid content of epithelial cells in frozen sections of this tumor.

Secondary Malignant Neoplasms

Histochemistry has limited value in the separation of primary and metastatic pulmonary neoplasms. One exception is that melanomas—which are virtually always secondary lesions in the lung—can often be separated from carcinomas by integrating the results of H&E, PAS-diastase, argyrophil, FM, and Schmorl's stains. The last two of these methods are expected to yield positivity in melanomas (91), whereas carcinomas would typically be nonreactive. In contrast, PAS-diastase or argyrophil stains may label poorly differentiated epithelial tumors. The histochemical profiling of other metastatic pulmonary neoplasms is beyond the scope of this chapter, and the reader is referred to other sections of this monograph for that information.

ILLUSTRATIVE CASE STUDIES

The following case studies are intended to present information which supplements that contained in the foregoing discussion.

CASE 1— ***Mucormycosis:*** A 56-year-old diabetic man was a homeless person who had not maintained adequate glycemic control for many years and was seen only rarely by a physician. He was admitted to the hospital after being found unresponsive in the street. Physical examination demonstrated a fever of 104 °F, with signs of marked dehydration. The patient was semicomatose, with findings suggesting a left-sided cerebrovascular accident; auscultation of the thorax demonstrated coarse rhonchi and rales at both lung bases. The blood glucose level was 850 mg/dL. Chest radiographs were obtained (Figure 3.36), but death occurred before active treatment could commence.

At autopsy, several hemorrhagic infarcts were present in the lungs (Figure 3.37); microscopically, these contained angioinvasive fungal organisms with branching nonseptate hyphae (Figure 3.38). They were weakly labeled by the Gomori's MS method (Figure 3.39), and cultures later grew *Mucor* species. Hemorrhagic infarcts also were found in both cerebral hemispheres.

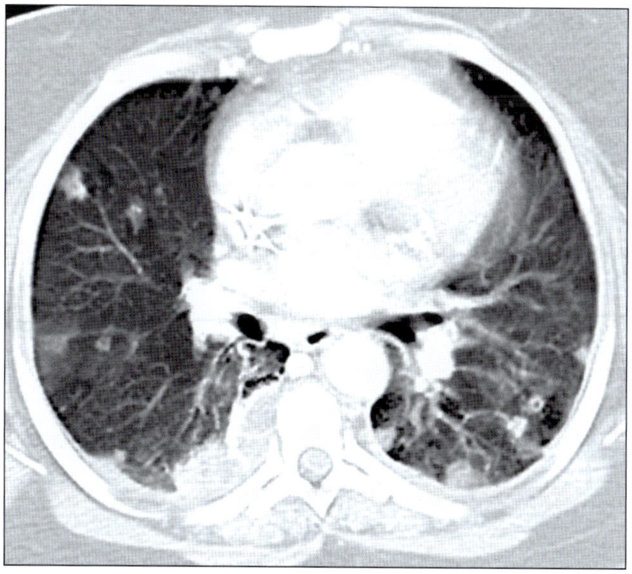

FIGURE 3.36: ***Computed tomogram of the thorax, showing multiple peripheral pulmonary infarcts and zones of parenchymal consolidation.***

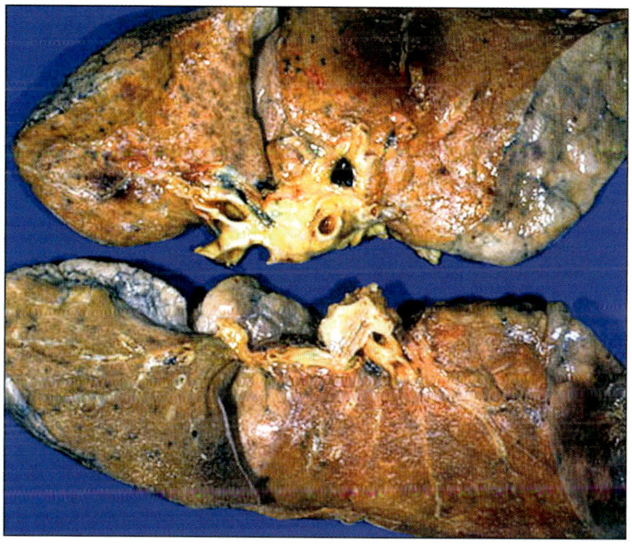

FIGURE 3.37: ***Autopsy photograph of the lungs, showing multiple hemorrhagic lesions throughout both lung fields.***

Mucorales is an order in the class of fungi known as zygomycetes, together with *Entomophthorales*. Genera in those groups of organisms include *Rhizopus*, *Mucor*, *Absidia*, and *Cunninghamella*, all of which have a similar appearance in tissue sections and comparable reactivity with histochemical methods (92–95). The zygomycetes are important to distinguish from *Aspergillus* or *Fusarium*, both of which are close morphologic mimics but contain septations in their hyphae. *Mucor* and related organisms often require substantial surgical debridement for successful therapy, in contrast to non-zygomycetes (92). ■

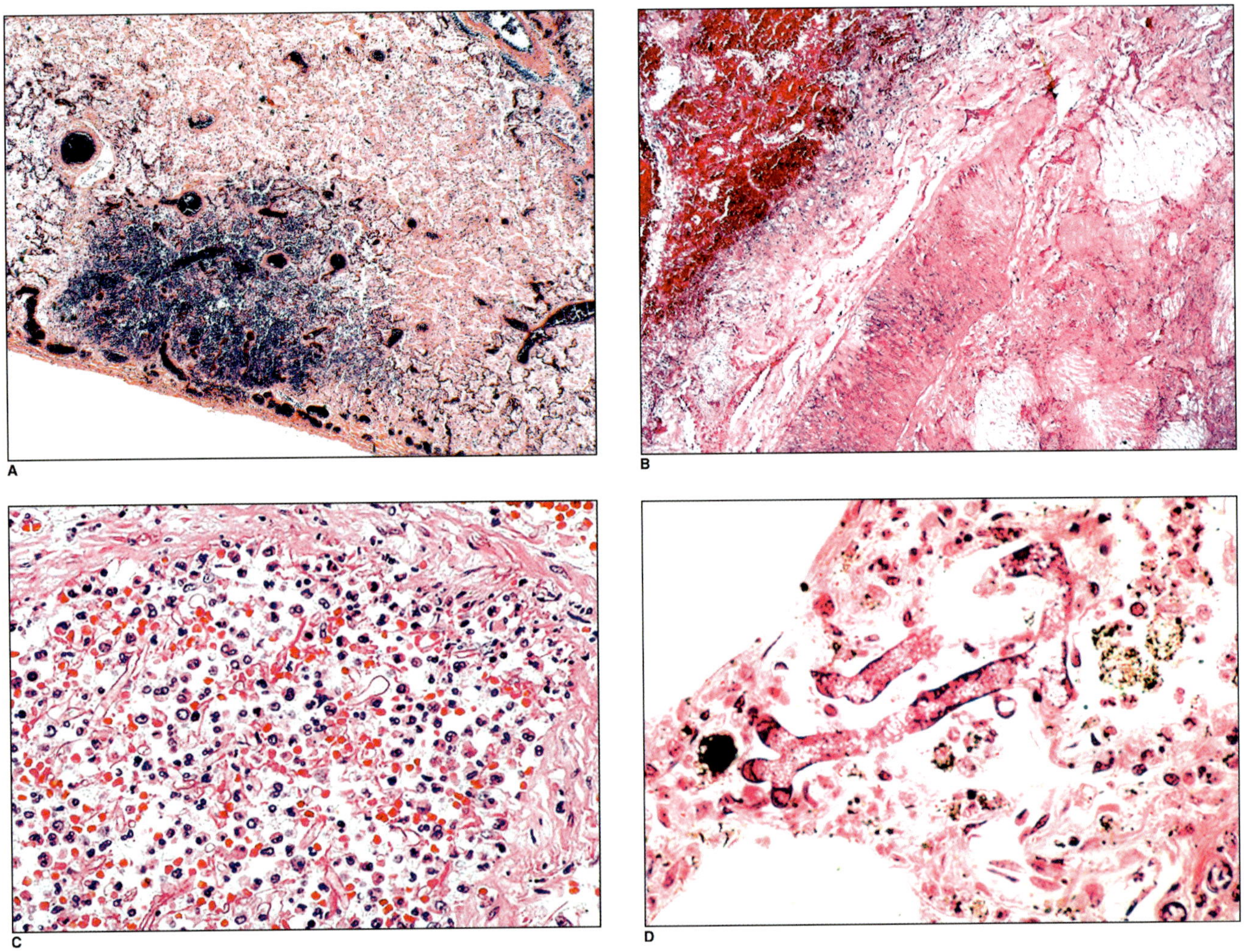

FIGURE 3.38: Histologically, peripheral hemorrhagic infarcts are evident (A). These contain neutrophilic infiltrates as well as pauci-septated hyphal fungal organisms (B–D) (A: H&E, 100×; B: H&E, 200×; C: H&E, 150×; D: H&E, 400×).

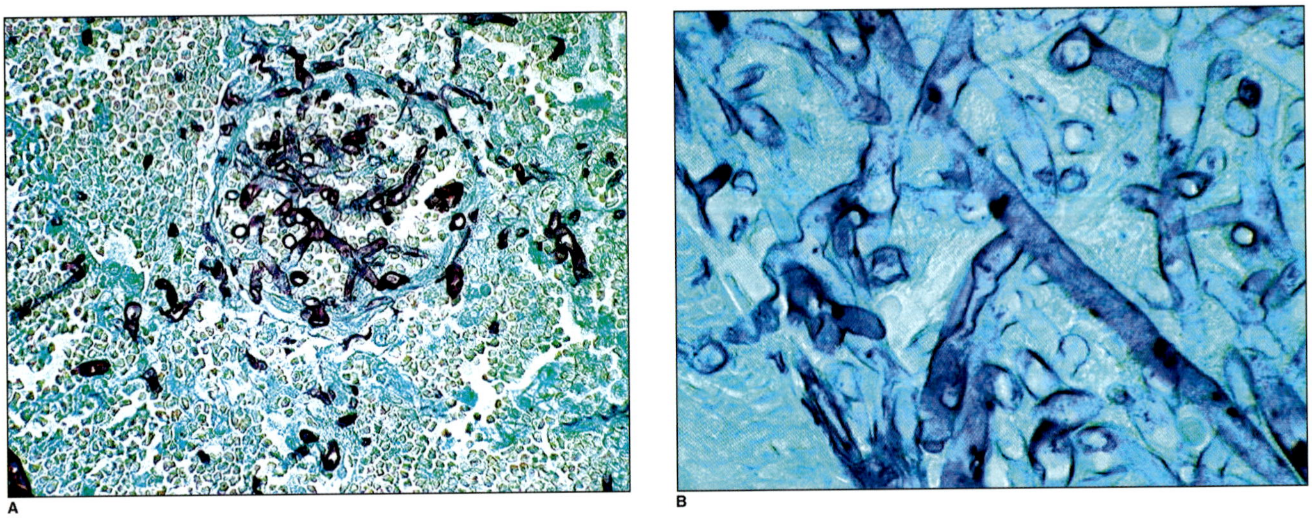

FIGURE 3.39: The hyphae are labeled with the MS method (A: 200×; B: 400×).

CASE 2— *Pulmonary Fat Embolism:* A 51-year-old man was involved in a motor vehicle accident, in which he suffered fractures of several long bones and ribs. These were stabilized surgically, and the patient initially did well postoperatively. However, 1 day later, he began to complain of rapidly worsening dyspnea, and a chest radiograph demonstrated bilateral interstitial and alveolar infiltrates (Figure 3.40). Progressive hypoxemia led to adult respiratory distress syndrome, multisystem failure, and death 1 week after surgery.

Because of the clinical context, it was suspected that the patient had fat microemboli in the lungs. Fresh pulmonary tissue obtained at autopsy was frozen and stained with the ORO method (Figure 3.41), yielding positive results in interalveolar septal capillaries.

The ORO, Sudan-III, Sudan-black-B, Sudan-IV, and osmic acid methods are capable of labeling triglycerides, owing to their properties as lysochromes. Although "wet" formalin-fixed tissue can be stained by these procedures with modest success, optimal results are obtained when frozen sections of unfixed specimens are employed as substrates (96,97). Fat embolization syndrome should be suspected whenever bone fractures are followed by unexplained pulmonary dysfunction or neurologic compromise, or both. ■

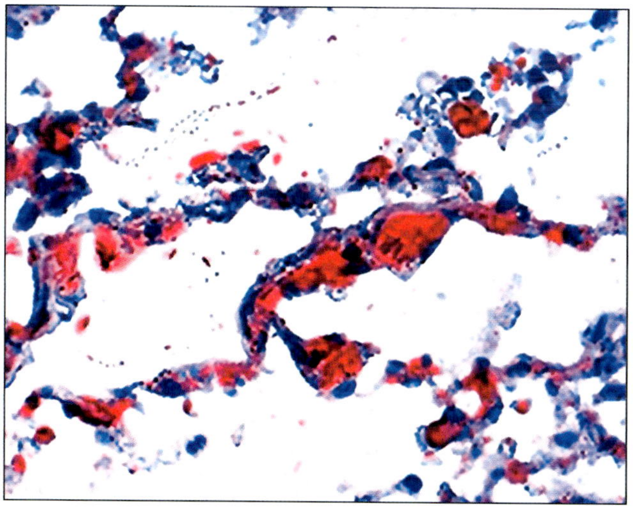

FIGURE 3.41: Fat emboli are visible in interalveolar capillary lumina, using the ORO method on frozen sections of autopsy lung (ORO method, 350×).

CASE 3— *Alveolar Proteinosis:* A 42-year-old man began to experience slowly progressive shortness of breath, accompanied by a low-grade fever and malaise. He consulted his internist, who noted the presence of bibasilar rales on auscultation of the thorax. Chest radiographs showed a bilateral alveolar-filling infiltrate with a "butterfly" pattern (Figure 3.42), and computed tomograms demonstrated a so-called crazy paving pattern in the lung fields. Bronchoalveolar lavage (BAL) was performed, with marked improvement in the patient's symptoms and signs. However, these recurred 5 months later and an open lung biopsy was obtained then. It demonstrated a granular eosinophilic exudate in the alveolar spaces, which was reactive with the PAS method (Figures 3.43 and 3.44). There was no evidence of microorganisms. A prior diagnosis of idiopathic alveolar proteinosis was confirmed. The patient again underwent BAL, and he remained disease-free thereafter.

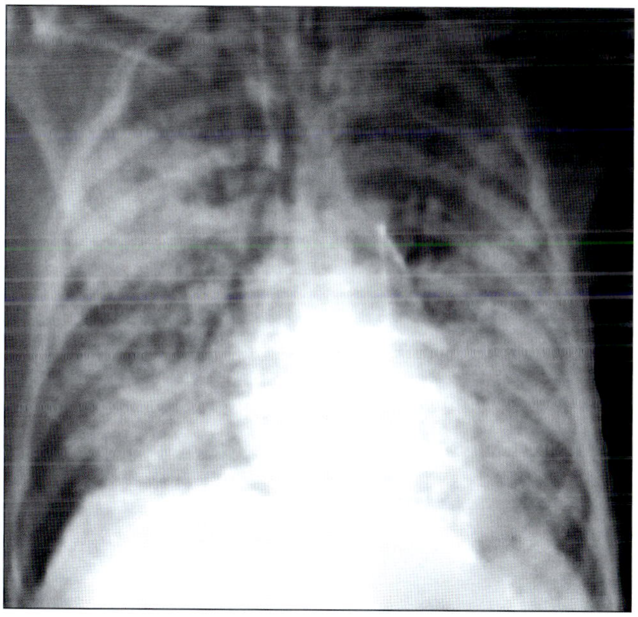

FIGURE 3.40: Diffuse alveolar and reticulonodular interstitial infiltrates are seen in this chest radiograph.

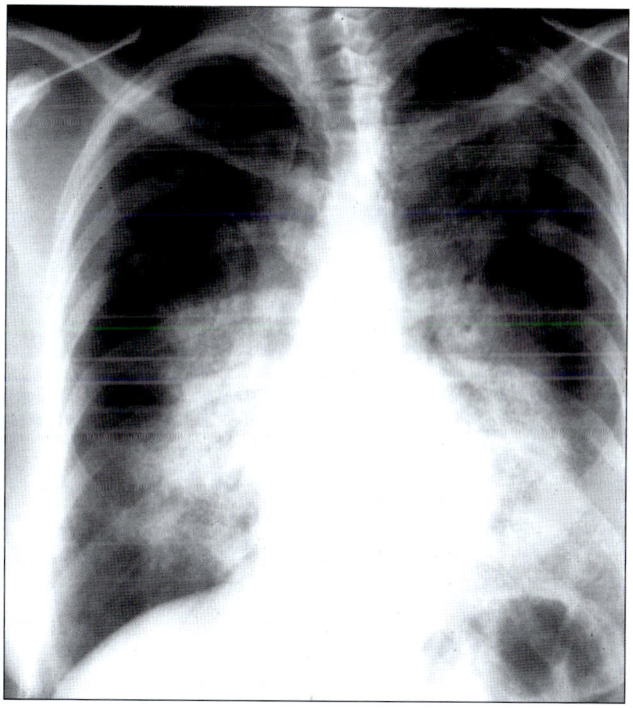

FIGURE 3.42: Pulmonary alveolar consolidation has a "butterfly" distribution in this chest radiograph.

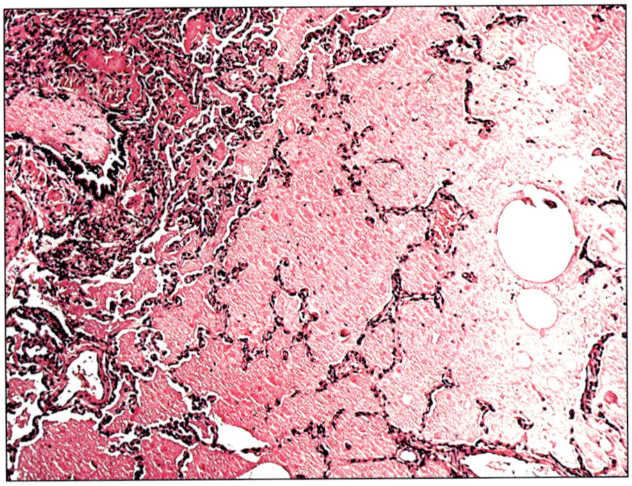

FIGURE 3.43: ***Homogeneous, faintly granular, and eosinophilic exudates fill the alveolar spaces (H&E, 150×).***

Alveolar proteinosis may be a primary disorder, or secondary to a pneumoconiosis, a malignancy, or pulmonary infection with *Nocardia* or other microorganisms (98–100). The intraalveolar exudate in this disorder contains phospholipoprotein derived from surfactant, accounting for its PAS positivity (99). ■

<u>CASE 4</u>— ***Pulmonary Malacoplakia:*** A 59-year-old man had had a cadaveric renal transplant for diabetic nephropathy and coronary revascularization for ischemic heart disease. He began to complain of low-grade fever and recalcitrant nonproductive cough, and chest radiographs showed fullness of the right hilum with a partially cavitary, pleural-based lesion in the right lower lung field (Figure 3.45). Because of concern over the possibility of posttransplant lymphoproliferative disease, a thoracotomy was performed and the lesion in the right lower lobe was excised. Cultures of this tissue subsequently grew *Rhodococcus equi*.

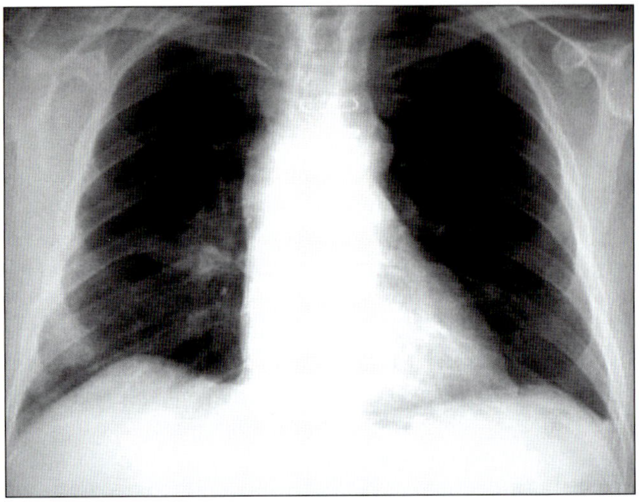

FIGURE 3.45: ***A right perihilar parenchymal infiltrate is accompanied by a peripheral, right pleural-based lesion in this chest radiograph.***

The specimen showed no evidence of an atypical lymphoid infiltrate but instead demonstrated sheets of histiocytes admixed with other inflammatory cells (Figures 3.46 and 3.47). The histiocytes contained internally laminated and rounded cytoplasmic inclusions that were labeled with the PAS, Perl's, and von Kossa techniques (Figures 3.48–3.50). These findings established the diagnosis of infection-associated pulmonary malacoplakia.

Malacoplakia results from defective killing of bacteria by macrophages that lack normal phagolysosomal activity. Partially digested organisms accumulate in the cells, producing deposits of calcium and iron on bacterial glycolipids (101). The resulting basophilic inclusions, known as Michaelis-Gutmann bodies, are characteristic of this disorder (102). ■

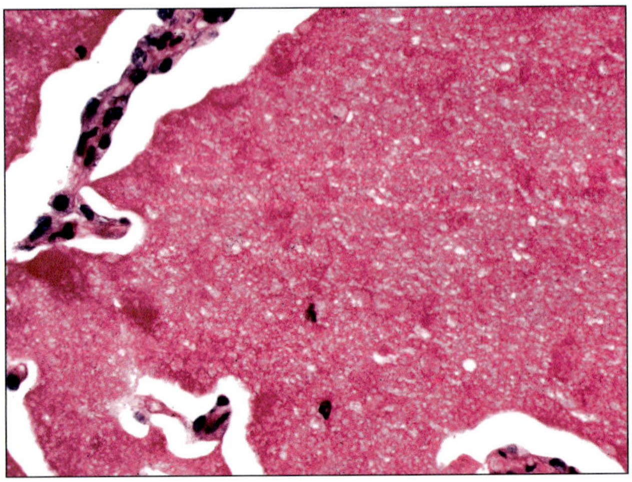

FIGURE 3.44: ***The exudates are labeled with the digested PAS method. (digested PAS technique, 400×).***

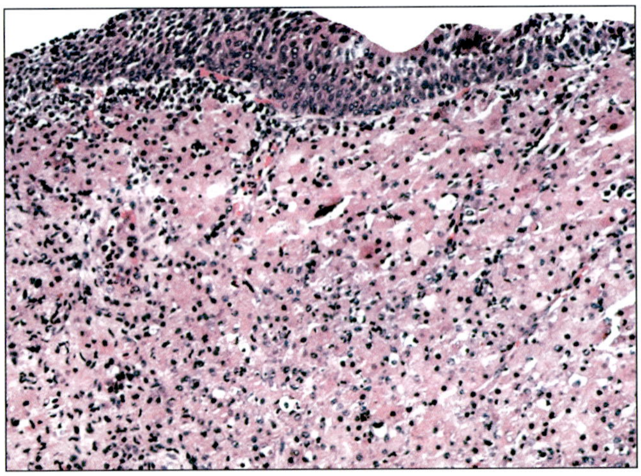

FIGURE 3.46: ***The bronchial mucosa is undermined by a mixed inflammatory infiltrate that is dominated by large histiocytes with eosinophilic cytoplasm (H&E, 200×).***

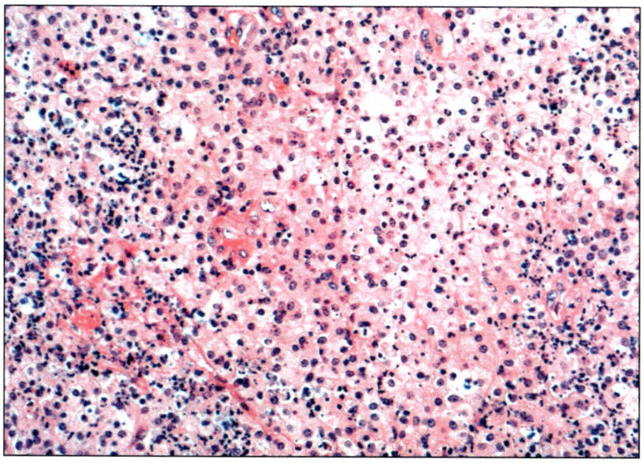

FIGURE 3.47: *Lymphocytes and scattered neutrophils are admixed with the histiocytes (H&E, 350×).*

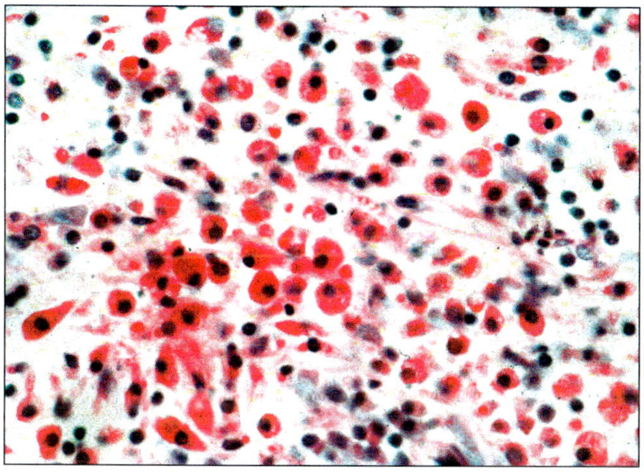

FIGURE 3.48: *The histiocytes are intensely reactive with the digested PAS method (digested PAS technique, 350×).*

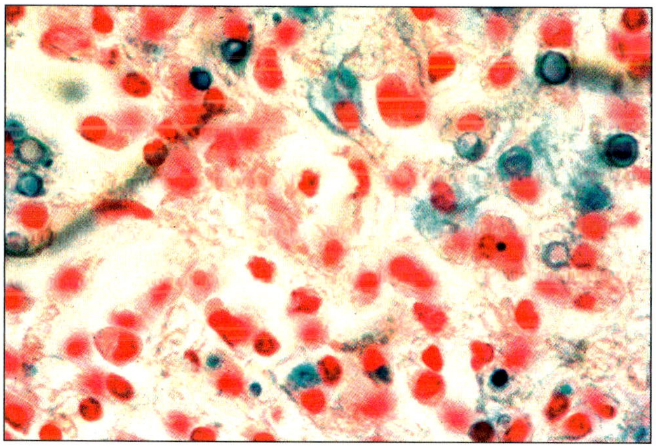

FIGURE 3.49: *Perl's method labels circular intracellular inclusions in the histiocytes (Perl's stain, 600×).*

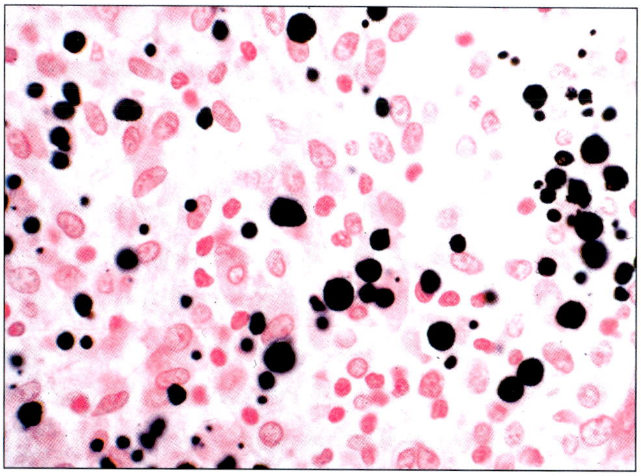

FIGURE 3.50: *The von Kossa stain similarly decorates the intracellular inclusions (von Kossa method, 400×).*

CASE 5— *Large Cell Undifferentiated Carcinoma:* A 62-year-old man developed a persistent dry cough and had lost 20 pounds involuntarily in 3 months. He had a 30 pack-year cigarette smoking history but had discontinued the habit 10 years previously. Physical examination disclosed dullness to percussion over the left anterior upper chest, and thoracic imaging studies showed a mass in the mid-left lung field (Figure 3.51). A percutaneous

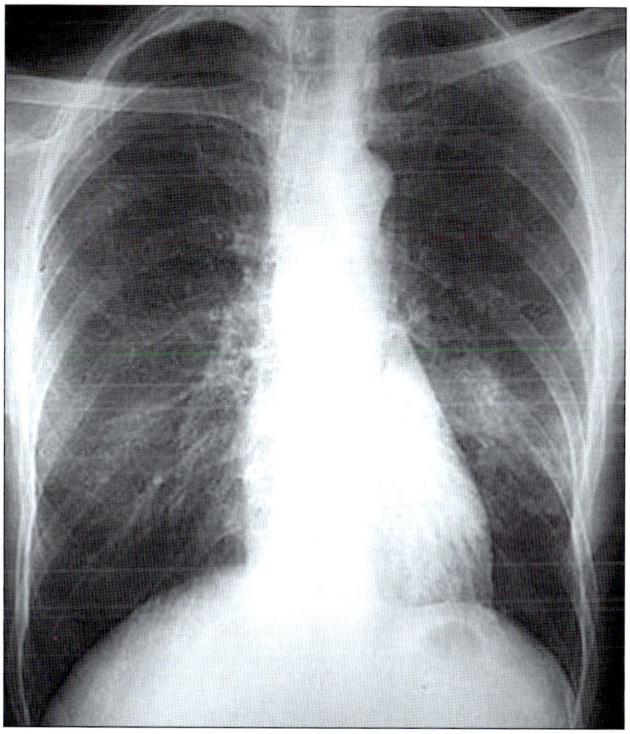

FIGURE 3.51: *An intrapulmonary mass is apparent in the left mid-lung field in this chest radiograph.*

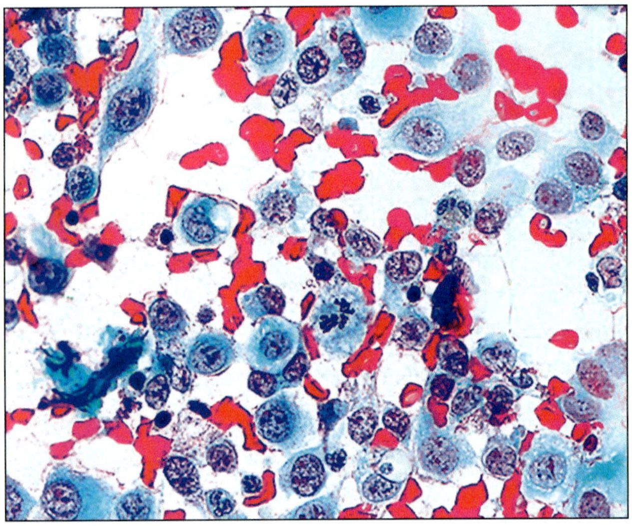

FIGURE 3.52: A fine-needle aspiration biopsy of the lesion yields a poorly cohesive population of undifferentiated polygonal cells (Papanicolaou stain, 300×).

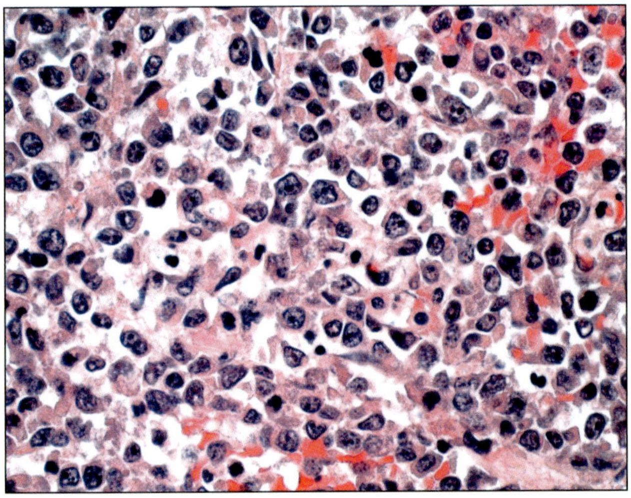

FIGURE 3.53: This photomicrograph of the resected tumor also demonstrates a composition by undifferentiated large neoplastic cells (H&E, 300×).

fine-needle aspiration biopsy (Figure 3.52) showed the presence of a large cell malignant neoplasm, but it could not be further characterized. Accordingly, a thoracotomy was done and a left upper lobectomy was performed.

Microscopic examination of the lesion showed a large cell anaplastic tumor, the principal differential diagnosis for which was a large cell malignant hematopoietic lesion versus large cell undifferentiated carcinoma (Figures 3.53 and 3.54). An MGP stain was moderately positive (Figure 3.55); an undigested PAS stain (Figure 3.56) and lectin histochemical stains with biotinylated *Ulex europaeus I* and *Griffonia simplicifonia* were also reactive (Figure 3.57). The von Leder stain was negative. An interpretation of large cell carcinoma was made.

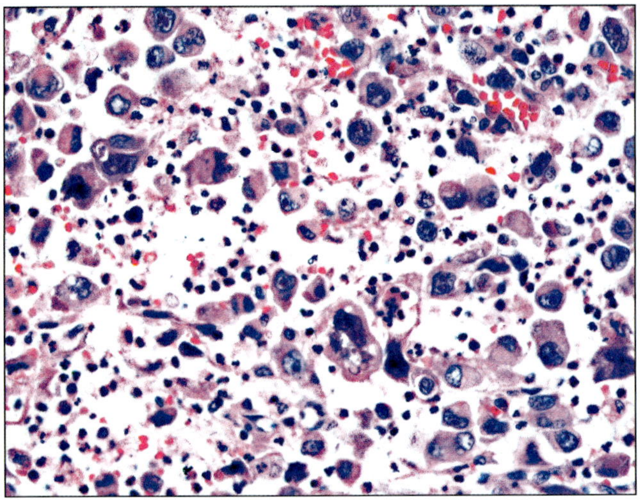

FIGURE 3.54: Pleomorphism is apparent among the tumor cell population (H&E, 400×).

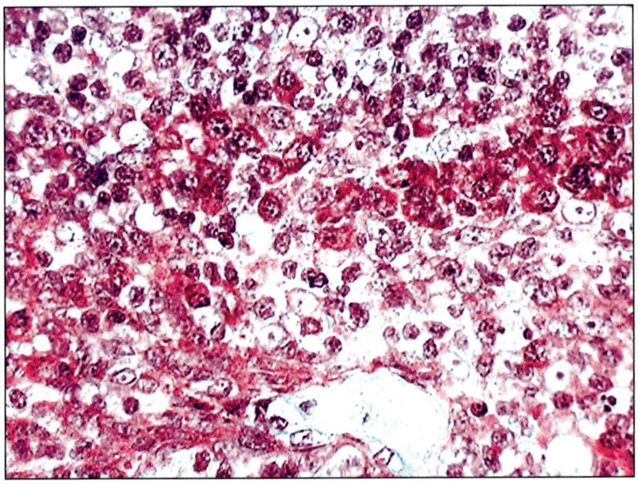

FIGURE 3.55: Histochemical reactivity is seen in the lesion with the MGP technique (MGP stain, 400×).

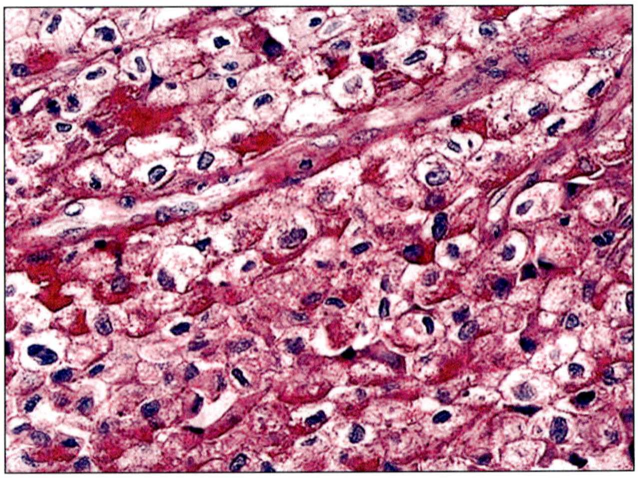

FIGURE 3.56: The tumor cells are also reactive with the undigested PAS method (undigested PAS stain, 350×).

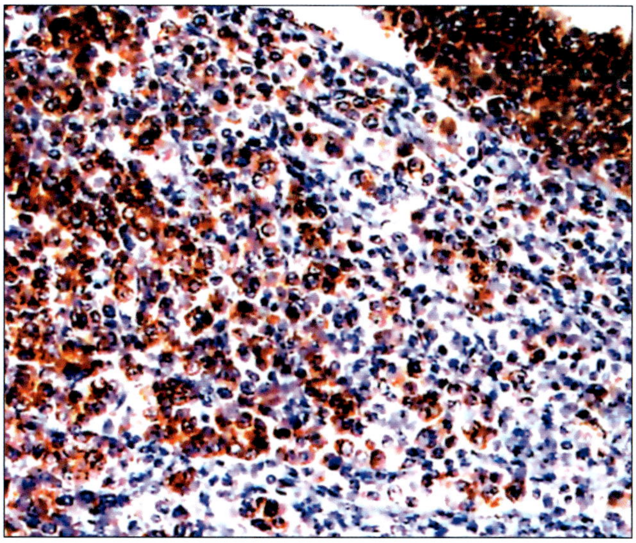

***FIGURE 3.57:** A histochemical stain with biotinylated* Ulex europaeus *I lectin shows positivity in the tumor cells (*Ulex europaeus *I: streptavidin-peroxidase method, 150×).*

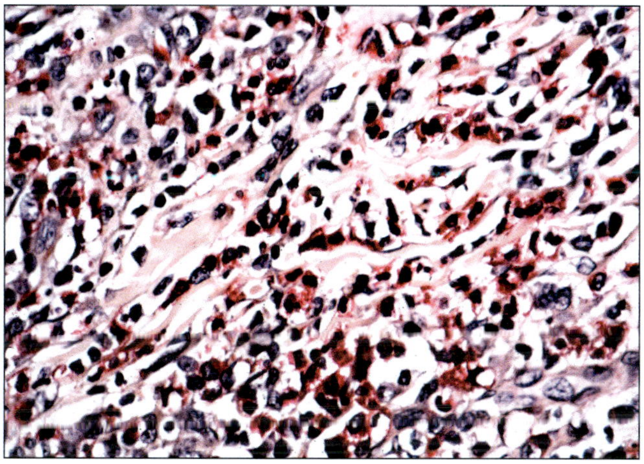

***FIGURE 3.58:** Histochemical reactivity for chloroacetate esterase is seen in this granulocytic sarcoma (extramedullary myeloid tumor) of the lung (von Leder method, 300×).*

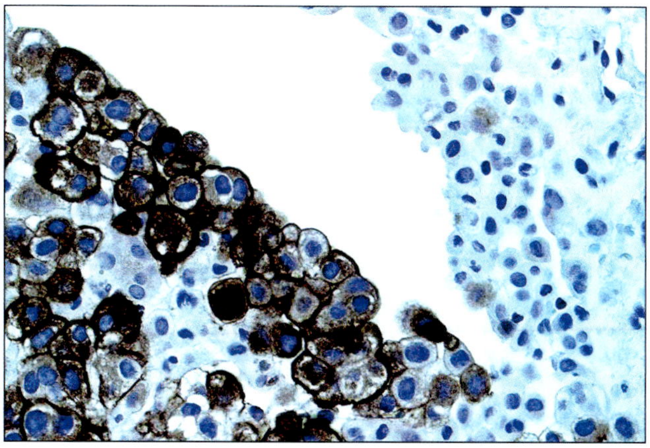

***FIGURE 3.59:** Histochemical positivity is shown in a pleural metastasis of adenocarcinoma, using biotinylated* Griffonia simplicifolia *lectin (*Griffonia simplicifolia*: streptavidin-peroxidase method, 400×).*

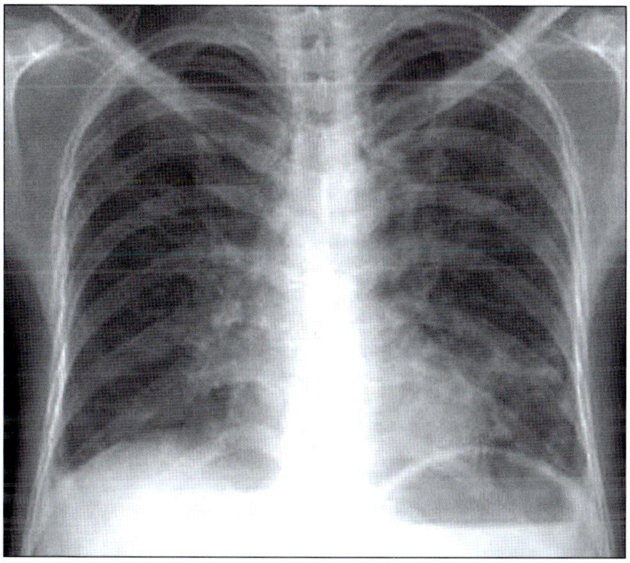

***FIGURE 3.60:** Chest radiograph showing bilateral alveolar and interstitial opacities in a 41-year-old intravenous drug abuser.*

Glycogen content in an undifferentiated large cell neoplasm—as demonstrated by the PAS technique—excludes the possibility of malignant lymphoma, despite concurrent reactivity with the MGP method (103,104). Conversely, positivity for chloroacetate esterase with the von Leder method (Figure 3.58) would have argued for a diagnosis of granulocytic sarcoma (extramedullary myeloid sarcoma) and eliminated carcinoma from consideration (105).

Lectins are plant proteins that bind to specific carbohydrate residues in tissue sections. In particular reference to the differential diagnosis in question in this case, it is important to note that lectins from *Ulex* and *Griffonia* do not label hematopoietic malignancies but instead are each associated with 60% positivity in lung carcinomas (106–110). Another potential application of lectin histochemistry is in the separation of epithelioid mesothelioma from pleural involvement by adenocarcinoma. Reactivity for *Griffonia* and *Ricinus communis* lectins is associated with the latter but not the former of those tumors (Figure 3.59) (108,111). ■

<u>CASE 6—</u> ***Pulmonary Kaposi's Sarcoma:*** A 41-year-old man had had progressive dyspnea, intermittent fever, and a recalcitrant cough for 6 months. He was known to be an intravenous drug abuser. Chest radiography showed irregular alveolar and interstitial densities throughout both lung fields (Figure 3.60). An open lung biopsy demonstrated

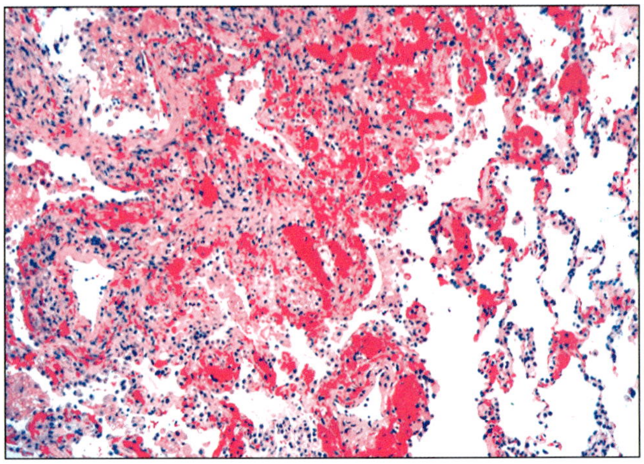

FIGURE 3.61: This open lung biopsy specimen demonstrates a spindle cell proliferation that is principally centered in the interstitium in this figure, in association with hemorrhage (H&E, 200×).

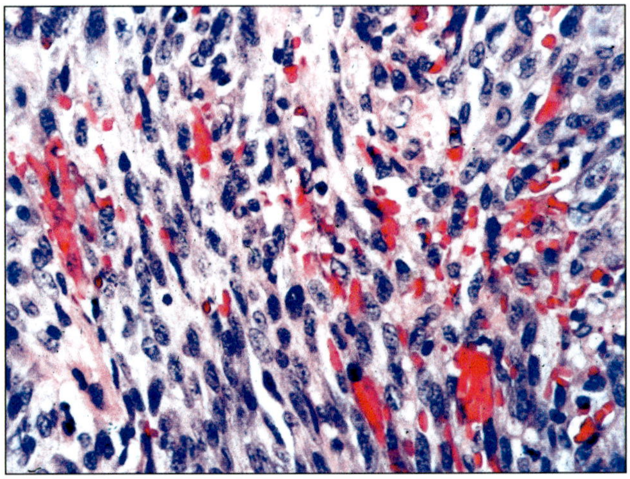

FIGURE 3.62: The lesion is typified by atypical fusiform cells, some of which contain cytoplasmic vacuoles (H&E, 350×).

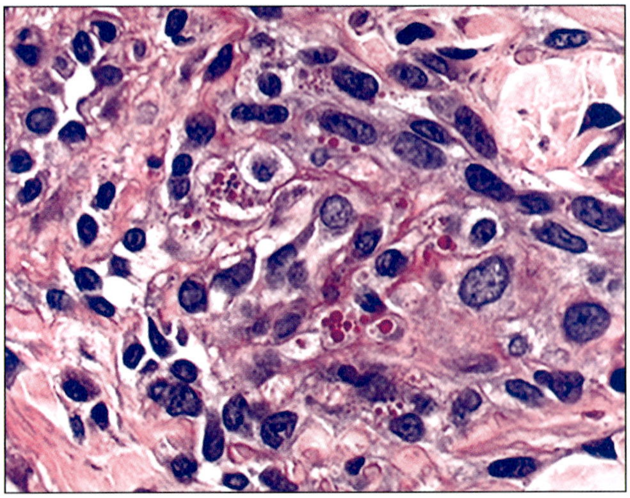

FIGURE 3.63: Cytoplasmic globules are present in the tumor cells in this digested PAS stain, supporting the diagnosis of Kaposi's sarcoma (digested PAS stain, 400×).

mass-forming proliferations of spindle cells in the parenchyma (Figures 3.61 and 3.62), including cells with intracytoplasmic vacuoles and eosinophilic globules. The latter were reactive with the digested-PAS method (Figure 3.63), as well as the phosphotungstic acid–hematoxylin (PTAH) technique.

Although it is not absolutely specific for Kaposi's sarcoma, the presence of eosinophilic intracellular globules is characteristic of the spindle cell variant of that tumor. Several studies have shown that these structures label with PAS, PTAH, and peroxidase procedures, supporting the premise that they represent partially digested erythrocytes (112–114). ■

REFERENCES

1. Taylor RD: Modification of the Brown and Brenn Gram stain for the differential staining of Gram-positive and Gram-negative bacteria in tissue sections. *Am J Clin Pathol* 1966; 46: 472–474.
2. Rudin JE, Wing EJ: A comparative study of *Legionella micdadeii* and other nosocomial acquired pneumonia. *Chest* 1984; 86: 675–680.
3. Genest DR, Choi-Hong SR, Tate JE, et al.: Diagnosis of congenital syphilis from placental examination: comparison of histopathology, Steiner stain, and polymerase chain reaction for *Treponema pallidum* DNA. *Hum Pathol* 1996; 27: 366–372.
4. Tome Y, Hayashi I, Matsuoka S, et al.: A simple and highly reproducible staining method for fungi and other polysaccharide-rich microorganisms in animal tissues. *Stain Technol* 1988; 63: 53–57.
5. Smith I, Nathan C, Peavy HH: Progress and new directions in the genetics of tuberculosis: an NHLBI Working Group report. *Am J Respir Crit Care Med* 2005; 172: 1491–1496.
6. Martins AB, Matos ED, Lemos AC: Infection with the *Mycobacterium avium* complex in patients without predisposing conditions: a case report and literature review. *Braz J Infect Dis* 2005; 9: 173–179.
7. Doucette K, Fishman JA: Nontuberculous mycobacterial infection in hematopoietic stem cell and solid organ transplant recipients. *Clin Infect Dis* 2004; 38: 1428–1439.
8. Olivier KN and the NTM in CF Study Group: The natural history of nontuberculous mycobacteria in patients with cystic fibrosis. *Paediatr Respir Rev* 2004; 5 (Suppl A): S213–S216.
9. Aksamit TR: *Mycobacterium avium* complex pulmonary disease in patients with preexisting lung disease. *Clin Chest Med* 2002; 23: 643–653.
10. Tang YW, Procop GW, Zheng X, Myers JL, Roberts GD: Histologic parameters predictive of mycobacterial infection. *Am J Clin Pathol* 1998; 109: 331–334.
11. Truant JP, Brett WA, Thomas W Jr.: Fluorescence microscopy of tubercle bacilli stained with auramine and rhodamine. *Henry Ford Hosp Med Bull* 1962; 10: 287–296.
12. Samuel KC, Chatterjee SN: Modification of Fite-Faraco staining for acid-fast bacilli. *Indian J Pathol Bacteriol* 1971; 14: 107–109.

13. Wade HW: A modification of the Fite formaldehyde (Fite-I) method for staining acid-fast bacilli in paraffin sections. *Stain Technol* 1957; 32: 287–292.
14. Poonwan N, Kusum M, Mikami Y, et al.: Pathogenic *Nocardia* isolated from clinical specimens including those of AIDS patients in Thailand. *Eur J Epidemiol* 1995; 11: 507–512.
15. Scott MA, Graham BS, Verrall R, et al.: *Rhodococcus equi*—an increasingly recognized opportunistic pathogen: report of 12 cases and review of 65 cases in the literature. *Am J Clin Pathol* 1995; 103: 649–655.
16. Magi B, Canocchi V, Tordini G, Cellesi C, Barberi A: *Cryptosporidium* infection: diagnostic techniques. *Parasitol Res* 2006; 98: 150–152.
17. Chappell CL, Okhuysen PC: Cryptosporidiosis. *Curr Opin Infect Dis* 2002; 15: 523–527.
18. Hilton E, Freedman RA, Cintron F, Isenberg HD, Singer C: Acid-fast bacilli in sputum: a case of *Legionella micdadeii* pneumonia. *J Clin Microbiol* 1986; 24: 1102–1103.
19. Blumenfeld W, Wagar E, Hadley WK: Use of the transbronchial biopsy for diagnosis of opportunistic pulmonary infections in acquired immunodeficiency syndrome (AIDS). *Am J Clin Pathol* 1984; 81: 1–5.
20. Musto L, Flanigan M, Elbadawi A: Ten-minute silver stain for *Pneumocystis carinii* and fungi in tissue sections. *Arch Pathol Lab Med* 1982; 106: 292–294.
21. Huppert M, Oliver DJ, Sun SH: Combined methenamine-silver nitrate and hematoxylin & eosin stain for fungi in tissues. *J Clin Microbiol* 1978; 8: 598–603.
22. La Manna S: The histochemical demonstration and silver impregnation of fungi. *Arch Venez Med Trop Parasitol Med* 1960; 3: 173–186.
23. Grocott RG: A stain for fungi in tissue sections and smears using Gomori's methenamine-silver nitrate technic. *Am J Clin Pathol* 1955; 25: 975–979.
24. Gridley ME: A stain for fungi in tissue sections. *Am J Clin Pathol* 1953; 23: 303–307.
25. Krane JF, Renshaw AA: Relative value and cost-effectiveness of culture and special stains in fine needle aspirates of the lung. *Acta Cytol* 1998; 42: 305–311.
26. Kimura M, McGinnis MR: Fontana-Masson-stained tissue from culture-proven mycoses. *Arch Pathol Lab Med* 1998; 122: 1107–1111.
27. Lazcano O, Speights VO Jr., Strickler JG, et al.: Combined histochemical stains in the differential diagnosis of *Cryptococcus neoformans*. *Mod Pathol* 1993; 6: 80–84.
28. Lazcano O, Speights VO Jr., Bilbao J, Becker J, Diaz J: Combined Fontana-Masson-mucin staining of *Cryptococcus neoformans*. *Arch Pathol Lab Med* 1991; 115: 1145–1149.
29. Dei-Cas E, Fleurisse L, Aliouat EM, et al.: Morphological and ultrastructural methods for *Pneumocystis*. *FEMS Immunol Med Microbiol* 1998; 22: 185–189.
30. Chien JW, Johnson JL: Viral pneumonias: infection in the immunocompromised host. *Postgrad Med* 2000; 107: 67–80.
31. Camazine B, Antkowiak JG, Nava ME, Lipman BJ, Takita H: *Herpes simplex* viral pneumonia in the postthoracotomy patient. *Chest* 1995; 108: 876–879.
32. Ebbert JO, Limper AH: Respiratory syncytial virus pneumonitis in immunocompromised adults: clinical features and outcome. *Respiration* 2005; 72: 263–269.
33. Caron F, Meurice JC, Ingrand P, et al.: Acute Q fever pneumonia: a review of 80 hospitalized patients. *Chest* 1998; 114: 808–813.
34. Johnson DH, Cunha BA: Atypical pneumonias: clinical and extrapulmonary features of *Chlamydia*, *Mycoplasma*, and *Legionella* infections. *Postgrad Med* 1993; 93: 69–82.
35. Lendrum AC: The phloxine-tartrazine method as a general histological stain and for the demonstration of inclusion bodies. *J Pathol Bacteriol* 1947; 59: 399–404.
36. Quezada M, Cayo L, Carrasco L, et al.: Characterization of lesions caused by a South American virulent isolate ("Quillota") of the hog cholera virus. *J Vet Med B Infect Dis Vet Public Health* 2000; 47: 411–422.
37. Roger F, Roger A: Stain affinities of *Rickettsiae* & large visible viruses (lymphogranuloma-psittacosis group): I. Mechanism of differentiation with Macchiavello stain and various practical applications. *Ann Inst Pasteur* 1958; 94: 126–128.
38. David-West TS: Cytological, cytochemical, and immunofluorescence studies with Dugbe virus—a new Nigerian tick-borne virus. *Acta Microbiol Acad Sci Hung* 1975; 22: 33–40.
39. Saiz E, Lubin J, Robinson MJ: The modified Steiner stain: a new use for an old stain? Staining cytomegalovirus-infected cells in gastrointestinal biopsies. *Histochem J* 1998; 30: 549–552.
40. Kasnic G Jr., Sayeed A, Azar HA: Nuclear and cytoplasmic inclusions in disseminated human cytomegalovirus infection. *Ultrastruct Pathol* 1982; 3: 229–235.
41. Tanyuksel M, Yilmaz H, Ulukanligil M, et al.: Comparison of two methods (microscopy and enzyme-linked immunosorbent assay) for the diagnosis of amebiasis. *Exp Parasitol* 2005; 110: 322–326.
42. Da Silva AV, Langoni H: The detection of *Toxoplasma gondii* by comparing cytology, histopathology, bioassay in mice, and the polymerase chain reaction (PCR). *Vet Parasitol* 2001; 97: 191–198.
43. Cornelissen AW, Overdulve JP, Van der Ploeg M: Cytochemical studies on nuclear DNA of four eucoccidian parasites, *Isospora (Toxoplasma) gondii*, *Eimeria tenella*, *Sarcocystis cruzi*, and *Plasmodium berghei*. *Parasitology* 1984; 88: 13–25.
44. Casemore DP, Armstrong M, Sands RL: Laboratory diagnosis of cryptosporidiosis. *J Clin Pathol* 1985; 38: 1337–1341.
45. Tracy JW, Mahmoud AA: Isolation of *Schistosoma japonicum* egg glycoprotein antigens which sensitize mice to lung granuloma formation and elicit an immediate hypersensitivity response. *Am J Trop Med Hyg* 1982; 31: 1201–1212.
46. Fukuda Y, Masuda Y, Ishizaki M, Masugi Y, Ferrans VJ: Morphogenesis of abnormal elastic fibers in lungs of patients with panacinar and centriacinar emphysema. *Hum Pathol* 1989; 20: 652–659.
47. Barbolini G, Bisetti A, Martinelli AM, Rivasi F, Saltini C: Diffuse interstitial lung diseases: a histochemical approach. *Respiration* 1984; 45: 339–346.
48. Negri EM, Montes GS, Saldiva PH, Capelozzi VL: Architectural remodelling in acute and chronic interstitial lung disease: fibrosis or fibroelastosis? *Histopathology* 2000; 37: 393–401.
49. Rozin GF, Gomes MM, Parra ER, et al.: Collagen and elastic system in the remodelling process of major types

of idiopathic interstitial pneumonias. *Histopathology* 2005; 46: 413–421.
50. Chung MJ, Lee KS, Franquet T, et al.: Metabolic lung disease: imaging and histopathologic findings. *Eur J Radiol* 2005; 54: 233–245.
51. Chan ED, Morales DV, Welsh CH, McDermott MT, Schwarz MI: Calcium deposition with or without bone formation in the lung. *Am J Respir Crit Care Med* 2002; 165: 1654–1669.
52. Pai U, McMahon J, Tomashefski JF Jr.: Mineralizing pulmonary elastosis in chronic cardiac failure: "endogenous pneumoconiosis" revisited. *Am J Clin Pathol* 1994; 101: 22–28.
53. Agresta F, Marin A, Della-Libera D, et al.: Primary pulmonary nodular amyloidosis. *Monaldi Arch Chest Dis* 2005; 63: 173–175.
54. Sharma OP: Unusual systemic disorders associated with interstitial lung disease. *Curr Opin Pulm Med* 2001; 7: 291–294.
55. Howard ME, Ireton J, Daniels F, et al.: Pulmonary presentations of amyloidosis. *Respirology* 2001; 6: 61–64.
56. Fujimura N: Pathology and pathophysiology of pneumoconiosis. *Curr Opin Pulm Med* 2000; 6: 140–144.
57. Honma K, Chiyotani K: Diffuse interstitial fibrosis in nonasbestos pneumoconiosis—a pathological study. *Respiration* 1993; 60: 120–126.
58. Raithel HJ, Stolte M, Otto H, Valentin H: Diagnostic and forensic problems in the simultaneous occurrence of asbestosis and pulmonary sideroelastosis (Ceelen disease). *Prax Klin Pneumonol* 1984; 38: 446–451.
59. Justo RN, Dare AJ, Whight CM, Radford DJ: Pulmonary veno-occlusive disease: diagnosis during life in four patients. *Arch Dis Child* 1993; 68: 97–100.
60. Palevsky HI, Schloo BL, Pietra GG, et al.: Primary pulmonary hypertension: vascular structure, morphometry, and responsiveness to vasodilator agents. *Circulation* 1989; 80: 1207–1221.
61. Loyd JE, Atkinson JB, Pietra GG, Virmani R, Newman JH: Heterogeneity of pathologic lesions in familial primary pulmonary hypertension. *Am Rev Respir Dis* 1988; 138: 952–957.
62. Wagenvoort CA, Wagenvoort N, Takahashi T: Pulmonary veno-occlusive disease: involvement of pulmonary arteries and review of the literature. *Hum Pathol* 1985; 16: 1033–1041.
63. Edwards WD, Edwards JE: Clinical primary pulmonary hypertension: three pathologic types. *Circulation* 1977; 56: 884–888.
64. Travis WD: Pathology of pulmonary vasculitis. *Semin Respir Crit Care Med* 2004; 25: 475–482.
65. Travis WD: Pathology of pulmonary granulomatous vasculitis. *Sarcoidosis Vasc Diffuse Lung Dis* 1996; 13: 14–27.
66. Travis WD, Hoffman GS, Leavitt RY, Pass HI, Fauci AS: Surgical pathology of the lung in Wegener's granulomatosis: review of 87 open lung biopsies from 67 patients. *Am J Surg Pathol* 1991; 15: 315–333.
67. Yousem SA, Lombard CM: The eosinophilic variant of Wegener's granulomatosis. *Hum Pathol* 1988; 19: 682–688.
68. Hammar SP: Granulomatous vasculitis. *Semin Respir Infect* 1995; 10: 107–120.
69. Niku S, Stark P, Levin DL, Friedman PJ: Lymphangioleiomyomatosis: clinical, pathologic, and radiologic manifestations. *J Thorac Imaging* 2005; 20: 98–102.
70. Sobonya RE, Quan SF, Fleishman JS: Pulmonary lymphangioleiomyomatosis: quantitative analysis of lesions producing airflow obstruction. *Hum Pathol* 1985; 16: 1122–1128.
71. Carrington CB, Cugell DW, Gaensler EA, et al.: Lymphangioleiomyomatosis: physiologic-pathologic-radiologic correlations. *Am Rev Respir Dis* 1977; 116: 977–995.
72. Ward C, Snell GI, Zheng L, et al.: Endobronchial biopsy and bronchoalveolar lavage in stable lung transplant recipients and chronic rejection. *Am J Respir Crit Care Med* 1998; 158: 84–91.
73. Cagle PT, Brown RW, Frost A, Kellar C, Yousem SA: Diagnosis of chronic lung transplant rejection by transbronchial biopsy. *Mod Pathol* 1995; 8: 137–142.
74. Marchevsky AM, Hartman G, Walts A, et al.: Lung transplantation: the pathologic diagnosis of pulmonary complications. *Mod Pathol* 1991; 4: 133–138.
75. Cerilli LA, Ritter JH, Mills SE, Wick MR: Neuroendocrine neoplasms of the lung. *Am J Clin Pathol* 2001; 116 (Suppl): S65–S96.
76. Corrin B: Lung endocrine tumors. *Invest Cell Pathol* 1980; 3: 195–206.
77. Vinik A: Carcinoid tumors. On: Endotext.com (http://www.endotext.org/guthormones/guthormone2/guthormone2.htm), accessed 12/10/2005.
78. Shibahara J, Goto A, Niki T, et al.: Primary pulmonary paraganglioma: report of a functioning case with immunohistochemical and ultrastructural study. *Am J Surg Pathol* 2004; 28: 825–829.
79. Wattanasirmkit V, Tanboon J, Shuangshoti S, Shuangshoti S: LacCur stain for detection of mucin in adenocarcinoma. *J Med Assoc Thai* 2004; 87: 87–91.
80. McGregor DH, Dixon AY, McGregor DK: Adenocarcinoma of the lung: a comparative diagnostic study using light and electron microscopy. *Hum Pathol* 1988; 19: 910–913.
81. Kennedy A, Burgin PD: A comparison of different methods of detecting mucin in adenocarcinomas of the lung. *Br J Dis Chest* 1975; 69: 137–143.
82. Wessely Z, Shapiro SH, Klavins JV, Tinberg HM: Identification of Mallory bodies with rhodamine-B fluorescence and other stains for keratin. *Stain Technol* 1981; 56: 169–176.
83. Waldrop FS, Puchtler H, Akamatsu Y: Staining of keratin and keratohyaline with the reactive dye Levafix-red violet E-2BL. *Stain Technol* 1976; 51: 219–225.
84. Koss MN, Fleming M, Przygodzki RM, et al.: Adenocarcinoma simulating mesothelioma: a clinicopathologic and immunohistochemical study of 29 cases. *Ann Diagn Pathol* 1998; 2: 93–102.
85. Hartmann CA, Schutze H: Mesothelioma-like tumors of the pleura: a review of 72 autopsy cases. *J Cancer Res Clin Oncol* 1994; 120: 331–347.
86. Wick MR, Loy T, Mills SE, Legier JF, Manivel JC: Malignant epithelioid pleural mesothelioma versus peripheral pulmonary adenocarcinoma: a histochemical, ultrastructural, and immunohistochemical study of 103 cases. *Hum Pathol* 1990; 21: 759–766.

87. Dewar A, Valente M, Ring NP, Corrin B: Pleural mesothelioma of epithelial type and pulmonary adenocarcinoma: an ultrastructural and cytochemical comparison. *J Pathol* 1987; 152: 309–316.
88. Lucas JG, Tuttle SE: Diagnostic histochemical and immunohistochemical studies in malignant mesothelioma. *J Surg Oncol* 1987; 35: 30–34.
89. Hammar SP, Bockus DE, Remington FL, Rohrbach KA: Mucin-positive epithelial mesotheliomas: a histochemical, immunohistochemical, and ultrastructural comparison with mucin-producing pulmonary adenocarcinomas. *Ultrastruct Pathol* 1996; 20: 293–325.
90. Borczuk AC, Sha KK, Hisler SE, Mann JM, Hajdu SI: Sebaceous carcinoma of the lung: histologic and immunohistochemical characterization of an unusual pulmonary neoplasm: report of a case and review of the literature. *Am J Surg Pathol* 2002; 26: 795–798.
91. Gupta SK, Rajwanshi AK, Das DK: Fine needle aspiration cytology smear patterns of malignant melanoma. *Acta Cytol* 1985; 29: 983–988.
92. Ribes JA, Vanover-Sams CL, Baker DJ: Zygomycetes in human disease. *Clin Microbiol Rev* 2000; 13: 236–301.
93. Prabhu RM, Patel R: Mucormycosis and entomorphthoramycosis: a review of the clinical manifestations, diagnosis, and treatment. *Clin Microbiol Infect* 2004; 10 (Suppl 1): 31–47.
94. Brown J: Zygomycosis: an emerging fungal infection. *Am J Health Syst Pharm* 2005; 62: 2593–2596.
95. Espinel-Ingroff A, Oakley LA, Kerkering TM: Opportunistic zygomycotic infections: a literature review. *Mycopathologia* 1987; 97: 33–41.
96. Armani A: Histochemical findings in fat embolism. *Pathologica* 1986; 78: 517–523.
97. Abramowsky CR, Pickett JP, Goodfellow BC, Bradford WD: Comparative demonstration of pulmonary fat emboli by 'en bloc' osmium tetroxide and oil red O methods. *Hum Pathol* 1981; 12: 753–755.
98. Mazzone PJ, Jane-Thomassen M, Kavuru MS: Pulmonary alveolar proteinosis: recent advances. *Semin Respir Crit Care Med* 2002; 23: 115–126.
99. Akin MR, Nguyen GK: Pulmonary alveolar proteinosis. *Pathol Res Pract* 2004; 200: 693–700.
100. De Blic J: Pulmonary alveolar proteinosis. *Paediatr Respir Rev* 2004; 5: 340–342.
101. Govender D, Essa AS: Malakoplakia and tuberculosis. *Pathology* 1999; 31: 280–283.
102. Stevens S, McClure J: The histochemical features of the Michaelis-Gutmann body and a consideration of the pathophysiological mechanisms of its formation. *J Pathol* 1982; 137: 119–127.
103. Oborne BM, Robboy SJ: Lymphomas or leukemia presenting as ovarian tumors: an analysis of 42 cases. *Cancer* 1983; 52: 1933–1943.
104. Grogan TM, Warnke RA, Kaplan HS: A comparative study of Burkitt's and non-Burkitt's "undifferentiated" malignant lymphoma: immunologic, cytochemical, ultrastructural, cytologic, histopathologic, clinical, and cell culture features. *Cancer* 1982; 49: 1817–1828.
105. Menasce LP, Banerjee SS, Beckett E, Harris M: Extramedullary myeloid tumor (granulocytic sarcoma) is often misdiagnosed: a study of 26 cases. *Histopathology* 1999; 34: 391–398.
106. Ree HJ, Raine L, Crowley JP: Lectin binding patterns in diffuse large cell lymphoma. *Cancer* 1983; 52: 2089–2099.
107. Kimura S, Saitou D, Kimura K, Yakata M, Hakomori S: Pathological study of carbohydrate antigens in human lung cancer with monoclonal antibodies. *Jpn J Clin Pathol* 1991; 39: 765–772.
108. Noguchi M, Nakajima T, Hirohashi S, Akiba T, Shimosato Y: Immunohistochemical distinction of malignant mesothelioma from pulmonary adenocarcinoma with anti-surfactant apoprotein, anti-Lewis(a), and anti-Tn antibodies. *Hum Pathol* 1989; 20: 53–57.
109. Zhang S, Zhang HS, Cordon-Cardo C, et al.: Selection of tumor antigens as targets for immune attack using immunohistochemistry: II. Blood group-related antigens. *Int J Cancer* 1997; 73: 50–56.
110. Sugiyama K, Kawai T, Nakanishi K, Suzuki M: Histochemical reactivities of lectins and surfactant apoprotein in pulmonary adenocarcinomas and their metastases. *Mod Pathol* 1992; 5: 273–276.
111. Kawai T, Greenberg SD, Truong LD, et al.: Differences in lectin binding of malignant pleural mesothelioma and adenocarcinoma of the lung. *Am J Pathol* 1988; 130: 401–410.
112. Senba M, Itakura H, Yamashita H, Huang HJ: Eosinophilic globules in Kaposi's sarcoma: a histochemical, immunohistochemical, and ultrastructural study. *Acta Pathol Jpn* 1986; 36: 1327–1333.
113. Kao GF, Johnson FB, Sulica VL: The nature of hyaline (eosinophilic) globules and vascular slits of Kaposi's sarcoma. *Am J Dermatopathol* 1990; 12: 256–267.
114. Fukunaga M, Silverberg SG: Hyaline globules in Kaposi's sarcoma: a light microscopic and immunohistochemical study. *Mod Pathol* 1991; 4: 187–190.

CHAPTER 4

Diagnostic Histochemistry of Mediastinal Diseases

Mark R. Wick, M.D., and
Nancy C. Mills, HT, QIHC (ASCP)

The mediastinum is a relatively small anatomical compartment, but many disease processes may affect it. These include a wide variety of developmental, inflammatory, and neoplastic disorders. Most of them are satisfactorily recognizable using routine histology and hematoxylin and eosin (H&E) stains, but histochemical methods are potentially helpful in delineating the nature of others. This chapter outlines an approach to the latter topic.

INFLAMMATORY/INFECTIOUS DISEASES OF THE MEDIASTINUM

Acute Mediastinitis

Most examples of acute mediastinal inflammation are caused by trauma, which may be penetrating in nature (including postsurgical) or secondary to perforations of the esophagus or gastric cardia as seen with ingestion of corrosive chemicals or endoscopic misadventures. Rarely, cervical abscesses also may rupture and track downward along the prevertebral fascia in the mediastinum (1).

Upper aerodigestive tract flora are principally represented in acute mediastinitis (2); hence, gram-positive cocci, gram-positive bacilli, and, occasionally, *Candida* may be identified with tissue Gram stains and silver methenamine or periodic acid–Schiff (PAS) stains (Figures 4.1 and 4.2).

Chronic Fibrosing Mediastinitis

Chronic fibrosing (sclerosing) mediastinitis (CFM) is much more common than the acute form, although neither one is frequently seen in general terms. CFM usually is idiopathic, but a distinct subgroup of patients with this disorder have underlying granulomatous infections of the mediastinal lymph nodes (3–5). These may be caused by mycobacteria or fungi in any of several genera. Typically, by the time a significant sclerotic response to the pathogenic organism has developed, it is poorly detectable histochemically. Nonetheless, Kinyoun/Ziehl-Neelsen, Fite, silver methenamine, and PAS stains are still worthwhile in the assessment of CFM. Another possibility is the application of lectin histochemistry in this context. Mycobacteria have an affinity

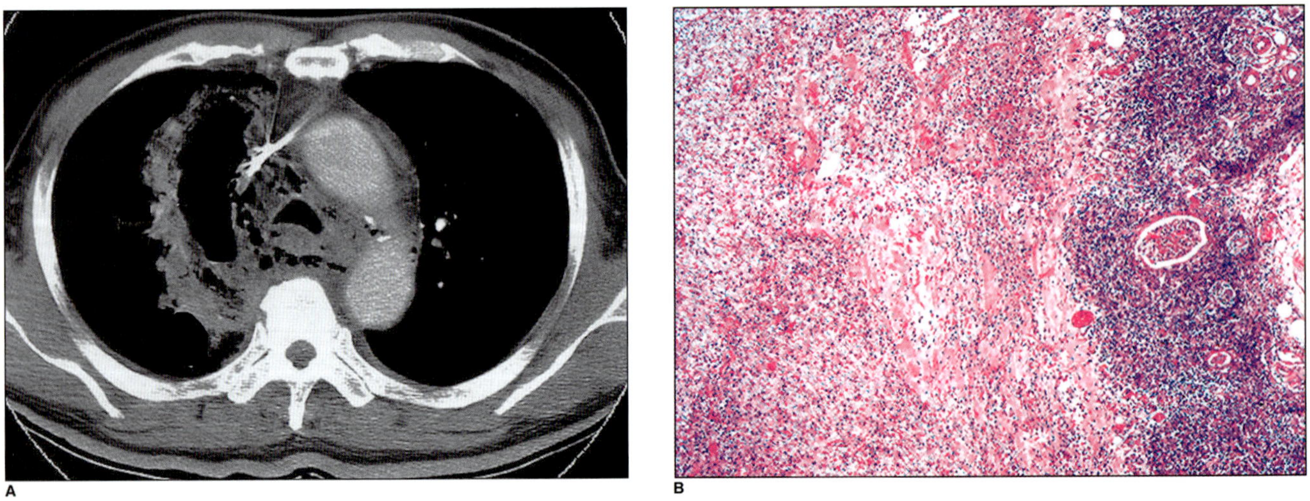

A B

FIGURE 4.1: (A) Irregular thickening of the mediastinum is seen in this thoracic computed tomogram. (B) Suppurative acute inflammation borders the thymus gland in this photomicrograph of the lesion.

for concanavalin-A (Con-A) generically, and various fungi have been shown to label with soybean agglutinin or wheat germ lectin, or both (6). However, the relative sensitivity of those reagents has not been formally compared with traditional histochemical procedures as yet. In cases where an infectious source can be identified, appropriate treatment may arrest progression of the process, which otherwise may proceed to enwrap and compress the great vessels and the tracheobronchial tree.

Connective tissue stains are not usually necessary to characterize CFM, but Masson trichrome and Verhoeff–Van Gieson methods can be employed to demonstrate the excess collagen and elastin deposition that typifies the disease (Figures 4.3 and 4.4). Lastly, it must be remembered that some cases of mediastinal Hodgkin lymphoma (HL), non-Hodgkin lymphoma, seminoma, or metastatic carcinoma may be accompanied by an idiosyncratic fibrous host response that simulates CFM (7). Careful examination of an adequate tissue sample with H&E stains is the best approach to this potentiality, but application of methyl green–pyronine (MGP) and PAS stains may assist in identifying atypical lymphoid cells and epithelial elements, respectively.

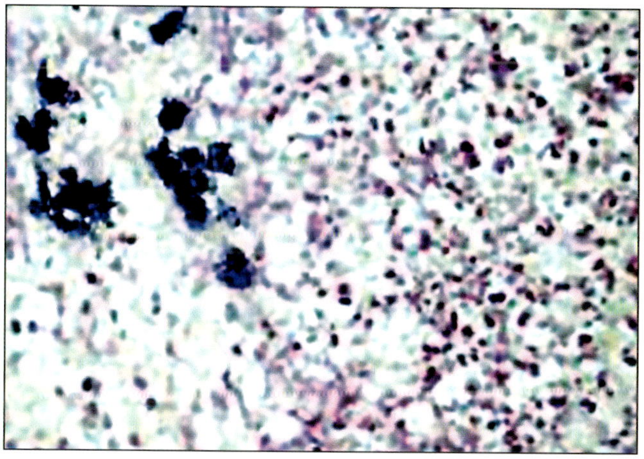

FIGURE 4.2: ***This tissue Gram (Brown-Hopps) stain shows colonies of gram-positive cocci, establishing the diagnosis of infectious acute mediastinitis.***

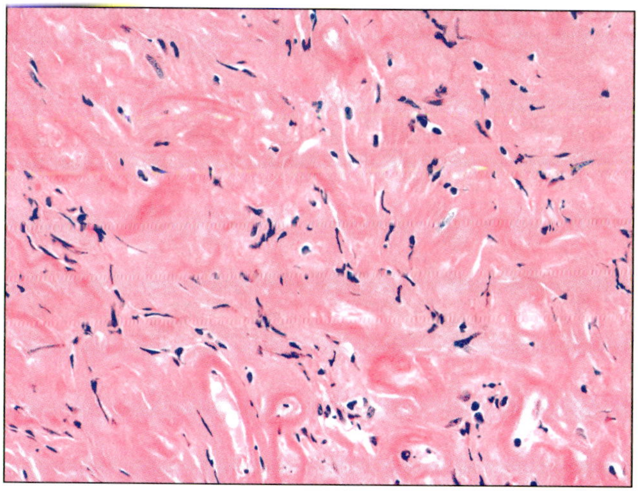

FIGURE 4.3: ***Dense collagen deposition in the anterior mediastinum is seen in FM.***

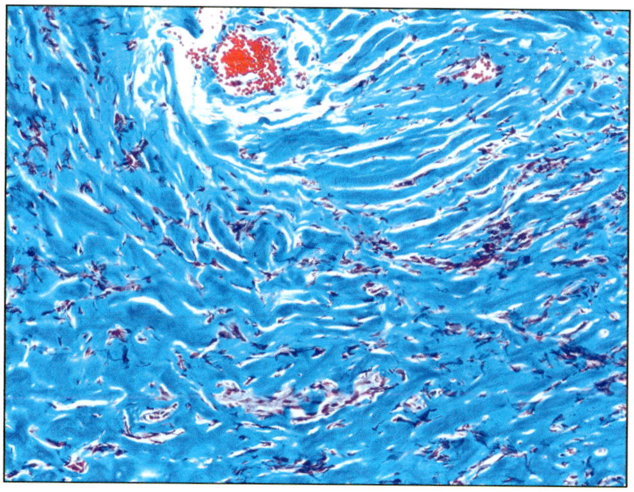

FIGURE 4.4: ***Masson trichrome staining accentuates the mature collagen in FM.***

DEVELOPMENTAL DISEASES

Developmental disorders of the mediastinum include thymic dysplasia, unilocular thymic cysts, pericardial cysts, ectopic parathyroid cysts, bronchogenic cysts, meningoceles, and gastroenteric cysts (8,9). The first four of these lesions preferentially affect the anterior compartment, whereas bronchogenic cysts are typically seen in the middle mediastinum and enteric cysts and meningoceles occur posteriorly. Again, H&E stains are sufficient for diagnosis in the vast majority of cases. PAS stains, done with and without diastase digestion, can be used to confirm the presence of epithelial mucin in bronchogenic or enteric cysts (Figure 4.5), or to demonstrate diastase-labile glycogen in the lining cells of parathyroid cysts (10). Smooth muscle elements are likewise expected in bronchogenic and enteric cysts, and trichrome stains may help to highlight this tissue component (Figure 4.6).

NONNEOPLASTIC ENDOCRINE PROLIFERATIONS

Hyperplastic proliferations of the thyroid and parathyroids may be seen in the anterosuperior mediastinum, even in the concurrent presence of normal glands in the neck. These two lesions can be difficult to distinguish from one another morphologically, as they also are in cervical locations, because of closely similar cytoarchitectural features. However, parathyroid tissue much more often contains abundant PAS-stainable glycogen than does thyroid parenchyma (11). Doi et al. (12) and Sasano et al. (13)

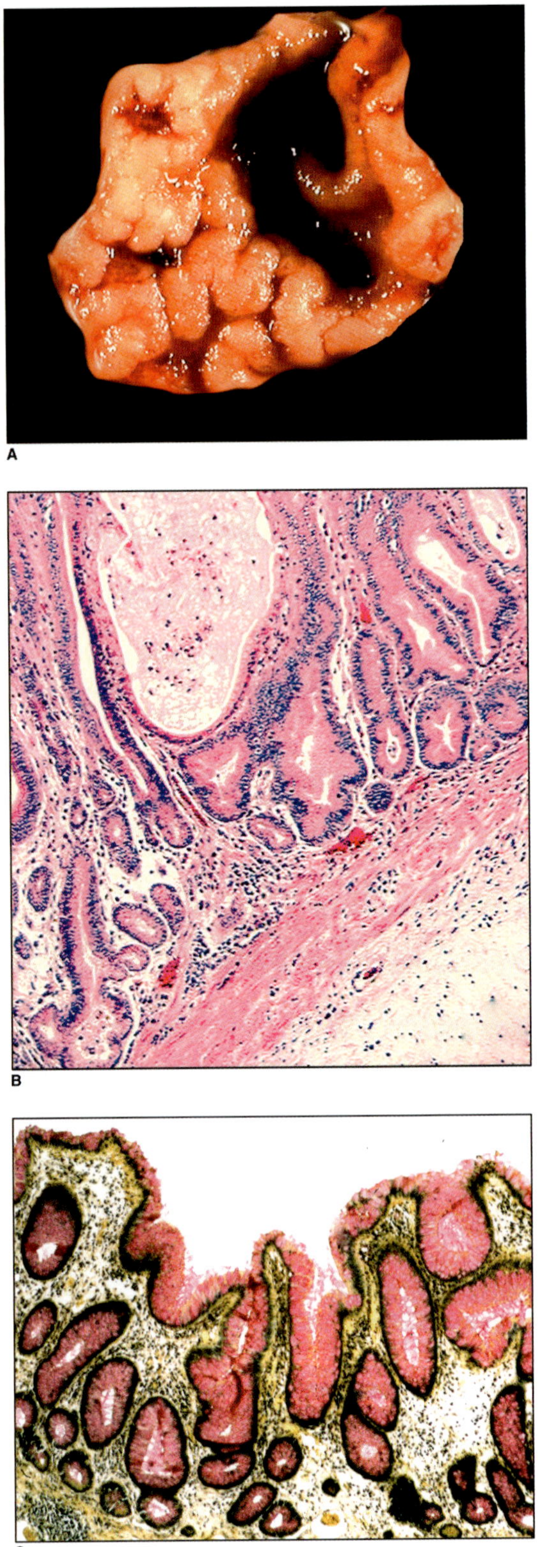

FIGURE 4.5: (A) Gross photograph of a gastroenteric duplication cyst of the posterior mediastinum. (B) The cyst lining resembles that of the stomach and intestine. (C) Mucicarmine staining of the cyst shows a pattern of positivity that is consistent with that of enteric metaplasia.

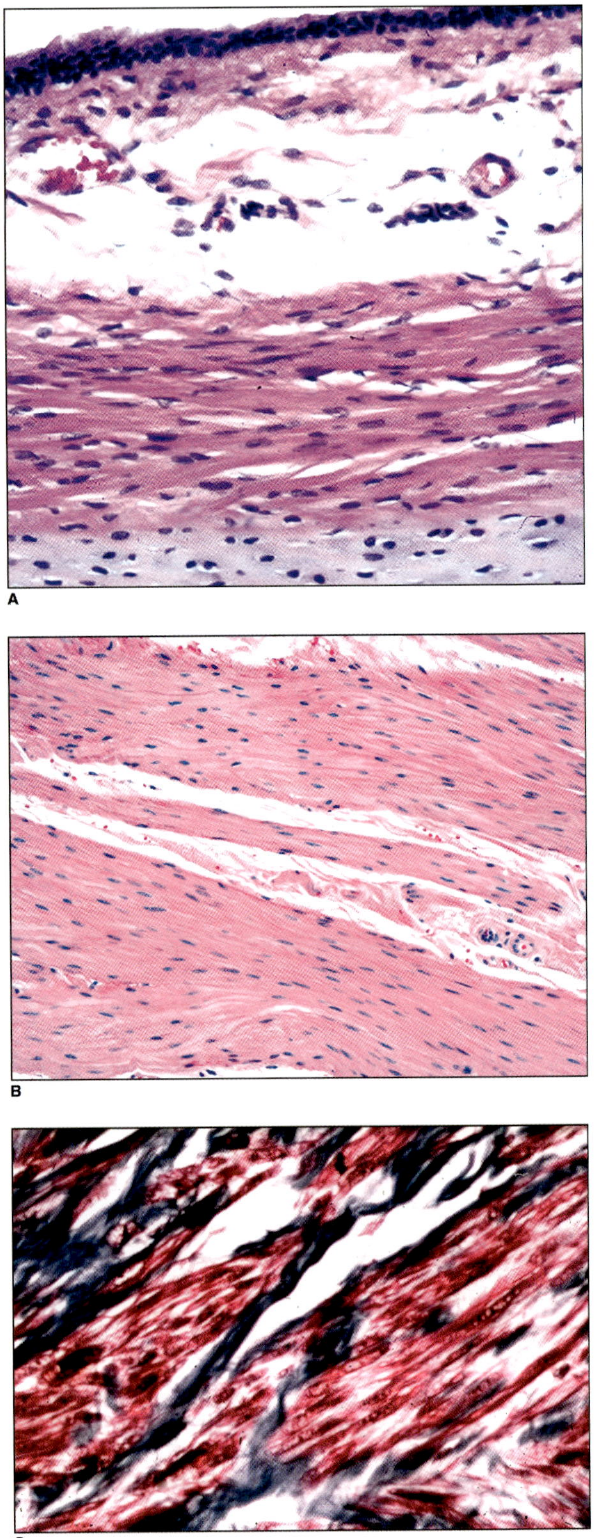

FIGURE 4.6: (A) The wall of this bronchogenic mediastinal cyst shows a pseudostratified columnar epithelial lining, as well as mural smooth muscle and hyaline cartilage. The smooth muscle component (B) is highlighted with a Masson trichrome stain, shown here (C).

examined the possible use of several commercial lectin preparations—including *Ulex europaeus* I (UEA-I), peanut agglutinin, wheat germ agglutinin, Con-A, soybean lectin, *Dolichos biflorus* agglutinin, *Ricinus communis* lectin, and *Helix pomatia* agglutinin—in attempts to diagnostically distinguish between thyroid and parathyroid proliferations. Their results suggest that this avenue of investigation is unfruitful because of a high level of shared reactivity with those reagents.

ACQUIRED CYSTIC LESIONS

Acquired cysts of the mediastinum are typically seen anteriorly; these include cystic parathyroid adenoma, cystic seminoma, cystic (degenerating) thymoma, cystic teratoma, cystic Hodgkin lymphoma (CHL) (actually secondary thymic cyst formation within lymphomatous masses), and multilocular thymic cyst (14). The constituent cells in all of these lesions except CHL potentially contain glycogen. Phosphotungstic acid–hematoxylin (PTAH) stains done on thin (2–3 μm) tissue sections may also assist in identifying thymomatous tissue because of its content of cytoplasmic tonofibrils (intermediate filament skeins). The latter are demonstrated by the PTAH method as delicate linear structures, often abutting cell membranes (15). Myofibrils also are PTAH positive; thus, so-called myoid cells in thymomas and smooth or striated muscle cells in cystic teratomas may be labeled as well (Figure 4.7).

THYMIC HYPERPLASIA

Thymic hyperplasia may be "true"—involving both thymic epithelium and lymphoid elements—or reactive, in which case it is typified only by the formation of lymphoid follicles with germinal centers (16). Reactive hyperplasia is most often seen in association with systemic diseases, such as thyrotoxicosis, rheumatic fever, autoimmune collagen vascular disorders, and myasthenia gravis (17). The lymphoid follicles in this condition are adequately visualized with H&E stains. However, they can sometimes be emphasized with Snook's or Sweet's procedure, which may show an increase in reticulin fibers surrounding central intrafollicular blood vessels, as well as external to the mantle zone of the follicles (Figure 4.8). Moreover, the MGP stain often demonstrates pyroninophilic (red staining) immunoblastic cells in the follicular interior, and plasma cells with cytoplasmic immunoglobulin content sometimes can be visualized there with the PAS method and the trichrome procedure, both of which also label the cells red (15).

Interestingly, the intrathymic germinal centers seen in connection with systemic diseases also may be juxtaposed to microcystic and metaplastic Hassall's corpuscles. These structures can facultatively produce epithelial mucin in such circumstances, as visualized with the digested PAS stain, the mucicarmine method, or the alcian blue technique (15).

MEDIASTINAL NEOPLASMS

Neoplasms in the mediastinum are usually grouped according to the compartment in which they are most commonly seen—anterior, middle, or posterior. They encompass a considerable spectrum of differentiation, including thymic and nonthymic epithelial, hematopoietic, germ cell, neural, and mesenchymal proliferations.

Anterior Mediastinal Tumors

All residents in training in surgery and pathology learn the four most common masses of the anterior mediastinum, those being thymomic epithelial neoplasms, lymphomas,

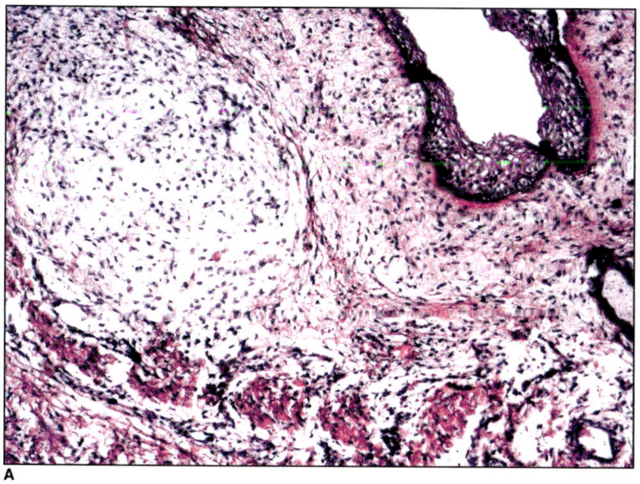

A

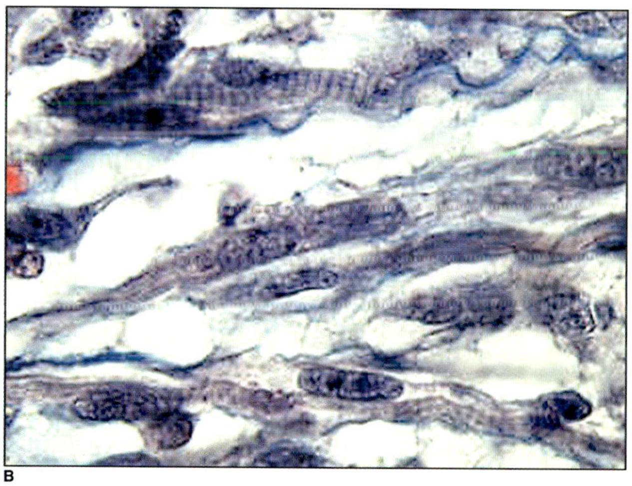

B

***FIGURE 4.7:** (A) Cystic mature teratoma of the anterior mediastinum, containing an admixture of several epithelial and mesenchymal tissues. (B) PTAH staining reveals cross-striations in striated muscle cells in the lesion.*

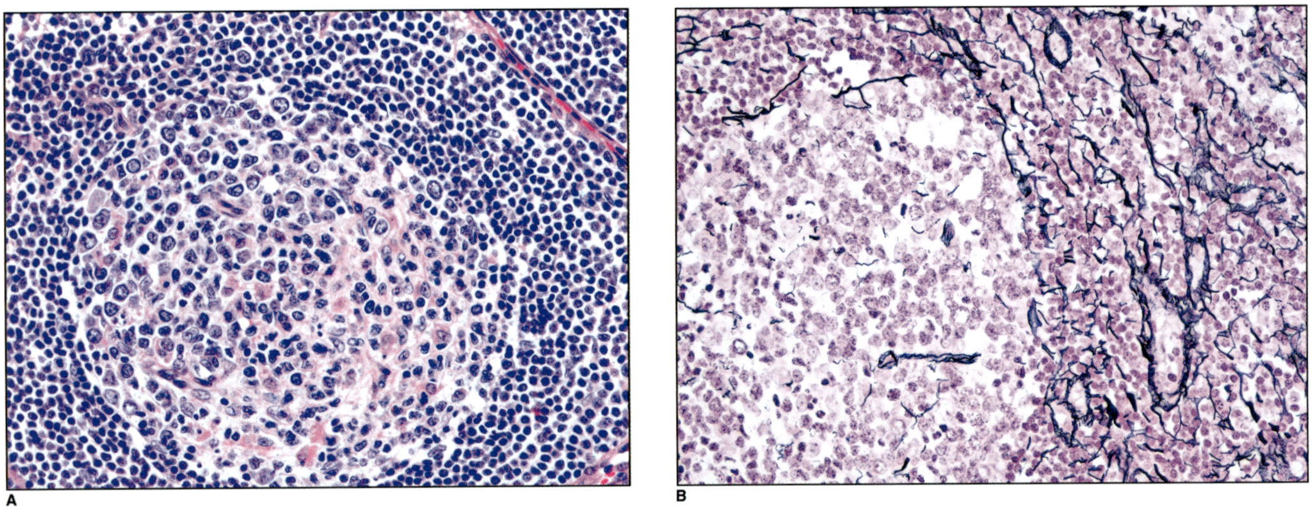
A B

FIGURE 4.8: (A) Germinal centers are present in this thymus from a patient with myasthenia gravis. (B) Snook's stain shows an increased number of reticulin fibers in and around the germinal centers. The diagnosis is that of thymic lymphoid hyperplasia.

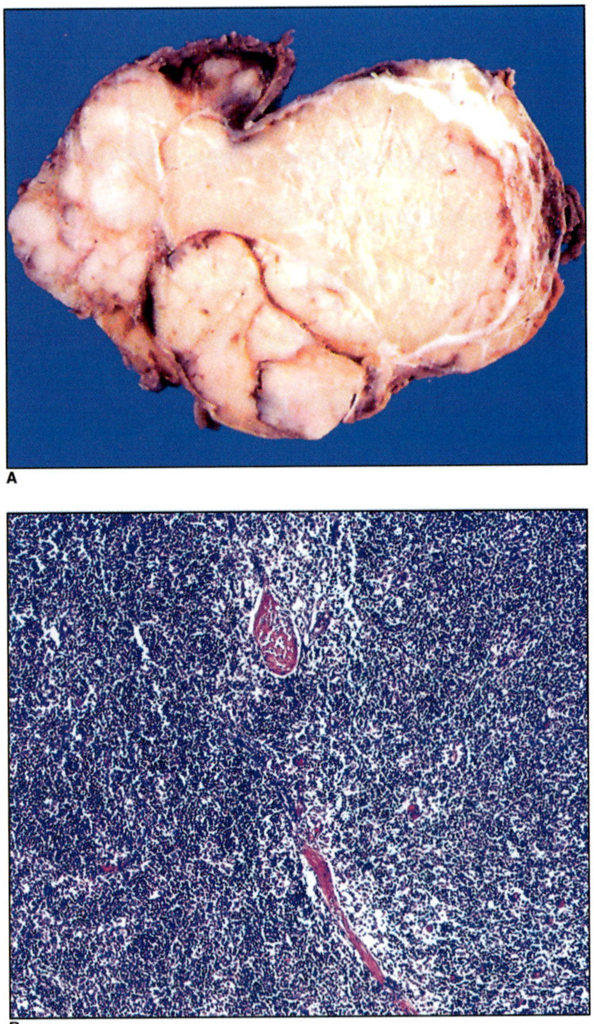
A B

FIGURE 4.9: [A (gross) and B (microscopic)] LPT may be strikingly similar to LBL morphologically, especially in limited biopsy samples.

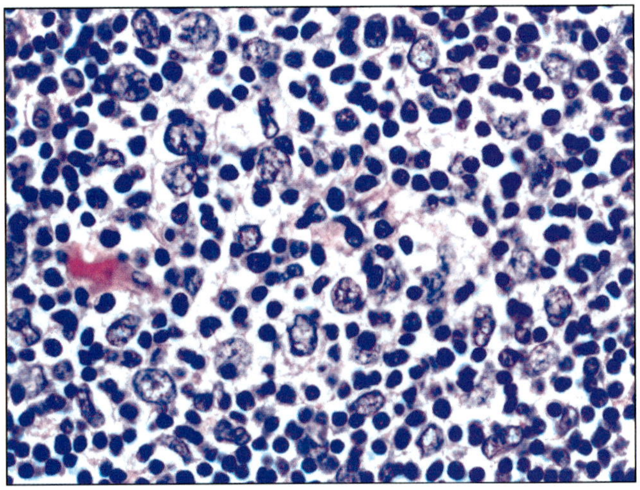

FIGURE 4.10: Other non-Hodgkin lymphomas of the thymic region also may be simulated microscopically by thymomas. A mixed lymphocytic-epithelial thymoma is seen here.

thyroid lesions, and teratoid tumors, in relative order of frequency (18). These lesions are amenable to investigation with histochemical methods, although most of them can be recognized satisfactorily by routine morphological study.

Thymoma and Thymic Carcinoma

Thymoma is defined as a cytologically bland or cytologically indeterminate neoplasm of thymic epithelial cells, with a variable content of reactive thymocytic lymphocytes. In those tumors where one sees a predominance of lymphoid cells, or a roughly equal admixture of epithelium and thymocytes, a resemblance to lymphoma may be striking (18) (Figures 4.9 and 4.10). One helpful stain in addressing the latter differential diagnosis is the trichrome method (15,19). It shows variably thick internal fibrous bands in thymoma, and, more importantly, these can be seen to intersect one another at acute angles (Figure 4.11). This characteristic is not observed in lymphomas. A second technique that is potentially helpful is the chloroacetate esterase (von Leder) procedure, which labels granulocytic leukocytes and mast cells. Thymomas often contain groups of mastocytes—typically arranged around blood vessels (Figure 4.12)—whereas non-Hodgkin lymphomas do not (19). It must be acknowledged that HL *does* contain mast cells, sometimes abundantly so, but the form of this neoplasm that might imitate lymphocyte-rich thymoma (lymphocyte-predominant HL) is extraordinarily rare in the mediastinum (20).

Lymphoblastic lymphoma (LBL) (21–27) is the lesion that most often enters differential diagnosis with lymphocyte-predominant thymoma (LPT). The former of these two tumors lacks acutely intersecting internal fibrous septations, and it also may demonstrate an increase in the number of delicate reticulin fibers, around groups of lymphocytes in the lesion, with Snook's method (15) (Figure 4.13). This pattern is not seen in thymomas; indeed, most of them lack reticulin altogether, except for that which surrounds intratumoral blood vessels (19). Other stains that label the cells of LBL in many cases (albeit sparsely) are the PAS technique (28–30) (Figure 4.14) and the MGP method, contrasting with the negativity of lymphoid cells in LPT. In contrast, lectin binding profiles of the two tumors are comparable, at least with peanut agglutinin and Con-A (31–34) (Figure 4.15). Lymphoblasts also show an affinity for lentil lectin (34), but this marker has not been evaluated with respect to the lymphoid constituents of LPT.

Indeed, the lectin binding profiles of thymoma have not been studied extensively. Several authors have reported that intratumoral lymphocytes variably bind peanut agglutinin and Con-A, as expected of immature T-cells (31,32).

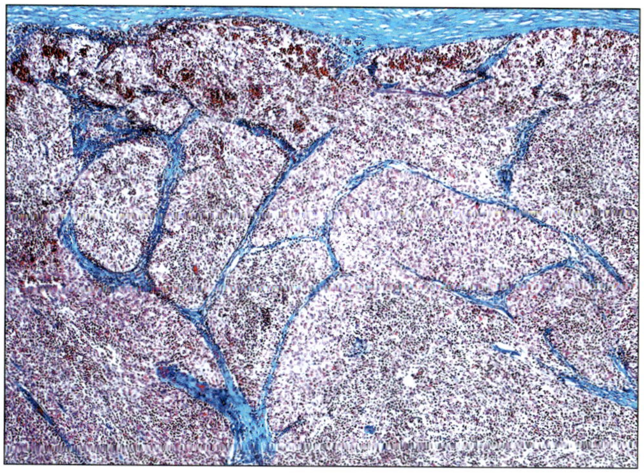

FIGURE 4.11: ***Masson trichrome staining accentuates the angular internal fibrous septa that typify thymomas but are not expected in lymphomas.***

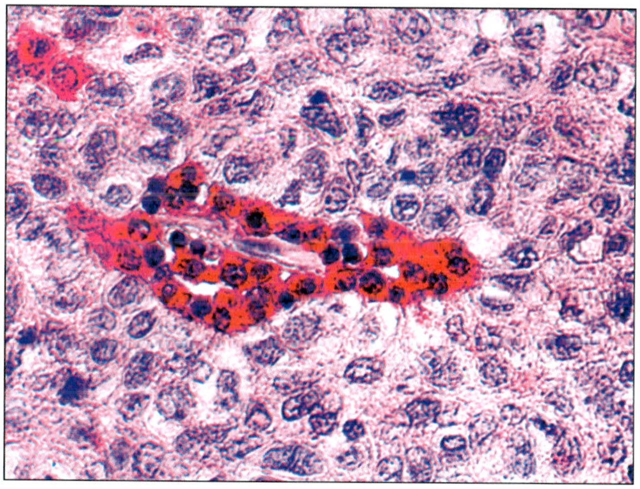

FIGURE 4.12: ***Perivascular mast cell aggregates can be highlighted with the Leder stain in thymomas. They are not present in non-Hodgkin lymphomas.***

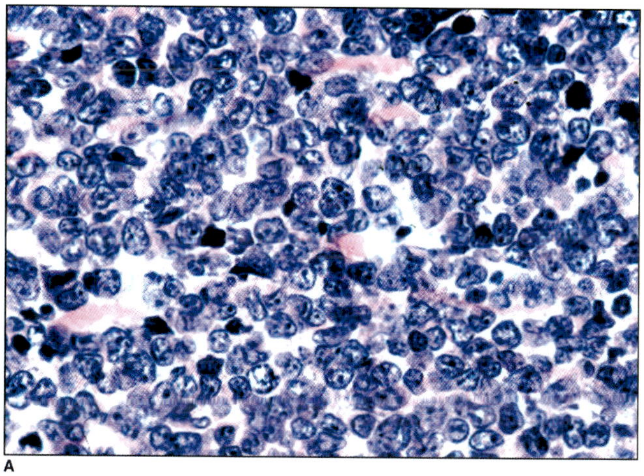

A

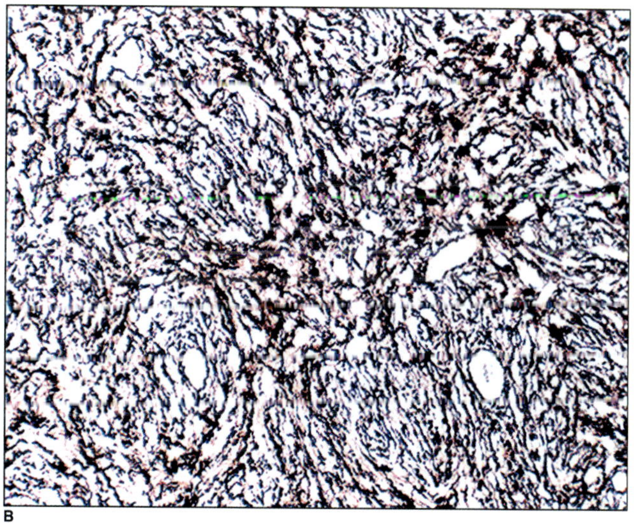

B

FIGURE 4.13: ***(A) LBL of the thymic region, characterized by sheets of intermediate-sized lymphoid cells with convoluted nuclear contours. (B) Snook's method shows abundant intercellular reticulin deposition in this LBL.***

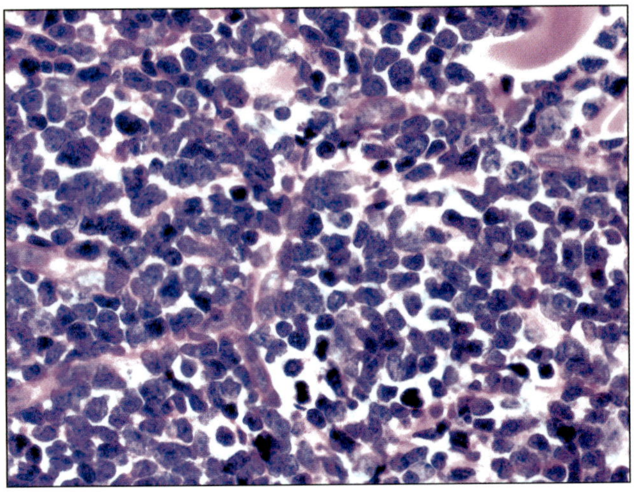

***FIGURE 4.14:* Scant but definite positivity is seen in this LBL with the undigested PAS technique. This reaction in LBL is singular among non-Hodgkin lymphomas and is abrogated by prior diastase digestion.**

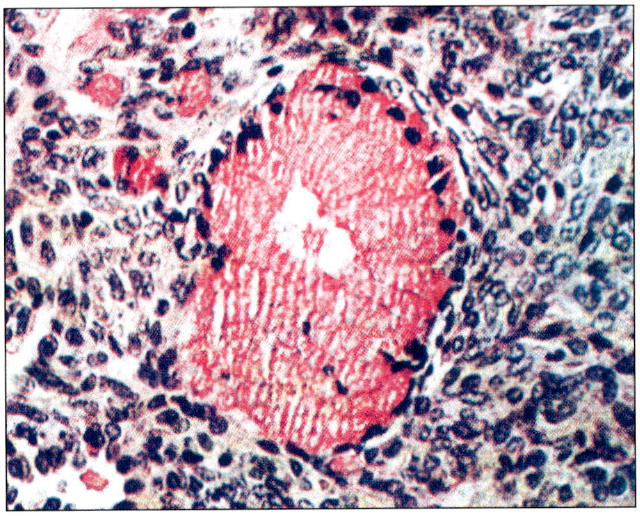

***FIGURE 4.16:* Divergent glandular differentiation in thymoma, as accentuated with a mucicarmine stain.**

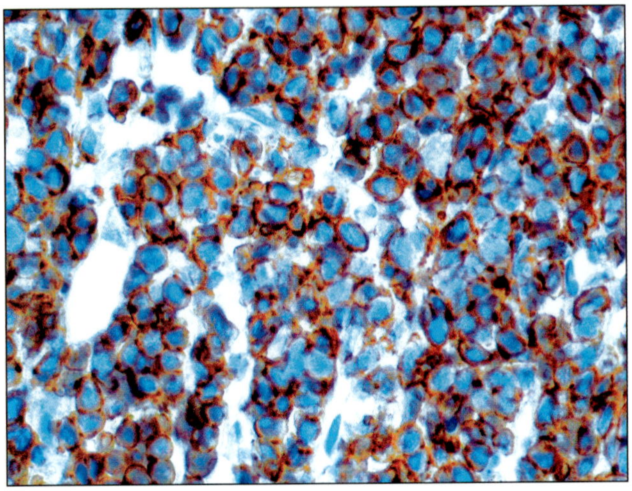

***FIGURE 4.15:* Diffuse binding of Con-A by the cells of LBL.**

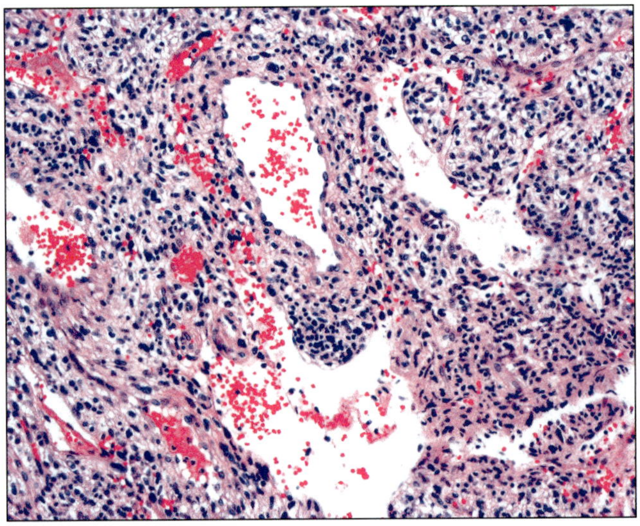

***FIGURE 4.17:* Some thymomas histologically simulate the appearance of HPC, with bluntly spindled tumor cells and blood vessels that resemble moose antlers.**

Tateyama et al. (35) found that epithelial components of thymomas reacted with Con-A, *Lens culinaris*, peanut, *Ricinus communis*, wheat germ, and *Vicia villosa* (VV) agglutinins in most cases. Interestingly, UEA-I did not label any of the lesions. Moreover, VV binding demonstrated an intriguing pattern in relation to the histologic type of thymoma according to the World Health Organization classification scheme. Type A and AB thymomas largely failed to bind VV, whereas tumors in the "B" subgroup progressively did so more avidly among categories B1 through B3. Thymic carcinomas were not evaluated in that analysis.

Occasional examples of epithelial-predominant thymoma may demonstrate divergent differentiation, containing glandular structures that are labeled with the mucicarmine, digested PAS, alcian blue, and colloidal iron techniques (19,21,22) (Figure 4.16). Generally speaking, however, the constituent tumor cells in such lesions lack the level of cytologic atypia that one usually associates with adenocarcinoma, and therefore, confusion between that neoplasm and glandular thymoma is unlikely.

Another potential problem presented by thymoma is its ability to resemble either hemangiopericytoma (HPC) or fibrous histiocytoma (23,36) (Figure 4.17). Snook reticulin stains in thymoma and HPC give similar results, with individual cells or small groups of them being surrounded by reticulin fibers (Figure 4.18), but fibrous histiocytomas are essentially devoid of them. The PTAH procedure, done on 2- to 3-μm sections, assists in separating HPC from thymoma. The latter neoplasm shows delicate PTAH-positive tonofibrils in the cytoplasm, often abutting cell

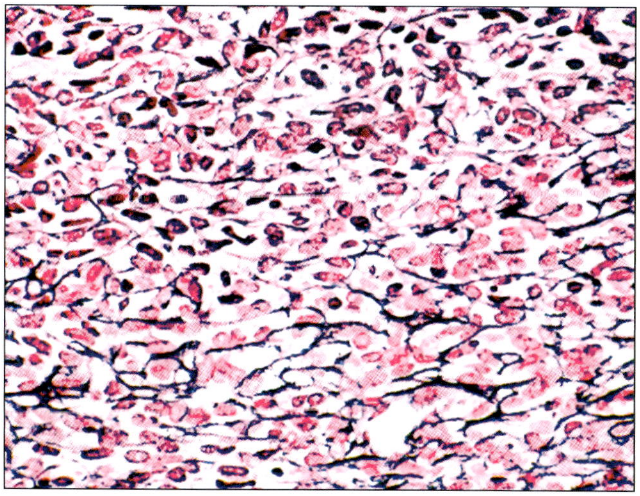

FIGURE 4.18: Snook's staining further imitates the profile of HPC in hemangiopericytoid thymoma, showing investment of individual tumor cells by reticulin.

membranes (15,19), but those structures are not apparent in either HPC or fibrous histiocytoma. So-called myoid cells, which are markedly PTAH reactive, can also be highlighted in some epithelial-predominant thymomas (37–39). *Ricinus communis* lectin binding has been reported as a sensitive marker of fibrohistiocytic neoplasms (40), but it is potentially seen in thymomas as well (35).

Primary thymic carcinomas (PTCs) differ morphologically from thymomas because of their overt cytologic malignancy (41–43). However, the histochemical properties of the two tumor types are virtually identical. A proportion of PTCs also demonstrates neuroendocrine differentiation, which may be either overt ("thymic carcinoid" and "small cell carcinoma") or occult (41). These tumors may exhibit granular cytoplasmic reactivity with argyrophilic methods such as the Grimelius, Churukian-Schenk, or Sevier-Munger techniques, especially if the tissue has been fixed in Bouin's solution (Figure 4.19).

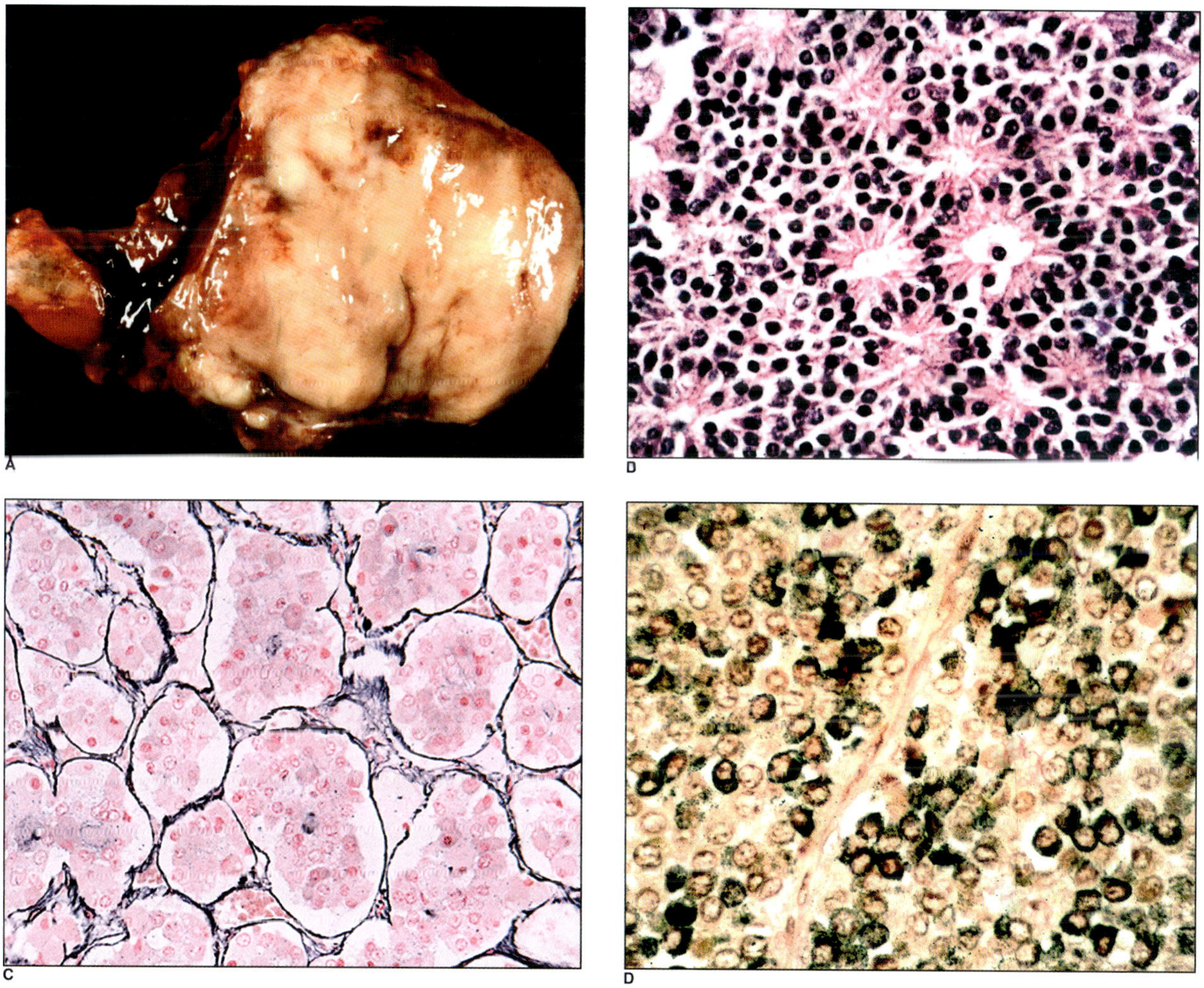

FIGURE 4.19: "Carcinoid tumor" [neuroendocrine carcinoma] (NEC) of the thymus (A) often shows microscopic rosette formation by the tumor cells (B), but this feature can also be seen in thymoma. Snook's reticulin stain (C) emphasizes the organoid growth pattern of this tumor. The Sevier-Munger argyrophil method (D) demonstrates positivity in the cells of thymic NEC, and this result is not expected in epithelial thymomas.

Germ Cell Tumors

Seminoma may arise within the thymus, and its morphologic features are identical to those of analogous neoplasms in the gonads, retroperitoneum, and pineal region (44–47). The problem in evaluating this tumor lies in separating it not from thymoma but from other cytologically malignant mediastinal lesions such as large cell non-Hodgkin lymphoma (LCNHL), metastatic carcinoma, and PTC (Figures 4.20 and 4.21).

The two most useful techniques for identifying this neoplasm are the PAS and reticulin methods. The first demonstrates intense and diffuse positivity in the tumor cells, owing to their uniform glycogen content; the second shows that reticulin fibers invest cell groups (45,46) rather than individual cells, as would be expected in lymphomas.

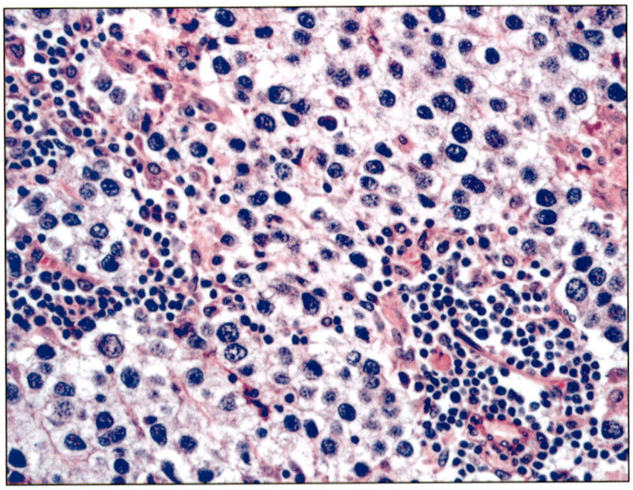

FIGURE 4.20: ***Seminoma of the thymus, showing sheets of atypical cells with clear cytoplasm and a lymphocyte-rich stroma.***

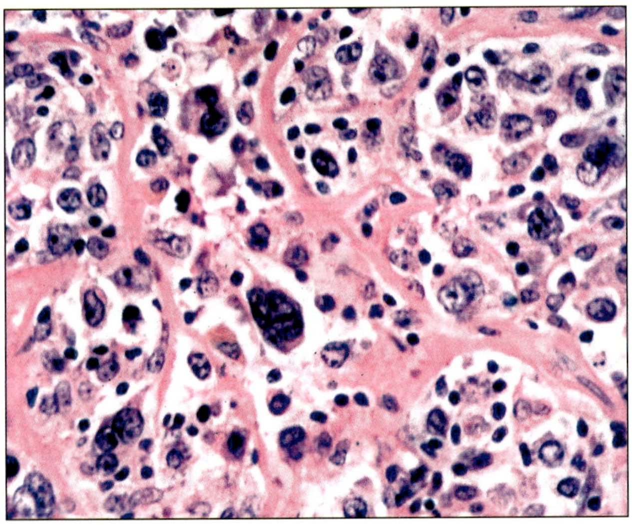

FIGURE 4.21: ***Large cell lymphoma of the thymus, comprising sheets of morphologically ambiguous polygonal tumor cells.***

Azzopardi and Hoffbrand (48) were the first to emphasize the use of the Masson trichrome stain in necrotic seminoma (48), to improve visualization of nuclear detail in the degenerative foci (Figure 4.22). This approach was recently reiterated by Florentine et al. (49). PAS positivity is lost in areas of necrosis, making the trichrome staining results especially valuable.

With respect to differential diagnosis with primary or secondary carcinoma in the anterior mediastinum, the documentation of significant glycogen content is likewise helpful in supporting a diagnosis of seminoma. Aside from clear cell PTC and rare examples of metastatic clear cell carcinoma of the lung or kidney, non–germ cell epithelial tumors are usually PAS negative or only feebly reactive (Table 4.1) (15,19,43).

Embryonal carcinoma of the thymic region similarly can exhibit lesser degrees of glycogen content, but the PAS method also may—in approximately 33% of cases—show the presence of diastase-resistant cytoplasmic globules

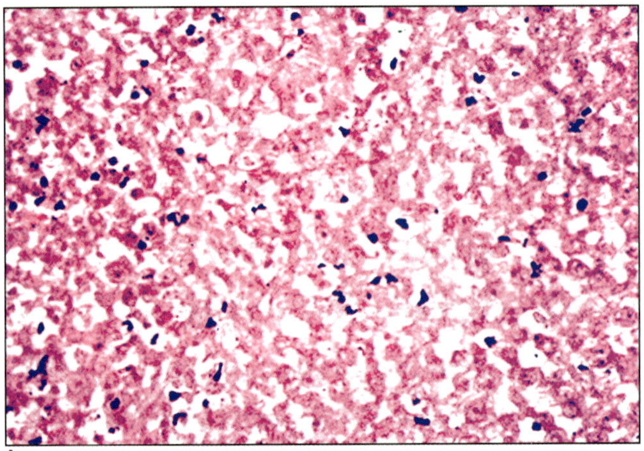

A

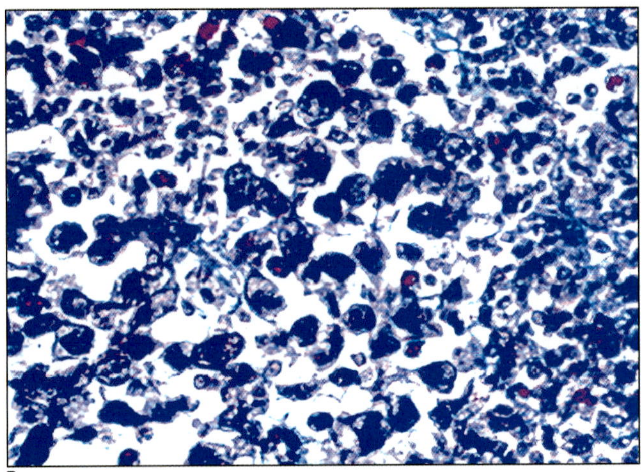

B

FIGURE 4.22: ***(A) This necrotic tumor of the thymus is of indeterminate lineage, based on this H&E stain. (B) Masson trichrome staining of the lesion "rehabilitates" its nuclear features, allowing for a diagnosis of necrotic seminoma.***

TABLE 4.1: Histochemical Differential Diagnosis of Cytologically Malignant Anterior Mediastinal Neoplasms

Stain	GCT	LCNHL	CARC	MEM
PAS	++–+++	0	+/–	+/–
PAS-D	++*	0	+/–	0**
RETIC	+***	+	+	+
CON-A	+	0	+/–	+/–
WGA	+	+/–	+/–	+/–
PNA	0	+/–****	+/–	+/–
MGP	+/– (weak)	+	+/– (weak)	+/– (weak)

CARC = primary or metastatic carcinomas, GCT = germ cell tumors MEM = malignant epithelioid mesotheliomas, PAS-D = PAS stain with prior diastase digestion, PNA = peanut agglutinin, RETIC = reticulin WGA = wheat germ agglutinin.

* Only embryonal carcinomas and yolk sac tumors show PAS-D-positivity among germ cell tumors.

** Rare epithelioid mesotheliomas are PAS-D-positive; reactivity is abrogated by pretreatment with hyaluronidase.

*** Reticulin staining in germ cell tumors, carcinomas, and mesotheliomas demonstrates investment of cell groups, whereas large cell lymphoma shows investment of individual cells.

**** Approximately 40% of large cell lymphomas are labeled by peanut agglutinin.

(50–52). These reflect a common admixture of yolk sac carcinomatous components with embryonal carcinoma, and they comprise a mixture of alpha-1-antitrypsin and alpha-fetoprotein (51). Such inclusions are not expected in other epithelial malignancies of the thymic region.

Lectin histochemical studies of germ cell tumors have demonstrated shared reactivity between seminoma, embryonal carcinoma, and yolk sac tumor, all of which bind Con-A and wheat germ agglutinin (53–55). UEA-I fails to label seminoma but does bind to embryonal and yolk sac carcinoma (55). This piece of information obviously is helpful in diagnostically separating seminomatous from nonseminomatous germ cell neoplasms. It also may be beneficial in distinguishing embryonal carcinoma of the diffuse type from large cell lymphomas, which lack UEA-I receptors (56). However, UEA-I affinity is shared by many metastatic somatic carcinomas in the mediastinum (56), abrogating its usefulness in their differential diagnosis with embryonal carcinoma.

A rare but important development in some mature teratomatous tumors of the anterior mediastinum is the appearance of malignant somatic tissues. The latter may include squamous carcinoma, adenocarcinoma, or neuroendocrine carcinoma (57)—each of which is typically obvious morphologically—hematopoietic tumors (58,59) or primitive sarcomas. Osteosarcoma and RMS are the most common mesenchymal malignancies in this context (58–62). In reference to the first of these entities, osteoid matrix may be difficult to separate histologically from hyalinized fibrous tissue or congealed fibrin. Li et al. (63) have shown that the modified Ponceau trichrome technique and Picrosirius red (PR) staining are helpful in making such distinctions. Osteoid assumes a reddish-blue hue with the Ponceau method, whereas other stromal tissues are blue. The PR procedure labels osteoid red purple, and fibrous tissue or fibrin is red. When sections stained with PR were also viewed by polarization microscopy, Junqueira et al. (64) found that osteoid took the form of weakly birefringent, randomly arranged, three-dimensional fibrillar material, unlike other stromal constituents.

RMS often shows PAS positivity that is diastase labile, owing to its content of glycogen (65,66) (Figure 4.23). In particular, mediastinal examples of this tumor—either in sporadic form or as part of a teratoma—may demonstrate cytoplasmic clarity in conventional microscopic sections because of this feature (65). PTAH or trichrome methods occasionally reveal cytoplasmic myofibrillary cross-striations

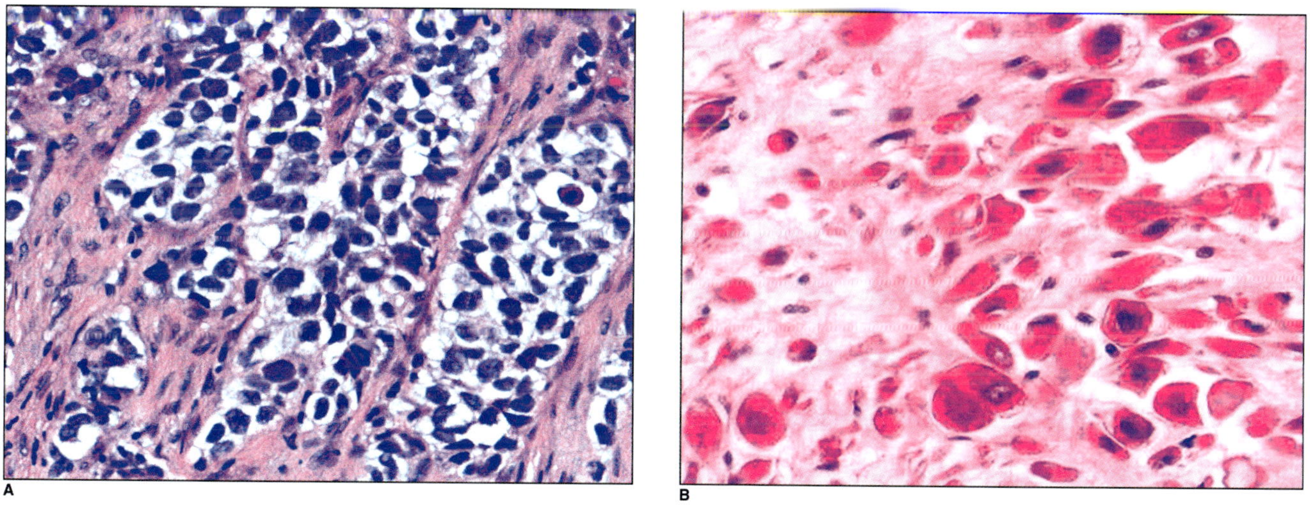

A B

FIGURE 4.23: (A) This malignant small round-cell tumor of the anterior mediastinum also has an indeterminate lineage based on morphologic findings alone. (B) Intense diastase-labile positivity is seen in the tumor cells with the PAS stain, supporting a diagnosis of RMS.

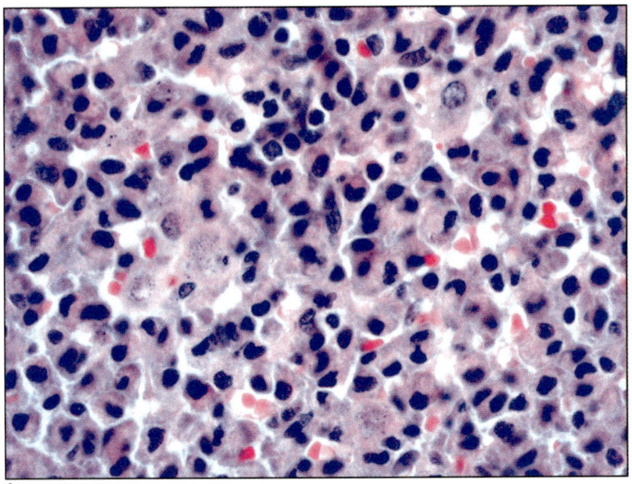

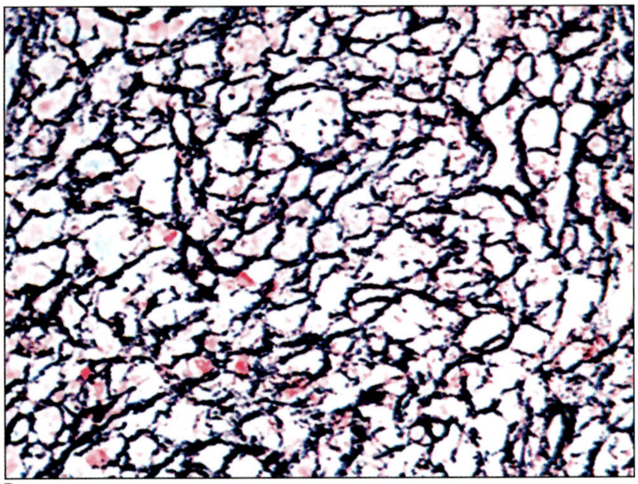

FIGURE 4.24: Individual tumor cells in this large cell non-Hodgkin lymphoma of the thymus (A) are invested by reticulin (B), as seen with Snook's method.

in this tumor type (in approximately 20% of cases), but this characteristic is only seen in thin, optimally stained sections of differentiating RMS (66). Embryonal and alveolar variants of RMS virtually never manifest a myofibrillary structure by light microscopy. Wheat germ lectin also binds to the majority of RMSs, but other small cell malignant mesenchymal tumors are typically nonreactive with this marker (67).

Hematolymphoid Malignancies

The histochemical characteristics of LCNHL are largely superimposable with those of LBL of the thymic region, as described above, with the exception that LCNHL only rarely manifests PAS positivity (68,69). The latter observation corresponds to the presence of abundant cytoplasmic immunoglobulin, which may take the form of Golgi-zonal PAS-positive, diastase-resistant globules or randomly dispersed crystalloids (15). Staining attributes of LCNHL with reticulin and MGP stains are comparable to those of LBL (70) (Figures 4.24 and 4.25).

A subtype of B-cell LCNHL in the thymic region shows marked sclerosis, such that small groups of tumor cells are partitioned by fibrous stroma (compartmentalization) (71–74). Trichrome methods are helpful for delineating this microscopic feature.

A particular differential diagnostic concern regarding LCNHLs is their possible confusion with extramedullary myeloid tumors [granulocytic sarcomas (GSs)] (75–77). The latter are potentially histochemically recognizable with the von Leder (chloroacetate esterase) procedure, which does not label lymphomas (78,79). Touch preparations stained with Romanowsky methods also may show primary cytoplasmic myeloid granules in myelogenous leukemia but not in LCNHL.

As mentioned earlier, seminoma and embryonal carcinoma of the mediastinum may resemble LCNHL histologically. These two germ cell tumors are both labeled by Con-A, and embryonal carcinoma shows UEA-I affinity, whereas LCNHL lacks receptors for both of those lectins (50–52,80). Conversely, peanut agglutinin binds to approximately 40% of LCNHL cases but not to germ cell tumors (80).

Most examples of HL are recognizable as such using H&E stains. Nevertheless, occasional cases of this cytologically heterogeneous tumor may cause diagnostic confusion with such alternatives as thymoma or sclerosing mediastinitis (81). Trichrome staining is useful in reference to the first of these lesions; as mentioned earlier in this chapter, thymoma contains internal fibrous septations that distinguish it from HL, which exhibits obtusely intersecting fibrous bands that encompass cellular nodules. Polarization microscopy further emphasizes this difference, inasmuch as the collagenous stroma in both thymoma and HL is somewhat birefringent. MGP stains also may prove useful in

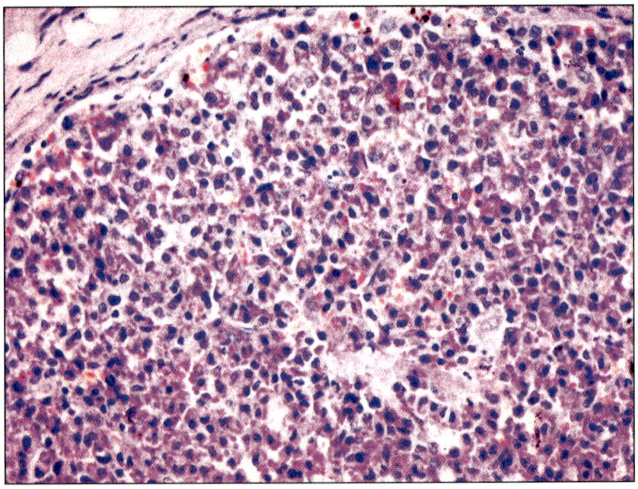

FIGURE 4.25: The tumor shown in the previous figure also shows intense nucleocytoplasmic positivity with the MGP procedure.

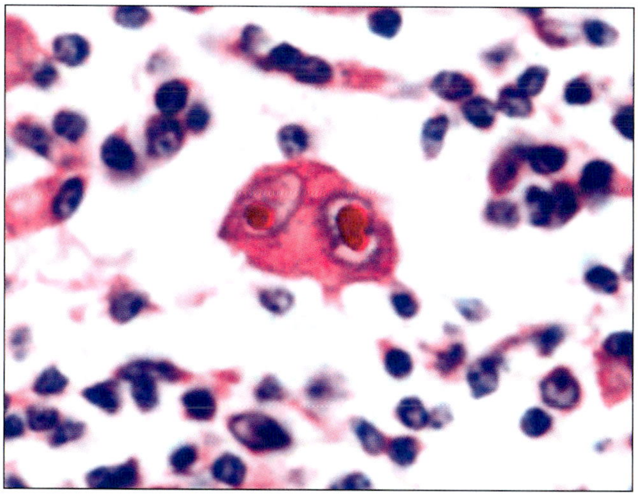

FIGURE 4.26: A classic Reed-Sternberg cell in thymic HL is also positive with the MGP technique.

highlighting the presence of Reed-Sternberg cells and Reed variants in HL (Figure 4.26). Lectin histochemistry with peanut agglutinin serves a comparable purpose (82).

Other Anterior Mediastinal Neoplasms

Parathyroid Adenoma and Carcinoma: Parathyroid tumors are not rare in the anterior mediastinum, even in patients who have four parathyroid glands in the neck (83–85). They are typified by cytologic homogeneity, often with cytoplasmic lucency. The latter characteristic reflects the consistent presence of glycogen in the neoplastic cells, manifested by PAS positivity and diastase lability. A majority—but not all—of mediastinal parathyroid tumors also show argyrophilia with silver impregnation stains (86–88). Malignancy in such tumors is usually recognized through their multinodularity and locally invasive growth, rather than by nuclear anaplasia (89). An often-cited shibboleth regarding parathyroid carcinomas is that they characteristically contain internal fibrous septa—as demonstrated with trichrome stains—as opposed to parathyroid adenomas (70). However, this has not been my experience or that of others (89–91).

Kanematsu et al. (92) found a statistical difference between the numbers of intranuclear silver colloid–stained nucleolar organizer regions (AgNORs), in a comparison of parathyroid hyperplasia-adenoma and parathyroid carcinoma. Carcinomas demonstrated a mean AgNOR value of 3.18, compared with 1.85 in hyperplasia and 1.67 in adenomas. It is possible that this information may supplement morphologic findings in making the specified diagnostic distinctions.

Aorticopulmonary Paraganglioma: Paragangliomas arising near the aortic arch (93–95) may be very difficult to distinguish from low-grade neuroendocrine carcinomas of the thymus (96) using H&E sections alone. Indeed, even with histochemical methods, a diagnostic separation between these tumor types cannot always be secured. Both of them are argyrophilic neoplasms, but if argentaffinity is observed with the Fontana-Masson procedure, an interpretation of paraganglioma would be preferred (97). Reticulin stains may be helpful in elucidating the organoid growth pattern of this neoplasm, as is true of any neuroendocrine tumor (98).

Malignant Mesothelioma: Malignant mesothelioma (MM) may potentially affect *any* of the three mediastinal compartments but is probably most commonly seen anteriorly (99–101). The principal differential diagnosis of epithelioid MM is that of metastatic adenocarcinoma, and these lesions differ in their production of mucins. MM synthesizes "stromal" mucin, which is only rarely labeled by the mucicarmine or digested PAS procedures; instead, it has affinity for the colloidal iron or alcian blue stains, positivity with which is abrogated by pretreatment of tissue sections with hyaluronidase (102–104). "Epithelial" mucin in adenocarcinomas has the converse of those histochemical properties (103–106) (Figure 4.27). The epithelial components of biphasic MMs show a comparable profile, but the staining characteristics of *sarcomatoid* or *desmoplastic* MMs are highly variable and may be shared by true mesenchymal tumors as well as primary or metastatic sarcomatoid carcinomas (107).

Castleman's Disease: Castleman initially described "angiofollicular lymphoid hyperplasia" (AFLH) of the anterior mediastinum as a condition with which thymoma could be confused (108). This is now more possible than ever, in light of the fact that one thymoma subtype known as "micronodular" thymoma (MNT)—described in 1999 by Suster and Moran (109)—has also been called "Castlemanoid" thymoma. AFLH and MNT are very closely comparable morphologically, demonstrating abnormal germinal center–like structures that contain PAS-positive basement membrane material (110). Around them, lymphocytic arrays are disposed in an "onion-skin" configuration, separated by stromal material that can be labeled with the reticulin method. Unless PTAH stains demonstrate the presence of thymic epithelial cell groups that contain tonofibrils (15), other nonhistochemical adjunctive studies must be used to distinguish AFLH from MNT.

Metastatic Tumors: Metastatic malignancies in the anterior mediastinum usually derive from other structures in the thorax, particularly the lungs (111–113). Because the thymus may also produce carcinomas that are histologically identical to pulmonary tumors, it stands to reason that the histochemical properties of squamous carcinoma variants, adenocarcinoma subtypes, and neuroendocrine carcinomas in those two locations are also essentially the

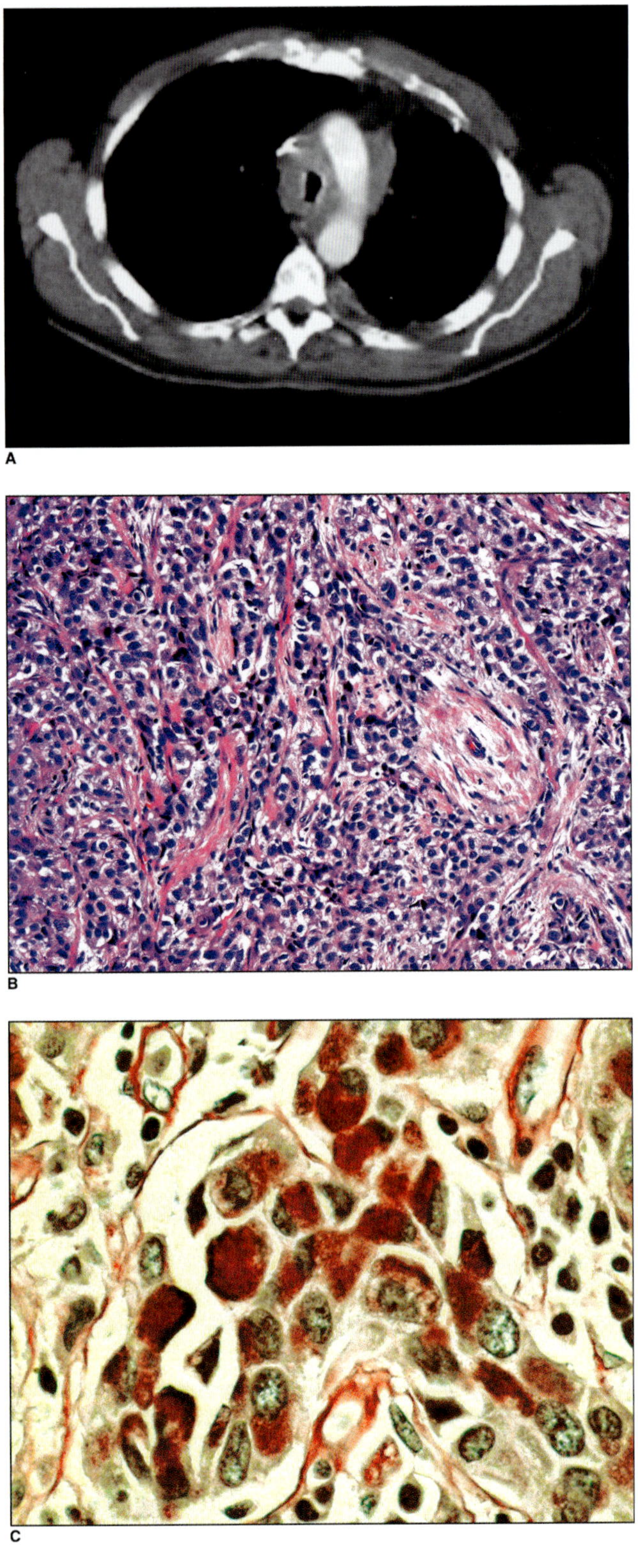

FIGURE 4.27: (A) Bilateral thickening of the mediastinal pleura is seen in this computed tomogram. (B) This biopsy of the pleura shows a malignant epithelioid neoplasm, and the differential diagnosis centers on epithelioid mesothelioma versus metastatic adenocarcinoma. (C) Mucicarmine staining of the neoplasm shows intracytoplasmic positivity, supporting interpretation of the lesion as an adenocarcinoma.

same. In that context, one must remember the diagnostic adage that any carcinoma in the mediastinum is much more likely to be secondary than primary.

Another consideration in the family of mediastinal metastatic lesions is that of melanoma (114). I have seen occasional examples of secondary melanoma in the thymic region, in patients with no known mucocutaneous primary tumors. Thus, all anaplastic small cell and epithelioid neoplasms of the mediastinum should be studied with the Fontana-Masson method or Schmorl's procedure, both of which label melanocytic proliferations.

Tumors of the Middle Mediastinum

Neoplasms in the middle mediastinum are less frequently seen than those of the anterior or posterior compartment. They are represented overwhelmingly by lymphomas—usually LCNHL or Hodgkin type—as well as benign mesenchymal tumors and metastatic carcinomas (115,116). The last of those lesions typically originate in the airways, lungs, or esophagus. Histochemical approaches to the nonmesenchymal lesions are as described above; mesenchymal tumors will be considered in the next section.

Posterior Mediastinal Neoplasms

Neural and neuroendocrine tumors predominate in the posterior mediastinum, and one may also encounter a range of other soft tissue neoplasms. Some are related to the esophagus, whereas others emanate from the mesenchymal investment of mediastinal structures. The following discussion will cover only those nonepithelial tumors that are characteristic of the posterior mediastinum; others are discussed in the chapter on soft tissue pathology in this monograph.

Neuroectodermal Tumors: Paragangliomas of the posterior mediastinum (PGPMs) are usually situated paravertebrally, and their basic histochemical attributes are as described earlier in reference to aorticopulmonary tumors. However, PGPMs are much more often functional clinically than the latter lesions are, producing symptoms and signs like those of adrenal pheochromocytoma in 50% of cases (93). Accordingly, the fact that argentaffinity is often seen in posterior mediastinal paragangliomas would be expected because this feature correlates with the synthesis of bioactive amines. In addition, melanin can also be detected in a proportion of PGPMs, with the Fontana-Masson or Schmorl method (117) (Figure 4.28).

Neuroblastic neoplasms also are seen next to the vertebral column in the mediastinum, typically in children. These small cell tumors demonstrate variable degrees of neurofibrillary maturation, and this attribute parallels their capacity for labeling with silver impregnation methods such as the

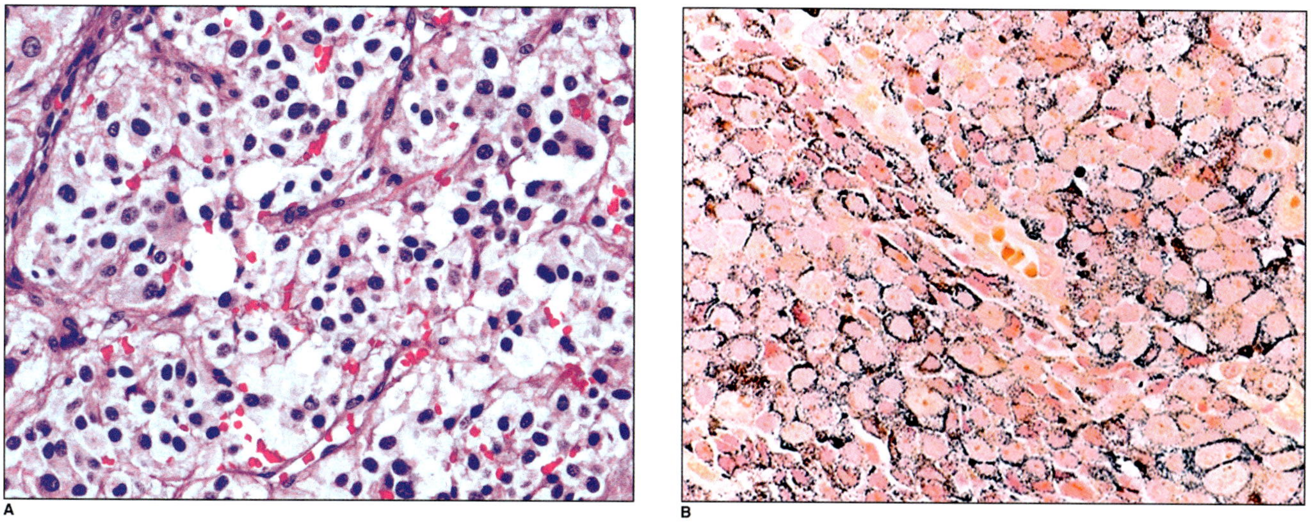

FIGURE 4.28: (A) PGPM, showing a nested growth pattern. (B) The lesion demonstrates reactivity with the Grimelius technique, shown here.

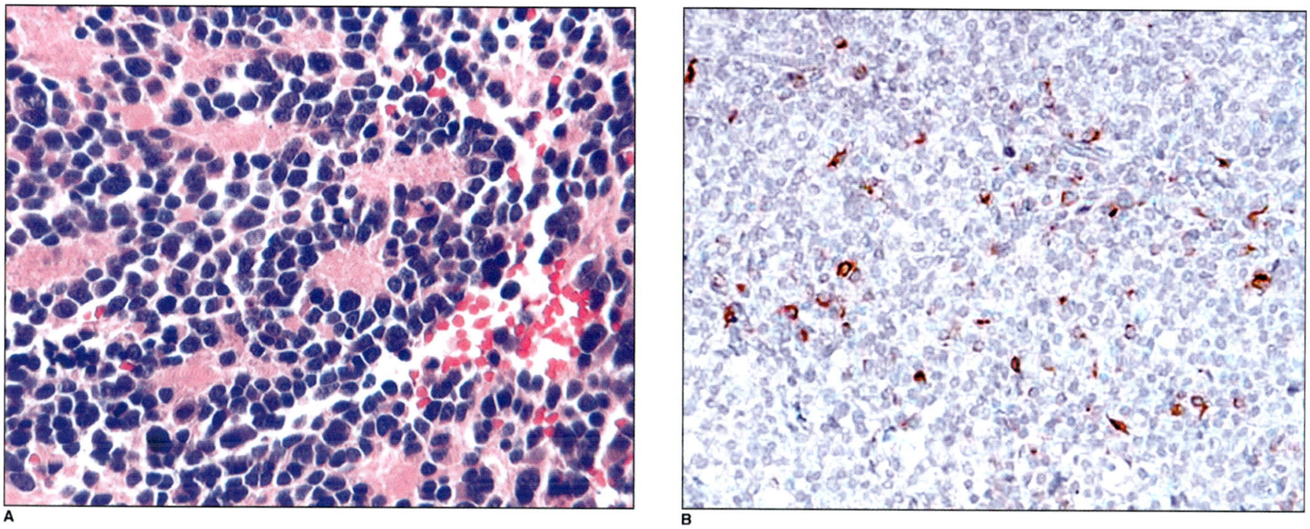

FIGURE 4.29: (A) NDL of the posterior mediastinum, comprising sheets of undifferentiated small tumor cells. (B) Multifocal binding of peanut agglutinin (Arachis hypogaea) is present in the tumor.

Rio-de-Hortega, Bodian, Bielschowsky, Gallyas, or Holmes procedures (118–120). Undifferentiated neuroblastomas (NBLs) fail to stain with any of them. Interestingly, Egan et al. (121) have suggested that stains for AgNORs demonstrate differential labeling of NBL as compared with other small cell malignancies, potentially aiding in differential diagnosis. The same group of authors further related the number of AgNORs in NBL to prognosis, with low scores being associated with more favorable clinical outcomes (122).

A lectin histochemical assessment of small round-cell tumors by Swanson et al. (67) showed that NBLs typically failed to bind wheat germ, *R. communis*, soybean, *D. biflorus*, *L. culinaris*, *Pisum sativum*, *Bandeiraea simplicifolia*, *H. pomatia*, or peanut agglutinin. Nonetheless, Kahn et al. (123) did observe labeling of differentiating neuronal cells in such lesions by peanut lectin (Figure 4.29). In comparison, RMS shows an isolated affinity for wheat germ agglutinin, and primitive neuroectodermal tumor (peripheral neuroepithelioma) manifests concomitant binding of *L. culinaris* and *P. sativum* lectins (67) (Figure 4.30).

Somatic Peripheral Nerve Sheath Tumors: Neurofibromas and neurilemmomas comprise the somatic peripheral nerve sheath tumors of the posterior mediastinum (124–126). These lesions each have a characteristic range of morphologic images in H&E stains, but, in small biopsies,

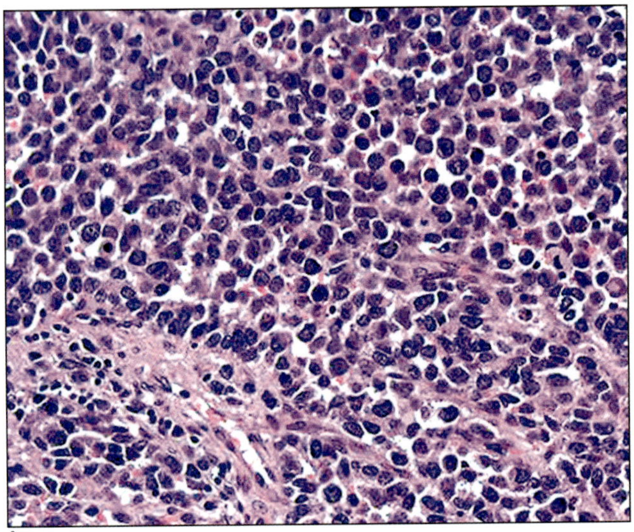
A

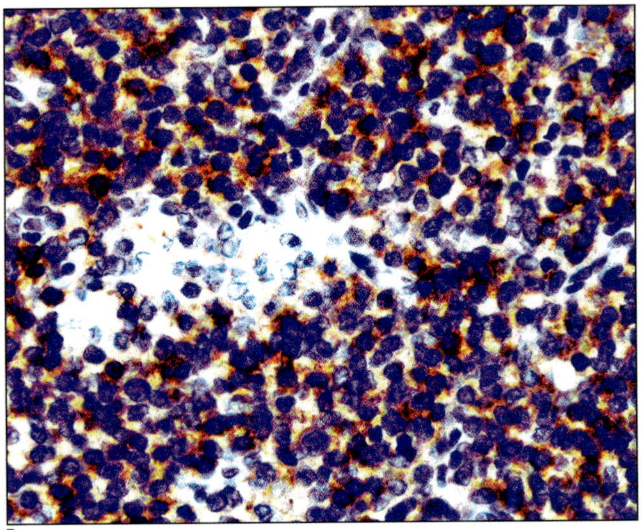
B

FIGURE 4.30: (A) Primitive neuroectodermal tumor (PNET) of the posterior mediastinum, exhibiting a morphologic profile that is similar to that seen in the previous figure. (B) Unlike NBL, however, PNET binds Lens culinaris **lectin diffusely.**

one may wish to apply histochemical methods to separate them from one another. The two techniques that are most useful in this regard are the Bodian and trichrome procedures. The first of them reveals the presence of entrapped neuritic processes throughout the substance of neurofibromas (Figure 4.31), but not neurilemmomas (schwannomas) (127). Similarly, internal collagenous stroma is more abundant and more evenly distributed in neurofibromas studied with trichrome stains. Johnson et al. (128) found that they also contained more mast cells—as demonstrated with von Leder or toluidine blue stain—than did neurilemmomas.

Potential malignancy is restricted to neurofibromatous lesions ["neurofibrosarcoma"; "malignant peripheral nerve sheath tumor" (MPNST)], and it may be associated with a lessening of Bodian positivity. In contrast, AgNOR labeling may be increased in MPNSTs as compared with neurofibroma variants (129).

ILLUSTRATIVE CASE STUDIES

CASE 1— ***Fibrosing Mediastinitis:*** A 55-year-old man developed progressive shortness of breath on exertion over the course of several months, culminating with an episode of hemoptysis. Physical examination showed mild fullness of the neck veins, as well as prominence of those in the skin of the chest. Computed tomograms of the thorax demonstrated an anterior mediastinal mass with distortion of the great vessels (Figure 4.32). Thoracoscopic biopsies of the mediastinum were done, showing dense fibrosis and admixed chronic inflammatory cells (Figures 4.33 and 4.34). An entrapped lymph node also contained epithelioid microgranulomas, which, on silver methenamine staining, harbored yeast-form fungal organisms with the morphologic features of *Histoplasma* (Figure 4.35). Multiple tissue sections—including some stained with MGP and PAS—showed no evidence of malignancy.

Only a minority of cases of fibrosing mediastinitis (FM) are associated with definable infectious causes upon histologic assessment. As mentioned earlier, both mycobacteria and fungi may be the inciting organisms for this condition; they are detectable with acid-fast, silver impregnation, or lectin histochemical method. However, in one series, microscopic studies failed to document any microbes in any case of FM, even though 61% of the patients had a history of pulmonary tuberculosis or positive serologic tests for prior *Histoplasma* infection (130). An alternative etiology for FM is that of autoimmune disease, such as rheumatoid arthritis or lupus erythematosus. Also, one must remember that FM-like reactions may occur to malignancies, including HL and non-Hodgkin lymphoma, seminoma, and metastatic carcinoma. MGP and PAS stains are potentially helpful in addressing this possibility. Finally, a sizable proportion of FM cases have no definable cause and are classified as idiopathic (3–5,130). ■

CASE 2— ***Seminoma:*** A 42-year-old man presented with vague and intermittent anterior chest pain, which steadily worsened over a period of 6 months. He also experienced an involuntary weight loss of 20 lb during the same time, but had no fever or night sweats. Chest imaging showed a 6-cm anterior mediastinal mass (Figure 4.36), from which a thoracoscopic biopsy was obtained.

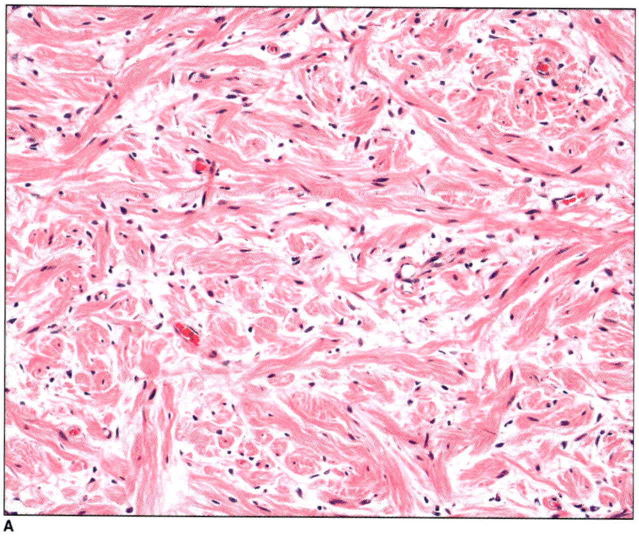

A

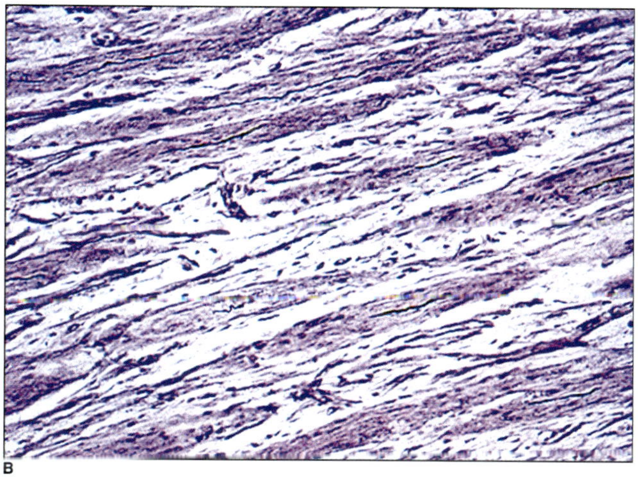

B

FIGURE 4.31: (A) Posterior mediastinal neurofibroma, showing a proliferation of bland spindle cells with serpiginous nuclear contours. (B) Multifocal linear staining is present in the tumor with the Bodian technique.

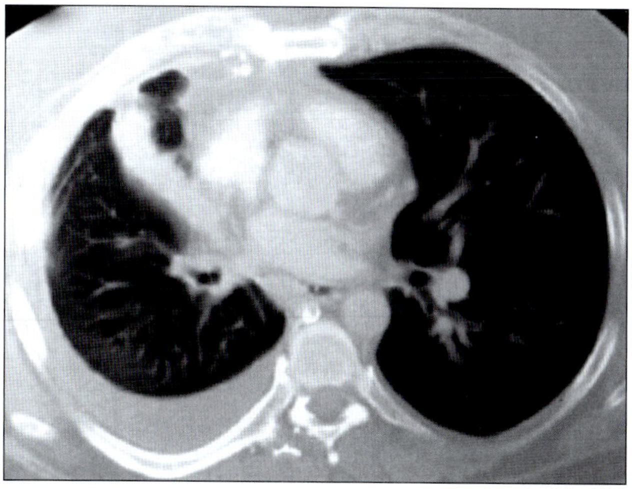

FIGURE 4.32: This computed tomogram of the thorax in a 55-year-old man shows an anterior mediastinal mass with distortion of the great vessels.

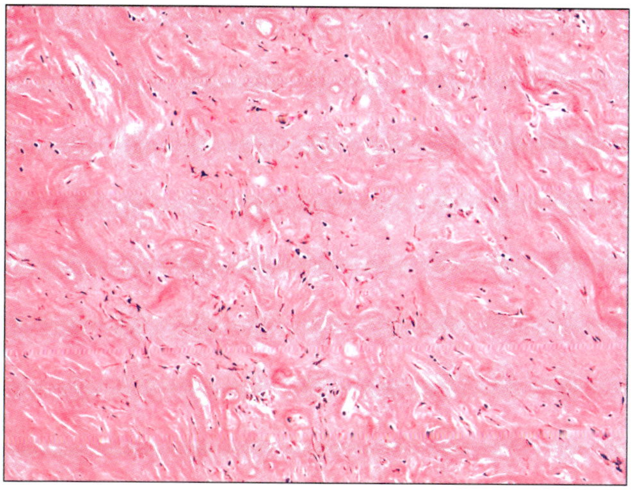

FIGURE 4.33: A biopsy of the mass reveals dense collagen deposition in the anterior mediastinum.

Sections showed a large cell malignant neoplasm with prominent nucleoli, growing in sheets and nests with interspersed lymphocytes. The supporting stroma was variably dense and fibrovascular (Figure 4.37). Principal differential diagnostic considerations included seminoma (germinoma) of the anterior mediastinum versus LCNHL of the thymic region. A PAS stain revealed strong positivity in the tumor cells (Figure 4.38), which was abrogated by prior digestion with diastase. There was also binding of Con-A lectin (Figure 4.39). Sweet's method (Figure 4.40) showed investment of groups of neoplastic cells by reticulin, but it did not envelop individual cells. A final diagnosis of seminoma was made.

In small biopsy specimens, the morphologic images of seminoma and LCNHL may be very similar. However, the above-cited histochemical profile definitively separates the two because large cell lymphomas of the mediastinum do not contain glycogen, have a different reticulin pattern, and fail to be labeled by Con-A (45,46,53–55). ■

CASE 3— ***Mediastinal Granulocytic Sarcoma:*** A 28-year-old woman complained of rapidly worsening dyspnea, and physical examination disclosed the signs of superior vena cava syndrome. Chest radiographs demonstrated a 10-cm mass in the anterior mediastinum, which was biopsied via thoracoscopy. Results of biochemical surveys of the blood were normal, as was a hemogram and peripheral blood smear.

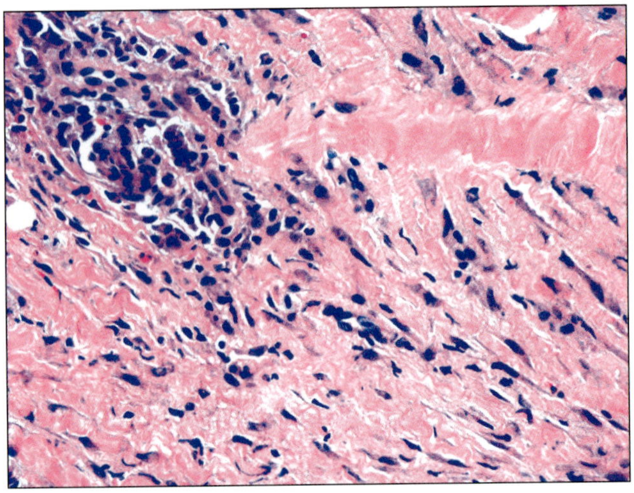

FIGURE 4.34: ***Lymphocytic infiltrates are also present multifocally in the lesion.***

Microscopic examination of the lesion showed sheets of large, modestly pleomorphic tumor cells, with frequent mitotic figures and a delicate fibrovascular stroma (Figures 4.41 and 4.42). There were no admixed mature lymphocytes, and zonal necrosis was likewise absent. Sweet's stain showed investment of individual neoplastic cells by reticulin, and the PAS method yielded no reactivity. However, chloroacetate esterase and MGP stains revealed multifocal positivity in the lesion (Figures 4.43 and 4.44), supporting the final diagnosis of aleukemic GS (extramedullary myeloid tumor).

GS can easily be confused with LCNHL on conventional morphologic studies, particularly in cases where the peripheral blood and bone marrow show no abnormalities. Parenthetically, the latter situation is not uncommon. Thus, it is prudent to include a Leder stain in evaluations of all presumed hematolymphoid malignancies. Another difference between GS and LCNHL is represented by their lectin binding profiles, which feature potential reactivity with Con-A or UEA-I, and peanut agglutinin, respectively (80,131,132). At the same time, this pattern of lectin affinity raises the possibility that embryonal carcinoma—which shares receptors for Con-A and UEA-I (53–56)—might be mistaken for GS if von Leder and PAS stains are not done as well. The treatment of aleukemic GS is similar to that of conventional myelogenous leukemia, being predicated on the use of multiagent chemotherapy or bone marrow transplantation (133–135). ▪

CASE 4— *Micronodular (Castlemanoid) Thymoma:* A 51-year-old woman developed weakness in her girdle muscles as well as ptosis, both of which were accentuated at the end of each day. Consultation with a neurologist and electromyographic testing resulted in a diagnosis of myasthenia gravis. Chest radiographs showed the presence of a 4-cm anterosuperior mediastinal mass, which was biopsied thoracoscopically.

Sections of the lesion showed a lymphocyte-rich proliferation in which the lymphoid cells were arranged in concentric reticulin-rich follicular profiles around distorted, vascularized germinal centers (Figure 4.45). Interfollicular areas of the mass were likewise primarily populated by large and small lymphoreticular elements; however, small zones of the tumor comprised nests of cohesive epithelioid cells that contained PTAH-positive fibrils (Figure 4.46). In light

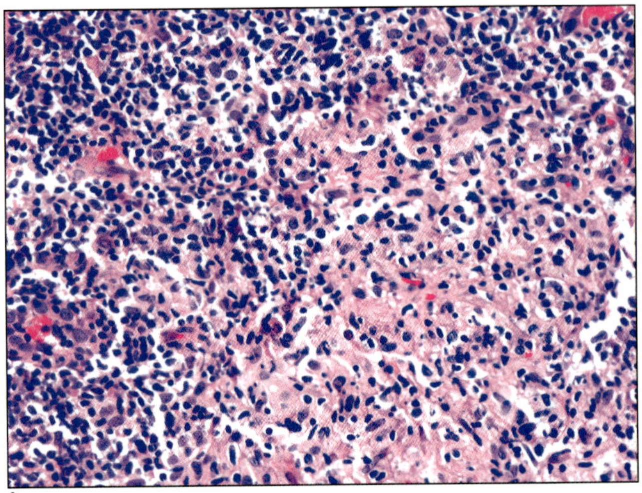

A

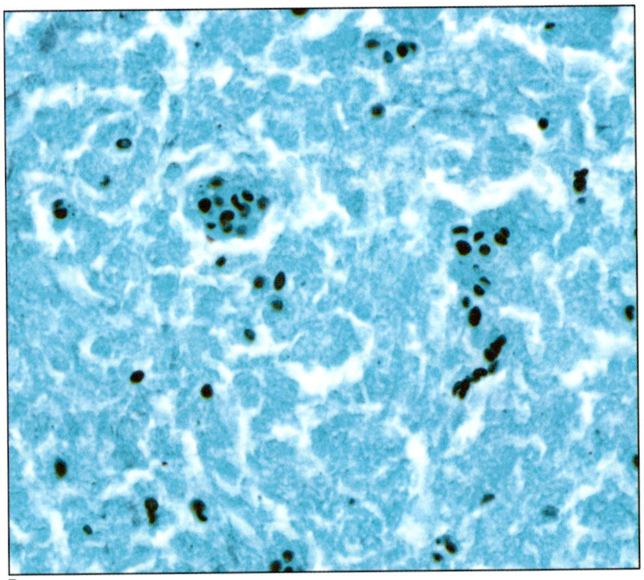

B

FIGURE 4.35: ***(A) An entrapped lymph node in the mass shows granulomatous inflammation. (B) Yeast-form fungal organisms are present in the granulomas. The final diagnosis was that of FM, associated with histoplasmosis of the mediastinum.***

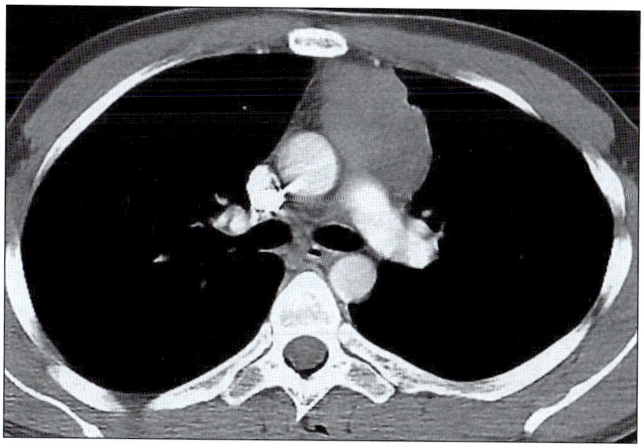

FIGURE 4.36: *An anterior mediastinal mass is seen in this computed tomogram from a 42-year-old man.*

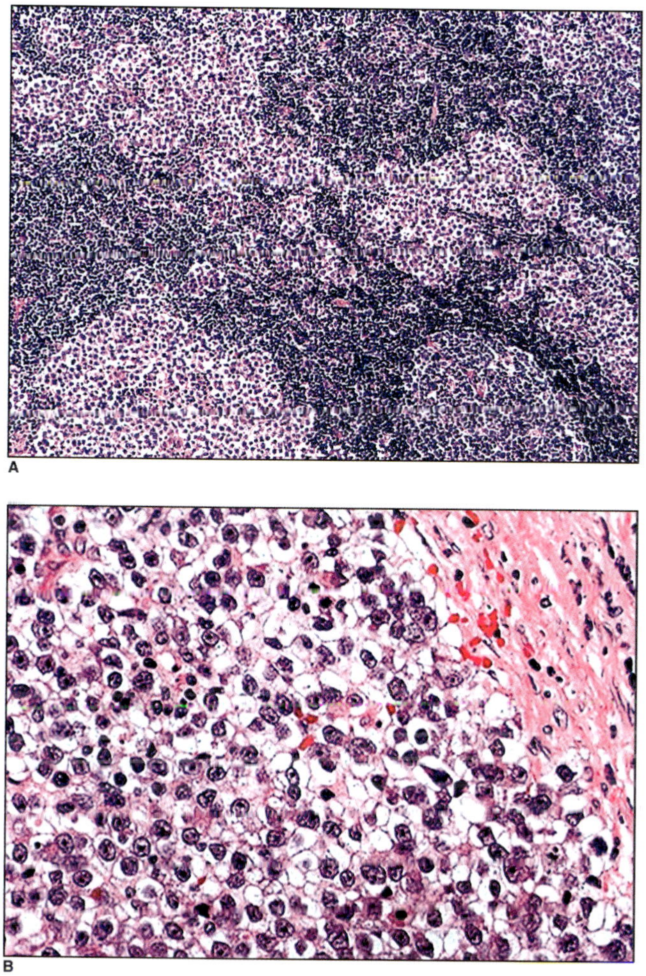

FIGURE 4.37: *(A and B) Nests of atypical polygonal cells with clear cytoplasm are separated by lymphocyte-rich stroma.*

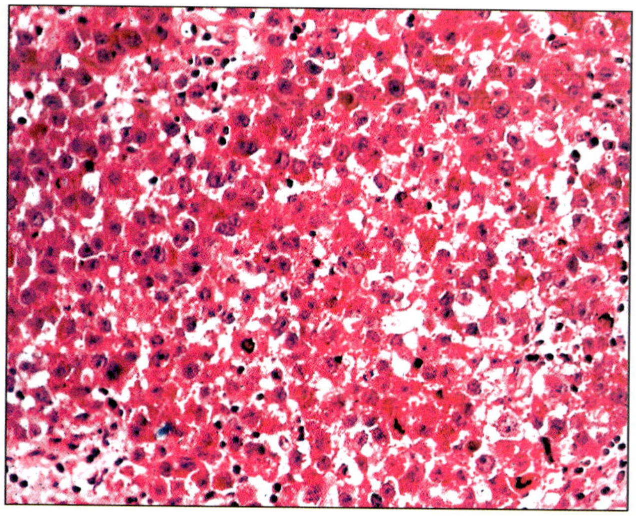

FIGURE 4.38: *Diffuse diastase-labile reactivity with the PAS procedure is seen in the tumor cells.*

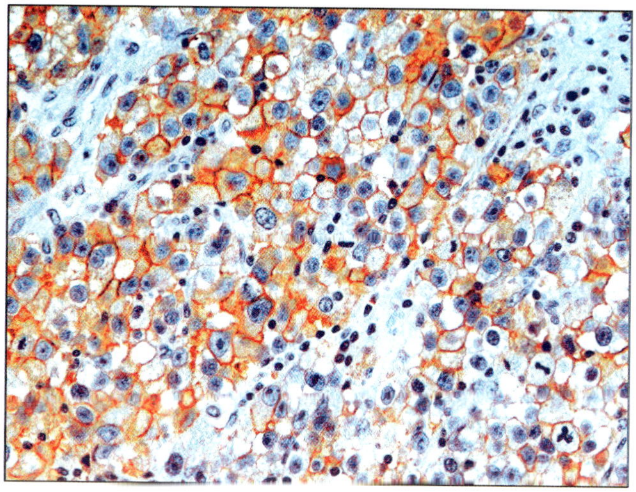

FIGURE 4.39: *The neoplasm also binds Con-A lectin diffusely.*

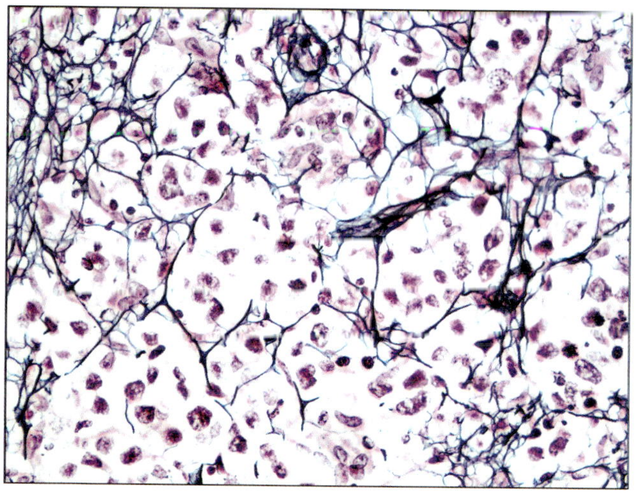

FIGURE 4.40: *Snook's stain shows investment of groups of tumor cells by reticulin. The final diagnosis was that of thymic seminoma.*

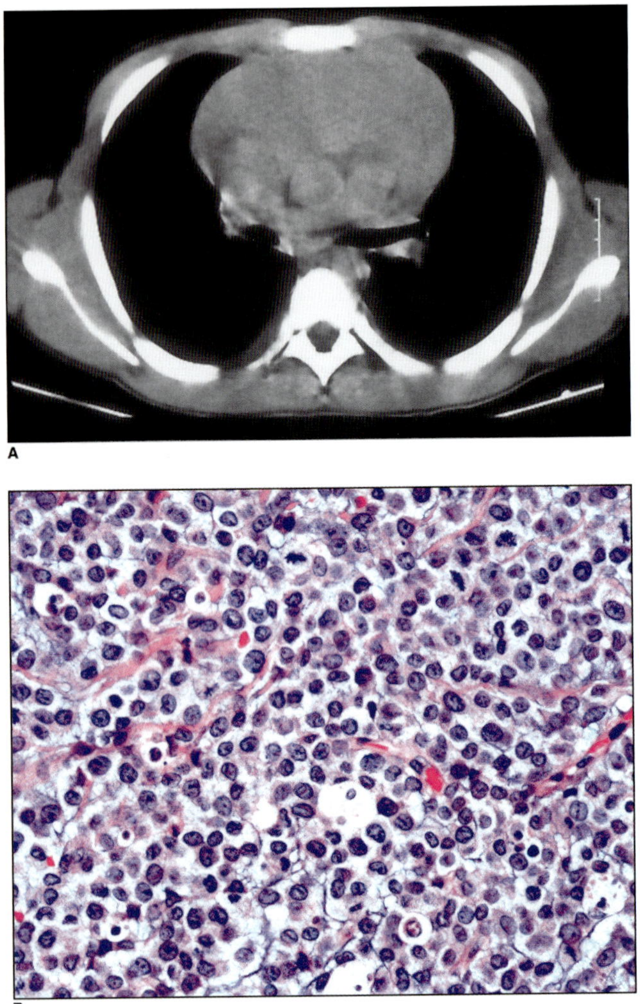

FIGURE 4.41: (A and B) A malignant large cell tumor is present in the mediastinum in a 28-year-old woman.

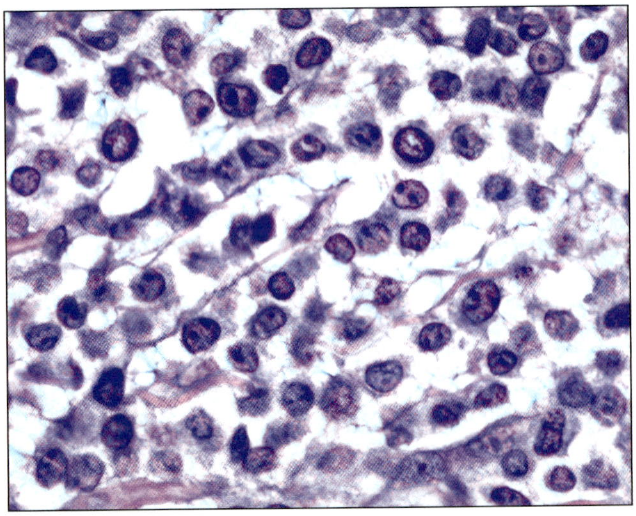

FIGURE 4.42: The neoplastic cells are polygonal in shape but indeterminate with regard to their lineage.

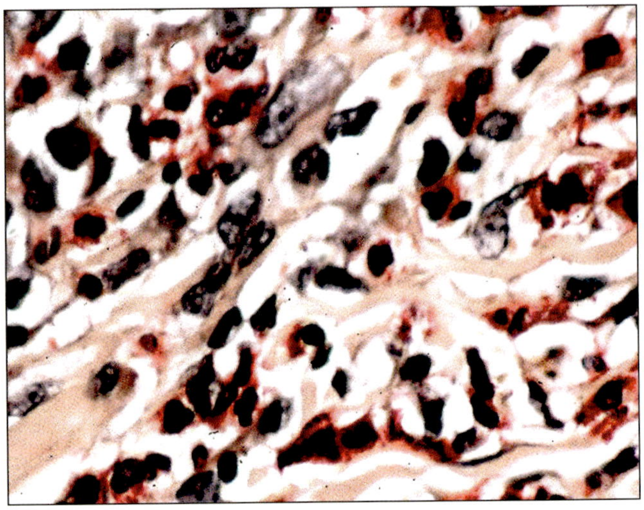

FIGURE 4.43: Cytoplasmic positivity for chloroacetate esterase is seen in the tumor cells with the Leder stain.

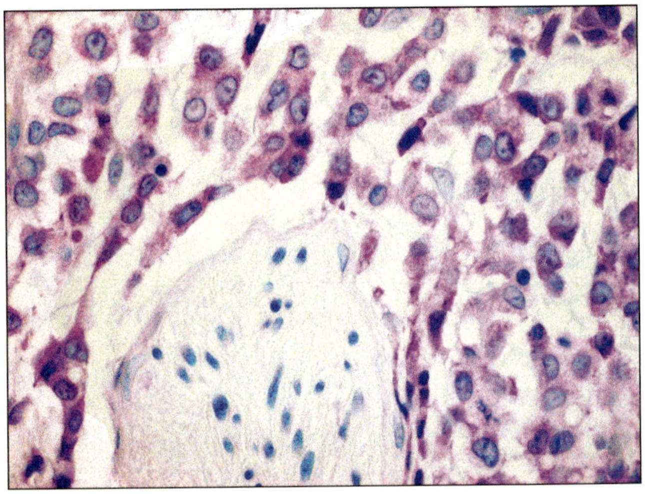

FIGURE 4.44: The MGP method also shows nucleocytoplasmic reactivity in the neoplasm. The final diagnosis was that of GS (extramedullary myeloid tumor).

of the latter observation and the clinical context, a diagnosis of MNT (Castlemanoid) was rendered.

MNT is a rare variant of thymoma; its histologic resemblance to Castleman's disease is complicated by a similar histochemical profile in the two lesions (109,110,136,137), and also by the fact that patients with MNT may also have a higher incidence of true lymphoproliferative disorders (138). Indeed, we have seen one remarkable case in which *both* MNT and true Castleman's disease were present. In those instances where thymoma-related paraneoplasias are absent and no PTAH-positive cell groups can be demonstrated, immunohistochemical and ultrastructural studies are required to distinguish MNT from Castleman's disease. ■

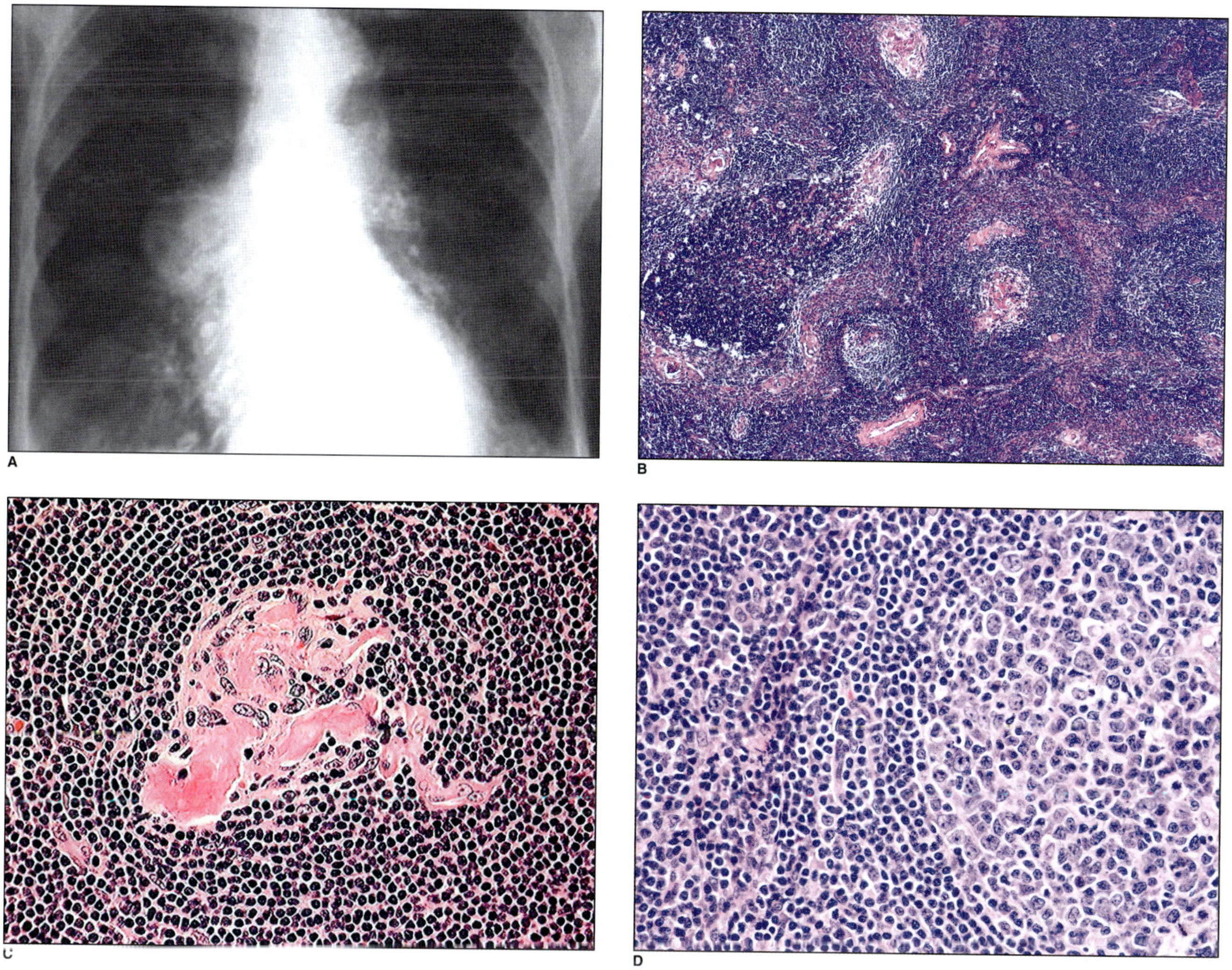

FIGURE 4.45: (A–D) A micronodular lymphocyte-predominant mass is seen in the anterior mediastinum, in a 51-year-old woman. The microscopic image suggests a diagnosis of Castleman's disease (AFLH).

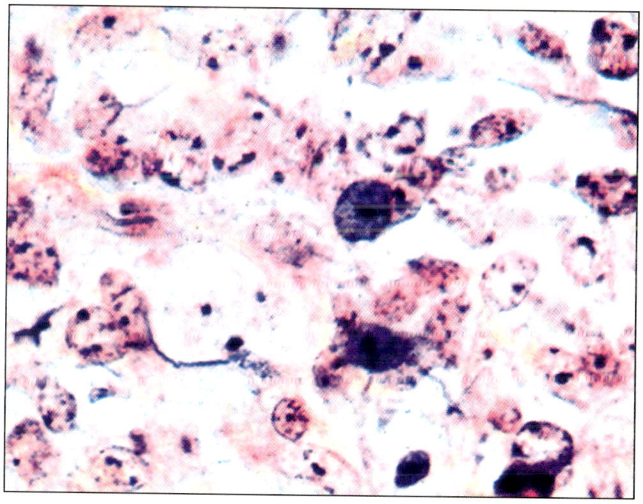

FIGURE 4.46: Linear and arcuate positivity is present in the tumor cell cytoplasm with the PTAH method, representing the presence of tonofibrils. The final diagnosis was that of micronodular (Castlemanoid) thymoma.

CASE 5— *Aorticopulmonary Paraganglioma*: A 38-year-old woman complained of nearly constant dull anterior chest pain. Computed tomograms of the thorax showed a 5-cm ill-defined anterior mediastinal mass, which partially surrounded the aorticopulmonary vascular trunk. A biopsy of the lesion was obtained during a thoracotomy, but the mass was judged to be unresectable.

Histologic examination of the tumor showed an organoid proliferation of large epithelioid cells with moderate nuclear pleomorphism, occasional intranuclear cytoplasmic invaginations, and no discernible mitotic activity or necrosis (Figures 4.47 and 4.48). Positivity was seen in the neoplastic cells with both the Churukian-Schenk and Fontana-Masson procedures (Figure 4.49), and a diagnosis of aorticopulmonary paraganglioma (APPG) was chosen.

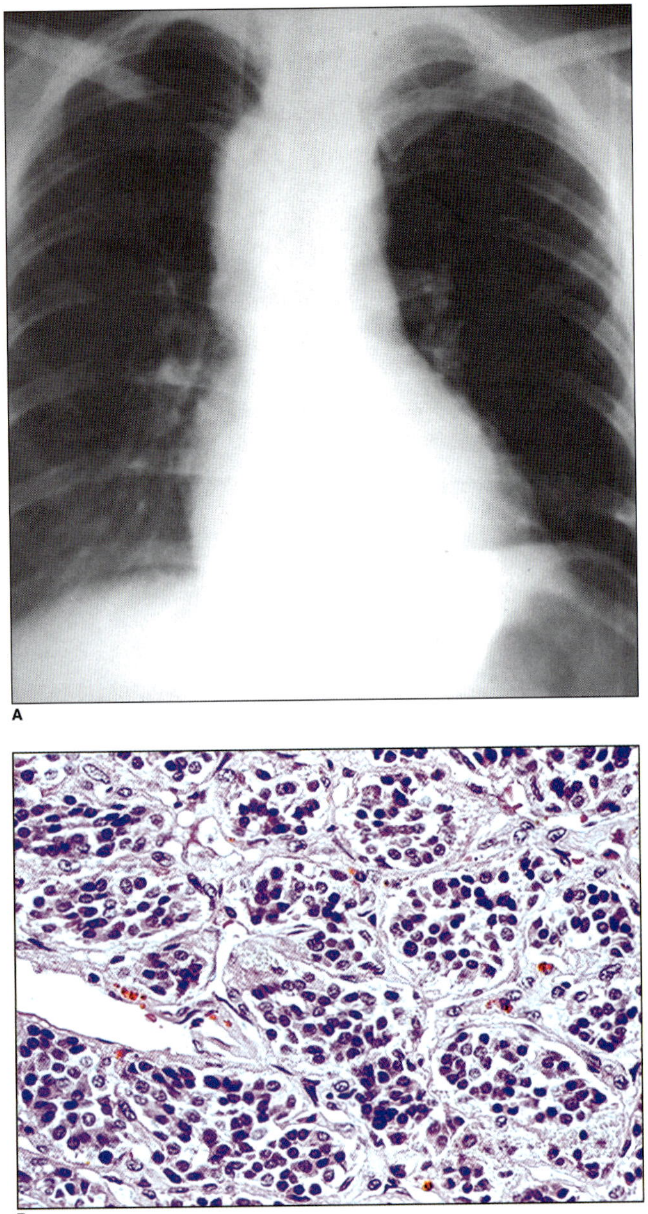

FIGURE 4.47: (A) An anterior mediastinal mass is seen in this chest radiograph of a 38-year-old woman. (B) A biopsy of the lesion showed an epithelioid neoplasm with a nested growth pattern.

The main diagnostic alternative to APPG is that of thymic neuroendocrine carcinoma (TNEC; thymic "carcinoid") (93,96) (Figure 4.50). These two neoplasms are potentially separable through the use of silver impregnation stains, as in this case, because TNEC typically lacks Fontana-Masson positivity, whereas APPG manifests it. In addition, TNEC characteristically contains easily seen mitotic activity and variably sized areas of geographic necrosis, both of which are rare in APPG (117).

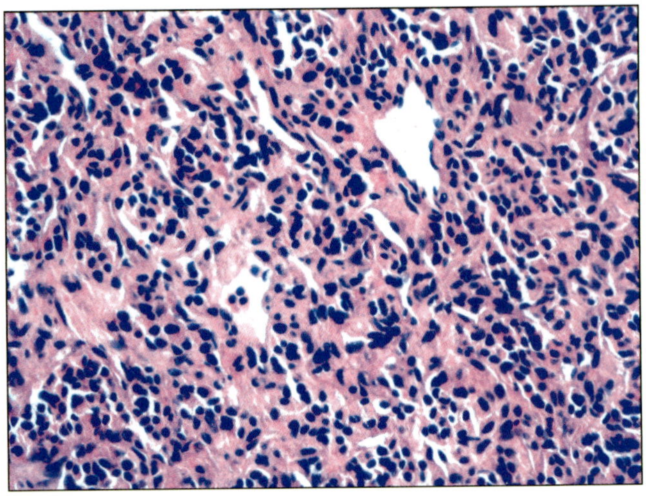

FIGURE 4.48: A richly vascular stroma was also present in the lesion. The differential diagnosis centered on TNEC (carcinoid) versus APPG.

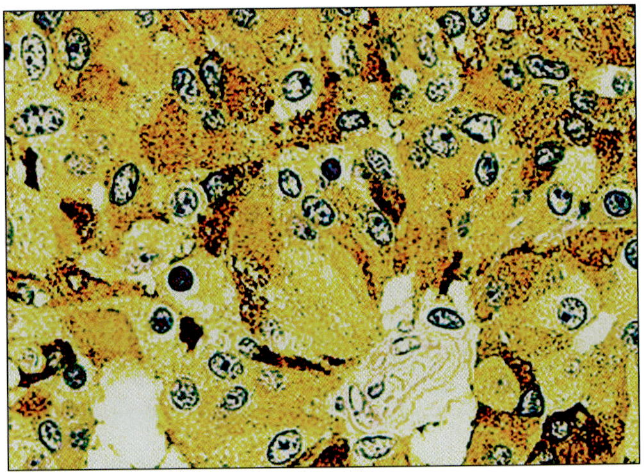

FIGURE 4.49: Fontana-Masson staining of the tumor showed multifocal positivity in the neoplastic cells, supporting a final diagnosis of APPG.

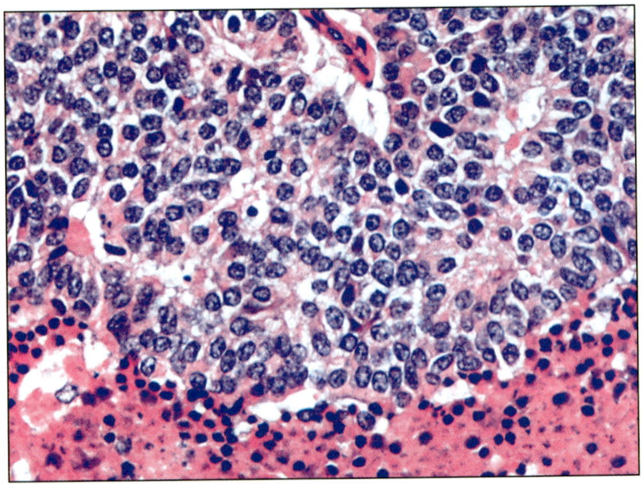

FIGURE 4.50: Thymic neuroendocrine carcinoma shows necrosis and mitotic activity, unlike paraganglioma. It is also nonreactive with the Fontana-Masson stain, whereas paraganglioma is often positive.

REFERENCES

1. Pierce TB, Razzuk MA, Razzuk LM, Luterman DL, Sutker WL: Acute mediastinitis. *Proc Baylor Univ Med Center* 2000; 13: 31–33.
2. Enquist RW, Blanck RR, Butler RH: Non-traumatic mediastinitis. *JAMA* 1976; 236: 1048–1049.
3. Wheat LJ, Conces D, Allen SD, Blue-Hindy D, Loyd J: Pulmonary histoplasmosis syndromes: recognition, diagnosis, and management. *Semin Respir Crit Care Med* 2004; 25: 129–144.
4. Rossi SE, McAdams HP, Rosado-de-Christenson ML, Franks TJ, Galvin JR: Fibrosing mediastinitis. *Radiographics* 2001; 21: 737–757.
5. Flieder DB, Suster S, Moran CA: Idiopathic fibroinflammatory (fibrosing/sclerosing) lesions of the mediastinum: a study of 30 cases with emphasis on morphologic heterogeneity. *Mod Pathol* 1999; 12: 257–264.
6. Slifkin M, Doyle RJ: Lectins and their application to clinical microbiology. *Clin Microbiol Rev* 1990; 3: 197–218.
7. Ritter JH, Humphrey PA, Wick MR: Malignant neoplasms capable of simulating inflammatory (myofibroblastic) pseudotumors and tumefactive fibroinflammatory lesions: pseudopseudotumors. *Semin Diagn Pathol* 1998; 15: 111–132.
8. Priola AM, Priola SM, Cardinale LO, Cataldi A, Fava C: The anterior mediastinum: diseases. *Radiol Med* 2006; 111: 312–342.
9. Salyer DC, Salyer WR, Eggleston JC: Benign developmental cysts of the mediastinum. *Arch Pathol Lab Med* 1977; 101: 136–139.
10. Marco V, Carrasco MA, Marco C, Bauza A: Cytomorphology of a mediastinal parathyroid cyst: report of a case mimicking malignancy. *Acta Cytol* 1983; 27: 688–692.
11. Welsch U: Development of C-cells and of the follicular epithelium in the mammalian thyroid gland: electron microscopy and histochemical studies. *Ergeb Anat Entwicklungsgesch [Res Anat Devel]* 1972; 46: 7–52.
12. Doi N, Moriyama N, Hosaka Y, et al.: Lectin expression of parathyroid glands with primary hyperparathyroidism. *Nippon Hinyokika Gakka Zasshi [Jpn J Urol]* 1991; 82: 572–578.
13. Sasano H, Rojas M, Silverberg SG: Analysis of lectin-binding in benign and malignant thyroid nodules. *Arch Pathol Lab Med* 1989; 113: 186–189.
14. Wick MR: Cystic lesions of the mediastinum. *Semin Diagn Pathol* 2005; 22: 241–253.
15. Henry K, Farrer-Brown G: *A Colour Atlas of Thymus & Lymph Node Histopathology*, Wolfe Medical Publishers, London, UK, 1981; pp. 15—74.
16. Levine GD, Rosai J: Thymic hyperplasia and neoplasia: a review of current concepts. *Hum Pathol* 1978; 9: 495–515.
17. Hofmann WJ, Moller P, Otto HF: Hyperplasia. In: *Surgery of the Thymus* (Givel JC, Ed.), Springer-Verlag, Berlin, Germany, 1990; pp. 59–70.
18. Wick MR: The mediastinum. In: *Sternberg's Diagnostic Surgical Pathology*, 4th ed., Lippincott-Williams & Wilkins, Philadelphia, 2004; pp. 1253–1321.
19. Rosai J, Levine GD: Tumors of the thymus. In: *Atlas of Tumor Pathology*, Series 2, Fascicle 13, Armed Forces Institute of Pathology, Washington, D.C., 1976; pp. 5–150.
20. Crocker J, Smith PJ: A quantitative study of mast cells in Hodgkin's disease. *J Clin Pathol* 1984; 37: 519–522.
21. Suster S, Moran CA: Primary thymic epithelial neoplasms: spectrum of differentiation and histological features. *Semin Diagn Pathol* 1999; 16: 2–17.
22. Engel P, Pilsgaard B, Francis D: Thymomas and thymic carcinomas: a retrospective investigation with histological reclassification. *APMIS* 1995; 103: 671–678.
23. Lewis JE, Wick MR, Scheithauer BW, Bernatz PE, Taylor WF: Thymoma: a clinicopathologic review. *Cancer* 1987; 60: 2727–2743.
24. Thomas DA, Kantarjian HM: Lymphoblastic lymphoma. *Hematol Oncol Clin North Am* 2001; 15: 51–95.
25. Loffler H, Gassmann W: Morphology and cytochemistry of acute lymphoblastic leukemia. *Baillieres Clin Haematol* 1994; 7: 263–272.
26. Jain U, Frable WJ: Thymoma: analysis of benign and malignant criteria. *J Thorac Cardiovasc Surg* 1974; 67: 310–321.
27. Wakely PE Jr.: Cytopathology-histopathology of the mediastinum: epithelial, lymphoproliferative, and germ cell neoplasms. *Ann Diagn Pathol* 2002; 6: 30–43.
28. Snower DP, Smith BR, Munz UJ, McPhedran P: Reevaluation of the periodic acid–Schiff stain in acute leukemia with immunophenotypic analysis. *Arch Pathol Lab Med* 1991; 115: 346–350.
29. Lilleyman JS, Britton JA, Sugden PJ, Hoyle BJ: Periodic acid–Schiff reaction changes during the course of relapsing lymphoblastic leukemia. *Scand J Haematol* 1979; 23: 30–32.
30. McKenna RW, Brynes RK, Nesbit ME, et al.: Cytochemical profiles in acute lymphoblastic leukemia. *Am J Pediatr Hematol Oncol* 1979; 1: 263–275.
31. Lauriola L, Maggiano N, Marino M, et al.: Human thymoma: immunologic characteristics of the lymphocytic component. *Cancer* 1981; 48: 1992–1995.
32. Piantelli M, Ranelletti FO, Musiani P, Lauriola L, Maggiano N: A human thymoma with prothymocyte-like infiltration. *Clin Immunol Immunopathol* 1983; 28: 350–360.
33. Schwenk HU, Schneider U, Herzog KH: Binding of lectins to leukemic cell lines. *Blut* 1980; 40: 7–15.
34. Slesak B, Boguslawska-Jaworska J, Pejcz J, Harlozinska A: Binding of lectins to human leukemic cells. *Folia Haematol Int Mag Klin Morphol Blutforsch* 1989; 116: 251–259.
35. Tateyama H, Matsuda H, Takahashi E, Eimoto T: Prognostic value of lectin VVA in thymoma (Abstract). *Acta Histochem Cytochem* 2002; 35: 257.
36. Nappi O, Ritter JH, Pettinato G, Wick MR: Hemangiopericytoma: histopathological pattern or clinicopathologic entity? *Semin Diagn Pathol* 1995; 12: 221–232.
37. Moran CA, Koss MN: Rhabdomyomatous thymoma. *Am J Surg Pathol* 1993; 17: 633–636.
38. Sato T, Tamaoki N: Myoid cells in the human thymus and thymoma revealed by three different immunohistochemical markers for striated muscle. *Acta Pathol Jpn* 1989; 39: 509–519.
39. Muller-Hermelink HK, Kirchner T, Hoppe F, Demel S: Thymic myoid cells—simulants and/or targets of autoimmune myasthenia gravis. *Monogr Allergy* 1988; 25: 68–74.

40. Ueda T, Aozasa K, Yamamura T, et al.: Lectin histochemistry of malignant fibrohistiocytic tumors. *Am J Surg Pathol* 1987; 11: 257–262.
41. Suster S: Thymic carcinoma: update of current diagnostic criteria and histologic types. *Semin Diagn Pathol* 2005; 22: 198–212.
42. Chalabreysse L, Etienne-Mastroianni B, Adeleine P, et al.: Thymic carcinoma: a clinicopathological and immunohistological study of 19 cases. *Histopathology* 2004; 44: 367–374.
43. Ritter JH, Wick MR: Primary carcinomas of the thymus gland. *Semin Diagn Pathol* 1999; 16: 18–31.
44. Chhieng DC, Lin O, Moran CA, et al.: Fine-needle aspiration biopsy of nonteratomatous germ cell tumors of the mediastinum. *Am J Clin Pathol* 2002; 118: 418–424.
45. Dehner LP: Germ cell tumors of the mediastinum. *Semin Diagn Pathol* 1990; 7: 266–284.
46. Luna MA, Valenzuela-Tamariz J: Germ-cell tumors of the mediastinum: post-mortem findings. *Am J Clin Pathol* 1976; 65: 450–454.
47. Geisinger KR: Differential diagnostic considerations and potential pitfalls in fine-needle aspiration biopsies of the mediastinum. *Diagn Cytopathol* 1995; 13: 436–442.
48. Azzopardi JG, Hoffbrand AV: Retrogression in testicular seminoma with viable metastases. *J Clin Pathol* 1965; 18: 135–141.
49. Florentine BD, Roscher AA, Garrett J, Warner NE: Necrotic seminoma of the testis: establishing the diagnosis with Masson trichrome stain and immunostains. *Arch Pathol Lab Med* 2002; 126: 205–206.
50. El-Khatib M, Chew FS: Embryonal carcinoma of the anterior mediastinum. *Am J Roentgenol* 1998; 170: 722.
51. Moran CA, Suster S, Koss MN: Primary germ cell tumors of the mediastinum: III. Yolk sac tumor, embryonal carcinoma, choriocarcinoma, and combined nonteratomatous germ cell tumors of the mediastinum—a clinicopathologic and immunohistochemical study of 64 cases. *Cancer* 1997; 80: 699–707.
52. Wick MR, Siegal GP: Primary mediastinal embryonal carcinoma. *Minn Med* 1980; 63: 723–726.
53. Malmi R, Soderstrom KO: Lectin histochemistry of embryonal carcinoma. *APMIS* 1991; 99: 233–243.
54. Malmi R, Kellokumpu-Lehtinen PL, Soderstrom KO: Correlation between lectin binding and clinical factors in seminoma patients. *J Cancer Res Clin Oncol* 1989; 115: 96–100.
55. Motzer RJ, Reuter VE, Cordon-Cardo C, Bosl GJ: Blood group-related antigens in human germ cell tumors. *Cancer Res* 1988; 48: 5342–5347.
56. Shield PW: Lectin binding properties of cells from serous effusion and peritoneal washing specimens. *J Clin Pathol* 1989; 42: 1178–1183.
57. Morinaga S, Nomori H, Kobayashi R, Atsumi Y: Well-differentiated adenocarcinoma arising from mature cystic teratoma of the mediastinum (teratoma with malignant transformation): report of a surgical case. *Am J Clin Pathol* 1994; 101: 531–534.
58. Saito A, Watanabe K, Kusakabe T, Abe M, Suzuki T: Mediastinal mature teratoma with coexistence of angiosarcoma, granulocytic sarcoma, and a hematopoietic region of the tumor: a rare case of association between hematological malignancy and mediastinal germ cell tumor. *Pathol Int* 1998; 48: 749–753.
59. Neiman RS, Orazi A: Mediastinal non-seminomatous germ cell tumors: their association with non-germ cell malignancies. *Pathol Res Pract* 1999; 195: 589–594.
60. Omezzine N, Khouatra C, Larive S, et al.: Rhabdomyosarcoma arising in mediastinal teratoma in an adult man: a case report. *Ann Oncol* 2002; 13: 323–326.
61. Manivel C, Wick MR, Abenoza P, Rosai J: The occurrence of sarcomatous components in primary mediastinal germ cell tumors. *Am J Surg Pathol* 1986; 10: 711–717.
62. Ulbright TM, Loehrer PJ, Roth LM, et al.: The development of non-germ cell malignancies within germ cell tumors: a clinicopathologic study of 11 cases. *Cancer* 1984; 54: 1824–1833.
63. Li Q, Gong XQ, Ma FC, Zhao YL, Zhu XH: Application of histochemical staining in diagnosis of osteosarcomas. *Zhonghua Zhong Liu Za Zhi [Chin J Oncol]* 2005; 27: 489–491.
64. Junqueira LC, Assis-Figeuiredo MT, Torloni H, Montes GS: Differential histologic diagnosis of osteoid: a study on human osteosarcoma collagen by the histochemical picrosirius-polarization method. *J Pathol* 1986; 148: 189–196.
65. Begin LR, Schurch W, Lacoste J, Hiscott J, Melnychuk DA: Glycogen-rich clear-cell rhabdomyosarcoma of the mediastinum: potential diagnostic pitfall. *Am J Surg Pathol* 1994; 18: 302–308.
66. Sarnat HB, DeMello DE, Siddiqui SY: Diagnostic value of histochemistry in embryonal rhabdomyosarcoma. *Am J Surg Pathol* 1979; 3: 177–183.
67. Swanson PE, Dehner LP, Wick MR: Small round cell tumors in children: a lectin-histochemical and immunohistochemical study of 30 cases. *Surg Pathol* 1993; 5: 5–16.
68. Morris MW, Davey FR: Immunologic and cytochemical properties of histiocytic and mixed histiocytic-lymphocytic lymphomas. *Am J Clin Pathol* 1975; 63: 403–414.
69. Davey RF, Goldberg J, Stockman J, Gottlieb AJ: Immunologic and cytochemical cell markers in non-Hodgkin's lymphomas. *Lab Invest* 1976; 35: 430–438.
70. Osborne BM, Robboy SJ: Lymphomas or leukemia presenting as ovarian tumors: an analysis of 42 cases. *Cancer* 1983; 52: 1933–1943.
71. Suster S: Primary large-cell lymphomas of the mediastinum. *Semin Diagn Pathol* 1999; 16: 51–64.
72. Strickler JG, Kurtin PJ: Mediastinal lymphoma. *Semin Diagn Pathol* 1991; 8: 2–13.
73. Lamarre L, Jacobson JO, Aisenberg AC, Harris NL: Primary large-cell lymphoma of the mediastinum: a histologic and immunophenotypic study of 29 cases. *Am J Surg Pathol* 1989; 13: 730–739.
74. Perrone T, Frizzera G, Rosai J: Mediastinal diffuse large-cell lymphoma with sclerosis: a clinicopathologic study of 60 cases. *Am J Surg Pathol* 1986; 10: 176–191.
75. Wong WS, Loong F, Ooi GC, Tse TC, Chim CS: Primary granulocytic sarcoma of the mediastinum. *Leuk Lymphoma* 2004; 45: 1931–1933.

76. Toback A, Hasbrouck DJ, Blaustein J, Ershler WB: Granulocytic sarcoma of the anterior mediastinum. *Am J Med Sci* 1985; 290: 206–208.
77. Gariepy G, Dumont A, Monte M, Noel D: Granulocytic sarcoma of the mediastinum. *Ann Pathol* 1983; 3: 167–170.
78. Chen YS, Lu DW: The diagnostic value of chloroacetate esterase staining in acute leukemias—observation of 70 cases. *Zhonghua Nei Ke Za Zhi [Chin J Int Med]* 1983; 22: 164–166.
79. Davey FR, Olson S, Kurec AS, et al.: The immunophenotyping of extramedullary myeloid cell tumors in paraffin-embedded tissue sections. *Am J Surg Pathol* 1988; 12: 699–707.
80. Ree HJ, Raine L, Crowley JP: Lectin binding patterns in diffuse large cell lymphoma. *Cancer* 1983; 52: 2089–2099.
81. Flannery MT, Espino M, Altus P, Messina J, Wallach PM: Hodgkin's disease masquerading as sclerosing mediastinitis. *South Med J* 1994; 87: 921–923.
82. Ree HJ, Neiman RS, Martin AW, Dallenbach F, Stein H: Paraffin section markers for Reed-Sternberg cells: a comparative study of peanut agglutinin, Leu-M1, LN-2, and Ber-H2. *Cancer* 1989; 63: 2030–2036.
83. Sicard GA: Parathyroid hyperplasia versus adenoma—clinical or histologic diagnosis. *Surgery* 1981; 90: 125–126.
84. McGarity WC, Boehm G: Carcinoma of the parathyroid. *South Med J* 1975; 68: 166–168.
85. Murphy MN, Glennon PG, Diocee MS, Wick MR, Cavers DJ: Nonsecretory parathyroid carcinoma of the mediastinum: light microscopic, immunocytochemical, and ultrastructural features of a case, and review of the literature. *Cancer* 1986; 58: 2468–2476.
86. Grimelius L, Ejerblad S, Johansson H, Werner I: Parathyroid adenomas and glands in normocalcemic hyperparathyroidism: a light microscopic study. *Am J Pathol* 1976; 83: 475–484.
87. Frigerio B, Capella C, Wilander E, Grimelius L: Argyrophil reaction in parathyroid glands: a light and electron microscopic study. *APMIS* 1982; 90: 323–326.
88. Pesce C, Tobia F, Carli F, Antoniotti GV: The sites of hormone storage in normal and diseased parathyroid glands: a silver-impregnation and immunohistochemical study. *Histopathology* 1989; 15: 157–166.
89. Evans HL: Criteria for diagnosis of parathyroid carcinoma: a critical study. *Surg Pathol* 1991; 4: 244–265.
90. Schantz A, Castleman B: Parathyroid carcinoma: a study of 70 cases. *Cancer* 1973; 31: 600–605.
91. Kay S, Hume DM: Carcinoma of the parathyroid gland. how reliable are the clinical and histologic features? *Arch Pathol* 1973; 96: 316–319.
92. Kanematsu E, Matsui H, Deguchi T, et al.: Significance of AgNOR counts for distinguishing carcinoma from adenoma and hyperplasia in parathyroid gland. *Hum Pathol* 1997; 28: 421–427.
93. Moran CA, Suster S, Fishback N, Koss MN: Mediastinal paragangliomas: a clinicopathologic and immunohistochemical study of 16 cases. *Cancer* 1993; 72: 2358–2364.
94. Assaf HM, Al-Momen AA, Martin JG: Aorticopulmonary paraganglioma: a case report with immunohistochemical and literature review. *Arch Pathol Lab Med* 1992; 116: 1085–1087.
95. Mauri MF, Mingazzini P, Sisti S, Montironi R, Scarpelli M: Histomorphometric and morphologic studies of the carotid body and aortic paragangliomas. *Appl Pathol* 1989; 7: 310–317.
96. Suster S, Moran CA: Neuroendocrine neoplasms of the mediastinum. *Am J Clin Pathol* 2001; 115 (Suppl): S17–S27.
97. Lack EE, Kim H, Reed K: Pigmented ("black") extraadrenal paraganglioma. *Am J Surg Pathol* 1998; 22: 265–269.
98. Meitner ER: Architectonics of the argyrophil fibers of some endocrine organs. *Folia Morphol* 1980; 28: 351–353.
99. Shrivastava CP, Devgarha S, Ahlawat V: Mediastinal tumors: a clinicopathological analysis. *Asian Cardiovasc Thorac Ann* 2006; 14: 102–104.
100. Henley JD, Loehrer PJ Sr., Ulbright TM: Deciduoid mesothelioma of the pleura after radiation therapy for Hodgkin's disease, presenting as a mediastinal mass. *Am J Surg Pathol* 2001; 25: 547–548.
101. Tagliamonti JA, Yannopoulos K, Kryle LS: Malignant epithelial mesothelioma of the anterior mediastinum. *NY State J Med* 1984; 84: 127–129.
102. Lucas JG, Tuttle SE: Diagnostic histochemical and immunohistochemical studies in malignant mesothelioma. *J Surg Oncol* 1987; 35: 30–34.
103. Wick MR, Loy T, Mills SE, Legier JF, Manivel JC: Malignant epithelioid pleural mesothelioma versus peripheral pulmonary adenocarcinoma: a histochemical, ultrastructural, and immunohistologic study of 103 cases. *Hum Pathol* 1990; 21: 759–766.
104. Hammar SP, Bockus DE, Remington FL, Rohrbach KA: Mucin-positive epithelial mesotheliomas: a histochemical, immunohistochemical, and ultrastructural comparison with mucin-producing pulmonary adenocarcinomas. *Ultrastruct Pathol* 1996; 20: 293–325.
105. Harwood TR, Gracey DR, Yokoo H: Pseudomesotheliomatous carcinoma of the lung: a variant of peripheral lung cancer. *Am J Clin Pathol* 1976; 65: 159–167.
106. Koss MN, Travis WA, Moran CA, Hochholzer L: Pseudomesotheliomatous adenocarcinoma: a reappraisal. *Semin Diagn Pathol* 1992; 9: 117–123.
107. Bolen JW: Tumors of serosal origin. *Clin Lab Med* 1987; 7: 31–50.
108. Castleman B, Iverson L, Menendez VP: Localized mediastinal lymph node hyperplasia resembling thymoma. *Cancer* 1956; 9: 822–830.
109. Suster S, Moran CA: Micronodular thymoma with lymphoid B-cell hyperplasia: clinicopathologic and immunohistochemical study of eighteen cases of a distinctive morphologic variant of thymic epithelial neoplasm. *Am J Surg Pathol* 1999; 23: 955–962.
110. Palestro G, Turrini F, Pagano M, Chiusa L: Castleman's disease. *Adv Clin Pathol* 1999; 3: 11–22.
111. Blanco N, Kirgan DM, Little AG: Metastatic squamous cell carcinoma of the mediastinum with unknown primary tumor. *Chest* 1998; 114: 938–940.

112. Kohdono S, Ishida T, Fukuyama Y, et al.: Lymph node cancer of the mediastinal or hilar region with an unknown primary site. *J Surg Oncol* 1995; 58: 196–200.
113. Tanaka F, Takenaka K, Oyanagi H, et al.: Skip mediastinal nodal metastases in non-small-cell lung cancer. *Eur J Cardiothorac Surg* 2004; 25: 1114–1120.
114. McCloud TC, Meyer JE: Mediastinal metastases. *Radiol Clin North Am* 1982; 20: 453–468.
115. Azarow KS, Pearl RH, Zurcher R, Edwards FH, Cohen AJ: Primary mediastinal masses: a comparison of adult and pediatric populations. *J Thorac Cardiovasc Surg* 1993; 106: 67–72.
116. Blegvad S, Lippert H, Simper LB, Dybdahl H: Mediastinal tumors: a report of 129 cases. *Scand J Thorac Cardiovasc Surg* 1990; 24: 39–42.
117. Wick MR, Rosai J: Neuroendocrine neoplasms of the mediastinum. *Semin Diagn Pathol* 1991; 8: 35–51.
118. Lonergan GJ, Schwab CM, Suarez ES, Carlson CL: Neuroblastoma, ganglioneuroblastoma, and ganglioneuroma: radiologic-pathologic correlation. *Radiographics* 2002; 22: 911–934.
119. Wakely PE Jr., Favara BE, Galliani CA: Diagnostic histopathologic criteria in pediatric tumors: an eclectic review. *Crit Rev Oncol Hematol* 1986; 6: 261–286.
120. Anthony PP, Coles HM, Pryse-Davies J, Sinclair L: Biochemical, pharmacological, and histochemical studies in neuroblastoma. *Arch Dis Child* 1965; 40: 411–417.
121. Egan MJ, Raafat F, Crocker J, Smith K: Nucleolar organizer regions in small cell tumors of childhood. *J Pathol* 1987; 153: 275–280.
122. Egan MJ, Raafat F, Crocker J, Williams D: Comparative study of the degree of differentiation of neuroblastoma and mean numbers of nucleolar organizer regions. *J Clin Pathol* 1988; 41: 527–531.
123. Kahn HJ, Baumal R, Thorner PS, Chan H: Binding of peanut agglutinin to neuroblastomas and ganglioneuromas: a marker for differentiation of neuroblasts into ganglion cells. *Pediatr Pathol* 1988; 8: 83–93.
124. Swanson PE: Soft tissue neoplasms of the mediastinum. *Semin Diagn Pathol* 1991; 8: 14–34.
125. Marchevsky AM: Mediastinal tumors of peripheral nervous system origin. *Semin Diagn Pathol* 1999; 16: 65–78.
126. Reeder LB: Neurogenic tumors of the mediastinum. *Semin Thorac Cardiovasc Surg* 2000; 12: 261–267.
127. Kao GF, Laskin WB, Olsen TG: Solitary cutaneous plexiform neurilemmoma (schwannoma): a clinicopathologic, immunohistochemical, and ultrastructural study of 11 cases. *Mod Pathol* 1989; 2: 20–26.
128. Johnson MD, Kamso-Pratt J, Federspiel CF, Whetsell WO Jr.: Mast cell and lymphoreticular infiltrates in neurofibromas: comparison with nerve sheath tumors. *Arch Pathol Lab Med* 1989; 113: 1263–1270.
129. Kuratsu S, Ohsawa M, Naka N, et al.: Usefulness of argyrophilic nucleolar organizer staining for predicting prognosis of patients with recurrent soft tissue sarcoma. *Oncology* 1994; 51: 244–250.
130. Mole TM, Glover J, Sheppard MN: Sclerosing mediastinitis: a report on 18 cases. *Thorax* 1995; 50: 280–283.
131. Stoddart RW, Collins RD, Jacobson W: Lectin staining of carbohydrates of haemic cells: the cells of normal blood and bone marrow and of the myeloid leukaemias. *J Pathol* 1980; 131: 321–332.
132. Delwel R, Touw I, Bot F, Lowenberg B: Fucose binding lectin for characterizing acute myeloid leukemia progenitor cells. *Blood* 1986; 68: 41–45.
133. Tsimberidou AM, Kantarjian HM, Estey E, et al.: Outcome in patients with nonleukemic granulocytic sarcoma treated with chemotherapy with or without radiotherapy. *Leukemia* 2003; 17: 1100–1103.
134. Finnegan DP, Jones FG, McMullin MF: Acute myeloid leukemia with concurrent myeloid sarcoma treated with autologous bone marrow transplantation: two illustrative cases and a literature review. *Hematol Oncol* 2005; 23: 133–135.
135. Tan D, Wong GC, Koh LP, et al.: Successful treatment of primary granulocytic sarcoma by non-myeloablative stem cell transplant. *Leuk Lymphoma* 2006; 47: 159–162.
136. Cioffi U, DeSimone M, Nosotti M, et al.: Hyaline vascular Castleman's disease of the mediastinum. *Int Surg* 1999; 84: 115–117.
137. Tateyama H, Saito Y, Fujii Y, et al.: The spectrum of micronodular thymic epithelial tumors with lymphoid B-cell hyperplasia. *Histopathology* 2001; 38: 519–527.
138. Strobel P, Marino M, Feuchtenberger M, et al.: Micronodular thymoma: an epithelial tumor with abnormal chemokine expression setting the stage for lymphoma development. *J Pathol* 2005; 207: 72–82.

CHAPTER 5

Diagnostic Histochemistry in Cardiac Diseases

Philip G. Robinson, M.D., and
Azorides R. Morales, M.D.

Assessment of cardiac pathology in autopsies or after cardiac transplantation requires careful gross examination of the organ; findings made during this process dictate the selection of appropriate histologic sections, usually of the coronary arteries and myocardium, and, less frequently, one or more of the valves and the cardiac conduction system. As is usual in the practice of anatomic pathology, these cardiac structures and their pathologic changes are studied with a routine hematoxylin and eosin stain (H&E).

There are definite indications, however, to supplement H&E with other staining techniques, either to highlight morphologic abnormalities such as fibrosis or to demonstrate alterations that are not particularly well seen with H&E such as the accumulation of mucopolysaccharides. Histochemical methods are particularly useful and practically essential in cardiac pathology with regard to a variety of diseases. These include infections and relatively uncommon conditions such as infiltrative diseases (amyloid, Gaucher's disease, and Hurler's disease) and storage disorders (hemochromatosis, Fabry's disease, and Pompe's disease). The scope of histochemical procedures that are used in addition to H&E in the study of endomyocardial biopsies is relatively narrow. Applicable stains may be used to demonstrate the nature of cardiomyopathies; some of them include the Masson trichrome technique, the Congo red method, and the periodic acid–Schiff (PAS) and Prussian blue procedures (1–3).

Histochemical stains that are frequently used in the study of medical conditions of the heart are discussed in this chapter. Methods for the detection of cardiac enzymes, such as phosphatases or dehydrogenases, are not included because they are seldom used. Also excluded from this review are intracardiac neoplasms (both primary and metastatic). This is because the histochemical features of most of the tumor types that are seen in the heart (solitary fibrous tumor, angiosarcoma, rhabdomyosarcoma, leiomyosarcoma) are covered in other chapters in this monograph.

MASSON TRICHROME STAINING

Masson's trichrome is one of the most common connective tissue stains used to identify collagenized tissue, such as myocardial fibrosis, in histologic sections. Fibrosis is potentially present in virtually any cardiac disease, but its extent, distribution, and microscopic characteristics vary according to the etiology and severity of the process. Evaluation of the gross and microscopic features of fibrosis often provides further insight into the nature of the disease. In 1979, a study attempting to quantitate and correlate myocardial fibrosis with causative conditions led to a still widely quoted system concerning the histopathologic types of that process. These center on "interfiber," "perivascular," and "plexiform" fibrosis (4).

In diseases such as myocardial infarction and myocarditis, scarring is the final step of the healing process after respective ischemic or inflammatory injury. The scarring is distributed randomly in cases of myocarditis (Figure 5.1) or in the area supplied by an occluded vessel in infarcts, most often due to arteriosclerosis (Figure 5.2). In addition, other conditions such as pathologic myocardial bridges may demonstrate fibrosis (Figure 5.3) (5).

The most common example of myocardial scarring would be that which follows myocardial infarction (6), because of the commonality of coronary artery disease. In addition, myocardial fibrosis is a frequently seen feature in the major types of cardiomyopathy (dilated, hypertrophic, restrictive, and arrhythmogenic right ventricular) (Figures 5.4–5.9) (7). Healed myocardial infarcts, typified by "replacement" fibrosis in the left ventricle and beginning in the subendocardial region, differ from idiopathic dilated cardiomyopathy. The latter condition principally exhibits "interstitial" fibrosis that is situated mainly in the mid-layers of the left ventricle (8) and may lead to as much as four times the normal quantity of intramyocardial collagen (9). In idiopathic hypertrophic cardiomyopathy and in hypertrophic subaortic stenosis, cardiocytes are arranged in a plexiform pattern with intervening fibrosis (10). In contrast, arrhythmogenic cardiomyopathy

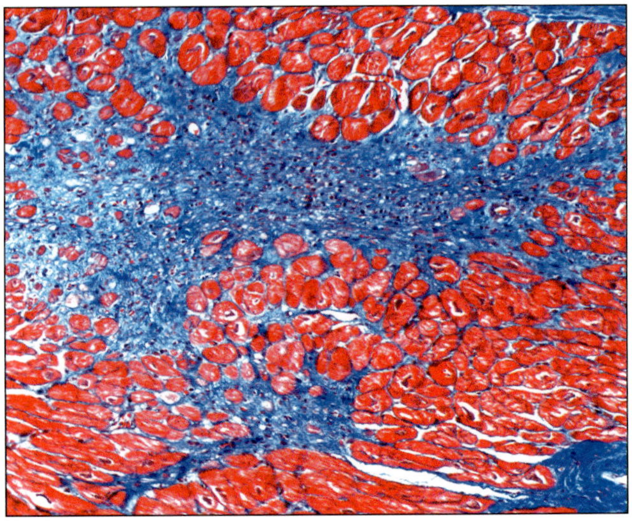

FIGURE 5.1: Irregular "replacement"-type fibrosis is associated with viral myocarditis, as shown in this trichrome stain.

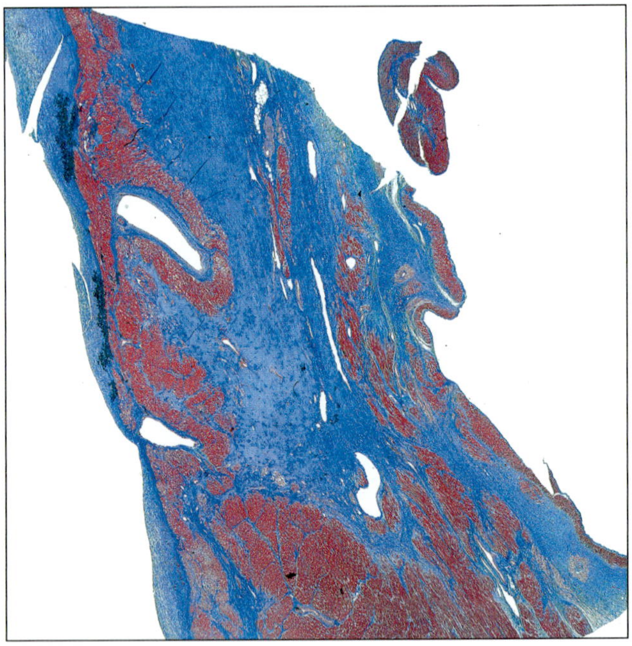

FIGURE 5.2: This Masson trichrome–stained section of the interventricular septum shows a dense scar, following myocardial infarction. Only a few bands of myocytes, primarily in the right subendocardium, remain.

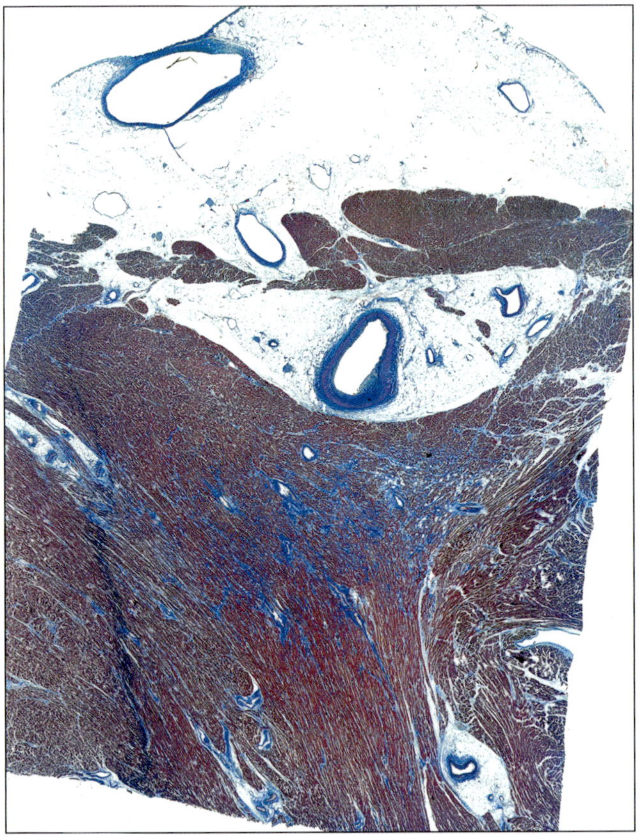

FIGURE 5.3: A focus of interstitial fibrosis, underlying a "bridged" left anterior descending coronary artery, is outlined by the trichrome stain.

principally affects the right ventricle and is characterized by irregular fibroadipose replacement of the myocardial wall (11). Involvement of the left ventricle is not uncommon in this condition as well, and, in such instances, fibrofatty tissue assumes a characteristic linear subepicardial arrangement (12). Restrictive cardiomyopathy features the presence of endomyocardial fibrosis (Figure 5.7A), in which trichrome staining is effective in highlighting the deposition of collagen (Figure 5.7B,C). Indirectly, trichrome stains may contribute to the recognition of cardiac amyloidosis and hemochromatosis as well, through their failure to label the increased interstitial material in those diseases.

Trichrome labeling is indispensable in studying the cardiac conduction system (Figures 5.10–5.14). The framework of the sinus node (Figure 5.10), and, to a lesser extent, that of the atrioventricular node (Figure 5.11), is composed of connective tissue that is abundantly interspersed in the middle of a network of specialized muscle fibers. Because of this close anatomic relationship between the two components, they are practically indistinguishable from one another in H&E stains. An increase in collagen is an element of normal aging of the sinus node; one may also see augmented reticulin fibers and surrounding fat (13). The bundle of His, as it penetrates the annulus fibrosus, is also best assessed with the trichrome method, as is the origin of the cardiac bundle branches. These are illustrated in Figures 5.12 through 5.14.

In, Figure 5.12, the sinus node is affected by lipomatous hypertrophy of the atrial septum. Although it is usually an incidental finding at autopsies, fatty infiltration may extend to the crista terminalis and replace the nodal fibers (14) or isolate the node from the surrounding myocardium. Such changes can result in arrhythmias or even sudden death (15).

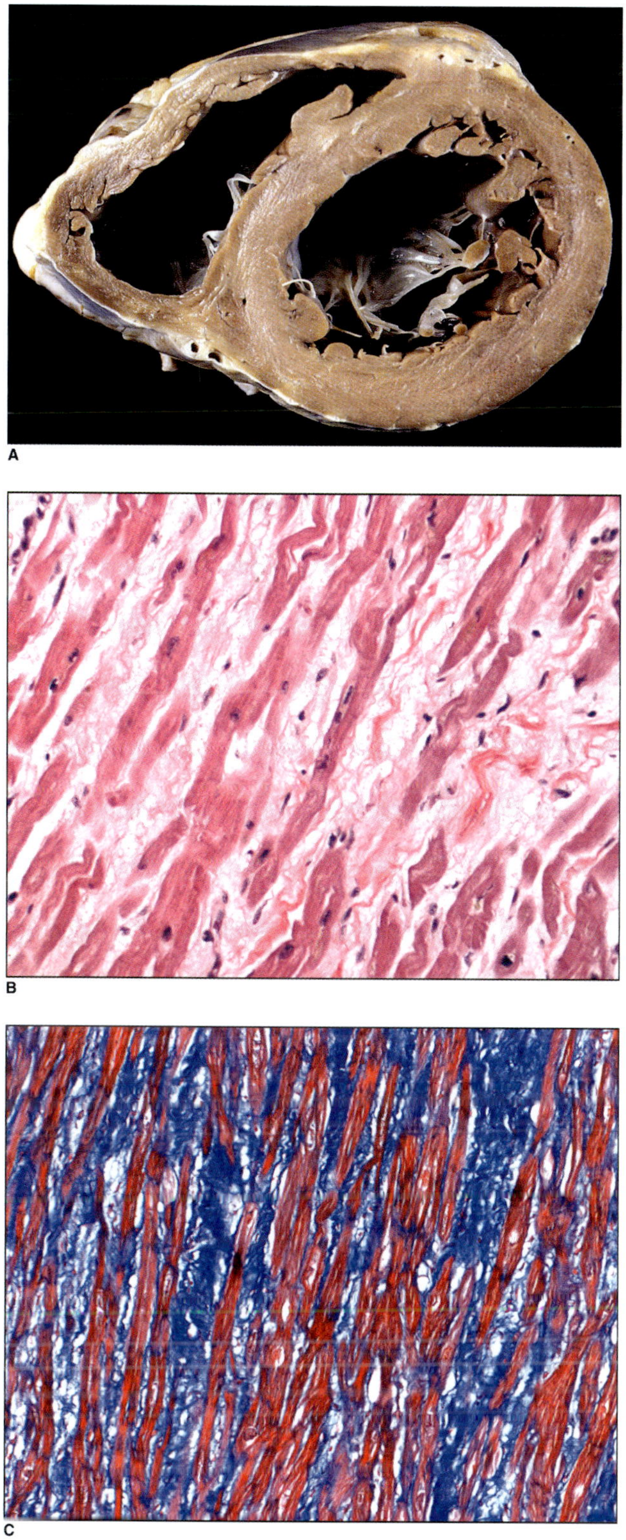

A

B

C

FIGURE 5.4: Idiopathic dilated cardiomyopathy. (A) This transverse section of the heart displays marked dilatation of the ventricles, particularly on the left. (B) Although not readily apparent on gross examination, microscopic sectioning showed interstitial fibrosis, as shown in this H&E section of the interventricular septum. (C) Unlike H&E, trichrome staining distinctly differentiates between myocardial fibers and fibrotic ones.

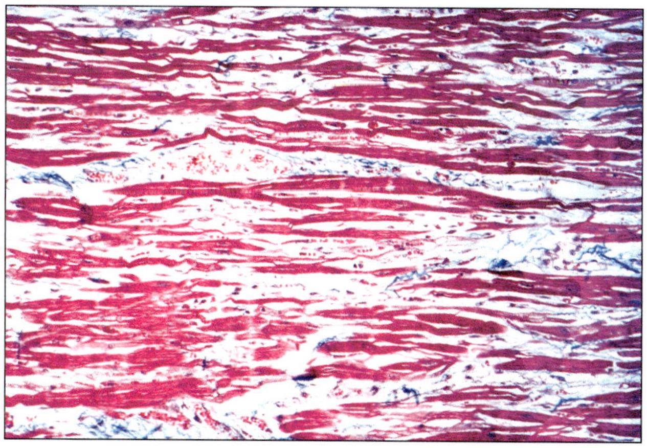

FIGURE 5.5: This illustration of a trichrome-stained section of a heart with adriamycin cardiomyopathy shows that, unlike idiopathic dilated cardiomyopathy, the myocardial fibers are thinned by myofibrillar loss and there is just slight increase, if any, of interstitial collagen.

Figure 5.13 demonstrates fibromuscular dysplasia, a developmental anomaly that also may involve large- and medium-sized arteries such as the carotid, renal, and sinus node arteries. Sudden death may again be associated with this condition.

Figure 5.14 shows a case of congenital heart block resulting from absence of the bundle of His. It demonstrates an empty space at the anatomical location normally occupied by that structure in the conduction system.

CONGO RED STAINING

Congo red was first introduced in 1884 as a dye for cotton, but it has subsequently become the stain of choice for demonstrating amyloid in tissue. In 1922, Benhold described the selective affinity of the dye for amyloid, which was labeled red (16). Romhanyi subsequently described apple-green birefringence of amyloid in tissue after staining with Congo red, when viewed with polarized light (17). There are a variety of ways by which the Congo red method can be performed; these include Highman's technique, the alkaline Congo red method, the high-pH Congo red procedure, and the Sirius red technique. Other histochemical procedures for the detection of amyloid include toluidine blue, methyl violet, crystal violet, and thioflavine-T.

Irrespective of the types of amyloidosis—with those being primary, secondary, senile, hereditary, or isolated-cardiac in nature—the proteinaceous deposits relating to this disease in the heart may involve small intracardiac blood vessels, valves, the atrioventricular interstitium, and the conduction system (18–20). Cardiac amyloidosis is usually recognized easily by gross examination of the heart because of the glistening, glassy appearance of the endocardium, which also contains small nodular deposits

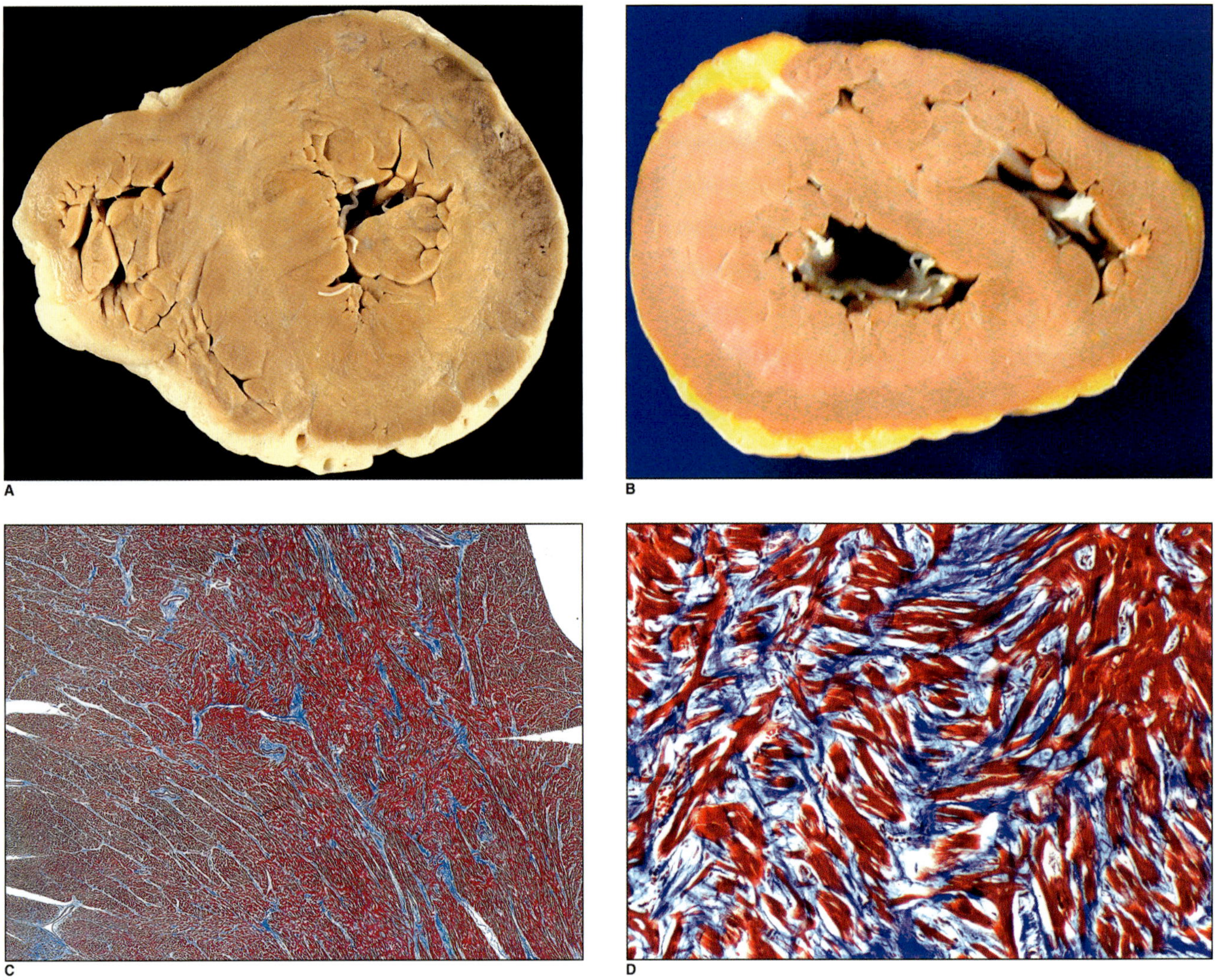

FIGURE 5.6: *Idiopathic hypertrophic cardiomyopathy is exemplified in these two cases: (A) thickening of the ventricular walls with asymmetric hypertrophy of the ventricular septum and small ventricular chambers and (B) uniformly hypertrophied ventricular walls with small left and right ventricular chambers. (C) Trichrome staining of the interventricular septum of the specimen shown in (B) shows disarray of the myocardial fibers. It is particularly pronounced in the mid-portion of the septum. (D) The disarray of cardiocytes and interstitial fibrosis is further illustrated in this higher magnification of the previous figure. The histologic appearance is characteristic of hypertrophic cardiomyopathy, restrictive type.*

(Figure 5.15A). It also is manifested by "rubbery" firmness of hypertrophic ventricular walls. In H&E sections of autopsy hearts or endomyocardial biopsies, the presence of interstitial deposits of homogeneous, hyaline eosinophilic material around cardiomyocytes is characteristic of amyloid (Figure 5.15B). However, definitive diagnosis requires confirmation with the stains listed above (Figure 5.15C,D).

VERHOEFF–VAN GIESON STAINING

The Verhoeff–Van Gieson (VVG) stain is another very useful stain in the evaluation of cardiovascular lesions. It is actually a combination of two methods for labeling connective tissue fibers. The Verhoeff stain is specific for elastic fibers, whereas the van Gieson procedure stains collagen.

The VVG technique is helpful in evaluating pathologic changes in the aorta, such as cystic medial degeneration. The condition is represented microscopically by fragmentation and loss of elastic fibers, with or without the deposition of glycoproteins in the arterial media (21). Accordingly, stains for stromal mucin—such as the colloidal iron or alcian blue (AB) method—may also be used diagnostically in this setting. Cystic medial degeneration may be seen in a variety of disorders including hypertension, coronary arterial atherosclerosis, congestive heart disease, bicuspid aortic valvular disease, and inherited or acquired connective tissue disease. The VVG stain can also delineate areas of previous aortic dissection.

An elastic stain may also contribute to evaluations of various aortitides, including syphilitic aortitis, giant cell

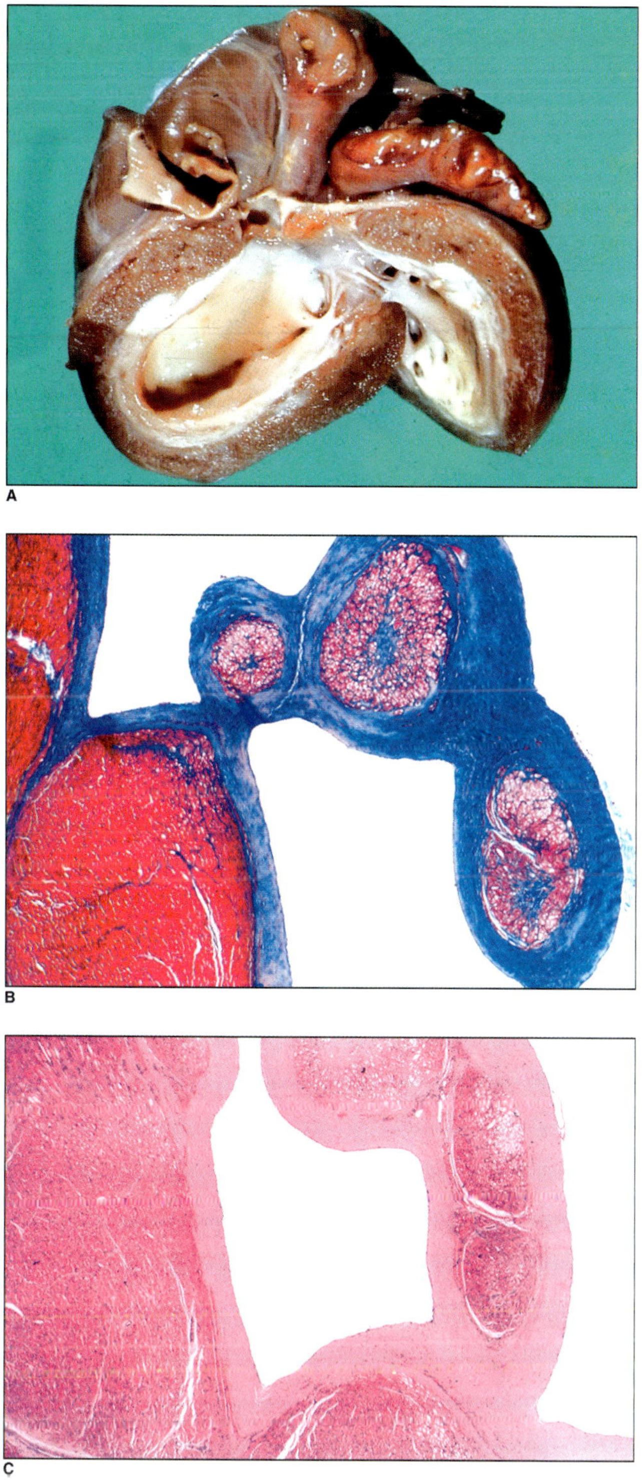

A

B

C

FIGURE 5.7: *(A) Diffuse thickening of the entire left ventricular endocardium is shown in this heart from a patient with endocardial fibroelastosis. (B) Trichrome staining distinctly outlines the endocardium from the myocardium; these two layers are not as clearly seen as with the routine H&E section (C).*

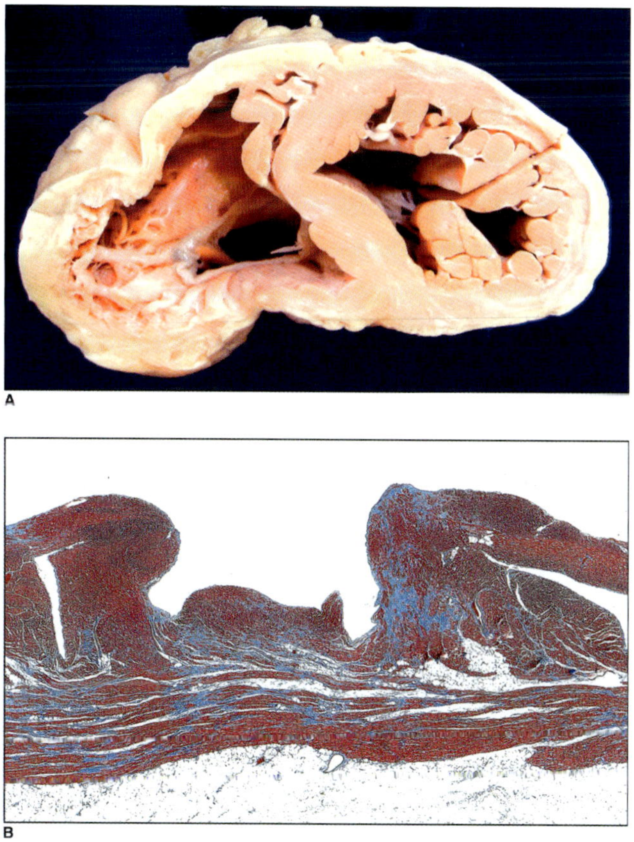

A

B

FIGURE 5.8: *(A) This gross photograph illustrates the features of arrhythmogenic right ventricular cardiomyopathy, namely, right ventricular dilatation and an abnormal right ventricular wall. Histologic examination of the right ventricle (B) demonstrates significant fibroadipose replacement of the myocardium.*

arteritis, Takayasu's aortitis, and aortitides associated with acquired collagen vascular diseases (22,23). In such cases, one may see perivascular lymphoplasmacytic infiltrates, giant cells, granulomas, medial cystic degeneration, and laminar medial necrosis. All of these lesions, as well as other types of vasculitis such as sarcoidal arteritis, are well shown with the VVG procedure (Figure 5.16) (24). An elastic stain may also be productive in characterizing Buerger's disease (thromboangiitis obliterans). In this disorder, small arteries show thrombosis and transmural neutrophilic inflammation. Subsequently, the areas of thrombosis are recanalized by small capillaries that are scattered throughout a luminal connective tissue plug. VVG stains show that the media and the internal elastic membrane are well preserved in Buerger's disease. The internal elastic membrane demonstrates marked undulation and splitting; those alterations are probably artifactual, as a consequence of tissue retraction (25).

H&E staining is typically sufficient to demonstrate the degree of luminal occlusion of coronary arteries by atherosclerosis. Calcifications are also well seen in conventionally

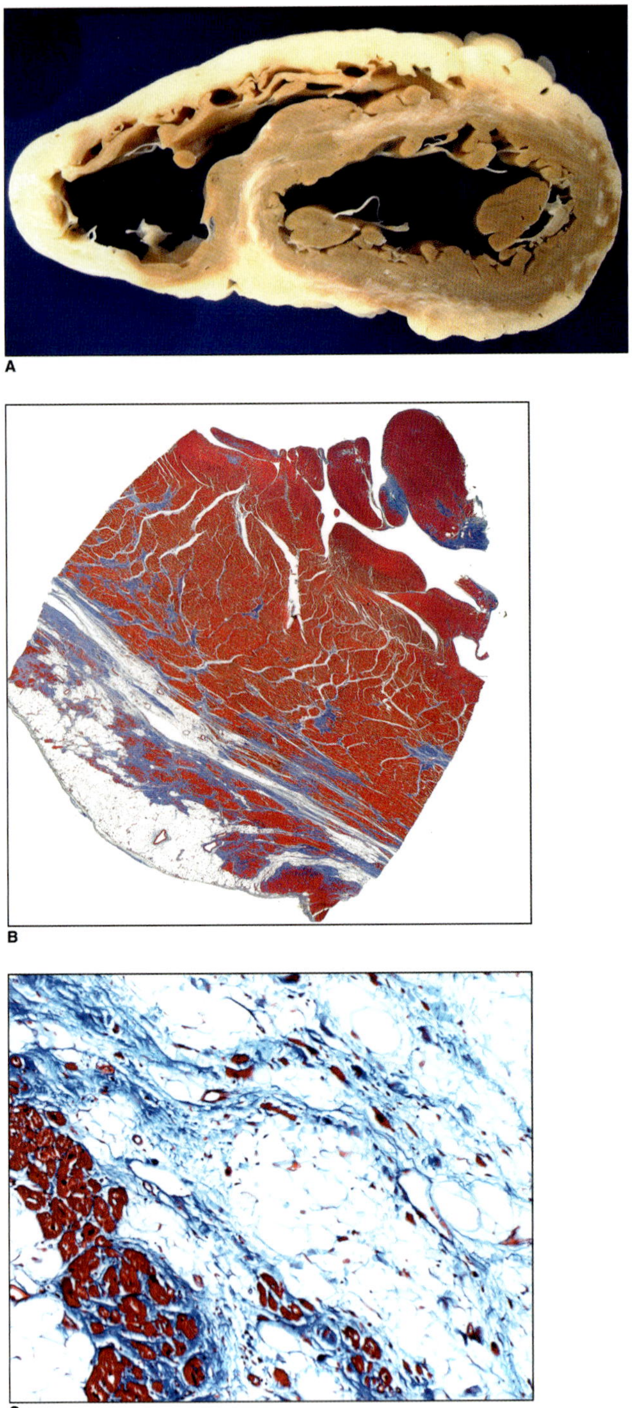

A

B

C

FIGURE 5.9: (A) Extensive left ventricular involvement by arrhythmogenic cardiomyopathy is seen here as well, including the interventricular septum. (B) The linear subepicardial pattern of this disease in the free left ventricular wall is demonstrated by the trichrome stain, illustrated in (C). It clearly shows the fibroadipose replacement of the myocardium and the remaining fragmented cardiocytes.

stained sections (Figure 5.17A–C). Nonetheless, the VVG stain does contribute to assessments of the coronary arteries in cardiac allografts. In transplanted hearts, the coronary vessels may develop apparent atherosclerosis, although this is actually felt to reflect the presence of chronic vascular rejection. One sees a concentric intimal proliferation that narrows the lumen of the transplanted coronary arteries throughout their length, in both epicardial and intramural locations (Figure 5.17D–G) (26–28).

ALCIAN BLUE STAINING

AB is a copper phthalocyanine dye; it specifically labels acid mucopolysaccharides. At a pH of 2.5, AB binds to carboxyl and sulfate-ester groups, but, at a pH of 1.0, it will only label sulfate-ester groups. This stain is particularly suitable for demonstrating alterations in the collagenous framework of the heart; in particular, it is used to study the cardiac valves in such conditions as mitral valve prolapse.

Myxomatous degeneration is the pathologic change that ultimately causes mitral valve prolapse. The latter disorder is the most common acquired abnormality of the cardiac valves, affecting up to 10% of the general population. Most patients are asymptomatic (29–31). Proliferation of fibroblasts in the spongiosa layer of the valve, with accumulation of mucopolysaccharides, leads to enlargement and thickening of the mitral leaflets. These can then prolapse if they fail to coapt (Figure 5.18). Mitral regurgitation and secondary structural changes in the valve are the events that result from this idiopathic condition. In selected cases, it may be associated with sudden death. Although the mechanism for this catastrophe is unknown, concurrent myxoid change in the sinus node and cardiac nerves has been advanced as a possible explanation (32).

AB staining is also used to demonstrate the valvular, vascular, and myocardial changes in hereditary mucopolysaccharidoses, such as Hurler's and Hunter's syndromes.

In these conditions, acid mucopolysaccharides accumulate in mesenchymal cells in several organ sites. Many of these elements are likely fibroblasts (33). The most prominent finding in the heart is diffuse thickening of the leaflets of all four valves, potentially causing valvular stenosis. In H&E sections, the affected leaflets contain a large number of clear cells, and the AB stain demonstrates that this vacuolization is caused by accumulation of mucopolysaccharides (34). The next most prominent finding is marked thickening of the coronary arterial intima, resulting from accrual of similar cells. They also may be present in the myocardial interstitium (35).

Other stains for acid mucopolysaccharide deposition include toluidine blue, PAS, colloidal iron, and azure A. These can be used with similar efficacy to that associated with AB.

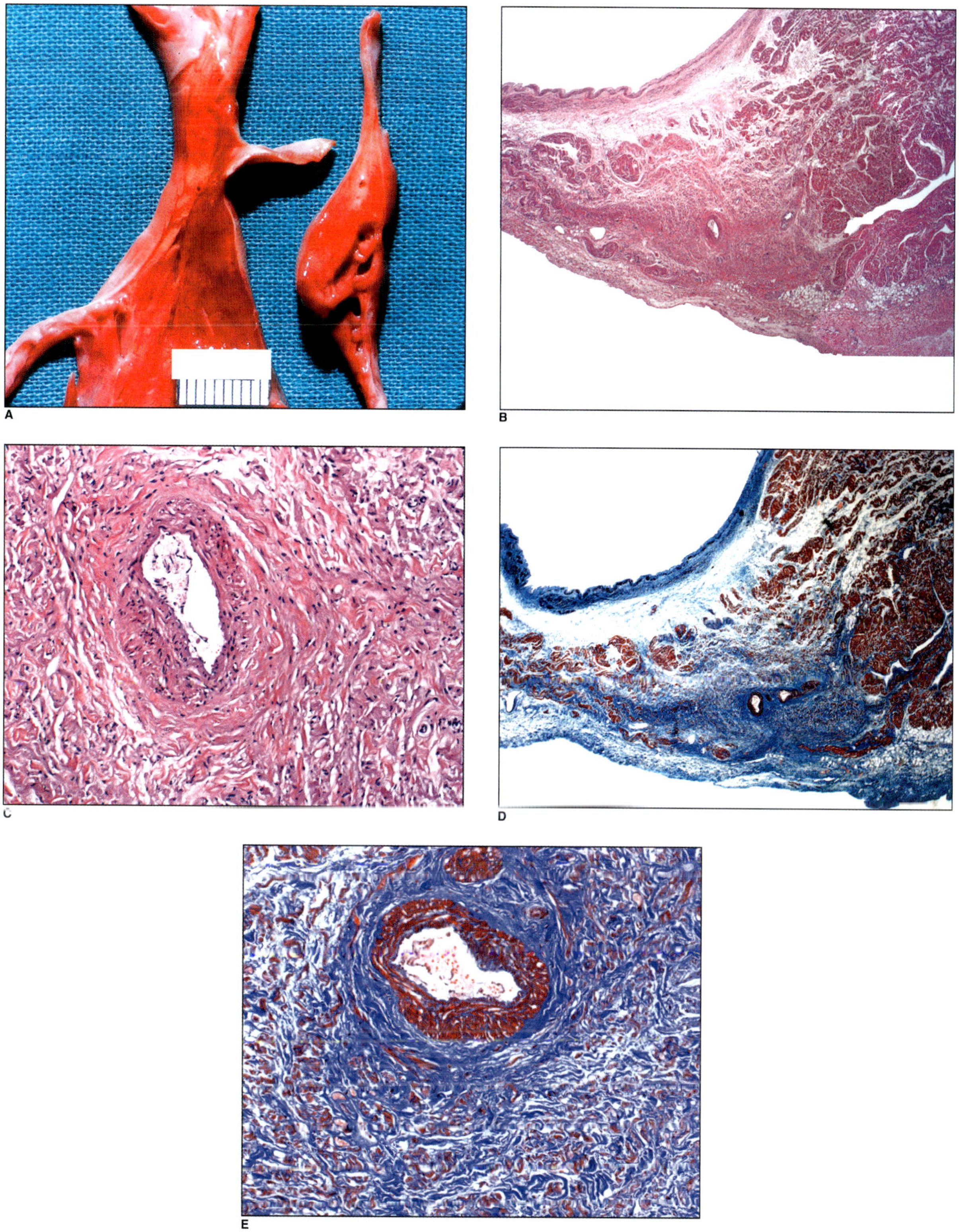

FIGURE 5.10: (A) Characteristic samples of tissue, one each from the sinus and atrioventricular nodes, that are required for histologic assessment of those structures. The sample on the right shows the location of the sinus node, with its centrally located artery, at the junction of the superior vena cava and the crista terminalis. The atrioventricular node is situated in the lowermost portion of the atrial septum, just above the septal leaflet of the tricuspid valve. (B) This H&E-stained section shows the sinus node as a dense aggregate of tissue, which, at higher magnification (C), is still difficult to identify. The nodal fibers merge with a collagenous framework around an artery. (D) Trichrome staining clearly delineates the sinus node from the adjacent myocardium, and its structure is also well demarcated at higher magnification (E).

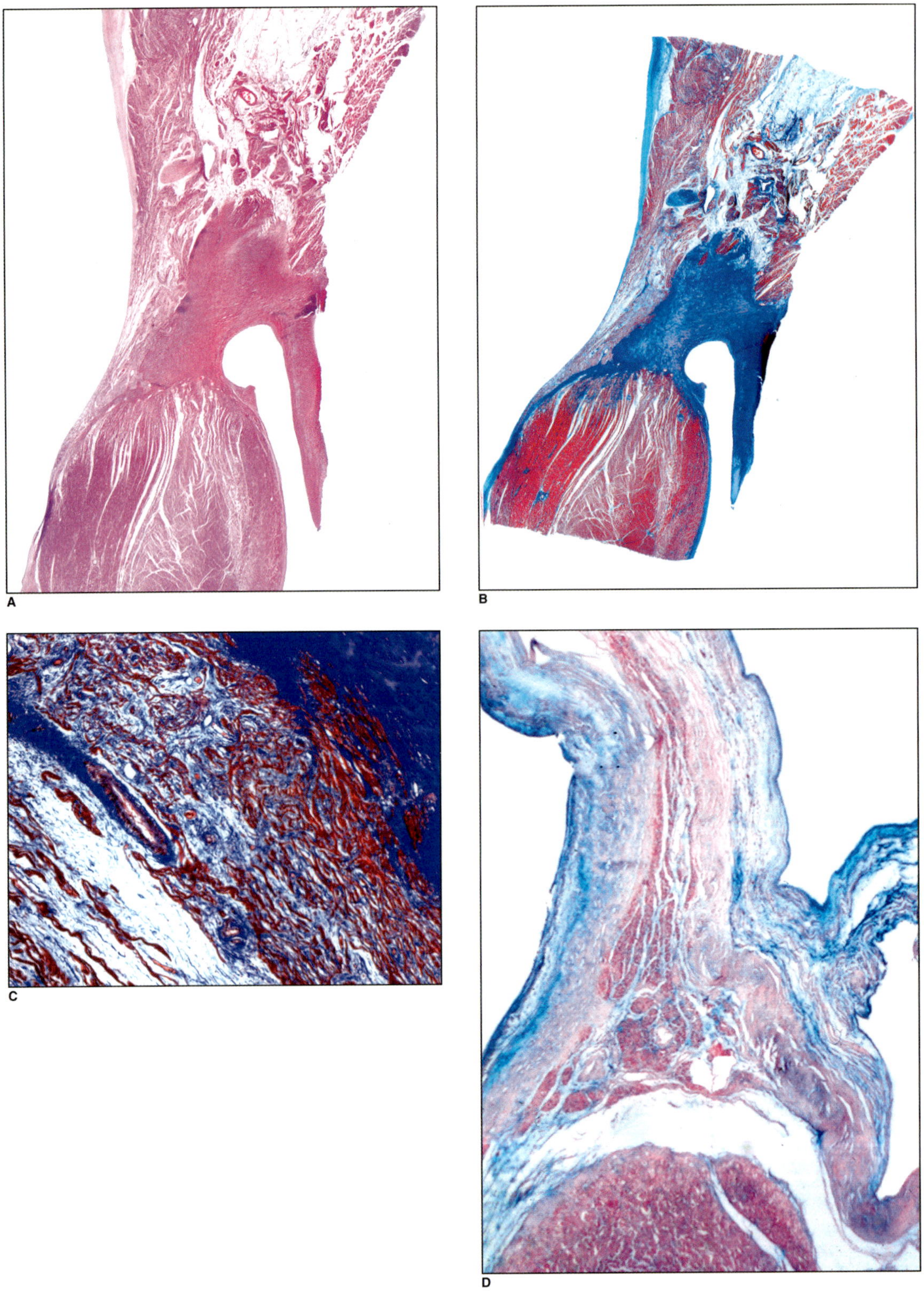

FIGURE 5.11: The advantage of trichrome staining in the microscopic evaluation of the atrioventricular node, bundle of His, and bundle branches of the conduction system is exemplified in this set of illustrations. (A) At this low magnification, the atrioventricular node, central fibrous body, lower portion of the atrial septum, and summit of the interventricular septum are not well delineated from each other by H&E staining. (B) This is the same view as that shown in (A) but the cited structures are better outlined with trichrome in this higher magnification image of the normal atrioventricular node. (C) One sees a plexiform arrangement of the specialized nodal fibers, supported by a collagenous framework, although not as dense as that of the sinus node. (D) The bundle of His is illustrated in this trichrome stain as it pierces the annulus fibrosus to reach the summit of the interventricular septum and give off branches.

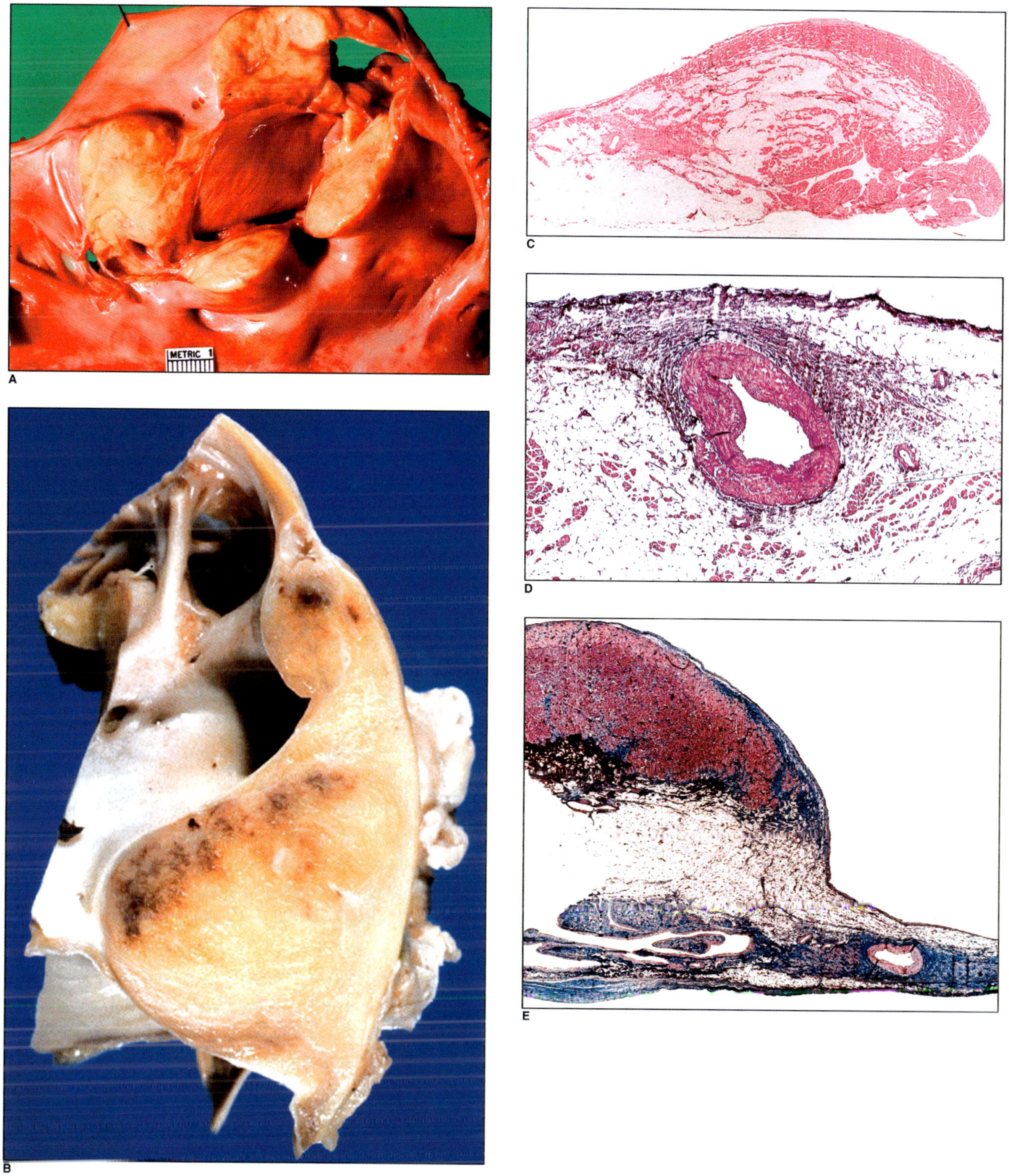

FIGURE 5.12: (A) Lipomatous hypertrophy of the atrial septum. (B) Another example of lipomatous hypertrophy of the atrial septum that has been sectioned to illustrate fatty infiltration of the crista terminalis, where the sinus node resides. A corresponding H&E section (C) is also shown of this structure. The pathology of the sinus node associated in lipomatous hypertrophy is well demonstrated with trichrome stains. In abnormal cases, one sees fatty replacement of the node (D) or isolation of that structure from the surrounding myocardium by adipose tissue (E).

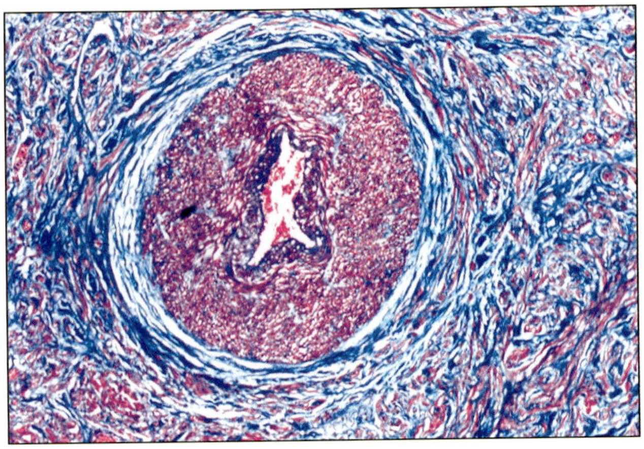

FIGURE 5.13: *Fibromuscular hyperplasia of the sinus node artery in a case of sudden death.*

PERIODIC ACID-SCHIFF WITH AND WITHOUT DIASTASE

The PAS stain, with or without diastase, is used to identify glycogen accumulation in the myocardium in a variety of genetic diseases. One of them, Pompe disease (glycogen storage disease type II), is an inherited condition in which there is a deficiency of acid alpha-glucosidase (36). Without this enzyme, lysosomes cannot hydrolyze glycogen and it accumulates in the cytoplasm. The infantile form of Pompe disease is the most severe, featuring cardiomegaly. In H&E stains, cardiac myocytes are obviously vacuolated (Figure 5.19).

Gaucher's disease results from a deficiency of beta-glucocerebrosidase, leading to accrual of glucocerebroside in macrophages. The left ventricle of the heart may

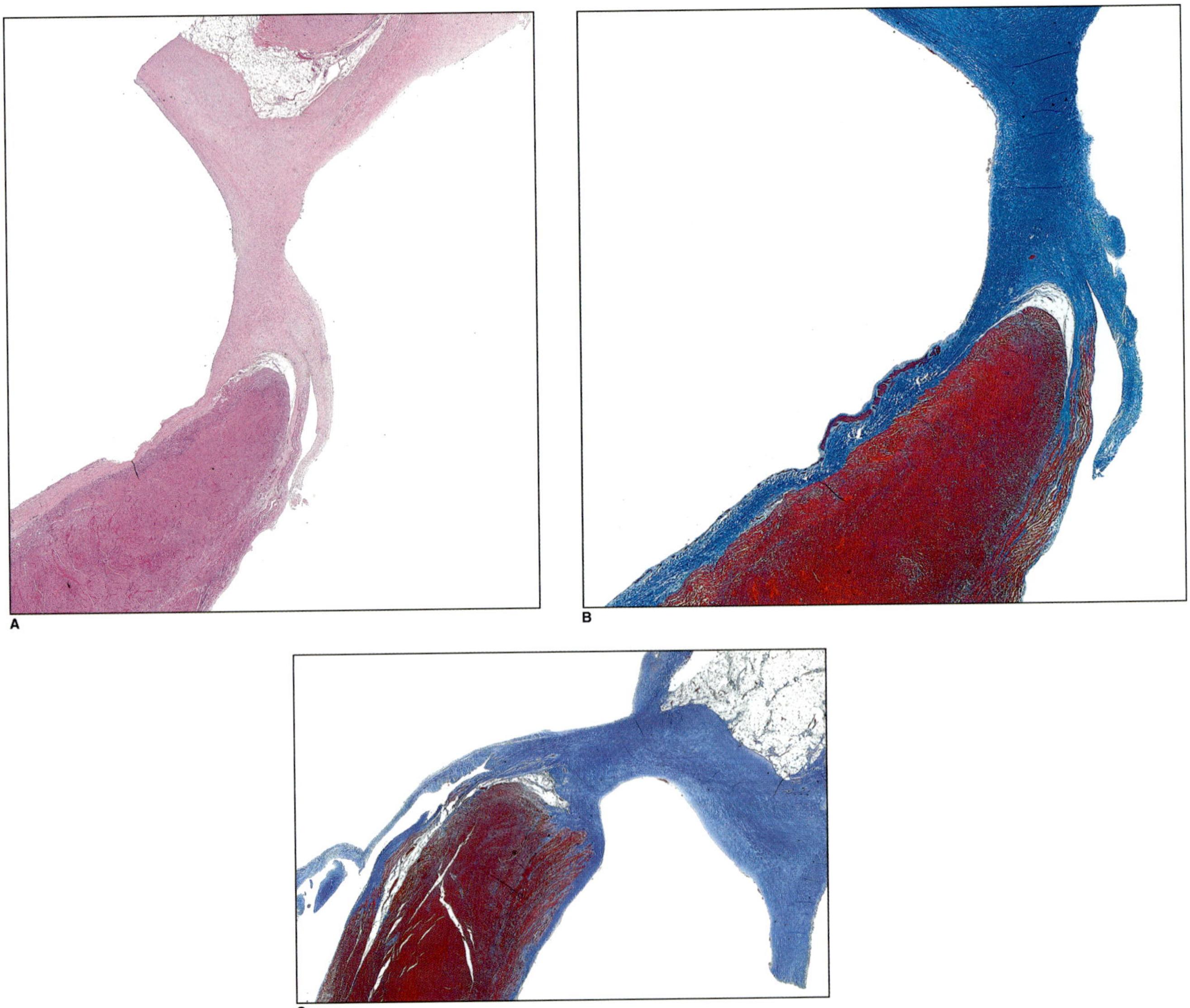

A

B

C

FIGURE 5.14: *This group of photographs shows an absence of the bundle of His in a case of congenital heart block. Although an empty space that normally harbors the bundle of His can be seen in an H&E-stained section (A), it is better demonstrated with trichrome staining immediately above the summit of the interventricular septum (B and C).*

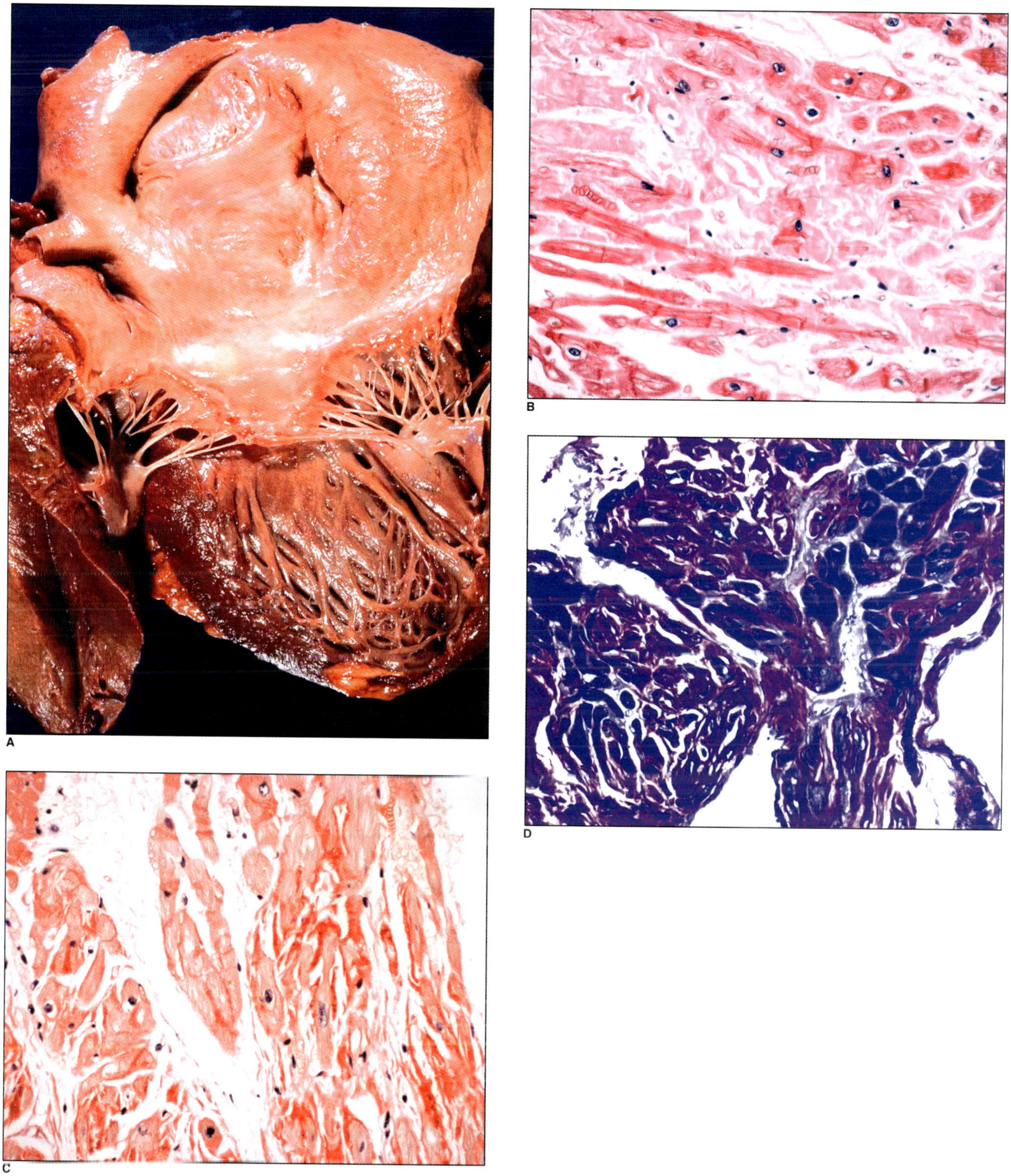

FIGURE 5.15: Cardiac amyloidosis. (A) The dilated left atrium of this heart shows translucent endocardium with finely nodular deposits, in a case of senile amyloidosis. (B) Endomyocardial biopsy demonstrating amyloidosis, which, in H&E sections, appears as interstitial hyalinized deposits. Their nature is clearly established by Congo red staining; it shows red amyloid embracing the myocardial fibers (C). The presence of amyloid is further confirmed by crystal violet staining (D).

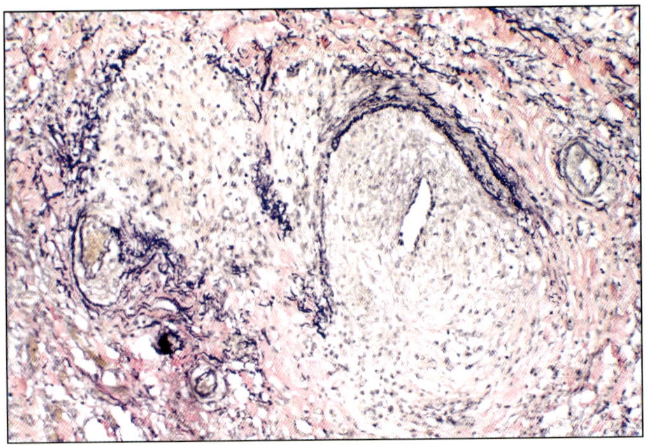

FIGURE 5.16: VVG staining shows marked fragmentation of the internal elastic membrane and fibrous intimal proliferation with luminal narrowing of the sinus node artery in a case of cardiac sarcoidosis.

become infiltrated by such macrophages, resulting in ventricular dysfunction (37). In H&E stains, the cells contain gray fibrillary cytoplasm that is sometimes likened to "crumpled tissue paper." They stain positively with the PAS method, even after diastase digestion, indicating that the cytoplasmic material they contain is not glycogen (38).

Fabry's disease is an X-linked recessive condition in which patients have a deficiency of alpha-galactosidase. One observes an accumulation of glycosphingolipids—especially globotriaosylceramide—in various organs including the heart (39,40). This results in left ventricular hypertrophy, and the affected myocardiocytes have a vacuolated appearance, with myofibrils and nuclei being displaced to the periphery of the cells (41). The storage material in the cells is water soluble, and therefore, histochemical stains for its detection need to be done on frozen or plastic-embedded tissue. Toluidine blue stains reveal metachromatic rounded structures in the vacuolated cardiocytes of Fabry's disease. In frozen-section preparations, the vacuoles again label with the PAS method, both before and after diastase digestion. The storage material also takes up Sudan black B (a stain for fat) (42) in frozen sections.

Admittedly, PAS is less than an optimal stain for delineation of this disease. As is true of other enzyme deficiencies, the most definitive diagnostic method is actual measurement of specific enzyme levels in fresh tissue.

Cardiac "rhabdomyoma" is actually a hamartoma that is composed of one or more circumscribed, yellow-tan myocardial nodules. It comprises large round or oval myocardiocytes with central nuclei and a rim of clear or amphophilic cytoplasm. Eosinophilic or amphophilic cytoplasmic strands often radiate outward from the nuclei and form clear vacuoles in what are known as "spider cells." They contain abundant glycogen that can be demonstrated with undigested PAS stains (43).

HISTOCHEMICAL STAINING FOR CALCIUM

H&E-stained sections can readily delineate several forms of calcium deposition in the heart, whether it is in atheromatous plaques, calcified valves, healed infarcts, or metastatic calcification as seen in hyperparathyroidism or uremia. Intracellular calcium in cardiomyocytes, however, may require special histochemical staining for detection. Applicable techniques principally include the von Kossa method and the alizarin red procedure. Mitochondrial calcium deposition is not uncommon in contraction band necrosis, primarily affecting actively contracting myocardium or, less frequently, the conduction system (Figure 5.20) (44).

IRON STAINS

Several stains are available to identify iron in tissue. Perl's Prussian blue reaction is commonly used for this purpose. In this method, ferric iron is unmasked with an acid ferrocyanide solution that converts it to a hydroxide [$Fe(OH)_3$]; this moiety in turn reacts with dilute potassium ferrocyanide to produce insoluble blue ferric ferrocyanide (1). Other stains that can be used to label iron include Lillie's procedure and the Hukill-Putt method.

Deposition of iron in the heart may result from hereditary hemochromatosis or hemosiderosis occurring after chronic blood transfusion. In both instances, hemosiderin accumulates in ventricular (and, to a lesser extent, atrial) myocytes but not in the cardiac interstitium (Figure 5.21) (45). The densest deposition of iron occurs in the subepicardial myocardium, with the subendocardium showing an intermediate amount and the middle third of the myocardium containing the least (46). Iron stains of endomyocardial biopsies are helpful in establishing the diagnosis of hemochromatosis, but they may not be equally beneficial in demonstrating early transfusion-related iron overload (47).

HISTOCHEMICAL STAINS FOR CARDIAC INFECTIONS

Approximately 15,000 patients per year develop infectious endocarditis in the United States (48). Broadly speaking, microbes that may affect the heart include various viruses, bacteria, mycobacteria, fungi, rickettsiae, spirochetes, helminths, and protozoa. Bacterial infections account for most cases of endocarditis and also those of "mycotic" aneurysms, as exemplified in, Figure 5.22. In some cases, H&E findings are nonspecific, showing only nondescript inflammation. In others examples of cardiac infection, typically caused by helminths and protozoa, morphological features can be diagnostic. Infectious disease of the heart is generally not recognized primarily through histologic examination of tissue specimens but rather from cultures of surgically procured samples or blood.

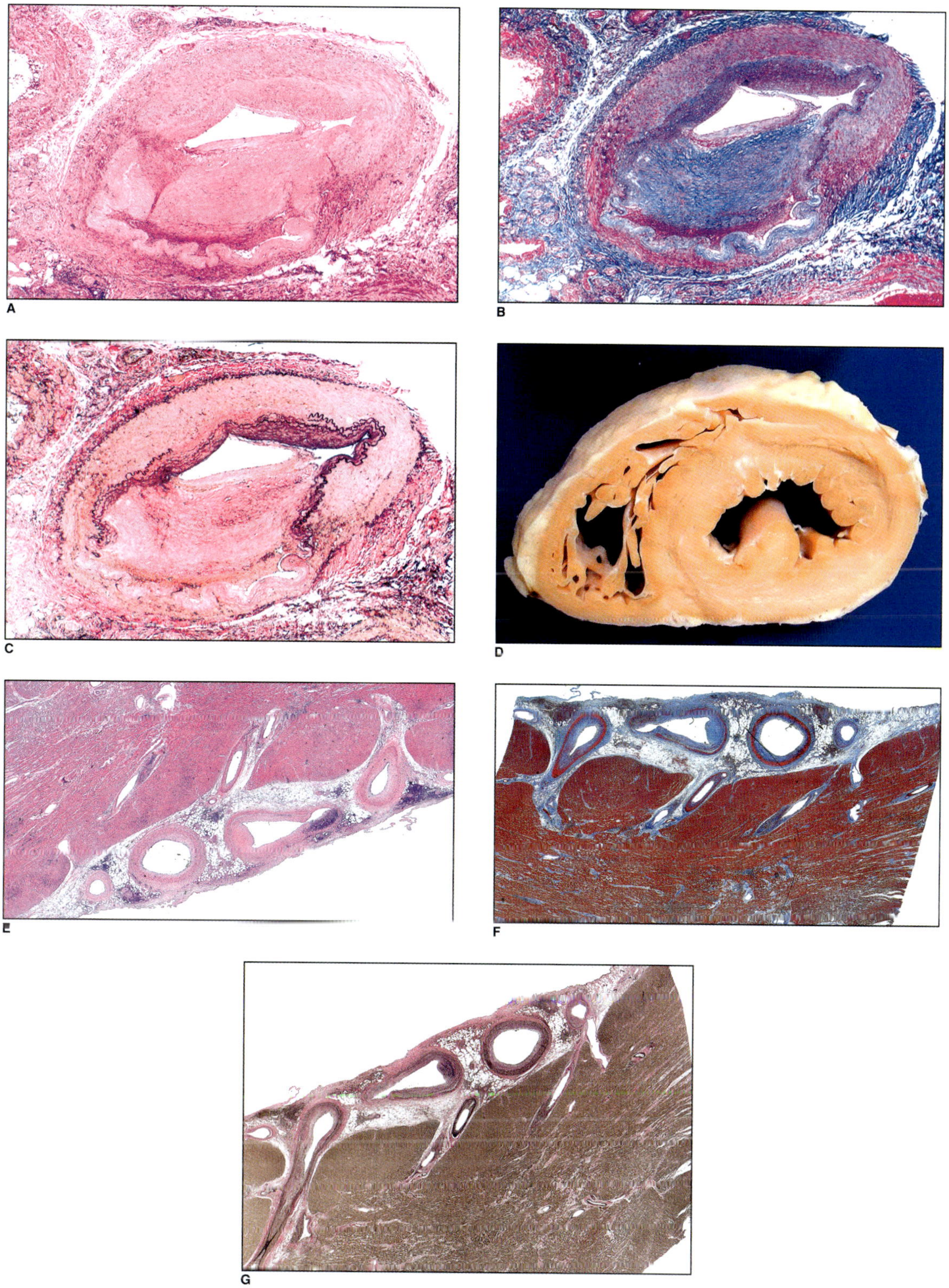

FIGURE 5.17: This group of photographs of the coronary arteries illustrates the value of histochemical stains that are used to evaluate vascular lesions. (A) Severe atherosclerosis of the left anterior descending coronary artery (LAD). H&E suffices to demonstrate the nature of the process. (B) Masson trichrome staining highlights the fibrous component of the atherosclerotic plaque. (C) VVG elastic staining outlines the layers of the vessel, clearly showing that atherosclerosis is centered in the intima. (D) This is a transverse section of the ventricles in chronic allograft rejection. Patchy, minute scars are present throughout the myocardium. (E) In addition to foci of lymphoid infiltrates in the myocardium and coronary arteries, this H&E section illustrates arteriopathy that is further highlighted by trichrome staining (F). The elastic stain demonstrates concentric intimal proliferation in the epicardial and intramyocardial arteries (G).

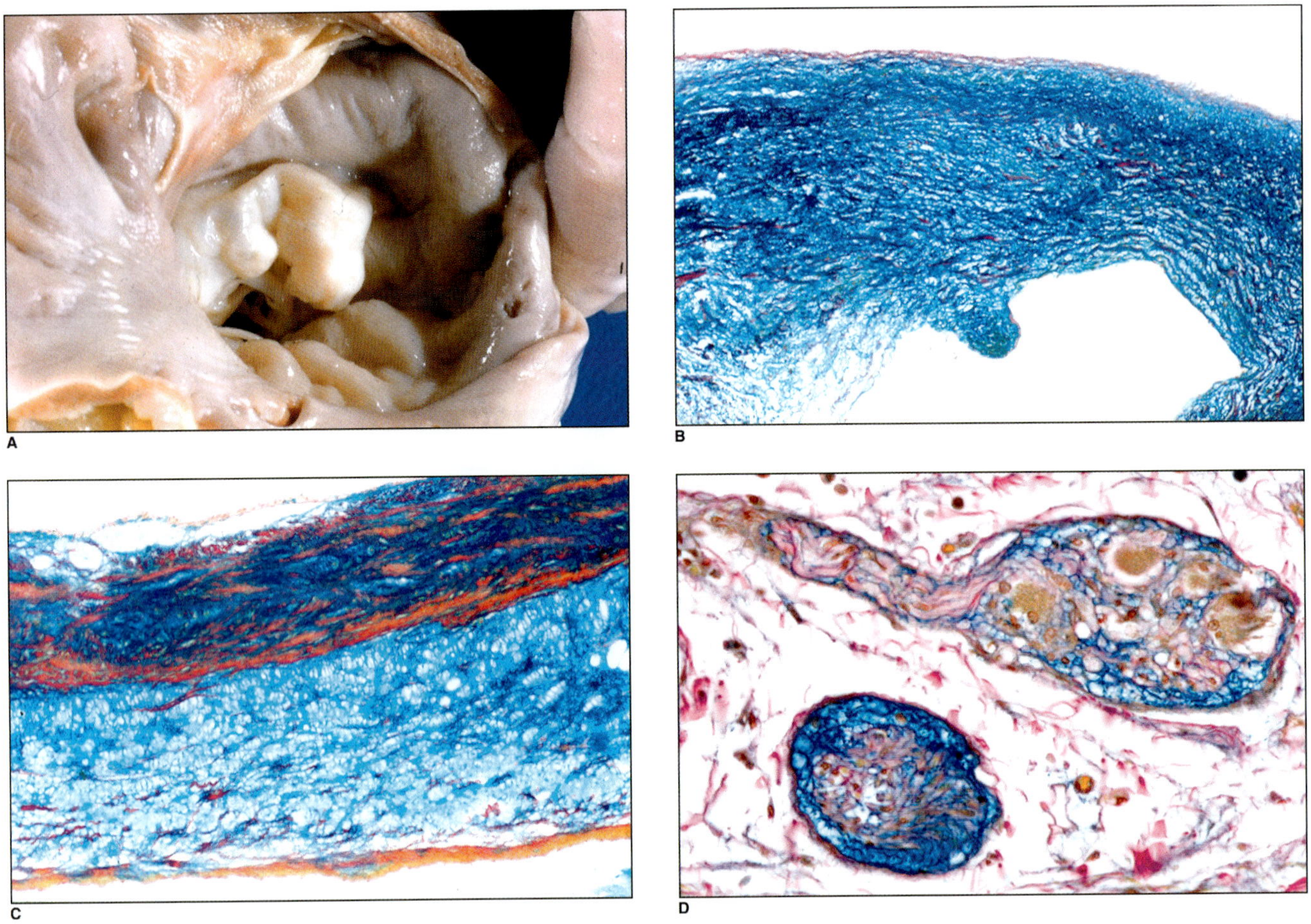

FIGURE 5.18: (A) Mitral valve prolapse, manifested in this photograph by herniation of the posterior valve leaflet into the left atrial cavity. (B) Histologic demonstration of acid mucopolysaccharide accumulation is shown by this AB stain. It demonstrates severe myxoid degeneration of the valve leaflet, erasing the normal architecture of the valve (C). AB staining also demonstrates accumulation of mucopolysaccharide in other cardiac structures, such as a ganglion and nerve trunk (D).

Bacterial Infections

The presence of structurally abnormal heart valves is the most common factor that predisposes to infectious endocarditis. A few virulent bacteria are capable of infecting morphologically normal valves, but, in most instances, prior endocardial damage is necessary for pathogenic bacteria to gain a foothold. The organisms may be identified histologically in endomyocardial biopsies, but, more often, this eventuality requires a greater volume of tissue such as that represented by resected infected valves or autopsy specimens. Staphylococci and streptococci account for 75% of endocarditis cases; enterococci are the pathogens in most of the remaining examples (48). The most important histochemical stains for detection of such organisms are the tissue Gram method, the Gram-Twort procedure, and the Churukian-Schenk modification of the Gram stain. All three methods label gram-positive organisms blue black, whereas gram-negative organisms are pink red (1). Occasional patients may also develop bacterial myocarditis, for which the most common causative organism is *Staphylococcus aureus* (49).

Mycobacterial Infections

Mycobacterial infections may manifest themselves in the heart with the formation of caseating granulomas. These lesions often contain red, rod-shaped organisms by Ziehl-Neelsen (acid-fast) staining. The Fite-Faraco (FF) method can also be used to detect "atypical" mycobacteria. The acid-alcoholic decolorizing agent in the latter procedure is not as strong as that used in the Ziehl-Neelsen technique, and the FF stain better visualizes *Mycobacterium leprae* and other nontuberculous mycobacteria.

Myocardial and pericardial tuberculosis is extremely rare in the United States at the present time. Tuberculosis of the myocardium has two forms—miliary and nodular. Miliary lesions predominantly manifest noncaseating granulomas, whereas nodular lesions also show caseation.

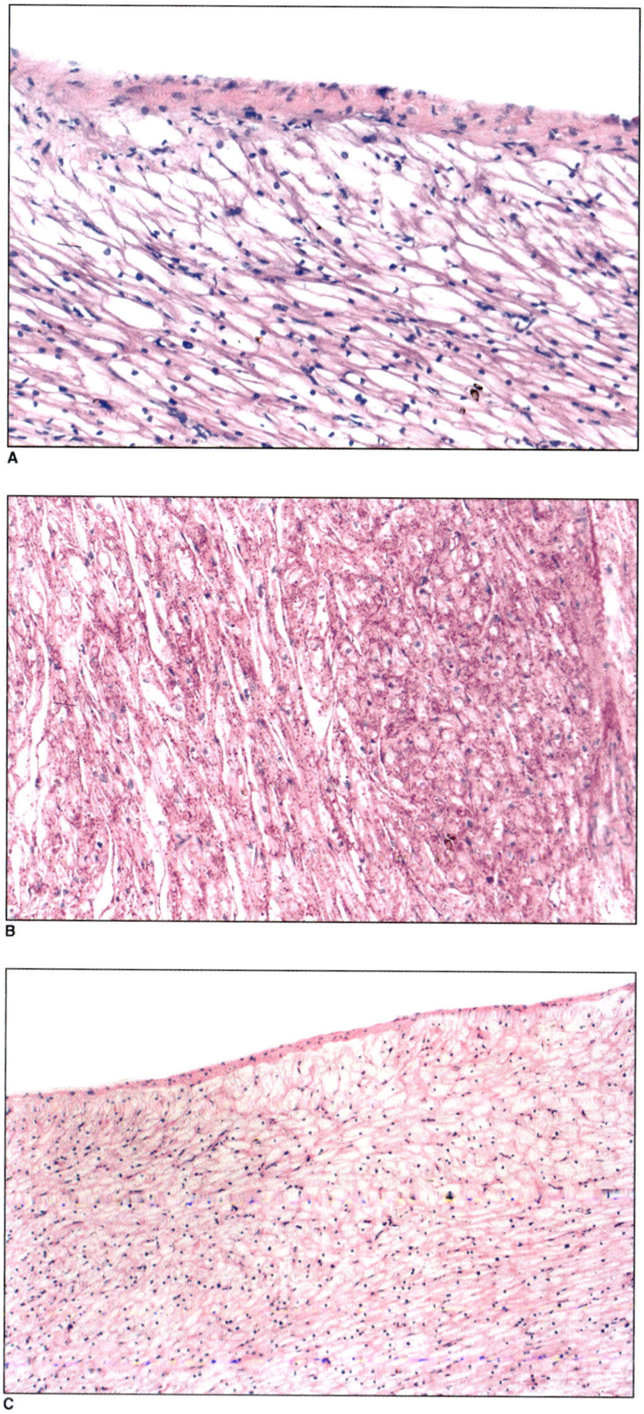

FIGURE 5.19: Glycogen storage disease. (A) H&E staining showing vacuolization of the myocardial fibers. (B) Glycogen accumulation is demonstrated with the PAS stain and further confirmed by abrogation of labeling after diastase digestion (C).

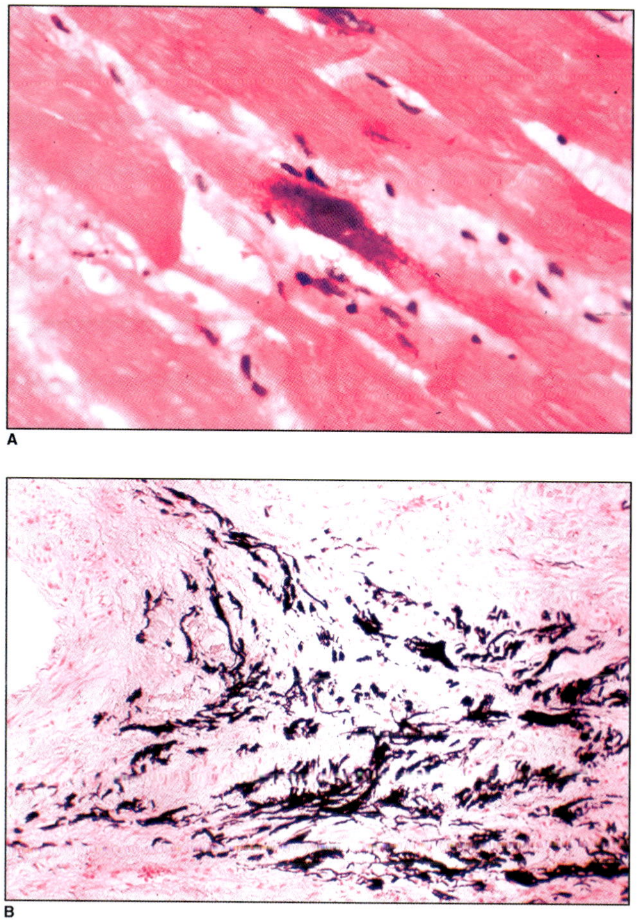

FIGURE 5.20: (A) H&E stain showing calcium deposition in a myocyte associated with contraction band necrosis. (B) von Kossa staining demonstrates multiple deposits of calcium in sinus node fibers.

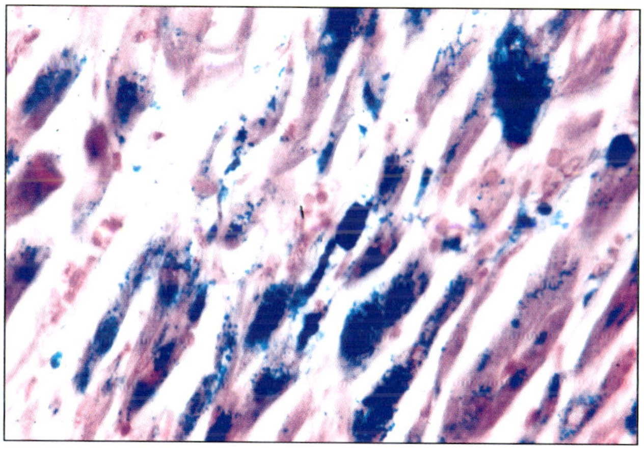

FIGURE 5.21: Hemochromatosis. Hemosiderin deposition in cardiocytes is demonstrated by the Prussian blue stain.

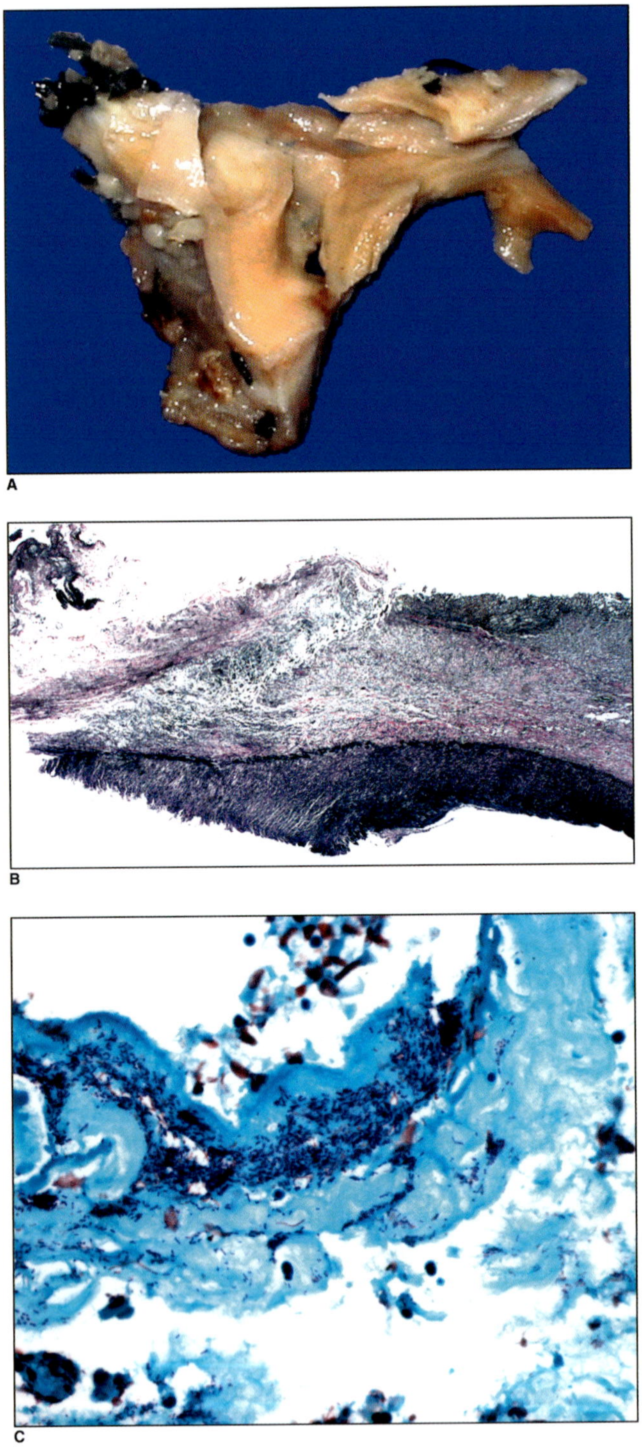

FIGURE 5.22: (A) "Mycotic" aneurysm at the anastomotic site of the hepatic artery, following liver transplantation. (B) Elastic staining shows perforation of the artery. (C) Tissue Gram staining demonstrates gram-positive bacteria.

Acid-fast organisms are typically difficult to find microscopically in myocardial granulomas (50), whereas this is not necessarily true in the lesions of mycobacterial pericarditis (51).

Myopericardial tuberculosis has occasionally been associated with instances of sudden death (52,53). In rare instances, atypical mycobacteria also can cause endocarditis of prosthetic or native cardiac valves (54).

Fungal Infections

The most common histochemical methods for visualization of fungi in tissue are the Grocott methenamine silver (GMS) stain and the PAS procedure. In the GMS technique, the walls of fungal organisms are black and the tissue background is light green. We believe that the latter method is the superior one because it will label fungi whether they are viable or not; in contrast, the PAS stain may fail to recognize devitalized organisms. In positive PAS stains, fungi are labeled magenta.

Candida spp. are the most common cause of fungal endocarditis and myocarditis (48,55). *Candida* infection is associated with intravenous drug abuse or immunosuppression, or both. *Aspergillus* is another relatively common cause of fungal myocarditis, usually in immunocompromised individuals, and *Cryptococcus neoformans* also can cause myocarditis. The latter organism is well visualized with the GMS stain, but its outer capsule will also label with the mucicarmine method; the capsule stains pink red against a yellow background (55).

Protozoa

Toxoplasmosis is caused by *Toxoplasma gondii*; it may affect the heart, particularly in immunosuppressed patients. The organisms of this disease have three forms—tachyzoites, bradyzoites, and oocysts. They may be seen on H&E- or Giemsa-stained sections. Another protozoan organism that can infect the heart is *Trypanosoma cruzi*; it does so in the form of trypomastigotes, which become amastigotes once they have entered the myocardiocytes (56).

REFERENCES

1. Cunningham KS, Veinot JP, Butany J: An approach to endomyocardial biopsy interpretation. *J Clin Pathol* 2006; 59: 121–129.
2. Winters GL, Costanzo-Nordin MR: Pathological findings in 2300 consecutive endomyocardial biopsies. *Mod Pathol* 1991; 4: 441–448.
3. Ardhali H, Qasim A, Cappola T, Howard D, Hruban R, Hare JM, Baughman KL, Kasper EK: Endomyocardial biopsy plays a role in diagnosing patients with unexplained cardiomyopathy. *Am Heart J* 2004; 147: 919–923.
4. Anderson KR, Sutton MG, Lie JT: Histopathological types of cardiac fibrosis in myocardial disease. *J Pathol* 1979; 128: 79–85.
5. Morales AR, Romanelli R, Tate LG, Boucek RJ, De Marchena E: Intramural left anterior descending coronary

artery: significance of the depth of the muscular tunnel. *Hum Pathol* 1993; 24: 693–701.
6. Kumar V, Abbas AK, Fausto N: Robbins and Cotran Pathologic Basis of Disease (7th Ed.), W.B. Saunders, Philadelphia, 2005.
7. Anonymous: Report of the WHO/ISFC Task Force on the definition and classification of cardiomyopathics. Circulation 1996; 93: 841–842.
8. Crossland DS, Edmonds K, Rassl D, et al.: Histology of the explanted hearts of children transplanted for dilated cardiomyopathy. *Pediatr Transplant* 2008; 12: 85–90.
9. Gunja-Smith Z, Morales AR, Romanelli R, Woessner JF: Remodeling of human myocardial collagen in idiopathic dilated cardiomyopathy. *Am J Pathol* 1996; 148: 1639–1648.
10. Davies MJ, McKenna WJ: Hypertrophic cardiomyopathy: pathology and pathogenesis. *Histopathology* 1995; 26: 493–500.
11. Thiene G, Nava A, Corrado D, et al.: Right ventricular cardiomyopathy and sudden death in young people. *N Engl J Med* 1988; 318: 129–133.
12. Fletcher A, Ho SY, McCarthy KP, et al.: Spectrum of pathological changes in both ventricles of patients dying suddenly with arrhythmogenic right ventricular dysplasia. Relation of changes to age. *Histopathology* 2006; 48: 445–452.
13. Lev M: Aging changes in the human sinoatrial node. *J Gerontol* 1954; 9: 1–9.
14. Balsaver AM, Morales AR, Whitehouse FW: Fat infiltration of myocardium as a cause of cardiac conduction defect. *Am J Cardiol* 1967; 19: 261–265.
15. Shirani WC Jr.: Clinical, electrocardiographic and morphologic features of massive fatty deposits ("lipomatous hypertrophy") in the atrial septum. *J Am Coll Cardiol* 1993; 22: 226–238.
16. Benhold H. Eine specifische amyloid-färbrung mit kangorot. *München Med Wschr* 1922; 33: 1537–1541.
17. Romhanyi G: Selective differentiation between amyloid and connective tissue structures based on collagen specific topo-optical staining reaction with Congo red. *Virchows Arch* 1971; 354: 209–222.
18. Falk RH: Diagnosis and management of the cardiac amyloidoses. *Circulation* 2005; 112: 2047–2060.
19. Shah KB, Inoue Y, Mehra MR: Amyloidosis and the heart: a comprehensive review. *Arch Intern Med* 2006; 166: 1805–1813.
20. Hassan W, Al-Sergani H, Mourad W, Tabbaa R: Amyloid heart disease: new frontiers and insights in pathophysiology, diagnosis and management. *Tex Heart Inst J* 2005; 32: 178–184.
21. Homme JL, Aubry M, Edwards WD, Bagniewski SM, Pankratz VS, Kral CA, Tazellar HD: Surgical pathology of the ascending aorta: a clinicopathologic study of 513 cases. *Am J Surg Pathol* 2006; 30: 1159–1168.
22. Miller DV, Isotalo PA, Weyand CM, Edwards WD, Aubry M, Tazellar HD: Surgical pathology of noninfectious ascending aortitis: a study of 45 cases with emphasis on an isolated variant. *Am J Surg Pathol* 2006; 30: 1150–1158.
23. Tavora F, Burke A: Review of isolated aortitis: differential diagnosis, including syphilitic, Takayasu's and giant cell aortitis. *Pathology* 2006; 38: 302–308.
24. Morales AR, Davis J, Fine G, Levy S: Sarcoidosis and the heart. *Pathol Annu* 1974; 174: 139–155.
25. Kim E, Cho B, Lee TS, Kim S, Seo J: Morphologic change of the internal elastic lamina in Buerger's disease. *J Korean Med Sci* 2000; 15: 44–48.
26. Winters GL: The pathology of allograft rejection. *Arch Pathol Lab Med* 1991; 115: 266–272.
27. Billingham ME: The pathologic changes in long-term heart and lung transplant survivors. *J Heart Lung Transplant* 1992; 11: S252–S257.
28. Billingham ME: Cardiac transplant atherosclerosis. *Transplant Proc* 1987; 19: 19–25.
29. Davies MJ, Moore BP, Braimbridge MV: The floppy mitral valve. Study of incidence, pathology and complications in surgical, necropsy and forensic material. *Br Heart J* 1978; 40: 468–481.
30. Gravanis MB, Campbell WG: The syndrome of prolapse of the mitral valve: an etiologic and pathogenic enigma. *Arch Pathol Lab Med* 1982; 106: 369–374.
31. Read RC, Thal AP, Wendt VE: Symptomatic valvular myxomatous transformation (the floppy valve syndrome). A possible forme fruste of the Marfan syndrome. *Circulation* 1964; 32: 897–910.
32. Morales AR, Romanelli R, Boucek, et al.: Myxoid heart disease. An assessment of extravalvular cardiac pathology in severe mitral valve prolapse. *Hum Pathol* 1992; 23: 129–137.
33. Wolfe HJ, Blennerhasset JB, Young GF, Cohen RB: Hurler's syndrome a histochemical study: new techniques for localization of very water-soluble acid mucopolysaccharides. *Am J Pathol* 1964; 45: 1007–1017.
34. Lagunoff D, Gritzka TL: The site of mucopolysaccharides accumulation in Hurler's syndrome: an electron microscopic and histochemical study. *Lab Invest* 1966; 15: 1578–1588.
35. Renteria VG, Ferrans VJ, Roberts WC: The heart in Hurler syndrome: gross, histologic and ultrastructural observations in five necropsy cases. *Am J Cardiol* 1976; 38: 487–500.
36. Kishnani P, Steiner RD, Bali D, et al.: Pompe disease diagnosis and management guideline. *Genet Med* 2006; 8: 267–288.
37. Smith RRL, Grover HM, Sack GH, et al.: Unusual cardiac, renal and pulmonary involvement in Gaucher's disease: interstitial glucocerebroside accumulation, pulmonary hypertension and fatal bone marrow embolization. *Am J Med* 1978; 65: 352–360.
38. Edwards WD, Hurdey HP, Partin JR: Cardiac involvement by Gaucher's disease documented by right ventricular endomyocardial biopsy. *Am J Cardiol* 1983; 52: 654.
39. Nakao S, Takenaka T, Maeda M, et al.: An atypical variant of Fabry's disease in men with left ventricular hypertrophy. *N Engl J Med* 1995; 333: 288–293.
40. Broadbent JC, Edwards WD, Gordon H, et al.: Fabry cardiomyopathy in the female confirmed by endomyocardial biopsy. *Mayo Clin Proc* 1981; 56: 623–628.
41. Owens CL, Russell SD, Halushka MK: Histologic and electron microscopy findings in myocardium of treated Fabry disease. *Hum Path* 2006; 37: 764–768.
42. Ferrans VJ, Hibbs RG, Burda CD: The heart in Fabry's disease: a histochemical and electron microscopic study. *Am J Cardiol* 1969; 24: 95–110.
43. Isaacs H: Fetal and neonatal cardiac tumors. *Pediatr Cardiol* 2004; 25: 252–273.

44. Morales AR: Cardiomyopathies: congenital and acquired: Chapter 10. In "The Heart," International Academy of Pathology Monograph #15 (Edwards JE, Lev M, MR, Abell Eds), Williams & Wilkins, Co., Baltimore, 1974; pp. 211–231.
45. Olson LJ, Edwards WD, McCall JT, Ilstrup DM, Gersh BJ: Cardiac iron deposition in idiopathic hemochromatosis: histologic and analytic assessment of 14 hearts from autopsy. *J Am Coll Cardiol* 1987; 10: 1239–1243.
46. Buja LM, Roberts WC: Iron in the heart: etiology and clinical significance. *Am J Med* 1971; 51: 209–221.
47. Olson LJ, Edwards WD, Holmes DR, Miller FA, Norodstrom LA, Baldus WP: Endomyocardial biopsy in hemochromatosis: clinicopathologic correlates in six cases. *J Am Coll Cardiol* 1989; 13: 116–120.
48. Bashore TM, Cabell C, Fowler V: Update on infective endocarditis. *Curr Probl Cardiol* 2006; 31: 274–352.
49. Wasi F, Shuter J: Primary bacterial infection of the myocardium. *Front Biosci* 2003; 8: S228–S231.
50. Rose AG: Cardiac tuberculosis: a study of 19 patients. *Arch Pathol Lab Med* 1987; 111: 422–426.
51. Cherian G, Uthaman B, Salama A, et al.: Tuberculous pericardial effusion: features, tamponade, and computed tomography. *Angiology* 2004; 55: 431–440.
52. Dada MA, Lazarus NG, Kharsany AB, et al.: Sudden death cause by myocardial tuberculosis: case report and review of the literature. *Am J Forensic Med Pathol* 2000; 21: 385–388.
53. Biedrzycki OJ, Baithun SI: TB-related sudden death (TBRSD) due to myocarditis complicating miliary TB: a case report and review of the literature. *Am J Forensic Med Pathol* 2006; 27: 335–336.
54. Olalla J, Pombo M, Aguado JM, et al.: *Mycobacterium fortuitum* complex endocarditis—case report and literature review. *Clin Microbiol Infect* 2002; 8: 125–129.
55. Nosanchuk JD: Fungal myocarditis. *Front Biosci* 2002; 7: d1423–d1438.
56. Kirchhoff LV, Weiss LM, Wittner M, et al.: Parasitic diseases of the heart. *Front Biosci* 2004; 9: 706–723.

CHAPTER 6

Diagnostic Histochemistry in Gastrointestinal Disorders

Patrick L. Fitzgibbons, M.D., and

Jay M. Packer, M.D.

Disorders of the gastrointestinal (GI) tract tend to involve the mucosa and submucosa more than the muscularis or subserosa. The mucosa lining the gut serves as a large interface with the "external" environment, and many disorders are caused by elements within the gut lumen such as microorganisms or toxins. While organisms such as *Mycobacterium tuberculosis* are considered always pathogenic, many others represent part of the normal gut flora and are associated with disease only under certain circumstances, for example, after mucosal damage or in patients with impaired immune surveillance. While infections are by far the most common cause of GI tract disease worldwide, important ailments are caused by structural damage to the mucosa from ischemia or toxic mucosal injury.

Neoplasms of the gut generally do not show the same degree of morphologic heterogeneity as is seen in some other organ systems, but tumors occur at all levels of the GI tract and can be associated with a variety of clinical presentations; in some cases, histochemical analysis can represent a useful tool in the differential diagnosis of these tumors.

INFECTIOUS DISORDERS

The GI tract contains a highly complex flora that varies based on organ site, patient age, genetics, nutrition, medications, and many other factors. Organisms are relatively sparse in the upper part of the GI tract—partly because of rapid passage of contents in the esophagus and a highly acidic environment in the stomach—but a large number and variety of microorganisms are present in the colon. While small numbers of fungi and protozoa are also normally present in the colon, bacteria are by far the largest component of the colonic flora; these organisms comprise at least 25% of the dry weight of stool (1). Bacteria that are almost always present in healthy individuals include *Bacteroides* spp., *Clostridium* spp., Enterobacteriaceae, *Enterococcus faecalis*, *Escherichia coli*, and *Lactobacillus* spp. Other organisms that may be part of the colonic flora in the absence of disease include *Staphylococcus aureus*, *Pseudomonas aeruginosa*, spirochetes, and some mycobacteria.

The normal gut flora contributes to good health in various ways, mainly by competing for essential nutrients and preventing overcolonization by opportunistic pathogens. Clinical infections of the GI tract are often due to relative overgrowth of one or more of these organisms because of antibiotic use, but some infections arise as a consequence of mucosal injury. Infections related to bacterial overgrowth can involve many different organisms, but a number of important infectious diseases of the gut are caused by specific organisms, which tend to present as clinicopathologically distinct syndromes. Examples of the latter include *Helicobacter*-associated gastritis, intestinal tuberculosis, and amoebic colitis. Because large numbers of bacteria are normally present in the lower GI tract, tissue Gram stains such as Brown-Brenn or Brown-Hopps have limited diagnostic utility, but special stains can be essential for diagnosing specific infections.

Bacterial Disorders

Helicobacter pylori is a curved bacillus that colonizes the gastric antral mucosa and causes active chronic gastritis. The organisms are gram-negative rods that are easily detected with routine hematoxylin and eosin (H&E) stain when present in large numbers (2) but may be more clearly visualized by stains such as modified Giemsa, Diff-Quik®, carbol fuchsin, and silver stains. Immunohistochemical stains are also commonly used (Figure 6.1) (3).

A number of studies have shown that special stains have higher sensitivity for identifying *H. pylori* compared with H&E when the bacterial count is low (1,4). The Giemsa stain is rapid, inexpensive, and readily available but tends to lack contrast. Diff-Quik®, which is a modified Romanowsky stain, is a very simple and effective stain for *H. pylori* using only the Diff Quik II® solution (5–7). The alcian yellow–toluidine blue (Leung) stain is an inexpensive and rapid method that results in high contrast between the blue-staining organisms and yellow mucus; this stain may also be useful in identifying *Giardia* and *Cryptosporidium* (8). Carbol fuchsin stains such as the modified Gimenez

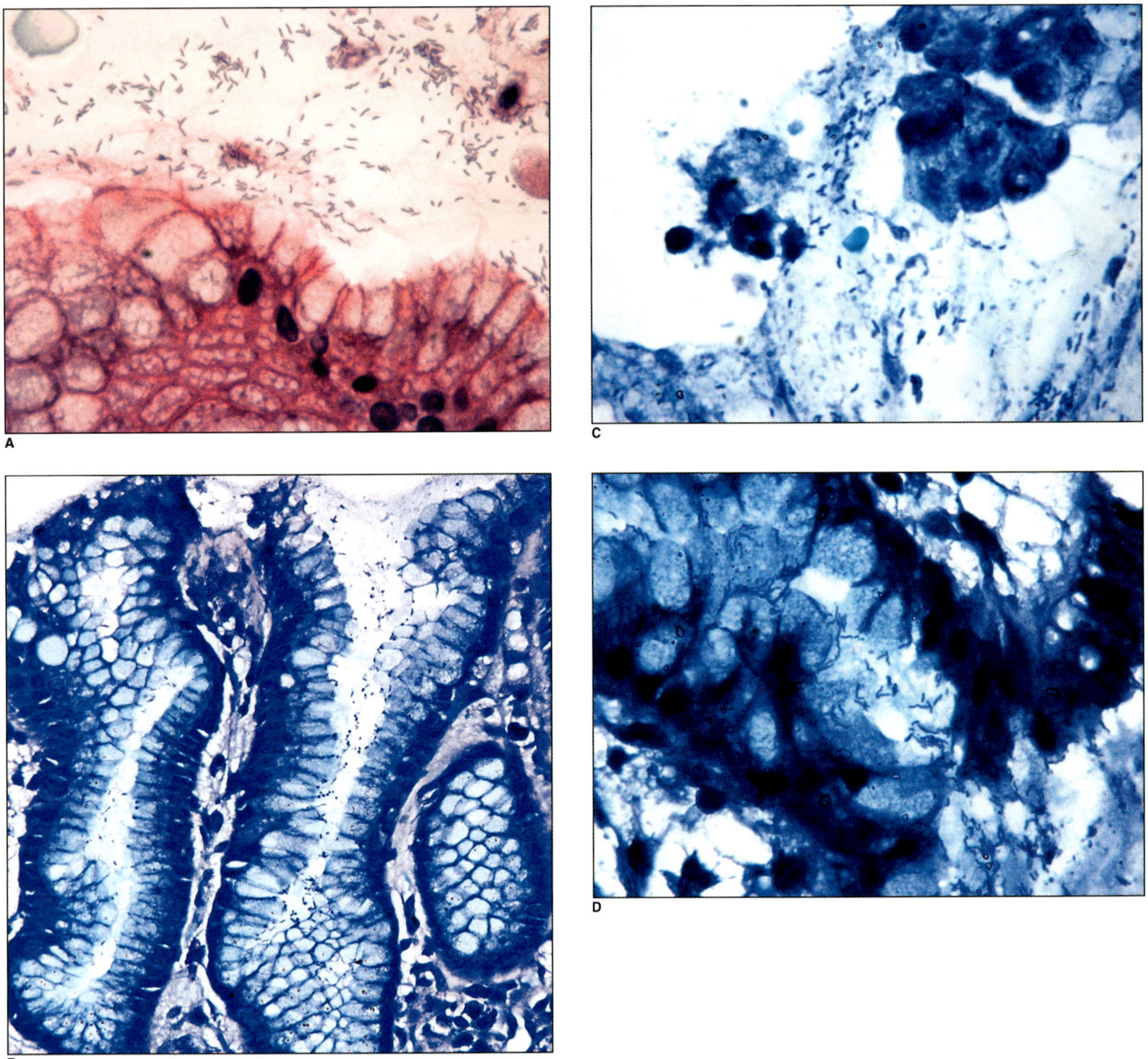

FIGURE 6.1: **Helicobacter pylori** ***organisms in gastric mucosa can be visualized with H&E as well as other stains. The organisms have a characteristic curved morphology that is usually easily distinguished from other rod-shaped bacteria (A–H).*** **H. pylori** ***may appear larger and therefore easier to identify with silver-based methods such as Steiner and Genta. In some cases, the organisms exhibit a coccoid morphology (I and J) (A—H&E, 1000×; B—Giemsa, 400×; C—Giemsa, 1000×; D—Diff-Quik®, 1000×; E*—Steiner, 1000×; F—Genta, 400×; H*—immunohistochemistry, 400×; I—Giemsa, 1000×; J—immunohistochemistry, 400×). *(Figures courtesy of Dr. Mark Wick.)***

stain are also simple and inexpensive (9,10). Modified Romanowsky Diff-Quik® and Giemsa are the stains in this group that are most commonly used to detect *H. pylori* (11).

Silver stains are very effective in demonstrating *H. pylori* and have the advantage of making organisms larger and therefore easier to see, but these stains are labor intensive and associated with much greater staining variability, particularly when done in inexperienced laboratories Among laboratories using silver stains for *H. pylori*, the Warthin-Starry and modified Steiner methods are most common (12). The modified Steiner stain has been recommended as a simpler technique than Warthin-Starry in detecting *H. pylori*, but both Steiner and Dieterle silver stains require uranyl nitrate, which is potentially toxic and expensive to dispose. Warthin-Starry does not require uranyl nitrate when used for *Helicobacter* sp. but is associated with higher background staining. The *Helicobacter pylori* silver stain, or HpSS, is a silver-based stain that targets the nucleolar organizing region (13). This stain has been reported to be simple and more sensitive than Giemsa.

Several combined stains have been described that allow simultaneous assessment of gastric mucosal histology and

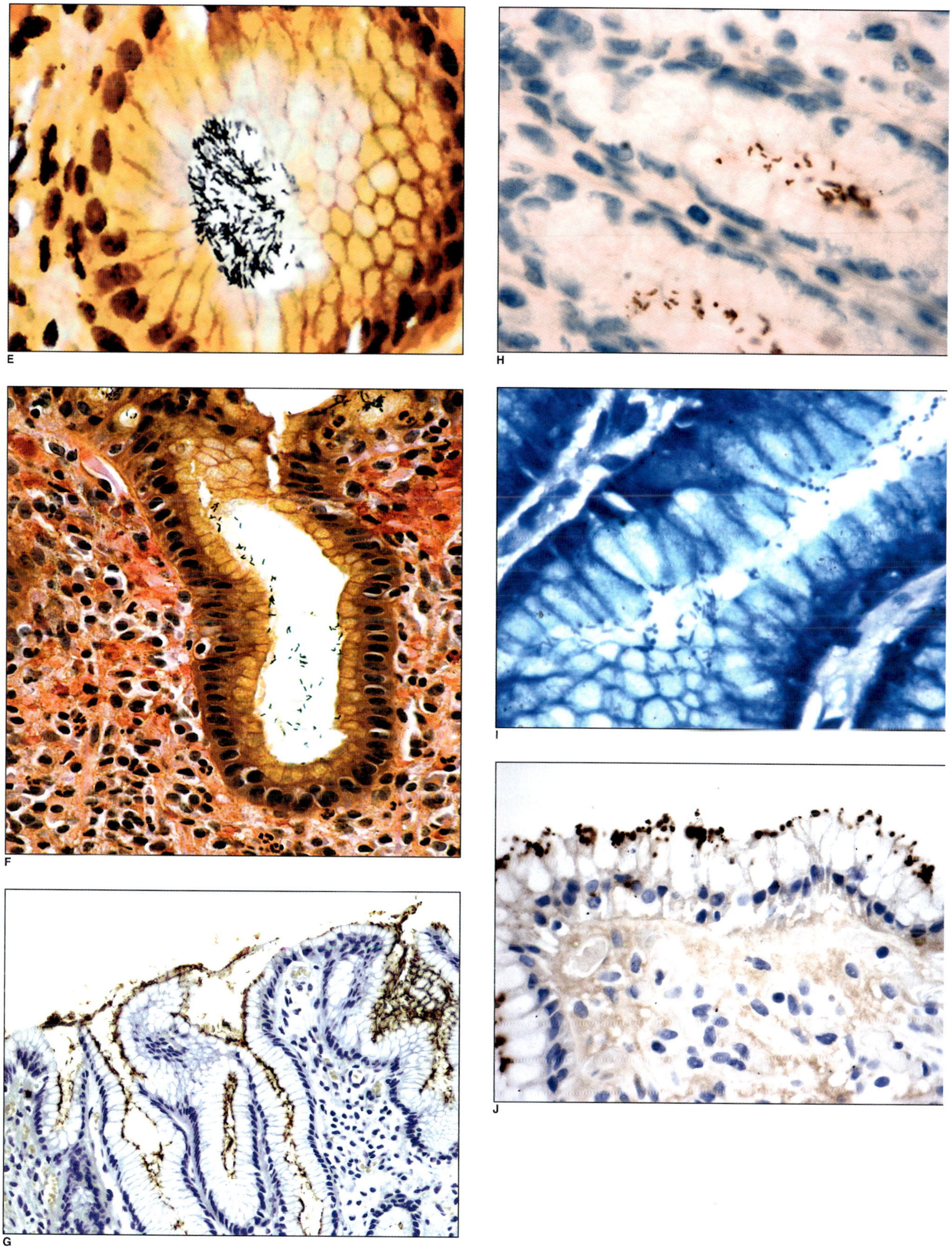

FIGURE 6.1: (Continued)

visualization of the organisms (14–16). The Genta and El-Zimaity stains both combine Steiner, H&E, and alcian blue at pH 2.5, although the El-Zimaity triple stain does not use uranyl nitrate. *Helicobacter heilmannii*, previously called *Gastrospirillum hominis*, is occasionally found in gastric biopsies in patients with gastritis. Initially thought to be nonpathogenic, the organism is now known to be an infrequent cause of active chronic gastritis, although this is generally milder and more limited than that seen with *H. pylori* (17,18). *H. heilmannii* is longer than *H. pylori* and has a distinctive, tightly coiled morphology (Figure 6.2). The number of organisms present in gastric biopsies is often low and infection may not be recognized initially, but the characteristic corkscrew-like appearance can be readily visualized with the same stains used for *H. pylori* (19).

Mycobacterial infections involve the gut in several distinctive ways depending on the organism. GI infection with *M. tuberculosis* (intestinal tuberculosis) is rare in the United States but is not uncommon in developing countries and is most often found in the distal ileum, ileocecal valve, or appendix (20,21). Other sites include the jejunum, ascending colon, and sigmoid colon. Intestinal tuberculosis presents in various ways but transverse ulcers, strictures, and small sessile polyps are the most common findings. The disease can have a segmental distribution (22,23) and as a result may be initially mistaken for Crohn's disease, particularly when granulomas are found on biopsy (24). In some cases, patients have large inflammatory masses with obstruction. Histologic examination usually reveals granulomas with caseous necrosis, and the Ziehl-Neelsen stain is the most useful stain for identifying acid-fast bacteria in these granulomas (Figure 6.3). This stain employs carbol-fuchsin and a decolorization step with acid-alcohol, and the rod-shaped organisms stain red. Mycobacteria are generally not visualized with tissue Gram stains, although *M. tuberculosis* may be very weakly gram positive.

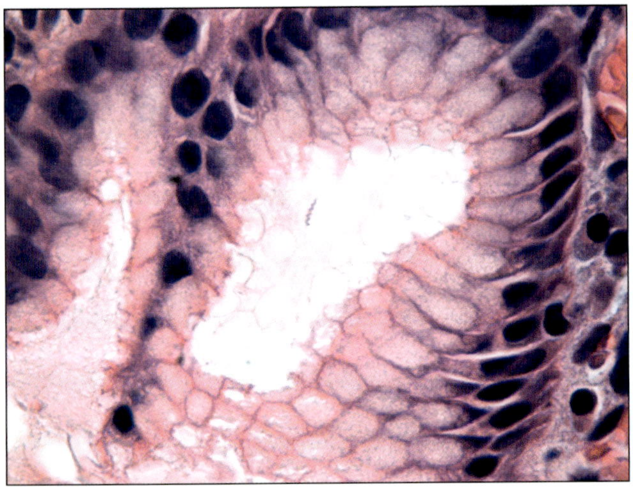

FIGURE 6.2: **Helicobacter heilmannii** ***has a distinctive spiral, tightly coiled shape that differs from*** **H. pylori** ***(H&E, 1000×).***

Mycobacterium avium-intracellulare (MAI) infection is usually seen in immunosuppressed patients, particularly those with the acquired immune deficiency syndrome (AIDS). GI involvement is most common in the proximal small intestine (25). The endoscopic appearance of the

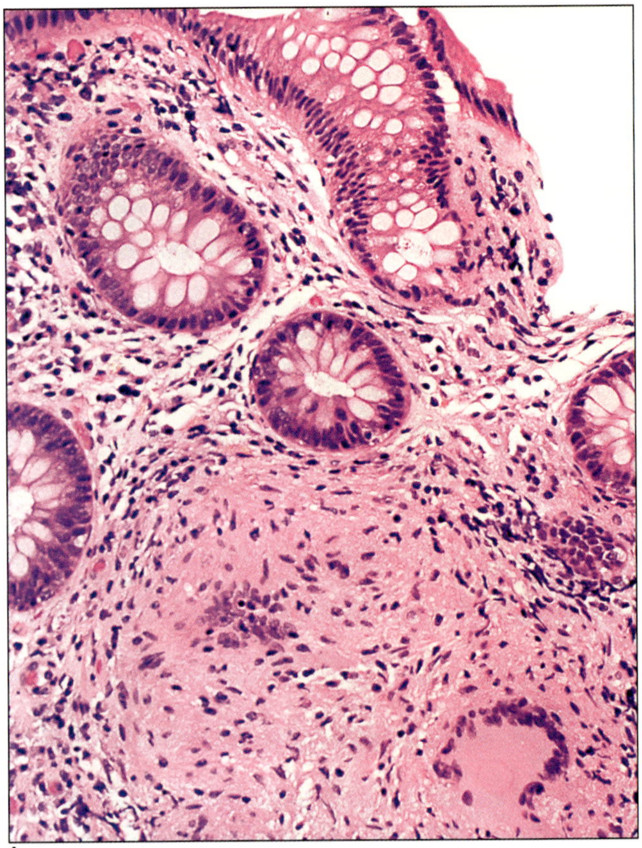

A

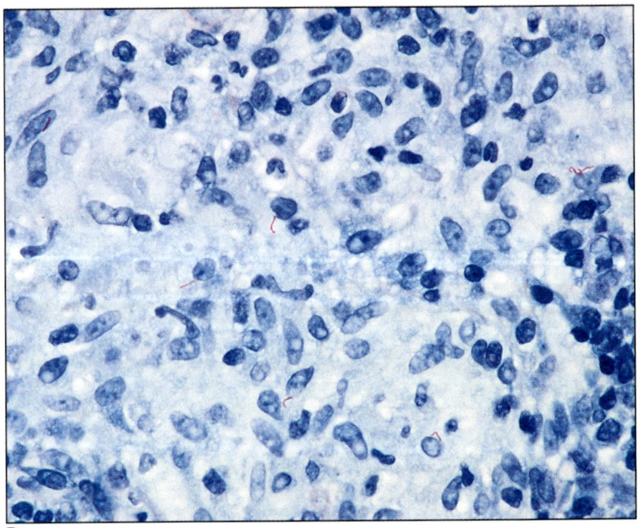

B

FIGURE 6.3: Granulomas in colonic mucosa (A). Multiple red acid-fast bacilli are seen within granulomas (B) (A—H&E, 1000×; B—Ziehl-Neelsen stain, 1000×).

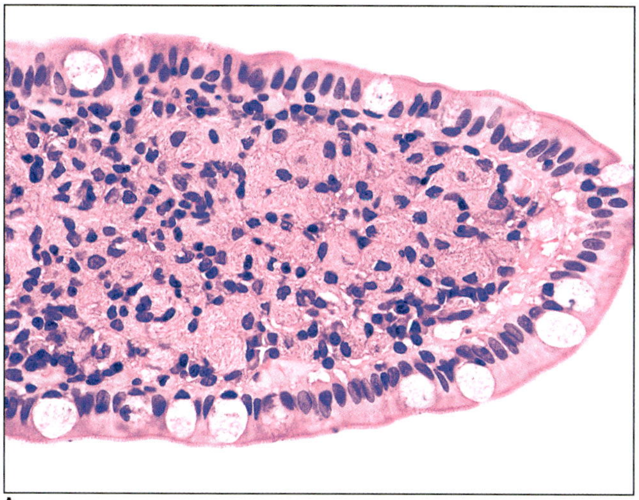

A

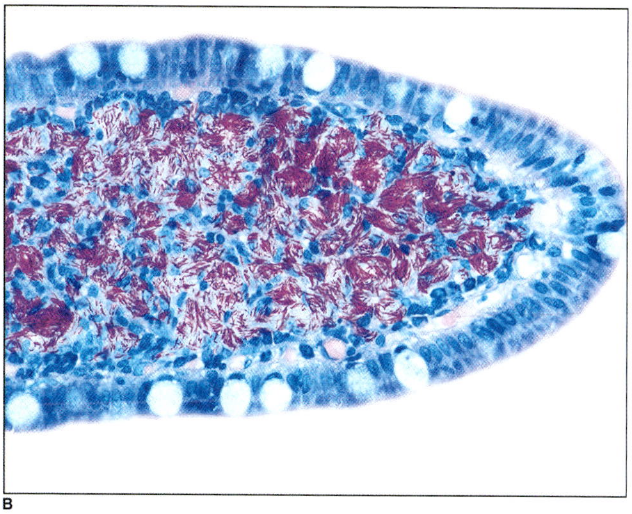

B

FIGURE 6.4: Numerous foamy macrophages fill the mucosa in MAI (A). These cells are stuffed with acid-fast bacilli (B) (A—H&E, 400×; B—Ziehl-Neelsen stain, 400×).

mucosa is often normal or only slightly abnormal, but nodules or even large masses may sometimes be seen. In patients with AIDS, biopsies of MAI lesions usually reveal sheets of large foamy macrophages in the mucosa and submucosa, often with little accompanying inflammation. Large numbers of bacillary organisms are present within the cytoplasm of the histiocytes, and the bacteria can usually be readily identified with a Ziehl-Neelsen stain (Figure 6.4) (26). The cells may also be periodic acid-Schiff (PAS) positive, which could lead to confusion with Whipple's disease.

In some cases of MAI, aggregates of phagocytes exhibit a spindle cell appearance morphologically suggestive of a mesenchymal neoplasm, particularly Kaposi sarcoma. These so-called mycobacterial spindle cell pseudotumors are usually due to MAI and may be found within mesenteric lymph nodes, skin, and elsewhere. Despite the unusual morphologic appearance, acid-fast organisms can be identified in these spindled cells (27,28).

Whipple's disease is a very rare bacterial infection typically found in Caucasian men older than 40 years (29). The disease was previously termed "intestinal lipodystrophy" because of prominent deposition of fat within intestinal mucosa and mesenteric lymph nodes (30). After numerous unsuccessful attempts to culture a causative agent, a gram-positive, acid-fast–negative bacterium related to actinomycetes was eventually characterized by 16S rRNA gene analysis and termed *Tropheryma whippelii* (31). The infection most commonly involves the small intestine but can involve almost any organ including the heart, joints, and central nervous system. Patients with intestinal involvement typically present with abdominal pain, weight loss, and/or diarrhea. The characteristic histologic finding is the presence of large PAS-positive macrophages in the lamina propria (Figure 6.5) (32).

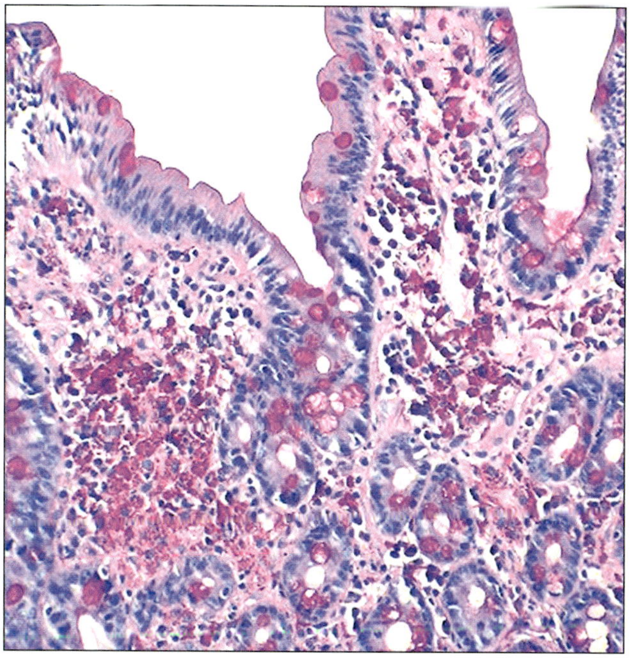

FIGURE 6.5: Whipple's disease in small intestinal mucosa. Large PAS-positive macrophages that are strongly positive for PAS fill the lamina propria (PAS, 200×). (Figure courtesy of Dr. Mark Wick.)

Malacoplakia of the GI tract is rare, but the colon is the second most common site after the urinary tract. Involvement of the stomach, small intestine, and appendix is less common. Colonic malacoplakia often occurs in association with tumors, most commonly invasive adenocarcinoma (33,34). In these cases, foci of malacoplakia are found in close proximity to the tumor, but the reason for this

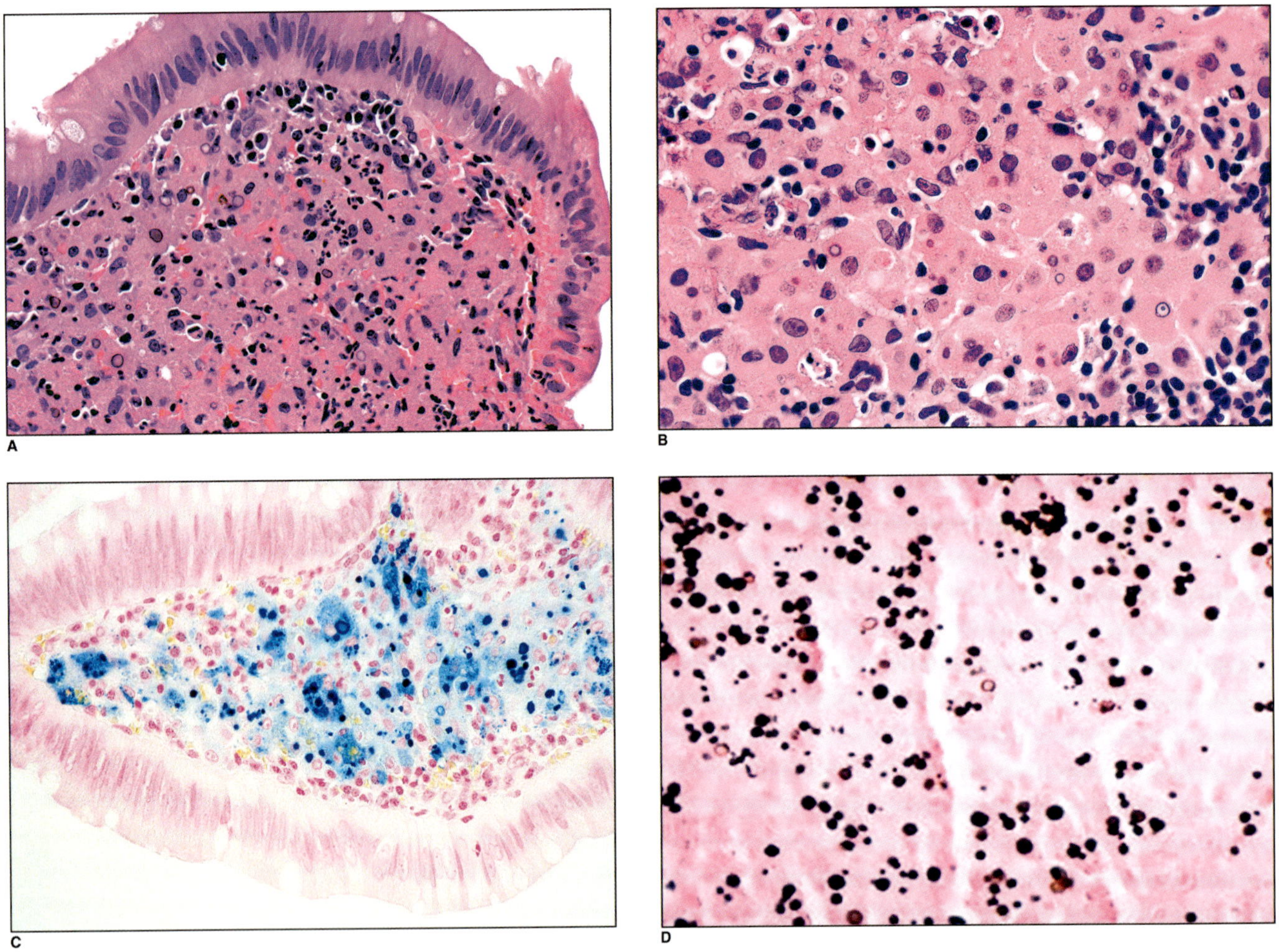

FIGURE 6.6: Malacoplakia is characterized by aggregates of histiocytes that contain intracytoplasmic concretions called Michaelis-Gutmann bodies (A and B). Note the distinctive targetoid appearance. These structures stain with iron (C) and calcium (D) stains (A—H&E, 400×; B—H&E, 1000×; C—Prussian blue, 400×; D—von Kossa, 400×). *(Figure courtesy of Dr. Mark Wick.)*

proximity is not known. Although most cases are discovered incidentally, endoscopy may reveal small yellow foci.

Malacoplakia appears to be due to a defect in bacterial killing by monocytes and macrophages (35). Partially digested bacteria such as *E. coli* accumulate in phagolysosomes, which become encrusted with calcium and iron salts. Histologically, malacoplakia is characterized by sheets of histiocytes (von Hansemann cells) that contain intracytoplasmic concretions called Michaelis-Gutmann bodies. These structures have a targetoid appearance and are considered pathognomonic for malacoplakia. Michaelis-Gutmann bodies range from 2 to 10 μm and stain with PAS, iron (Perl's and Prussian blue), and calcium (von Kossa and alizarin red) stains (Figure 6.6). The von Kossa is the most commonly used stain to confirm the diagnosis of malacoplakia, but alizarin red may be useful when only small amounts of calcium are present, as is often the case.

Other unusual infectious disorders of the GI tract include intestinal spirochetosis (36,37). In this condition, biopsies of the colonic mucosa reveal a uniform basophilic layer on the surface of the colonic mucosa that resembles an unusually prominent brush border. There is usually no associated inflammatory infiltrate. Intestinal spirochetosis has been associated with watery diarrhea, but most patients have few distinct symptoms and it remains unclear whether the presence of the organisms indicates an infectious disease (38). The organisms can be identified with a wide variety of stains including Giemsa and silver stains such as Steiner, Dieterle, and Warthin-Starry (Figure 6.7) (39). Similar findings have been described with enteroadherent *E. coli*, but the latter are gram negative and should be easily distinguished from spirochetosis.

Fungal Disorders

Candida esophagitis is the most common fungal infection of the GI tract, especially in immunocompromised patients (40,41). The organism is typically identified within the

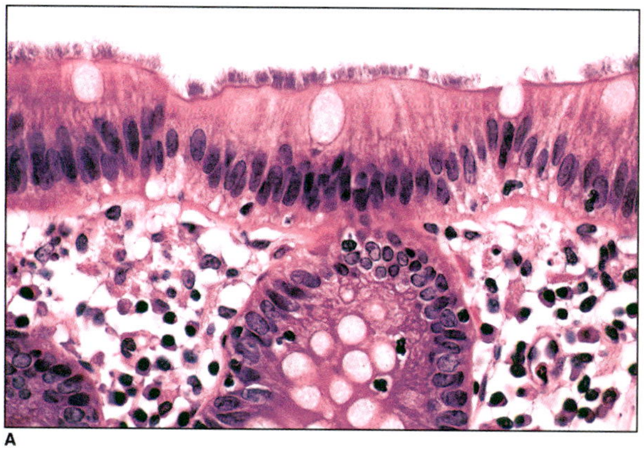

A

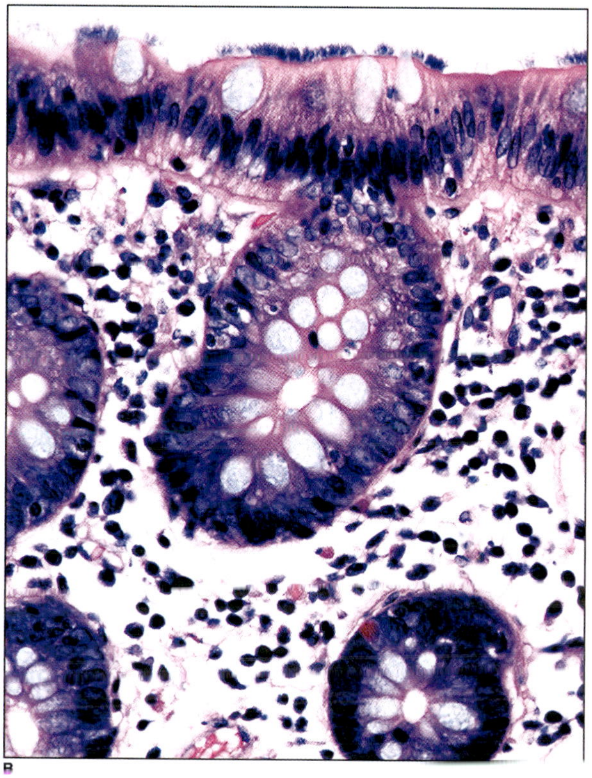

B

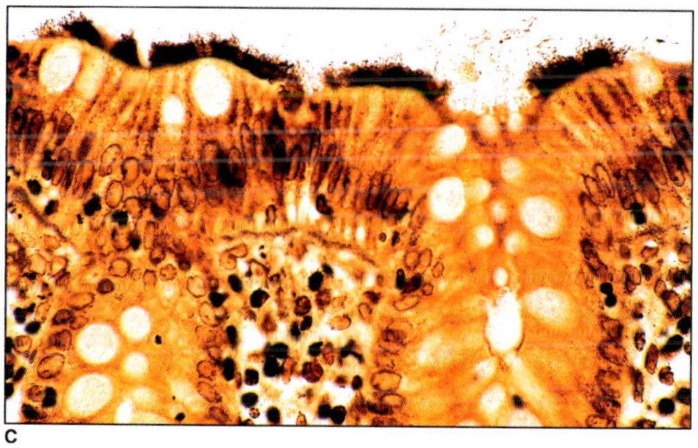

C

FIGURE 6.7: ***Colonic spirochetosis is characterized by organisms along the luminal surface resembling a prominent brush border (A—H&E, 1000×; B—Giemsa, 1000×; C—Steiner, 1000×).***

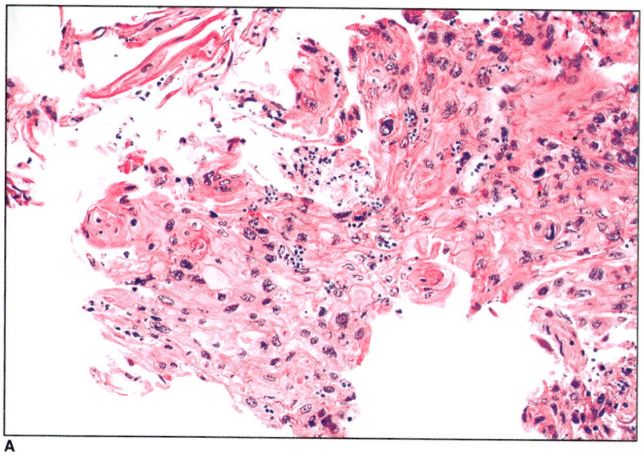

A

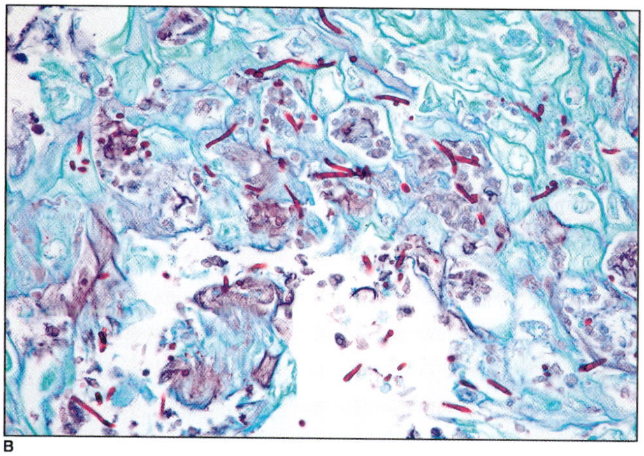

B

FIGURE 6.8: ***Candida species can be seen in the squamous epithelium of the esophagus with H&E and PAS stains (A—H&E, 400×; B—PAS, 400×).***

superficial layers of the esophageal mucosa and is identified clinically by the presence of white mucosal patches or plaques. Either PAS or methenamine silver stain can confirm the presence of pseudohyphae of *Candida* spp. (Figure 6.8).

Fungal infection of the intestine is uncommon, with one study finding that *Candida* and *Aspergillus* accounted for all fungal infections of the small and large bowel (42). *Aspergillus* infections of the gut are almost always seen in immunocompromised patients. Other fungal infections that can involve the GI tract include *Histoplasma*, *Zygomycetes*, and *Cryptococcus* (43,44). Most of these infections are seen in patients with either disseminated or pulmonary involvement.

Parasitic Disorders

Amoebic colitis is caused by the protozoan organism *Entamoeba histolytica*. The disease is acquired by ingesting cysts of *E. histolytica*, which are found in food or water contaminated by feces. Although rare in developed countries, the disease is quite common in Central and South America and Southeast Asia. Most patients are asymptomatic, but

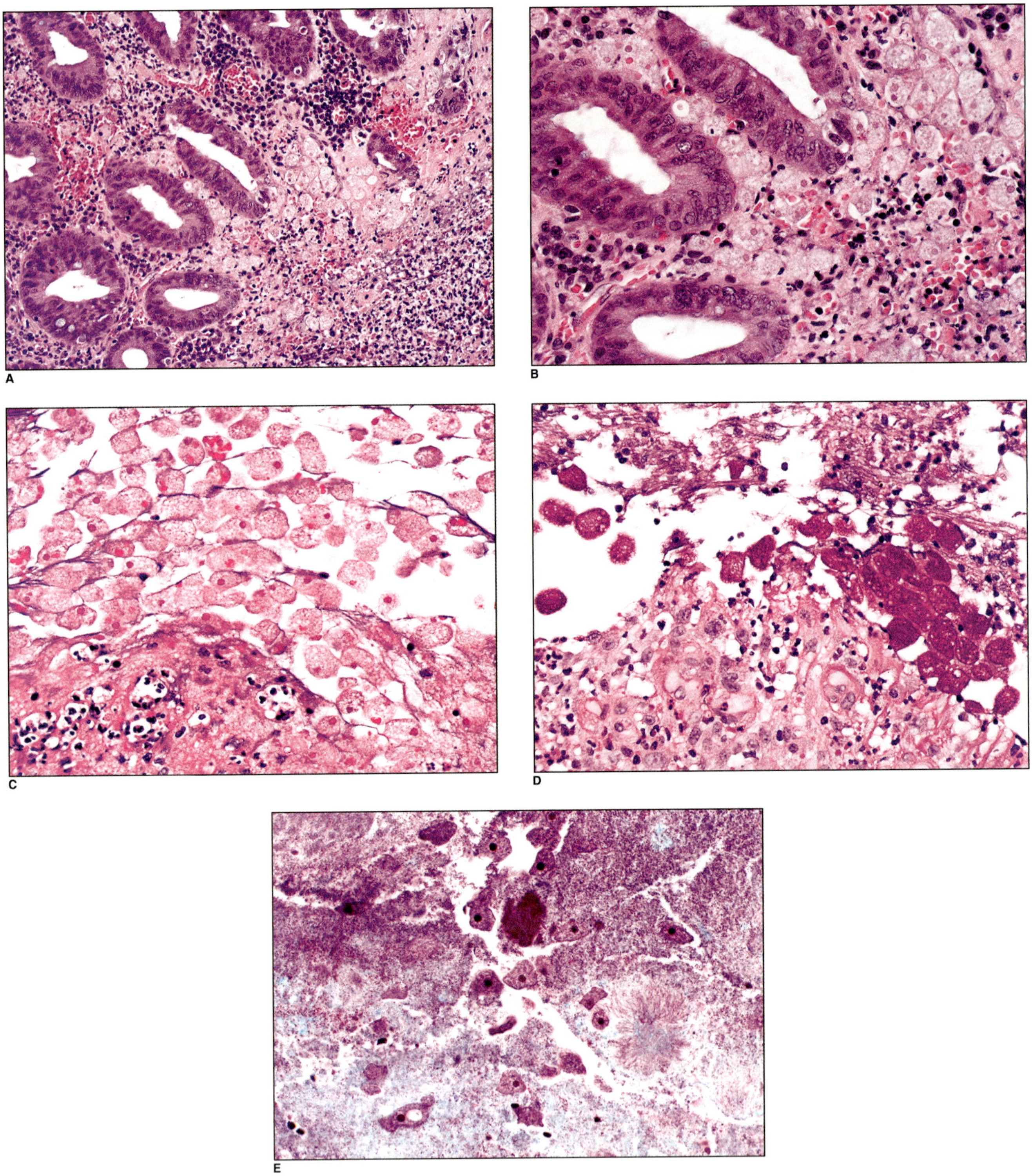

FIGURE 6.9: Amebic colitis is caused by Entamoeba histolytica, *which invade into the intestinal wall and result in flask-shaped ulcers with submucosa abscesses. Note large numbers of organisms adjacent to an abscess (A and B). In tissue sections, the organisms are normally visualized with H&E (C) and PAS (D) stains. In fecal smears, a trichrome stain (E) is most common (A—H&E, 200×; B—H&E, 400×; C—H&E, 1000×; D—PAS, 1000×; E—trichrome, 1000×).*

bloody diarrhea and colicky pain occur if trophozoites invade the intestinal mucosa. The classic histologic finding in such patients is a flask-shaped mucosal ulcer that contains numerous trophozoites, which are PAS positive (Figure 6.9). The organisms often contain red blood cells, which is a helpful diagnostic clue.

Cryptosporidiosis is caused by infection with protozoans of the genus *Cryptosporidium*, most commonly

Cryptosporidium parvum. Healthy individuals with cryptosporidiosis have mild, self-limited diarrhea, but severe, persistent watery diarrhea can be seen in children and in immunocompromised patients. Transmission is usually via the fecal-oral route, but animal-to-person and waterborne transmissions also occur (45). Involvement of the duodenum is most common; the colon and stomach are less frequently involved (46). Histologic examination shows round basophilic organisms 2–4 μm in diameter attached to the brush border of the small intestine (Figure 6.10). The parasites may also be identified in the biliary tract or in colon biopsies. The underlying mucosa may be normal or show only mild villous blunting, and inflammatory infiltrates are generally sparse. The organisms are easily identified by H&E stain but will also stain with PAS, Giemsa, Gram, and silver stains. The organisms are acid fast, and the diagnosis is most often made by identifying oocysts in stool specimens with Fite or Ziehl-Neelsen stain.

Isosporiasis (*Isospora belli*) is another coccidian infection seen in immunocompromised patients; most patients have AIDS, but the infection is also found in patients with lymphoma and other malignancies (47). Signs and symptoms are similar to cryptosporidiosis, with cramping abdominal pain and severe watery diarrhea (48). Like cryptosporidiosis, infection is usually from fecal contamination. Small bowel biopsy reveals relatively large intraepithelial inclusions (Figure 6.11). *Isospora* cysts are found within the cytoplasm of enterocytes and are much larger than those of cryptosporidia, measuring up to 25 μm. An acid-fast stain reveals oocytes in stool, but special stains are of limited utility in biopsies.

Microsporidiosis is caused by many organisms in the order Microsporidia, most commonly *Enterocytozoon bieneusi*. Chronic diarrhea in HIV-positive patients is the most common clinical picture, but infection also occurs in renal transplant and other immunosuppressed patients (49,50). The small intestinal mucosa may appear normal or show

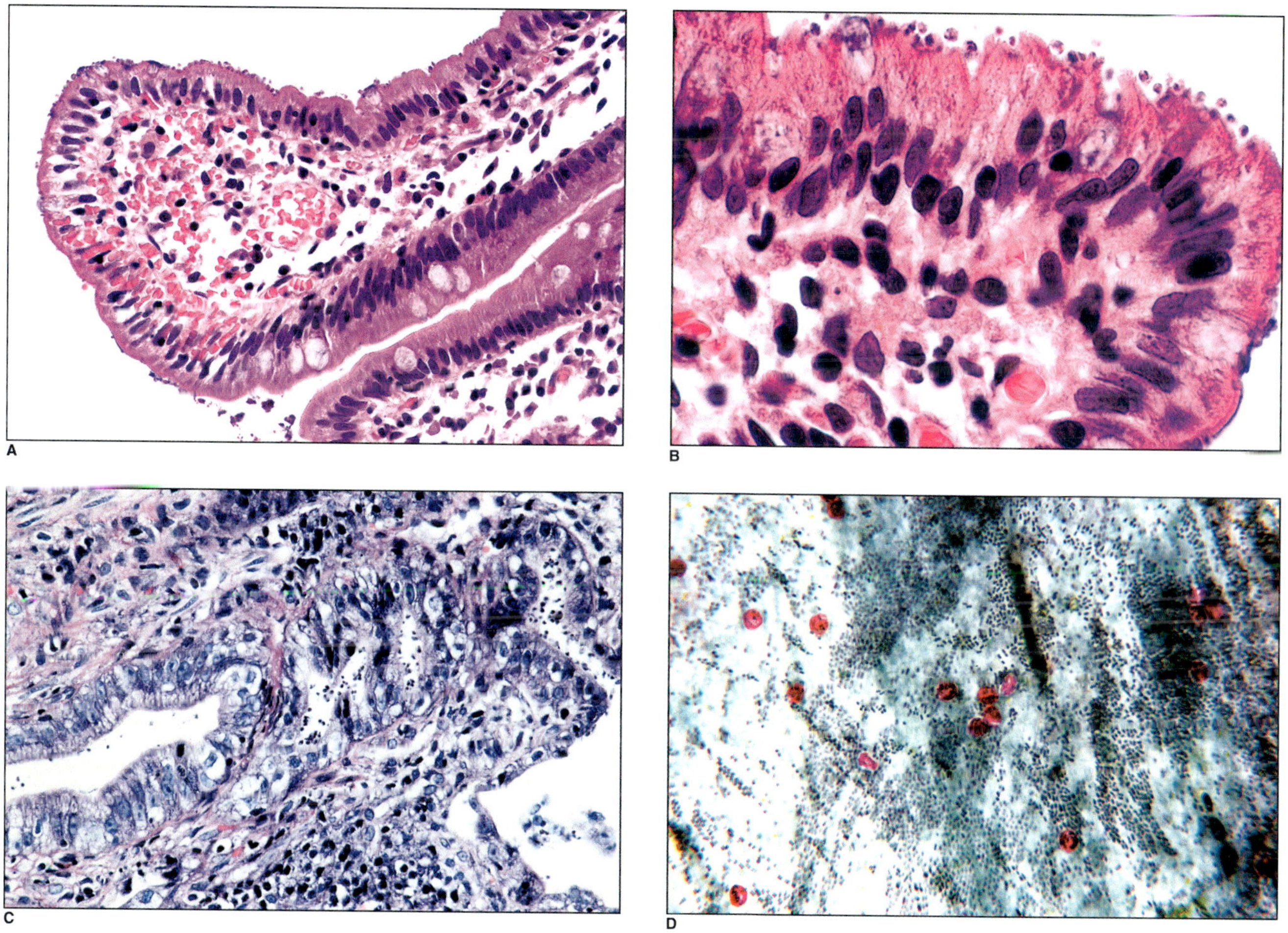

A B C D

FIGURE 6.10: Cryptosporidiosis is characterized by many tiny round basophilic organisms attached to the epithelial surface of the small intestine (A and B). Giemsa is useful in tissue sections to identify the organism (C), but acid-fast stains are most useful in stool. Note the brightly acid-fast organisms (D) (A—H&E, 200×; B—H, 1000×; C—Giemsa, 400×; D—Ziehl-Neelsen, 1000×).

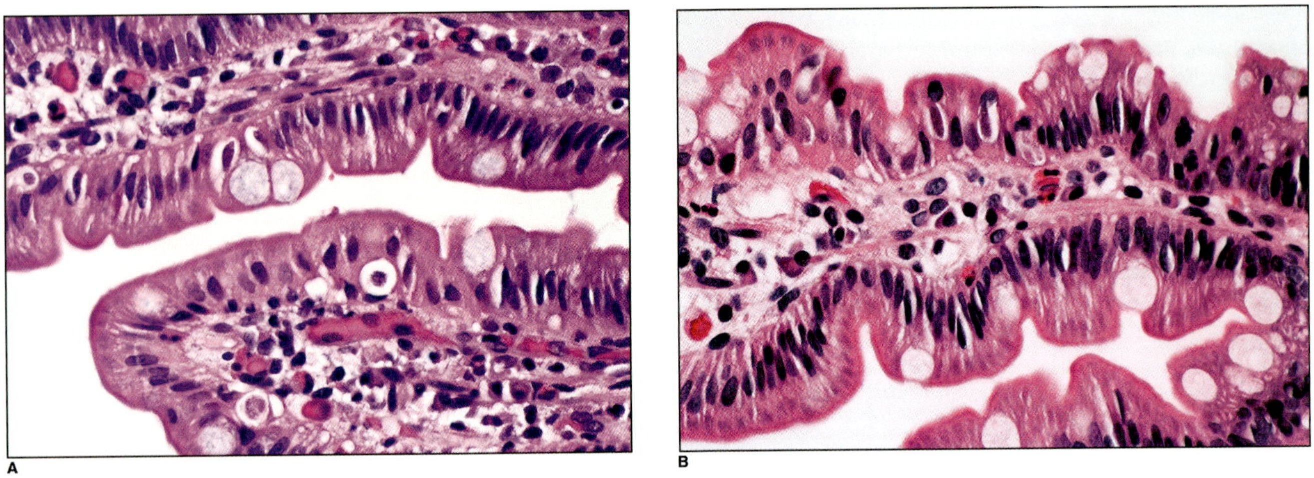

A B

FIGURE 6.11: (A and B) Immunosuppressed patients develop severe diarrhea due to infection with Isospora belli. *Large ovoid intracellular inclusions are seen in the small intestinal epithelium (H&E, 1000×).*

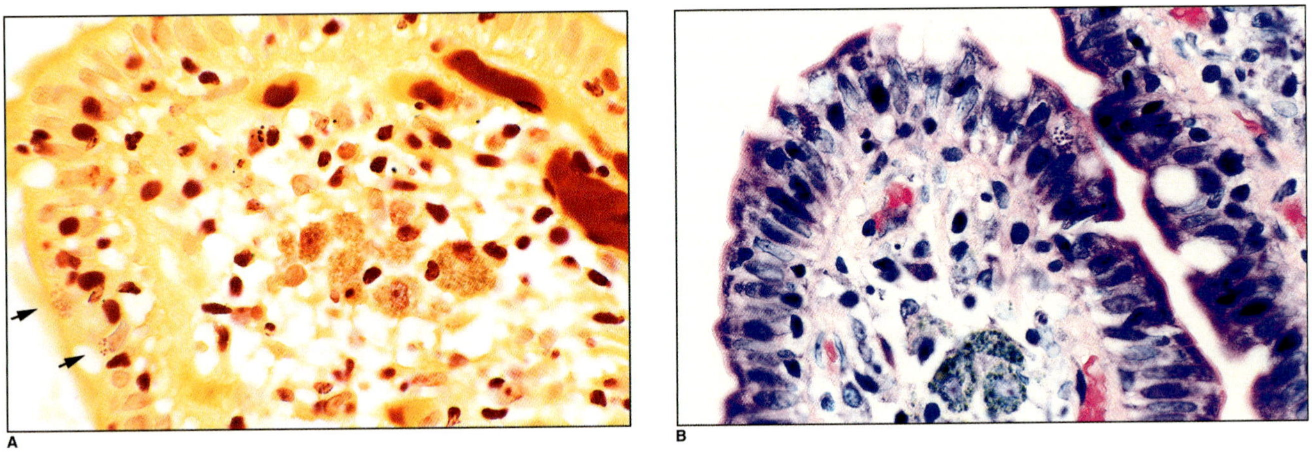

A B

FIGURE 6.12: Tiny spherical intracellular organisms characterize microsporidiosis (A—Brown-Brenn, 1000×; B—Giemsa, 1000×).

mild villous blunting and other features of epithelial injury. The organisms are found within the cytoplasm of enterocytes but are very small and difficult to detect in H&E-stained sections. They stain variably with histochemical stains such as Brown-Brenn, Brown-Hopps, Giemsa, modified trichrome, and silver impregnation methods (Figure 6.12) (51,52).

NONNEOPLASTIC STRUCTURAL DISORDERS

As in the case of infections, the mucosa is the site of most nonneoplastic structural disorders of the GI tract.

Metaplasias

Intestinal metaplasia of the stomach, defined as replacement of the normal gastric epithelium by intestinalized epithelium, is associated with an increased risk for gastric adenocarcinoma and has long been thought to represent an important step in gastric carcinogenesis. Complete intestinal metaplasia reflects a small intestinal phenotype, with a mixture of goblet cells, Paneth cells, and mature absorptive cells with a distinct brush border. Incomplete intestinal metaplasia is composed of goblet cells and intervening columnar cells that lack a brush border.

Mucin histochemistry can be used to assess intestinal metaplasia (53). Mucins are glycoproteins that represent the major component of the mucus layer covering the surface of the gastric mucosa. Epithelial mucins are generally classified as neutral or acidic. Neutral mucins are present in normal gastric foveolar–type surface epithelium, while acid mucins, which can be either sulfated or nonsulfated (carboxylated), are characteristic of normal goblet cells. In complete (type I) intestinal metaplasia, the goblet cells secrete sialomucin (a nonsulfated acidic mucin), while the columnar cells generally do not contain mucin. The goblet cells in incomplete intestinal metaplasia type II (or IIa) secrete sialomucin predominantly, while those in incomplete intestinal metaplasia type III (or IIb) secrete sulfated mucin (sulfomucin). The columnar cells of incomplete intestinal metaplasia can exhibit variable amounts of either neutral and/or acidic mucins in the apical cytoplasm.

These mucins can be identified by histochemical stains (Table 6.1). PAS is the only stain that identifies neutral mucin, and removing glycogen with diastase predigestion can help in confirming it. Both forms of acidic mucin stain with Mayer mucicarmine (red) and alcian blue at pH 2.5 (blue), but only alcian blue at pH 0.5–1.0 stains sulfated mucins. Sulfomucin also stains brown black with high-iron diamine (HID).

Because of these differences, combined stains have proven useful in assessing intestinal metaplasia in gastric specimens (Figure 6.13). A combined PAS/alcian blue (pH 2.5) stain is a simple means of identifying both acidic mucins (blue) and neutral mucins (magenta) in routine specimens. In complete (type I) intestinal metaplasia, the goblet cells will stain dark blue but the columnar cells are negative. In incomplete intestinal metaplasia, the goblet cells will be dark blue and the columnar cells magenta or purple (depending on the amount of acidic and/or neutral mucin present). A combined HID/alcian blue (pH 2.5) stain can distinguish sialomucin from sulfomucin and is therefore useful in distinguishing type II from type III intestinal metaplasia (54). Since this stain requires prolonged staining time and involves the use of potentially toxic reagents, some authors have recommended using Gomori's aldehyde fuchsin as a substitute to identify sulfomucin (55).

While the extent of intestinal metaplasia is clearly associated with an increased risk for gastric cancer, there are conflicting data on whether subtyping is of any utility in stratifying patients into clinically significant risk groups. This is partly because patients very often exhibit mixtures of different metaplasia types (56–58). A number of studies have suggested that type III intestinal metaplasia is most clearly associated with an increased risk of malignant transformation (59–61), but this could be simply related to the increased likelihood of its presence when intestinal metaplasia is extensive (62). As a result, gastric cancer surveillance by mucin histochemistry has not achieved wide acceptance (63,64).

Barrett's Esophagus

Barrett's esophagus is identified when portions of the normal stratified squamous epithelium of the lower esophagus are replaced by a specialized form of metaplastic columnar epithelium. This metaplastic epithelium appears to arise partly as a response to acid reflux and is associated with a significantly increased risk for adenocarcinoma. The diagnosis requires histologic identification of intestinal metaplasia in the esophagus; in practical terms, goblet cells must be identified in biopsies from the esophagus (65). Goblet cells are usually easily recognized by routine H&E stains, but some cases present diagnostic difficulties. While stains for acid mucin, most often alcian blue, may help identify goblet cells in such cases, it is important to understand that not all alcian blue–positive cells are goblet cells (66,67). Non-Barrett columnar cells (pseudogoblet cells) can also have blue-staining apical cytoplasmic mucin, and in some cases may be difficult to distinguish from true goblet cells (Figure 6.14). In general, goblet cells are dark blue with alcian blue at pH 2.5, while pseudogoblet cells tend to show lighter staining, but because of the ease with which goblet cells can usually be recognized in H&E stains and the low yield of mucin stains, some authors recommend routine alcian blue stains against Barrett's esophagus (68). Type III intestinal metaplasia has been identified in Barrett's esophagus and some authors (69) have suggested an association with an increased risk for adenocarcinoma, but other studies have not supported this finding (70).

Other Structural Disorders

Amyloid deposition is commonly found in the GI tract in patients with systemic or localized amyloidosis (71). Amyloid is most often identified in biopsies from the rectum or stomach and most commonly appears as perivascular deposits; rare cases present as globular deposits or large nodular masses (72). Congo red is the stain most

TABLE 6.1: Differentiation of Epithelial Mucins by Histochemical Stains

	Neutral Mucin	Nonsulfated Acid Mucin (sialomucin)	Sulfated Acid Mucin (sulfomucin)
PAS	Magenta	Negative	Negative
Alcian blue (pH 2.5)	Negative	Dark blue	Negative or light blue
Mayer mucicarmine	Negative	Red	Red
HID/alcian blue (pH 2.5)	Negative	Dark blue	Brown-black
PAS/alcian blue (pH 2.5)	Magenta	Dark blue	Negative or light blue

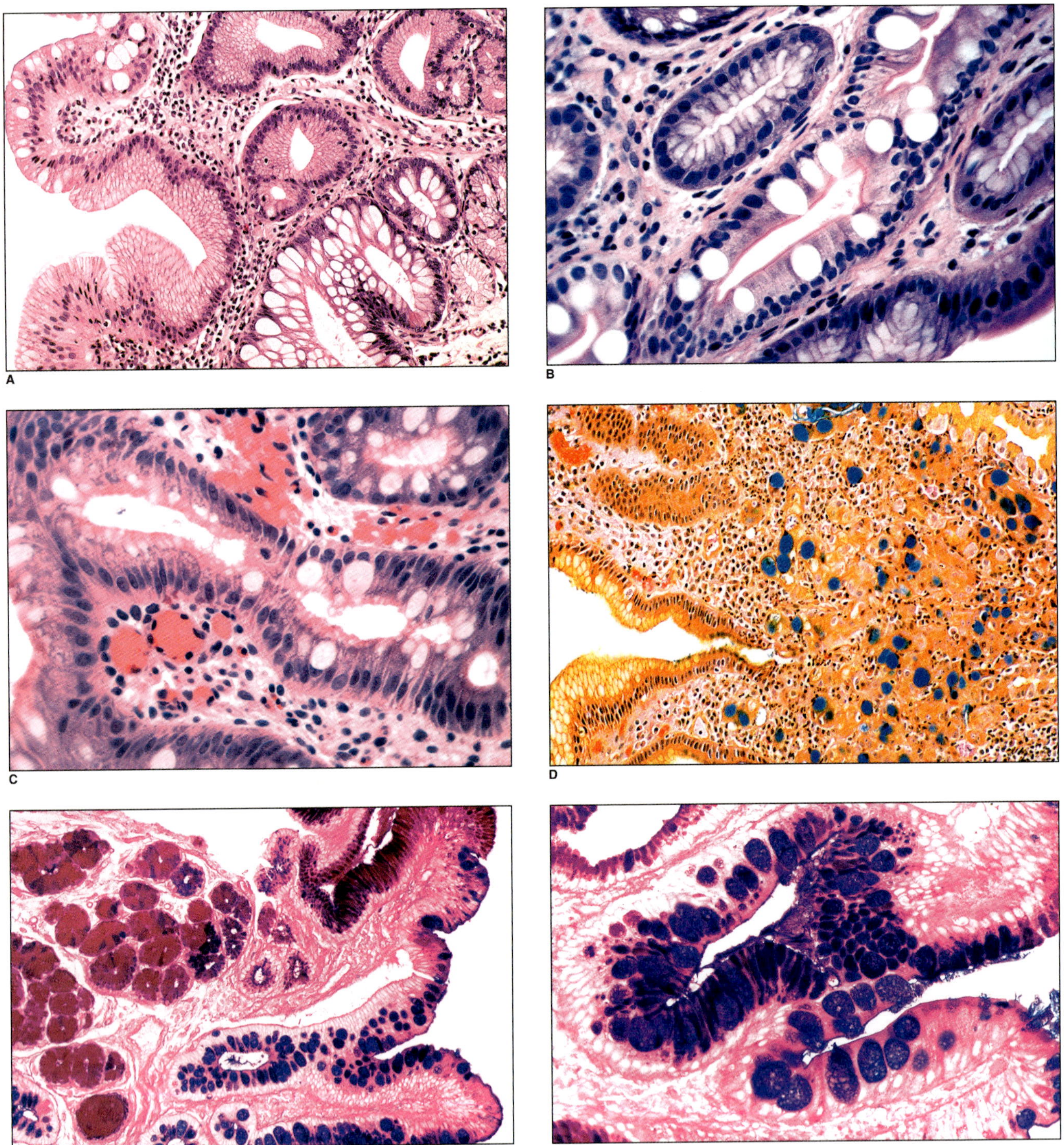

A B C D E F

FIGURE 6.13: Intestinal metaplasia of gastric mucosa (A). A distinct brush border is seen in complete intestinal metaplasia (B) but not in incomplete intestinal metaplasia (C). Goblet cells contain acid mucin and stain blue with the Genta stain (D). A combined PAS/alcian blue at pH 2.5 simultaneously demonstrates the acid mucin–containing goblet cells, which stain blue, and cells with neutral mucin, which stain red (E and F). Note that some of the columnar cells have a purple color due to mixtures of neutral and acid mucins (A—H&E, 200×; B—H, 400×; C—H&E, 400×; D—Genta, 200×; E and F—PAS/alcian blue, 200×).

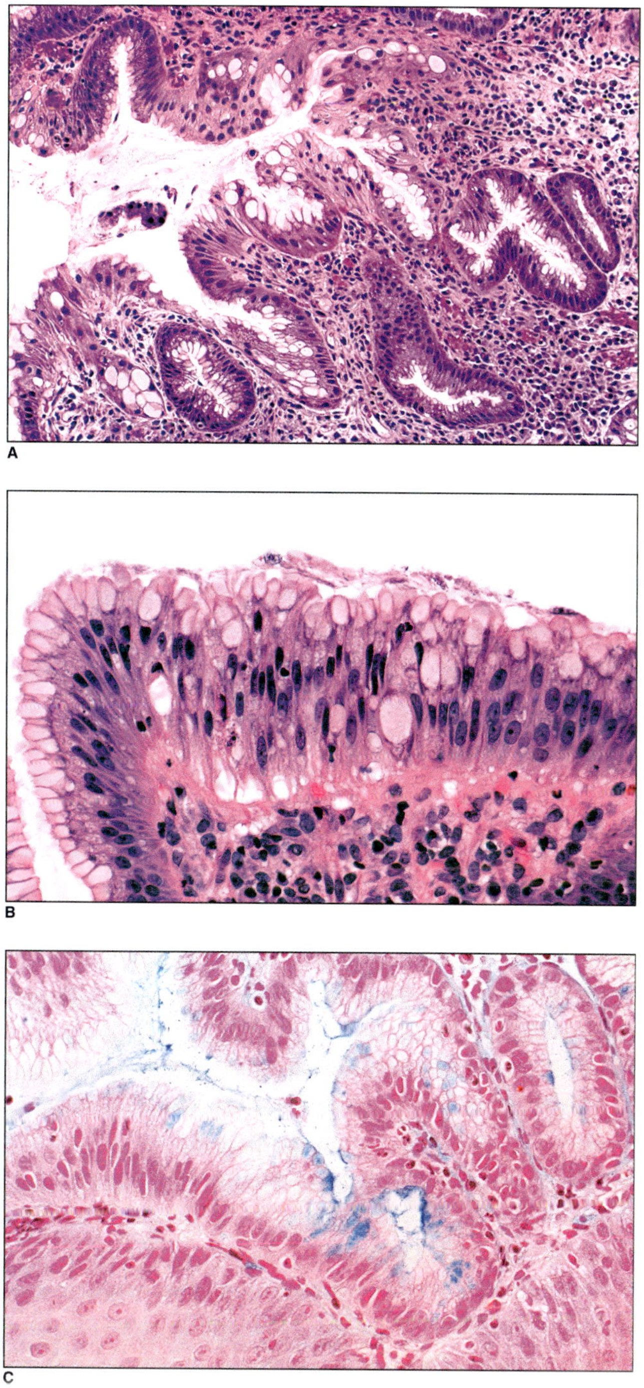

A

B

C

FIGURE 6.14: Columnar cells in the gastric cardia can have abundant apical mucus that results in an appearance resembling goblet cells (pseudogoblet cells), which can be difficult to distinguish from Barrett's mucosa (A and B). An alcian blue stain may show faint staining of these cells (C), which should not be misinterpreted as intestinal metaplasia (A—H&E, 200×; B—H&E, 400×; C—alcian blue, 200×).

commonly used to identify amyloid protein, but other methods include the fluorescent dyes thioflavin-T and thioflavin-S and metachromatic stains such as crystal violet, methyl violet, and toluidine blue (Figure 6.15). Thioflavin-T

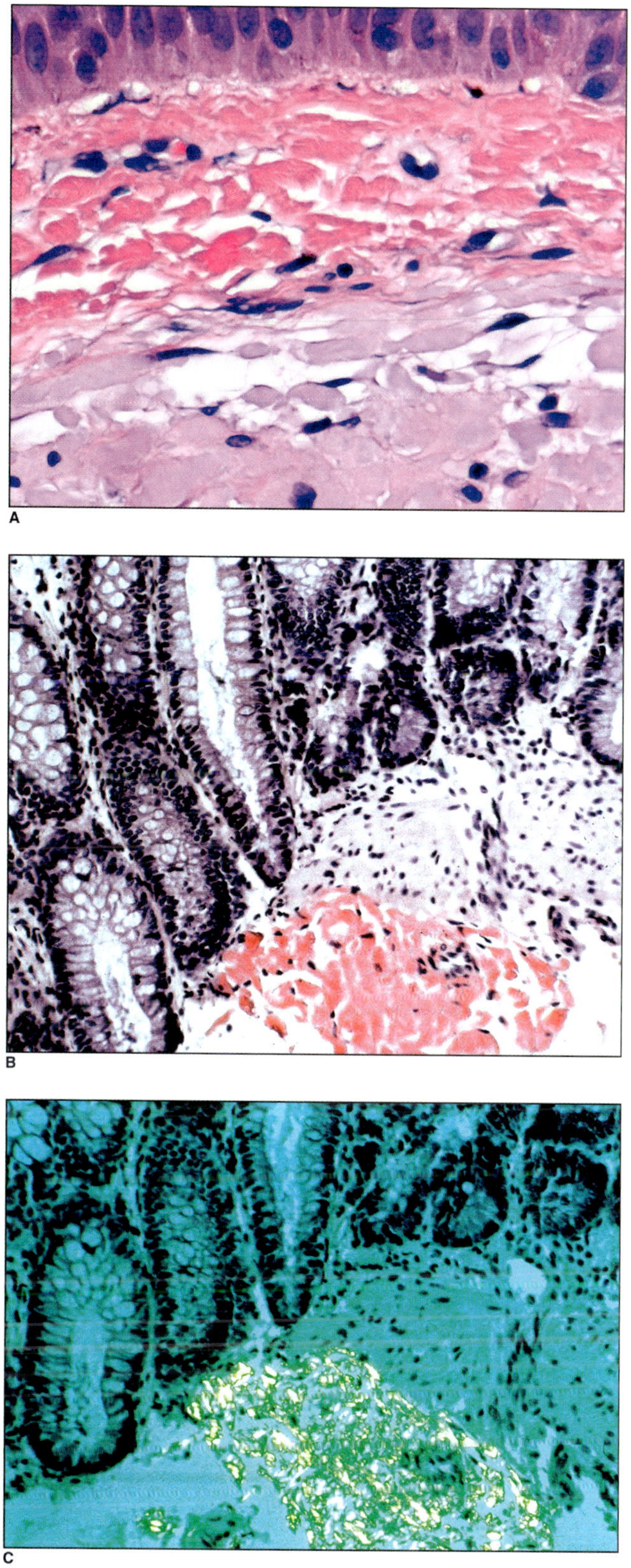

A

B

C

FIGURE 6.15: GI amyloidosis is most often identified in rectal biopsies, which can show submucosal or perivascular deposits of pale eosinophilic material (A). Congo red shows a characteristic red color with ordinary transmitted light (B) and apple-green birefringence with polarized light (C) (A—H&E, 200×; B and C—Congo red, 200×). (Figures courtesy of Dr. Mark Wick.)

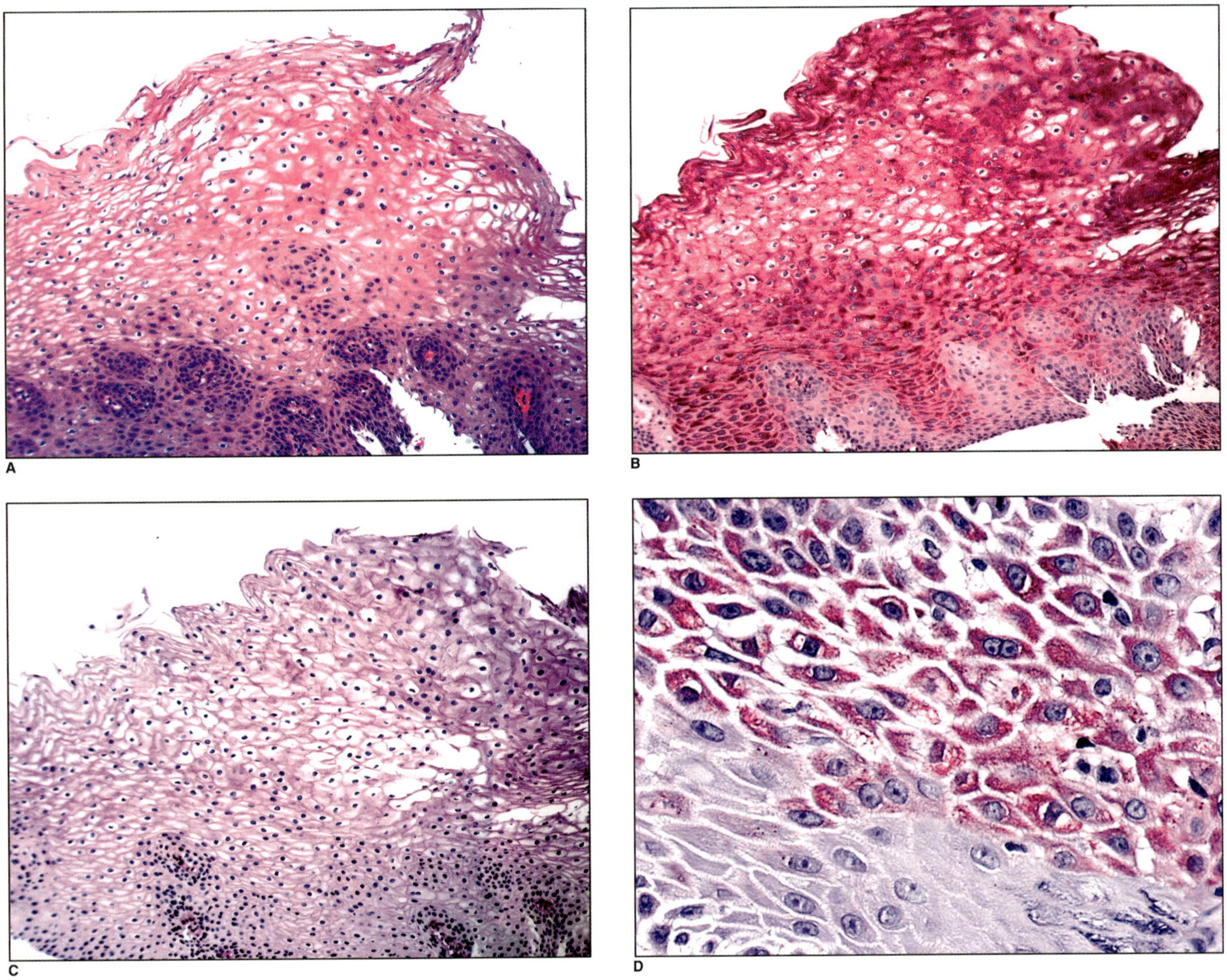
A

B

C

D

FIGURE 6.16: A focus of glycogenic acanthosis is seen in an esophageal biopsy. Glycogen-filled cells are identified by PAS stains (A—H&E, 100×; B—PAS, 100×; C—digested PAS, 100×; D*—PAS, 400×). *(Figures courtesy of Dr. Mark Wick.)

is reported to be more sensitive than Congo red, while the metachromatic stains are significantly less sensitive.

Glycogenic acanthosis of the esophagus is a relatively common finding in mucosal biopsies, occurring as small (<1 cm) mucosal plaques in the lower esophagus that resemble leukoplakia. Histologic examination reveals hyperplasia of the squamous epithelium accompanied by an accumulation of glycogen within squamous cells (73). PAS stains with and without diastase predigestion may be helpful in demonstrating abundant glycogen (Figure 6.16). Although the cause of this condition is not known, one study showed that most patients have gastroesophageal acid reflux (74). Some patients with Cowden's disease have been reported to have extensive esophageal glycogenic acanthosis (75).

Xanthelasma is a small (<0.5 cm) cluster of foamy macrophages in the superficial mucosa, most often in the stomach (76). These cells are positive with lipid stains such as oil red O and Sudan III, but reported results of staining with PAS, mucicarmine, and alcian blue have been variable (77).

Progressive systemic sclerosis (or generalized scleroderma) is a systemic disorder of collagen deposition that nearly always involves the GI tract, usually the esophagus (78). Trichrome stains highlight diffuse fibrosis of the submucosa and inner layer of the muscularis propria.

Involvement of the digestive tract by vasculitis is rare. Intestinal involvement usually occurs in association with systemic vasculitis and is a poor prognostic finding (79,80). Polyarteritis nodosa, systemic lupus erythematosus, and Henoch-Schonlein purpura are the systemic vasculitides that most commonly involve the GI tract (79,81,82). Vasculitis is usually not suspected as a cause of intestinal ischemia but should be considered in young patients because of the significant consequence of inaccurate diagnosis. Histologic examination reveals an inflammatory infiltrate involving the wall of small- and medium-sized blood vessels. Fibrinoid necrosis and thrombosis may also be seen. Histochemical stains are unlikely to be of diagnostic utility in superficial mucosal biopsies, but for resection specimens, stains to highlight inflamed blood vessels may be useful. Elastic stains are used most frequently for this purpose. The most commonly used elastic stain is Verhoeff's method (Verhoeff's hematoxylin with Van Gieson counterstain). Other elastic stains include orcein, Weigert's resorcin fuchsin, and the aldehyde fuchsin methods of Biot and of Gomori.

Hirschsprung's disease (aganglionosis) is caused by the congenital absence of ganglion cells in the distal large intestine and is usually diagnosed following rectal biopsy. In addition to the absence of ganglion cells in the submucosal or myenteric plexus, acetylcholinesterase stains performed on frozen sections reveal hyperplasia of cholinergic fibers and can be used to increase the accuracy of histologic assessment (Figure 6.17) (83–86).

Finally, crystalline deposits are sometimes seen in mucosal biopsies from the GI tract, and these can be identified by special stains. Crystalline iron deposits are occasionally seen in patients taking therapeutic dosage of iron tablets, especially ferrous sulfate (Figure 6.18) (87–89).

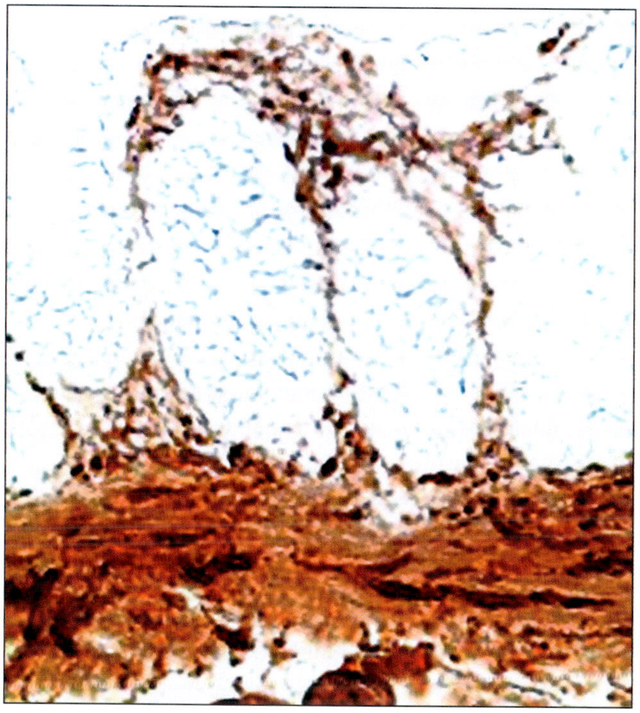

***FIGURE 6.17:** Hyperplasia of cholinergic fibers can be seen in Hirschsprung's disease (acetylcholinesterase, 200×). (Figure courtesy of Dr. Mark Wick.)*

NEOPLASTIC DISORDERS

Histochemical stains are less important than immunohistochemical and molecular methods in assessing poorly differentiated neoplasms, but histochemistry is useful in several selected situations.

Adenocarcinoma

Adenocarcinomas are the most common malignant neoplasms of the stomach and large intestine. An unequivocal diagnosis is easily made from the H&E stains alone in the great majority of cases, particularly in tumors of the colon and rectum. The primary utility of histochemistry in GI adenocarcinomas is in helping identify neutral or acidic mucins in poorly differentiated tumors (Figure 6.19). One example of this utility is in distinguishing diffuse-type gastric adenocarcinoma from gastric xanthelasma. Alcian blue will stain acid mucins in gastric adenocarcinoma but not in xanthelasma. Identifying acidic mucin with alcian

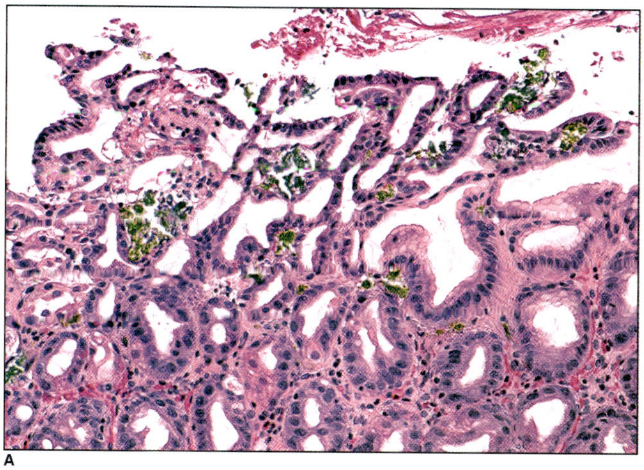

A

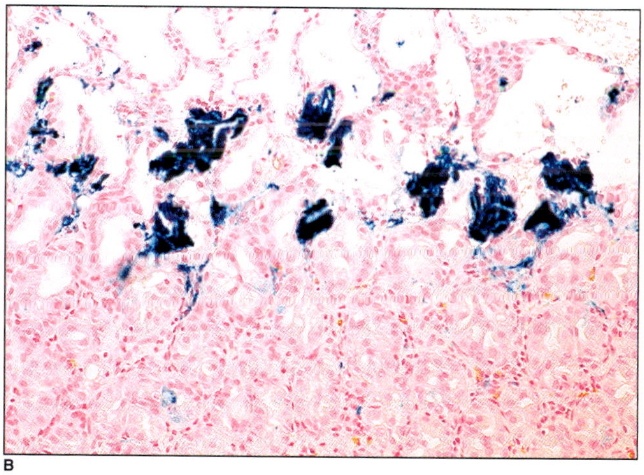

B

***FIGURE 6.18:** Golden brown, slightly refractile crystalline material is present on the surface of the gastric mucosa in a patient taking iron sulfate tablets (A). These crystalline iron deposits can be confirmed by iron stains (A—H&E, 100×; B—Prussian blue, 100×).*

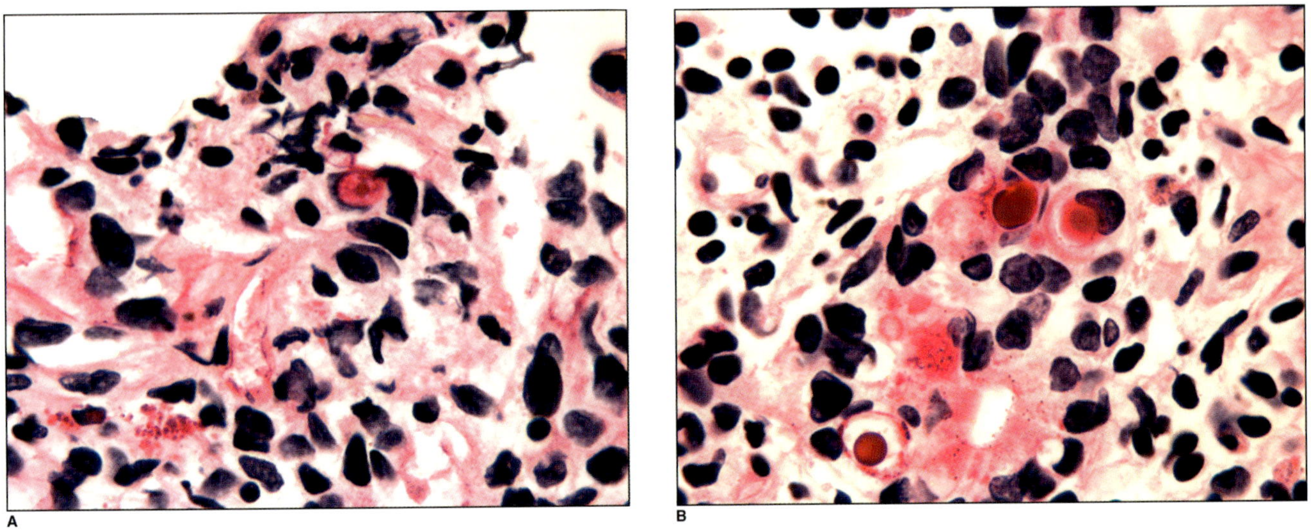

FIGURE 6.19: Poorly differentiated adenocarcinoma of the stomach. Intracytoplasmic mucin droplets help confirm the diagnosis (A—digested PAS, 400×; B—digested PAS, 1000×).

blue or mucicarmine stain may also be helpful in identifying mucoepidermoid carcinomas. Mucin histochemistry has not been shown to be of value in identifying patients with ulcerative colitis who are at increased risk for carcinoma (90).

Neuroendocrine Tumors

Carcinoid tumor is the most common neuroendocrine tumor of the GI tract. The appendix is the most common site of GI carcinoids, followed by the rectum and small bowel (91,92). Most are clinically silent and discovered incidentally, but the presentation varies to some extent depending on tumor location (93). Carcinoid tumors of the digestive tract have traditionally been subdivided into foregut, midgut, and hindgut types based on similarities in histology and histochemistry (Table 6.2) (94). Midgut carcinoids are the most frequent GI carcinoid tumors and are the tumors most commonly associated with the carcinoid syndrome (95,96). These tumors are found in the small bowel, appendix, and proximal colon, most occurring within 1 m of the ileocecal valve.

In addition to location and histologic pattern, carcinoid tumors have also been subclassified by the results of argentaffin and argyrophil staining. These stains depend on differences in the cell's ability to form intracellular silver deposits by reducing ammoniacal silver nitrate. Most carcinoid tumors are argyrophilic and exhibit intracytoplasmic brown-black silver granules by argyrophil stains, but only midgut carcinoids consistently stain with argentaffin stains (97). Argentaffin stains include Fontana-Masson, Gomori alkaline diazo technique, and Schmorl's methods. Argyrophil stains differ from argentaffin stains by the addition of a reduction step and include Grimelius, Churukian-Schenk, Pascual's, and Sevier-Munger stains (Figure 6.20). One study suggested that Churukian-Schenk was the most sensitive histochemical method in identifying carcinoid tumors (98).

TABLE 6.2. GI Carcinoid Tumors

Foregut tumors
- ➢ Stomach and proximal duodenum
- ➢ Trabecular growth pattern (anastomosing ribbons)
- ➢ Positive with Grimelius stain, rarely positive with Fontana-Masson
- ➢ Rarely associated with carcinoid syndrome

Midgut tumors
- ➢ Jejunum, ileum, appendix, proximal colon
- ➢ Insular pattern (solid nests)
- ➢ Positive with Fontana-Masson and Grimelius
- ➢ Carcinoid syndrome seen in 5–39%

Hindgut tumors
- ➢ Distal colon and rectum
- ➢ Mixed growth patterns
- ➢ Positive with Grimelius stain, negative with Fontana-Masson

Other Tumors

Histochemical stains may be of value in selected mesenchymal neoplasms, but for most of the common nonepithelial tumors in the GI tract, such as malignant lymphoma and mucosal melanoma, immunohistochemical evaluation has essentially supplanted traditional histochemical stains.

ILLUSTRATIVE CASE STUDIES

<u>CASE 1—</u> ***Granulocytic Sarcoma***: A 42-year-old woman presented with vague abdominal pain. Her medical history

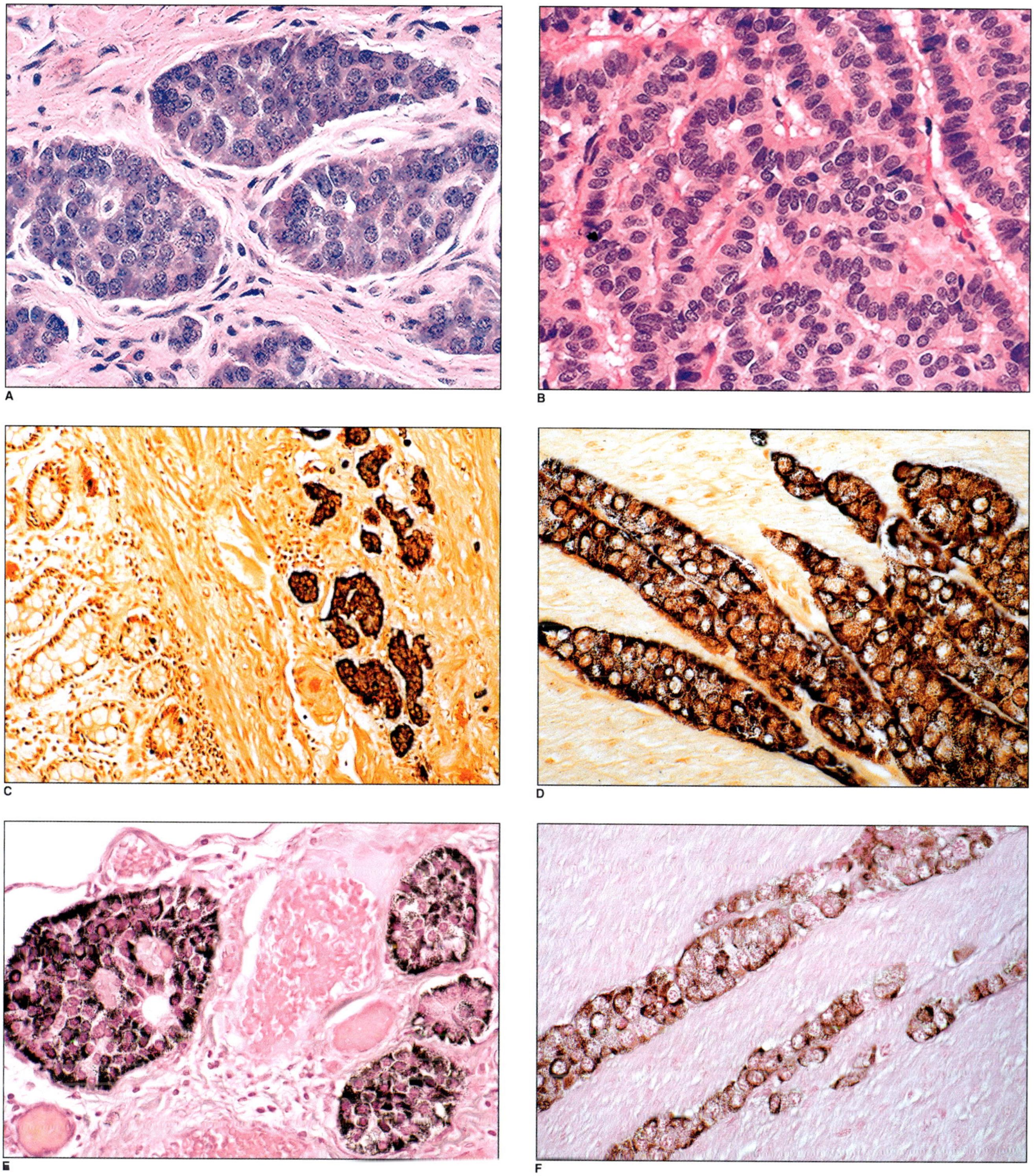

FIGURE 6.20: *Carcinoid tumor with typical nested (A) and trabecular (B) growth patterns. Most GI carcinoids will stain with an argyrophil stain (C and D). Only midgut carcinoids are argentaffin positive (E and F) (A and B*—H&E, 200×; C—Grimelius, 100×; D—Grimelius, 100×; E—Fontana-Masson, 100×; F—Fontana-Masson, 100×). *(Figures courtesy of Dr. Mark Wick.)*

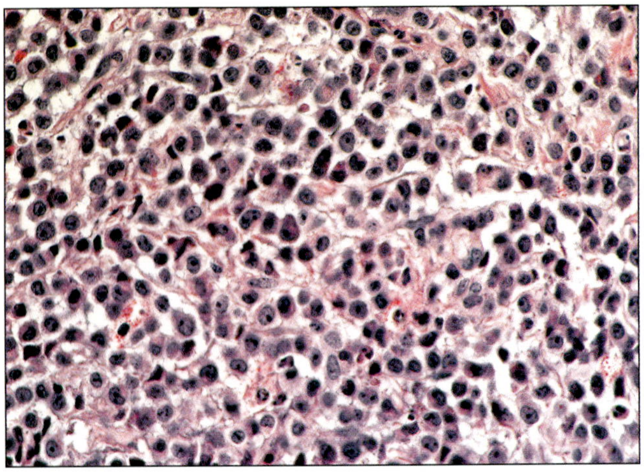

FIGURE 6.21: Undifferentiated tumor composed of primitive cells arranged in patternless sheets. The cells have dispersed chromatin, prominent nucleoli, and small amounts of cytoplasm (H&E, 400×).

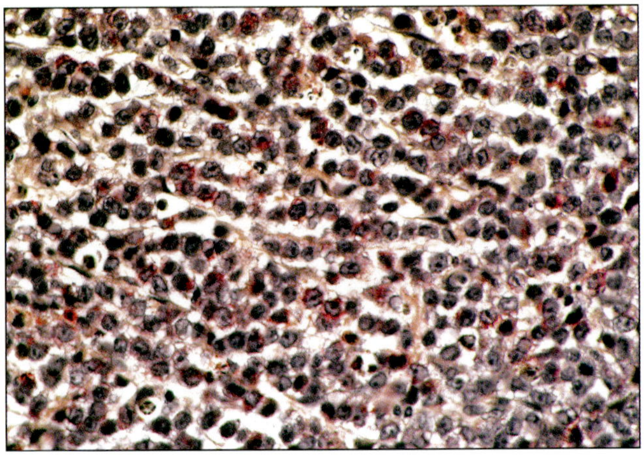

FIGURE 6.22: Some of the cells exhibit reaction with α-naphthol AS-D chloroacetate esterase indicating myeloid differentiation (Leder stain, 400×).

was negative. Physical examination did not disclose any specific findings, but an abdominal computed tomography (CT) scan revealed a mass in the ileum with evidence of partial bowel obstruction. A segmental resection of the small bowel was done and a 4.5-cm submucosal mass was found in the distal small bowel.

Microscopic examination revealed an undifferentiated malignant tumor composed of uniform cells with scant cytoplasm (Figure 6.21). The differential diagnosis included lymphoma, melanoma, anaplastic myeloma, and granulocytic sarcoma. The cells were strongly positive for α-naphthol AS-D chloroacetate esterase (Figure 6.22) and were also immunoreactive for lysozyme.

Granulocytic sarcoma (myeloid sarcoma, chloroma) is an extramedullary tumor of primitive myeloid cells that often precedes a diagnosis of acute myelogenous leukemia. The tumor can arise in any anatomic site but most often involves bone, soft tissue, skin, lymph nodes, or the central nervous system. Granulocytic sarcoma of the GI tract is rare and most frequently involves the small intestine (99). Because these tumors are morphologically undifferentiated

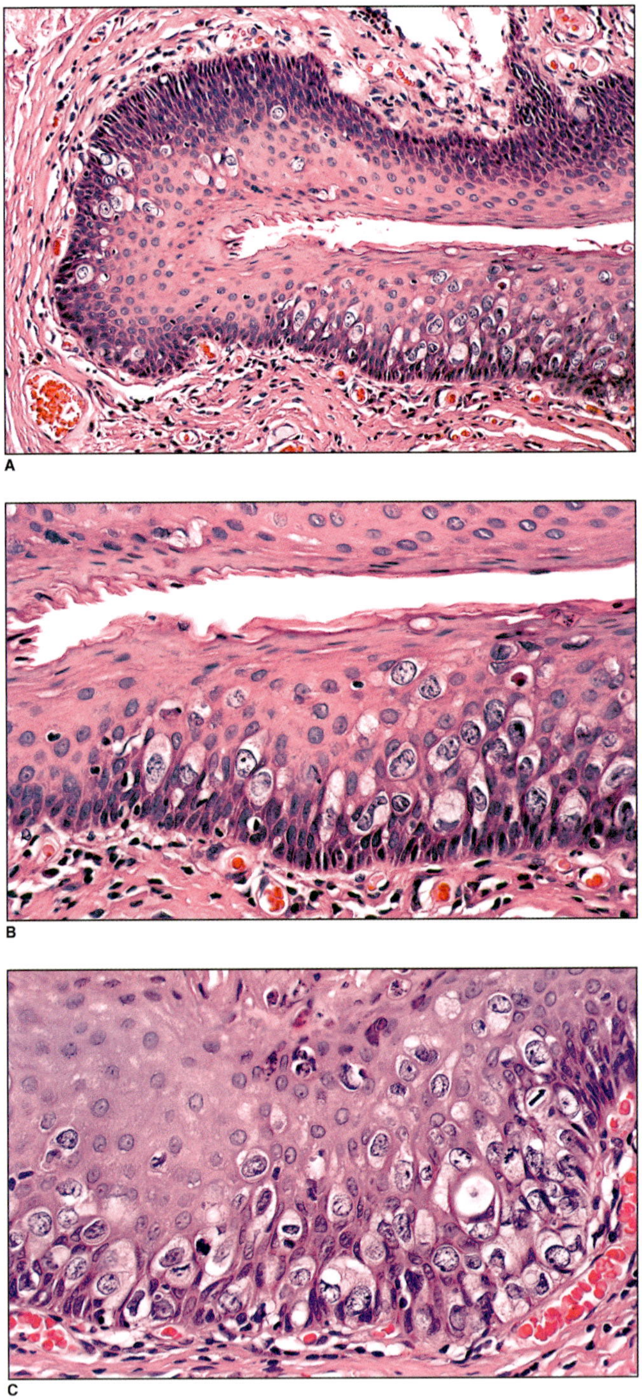

FIGURE 6.23: Large polygonal cells with abundant cytoplasm in the anal squamous epithelium (A—H&E, 100×; B—H&E, 200×; C—H&E, 400×).

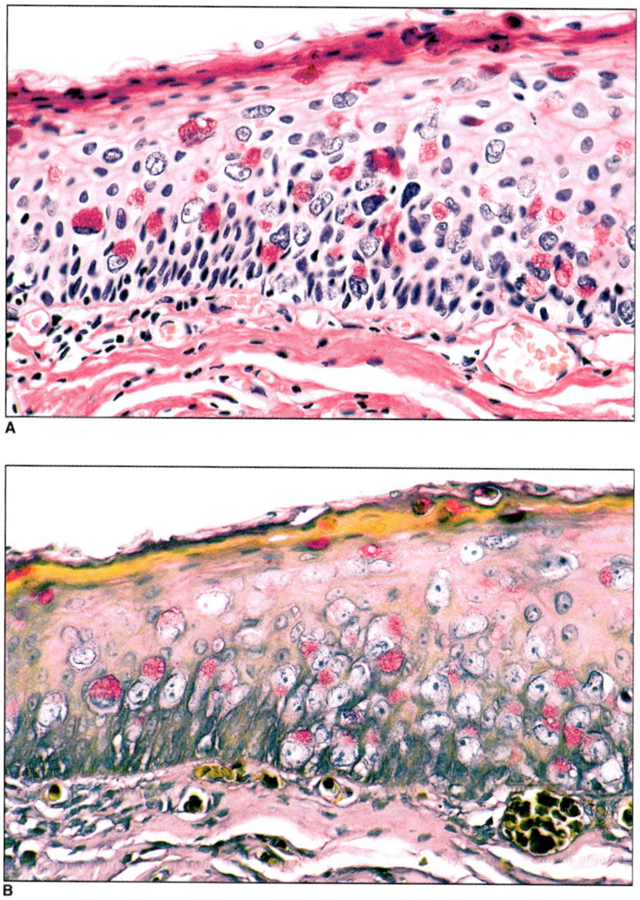

FIGURE 6.24: Paget disease cells stain brightly with both digested PAS (A) and mucicarmine (B) (A—digested PAS, 400×; B—Mayer mucicarmine, 400×).

and arise in patients not known to have leukemia, the tumor is often initially mistaken for lymphoma. Stains helpful in identifying myeloid lineage include α-naphthol AS-D chloroacetate esterase (Leder stain) and myeloperoxidase.

Chloroacetate esterase is present in granulocytes and mast cells but not monocytes or megakaryocytes. The stain identifies cells in the myeloid series from early promyelocytes to mature segmented neutrophils (100). Myeloblasts may be negative or only weakly positive for chloroacetate esterase but are usually positive for myeloperoxidase. As such, chloroacetate esterase is less sensitive than myeloperoxidase but is more specific for myeloid differentiation. PAS is not specific but mature myeloid cells and monocytes exhibit weak cytoplasmic positivity for PAS. ▪

CASE 2— ***Perianal Paget's Disease:*** A 51-year-old man presented with anal bleeding and pruritis. Clinical examination revealed scaling erythema of the anal mucosa, but no discrete masses were found. An anal biopsy revealed large polygonal cells in the anal epithelium (Figure 6.23). These cells had abundant pale cytoplasm and vesicular nuclei. No cytoplasmic pigment was found. The cells stained strongly with PAS and mucicarmine, typical of Paget's disease (Figure 6.24).

Extramammary Paget's disease is an intraepithelial malignancy most often found in the anogenital region. Large polygonal tumor cells confined to the squamous epithelium characterize the disease. In contrast to Paget's disease of the breast, in which nearly all patients have an underlying breast carcinoma, a clinically evident underlying carcinoma is frequently lacking in patients with extramammary Paget's disease. About one third of patients with perianal Paget's disease are found to have an associated carcinoma, usually in the anorectal region (101). The histologic differential diagnosis includes pagetoid actinic keratosis, squamous carcinoma in situ, and melanoma in situ. Histochemical stains can be very helpful in distinguishing among these lesions. The neoplastic cells stain with alcian blue, digested PAS, and mucicarmine; however, focal staining for mucin can be a potential diagnostic pitfall in small biopsy specimens (102). ▪

CASE 3— ***Abdominal Actinomycosis:*** A 23-year-old man presented with an acute onset of right lower quadrant abdominal pain. The white blood cell count was 18,600 and a CT scan revealed changes compatible with appendicitis. An appendectomy was performed. The appendix was covered by thick purulent exudate and the mesoappendix was firm and hemorrhagic. Sectioning revealed a focal perforation. Microscopic examination showed extensive suppurative inflammation and a small intramural abscess that contained several small sulfur granules (Figure 6.25). Brown-Brenn stain revealed thin filamentous bacteria typical of actinomyces organisms (Figure 6.26).

Abdominal actinomycosis is most often seen in patients with appendicitis or diverticulitis (103,104). The appendix and ileocecal region are the most common sites of GI actinomycosis; the majority of cases in one study were seen in patients with acute appendicitis with rupture (103). The infection is usually due to *Actinomyces israelii* (105,106). *Actinomyces* are anaerobic, gram-positive, non–acid-fast, filamentous bacteria that represent a normal component of the GI flora, especially the upper aerodigestive tract. The organism tends to form large colonies called sulfur granules surrounded by many neutrophils. Brightly eosinophilic club-like deposits are often seen radiating from the periphery of the sulfur granules (Splendore-Hoeppli phenomenon). These deposits are composed of fibrin and immunoglobulins (107). The sulfur granules are PAS positive, but the diagnosis of actinomycosis requires identifying the filamentous organisms, most often with a tissue Gram stain such as Brown-Brenn, Brown-Hopps, or Gram-Weigert stain. The bacteria will also stain with methenamine silver (Figure 6.27). ▪

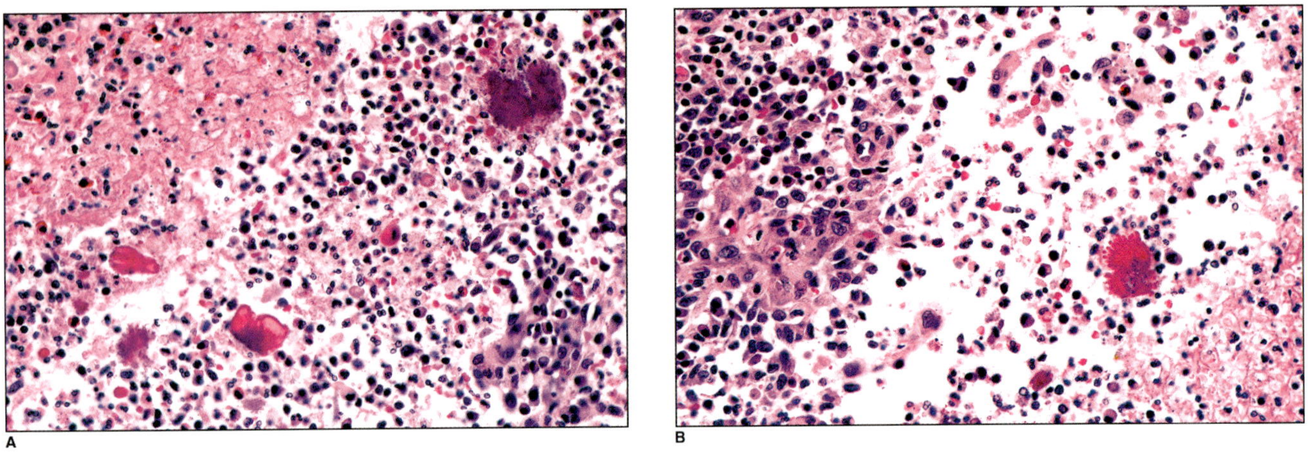

FIGURE 6.25: Small sulfur granules are found in an appendiceal abscess (A and B—H&E, 200×).

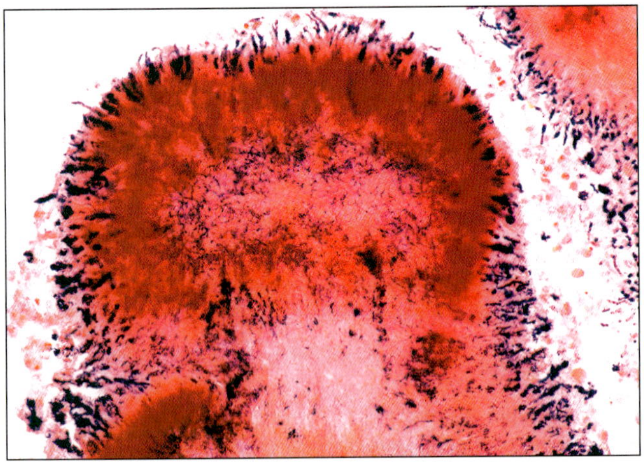

FIGURE 6.26: Actinomyces *organisms are gram positive (Brown-Brenn, 400×).*

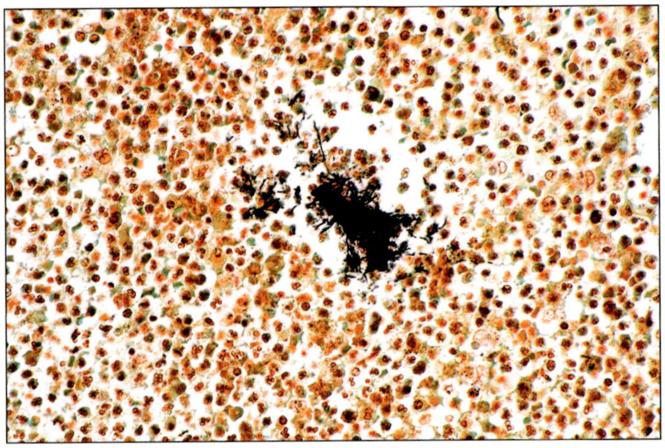

FIGURE 6.27: Actinomyces *will also stain with silver stains (Warthin-Starry, 400×).*

CASE 4— *Collagenous Colitis*: A 43-year-old woman was evaluated for persistent watery diarrhea. Colonoscopy revealed no significant endoscopic abnormalities. Random biopsies of the right and left colon were obtained. Histologic examination revealed a moderate increase in the number of lymphocytes in the lamina propria and epithelium, and marked thickening of the subepithelial collagen table (Figure 6.28).

Microscopic colitis, which is characterized by chronic watery (secretory) diarrhea and an endoscopically normal colonic mucosa, is divided into collagenous colitis and lymphocytic colitis. In collagenous colitis, biopsies show thickening of the subepithelial collagen table (at least 10 μm) (108). This is most easily assessed with a trichrome stain. In some cases, other eosinophilic fibrillary deposits such as amyloid must be excluded by Congo red or other stains.

The diagnosis of collagenous colitis requires an increased number of inflammatory cells in the lamina propria and intraepithelial lymphocytosis; thickening of the collagen table alone is insufficient for the diagnosis (109). A recent study suggests collagenous colitis may be underreported (110). Inflammatory bowel disease–like changes including crypt distortion and crypt abscesses can be seen in collagenous colitis and may result in the correct diagnosis being overlooked (111). ■

CASE 5— *Endocrine Cell Hyperplasia*: A 57-year-old man with a long history of peptic ulcer disease had been taking the proton pump inhibitor omeprazole for the past 3 years. Because of persistent upper abdominal pain, upper GI endoscopy was performed. The gastric mucosa showed mild hyperemia, and an antral biopsy was performed. The gastric mucosa showed mild chronic gastritis and patchy foci of incomplete intestinal metaplasia. Stains for neuroendocrine cells reveal increased numbers of neuroendocrine cells within the gastric antral glands (Figure 6.29).

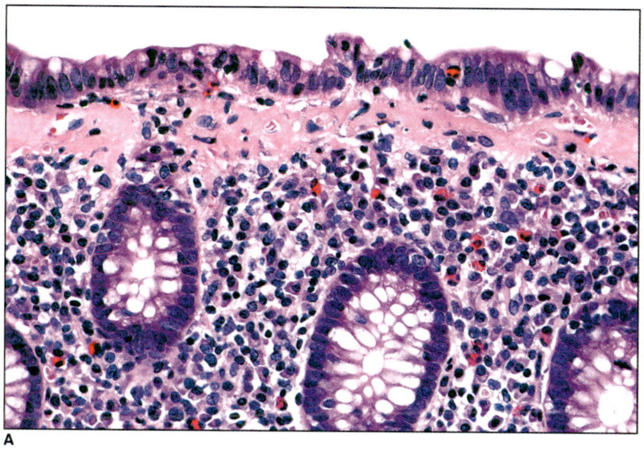

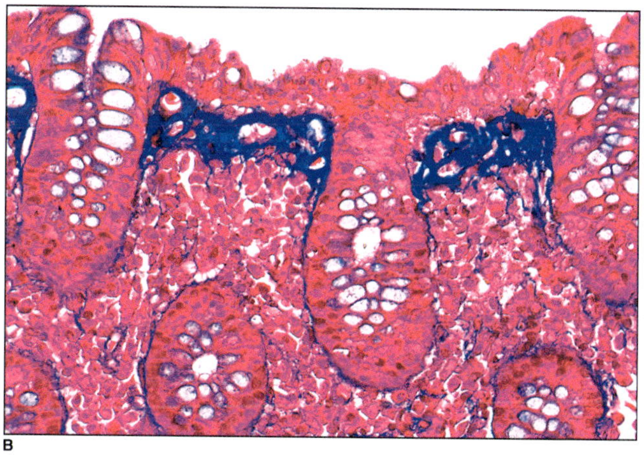

FIGURE 6.28: The subepithelial collagen table is markedly thickened in collagenous colitis (A—H&E, 200×; B—Mallory trichrome, 200×).

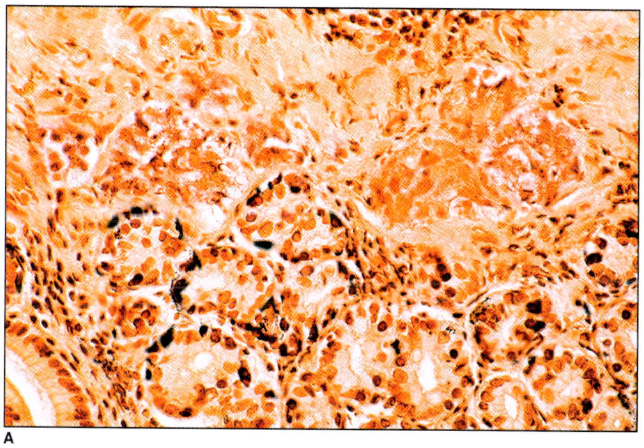

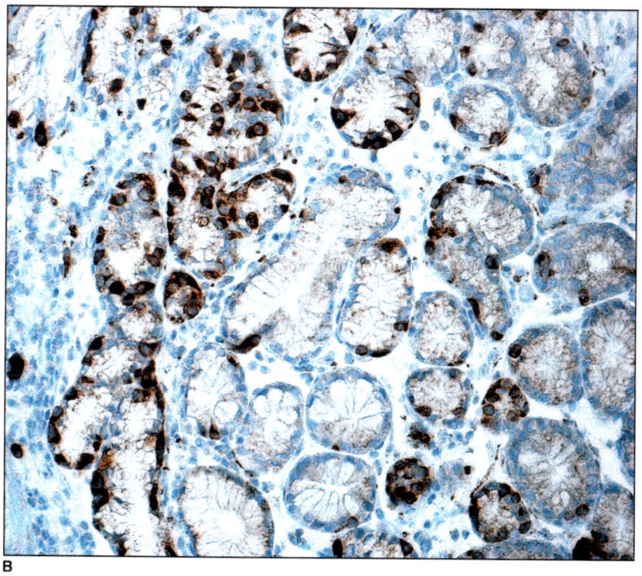

FIGURE 6.29: Stains for neuroendocrine markers demonstrate gastric endocrine cell hyperplasia (A—Grimelius, 200×; B—chromogranin A, 200×).

Proton pump inhibitors, which are commonly used to treat symptomatic gastroesophageal reflux disease and erosive esophagitis, are potent inhibitors of gastric acid secretion. In response to hypochlorhydria, most patients undergoing treatment with these drugs have moderate elevation of serum gastrin (112,113). Gastrin has a trophic influence on the enterochromaffin-like cells of the stomach (114), and mild gastric endocrine cell hyperplasia is relatively common after long-term acid suppression (113,115). Endocrine cells are difficult to identify by routine H&E stains, and an argyrophil stain such as Grimelius can be used to assess endocrine cell hyperplasia. Although hypergastrinemia clearly can cause significant endocrine cell hyperplasia, the risk of developing gastric carcinoids in these patients appears to be low (116). ■

CASE 6— ***Disseminated Histoplasmosis***: A 35-year-old man with a history of AIDS presented to the AIDS clinic with fever and abdominal pain. Multiple, small, sessile mucosal nodules were found in the ileocecal valve and colon during colonoscopy. Biopsies showed marked acute and chronic inflammation with large numbers of histiocytes in the mucosa and submucosa. Numerous small rounded organisms are seen in the cytoplasm of these cells (Figure 6.30).

Histoplasma capsulatum is a dimorphic fungus that is endemic to the river valleys of the central United States. The organism is a common cause of respiratory infection in these areas, which in immunocompetent patients is usually mild and self limiting. Those who are immunocompromised, by contrast, can develop disseminated histoplasmosis, which is invariably fatal if left untreated (117,118). GI involvement is common in patients with disseminated histoplasmosis, most of whom have AIDS. The most commonly involved sites in the gut are the terminal ileum and cecum (119). A wide variety of signs and symptoms accompany intestinal histoplasmosis. Patients can present with diarrhea, lower

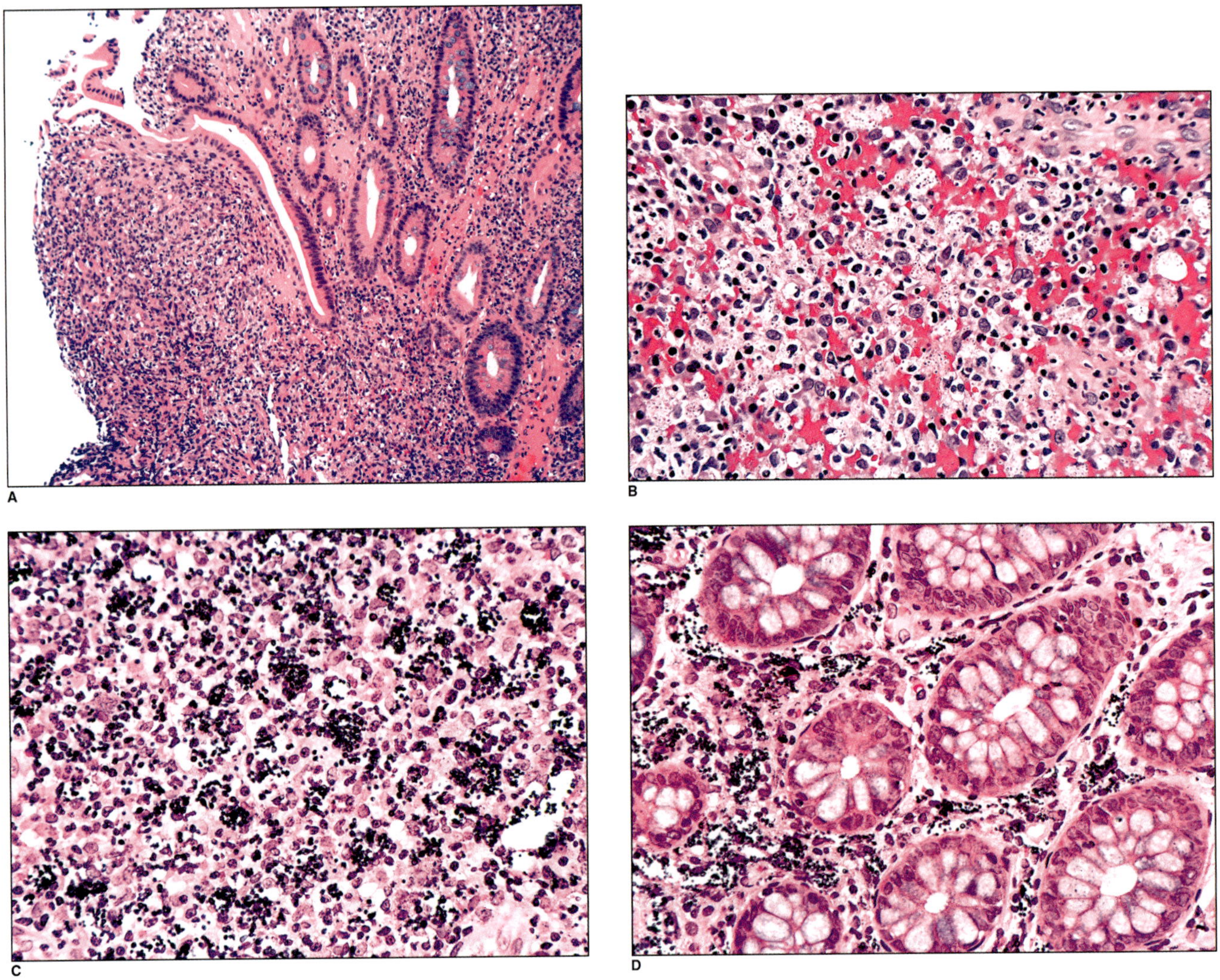

FIGURE 6.30: *Biopsies of the ileocecal valve and colon contain sheets of histiocytes, many of which contain small 3- to 5-μm intracytoplasmic organisms (A—H&E, 100×; B—H&E, 200×; C and D—combined methenamine silver/H&E stain, 200×).*

GI bleeding, abdominal pain, and/or obstruction. The diagnosis is rarely considered in the absence of known disseminated disease, and patients are commonly misdiagnosed with inflammatory bowel disease or cancer. ■

REFERENCES

1. Guarner F, Malagelada JR. Gut flora in health and disease. *Lancet* 2003;361:512–519.
2. Laine L, Lewin DN, Naritoku W, Cohen H. Prospective comparison of H&E, Giemsa, and Genta stains for the diagnosis of *Helicobacter pylori*. *Gastrointest Endosc* 1997;45:463–467.
3. Young DG. A stain for demonstrating *Helicobacter pylori* in gastric biopsies. *Biotechnic Histochem* 2001;76:31–34.
4. Rotimi O, Cairns A, Gray S, Moayyedi P, Dixon MF. Histological identification of *Helicobacter pylori*: comparison of staining methods. *J Clin Pathol.* 2000;53:756–759.
5. Skipper R, DeStephano DB. A rapid stain for *Campylobacter pylori* in gastrointestinal sections using Diff-Quik. *J Histotechnol* 1989;12:303–304.
6. El-Zimaity HM, Segura AM, Genta RM, Graham DY. Histologic assessment of *Helicobacter pylori* status after therapy: comparison of Giemsa, Diff-Quik, and Genta stains. *Mod Pathol* 1998;11:288–291.
7. Potvin CA. A modified Diff-Quik stain for *Helicobacter pylori* in gastrointestinal biopsies. *Lab Med* 1994;25:389–391.
8. Vartanian RK, Leung JK, Davis JE, Kim YB, Owen DA. A novel Alcian yellow–toluidine blue (Leung) stain for Helicobacter species: comparison with standard stains, a cost-effectiveness analysis, and supplemental utilities. *Mod Pathol* 1998;11:72–78.
9. Iwaki H, Sugiyama T, Asaka M. A modified McMullen's staining for *Helicobacter pylori*: a high-contrast, visibly prominent method. *Helicobacter* 1998;3:45–48.
10. McMullen L, Walker MM, Bain LA, Karim QN, Baron JH. Histological identification of *Campylobacter* using Gimenez

technique in gastric antral mucosa. *J Clin Pathol* 1987; 40:464–465.
11. Brown RW, Carson FL, Della Speranza V, Lewis SE, Fitzgibbons PL, Spencer LT. NSH/CAP HistoQIP Program. HQIP-B Final Critique. College of American Pathologists; Northfield, IL 2004.
12. Brown RW, Carson FL, Mohan D, Lewis SE, Spencer LT, Lott RL. NSH/CAP HistoQIP Program. HQIP-B Final Critique. College of American Pathologists; Northfield, IL 2006.
13. Doglioni C, Turrin M, Macri E, Chiarelli C, Germana B, Barbareschi M. HpSS: a new silver staining method for *Helicobacter pylori*. *J Clin Pathol* 1997;50:461–464.
14. Genta RM, Robason GO, Graham DY. Simultaneous visualization of *Helicobacter pylori* and gastric morphology: a new stain. *Hum Pathol* 1994;25:221–226.
15. El-Zimaity HMT, Ota H, Scott S, Killen DE, Graham DY. A new triple stain for *Helicobacter pylori* suitable for the autostainer: Carbol fuchsin/alcian blue/hematoxylin-eosin. *Arch Pathol Lab Med* 1998;122:732—736.
16. El-Zimaity HM. Modified triple stain (carbol fuchsin/alcian blue/hematoxylin-eosin) for the identification of *Helicobacter pylori*. *Arch Pathol Lab Med* 2000;124:1416–1417.
17. Heilman KL, Borchard F. Gastritis due to spiral shaped bacteria other than *Helicobacter pylori*: clinical, histological and ultrastructural findings. *Gut* 1991;32:137–140.
18. Okiyama Y, Matsuzawa K, Hidaka E, Sano K, Akamatsu T, Ota H. *Helicobacter heilmannii* infection: Clinical, endoscopic and histopathological features in Japanese patients. *Pathol Int* 2005;55:398–404.
19. Hilzenrat N, Lamoureux E, Weintrub I, et al. *Helicobacter heilmannii*-like spiral bacteria in gastric mucosal biopsies. Prevalence and clinical significance. *Arch Pathol Lab Med* 1995;119:1149–1153.
20. Leung VK, Law ST, Lam CW, et al. Intestinal tuberculosis in a regional hospital in Hong Kong: a 10-year experience. *Hong Kong Med J* 2006;12:264–271.
21. Horvath KD, Whelan RL. Intestinal tuberculosis: return of an old disease. *Am J Gastroenterol* 1998;93:692–696.
22. Naga MI, Okasha HH, Ismail Z, El-Fatatry M, Hassan S, Monir BE. Endoscopic diagnosis of colonic tuberculosis. *Gastrointest Endosc* 2001;53:789–793.
23. Misra SP, Misra V, Dwivedi M, Gupta SC. Colonic tuberculosis: clinical features, endoscopic appearance and management. *J Gastroenterol Hepatol* 1999;14:723–729.
24. Gan HT, Chen YQ, Ouyang Q, Bu H, Yang XY. Differentiation between intestinal tuberculosis and Crohn's disease in endoscopic biopsy specimens by polymerase chain reaction. *Am J Gastroenterol* 2002;97:1446–1451.
25. Gray JR, Rabeneck L. Atypical mycobacterial infection of the gastrointestinal tract in AIDS patients. *Am J Gastroenterol* 1989;84:1521–1524.
26. Roth RI, Owen RL, Keren DF, Volberding PA. Intestinal infection with *Mycobacterium avium* in acquired immune deficiency syndrome (AIDS). Histological and clinical comparison with Whipple's disease. *Dig Dis Sci* 1985;30:497–504.
27. Brandwein M, Choi HSH 2, Strauchen J, Stoler M, Jagirdar J. Spindle cell reaction to nontuberculous mycobacteriosis in AIDS mimicking a spindle cell neoplasm. Evidence for dual histiocytic and fibroblast-like characteristics of spindle cells. *Virchows Arch* 1990;416:281–286.
28. Wood C, Nickoloff BJ, Todes-Taylor NR. Pseudotumor resulting from atypical mycobacterial infection: a "histoid" variety of *Mycobacterium avium-intracellulare* complex infection. *Am J Clin Pathol* 1985;83:524–527.
29. Dutly F, Altwegg M. Whipple's disease and "*Tropheryma whippelii*." *Clin Microbiol Rev* 2001;14:561–583.
30. Swartz MN. Whipple's disease—past, present, and future. *N Engl J Med* 2000;342:648–650.
31. Relman DA, Schmidt TM, MacDermott RP, Falkow S. Identification of the uncultured bacillus of Whipple's disease. *N Engl J Med* 1992; 327:293–301.
32. von Herbay A, Maiwald M, Ditton HJ, Otto HF. Histology of intestinal Whipple's disease revisited: a study of 48 patients. *Virchows Arch* 1996;429:335–343.
33. Sanmeier D, Guillou L, Malakoplakia and adenocarcinoma of the caecum: a rare association. *J Clin Pathol* 1993;46:959–960.
34. Pillay K, Chetty R. Malakoplakia in association with colorectal carcinoma: a series of four cases. *Pathology* 2002;34:332–335.
35. Abdou NI, NaPombejara C, Sagawa A, et al. Malakoplakia: evidence for monocyte lysosomal abnormality correctable by cholinergic agonist in vitro and in vivo. *N Engl J Med* 1977;297:1413–1419.
36. Korner M, Gebbers JO. Clinical significance of human intestinal spirochetosis—a morphologic approach. *Infection* 2003; 31:341–349.
37. Koteish A, Kannangai R, Abraham SC, Torbenson M. Colonic spirochetosis in children and adults. *Am J Clin Pathol* 2003; 120:828–832.
38. Surawicz CM, Roberts PL, Rompalo A, Quinn TC, Holmes KK, Stamm WE. Intestinal spirochetosis in homosexual men. *Am J Med* 1987,82.587–592.
39. Alsaigh N, Fogt F. Intestinal spirochetosis: clinicopathological features with review of the literature. *Colorectal Dis* 2002;4:97–100.
40. Wilcox CM, Karowe MW. Esophageal infections: etiology, diagnosis, and management. *Gastroenterologist* 1994;2: 188–206.
41. Thom K, Forrest G. Gastrointestinal infections in immunocompromised hosts. *Curr Opin Gastroenterol* 2006;22:18–23.
42. Prescott RJ, Harris M, Banerjee SS. Fungal infections of the small and large intestine. *J Clin Pathol* 1992;45:806–811.
43. Lamps LW, Molina CP, West AB, Haggitt RC, Scott MA. The pathologic spectrum of gastrointestinal and hepatic histoplasmosis. *Am J Clin Pathol* 2000;113:64–72.
44. Cappell MS, Mandell W, Grimes MM, Neu HC. Gastrointestinal histoplasmosis. *Dig Dis Sci* 1988;33:353–360.
45. Schuster CJ, Ellis AG, Robertson WJ, et al. Infectious disease outbreaks related to drinking water in Canada, 1974–2001. *Can J Public Health* 2005;96:254–258.
46. Lumadue JA, Manabe YC, Moore RD, Belitsos PC, Sears CL, Clark DP. A clinicopathologic analysis of AIDS-related cryptosporidiosis. *AIDS* 1998;12:2459–2466.
47. Lindsay DS, Dubey JP, Blagburn BL. Biology of *Isospora* spp. from humans, nonhuman primates and domestic animals. *Clin Microbiol Rev* 1997;10:19–34.

48. DeHovitz JA, Pape JW, Boncy M, Johnson WD Jr. Clinical manifestations and therapy of *Isospora belli* infection in patients with acquired immunodeficiency syndrome. *N Engl J Med* 1986;315:87–90.
49. Mannheimer SB, Soave R. Protozoal infections in patients with AIDS. Cryptosporidiosis, isosporiasis, cyclosporiasis, and microsporidiosis. *Infect Dis Clin North Am* 1994;8: 483–498.
50. van Gool T, Dankert J. Human microsporidiosis: clinical, diagnostic and therapeutic aspects of an increasing infection. *Clin Microbiol Infect* 1995;1:75–85.
51. Lamps LW, Bronner MP, Vnencak-Jones CL, Tham KT, Mertz HR, Scott MA. Optimal screening and diagnosis of microsporidia in tissue sections: a comparison of polarization, special stains, and molecular techniques. *Am J Clin Pathol* 1998;109:404–410.
52. Sun T, Kaplan MH, Teichberg S, Weissman G, Smilari T, Urmacher C. Intestinal microsporidiosis. Report of five cases. *Ann Clin Lab Sci* 1994;24:521–532.
53. Jass JR, Filipe MI. The mucin profiles of normal gastric mucosa, intestinal metaplasia and its variants and gastric carcinoma. *Histochem J* 1981;13:931–939.
54. McFadden DE, Owen DA, Reid PE, Jones EA. The histochemical assessment of sulphated and non-sulphated sialomucin in intestinal epithelium. *Histopathology* 1985;9: 1129–1137.
55. Shah KA, Deacon AJ, Dunscombe P, Price AB. Intestinal metaplasia subtyping: evaluation of Gomori's aldehyde fuchsin for routine diagnostic use. *Histopathology* 1997;31:277–283.
56. Sipponen P, Seppala K, Varis K, et al. Intestinal metaplasia with colonic-type sulphomucins in the gastric mucosa; its association with gastric carcinoma. *Acta Pathol Microbiol Scand [A]* 1980;88:217–224.
57. El-Zimaity HM, Ramchatesingh J, Saeed MA, Graham DY. Gastric intestinal metaplasia: subtypes and natural history. *J Clin Pathol* 2001;54:679–683.
58. Baracchini P, Fulcheri E, Lapertosa G. Patterns of intestinal metaplasia in gastric biopsies. A comparison of different histochemical classifications. *Histochem J* 2005;23:1–9.
59. Jass JR. Role of intestinal metaplasia in the histogenesis of gastric carcinoma. *J Clin Pathol* 1980;33:801–810.
60. Tosi P, Filipe MI, Luzi P, et al. Gastric intestinal metaplasia type III cases are classified as low-grade dysplasia on the basis of morphometry. *J Pathol* 1993;169:73–78.
61. Filipe MI, Munoz N, Matko I, et al. Intestinal metaplasia types and the risk of gastric cancer: a cohort study in Slovenia. *Int J Cancer* 1994;57:324–329.
62. Cassaro M, Rugge M, Gutierrez O, Leandro G, Graham DY, Genta RM. Topographic patterns of intestinal metaplasia and gastric cancer. *Am J Gastroenterol* 2000;95: 1431–1438.
63. Ramesar KC, Sanders DS, Hopwood D. Limited value of type III intestinal metaplasia in predicting risk of gastric carcinoma. *J Clin Pathol* 1987;40:1287–1290.
64. Matsukuma A, Mori M, Enjoji M. Sulphomucin-secreting intestinal metaplasia in the human gastric mucosa. An association with intestinal-type gastric carcinoma. *Cancer* 1990;66:689–694.
65. Haggitt RC. Barrett's esophagus, dysplasia, and adenocarcinoma. *Hum Pathol* 1994;25:982–993.
66. Odze RD. Pathology of the gastroesophageal junction. *Semin Diagn Pathol* 2005;22:256–265.
67. Younes M, Ertan A, Ergun G, et al. Goblet cell mimickers in esophageal biopsies are not associated with an increased risk for dysplasia. *Arch Pathol Lab Med* 2007; 131:571–575.
68. Wright CL, Kelly JK. The use of routine special stains for upper gastrointestinal biopsies. *Am J Surg Pathol* 2006;30: 357–361.
69. Peuchmaur M, Potet F, Goldfain D. Mucin histochemistry of the columnar epithelium of the oesophagus (Barrett's oesophagus): a prospective biopsy study. *J Clin Pathol* 1984; 37:607–610.
70. Jauregui HO, Davessar K, Hale JH, et al. Mucin histochemistry of intestinal metaplasia in Barrett's esophagus. *Mod Pathol* 1988;1:188–192.
71. Ohashi K, Takagawa R, Hara M. Visceral organ involvement and extracellular matrix changes in beta 2-microglobulin amyloidosis—a comparative study with systemic AA and AL amyloidosis. *Virchows Arch* 1997;430:479–487.
72. Hemmer PR, Topazian MD, Gertz MA, Abraham SC. Globular amyloid deposits isolated to the small bowel: a rare association with AL amyloidosis. *Am J Surg Pathol* 2007;31: 141–145.
73. Bender MD, Allison J, Cuartas F, Montgomery C. Glycogenic acanthosis of the esophagus: a form of benign epithelial hyperplasia. *Gastroenterology* 1973;65:373–380.
74. Vadva MD, Triadafilopoulos G. Glycogenic acanthosis of the esophagus and gastroesophageal reflux. *J Clin Gastroenterol* 1993;17:79–83.
75. Kay PS, Soetikno RM, Mindelzun R, Young HS. Diffuse esophageal glycogenic acanthosis: an endoscopic marker of Cowden's disease. *Am J Gastroenterol* 1997;92:1038–1040.
76. Gencosmanoglu R, Sen-Oran E, Kurtkaya-Yapicier O, Tozun N. Xanthelasmas of the upper gastrointestinal tract. *J Gastroenterol* 2004;39:215–219.
77. Bejarano PA, Aranda-Michel J, Fenoglio-Preiser C. Histochemical and immunohistochemical characterization of foamy histiocytes (muciphages and xanthelasma) of the rectum. *Am J Surg Pathol* 2000;24:1009–1015.
78. Lock G, Holstege A, Lang B, Scholmerich J. Gastrointestinal manifestations of progressive systemic sclerosis. *Am J Gastroenterol* 1997;92:763–771.
79. Scott DG, Bacon PA, Elliott PJ, Tribe CR, Wallington TB. Systemic vasculitis in a district general hospital 1972–1980: clinical and laboratory features, classification and prognosis of 80 cases. *Q J Med* 1982;51:292–311.
80. Camilleri M, Pusey CD, Chadwick VS, Rees AJ. Gastrointestinal manifestations of systemic vasculitis. *Q J Med* 1983;52:141–149.
81. Passam F, Diamantis I, Perisinaki G, et al. Intestinal ischemia as the first manifestation of vasculitis. *Semin Arthritis Rheum* 2004;34:431–441.
82. Hallegua DS, Wallace DJ. Gastrointestinal manifestations of systemic lupus erythematosus. *Curr Opin Rheum* 2000; 12:379–385.

83. Hamoudi AB, Reiner CB, Boles ET Jr., McClung HJ, Kerzner B. Acetylthiocholinesterase staining activity of rectal mucosa. Its use in the diagnosis of Hirschsprung's disease. *Arch Pathol Lab Med* 1982;106:670–672.
84. Qualman SJ, Murray R. Aganglionosis and related disorders. *Hum Pathol* 1994;25:1141–1149.
85. Galvis DA, Yunis EJ. Comparison of neuropeptide Y, protein gene product 9.5, and acetylcholinesterase in the diagnosis of Hirschsprung's disease. *Pediatr Pathol Lab Med* 1997;17:413–425.
86. Nakao M, Suita S, Taguchi T, Hirose R, Shima Y. Fourteen-year experience of acetylcholinesterase staining for rectal mucosal biopsy in neonatal Hirschsprung's disease. *J Pediatr Surg* 2001;36:1357–1363.
87. Abraham SC, Yardley JH, Wu TT. Erosive injury to the upper gastrointestinal tract in patients receiving iron medication: an underrecognized entity. *Am J Surg Pathol* 1999;23:1241–1247.
88. Haig A, Driman DK. Iron-induced mucosal injury to the upper gastrointestinal tract. *Histopathology* 2006;48:808–812.
89. Ji H, Yardley JH. Iron medication–associated gastric mucosal injury. *Arch Pathol Lab Med* 2004;128:821–822.
90. Jass JR, England J, Miller K. Value of mucin histochemistry in follow up surveillance of patients with long standing ulcerative colitis. *J Clin Pathol* 1986;39:393–398.
91. Godwin JD. Carcinoid tumors. An analysis of 2837 cases. *Cancer* 1975;36:560–569.
92. Hajdu SI, Winawer SJ, Myers WPL. Carcinoid tumors. A study of 204 cases. *Am J Clin Pathol* 1974;61:521–528.
93. Lechago J. Gastrointestinal neuroendocrine cell proliferations. *Hum Pathol* 1994;25:1114–1122.
94. Williams ED, Sandler M. The classification of carcinoid tumors. *Lancet* 1963;1:238–243.
95. Burke AP, Thomas RM, Elsayed AM, Sobin LH. Carcinoids of the jejunum and ileum. An immunohistochemical and clinicopathologic study of 167 cases. *Cancer* 1997;79:1086–1093.
96. Cunningham JD, Aleali R, Aleali M, Brower ST, Aufses AH. Malignant small bowel neoplasms. Histopathologic determinants of recurrence and survival. *Ann Surg* 1997;225:300–306.
97. Wilander E, Portela-Gomes G, Grimelius L, Westermark P. Argentaffin and argyrophil reactions of human gastrointestinal carcinoids. *Gastroenterology* 1977;73:733–736.
98. Smith DM Jr., Haggitt RC. A comparative study of generic stains for carcinoid secretory granules. *Am J Surg Pathol* 1983;7:61–68.
99. Kohl SK, Aoun P. Granulocytic sarcoma of the small intestine. *Arch Pathol Lab Med* 2006;130:1570–1574.
100. Leder LD. Diagnostic experiences with the naphthol AS-D chloroacetate esterase reaction. *Ann Hematol* 1970;21:1–8.
101. Goldman S, Ihre T, Lagerstedt U, Svensson C. Perianal Paget's disease: report of five cases. *Int J Colorect Dis* 1992;7:167–169.
102. Battles OE, Page DL, Johnson JE. Cytokeratins, CEA, and mucin histochemistry in the diagnosis and characterization of extramammary Paget's disease. *Am J Clin Pathol* 1997;108:6–12.
103. Deshmukh N, Heaney S. Actinomycosis at multiple colonic sites. *Am J Gastroenterol* 1986;81:1212–1214.
104. Weese WC, Smith IM. A study of 57 cases of actinomycosis over a 36-year period. *Arch Intern Med* 1975;135:1562–1568.
105. Russo T. Agents of Actinomycosis. In: Mandell G, Bennett J, Dolin R, eds. *Principles and Practice of Infectious Diseases.* 4th ed. New York, NY: Churchill Livingston; 1995: 2280–2288.
106. Cintron JR, Del Pino A, Duarte B, Wood D. Abdominal actinomycosis. *Dis Colon Rectum* 1996;39:105–108.
107. Rodig SJ, Dorfman DM. Splendore-Hoeppli phenomenon. *Arch Pathol Lab Med* 2001;125:1515–1516.
108. Offner FA, Jao RV, Lewin KJ, Havelec L, Weinstein WM. Collagenous colitis: a study of the distribution of morphological abnormalities and their histological detection. *Hum Pathol* 1999;30:451–457.
109. Lazenby AJ. Collagenous and lymphocytic colitis. *Semin Diagn Pathol* 2005;22:295–300.
110. Jaskiewicz K, Rzepko R, Adrych K, Smoczynski M. Microscopic colitis in routine colonoscopies. *Dig Dis Sci* 2006,51.241–244.
111. Ayata G, Ithamukkala S, Sapp H, et al. Prevalence and significance of inflammatory bowel disease-like morphologic features in collagenous and lymphocytic colitis. *Am J Surg Pathol* 2002;26:1414–1423.
112. Bordi C, D'Adda T, Azzoni C, Pilato FP, Caruana P. Hypergastrinemia and gastric enterochromaffin-like cells. *Am J Surg Pathol* 1995;19(Suppl 1):S8–S19.
113. Eissele R, Brunner G, Simon B, Solcia E, Arnold R. Gastric mucosa during treatment with lansoprazole: *Helicobacter pylori* is a risk factor for argyrophil hyperplasia. *Gastroenterology* 1997;112:707–717.
114. Tielemans Y, Hakanson R, Sundler F, Willems G. Gastrin stimulates the self-replication rate of enterochromaffinlike cells in the rat stomach. *Gastroenterology* 1989;96:723–729.
115. Lamberts R, Creutzfeldt W, Strüber HG, Brunner G, Solcia E. Long-term omeprazole therapy in peptic ulcer disease: gastrin, endocrine cell growth, and gastritis. *Gastroenterology* 1993;104:1356–1370.
116. Peghini PL, Annibale B, Azzoni C, et al. Effect of chronic hypergastrinemia on human enterochromaffin-like cells: insights from patients with sporadic gastrinomas. *Gastroenterology* 2002;123:68–85.
117. Kahi CJ, Wheat LJ, Allen SD, Sarosi GA. Gastrointestinal histoplasmosis. *Am J Gastroenterol* 2005;100:220–231.
118. Lee JT, Dixon MR, Murrell Z, et al. Colonic histoplasmosis presenting as colon cancer in the nonimmunocompromised patient: report of a case and review of the literature. *Am Surg* 2004;70:959–963.
119. Assi M, McKinsey DS, Driks MR, et al. Gastrointestinal histoplasmosis in the acquired immunodeficiency syndrome: report of 18 cases and literature review. *Diagn Microbiol Infect Dis* 2006;55:195–201.

CHAPTER 7

Diagnostic Histochemistry in Hepatic Pathology

Julia C. Iezzoni, M.D.

Both historically and in current practice, diagnostic hepatopathology is based on the interpretation of the morphologic features demonstrated on hematoxylin and eosin (H&E)–stained tissue sections in conjunction with findings in a series of histochemical stains. Because of their selective biochemical properties, histochemical procedures contribute useful information by the delineation of specific architectural and cellular features. Although histochemistry remains an important part of the evaluation of liver tissue, its specific role has changed because of other advances in diagnostic surgical pathology, as well as in medicine as a whole. For example, the widespread availability of sensitive and specific immunohistochemical stains has, in selected instances, replaced some of the functions previously served by histochemical methods. Specific enzymatic and genetic tests have modified and, in some cases, supplanted, the role of histochemistry in the diagnosis of many metabolic disorders. Moreover, the role of liver biopsy itself is changing (1–3). Emphasis was previously given to an etiologic diagnosis of the hepatic diseases, but liver biopsy now has important additional roles in the assessment of cases for various therapeutic modalities and in prognosis. In addition, the advent and widespread use of liver transplantation has added a new group of issues that must be addressed in diagnostic hepatopathology.

Because of these advances, many of the questions attached to liver biopsies—in both H&E and special histochemical sections—have been altered as well. This chapter focuses on the practical applications of histochemical techniques in contemporary hepatopathology. In addition, in view of the fact that most diagnostic liver samples are needle core biopsies, the utility of histochemistry in evaluating those particular specimens will be emphasized. For technical details of specific histochemical procedures and biochemical formulations, the reader is referred to other manuals and texts (4–6).

CONNECTIVE TISSUE STAINS IN NONNEOPLASTIC LIVER DISEASE

Connective tissue stains are an integral part of the routine evaluation of liver biopsies because of their ability to delineate specific architectural features of the hepatic lobules. By doing so, such techniques assist in the identification of morphologic injury patterns. In turn, because each pattern is associated with a well-recognized group of possible etiologies, the mode of injury provides clues to the specific cause of the liver disease. In addition, certain architectural features may provide information about the response to treatment as well as disease progression and stage. Accordingly, in current practice, connective tissue stains are utilized for diagnostic, therapeutic, and prognostic purposes.

By way of review, three main types of connective tissue fibers are present in the normal liver (5,6). *Type I collagen* is located predominantly in the portal tracts and the walls of efferent veins. It is composed of tightly packed thick fibrils that form broad fibers. Type I collagen can be demonstrated readily with a variety of histochemical methods, but the Masson trichrome technique is probably most commonly used for this purpose (Figure 7.1A). *Type III collagen* is the principal component of reticulin fibers in the space of Disse, and it comprises loosely aggregated fibrils. The reticulin fibers mantle the hepatic plates, thereby forming a delicate network that extends in a regular, radial pattern between the portal tracts and the terminal venules. The reticulin fibers delineate the integrity of the hepatic plates and the overall acinar structure. They can be demonstrated with silver impregnation methods such as Snook's reticulin stain (Figure 7.1B). *Elastic fibers* are prominent in the portal tracts and can be labeled with several histochemical procedures; these include the Verhoeff–Van Gieson (VVG), orcein, and Victoria blue methods (Figure 7.1C).

In several disease states, connective tissue stains highlight a variety of structural alterations (6–8). Because of its ability to decorate type I collagen fibers, the trichrome stain is effective for demonstration of fibrosis with respect to its presence and its pattern of distribution. A silver stain for reticulin outlines abnormalities in the acinar architecture and the hepatic plates, such as hepatocyte dropout, regeneration-related thickening of the hepatic plates, loss of the normal spatial relationship between portal tracts and terminal venules, and lobular collapse owing to zonal or

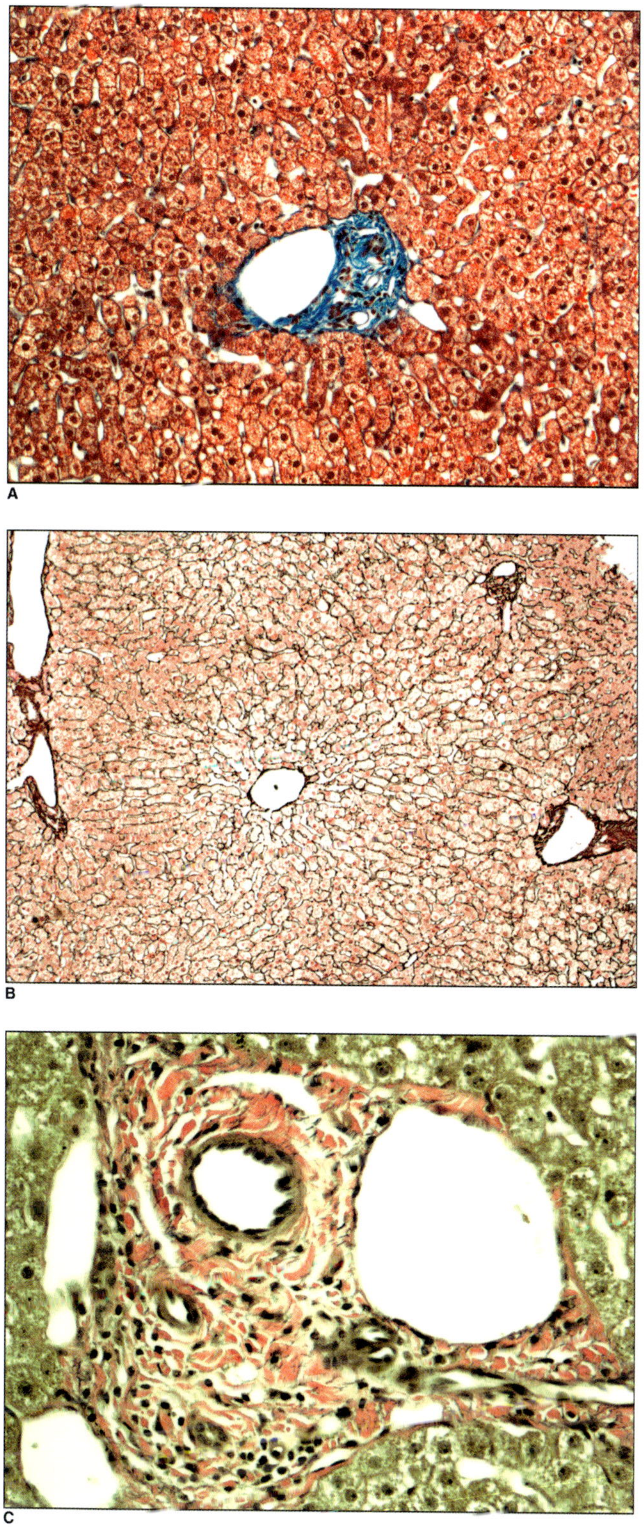

A B C

FIGURE 7.1: *Normal liver: (A). Type I collagen, which is readily demonstrated with the trichrome stain, is located predominantly in portal tracts and the wall of efferent veins. (B). The reticulin stain, which stains type III collagen, delineates the reticulin fibers that line the hepatic plates. (C). Elastic fibers, as identified with the Verhoeff–Van Gieson stain, are prominent in the portal tracts (A—Masson trichrome, 100×; B—Snook's reticulin, 40×; C—Verhoeff–Van Gieson, 200×)*

panacinar hepatocellular necrosis. In addition, stains for elastic fibers can distinguish areas of recent lobular collapse from the bridging interportal fibrous septa of chronic liver disease.

Each morphologic injury pattern has a constellation of cellular and architectural features. The following examples demonstrate how connective tissue stains help to identify those morphologic features:

Acute Hepatitis

Acute hepatitis is characterized morphologically by acinar inflammation (with or without portal inflammation), with variable degrees of "ballooning" hepatocellular degeneration, apoptosis, Kupffer cell hypertrophy, hepatocyte loss, and focal hepatocytic necrosis. More severe injury is typified by zonal or panacinar hepatocellular necrosis. The histologic features of the hepatitic recovery phase (after acute disease has passed) include regenerative changes such as hepatocytic mitosis, hepatocellular rosettes, and thickened hepatic plates. Because of their ability to delineate the hepatic plates, reticulin stains demonstrate the characteristic irregularities in these structures caused by hepatocyte dropout, inflammation, and focal regenerative changes, known in aggregates as lobular "disarray" (Figure 7.2A). When the extent of hepatocyte necrosis is severe, as in zonal or panacinar necrosis, Snook's or Wilder's stain shows the subsequent coalescence of reticulin fibers in so-called lobular "collapse" (Figure 7.2B).

Chronic Hepatitis

Morphologically, chronic hepatitis is characterized by expansion of the portal tracts by a chronic inflammatory infiltrate and variable degrees of lobular and periportal inflammation and hepatocellular damage. Fibrosis may or may not be present, but, with progressive disease, one sees portal and periportal fibrosis, bridging interportal fibrous bands, and, eventually, cirrhosis with regenerative hepatocellular nodules. Because of its ability to label type I collagen fibers, the trichrome stain is effective for the elucidation of fibrosis as well as its pattern of distribution. Predominantly, portal-based fibrosis is characteristic of the chronic hepatitic injury pattern (Figure 7.3).

Cirrhosis

At a histologic level, cirrhosis is a diffuse process in which the normal hepatic parenchyma is replaced by rounded regenerative nodules that are separated by thick fibrous septa. The latter structures are again highlighted with the trichrome stain (Figure 7.4A). On occasion, it may be difficult to distinguish between fibrous septa and areas of recent bridging necrosis with lobular collapse, but this

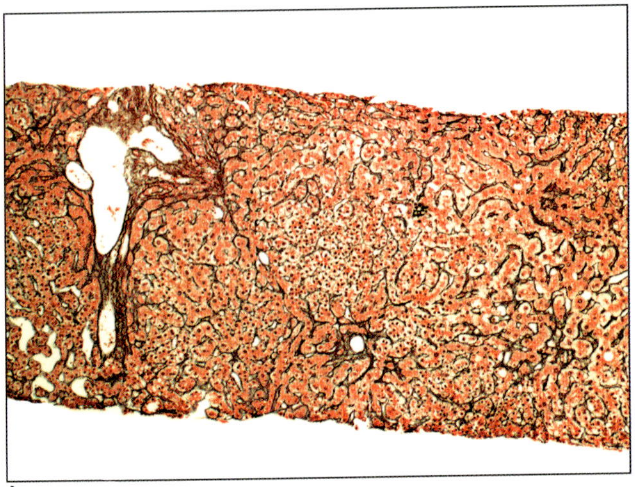

A

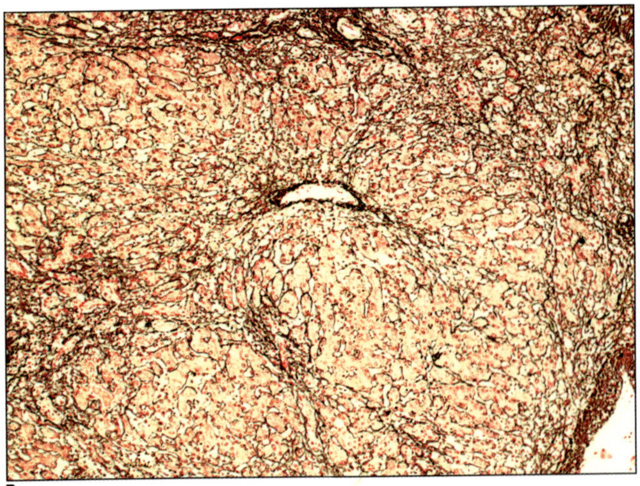

B

FIGURE 7.2: Acute hepatitis: (A). The reticulin stain highlights the characteristic irregularities in the hepatic plates due to hepatocyte dropout, inflammation, and focal regenerative hyperplasia, together known as "lobular disarray." (B). When the hepatocyte necrosis is more extensive, the reticulin stain demonstrates the subsequent coalescence of reticulin fibers—so-called lobular collapse (A and B—Snook's reticulin, 40×).

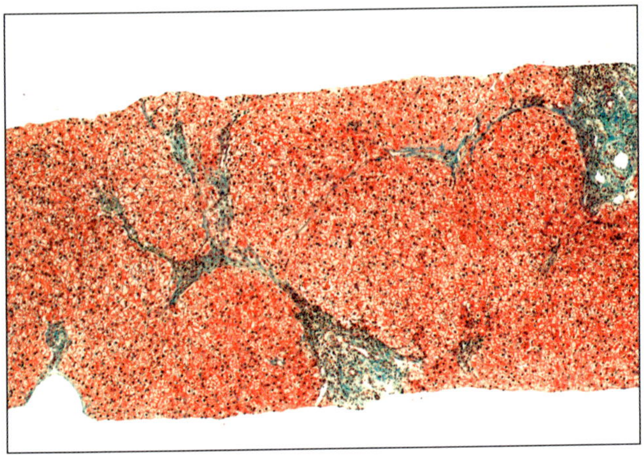

FIGURE 7.3: Chronic hepatitis: The trichrome stain demonstrates both the extent of fibrosis and its pattern of distribution, which in chronic hepatitis, is predominantly portal based (Masson trichrome, 40×).

interpretative separation can be made with stains for elastic fibers (6).

Although cirrhosis can be recognized readily on a liver biopsy specimen that is sufficiently large and intact, and in which regenerative nodules are small, this diagnosis can be challenging when faced with a slender core biopsy from a large cirrhotic nodule or a specimen that is very fragmented. In such instances, the reticulin stain is useful because it delineates abnormalities of acinar structure and disturbances in the hepatic plates that suggest the presence of a regenerative nodule (6,8). Specifically, thickened plates, loss of their radial arrangement, and an absence of the normal spatial relationship between portal and blood vessels allow for the identification of regenerative nodules (6) (Figure 7.4B).

Submassive or Massive Necrosis

"Submassive" and "massive" necrosis are terms for multiacinar, confluent hepatocytic necrosis. They correlate with clinical subfulminant or fulminant hepatic failure. The areas of bridging or confluent necrosis with multiacinar collapse can be demonstrated with reticulin stains, confirming the presence of extensive hepatocytic dropout, lobular collapse, and coalescence of the reticulin network around hepatic plates (Figure 7.5A).

Confluent areas of acinar collapse may be mistaken for fibrous septa of chronic liver disease, as seen in chronic hepatitis or cirrhosis (6). VVG stains are helpful in making this important distinction. Specifically, the hepatic acinae normally lack elastic fibers, and after injury, several months must elapse for elastic fibers to accumulate in fibrous septa. As such, areas of recent collapse are devoid of elastic fibers and well-developed connective tissue septa contain it (Figure 7.5B,C). Obviously, then, the presence of elastic fibers indicates chronicity.

Venous Outflow Obstruction

Two hepatic vein disorders result in hepatic venous obstruction. These are the Budd-Chiari syndrome and hepatic venoocclusive disease (VOD). The Budd-Chiari syndrome is caused by impedance of circulatory outflow at the level of the large hepatic veins or the inferior vena cava; some definitions also extend to include relative

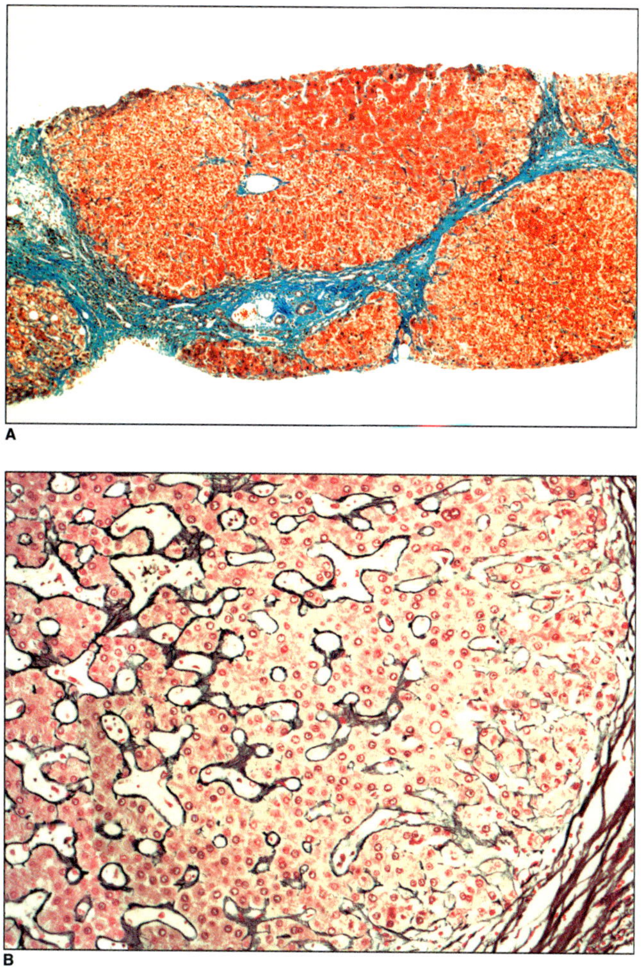

A

B

FIGURE 7.4: Cirrhosis: A. The trichrome stain demonstrates the characteristic bridging fibrosis that surrounds the regenerative nodules. B. Thickened hepatic plates and loss of their normal radial arrangement, as delineated on the reticulin stain, are features that characterize a regenerative nodule (A—Masson trichrome, 40×; B—Snook's reticulin, 100×).

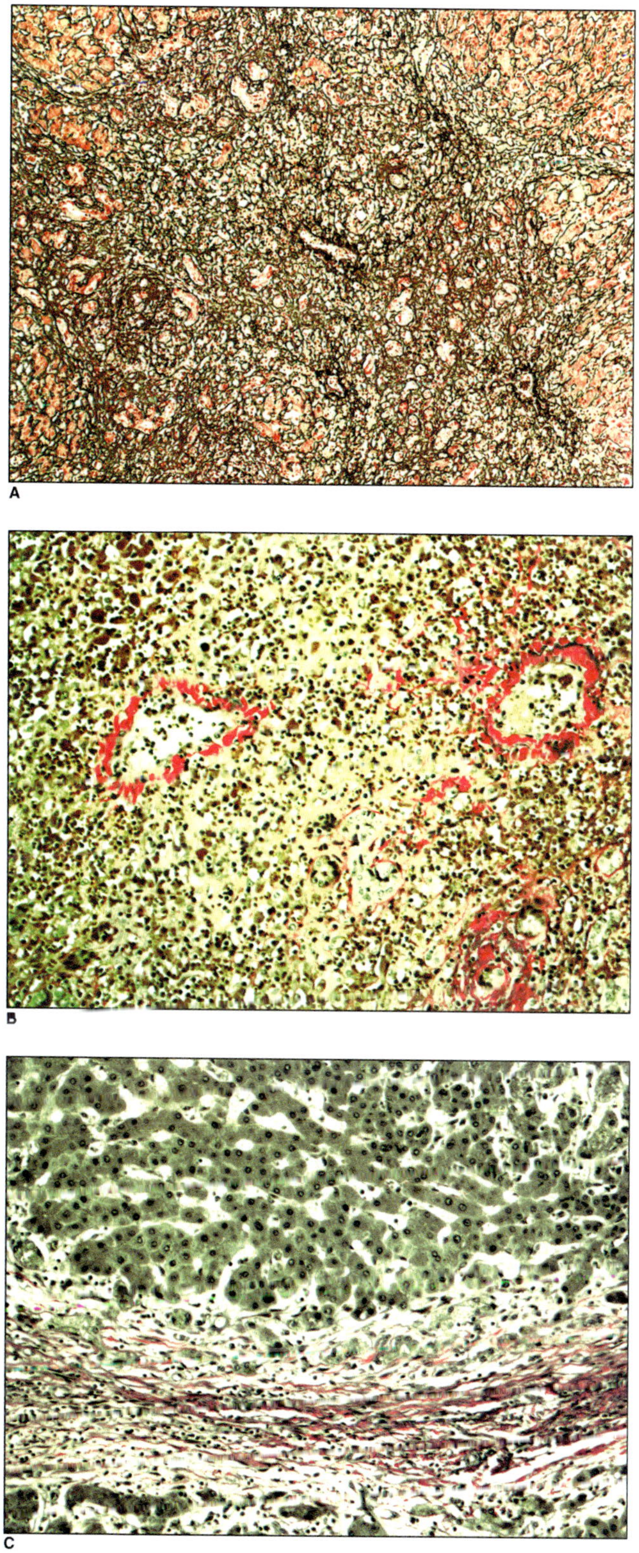

A

B

C

FIGURE 7.5: Submassive necrosis: A. The reticulin stain demonstrates extensive multiacinar collapse. B. Areas of recent collapse, which characterize submassive necrosis, are negative for elastic fibers. C. In contrast, older fibrous septa, such as seen with chronic hepatitis or cirrhosis, contain elastic fibers (A—Snook's reticulin, 40×; B and C—Verhoeff–Van Gieson, 100×).

obstruction at the cardiac level. With acute obstruction, one observes marked sinusoidal dilatation in "zone 3" (around the central hepatic venules) of the hepatic lobules, with lobular congestion. In chronic obstruction, perivenular and sinusoidal fibrosis appears and partially replaces the hepatic parenchyma. This process may progress to result in bridging fibrosis between adjacent centrilobular areas, producing a pattern of "reverse lobulation" as well demonstrated with trichrome stains.

Hepatic VOD results from endothelial injury and fibrous occlusion of small tributaries of the hepatic veins. Specifically, these include terminal hepatic venules and the sublobular outflow veins. One sees partial or complete luminal sclerosis of those vessels, together with sinusoidal congestion and dilatation, and perivenular fibrosis. The obliterative lesions in terminal hepatic venules in VOD may be difficult to see in H&E sections, but they are easily documented with trichrome stains (Figure 7.6A,B).

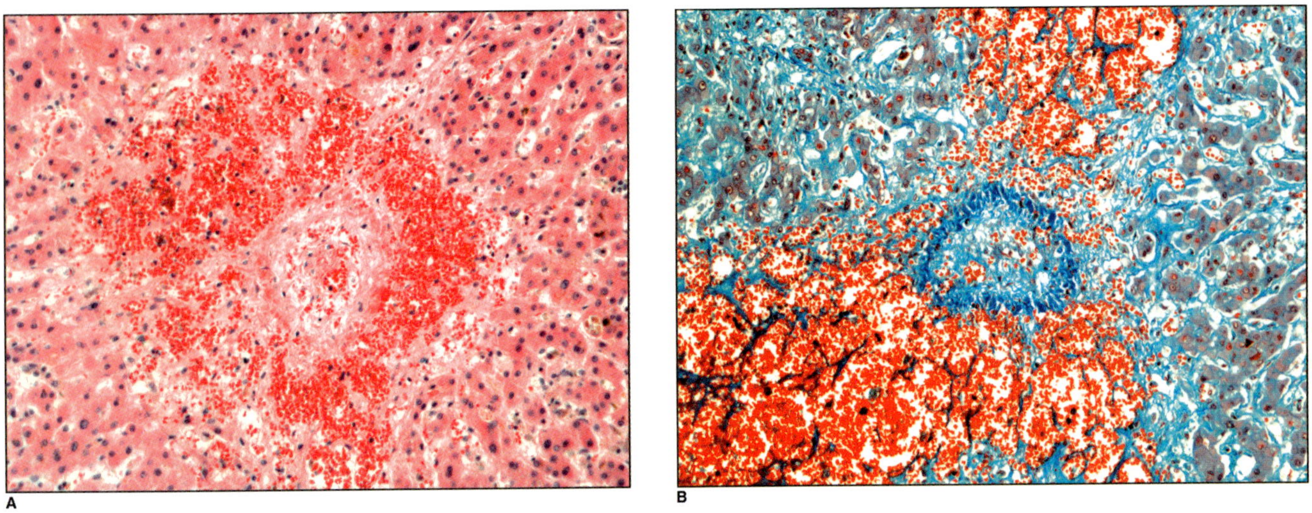

A B

FIGURE 7.6: VOD: A. The characteristic fibrous obliteration of the small, terminal hepatic venules may be difficult to see on the H&E stain. B. However, it is well documented on the trichrome stain (A—H&E, 100×; B—Masson trichrome, 100×).

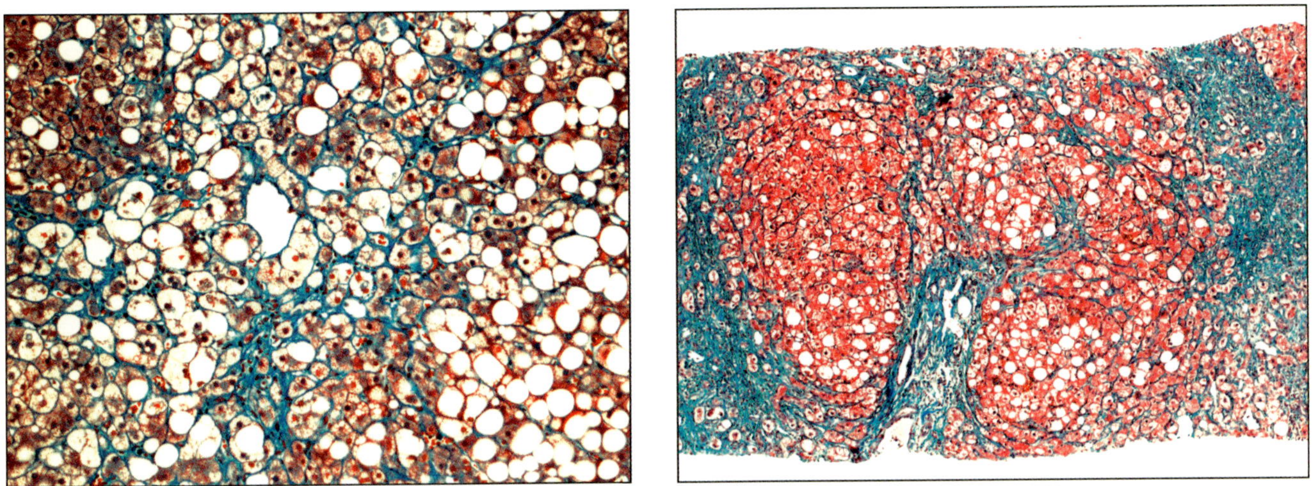

A B

FIGURE 7.7: Steatohepatitis: A. As highlighted on the trichrome stain, the fibrosis occurs initially in the perivenular area of acinar zone 3; the collagen is deposited along the sinusoids and characteristically appears to invest single or small groups of hepatocytes in a pattern referred to as "perisinusoidal," "pericellular," or "chicken-wire" fibrosis. B. This perisinusoidal/pericellular quality of the fibrosis is often retained and can be identified even in disease that has progressed to cirrhosis (A—Masson trichrome, 100×; B—Masson trichrome, 40×).

Steatohepatitis

Steatohepatitis is a distinctive pattern of hepatic injury characterized by steatosis, predominantly lobular inflammation, hepatocytic ballooning degeneration, and fibrosis that has a characteristic perivenular, perisinusoidal-pericelluar distribution (9). Early pathologic alterations are concentrated in zone 3 with a perivenular-acinar localization, and, as the disease progresses, these changes become panacinar and also involve portal tracts. Even though the chronic nonsteatotic hepatitides and the steatohepatitides are forms of chronic necroinflammatory hepatocellular injury, their etiologies, prognosis, and treatment are different. Hence, this distinction is an important one for pathologists. The pattern of fibrosis, as demonstrated on trichrome staining, is helpful in this separation. In chronic nonsteatotic hepatitis, fibrosis is, at least in part, portal based. In contrast, steatohepatitis manifests fibrosis that begins in the perivenular area of acinar zone 3, usually in association with hepatocytes that show steatosis or ballooning degeneration (Figure 7.7A). Collagen is deposited along the sinusoids and characteristically appears to invest single cells or small groups of hepatocytes. This pattern has been referred to as "perisinusoidal," "pericellular," or "chicken-wire" fibrosis. Collagen deposition in portal and periportal regions is generally a minor feature during the

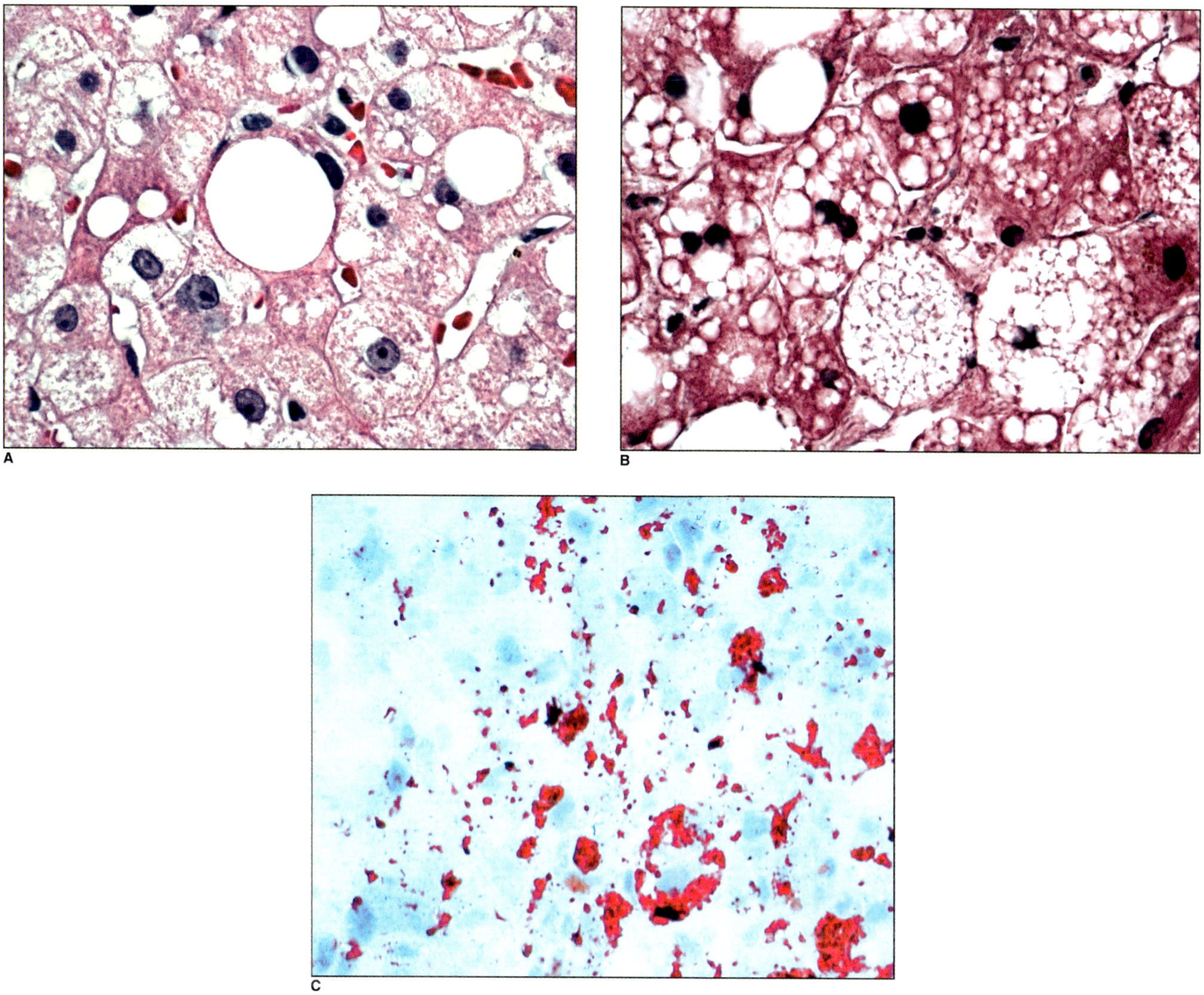

FIGURE 7.8: Steatosis: A. Macrovesicular steatosis is characterized by a single, large droplet of fat, which often distends the hepatocyte and compresses and displaces the nucleus to the edge of the cell. B. In contrast, microvesicular steatosis is characterized by numerous, small, cytoplasmic fat droplets, and the hepatocyte nucleus remains centrally located. C. In cases with suspected microvesicular steatosis, such as this case of acute fatty liver disease of pregnancy, an ORO stain may be performed on a frozen section to confirm this finding (A and B—H&E, 400×; C—ORO, 400×, photomicrograph kindly provided by Elizabeth M. Brunt, M.D.).

early stages of steatohepatitis. Even though progressive injury in steatohepatitis results in portal, periportal, and perivenular fibrosis, the perisinusoidal-pericellular quality of the collagenization is often retained and can be identified even in disease that has progressed to cirrhosis (Figure 7.7B).

In addition to their important contribution to the etiologic diagnosis of the liver disease, connective tissue stains—especially the trichrome method and its equivalents—play a central role in the prognostication of patients with liver disease (1–3). Fibrosis is a marker of cellular damage that often is irreversible. Accordingly, determinations of the extent of fibrosis are used to evaluate therapeutic responses and disease progression. In addition, the outlook for patients with many forms of nonneoplastic liver disease is, in large part, based on the severity of fibrosis (the so-called histologic stage) (10–13). The presence of extensive hepatic fibrosis or cirrhosis may eliminate the possibility of using other treatments or procedures, such as renal transplantation or bariatric surgery for morbid obesity.

FAT STAINS IN LIVER DISEASE

Lipid accumulates in hepatocytes in two basic morphologic and pathogenetic forms (14,15), that is, macrovesicular steatosis (large-droplet fat) and microvesicular steatosis (small-droplet fat). Histologically, the lipid in macrovesicular steatosis distends the hepatocytes, compressing and displacing nuclei peripherally (Figure 7.8A). In small amounts, it is a common and etiologically nonspecific finding; more extensive macrovesicular steatosis, however, characterizes fatty liver disease that is associated with alcohol ingestion, extreme obesity, and the hepatopathy of type II diabetes mellitus. In contrast, microvesicular steatosis demonstrates numerous small cytoplasmic fat droplets, and the hepatocyte nucleus remains centrally located (Figure 7.8B). Microvesicular lipid accumulation is caused by disturbed mitochondrial beta-oxidation of fatty acids in hepatocytes. It is seen in Reye's syndrome, acute fatty liver of pregnancy, tetracycline or valproate toxicity, alcoholic "foamy degeneration," and some other metabolic disorders, such as Alpers' disease.

Several histochemical methods, such as the oil red O (ORO) stain, are specific for fat. Because dehydrating and clearing agents that are used during standard tissue processing dissolve lipids, stains for them must usually be performed on frozen sections. This may require advance planning on the part of clinicians, in consultation with the pathologist. In practical terms, however, affirmative demonstration of intrahepatic lipid is not always necessary. Macrovesicular steatosis is easily recognized on H&E sections—both frozen and permanent—and an ORO preparation is not mandatory to make this diagnosis. In contrast, lipid stains are indeed useful for the identification of *microvesicular* steatosis because it results in finely reticulated hepatocellular cytoplasm that simulates ballooning degeneration or rarification caused by glycogen accumulation. The diseases associated with microvesicular steatosis usually present with dramatic clinical features (e.g., acute hepatic failure, coma, or death). Therefore, clinicians and pathologists can anticipate the need to submit fresh tissue in such clinical settings. The ORO method is usually diagnostic (Figure 7.8C).

In the setting of hepatic transplantation, it has been proposed that the extent of steatosis in a potential allograft should be evaluated with an ORO stain on a frozen section of the donor liver. Extensive macrovesicular steatosis of the potential allograft is associated with an increased risk of poor posttransplant function; as a general rule, donor livers with >50% macrovesicular fat content are considered unsuitable for transplantation (16–18). Thus, pathologists are currently asked to do ORO evaluations on frozen sections of the potential allograft. In this context, it is critical to evaluate macrovesicular steatosis only. Microvesicular steatosis is a common finding in hepatic allografts, and it is considered to be a reversible change related to organ harvest–related ischemia. Although macrovesicular steatosis reproducibly affects allograft function, microvesicular fat accumulation, even if it is severe, does not (19,20). Macrovesicular steatosis can be identified readily in H&E stains of frozen sections, but the potential exists for confusion of that condition with *microvesicular* steatosis in ORO stains. Thus, we recommend that the latter preparations should *not* be used in evaluations of allograft organs.

IRON STAINS IN HEPATIC DISEASE

The histochemical stain that is usually employed to evaluate the iron content of liver tissue is Perl's Prussian blue method or a modification thereof. This technique demonstrates only ferric compounds, and hepatic iron stores usually are represented by this chemical class. Perl's stain identifies both ferritin and hemosiderin (4,5), and therefore, it is an effective measure of hepatic iron stores. Moreover, this procedure distinguishes iron from other hepatocellular cytoplasmic pigments such as bile and lipofuscin.

The role of the liver biopsy for histochemical evaluation of iron content has changed in the past decade because of the discovery of the gene mutations that are linked to most examples of hereditary hemochromatosis (HH), in the *HFE* gene (21). The development of commercially available tests for such mutations has essentially supplanted tissue biopsy as the preferred means of diagnosing HH (22). Nonetheless, liver biopsy and histochemical stains are still valuable in the evaluation and management of patients with iron overload, whether they have HH or another underlying condition (23,24). In individuals with HH, the risk of hepatocellular carcinoma (HCC) is significantly increased. So-called iron-free foci, which are discrete lobular hepatocellular clusters with a low iron content, are considered to be preneoplastic in this context. They are felt to represent an early lesion in the development of HCC in patients with HH (25) (Figure 7.9). Thus, all liver biopsies from patients with HH should be studied with Perl's stain. Individuals who are not homozygous for gene variants associated with HH may have other causes of iron overload that are suggested by histochemical results because selected histologic patterns of iron accumulation can provide clues regarding etiology.

Unsuspected hepatic iron overload also may be an important cofactor in the morbidity of several chronic hepatic disorders, such as viral hepatitis C, alcohol-related hepatopathy, nonalcoholic fatty liver disease, and porphyria (24,26). In these settings, it has been hypothesized that increased iron stores may promote progression of the primary diseases (27).

Microscopic evaluation of liver biopsies from patients with possible iron overload should include two steps. The first is a semiquantitative assessment of the amount of iron, and the second is a qualitative analysis of iron distribution in various intrahepatic cell types and lobular zones.

Several schemes have been proposed for histologic semiquantitation of hepatic iron stores. One of the most widely used is that of Searle et al. (28) (Table 7.1). With this system, the amount of iron is assigned a grade reflecting the ease with which it can be discerned; no attention is paid to the cell types that harbor the iron.

The qualitative assessment of iron overload does identify specific cell types and zones in the lobule in which iron is seen. It may accumulate in parenchymal cells, that is,

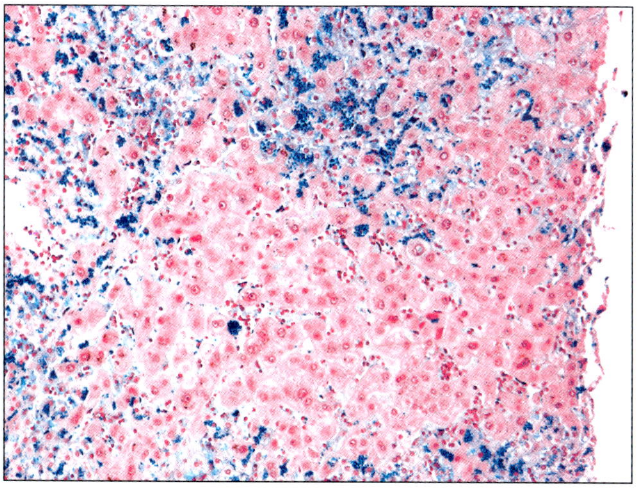

FIGURE 7.9: ***Iron: In patients with HH, so-called iron-free foci, small discrete sublobular nodular clusters of hepatocytes devoid of iron or with a low iron content within an otherwise iron-overloaded liver are considered to be preneoplastic and an early step in the development of HCC (Prussian blue, 100×, photomicrograph kindly provided by Maria Isabel Fiel, M.D.).***

hepatocytes and bilary epithelium, or so-called mesenchymal cells of the reticuloendothelial system, including portal macrophages, Kupffer cells, stellate cells, and endothelial cells. Based on the predominant pattern of cellular localization, iron distribution is therefore categorized as "parenchymal," "mesenchymal," or "mixed." Such patterns may provide etiologic clues, as follows:

A parenchymal pattern of iron overload is characteristic of HH. This pattern is typified by deposition of iron predominantly in hepatocytes, with a distinct pericanalicular location (biliary poles), and an obvious gradient of iron distribution in the lobules that is greatest in the periportal acinus (zone 1) and least in zone 3 (Figure 7.10A). Early in the disease, iron is seen only in zone 1. As it continues to accumulate, hepatocytes throughout the lobule acquire iron, although a periportal-to-centrilobular gradient is maintained. In later stages, iron also accumulates in bile duct epithelium as well as reticuloendothelial cells, producing a mixed distribution pattern. Although the parenchymal pattern of iron distribution is characteristic of HH, it is not specific for this condition (29).

TABLE 7.1: Histologic Grading of Iron Stores (Searle et al.)

Grade	Ease of Observation; Magnification (eyepiece × objective) Required for Identification
0	Granules absent or barely discernible at 400×
1+	Granules barely discernible at 250× and easily confirmed at 400×
2+	Discrete granules resolved at 100×
3+	Discrete granules resolved at 25×
4+	Masses visible at 10× or naked eye

In the mesenchymal pattern of iron overload, deposition occurs predominantly in the intrahepatic reticuloendothelial system (Figure 7.10B). As the process progresses, large aggregates of iron-laden Kupffer cells (siderotic nodules) may form (Figure 7.10C). As the reticuloendothelial system is saturated, hepatocytes also accumulate iron, usually in foci that are adjacent to clusters of iron-laden macrophages. The mesenchymal iron-storage pattern is characteristic of acquired iron overload, as seen with chronic blood transfusion, long-term hemodialysis, and alcohol-related hepatopathy.

The mixed pattern of iron storage has histologic characteristics of both the parenchymal and the mesenchymal profiles. Pathogenetically, it reflects complex mechanisms associated with diverse sources of iron overload, such as concomitant hemochromatosis and chronic alcohol abuse.

COPPER STAINS IN LIVER DISEASE

Copper accumulates in hepatocytes in a variety of pathologic conditions. These include primary metabolic disorders of copper metabolism, most notably Wilson's disease, and nonspecific chronic cholestatic abnormalities. It should also be noted that normal neonatal liver is rich in copper (30). Several histochemical stains are available for the detection of this mineral, including the rhodanine, orcein (labeling copper-binding protein), and Timm procedures. Of these, the modified Timm's method is the most sensitive (31) but the rhodanine technique is the most commonly used.

Histochemical evaluation of hepatic copper content is principally undertaken in cases of suspected Wilson's disease. Nevertheless, it may yield seemingly normal results even in documented cases of this disorder; this is because intracellular distribution of copper is highly variable in different stages of the disease (32). Early on, an excess of the mineral is seen diffusely in the hepatocytic cytoplasm. It is difficult to identify histochemically because the copper is not yet concentrated in lysosomes. Later, copper is indeed lysosome centered and the rhodanine stain is typically positive (Figure 7.11A). In the precirrhotic stages of Wilson's disease, histochemically demonstrable copper predominates in periportal hepatocyes (33). Once cirrhosis has eventuated, the copper distribution becomes uneven, with some regenerative nodules staining intensely and others being essentially negative (7). Thus, negative copper stains in limited biopsy specimen, even those showing cirrhosis, do not exclude the diagnosis of Wilson's disease.

Both copper and copper-binding protein may be found in chronic liver diseases other than those concerning disorders of copper metabolism. In chronic cholestatic processes, such as primary biliary cirrhosis and primary sclerosing cholangitis, copper accumulation may occur and it preferentially involves periportal or periseptal hepatocytes (6,34) (Figure 7.11B). Also, copper may be

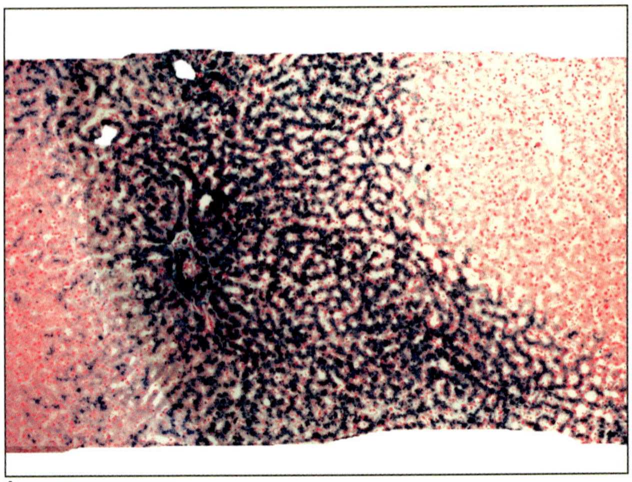

A

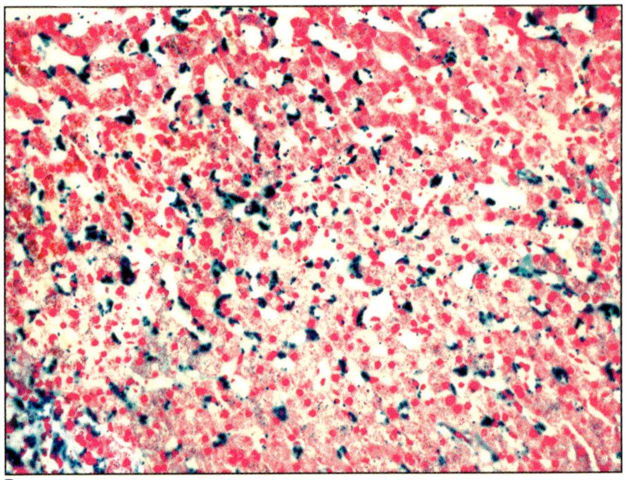

B

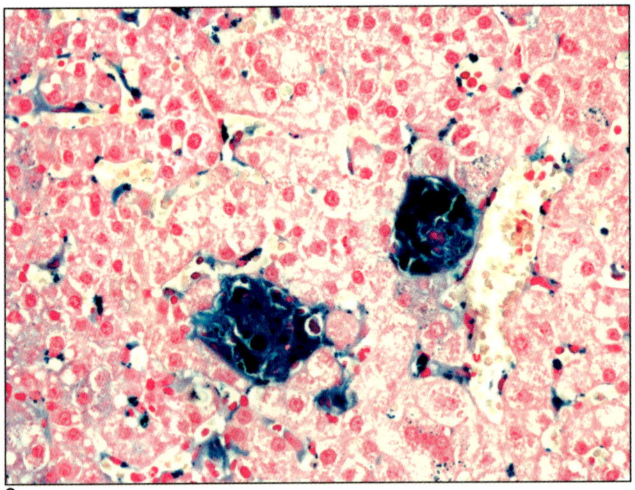

C

FIGURE 7.10: Iron: A. In the parenchymal pattern of iron overload, which is characteristic of HH, the deposition of iron is predominantly within hepatocytes and demonstrates an obvious gradient of iron distribution within the hepatic lobules that is greatest in the periportal acinar zone 1 hepatocytes and decreases toward the acinar zone 3 hepatocytes. B. In the mesenchymal pattern of iron overload, iron deposition occurs predominately in the cells of the reticuloendothelial system, such as the sinusoidal lining cells and Kupffer cells. C. As mesenchymal iron overload progresses, large aggregates of iron-laden Kupffer cells, so-called siderotic nodules, may form (A—Prussian blue, 40×; B—Prussian blue, 100×; C—Prussian blue, 200×).

identified in selected cases of chronic viral hepatitis and alcohol-related liver disease (35). In these settings, the mineral also tends to accrue in periportal or periseptal hepatocytes, and it has been speculated that subclinical cholestasis or local disturbances of bile flow are the causative factors.

CARBOHYDRATE STAINS IN HEPATIC DISORDERS

The most widely used carbohydrate stain in hepatopathology is the periodic acid-Schiff (PAS) method. It is frequently employed before and after diastase digestion (PAS-d), with the latter procedure being used to remove glycogen from tissue sections. This process reveals other complex carbohydrate–containing macromolecules, including metabolic products of hepatocytes as well as intrahepatic structural components such as basal lamina (8). The PAS-d stain is useful for labeling PAS-d–positive lipofuscin and cellular debris in activated Kupffer cells, indicating hepatocellular injury with subsequent phagocytosis of necrotic hepatocytes (Figure 7.12A); disruption of the basement membrane that normally encircles interlobular bile ducts, as seen in bile duct injury (Figure 7.12B,C); and PAS-d–positive intracytoplasmic globules in zone 1 hepatocytes, typical of α_1-antitrypsin (A1AT) deficiency (7) (Figure 7.12D). Cytoplasmic PAS-d–positive globules also may be seen in passive hepatic congestion, in the absence of A1AT deficiency (36). The globules are perivenular in location, adjacent to areas of sinusoidal congestion, and they usually contain fibrinogen as well. PAS-d–positive intrahepatocellular globules may be identified in cirrhosis

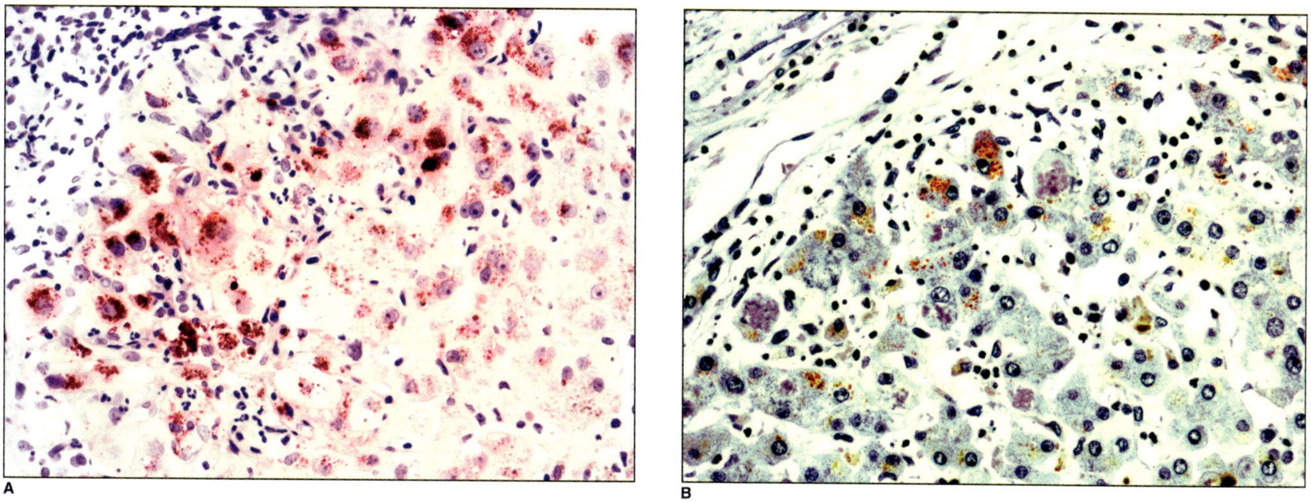

FIGURE 7.11: Copper: A. In Wilson's disease, copper is more readily identified histochemically in advanced stages of the disease because the copper is concentrated in lysosomes. B. Copper accumulation may also occur in chronic cholestatic diseases, such as primary sclerosing cholangitis, and it preferentially involves periportal or periseptal hepatocytes (A and B—rhodanine, 200×).

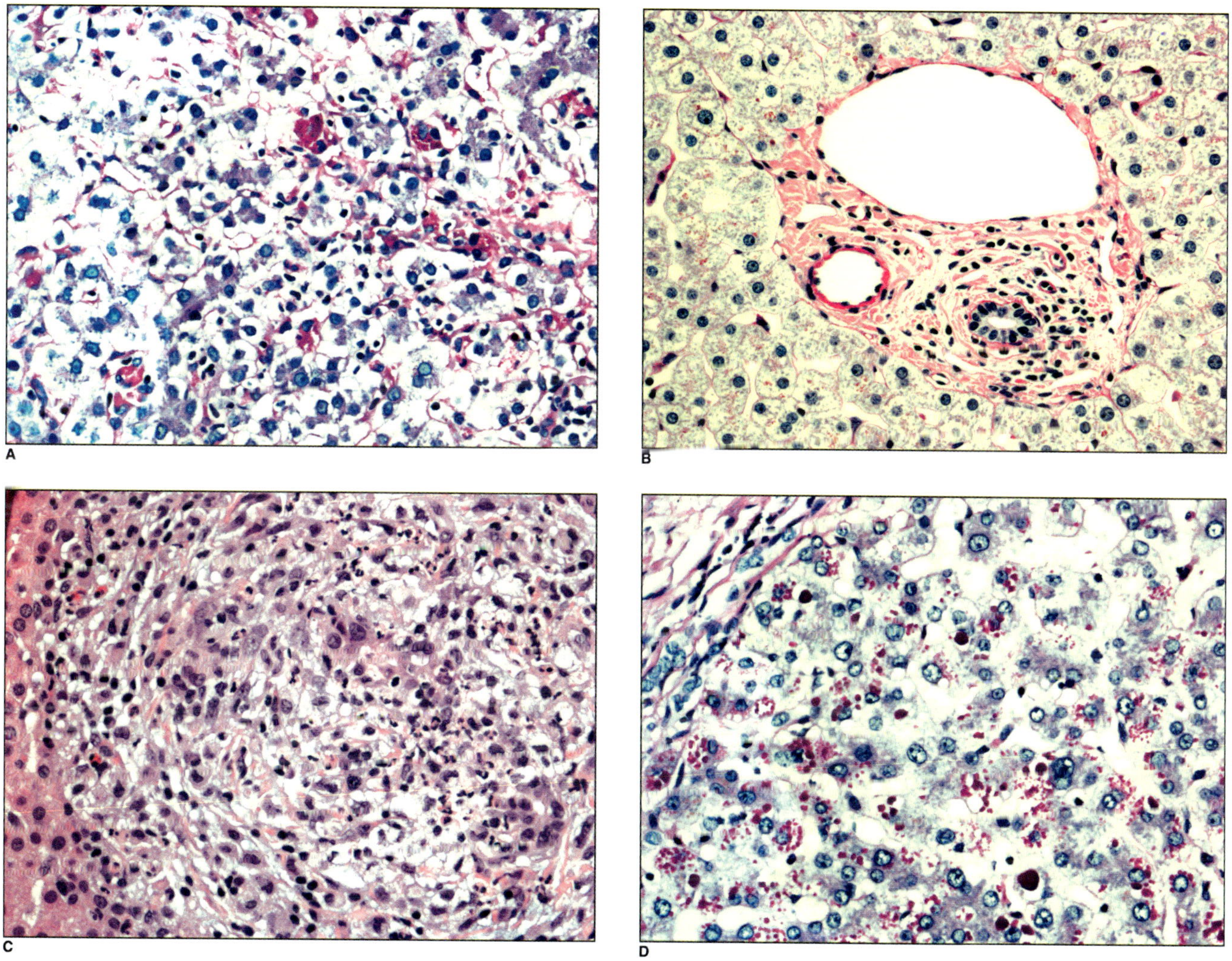

FIGURE 7.12: The PAS-d stain is useful for assessing a variety of processes: A. PAS-d–positive lipofuscin and cellular debris in activated Kupffer cells is indicative of hepatocellular injury with subsequent phagocytosis of the necrotic hepatocytes. B. The interlobular bile ducts are normally encircled by a basement membrane, which is PAS-d positive. C. In conditions that result in bile duct injury, such as primary biliary cirrhosis, there is disruption of this PAS-d–positive basement membrane. D. PAS-d–positive intracytoplasmic globules predominantly in the zone 1 hepatocytes is characteristic of A1AT deficiency (A–D—PAS-d, 200×).

caused by diverse etiologies, in patients with a normal A1AT phenotype (37). Hence, although one may consider the possibility of A1AT deficiency if PAS-d–positive globules are seen in liver biopsy specimens, genotypic analysis is required for definitive diagnosis.

Biochemical analysis of frozen tissue has largely supplanted morphologic and histochemical evaluations of liver tissue for suspected glycogen-storage disease. Nevertheless, the PAS stain continues to be an important screening test in this setting. Glycogenic hepatopathy represents the pathologic overloading of hepatic epithelium with glycogen, often associated with suboptimally controlled diabetes mellitus (38). The liver shows pale, swollen hepatocytes on H&E-stained sections (Figure 7.13A), and the PAS stain demonstrates an excess of cytoplasmic glycogen that disappears after diastase digestion (Figure 7.13B,C).

AMYLOID STAINS IN LIVER DISEASE

The liver is commonly involved in systemic amyloidosis, in "AL" disease (primary amyloidosis) or the "AA" form of the disease (secondary amyloidosis). Deposits of amyloid occur along preexisting structures, such as the perisinusoidal spaces of Disse, vascular walls, and perivascular areas in portal tracts (39) (Figure 7.14A). Extracellular globules of this protein are typical, although rare intracytoplasmic deposits may be observed as well. Amyloid can be seen in H&E-stained sections, as well as trichrome-stained preparations. The identification of amyloid is confirmed by characteristic green birefringence on examination of Congo red–stained sections with polarized light (Figure 7.14B). A histochemical distinction between AL and AA amyloid can be made by the loss of labeling of AL deposits with Congo red, after permanganate pretreatment. This separation continues to be valuable, even in the current climate of immunohistochemical evaluation. Rare involvement of the liver by light chain immunoglobulin deposits may resemble hepatic amyloidosis morphologically (40), but the light chain aggregates fail to label with the Congo red method (6).

HEPATOCELLULAR NEOPLASMS AND OTHER MASS LESIONS

Virtually all types of neoplasms may present in the liver, either as primary tumors or metastases. If the lineage of a neoplastic lesion is not apparent based on its morphologic features, additional adjunctive studies will be necessary to determine it. Histochemical stains for mucin, using the PAS-d or mucicarmine techniques, are often employed in this setting. HCC is mucin negative, but combined HCC-cholangiocarcinoma (CC), CC, and many metastatic adenocarcinomas do contain epithelial mucins. Some HCCs also retain their ability to produce bile (Figure 7.15A), even in extrahepatic metastases of these tumors, and Fouchet's or

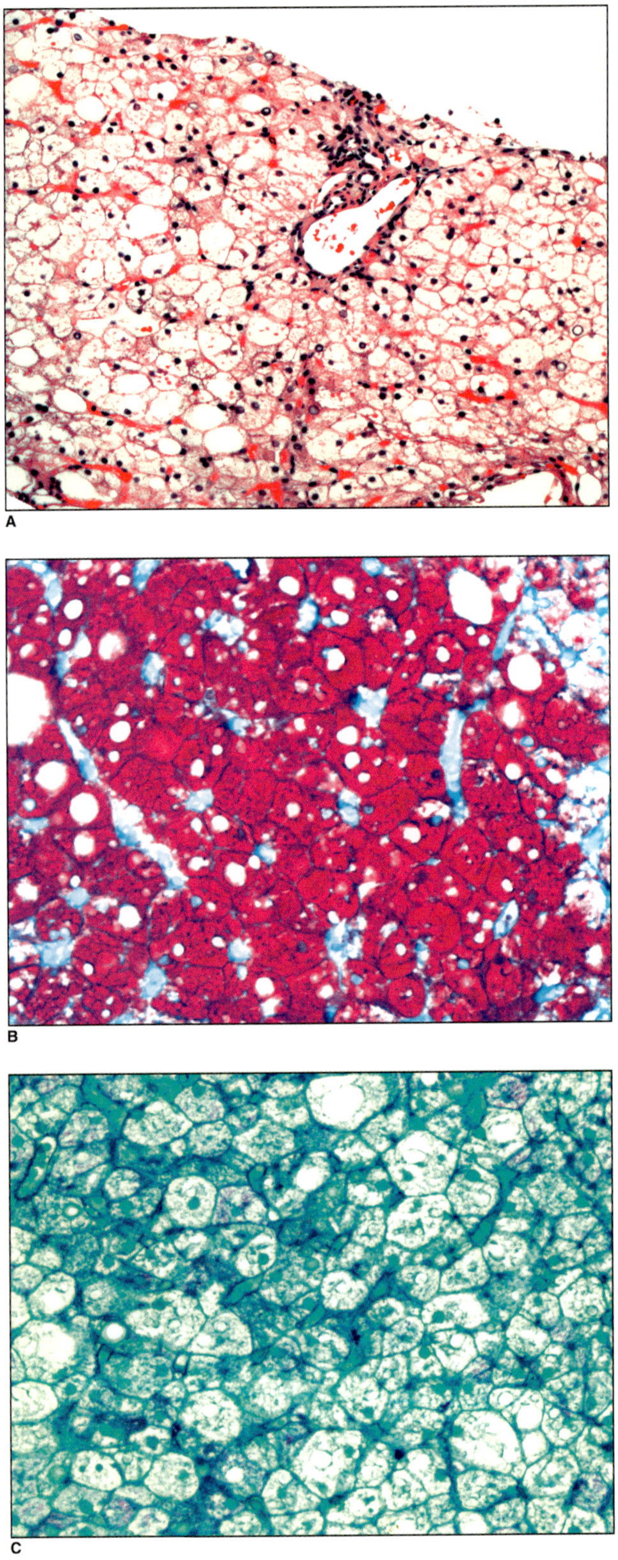

FIGURE 7.13: A. Glycogenic hepatopathy morphologically is characterized by diffuse, pale staining, swollen-appearing hepatocytes. B and C. The PAS stain demonstrates that these hepatocellular changes are due to excessive cytoplasmic glycogen accumulation, which disappears after diastase digestion (A—H&E, 100×; B—PAS, 200×; C—PAS-d, ×200).

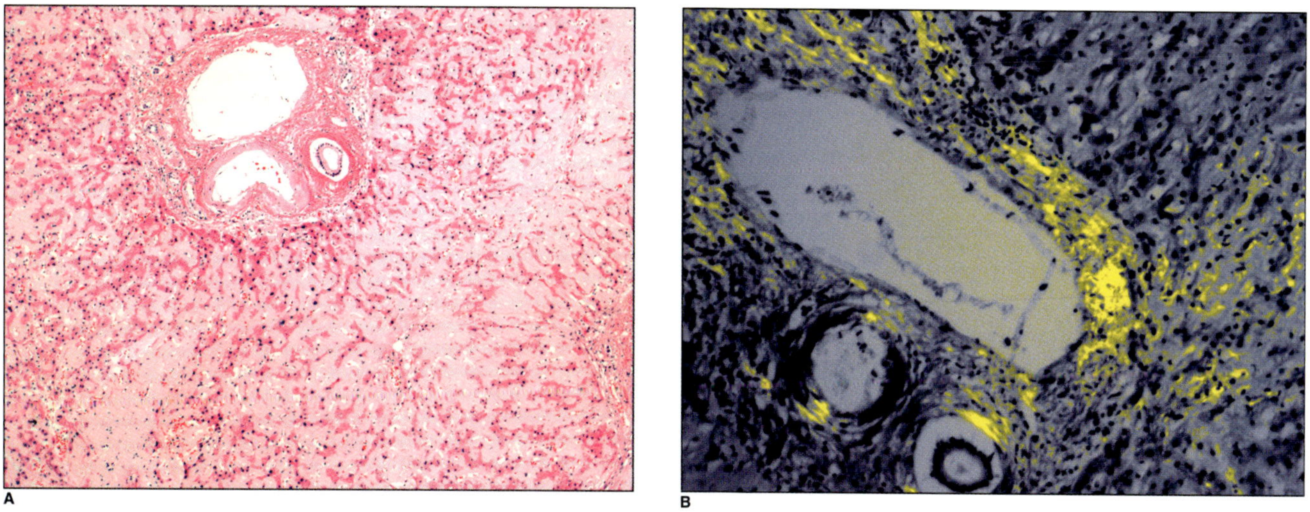

FIGURE 7.14: Amyloid: A. In the liver, the amyloid deposits occur along preformed architectural structures, specifically the perisinusoidal space of Disse, within vascular walls or in the perivascular area within portal tracts. B. The identification of this material as amyloid is confirmed by the characteristic green birefringence seen upon examination of the Congo red stain when examined under polarized light (A—H&E, 40×; B—Congo red, 100×).

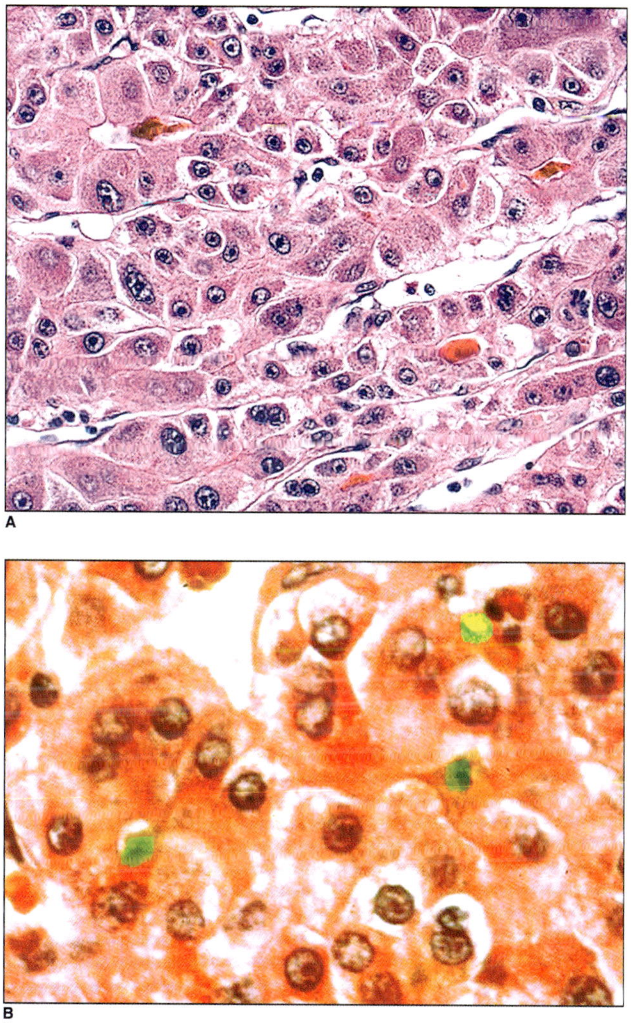

FIGURE 7.15: This HCC retains the capacity for bile production, manifested by small greenish intercellular plugs of pigment (A). A Fouchet stain accentuates the bile, labeling it bright yellow green (B).

Hall's techniques for that pigment (Figure 7.15B) can be used to confirm its presence in such neoplasms.

Because of the ability to delineate the architecture of the hepatic plates, the reticulin stain is also helpful in the evaluation of cytologically well-differentiated liver masses with hepatocellular differentiation (41). The distinction between macroregenerative nodules, focal nodular hyperplasia, hepatic adenomas, and well-differentiated HCCs may be difficult in that context, especially with limited biopsy specimens. Reticulin stains help to separate the three benign entities just mentioned from well-differentiated HCC. Specifically, in the benign lesions, the hepatic plates are no more than three cells thick, and the reticulin framework is intact or only focally decreased (Figure 7.16A). In contrast, HCC contains cell plates which are ⩾3 cells thick, with common formation of trabecula. The reticulin framework is often absent or markedly distorted, with irregular or absent labeling at the edges of the trabecula (42,43) (Figure 7.16B).

In addition, reticulin stains are useful for the identification of nodular regenerative hyperplasia, another lesion that is difficult to diagnose in limited biopsy specimens (7). Reticulin labeling delineates the characteristics of this mass-forming lesion, represented by nodules comprising hypertrophic cords of hepatocytes that are surrounded by compressed, atrophic hepatic plates (Figure 7.17).

LECTIN HISTOCHEMISTRY OF HEPATIC TUMORS

Lectins have not been utilized extensively in the diagnostic study of hepatic neoplasms, except in the nonmorphologic identification of a subfraction of alpha-fetoprotein through its binding to *Lens culinaris*. The latter is felt to possibly have a diagnostic and prognostic use in regard to small HCCs

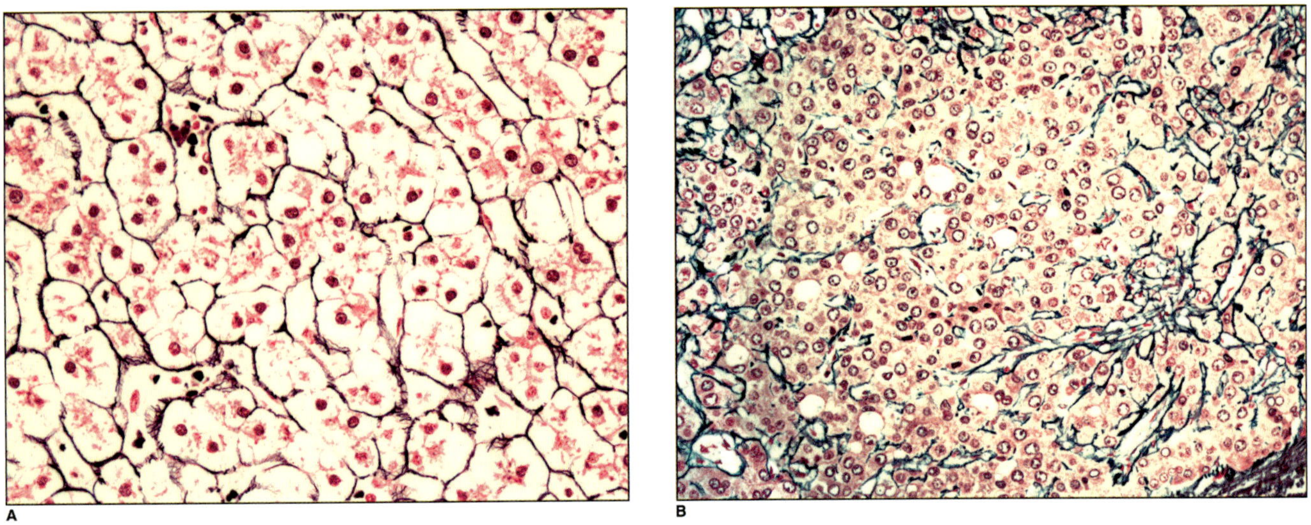

FIGURE 7.16: The reticulin stain helps to distinguish benign hepatocellular mass lesions from well-differentiated HCC. A. In the benign lesions, such as hepatic adenoma, the cell plates are no more than three cells thick, and the reticulin framework is intact or only focally decreased. B. In contrast, in HCC, the cell plates are three cells thick or greater, with trabeculae commonly seen, and the reticulin framework is often absent in areas (A and B—Snook's reticulin).

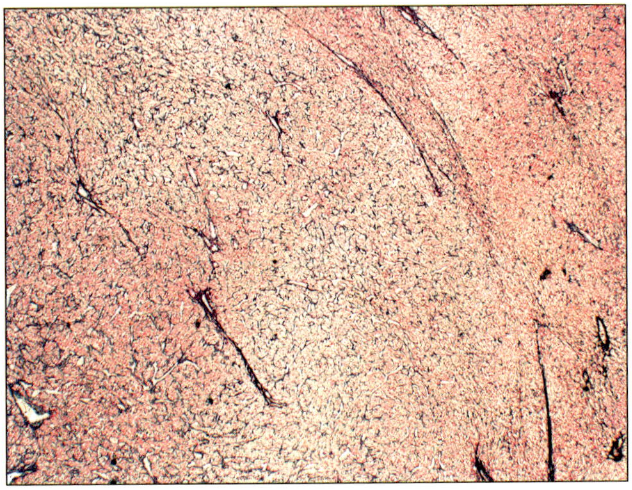

FIGURE 7.17: The reticulin stain is useful for the identification of nodular regenerative hyperplasia because it delineates this mass-forming lesion's characteristic findings of nodularity composed of hypertrophic cords of hepatocytes surrounded by compressed, atrophic hepatic plates (Snook's reticulin, 20×).

(44). Otherwise, most analyses of plant agglutinin binding to human liver tumors have concerned lectins that recognize blood group antigens (45–49), usually in comparisons of HCC and CC. Blood groups A, B, and H—recognized respectively, by *Dolichos biflorus*, *Bandeiraea simplicifolia*, and *Ulex europaeus* I—are more typically expressed by CC than by HCC; however, there is significant overlap in this regard between these lesions. Conversely, concanavalin-A, wheat germ agglutinin, and *Ricinus communis* often label HCC, but other tumors (both CCs and metastatic carcinomas) may bind to such lectins also. Lastly, epithelioid hemangioendothelioma (EHE)—a "borderline" hepatic vascular neoplasm (Figure 7.18A)—typically is labeled with *U. europaeus* I agglutinin (50) (Figure 7.18B). Nevertheless, because this lectin can bind to epithelial proliferations as well, it cannot be used in isolation to establish the diagnosis of EHE definitively.

ILLUSTRATIVE CASE STUDIES

CASE 1— A 46-year-old woman presented with a 7-month history of intermittent lethargy and malaise. She denied ethanol abuse and was not taking any medications or herbal supplements. Liver function tests demonstrated moderately elevated aminotransferase levels. Viral serologic studies were positive for anti–hepatitis-C; hepatitis C virus (HCV) RNA was also detectable, and genomic testing identified the presence of HCV genotype-1. A liver biopsy was performed, showing minimal necroinflammatory activity and no fibrosis (grade 1). No treatment was given initially, and the patient was to be reassessed in 3 years. At that point, a liver biopsy was performed, revealing mild necroinflammatory activity and periportal fibrosis (Figure 7.19). Based on the progression of disease, especially the appearance of fibrosis, antiviral treatment was begun. ■

CASE 2— A 28-year-old man was noted to have mildly elevated aminotransferase levels and mildly elevated serum iron levels during a routine health assessment. Otherwise, he was asymptomatic. The patient denied alcohol abuse and was not taking any medications or herbal supplements. Physical examination showed obesity and hepatomegaly. Viral serologic studies and analyses for autoantibodies were negative. A liver biopsy was performed (Figures 7.20 and 7.21), demonstrating steatohepatitis as well as increased iron

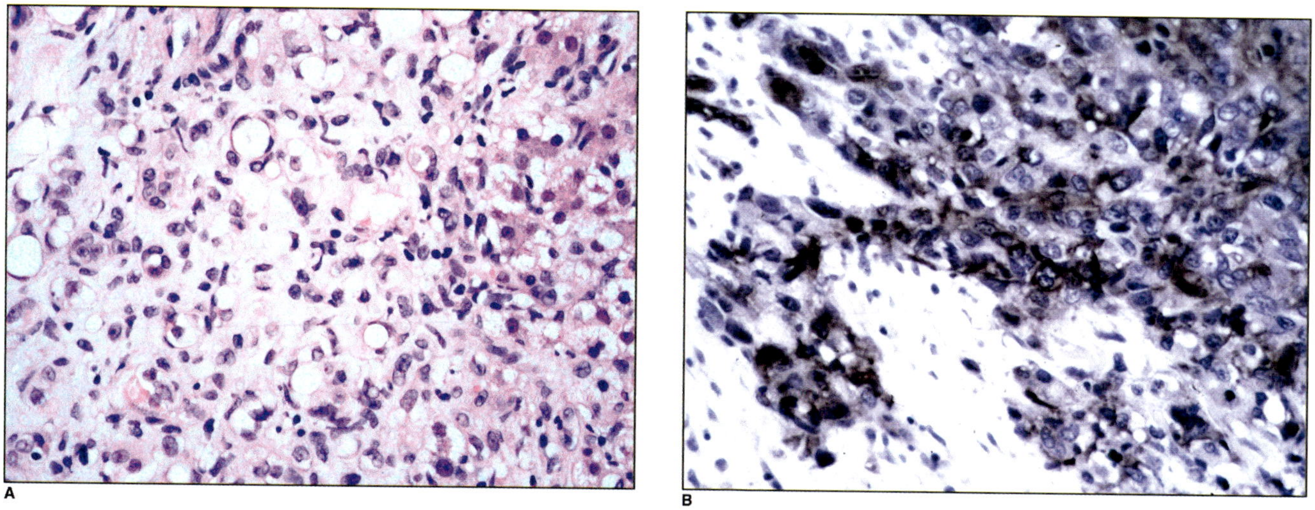

FIGURE 7.18: EHE of the liver, represented by randomly disposed aggregates of polygonal cells that efface a portion of the hepatic parenchyma. A fraction of the neoplastic cells contain intracytoplasmic vacuoles (A). Lectin histochemistry with Ulex europaeus *I agglutinin shows labeling of the tumor cells in EHE (B).*

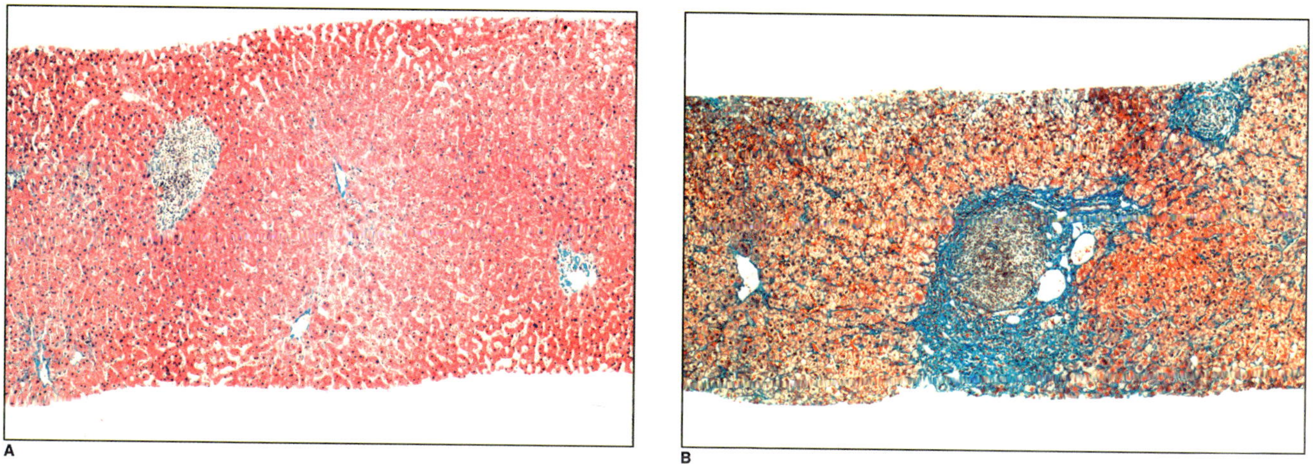

FIGURE 7.19. The initial liver biopsy in case 1 demonstrates no fibrosis (A—Masson trichrome stain). Another liver biopsy in case 1, performed 3 years later, shows periportal fibrosis (B—Masson trichrome stain).

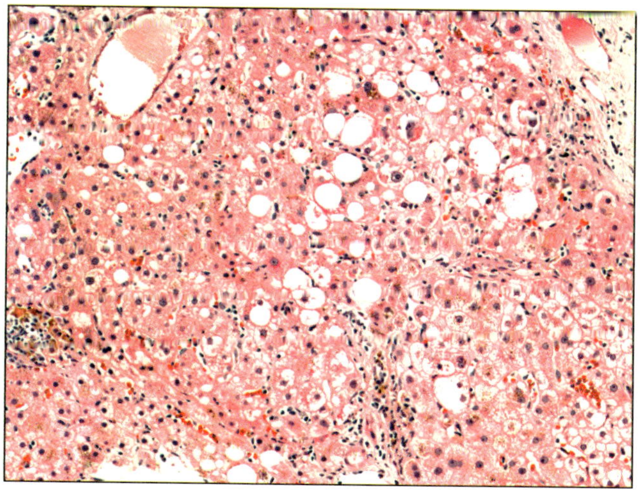

FIGURE 7.20: The liver biopsy in case 2 shows moderate macrovesicular steatosis, predominantly in acinar zone 3 with extension into zone 2. Scattered hepatocytes, also within zone 3, demonstrate ballooning degeneration and Mallory's hyaline, and there is a sparse lobular inflammatory infiltrate.

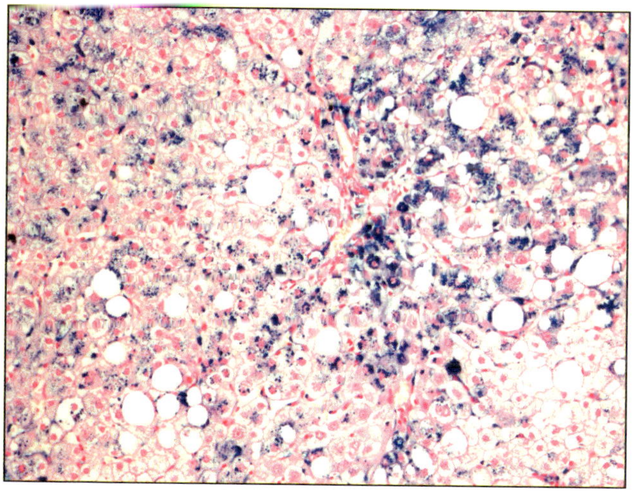

FIGURE 7.21: With the iron stain, increased iron stores are identified in the biopsy from case 2, predominantly in periportal hepatocytes (Perl's Prussian blue stain).

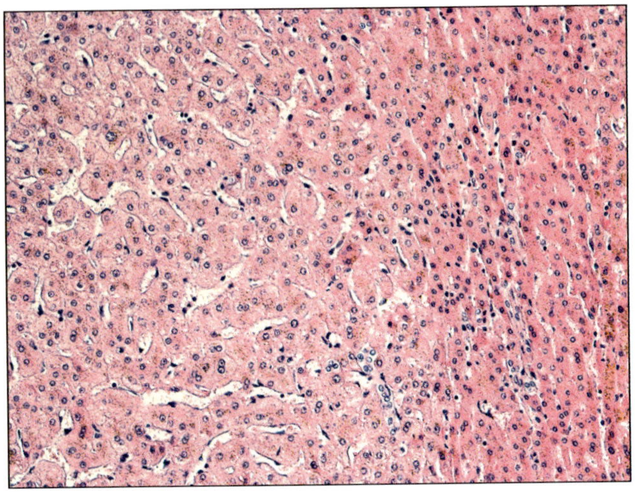

FIGURE 7.22: The liver biopsy in case 3 demonstrates vague nodularity and is devoid of fibrosis.

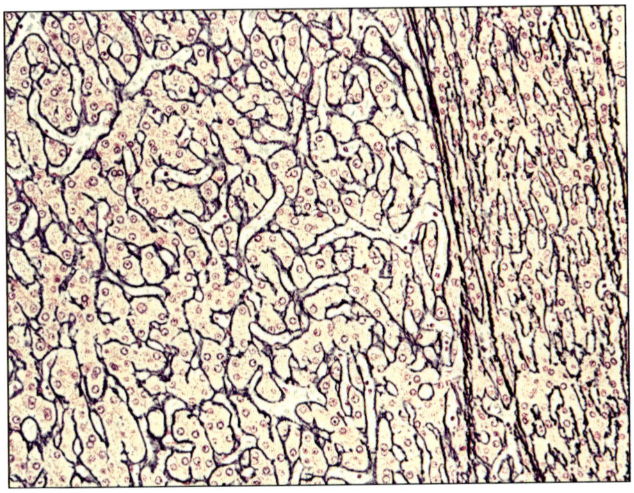

FIGURE 7.23: Reticulin staining in case 3 shows parenchymal nodules formed by hypertophic cords of hepatocytes, surrounded by compressed, atrophic hepatic plates (Snook's reticulin stain).

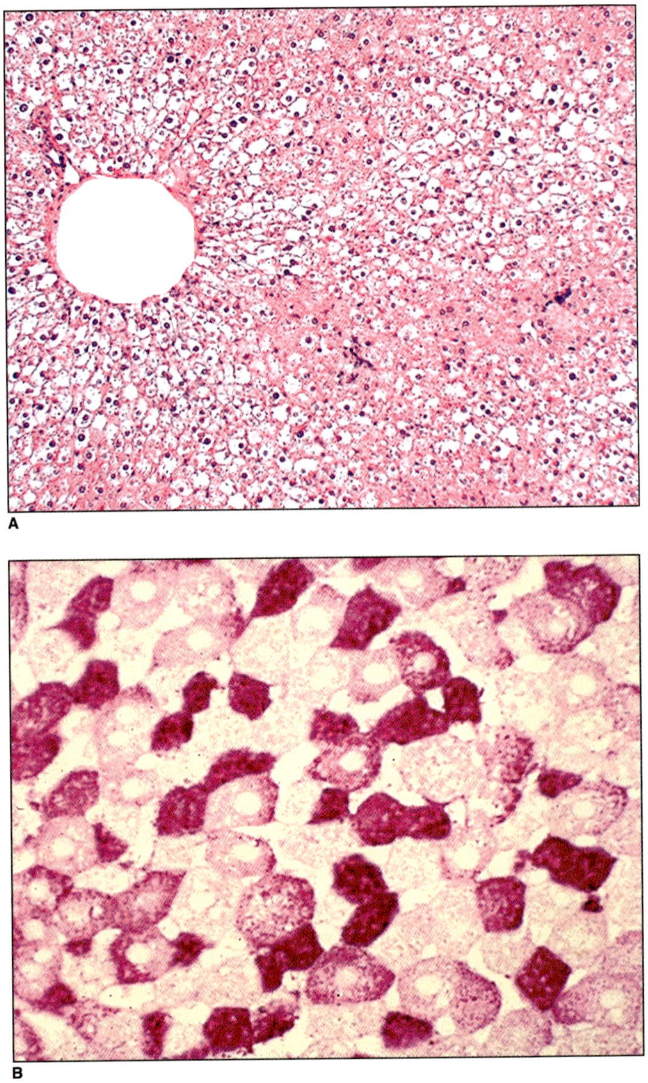

FIGURE 7.24: The liver biopsy in case 4 shows diffuse hepatocytic rarefication, with intranuclear glycogenosis (A). Staining with the PAS method demonstrates intense reactivity in many of the hepatocytes. It was removed with prior diastase digestion (B).

stores with a parenchymal pattern. Subsequent genetic testing revealed C282Y homozygosity for the *HFE* gene. Thus, the liver disease in this case was attributed to both non–alcohol-related steatohepatitis and HH. ▪

<u>CASE 3</u>— A 52-year-old woman, with a long-standing history of mixed connective tissue disease, presented with stigmata of portal hypertension. She denied ethanol abuse or the use of any medications. Viral serologic assessments and evaluations for autoantibodies were negative; levels of urinary copper, serum iron, and serum A1AT were normal. A liver biopsy was performed, showing vague nodularity and an absence of fibrosis (Figure 7.22). A reticulin stain revealed that the nodularity was related to hypertophic cords of hepatocytes that were surrounded by compressed and atrophic heptocytic plates (Figure 7.23). The diagnosis of nodular regenerative hyperplasia was made. ▪

<u>CASE 4</u>— A 3-day-old full-term baby boy developed apneic "spells" that progressed to generalized seizure activity. The first of these occurred after the child's normal feeding time was delayed. After admission to the hospital, it was noted that the baby was irritable and had tremors of the extremities. During one of the spells, blood was drawn for metabolic evaluation. It showed the presence of marked hypoglycemia, with a blood glucose level of 25 mg/dL, as well as hypercholesterolemia, lactatemia, and hypertriglyceridemia. Continuous administration of intravenous glucose solution prevented the episodes thereafter. A liver biopsy was done, showing marked rarefaction of the hepatocytic cytoplasm; a PAS stain was strikingly positive, with lability to diastase (Figure 7.24A,B). A diagnosis of glycogen storage disease type Ib was made subsequently. ▪

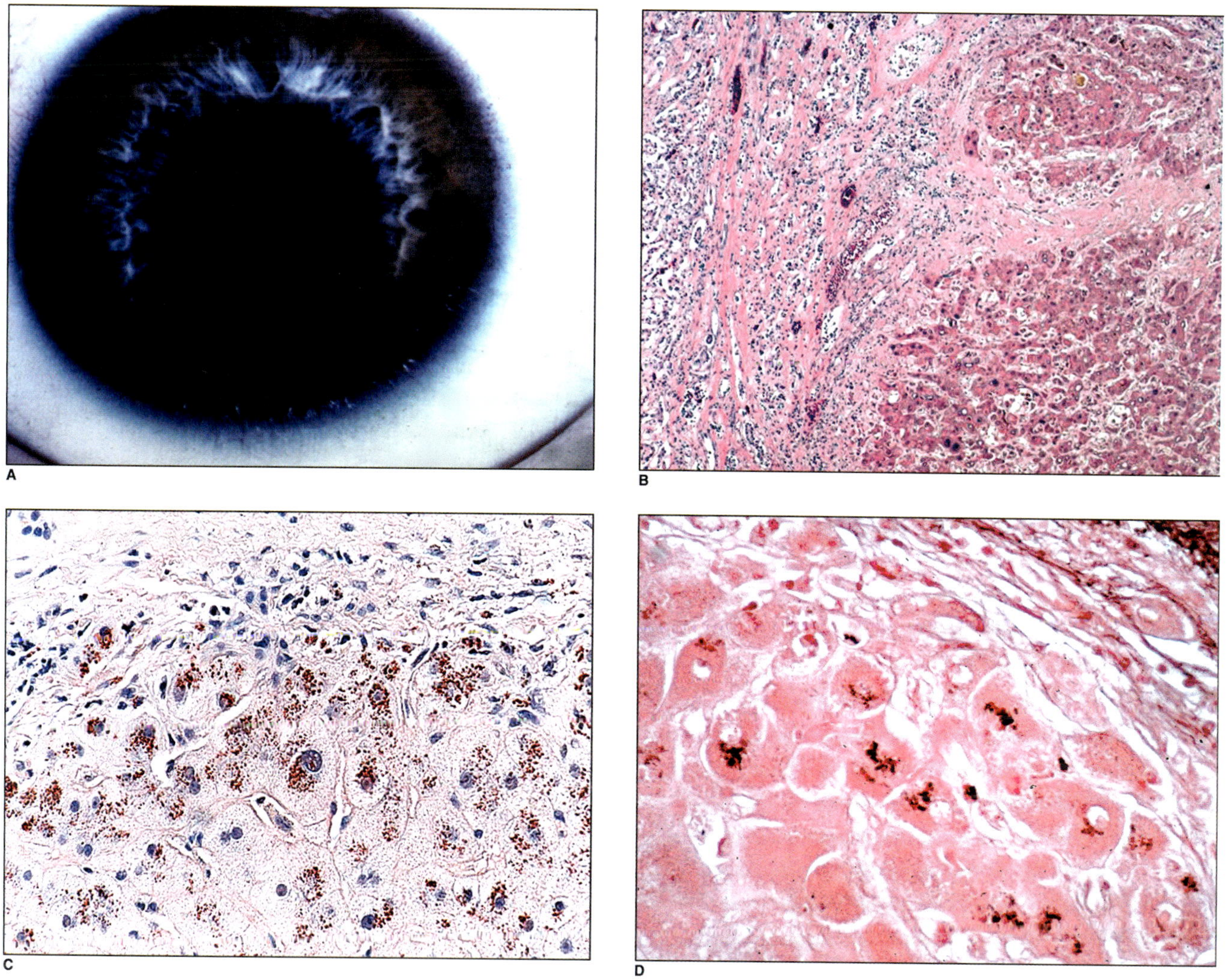

FIGURE 7.25: The patient in case 5 had peripheral intracorneal "rings" by slit-lamp examination of the eyes, representing Kayser-Fleischer deposits (A). Liver biopsy showed lobular inflammation with bridging portal fibrosis. Peripherolobular Mallory's hyaline was also apparent (B). Rhodanine staining (C) demonstrated easily seen intrahepatocellular deposits of copper, which were confirmed with the orcein (Shikata) method (D).

CASE 5— A 20-year-old man told his college roommates that he was "hearing voices," and they also noted the appearance of paranoid, delusional ideation on his part. Assessment by a psychiatrist confirmed the presence of a schizophreniform illness, and she also found the neurologic examination to be abnormal. It showed generalized hyperreflexia and the presence of an intermittent "pill-rolling" tremor of the hands. The patient was admitted to the hospital for further evaluation. A metabolic survey demonstrated a serum ceruloplasmin level of 10 mg/dL (with the normal being >20), as well as a 24-hour urine copper excretion of 200 μg and the presence of aminotransferasemia. Ophthalmologic examination disclosed bilateral corneal Kayser-Fleischer rings (Figure 7.25A); a hepatic biopsy showed chronic lobular hepatitis with portal fibrosis (Figure 7.25B). Histochemical studies of the liver tissue revealed a marked excess of hepatocellular copper content with the rhodanine and orcein methods (Figure 7.25C,D). The diagnosis of Wilson's disease was made and treatment with penicillamine was instituted. ■

REFERENCES

1. Sheela H, Seela S, Caldwell C, Boyer JL, Jain D. Liver biopsy: evolving role in the new millennium. *J Clin Gastroenterol* 2005;39:603–610.
2. Campbell MS, Reddy KR. Review article: the evolving role of the liver biopsy. *Aliment Pharmacol Ther* 2004;20:249–259.
3. Siegel CA, Silas AM, Suriawinata AA, van Leeuwen DJ. Liver biopsy 2005: when and how? *Clev Clin J Med* 2005;72:199–224.
4. FL Carson. *Histotechnology: A Self-Instructional Text* (2nd ed.). Chicago, ASCP Press, 1997; pp. 1–110.
5. JD Bancroft, M Gamble (Eds.). *Theory and Practice of Histologic Techniques* (5th ed.). New York, Churchill Livingstone, 2002.

6. PJ Scheuer, JH Lefkowitch. *Liver Biopsy Interpretation* (6th ed.). New York, W.B. Saunders, 2000.
7. KE Brown, CG Janney, Brunt EM. Liver biopsy: indications, technique, complications, and interpretation. In *Liver Disease: Diagnosis and Management*. BR Bacon, Adrian M. Di Bisceglie (Eds.). New York, Churchill Livingstone, 2000; pp. 47–75.
8. ZD Goodman. Basics of diagnostic hepatic histopathology. In: *Clinical and Pathologic Correlations in Liver Disease: Approaching the Next Millennium*. JR Bloomer, ZD Goodman, Ishak KG (Eds.). Washington, D.C., The American Association for the Study of Liver Diseases, 1998; pp. 3–19.
9. Brunt EM. Nonalcoholic steatohepatitis. *Semin Liver Dis* 2004;24:3–20.
10. Knodell RG, Ishak KG, Black WC, et al. Formulation and application of a numerical scoring system for assessing histological activity in asymptomatic chronic active hepatitis. *Hepatology* 1981;1:431–435.
11. Ishak K, Baptista A, Bianchi L, et al. Histologic grading and staging of chronic hepatitis. *J Hepatol* 1995;22:696–699.
12. Batts KP, Ludwig J. Chronic hepatitis: an update on terminology and reporting. *Am J Surg Pathol* 1995;19:1409–1417.
13. Kleiner DE, Brunt EM, Van Natta M, Behling C, Contos MJ, Cummings OW, Ferrell LD, Liu Y-C, Torbenson MS, Unalp-Arida A, Yeh M, McCullough AJ, Sanyal AJ. Design and validation of a histologic scoring system for nonalcoholic fatty liver disease. *Hepatology* 2005;41:1313–1321.
14. Bradbury MW, Berk PD. Lipid metabolism in hepatic steatosis. *Clin Liver Dis* 2004;8:639–671.
15. Pessayre D, Mansouri A, Haouzi D, Fromenty B. Heptotoxicity due to mitochondrial dysfunction. *Cell Biol Toxicol* 1999;15:367–373.
16. Markin RS, Wisecarver JL, Radio SJ, Stratta RJ, Langas AN, Hirst K, Shaw BW Jr. Frozen section evaluation of donor liver before transplantation. *Transplant* 1993;56:1403–1409.
17. Selzner M, Clavien P-A. Fatty liver in liver transplantation and surgery. *Sem Liver Dis* 2001;21:105–113.
18. Perez-Daga JA, Santoyo J, Suarez MA, Fernandez-Aguilar JA, Rameriz C, Rodriquez-Canete A, Aranda JM, Sanchez-Perez B, Montiel C, Palomo D, Ruiz M, Mate A. Influence of the degree of hepatic steatosis on graft function and postoperative complications of liver transplantation. *Transplant Proc* 2006;38:2468–2470.
19. Crowley H, Lewis WD, Gordon F, Jenkins R, Khettry. Steatosis in donor and transplant liver biopsies. *Hum Pathol* 2000;31:1209–1213.
20. Zamboni F, Franchello A, David E, Rocca G, Ricchiuti A, Lavezzo B, Rizzetto M, Salizzoni M. Effect of macrovesicular and other donor and recipient characteristics on the outcome of liver transplantation. *Transplant* 2001;15:53–57.
21. Feder JN, Gnirke A, Thomas W, Tsuchihashi Z, Ruddy DA, Basava A, Dormishian F, Domingo R Jr., Ellis MC, Fullan A, Hinton LM, Jones NL, Kimmel BE, Dronmal IGS, Lauer P, Lee VK, Loeb DB, Mapa FA, McClelland E, Meyer MC, Mintier GA, Moeller N, Moore T, Morikang E, Prass CE, Quintana L, Starnes SM, Schatzman RC, Brunke KJ, Drayna DT, Risch NJ, Bacon BR, Wolff RK. A novel MHC class I-like gene is mutated in patients with hereditary hemochromatosis. *Nat Genet* 1996;13:399–408.
22. Bacon BR. Hemochromatosis: diagnosis and management. *Gastroenterology* 2001;120:718–725.
23. Turlin B, Deugnier Y. Iron overload disorders. *Clin Liver Dis* 2002;6:481–496.
24. Brunt EM. Pathology of hepatic iron overload. *Sem Liver Dis* 2005;25:392–401.
25. Deugnier YM, Guyader D, Crantock L, Lopez JM, Turlin B, Yaouanq J, Jouanolle H, Campion JP, Launois B, Halliday JW, Powell LW, Pierre Brissot. Primary liver cancer in genetic hemochromatosis: a clinical, pathological, and pathogenic study of 54 cases. *Gastroenterology* 1993;104:228–234.
26. Alla V, Bonkovsky HL. Iron in nonhemochromatotic liver disorders. *Sem Liver Dis* 2005;25:461–472.
27. Ramm A, Ruddell RG. Hepatotoxicity of iron overload: mechanisms of iron-induced hepatic fibrogenesis. *Sem Liver Dis* 2005;25:433–449.
28. Searle J, Leggett BA, Crawford DHG, Powell LW. Iron storage disease. In *Pathology of the Liver* (4th ed.). MacSween RNM, Burt AD, Portmann BC, Ishak KG, Scheuer PJ, Anthony PP (Eds.). London, Churchill Livingstone, 2002; pp. 257–272.
29. Brunt EM, Olynyk JK, Britton RS, Janney CG, Di Bisceglie AM, Bacon BR. Histological evaluation of iron in liver biopsies: relationship to HFE mutations. *Am J Gastroenterol* 2000;95:1788–1793.
30. Faa G, Liguori C, Columbano A, Diaz G. Uneven copper distribution in the human newborn liver. *Hepatology* 1995;22:303–308.
31. Pilloni L, Lecca S, Van Eyken P, Flore C, Demelia L, Pilleri G, Nurchi AM, Farci AMG, Ambu R, Callea F, Faa G. Value of histochemical stains for copper in the diagnosis of Wilson's disease. *Histopathology* 1998;33:28–33.
32. Goldfischer S, Sternlieb I. Changes in the distribution of hepatic copper in relation to the progression of Wilson's disease (hepatolenticular degeneration). *Am J Pathol* 1986; 53:883–901.
33. Ishak KG. Inherited metabolic diseases of the liver. *Clin Liver Dis* 2002;6:455–479.
34. Guarascio P, Yentis F, Cevikbas U, Portman B, Williams R. Value of copper-associated protein in diagnostic assessment of liver biopsy. *J Clin Pathol* 1983;36:18–23.
35. Miyamura H, Nakanuma Y, Kono N. Survey of copper granules in liver biopsy specimens from various liver abnormalities other than Wilson's disease and biliary diseases. *Gastroenterol Jpn* 1988;23:633–638.
36. Klatt EC, Koss MN, Young TS, Macauley L, Martin SE. Hepatic hyaline globules associated with passive congestion. *Arch Pathol Lab Med* 1988;112:510–513.
37. Iezzoni JC, Gaffey MJ, Stacy EK, Normansell DE. Hepatocytic globules in end-stage hepatic disease: relationship to alpha$_1$-antitrypsin phenotype. *Am J Clin Pathol* 1997;107:692–697.
38. Torbenson M, Chen Y-Y, Brunt E, Cummings OW, Gottfried M, Jakate S, Liu Y-C, Yeh MM, Ferrell L. Glycogenic hepatopathy: an underrecognized hepatic complication of diabetes mellitus. *Am J Surg Pathol* 2006;30:508–513.
39. Brunt EM, Tiniakos DG. Metabolic storage diseases: amyloidosis. *Clin Liver Dis* 2004;8:915–930.

40. Droz D, Noel LH, Carnot F, Degos F, Ganeval D, Grunfeld JP. Liver involvement in nonamyloid light chain deposits disease. *Lab Invest* 1984;50:683–689.
41. Ferrell LD, Frankel WL. Tumors in the liver—diagnostic problems. Available at: http://test.pathologyportal.org/newindex.htm?96th/education.htm (accessed 6/29/2007).
42. Bergman S, Graeme-Cook F, Pitman MB. The usefulness of the reticulin stain in the differential diagnosis of liver nodules on fine-needle aspiration biopsy cell block preparations. *Mod Pathol* 1997;10:1258–1264.
43. Kuo F-Y, Chen W-J, Lu S-N, Wang J-H, Eng H-L. Fine needle aspiration cytodiagnosis of liver tumors. *Acta Cytol* 2004;48:142–148.
44. Leerapun A, Suravarapu SV, Bida JP, et al. The utility of *Lens culinaris* agglutinin-reactive alpha-fetoprotein in the diagnosis of hepatocellular carcinoma: evaluation in a United States referral population. *Clin Gastroenterol Hepatol* 2007;5:394–402.
45. Machinami R, Oono Y. Carcinoembryonic antigen and lectin binding in the bile canalicular structures of hepatocellular carcinoma. *Virchows Arch A Pathol Anat Histopathol* 1987;412:111–118.
46. Zhang SM, Wu M, Chen H, Zhang X. Changes of glycoconjugates in human hepatocellular carcinoma. *Histochemistry* 1989;92:171–175.
47. Lin XS. The expression and significance of blood group antigens (BGA) A, B, H, Le(a), and Le(b) in hepatocellular carcinoma and chronic hepatitis. *Chin J Pathol* 1992;21:24–26.
48. Okada Y, Jinno K, Moriwaki S, et al. Expression of ABH and Lewis blood group antigens in combined hepatocellular-cholangiocarcinoma. Possible evidence for the hepatocellular origin of combined hepatocellular-cholangiocarcinoma. *Cancer* 1987;60.345–352.
49. Zhang SM, Wu MC, Chen H, et al. Characteristics of the distribution of lectin receptors in intrahepatic cholangiocellular carcinoma. *Histochem J* 1989;21:296–300.
50. Terg R, Bruguera M, Campo E, et al. Epithelioid hemangioendothelioma of the liver: report of two cases. *Liver* 1988;8:106–110.

CHAPTER 8

Diagnostic Histochemistry of Renal and Urologic Diseases

Helen P. Cathro, M.B., Ch.B., M.P.H.,
Steven S. Shen, M.D., Ph.D., and
Luan D. Truong, M.D.

The use of special stains is indispensable in the diagnosis of medical renal diseases. For example, the periodic acid-Schiff (PAS) stain and the Jones methenamine silver (Jones) method are both fundamental to the recognition of glomerular disorders. The Masson trichrome (trichrome) procedure, together with standard hematoxylin and eosin (H&E) staining, completes the conventional tetrad of histochemical methods that are used to examine kidney biopsies with light microscopy (Figure 8.1) (1). Because the kidney can be involved in a variety of systemic immune complex–related, autoimmune, and vascular diseases, as well as diabetes and a host of kidney-specific conditions, a broad array of adjunctive pathologic procedures are often needed to achieve a meaningful diagnostic interpretation. The recognition of some relatively common renal disorders, such as amyloidosis and light chain deposition disease, is predicated almost entirely on the use of special stains. Although the kidney is infrequently biopsied in suspected infectious disease, various bacteria in that organ can cause malacoplakia, a renal mass that simulates the appearance of renal cell carcinoma (RCC). Sarcoidosis involving the kidney must be distinguished from infectious granulomatous inflammation, and histochemistry is important in that regard. Viral infections of the kidney have become more prevalent in the era of solid organ transplantation, and some viral inclusions are accentuated by special staining (2). Among all the areas of surgical pathology in which histochemistry can be used, it is in nephropathology that this low-cost technology continues to have one of the greatest values.

H&E sections provide an overview of a disease process on initial examination of renal biopsies; they are particularly helpful for identifying vascular hyalinization and different types of inflammatory cells (Figure 8.2 and Table 8.1). Hyaline is distinguished by its smooth, glassy, pink appearance. The Jones and PAS methods allow one to determine which renal compartments (i.e., glomeruli, tubules, interstitium, or blood vessels) are involved in the process at hand. Both those procedures demarcate the basement membrane and mesangial matrix clearly, whereas H&E does not. The distinction between tubulitis and purely interstitial inflammation can be made using the PAS method because it labels the tubular basement membrane, as in the setting of acute cellular allograft rejection (Figure 8.3). Likewise the Jones and PAS stains allow for evaluation of mesangial expansion, as one might see in IgA nephropathy (Figure 8.4). The Jones method highlights adhesions involving Bowman's capsule, as well as breaks in the glomerular basement membrane (GBM), as in focal segmental glomerulosclerosis (FSGS) or crescentic glomerulonephritis (GN), respectively. The Jones and trichrome stains can also demonstrate immune complexes in the GBM, which are seen as "holes" or irregularities in the membrane (Jones) or bright red deposits (trichrome) (Figure 8.5). Neither procedure is particularly effective at demonstrating mesangial deposits. Renal tissue is frequently submitted for ultrastructural examination as well, during which 0.5- to 1-μm "pilot" sections are stained with cationic dyes—most frequently toluidine blue—for light microscopic screening. Although the latter stain is monochromatic, it yields a gradation in tissue color, which, together with the thinness of the sections, enables one to visualize structures that are not clearly seen with other stains (3).

The Jones method is capricious technically and particularly so when it is done with automated platforms. The trichrome stain performs more consistently in demonstrating immune deposits. This technique is particularly valuable in labeling fibrin because it is characteristically "ropey" or "stringy" orange-red appearance is well seen in contrast to erythrocytes (Figure 8.6). The classic use of the trichrome method is to delineate the degree of interstitial fibrosis, but collagenized tissue also is well seen with the Jones technique as fibrillary black material (Figure 8.7). Another useful application of the trichrome stain is to distinguish interstitial edema from fibrosis. This step is necessary to separate acute tubulointerstitial nephritis (TIN) or acute cellular allograft rejection from chronic interstitial

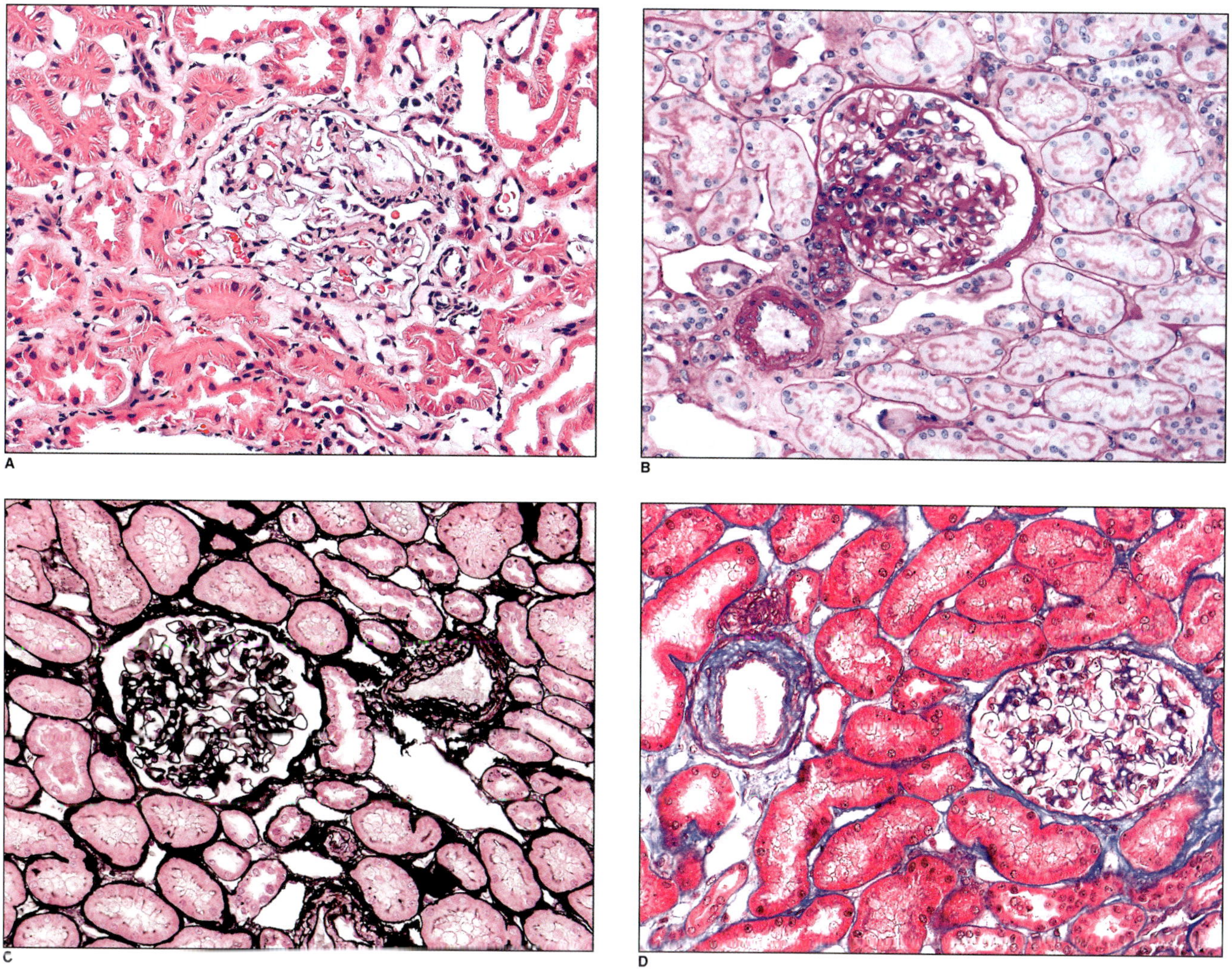

A B C D

FIGURE 8.1: Normal kidney stained with H&E (A), PAS (B), Jones (C), and trichrome (D), illustrating the enhancement of the glomerular and tubular basement membranes on the PAS and Jones stains (all 200×).

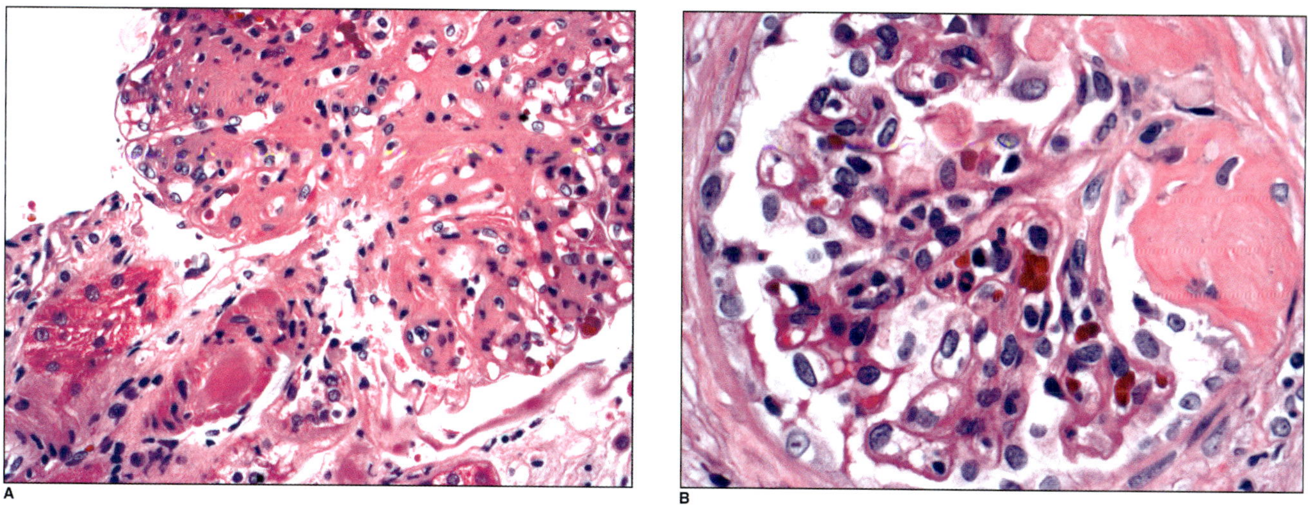

A B

FIGURE 8.2: Arteriolar hyalinization (A) and a mild increase in glomerular neutrophils (B) both demonstrated beautifully on H&E (both 200×).

fibrosis. Another mainstay in nephropathology is the Congo red stain, yielding a typical red-orange result with characteristic "apple-green" birefringence when amyloid deposits are present (Figure 8.8) (4,5).

Many other special stains have been used historically in the diagnosis of medical renal disease. However, those that have been discussed above are the most consistently useful and reliable. Less commonly employed histochemical

TABLE 8.1: Staining of Various Renal Compartments and Pathologic Structures

	H&E	PAS	Trichrome	Jones
Mesangial matrix	Poor	Good	Good	Good
Glomerular sclerosis (global or segmental)	Poor	Excellent	Excellent	Good
Immune complexes	Not visible	Poor	Variable	Negative relief
Basement membrane (glomerular or tubular)	Poor	Excellent	Good	Excellent
Interstitial fibrosis	Poor	Poor	Excellent	Excellent
Arteriolar and capillary loop hyaline	Good	Poor	Good	Negative relief
Microthrombi	Good	Poor	Variable	Variable

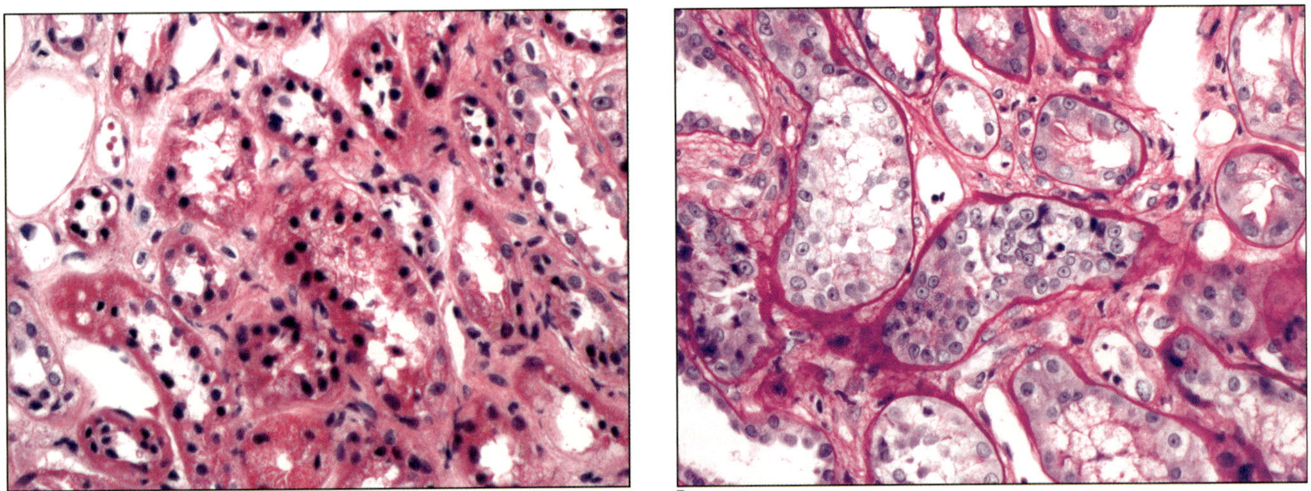

A B

***FIGURE 8.3:** Adjacent levels demonstrating the difficulty of distinguishing renal compartments on H&E (A), as opposed to PAS (B), when evaluating an allograft biopsy for tubulitis (both 200×).*

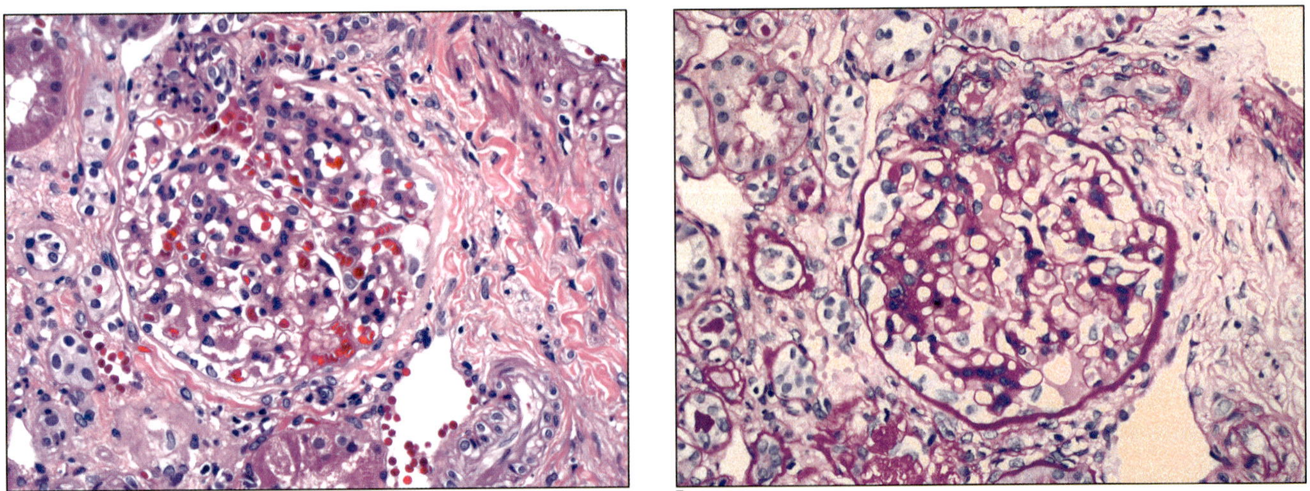

A B

***FIGURE 8.4:** Segmental mesangial expansion in IgA nephropathy that is difficult to identify on H&E (A), but obvious on PAS (B) staining (both 200×).*

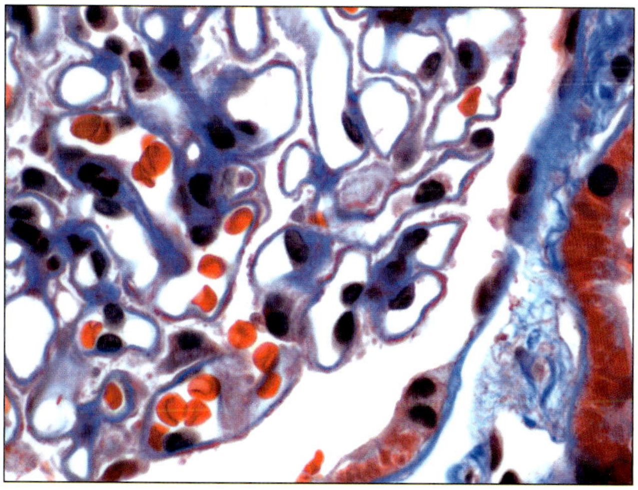

FIGURE 8.5: Subepithelial capillary loop basement membrane deposits seen as red granules on trichrome staining (400×).

procedures will be mentioned in subsequent sections of this chapter in reference to specific diseases.

Arterionephrosclerosis

Arterionephrosclerosis is a very common disease, characterized by aging-related or hypertension-induced fibrosis and hyalinization of blood vessel walls in the kidney. In its early phase, affected glomeruli may show partial collapse and corrugated GBMs, best demonstrated with the PAS and Jones stains (Figure 8.9). Later, the glomeruli become obsolescent (global sclerosis), as characterized by a loss of cellularity, marked corrugation of the GBM, and expanded mesangial matrix. The glomeruli tend to "fade into the background" of the interstitium in H&E stains, leading one to underestimate their density. However, obsolescent glomeruli can be made clearly visible with the PAS or Jones method.

Although global sclerosis is a common glomerular response to injury, regardless of etiology, sclerotic glomeruli in chronic ischemic disease often also show deposition of fibrous tissue beneath Bowman's capsule. This subcapsular fibrosis is PAS negative, whereas underlying sclerotic glomerular tufts are PAS positive (Figure 8.10). This differential labeling is probably caused by deposition of dissimilar collagen types. Subcapsular fibrosis must be separated from the formation of fibrous glomerular crescents, and the PAS stain is again helpful in that regard (see below). Vascular changes are an integral element of arterionephrosclerosis, and histochemical methods may be helpful in evaluating such alterations (see subsequent section on vascular lesions).

Focal Segmental Glomerulosclerosis

Focal segmental sclerosis of glomerular capillaries is the diagnostic lesion of FSGS, one of the most frequently encountered disorders in medical renal pathology. In the

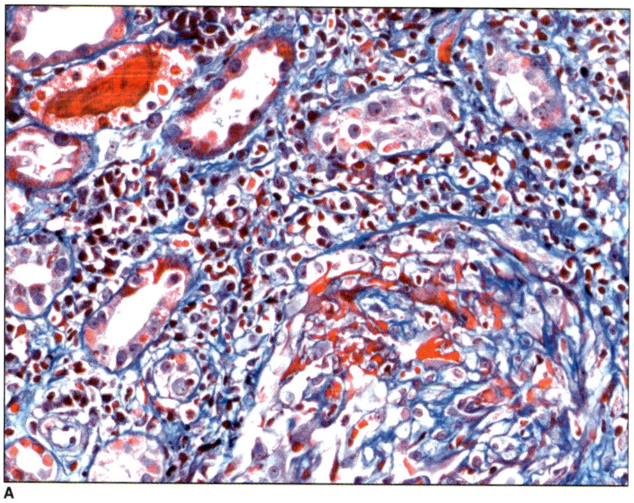

A

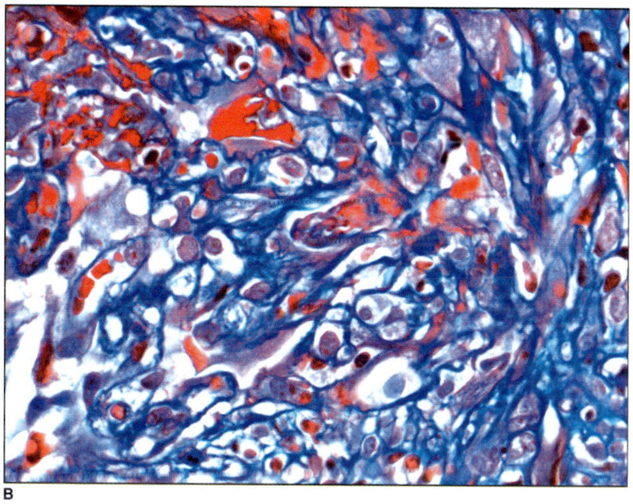

B

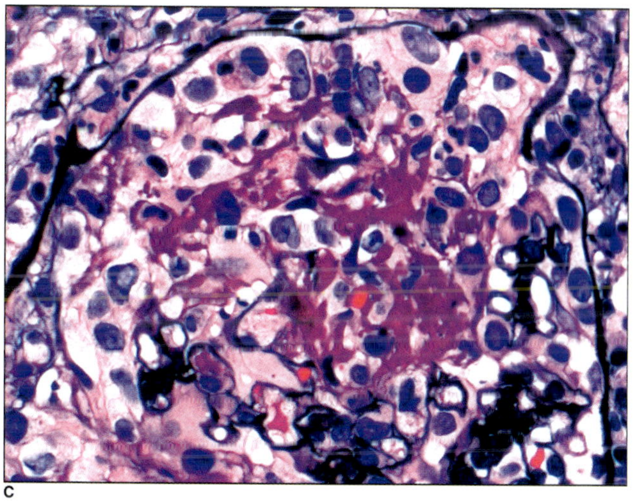

C

FIGURE 8.6: Trichrome (A and B) and Jones (C) stains highlighting glomerular fibrin in a case of pauci-immune necrotizing GN. The low-power trichrome image illustrates the subtle difference in color between superimposed erythrocytes and fibrin, while the high-power images illustrate the textural qualities of fibrin (200× for A, 400× for B and C).

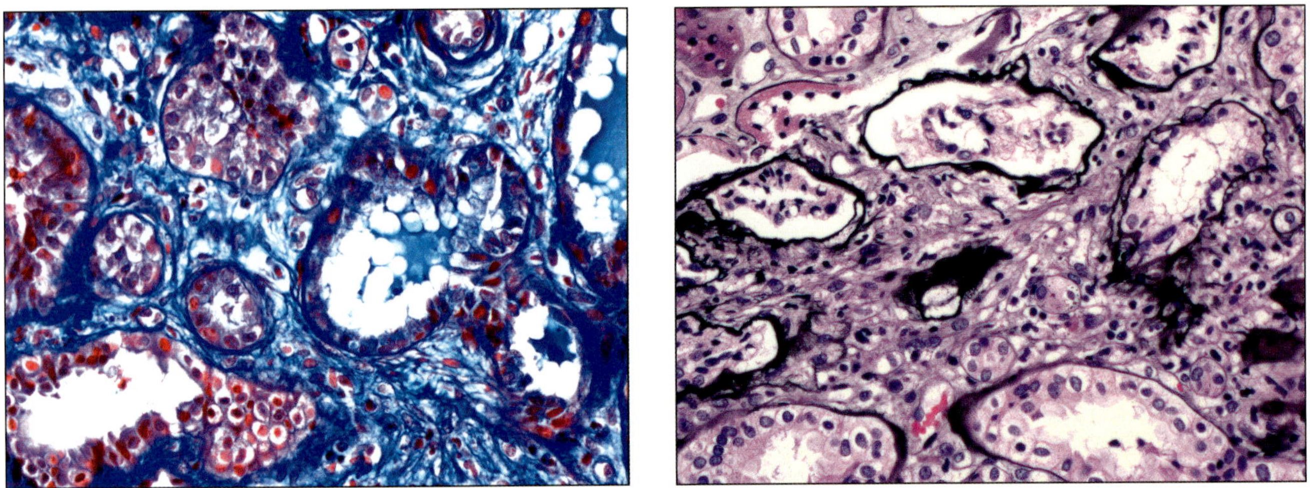

A B

FIGURE 8.7: Interstitial fibrosis compared on trichrome (A) and Jones (B). The collagen has a more solid appearance on the former stain and a wispy appearance on the latter (both 200×).

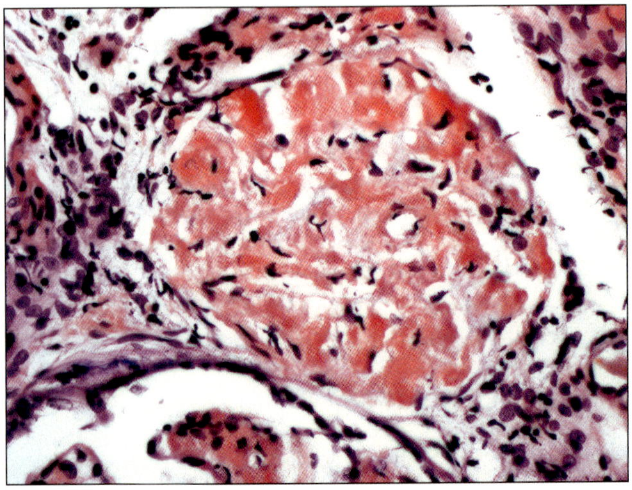

FIGURE 8.8: Amyloid stained with Congo red, showing the typical pale, salmon-red nodular mesangial expansion of the glomerulus. See Figure 8.23 also (200×).

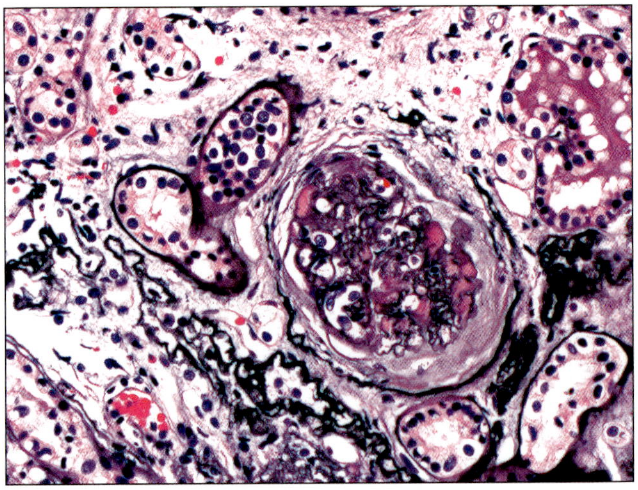

FIGURE 8.9: Corrugated basement membrane and subcapsular fibrosis in a glomerulus damaged by arterial sclerosis on Jones stain (200×).

early phase of the disease, lesions may be very limited and involve only rare glomeruli, thus being easily missed on H&E stains. FSGS is most visible with the PAS and Jones stains, although the trichrome stain is helpful as well. Specifically, one looks for areas of condensation or collapse of glomerular tufts, often seen in only one or two glomeruli in a needle biopsy specimen. As in glomerular sclerosis from any cause, the sclerotic areas in FSGS are densely black in Jones preparations, deep purple-pink in PAS stains, and blue with the trichrome method (Figure 8.11). Because glomeruli that are not affected by segmental sclerosis are normal by light microscopy, FSGS can easily be confused with other glomerulopathies, especially minimal-change disease. The PAS stain is valuable in that context because it also highlights tubulointerstitial injury; this process is usually absent in minimal-change disease but pronounced in FSGS, even when glomerular sclerosis is not obvious. Sclerotic foci in FSGS often show attachment to Bowman's capsule, with hyalinosis. Hyaline foci are easily seen with H&E, trichrome, PAS, and Jones stains; synechiae are best identified with the PAS method. The Jones technique demonstrates the shrunken glomerular tufts that are characteristic of the "collapsing" variant of FSGS. They stand out in dark relief against overlying visceral epithelial cells (Figure 8.12).

Glomerulonephritis with Crescent Formation

Glomerular epithelial crescents, defined as hyperplastic proliferations of parietal Bowman's epithelial cells, are a frequent lesion in a variety of GNs. These are typified by crescentic GN, in which crescents involve >50% of glomeruli in the sample. GN with crescents may have several underlying etiologies, which cannot necessarily be separated from one another with special stains. However, histochemical methods do play a role in this setting.

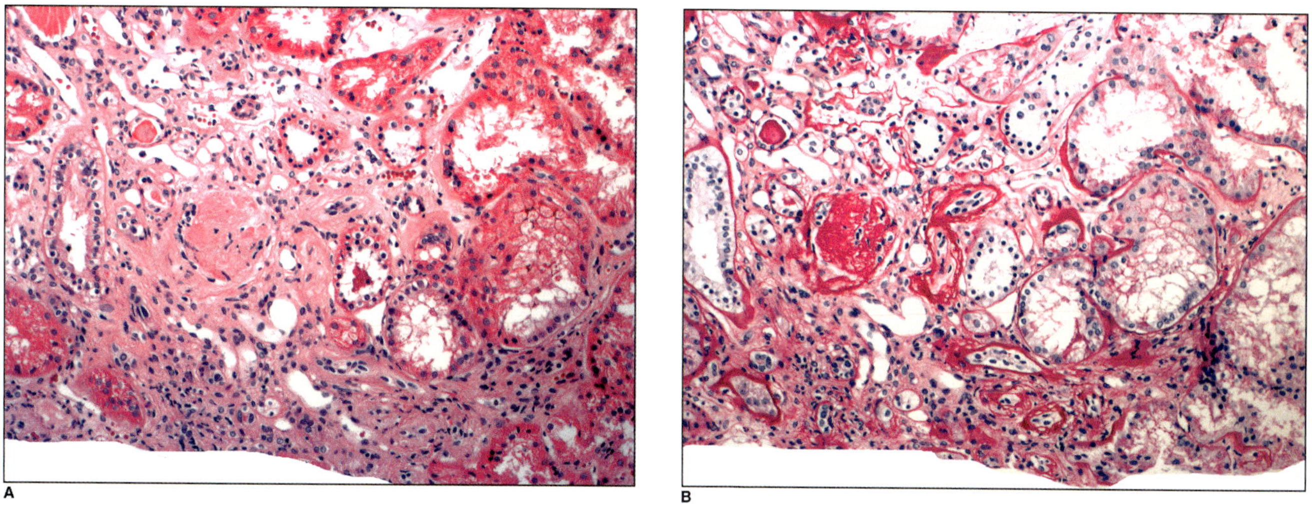

FIGURE 8.10: Low-power images demonstrating the difficulty of visualizing sclerotic glomerular tufts on H&E (A) staining, contrasted with the ease of identification on PAS (B) (both 100×).

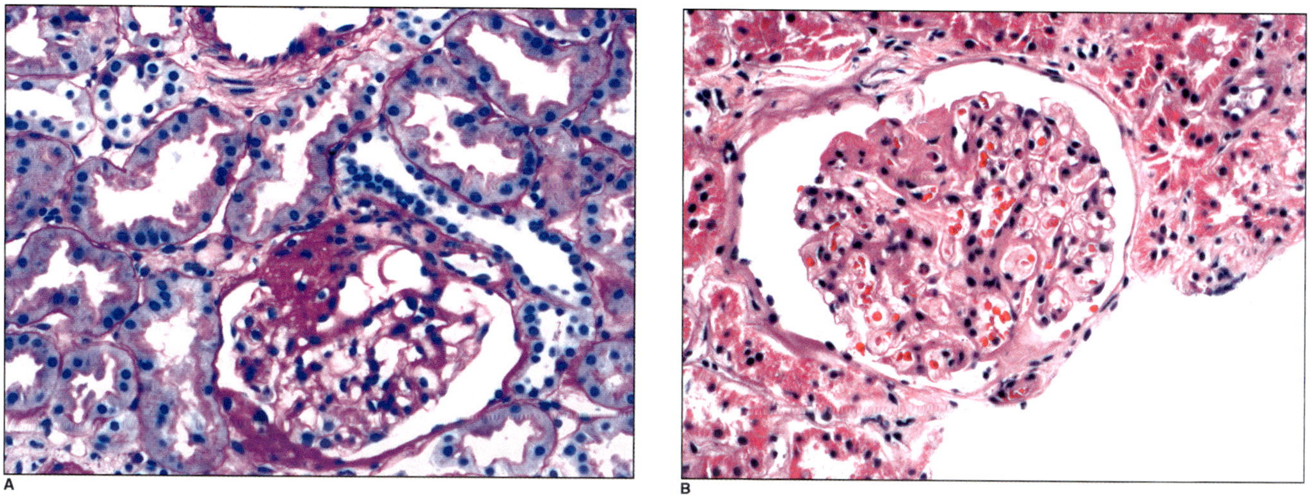

FIGURE 8.11: Segmental glomerulosclerosis delineated much more clearly on PAS (A) than on H&E (B) (both 200×).

Cellular crescents may be so pronounced that they essentially destroy all native glomerular structure. PAS stains are useful in identifying remaining GBM or basement membranes of Bowman's capsule. Fibrous crescents must be distinguished from subcapsular fibrosis because those conditions have different causes. PAS labeling is likewise productive in that regard (Figure 8.13). Healed segmental proliferative GN, in which glomerular lesions are composed of basement membrane matrix, can be confused with FSGS in H&E sections. The former of those diseases is often associated with segmental crescents, which can be outlined with PAS stains. This finding effectively excludes the alternative possibility of FSGS.

Diabetic Nephropathy

Lobular mesangial glomerulosclerosis (so-called Kimmelstiel-Wilson nodules) is characteristic of advanced diabetic nephropathy. The mesangial micronodules in that condition stand out in PAS stains as dark purple-pink structures. They are dark blue with the trichrome technique and brown with the Jones procedure (Figure 8.14) (7). Earlier in the course of the disease, the GBM in diabetic nephropathy is thickened and the mesangial matrix expands, as shown with PAS and trichrome stains (Figure 8.15). Other changes that are easily demonstrated with special stains are mesangiolysis and microaneurysm formation in glomerular capillary walls. The Jones method reveals lamination of neobasement membrane material in the mesangium, which may be related to microaneurysms (Figure 8.16). It should be noted, however, that the same lamellated appearance can be seen in glomerular nodules in light chain immunoglobulin deposition disease. Insudative lesions in diabetes—that is, fibrin caps in capillary loops and capsular "gumdrops" between Bowman's capsule and parietal epithelial cells—are easily visualized with standard H&E stains. The

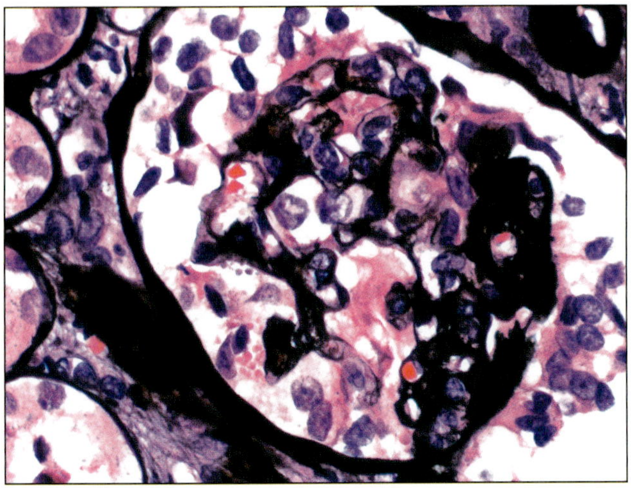

FIGURE 8.12: The collapsing variant of FSGS illustrated on Jones stain with shrunken tufts and associated protein droplets (400×).

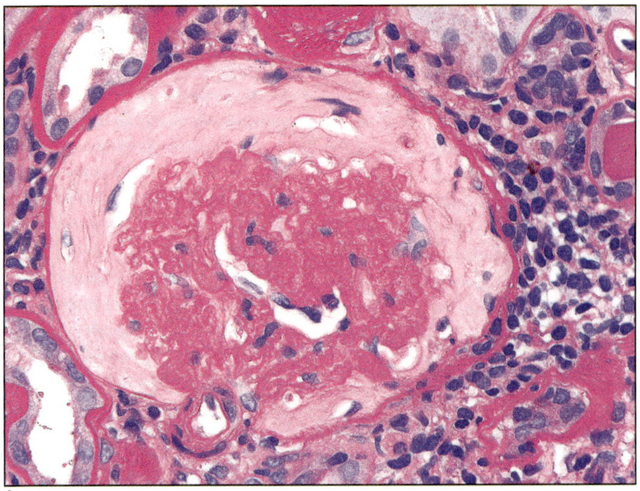

A

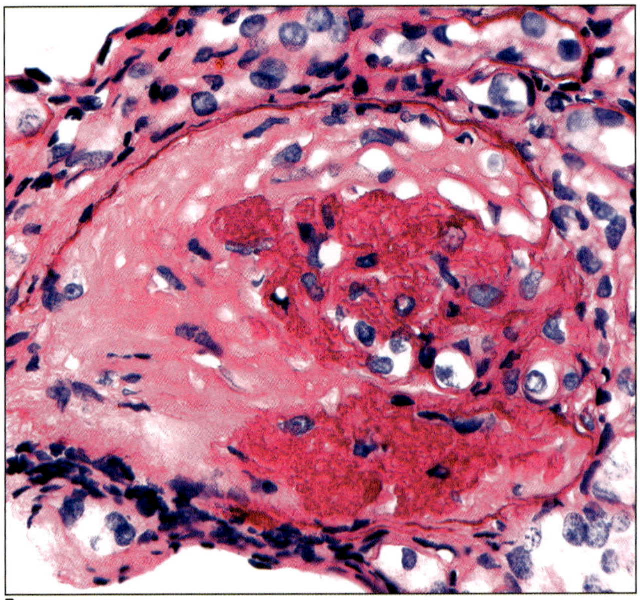

B

FIGURE 8.13: Subcapsular fibrosis (not stained) in arterionephrosclerosis underneath an intact Bowman's capsule highlighted on PAS (A), in contrast to the disrupted Bowman's capsule overlying a fibrous crescent (B) (both 200×).

simultaneous hyalinization of both efferent and afferent arterioles is likewise recognizable. Nevertheless, concern may arise over the possible presence of glomerular amyloid, as a diagnostic alternative to diabetic glomerulosclerosis. PAS and Congo red staining should be done in those circumstances (see subsequent section on amyloidosis).

Membranous Glomerulonephropathy

The classic finding of GBM "spikes" in Jones preparations is expected in well-developed examples of membranous glomerulonephropathy (MGN). It represents the formation of neobasement membrane material surrounding subepithelial immune complexes (Figure 8.17). As the basement membrane completely surrounds immune deposits, the spikes may instead become club shaped, although this eventuality is rare. The immune complexes in MGN also can be demonstrated with Jones and trichrome stains as rounded defects in the GBM and bright red deposits, respectively. In our experience, these findings are far more common than the aforementioned spikes (Figure 8.18). A reticular pattern can also be seen in *en-face* Jones-stained sections of the GBM. Basement membrane thickening in later stages of MGN is nonspecific. It can also be seen in diabetes and is best visualized with PAS and trichrome stains.

Membranoproliferative Glomerulonephropathy

The most common feature of membranoproliferative GN (MPGN; "mesangiocapillary" GN) is that of reduplication of the GBM. This is caused by a new basement membrane that is laid down beneath the capillary loop endothelium. PAS and Jones methods are most useful for detecting basement membrane reduplication (Figure 8.19). "Chunky" deposits of cryoglobulin in capillary lumina in hepatitis C–related MPGN likewise are well seen with the PAS procedure (Figure 8.20). By ultrastructural examination, one sees curved microtubular aggregates in the latter condition. Thick, ribbon-like GBM in cases of dense deposit disease (DDD; MPGN type II) is best visualized as fuchsia colored with the PAS method or crimson with the trichrome procedure. It is also obvious in toluidine blue–stained thick sections that are prepared for electron microscopy (3).

Nephritis of Lupus Erythematosus

Epithelial crescents that are typical of proliferative lupus nephritis stand out in relief with Jones method and are also crisply demonstrated with PAS staining (Figure 8.21). The trichrome and Jones procedures are also helpful for characterizing the degree of fibrosis associated with the

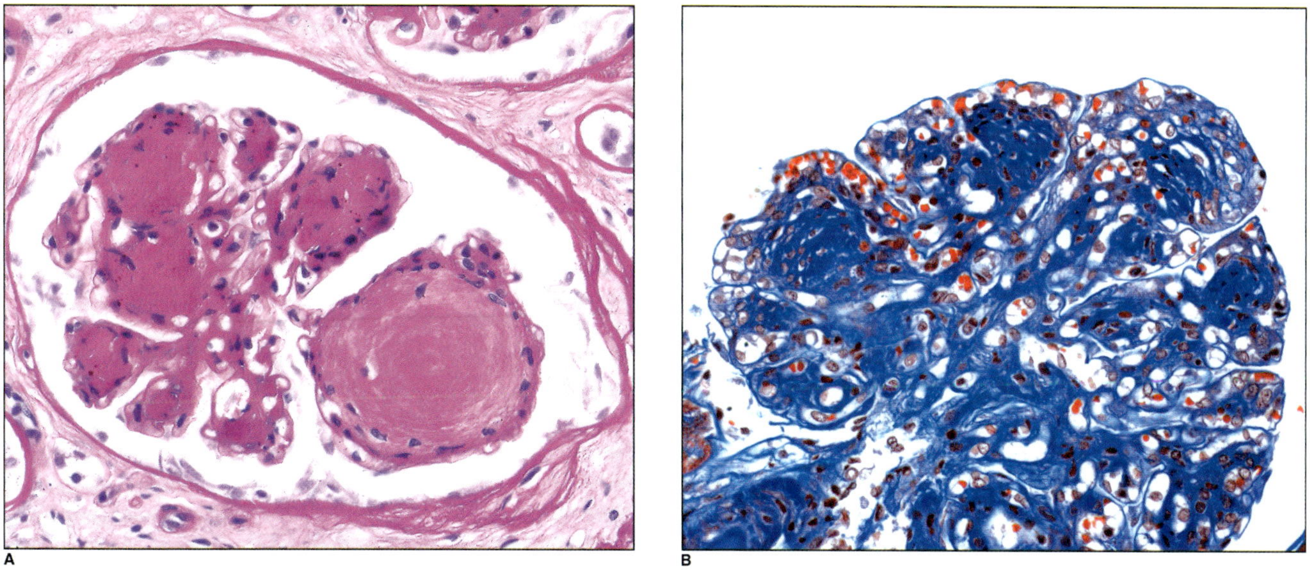

FIGURE 8.14: The Kimmelstiel-Wilson nodules of advanced diabetic glomerulosclerosis on PAS (A) and Jones (B) stains (both 200×).

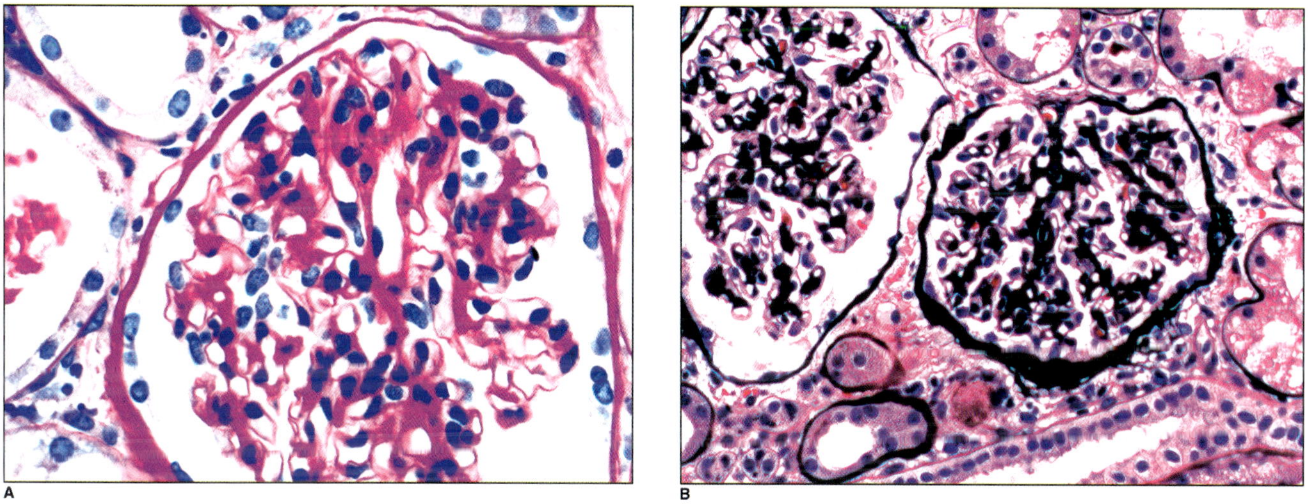

FIGURE 8.15: Subtle mesangial expansion characteristic of an early stage of diabetic damage on PAS (A) and trichrome (B) staining (both 200×).

crescents. Fibrin in necrotic glomerular tufts or crescents is obvious in trichrome stains; so-called "wire loops" and hyaline thrombi in the capillary loops are densely purple pink with PAS and pale pink in Jones preparations in the nephritis of systemic lupus erythematosus (SLE) (Figure 8.22). Mesangial immune complexes are used to distinguish primary MGN from the membranous form of SLE nephritis. However, they cannot be reliably visualized with special stains. Subepithelial immune complexes in membranous lupus nephritis are more easily demonstrated with Jones and trichrome procedures, as compared with subendothelial deposits as seen in proliferative lupus nephritis.

Amyloidosis

"Amyloid" is not a single entity but rather a set of proteins that features the formation of disordered β-pleated sheets at a biophysical level. The two types of amyloid that are most commonly encountered in renal biopsies are "AL" amyloid, usually deriving from immunoglobulin light chains in the setting of myeloma, and secondary "AA" amyloid, originating from serum AA protein in chronic inflammatory diseases. Amyloid can be deposited in the mesangium, the basement membrane of the capillary loops, Bowman's capsule, arterial walls, tubular basement membranes, and the renal interstitium. Because it can be only subtly present in H&E sections, the use of Congo red staining is preferable in any case where amyloid deposition is a clinical consideration. In conventional sections, amyloid is pale pink, amorphous, and "fluffy" (Figure 8.23). PAS and Jones methods label amyloid pale pink and pink brown, respectively. Sirius-red has been used to recognize amyloid, but general consensus holds that Congo red is more sensitive and specific (8). Sections for the Congo

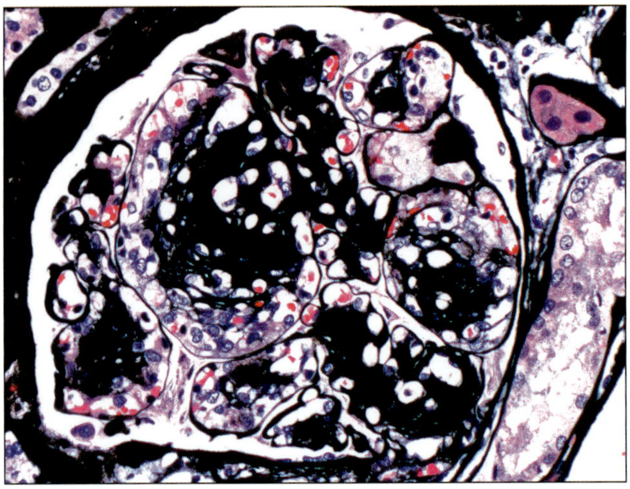

FIGURE 8.16: Early lamination of one of the mesangial nodules in advanced diabetic glomerulosclerosis on Jones (200×).

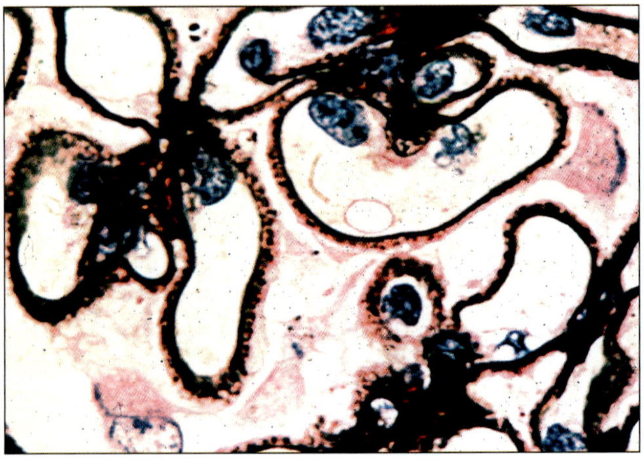

FIGURE 8.17: "Spikes" formed by neobasement membrane surrounding glomerular capillary deposits in membranous GN on Jones (400×).

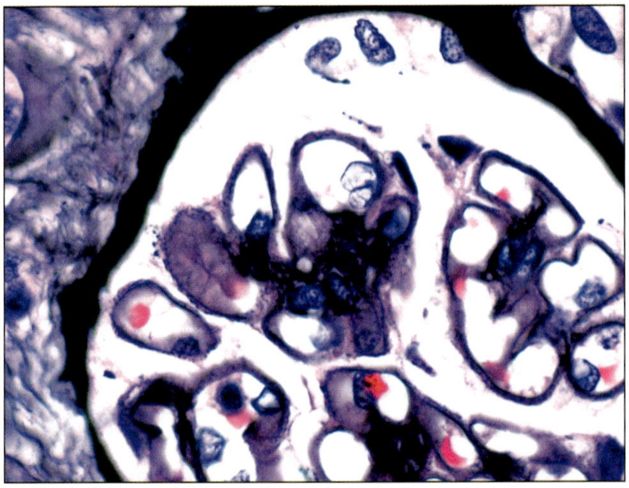

FIGURE 8.18: The subepithelial deposits of membranous GN seen as holes or defects in the capillary loop basement membrane on Jones (400×).

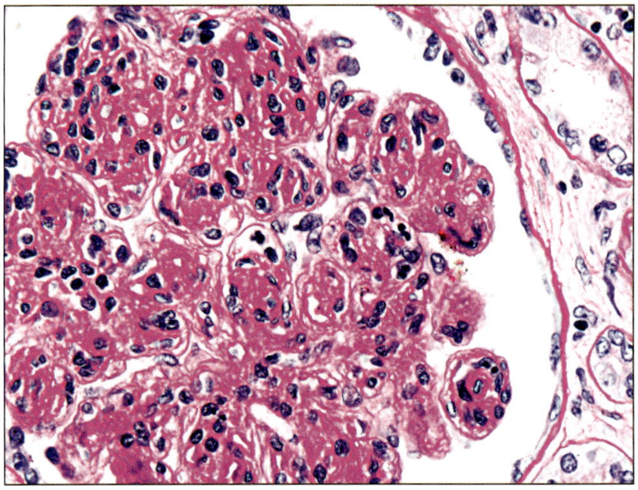

FIGURE 8.19: Reduplication of the capillary loop basement membrane and endocapillary hypercellularity in MPGN on PAS (400×).

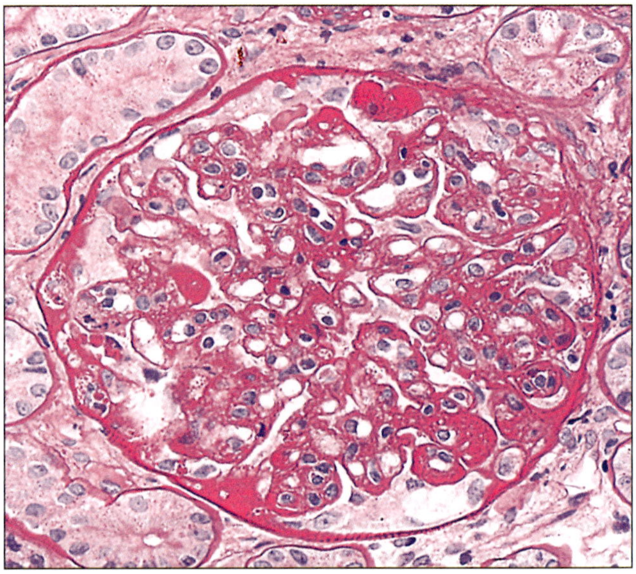

FIGURE 8.20: Chunky deposits of cryoglobulin on PAS of hepatitis C–related membranoproliferative GN type 1 (200×).

red procedure must be cut at 9–10 μm, rather than the usual 2–4 μm. The method of Puchtler et al., using an alkaline Congo red solution, seems to work best. One obtains optimal results if the solution is fully saturated with NaCl (5,9). Amyloid is a dull red orange by light microscopy, but polarization produces specific apple-green refringence that alternates with yellow-orange refringence (Figure 8.24) (4,10). Collagen is also dull orange in Congo red sections, but it is exclusively yellow-orange with polarization. Use of the correct conditions for polarization is crucial to avoid false-negative results; maximum microscope light levels should be used in a dark ambient environment, with all color filters having been removed (11).

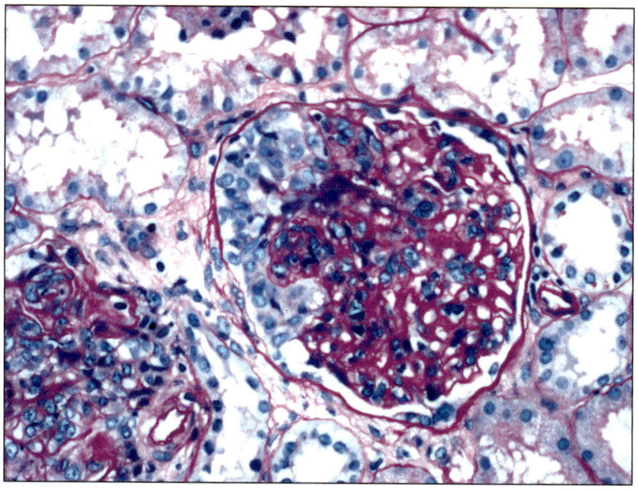

FIGURE 8.21: A cellular crescent seen in proliferative lupus GN on PAS (100×).

Differential staining, with and without potassium permanganate, is useful in increasing the sensitivity of the method. It has also been proposed as a way of separating the AL and AA types of amyloid. If staining is weakened by the addition of potassium permanganate, AA-type amyloid is said to be more likely (9,12). However, we have found that immunostaining—usually performed in reference laboratories—provides more reliable typing of amyloid. Crystal violet, methyl violet, "cotton" dyes, and sulfated alcian blue stains can also be used to recognize amyloid, but the first two of those preparations fade rapidly. Thioflavin T and S methods necessitate the use of immunofluorescence microscopy. Congo red stains can also be subjected to fluorescence to accentuate the presence of small deposits; this effect is enhanced with the Eastwood modification of the Congo red method (13).

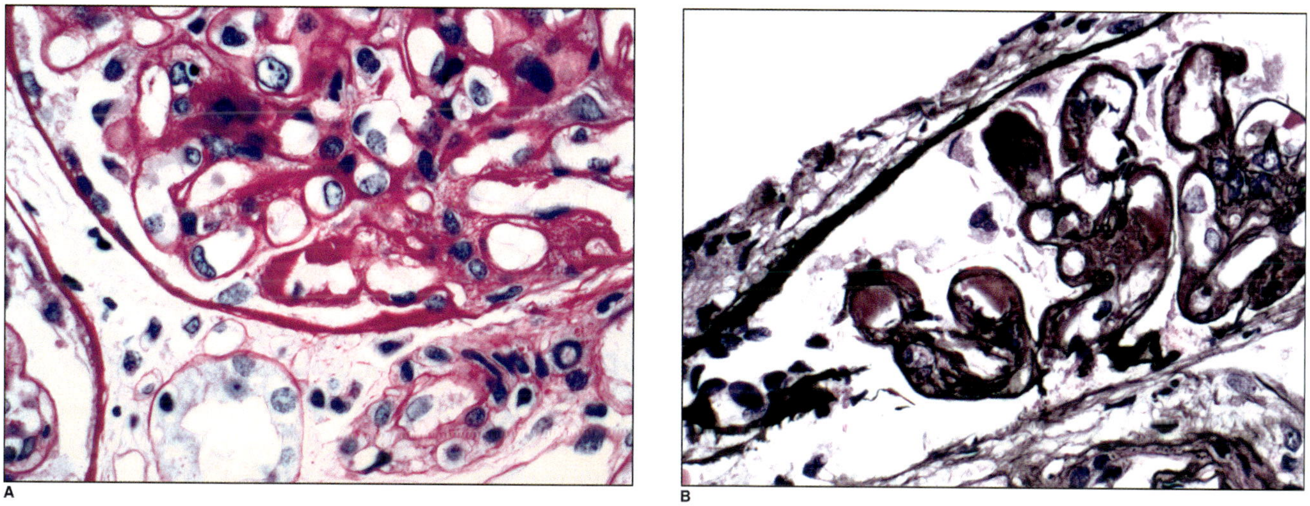

A B

FIGURE 8.22: "Wireloops" on PAS (A) and Jones (B) in proliferative lupus GN (both 400×).

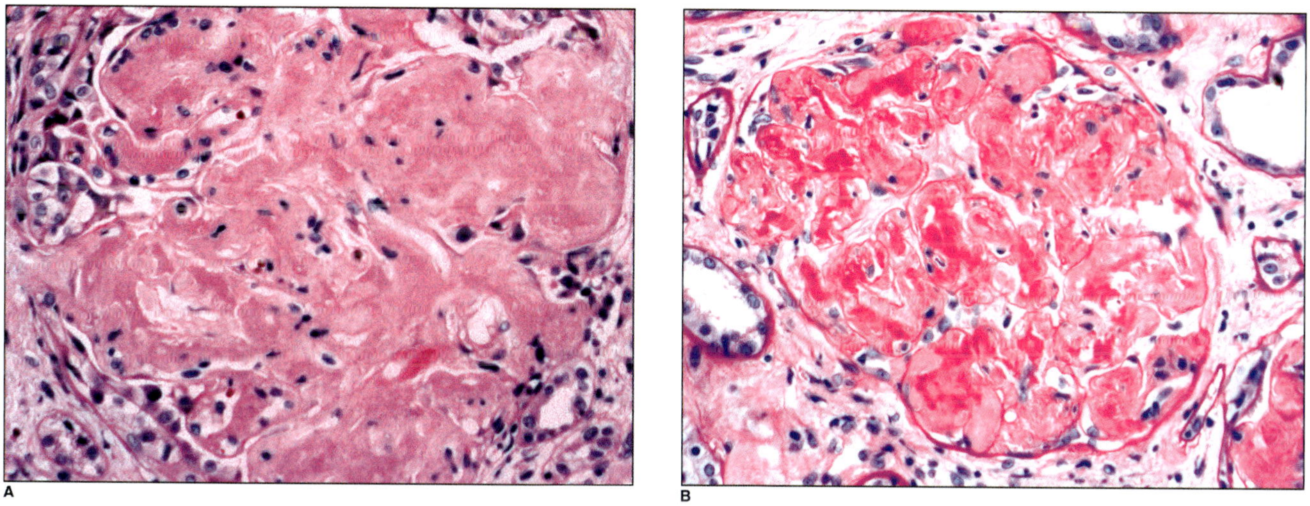

A B

FIGURE 8.23: Amyloid stained with H&E (A) and PAS (B). Note the negative stain of amyloid (pale pink) in PAS, in comparison to the positive stain (purple) of sclerotic glomerular capillaries and Bowman's capsule, and the basement membrane of adjacent tubules (both 200×).

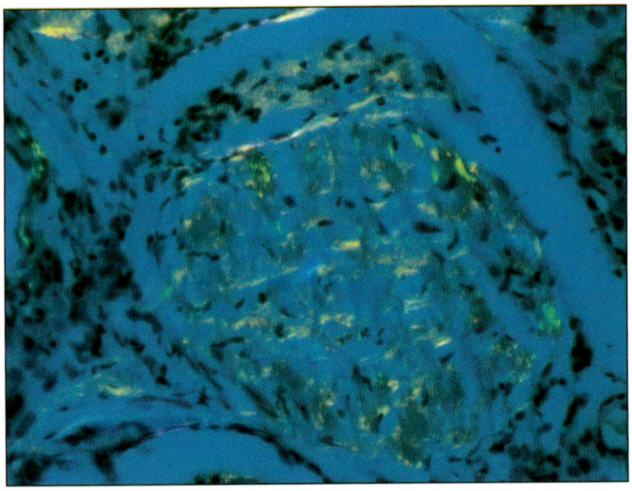

FIGURE 8.24: Amyloid staining with Congo red under conventional (see Figure 8.8) and polarization microscopy (200×).

By H&E staining, glomerular amyloidosis may be confused with other morphologically similar lesions such as advanced MPGN, light chain immunoglobulin deposition disease, amyloid-like nephropathy-fibrillary GN, and idiopathic lobular GN glomerulosclerosis. However, glomerular deposits in those conditions are Congo red negative and variably positive with the PAS and Jones procedures.

Monoclonal Immunoglobulin Deposition Disease

Monoclonal immunoglobulin deposition diseases (MIDD), including those involving light chains, heavy chains, and mixed light and heavy chains are suggested by suitable light microscopic images in the correct clinical settings. However, histochemical methods are necessary to make such diagnoses with certainty. Typical immunoglobulin deposits, if they are large enough, will stain with the PAS method but not with the Jones technique (Table 8.2). (Figure 8.25). Collagen is the main simulant of amyloid in this setting, but this distinction is easily resolved by polarization microscopy, as discussed above. Hence, histochemistry is indeed helpful in the recognition of MIDD (Table 8.1).

Tubulointerstitial Nephritis

Histochemical methods also assist in recognizing the acute and chronic forms of TIN. Chronic TIN is characterized by interstitial fibrosis and tubular atrophy (Figure 8.26); it shows an increased volume of dense blue matrix between the tubules in trichrome stains or black fibrillary material in Jones preparations. The degree of interstitial fibrosis provides independent prognostic information in GN cases. The PAS stain labels tubular basement membranes, facilitating the interpretation of acute pyelonephritis when seen in association with intratubular neutrophilia. Lymphocytic tubulitis in acute cellular allograft rejection can be accentuated with the PAS technique as well.

Acute TIN is characterized by interstitial edema and an absence of tubular atrophy. The trichrome stain shows a wispy pale blue interstitium in that condition, as opposed to the dense blue matrix seen in chronic TIN. Subtle interstitial edema is also a diagnostic feature of acute cellular rejection (Figure 8.27). Because these distinctions depend on the staining intensity of the trichrome method, one must assess the intensity of staining in control sections that are included in each "run" of automated histochemical instruments. Perivascular or renal capsular collagen also serves as an internal control.

Tubular protein resorption droplets are frequently seen in renal biopsies. They may be particularly pronounced in human immunodeficiency virus–related nephropathy, as delineated with the PAS stain. Isometric tubular vacuolization, a common finding in diseases causing nephrotic-range

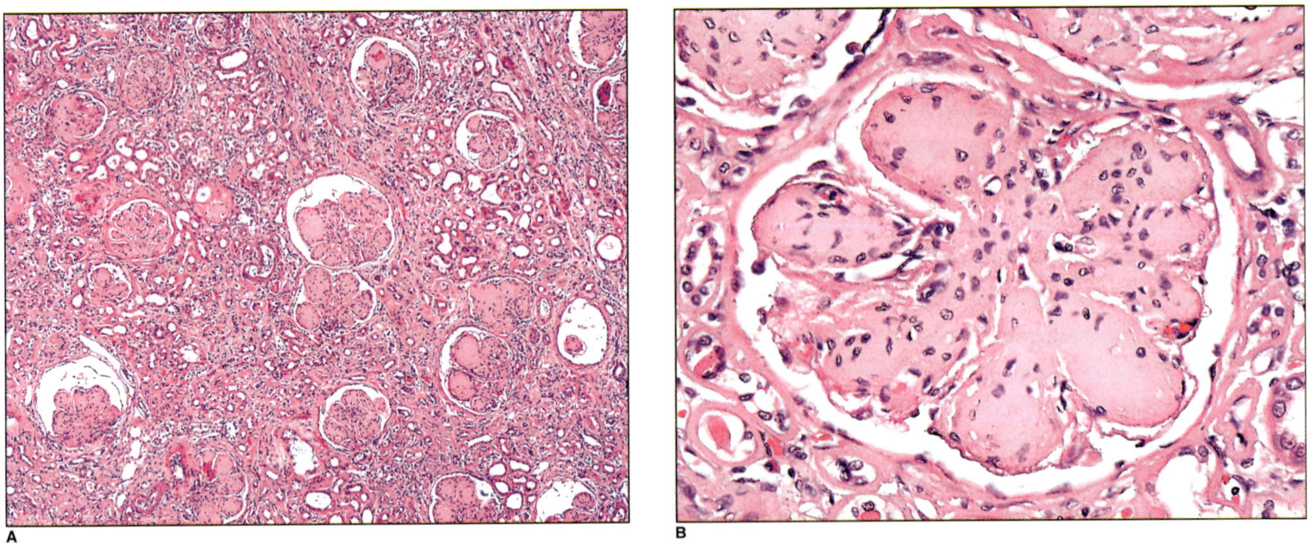

A B

FIGURE 8.25: Light chain glomerular deposits in the lobular pattern characteristic of monoclonal immunoglobulin deposition disease at low (A) and high power (B) by H&E (100× and 200×).

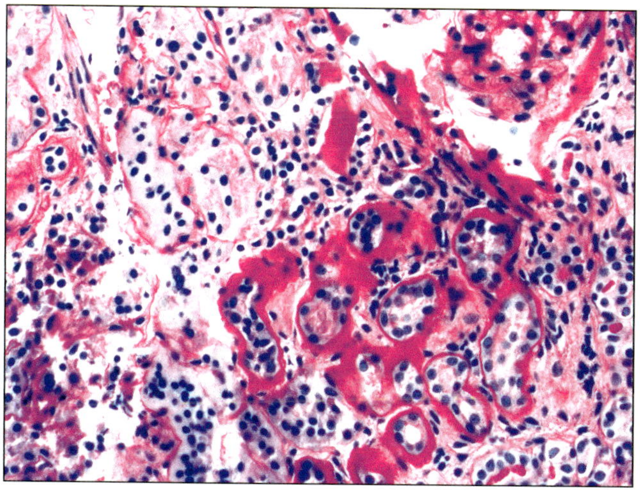

***FIGURE 8.26:** Tubular atrophy associated with thickened tubular basement membrane on PAS (200×).*

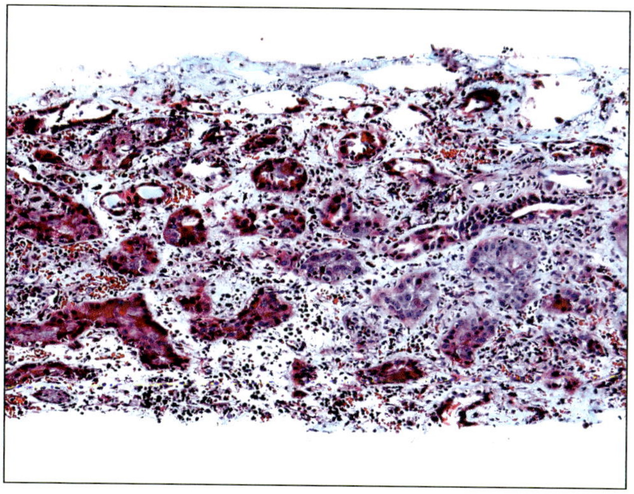

***FIGURE 8.27:** Interstitial edema as seen on trichrome, supporting a diagnosis of acute cellular renal allograft rejection (200×).*

proteinuria (and others) can be accentuated with trichrome, Jones, or toluidine blue preparations. Hyaline casts, as seen in tubular atrophy regardless of cause, have a characteristic staining profile. They are PAS positive and variably labeled with Jones and trichrome stains, helping to separate them from myeloma casts (see below). That is because Tamm-Horsfall protein (THP), a protein synthesized in the distal nephron, is a constant glycoprotein-containing constituent of tubular casts. It is uniformly pale blue in H&E sections, intensely red in PAS preparations, light blue with the trichrome method, and dull purple brown in Jones stains.

Granulomatous Interstitial Nephritis

Occasionally TIN assumes a granulomatous form, as seen in association with underlying infections, allergic reactions to medications, and some connective tissue diseases. However, renal interstitial granulomas are rarely found in Wegener's granulomatosis despite their common presence in sinonasal and lung biopsies in that condition. Special stains for fungi, such as Gomori's methenamine silver (GMS) or the PAS–light green methods, and others for mycobacteria (e.g., Ziehl-Neelsen and Fite stains) help to exclude underlying infections but have limited sensitivity. Another problem is that granulomas may not be present in sections cut for histochemical studies, given the small amount of tissue in renal biopsies. If the associated inflammatory infiltrate contains eosinophils, a drug-related granulomatous nephritis is a possibility. In contrast, GN usually accompanies connective tissue disease–related granulomas. Rarely, granulomas represent a response to primary tubular basement membrane diseases. The PAS stain may also help to highlight subtle granulomas that are associated with arteries or glomeruli.

Malacoplakia

Malacoplakia is a potentially tumefactive interstitial inflammatory process that is characterized by a mixture of inflammatory cells. These include lymphocytes and plasma cells, but the predominant elements are so-called von Hansemann cells, a special type of macrophages with abundant eosinophilic cytoplasm (14,15). The diagnosis of malacoplakia requires the presence of Michaelis-Gutmann bodies in the cytoplasm of von Hansemann cells or in the interstitium. These structures are 2- to 5-µm, round, concentrically multilayered, basophilic calcospherites; they are thought to represent phagolysosomes, reflecting an impaired capacity to digest phagocytosed bacteria. Partially digested glycolipids remain as the nidus for secondary deposition of calcium and iron salts.

Histochemical methods facilitate the diagnosis of malacoplakia when morphologic image is indeterminate or obscured by fibroinflammatory changes. Michaelis-Gutmann bodies are reactive with the PAS method—both before and after diastase digestion—as well as Perl's stain, the von Kossa and alizarin red procedures, GMS, mucicarmine, alcian blue, and toluidine blue (Figure 8.28). The PAS, iron, and von Kossa stains are the most consistently positive (14,16,17).

Malacoplakia must be distinguished from RCC, xanthogranulomatous nephritis, and megalocytic interstitial nephritis. The last of these disorders shares many features of renal malacoplakia and may be an early form lacking Michaelis-Gutmann bodies.

Renal Calcium Deposits

Deposition of calcium salts (phosphate, carbonate, or oxalate) may be seen in the kidney. Calcium phosphate and calcium carbonate are associated with tubular necrosis, interstitial fibrosis, and nephrocalcinosis. When they are abundant, deposits of either salt have a characteristic appearance; nonetheless, the von Kossa and alizarin red methods may be needed to detect small amounts of calcium. The first of those stains is sensitive, but it detects the

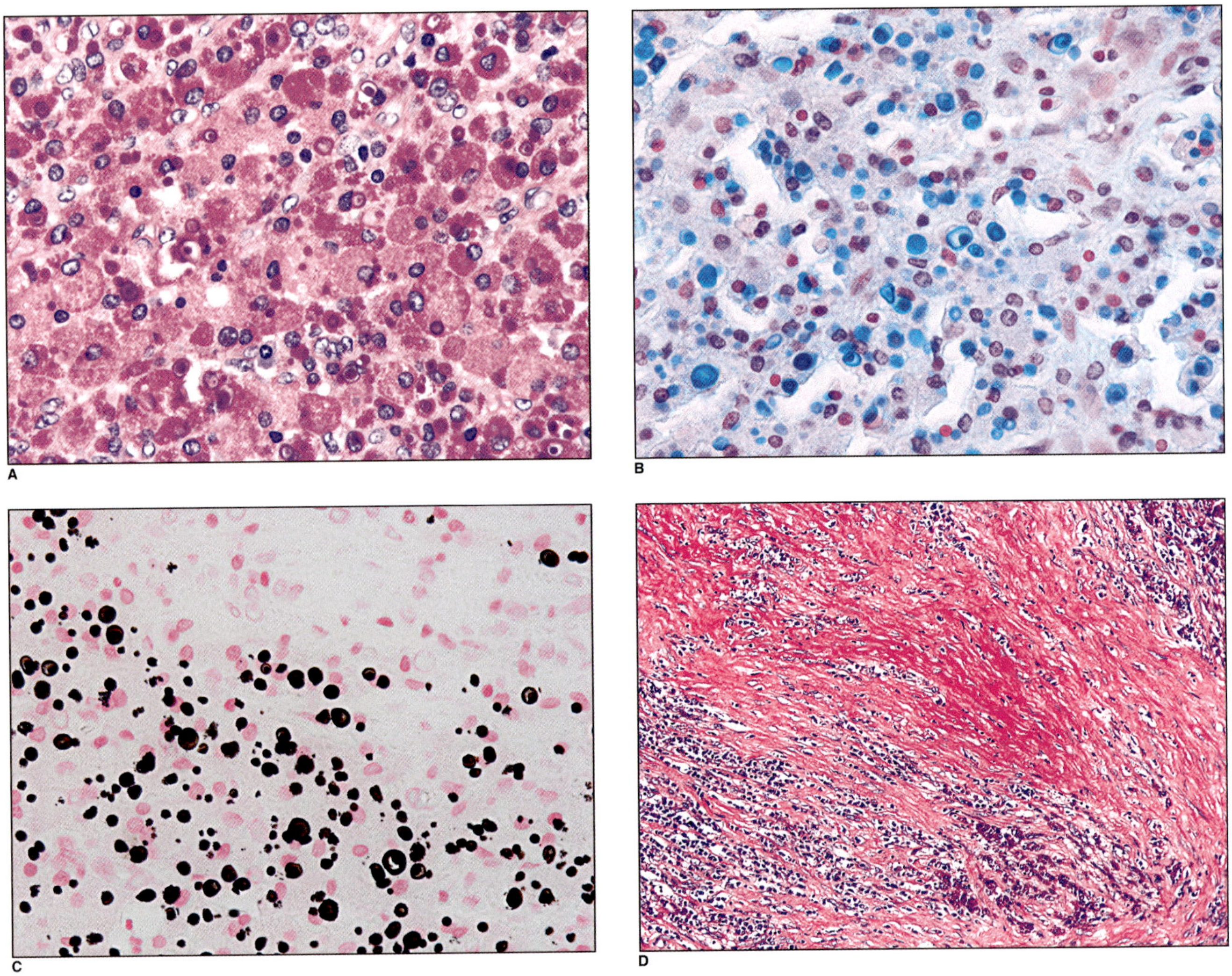

FIGURE 8.28: Aggregations of diagnostic von Hansemann cells with abundant PAS-positive cytoplasmic granules and the diagnostic Michaelis-Gutmann bodies (A). These bodies are strongly positive with alcian blue (B) and von Kossa stains (C). The von Hansemann cells are clearly demonstrated by the PAS stain, even when they are present in small numbers against the background of nonspecific fibroinflammatory changes, as seen in the late phase of malacoplakia or at the periphery of this lesion (D) (400× for A and C, 200× for B, and 100× for D).

phosphate or carbonate congener of calcium rather than that element itself. In contrast, the alizarin red procedure is less sensitive but it is specific for calcium ions.

Calcium oxalate (CO) deposition is potentially seen in any form of chronic renal damage, but it is diagnostic for CO nephropathy. CO is represented by colorless refractile crystals that polarize in multiple hues. They are present in tubular lumina and in the interstitium (Figure 8.29). The crystals are labeled with the alizarin red procedure but not with the von Kossa stain. PAS and trichrome stains are likewise negative.

Diseases Featuring Acute Tubular Damage

Histochemical stains help to recognize the presence of acute tubular necrosis (ATN), as well as specific intratubular substances that may be damaging to the kidney. The PAS stain can be used to demonstrate the early loss of brush borders on the luminal surfaces of proximal tubular epithelial cells in ATN; this change may also be seen with the Jones procedure (Figure 8.30). The tubular casts in hemolytic diseases and rhabdomyolysis comprise coarsely granular material that is usually brown on H&E and crimson with the trichrome method. In severe or long-standing chronic hemolysis, intratubular iron deposition can be confirmed with the Prussian blue stain (Figure 8.31). Bile casts are labeled bright green by Hall's or Fouchet's procedure (Figure 8.32). The casts that are diagnostic of myeloma nephropathy are typically red to pink with trichrome stains, but they are unlabeled or very pale pink with the PAS technique. Another rare cause of tubular epithelial damage is lead poisoning; if this is suspected clinically, Ziehl-Neelsen stains can be employed to color the diagnostic intranuclear inclusions in that condition bright red.

Viral Infections

Another cause of severe tubular damage is viral infection, often seen in the context of immunosuppression for renal transplantation. The most common virus that is observed in that context is the polyomavirus, specifically the "BK" virus (given the initials of the first patient in whom it was recognized). The most common type of nuclear change in BK viral infection is a basophilic glassy alteration (Figure 8.33). Other BK inclusions are characterized by focal clearing of chromatin, vesicular change, and eosinophilic bodies (2). Those structures are accentuated with PAS, Jones, and trichrome stains, often being seen in relief or in a different color from the rest of the nucleus (Figure 8.34). Renal inclusions of cytomegalovirus and adenovirus may also be identified histochemically with the Jones or PAS procedure (Figure 8.35).

Nonvasculitic Vascular Damage in the Kidneys

Histochemical stains can be utilized to help delineate the intrarenal vascular lesions of severe essential hypertension or aging. Reduplicative hyperplasia of the vascular intima and subintima—causing so-called onion-skin layering—typifies the arterial and arteriolar damage of "malignant" hypertension. This change is apparent with PAS stains, and reduplication or fragmentation of the internal arterial elastica is well demonstrated with the Verhoeff–Van Gieson (VVG) technique (Figure 8.36). PAS-positive duplication of the basement membranes of peritubular capillaries also is a sequela of chronic antibody-mediated rejection (18,19).

Renal Vasculitides

Acute renal arteritides are characterized by fibrin deposits in vascular walls, which can be accentuated with

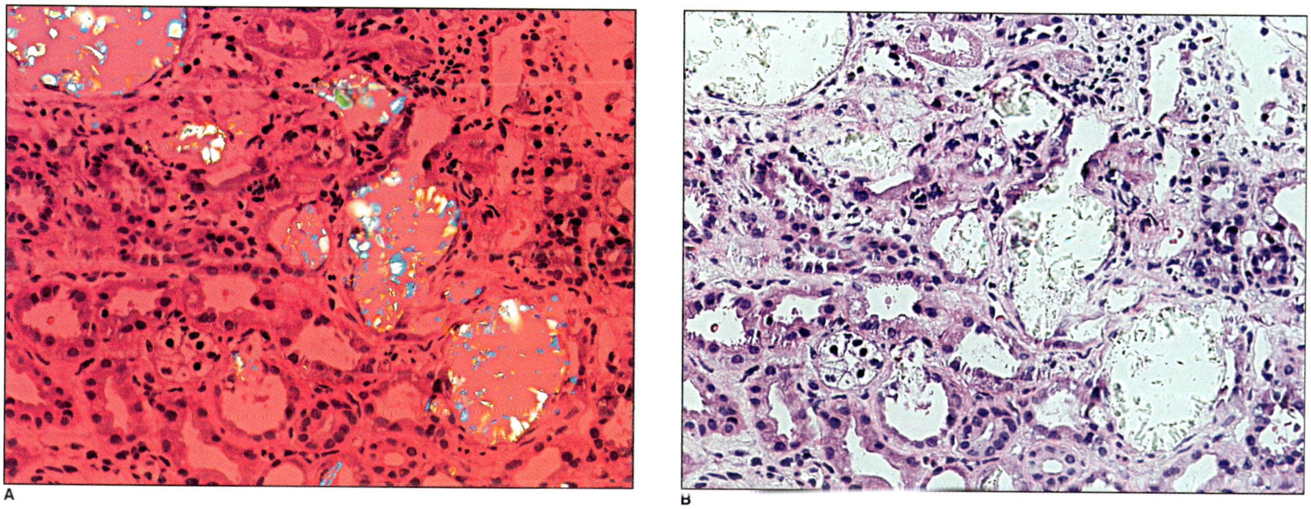

A B

FIGURE 8.29: Polarized (A) predominantly intraluminal CO crystals in chronic renal damage, much less obvious on H&E (B) (both 200×).

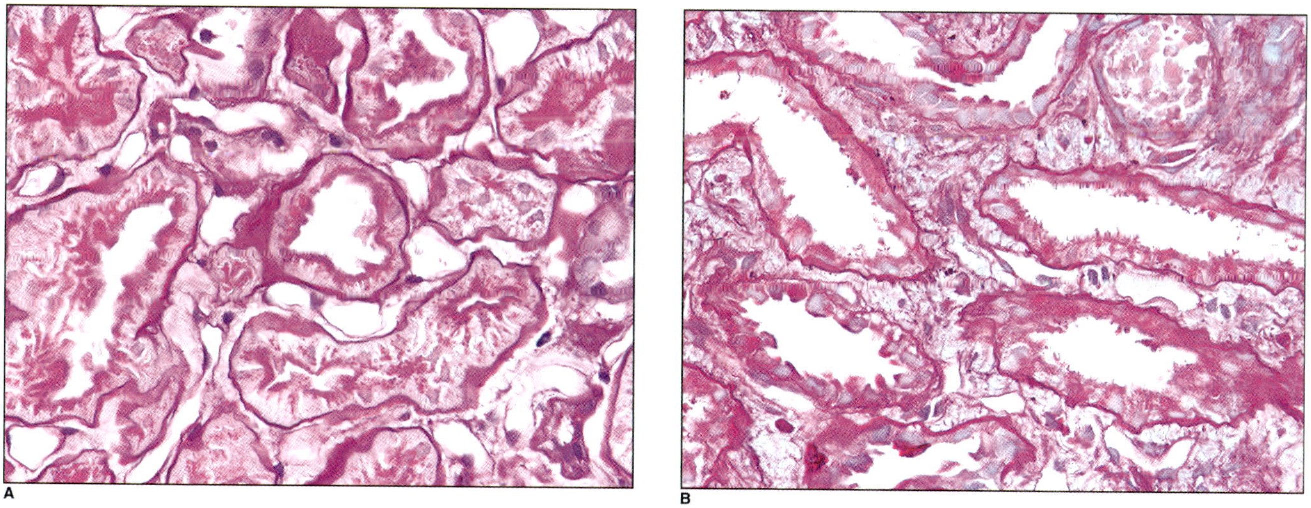

A B

FIGURE 8.30: Normal brush border (A), which is lost in ATN (B) as shown by PAS (both 400×).

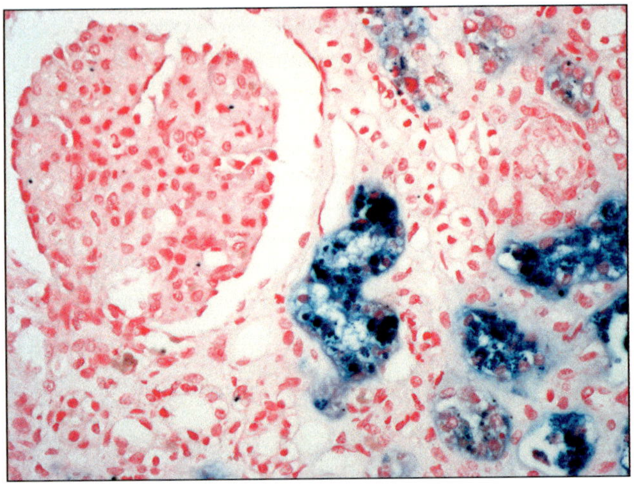

FIGURE 8.31: Abundant iron in renal tubular cell cytoplasm in a patient with hemolysis, on Prussian blue staining (200×).

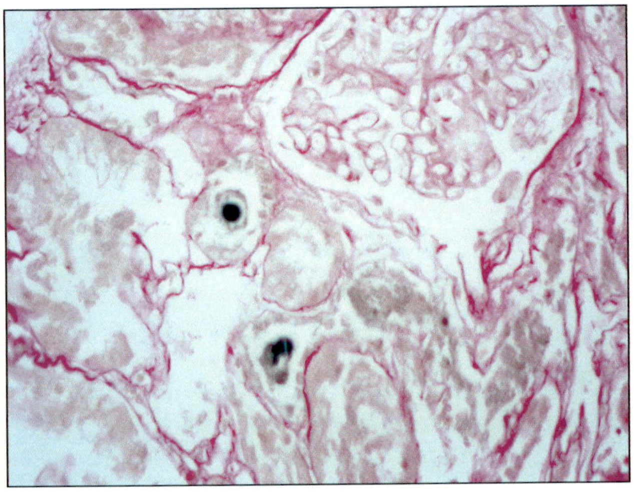

FIGURE 8.32: Intraluminal bilirubin confirmed on a Hall's stain associated with acute tubular epithelial damage (200×).

phosphotungstic acid–hematoxylin or trichrome stains. In cases of "burned-out" vasculitis, a VVG stain allows disrupted elastica to be recognized. That is not always seen, however, because in the most common renal vasculitides (Wegner's granulomatosis and microscopic polyangiitis), the primary target of disease is the glomeruli rather than larger blood vessels. Histochemical stains may suggest the presence of immune complexes in rare examples of SLE or other vasculitides, but this possibility would always require validation using immunofluorescence techniques.

Thrombotic Microangiopathy

Glomerular hilar arterioles and glomerular capillaries are most often affected by the vascular lesions of thrombotic microangiopathy (TMA). These manifest minute tangled complexes of fibrin and platelets, and they can be visualized

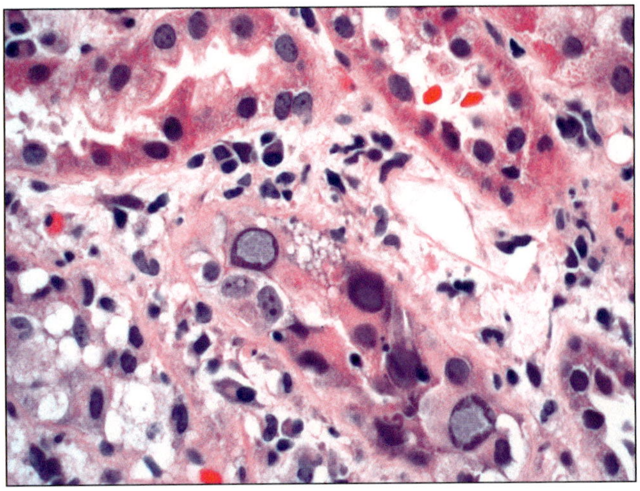

FIGURE 8.33: The most common type of tubular cell nuclear inclusions characteristic for BK virus nephropathy in renal allograft illustrated nicely on H&E (400×).

with trichrome stains (Figure 8.37). Fragmented erythrocytes also can be seen.

TMA may be a component of malignant hypertension, as described above. Other possible causes include connective tissue diseases such as SLE. In cases of *prior* TMA, mesangiolysis (dissolution of the glomerular substructure) allows the glomerular capillaries to fuse. This process can be seen well with Jones, trichrome, or PAS preparations.

RENAL NEOPLASMS

Renal neoplasms account for approximately 2–3% of all new cancer cases worldwide, and this incidence appears to be increasing. The vast majority of such lesions are of renal tubular origin and they encompass a spectrum of histologic types and biologic behaviors. RCCs have been categorized based on their morphologic characteristics, including growth patterns and cytologic features. The recent World Health Organization classification system also takes cytogenetic findings into consideration (20). Histochemical stains have figured significantly in the past in the diagnosis and differential diagnosis of renal tumors, and, although their use has been attenuated by immunohistochemistry, they can still be useful in selected situations. We herein summarize the role of histochemistry in this context, both historically and in current practice. The use of histochemical analysis in studying renal neoplasms is aimed at the identification of specialized cytoplasmic or stromal contents in those tumors. A brief summary of pertinent histochemical stains is shown in Table 8.3.

Renal Cell Carcinoma

PAS Staining with and without Diastase Digestion: In clear cell RCC, abundant cytoplasmic glycogen is

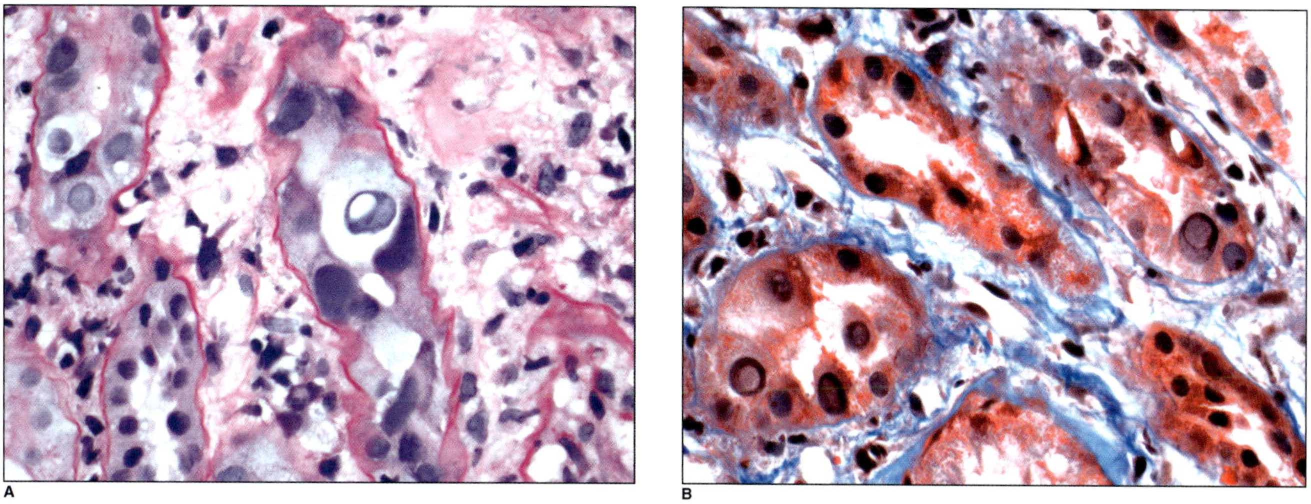

A B

FIGURE 8.34: BK viral nuclear inclusions can also be well visualized on PAS (A) and trichrome (B) (both 400×).

a prominent feature and can be demonstrated by PAS staining with and without diastase. However, glycogen content is variable among individual tumors, especially among high-grade lesions. Glycogen is usually scant in other adult renal neoplasms, including other types of RCC, oncocytoma, and urothelial carcinoma.

Neoplastic cells with granular eosinophilic cytoplasm are also common to several types of renal tumors including "conventional" RCC, type 2 papillary RCC, chromophobe-type RCC, and oncocytoma. Indeed, some "clear cell" RCCs are actually composed predominantly of granular cells. Hes et al. studied the frequency with which glassy hyaline globules were found in all types of renal cell neoplasms using the PAS stain (21). They found that such inclusions were strongly PAS positive both with and without diastase digestion. The globules were seen in 77% of clear cell RCCs containing granular cells (Figure 8.38), but in only 15% of papillary RCCs and no examples of chromophobic RCC or oncocytoma. Therefore, PAS staining may be useful in differential diagnosis of clear cell RCC from other potential granular cell neoplasms. The presence of glassy hyaline globules helps to exclude chromophobe RCC or oncocytoma as possibilities.

Advanced fibroinflammatory change in the renal parenchyma, in which atrophic renal tubules and sclerotic glomeruli may be inconspicuous in H&E stains, can closely simulate other pathologic lesions. A renal origin for fibroinflammatory masses may be diagnostically important, as in separating xanthogranulomatous pyelonephritis from granulomatous pyelitis, or segmental renal cystic disease from cystic nephroma. PAS stains can help to identify glomeruli and tubular basement membranes, even in markedly damaged renal tissue.

Colloidal Iron Staining: The colloidal iron stain, which was first introduced in 1946 by Hale, relies on the affinity of free acidic groups in tissue for positively charged colloidal iron (Fe^{3+}) at pH <2. Bound ferric iron is demonstrated using the Prussian blue reaction (4). Although the latter method is still the most popular, another one described by Mowry and modified by Lillie and Follmer is also used clinically. Both of them use a working pH of 1.9 with different pretreatment conditions. A study by Bonsib suggested that colloidal iron reactivity resides in microvescicles, which are characteristically prominent in chromophobic RCC (22). Hale's method is therefore useful in the diagnosis of that tumor type, and its staining pattern is sufficiently distinctive to separate it from oncocytoma. In chromophobic RCC, diffuse cytoplasmic labeling with colloidal iron is present in >75% of the tumor cells. In contrast, oncocytoma demonstrates a perimembranous, apical, or perinuclear pattern involving only some areas of the tumor (Figure 8.39) (23). Typically, other types of RCCs are either colloidal iron negative or they show a focal and coarse staining pattern.

Mucicarmine and Alcian Blue Stains: The histochemical classification of mucin types is complex with available staining procedures (24). Mucin may be of epithelial or stromal types; the former of those includes neutral mucin and acidic mucin in which the acidity is due to components of sulfuric acid (sulfated mucin) or sialic acid (sialated mucin). Stromal mucin is only acidic in type. Three stains have traditionally been used to detect tissue mucins: PAS with and without diastase digestion, alcian blue, and Mayer's or Best's mucicarmine. The PAS procedure consistently decorates neutral mucin after diastase digestion of the specimen; it also variably detects epithelial acidic mucin but fails to label stromal mucin. The alcian blue stain reliably detects all types of acidic mucin but is nonreactive with glycogen and neutral mucin. Mucicarmine is specific for epithelial mucins, but it lacks sensitivity. Neutral

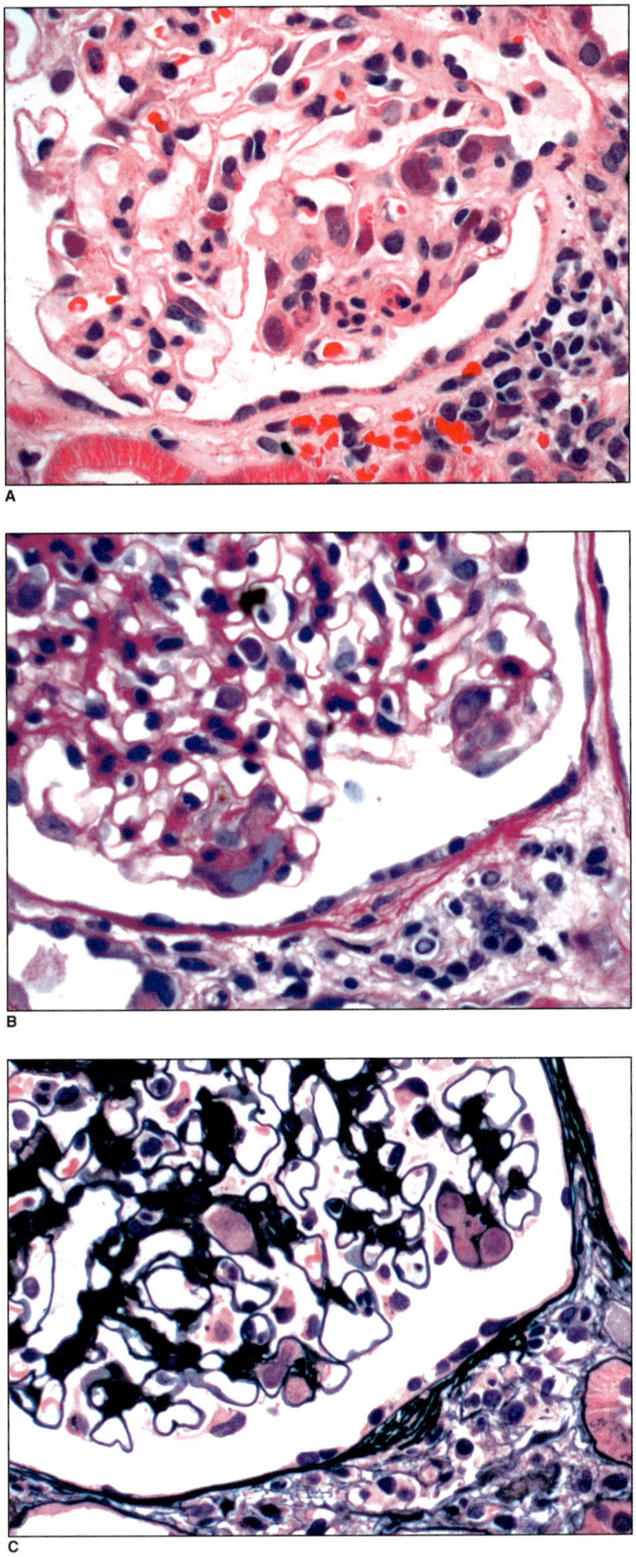

FIGURE 8.35: An unusual case of glomerular involvement by cytomegalovirus (CMV) infection in a renal allograft with nuclear and cytoplasmic inclusions on H&E (A), PAS (B), and Jones (C) stains (all 400×).

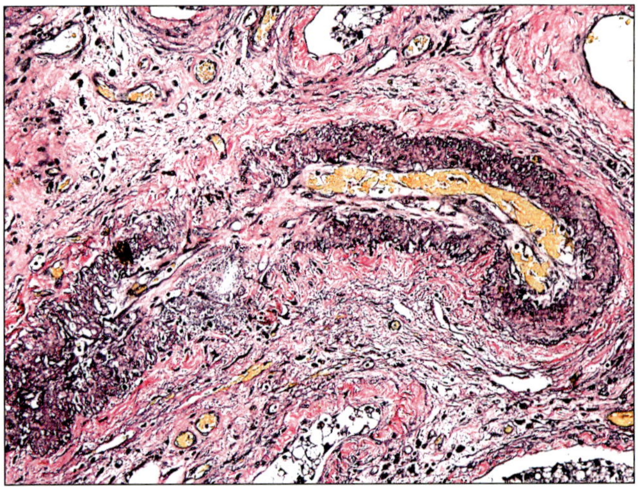

FIGURE 8.36: Myoepithelial hypertrophy and disruption of the elastic in malignant hypertension on a VVG stain (400×).

mucin and some forms of epithelial acidic mucin fail to label. Moreover, all of the aforementioned methods may fail to detect mucin when it is present in only minute amounts.

In the context of neoplasia, mucin may be present in variable quantities and more than one type may be present. This necessitates the use of a panel of stains for optimal detection.

The presence of mucin is an uncommon finding in RCC in general, but, when present in that tumor, it can be either of stromal or epithelial type. Stromal mucin can be seen in papillary RCC, papillary adenoma, and collecting duct carcinoma, as well as in mucinous tubular spindle cell carcinoma. It is rarely seen in clear cell or chromophobic RCCs, or in oncocytomas. A publication by Val-Bernal et al. showed that acid mucin was detected in 15% of papillary RCCs, in contrast to 4.3% of papillary adenomas (25). Either cytoplasmic or luminal mucin was highlighted by mucicarmine and alcian blue stains. An absence of labeling with the PAS method indicated that the mucins in such lesions were sulfated. Mucin-secreting RCC has also been reported by Grignon et al. (26); luminal and cytoplasmic mucin was demonstrated with the mucicarmine and alcian blue techniques at a pH of 0.9 and 2.5, respectively. Mucicarmine-positive mucinous metaplasia is possible in collecting duct or renal pelvic epithelium, as well as collecting duct carcinoma and mucin-secreting adenocarcinoma of the renal pelvis (27,28).

Lectins and Renal Epithelial Tumors: Biotinylated lectins were shown in the 1970s to bind to normal renal tubular epithelial cells in different nephron segments. Truong et al. performed a comprehensive histochemical study with 13 biotinylated lectins and showed that some of them, such as *Triticum vulgaris* and *Concanavalin ensiformis*, stained all tubules. Others predominantly bound to proximal or distal

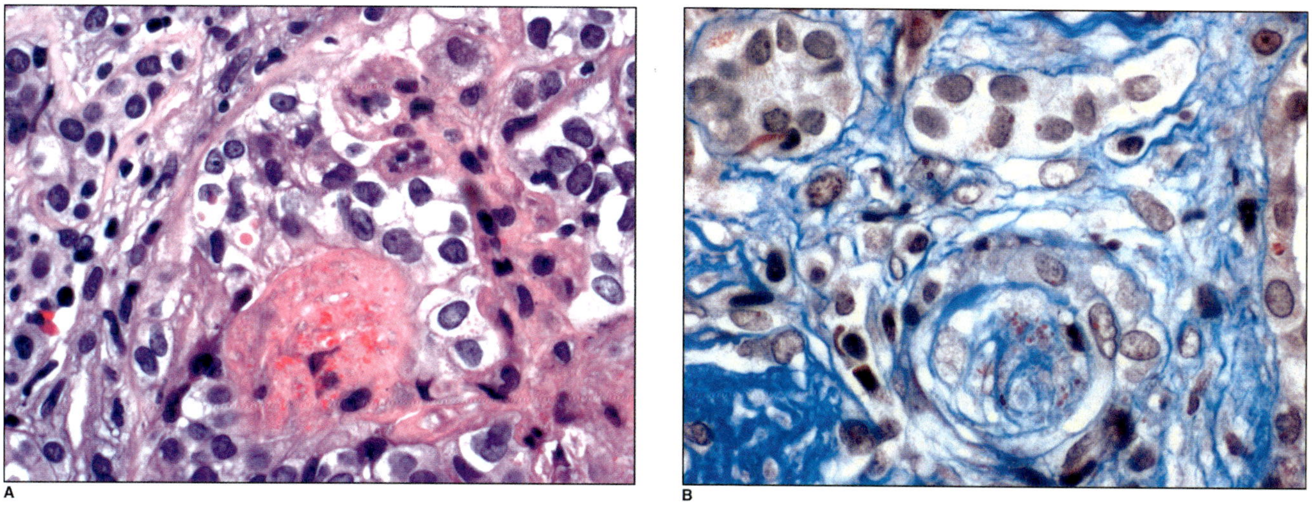

FIGURE 8.37: Glomerular capillary thrombi and erythrocytic fragments in TMA, seen particularly well on H&E (A) and trichrome (B) (both 400×).

TABLE 8.2: Characteristic Histochemical Profiles of Selected Nodular Glomerular Lesions

	H&E	PAS	Jones	Trichrome	Congo red
Light chain	Pink	Pale pink	Pale pink	Dark blue	Negative
Amyloid	Pink	Pale pink	Pink/brown	Dark blue	Positive*
Collagen	Pink	Pale pink	Pale pink	Dark blue	Orange

* Orange with apple-green birefringence on polarization.

TABLE 8.3: Selective Histochemical and Immunohistochemical Staining of Renal Neoplasms

	Glycogen	Lipid	Mucin	Colloidal iron	Cytokeratin	Vimentin
Oncocytoma	Scant	No	No	Pos/neg*	LMWCK +	Neg
Clear RCC	Abundant	Yes	No	Neg	LMWCK +	Pos/neg
Papillary RCC	Scant	No	Yes or no	Neg	LMWCK +	Pos/neg
Chromophobe RCC	Scant	No	No	Pos	LMWCK +	Neg
Collecting duct RCC	Scant	No	Yes, focal	Neg	HMWCK +	Pos/neg
Urothelial carcinoma	Absent	No	Rare	Neg	HMWCK +	Neg

HMWCK = high-molecular weight cytokeratin, LMWCK = low-molecular weight cytokeratin, neg = negative, pos = positive, RCC = renal cell carcinoma. *If positive in a perimembranous, apical, or perinuclear pattern.

tubules, or to collecting ducts (29). These properties may be helpful in the differential diagnosis of renal epithelial neoplasms, particularly collecting duct carcinoma, oncocytoma, and chromophobic RCC (30). Collecting duct carcinoma has a strong affinity for peanut lectin, *Dolichos biflorus* agglutinin, and soybean lectin. Oncocytoma binds only to *Dolichos biflorus* agglutinin. Although renal epithelial neoplasms do show restricted lectin-binding profiles, significant overlap exists (31). Therefore, definitive determinations of proximal or distal-medullary tubular differentiation in such tumors may not be possible with lectin histochemistry, and this method has limited utility in clinical practice in this context.

Other Intrarenal Tumors

Primary neuroendocrine tumors of the kidney are extremely rare, but they range from well-differentiated neuroendocrine carcinoma ("classic" carcinoid tumor) to poorly differentiated neuroendocrine carcinoma (small cell carcinoma). In similarity to neuroendocrine proliferations in other sites, the presence of cytoplasmic argyrophilic neurosecretory granules in such tumors can be demonstrated with the Grimelius, Churukian-Schenk, and Sevier-Munger procedures.

Primitive neuroectodermal tumor/Ewing sarcoma may rarely present as a primary renal neoplasm. PAS staining, with and without digestion, can be used to demonstrate the glycogen content that usually typifies that tumor. In

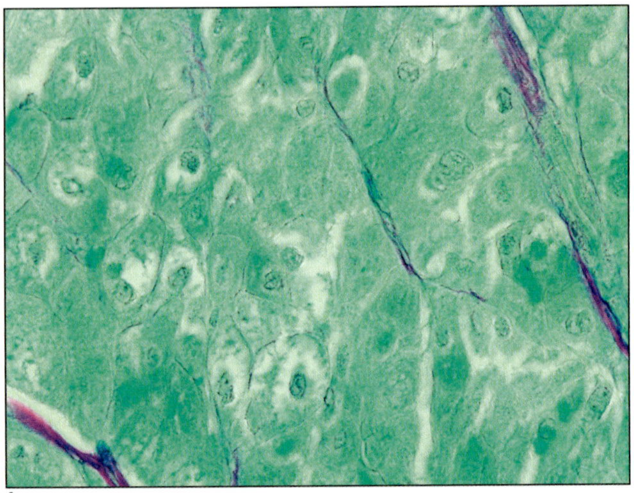

A

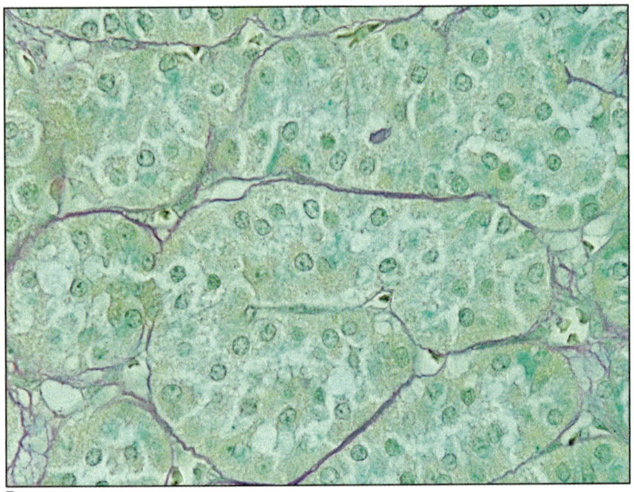

B

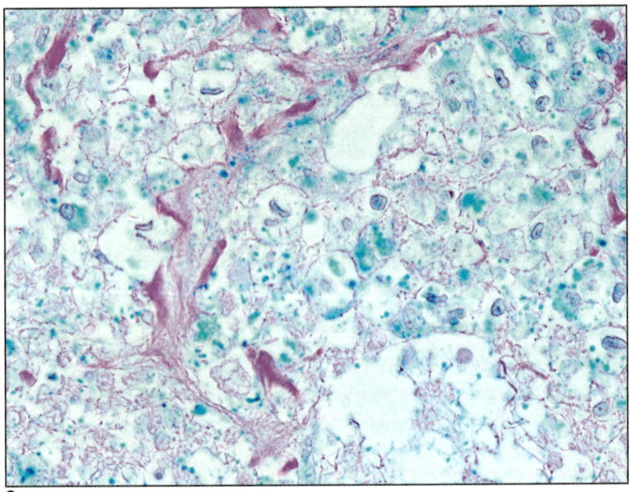

C

FIGURE 8.38: ***Colloidal iron staining for renal neoplasms. Chromophobe renal cell carcinoma with diffuse cytoplasmic staining pattern (A), oncocytoma with weak perimembranous pattern (B), clear cell RCC with focal coarse staining pattern (C) (all 200×).***

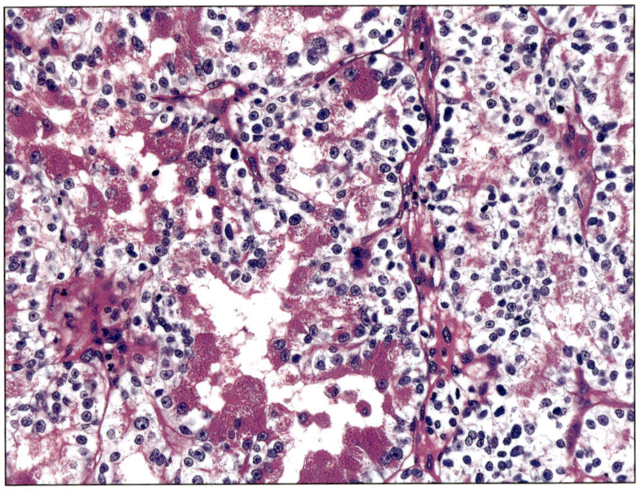

FIGURE 8.39: ***PAS with diastase staining of clear cell RCC: tumor cells showing strong positively stained fine granular and globules, a feature of clear cell RCC with granular cells (200×).***

juxtaglomerular cell tumors, which are rare benign renin-producing lesions that cause hypertension, characteristic cytoplasmic granules can be demonstrated with Bowie's stain (32).

Renal Pelvic Urothelial Carcinomas

The group of neoplasms arising in the pelvic urothelium shows essential identity to that of the urinary bladder mucosa. Those lesions also have similar histochemical properties, as discussed subsequently in this chapter (see section on urinary bladder neoplasms).

MEDICAL DISEASES OF THE URINARY TRACT

Histochemical stains play a limited role in the diagnosis of most nonneoplastic diseases of the urinary tract. However, there are some conditions in which histochemistry facilitates such diagnoses.

Interstitial Cystitis

Interstitial cystitis (IC), recently renamed "painful bladder syndrome/interstitial cystitis," is a chronic disease that predominantly affects middle-aged women. They present with chronic recurrent dysuria, frequency, urgency, and suprapubic pain (33,34). The etiology for IC is unknown but it may be associated with infection, autoimmunity, insufficiency of the urothelial barrier, inflammatory cytokines, or mast cell degranulation in the bladder wall (35). The diagnosis is made from the characteristic clinical presentation and cystoscopic findings. IC has two distinct forms, with one featuring mucosal ulcers (Hunner's ulcers) and the other being a nonulcerating variant with "strawberry-like" mucosal hemorrhage (glomerulation) (34,35). Bladder biopsies show nonspecific changes including ulceration, submucosal edema, congestion, and chronic inflammation (34,36). The muscularis propria is unaffected in the early stage but fibrotic later (34). Biopsy is not needed for the diagnosis, but it helps to eliminate clinically similar conditions such as urothelial neoplasia, other forms of sterile cystitis, and infection (34,35).

Mast cell infiltration of bladder wall has been stressed as a diagnostic feature of IC in the past, but this assertion remains controversial (34,36,37). Mast cells are best visualized with histochemical stains (38), the sensitivity of which is variable. The best of them are the Romanowsky, May-Grunwald-Giemsa, and toluidine blue methods, in which mast cells have reddish-purple metachromatic granules against a blue background (Figure 8.40) (24,38). The Leder stain is also effective, yielding red positivity in mast cells.

Although several studies have claimed that the bladder mast cell density per mm^2 in IC is higher than in controls, other analyses have not confirmed that observation (36,37,39). Furthermore, the former group of those publications reported a wide range of values in both IC and controls (Table 8.4), precluding accurate definition of a diagnostic threshold. These discrepancies partly reflect vagaries in the histochemical detection of mast cells, relating to sensitivity of the various stains, tissue preservation (38), and the type of fixative used. In particular, formalin blocks binding of some dyes to mast cell granules (34,40).

Because activated mast cells may lose their cytoplasmic granules, they also may not be detected histochemically (41). Indeed, the density of mast cells in immunostains for tryptase, chymase, or CD117 is approximately 30% higher than that detected with histochemical methods (42). All of these considerations complicate the use of mast cell stains for the diagnosis or prognostication of IC.

Fibrosis of the muscularis propia may supervene in advanced IC, as shown with trichrome stains. A suggestion has not been validated that THP in the bladder mucosa supports the diagnosis of IC (43). Nevertheless, recognition of this protein is facilitated by histochemical staining (see below).

Bacterial Cystitis

Bacterial infection is, by far, the most common cause of cystitis. The causal organisms are *Escherichia coli*, *Staphylococcus saprophyticus*, *Klebsiella* species, *Proteus mirabilis*, or enterococci. Some bacteria possess mucosa-specific proteins on their fimbria that facilitate attachment to the bladder mucosa. Thus, they can be detected with tissue Gram stains in some cases (44). However, bacterial cystitis is usually clinically apparent and biopsies are rarely needed.

Viral Cystitis

Viral cystitis is potentially caused by adenovirus, the BK subtype of polyoma virus, cytomegalovirus, and *Herpes* virus, among other agents (45–48). It manifests clinically as a hemorrhagic process and is seen almost exclusively in patients with severe immunosuppression. Viral cystitis is usually confirmed with urine viral cultures or by the polymerase chain reaction (46,47). In rare cases where a tissue diagnosis of that condition is made, cytopathic effects seen in H&E sections are usually diagnostic, obviating the need for histochemical stains (45,49–51). However, such methods as the Feulgen stain, the methyl green–pyronine technique, and the Steiner procedure can be used to highlight viral inclusions.

TABLE 8.4: Mast Cell Density/mm^2 in IC*

	Mucosa	Muscularis Propria
Control	15–95	3.3–46
Nonulcerated form	34–146	18–170
Ulcerated form	99–270	120–190

* Adapted from Theoharides et al. (35).

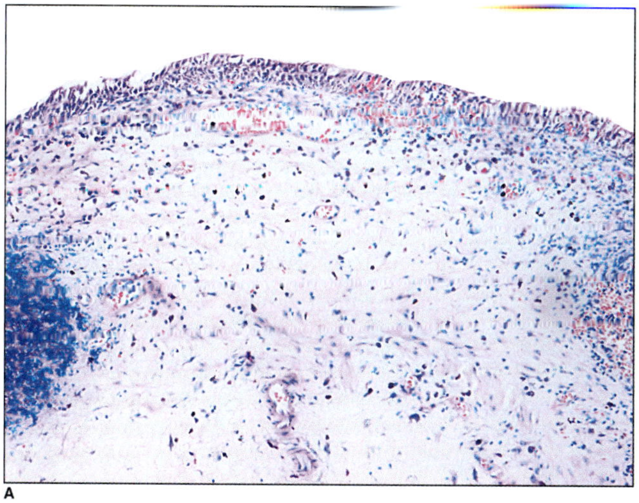

A

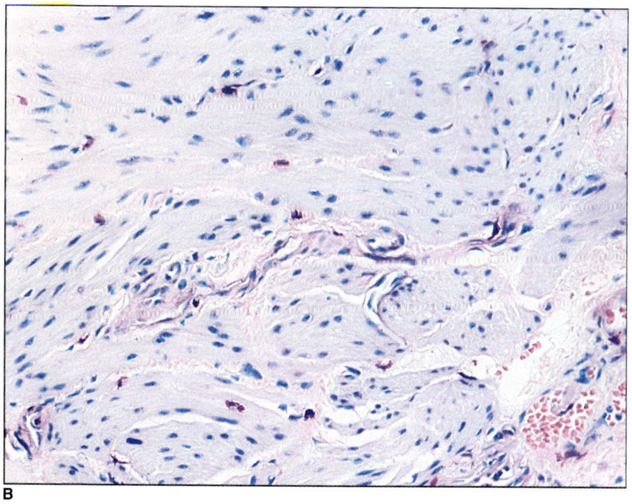

B

FIGURE 8.40: ***IC: several mast cells with cytoplasmic metachromatic staining by the Giemsa stain (A). Mast cells are also noted in the muscularis propria (B) (200× for A, 400× for B).***

Encrusted Cystitis

Encrusted cystitis is characterized by the deposition of calcium-containing salts in the bladder mucosa, caused by urinary infection with urea-splitting bacteria. von Kossa or alizarin red stain may help to better delineate the areas of calcification.

Granulomatous Cystitis

Granulomatous cystitides are rare, but recognizing their causes (Table 8.5) is essential for proper treatment. The morphologic features of granulomatous cystitis are potentially similar regardless of etiology, and therefore, histochemical procedures are important for accurate classification. Stains to detect mycobacteria, conventional bacteria, and fungi are needed in this context.

Fungal Cystitis

Fungal cystitis is rare; it is most frequently caused by *Candida albicans* or extraordinarily by other organisms including *Aspergillus*, *Cryptococcus*, *Torulopsis*, and *Mucor*. Other portions of the urinary tract or kidney may be affected as well. Predisposing factors include diabetes mellitus, prolonged antibiotic treatment, indwelling catheters, and organ transplantation. Biopsies usually show nonspecific acute and chronic inflammation rather than granulomas (52–55).

Aspergillus or *Zygomyces* organisms may be visible in H&E sections, but other fungi are usually not; they require silver (GMS) or PAS stain for adequate identification (44). In general, the GMS method is more sensitive than the PAS stain, but nuclear debris, erythrocytes, and mucin granules can also be GMS positive. Rare examples of fungal cystitis will yield reactivity with the GMS but not the PAS procedures, or vice versa, making both of those methods desirable in each case.

Tuberculous Cystitis

Tuberculous cystitis is unusual in Western countries but remains prevalent in some parts of the world (56). Most cases are part of a systemic infection (57,58).

TABLE 8.5: Causes of Granulomatous Cystitis

Fungal
Candida
Parasitic
Schistosomiasis
Bacterial
Tuberculosis
BCG
Others
Postsurgical (transurethral resection of bladder tumor)
Childhood granulomatous diseases
Crohn's disease

As true in other anatomic locations, definite histologic diagnosis requires the demonstration of mycobacteria (44,59). The organisms are red with the Ziehl-Neelsen or Fite method, against a methylene blue counterstain. Degenerated mycobacteria may lose their fatty acid capsules and are not well detected by the Ziehl-Neelsen stain, but their carbohydrate components may still be labeled with the GMS method (44).

Bacilli Calmette-Guerin–Induced Cystitis

Intravesical administration of bacilli Calmette-Guerin (BCG), as a treatment for in situ urothelial carcinoma, causes granulomatous inflammation in 90% of cases (60,61). Rarely (<1%) granulomas are seen in other organs as well, including prostate, kidney, lung, liver, and bone (62,63). BCG treatment is currently the most common cause of granulomatous cystitis. Acid-fast bacilli are rarely seen in this context, and, even if they are, this finding does not usually precipitate the use of antituberculous treatment (64). Thus, routine histochemical stains are not needed if it is known that the patient has received BCG (65,66). One study reported positive results for acid-fast bacilli in 77% of granulomatous prostatitis cases after BCG administration (67). Positive blood cultures or results of polymerase chain reaction testing have rarely been reported (65,68,69), and those eventualities are not necessarily associated with detectable organisms in the granulomas (63,69).

Schistosomal Cystitis

Schistosomal cystitis is principally seen in the Middle East and Africa. It is characterized by localization of viable schistosomes and their eggs in the pelvic veins (including those in the bladder wall), leading to granulomatous inflammation and progressive fibrosis. As the disease evolves, the parasite eggs are destroyed and calcified. Special stains are not needed for diagnosis, but the von Kossa or alizarin red methods may help to highlight the calcified eggs. The Ziehl-Neelsen method is also useful for speciation, in that the shells of *Schistosoma haematobium* organisms are acidfast but those of *S. japonicum* and *S. mansoni* are not (70,71).

Amyloidosis

Bladder amyloidosis may be "primary" (localized) or "secondary" (part of systemic disease). Primary amyloidosis of the bladder is rare (72), but, in a series of 32 cases from the Mayo Clinic, all of them were of the primary type (73). Most examples derive from light chain immunoglobulin (AL-type amyloidosis) (73–75). Nonetheless, the development of systemic amyloidosis or multiple myeloma has not been reported in patients with primary bladder amyloidosis (73,76,77). The frequency of secondary amyloidosis in

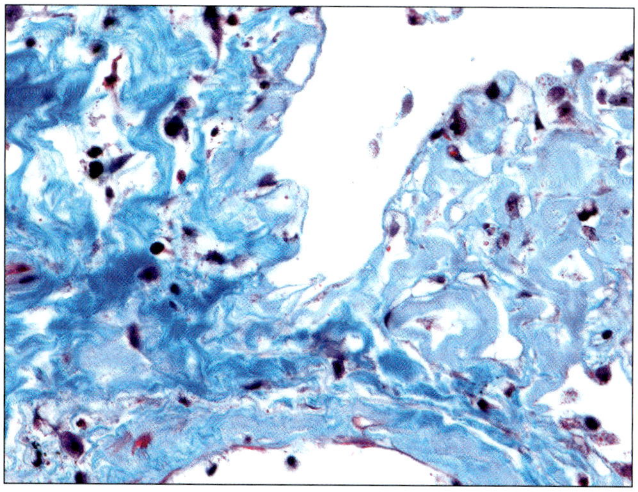

FIGURE 8.41: ***Bladder amyloidosis: the trichrome stain helps differentiate amyloid (glassy, "lilac blue") from dense fibrous tissue (fibrillary, deep blue) (200×).***

the bladder is not known, but it is probably extraordinarily rare. Most examples are of the AA type (78–80).

Painless hematuria is the most frequent manifestation of bladder amyloidosis; the lesions can be unifocal, multifocal, or diffuse, and may closely simulate neoplasms endoscopically. The amyloid deposits are glassy, markedly hypocellular, and eosinophilic, potentially affecting all layers of the bladder wall (81–83).

A definite diagnosis of bladder amyloidosis requires histochemical confirmation. The most reliable stain for this purpose is Congo red, yielding apple-green birefringence on polarization microscopy. Nevertheless, it must be remembered that Congo red can bind to other tissues including collagen, red blood cells, cartilage, and elastic tissue, albeit with different polarization microscopic features. As mentioned earlier in this chapter, optimal specimen preparation and staining is crucial to interpretation of Congo red stains (84).

Other stains for amyloid, including methyl violet, toluidine blue, and sulfated alcian blue "metachromatically" stain amyloid, producing a red color despite the blue colors of the dyes. Compared with Congo red, they are less specific and sensitive. We have noted that amyloid has a characteristic lilac-blue color in trichrome-stained sections; this contrasts markedly with the dark blue hue of hyalinized collagen (Figure 8.41).

Histochemical stains are limited in their ability to distinguish between the different types of amyloid. The only method applicable to that goal is potassium permanganate pretreatment followed by Congo red staining. AA-type amyloid is said to lose its affinity for Congo red in that setting. This approach has now generally been supplanted by the use of immunohistochemical amyloid typing.

Bladder amyloidosis must be distinguished from fibrin deposition (Figure 8.42), densely hyalinized fibrous tissue,

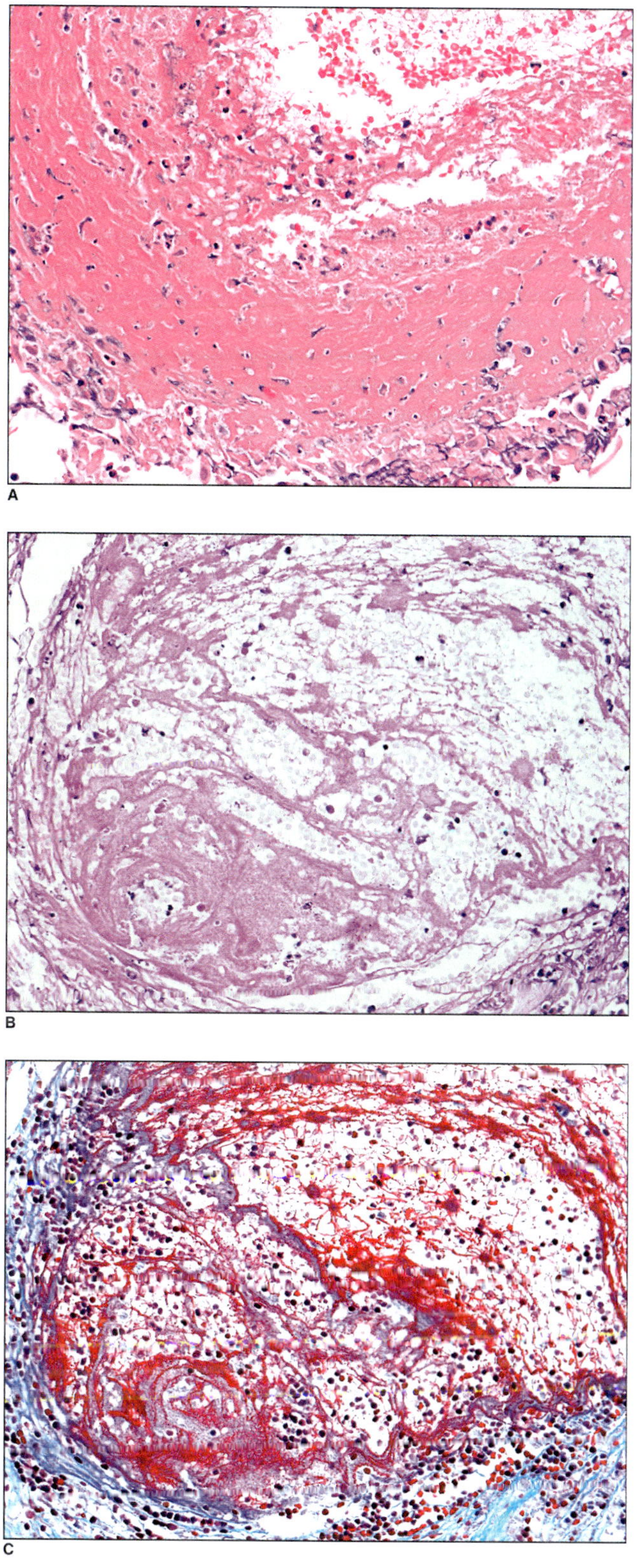

A

B

C

FIGURE 8.42: ***Fibrin deposition: fibrin appears as eosinophilic material on H&E (A), which is PAS negative (B), but is bright red on trichrome (C) (all 200×).***

TABLE 8.6: Urinary Bladder Eosinophilic Deposits

	Congo Red	PAS	Trichrome	VVG
Amyloid	Positive	Weak or 1+	"Lilac" blue	Negative
Elastic tissue	Negative or weakly positive*	Negative	Negative	Positive
Hyalinized fibrous tissue	Negative or weakly positive*	Negative	Strongly blue	Negative
THP	Negative	Positive	Pale blue	Negative
Necrotic tissue	Negative	Negative	Negative	Negative
Fibrin	Negative	Negative	Bright red	Negative

* Due to nonspecific binding of Congo red dye; can be improved with technical modifications.

necrotic debris, edema fluid, elastic tissue deposition, and THP deposits. Histochemical stains are helpful in that respect (Table 8.6). In particular, the VVG method labels elastin black, whereas amyloid is unstained (Figure 8.43).

Malacoplakia

Malacoplakia is a rare but distinctive form of localized chronic inflammation that can involve several organs, the most common of which is the urinary bladder (15,17,85,86). It is an isolated or multifocal lesion of the mucosa. The microscopic and histochemical attributes of urinary bladder malacoplakia are similar to those of the same process in the kidney, as described earlier in this chapter.

Xanthogranulomatous Cystitis

Xanthogranulomatous cystitis (XC) is rare, with approximately 20 reported cases (87). It is characterized by a mixture of xanthomatized histiocytes, other chronic inflammatory elements, and fibroblasts. XC can be distinguished from malacoplakia by the microscopic appearance of the constituent macrophages, and the absence of Michaelis-Gutmann bodies in the former of those conditions. Histochemistry also plays a role in differential diagnosis because the foamy macrophages in XC are PAS negative or have far fewer PAS-positive granules than von Hansemann cells do (Figure 8.44).

Tamm-Horsfall Protein Deposition

As described above, THP is a glycoprotein that is synthesized exclusively by the thick descending loop of Henle and collecting ducts. Its function is unclear but may serve to prevent urinary tract infection (88). THP is seen in bladder biopsies or cystectomy specimens in about 7% of cases (89). Its deposition is frequently observed in necrotic or inflamed tissue, as waxy, pale, or weakly eosinophilic and homogeneous extracellular material. THP may also be seen in the kidney and pelvic lymph nodes (89).

This protein must be differentiated from other eosinophilic deposits in the bladder, including amyloid, and histochemical stains are helpful in this respect (Table 8.6). Regardless of its localization, THP is PAS positive (before and after diastase digestion), pale blue on trichrome staining, and negative with the mucicarmine, alcian blue, and Congo red procedures (Figure 8.45).

NEOPLASMS OF THE URINARY TRACT

Generally, the diagnosis of most urinary tract neoplasms does not require the use of special stains. However, histochemical methods may contribute to the delineation of glandular differentiation and the evaluation of tumor invasion.

Transitional Cell Carcinoma

Normal urothelial cells contain glycogen (24). Most transitional cell carcinoma (TCC) also show focal PAS positivity, but they lack mucin (90). Indeed, those neoplasms that demonstrate clear cell change may have extensive PAS reactivity (91).

Divergent glandular differentiation in TCC is most frequently characterized by the presence of tubular acini containing cytoplasmic and luminal mucin. Rarely, one may also encounter an overtly mucinous (colloid) or signet-ring cell pattern. Those phenotypes must be differentiated from the simple presence of intracytoplasmic mucin in otherwise typical urothelial carcinoma (92,93) or microcystic change in TCC (94,95). Mucin stains such as the PAS-diastase (PAS-D), mucicarmine, and digested colloidal iron methods are valuable in better defining mucinous components. The clinical significance of glandular differentiation in TCCs is uncertain, but it is thought that the prognosis of such tumors is more related to stage than to cytomorphology (91).

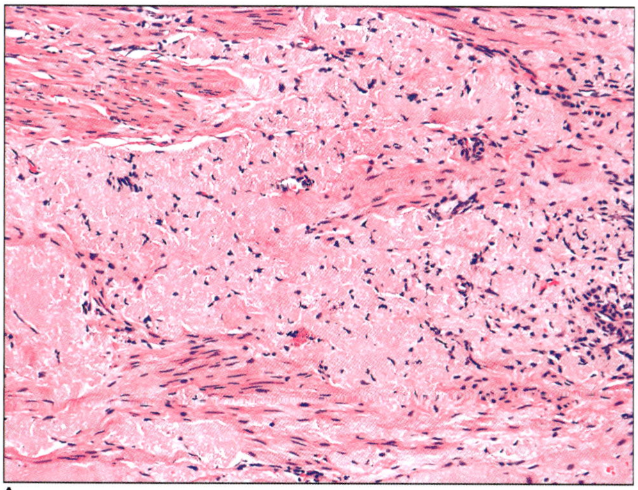

A

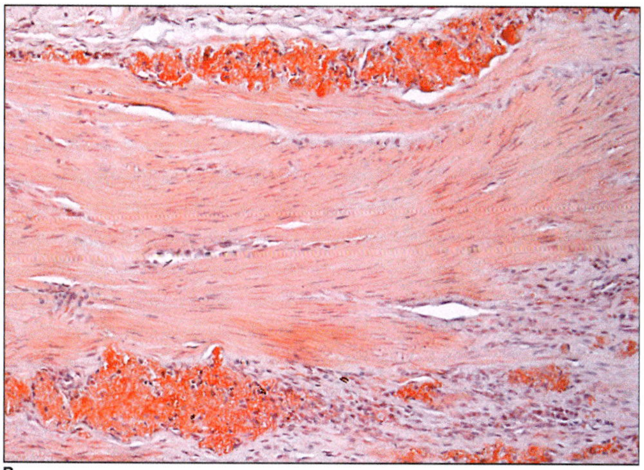

B

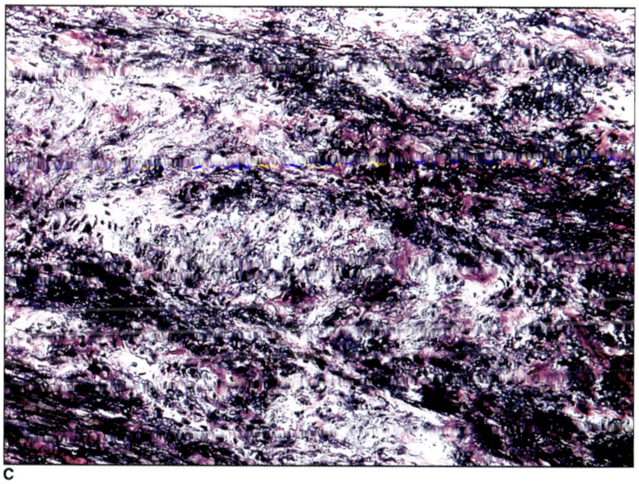

C

FIGURE 8.43:* *Bladder elastosis: the elastic tissue appears as homogeneous eosinophilic material, reminiscent of amyloid (A); this confusion is enhanced by a Congo red positivity often seen for elastic tissue (B); it is however strongly positive when using the elastin stain VVG (C) (all 200×).

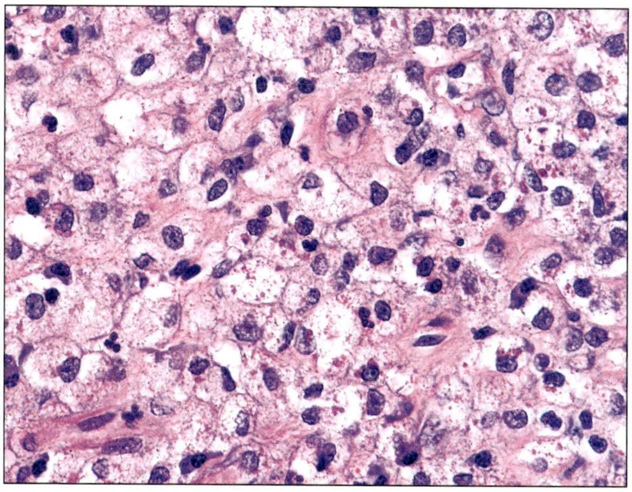

FIGURE 8.44:* *XC: small PAS-positive cytoplasmic inclusions are noted in the abundant foamy cytoplasm of xanthoma cells (200×).

The microcystic variant of TCC can potentially simulate nephrogenic adenomas or large von Brunn's nests. Our limited experience suggests that PAS-D stains demonstrate basement membrane material around the tumor cell clusters in benign lesions but not in microcystic TCCs (Figure 8.46).

Invasion by TCC is an important prognostic feature. Disruption of the urothelial basement membrane can again be shown with the PAS-D method; invasion into the muscularis propia may be demonstrated with trichrome stains. These help to identify large bundles of smooth muscle as opposed to myofibroblastic tissue reactions (Figure 8.47). Fibrous stroma within the muscularis is also well seen with the trichrome procedure.

Glandular Lesions

Malignant glandular lesions of the bladder are represented by primary and metastatic adenocarcinomas. Histochemical stains do not play a meaningful role in the differential diagnosis of those two categories because the same profiles of mucin production (and other biochemical analytes) are shared by them (Figure 8.48) (90,96–98).

Benign glandular proliferations include cystitis glandularis, intestinal metaplasia, villous adenoma, endometriosis, endocervicosis, and nephrogenic adenoma. The mucins seen in cystitis glandularis, intestinal metaplasia, and villous adenoma are similar to those of the colonic epithelium (99) and primary adenocarcinomas of the intestine or the bladder (90,96,100,101). Histochemical analysis may, however, be diagnostically helpful in selected instances. Endocervicosis has a mucin profile that is similar to that of normal endocervical epithelium (102). In addition, nephrogenic adenomas may contain arrays of

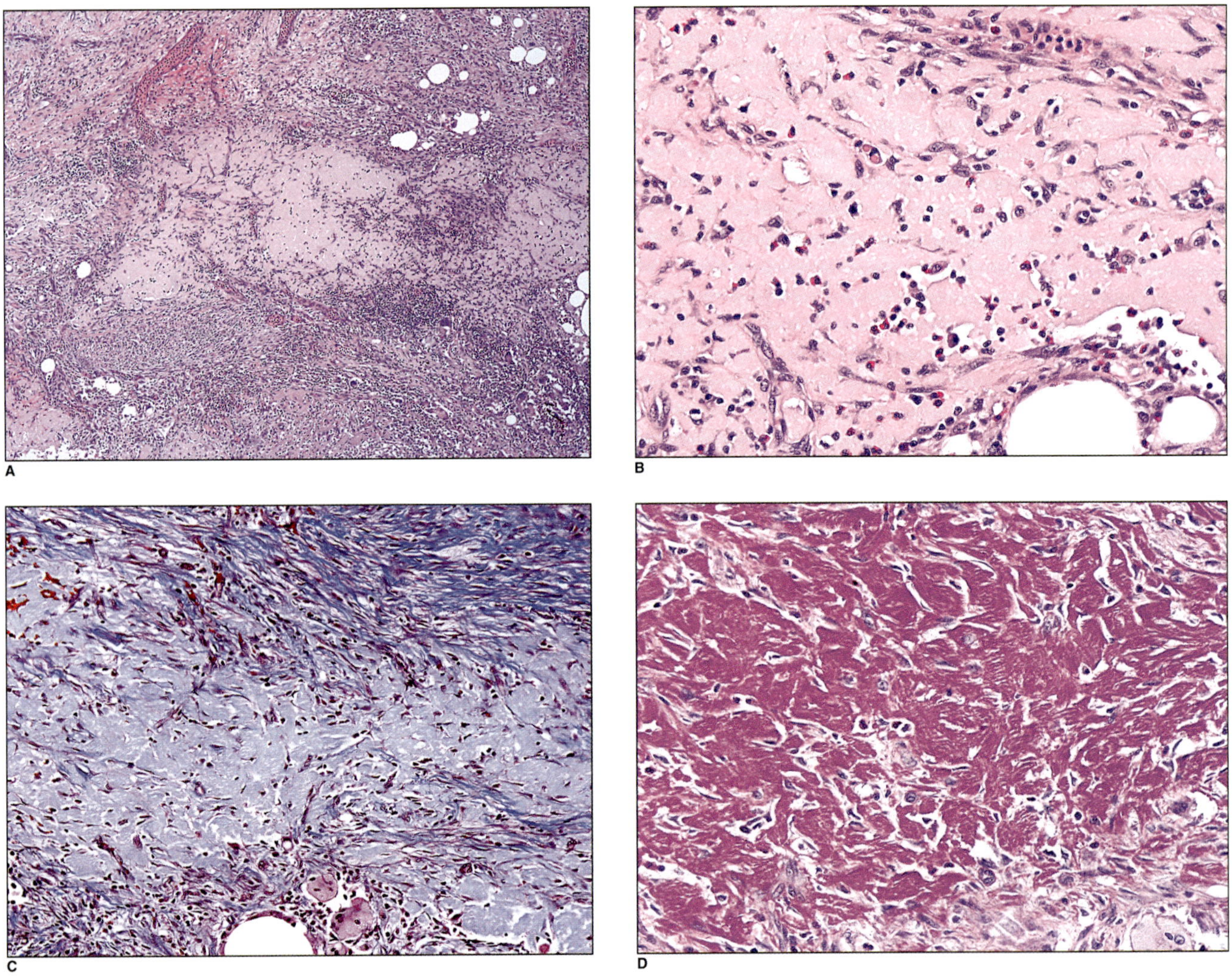

FIGURE 8.45: Bladder THP deposition: this protein appears as a glassy eosinophilic material, surrounding inflammatory cells and reactive fibroblasts (A and B), and is stained a "lilac" blue color on trichrome (C). These features are reminiscent of amyloid, but this protein is strongly PAS positive (D), in contrast to a faint PAS positivity for amyloid (100× for A, 200× for B, 400× for B and D).

signet-ring–like cells, which may focally label weakly with mucin stains. However, those structures are also surrounded by basement membrane material that is PAS positive (Figure 8.46), unlike the attributes of adenocarcinomas (103–106). Signet-ring cell carcinomas (SRCCs) of the bladder do exist, but they manifest a solid growth pattern (91,107) with no basement membrane deposition. SRCCs also can be confused with lymphomas and the plasmacytoid variant of TCC. PAS-D, mucicarmine, and digested colloidal iron stains are effective at separating those diagnostic alternatives (108–112). Positivity with such procedures is not limited to *primary* SRCCs of the bladder, however, because it is also possibly seen in rare cases of metastatic lobular breast carcinoma in the bladder with signet-ring cell differentiation (108).

Clear Cell Tumors

Clear cell neoplasms of the bladder potentially include TCCs with clear cell change, clear cell adenocarcinoma (111,112), nephrogenic adenoma, metastatic clear cell RCC, clear cell myelomelanocytic (sugar) tumors (113), and TCCs with lipidized cells (91). The clear cell appearance of these tumors is variable in scope. The first five lesions just listed all contain glycogen; in addition, mucin can be seen in gland-like structures in nephrogenic adenomas and clear cell adenocarcinomas (103–106,112,114). As implied by its name, TCC with lipidized cells shows the presence of neutral cytoplasmic fat, which is undetectable in routinely processed specimens. Mucin stains have limited application in the differential diagnosis of these tumors.

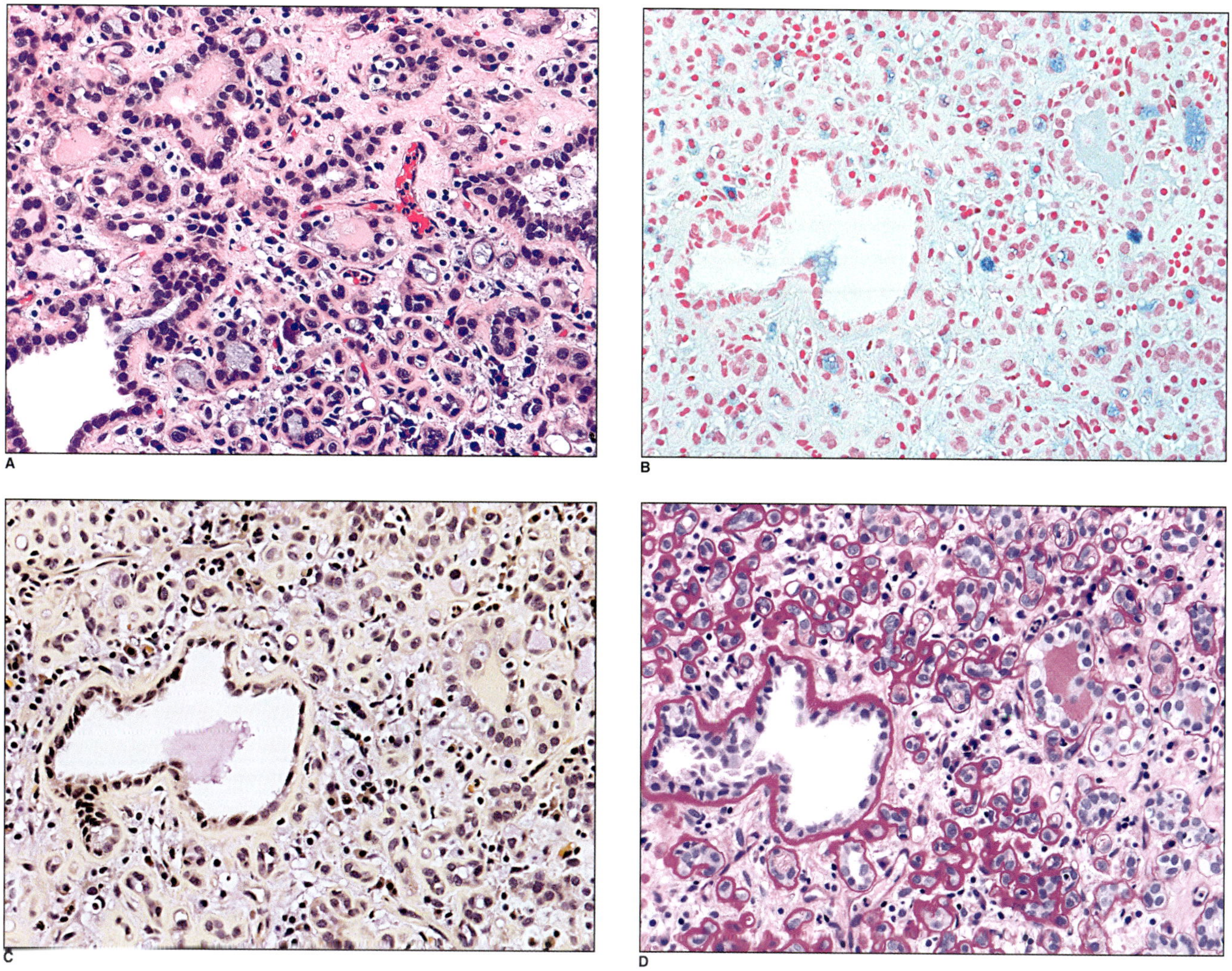

FIGURE 8.46: Nephrogenic adenoma: characterized by gland-like structures or individual cells (A). These structures are surrounded by PAS-positive basement membrane (B). Scanty mucin is noted in the lumens of some gland-like structures (A), which is negative using both PAS (B) and mucicarmine (C) but is weakly positive when using alcian blue (D) (all 200×).

Only a single case has been reported of clear cell myelomelanocytic tumor in the bladder (113). That lesion is a member of the "PEcoma" family, reputed to show partial melanocytic differentiation. However, the aforementioned case showed no reactivity with a Fontana-Masson stain.

Spindle Cell Tumors

The differential diagnosis of spindle cell proliferations in the bladder is broad. It includes sarcomatoid TCC, true sarcomas, TCC with pseudosarcomatous stroma, solitary fibrous tumor, inflammatory myofibroblastic tumor, schwannoma, neurofibroma, fibroma, postsurgical spindle cell lesions, and leiomyoma. Trichrome staining may be helpful in this group because the bright red fibrillary cytoplasmic staining is strongly suggestive of smooth muscle differentiation. Basement membrane material can also be seen around individual smooth muscle cells with the PAS-D method. The Bodian or Bielschowsky stain can be used to demonstrate delicate nerve processes that are present among the tumor cells in neurofibromas and are adjacent to neurilemmomas.

Other Neoplasms

Histochemical stains may contribute to the recognition of other bladder neoplasms as well. PAS-D–positive granules are seen in granular cell tumors (115); glycogen is present in primitive neuroectodermal tumors; melanin is often demonstrable in primary or metastatic malignant melanomas with the Fontana-Masson or Schmorl procedure (116);

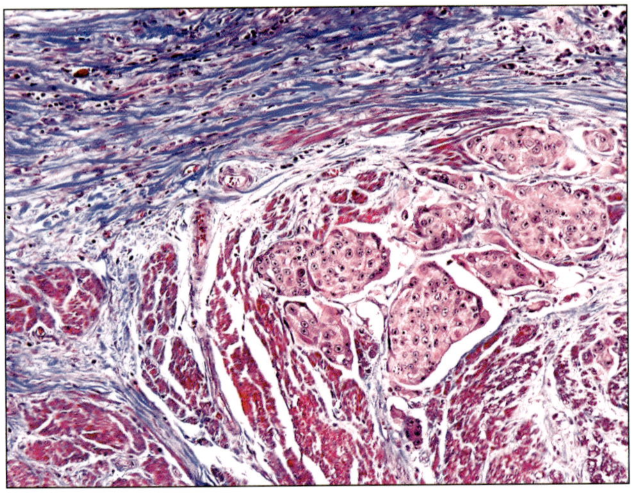

FIGURE 8.47: Muscularis propia invasion by urothelial carcinoma: the smooth muscle fiber cytoplasm (lower left) shows strong staining (red) with trichrome, contrasting with cellular desmoplastic stroma (blue, upper) (200×).

and neuroendocrine carcinomas manifest positivity with argyrophil or argentaffin stains (e.g., Grimelius, Sevier-Munger, Churukian-Schenk, Fontana-Masson methods), or both (117).

Lectin Histochemistry of Bladder Tumors

Lectin binding has been noted to TCCs, including peanut lectin (which binds to the T blood group antigen) (118). Both cytoplasmic and membranous patterns of labeling can be seen, and they tend to show an inverse relationship with tumor grade. Binding of several other lectins has been reported in the nonneoplastic urothelium and in TCCs; again, their avidity has correlated with tumor stage and grade (119) The lectin affinity pattern of bladder mucosa that shows intestinal metaplasia is similar to that of the normal colonic epithelium (100) Other diagnostic and prognostic roles for lectin histochemistry are still largely unstudied with regard to tumors of the bladder (120,121).

ILLUSTRATIVE CASE STUDIES

CASE 1— A 70-year-old man with multiple myeloma presented with renal insufficiency. His medical history included essential hypertension, prostatic adenocarcinoma, and hyperlipidemia.

A bone marrow biopsy revealed 10% plasma cells and a monoclonal IgG-kappa paraprotein was found in electrophorograms of the serum and urine. The level of serum creatinine was 2.7 mg/dL with a 24-hour urine protein excretion of 6 gm. Urinalysis showed no casts, but there

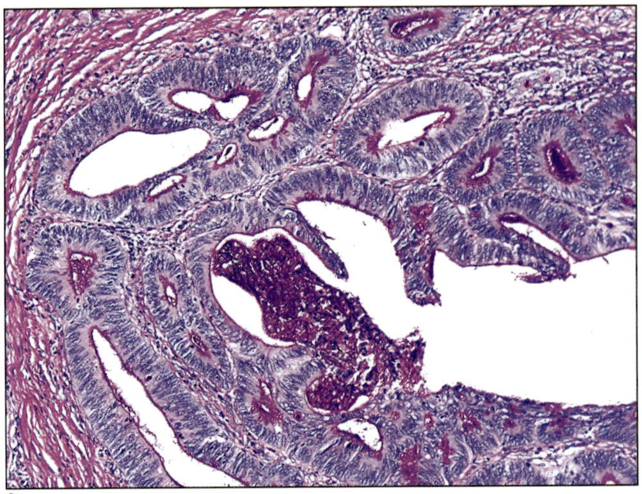

A

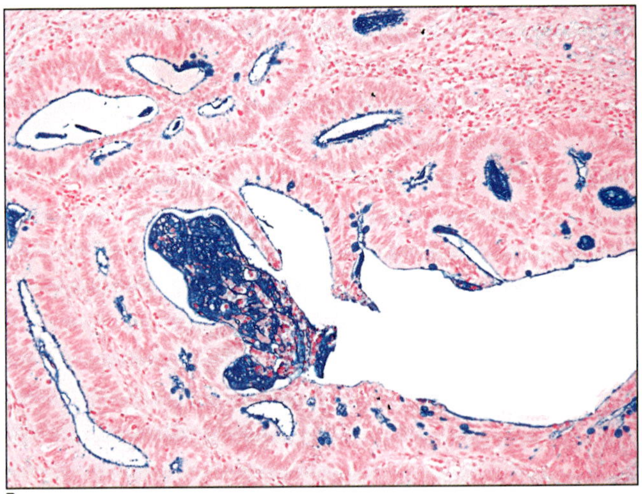

B

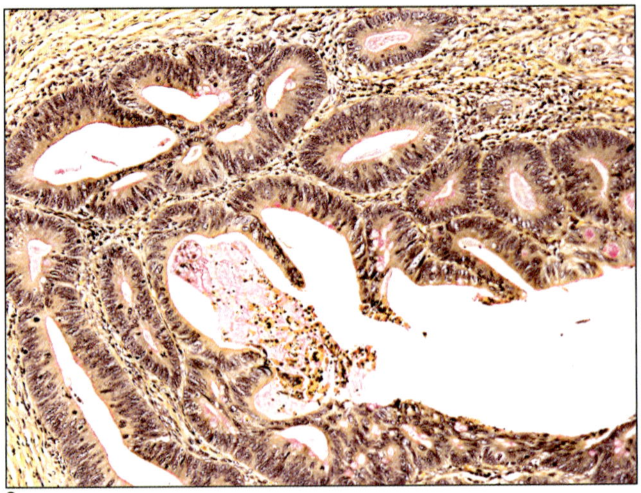

C

FIGURE 8.48: Primary bladder adenocarcinoma: mucin is clearly shown by the PAS with digestion (A), and alcian blue (B), whereas much less mucin is noted with the mucicarmine (C) (all 200×).

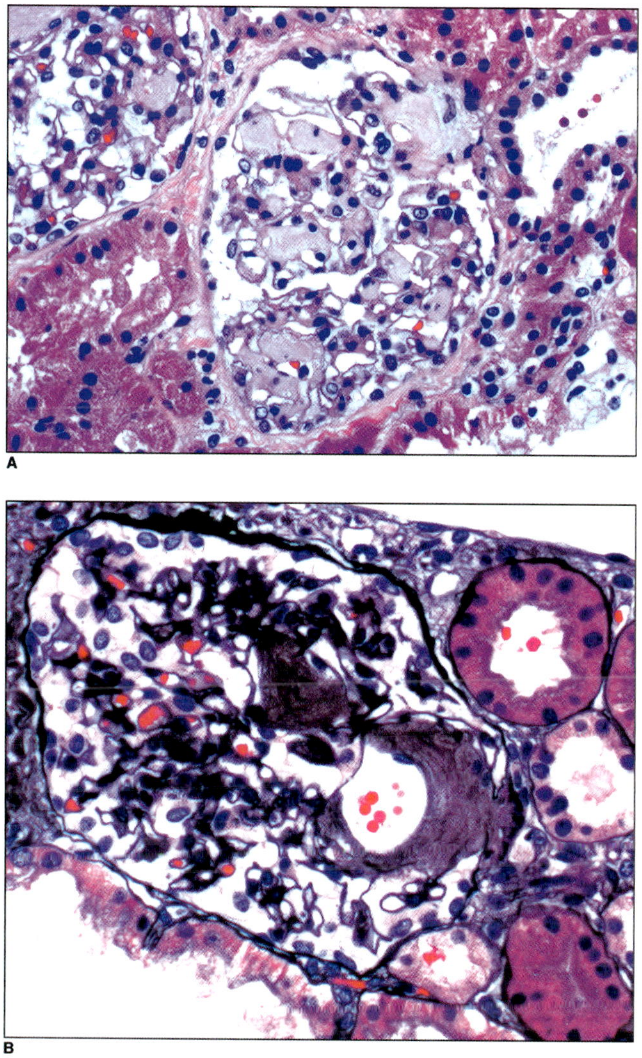

FIGURE 8.49: This glomerulus demonstrates diffuse expansion of the mesangium by amorphous red-lavender material (A). It shows perivascular accentuation and is not labeled with the Jones method (B).

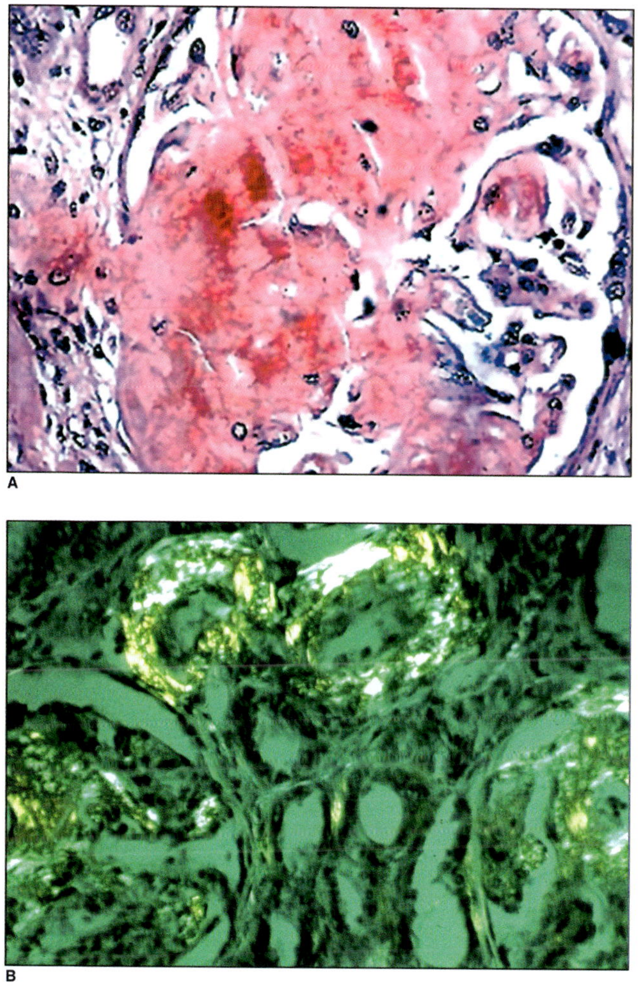

FIGURE 8.50: Congo red staining of the specimen shown in the previous figure produced intense reactivity of the interstitial material (A), and polarization microscopy showed that it had green birefringence (B), establishing the diagnosis of amyloidosis.

were 20 erythrocytes per high-power microscopic field; none of the cells was dysmorphic.

Light microscopic examination of a renal biopsy revealed diffuse lobular glomerular expansion by acellular, amorphous eosinophilic material involving the mesangium and partially obliterating the capillary loops (Figure 8.49). A mild, predominantly lymphocyte interstitial infiltrate was present, accompanied by tubular atrophy and interstitial fibrosis. Intramural eosinophilic material was also seen in arteriolar walls. Congo red staining showed salmon-pink staining of the mesangial and vascular deposits, with apple-green birefringence by polarization microscopy (Figure 8.50).

Immunofluorescence studies demonstrated only smudgy mesangial 1+ labeling for lambda light chain immunoglobulin. Ultrastructural analysis confirmed the presence of amyloid fibrils, and the final diagnosis was that of renal amyloidosis. ■

CASE 2— An 8-year-old boy had a history of hemolytic-uremic syndrome following a pneumococcal infection. He had received a second cadaveric renal allograft 3 years previously. The immunosuppressive regime consisted of prednisone, mycophenolate mofetil, and tacrolimus. The patient developed increasing creatininemia and proteinuria that was unresponsive to hydration, and he also had hypertension of recent onset.

The serum creatinine level was 6.3 mg/dL (with a baseline of 1.2) and the blood urea nitrogen (BUN) level was 111 mg/dL. The 24-hour urinary protein excretion was 4 gm and hematuria was present.

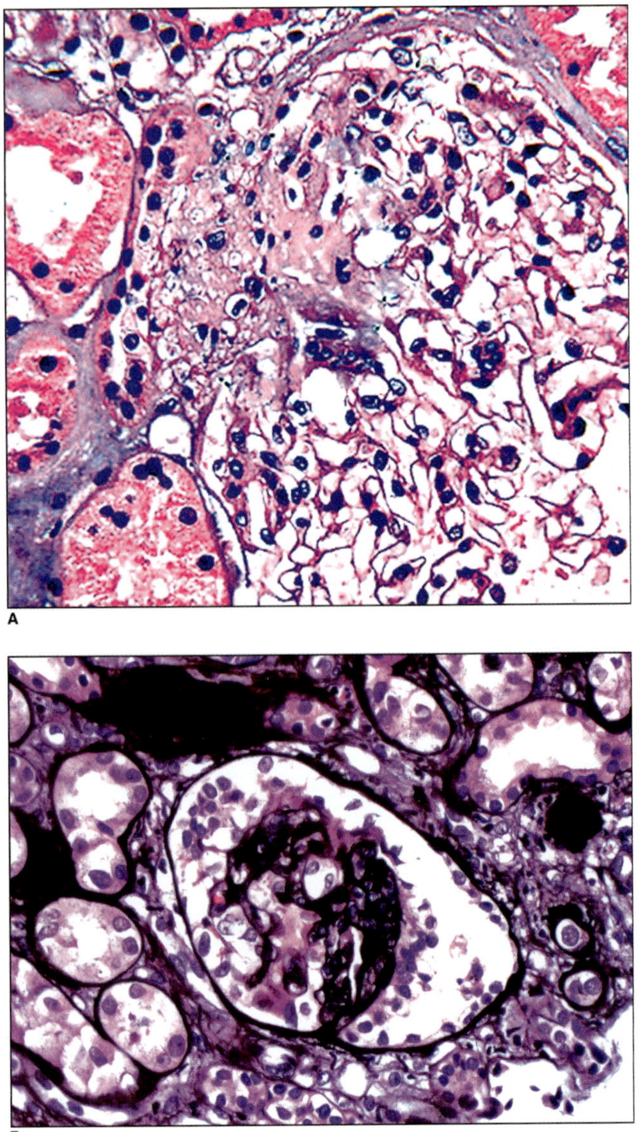

A

B

FIGURE 8.51: This renal biopsy shows focal mesangial sclerosis (A), corresponding to an area of silver-labeling with Jones' method (B).

A renal biopsy was obtained, showing that only 5 of 13 glomeruli in the specimen were viable. They contained collapsed tufts with reactive visceral epithelial cell proliferation, areas of segmental sclerosis, and protein resorption droplets (Figure 8.51). A marked interstitial inflammatory infiltrate was accompanied by widespread tubular atrophy, interstitial fibrosis, and lymphocytic tubulitis (Figure 8.52). Intrarenal blood vessels showed mild myoepithelial hypertrophy, but there was no microthrombosis, nodular arteriolar hyalinization, or vasculitis.

Immunofluorescence studies demonstrated only fine (1+) mesangial granular deposits of IgM. The peritubular capillaries did not stain for C4d. Those findings were interpreted as representing FSGS of the collapsing type, with severe interstitial fibrosis and acute cell–mediated rejection of Banff grade 1A. ■

CASE 3— A 63-year-old woman presented with dyspnea, hematuria, pulmonary infiltrates, proteinuria, and renal insufficiency. The presumptive clinical diagnosis was that of systemic vasculitis. The patient had been taking moxifloxacin for approximately 1 month.

The serum creatinine level was 2.1 mg/dL with a BUN level of 30 mg/dL. Urinalysis demonstrated 1+ proteinuria with erythrocyte casts. Serologic testing for hepatitis virus infection was negative, and serum complement levels were normal. Antineutrophilic cytoplasmic antibody (ANCA) was present in serum at a titer of 1:320, with a perinuclear pattern.

A renal biopsy showed that several glomeruli contained predominantly cellular crescents with less common fibrocellular forms; some disrupted Bowman's capsules (Figure 8.53). At least four glomeruli showed fibrinoid necrosis with karyorrhectic debris. A marked mixed chronic inflammatory infiltrate was accompanied by tubular atrophy and interstitial fibrosis. Intrarenal arteries were moderately sclerotic with no inflammatory lesions. No specific abnormalities were noted in immunofluorescence studies, and ultrastructural analysis showed no electron-dense deposits. The final diagnosis was that of ANCA-associated necrotizing GN. ■

CASE 4— Radical cystectomy was performed in the case of a 62-year-old man who had had a grade III urothelial carcinoma in a previous excisional biopsy. The bladder wall contained a firm 3-cm area with a shaggy, hemorrhagic surface but no obvious mass. The rest of the mucosa contained focal hemorrhages.

Sections of the firm lesion revealed no residual carcinoma, but there were changes related to the previous biopsy. These included fibrosis, hemorrhage, necrosis, fibrin deposition, and chronic inflammation with eosinophilia. Large multifocal deposits of glassy eosinophilic material were also present, within which reactive fibroblastic proliferation, inflammation, and numerous capillaries were noted (Figure 8.54A,B). The differential diagnosis included necrosis, fibrin deposition, fibrosis, amyloidosis, and deposition of THP. Histochemical findings included "lilac" staining of the deposits with trichrome, negativity with the Congo red method, and strong PAS reactivity (Figure 8.54C,D). They were considered diagnostic of THP deposits; this impression was confirmed immunohistochemically. ■

CASE 5— A 32-year-old woman had a cadaveric renal allograft, which she had received for end-stage IgA nephropathy. Currently, she presented with dysuria.

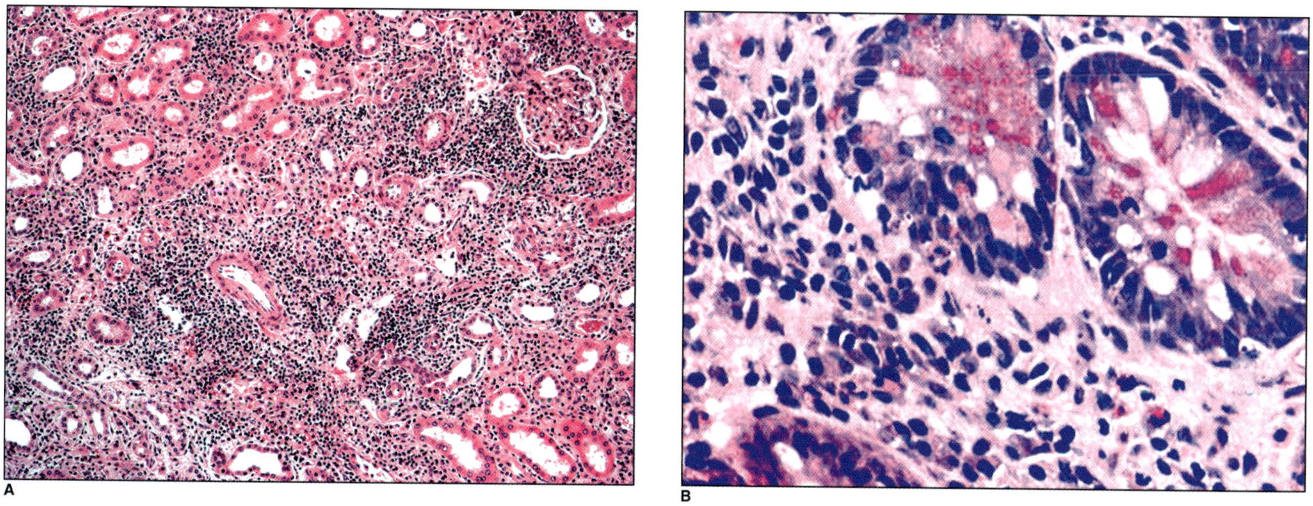

A B

FIGURE 8.52: *The interstitium and tubules in the biopsy shown in the previous figure were chronically inflamed (A); tubulitis was focally present as well (B), and the latter feature was accentuated by staining with the periodic acid-Schiff method.*

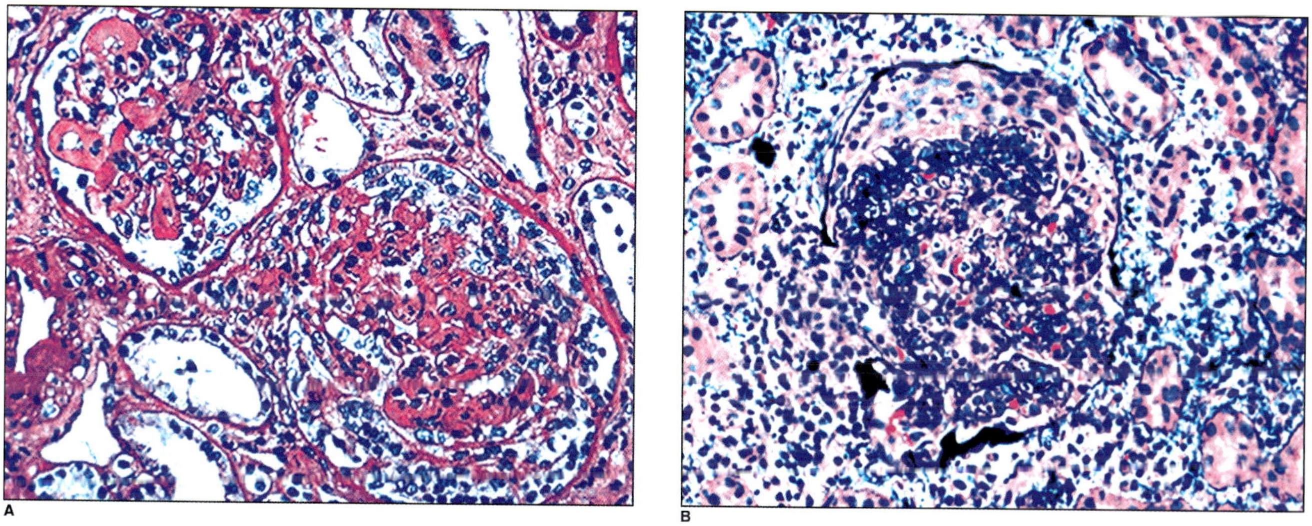

A B

FIGURE 8.53: *Epithelial crescent formation is present in this renal biopsy specimen (A); a Jones silver stain demonstrates focal-disruption of Bowman's capsule and mesangial sclerosis (B).*

Bladder biopsies showed "nonspecific cystitis" and the patient was treated with antibiotics without improvement. Subsequent cystoscopic examination showed a 1.5-cm mucosal area of hemorrhage and ulceration, which was biopsied.

The specimen showed aggregates of variably sized gland-like structures and single epithelioid elements. The constituent cells exhibited slight nuclear atypia and eosinophilic or clear cytoplasm (Figure 8.55A). Scanty mucin or weakly eosinophilic material was present in glandular lumina. The stroma was poorly vascularized with alternating fibrosis, edema, and chronic inflammation. Histochemical staining showed that the clear cytoplasm in the lesional cells contained glycogen. The presence of limited luminal mucin was confirmed with the PAS-D and alcian blue procedures, but the mucicarmine stain was negative (Figure 8.55B–D). Basement membranes were clearly demonstrated around groups of cells with the PAS-D stain (Figure 8.55B). The diagnosis was that of nephrogenic adenoma. ■

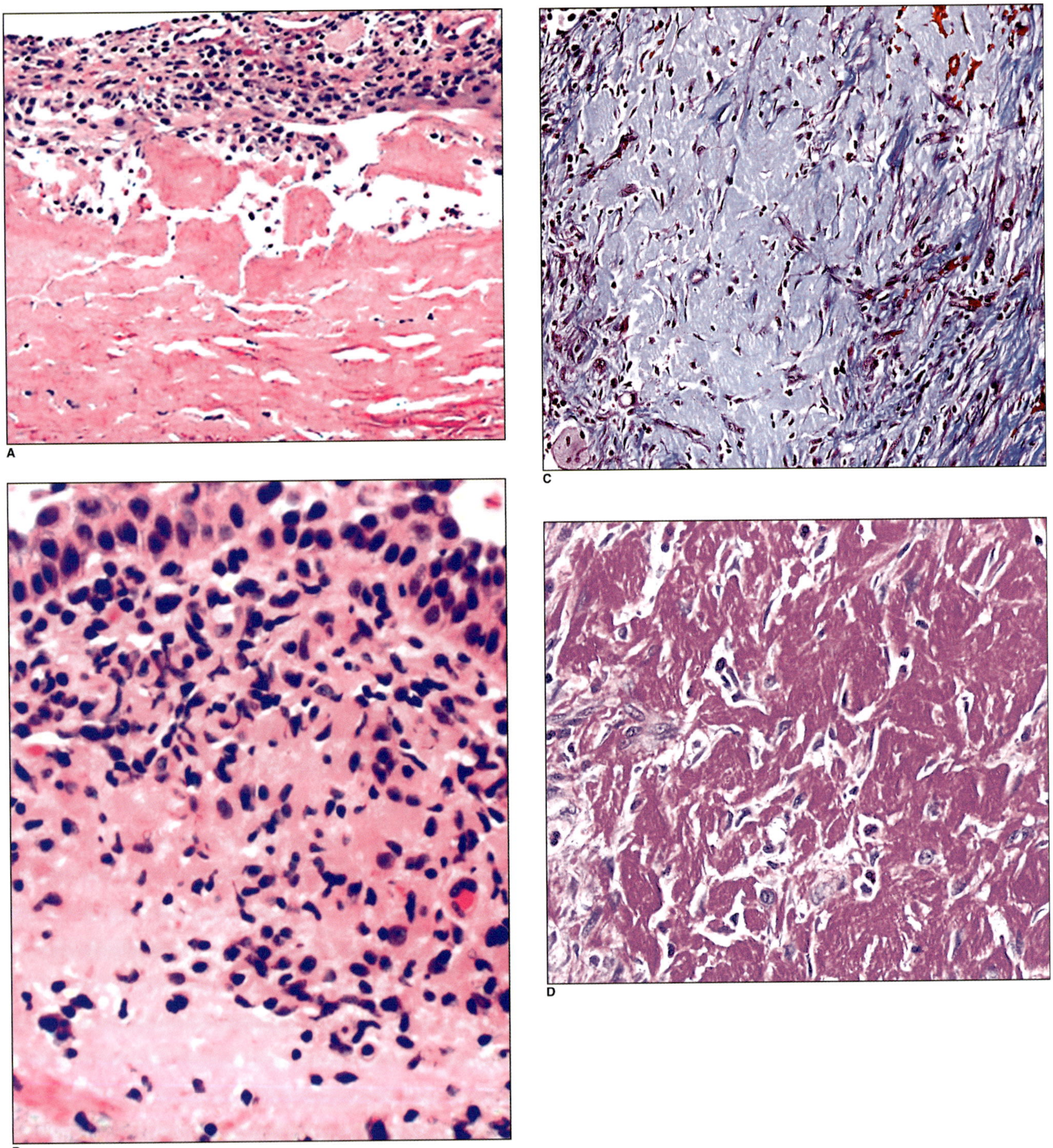

FIGURE 8.54: Amorphous eosinophilic material is admixed with chronic inflammatory cells in this section of the urinary bladder (A & B). The interstitial material is lilac-colored in a Masson trichrome stain (C) and intensely reactive with the periodic acid-Schiff technique (D), establishing its identity as Tamm-Horsfall protein.

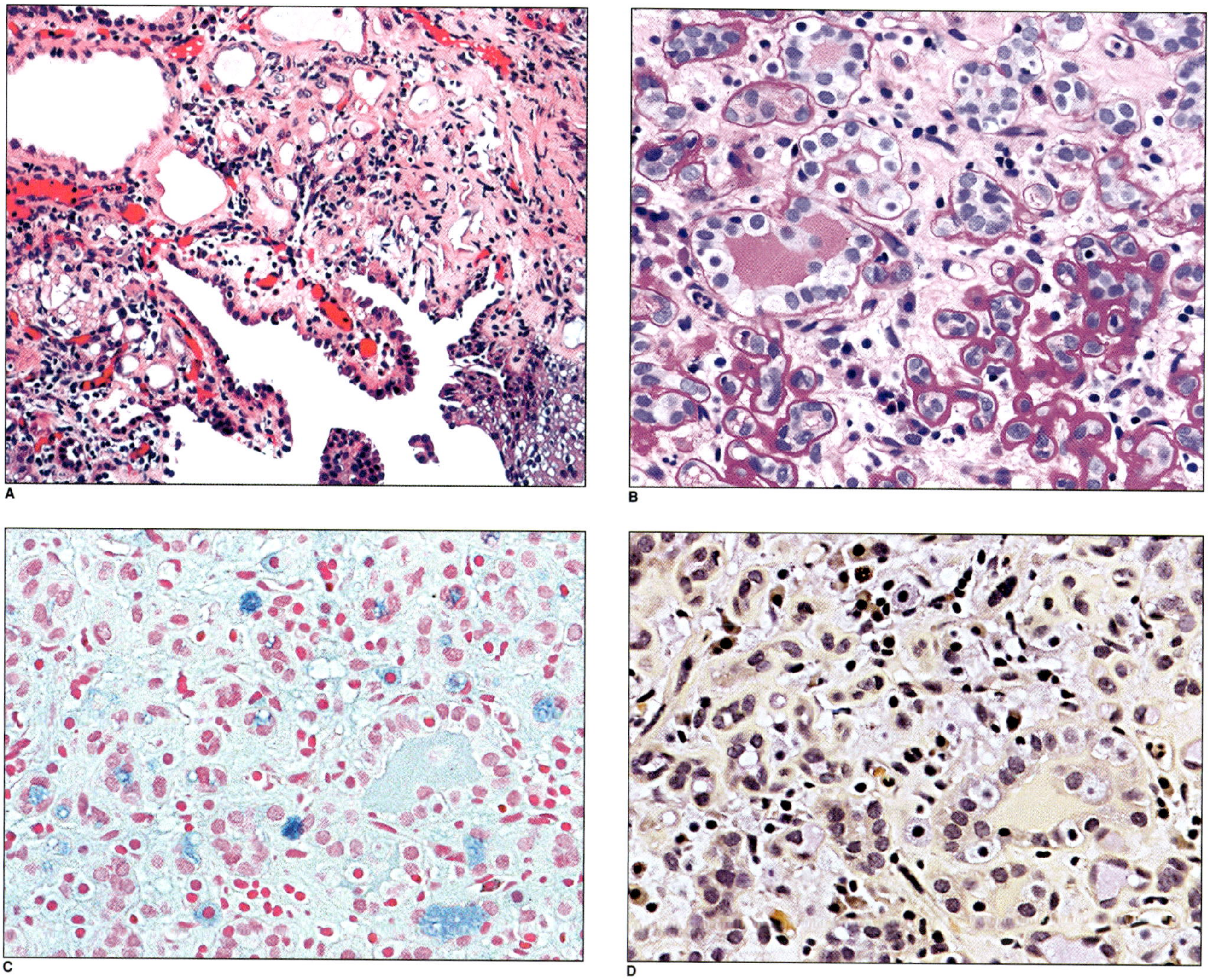

FIGURE 8.55: ***This micropapillary and gland-forming lesion of the bladder mucosa has a partial clear-cell constituency and modest nuclear atypia (A). A digested periodic acid-Schiff stain reveals intraluminal mucin and investment of tubules by basement membrane (B). Stromal mucin is also shown with the alcian blue procedure (C), but a mucicarmine stain is negative (D). The final diagnosis was nephrogenic adenoma.***

REFERENCES

1. Walker PD, Cavallo T, Bonsib SM. Practice guidelines for the renal biopsy. *Mod Pathol* 2004; 17:1555–63.
2. Nickeleit V, Hirsch HH, Zeiler M, et al. BK-virus nephropathy in renal transplants-tubular necrosis, MHC-class II expression and rejection in a puzzling game. *Nephrol Dial Transplant* 2000; 15:324–32.
3. Walker PD, Ferrario F, Joh K, Bonsib SM. Dense deposit disease is not a membranoproliferative glomerulonephritis. *Mod Pathol* 2007; 20:605–16.
4. Pearse AGE. Histochemistry: Theoretical and Applied. London: Churchill, 1985.
5. Puchtler H, Sweat F. Congo red as a stain for fluorescence microscopy of amyloid. *J Histochem Cytochem* 1965; 13:693–4.
6. Rossini M. Interpreting segmental glomerular sclerosis. *Curr Diag Pathol* 2004; 10:1–10.
7. Kimmelstiel P. Intercapillary lesions in glomeruli of the kidney. *Am J Pathol* 1936; 12:83.
8. Brigger D, Muckle RJ. Comparison of Sirius red and Congo red as stains for amyloid in animal tissues. *J Histochem Cytochem* 1975; 23:84–8.
9. Elghetany MT, Saleem A, Barr K. The congo red stain revisited. *Ann Clin Lab Sci* 1989; 19:190–5.
10. Elghetany MT, Saleem A. Methods for staining amyloid in tissues: a review. *Stain Technol* 1988; 63:201–12.
11. Rocken C, Sletten K. Amyloid in surgical pathology. *Virchows Arch* 2003; 443:3–16.
12. Wright JR, Calkins E, Humphrey RL. Potassium permanganate reaction in amyloidosis. A histologic method to

assist in differentiating forms of this disease. *Lab Invest* 1977; 36:274–81.

13. Churukian CJ, Shenck EA. Eastwood Congo red method for demonstrating amyloid. *J Histotechnol* 1988; 11:105.
14. Long JP Jr., Althausen AF. Malacoplakia: a 25-year experience with a review of the literature. *J Urol* 1989; 141:1328–31.
15. Matter MJ, Gygi C, Gillet M, Gebhard S, Bouzourene H. Malacoplakia simulating organ invasion in a rectosigmoid adenocarcinoma: report of a case. *Dis Colon Rectum* 2001; 44:1371–5.
16. Esparza AR, McKay DB, Cronan JJ, Chazan JA. Renal parenchymal malakoplakia. Histologic spectrum and its relationship to megalocytic interstitial nephritis and xanthogranulomatous pyelonephritis. *Am J Surg Pathol* 1989; 13:225–36.
17. Colvin RB. Antibody-mediated renal allograft rejection: diagnosis and pathogenesis. *J Am Soc Nephrol* 2007; 18:1046–56.
18. Curran FT. Malakoplakia of the bladder. *Br J Urol* 1987; 59:559–63.
19. Solez K, Colvin RB, Racusen LC, et al. Banff' 05 meeting report: differential diagnosis of chronic allograft injury and elimination of chronic allograft nephropathy ('CAN'). *Am J Transplant* 2007; 7:518–26.
20. Eble JN, Sauter G, Epstein JI, Sesterhenn IA. World Health Organization Classification of Tumours. Pathology and Genetics of the Urinary System and Male Genital Organs. Lyon, France: IARC Press, 2004.
21. Hes O, Michal M, Sulc M, Kocova L, Hora M, Rousarova M. Glassy hyaline globules in granular cell carcinoma, chromophobe cell carcinoma, and oncocytoma of the kidney. *Ann Diagn Pathol* 1998; 2:12–8.
22. Bonsib SM. Renal chromophobe cell carcinoma: the relationship between cytoplasmic vesicles and colloidal iron stain. *J Urol Pathol* 1996; 4:9–14.
23. Tickoo SK, Amin MB, Zarbo RJ. Colloidal iron staining in renal epithelial neoplasms, including chromophobe renal cell carcinoma: emphasis on technique and patterns of staining. *Am J Surg Pathol* 1998; 22:419–24.
24. Totty BA. Mucin. In: M. BJDaG, ed. Theory and Practice of Histological Techniques. Edinburgh: Churchill Livingstone, 2002:163–200.
25. Val-Bernal JF, Gomez-Roman JJ, Vallina T, Villoria F, Mayorga M, Garcia-Arranz P. Papillary (chromophil) renal cell carcinoma with mucinous secretion. *Pathol Res Pract* 1999; 195:11–7.
26. Grignon DJ, Ro JY, Ayala AG. Primary mucin-secreting adenocarcinoma of the kidney. *Arch Pathol Lab Med* 1988; 112:847–9.
27. Kutscher HA, Trainer TD, Fagan WT Jr. Mucinous adenocarcinoma of renal pelvis. *Urology* 1982; 20:94–5.
28. Verani RR. Transitional and mucous metaplasia of the renal collecting ducts and renal papilla. *Am J Surg Pathol* 1987; 11:83–4.
29. Truong LD, Phung VT, Yoshikawa Y, Mattioli CA. Glycoconjugates in normal human kidney. A histochemical study using 13 biotinylated lectins. *Histochemistry* 1988; 90:51–60.
30. Ulrich W, Horvat R, Krisch K. Lectin histochemistry of kidney tumours and its pathomorphological relevance. *Histopathology* 1985; 9:1037–50.
31. Cohen C, McCue PA, Derose PB. Histogenesis of renal cell carcinoma and renal oncocytoma. An immunohistochemical study. *Cancer* 1988; 62:1946–51.
32. Abbi RK, McVicar M, Teichberg S, Fish L, Kahn E. Pathologic characterization of a renin-secreting juxtaglomerular cell tumor in a child and review of the pediatric literature. *Pediatr Pathol* 1993; 13:443–51.
33. Kelada E, Jones A. Interstitial cystitis. *Arch Gynecol Obstet* 2007; 275:223–9.
34. Johansson SL. In consultation. Interstitial cystitis. *Mod Pathol* 1993; 6:738–42.
35. Nickel JC. Interstitial cystitis: the paradigm shift. *Rev Urol* 2004; 6:200–2.
36. Lynes WL, Flynn SD, Shortliffe LD, Stamey TA. The histology of interstitial cystitis. *Am J Surg Pathol* 1990; 14:969–76.
37. Theoharides TC, Kempuraj D, Sant GR. Mast cell involvement in interstitial cystitis: a review of human and experimental evidence. *Urology* 2001; 57:47–55.
38. Xaubet A, Moises JA, Agusti C, Martos JA, Picado C. Identification of mast cells in bronchoalveolar lavage fluid. Comparison between different fixation and staining methods. *Allergy* 1991; 46:222–7.
39. Dundore PA, Schwartz AM, Semerjian H. Mast cell counts are not useful in the diagnosis of nonulcerative interstitial cystitis. *J Urol* 1996; 155:885–7.
40. Aldenborg F, Peeker R, Fall M, Olofsson A, Enerback L. Metaplastic transformation of urinary bladder epithelium: effect on mast cell recruitment, distribution, and phenotype expression. *Am J Pathol* 1998; 153:149–57.
41. Claman HN, Choi KL, Sujansky W, Vatter AE. Mast cell "disappearance" in chronic murine graft-vs-host disease (GVHD)-ultrastructural demonstration of "phantom mast cells." *J Immunol* 1986; 137:2009–13.
42. Yamada T, Murayama T, Mita H, Akiyama K. Subtypes of bladder mast cells in interstitial cystitis. *Int J Urol* 2000; 7:292–7.
43. Stein PC, Santamaria PJ, Kurtz SB, Parsons CL. Evaluation of urothelial Tamm-Horsfall protein and serum antibody as a potential diagnostic marker for interstitial cystitis. *J Urol* 1993; 150:1405–8.
44. Swisher B. Microorganisms. In: Bancroft JD, Gamble M, eds. Theory and Practice of Histological Techniques. Edinburgh: Churchill Livingstone, 2002:325–38.
45. Tutuncuoglu SO, Yanovich S, Ozdemirli M. CMV-induced hemorrhagic cystitis as a complication of peripheral blood stem cell transplantation: case report. *Bone Marrow Transplant* 2005; 36:265–6.
46. Gorczynska E, Turkiewicz D, Rybka K, et al. Incidence, clinical outcome, and management of virus-induced hemorrhagic cystitis in children and adolescents after allogeneic hematopoietic cell transplantation. *Biol Blood Marrow Transplant* 2005; 11:797–804.
47. Hofland CA, Eron LJ, Washecka RM. Hemorrhagic adenovirus cystitis after renal transplantation. *Transplant Proc* 2004; 36:3025–7.
48. Hoffman JA, Shah AJ, Ross LA, Kapoor N. Adenoviral infections and a prospective trial of cidofovir in pediatric hematopoietic stem cell transplantation. *Biol Blood Marrow Transplant* 2001; 7:388–94.

49. Spach DH, Bauwens JE, Myerson D, Mustafa MM, Bowden RA. Cytomegalovirus-induced hemorrhagic cystitis following bone marrow transplantation. *Clin Infect Dis* 1993; 16:142–4.
50. McClanahan C, Grimes MM, Callaghan E, Stewart J. Hemorrhagic cystitis associated with herpes simplex virus. *J Urol* 1994; 151:152–3.
51. Nguyen ML, Borochovitz D, Thomas G, McClure T, Ruben FL. Hemorrhagic cystitis with herpes simplex virus type 2 in the bladder mucosa. *Clin Infect Dis* 1992; 14:767–8.
52. Traboulsi R, Kanafani ZA, Kanj SS. Fungal infections of the genitourinary tract. *J Med Liban* 2004; 52: 202–9.
53. Mlinaric-Missoni E, Lipozencic J, Marinovic-Kulisic S, Mlinaric-Dzepina A. Fungal infections of urogenital system. *Acta Dermatovenerol Croat* 2004; 12:77–83.
54. Florentine BD, Carriere C, Abdul-Karim FW. Mucor pyelonephritis. Report of a case diagnosed by urine cytology, with diagnostic considerations in the workup of funguria. *Acta Cytol* 1997; 41:1797–800.
55. Siminovitch JM, Herman GP. Torulopsis glabrata fungal cystitis. *Urology* 1984; 24:343–4.
56. Kul'chavenia EV, Brizhatiuk EV. [Tuberculosis of the urinary bladder: diagnosis and treatment]. *Urologiia* 2006: 61–4, 66.
57. Wise CJ, Marella VK. Genitourinary manifestations of tuberculosis. *Urol Clin North Am* 2003; 30:111–21.
58. el Khader K, el Fassi J, Karmouni T, et al. [Urogenital tuberculosis. Apropos of 40 cases]. *Ann Urol (Paris)* 1997; 31:339–43.
59. Lewis KE, Lucas MG, Smith R, Harrison NK. Urogenital infection by *Mycobacterium bovis* relapsing after 50 years. *J Infect* 2003; 46:246–8.
60. Lamm DL. Complications of bacillus Calmette-Guerin immunotherapy. *Urol Clin North Am* 1992; 19:565–72.
61. Steg A, Adjiman S, Debre B. BCG therapy in superficial bladder tumours—complications and precautions. *Eur Urol* 1992; 21 (Suppl 2):35–40.
62. Rischmann P, Desgrandchamps F, Malavaud B, Chopin DK. BCG intravesical instillations: recommendations for side-effects management. *Eur Urol* 2000; 37 (Suppl 1):33–6.
63. Trevenzoli M, Cattelan AM, Marino F, Sasset L, Dona S, Meneghetti F. Sepsis and granulomatous hepatitis after bacillus Calmette-Guerin intravesical installation. *J Infect* 2004; 48:363–4.
64. Huang TC. Management of complications of bacillus Calmette-Guerin immunotherapy in the treatment of bladder cancer. *Ann Pharmacother* 2000; 34:529–32.
65. Proctor DD, Chopra S, Rubenstein SC, Jokela JA, Uhl L. Mycobacteremia and granulomatous hepatitis following initial intravesical bacillus Calmette-Guerin instillation for bladder carcinoma. *Am J Gastroenterol* 1993; 88:1112–5.
66. Leebeek FW, Ouwendijk RJ, Kolk AH, et al. Granulomatous hepatitis caused by bacillus Calmette-Guerin (BCG) infection after BCG bladder instillation. *Gut* 1996; 38:616–8.
67. LaFontaine PD, Middleman BR, Graham SD Jr., Sanders WH. Incidence of granulomatous prostatitis and acid-fast bacilli after intravesical BCG therapy. *Urology* 1997; 49: 363–6.
68. Elkabani M, Greene JN, Vincent AL, VanHook S, Sandin RL. Disseminated *Mycobacterium bovis* after intravesicular bacillus calmette-Guerin treatments for bladder cancer. *Cancer Control* 2000; 7:476–81.
69. Rawls WH, Lamm DL, Lowe BA, et al. Fatal sepsis following intravesical bacillus Calmette-Guerin administration for bladder cancer. *J Urol* 1990; 144:1328–30.
70. Scully RE, Mark EJ, McNeely WF, et al. Case records of the Massachusetts General Hospital. Weekly clinicopathological exercises. Case 31-2000: 32-year old man was evaluated for a lesion of the urinary bladder. *N Engl J Med* 2000; 343:1105–11.
71. Zhou M, Magi-Galuzzi C, Epstein JI. Non-neoplastic diseases of the urinary bladder. In: Zhou M, Magi-Galuzzi C, eds. Genitourinary Pathology. Philadelphia: Churchill Livingstone, 2007:1105–11.
72. Livneh A, Shtrasburg S, Martin BM, Baniel J, Gal R, Pras M. Light chain amyloidosis of the urinary bladder. A site restricted deposition of an externally produced immunoglobulin. *J Clin Pathol* 2001; 54:920–3.
73. Tirzaman O, Wahner-Roedler DL, Malek RS, Sebo TJ, Li CY, Kyle RA. Primary localized amyloidosis of the urinary bladder: a case series of 31 patients. *Mayo Clin Proc* 2000; 75:1264–8.
74. Boorjian S, Choi BB, Loo MH, Kim P, Sandhu J. A rare case of painless gross hematuria: primary localized AA-type amyloidosis of the urinary bladder. *Urology* 2002; 59:137.
75. Akram CM, Al-Marhoon MS, Mathew J, Grant CS, Rao TV. Primary localized AA type amyloidosis of urinary bladder: case report of rare cause of episodic painless hematuria. *Urology* 2006; 68:1343 e15–7.
76. Lipper S, Kahn LB. Amyloid tumor. A clinicopathologic study of four cases. *Am J Surg Pathol* 1978; 2:141–5.
77. Merrimen JL, Alkhudair WK, Gupta R. Localized amyloidosis of the urinary tract: case series of nine patients. *Urology* 2006; 67:904–9.
78. Hajji K, Martin L, Devevey JM, et al. Rheumatoid arthritis-induced pseudotumoral AA amyloidosis of the bladder with vesico-peritoneal fistula. *Clin Nephrol* 2007; 67:38–43.
79. Treaba DO, Benson MD, Assad LW, Dainauskas JR. Sarcoidosis and immunoglobulin lambda II light-chain amyloidosis diagnosed after orthotopic heart transplantation: a case report and review of the literature. *Mod Pathol* 2005; 18:451–5.
80. Kanat O, Evrensel T, Filiz G, et al. Systemic AA amyloidosis and nephrotic syndrome associated with small cell carcinoma of the bladder. *Nephrol Dial Transplant* 2003; 18:2453–4.
81. Huang CY, Shun CT, Huang KH, Chen J, Pu YS. Primary amyloidosis of the urinary bladder. *J Formos Med Assoc* 2006; 105:164–7.
82. Anavi BL, Poriazova EG, Staikova ND, Staikov DS. Primary localized amyloidosis of the urinary bladder. A case report. *Folia Med (Plovdiv)* 2001; 43:47–52.

83. Khan SM, Birch PJ, Bass PS, Williams JH, Theaker JM. Localized amyloidosis of the lower genitourinary tract: a clinicopathological and immunohistochemical study of nine cases. *Histopathology* 1992; 21:143–7.
84. Vowles GH, Frincis RJ. In: Bancroft JD, Gamble M, eds. Theory and Practice of Histologic Techniques. Edinburgh: Churchill Livingstone, 2002:303–20.
85. Stanton MJ, Lynch JH, Maxted WC, Chun BK. Malacoplakia of the bladder: a case report of resolution with bethanechol, trimethoprim-sulfamethoxazole and ascorbic acid. *J Urol* 1983; 130:1174–5.
86. Darvishian F, Teichberg S, Meyersfield S, Urmacher CD. Concurrent malakoplakia and papillary urothelial carcinoma of the urinary bladder. *Ann Clin Lab Sci* 2001; 31:147–50.
87. Fornari A, Dambros M, Teloken C, Hartmann AA, Kolling J, Seben R. A case of xanthogranulomatous cystitis. *Int Urogynecol J Pelvic Floor Dysfunct* 2007; 18:1233–1235.
88. Raffi HS, Bates JM Jr., Laszik Z, Kumar S. Tamm-Horsfall protein acts as a general host-defense factor against bacterial cystitis. *Am J Nephrol* 2005; 25:570–8.
89. Truong LD, Ostrowski ML, Wheeler TM. Tamm-Horsfall protein in bladder tissue. Morphologic spectrum and clinical significance. *Am J Surg Pathol* 1994; 18: 615–22.
90. Newbould M, McWilliam LJ. A study of vesical adenocarcinoma, intestinal metaplasia and related lesions using mucin histochemistry. *Histopathology* 1990; 17:225–30.
91. Lopez-Beltran A. Infiltrating urothelial carcinoma. In: Eble JN, Sauter G, Epstein JI, Sesterhenn IA, eds. Pathology & Genetics: Tumor of the Urinary System and Male Genital Organs. Lyon, France: IARC Press, 2004:93–109.
92. Donhuijsen K, Schmidt U, Richter HJ, Leder LD. Mucoid cytoplasmic inclusions in urothelial carcinomas. *Hum Pathol* 1992; 23:860–4.
93. Yang GC, Campbell WG Jr. Morphogenesis of inclusion bodies of urothelial carcinoma: a case study. *Mod Pathol* 1996; 9:566–70.
94. Young RH, Eble JN. Unusual forms of carcinoma of the urinary bladder. *Hum Pathol* 1991; 22:948–65.
95. Eble JN, Young RH. Carcinoma of the urinary bladder: a review of its diverse morphology. *Semin Diagn Pathol* 1997; 14:98–108.
96. Hasegawa R, Fukushima S, Hirose M, et al. Histochemical demonstration of colonic type mucin in glandular metaplasia and adenocarcinoma of the human urinary bladder. *Acta Pathol Jpn* 1987; 37:1097–103.
97. Nakanishi K, Tominaga S, Kawai T, Torikata C, Aurues T, Ikeda T. Mucin histochemistry in primary adenocarcinoma of the urinary bladder (of urachal or vesicular origin) and metastatic adenocarcinoma originating in the colorectum. *Pathol Int* 2000; 50:297–303.
98. Grignon DJ, Ro JY, Ayala AG, Johnson DE, Ordonez NG. Primary adenocarcinoma of the urinary bladder. A clinicopathologic analysis of 72 cases. *Cancer* 1991; 67: 2165–72.
99. Cheng L, Montironi R, Bostwick DG. Villous adenoma of the urinary tract: a report of 23 cases, including 8 with coexistent adenocarcinoma. *Am J Surg Pathol* 1999; 23: 764–71.
100. Barresi G, Marafioti T. Mucin histochemistry and lectin binding sites in intestinal metaplasia of the urinary bladder. *Histopathology* 1990; 17:219–23.
101. Wells M, Anderson K. Mucin histochemistry of cystitis glandularis and primary adenocarcinoma of the urinary bladder. *Arch Pathol Lab Med* 1985; 109:59–61.
102. Clement PB, Young RH. Endocervicosis of the urinary bladder. A report of six cases of a benign mullerian lesion that may mimic adenocarcinoma. *Am J Surg Pathol* 1992; 16:533–42.
103. Allan CH, Epstein JI. Nephrogenic adenoma of the prostatic urethra: a mimicker of prostate adenocarcinoma. *Am J Surg Pathol* 2001; 25:802–8.
104. Oliva E, Young RH. Nephrogenic adenoma of the urinary tract: a review of the microscopic appearance of 80 cases with emphasis on unusual features. *Mod Pathol* 1995; 8: 722–30.
105. Young RH. Pseudoneoplastic lesions of the urinary bladder and urethra: a selective review with emphasis on recent information. *Semin Diagn Pathol* 1997; 14:133–46.
106. Young RH, Scully RE. Nephrogenic adenoma. A report of 15 cases, review of the literature, and comparison with clear cell adenocarcinoma of the urinary tract. *Am J Surg Pathol* 1986; 10:268–75.
107. Reuter VE. Urinary tract: renal pelvis, ureter, urinary bladder, and urethra. In: Mills SE, ed. Sternberg's Diagnostic Surgical Pathology. Philadelphia: Lippincott Williams & Wilkin, 2004.
108. Baldwin L, Lee AH, Al-Talib RK, Theaker JM. Transitional cell carcinoma of the bladder mimicking lobular carcinoma of the breast: a discohesive variant of urothelial carcinoma. *Histopathology* 2005; 46:50–6.
109. Azadeh B, Vijayan P, Chejfec G. Linitis plastica-like carcinoma of the urinary bladder. *Br J Urol* 1989; 63: 479–82.
110. Torenbeek R, Koot RA, Blomjous CE, De Bruin PC, Newling DW, Meijer CJ. Primary signet-ring cell carcinoma of the urinary bladder. *Histopathology* 1996; 28: 33–40.
111. Humphrey PA. Clear cell neoplasms of the urinary tract and male reproductive system. *Semin Diagn Pathol* 1997; 14:240–52.
112. Drew PA, Murphy WM, Civantos F, Speights VO. The histogenesis of clear cell adenocarcinoma of the lower urinary tract. Case series and review of the literature. *Hum Pathol* 1996; 27:248–52.
113. Pan CC, Yu IT, Yang AH, Chiang H. Clear cell myomelanocytic tumor of the urinary bladder. *Am J Surg Pathol* 2003; 27:689–92.
114. Cheng L, Cheville JC, Sebo TJ, Eble JN, Bostwick DG. Atypical nephrogenic metaplasia of the urinary tract: a precursor lesion? *Cancer* 2000; 88:853–61.
115. Fletcher MS, Aker M, Hill JT, Pryor JP, Whimster WF. Granular cell myoblastoma of the bladder. *Br J Urol* 1985; 57:109–10.
116. Rosai J. Rosai and Ackerman's Surgical Pathology. Edinburgh: Mosby, 2004.

117. Cheng L. Carcinoid. In: Eble JN, Sauter G, Epstein JI, Sesterhenn IA, eds. Pathology & Genetics: Tumor the Urinary System and Male Genital Organs. Lyon, France: IARC Press, 2004:138.
118. Lehman TP, Cooper HS, Mulholland SG. Peanut lectin binding sites in transitional cell carcinoma of the urinary bladder. *Cancer* 1984; 53:272–7.
119. Neal DE, Charlton RG, Bennett MK. Histochemical study of lectin binding in neoplastic and non-neoplastic urothelium. *Br J Urol* 1987; 60:399–404.
120. Okamura T, Ueda K, Ohtaguro K, et al. Lectin immunohistochemical evaluation of human bladder carcinomas. A comparison of Carnoy's and formalin fixation. *Hinyokika Kiyo* 1993; 39:899–905.
121. Langkilde NC, Hastrup J, Olsen S, Wolf H, Orntoft TF. Immunohistochemistry and cytochemistry of experimental rat bladder cancer: binding of the lectins PNA and WGA and of a Le(Y) mouse monoclonal antibody. *J Urol* 1989; 141:981–6.

CHAPTER 9

Diagnostic Histochemistry of the Gynecologic Tract

Joseph T. Rabban, M.D., M.P.H., and
Charles J. Zaloudek, M.D.

Tissues from the female genital tract are among the most common specimens that surgical pathologists are called upon to evaluate. In most instances, one can make a diagnosis after examining slides that have been stained routinely with hematoxylin and eosin (H&E). On occasion, however, special stains must be used to clarify the interpretation. Currently, pathologists tend to rely mainly, and, perhaps, excessively, on immunohistochemistry to help them resolve difficult diagnostic issues. Immunostains employ antigen-antibody reactions to identify specific proteins in cells or organisms. Although they are extremely valuable, such studies are not always the best or most cost-effective way to solve diagnostic problems. An appropriate antibody may not be available, or technical problems with tissue processing may cause the specimen to be antigenically "inert." In contrast, histochemical stains have been used in diagnostic work for many years and are not affected much by fixation-related problems with tissue samples. If an appropriate histochemical stain is available for application, it is almost always preferable to use it rather than other diagnostic methods.

The histochemical stains that are most useful in gynecologic pathology are listed in Table 9.1. They are less expensive than immunohistologic methods, can generally be performed more rapidly, and are available in virtually all laboratories. One does not need to have complicated instruments or a large library of biologic reagents on hand in order to do histochemical procedures. They employ specific chemical reactions that yield visible, multicolored products to signal a positive result. One of the principal uses of such techniques in gynecologic pathology is to evaluate inflammatory processes because a variety of bacterial and fungal organisms can be labeled with histochemical stains. Positive identification of a particular microbe can help in recognizing a specific infection and allow for rapid institution of corresponding treatment.

Another common use of histochemistry is in the recognition of tumors and tumorlike conditions of the female genital tract. Some of those lesions have distinctive cellular or stromal characteristics that can be identified biochemically. In this chapter, we survey the spectrum of gynecologic conditions in which histochemical staining is helpful diagnostically.

VULVAR LESIONS

Nonneoplastic Disorders

Inflammatory disorders of the vulva fall into two categories—infectious lesions and others. Infections of the vulva are caused by a wide range of organisms, including viruses, fungi, bacteria, and parasites. Histochemical methods can be used to identify some of those agents, but others require immunohistochemical stains, in situ hybridization, or standard culture techniques. Histochemistry has limited value in the evaluation of noninfectious inflammatory processes, which are recognized mainly by their clinical appearances and morphologic features in H&E-stained slides. Nonetheless, histochemical stains can be used in selected situations to identify processes that simulate noninfectious vulvar inflammation but which have a different biologic nature.

Bartholin Cyst and Abscess: The Bartholin glands are paired mucin-producing structures that are located deep in the posterior vulva. They drain into the vulvar vestibule via the Bartholin duct, which is lined by mucinous, transitional, or squamous epithelium. Inflammation of the duct is common and can result in ulceration of the ductal mucosa with fibrosis, producing ductal luminal obstruction. This, in turn, can lead to development of a Bartholin cyst, as mucinous secretions accumulate behind the obstruction. Histologically, Bartholin cysts have varied appearances. In some instances, they have squamous, transitional, or mucinous epithelial linings. In others, the epithelium is completely eroded and the cyst contains only inflamed granulation tissue. Staining for mucin in normal Bartholin acini, proximal Bartholin ducts, and Bartholin duct cysts with the mucicarmine or diastase-digested periodic acid-Schiff (PAS-D) methods can help to confirm that a cyst

TABLE 9.1: Useful Histochemical Stains in Gynecologic Pathology

Stain	Target	Common Diagnostic Uses
PAS, GMS	Fungi	Vulvar fungal infection
PAS-D	Glycogen	Clear cell carcinoma, dysgerminoma, and ASR
PAS-D	Hyaline globules or hyaline stroma	Clear cell carcinoma, yolk sac tumor
Methyl green–pyronine	Plasma cells	Chronic endometritis
Warthin-Starry	Organisms	BA, syphilis, actinomycosis, GI
Mucicarmine	Mucin	Mucinous carcinoma, vulvar Paget disease
Reticulin	Pericellular fibrils	Thecoma versus GCT
Oil red O	Lipid	Thecoma, steroid cell tumor
Alcian blue	Mucopolysaccharides	Myxoid smooth muscle tumor versus hydropic degeneration
PTAH	Skeletal muscle cross-striations	Rhabdomyomatous tumors
Prussian blue iron stain	Iron	Meconium exposure versus placental hemorrhage

or inflammatory process involves the Bartholin gland (Figures 9.1 and 9.2).

Bartholin ducts and Bartholin cysts also can become secondarily infected. In such cases, a suppurative exudate comprising neutrophils, macrophages, red blood cells, fibrin, and necrotic debris is present in the duct or cyst lumen, and the inflammation frequently extends into the surrounding connective tissue. Bacteria are typically identified in cultures of such lesions, and the most common organisms in this setting are *Bacteroides* species, *Peptostreptococcus* species, *Escherichia coli*, and *Neisseria gonorrhoeae* (1–3). If lesional tissue is debrided or the abscess contents are smeared on glass slides, one may be able to identify the bacteria on Gram-stained slides, thereby providing the clinician with preliminary information about the microbes that are visualized. Similar studies of permanent (paraffin embedded) sections serve a comparable purpose, using tissue Gram, Grocott's silver methenamine (GMS), and acid-fast staining procedures.

Vulvar Vestibulitis: Vulvar vestibulitis is a chronic condition of unknown etiology, characterized by dyspareunia, point tenderness, and varying degrees of erythema. Pain is the most diagnostic clinical feature (4). Proposed etiologies have included infections, especially recurrent *Candida* infections, viral infections with *Herpes* or human papillomavirus (HPV), trauma, dermatitis, and vasculitis.

Microscopically, the main findings are a lymphocytic infiltration of the lamina propria and of the periglandular stroma around the minor vestibular glands (5,6). Lymphocytes—most of which are T-cells—focally infiltrate both the surface epithelium and the ductal epithelium. Plasma cells are also present. There is no inflammation of the vestibular lumen, nerves, or blood vessels. Squamous metaplasia of the vestibular gland ducts is typically present. Neither bacterial nor fungal infection has been shown to cause vulvar vestibulitis. Comparative studies have shown that the inflammatory changes that are seen in patients with vestibulitis are similar to findings in controls (7,8). The disease poses a difficult clinical problem and one

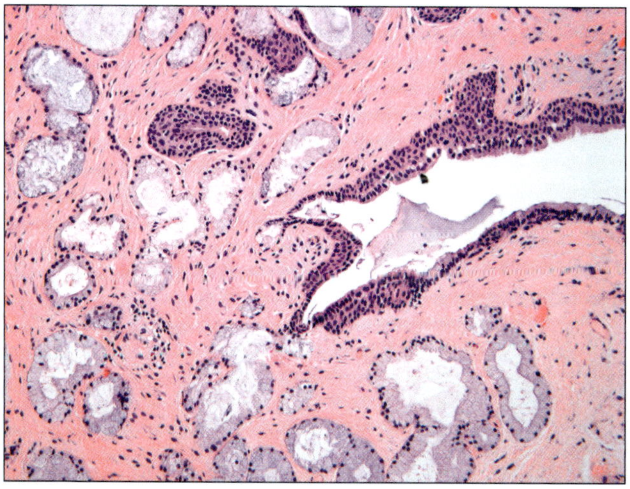

FIGURE 9.1: *Bartholin glands secrete mucus into the vaginal orifice via a duct lined by stratified columnar epithelium with mucinous, transitional, and/or squamous features.*

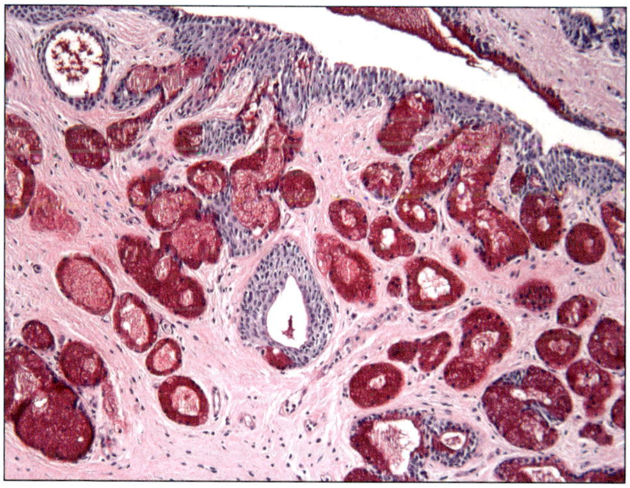

FIGURE 9.2: *PAS highlights the mucin contents of Bartholin glands.*

is typically called on to exclude an infectious etiology. However, Gram and GMS stains are negative for organisms. Although some reports indicated that HPV was detectable in cases of vulvar vestibulitis, koilocytosis is rare in typical cases (9) and most recent studies have found few if any patients to have vulvar HPV infection (5,6,10). Treatment is difficult, and a wide variety of modalities have been used with variable results (11).

Bacillary Angiomatosis: Bacillary angiomatosis (BA) is a vascular proliferative disease caused by *Bartonella* organisms, usually *Bartonella henselae*. It occurs in immunosuppressed patients, most of whom have acquired immunodeficiency syndrome. However, BA also has been reported in patients who are immunocompromised for other reasons. A variety of body sites can be involved, including the female genital tract. In the latter location, BA has typically been observed in the vulva and the cervix (12). This disease is most readily recognized when the skin is involved. Patients develop red-purple nodules that are clinically thought to represent Kaposi sarcoma. Microscopically, the process suggests a vascular neoplasm, with multilobular proliferation of small blood vessels lined by epithelioid endothelial cells. They have vesicular nuclei, prominent nucleoli, and eosinophilic cytoplasm. The key diagnostic finding in H&E sections is the presence of hazy purple clumps of bacteria in edematous stroma between the blood vessels. *Bartonella henselae* is a short gram-negative bacillus that can be identified in the purple clumps in sections stained with the Warthin-Starry silver method. Blood cultures are typically positive as well. BA is an infrequent finding now that patients with human immunodeficiency virus infection are routinely treated with antiretroviral therapy. Nonetheless, awareness of this disorder is still mandatory.

Syphilis: Syphilis (lues) is caused by infection with a spirochete, *Treponema pallidum*. Untreated disease may evolve through three clinical phases, featuring mucocutaneous chancres, condylomata lata, and gummas. Microscopically, the condylomata lata show hyperkeratosis, psoriasiform epidermal hyperplasia, neutrophils in the epidermis, and a dermal inflammatory cell infiltrate containing numerous plasma cells (13). Spirochetes are numerous in condylomata lata, and less so in primary chancres. Tertiary (gummatous) syphilis generally does not involve the female genital tract. It is most often centered in the cardiovascular or nervous systems, but may affect other viscera as well.

The diagnosis of lues is most often made by clinical examination together with laboratory testing. Spirochetes can be identified in scrapings from the lesions that are examined with dark-field microscopy or a fluorescent antibody technique. Alternatively, serologic testing for reagin can be performed with the Venereal Disease Research Laboratory or Rapid Plasma Reagin test. Tissue biopsies are most likely when the clinician does not suspect syphilis or the presentation is atypical. Histochemical stains for spirochetes should be performed in any case where histologic findings raise a question of syphilis, remembering that the microscopic images of lues are potentially several. The Warthin-Starry, Dieterle, and Steiner procedures can all be used effectively to visualize *T. pallidum* in tissue sections. With those stains, black-brown pigment is deposited on the spirochetes, which are tightly coiled organisms measuring 6–15 μm in length (Figure 9.3) (14). The bacteria are most likely to be found along the dermal-epidermal junction and in the superficial epidermis (15).

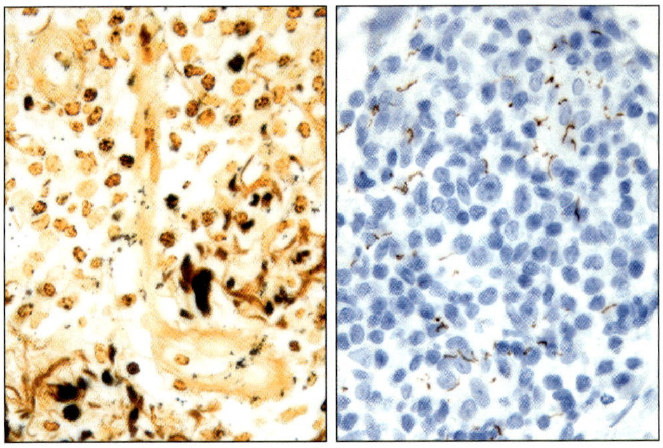

FIGURE 9.3: *Tightly coiled* **Treponema pallidum**, ***the spirochetes causing syphilis, are seen on Warthin-Starry stain (left). Immunohistochemical staining for the spirochetes can also be performed (right).***

Unfortunately, the aforementioned stains are somewhat difficult to perform and interpret. Some authors have reported that silver impregnation procedures are highly effective for spirochete identification, but others have realized a success rate approximating 40% in cases of lues (15,16). Background staining and precipitated silver debris can make interpretation difficult; thus, it is important to report a positive result only when typical spirilliform organisms are clearly identified. We find it easier to perform and interpret immunostains for *Treponema*, as compared with silver-based histochemical techniques (16,17).

Granuloma Inguinale: Granuloma inguinale (GI; donovanosis) is caused by the gram-negative coccobacillus *Calymmatobacterium granulomatis* (18). Fewer than 100 cases are reported annually in the United States, but this may be an underestimate. GI is a chronic progressive inflammatory disease that predominantly affects the vulva but also may involve the cervix, vagina, and perineum.

The earliest lesions are papules, which subsequently ulcerate. Granulation tissue is present at the base of the ulcers, which also manifest a mixed inflammatory cell infiltrate that usually contains many plasma cells. Microabscesses are noted in some cases, and foamy histiocytes are present in the inflammatory infiltrate. Pseudoepitheliomatous hyperplasia of the epidermis can be present as well.

As the disease progresses, the ulcers may enlarge and incite a secondary fibrotic response that causes distortion of the tissues and may simulate malignancy (19,20). The differential diagnosis also includes mycobacterial infection, malacoplakia, syphilis, and granular cell tumor.

The bacteria that cause GI generally cannot be cultured, so the diagnosis depends on identification of organisms in smears or tissue samples. The lesional bacilli are present within foamy (parasitized) histiocytes (Donovan bodies) but they cannot be seen in H&E-stained sections. Slides labeled with the Giemsa or Warthin-Starry technique will show single or clustered bacilli within histiocytes. The organisms may be straight, curved, or dumbbell shaped.

Chancroid: Chancroid is an ulcerative inflammatory disease that involves the anogenital region. In women, lesions can be located in the vulva, the perianal region, or the cervix (21). Chancroid is rare in the United States and Europe (22), but it is common in tropical countries. The initial lesions of this disorder are papules that progress to pustules. Rupture leads to the formation of shallow ulcers with undermined edges. Subsequently, there is unilateral enlargement of the inguinal lymph nodes in roughly 50% of cases. Chancroid is caused by a gram-negative bacterium, *Haemophilus ducreyi*.

Microscopically, the shallow ulcers of this disease are bordered by psoriasiform epidermal hyperplasia, spongiosis, and a neutrophilic infiltrate (23). One may see dermal vasculitis, with lymphocytes and histiocytes surrounding and infiltrating blood vessels, endothelial cell swelling, mural edema, and, in some instances, vascular fibrin deposition. A dermal chronic inflammatory infiltrate may also be seen, in which plasma cells predominate (13). Some reports have posited that the histologic image of chancroid suggests the presence of a cell-mediated immune response (23).

The causative organisms can be visualized with histochemical methods. Giemsa staining or silver impregnation methods reveal metachromatic 1- to 2-μm coccobacilli in intercellular spaces of the epidermis and in the superficial dermis. The organisms are weakly gram negative; therefore, tissue Gram stains such as the Brown-Brenn or Brown-Hopps procedure is typically false negative in cases of chancroid. *Haemophilus ducreyi* can be cultured, but special media are required to do so and these are typically available only in selected laboratories.

Fungal Infections: *Candida* or dermatophyte infections are frequent causes of chronic vulvar inflammation. Symptoms include itching and burning, and the involved areas of the vulva are red and edematous. Simultaneous vaginal infection is frequently present in women with vulvar candidiasis. The diagnosis is generally made by identifying fungal yeast forms or pseudohyphae in scrapings from the lesions, using saline suspension microscopy with or without addition of 10% potassium hydroxide.

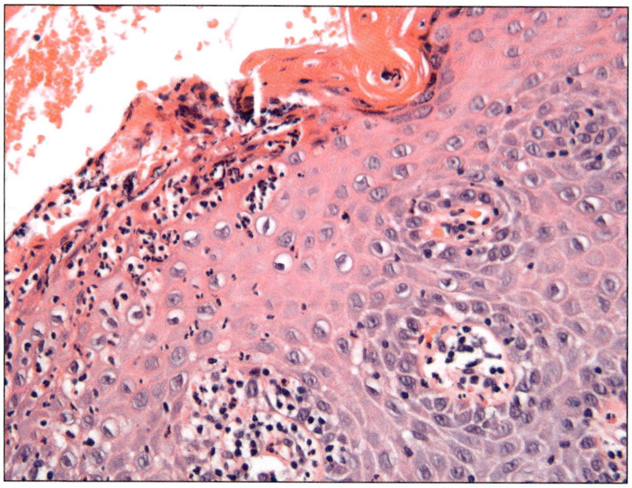

FIGURE 9.4:* *The acute inflammatory infiltrate accompanying vulvovaginal candidiasis may obscure recognition of the fungal organisms present.

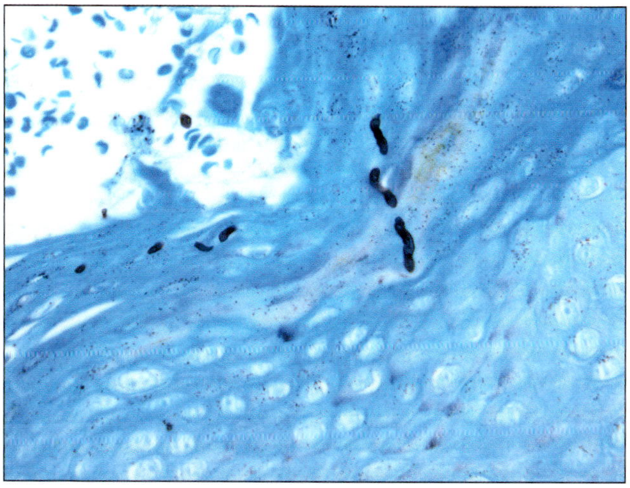

FIGURE 9.5:* *Fungal forms are highlighted by GMS staining in the parakeratotic zone of this vulvar biopsy.

Biopsy specimens are rarely obtained in clinically typical cases, but they may be procured in cases with atypical presentations or when response to therapy is poor. Microscopically, the epidermis tends to be thickened with parakeratosis or hyperkeratosis, and a lymphocytic infiltrate is seen in the underlying dermis. Small aggregates of neutrophils may form microabscesses in the surface epithelium. Fungal organisms can be identified with the GMS or PAS-D methods (Figures 9.4 and 9.5). Yeast forms or pseudohyphae are most conspicuous in the parakeratotic zone just beneath the epidermal surface.

Other Infections: Diagnosis of other vulvar infectious diseases, such as *Herpes simplex* virus infection, condyloma acuminatum, molluscum contagiosum, or lymphogranuloma venereum, relies on their morphologic characteristics.

Adjunctive nonmorphologic laboratory testing is also contributory.

Malacoplakia: Malacoplakia is a histocytic reaction that is associated with persistent bacterial infection. It typically affects the urinary tract not only has been reported in other sites including the gynecologic tract. This disorder may present as soft plaques, nodules, or masses in the ovary, endometrium, cervix, vagina, or vulva (24–28). Microscopically, it comprises aggregates of vacuolated histiocytes, known as von Hansemann's cells, some of which contain Michaelis-Gutmann bodies (Figure 9.6) (29). The latter structures are intracytoplasmic rounded inclusions, 5–10 μm in diameter. In H&E stains, they have a targetoid or concentric laminar appearance but are much better demonstrated with von Kossa, Prussian blue, PAS-D, or alcian blue method (Figure 9.7). Michaelis-Gutmann bodies are not labeled with tissue Gram stains.

Malacoplakia is thought to reflect a defect in the lysosomal degradation of bacteria. At the ultrastructural level, Michaelis-Gutmann bodies contain calcium and bacterial fragments (29). The differential diagnosis of this disease includes foreign body reactions, sarcoidosis (Figure 9.8) (30–32), and infections with treponemes, fungi, or mycobacteria (33,34). Thus, Ziehl-Neelsen, Fite, silver impregnation, GMS, and PAS-D stains are indicated when considering a diagnosis of malacoplakia.

Lichen Sclerosus et Atrophicus: Lichen sclerosus et atrophicus (LSEA; "kraurosis vulvae") is a common noninfectious dermatosis that can involve the vulva and the perineum. Although it may present at any age, LSEA typically arises in postmenopausal women. The etiology for this disorder is unknown. It presents as irregular white patches on the vulva although red-brown discoloration may be seen as well. Advanced LSEA is characterized by scarring, wrinkling, and atrophy of the affected skin (35).

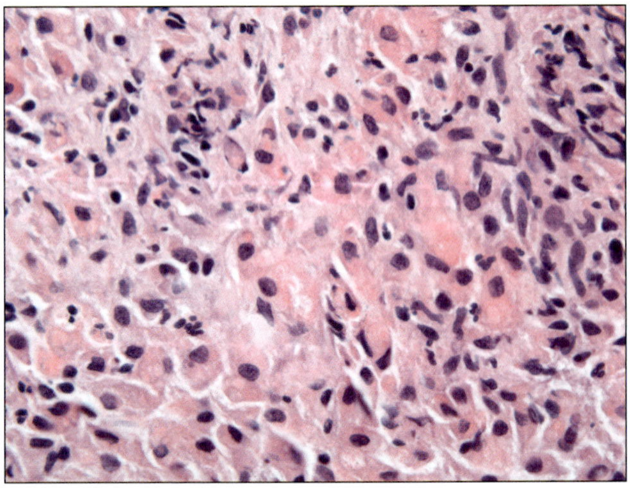

FIGURE 9.6: *Malacoplakia consists of aggregates of foamy histiocytes, so-called von Hansemann's cells. Intracytoplasmic Michaelis-Gutmann bodies are not easily visible on H&E stain.*

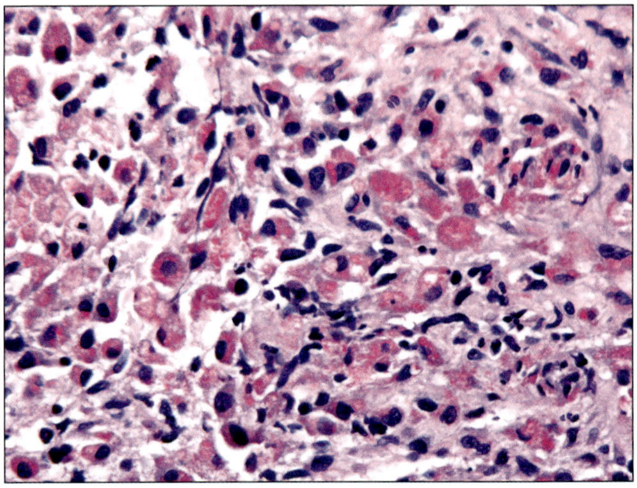

FIGURE 9.7: *Intracytoplasmic Michaelis-Gutmann bodies (calcium and bacterial fragments), highlighted on PAS stain, accumulate in the histiocytes of malacoplakia.*

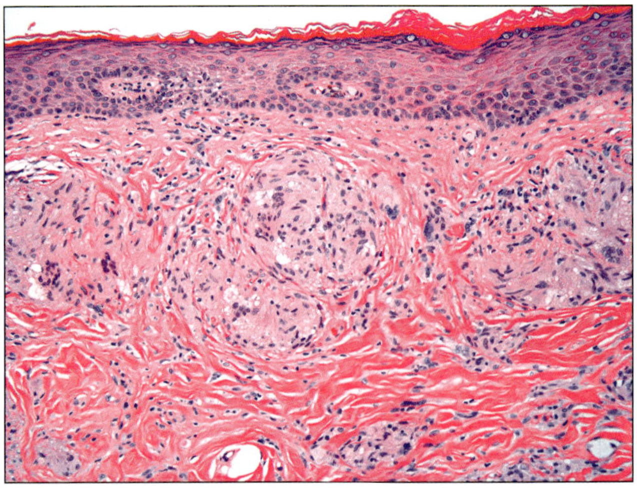

FIGURE 9.8: *Sarcoidosis of the vulva is a rare entity but should be considered in the differential diagnosis of granulomatous lesions.*

Microscopically, the characteristic finding is sclerosis and homogenization of the subepithelial dermis due to loss of papillary dermal elastin fibers, which can be demonstrated with an elastin stain (Figures 9.9 and 9.10). Associated findings include epidermal atrophy, basement membrane thickening, a psoriasiform lichenoid pattern, basilar lymphocytosis, hyperkeratosis, parakeratosis, and spongiosis (36,37).

Cutaneous Amyloidosis: The dermal hyalinization seen in LSEA can be mimicked to some degree by cutaneous amyloidosis, an entity that occasionally involves the vulva (38). Primary nodular amyloidosis of the skin can be associated with autoimmune disorders, such as Sjogren's syndrome (39), and vulvar amyloidosis has also been reported

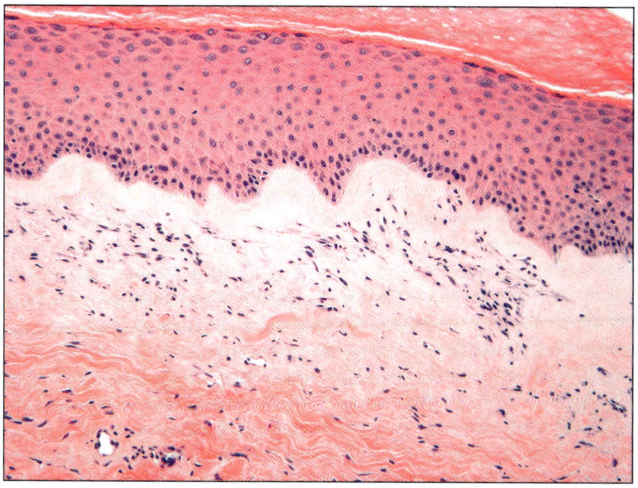

FIGURE 9.9: Lichen sclerosus of the vulva/perineum is characterized by homogenization of the subepithelial dermis.

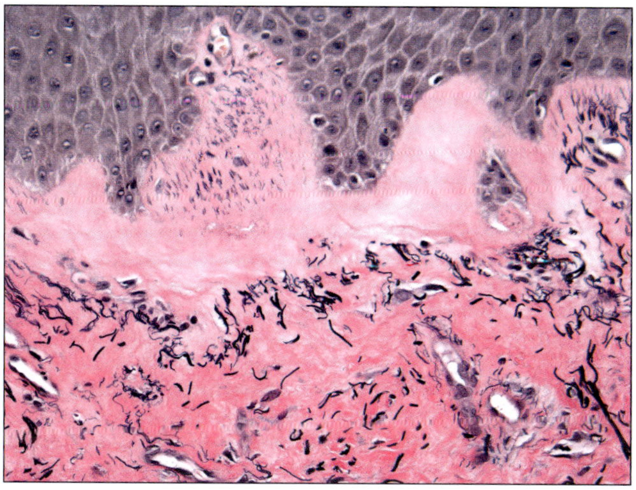

FIGURE 9.10: Superficial dermal homogenization in lichen sclerosus is due to loss of papillary dermal elastin fibers, as demonstrated by loss of elastin staining in this zone, compared to normal fiber distribution in the lower half of the image.

as a paraneoplastic complication of vulvar carcinoma (40). Amyloid deposition results in dermal hyalinization and eosinophilia in H&E-stained sections (Figure 9.11). In the nodular form of this disease, the lesional configuration helps to distinguish it from LSEA. Congo red staining confirms the presence of amyloid (41,42), labeling it red orange. With polarized light, the amyloid deposits are birefringent and apple green (Figure 9.12). Alternative histochemical methods for amyloid detection include the crystal violet stain and the thioflavin-T procedure. However, the Congo red stain is felt to be the most reliable and is probably the most widely used.

Tumors of the Vulva

Squamous cell carcinoma, and its precursor, vulvar intraepithelial neoplasia (VIN) are the most common neoplasms

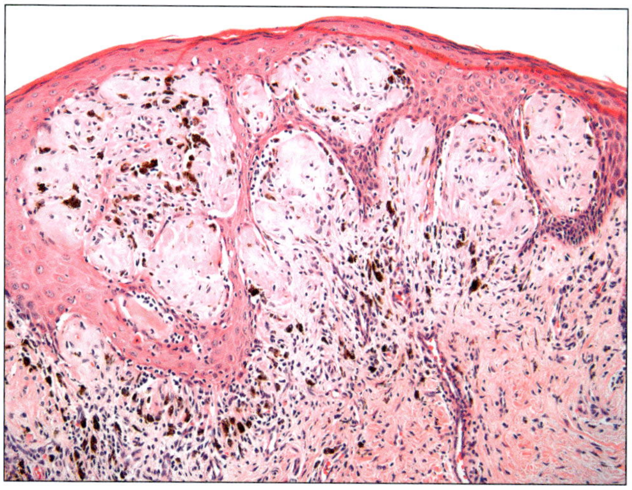

FIGURE 9.11: The superficial dermal nodules of amyloid in vulvar cutaneous amyloidosis, a rare entity, can mimic lichen sclerosus.

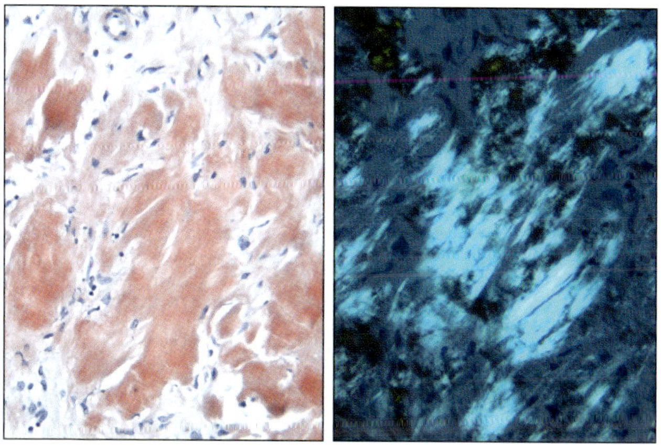

FIGURE 9.12: Congo red staining of amyloid (left) shifts to apple-green under polarized light (right).

of the vulva. Generally, these lesions are diagnosable with H&E-stained slides. When ancillary preparations are needed for their recognition, one usually resorts to immunostaining because histochemical methods are generally not helpful in evaluating squamous proliferations. Nonetheless, many other neoplasms occur in the vulva, and histochemical stains contribute to assessments of some of them. These tumors include Paget disease, adenocarcinoma of the Bartholin gland, some mesenchymal tumors, and melanoma.

Paget Disease: Paget disease of the vulva is a form of intraepidermal adenocarcinoma. Unlike its morphologic counterpart in the breast, primary vulvar Paget disease is typically not associated with an underlying malignancy. Adenocarcinomas of the vulvar sweat glands are present in only a few cases; vulvar Paget disease may also be related to adenocarcinomas of colonic or urologic origin in rare instances. Paget disease occurs mainly between the ages of 50 and 90, with the average being in the mid-60s

(43–45). Symptoms are often present for a lengthy period before the patient consults a physician. The most common presentation is with pruritis or burning, with one or more eczematous erythematous plaques on the vulva, perineum, or perianal skin (46,47). When epidermal hyperplasia is present, the lesions may be white rather than red.

Microscopically, intraepidermal Paget cells are large and polygonal with abundant amphophilic and vacuolated cytoplasm (Figure 9.13). Occasionally, nuclei are displaced to the periphery of the cells by large mucin vacuoles, yielding a signet-ring cell appearance. Paget cells are most numerous in the basal epidermis, but they are scattered throughout other layers of the integument as well. They may be dispersed singly, in clusters, or with gland formation, extending along epidermal appendages, and potentially being present in the lumina of sweat glands. In approximately 1% of cases, the Paget cells invade the dermis, where they are surrounded by lymphocytes and a desmoplastic stromal reaction. The differential diagnosis of Paget disease includes Pagetoid (superficial spreading) melanoma and the Pagetoid variant of high-grade VIN. The diagnosis of Paget disease can be confirmed with the mucicarmine, PAS-D, colloidal iron, zirconyl-hematoxylin, or alcian blue method, all of which label epithelial mucin (Figure 9.14). Unlike Paget disease of the breast, mucin stains are almost always positive in vulvar lesions, at least focally. These are helpful in evaluating surgical margins, which can be difficult to define clinically and histologically (45,46).

Mimics of Paget disease, such as melanoma and pagetoid VIN, do not contain intracellular mucin. The identification of perianal Paget disease that is associated with colorectal neoplasia is best accomplished with immunostains (48), although the high-iron diamine–alcian blue method may also be helpful. Although it is not absolutely specific, positivity with the latter technique supports the presence of enteric-type epithelial differentiation (48a).

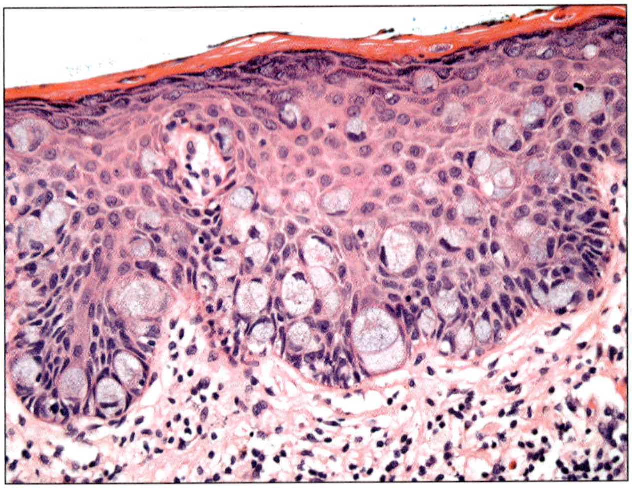

FIGURE 9.13: ***Tumor cells of vulvar Paget disease are polygonal and contain abundant amphophilic foamy cytoplasm, sometimes resembling signet-ring cells.***

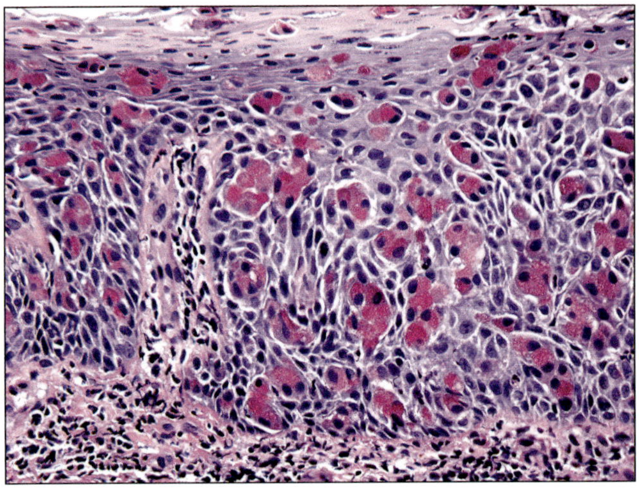

FIGURE 9.14: ***PAS-D marks Paget tumor cells, as does mucicarmine and alcian blue. Melanoma and pagetoid VIN, the main mimics of Paget disease, are negative for these stains.***

Other Vulvar Adenocarcinomas: Other invasive adenocarcinomas of the vulva are rare. Most are of Bartholin gland origin, but such tumors occasionally arise from other adnexal glands. Bartholin gland carcinomas are most common in middle-aged and elderly women, and they present as masses in the Bartholin gland region. Reported survival rates range from 20% to 80% (49), with stage being the most important prognostic variable.

Pathologically, the diagnosis of Bartholin gland carcinoma is predicated on proof that the tumor has indeed arisen in the region of the Bartholin gland. The histologic image may be that of a micropapillary adenocarcinoma, adenoid cystic carcinoma, mucinous carcinoma, adenocarcinoma not further specified, transitional carcinoma, squamous carcinoma, or undifferentiated carcinoma. Histochemical studies can be helpful in this context. The Bartholin gland is a mucin-producing structure, and special stains for epithelial mucin, such as mucicarmine or PAS-D, can delineate residual Bartholin gland acini or ducts. This finding supports the conclusion that the carcinoma has arisen from the Bartholin gland and does not represent a metastasis. Mucin stains can also label mucin within the tumor cells themselves, confirming a diagnosis of adenocarcinoma. In adenoid cystic carcinomas, cribriform spaces or pseudolumina contain mucoid secretions or dense cylinders of hyalinized basement membrane material (Figure 9.15). The cylinders tend to stain strongly with the PAS-D technique and tumor-related mucin labels with mucicarmine, colloidal iron, and alcian blue. Adenoid cystic carcinoma is reported to have a more favorable prognosis than other forms of Bartholin gland carcinoma (50,51).

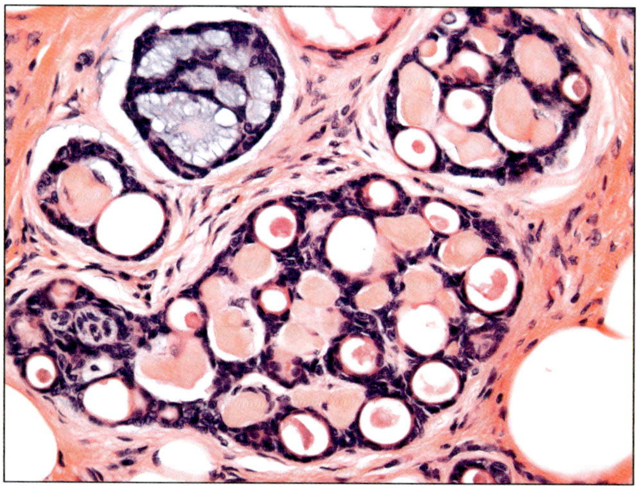

FIGURE 9.15: ***Hyaline cylinders of basement membrane material are characteristic of vulvar adenoid cystic carcinoma, as seen in other anatomic sites, and can be highlighted by PAS stain.***

Granular Cell Tumor: Granular cell tumors can arise in the skin, subcutaneous tissue, or submucosa in nearly any anatomic site, including the vulva (52,53). It typically presents as a small painless mass.

Microscopically, the tumor cells in this lesion grow in nests, sheets, or trabecula. The overlying epidermis frequently shows pseudoepitheliomatous hyperplasia. The polygonal or fusiform neoplastic cells have small bland nuclei and abundant granular eosinophilic cytoplasm (Figure 9.16). Electron microscopic studies have shown that the latter feature reflects the presence of innumerable phagolysosomes. The cytoplasm stains strongly with the PAS-D technique and also may be labeled bright red with the Masson trichrome stain (Figure 9.17). Adjunctive confirmatory diagnostic findings include immunoreactivity for S100 protein, neuron-specific enolase, inhibin, calretinin, and NK1-C3 (54,55).

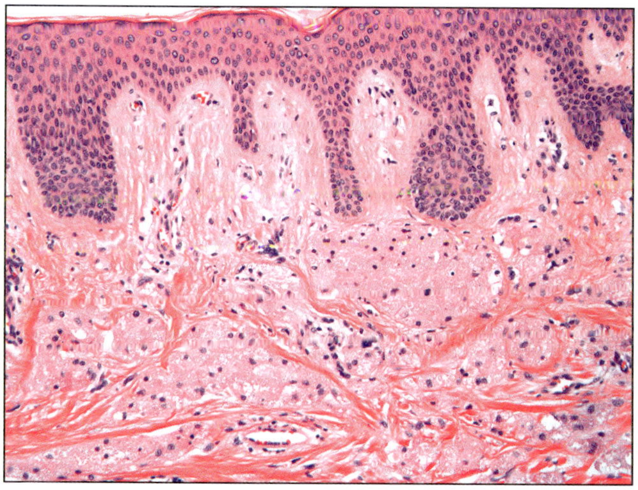

FIGURE 9.16: ***Granular cell tumor of the vulva consists of submucosal aggregates of bland polygonal cells with granular pink cytoplasm.***

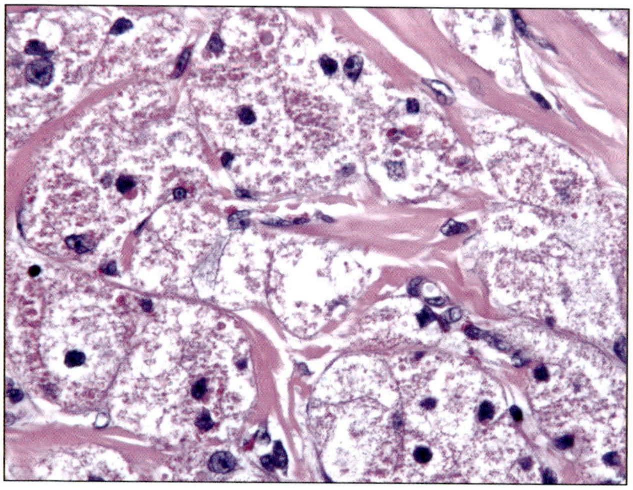

FIGURE 9.17: ***The granular cytoplasm of granular cell tumor is highlighted by PAS-D stain.***

The differential diagnosis includes various histiocytic and granulomatous processes, including infections, malacoplakia, and sarcoidosis. Associated pseudoepitheliomatous epidermal hyperplasia can be mistaken for carcinoma if the biopsy is superficial.

VAGINAL LESIONS

Vaginal biopsies are relatively rarely submitted for histopathologic evaluation. Infectious diseases of the vagina are common, including bacterial infections, fungal infections with *Candida*, and infestations with *Trichomonas*, but these are almost invariably diagnosed clinically. The most common conditions for which biopsies are obtained are the developmental abnormality known as adenosis, suspected neoplasms, and nonneoplastic polyps. Most tumors of the vagina are squamous cell neoplasms, namely, squamous cell carcinoma and its precursor, vaginal intraepithelial neoplasia. Histochemical stains are again relatively unhelpful in characterizing squamous cell lesions, but they do play a role in evaluating selected nonsquamous tumors and tumorlike conditions.

Vaginal Adenosis and Clear Cell Carcinoma

The normal vaginal mucosa comprises a stratified squamous epithelium and a lamina propria that contains blood and lymphatic vessels, nerves, and connective tissue. In vaginal adenosis, the vaginal epithelium is focally or extensively altered by the presence of columnar epithelium or its mucinous products (56). Adenosis has been seen most frequently in women who had in utero exposure to diethylstilbestrol (DES), and, in those individuals, it appears to represent a form of malformation. Adenosis is also occasionally found in women with no history of DES exposure. Clinically, vaginal adenosis may appear as red velvety mucosal patches or as white areas, typically involving the upper vagina.

Microscopically, the vaginal mucosa in adenosis focally or extensively comprises benign glandular epithelium. The constituent cells in such foci generally resemble those of the endocervix. They are columnar with basal round to oval nuclei and abundant mucinous cytoplasm (Figure 9.18). Glands are also present in the lamina propria in some cases. Although the glandular cells are typically of the endocervical type, they can also be endometrial or ciliated tubal-type epithelium, or mixtures thereof. As in the cervical transformation zone, the glandular epithelium in adenosis undergoes squamous metaplasia, wherein the number of glandular cells progressively diminishes.

Microglandular hyperplasia (MGH) may also occur in areas involved by adenosis. In those instances, histochemical stains for mucin help to identify remaining patches of glandular epithelium or mucin droplets within metaplastic or hyperplastic epithelium. Mucicarmine or PAS-D methods are useful in that regard (Figure 9.19), and positivity with either of them confirms the presence of adenosis. Other glandular lesions that must be considered in differential diagnosis include simple Müllerian cysts, mesonephric cysts (Gartner's duct cysts), Bartholin duct cysts, and endometriosis. With the exception of Bartholin duct cysts, these conditions generally lack the intense staining for mucin that characterizes adenosis. In this regard, it is pertinent to know that Bartholin duct cysts do not involve the vaginal mucosa or lamina propria.

A rare form of vaginal adenocarcinoma—clear cell carcinoma—occurs in young women who have adenosis and prior DES exposure. Clear cell carcinoma of the vagina is also seen, rarely, in non-DES–exposed patients. This tumor is histologically similar to clear cell adenocarcinoma arising in other gynecologic sites. Clear "hobnail"-shaped

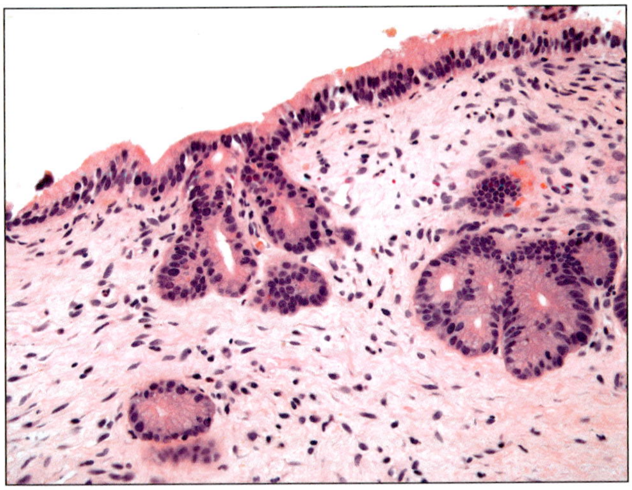

***FIGURE 9.18:** In vaginal adenosis, the mucosa is replaced by benign columnar cells containing mucin; superficial glands may also be present.*

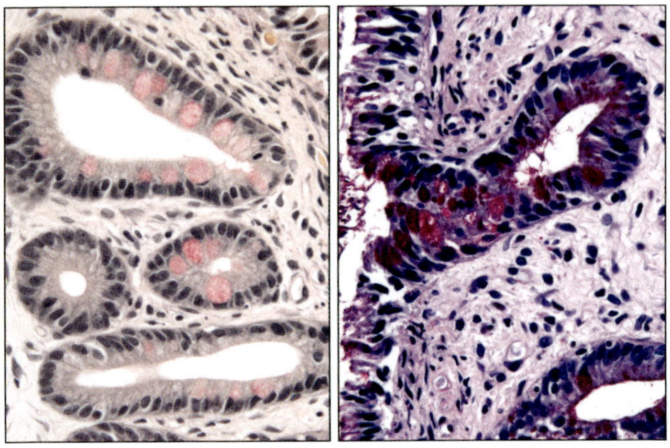

***FIGURE 9.19:** The glandular epithelium of vaginal adenosis resembles that of the endocervix and the mucinous contents are demonstrated by mucicarmine (left) and PAS (right) stains.*

tumor cells grow in sheets, tubules, or small cysts, or mantle micropapillae. Although clear cell carcinoma is linked to adenosis, histologic transitions between the two are rarely demonstrated. Also, the cells of adenosis are typically mucin positive, but those in clear cell carcinoma are not. The tumor cells in clear cell carcinoma have lucent cytoplasm because of their content of glycogen. Thus, positivity with the undigested PAS method is typical of the latter tumor. The extent and intensity of PAS staining depend on optimal fixation of the tissue.

Polypoid Mesenchymal Lesions

A variety of polypoid and papillary mesenchymal proliferations occur in the vulva, vagina, and cervix (57). Vulvovaginal fibroepithelial polyps are the most common of them. They have a loose connective tissue core and are covered by bland but sometimes reactive stratified squamous epithelium. The core of the polyps contains fibroblasts, collagen, amorphous ground substance, and prominent blood vessels. In some polyps, the constituent spindle cells have rounded or stellate nuclei. Polyps pose a diagnostic problem only when the stroma is hypercellular or it contains atypical fibroblasts or numerous mitotic figures. Such polyps can be difficult to distinguish from rhabdomyosarcoma (RMS) or other mesenchymal tumors. Features that help to separate atypical polyps from sarcomas include the circumscribed nature of the former lesions, the absence of a subsurface "cambium" layer, the lack of tumor cells with cytoplasmic cross-striations, the absence of cells with prominent eosinophilic cytoplasm, a paucity of mitotic figures, and the fibroblastic or myofibroblastic appearance of stromal cells.

Rarely, vaginal polyps demonstrate skeletal muscle differentiation. The two types of skeletal muscle tumors that occur in the vagina are rhabdomyoma and RMS. Rhabdomyoma is a benign polypoid vaginal or vulvar tumor that is seen in middle-aged women (58–60). It may be asymptomatic, or present with vaginal bleeding or a mass. Microscopically, a covering of stratified squamous epithelium is observed over aggregates of round, strap-, or racquet-shaped cells with abundant eosinophilic cytoplasm (61). One can sometimes identify longitudinal fibrils or cross-striations in such cells (Figure 9.20) (62). A phosphotungstic acid–hematoxylin (PTAH) stain can be helpful in delineating cross-striations that are not readily seen in H&E-stained sections of rhabdomyoma (Figure 9.21) (63,64). The striations are dark blue and the background cytoplasm stains dark red. Electron microscopic studies show that the cells are surrounded by a basal lamina and the cytoplasm contains actin and myosin filaments with occasional well-formed sarcomeres. The latter structures correspond with the cross-striations seen in H&E- or PTAH-stained sections (62,65) Large numbers of differentiated muscle cells are seen in genital rhabdomyoma, accounting for a level of PTAH labeling, which far exceeds that expected in RMSs.

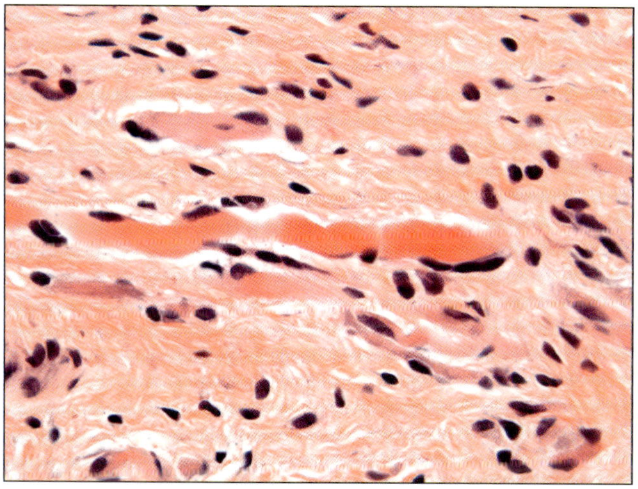

FIGURE 9.20: ***Cross-striations can sometimes be identified by H&E staining in skeletal muscle strap cells of rhabdomyoma and RMS.***

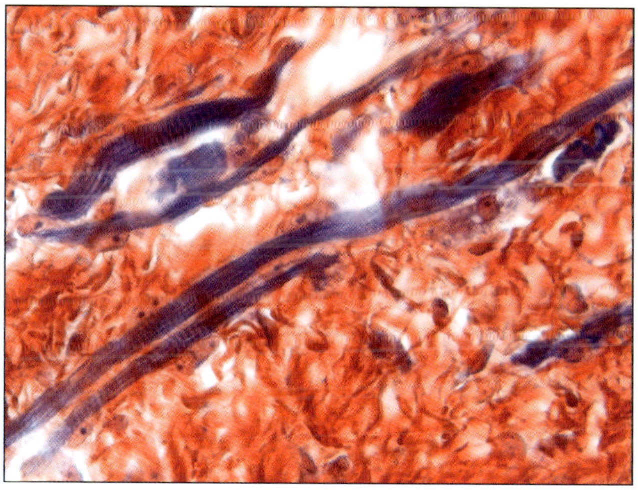

FIGURE 9.21: ***PTAH staining of rhabdomyomatous lesions facilitates recognition of cross-striations, indicating skeletal muscle differentiation.***

Approximately 3% of all RMSs arise in the lower female genital tract, with the vagina being the single most common site. Most often, RMS presents as a vaginal tumor in a young child, with an average age of around 2 years (66,67). The clinical presentation is with vaginal bleeding or discharge, or with a mass protruding from the introitus (sarcoma botryoides). RMS can occur in older patients as well (68). Vulvar RMS may be either embryonal or alveolar in nature, with the latter variant predominating (66,67,69,70) Cross-striations are occasionally seen in the cytoplasm of the differentiated tumor cells in RMSs, regardless of subtype, and staining with PTAH can make them more obvious. A well-developed alveolar pattern is often difficult to identify in alveolar RMS, and tumors with only focal alveolar differentiation, gland-like spaces, or solid sheets of small round tumor cells also qualify as alveolar neoplasms. Reticulin or Masson trichrome stains may be helpful in accentuating the alveolar architecture. Most RMSs also contain glycogen, as reflected by PAS positivity and diastase sensitivity.

The presence of *any* alveolar component in RMS mandates classification of the entire lesion as an alveolar RMS. A distinctive chromosomal translocation [t(2:13)] has been identified in this neoplasm, and its identification by cytogenetic or molecular testing can facilitate the diagnosis. Immunostains for muscle-related proteins are necessary for the definitive diagnosis of most RMSs unless cross-striations are present (71–73).

UTERINE CERVICAL LESIONS

As true in the vulva and vagina, most cervical biopsies are performed for squamous proliferations, including squamous intraepithelial lesions [cervical intraepithelial neoplasia (dysplasia)] or squamous cell carcinoma, and these do not usually require special histochemical stains for diagnosis. However, glandular epithelia are also present in the cervix, and adjunctive stains are indeed helpful in evaluating proliferations of these elements.

Microglandular Hyperplasia

MGH is a benign proliferative endocervical lesion. It usually occurs in premenopausal women, often as an incidental microscopic finding, but sometimes as an endocervical polyp. The hyperplastic glands are crowded, forming an irregular cribriform or mesh-like pattern with variably sized lumina; microcysts may be seen as well (Figure 9.22) (74). Intraluminal mucin and inflammatory cells, especially neutrophils, are typically present. The differential diagnosis includes clear cell carcinoma and mucinous

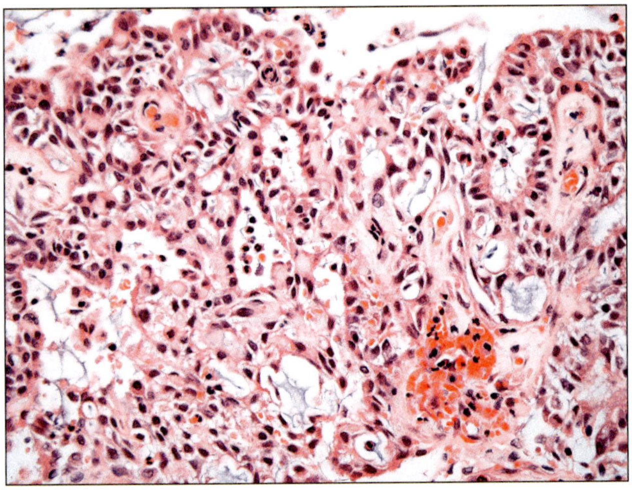

FIGURE 9.22: ***A mesh-like pattern of microlumens characterizes endocervical MGH. Squamous metaplasia and neutrophils are present to varying degrees. Residual patchy mucin, both intraluminal and intracytoplasmic, can be demonstrated by mucicarmine.***

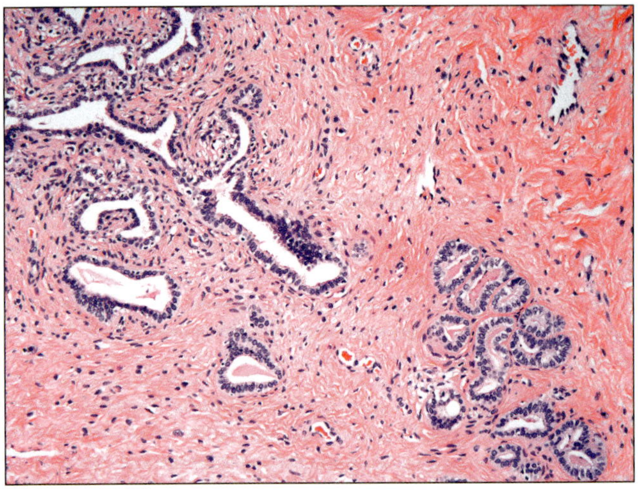

FIGURE 9.23: ***Remnant mesonephric glands (lower right) and mesonephric duct (upper left) can be mistaken for adenocarcinoma in the deep cervical wall.***

endocervical adenocarcinoma. In regard to both of these possibilities, mucin stains are helpful. The mucicarmine and PAS-D methods label intraluminal mucin and intracellular mucin in some of the hyperplastic glands of MGH. Typically, the staining is focal and globular. In endocervical adenocarcinoma, in contrast, mucin reactivity is more diffuse because it is not interrupted by areas of squamous metaplasia as seen in MGH.

Atypical forms of MGH show a sheet-like cellular growth pattern, or contain signet-ring–like cells, or have a pseudoinfiltrative appearance, or hyalinized stroma, or show nuclear atypia and mitotic activity (75). Hence, such variants are apt to be mistaken for clear cell carcinoma. PAS stains are particularly helpful in this specific differential diagnosis. Undigested PAS highlights the glycogen-rich cytoplasm in clear cell carcinoma, and PAS-D labels the intracytoplasmic globules that are often present in MGH (76). Thus, positivity with PAS and a lack of staining with PAS-D favors the diagnosis of clear cell carcinoma, whereas reactivity with both PAS and PAS-D supports the interpretation of MGH.

Mesonephric Remnants and Hyperplasia

Mesonephric remnants are occasionally seen in the endocervical stroma. They are generally located in the lateral cervix, at the 3 and 9 o'clock positions, and are most likely to be found in large conization specimens, but not in samples that are more limited in volume. Findings range from a few mesonephric tubules, with or without accompanying ducts, to striking lobular or diffuse tubular hyperplasia. Mesonephric elements comprise collections of small tubules that contain distinctive eosinophilic, colloid-like material (Figure 9.23). Constituent epithelium is cuboidal and cytologically bland, with scanty cytoplasm.

Histochemical stains for mucin assist in the recognition of mesonephric elements. The eosinophilic intraluminal material is positive with the PAS-D and mucicarmine techniques, but the cytoplasm is negative (Figure 9.24). This finding helps to distinguish mesonephric tubular cells from those of endocervical glands, which show cytoplasmic labeling with both mucicarmine and PAS-D.

Another common diagnostic problem is to separate hyperplastic mesonephric remnants from "minimal deviation" adenocarcinoma (MDA) of the cervix. Histochemical stains for mucin also resolve this problem because tumor cells in MDA almost invariably show cytoplasmic mucin positivity. PAS also helps to distinguish mesonephric tubules from clear cell carcinoma, in which the glycogen-rich cytoplasm is diastase sensitive (76,77). Moreover, mesonephric remnants are cytologically bland and contain no mitoses.

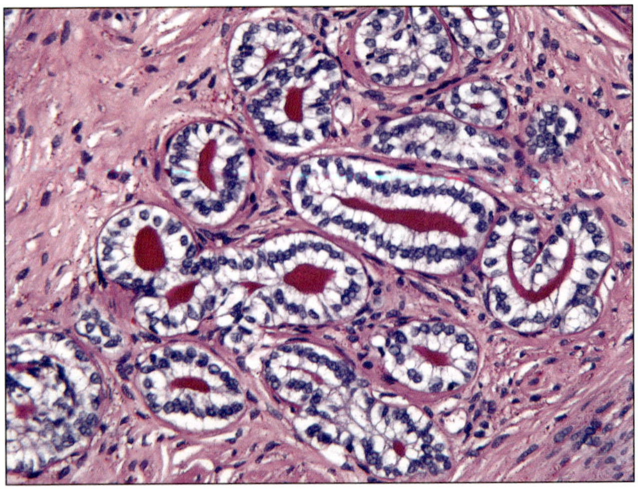

FIGURE 9.24: ***Mesonephric glands contain a brightly eosinophilic intraluminal deposit that stains strongly with PAS.***

Intraepithelial Lesions

Several intraepithelial proliferations occur in the cervix. Squamous intraepithelial lesions—also known as cervical intraepithelial neoplasia or squamous dysplasia—are by far the most common. Special histochemical methods do not play a role in the diagnosis of squamous intraepithelial lesions; if adjunctive stains are needed, they are typically immunohistochemical in nature. Glandular intraepithelial lesions are less well defined in the cervix, except, perhaps, for adenocarcinoma in situ. Lesions that exhibit lesser degrees of atypia have been classified as "glandular atypia," but diagnostic criteria for this category are vague. Adenocarcinoma in situ and glandular dysplasia show a variable loss of cytoplasmic mucin; however, stains for this substance are not helpful diagnostically because the decrement in mucin content is so variable. Recently, an intermediate type of intraepithelial neoplasia, termed "stratified mucin-producing intraepithelial lesion" (SMILE), has been described (78). In this disorder, the cells are stratified, as in squamous intraepithelial lesions, but some have mucinous cytoplasm and replace the full thickness of the epithelium. The mucicarmine and PAS-D procedures are essential to identifying the mucinous differentiation in SMILE. Some reports have suggested that SMILE may be a precursor of invasive cervical adenosquamous carcinoma.

Invasive Cervical Carcinoma

A variety of invasive carcinomas may be encountered in the cervix, including squamous cell carcinoma, adenocarcinoma, adenosquamous carcinoma, and neuroendocrine carcinoma. The diagnosis of such lesions is usually made definitively with H&E-stained slides, but special stains may occasionally be required to confirm a diagnosis of neuroendocrine carcinoma. Histochemical procedures for mucin staining are also helpful in selected cases. Some poorly differentiated adenocarcinomas show extensively solid growth, and, as a result, they can be difficult to separate from squamous cell carcinomas. Positive labeling of the cytoplasm or luminal contents with the mucicarmine or PAS-D procedure supports the presence of glandular differentiation. The same comments apply to adenosquamous carcinomas. Some predominantly squamous cell carcinomas contain scattered mucin-positive cells as well. These tumors have been variably interpreted as adenosquamous carcinomas or squamous cell neoplasms with mucin secretion (79). They are more prevalent than classical adenosquamous carcinomas; in one series of 196 invasive cervical carcinomas, there were 4% adenosquamous carcinomas and 20% squamous carcinomas with mucin secretion (80). Squamous cell carcinomas with mucin secretion may occur in younger women and have a worse prognosis than "pure" squamous cell carcinomas (81), although some studies have demurred on these points (80,82,83). In our opinion, mucin-positive areas of such tumors should show definite glandular differentiation and cytologic atypia to qualify for the diagnosis of adenosquamous carcinoma. If these stipulations are not met, it is best to classify the neoplasm as a squamous cell carcinoma.

LESIONS OF THE UTERINE CORPUS

A wide range of pathologic processes affect the endometrium, but histochemical stains are useful in evaluating only a few of them. These include inflammatory and infectious conditions, certain forms of endometrial metaplasia, and some types of endometrial carcinoma. Histochemical evaluations likewise have limited use in evaluating myometrial disorders.

Chronic Endometritis

Chronic plasma cell endometritis is often associated with pelvic inflammatory disease, particularly infections with *Chlamydia trachomatis* (84,85). Polyps, leiomyomas, and intrauterine devices (IUDs) are found in some patients with this disorder in the absence of infection. Histologic features of chronic endometritis include disordered glandular and stromal maturation, an edematous spindle cell stroma, endometrial breakdown, and plasma cells in the stroma. Lymphocytes and rare neutrophils may also be seen. Endometrial plasma cells can be difficult to identify because they are often present in small numbers and may resemble endometrial stromal cells (Figure 9.25). The methyl green-pyronine (MGP) stain can be helpful in accentuating plasma cells, by producing bright magenta cytoplasmic staining in those elements (Figure 9.26) (86). PAS-D stains also may label plasma cells in some cases.

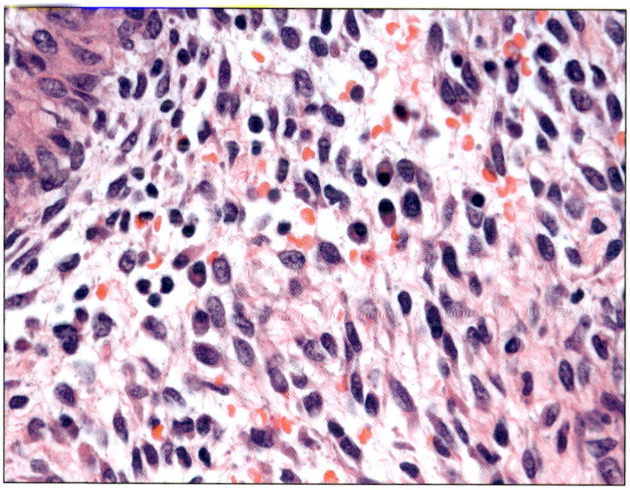

FIGURE 9.25: ***In chronic endometritis, plasma cells may be difficult to find because they may be infrequent, inconspicuous, or resemble endometrial stroma cells and lymphocytes.***

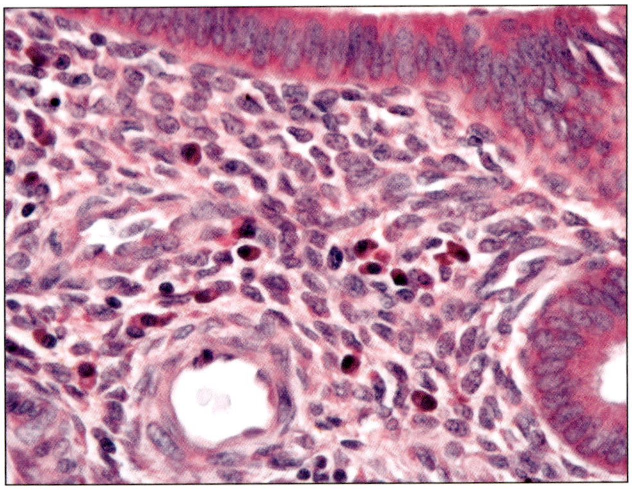

***FIGURE 9.26:** MGP stains the cytoplasm of plasma cells bright magenta, facilitating their distinction from endometrial stromal cells and lymphocytes.*

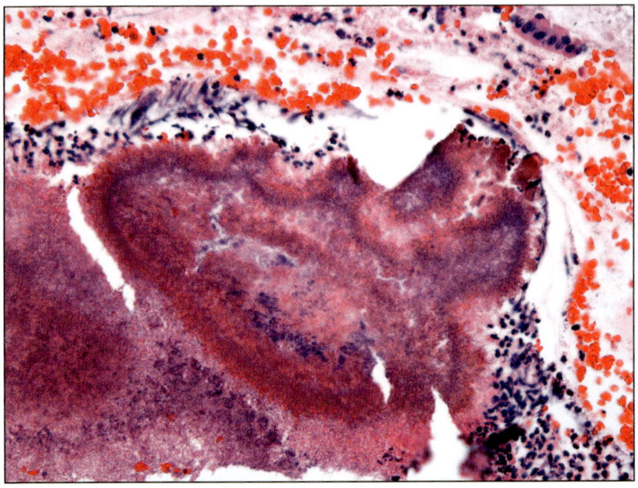

***FIGURE 9.27:** IUD-related endometrial actinomycosis is characterized by so-called sulfur granules, formed by protein precipitates on the filamentous bacteria, which are only seen on Gram or silver stain.*

Actinomycosis

Actinomyces are gram-positive filamentous bacteria. Long-term use of an IUD can lead to colonization or infection by *Actinomyces* (87–89). In some instances, the endometrium has a normal appearance, but tissue taken from the IUD itself contains aggregates of *Actinomyces* organisms that sometimes form classic "sulfur granules." In other cases, such granules are present in endometrial curettings, associated with acute or chronic endometritis. Histochemical stains show filamentous beaded organisms in a background of amorphous material. The organisms can be identified well with tissue Gram stains and the GMS method. Other silver impregnation procedures, including the Warthin-Starry stain, the Dieterle stain, and the Steiner procedure, can also be used. The organisms do not stain with PAS. Protein often precipitates on the filaments at the edges of the sulfur granules, resulting in thick eosinophilic radiations known as the "Splendore-Hoeppli phenomenon" (Figure 9.27). This proteinaceous material is PAS positive.

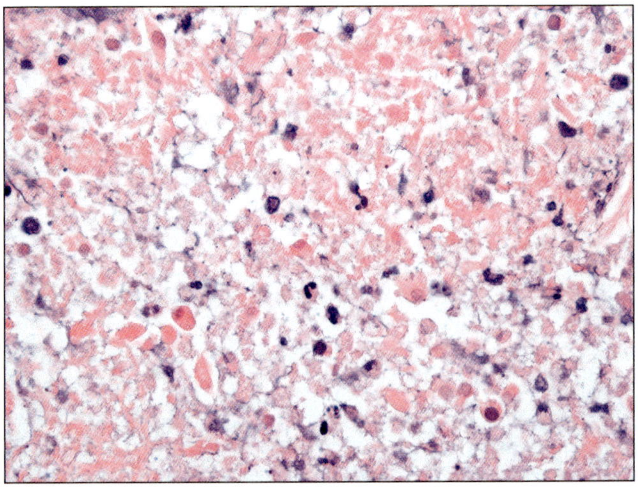

***FIGURE 9.28:** Endometrial TB, a rare entity in the United States, can sometimes present with pyometria in the absence of granulomatous inflammation.*

Tuberculosis

Granulomatous inflammation of the endometrium is usually a manifestation of systemic tuberculosis (TB). It is rare in the United States and Europe, and is usually encountered in developing countries (90). The endometrium is the second most common site of involvement for TB in the female genital tract, after the fallopian tubes (91). Indeed, endometrial involvement is felt to result from secondary infection by drainage from an infected tube. Cultures are often not initially performed in such cases because the clinician does not anticipate the diagnosis. Thus, pathologic detection is crucial. Endometrial involvement by TB can be focal or extensive. In some cases, well-formed epithelioid granulomas are present, but others show only vague granulomas or the image of chronic endometritis or pyometria without granulomatous inflammation (Figure 9.28). All of these histologic scenarios justify the application of histochemical stains for acid-fast bacteria (Figure 9.29), especially in immunocompromised patients. Organisms are most likely to be identified in cases where numerous caseating granulomas are present. Standard methods such as the Ziehl-Neelsen stain label the bacilli red with a blue background. A Fite stain is recommended if an "atypical" tuberculous organism is a possibility.

Endometrial Metaplasias

Endometrial metaplasias feature the development of epithelium that is not native to the uterine corpus, within

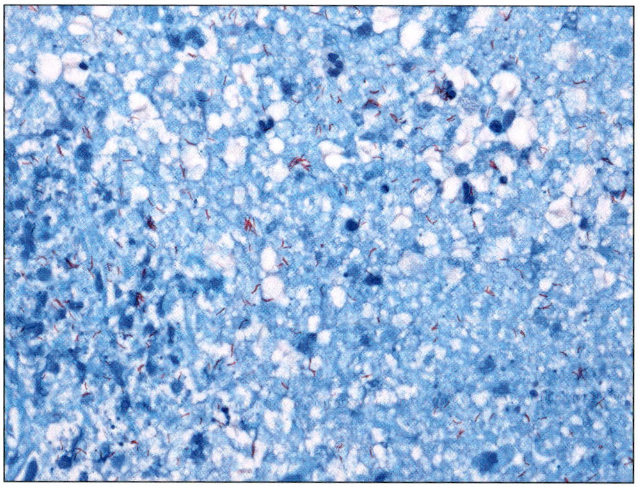

FIGURE 9.29: Acid-fast staining demonstrates numerous mycobacterial organisms in endometrial necroinflammatory debris, despite the absence of granulomatous inflammation.

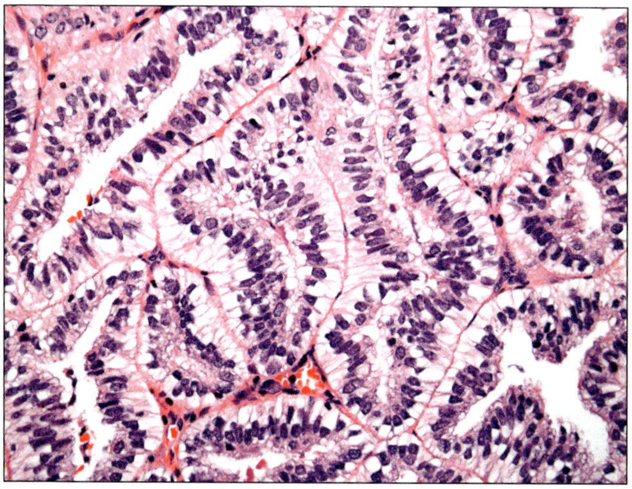

FIGURE 9.30: Abundant subnuclear and supranuclear vacuoles characterize secretory change in endometrioid adenocarcinoma.

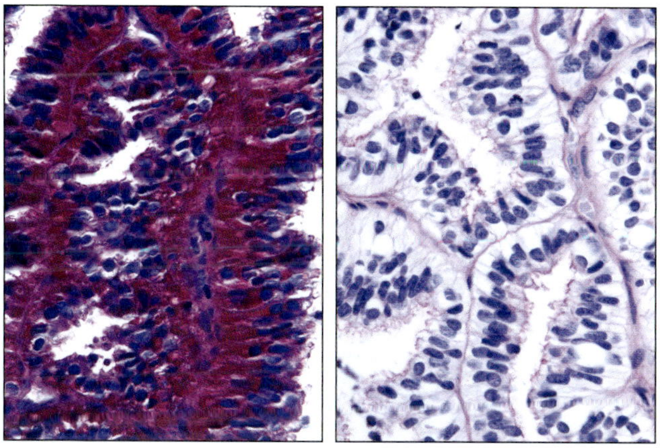

FIGURE 9.31: Secretory change in endometrioid adenocarcinoma is due to glycogen, which stains with PAS (left) but is diastase sensitive (right). In contrast to mucinous adenocarcinoma, mucicarmine and alcian blue stains are negative.

endometrial glands. Metaplasias are most often seen in proliferative and hyperplastic endometrium; they include squamous metaplasia, eosinophilic metaplasia, ciliated cell metaplasia, and mucinous metaplasia. Among these, histochemical methods are most helpful in the recognition of mucinous metaplasia. The constituent cells are columnar with basal nuclei and basophilic or clear cytoplasm, like endocervical epithelium. The mucicarmine and PAS-D procedures strongly label the cytoplasm of the mucinous elements. Differential diagnosis centers on secretory change in the endometrium, which is mucin negative. Mucinous changes also can be observed in endometrial hyperplasia and in endometrioid adenocarcinoma. Mucinous metaplasia is, of course, nonmalignant, but complex mucinous proliferations of the endometrium are felt to have an association with carcinoma (92,93). When >50% of an endometrioid endometrial adenocarcinoma shows mucinous differentiation, the tumor can be classified as mucinous adenocarcinoma (94); the prognosis of this lesion is felt to be similar to that of nonmucinous endometrioid carcinoma (95).

Endometrial Adenocarcinoma

For the most part, special histochemical stains are not necessary to diagnose endometrial carcinomas. Endometrioid adenocarcinoma is the most common of these lesions, and serous carcinoma is the second most frequent. As mentioned above, histochemical stains contribute to the recognition of mucinous adenocarcinoma, clear cell adenocarcinoma, and some cases of neuroendocrine carcinoma.

As discussed above, mucinous differentiation is common in cases of endometrioid adenocarcinoma. The diagnosis can be confirmed with the mucicarmine, PAS-D, or alcian blue method (95). Differential diagnosis centers on the possibilities of clear cell and secretory adenocarcinomas of the endometrium. The cytoplasm of these tumors is mucin negative (Figures 9.30 and 9.31). The distinction of mucinous endometrial carcinoma from endocervical adenocarcinoma is more difficult; the tumor cells in both of these tumor types exhibit similar mucin profiles, with comparable histochemical staining characteristics (96).

Clear cell endometrial carcinomas resemble tumors with the same name that are seen elsewhere in the female genital tract. It occurs in elderly women and is generally viewed as an unfavorable histologic type (97). PAS staining tends to be positive, with loss of reactivity after prior diastase treatment. The PAS-D stain is also helpful in identifying two additional characteristics—hyaline bodies, which are reactive using both the PAS and PAS-D procedures (Figures 9.32 and 9.33), and waxy eosinophilic stroma between tumoral glands and papillary stromal cores, which can also be labeled by the same two methods (Figures 9.34 and 9.35).

Neuroendocrine carcinomas are among the rarest of endometrial malignancies, but they are important because

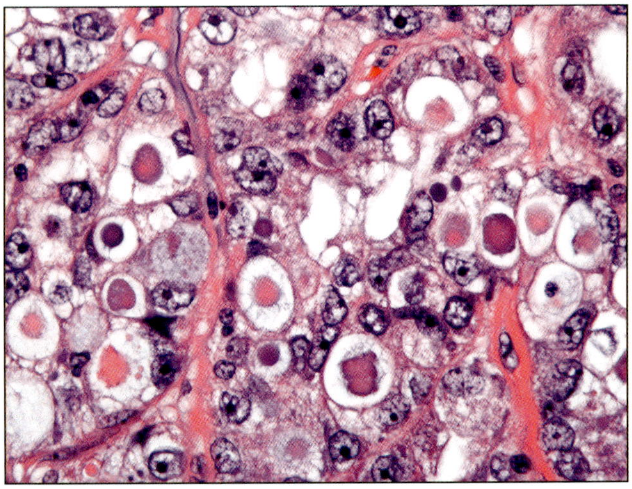

FIGURE 9.32: *Intracytoplasmic and extracellular hyaline globules can be found in endometrial and ovarian clear cell carcinoma (pictured) as well as in yolk sac tumor.*

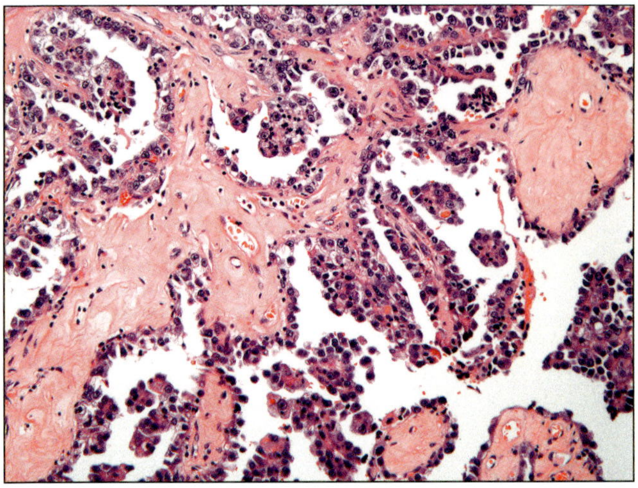

FIGURE 9.34: *Stromal hyalinization is a helpful feature seen in the papillary variant of clear cell carcinoma.*

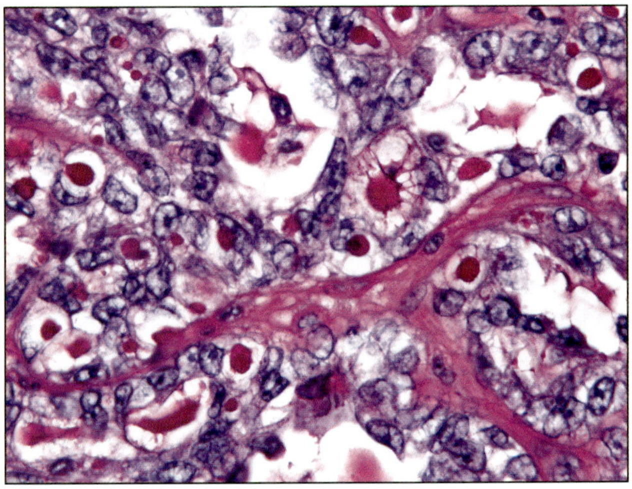

FIGURE 9.33: *Hyaline globules of clear cell carcinoma are PAS-D positive, as are those in yolk sac tumor.*

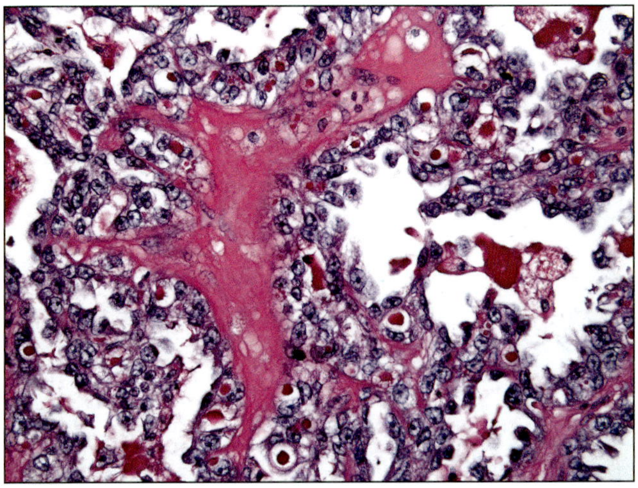

FIGURE 9.35: *PAS highlights stromal hyalinization of clear cell carcinoma.*

of their aggressivity (98). Some of these tumors resemble small cell carcinoma of the lung (99,100). In these lesions, the tumor cells are small to medium sized with hyperchromatic molded nuclei, inconspicuous nucleoli, and scanty cytoplasm. Mitotic activity is brisk with frequent apoptotic bodies and necrosis. Other neuroendocrine carcinomas comprise larger cells with more vesicular nuclei and discernible nucleoli (large cell neuroendocrine carcinoma). In many cases, cytoplasmic neurosecretory granules in neuroendocrine carcinomas can be labeled with histochemical argyrophil stains, such as the Grimelius, Sevier-Munger, or Churukian-Schenk procedure (100).

Mesenchymal Tumors

Smooth muscle tumors are the commonest mesenchymal neoplasms of the uterus and most are leiomyomas. They undergo various degenerative alterations, including hyaline degeneration, edema, and hemorrhage. One unusual secondary event, termed diffuse or perinodular hydropic degeneration, results in intratumoral nodules of cellular smooth muscle that are surrounded by pale stroma containing collagen fibers and scattered smooth muscle cells (101). This pattern can cause diagnostic difficulties because it can mimic intravenous leiomyomatosis, an endometrial stromal neoplasm, or myxoid change (Figures 9.36 and 9.37). Myxoid smooth muscle tumors are uncommon, but some of them behave as leiomyosarcomas (102). Histochemical stains for acid mucopolysaccharides—including the alcian blue method and the colloidal iron procedure—are helpful in this setting. The mucoid material in the matrix of myxoid smooth muscle tumors is labeled by such techniques, but areas of hydropic degeneration are not.

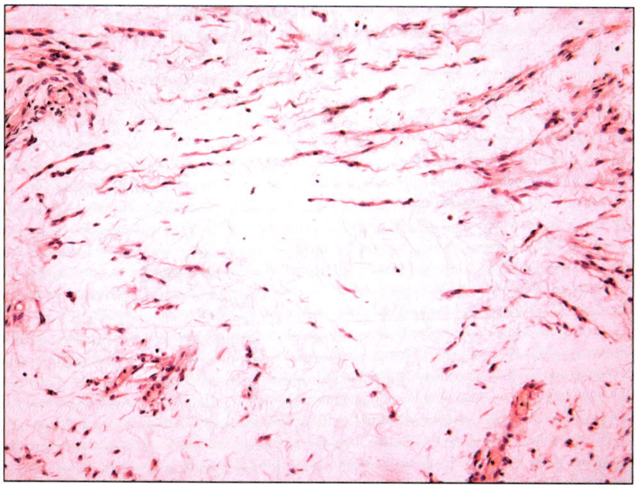

FIGURE 9.36: ***Hydropic change in a uterine leiomyoma can lead to edematous stromal changes that can mimic the stroma found in myxoid smooth muscle tumors.***

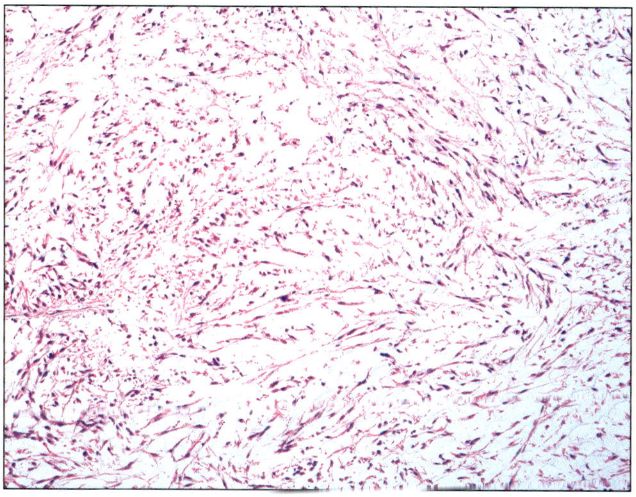

FIGURE 9.37: ***Myxoid leiomyosarcoma is a deceptively bland tumor cytologically but typically has infiltrative margins. Alcian blue staining (not shown) can highlight the myxoid stroma and help distinguish from the stroma of a hydropic leiomyoma, which will not stain.***

OVARIAN LESIONS

Somatic Ovarian Carcinomas

The major types of surface epithelial ovarian malignancies include serous, endometrioid, clear cell, mucinous, and transitional cell carcinomas and mixtures thereof. Histochemical stains are not commonly employed in evaluating such lesions, with the exception of the PAS and PAS-D methods in the diagnosis of clear cell carcinoma. These have been discussed above in the section on endometrial adenocarcinomas.

Germ Cell Tumors

The principal forms of ovarian germ cell tumor include teratoma, dysgerminoma, yolk sac carcinoma (YSC), embryonal carcinoma, and choriocarcinoma. As true of surface epithelial neoplasms, the diagnosis of most germ cell tumors is predicated primarily on morphologic findings. However, selected histochemical procedures merit consideration in this context.

Dysgerminoma is the most common nonteratomatous ovarian germ cell tumor, typically being seen in young women and children (103–105). This tumor usually demonstrates nests and sheets of large polygonal tumor cells with clear cytoplasm, round nuclei, and macronucleoli. The cellular nests are bounded by thin fibrous septa and a lymphoplasmacytic infiltrate that may percolate among the tumor cells. The lucent cytoplasm reflects the presence of abundant glycogen, which can be labeled by the PAS procedure before diastase digestion. PAS-D stains are negative (Figure 9.38).

YSC is likewise encountered in young women and children (106–108). Most cases are associated with elevated serum alpha-fetoprotein levels (109). The microscopic growth patterns of YSC are diverse and often mixed, including reticular (microcystic), endodermal sinus like, "solid," polyvesicular-vitelline, and glandular forms. Schiller-Duval bodies, which are micropapillary structures surrounding central blood vessels, are virtually diagnostic of YSC. Histochemical stains can help to identify two additional features of this tumor—hyaline globules and hyalinized stroma. The globules are rounded, eosinophilic, glassy deposits that may be intracellular or extracellular; they are positive with both the PAS and PAS-D procedures (Figures 9.32 and 9.33). As mentioned previously, the cells of clear cell carcinomas may also contain hyaline globules with similar biochemical features; thus, caution is warranted in interpreting such findings. Hyalinized stroma is seen in the reticular and solid patterns of YSC; it is also PAS positive and can likewise be seen in clear cell carcinomas.

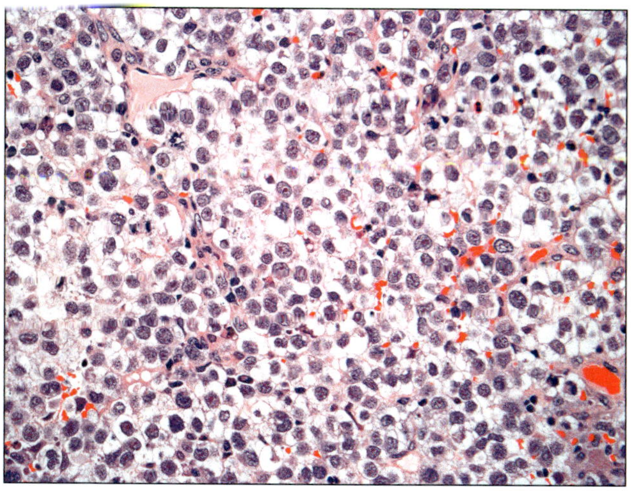

FIGURE 9.38: ***The clear cytoplasm of tumor cells in ovarian dysgerminoma is due to glycogen, confirmed by PAS-positive, diastase-sensitive staining (not shown).***

Sex Cord Stromal Tumors of the Ovary

Tumors that differentiate toward the ovarian sex cords or ovarian stroma are uncommon. Within this family of neoplasms, granulosa cell tumor (GCT) is the principal potentially malignant representative and fibroma-thecoma is the leading benign tumor (103,110). Histochemical methods can be applied to good advantage in selected cases.

GCTs have varied architectural patterns that may overlap with those of other sex cord neoplasms, epithelial tumors, neuroendocrine lesions, and even lymphomas or melanomas. Nuclear grooves and rosette-like "Call-Exner" bodies are helpful features in the diagnosis of GCTs, but their distinction from thecoma can sometimes be challenging because some GCTs have a prominent fibrothecomatous background. By convention, tumors containing >10% granulosa cells are considered to be GCTs (111,112). Reticulin (Wilder's, Snook's) stains can help to distinguish GCT from thecoma (Figure 9.39). Individual tumor cells are invested by reticulin-positive fibrils only in thecomas. In GCT, in contrast, reticulin may outline cellular nests but not individual cells. Cytoplasmic lipid in thecoma cells can be demonstrated with the oil red O method or the Sudan IV procedure, using fresh tissue. Generally, thecomas are distinguished from fibromas by the prominence of clear, lipid-rich tumor cells in the former of those lesions. Intracellular lipid is also seen in ovarian steroid cell tumors and in sclerosing stromal tumors, which have dissimilar morphologic images. Vacuolated cells in sclerosing stromal tumors can also be mistaken for metastatic adenocarcinoma cells, as seen in "Krukenberg" tumors. Nonetheless, neoplastic elements in metastatic adenocarcinomas in the ovaries are typically mucin positive, in contrast to those of sclerosing stromal tumors (113).

Ovarian sex cord tumors with annular tubules are rare; they may be incidental or found in patients with the Peutz-Jeghers syndrome. These tumors comprise arrays of annular tubules that are composed of sex cord–like cells. The

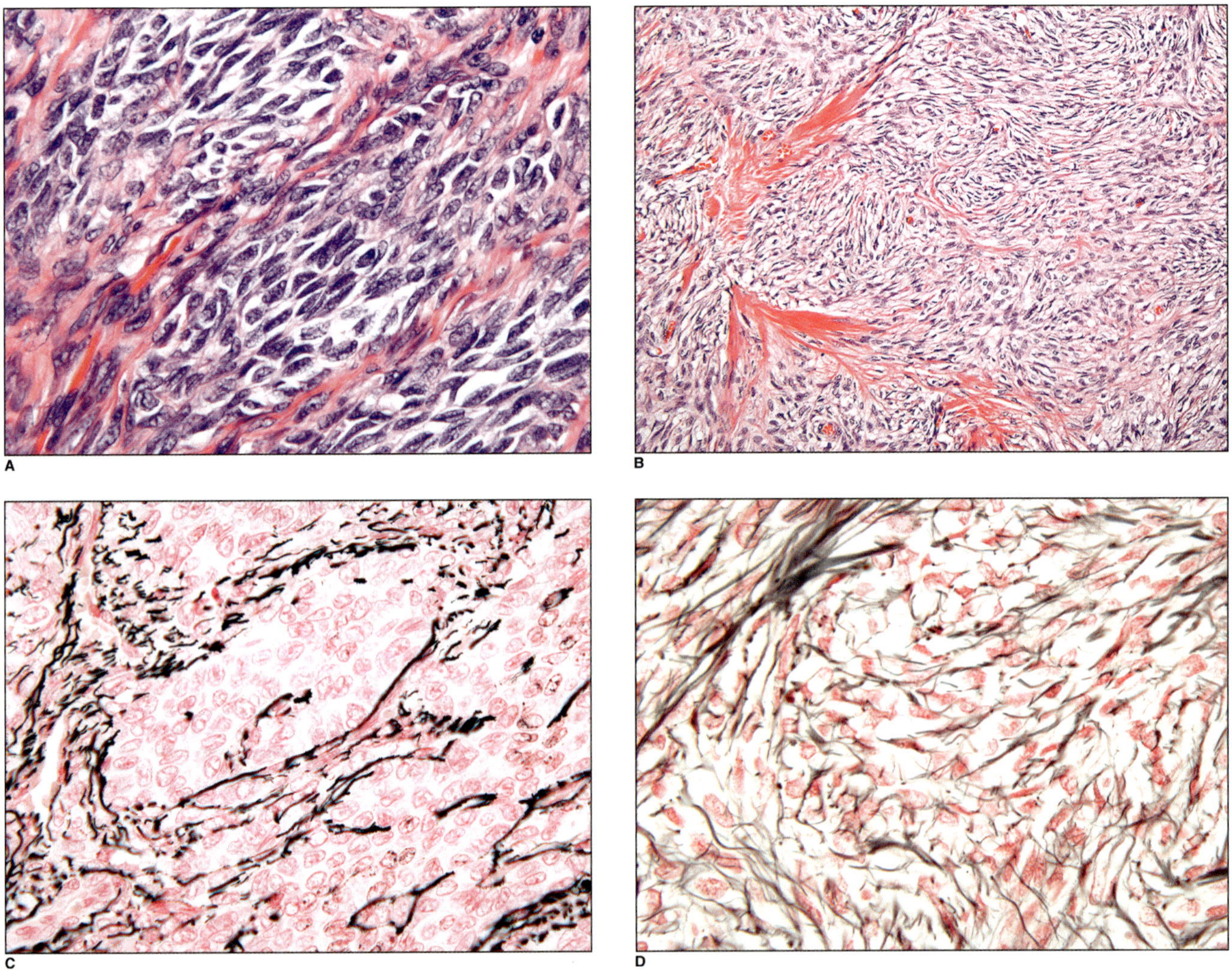

***FIGURE 9.39:** Ovarian GCTs (A) and thecomas (B) may share overlapping morphologic features. Reticulin staining demonstrates fibrils surrounding nests of tumor cells in a GCT (C), whereas there is pericellular staining around individual tumor cells of a thecoma (D).*

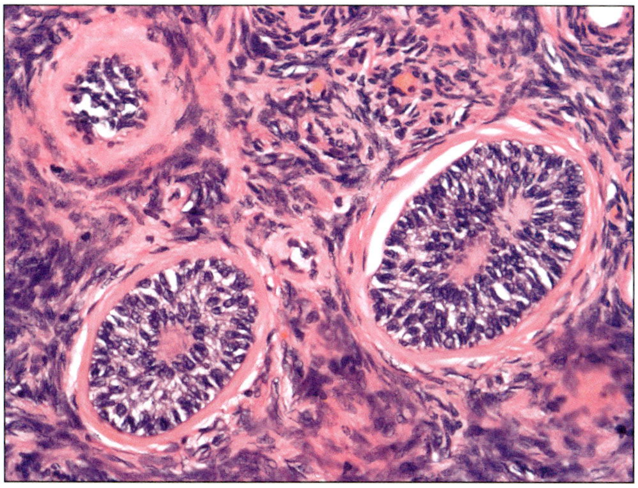

FIGURE 9.40: ***In this incidental microscopic ovarian sex cord tumor with annular tubules, the tubules contain an intraluminal eosinophilic deposit.***

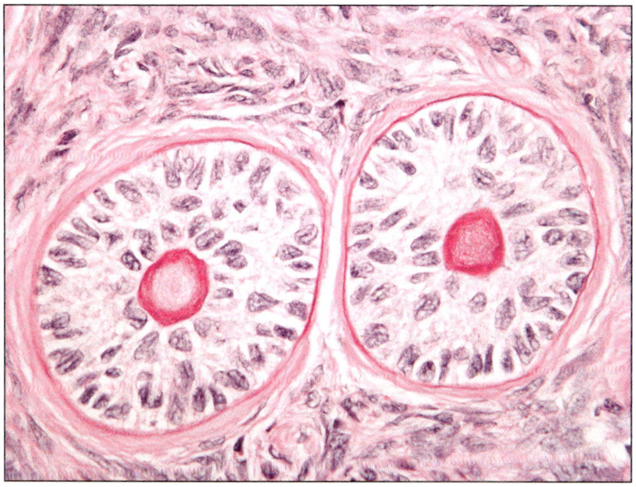

FIGURE 9.41: ***PAS highlights the intraluminal basement membrane deposits in an ovarian sex cord tumor with annular tubules.***

tubules contain intraluminal deposits of eosinophilic acellular basement membrane material, which is PAS positive and diastase resistant (Figures 9.40 and 9.41).

LESIONS OF THE FALLOPIAN TUBES

The fallopian tubes can be affected by inflammatory processes, infections, endometriosis, benign cysts, and neoplasms. Most of those entities have features which parallel those of their counterparts elsewhere in the gynecologic tract.

Salpingitis and Tuboovarian Abscesses

The organisms that most often cause acute salpingitis are *N. gonorrhoeae* and *C. trachomatis*, but a wide range of other microbes are potential pathogens as well. Their identification is best achieved through standard microbiologic culture procedures rather than with special stains. However, two exceptions concern granulomatous salpingitis and actinomycosis. Though it is uncommon in the United States, tuberculous salpingitis is a possibility when granulomatous salpingitis is present histologically. Ziehl-Neelsen and Fite stains are appropriate in that context. Actinomycosis can involve the ovarian adnexa and peritoneum, and, as stated earlier, it is often associated with the use of IUDs (114–117). *Actinomyces israelii*, a gram-positive filamentous bacterium, can cause tuboovarian abscesses in which "sulfur" granules are present (Figure 9.27). The histochemical characteristics of such structures have been discussed before in reference to endometrial infections.

Pregnancy-Associated Changes

The Arias-Stella reaction (ASR) is seen in the endometrium or endocervix of some pregnant women, especially with extrauterine pregnancies. It can also be identified sometimes in the fallopian tubes. The ASR is represented by the proliferation of polygonal cells with clear to eosinophilic cytoplasm (Figure 9.42). Their nuclei are bland but may be enlarged and contain smudged chromatin, occasionally leading to concern over malignancy, but there is typically no nuclear atypia or mitotic activity. The glycogen content of the cells in ASRs is confirmed with the PAS and PAS-D methods, which are positive and negative, respectively. The main differential diagnosis for these lesions centers on clear cell carcinoma. As discussed above, the histochemical attributes of this tumor are dissimilar; furthermore, the lack of nuclear atypia and mitotic activity in ASR also is discriminatory.

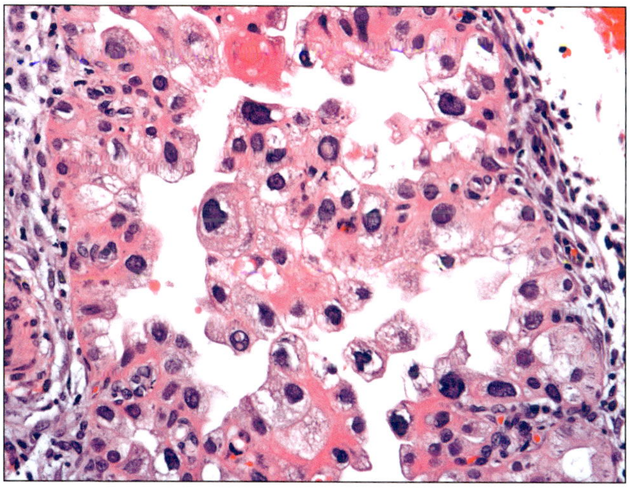

FIGURE 9.42: ***The ASR of pregnancy consists of a proliferation of large polygonal foamy or clear cells with abundant cytoplasm and enlarged nuclei but no atypia or conspicuous mitotic activity. The glycogen contents of the cytoplasm can be confirmed by PAS-positive, diastase-sensitive staining (not shown).***

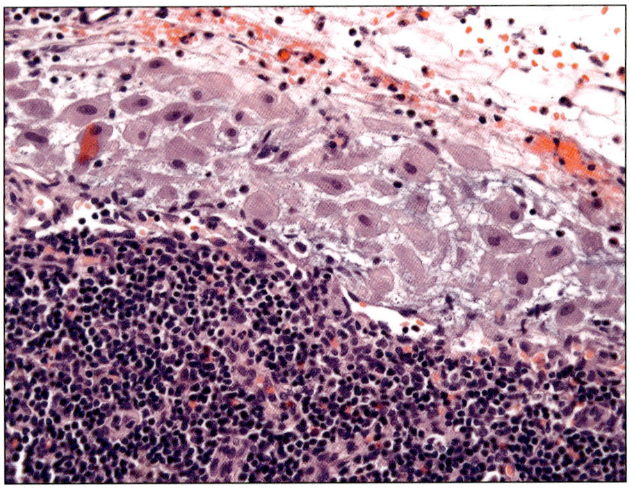

***FIGURE 9.43:** Ectopic decidua in pregnant or postpartum women can be found within lymph nodes (shown) or in the submesothelial layer of the peritoneum. Unlike its mimic, signet-ring cell carcinoma, the cells of this entity do not get marked by PAS stain.*

Ectopic decidual tissue in the fallopian tubes can also be associated with pregnancy. Like its counterparts in other gynecologic sites, as well as lymph node sinuses and the peritoneum, ectopic salpingeal decidua may be located in the submucosa or in subperitoneal tissue. It features large polygonal cells that usually assume a syncytial growth pattern but also may be represented by single or clustered cells (Figure 9.43). Rarely, mucinous cytoplasm can be seen, even with a pseudosignet-ring cell appearance. Unlike the case in true signet-ring cell adenocarcinomas, the PAS and PAS-D stains are negative in ectopic decidual tissue (113).

LESIONS OF THE PELVIC PERITONEUM

Mucicarminophilic Histiocytosis

An unusual reaction in the pelvic peritoneum is composed of histiocytes with abundant vacuolated mucicarminophilic cytoplasm. It also may involve lymph nodes. This entity, known as mucicarminophilic histiocytosis (MH), occurs in two particular clinical settings—patients who have received polyvinylpyrrolidine (an intravenous blood substitute) or have gotten applications of oxidized regenerated cellulose, a hemostatic agent used intraoperatively. Aggregates of the vacuolated histiocytes in MH may coalesce in the submesothelium or in lymph nodes (Figure 9.44). They are reactive with Best's mucicarmine technique, simulating the histochemical profile of signet-ring cell adenocarcinoma (Figure 9.45). Immunostaining is often necessary to exclude the latter possibility (118,119).

Ectopic Decidual Tissue

Ectopic decidua also may be encountered in the peritoneum and sinuses of pelvic lymph nodes. The pathologic

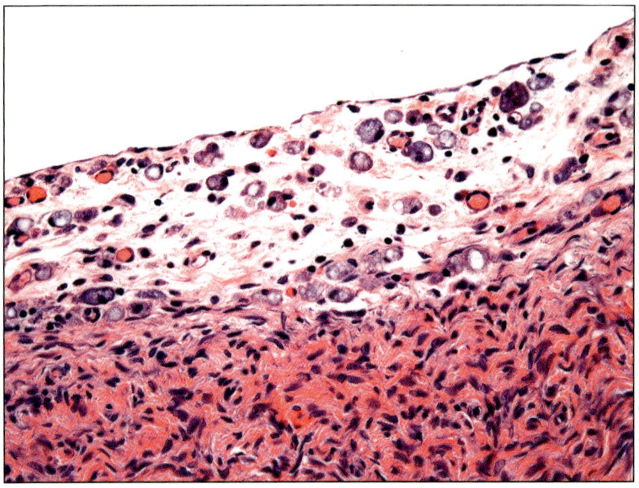

***FIGURE 9.44:** MH is a rare benign reaction in the peritoneal lining or ovarian surface that mimics signet-ring cell carcinoma.*

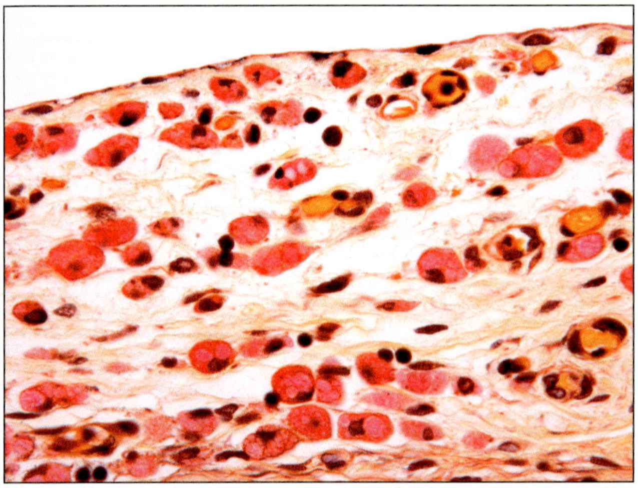

***FIGURE 9.45:** Staining of histiocytes with mucicarmine is a pitfall that can lead to misinterpretation as signet-ring cell carcinoma. Immunohistochemical stains may be needed to arrive at the correct diagnosis.*

features of this tissue parallel those described earlier in regard to other anatomic locations.

PLACENTAL LESIONS

Special histochemical procedures are only uncommonly needed in the interpretation of placental diseases. However, adjunctive stains can be useful in recognizing placental infections and meconium exposure.

Chorioamnionitis and Funisitis

Intrauterine infection can lead to placental inflammation. Most commonly, this process involves the placental membranes (so-called chorioamnionitis), but inflammation of the umbilical cord (funisitis), decidua, and chorionic villi is also possible. Acute chorioamnionitis is thought to be a major cause of preterm delivery and neonatal morbidity. It most often originates via ascending infection from the vagina and endocervical canal. The spectrum of potential infectious agents is broad, but the most common pathogens are group B streptococci, *Chlamydia*, *Mycoplasma*, *Clostridia*, *Listeria*, *Escherichia coli*, and *Fusobacteria* (120–123). The placental membranes become opaque and white, yellow, or green, and neutrophils are present in them microscopically. In funisitis, fetal neutrophils and eosinophils aggregate along the umbilical vascular endothelium and also involve the substance of the cord. Necrosis may occur with fungal infections. Identification of the causative organisms can sometimes be accomplished with tissue Gram stains, the GMS technique, or PAS-D staining.

Meconium Exposure

The bile-stained intestinal content of the fetus is called meconium. Generally, it is not discharged in utero, but, in some instances including fetal stress, the passage of meconium occurs into the amniotic fluid. This causes placental membranes to appear dark green. There may also be microscopic evidence of meconium staining of the placenta. Pigmented histiocytes that contain ingested meconium can be seen in the subamnionic space (Figure 9.46). The meconium pigment is not easily labeled by any commonly available histochemical method, though the Luna-Ishak stain has been described for that purpose (124). The differential diagnostic alternative to subamnionic meconium pigmentation is that of hemosiderin deposition from prior trauma, abruption, a circumvallate placental configuration, and intervillous thrombosis. Hemosiderin tends to

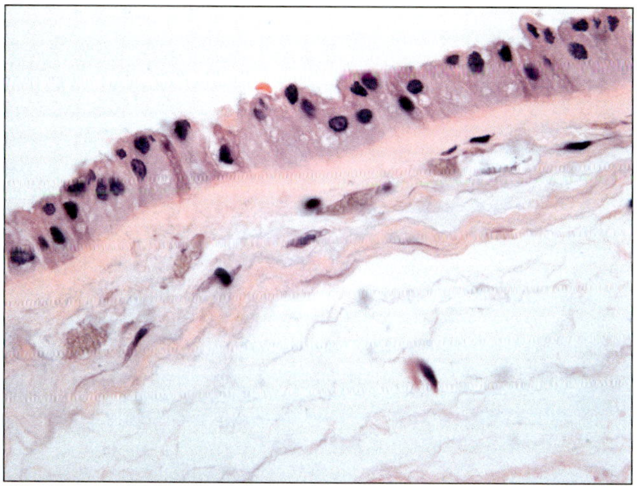

***FIGURE 9.46:** Intrauterine meconium passage can lead to columnar transformation of the amnion and subamnionic pigmented histiocytes, which sometimes can be difficult to distinguish from pigmented histiocytes due to hemosiderin accumulation in the setting of hemorrhage (Figure 9.47).*

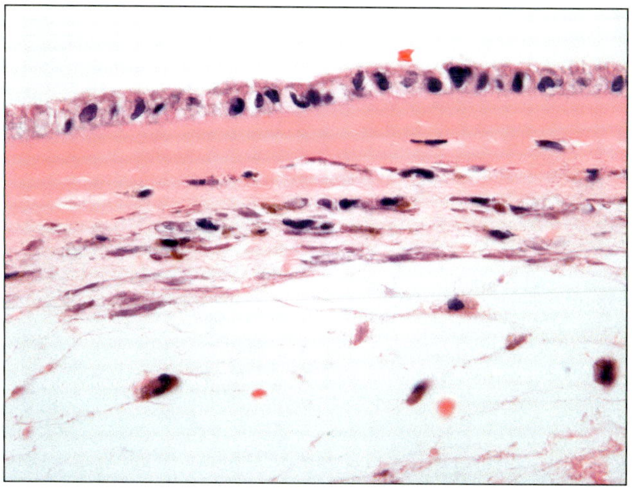

***FIGURE 9.47:** Placental hemorrhage can lead to accumulation of hemosiderin-laden histiocytes in the subamnion.*

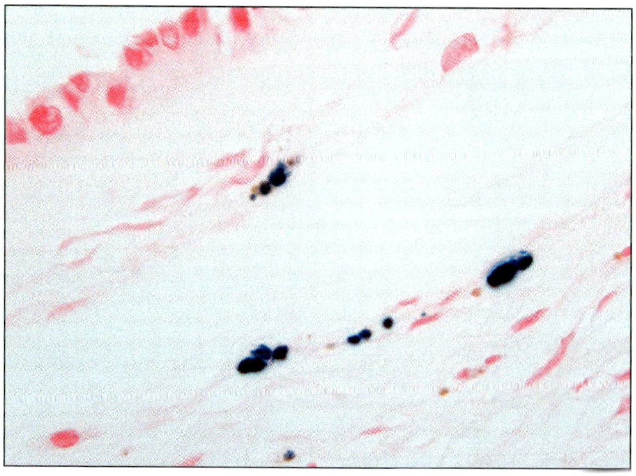

***FIGURE 9.48:** Iron staining will mark the subamnionic hemosiderin-laden histiocytes seen in placental hemorrhage but will not mark pigmented histiocytes seen in meconium exposure (not shown).*

have a granular yellow-brown appearance in H&E-stained sections, and it is refractile (Figure 9.47). It can be confirmed with Perl's method or Prussian blue stains (Figure 9.48). In contrast, pigmented histiocytes in meconium exposure are negative for iron (125).

REFERENCES

1. Brook I. Aerobic and anaerobic microbiology of Bartholin's abscess. *Surg Gynecol Obstet* 1989;169:32–4.
2. Tanaka K, Mikamo H, Ninomiya M, Tamaya T, Izumi K, Ito K, Yamaoka K, Watanabe K. Microbiology of Bartholin's gland abscess in Japan. *J Clin Microbiol* 2005;43:4258–61.
3. Wren MW. Bacteriological findings in cultures of clinical material from Bartholin's abscess. *J Clin Pathol* 1977;30:1025–7.
4. Bergeron S, Binik YM, Khalife S, Pagidas K, Glazer HI. Vulvar vestibulitis syndrome: reliability of diagnosis and

evaluation of current diagnostic criteria. *Obstet Gynecol* 2001;98:45–51.
5. Prayson RA, Stoler MH, Hart WR. Vulvar vestibulitis: a histopathologic study of 36 cases, including human papillomavirus in situ hybridization analysis. *Am J Surg Pathol* 1995;19:154–60.
6. Chadha S, Gianotten WL, Drogendijk AC, Weijmar SW, Blindeman LA, Van Der Meijden WI. Histopathologic features of vulvar vestibulitis. *Int J Gynecol Pathol* 1998;17:7–11.
7. Lundqvist EN, Hofer PA, Olofsson JI, Sjoberg I. Is vulvar vestibulitis an inflammatory condition? A comparison of histological findings in affected and healthy women. *Acta Derm Venereol* 1997;77:319–22.
8. Halperin R, Zehavi S, Vaknin Z, Ben-Ami I, Pansky M, Schneider D. The major histopathologic characteristics in the vulvar vestibulitis syndrome. *Gynecol Obstet Invest* 2005; 59:75–9.
9. Wilkinson EJ, Guerrero E, Daniel R, Shah K, Stone IK, Hardt NS, Friedrich EG Jr. Vulvar vestibulitis is rarely associated with human papillomavirus infection types 6, 11, 16, or 18. *Int J Gynecol Pathol* 1993;12:344–9.
10. Morin C, Bouchard C, Brisson J, Fortier M, Blanchette C, Meisels A. Human papillomaviruses and vulvar vestibulitis. *Obstet Gynecol* 2000;95:683–7.
11. Chaim W, Meriwether C, Gonik B, Qureshi F, Sobel JD. Vulvar vestibulitis subjects undergoing surgical intervention: a descriptive analysis and histopathological correlates. *Eur J Obstet Gynecol Reprod Biol* 1996;68:165–8.
12. Long SR, Whitfeld MJ, Eades C, Koehler JE, Korn AP, Zaloudek CJ. Bacillary angiomatosis of the cervix and vulva in a patient with aids. *Obstet Gynecol* 1996;88(4 Pt 2):709–11.
13. Freinkel AL. Histological aspects of sexually transmitted genital lesions. *Histopathology* 1987;11:819–31.
14. Fu YS. Infectious and inflammatory diseases of the lower female genital tract. Pathology of the uterine cervix, vagina, and vulva. Philadelphia: Saunders; 2002. pp. 88–9.
15. Engelkens HJ, Ten Kate FJ, Judanarso J, Vuzevski VD, Van Lier JB, Godschalk JC, Van Der Sluis JJ, Stolz E. The localisation of treponemes and characterisation of the inflammatory infiltrate in skin biopsies from patients with primary or secondary syphilis, or early infectious yaws. *Genitourin Med* 1993;69:102–7.
16. Hoang MP, High WA, Molberg KH. Secondary syphilis: a histologic and immunohistochemical evaluation. *J Cutan Pathol* 2004;31:595–9.
17. Phelps RG, Knispel J, Tu ES, Cernainu G, Saruk M. Immunoperoxidase technique for detecting spirochetes in tissue sections: comparison with other methods. *Int J Dermatol* 2000;39:609–13.
18. Rashid RM, Janjua SA, Khachemoune A. Granuloma inguinale: a case report. *Dermatol Online J* 2006;12:14.
19. Majmudar B. Granuloma inguinale (donovanosis). In: Connor HD, Chandler FW, Schwartz DA, Manz HJ, Lack EE, Editors. Pathology of Infectious Diseases. Stamford, CT: Appleton & Lange;1997. pp. 565–570.
20. Richens J. The diagnosis and treatment of donovanosis (granuloma inguinale). *Genitourin Med* 1991;67:441–52.
21. Lewis DA. Chancroid: clinical manifestations, diagnosis, and management. *Sex Transm Infect* 2003;79:68–71.
22. Trees DL, Morse SA. Chancroid and Haemophilus ducreyi: an update. *Clin Microbiol Rev* 1995;8:357–75.
23. Magro CM, Crowson AN, Alfa M, Nath A, Ronald A, Ndinya-Achola JO, Nasio J. A morphological study of penile chancroid lesions in human immunodeficiency virus (HIV)-positive and -negative African men with a hypothesis concerning the role of chancroid in HIV transmission. *Hum Pathol* 1996;27:1066–70.
24. Chalvardjian A, Picard L, Shaw R, Davey R, Cairns JD. Malacoplakia of the female genital tract. *Am J Obstet Gynecol* 1980;138:391–4.
25. Chou SC, Wang JS, Tseng HH. Malacoplakia of the ovary, fallopian tube and uterus: a case associated with diabetes mellitus. *Pathol Int* 2002;52:789–93.
26. Falcon-Escobedo R, Mora-Tiscareno A, Pueblitz-Peredo S. Malacoplakia of the uterine cervix. histologic, cytologic and ultrastructural study of a case. *Acta Cytol* 1986;30:281–4.
27. Paquin ML, Davis JR, Weiner S. Malacoplakia of Bartholin's gland. *Arch Pathol Lab Med* 1986;110:757–8.
28. Willen R, Stendahl U, Willen H, Trope. Malacoplakia of the cervix and corpus uteri: a light microscopic, electron microscopic, and X-ray microprobe analysis of a case. *Int J Gynecol Pathol* 1983;2:201–8.
29. Dasgupta P, Womack C, Turner AG, Blackford HN. Malacoplakia: Von Hansemann's disease. *BJU Int* 1999;84:464–9.
30. Decavalas G, Adonakis G, Androutsopoulos G, Gkermpesi M, Kourounis G. Sarcoidosis of the vulva: a case report. *Arch Gynecol Obstet* 2007;275:203–205.
31. Dicarlo FJ Jr., Dicarlo JP, Robboy SJ, Lyons MM. Sarcoidosis of the uterus. *Arch Pathol Lab Med* 1989;113:941–3.
32. Tatnall FM, Barnes HM, Sarkany I. Sarcoidosis of the vulva. *Clin Exp Dermatol* 1985;10:384–5.
33. Namavar JB, Parsanezhad ME, Ghane-Shirazi R. Female genital tuberculosis and infertility. *Int J Gynaecol Obstet* 2001; 75:269–72.
34. Tripathy SN, Tripathy SN. Genital manifestation of pulmonary tuberculosis. *Int J Gynaecol Obstet* 1981;19:319–26.
35. Smith YR, Haefner HK. Vulvar lichen sclerosus: pathophysiology and treatment. *Am J Clin Dermatol* 2004;5:105–25.
36. Carlson JA, Lamb P, Malfetano J, Ambros RA, Mihm MCJ. Clinicopathologic comparison of vulvar and extragenital lichen sclerosus: histologic variants, evolving lesions, and etiology of 141 cases. *Mod Pathol* 1998;11:844–54.
37. Fung MA, Leboit PE. Light microscopic criteria for the diagnosis of early vulvar lichen sclerosus—a comparison with lichen planus. *Am J Surg Pathol* 1998;22:473–8.
38. Northcutt AD, Vanover MJ. Nodular cutaneous amyloidosis involving the vulva. Case report and literature review. *Arch Dermatol* 1985;121:518–21.
39. Yoneyama K, Tochigi N, Oikawa A, Shinkai H, Utani A. Primary localized cutaneous nodular amyloidosis in a patient with Sjogren's syndrome: a review of the literature. *J Dermatol* 2005;32:120–3.
40. Persoons JH, Sutorius FJ, Koopman RJ, Schaafsma MR, Van Doorn GA. Vulvar paraneoplastic amyloidosis with the appearance of a vulvar carcinoma. *Am J Obstet Gynecol* 1999;180:1041–4.
41. Breathnach SM. Amyloid and amyloidosis. *J Am Acad Dermatol* 1988;18(1 Pt 1):1–16.

42. Cheung ST, Maheshwari MB, Tan CY. A comparative study of two Congo red stains for the detection of primary cutaneous amyloidosis. *J Am Acad Dermatol* 2006;55:363–4.
43. Goldblum JR, Hart WR. Vulvar Paget's disease: a clinicopathologic and immunohistochemical study of 19 cases. *Am J Surg Pathol* 1997;21:1178–87.
44. Fanning J, Lambert L, Hale TM, Morris PC, Schuerch C. Paget's disease of the vulva: prevalence of associated vulvar adenocarcinoma, invasive Paget's disease, and recurrence after surgical excision. *Am J Obstet Gynecol* 1999;180(1 Pt 1): 24–7.
45. Parker LP, Parker JR, Bodurka-Bevers D, Deavers M, Bevers MW, Shen-Gunther J, Gershenson DM. Paget's disease of the vulva: pathology, pattern of involvement, and prognosis. *Gynecol Oncol* 2000;77:183–9.
46. Zollo JD, Zeitouni NC. The Roswell Park Cancer Institute experience with extramammary Paget's disease. *Br J Dermatol* 2000;142:59–65.
47. Baehrendtz H, Einhorn N, Pettersson F, Silfverswaerd C. Paget's disease of the vulva: the Radiumhemmet series 1975–1990. *Int J Gynecol Cancer* 2001;4:1–6.
48. Rabban J, Zaloudek C. Vulvar Paget disease. *Pathol Case Rev* 2005;10:41–5.
48a. Reid PE, Owen DA, Fletcher K, et al. The histochemical specificity of high iron diamine–alcian blue. *Histochem J* 1989; 21: 501–7.
49. Copeland LJ, Sneige N, Gershenson DM, Mcguffee VB, Abdul-Karim F, Rutledge FN. Bartholin gland carcinoma. *Obstet Gynecol* 1986;67:794–801.
50. Rosenberg P, Simonsen E, Risberg B. Adenoid cystic carcinoma of Bartholin's gland: a report of five new cases treated with surgery and radiotherapy. *Gynecol Oncol* 1989;34:145–7.
51. Copeland LJ, Sneige N, Gershenson DM, Saul PB, Stringer CA, Seski JC. Adenoid cystic carcinoma of the Bartholin gland. *Obstet Gynecol* 1986;67:115–20.
52. Birch HW, Sondag DR. Granular cell myoblastoma of the vulva. *Obstet Gynecol* 1961;18:443.
53. Levavi H, Sabah G, Kaplan B, Tytiun Y, Braslavsky D, Gutman H. Granular cell tumor of the vulva: six new cases. *Arch Gynecol Obstet* 2006;273:246–9.
54. Mahalingam M, Lopiccolo D, Byers HR. Expression of PGP 9.5 in granular cell nerve sheath tumors: an immunohistochemical study of six cases. *J Cutan Pathol* 2001;28: 282–6.
55. Mazur MT, Shultz JJ, Myers JL. Granular cell tumor. Immunohistochemical analysis of 21 benign tumors and one malignant tumor. *Arch Pathol Lab Med* 1990;114:692–6.
56. Scullly RE, Robboy SJ, Welch WR. Pathology and pathogenesis of diethylstilbestrol related disorders of the female genital tract. Intrauterine exposure to diethylstilbesterol in the human. Chicago: American College of Obstetrics and Gynecology; 1978. p. 8.
57. Nucci MR, Fletcher CDM. Vulvovaginal soft tissue tumours: update and review. *Histopathology* 2000;36:97–108.
58. Di Sant'Agnese PA, Knowles DM. Extracardiac rhabdomyoma: a clinicopathologic study and review of the literature. *Cancer* 1980;46:780–9.
59. Willis J, Abdul-Karim F, Di Sant'Agnese PA. Extracardiac rhabdomyomas. *Semin Diagn Pathol* 1994;11:15–25.
60. Chabrel CM, Beilby JO. Vaginal rhabdomyoma. *Histopathology* 1980;4:645–51.
61. Gee DC, Finckh ES. Benign vaginal rhabdomyoma. *Pathology* 1977;9:263–7.
62. Gold JH, Bossen EH. Benign vaginal rhabdomyoma. a light and electron microscopic study. *Cancer* 1976;2283: 2294–91.
63. Bjorndal SK, Godballe C, Ostergaard B, Krogdahl A. Adult extracardiac rhabdomyoma: light and immunohistochemical studies of two cases in the parapharyngeal space. *Head Neck* 2006;28:275–9.
64. Gale N, Rott T, Kambic V. Nasopharyngeal rhabdomyoma. Report of case (light and electron microscopic studies) and review of the literature. *Pathol Res Pract* 1984;178:454–60.
65. Leone PG, Taylor HB. Ultrastructure of a benign polypoid rhabdomyoma of the vagina. *Cancer* 1973;31:1414–7.
66. Copeland LJ, Gershenson DM, Saul PB, Sneige N, Stringer CA, Edwards CL. Sarcoma botryoides of the female genital tract. *Obstet Gynecol* 1985;66:262–6.
67. Andrassy RJ, Hays DM, Raney RB, Wiener ES, Lawrence W, Lobe TE, Corpron CA, Smith M, Maurer HM. Conservative surgical management of vaginal and vulvar pediatric rhabdomyosarcoma: a report from the Intergroup Rhabdomyosarcoma Study III. *J Pediatr Surg* 1995;30: 1034–7.
68. Lloyd RV, Hajdu SI, Knapper WH. Embryonal rhabdomyosarcoma in adults. *Cancer* 1983;51:557–65.
69. Copeland LJ, Sneige N, Stringer CA, Gershenson DM, Saul PB, Kavanagh JJ. Alveolar rhabdomyosarcoma of the female genitalia. *Cancer* 1985;56:849–55.
70. Imachi M, Tsukamoto N, Kamura T, Shigematsu T, Funakoshi K, Nakano H. Alveolar rhabdomyosarcoma of the vulva. Report of two cases. *Acta Cytol* 1991;35:345–9.
71. Parham DM, Webber B, Holt H, Williams WK, Maurer H. Immunohistochemical study of childhood rhabdomyosarcomas and related neoplasms. Results of an intergroup rhabdomyosarcoma study project. *Cancer* 1991;67:3072–80.
72. Parham DM. Immunohistochemistry of childhood sarcomas: old and new markers. *Mod Pathol* 1993;6:133–8.
73. Tsokos M. The diagnosis and classification of childhood rhabdomyosarcoma. *Semin Diagn Pathol* 1994;11:26–38.
74. Chumas JC, Nelson B, Mann WJ, Chalas E, Kaplan CG. Microglandular hyperplasia of the uterine cervix. *Obstet Gynecol* 1985;66:406–9.
75. Young RH, Scully RE. Atypical forms of microglandular hyperplasia of the cervix simulating carcinoma: a report of five cases and review of the literature. *Am J Surg Pathol* 1989;13:50–6.
76. Clement PB, Young RH. Non-endometrioid carcinomas of the uterine corpus: a review of their pathology with emphasis on recent advances and problematic aspects. *Adv Anat Pathol* 2004;11:117–42.
77. Ferry JA, Scully RE. Mesonephric remnants, hyperplasia, and neoplasia in the uterine cervix: a study of 49 cases. *Am J Surg Pathol* 1990;14:1100–11.
78. Park JJ, Sun DQ, Quade BJ, Flynn C, Sheets EE, Yang A, Mckeon F, Crum CP. Stratified mucin-producing intraepithelial lesions of the cervix—adenosquamous or columnar cell neoplasia? *Am J Surg Pathol* 2000;24:1414–9.

79. Bethwaite P, Yeong ML, Holloway L, Robson B, Duncan G, Lamb D. The prognosis of adenosquamous carcinomas of the uterine cervix. *Br J Obstet Gynaecol* 1992;99:745–50.
80. Colgan TJ, Auger M, Mclaughlin JR. Histopathologic classification of cervical carcinomas and recognition of mucin-secreting squamous carcinomas. *Int J Gynecol Pathol* 1993; 12:64–9.
81. Buckley CH, Beards CS, Fox H. Pathological prognostic indicators in cervical cancer with particular reference to patients under the age of 40 years. *Br J Obstet Gynaecol* 1988; 95:47–56.
82. Langlois NEI, Ellul B, Miller ID. A study of the value and prognostic significance of mucin staining in squamous cell carcinoma of the uterine cervix. *Histopathology* 1996;28: 175–8.
83. Samlal RA, Ten Kate FJ, Hart AA, Lammes FB. Do mucin-secreting squamous cell carcinomas of the uterine cervix metastasize more frequently to pelvic lymph nodes? A case-control study. *Int J Gynecol Pathol* 1998;17:201–4.
84. Mount S, Mead P, Cooper K. Chlamydia trachomatis in the endometrium: can surgical pathologists identify plasma cells? *Adv Anat Pathol* 2001;8:327–9.
85. Paukku M, Puolakkainen M, Paavonen T, Paavonen J. Plasma cell endometritis is associated with Chlamydia trachomatis infection. *Am J Clin Pathol* 1999;112:211–5.
86. Gilmore H, Fleischhacker D, Hecht JL. Diagnosis of chronic endometritis in biopsies with stromal breakdown. *Hum Pathol* 2007;38:581–584.
87. Chatwani A, Amin-Hanjani S. Incidence of actinomycosis associated with intrauterine devices. *J Reprod Med* 1994; 39:585–7.
88. Kayikcioglu F, Akif AM, Haberal A, Faruk DO. Actinomyces infection in female genital tract. *Eur J Obstet Gynecol Reprod Biol* 2005;118:77–80.
89. Schmidt WA, Bedrossian CW, Ali V, Webb JA, Bastian FO. Actinomycins and intrauterine contraceptive devices: the clinicopathologic study. *Diagn Gynecol Obstet* 1980;2: 165–77.
90. Namavar JB, Parsanezhad ME, Ghane-Shirazi R. Female genital tuberculosis and infertility. *Int J Gynaecol Obstet* 2001;75:269–72.
91. Nogales-Ortiz F, Tarancon I, Nogales FF Jr. The pathology of female genital tuberculosis. A 31-year study of 1436 cases. *Obstet Gynecol* 1979;53:422–8.
92. Nucci MR, Prasad CJ, Crum CP, Mutter GL. Mucinous endometrial epithelial proliferations: a morphologic spectrum of changes with diverse clinical significance. *Mod Pathol* 1999;12:1137–42.
93. Vang R, Tavassoli FA. Proliferative mucinous lesions of the endometrium: analysis of existing criteria for diagnosing carcinoma in biopsies and curettings. *Int J Surg Pathol* 2003; 11:261–70.
94. Tiltman AJ. Mucinous carcinoma of the endometrium. *Obstet Gynecol* 1980;55:244–7.
95. Melhem MF, Tobon H. Mucinous adenocarcinoma of the endometrium: a clinico-pathological review of 18 cases. *Int J Gynecol Pathol* 1987;6:347–55.
96. Ross JC, Eifel PJ, Cox RS, Kempson RL, Hendrickson MR. Primary mucinous adenocarcinoma of the endometrium. a clinicopathologic and histochemical study. *Am J Surg Pathol* 1983;7:715–29.
97. Abeler VM, Vergote IB, Kjorstad KE, Tropé CG. Clear cell carcinoma of the endometrium—prognosis and metastatic pattern. *Cancer* 1996;78:1740–7.
98. Eichhorn JH, Young RH. Neuroendocrine tumors of the genital tract. *Am J Clin Pathol* 2001;115 (Suppl): S94–112.
99. Abeler VM, Kjorstad KE, Nesland JM. Undifferentiated carcinoma of the endometrium: a histopathologic and clinical study of 31 cases. *Cancer* 1991;68:98–105.
100. Huntsman DG, Clement PB, Gilks CB, Scully RE. Small-cell carcinoma of the endometrium: a clinicopathological study of sixteen cases. *Am J Surg Pathol* 1994;18:364–75.
101. Clement PB, Young RH, Scully RE. Diffuse, perinodular, and other patterns of hydropic degeneration within and adjacent to uterine leiomyomas. problems in differential diagnosis. *Am J Surg Pathol* 1992;16:26–32.
102. King ME, Dickersin GR, Scully RE. Myxoid leiomyosarcoma of the uterus: a report of six cases. *Am J Surg Pathol* 1982;6:589–98.
103. Koonings PP, Campbell K, Mishell DR Jr., Grimes DA. Relative frequency of primary ovarian neoplasms: a 10-year review. *Obstet Gynecol* 1989;74:921–6.
104. Bjorkholm E, Lundell M, Gyftodimos A, Silfversward C. Dysgerminoma. The Radiumhemmet series 1927–1984. *Cancer* 1990;65:38–44.
105. Kurman RJ, Norris HJ. Malignant germ cell tumors of the ovary. *Hum Pathol* 1977;8:551–64.
106. Gershenson DM, Del Junco G, Herson J, Rutledge FN. Endodermal sinus tumor of the ovary: the M.D. Anderson experience. *Obstet Gynecol* 1983;61:194–202.
107. Kurman RJ, Norris HJ. Endodermal sinus tumor of the ovary. A clinical and pathologic analysis of 71 cases. *Cancer* 1976;38:2404–19.
108. Teilum G. Endodermal sinus tumors of the ovary and testis: comparative morphogenesis of the so-called mesonephroma ovarii (Schiller) and extraembryonic (yolk sac-allantoic) structures of the rat's placenta. *Cancer* 1959;12:1092–105.
109. Ramalingam P, Malpica A, Silva EG, Gershenson DM, Liu JL, Deavers MT. The use of cytokeratin 7 and EMA in differentiating ovarian yolk sac tumors from endometrioid and clear cell carcinomas. *Am J Surg Pathol* 2004;28:1499–505.
110. Katsube Y, Berg JW, Silverberg SG. Epidemiologic pathology of ovarian tumors: a histopathologic review of primary ovarian neoplasms diagnosed in the Denver Standard Metropolitan Statistical Area, 1 July–31 December 1969 and 1 July–31 December 1979. *Int J Gynecol Pathol* 1982;1:3–16.
111. Young RH, Scully RE. Ovarian stromal tumors with minor sex cord elements: a report of seven cases. *Int J Gynecol Pathol* 1983;2:227–34.
112. Young RH. Sex cord-stromal tumors of the ovary and testis: their similarities and differences with consideration of selected problems. *Mod Pathol* 2005;18 (Suppl 2):S81–98.
113. Scully RE, Young RH, Clement PB. Tumors of the ovary, maldeveloped gonads, fallopian tube, and broad ligament. Vol Fascicle 23. Washington, D.C.: Armed Forces Institute of Pathology; 1998.
114. Evans DT. Actinomyces israelii in the female genital tract: a review. *Genitourin Med* 1993;69:54–9.

115. Hsu CT, Roan CH, Rai SY, Jong HL, Chen TY, Lin YN, Lan CC. Actinomycosis affecting the fallopian tube and ovary: report of 3 cases, with special reference to 2 cases following IUD application. *Asia Oceania J Obstet Gynaecol* 1988;14:275–84.
116. Lininger JR, Frable WJ. Diagnosis of pelvic actinomycosis by fine needle aspiration. A case report. *Acta Cytol* 1984;28:601–4.
117. Asuncion CM, Cinti DC, Hawkins HB. Abdominal manifestations of actinomycosis in IUD users. *J Clin Gastroenterol* 1984;6:343–8.
118. Kershisnik MM, Ro JY, Cannon GH, Ordoñez NG, Ayala AG, Silva EG. Histiocytic reaction in pelvic peritoneum associated with oxidized regenerated cellulose. *Am J Clin Pathol* 1995;103:27–31.
119. Kuo TT, Hsueh S. Mucicarminophilic histiocytosis. A polyvinylpyrrolidone (PVP) storage disease simulating signet-ring cell carcinoma. *Am J Surg Pathol* 1984;8:419–28.
120. Garland SM, Ni CF, Satzke C, Robins-Browne R. Mechanisms, organisms and markers of infection in pregnancy. *J Reprod Immunol* 2002;57:169–83.
121. Newton ER. Chorioamnionitis and intraamniotic infection. *Clin Obstet Gynecol* 1993;36:795–808.
122. Gibbs RS, Duff P. Progress in pathogenesis and management of clinical intraamniotic infection. *Am J Obstet Gynecol* 1991;164(5 Pt 1):1317–26.
123. Madan E, Meyer MP, Amortequi A. Chorioamnionitis: a study of organisms isolated in perinatal autopsies. *Ann Clin Lab Sci* 1988;18:39–45.
124. Luna LG. Manual of Histologic Staining Methods of the Armed Forces Institute Of Pathology. 3rd Ed. New York: Mcgraw-Hill Book; 1968.
125. Henry JA, Baker RW, Yanowitz TD. The in utero passage of meconium by very low birth weight infants: a marker for adverse outcomes. *J Perinatol* 2006;26:125–9.

CHAPTER 10

Diseases of the Male Genital System

Helen P. Cathro, M.B., Ch.B., M.P.H.,
Mark A. Weiss, M.D., and
Stacey E. Mills, M.D.

PROSTATIC DISEASES

Histochemical staining is of limited usefulness in the common clinical setting of acute (and chronic) bacterial prostatitis, but it remains important in the differential diagnosis of granulomatous prostatitis, which can be related to a variety of specific organisms, have a noninfectious pathogenesis, or be caused by systemic or localized vasculitis. Special stains are also useful in confirming the presence of metaplasia, evaluating epithelial and stromal pigment deposits, and diagnosing certain benign conditions, such as amyloidosis.

Prostate-specific antigen (PSA) screening has dramatically increased the number of needle biopsies, with the aim of early detection of prostate cancer. As a consequence, the surgical pathologist is faced with the challenge of accurately diagnosing prostate carcinoma and its putative precursor lesion [high-grade prostatic intraepithelial neoplasia (HGPIN)] as well as distinguishing cancer from a variety of benign mimics, including adenosis and sclerosing adenosis. Although not recommended for diagnosis, mucin histochemistry has proven useful in characterizing this spectrum of neoplastic, preneoplastic, and pseudoneoplastic lesions, as have studies employing biological tumor markers, for example, AgNOR and lectins. Finally, variants of prostatic carcinoma have been further characterized with histochemical stains.

Nonneoplastic Conditions of the Prostate

Infections and Inflammatory Lesions: *Acute bacterial prostatitis* is most commonly caused by gram-negative organisms, with *Escherichia coli* strains identified in 80% of infections (1). Prostatic biopsy is contraindicated in acute bacterial prostatitis because of the risk of sepsis, and the diagnosis is principally based on clinical findings, including urine culture.

Prostatic abscess may complicate bacterial prostatitis, particularly in patients with diabetes and AIDS (2–6). It is generally not necessary to apply histochemical stains for organisms to tissue sections with acute inflammation, unless there is abscess formation with necrosis or the patient is immunocompromised. Abscess tissue may be submitted to pathology as a surgical specimen, either transurethral resection or suprapubic prostatectomy, or encountered at autopsy, and tissue Gram stains (Brown and Brenn or Brown and Hopps) can be utilized to identify bacterial organisms. Unfortunately, these stains are almost always negative.

Actinomycosis of the prostate is rare but has been reported to produce abscesses containing sulfur granules associated with marked acute and chronic inflammation, fibrosis, and granulation tissue (7). A Gram stain shows numerous gram-positive, rod-like, branching filamentous *Actinomyces* organisms.

Granulomatous prostatitis can have numerous causes (see Table 10.1) and histochemical stains can be crucial in establishing a definitive diagnosis, particularly for those related to infection (8–11).

Infectious granulomatous prostatitis is most often due to *Mycobacteria tuberculosis*, which causes necrotizing granulomata (Figure 10.1). Acid-fast bacilli can be demonstrated with the Ziehl-Neelsen/Kinyoun stain. Rare cases due to *Mycobacterium bovis*, *Mycobacterium kansasii*, and *Mycobacterium avium* have been reported (12); a Fite stain is preferable for atypical mycobacteria. Granulomatous prostatitis has been reported to occur in 1.3% of patients treated with intravesical bacillus Calmette-Guérin (BCG) (Figure 10.2) (13). In one study (10), BCG prostatitis was the second most frequent form of granulomatous prostatitis found on needle biopsy. In an individual with granulomatous prostatitis and a history of BCG therapy, it is not necessary to perform stains for acid-fast organisms.

A variety of fungi have been implicated in specific granulomatous prostatitis (14). Epidemiologic changes that include immune-compromised patients and drug-resistant fungi have caused an increase in nosocomial infections by *Candida albicans* and *non-albicans Candida* species. *Aspergilla* (15) and *Cryptococcus* (16–19) are opportunistic invaders of the immune-compromised (transplant, HIV) patient. The environmental fungi *Coccidioides immitis* (20), *Paracoccidioides* (21), *Histoplasma capsulatum* (22) (avian-infested areas), and *Blastomyces dermatitidis* (aquatic areas) (Figure 10.3) can cause infections in immunocompetent

TABLE 10.1: Classification of Granulomatous Prostatitis

Infectious
Bacterial
Tuberculosis
Brucellosis
Spirochetal
Syphilis
Fungal
Coccidiomycosis
Cryptococcosis
Blastomycosis
Histoplasmosis
Paracoccidiomycosis
Candidiasis
Aspergillosis
Parasitic
Schistosomiasis
Echinococcosis
Viral
Herpes zoster
Malacoplakia
Iatrogenic
Postsurgical
BCG-induced
Nonspecific
Systemic granulomatous disease
Allergic ("eosinophilic")
Sarcoidosis
Autoimmune-vascular
Wegener's granulomatosis
Polyarteritis nodosa

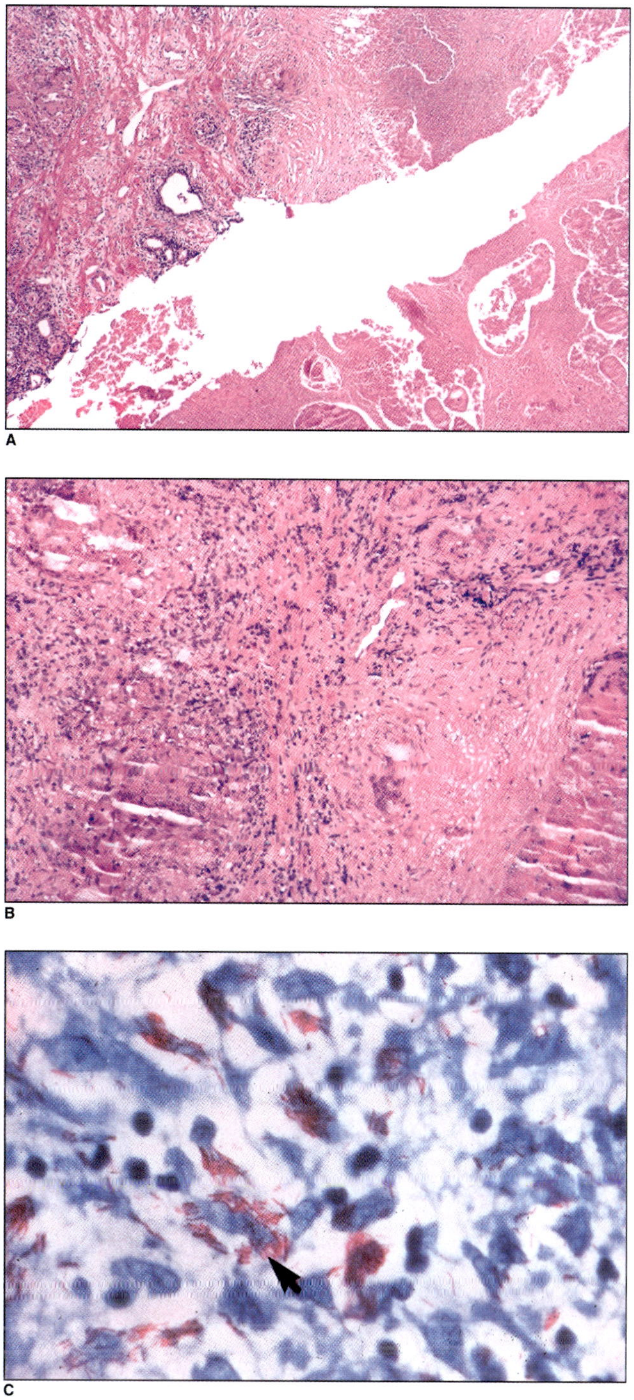

FIGURE 10.1: *Tuberculous prostatitis is characterized by zones of necrosis (A) and granulomatous inflammation with multinucleated giant cells (B). Numerous acid-fast mycobacteria are present (C) (A—H&E, 40×; B—H&E, 100×; C—Ziehl-Neelsen stain, 1000×).*

and immune-deficient patients. Periodic acid-Schiff (PAS) and Grocott-Gomori methenamine silver (MS) will stain fungal hyphae and spores (Figure 10.4). In addition, Mayer's mucicarmine will stain the capsule of *Cryptococcus neoformans* red (Figure 10.5). Of particular note, *C. neoformans* is found in up to 11% of AIDS cases, and it has been postulated that it is sequestered in the prostate from where it may cause relapse of systemic disease (17,18).

Herpes zoster has been implicated in a granulomatous prostatitis (23). *Cytomegalovirus* may produce a necrotizing prostatitis with associated intranuclear and intracytoplasmic inclusions, but the lesion is not typically granulomatous (24,25). Immunohistochemistry has generally replaced histochemical techniques for identifying inclusions and can identify specific viral antigens. Of historical interest, syphilis produces a granulomatous prostatitis (26) and *spirochetes* can be identified with a Warthin-Starry stain. The prostate is involved in approximately 20% of cases of schistosomiasis (27). The terminal spine eggs of *Schistosoma haematobium* may evoke a granulomatous reaction and can be highlighted with a PAS stain. Rare cases of cystic echinococcosis (hydatidosis) due to the tape worm *Echinococcus granulosus* have involved the prostate and seminal vesicles (28).

The most commonly diagnosed granulomatous entity within the prostate has been *nonspecific granulomatous prostatitis*, which is characterized by large, nodular, cellular infiltrates of epithelioid histiocytes, lymphocytes, and plasma cells with variable multinucleated giant cells (Figure 10.6) (8,9,11). *Posttransurethral resection granulomatous*

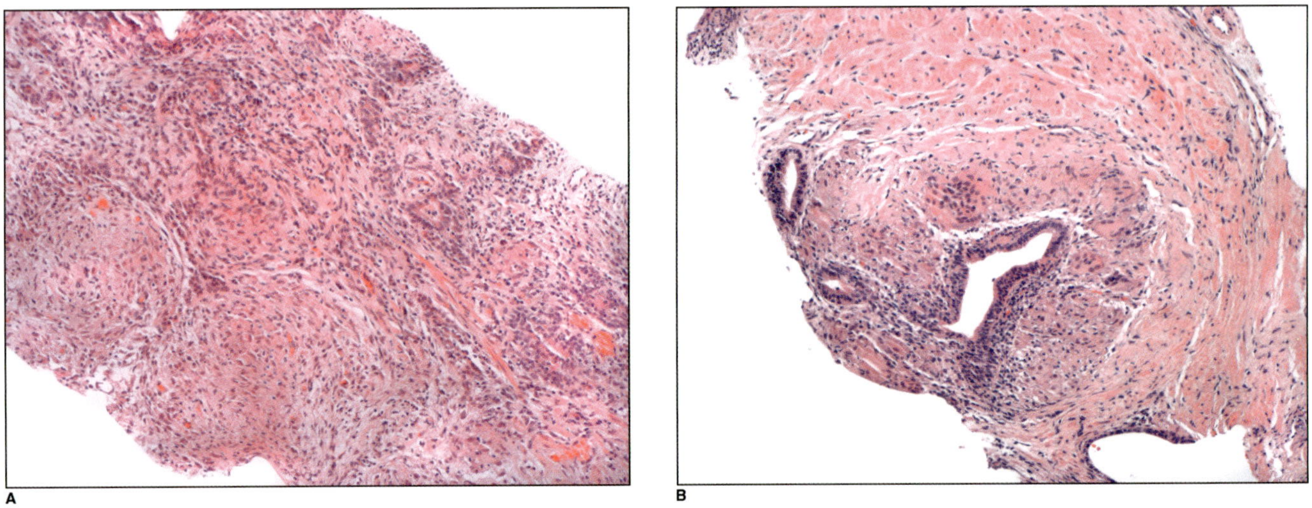

A B

FIGURE 10.2: Granulomatous prostatitis secondary to intravesical BCG therapy may produce multiple confluent granulomas (A) with foci of necrosis (not shown) or a discrete periductal granuloma (B). In the appropriate clinical setting, it is not necessary to order acid-fast stains (A—H&E, 100×; B—H&E, 200×).

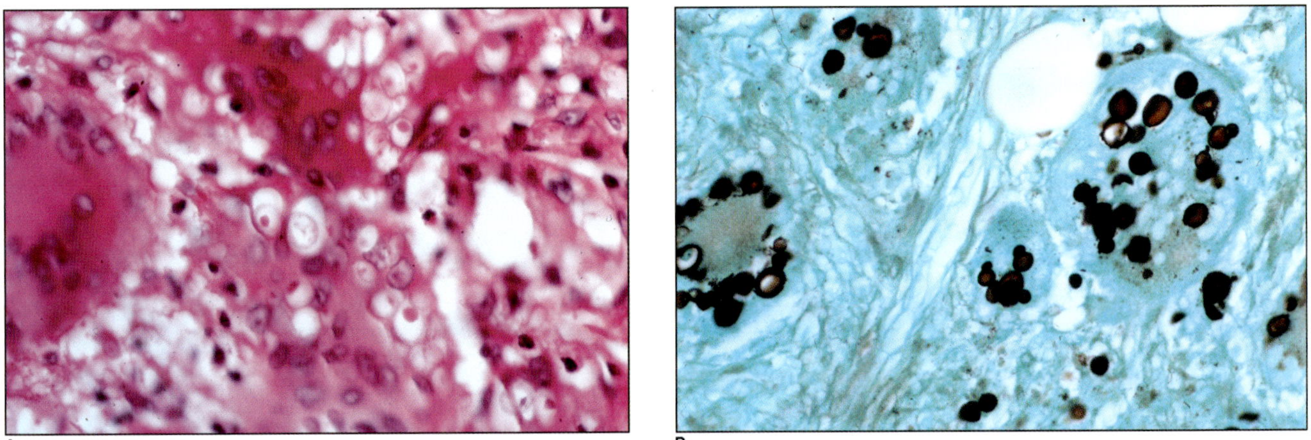

A B

FIGURE 10.3: Tissue infected with blastomycosis typically contains numerous organisms, which can be easily identified within giant cells (A). With silver stains, they appear as brown-staining yeast forms. Budding yeast have a characteristic broad base (B) (A—H&E, 400×; B—silver methenamine, 400×).

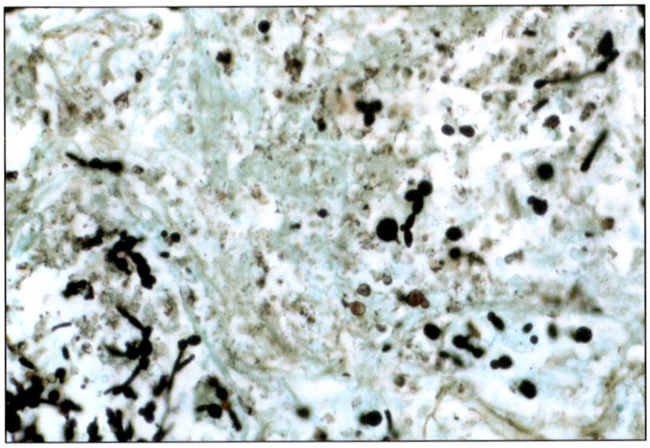

FIGURE 10.4: In candidiasis of the prostate, both yeast forms (upper right) and pseudohyphal forms (lower left) are typically present (silver methenamine, 400×).

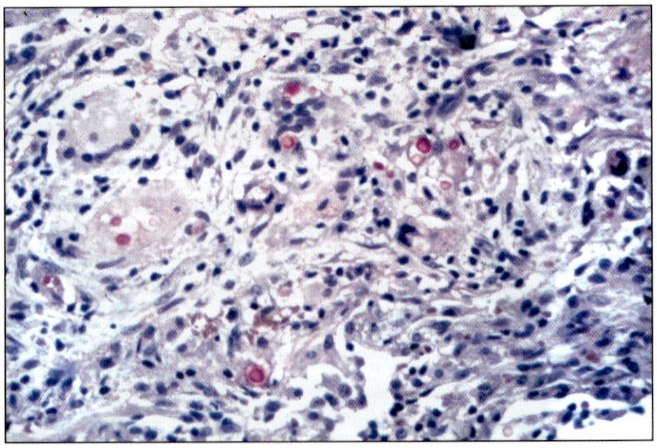

FIGURE 10.5: Special stains are virtually diagnostic of Cryptococcus neoformans *infection. Organisms have a characteristic mucicarmine-positive (red) capsule around them. A granulomatous inflammatory reaction is also present (mucicarmine stain, 400×).*

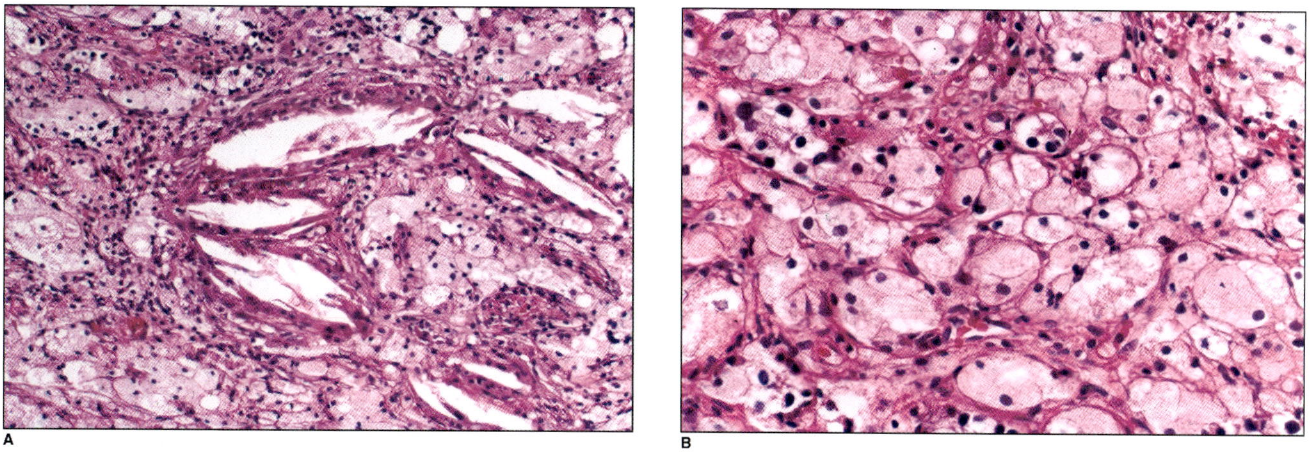

FIGURE 10.6: ***Xanthogranulomatous prostatitis is a variant of nonspecific prostatitis characterized by numerous lipid-laden histiocytes. The elliptical spaces are cholesterol clefts that were filled with lipid prior to tissue processing (A). Large lipid-laden histiocytes, which predominate in this condition (B), may be confused with the clear cell variant of prostatic carcinoma (H&E—A, 100×; B, 200×).***

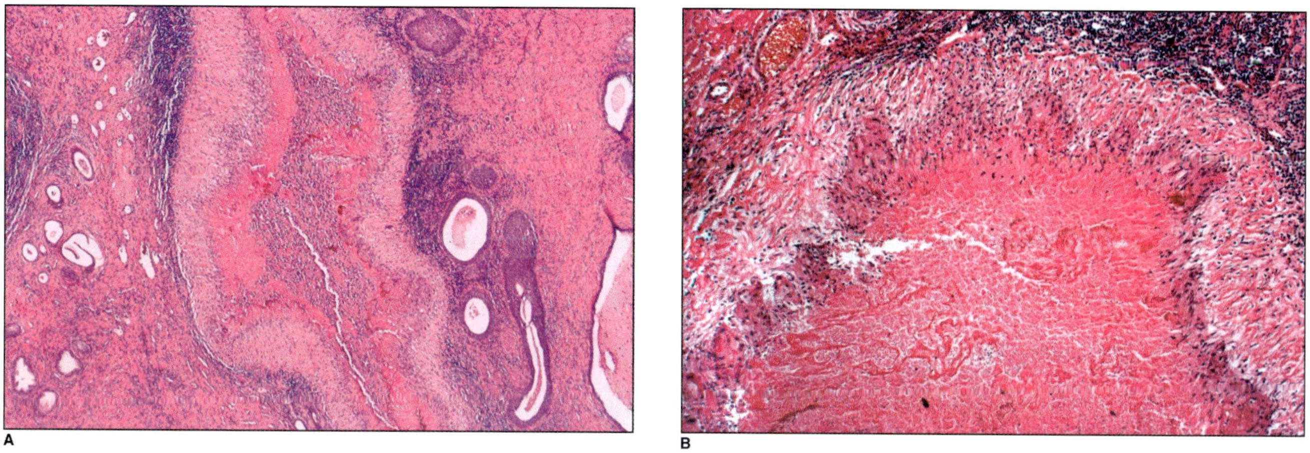

FIGURE 10.7: ***Posttransurethral resection granulomatous prostatitis consists of large, discrete necrobiotic granulomas (A). The central zone of amorphous fibrinoid material is rimmed by a band of palisaded fibrohistiocytic cells, indistinguishable from a rheumatoid nodule (B) (H&E—A, 40×; B, 100×).***

prostatitis is also commonly encountered (8). It resembles rheumatoid nodules (Figure 10.7) and consists of discrete necrobiotic granulomas containing amorphous fibrinoid material that does not stain red with Masson's trichrome. The palisading granulomas are causally related to superficial thermonecrosis, and in almost all cases, characteristic brownish, iron-negative carbonization inclusions can be detected within multinucleated giant cells (29). Nonspecific and posttransurethral resection granulomatous prostatitis may both have abundant infiltrates of eosinophils and appear histologically identical to allergic granulomatous prostatitis.

Prostatic *sarcoidosis* is rare (30). It produces characteristic noncaseating epithelioid granulomas and must be distinguished from other causes of infectious and noninfectious granulomatous prostatitis, which also may have negative organism stains.

Malacoplakia is characterized by dense aggregates of large mononuclear phagocytes (von Hansemann histiocytes) admixed with intracellular and extracellular Michaelis-Gutmann bodies (Figure 10.8). Similar to other sites, prostatic malacoplakia is most often associated with *E. coli* infections (31). Michaelis-Gutmann bodies have been found to be gram negative, stain positive with PAS, black with von Kossa, and positive with Prussian blue (iron), alizarin red (calcium salts), and Gomori iron stain (32).

Vasculitis is uncommon in the prostate and usually reflects systemic disease, such as *Wegener's granulomatosis* or *polyarteritis nodosa*. Wegener's granulomatosis produces a necrotizing and granulomatous vasculitis, and prostatitis rarely is the presenting manifestation (33,34). At autopsy, the prostate is involved in 7.4% of cases (35). Isolated prostatic vasculitis is rare. It may resemble polyarteritis nodosa (36,37) or giant cell arteritis (38). Fibrinoid

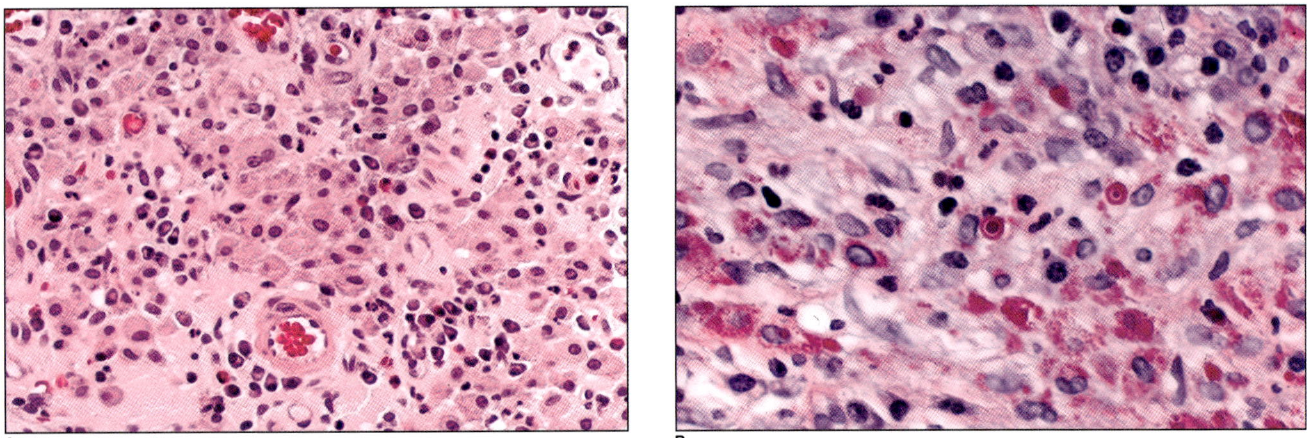

A B

FIGURE 10.8: Sheets of histiocytes with abundant eosinophilic cytoplasm characterize prostatic malacoplakia (A). Michaelis-Gutmann bodies can be difficult to see in H&E-stained sections but stain strongly with the PAS stain, which highlights their targetoid appearance (B) (A—H&E, 200×; B—PAS stain, 400×).

material can be stained with Masson's trichrome and Fraser-Lendrum. Destruction of the elastic laminae and scarring are highlighted with the Verhoeff–Van Gieson stain and Masson's trichrome.

Metaplasia: *Mucinous metaplasia* of the prostate has been identified in 8% of consecutive needle biopsy accession (39) in comparison to an incidence of <1% in autopsy cases. Following external beam radiation therapy with or without androgen deprivation therapy, it has been seen in 21% of biopsies (40). It is characterized by small groups of tall columnar cells with fine, foamy cytoplasm that stain intensely with PAS, mucicarmine, and alcian blue (pH 2.5) (Figure 10.9) (41,42).

Mucinous metaplasia should be distinguished from *Cowper's glands*, which are located in the urogenital diaphragm and situated laterally to the membranous urethra and may be sampled inadvertently by transurethral resection and rarely by needle biopsy. Cowper's (bulbourethral) glands consist of well-demarcated lobules of small compact tubuloalveolar glands radiating from a central excretory duct and entrapped within fascicles of striated muscle (Figure 10.10) (43). The acinar cells stain with digested PAS and mucicarmine (44,45). In the study by Cina et al. (45), the mucinous cytoplasm of Cowper's gland reacted with *Ulex europaeus* agglutinin I (UEA-1). In another study (46), Cowper's gland did not bind diaminobenzidine tetrahydrochloride (DBA) or UEA-1 and reacted with peanut agglutinin (PNA) in one-third of cases. In comparison, the acini of mucinous metaplasia reacted with DBA (45% cases), UEA-I (64% cases), and PNA (36% cases).

Paneth cell–like change is characterized by bright, eosinophilic cytoplasmic granules on routine hematoxylin and eosin (H&E) staining (Figures 10.11 and 10.12). In benign and hyperplastic glands, the granules are PAS positive and diastase resistant, as well as negative with Fontana-Masson and Grimelius silver stains (47). Ultrastructurally, the eosinophilic granules in benign epithelium represent exocrine-like or lysosomal-like vesicles (47), and the lesion represents Paneth cell–like change rather than Paneth cell metaplasia. Following external beam radiation, this change was found in 32% of biopsies (40). Paneth cell–like change in dysplastic [prostatic intraepithelial neoplasia (PIN)] glands and malignant prostatic epithelium represents neuroendocrine differentiation, and the granules are PAS negative and positive with the Fontana-Masson and Grimelius silver stains (47) (see section on prostatic adenocarcinoma).

Miscellaneous Benign Conditions: The prostate can be involved by primary (idiopathic or myeloma associated) *amyloidosis* consisting of monoclonal antibody light (AL) chain deposits or secondary amyloidosis associated with chronic inflammation and consisting of amyloid A protein (AA), but localized senile seminal vesicle amyloid is more common (48–50). In addition, prostatic corpora amylacea represent a localized amyloidosis of B2 microglobulin origin (51). Extensive vascular deposits, as well as stromal deposits, favor systemic amyloidosis. With localized senile seminal vesicle amyloid, the material is deposited in a linear and/or nodular subepithelial pattern (Figure 10.13) (48). All forms of amyloid share an affinity for the Congo red stain and demonstrate apple-green birefringence under polarized light. Primary (AL) amyloid is resistant to pretreatment with potassium permanganate (Wright-modified Rohmanyi method), whereas systemic AA amyloid and localized senile vesicle amyloid (48), as well as corpora amylacea amyloid (51), are potassium permanganate sensitive.

Prostatic *blue nevus*, a rare entity, is characterized by heavily pigmented round, polygonal, or elongated cells with dendritic cytoplasmic processes within the stroma. By electron microscopy, numerous melanosomes in all stages of differentiation or only mature stage IV

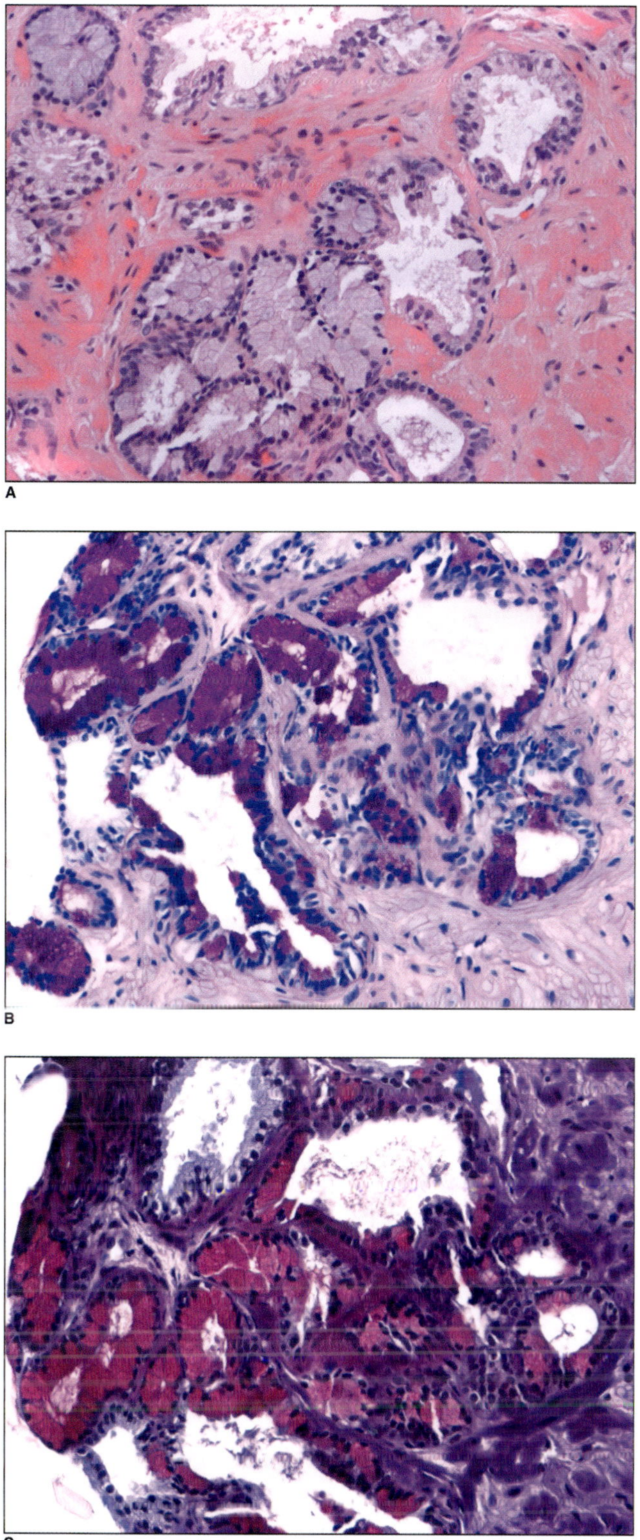

A

B

C

FIGURE 10.9: ***Mucinous metaplasia is characterized by glands lined by columnar cells with fine, foamy cytoplasm. This benign gland cluster is surrounded by prostatic fibromuscular stroma (A). The metaplastic cells stain intensely for neutral (B and C) and acid mucins (A—H&E, 200×; B—digested PAS, 200×; C—mucicarmine, 200×).***

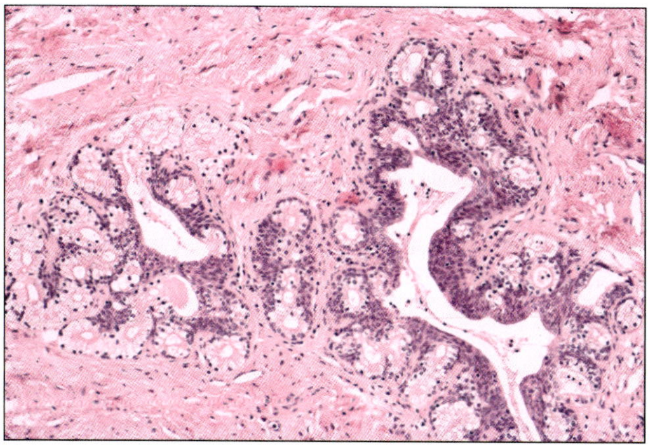

FIGURE 10.10: ***Cowper's glands have a lobular appearance with a central duct and surrounding compact tubuloalveolar glands associated with bundles of skeletal muscle (H&E, 100×).***

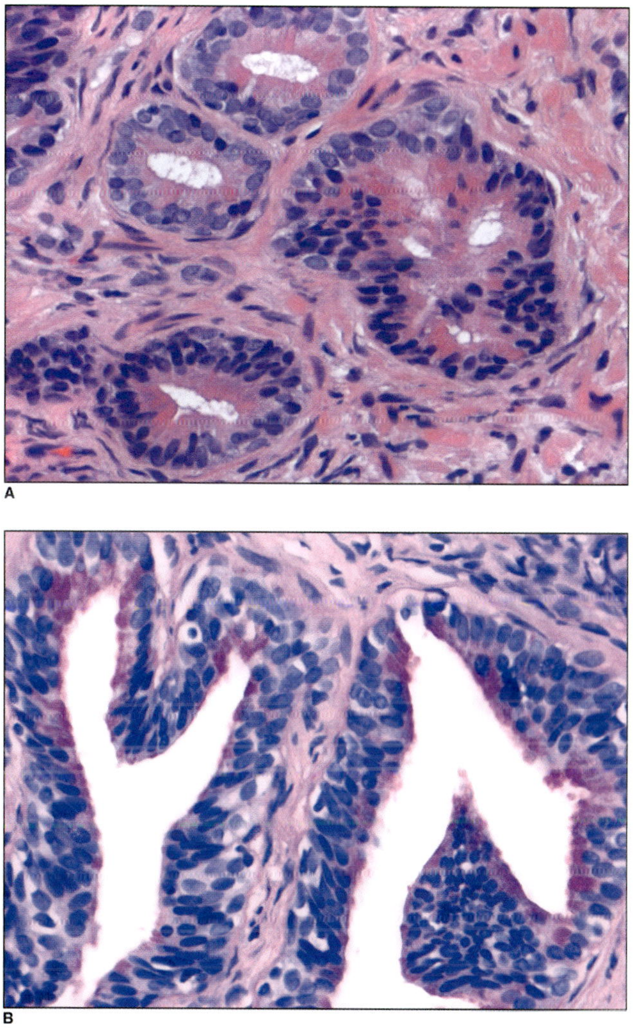

A

B

FIGURE 10.11: ***Paneth cell–like change in these benign glands is recognized by cells with bright, eosinophilic cytoplasmic granules (A). The exocrine- or lysosomal-like vesicles are PAS positive and diastase resistant (B) (A—H&E, 400×; B—digested PAS, 400×).***

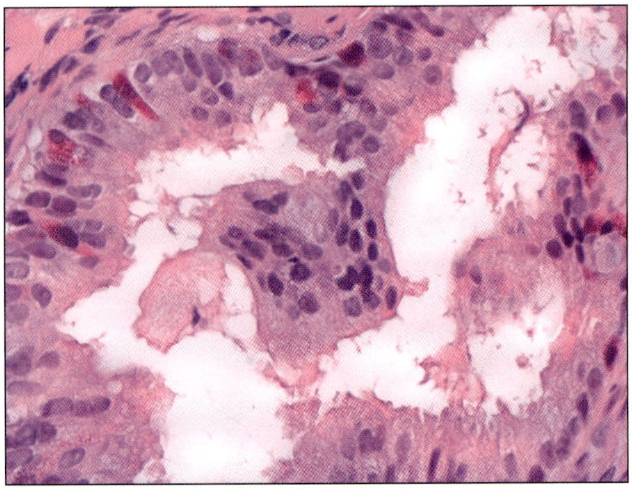

FIGURE 10.12: Paneth cell–like change in dysplastic (PIN) glands represents neuroendocrine differentiation (H&E, 400×).

melanosomes are present (52). In H&E sections, the cytoplasmic pigment is finely granular, brown or black. It stains black with the Fontana-Masson stain and disappears after melanin bleach with potassium permanganate, is iron negative, is strongly positive with Lillie's ferrous iron for melanin, and is negative with the acid-fast stain for ceroid and Schmorl's lipofuscin stain (52).

The term *melanosis* has been used when melanin-like pigment is present in the glandular epithelium with or without the presence of stromal melanocytes (Figure 10.14). The epithelial pigment has been described as 1–3 mm in diameter, predominantly subnuclear, yellow-brown to gray-brown granules with a dark blue rim (53) or golden yellow-brown (54) and tends to be less coarse and less refractile than seminal vesicle/ejaculatory duct pigment (Figure 10.15). It can be observed in all zones of the prostate (54,55) but had a preferential distribution in the central zone in the study by Leung and Srigley (55). In addition to benign and hyperplastic epithelium, the pigment can also be seen in preneoplastic (high-grade prostatic intraepithelial neoplasia) and carcinomatous glands (53,54). The pigment granules have melanin-like properties (Fontana-Masson positive and bleached by permanganate, positive for Lillie's ferrous iron and Lillie's sulfuric Nile blue for melanin, and negative for Mallory's hematoxylin-basic fuchsin and for Turnbull's stain for iron), but they also react like lipofuscin (pink coloration with the short Ziehl-Neelsen, strong affinity for basic dyes such as methylene blue and Nile blue, and some resistance to bleaching after Hueck's Nile blue stain) (56). The lipofuscin-type pigment also exhibits yellow autofluorescence and stains blue with luxol fast blue and with Giemsa, red to orange with oil red O, and focally orange-red with both Congo red and PAS with diastase (53). The Prussian blue stain and Perl's stain for hemosiderin are negative.

Benign Mimics of Cancer: *Adenosis* (atypical adenomatous hyperplasia) is a benign mimicker of Gleason grades 1 and 2 carcinoma. It has a predilection for the transition zone and is uncommon in needle biopsies (57) but has been identified in 1.5–19.6% of transurethral resectates and up to 33% of radical prostatectomy specimens (58). Adenosis is a proliferative lesion characterized by crowded small acini, usually forming well-circumscribed nodules (Figure 10.16). Unlike carcinoma, macronucleoli (>3 μm) are not present. Gaudin and Epstein (59) found blue-tinged mucinous secretions in 3% of cases of adenosis on needle biopsy compared to 30% of adenocarcinomas with Gleason scores 2–4 (Figure 10.17). Other studies have compared mucin expression using histochemistry: PAS-diastase (PAS-D) for neutral mucins, alcian blue, pH 2.7, for total

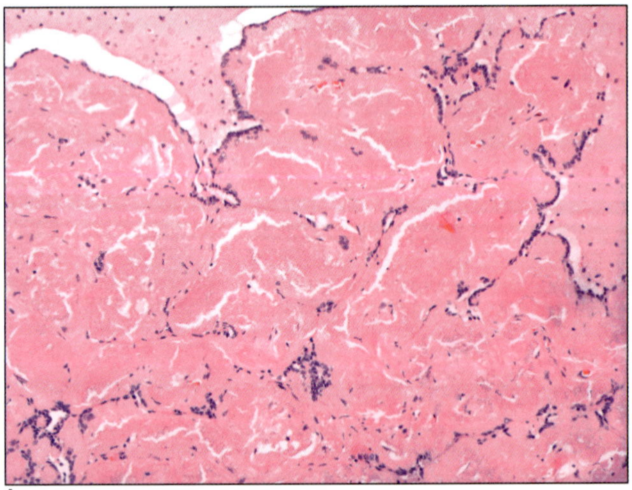

A

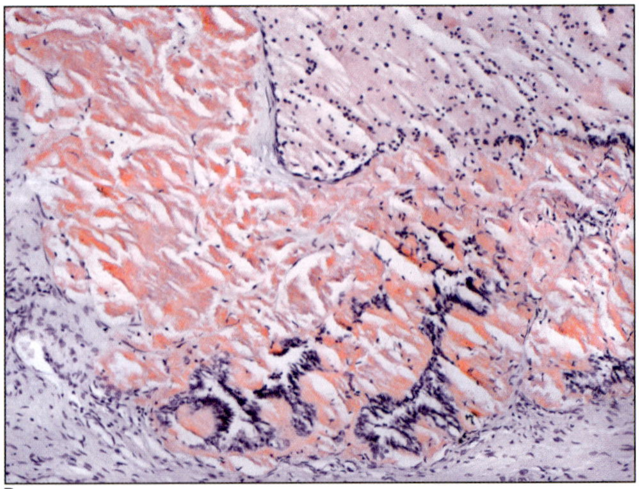

B

FIGURE 10.13: Localized senile seminal vesicle amyloid has a characteristic nodular subepithelial pattern of deposition (A). The hyaline material appears orangeophilic with the Congo red stain (B) (A—H&E, 100×; B—Congo red, 100×).

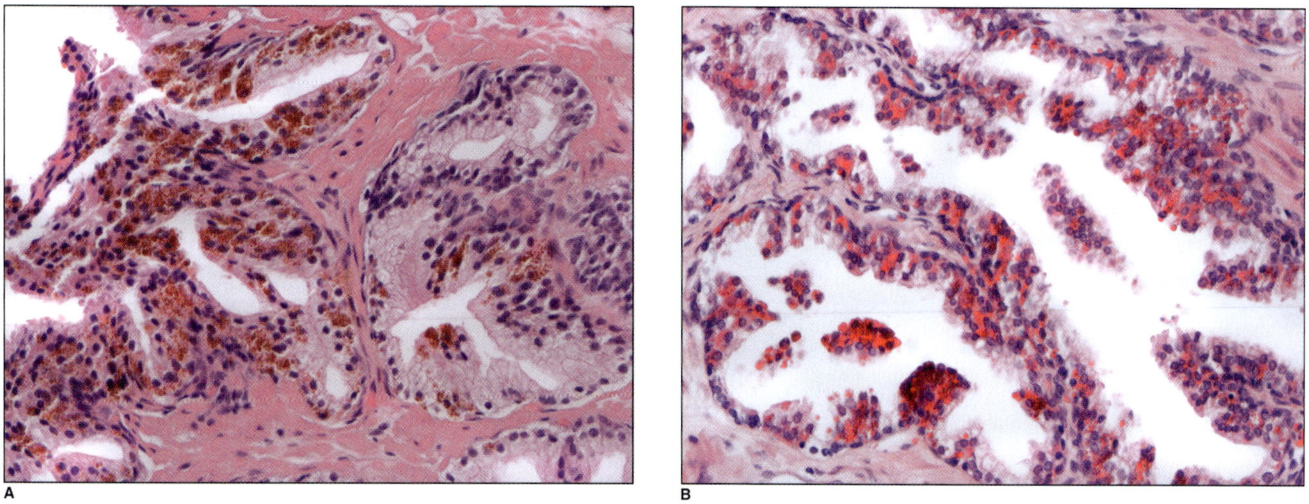

FIGURE 10.14: Prostatic glandular melanosis represents pigment granules that have both melanin-like and lipofuscin-like histochemical staining properties. The pigment can appear golden (A) or reddish (B) (H&E—A, 200×; B, 200×).

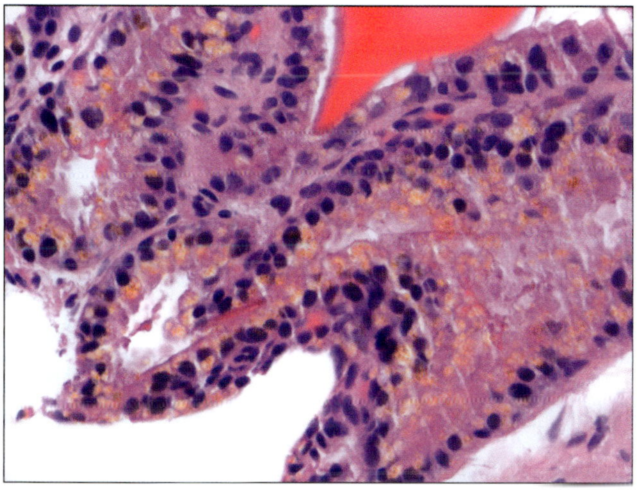

FIGURE 10.15: Lipofuscin pigment in the ejaculatory duct is coarser and more refractile than the pigment granules of prostatic granular melanosis (H&E, 400×).

(sulfated and nonsulfated) acidic mucin, and alcian blue, pH 0.7, for sulfated acidic mucin (60,61). Acid mucin expression was found in 63% of cases of adenosis and 60% of well-differentiated carcinoma (60). Luna-More et al. (61) found a 54.5% incidence of acidic mucin in adenosis, and an 80% incidence in carcinomas.

Other studies have utilized biological tissue markers to characterize adenosis. Helpap and Riede (62) studied the number of silver-stained nucleolar organizing regions (AgNOR) as a method for measuring proliferative activity and found the number of granules in adenosis (3.7) to be greater than that of usual hyperplasia (2.6) and less than in adenocarcinoma (5.0). Chastonay et al. (63) found that blood group antigen expression in adenosis (20% positive) is more like that of low-grade adenocarcinoma (15%) than normal epithelium (75%); PNA receptors were identified in 53% of adenocarcinoma and 50% of adenosis but were completely absent in benign epithelium. Shamsuddin and Beasly (64) reported expression of the carbohydrate D-galactose-*N*-acetyl-D-galactosamine in 95% of prostate cancers and no examples of adenosis studied.

Sclerosing adenosis is characterized by a more or less circumscribed proliferation of variably sized, often small glands embedded in a cellular and often edematous stroma (Figure 10.18) (65–67). Tiny microacini, cords, solid clusters, and single cells are seen. Sclerosing adenosis is unique in that the basal cells show myoepithelial differentiation. The lesion is further characterized using histochemical stains (Figure 10.19) (65–67). The luminal secretions stain strongly for PAS-D, which also highlights the distinct basement membrane surrounding the lesional glands and small clusters of epithelial cells. The alcian blue stain at pH 2.5 shows patchy but strong positivity with the luminal secretions as well as within the stroma. Cytoplasmic staining for either neutral or acid mucin is not present. The reticulin stain shows thin fibers surrounding the acini and cell clusters, and accentuates branching glands. Masson trichrome stain shows virtually no smooth muscle within the lesion; condensed collagen is present around lesional glands.

Distorted *rectal tissue* on prostate needle biopsy is a potential mimicker of prostate cancer, particularly since these glands lack a p63-stainable basal cell layer and can stain for alpha-methyl-acyl CoA racemase (AMACR) (68). Additional features mimicking prostate cancer include blue-tinged intraluminal mucinous secretions, extracellular mucin, and prominent nucleoli.

Benign *signet-ring cell change* can be observed in transurethral resection and needle biopsy specimens as a result of artifactual cytoplasmic vacuolation occurring in stromal cells and/or lymphocytes because of the effects of diathermy, trauma, or other cellular injuries (69). Exaggerated signet-ring cell change has also been reported in a benign,

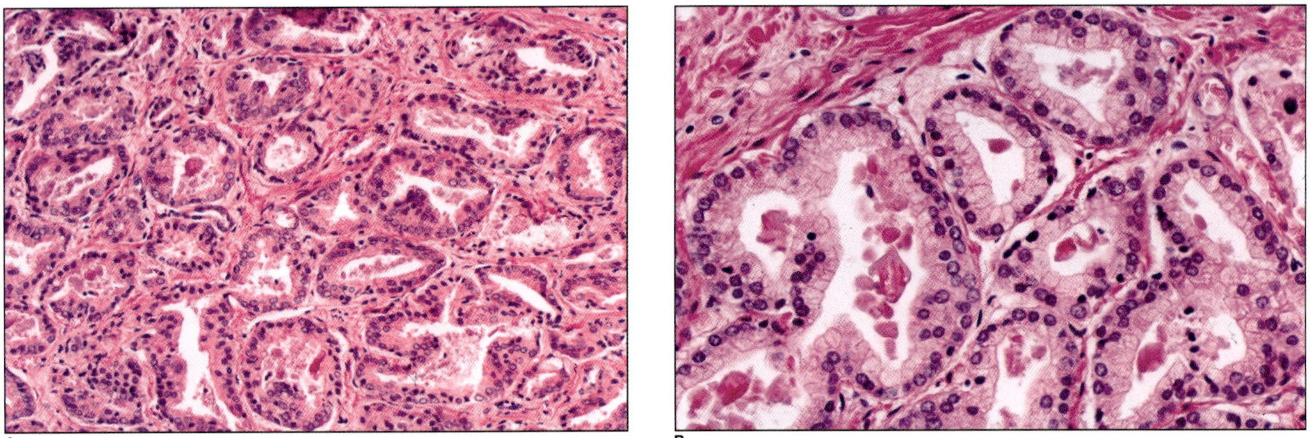

A B

FIGURE 10.16: Adenosis in this transurethral resection specimen is characterized by small, irregular, closely packed glands that form a circumscribed nodule (A). The focal presence of occasional basal cells and absence of macronucleoli (B) distinguish adenosis from low-grade carcinoma (H&E—A, 100×; B, 200×).

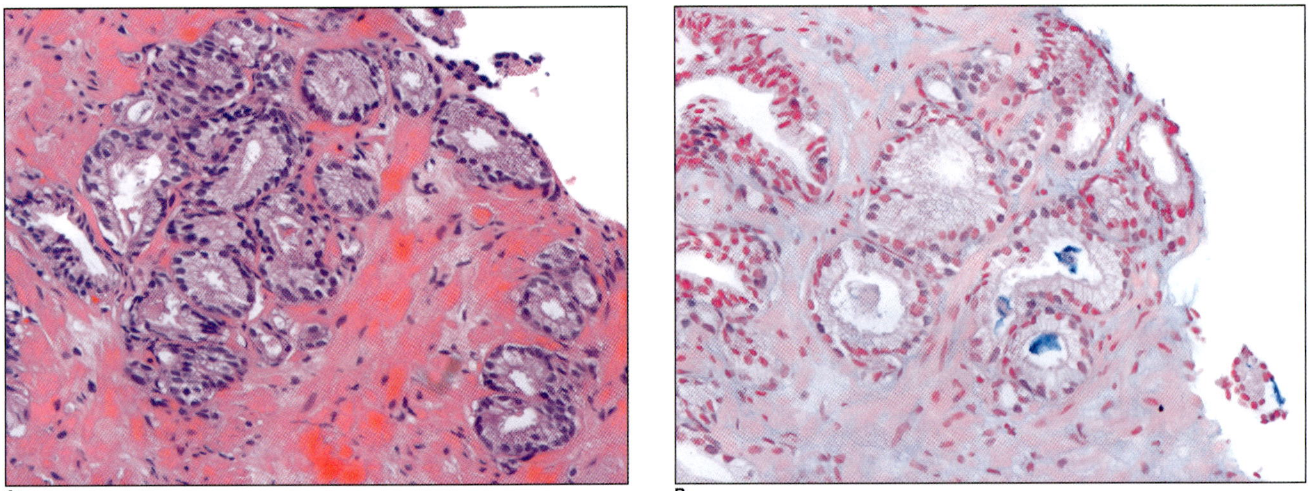

A B

FIGURE 10.17: Adenosis in this needle biopsy lacks nodular circumscription and mimics carcinoma (A). Although blue-tinged mucinous secretions are uncommon, intraluminal acid mucin is frequently found when histochemical stains are used (B) (A—H&E, 200×; B—alcian blue pH 2.5, 200×).

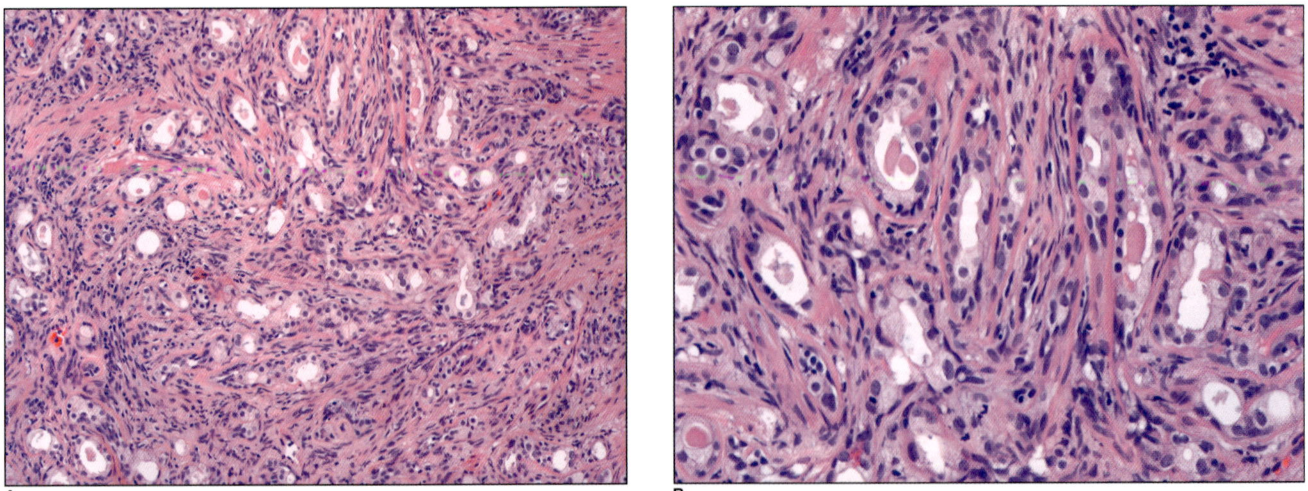

A B

FIGURE 10.18: Sclerosing adenosis is characterized by a proliferation of predominantly small glands embedded in a cellular stroma (A). Microacini, solid clusters, and single cells are present (B) (H&E—A, 100×; B, 200×).

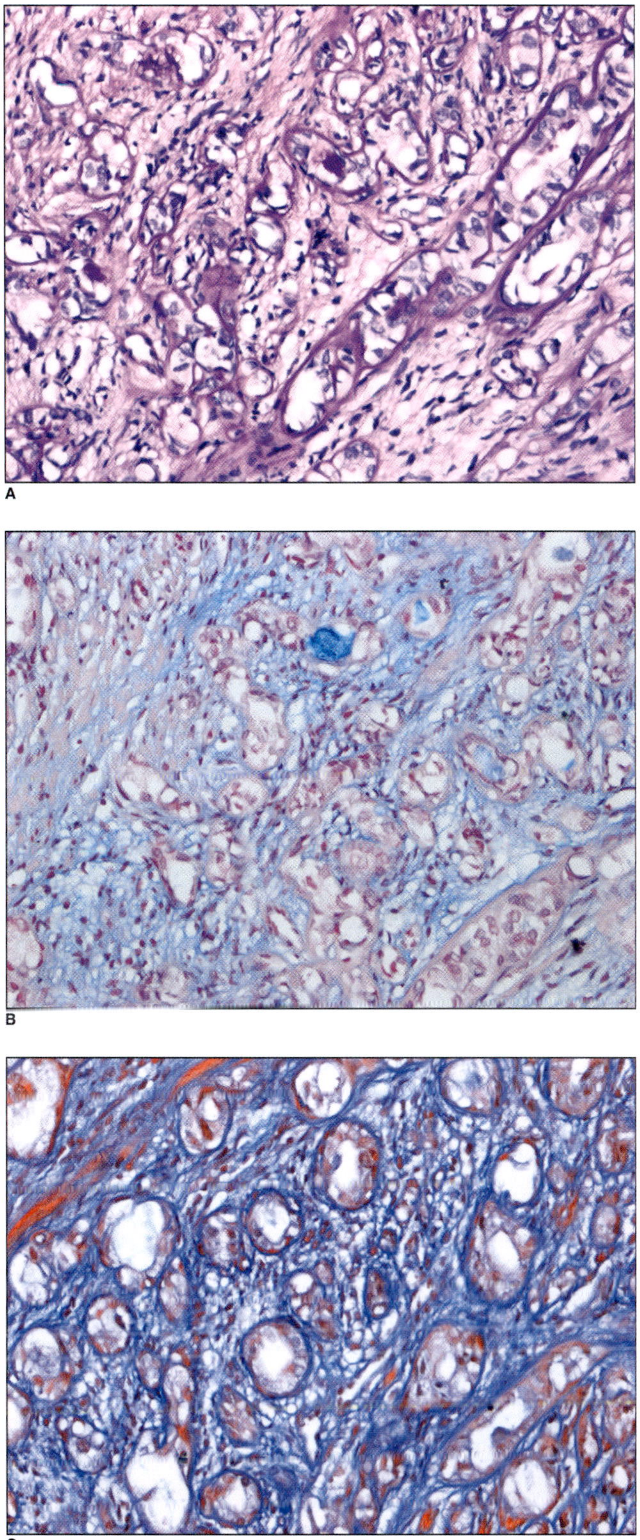

FIGURE 10.19: *The glands in sclerosing adenosis have a distinct basement membrane and luminal neutral mucin (A), acid mucins are focally present within luminal secretions and stroma (B), and smooth muscle is virtually absent (A—digested PAS, 200×; B—alcian blue pH 2.5, 200×; C—Masson trichrome, 200×).*

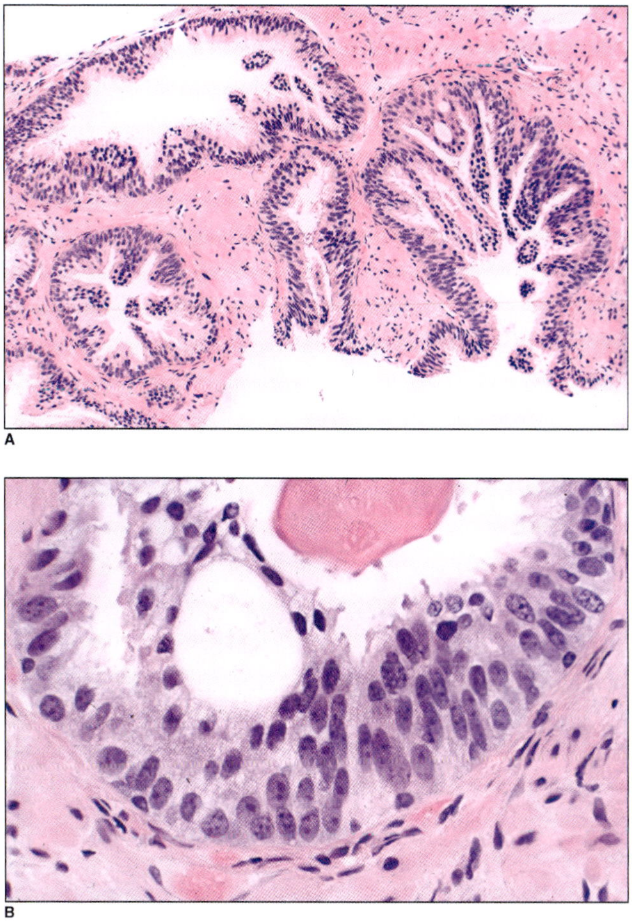

FIGURE 10.20: *High-grade PIN is recognized by the presence of normal-sized basophilic appearing glands and ducts having a micropapillary or tufted architecture (A). The stratified nuclei are enlarged and hyperchromatic with macronucleoli (B) (H&E—A, 100×; B, 400×).*

hyperplastic stromal nodule present in a simple prostatectomy specimen (70). The cytoplasmic vacuoles stain negative for mucin (PAS-D and mucicarmine). Signet-ring sinus histiocytes have been reported in 1.2% of lymph nodes from radical prostatectomy specimens (71). This reactive process can easily be confused with metastatic adenocarcinoma in H&E sections. Negative mucin stains (PAS-D, mucicarmine, and alcian blue), as well as lack of cytologic atypia and mitoses, are helpful in excluding malignancy.

High-Grade Prostatic Intraepithelial Neoplasia: Special stains are generally not needed in order to diagnose HGPIN, which has characteristic cytohistologic features in H&E sections (Figure 10.20). However, histochemistry has proven useful in demonstrating similarities between HGPIN and cancer and providing evidence linking HGPIN as a precursor lesion to peripheral zone carcinomas (72). A progressively increased content of acidic mucin (alcian blue pH 2.5) was reported in areas of PIN 2 (25%), PIN 3 (35%), and prostatic carcinoma (70%); sulfated acid mucin (alcian blue, pH 1) was secreted only in PIN 3 (5%) and carcinoma (10%) (73). Mucin stains have also helped to characterize unusual histologic types of HGPIN (Figure 10.21) (74). The signet-ring HGPIN (Figure 10.22) has a micropapillary and tufted architectural growth pattern; the prominent intracytoplasmic vacuoles have negative mucin staining with mucicarmine, alcian blue (pH 2.7), and PAS-D. In HGPIN with mucinous features, glands are distended and filled with intraluminal blue mucin (PAS and alcian blue positive), and the most frequent architectural pattern is the flat type.

Histochemical studies comparing AgNOR values have given conflicting results. Deschenes and Weidner (75), using a manual counting procedure, found AgNOR counts to increase progressively from benign nodular prostatic hyperplasia (BPH) through adenosis to carcinoma. HGPIN and carcinoma AgNOR counts were significantly different from those in BPH. Sakr et al. (76) found similar results. However, Cheville et al. (77), using image analysis, found no differences between AgNOR scores for hyperplasia, PIN, and adenocarcinoma.

Lectin binding studies in the prostate have given both variable and contradictory results. McNeal et al. (78) used a panel of lectins (see Table 10.2) and found reduced binding for all of them in PIN. Janssen et al. (79), however, reported that neoplastic prostate tissues and HGPIN exhibit a significantly higher number of peanut (*Arachis hypogea*) lectin (PNA) receptors than benign prostatic hyperplasias and low-grade PIN.

Prostatic Neoplasms

Prostatic Adenocarcinoma, Acinar (NOS): Early studies of mucin histochemistry of the prostate demonstrated neutral mucins in normal prostate and in prostatic carcinoma, whereas luminal alcian blue–positive acidic mucins containing sulfated sialic acid residues were primarily found in well- and moderately differentiated carcinomas and only rarely in nonmalignant prostatic tissue (80–82). In a more recent study, neutral (PAS positive) mucin was more frequently found in benign prostatic lesions (93%) as compared to carcinoma (36%), while acid (alcian blue positive) mucin was found more in carcinoma (68%) as compared to benign prostate lesions (16%) (83).

It is now apparent that the demonstration of acid mucins by histochemistry is not helpful in differentiating atypical acinar proliferations from well-differentiated prostatic carcinoma since it can be seen in basal cell hyperplasia sclerosing adenosis and adenosis (54–63%) (84). However, the finding of blue-tinged, stringy mucinous secretions in H&E-stained sections in a glandular proliferation suspicious for cancer can be useful for establishing a definitive diagnosis of malignancy (Figure 10.23). In studies outlining the diagnostic criteria of limited prostate adenocarcinoma, blue-tinged mucinous secretions were present in 18–52%

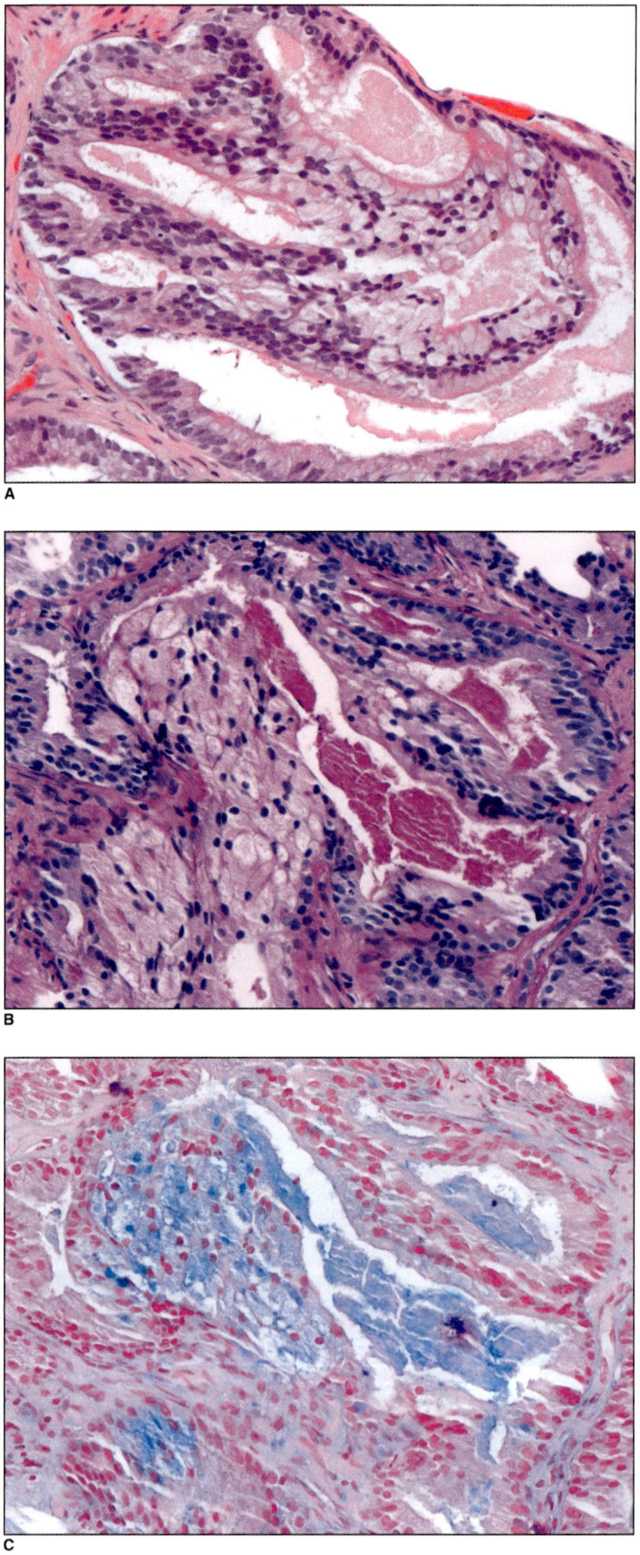

***FIGURE 10.21:** In this example of high-grade PIN, the cells have a foamy or secretory appearance (A). Intraluminal neutral mucin (B) and both cytoplasmic and luminal acid mucin (C) are present (A—H&E, 200×; B—digested PAS, 200×; C—alcian blue pH 2.5, 200×).*

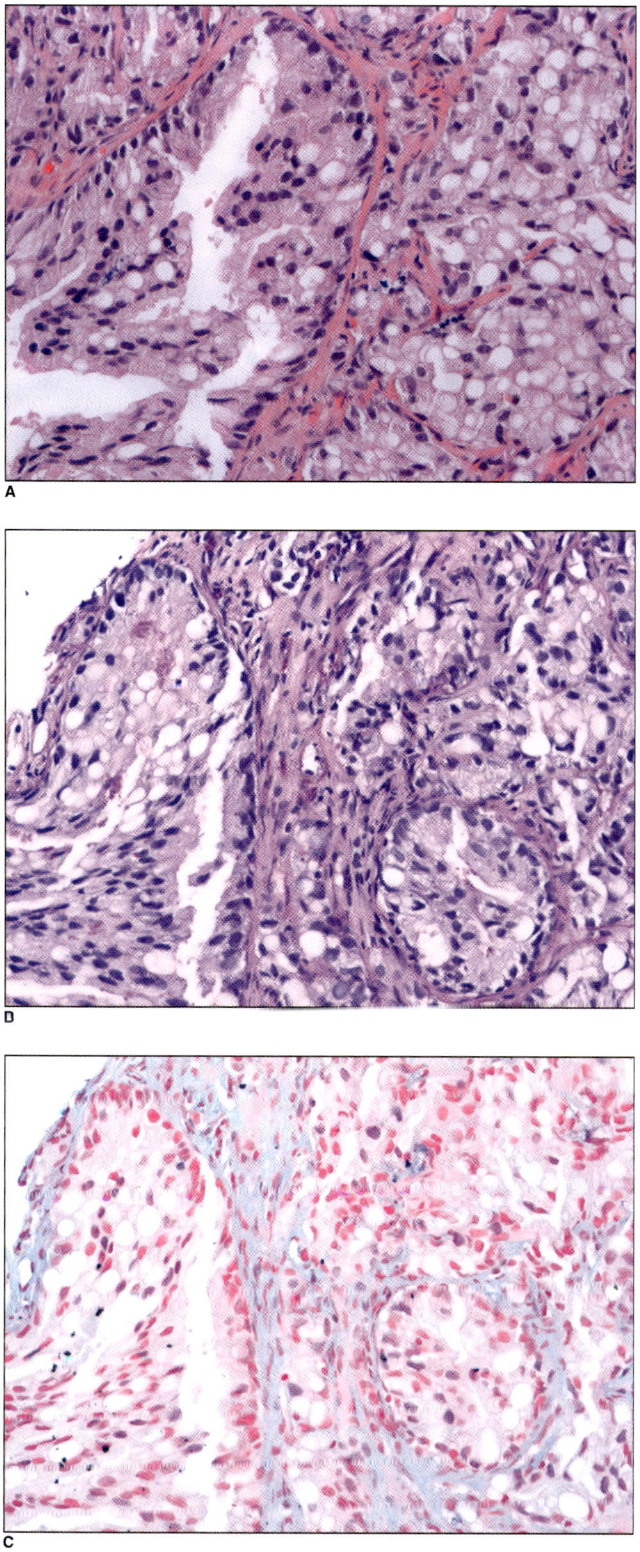

***FIGURE 10.22:** Signet-ring high-grade PIN is a rare variant. In this example, it is associated with infiltrating signet-ring carcinoma (A). Special stains for neutral (B) and acid (C) mucins are typically negative, and the appearance is due to intracytoplasmic lumina and vacuoles (A—H&E, 200×; B—digested PAS, 200×; C—alcian blue pH 2.5, 200×).*

of cases (84–86). In comparison, intraluminal amorphous pink material is present in 53–87% of cancers (84–86).

Collagenous micronodules, also termed mucinous fibroplasia, are microscopic nodular aggregates of paucicellular eosinophilic fibrillar stroma that are highly specific for carcinoma but are infrequent in needle biopsy specimens (0.1–2% cases) (Figures 10.24 and 10.25) (84,86,87,88) and 2–12.7% prostatectomies (87,89). McNeal et al. (90) first described collagenous micronodules as an incidental finding in mucin-producing adenocarcinoma. All stages of stromal micronodules were seen and ranged from lakes of pure mucin to areas in which mucinous lakes were obliterated by collagenous tissue. Bostwick et al. (87) also found that, in most cases, collagenous micronodules were associated with mucin-producing glands. The micronodules invariably stained with trichrome, indicating the presence of collagen, and alcian blue stain revealed at least focal luminal acidic mucin in glands with collagenous micronodules or adjacent to micronodules (87). Stains for elastin and glycogen were negative in the micronodules. Arangelovich et al. (89) could not demonstrate an association of the micronodules with mucin (negative staining with mucicarmine and alcian blue). The nodules stained weakly for PAS and blue with trichrome, and it was concluded that the stromal nodules are formed by the subepithelial accumulation of fragmented collagen fibers.

Morphologic features of *prostatic adenocarcinoma with atrophic features* include 1) loss of cytoplasmic volume such that tumor cell nuclei frequently touch the basal and apical cell membranes and 2) frequent flattening of tumor cell nuclei. Infiltrative growth pattern, nucleomegaly, and enlargement of nucleoli are key features in making the correct diagnosis. Intraluminal amorphous pink material, intraluminal wispy blue mucin, and collagenous micronodules are found less frequently than in usual carcinoma (91).

Prostatic foamy gland carcinoma is characterized by abundant foamy, xanthomatous cytoplasm and minimal cytologic atypia (Figure 10.26) (92,93). The foamy tumor cells are negative for mucin (PAS and mucicarmine) and lipid stains and are positive for colloidal iron and alcian blue (pH 2.5) consistent with the presence of numerous intracytoplasmic vesicles (93).

Focal individual cell *neuroendocrine differentiation* in conventional prostatic adenocarcinoma is the most common type of neuroendocrine differentiation in prostate malignancy and is present in approximately 10% of adenocarcinomas (94). The majority (and in some reports, 100%) of conventional prostatic adenocarcinomas show at least focal minimal neuroendocrine differentiation, although in approximately 10% of cases, this differentiation is quite extensive (94). In H&E sections, neuroendocrine differentiation can be recognized by cells having large eosinophilic cytoplasmic granules (positive for neuroendocrine markers) and does not represent true Paneth cell differentiation (Figure 10.27) (95). The prognostic significance of focal

TABLE 10.2: Lectins Used for Identifying Carbohydrate Residues*

Lectin Origin	Common Name	Acronym	Major Sugar Specification	Binding Inhibitor
Arachis hypogea	Peanut	PNA	Gal-α-(1-3)-GalNAc	Lactose
Bandeiraea simplicifolia	Griffonia	BS-I	α-D-Gal	Lactose
Canavalia ensiformis	Jack bean	Con-A	α-D-Glc, α-D-Man	α-D-Methyl-Man
Dolichos biflorus	Horse gram	DBA	α-D-GalNac	α-D-GalNAc
Glycine max	Soybean	SBA	α-D-GalNAc, α-D-Gal	α-D-GalNAc
Lens culinaris	Common lentil	LCA	α-D-Glc, α-D-Man	α-D-Methyl-Man
Ricinus communis	Castor bean	RCA-I	β-D-Gal	Lactose
Triticum vulgaris	Wheat germ	WGA	$(\beta\text{-}(1\text{-}4)\text{-D-GlcNAc})_2$ NeuNAc	NeuNAc
	Succinyl-WGA	S-WGA	$(\beta\text{-}(1\text{-}4)\text{-D-GlcNAc})_2$	β-D-GlcNAc
Ulex europaeus	Gorse	UEA-I	α-L-Fucose	α-L-Fucose

Gal = galactose, GalNAc = *N*-acetyl-galactosamine, GlcNAc = *N*-acetyl-glucosamine, NeuNAc = acetyl-neuraminic acid (sialic acid), Man = mannose
* From McNeal et al. (78) and Janssen et al. (79).

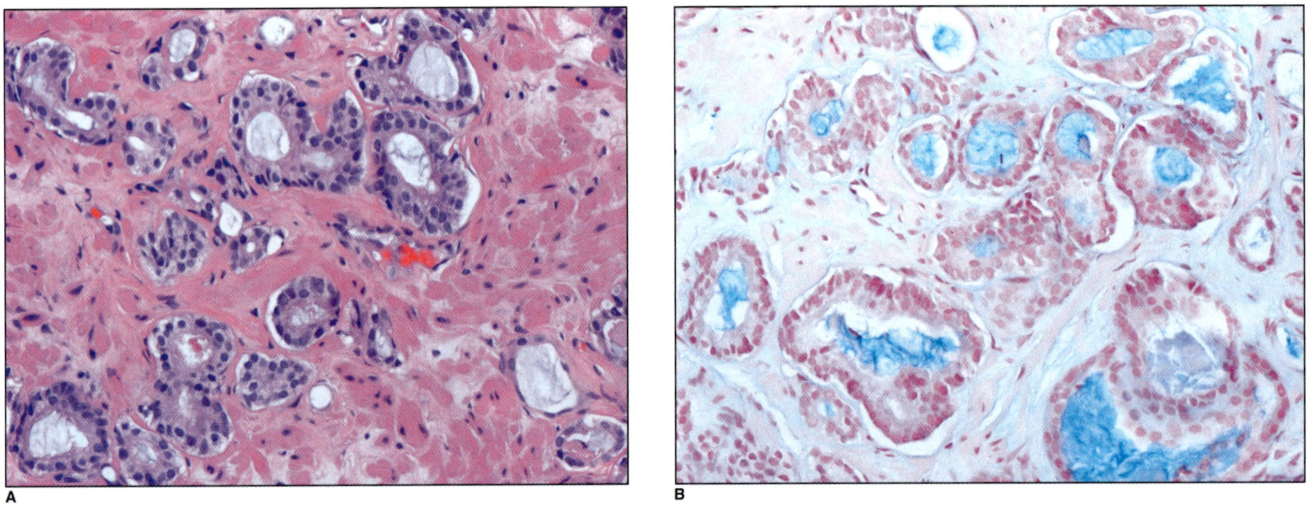

A B

***FIGURE 10.23:** The presence of pale-staining, blue-tinged, stringy mucinous secretions in an H&E section (A) can be useful in establishing a diagnosis of carcinoma. This alcian blue (+) acid mucin (B) is more frequent in carcinomas than benign prostatic lesions, but it is not a pathognomonic finding (A—H&E, 200×; B—alcian blue pH 2.5, 200×).*

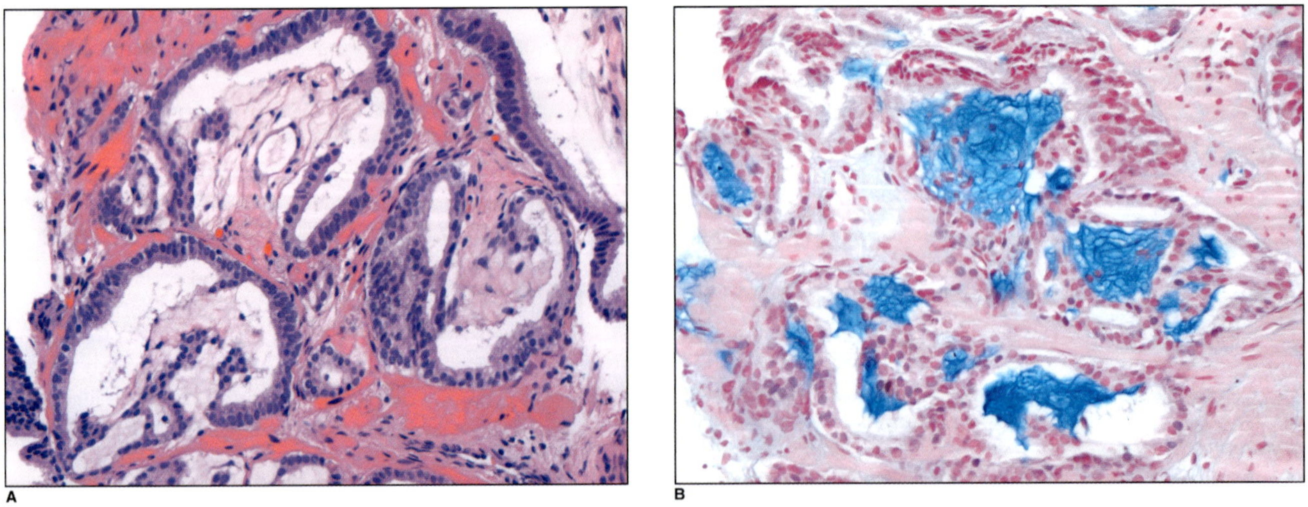

A B

***FIGURE 10.24:** Early collagenous micronodules are characterized by small paucicellular stromal mucinous lakes that are surrounded by neoplastic glands (A) and stain strongly for acid mucin B (A—H&E, 200×; B—alcian blue pH 2.5, 200×).*

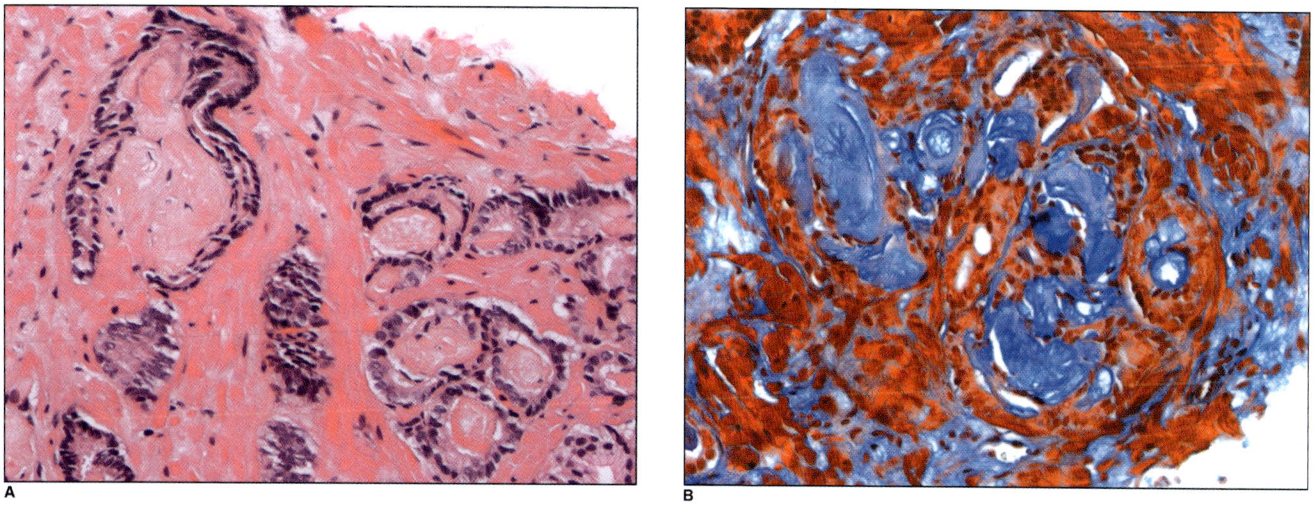

FIGURE 10.25: ***Collagenous micronodules typically appear as nodular aggregates of hyalinized stroma enveloped by infiltrating glands (A) that stain for collagen (B) (A—H, 200×; B—Masson trichrome, 200×).***

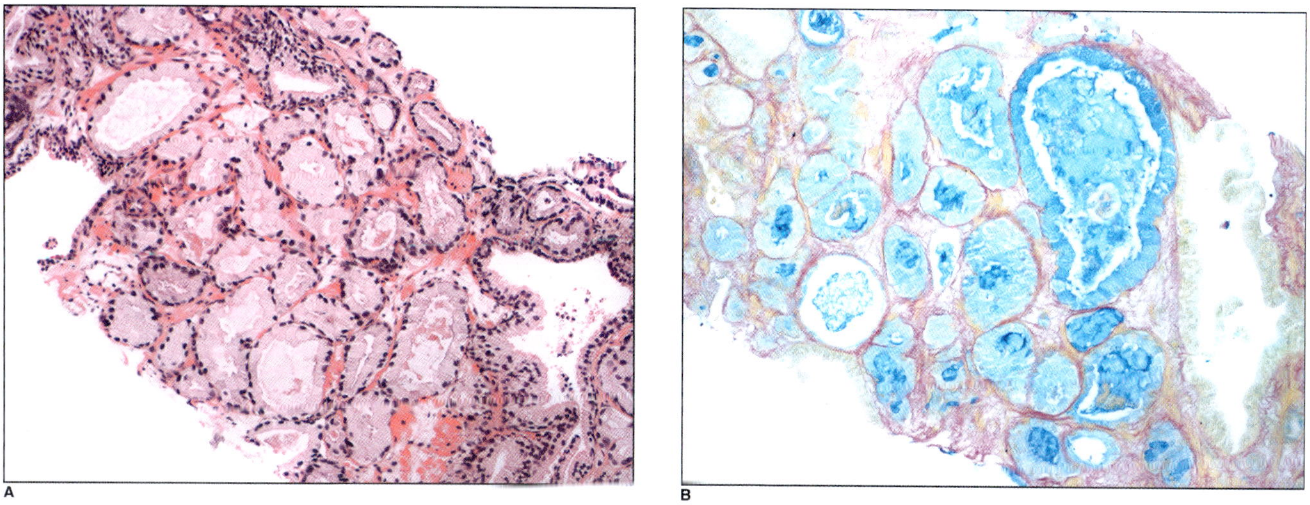

FIGURE 10.26: ***Prostatic foamy gland carcinoma can have a deceptively benign appearance with medium-sized glands, abundant xanthomatous cytoplasm, and minimal cytologic atypia (A). The foamy appearance is due to abundant intracytoplasmic vesicles, which stain with colloidal iron (B) (A—H&E, 100×; B—colloidal iron, 100×).***

neuroendocrine differentiation in prostate carcinoma has been controversial (94). The presence of neuroendocrine cells on needle biopsy or within radical prostatectomy specimens does not invariably indicate aggressive tumor (96).

Variants of Prostatic Carcinoma: Uncommon or rare variants of prostatic carcinoma can histogenetically be related to the secretory epithelium (mucinous and signet-ring cell adenocarcinoma), the basal cells (adenoid cystic carcinoma), and neuroendocrine cells (pure carcinoid tumor and small cell carcinoma) (97).

Criteria for the diagnosis of *mucinous adenocarcinoma* of the prostate include a total tumor volume more than 25% mucinous and single or clustered tumor cells floating in mucin lakes (Figure 10.28) (98). A cribriform pattern may be prominent; a pattern of tall columnar epithelium is absent and signet ring cells or goblet cells are rare (98,99). Mucin lakes and intraluminal spaces stain intensely with alcian blue (pH 2.7), mucicarmine, and PAS with and without diastase (Figure 10.29); staining intensity is diminished or absent altogether with alcian blue (pH 0.9), indicating the presence of predominantly nonsulfated acidic mucin (98). The tumor cells are negative with the mucin stains. The sialomucin in mucinous adenocarcinoma has been shown with histochemistry to be heavily O-acylated (100). Lectin histochemistry has also been used to characterize the composition of the acidic mucins (90). D-Galactosyl-*B*(1,3)-*N*-acetylgalactosamine and *B*-*N*-acetylgalactosamine residues are consistently present, and the secretions are

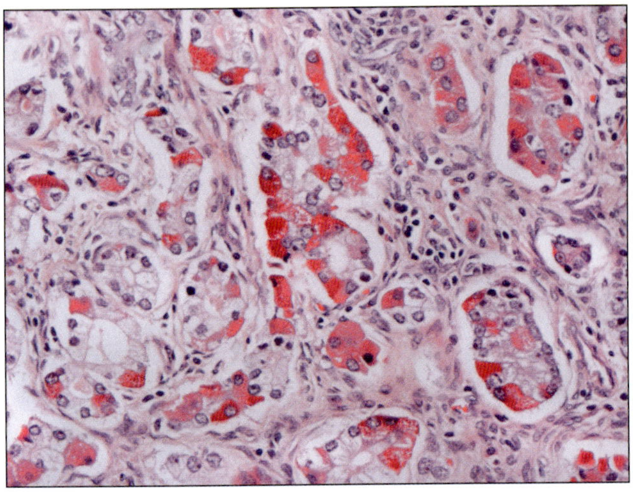

FIGURE 10.27: ***Focal neuroendocrine differentiation in conventional adenocarcinoma may be inapparent in H&E sections. In this example, there is a patchy distribution of neuroendocrine cells having red cytoplasmic granules (H&E, 200×).***

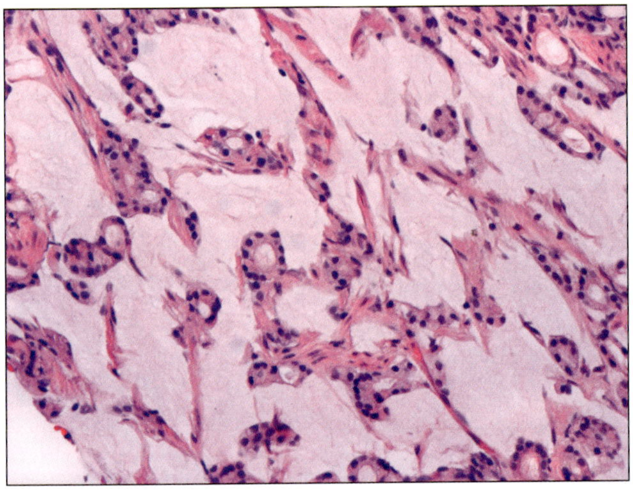

FIGURE 10.28: ***Prostatic mucinous adenocarcinoma consists of neoplastic glands floating in mucin lakes. The glands resemble conventional acinar carcinoma. Tall columnar cells, signet-ring cells, and goblet cells are not present (H&E, 200×).***

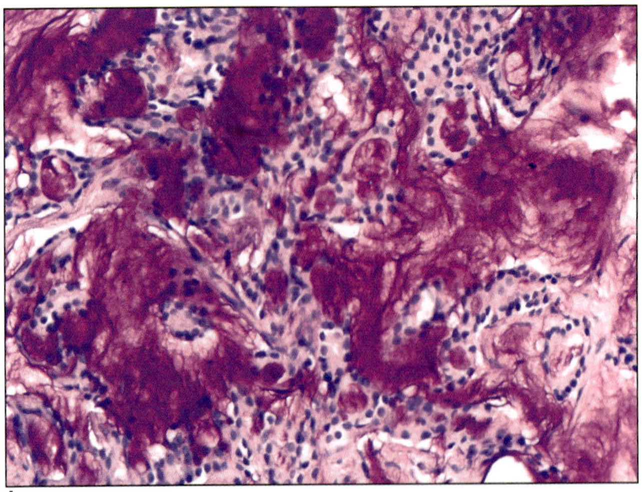

A

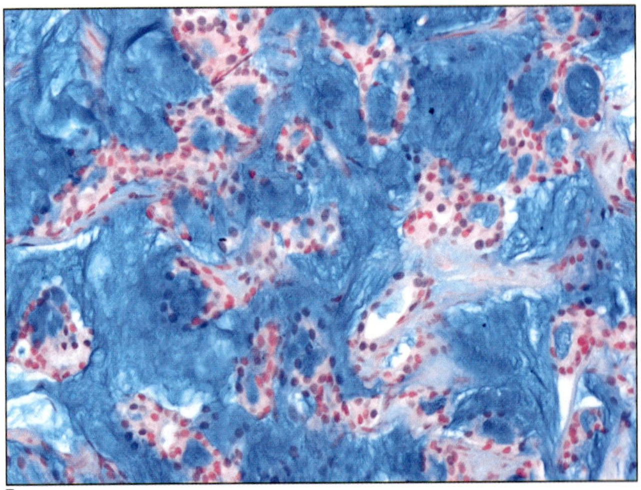

B

FIGURE 10.29: ***Stains for neutral (A) and acid (B) mucins are positive in prostatic mucinous adenocarcinoma. The tumor cells are negative (A—digested PAS, 200×; B—alcian blue pH 2.5, 200×).***

identified as O-linked by their staining with peanut (PNA), soybean (SBA), wheat germ (WBA), and absence of staining with jack bean (Con-A) and common lentil (LCA). Mucinous carcinoma of the prostate is regarded as a variant of high-grade adenocarcinoma (Gleason grade 4). The differential diagnosis includes secondary adenocarcinoma of colonic origin and urothelial-type adenocarcinoma arising from either the prostatic urethra or the proximal ducts, which may require immunohistochemistry for an accurate diagnosis (101).

Prostatic signet-ring cell carcinoma is typically negative for neutral (PAS-D) cell (see Figure 10.22) and acidic (alcian blue) mucins, and ultrastructurally, the signet-ring appearance is due to intracytoplasmic lumina and vacuoles (102). However, staining for mucin has been variable (Figure 10.30) with approximately 50% of reported cases weakly positive for neutral and/or acidic mucin (103–107). A rare case has shown positive for Sudan black staining, indicating the presence of intracellular fat (108). Signet-ring cell carcinoma is usually associated with other forms of high-grade prostatic carcinoma. Like the rare prostatic mucinous carcinoma with signet-ring cells, which responds poorly to endocrine therapy (109), it is an aggressive tumor with a poor prognosis (106).

Malignant neoplasms derived from prostatic basal cells have been variably termed *adenoid cystic carcinoma* (110,111) and *adenoid cystic/basal cell* (112) or basaloid (113) *carcinoma*. These carcinomas can have a predominant adenoid pattern (nests and trabeculae of cells with moderate cytoplasm and cribriform growth with microcystic spaces containing basement membrane material) *or* a basaloid pattern (nests

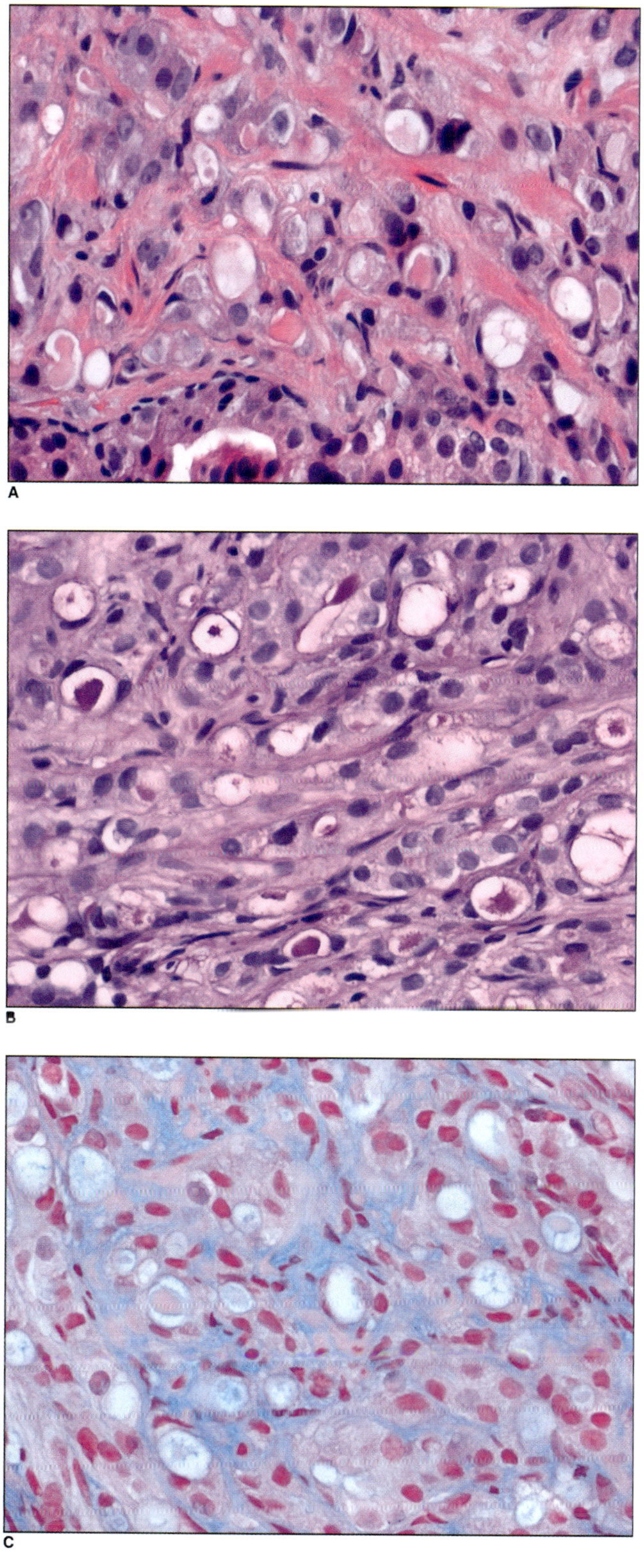

***FIGURE 10.30:** The infiltrating signet-ring cells in this case represented a focal finding in an otherwise typical, high-grade prostatic carcinoma. The cells are characterized by a clear cytoplasmic vacuole that displaces the nucleus to one side (A). The signet-ring cells contain neutral mucin (B) and show focal weak staining for acid mucin (C) (A—H&E, 400×; B—digested PAS, 400×; C—alcian blue pH 2.5, ×400).*

of basaloid cells with scant cytoplasm and peripheral palisading). Material within lumina stains positively with both PAS-D and mucicarmine (110). The stroma is desmoplastic or myxoid. A diffusely infiltrative growth pattern, perineural invasion, and extraprostatic extension are features of malignancy.

These carcinomas may be difficult to distinguish from forms of florid basal cell hyperplasia (Figure 10.31), termed adenoid cystic–like tumor (114)/hyperplasia (113) and adenoid basal cell tumor (111). These lesions can produce well-formed expansile nodules composed of basaloid nests. Cribriform lumina contain hyalinzed eosinophilic material or watery basophilic material that stains strongly with PAS-D and alcian blue (pH 2.5) and weakly with mucicarmine (111,114). The stroma contains hyaluronidase-sensitive, and alcian blue– (pH 2.5) and colloidal iron–positive material (114). Individual nests are surrounded by a thin PAS-positive basement membrane (111).

Carcinoid tumors of the prostate (Figure 10.32) are extremely rare and should be distinguished from "carcinoid-like" adenocarcinomas. The cells of pure prostatic carcinoid are argyrophilic, as are the mixed tumors composed of conventional adenocarcinoma with foci of carcinoid-like neuroendocrine differentiation (115). Pure carcinoids can have an insular, trabecular, glandular, or mixed architectural pattern and are composed of uniform rounded, polygonal, or spindle-shaped cells with round nuclei lacking conspicuous nucleoli. True prostatic carcinoid tumors appear to have a relatively indolent course in comparison to the carcinoid-like carcinomas, which are reported to have an aggressive behavior (115).

Small cell carcinoma of the prostate is rare and is an extremely aggressive tumor. It is morphologically similar to small cell carcinoma of the lung and must be distinguished from undifferentiated (Gleason grade 5) carcinoma of the prostate with small cell features (116,117) because of therapeutic differences, chemotherapy/radiation versus antiandrogen therapy, respectively. In about half the cases, small cell carcinoma coexists with the common adenocarcinoma (118). Routine histochemical studies with argentaffin (Fontana Masson method) and argyrophil (Grimelius method) stains are not useful because of low sensitivity—in 20% of cases, 35% of cells were argyrophil and 5% argentaffin positive (118).

Rare cases of large cell neuroendocrine carcinoma have been described (119). These tumors seem to have a predilection to arise from conventional prostatic adenocarcinoma following long-term antiandrogen therapy.

TESTIS AND PARATESTIS

Histochemical staining is no longer used on a routine basis in the diagnosis of most testicular and paratesticular lesions; however, it can still be extremely helpful in selected cases. The underlying organism in some cases of

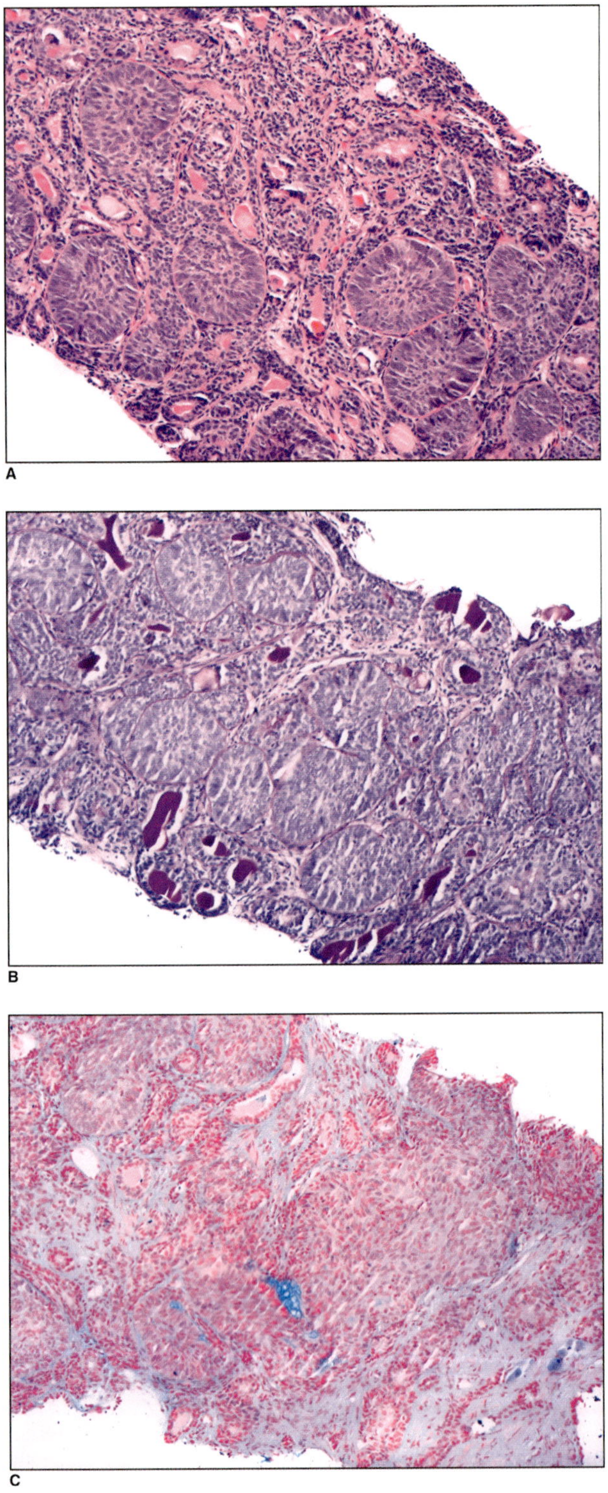

FIGURE 10.31: The adenoid basal form of florid basal cell hyperplasia has closely packed basaloid nests with peripheral palisading and foci of luminal differentiation, as well as focal interconnecting cords (A). The basaloid cells are two to three times larger than those of typical basal cell hyperplasia. In the adenoid pattern, the gland-like spaces contain variably dense, eosinophilic, and intensely PAS (+) concretions (B) as well as focal acid mucin (C) (A—H&E, 100×; B—digested PAS, 100×; C—alcian blue pH 2.5, 100×).

epididymo-orchitis can be diagnosed using special stains, which can also help exclude an infectious etiology in noninfectious granulomatous inflammation. Pre- and postpubertal hypogonadism can be separated, and regressed testicular tissue definitively diagnosed with the aid of histochemistry. Disruption of the elastic layer of arterial walls may be needed to make a diagnosis of vasculitis involving the testis. Although surgical pathologists now rely heavily on the use of immunohistochemical stains to classify germ cell neoplasms, histochemistry can have added value in this setting. The testis is the target of a broad range of metastatic tumors, many of which are amenable to histochemical analysis. Similarly, paratesticular tissues can be the site of cytologically confusing benign lesions, as well as a variety of benign and malignant neoplasms with characteristic histochemical profiles.

Nonneoplastic Conditions of the Testis

Infertility and Cryptorchism: The elastic fibers of the tunica propria surrounding seminiferous tubules are attenuated in cryptorchid testes from postpubertal patients, but are completely absent in undescended testes from prepubertal and pubertal patients (120). The attenuated fibers can be visualized using a Weigert resorcin-fuschin stain. Aldehyde fuschin, orcein, and Verhoeff's van Gieson (EVG) staining can be used for the same purpose. In the Sertoli cell–only syndrome, Sertoli cells can be confirmed as the only cells within seminiferous tubules on PAS and trichrome staining (Figure 10.33). Postpubertal cryptorchid testes are prone to intratubular germ cell neoplasia (ITGCN), indicating a significant risk for the development of all types of germ cell tumors except spermatic seminoma (121). ITGCN is reliably identified by the numerous PAS-positive intracytoplasmic granules (Figure 10.34).

Systemic Diseases: The testis can be involved in systemic diseases such as vasculitis and amyloidosis, both of which can present as a mass. Although an association with polyarteritis nodosa has been reported most frequently, the testes have also been involved by systemic Wegener's granulomatosis and Henoch-Schönlein purpura (122–124). Occasionally, vasculitis can present in the testis (125,126). Vessels in the testicular parenchyma, tunica vasculosa, and epididymis can be involved. EVG and Masson's trichrome stains can reveal disrupted elastic laminae and arterial aneurysms, even in the chronic phases of the disease. Martin's scarlet blue or trichrome histochemistry can confirm the presence of fibrin in the active phase.

Testicular amyloidosis is far less common, involving a similarly wide array of structures including blood vessel walls in the testis, the epididymis and the spermatic cord, and the tunica propria of the seminiferous tubules (127). Amyloidosis can not only present as a mass but can sometimes also displays rapid growth (128,129). Congo red

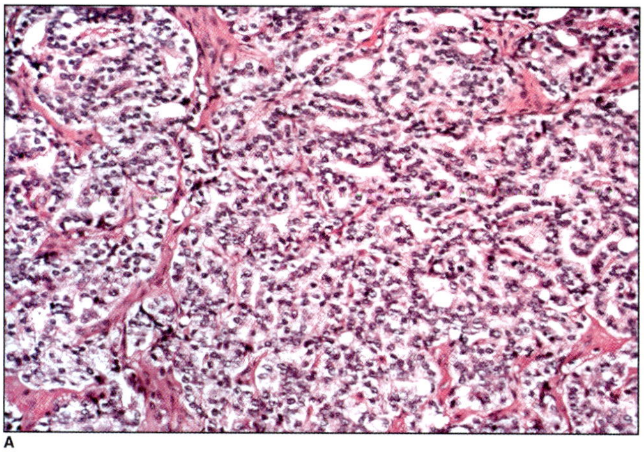

A

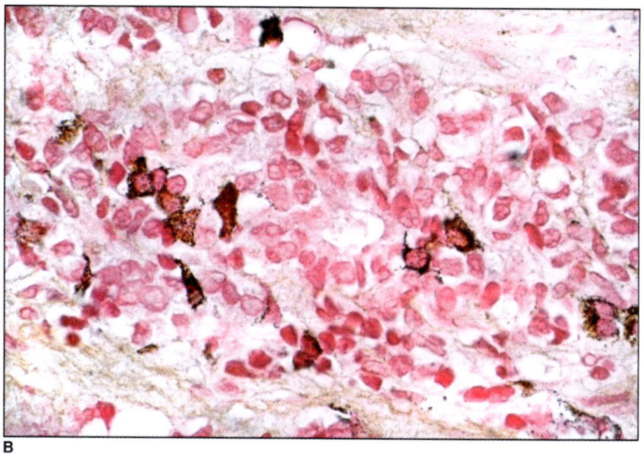

B

FIGURE 10.32: *This prostatic carcinoid tumor consists of nests of cells with uniform nuclei and a trabecular-glandular growth pattern (A). Argyrophil stain demonstrates numerous positive (brown) cells (B) (A—H&E, 100×; B—Churukian-Schenk, 400×).*

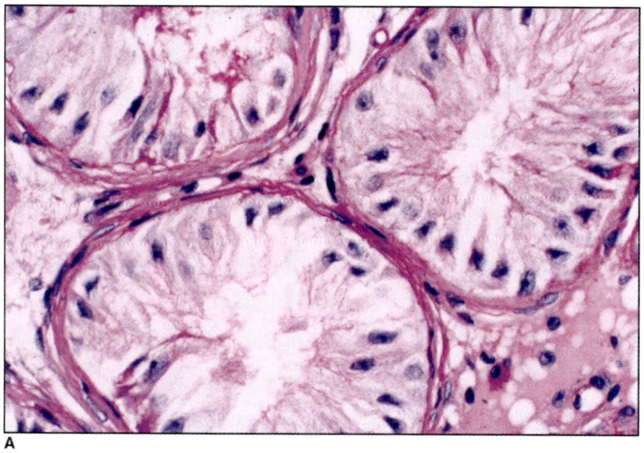

A

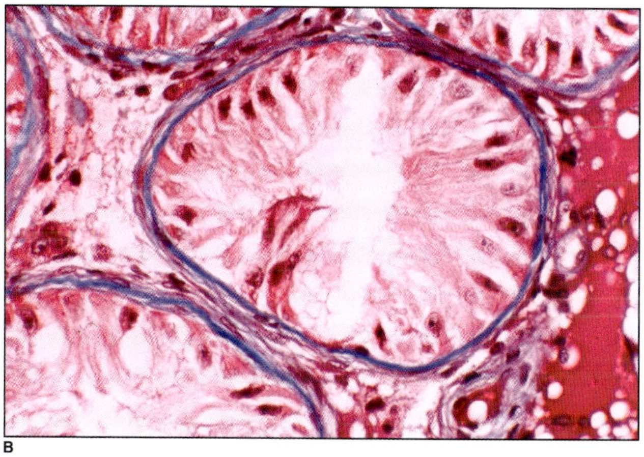

B

FIGURE 10.33: *Seminiferous tubules with a pure population of Sertoli cells in Sertoli cell–only syndrome highlighted on PAS (A) and trichrome staining (B) (PAS 400× and trichrome 400×).*

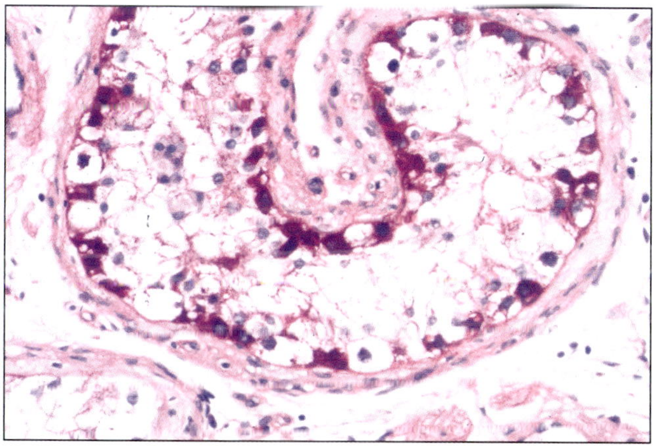

FIGURE 10.34: *ITGCN with intracytoplasmic granules highlighted on PAS staining (PAS, 400×).*

staining with specificity for amyloid confirmed by apple-green and orange-yellow birefringence is diagnostic; however, the crystal violet stain can be useful as well, as long as it is examined soon after preparation.

Infections: As infections of the testis are relatively uncommon in developed countries, the diagnosis may be missed in this setting. Histochemical staining should be considered in cases with unexplained orchitis, particularly if associated with granuloma formation. A whole range of bacterial infections have been documented at this site, including *E. coli*, *Pseudomonas aeruginosa*, staphylococci, streptococci, and various salmonellae and klebsiellae. Many cases of orchitis involve the epididymis as well, with *Chlamydia trachomatis* and *Neisseria gonorrhoeae* being the most common culprits (130,131). Both Brown and Brenn, and Brown and Hopps stains can demonstrate bacteria but have low sensitivity and will also stain some fungi particularly *Candida* and *Torulopsis*. Various bacteria may be aberrantly processed by histiocytes, leading to granulomatous inflammation with sheets of histiocytes containing the characteristic Michaelis-Gutmann cytoplasmic bodies of malacoplakia. *E. coli* is the most commonly implicated organism (132–135). The phagolysosomes contain incompletely destroyed bacteria coated with calcium and iron. Thus, the Prussian blue iron stain, PAS, and the von Kossa stain for calcium are most useful for demonstrating the lamellated inclusions (Figure 10.35).

Granulomatous inflammation of the testis may be attributable to syphilis, tuberculosis or other mycobacteria, leprosy, or brucellosis among other organisms (129,130,136–139). Acid-fast (Ziehl-Neelsen/Kinyoun) stains for mycobacteria, Fite or Fite-Faraco stains for leprosy, and Warthin-Starry stains for spirochetes may be helpful in identifying these agents. The sensitivity of these stains is low and organisms are difficult to detect unless the bacterial load is significant, as in immunosuppressed individuals. In the early stages of leprosy, lepra cells with large numbers of organisms contained in cytoplasmic vacuoles are present within vessel walls (140). Sarcoidosis rarely involves the testis, with the diagnosis being one of exclusion (141,142). Culture of tissue samples would be necessary to make this diagnosis, as infection cannot be confidently excluded based only on histochemistry.

Mycotic infections involve the testis uncommonly, producing variably tumefactive and/or multiple lesions, that have on occasion been mistaken for metastatic disease. A broad range of organisms have been implicated including *Actinomyces israelii*, *Candida* sp., and *H. capsulatum* (143–145). *Echinococcus granulosus*, *Echinococcus multiloculare*, and *Toxoplasma gondii* are among the protozoan infections that have been described in the testis (146,147). PAS, Gomori's, or Grocott's MS preparation can be useful in the identification of fungal hyphae and spores, while Giemsa, trichrome, and PAS stains may help visualize protozoa. Reliable speciation cannot, however, be performed using this method exclusively.

Testicular Neoplasms

Intratubular Germ Cell Neoplasia: As mentioned in the section on cryptorchid testes, identification of ITGCN is important because of its association with subsequent germ cell neoplasms. Only spermatocytic seminomas, unusual pure yolk sac tumors, and mature teratomas are not preceded by or associated with ITGCN. A whole variety of lectins can be used to stain ITGCN, *Datura stramomium*, *Lycopersicum esculentum*, *Phaseolus vulgaris*, and concavalin A lectins among them (148). ITGCN can involve the rete testis and the interstium between seminiferous tubules as well. PAS staining of the numerous intracytoplasmic glycogen granules, cleared by diastase digestion, may aid in distinguishing the process in these areas (121). Almost a third of patients with extragonadal germ cell tumors involving the mediastinum, and especially the retroperitoneum, demonstrate intratesticular ITGCN, indicating a higher risk of subsequent testicular tumors (149). Careful histopathologic examination of the testes for areas of scarring and ITGCN suggests that the majority of retroperitoneal germ cell tumors represent metastases from burnt-out testicular primaries (150–153).

Germ Cell Neoplasms: Although most types of germ cell neoplasms can be distinguished using conventional H&E morphology together with immunohistochemical staining, it should not be forgotten that histochemical stains can also aid in their diagnosis. Up to a third of germ cell tumors of the testis have a mixture of components, with various combinations of the different types occurring (153). Diastase-sensitive PAS-positive glycogen granules are found in the majority of usual-type seminomas, embryonal carcinomas, and choriocarcinomas, whereas teratomas and spermatocytic seminomas as well as testicular lymphomas all lack PAS staining (Figure 10.36). PAS can also be used to highlight the hyaline globules and intercellular basement membrane material (or so-called parietal yolk sac differentiation) of yolk sac tumors (Figure 10.37) (154,155). Alkaline phosphatase has a strong membranous reactivity in seminomas

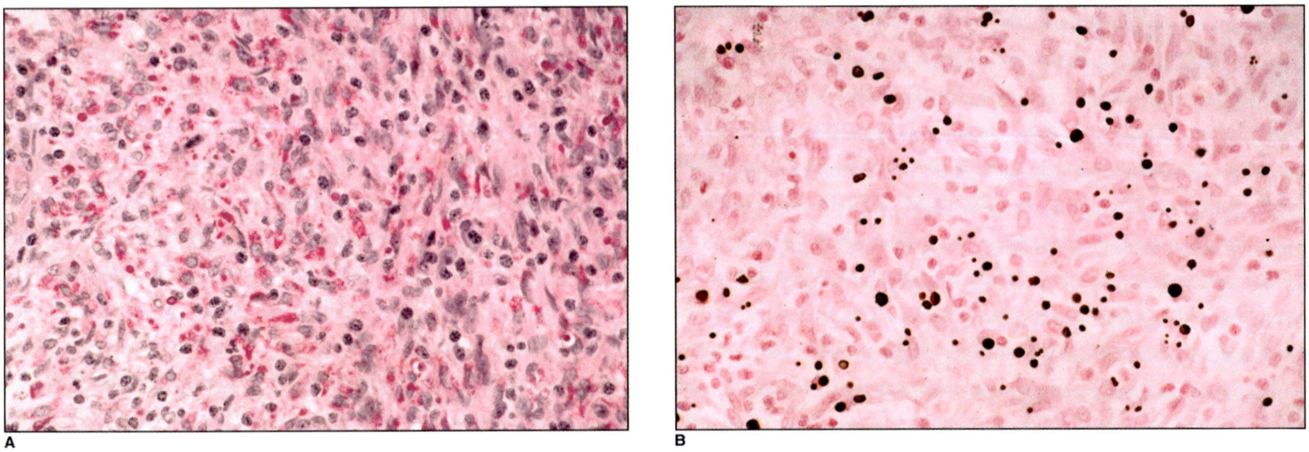

FIGURE 10.35: ***Testicular malacoplakia shows PAS-positive cytoplasmic granules (A). A von Kossa stain better highlights the Michaelis-Gutmann bodies in this lesion (B).***

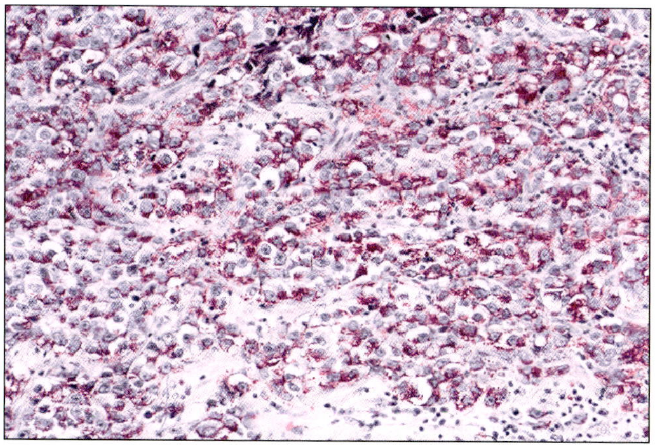

FIGURE 10.36: Usual-type seminoma demonstrating numerous diastase-sensitive PAS-positive intracytoplasmic granules (PAS without diastase 100×).

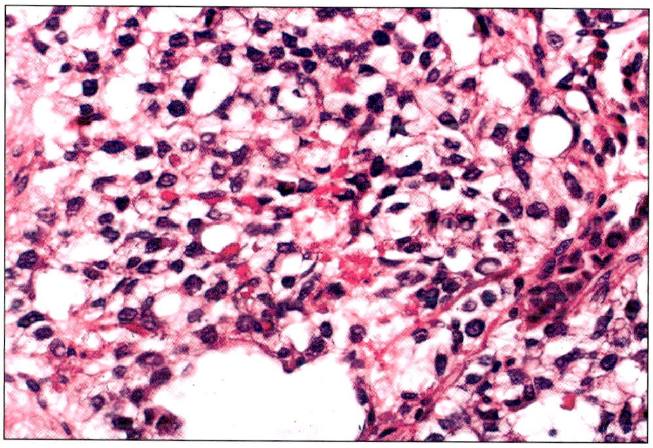

FIGURE 10.37: Hyaline globules highlighted on PAS staining of a yolk sac tumor (PAS, 200×).

and embryonal carcinomas but fails to stain choriocarcinomas and teratomas, apart from the brush border in glandular elements of the latter (156).

Carcinoid Tumors: Primary carcinoid tumors involving the testis are commonly associated with other teratomatous elements, but unusual monodermal variants of teratomas with a carcinoid phenotype do occur. They display both intracytoplasmic argyrophilic granules using silver based preparations such as Grimelius, Churukian-Schenk, or Sevier-Munger stains, as well as argentaffin granules, which appear black with the Masson-Fontana preparation, while the diazo method stains them orange (157–159). This dual pattern of staining can help distinguish these tumors from metastatic carcinoid tumors, with those from the lung, for example, being typically argyrophilic, but rarely argentaffinic.

Sex Cord Stromal Tumors: Leydig cell tumors are the most common in this group, but their multitudinous morphologic patterns can make diagnosis problematic. Reinke's

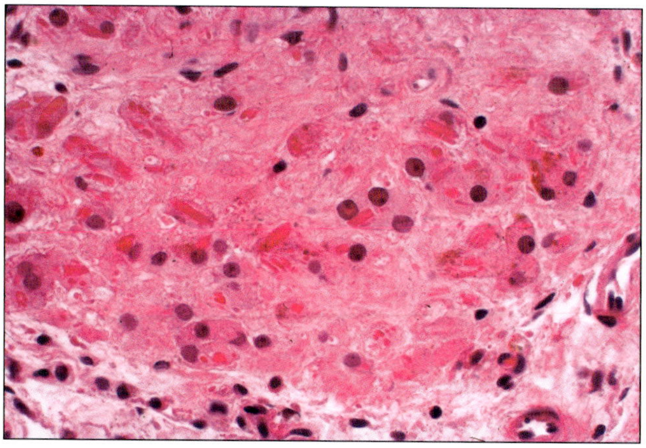

FIGURE 10.38: Testicular Leydig cells contain prominent Reinke crystals (H&E, 200×).

crystals within Leydig cell tumors are often visible on H&E stain (Figure 10.38) but can be more easily identified using the Masson trichrome stain. Frozen tissue must be used to perform any of the lipid stains, namely, Sudan III, Sudan IV, or oil red O, as formalin processing removes the cytoplasmic lipid (160).

Metastatic Tumors Involving the Testis: While lymphomas of various types may present with testicular, spermatic cord, or epididymal symptoms, most have nodal involvement on workup of the patient. The majority of testicular lymphomas in adults are of diffuse large B-cell type; however, virtually all types of lymphomas have been reported at this site. As lymphoma can involve the seminiferous tubules, its PAS negativity is helpful to rule out ITGCN (161). Methyl green-pyronine will label large cell lymphoma, as well as myeloid leukemia (granulocytic sarcoma); the latter displays granular cytoplasmic positivity with Leder, Giemsa, and toluidine blue stains, while lymphoma does not.

Although malignant melanoma metastatic to the testis is found in up to 15% of patients at autopsy, it is rarely diagnosed antemortem (162). The Fontana-Masson and Schmorl's methods will demonstrate black granular cytoplasmic staining in melanoma. Metastatic adenocarcinoma, sometimes in the differential diagnosis with metastatic melanoma, will usually have intracytoplasmic mucin with diastase-resistant PAS and mucicarmine positivity (Figure 10.39) (163). In differentiating metastatic adenocarcinoma from primary lesions, it should be remembered that the glandular elements within teratomas may of course contain mucin. The PAS positivity of the glycogen within seminomas will be removed with the addition of diastase (164).

Nonneoplastic Conditions of the Paratestis

Mullerian duct remnants can occur in association with inguinal hernias or congenital abnormalities of various

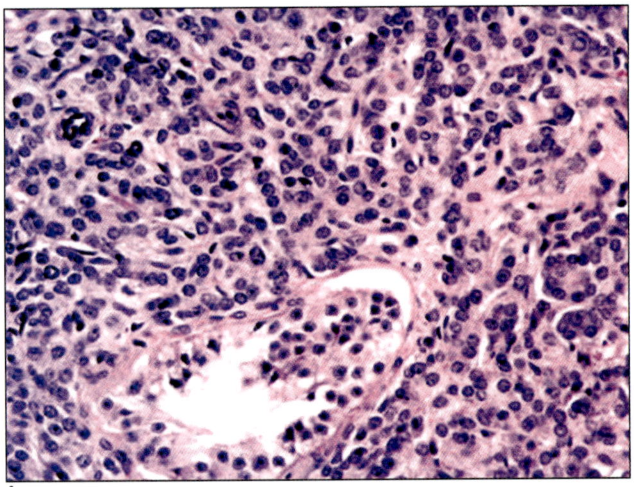
A

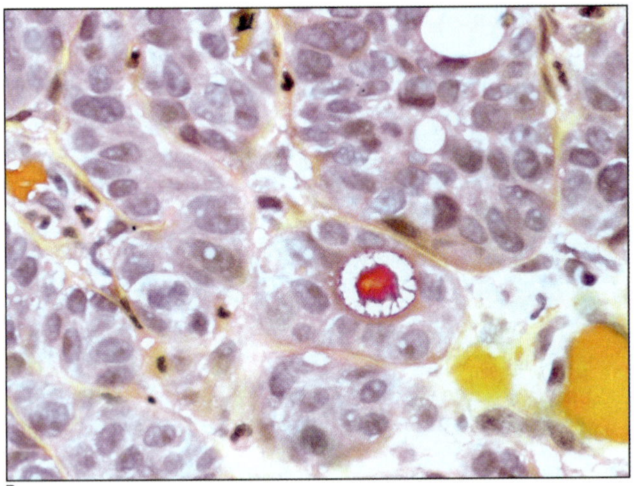
B

FIGURE 10.39: Poorly differentiated prostatic adenocarcinoma metastatic to the testis surrounds a residual seminal vesicle (A). At higher magnification, a mucicarmine stain shows positivity in a rare lumen (B) (A—H&E, 100×; B—mucicarmine, 400×).

urogenital organs (165,166). Their identification can be aided by the use of trichrome to delineate the surrounding smooth muscle coat. Sperm granulomas involving the epididymis are found most commonly postvasectomy but can also arise after infections in this region. The granulomatous histiocytes may contain PAS-positive granules; however, no mycobacteria, fungi, or spirochetes are detected on acid-fast, silver, or Warthin-Starry staining (167–169). Vasitis nodosa also occurs postvasectomy and is characterized by a fusiform nodular thickening of the vas deferens by a reactive proliferation of ductular epithelium. As this process can invade nerves, it represents a classic diagnostic pitfall, frequently being mistaken for an adenomatoid tumor or low-grade adenocarcinoma (170,171). Masson trichrome, EVG, luxol fast blue, and Bodrian's preparations can delineate nerve structures infiltrated by benign glands more clearly, with nerve appearing blue on the luxol fast blue stain and black on Bodrian's preparation (Figure 10.40) (172). Two other benign conditions in which PAS staining can be helpful are rete testis hyperplasia in which the characteristic hyaline globules within the epithelium are positive with diastase resistance and the so-called nodular "mesothelial" hyperplasia found within inguinal hernia sacs, in which the histiocytes contain diastase-sensitive cytoplasmic granules (173–175).

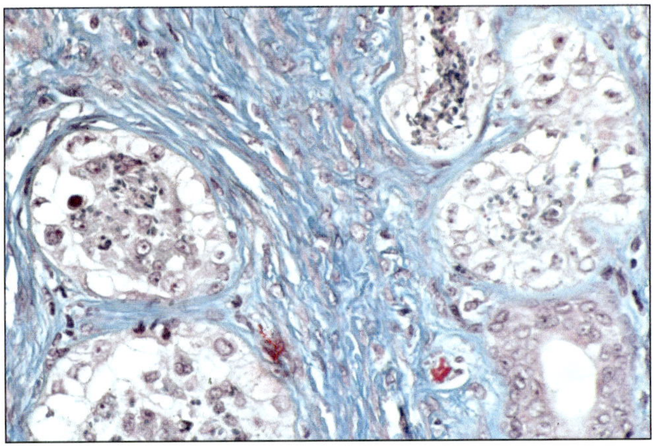

FIGURE 10.40: Infiltrated nerve highlighted between the proliferative ductular epithelium in vasitis nodosa (trichrome, 250×).

Infections: Epididymitis tends to be associated with *N. gonorrhoeae* or *C. trachomatis* infections in young males and *E. coli* or *P. aeruginosa* in males over 35 years of age (176). Other less frequently involved organisms include *Staphylococcus aureus* and *Aerobacter* sp. (177). As in the testis, urine culture is likely to be more sensitive and specific than tissue histochemistry (Brown and Brenn or Brown and Hopps). Other infections involving the epididymis include tuberculosis; leprosy; various mycoses such as blastomycosis, cryptococcosis, coccidiomycosis, and histoplasmosis; and rarely sportrichosis. Acid-fast, silver, and mucicarmine stains (for cryptococcus) can sometimes be supportive of the diagnosis (178–185). Care should be taken when attempting to speciate fungi on special stains as the morphology of various organisms can overlap and be misleading. Brucellosis can cause acid-fast–negative granulomatous lesions of the epididymis or the testis (186). Protozoal organisms implicated in the paratesticular region include *S. hematobium*, less commonly *S. mansonii*, and *Wuchereria bancrofti*. Giemsa staining may allow identification via the characteristic larval forms (187,188) Sarcoidosis can involve the epididymis, being characterized by typically well-delineated granulomas that are negative on acid-fast and silver staining (189).

Other Granulomatous Processes: As in the testis, xanthogranulomatous epididymitis can be separated from malacoplakia using PAS, von Kossa, and Prussian blue stains to visualize the characteristic Michaelis-Gutmann bodies of the latter (Figure 10.41) (190–192). Granulomas should

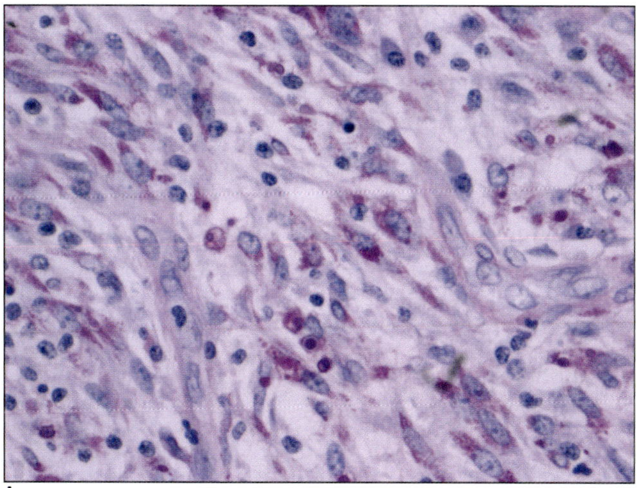

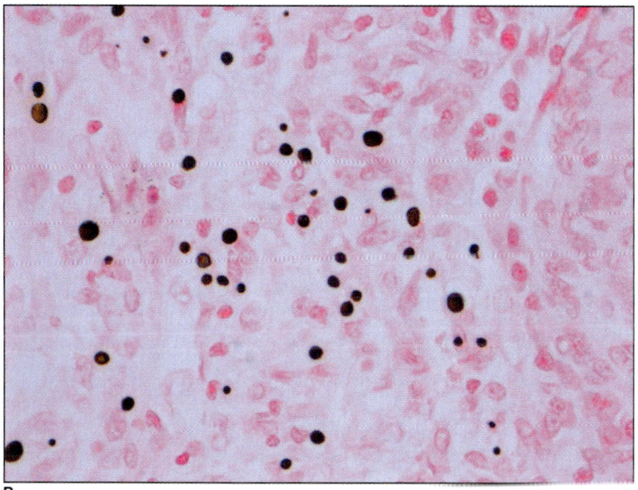

FIGURE 10.41: ***Malacoplakia of the epididymis with dense, round intracytoplasmic Michaelis-Gutmann bodies that are highlighted with a PAS stain (A) and a von Kossa stain (B) (A—PAS, 200×; B—von Kossa, 200×).***

also cause the pathologist to consider the possibility of involvement of the paratesticular area by vasculitis, with polyarteritis nodosa perhaps the best documented type. EVG histochemistry to demonstrate arterial wall damage to the internal and external elastic laminae and Martin's scarlet blue or trichrome histochemistry to confirm the presence of fibrin can be supportive of this diagnosis (193–195).

Paratesticular Neoplasms

Benign Neoplasms: Adenomatoid tumors can arise in any part of the paratestis but occur most commonly in the epididymis. Although these are benign nonpapillary mesotheliomas, they can display disconcerting cytologic features such as signet-ring cell change and intracytoplasmic bridging. Hyaluronidase-sensitive acid mucopolysaccharides are present in the cytoplasm on alcian blue staining, whereas papillary cystadenomas of the epididymis have PAS-positive diastase-resistant secretions (196,197). Alcian blue will highlight the brush border of the epithelium, while lipid stains performed on frozen tissue, such as Sudan III, Sudan IV, or oil red O, demonstrate intracytoplasmic lipid (198,199). One small study used soybean agglutinin to differentiate these tumors from renal cell carcinoma, another tumor containing both intracytoplasmic lipid and glycogen; the soybean agglutinin being present in adenomatoid tumors and absent in renal cell carcinoma (200).

Malignant Neoplasms: Adenocarcinomas of the rete testis can display a serous phenotype with corresponding diastase-sensitive PAS positivity (201). Cytoplasmic glycogen may not, however, be uniformly present in these tumors, and lipid is also sometimes found (202). Rare cystadenocarcinomas of the epididymis can display serous or mucinous phenotypes with dual diastase-resistant PAS and mucicarmine positivity in the latter (203,204). Paratesticular mesotheliomas, like benign mesothelial proliferations, contain intracytoplasmic glycogen and hyaluronic acid, with a resulting alcian blue hyaluronidase-sensitive and PAS-D–sensitive histochemical profile (196,205,206). A tumor of infancy occurring rarely in the epididymis, as opposed to the usual site in the head and neck, is the melanotic neuroectodermal tumor of infancy. Pointed oval, Fontana-positive granules in pigment-laden cuboidal cells and a neuroglia-like stroma devoid of collagen or reticulin fibers on Masson and reticulin stains are characteristic of these benign tumors. The melanin staining is of course susceptible to melanin bleach (207–210). The age of the patient and lesional architecture will allow distinction of these tumors from metastatic malignant melanoma; however, the histochemical pattern would be identical in the pigmented cells (211). Various types of metastatic carcinoma involving the epididymis have been reported, including renal cell carcinoma (intracytoplasmic glycogen and lipid), prostatic carcinoma (intracytoplasmic mucin in some cases), and colon carcinoma (intracytoplasmic mucin more common) (212–214). Lymphomas can involve the spermatic cord and epididymis, and as in the testis are most likely to fall into the diffuse large B-cell category (215,216). As in the testis, histochemical staining is of utility in this setting only insofar as it helps exclude other tumors on the differential diagnosis, like seminoma (PAS positive).

ILLUSTRATIVE CASE STUDIES

CASE 1— ***Granulomatous Fungal Prostatitis-Histoplasmosis:*** A 62-year-old immunosuppressed renal transplant patient developed obstructive voiding symptoms, and imaging studies suggested prostatic abscesses. A transurethral resection of the prostate was performed. The resected tissue had a severe destructive granulomatous prostatitis. Many of the granulomas had variably-sized central abscesses (Figure 10.42). Numerous fungal yeast about the size of erythrocytes were identified with a silver methenamine stain, including some with narrow-based buds (Figure 10.43).

Immunocompromised hosts are at high risk for developing disseminated histoplasmosis. GU involvement, which follows or is concurrent with disseminated disease, has a 6% incidence (14,22). Patients may be asymptomatic or develop hematuria and lower urinary tract symptoms. Culture is desirable to confirm the diagnosis. ■

CASE 2— ***Pseudohyperplastic Prostatic Adenocarcinoma:*** A 64-year-old with a suspicious prostatic nodule on digital rectal exam and a serum PSA of 8 underwent sextant biopsies. One of the cores contained a grouping of medium-sized glands with luminal undulations and intraluminal amorphous pink material (Figure 10.44). The glands were lined by columnar cells with enlarged basally oriented nuclei and occasional macronucleoli (Figure 10.45). The intraluminal material represented neutral mucin (Figure 10.46).

Pseudohyperplastic carcinoma can closely mimic nodular hyperplasia because it has a nodular growth pattern and consists of glands with cystic change, luminal undulations, and branching patterns as well as intermediate to large complex glands with a papillary architecture. Nuclear atypia, including nucleomegaly and nucleolomegaly, and complete absence of basal cells are features of malignancy. ■

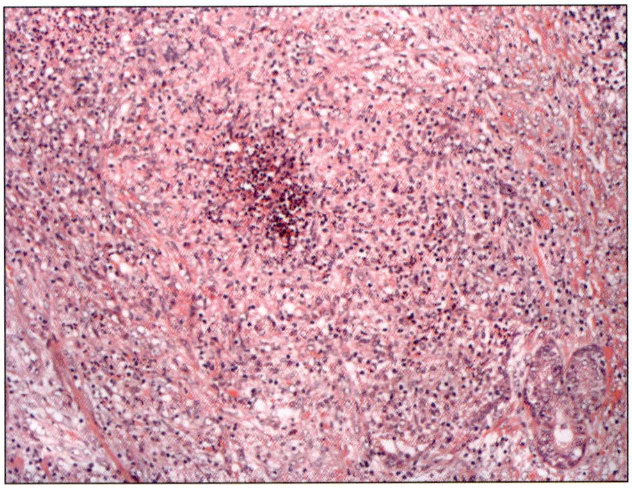

FIGURE 10.42: Necrotizing granuloma with punctate abscess (H&E, 100×).

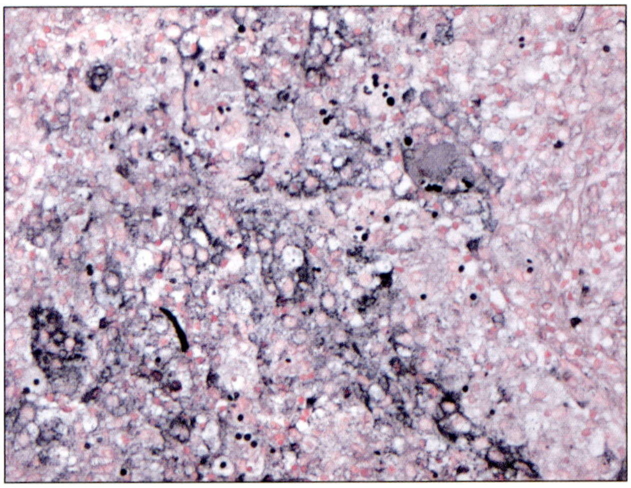

FIGURE 10.43: Budding yeast consistent with Histoplasma capsulatum *(silver methenamine, 200×).*

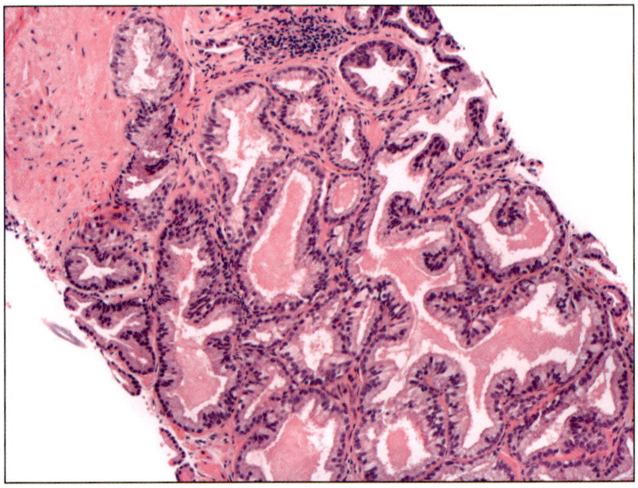

FIGURE 10.44: This prostatic adenocarcinoma lacks an invasive growth pattern and mimics nodular hyperplasia (H&E, 100×).

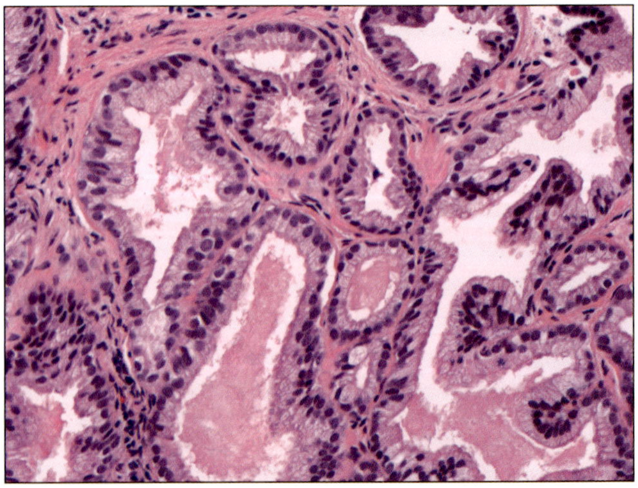

FIGURE 10.45: Cytologic features of malignancy are evident in the pseudohyperplastic glands (H&E, 200×).

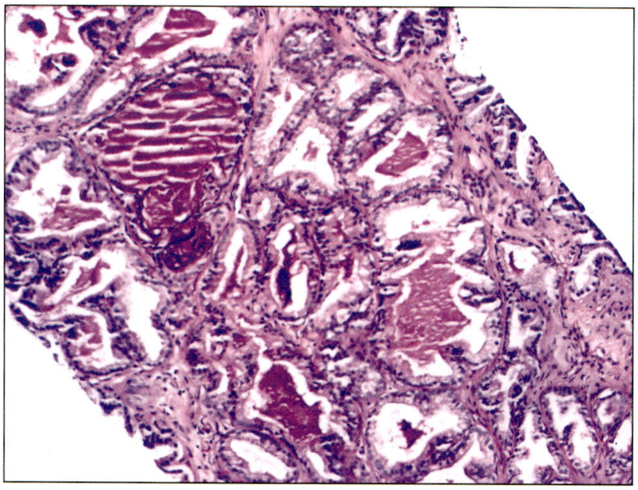

FIGURE 10.46: *The malignant glands contain neutral mucin and not acid mucin (digested PAS, 100×).*

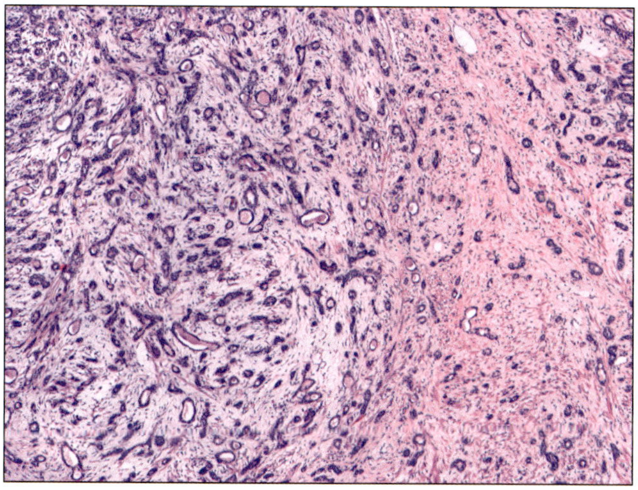

FIGURE 10.47: *Basal cell carcinoma of the prostate has a diffusely infiltrative pattern with associated myxoid and desmoplastic stroma (H&E, 40×).*

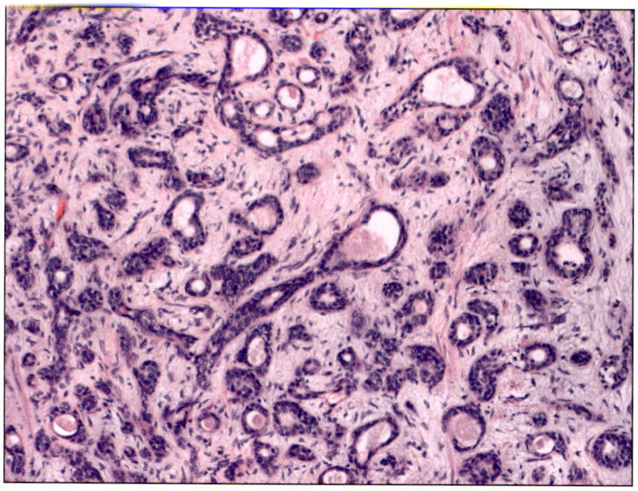

FIGURE 10.48: *A glandular and cord-like basaloid pattern is present (H&E, 100×).*

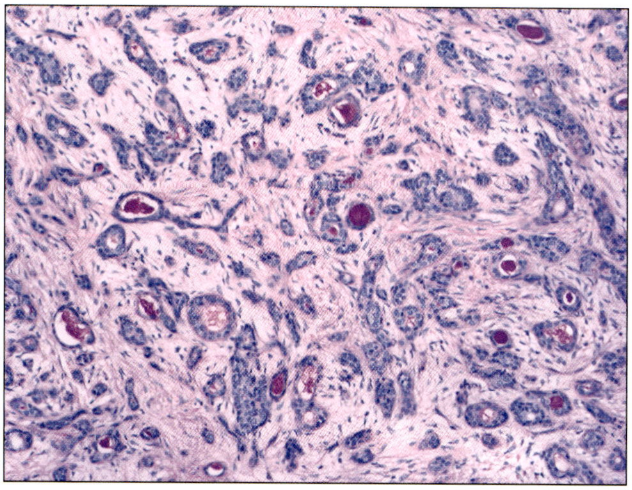

FIGURE 10.49: *Neutral mucin is present within lumina. The nests lack a stainable basement membrane (digested PAS, 100×).*

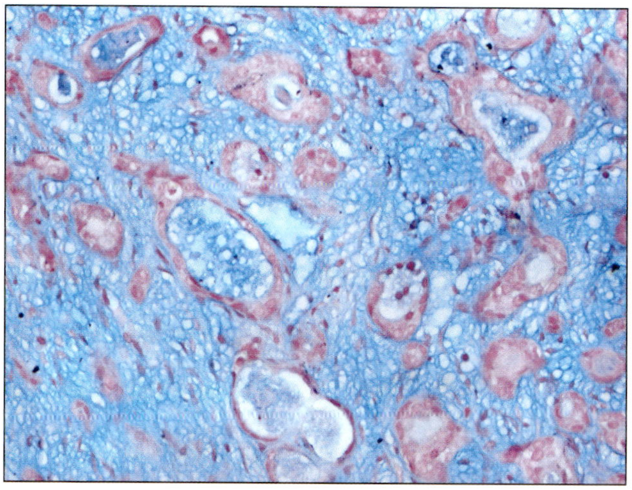

FIGURE 10.50: *Alcian blue (+) material is present within glandular lumina and the myxoid stroma (alcian blue pH 2.5, 100×).*

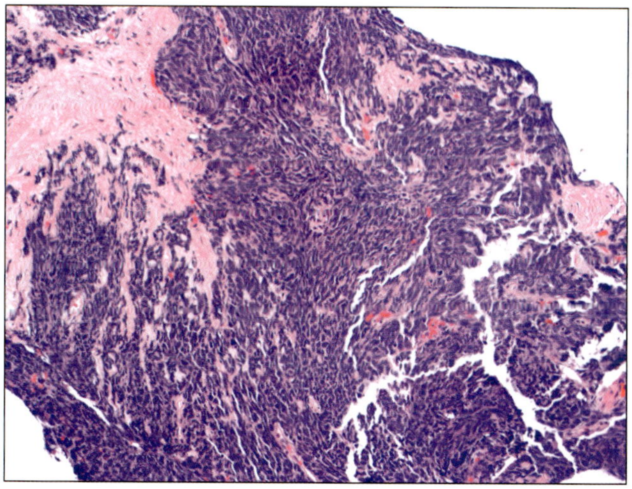

FIGURE 10.51: *Sheet-like arrangement of small blue tumor cells. Nuclei have a fusiform appearance (H&E, 100×).*

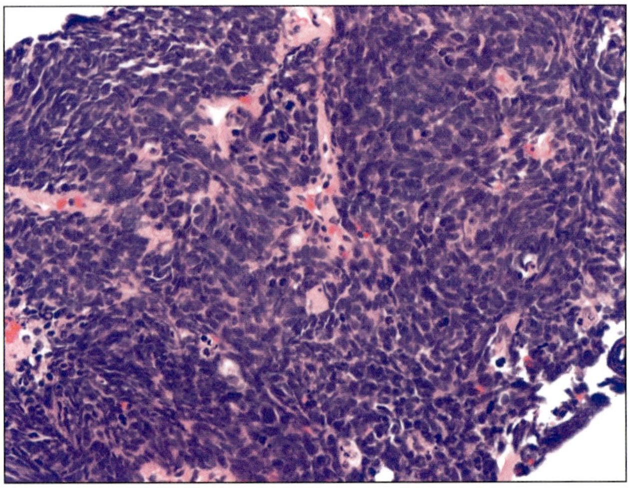

FIGURE 10.52: Sheet of small, polygonal to round neoplastic cells with scant cytoplasm, hyperchromatic nuclei with finely stippled chromatin and inconspicuous nucleoli. Mitotic figures are easily identified (H&E, 200×).

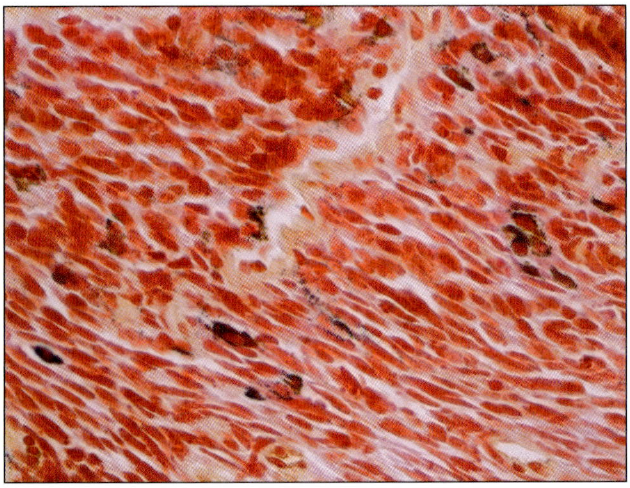

FIGURE 10.53: Occasional cells are positive with an argyrophil stain (Grimelius, 400×). (Note: tumor cells were more diffusely positive for neuron-specific enolase and synaptophysin.)

CASE 3— ***Basal Cell Carcinoma (Courtesy of David Grignon, Indianapolis, Indiana):*** A 58-year-old with an enlarged prostate and a clinical diagnosis of benign prostatic hyperplasia underwent a transurethral resection, which yielded 200 g of prostatic chips. A high percentage of the chips contained a diffusely infiltrative tumor composed of haphazardly dispersed nests and tubules (Figure 10.47). The stroma was partly myxoid and partly desmoplastic. Cord-like, glandular, and nested basaloid patterns were present (Figure 10.48). Glandular lumina contained neutral mucin (Figure 10.49) and acid mucin (Figure 10.50). The myxoid stroma contained abundant alcian blue (+) material, which would be expected to be hyaluronidase sensitive. ■

CASE 4— ***Small Cell Carcinoma:*** A 71-year-old presented with an enlarged, irregular, and hard prostate and symptoms of bladder outlet obstruction. The clinical diagnosis was prostatic cancer, and he underwent random prostate biopsies. All cores contained an undifferentiated carcinoma (Figure 10.51). There was no evidence of a mixture of prostatic acinar adenocarcinoma, and the cellular features (Figure 10.52) were typical of small cell neuroendocrine carcinoma. The Fontana-Masson (argentaffin) stain was negative. However, an argyrophil stain (Grimelius) was positive in a small percentage of cells (Figure 10.53). ■

REFERENCES

1. Roberts RO, Lieber MM, Bostwick DG, et al.: A review of clinical and pathological prostatitis syndromes. *Urology* 1997; 49:809–821.
2. Nickel JC: Prostatitis and related conditions. In: Walsh PC, Retik AB, Vaughan ED Jr., et al., eds. Campbell's Urology, 8th ed. Philadelphia: Saunders 2002; 603–630.
3. Ludwig M, Schroeder-Printzen I, Schiefer HG, et al.: Diagnosis and therapeutic management of 18 patients with prostatic abscess. *Urology* 1999; 53:340–345.
4. Weinberger M, Cytron S, Servadio C, et al.: Prostatic abscess in the antibiotic era. *Rev Infect Dis* 1988; 10:239–249.
5. Leport C, Rousseau F, Perronne C, et al.: Bacterial prostatitis in patients infected with the human immunodeficiency virus. *J Urol* 1989; 141:334–336.
6. Kwan DJ, Lowe FC: Genitourinary manifestations of the acquired immunodeficiency syndrome. *Urology* 1995; 45: 13–27.
7. de Souza E, Katz DA, Dworzack DL, et al.: Actinomycosis of the prostate. *J Urol* 1985; 133:290–291.
8. Epstein JI, Hutchins GM: Granulomatous prostatitis: distinction among allergic, nonspecific, and posttransurethral resection lesions. *Hum Pathol* 1984; 15:818–825.
9. Stillwell TJ, Engen DE, Farrow GM. The clinical spectrum of granulomatous prostatitis: a report of 200 cases. *J Urol* 1987; 138:320–323.
10. Oppenheimer JR, Kahane H, Epstein JI: Granulomatous prostatitis on needle biopsy. *Arch Pathol Lab Med* 1997; 121: 724–729.
11. Mohan H, Bal A, Punia RP, Bowa AS: Granulomatous prostatitis—an infrequent diagnosis. *Int J Urol* 2005; 12: 474–478.
12. Mikolich DJ, Mates SM: Granulomatous prostatitis due to *mycobacterium avium* complex. *Clin Infect Dis* 1992; 14: 589–591.
13. Lamm DL, Stogdill VD, Stogdill BJ, et al.: Complications of bacillus Calmette-Guerin immunotherapy in 1,278 patients with bladder cancer. *J Urol* 1986; 135:272–274.
14. Wise GJ, Shteynshluger A: How to diagnose and treat fungal infections in chronic prostatitis. *Curr Urol Rep* 2006; 7:320–328.
15. Abbas F, Kamal MK, Talati J: Prostatic aspergillosis. *J Urol* 1995; 153:748–750.

16. Milchgrub S, Visconti E, Avellini J: Granulomatous prostatitis induced by capsule-deficient cryptococcal infection. *J Urol* 1990; 143:365–366.
17. Adams JR, Mata Ja, Culkin DJ, et al.: Acquired immunodeficiency syndrome manifesting as prostate nodule secondary to cryptococcal infection. *Urology* 1992; 39:289–291.
18. Ndimbie OK, Dekker A, Martinez AJ, et al.: Prostatic sequestration of Cryptococcus neoformans in immunocompromised persons treated for cryptococcal meningoencephalitis. *Histol Histopathol* 1994; 9:643–648.
19. Caballes RL, Cabelles RA Jr.: Primary cryptococcal prostatitis in an apparently compromised host. *Prostate* 1999; 39:119–122.
20. Yurkanin JP, Ahmann F, Oalkin BL: Coccidioidomycosis of the prostate: a determination of incidence, report of 4 cases, and treatment recommendations. *J Infect* 2006; 52:e19–e25.
21. Melo CR, Melo JS, Cerski CT: Leukemic infiltration, paracoccidioidomycosis and nodular hyperplasia of the prostate. *Br J Urol* 1992; 70:329–330.
22. Mawhorter SD, Curley GU, Kursh ED, Farver CE: Prostatic and central nervous system histoplasmosis in an immunocompromised host: case report and review of the prostatic histoplasmosis literature. *Clin Infect Dis* 2000; 30:595–598.
23. Clason AE, McGeorge A, Garland C, et al.: Urinary retention and granulomatous prostatitis following sacral herpes zoster infection. A report of 2 cases with a review of the literature. *Br J Urol* 1982, 54:166–169.
24. McKay TC, Albala DM, Sendelbach K, et al.: Cytomegalovirus prostatitis case: report and review of the literature. *Int Urol Nephrol* 1994; 26:535–540.
25. Mastroianni A, Coronado O, Manfredi R, et al.: Acute cytomegalovirus prostatitis in AIDS. *Genitourin Med* 1996; 72:447–448.
26. Thompson L: Syphilis of the prostate. *Am J Syph* 1920; 4:323–341.
27. Gelfand M, Ross CMD, Blair DM, et al.: Schistosomiasis of the male pelvic organs. Severity of infection as determined by digestion of tissue and histologic methods in 300 cadavers. *Am J Trop Med Hyg* 1970; 19:779–784.
28. Vasileios R, Athanasios P, Stavros T: Echinococcal cyst of the seminal vesicles: a case report and literature review. *Int Urol Nephrol* 2002–2003; 34:527–530.
29. Helpap B, Vogel J.: TUR-prostatitis. Histological and immunohistochemical observations in a special type of granulomatous prostatitis. *Pathol Res Pract* 1986; 181:301–307.
30. Morris SB, Gordon EM, Corbishley CM: Prostatic sarcoidosis. A review of genitourinary sarcoidosis. *Br J Urol* 1993; 72:462–464.
31. Arena F, Fortunati C, di Stefano C, et al.: Prostatic malacoplakia associated with prostatic abscess: diagnosis and treatment. *Urol Int* 2001; 66:212–215.
32. Stanton MJ, Maxted W: Malacoplakia: A study of the literature and current concepts of pathogenesis, diagnosis and treatment. *J Urol* 1981; 125:139–146.
33. Murty GE, Powell PH: Wegner's granulomatosis presenting as prostatitis. *Br J Urol* 1991; 67:107–108.
34. Bray VJ, Hasbargen JA: Prostatic involvement in Wegener granulomatosis. *Am J Kidney Dis* 1991; 17:578–580.
35. Awadh B, Watson K, Adou NI: Wegener prostatitis presenting with acute urinary retention. *J Clin Rheumatol* 2006; 12:50–51.
36. Lopez-Beltran A: Vasculitis involving the prostate: report of two cases. *Pathol Case Rev* 1996; 1:70–73.
37. Val-Bernal JF, Gonzalez-Vela C, Mayorga M, et al.: Isolated fibrinoid arteritis of the prostate. *Int J Surg Pathol* 1997; 4:143–148.
38. Bretal-Laranga M, Insua-Vilarino S, Blanco-Rodriquez J, et al.: Giant cell arteritis limited to the prostate. *J Rheumatol* 1995; 22:566–568.
39. Hu JC, Palapattu GS, Kattan MW, et al.: The association of selected pathological features with prostate cancer in a single-needle biopsy accession. *Hum Pathol* 1998; 29:1536–1538.
40. Gaudin PB, Zelefsky MJ, Leibel SA, et al.: Histopathologic effects of three-dimensional conformal external beam radiation therapy on benign and malignant prostate tissue. *Am J Surg Pathol* 1999; 23:1021–1031.
41. Grignon DJ, O'Malley FP: Mucinous metaplasia in the prostate gland. *Am J Surg Pathol* 1993; 17:287–290.
42. Gal R, Koren R, Nofech-Mozes S, et al.: Evaluation of mucinous metaplasia of the prostate gland by mucin histochemistry. *Br J Urol* 1996; 77:113–117.
43. Chughtai B, Sawas A, O'Malley RL, et al.: A neglected gland: a review of Cowper's gland. *Int J Androl* 2005; 28:74–77.
44. Saboorian MH, Huffman H, Ashfaq R, et al.: Distinguishing Cowper's glands from neoplastic and pseudoneoplastic lesions of prostate. Immunohistochemical and ultrastructural studies. *Am J Surg Pathol* 1997; 21:1069–1074.
45. Cina SJ, Silberman MA, Kahane H, et al.: Diagnosis of Cowper's glands on prostate needle biopsy. *Am J Surg Pathol* 1997; 21:550–555.
46. Shiraishi T, Kusano I, Watanabe M, et al.: Mucous gland metaplasia of the prostate. *Am J Surg Pathol* 1993; 17:618–622.
47. Weaver MG, Abdul-Karim FW, Srigley J, et al.: Paneth cell-like change of the prostate gland. A histological, immunohistochemical, and electron microscopic study. *Am J Surg Pathol* 1992; 16:62–68.
48. Coyne JD, Kealy WF: Seminal vesicle amyloidosis: morphological, histochemical and immunohistochemical observations. *Histopathology* 1993; 22:173–176.
49. Maroun L, Jakobsen H, Kromann-Anderson B, et al.: Amyloidosis of the seminal vesicle—a case report and review of the literature. *Scand J Urol Nephol* 2003; 37:519–521.
50. Lawrentshuk N, Pan D, Stillwell R, Bolton DM. Implications of amyloidosis on prostate biopsy. *Int J Urol* 2004; 11:925–927.
51. Cross PA, Bartley CJ, McClure J: Amyloid in prostatic corpora amylacea: *J Clin Pathol* 1992; 45:894–897.
52. Ro JY, Grignon DJ, Ayala AG, et al.: Blue nevus and melanosis of the prostate. Electron-microscopic and immunohistochemical studies. *Am J Clin Pathol* 1998; 90:530–535.

53. Brennick JB, O'Connell JX, Dickersin GR, et al.: Lipofuscin pigmentation (so called "melanosis") of the prostate. *Am J Surg Pathol* 1994; 18:446–454.
54. Amin MB, Bostwick DG: Pigment in prostatic epithelium and adenocarcinoma: a potential source of diagnostic confusion with seminal vesicular epithelium. *Mod Pathol* 1996; 9:791–795.
55. Leung CS, Srigley JR: Distribution of lipochrome pigment in the prostate gland: biological and diagnostic implications. *Hum Pathol* 1995; 28:1302–1307.
56. Seman G, Gallager S, Johnson DE: Melanin-like pigment in the human prostate. *Prostate* 1982; 3:59–72.
57. Epstein JI: Adenosis (atypical adenomatous hyperplasia): histopathology and relationship to carcinoma. *Pathol Res Pract* 1995; 191:888–898.
58. Srigley JR: Benign mimickers of prostate adenocarcinoma. *Mod Pathol* 2004; 17:328–348.
59. Gaudin PB, Epstein JI: Adenosis of the prostate. Histologic features in need biopsy specimens. *Am J Surg Pathol* 1995; 19:737–747.
60. Goldstein NS, Qian J, Bostwick DG: Mucin expression in atypical adenomatous hyperplasia of the prostate. *Hum Pathol* 1995; 26:887–891.
61. Luna-More S, Florez P, Ayala A, et al.: Neutral and acid mucins and eosinophil and argyrophil crystalloids in carcinoma and atypical adenomatous hyperplasia of the prostate. *Pathol Res Pract* 1997; 193:291–298.
62. Helpap B, Riede C: Nucleolar and AgNOR-analysis of prostatic intraepithelial neoplasia (PIN), atypical adenomatous hyperplasia (AAH) and prostatic adenocarcinoma. *Pathol Res Pract* 1995; 191:381–390.
63. Chastonay P, Hurlimannn J, Gardiol D: Biological tissue markers in benign and malignant disease of the human prostate. *Virchows Arch A* 1986; 410:221–229.
64. Shamsuddin AM, Beasly WM: Expression of the tumor marker D-galactose-B-(1-3)-N-acetyl-D-galactosamine by premalignant and malignant prostate. *Arch Pathol Lab Med* 1994; 118:48–51.
65. Jones EC, Clement PB, Young RH: Sclerosing adenosis of the prostate gland. A clinicopathologic and immunohistochemical study of 11 cases. *Am J Surg Pathol* 1991; 15:1171–1180.
66. Sakamoto N, Tsuneyoshi M, Enjoji M: Sclerosing adenosis of the prostate. Histopathologic and immunohistochemical analysis. *Am J Surg Pathol* 1991; 15:660–667.
67. Grignon DJ, Ro JY, Srigley JR, et al.: Sclerosing adenosis of the prostate. A lesion showing myoepithelial differentiation. *Am J Surg Pathol* 1992; 16:383–391.
68. Schowinsky JT, Epstein JI: Distorted rectal tissue on prostate needle biopsy. A mimicker of prostate cancer. *Am J Surg Pathol* 2006; 30:866–870.
69. Alguacil-Garcia A: Artifactual changes mimicking signet ring cell carcinoma in transurethral prostatectomy specimens. *Am J Surg Pathol* 1986; 10:795–800.
70. Wang HL, Humphrey PA: Exaggerated signet-ring cell change in stomal nodule of prostate: a pseudoneoplastic proliferation. *Am J Surg Pathol* 2002; 26:1066–1070.
71. Guerrero-Medrano J, Delgado R, Albores-Saavedra J: Signet-ring histocytosis. A reactive disorder that mimics metastatic adenocarcinoma. *Cancer* 1997; 80:277–285.
72. Amin MB, Ro JY, Ayala AG: Putative precursor lesions of prostatic adenocarcinoma: fact or fiction. *Mod Pathol* 1993; 6:476–483.
73. Sentinelli S: Mucins in prostatic intra-epithelial neoplasia and prostatic carcinoma. *Histopathology* 1993; 22:271–274.
74. Reyes AO, Swanson PE, Carbone JM, Humphrey PA: Unusual histologic types of high-grade prostatic intraepithelial neoplasia. *Am J Surg Pathol* 1997; 21:1215–1222.
75. Deschenes J, Weidner N: Nucleolar organizer regions (NOR) in hyperplastic and neoplastic prostate disease. *Am J Surg Pathol* 1990; 14:1148–1155.
76. Sakr WA, Sarkar FH, Sreepathi P, et al.: Measurement of cellular proliferation in human prostate by AgNOR, PCNA, and SPF. *Prostate* 1993; 22:147–154.
77. Cheville JC, Clamon GH, Robinson RA: Silver-stained nucleolar organizer regions in the differentiation of prostatic hyperplasia, intraepithelial neoplasia, and adenocarcinoma. *Mod Pathol* 1990; 3:596–598.
78. McNeal JF, Leav I, Alroy J, et al.: Differential lectin staining of central and peripheral zones of the prostate and alterations in dysplasia. *Am J Clin Pathol* 1988; 89:41–48.
79. Janssen T, Petein M, Van Velthoven R, et al.: Differential histochemical peanut agglutinin stain in benign and malignant human prostate tumors: relationship with prostatic specific antigen immunostain and nuclear DNA content. *Hum Pathol* 1996; 27:1341–1347.
80. Franks LM, O'Shea JD, Thomson AER: Mucin in the prostate: a histochemical study in normal glands, latent, clinical, and *colloid cancers*. 1964; 17:983–991.
81. Hukill PB, Vidone RA: Histochemistry of mucus and other polysaccharides in tumors. II Carcinoma of the prostate. *Lab Investigation* 1967; 16:395–406.
82. Taylor NS: Histochemistry in the diagnosis of early prostatic carcinoma. *Hum Pathol* 1979; 10:513–520.
83. Mathur SK, Gupta S, Marwah N, et al.: Significance of mucin stain in differentiating benign and malignant lesions of prostate. *Indian J Pathol Microbiol* 2003; 46:593–595.
84. Varma M, Lin MW, Tamboli P, et al.: Morphologic criteria for the diagnosis of prostatic adenocarcinoma in needle biopsy specimens: a study of 250 consecutive cases in routine surgical pathology practice. *Arch Pathol Lab Med* 2002; 126:554–561.
85. Epstein JI: Diagnostic criteria of limited adenocarcinoma of the prostate on needle biopsy. *Hum Pathol* 1995; 26:223–229.
86. Thorson P, Vollmer RT, Arcangeli C, et al.: Minimal carcinoma in prostate needle biopsy specimens; diagnostic features and radical prostatectomy follow up. *Mod Pathol* 1998; 11:543–551.
87. Bostwick DG, Wollan P, Adlakha K: Collagenous micronodules in prostate cancer. A specific but infrequent diagnostic finding. *Arch Pathol Lab Med* 1995; 119:444–447.
88. Baisden BL, Kahane H, Epstein JI: Perineural invasion, mucinous fibroplasia, and glomerulations. Diagnostic features of limited cancer on prostate needle biopsy. *Am J Surg Pathol* 1999; 23:918–924.
89. Arangelovich V, Tretiakova M, Sengupta E, et al.: Pathogenesis and significance of collagenous micronodules of the prostate. *Appl Immunohistochem Mol Morphol* 2003; 11:15–19.

90. McNeal JE, Alroy J, Villers A, et al.: Mucinous differentiation in prostatic adenocarcinoma. *Hum Pathol* 1991; 22:979–988.
91. Kaleem Z, Swanson PE, Vollmer RT, et al.: Prostatic adenocarcinoma with atrophic features. A study of 202 consecutive completely-embedded radical prostatectomy specimens. *Am J Clin Pathol* 1998; 109:695–703.
92. Nelson RS, Epstein JI: Prostatic carcinoma with abundant xanthomatous cytoplasm. Foamy gland carcinoma. *Am J Surg Pathol* 1996; 20:419–426.
93. Tran TT, Sengupta E, Yang XY: Prostatic foamy gland carcinoma with aggressive behavior. Histologic, immunohistochemical, and ultrastructural analysis. *Am J Surg Pathol* 2001; 25:618–623.
94. di Sant' Agnese PA: Divergent neuroendocrine differentiation in prostatic carcinoma. *Semin Diagn Pathol* 2000; 17:149–161.
95. Adlakha H, Bostwick DG: Paneth cell-like in prostatic adenocarcinoma represents neuroendocrine differentiation. Report of 30 cases. *Hum Pathol* 1994; 25:135–139.
96. Tamas E, Epstein JI: Prognostic significance of Paneth cell-like neuroendocrine differentiation in adenocarcinoma of the prostate. *Am J Surg Pathol* 2006; 30:980–985.
97. Dhom G: Unusual prostatic carcinomas. *Pathol Res Pract* 1990; 186:28–36.
98. Ro JY, Grignon DJ, Ayala AG, et al.: Mucinous adenocarcinoma of the prostate: histochemical and immunohistochemical studies. *Hum Pathol* 1990; 21:593–600.
99. Epstein JI, Lieberman PH: Mucinous adenocarcinoma of the prostate gland. *Am J Surg Pathol* 1985; 9:299–308.
100. Saez C, Japon MA, Conde AF, et al.: Sialomucins are characteristically O-acylated in poorly differentiated and colloid prostatic adenocarcinomas. *Mod Pathol* 1998; 11:1193–1197.
101. Curtis MW, Evans AJ, Srigley JR: Mucin-producing urothelial-type adenocarcinoma of prostate: report of two cases of a rare and diagnostically challenging entity. *Mod Pathol* 2005; 18:585–590.
102. Ro JY, EL-Naggar A, Ayala AG, et al.: Signet-ring cell carcinoma of the prostate. Electron-microscopic and immunohistochemical study of eight cases. *Am J Surg Path* 1988; 12:453–460.
103. Hejka AG, England PM: Signet-ring cell carcinoma of prostate: immunohistochemical and ultrastructural study of a case. *Urology* 1989; 34:155–158.
104. Alline KM, Cohen MB: Signet-ring cell carcinoma of the prostate. *Arch Pathol Lab Med* 1992; 116:99–102.
105. Guerin D, Hosan N, Keen CE: Signet ring cell differentiation in adenocarcinoma of the prostate: a study of five cases. *Histopathology* 1993; 22:367–371.
106. Leong FJ, Leong AS, Swift J: Signet-ring carcinoma of the prostate. *Pathol Red Pract* 1996; 192:1232–1241.
107. Torbenson M, Dhir R, Nangia A, et al.: Prostatic carcinoma with signet ring cells: a clinicopathologic and immunohistochemical analysis of 12 cases, with review of the literature. *Mod Pathol* 1998; 11:552–559.
108. Kums JJM, Heldsdingen PJRO: Signet-ring-cell carcinoma of the bladder and prostate. *Urol Int* 1985; 40:116–119.
109. Saito S, Iwaki H: Mucin-producing carcinoma of the prostate: review of 88 cases. *Urology* 1999; 54:141–144.
110. Kuhajda FP, Mann RB: Adenoid cystic carcinoma of the prostate. A case report with immunoperoxidase staining for prostate-specific acid phosphatase and prostate-specific antigen. *Am J Clin Pathol* 1984; 81:257–260.
111. Grignon DJ, Ro JY, Ordonez NG, et al.: Basal cell hyperplasia, adenoid basal cell tumor, and adenoid cystic carcinoma of the prostate gland: an immunohistochemical study. *Hum Pathol* 1988; 19:1425–1433.
112. Iczkowski KA, Ferguson KL, Grier DD, et al.: Adenoid cystic/basal cell carcinoma of the prostate: clinicopathologic findings in 19 cases. *Am J Surg Pathol* 2003; 27:1523–1529.
113. McKenney JK, Amin MB, Srigley JR, et al.: Basal cell proliferations of the prostate other than usual basal cell hyperplasia: a clinicopathologic study of 23 cases, including four carcinomas, with a proposed classification. *Am J Surg Pathol* 2004; 28:1289–1298.
114. Young RH, Frierson HF Jr., Mills SE, et al.: Adenoid cystic-like tumor of the prostate gland. A report of two cases and review of the literature on "adenoid cystic carcinoma" of the prostate. *Am J Clin Pathol* 1988; 89: 49–56.
115. Murali R, Kneale K, Lalak N, Delprado W: Carcinoid tumors of the urinary tract and prostate. *Arch Pathol Lab Med* 2006; 130:1693 1706.
116. Helpap B, Kollerman J: Undifferentiated carcinoma of the prostate with small cell features: immunohistochemical subtyping and reflections on histogenesis. *Virchows Arch* 1999; 434:385–391.
117. Yao JL, Madeb R, Bourne P, et al.: Small cell carcinoma of the prostate: an immunohistochemical study. *Am J Surg Pathol* 2006; 30:705–712.
118. Abbas F, Civantos F, Benedetto P, et al.: Small cell carcinoma of the bladder and prostate. *Urology* 1995, 46:617–630.
119. Evans AJ, Humphrey PA, Belani J, et al.: Large cell neuroendocrine carcinoma of prostate. A clinicopathologic summary of 7 cases of a rare manifestation of advanced prostate cancer. *Am J Surg Pathol* 2006; 30:684–693.
120. Gotoh M, Miyake K, Mitsuya H, Elastic fibers in tunica propria of undescended and contralateral scrotal testes from cryptorchid patients. *Urology* 1987; 30:359–363.
121. Coffin CM, Ewing S, Dehner LP: Frequency of intratubular germ cell neoplasia with invasive testicular germ cell tumors. Histologic and immunocytochemical features. *Arch Pathol Lab Med* 1985; 109:555–559.
122. Dahl E: Testicular lesions of periarteritis nodosa, with special reference to diagnosis. *Am J Med* 1960; 28:222–228.
123. Shurbaji MS, Epstein JI: Testicular vasculitis: implications for systemic disease. *Hum Pathol* 1988; 19:186–189.
124. Teichman J: Polyarteritis nodosa presenting as acute orchitis: a case report and review of the literature. *J Urol* 1993; 149:1139–1140.
125. Huisman TK, Collins WT Jr., Voulgarakis GR: Polyarteritis nodosa masquerading as a primary testicular neoplasm: a case report and review of the literature. *J Urol* 1990; 144:1236–1238.
126. McGuire BB, et al.: Testicular vasculitis mimicking a testicular neoplasm. *Ir Med J* 2006: 99:27–28.

127. Nistal M, et al.: Secondary amyloidosis of the testis: an electron microscopic and histochemical study. *Appl Pathol* 1989; 7:2–7.
128. Corvino C, et al.: Testicular amyloidosis. *Urol Int* 2002; 69:162–163.
129. Casella R, et al.: Primary testicular amyloidosis mimicking tumor in a cryptorchid testis. *Urology* 2002; 59:445.
130. Ejlertsen T, Jensen HK:, Orchitis and testicular abscess formation caused by non-typhoid salmonellosis. A case report. *Apmis* 1990; 98:294–298.
131. Krieger JN: Epididymitis, orchitis, and related conditions. *Sex Transm Dis* 1984; 11:173–181.
132. Diaz Gonzalez R, et al.: Testicular malacoplakia. *J Urol* 1982; 127:325–328.
133. Saraf P et al.: An unusual case of malacoplakia involving the testis and prostate. *J Urol* 1983; 129:149–151.
134. Kostakopoulos A, et al.: Malakoplakia of the testis. *Int Urol Nephrol* 1997; 29:461–463.
135. McClure J: Malakoplakia of the testis and its relationship to granulomatous orchitis. *J Clin Pathol* 1980; 33:670–678.
136. Persaud V, Rao A: Gumma of the testis. *Br J Urol* 1977; 49:142.
137. Cabral DA, et al.: Tuberculous epididymitis as a cause of testicular pseudomalignancy in two young children. *Pediatr Infect Dis* 1985; 4:59–62.
138. Ibrahiem AA, et al.: Pathologic changes in testis and epididymis of infertile leprotic males. *Int J Lepr Other Mycobact Dis* 1979; 47:44–49.
139. Spjut HJ, Thorpe JD: Granulomatous orchitis. *Am J Clin Pathol* 1956; 26:136–145.
140. Akhtar M, Ali MA, Mackey DM: Lepromatous leprosy presenting as orchitis. *Am J Clin Pathol* 1980; 73:712–715.
141. Singer AJ, et al.: Genitourinary involvement of systemic sarcoidosis confined to testicle. *Urology* 1990; 35:442–424.
142. Haas GP, et al.: Testicular sarcoidosis: case report and review of the literature. *J Urol* 1986; 135:1254–1256.
143. Monroe M: Proceedings: granulomatous orchitis due to histoplasma capsulatum masquerading as sperm granuloma. *J Clin Pathol* 1974; 27:929–930.
144. Schuster TG, et al.: Testicular histoplasmosis. *J Urol* 2000; 164:1652.
145. Jenkin GA, et al.: Candidal epididymo-orchitis: case report and review. *Clin Infect Dis* 1998; 26:942–945.
146. Strohmaier WL, et al.: Alveolar echinococcosis with involvement of the ureter and testis. *J Urol* 1990; 144:733–734.
147. Nistal M, et al.: Testicular toxoplasmosis in two men with the acquired immunodeficiency syndrome (AIDS). *Arch Pathol Lab Med* 1986; 110:744–746.
148. Perry A: Pagetoid spread of intratubular germ cell neoplasia into rete testis: a morphologic and histochemical study of 100 orchiectomy specimens with invasive germ cell tumors. *Hum Pathol* 1994; 25:235–239.
149. Fossa S: Testicular carcinoma in situ in patients with extragonadal germ-cell tumors: the clinical role of pretreatment biopsy. *Annals Onc* 2003; 14:1412–1418.
150. Scholz M, et al.: Extragonadal retroperitoneal germ cell tumor: evidence of origin in the testis. *Ann Oncol* 2002; 13:121–124.
151. Fabre E et al.: 'Burned-out' primary testicular cancer. *BJU Int* 2004; 94:74–78.
152. Balzer BL, Ulbright TM: Spontaneous regression of testicular germ cell tumors: an analysis of 42 cases. *Am J Surg Pathol* 2006; 30:858–865.
153. Ulbright TM: Germ cell tumors of the gonads: a selective review emphasizing problems in differential diagnosis, newly appreciated, and controversial issues. *Mod Pathol* 2005; 18 (Suppl 2): S61–S79.
154. Ulbright TM, Roth LM: Recent developments in the pathology of germ cell tumors. *Semin Diagn Pathol* 1987; 4:304–319.
155. Ulbright TM, Roth LM, Brodhecker CA: Yolk sac differentiation in germ cell tumors. A morphologic study of 50 cases with emphasis on hepatic, enteric, and parietal yolk sac features. *Am J Surg Pathol* 1986; 10:151–164.
156. Beckstead J, Alkaline phosphatase histochemistry in human germ cell neoplasms. *Am J Surg Pathol* 1983; 7:341–349.
157. Zavala-Pompa A, et al.: Primary carcinoid tumor of testis. Immunohistochemical, ultrastructural, and DNA flow cytometric study of three cases with a review of the literature. *Cancer* 1993; 72:1726–1732.
158. Talerman A, et al.: Primary carcinoid tumor of the testis: case report, ultrastructure and review of the literature. *Cancer* 1978; 42:2696–2706.
159. Berdjis CC, Mostofi FK: Carcinoid tumors of the testis. *J Urol* 1977; 118:777–782.
160. Kurman RJ, et al.: An immunohistological study of steroid localization in Sertoli-Leydig tumors of the ovary and testis. *Cancer* 1978; 42:1772–1783.
161. Ferry JA, et al.: Malignant lymphoma of the testis, epididymis, and spermatic cord. A clinicopathologic study of 69 cases with immunophenotypic analysis. *Am J Surg Pathol* 1994; 18:376–390.
162. Richardson PG et al.: Metastatic melanoma to the testis simulating primary seminoma. *Br J Urol* 1992; 69:663–665.
163. Deshpande MS, Kulkarni JN: Metastatic adenocarcinoma in testis presenting as a testicular mass—a case report and review of literature. *Can J Urol* 2003; 10:1770–1771.
164. Haupt HM, et al.: Metastatic carcinoma involving the testis. Clinical and pathologic distinction from primary testicular neoplasms. *Cancer* 1984; 54:709–714.
165. Walker AN, Mills SE: Glandular inclusions in inguinal hernial sacs and spermatic cords. Mullerian-like remnants confused with functional reproductive structures. *Am J Clin Pathol* 1984; 82:85–89.
166. Cerilli LA, Sotelo-Avila C, Mills SE: Glandular inclusions in inguinal hernia sacs: morphologic and immunohistochemical distinction from epididymis and vas deferens. *Am J Surg Pathol* 2003: 27:469–476.
167. Dunner PS, Lipsit ER, Nochomovitz LE: Epididymal sperm granuloma simulating a testicular neoplasm. *J Clin Ultrasound* 1982; 10:353–355.
168. Cullen TH, Voss HJ: Sperm granulomata of the testis and epididymis. *Br J Urol* 1966; 38:202–207.
169. Silber SJ: Sperm granuloma and reversibility of vasectomy. *Lancet* 1977; 2:588–589.

170. Hirschowitz L, et al.: Vasitis nodosa and associated clinical findings. *J Clin Pathol* 1988; 41:419–423.
171. Deshpande RB et al.: Vasitis nodosa (a report of 7 cases). *J Postgrad Med* 1985; 31:105–108.
172. Balogh K, Travis WD: The frequency of perineurial ductules in vasitis nodosa. *Am J Clin Pathol* 1984; 82:710–713.
173. Chan JK, et al.: Nodular histiocytic/mesothelial hyperplasia: a lesion potentially mistaken for a neoplasm in transbronchial biopsy. *Am J Surg Pathol* 1997; 21:658–663.
174. Rosai J, Dehner LP: Nodular mesothelial hyperplasia in hernia sacs: a benign reactive condition simulating a neoplastic process. *Cancer* 1975; 35:165–175.
175. Ulbright TM, Gersell DJ: Rete testis hyperplasia with hyaline globule formation. A lesion simulating yolk sac tumor. *Am J Surg Pathol* 1991; 15:66–74.
176. Berger RE, et al.: Etiology, manifestations and therapy of acute epididymitis: prospective study of 50 cases. *J Urol* 1979; 121:750–754.
177. Mittemeyer BT, Lennox KW, Borski AA: Epididymitis: a review of 610 cases. *J Urol* 1966; 95:390–392.
178. Almagro UA, Tresp M, Sheth NK: Tuberculous epididymitis occurring 35 years after renal tuberculosis. *J Urol* 1989; 141:1204–1205.
179. Pareek SS, Tandon RC: Epididymal lesion in tuberculoid leprosy. Br Med J (Clin Res Ed) 1985; 291:313.
180. Orr WA, Mulholland SG, Walzak MP Jr, Genitourinary tract involvement with systemic mycosis. *J Urol* 1972; 107:1047–1050.
181. Eickenberg HU, Amin M, Lich R Jr: Blastomycosis of the genitourinary tract. *J Urol* 1975; 113:650–652.
182. Kauffman CA, Slama TJ, Wheat LJ: Histoplasma capsulatum epididymitis. *J Urol* 1981: 125:434–435.
183. Sohail MR, Andrews PE, Blair JE: Coccidioidomycosis of the male genital tract. *J Urol* 2005; 173:1978–1982.
184. James CL, Lomax-Smith JD: Cryptococcal epididymo-orchitis complicating steroid therapy for relapsing polychondritis. *Pathology* 1991; 23:256–258.
185. Selman SH, Hampel N: Systemic sporotrichosis: diagnosis through biopsy of epididymal mass. *Urology* 1982; 20:620–621.
186. Aklnci E, et al.: A complication of brucellosis: epididymoorchitis. *Int J Infect Dis* 2006; 10:171–177.
187. Jayaram G: Microfilariae in fine needle aspirates from epididymal lesions. *Acta Cytol* 1987; 31:59–62.
188. Gelfand M, et al.: Schistosomiasis of the male pelvic organs. Severity of infection as determined by digestion of tissue and histologic methods in 300 cadavers. *Am J Trop Med Hyg* 1970; 19:779–784
189. Yamamoto N, et al.: Bilateral epididymal sarcoidosis. Case report. *Scand J Urol Nephrol* 1992; 26:301–303.
190. Guccion JG, Thorgeirsson UP, Smith BH: Malacoplakia of epididymis. *Urology* 1978; 12:713–716.
191. Wiener LB, Riehl PA, Baum N: Xanthogranulomatous epididymitis: a case report. *J Urol* 1987; 138:621–622.
192. Dieckmann KP, Henke RP, Zimmer-Krolzig G: Malacoplakia of the epididymis. Report of a case and review of the literature. *Urol Int* 1995; 55:222–225.
193. Womack C, Ansell ID: Isolated arteritis of the epididymis. *J Clin Pathol* 1985; 38:797–800.
194. Persellin ST, Menke DM: Isolated polyarteritis nodosa of the male reproductive system. *J Rheumatol* 1992; 19:985–988.
195. Karnauchow PN, Steele AA: Isolated necrotizing granulomatous vasculitis of the spermatic cords. *J Urol* 1989; 141:379–381.
196. Srigley JR, Hartwick RW: Tumors and cysts of the paratesticular region. *Pathol Annu* 1990; 25(Pt 2): 51–108.
197. Price EB Jr: Papillary cystadenoma of the epididymis. A clinicopathologic analysis of 20 cases. *Arch Pathol* 1971; 91:456–470.
198. Axiotis CA: Intratesticular serous papillary cystadenoma of low malignant potential: an ultrastructural and immunohistochemical study suggesting mullerian differentiation. *Am J Surg Pathol* 1988; 12:56–63.
199. Torikata C: Papillary cystadenoma of the epididymis. An ultrastructural and immunohistochemical study. *J Submicrosc Cytol Pathol* 1994; 26:387–393.
200. Kragel PJ, et al.: Papillary cystadenoma of the epididymis. A report of three cases with lectin histochemistry. *Arch Pathol Lab Med* 1990; 114:672–675.
201. Crisp-Lindgren N, et al.: Papillary adenocarcinoma of rete testis. Autopsy findings, histochemistry, immunohistochemistry, ultrastructure, and clinical correlations. *Am J Surg Pathol* 1988; 12:492–501.
202. Jones EC, Murray SK, Young RH, Cysts and epithelial proliferations of the testicular collecting system (including rete testis). *Semin Diagn Pathol* 2000; 17:270–293.
203. Nistal M, Revestido R, Paniagua R, Bilateral mucinous cystadenocarcinoma of the testis and epididymis. *Arch Pathol Lab Med* 1992; 116:1360–1363.
204. Kurihara K, et al., Papillary adenocarcinoma of the epididymis. *Acta Pathol Jpn*, 1993. 43: p. 440–3.
205. McDonald RE, et al., Paratesticular mesotheliomas. *J Urol* 1983; 130:360–361
206. Japko L, et al., Malignant mesothelioma of the tunica vaginalis testis: report of first case with preoperative diagnosis. *Cancer* 1982; 49:119–127
207. Ricketts RR, Majmudarr B, Epididymal melanotic neuroectodermal tumor of infancy. *Hum Pathol* 1985; 16:416–420.
208. Frank GL, Koten JW, Melanotic hamartoma ("retinal anlage tumour") of the epididymis. *J Pathol Bacteriol* 1967; 93:549–554
209. Kanungo A, Chandi SM, Melanotic neuroectodermal tumor of the epididymis. A case report. *Indian J Cancer* 1994; 31:138–140.
210. Jurincic-Winkler C, Metz KA, Klippel KF: Melanotic neuroectodermal tumor of infancy (MNTI) in the epididymis. A case report with immunohistological studies and special consideration of malignant features. *Zentralbl Pathol* 1994; 140:181–185.
211. Hammad FA: Metastatic malignant melanoma of the epididymis. *Br J Urol* 1992; 69:661.
212. Kanno K, et al.: Epididymis metastasis from colon carcinoma: a case report and a review of the Japanese literature. *Jpn J Clin Oncol* 1994; 24:340–344.

213. de Riese W, Warmbold H, Aeikens B: Intrascrotal metastases from renal cell carcinoma. *Int Urol Nephrol* 1986; 18:449–452.
214. Wiebe B, et al.: Epididymal metastasis from prostatic carcinoma. *Scand J Urol Nephrol* 1993; 27:553–555.
215. Vega F, Medeiros LJ, Abruzzo LV: Primary paratesticular lymphoma: a report of 2 cases and review of literature. *Arch Pathol Lab Med* 2001; 125:428–432.
216. Moller MB: Non-Hodgkin's lymphoma of the spermatic cord. *Acta Haematol* 1994; 91:70–72.

CHAPTER 11

Diagnostic Histochemistry in Breast Pathology

Jamie Shutter, M.D., and
Mark R. Wick, M.D.

Although breast diseases are often studied nowadays with immunohistochemical staining techniques, there are several mammary lesions that have been the targets of histochemical analysis in the past. This method is still useful in contemporary clinical practice for diagnostic purposes, as well as for corroboration of results obtained through other methods of ancillary investigation.

SELECTED BENIGN LESIONS OF THE BREAST

Amyloid Deposits

Patients with or without plasma cell dyscrasias may develop tumefactive deposits of amyloid in the breast parenchyma (1–3). These are concentrated not only around ducts and blood vessels but they also occupy large portions of the mammary stroma (Figure 11.1). In likeness to amyloid elsewhere in the body, one observes histochemical reactivity with crystal violet, Congo red (Figure 11.2), thioflavine-T, sulfated alcian blue (AB), or cotton dye methods in breast lesions of this type.

Mammary Sarcoidosis

Sarcoidosis represents another systemic disease with possible manifestation in the breast. In 20% of cases with mammary involvement, however, the patients present with a mass and no evidence of disease elsewhere (4–6). Microscopically, one observes noncaseating granulomatous inflammation (Figure 11.3), sometimes with "asteroid" or "Schaumann" bodies in giant cells. The latter structures can be labeled, respectively, with the phosphotungstic acid–hematoxylin (PTAH) and von Kossa procedures (Figure 11.4). Stains for acid-fast bacteria and fungi should also be obtained, with negative results being expected in sarcoidosis.

Infectious Diseases of the Breast

Bacterial abscesses are potentially observed in the breast parenchyma, particularly after surgical procedures. In addition, "deep" fungal infections such as cryptococcosis, aspergillosis, chromomycosis, and coccidioidomycosis are possibilities as well (7–10). Tuberculosis of the breast is exceedingly rare, even in endemic areas of the world, but many well-documented cases are represented in the literature (11).

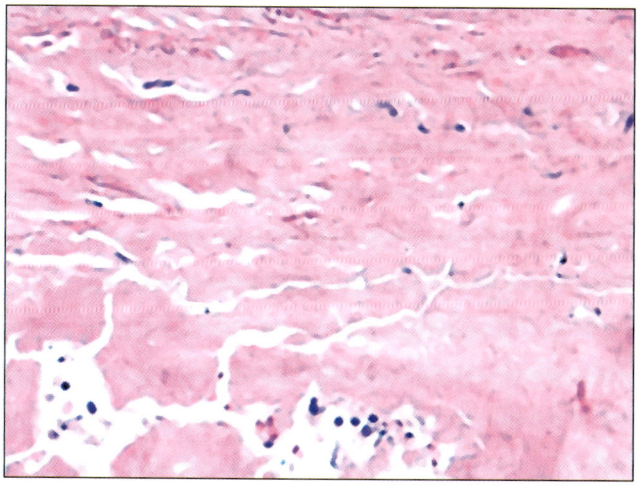

FIGURE 11.1: Amyloid tumor of the breast, demonstrating a deposit of amorphous eosinophilic material in the mammary stroma.

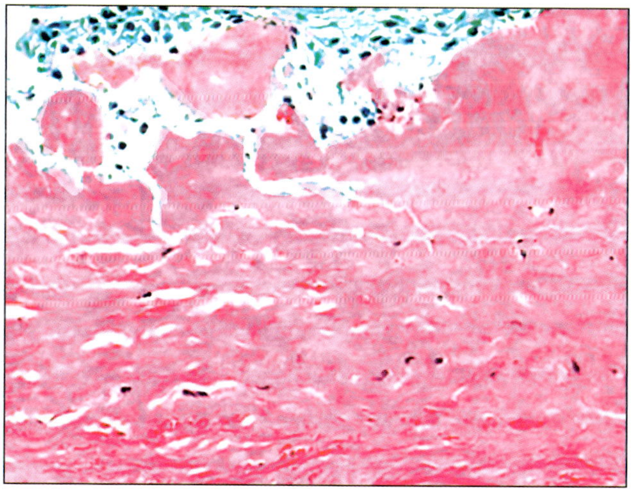

FIGURE 11.2: Congo red labeling of the lesion shown in Figure 11.1.

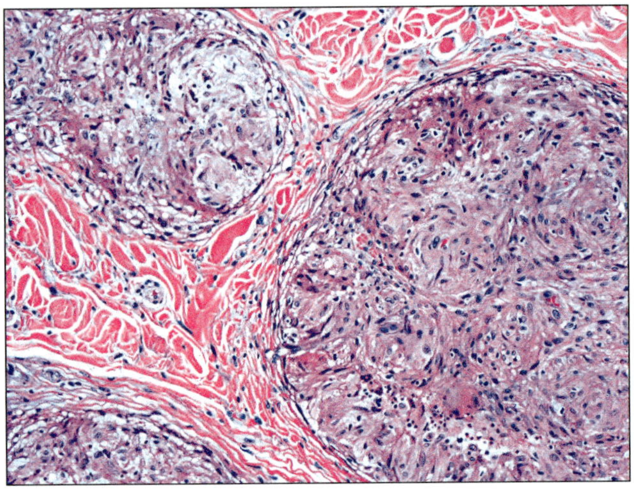

FIGURE 11.3: *Noncaseating granulomatous inflammation in mammary sarcoidosis.*

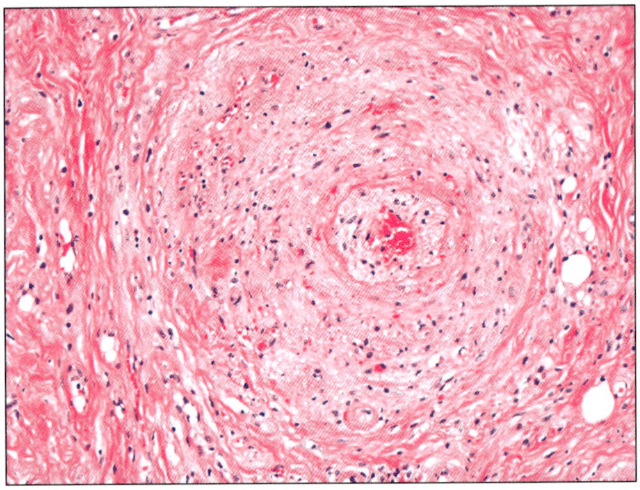

FIGURE 11.5: *Giant cell arteritis of the breast, showing intimal fibrosis and inflammatory involvement of the vascular media.*

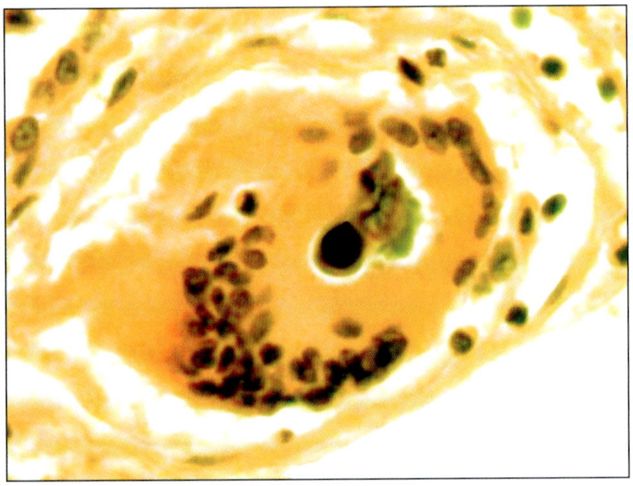

FIGURE 11.4: *The von Kossa stain labels Schaumann's bodies in multinucleated histiocytes of mammary sarcoidosis.*

As true of any other organ site, appropriate tissue stains for these microorganisms are indicated if an infectious process is suspected from the histologic image. Brown-Brenn staining for conventional bacteria, methenamine silver or periodic acid–Schiff (PAS) methods for fungi, and the Ziehl-Neelsen and Fite procedures for mycobacteria are the most frequently employed.

Large-Vessel Vasculitis

Isolated involvement of large intramammary blood vessels by vasculitic processes is a potential problem in giant cell arteritis and Wegener's granulomatosis (12–19) (Figure 11.5). Patients with such diseases present with tender masses of the breast; they may or may not have extramammary involvement of the lungs or other organs.

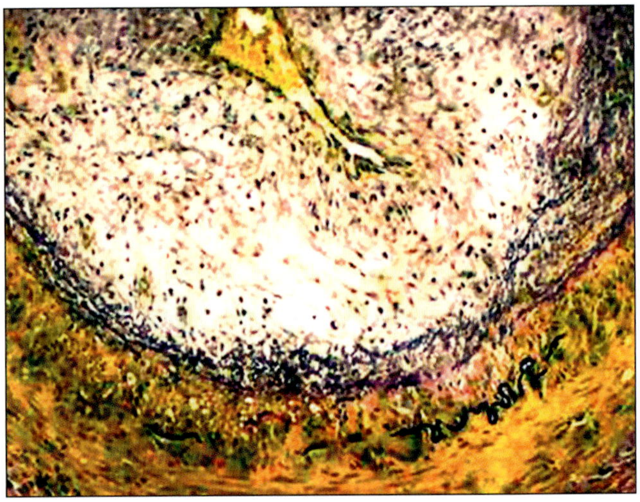

FIGURE 11.6: *Fragmented vascular elastic laminae are seen in giant cell arteritis of the breast, with the Verhoeff–Van Gieson method.*

Evidence for true mural destruction of affected blood vessels can be obtained with stains for elastin (Verhoeff–Van Gieson) and collagen (trichrome). Typically, the elastic laminae are disrupted (Figure 11.6) and replaced by zones of dense collagenosis.

Collagenous Spherulosis

Collagenous spherulosis (CS) of the breast was originally described by Clement et al. in 1987 as an incidental microscopic finding involving lobular acini and ductules. It comprises intraluminal clusters of eosinophilic spherules that often exhibit concentric and radiating fibrillar patterns (20). In the intervening period, CS has also been reported

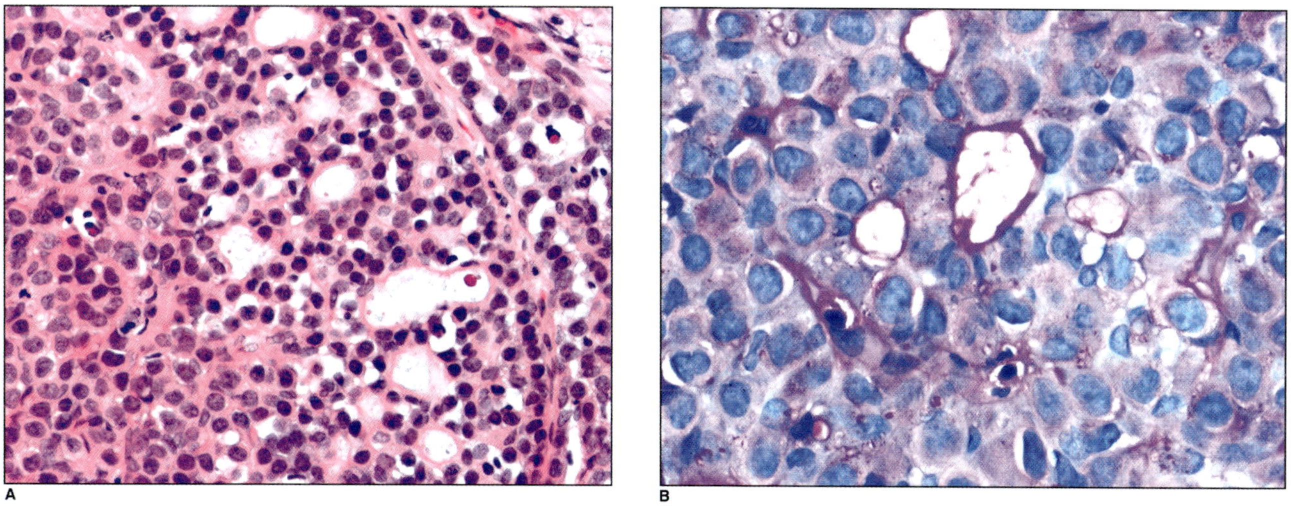

FIGURE 11.7: ***(A and B) CS of the breast shows labeling with the PAS procedure.***

as a mass-forming lesion, and it has demonstrated a potential association with mammographically detected microcalcifications (21,22). Even though the original series on CS demonstrated a linkage with various benign processes in the breasts, it has since been seen in conjunction with in situ carcinoma as well. In 2006, Resetkova et al. observed that CS was associated with atypical ductal hyperplasia (ADH) and lobular carcinoma in situ (LCIS) in 3 of 59 and 15 of 59 cases, respectively (22).

The main differential diagnosis for CS is adenoid cystic carcinoma (ACC) and cribiform-type ductal adenocarcinoma in situ (C-DCIS), with the latter two lesions having markedly different treatment regimens and prognostic implications. In those cases diagnosed in tertiary care centers as CS, the correct diagnosis of CS in made in only small proportion (approximately 15%) of cases; potential misinterpretations include ADH, C-DCIS, and even invasive carcinoma.

Histochemistry can provide assistance in the separation of CS from its morphologic simulants. The fibrillary pattern seen in the internal aspects of CS can be enhanced with the Masson trichrome stain, which yields a blue color in the fibrils. The van Gieson method labels them pink to red, and Snook's reticulin procedure stains them black. An AB or a PAS stain invariably stains the peripheral ductal cuticle of CS positively (red) (Figure 11.7), with the luminal contents being blue. The hyalinized globules in the spherules also stain for elastin (20). PAS-positive material in CS represents basement membrane–associated glycoprotein that is produced by myoepithelial cells, as supported by ultrastructural analysis (23).

Another study by Mooney et al. examined the spectrum of mammary spherulosis (24). Those authors coined the term "mucinous spherulosis" to represent examples of spherulosis with internally basophilic (rather than eosinophilic) luminal contents. All such cases were labeled with PAS and showed peripheral enhancement of the cuticle. Moderate to intense staining with AB was shown in all but one case, and the mucicarmine method yielded positivity in all examples (24).

Granular Cell Tumor

Granular cell tumor (GCT) (formerly "granular cell myoblastoma") is a benign, usually Schwannian, neoplasm that not uncommonly arises in the breast (25–30). It is a particularly good mimic of carcinoma at radiographic and macroscopic levels of examination because a stellate lesion is observed that distorts the surrounding mammary parenchyma (29).

Histochemically, particulate, diastase-resistant, and PAS-positive material is present in the cytoplasm of the tumor cells (Figure 11.8). In addition, one may see small intracellular spherules as well that are similarly labeled.

Adenosis

Mammary adenosis, especially the sclerosing type (Figure 11.9), can be frequently confused with invasive ductal adenocarcinoma, most specifically, with tubular carcinoma. Both these conditions feature a densely collagenized stroma in which proliferating ducts are embedded. Histochemical staining with reticulin or PAS methods usually demonstrates a continuous basement membrane surrounding the ductal structures in sclerosing adenosis (31) but not in tubular carcinoma. Microglandular adenosis (MGA) is another benign lesion in the breast that can produce a palpable mass (32). Because of the random disposition of the constituent gland-like elements—often involving mammary fat (Figure 11.10)—MGA is at particular risk for confusion with low-grade carcinoma. To complicate matters further, some reported cases of MGA have been concurrently associated with carcinoma (33,34). The lumina of

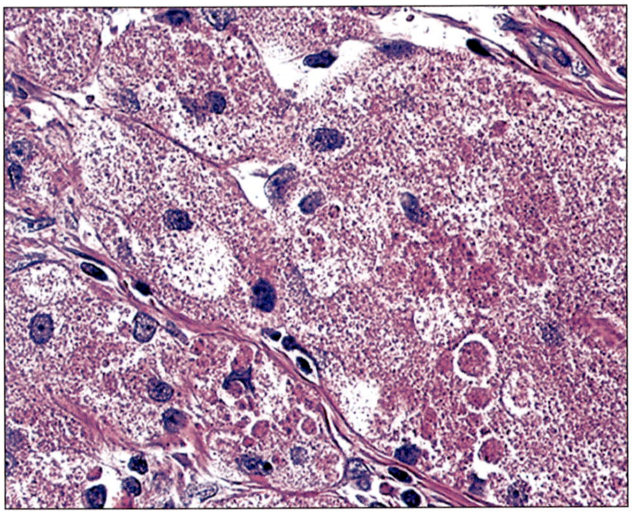

FIGURE 11.8: GCT of the breast demonstrates reactivity of intracytoplasmic granules with the digested PAS stain.

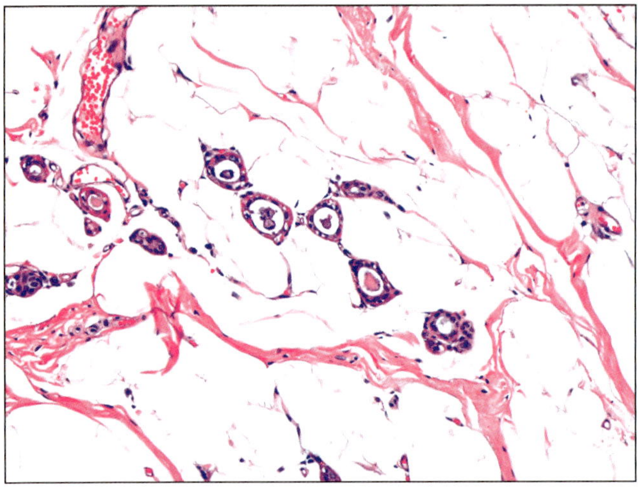

FIGURE 11.10: Microglandular adenosis, demonstrating irregular permeation of the breast stroma by glandular elements with secretory contents.

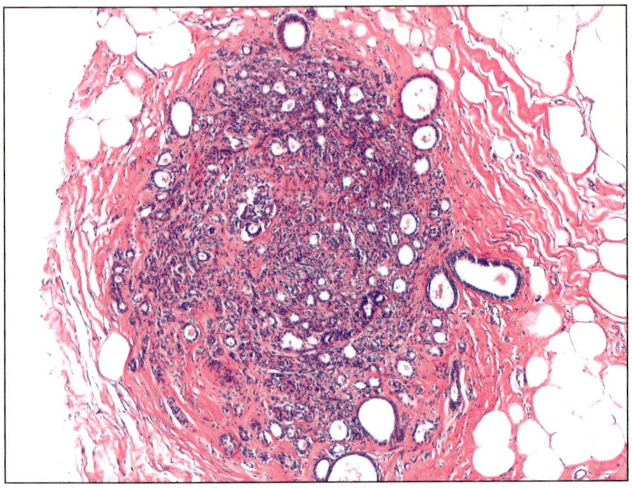

FIGURE 11.9: Sclerosing adenosis of the breast, showing replacement of mammary lobules by proliferating ductular units in a fibrotic stroma.

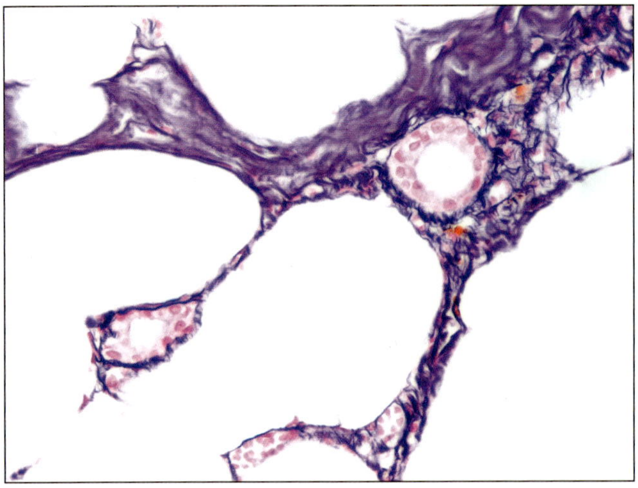

FIGURE 11.11: Snook's method shows investment of the glandular elements of MGA by reticulin.

glands in MGA contain eosinophilic secretions that are reactive with the PAS and mucicarmine techniques (35). Moreover, they are surrounded by completely circumferential basal lamina, as shown with reticulin (Figure 11.11) or PAS stains. The latter result differs from the characteristics of well-differentiated ductal adenocarcinomas.

Mucinous Lesions of the Breast

Mucinous lesions of the breast range from benign mucinous lesions to others that span the spectrum from benign to atypical to frankly malignant. Nodular mucinosis (NM) is a benign lesion seen in young women, sometimes seen in association with Carney's syndrome. NM typically stains for acidic mucins with the AB and colloidal iron (CI) methods. It is faintly positive with mucicarmine and negative with the digested PAS procedure (36). This staining profile contrasts to that of mucocele-like lesions (MLLs), as found in association with in situ and invasive carcinomas. MLLs stain strongly for neutral mucins with PAS and mucicarmine (37). The latter lesions must also be distinguished from colloid carcinomas (CCs) because they have identical mucin-staining characteristics. However, CCs often show neuroendocrine differentiation as well, as evidenced by argyrophilia with the Grimelius, Churukian-Schenk, or Sevier-Munger stain (38–40).

Sclerosing and Proliferative Epithelial Lesions

Radial scars are potentially misdiagnosed as invasive cancers of the breast. Radiographically and histologically, they exhibit a stellate configuration (Figure 11.12) that is often

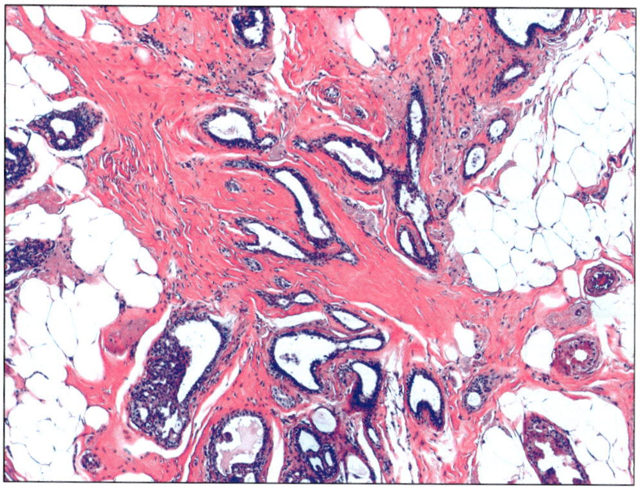

FIGURE 11.12: ***Radial scar of the breast, in which central stellate areas of fibrosis are surrounded by proliferative ductal epithelial units.***

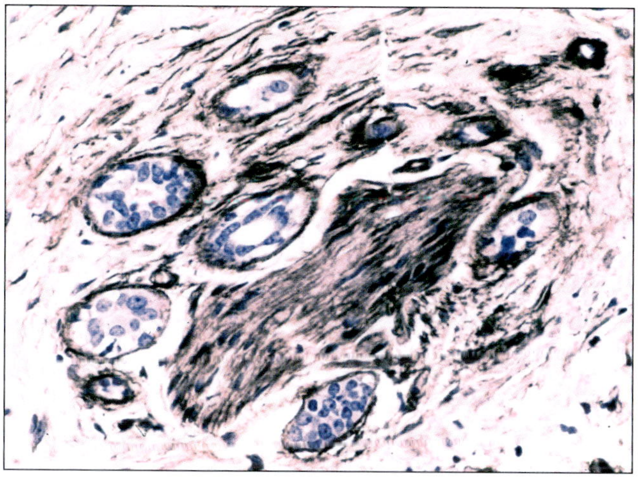

FIGURE 11.13: ***Snook's reticulin method shows complete investment of ductal units by basement membrane material in radial scar of the breast.***

shared by infiltrative adenocarcinomas. In radial scar, elastin stains show a density of that protein in the central stroma of the lesion, whereas it is concentrated around the ducts of entrapped nonneoplastic ducts in invasive carcinomas (31). Furthermore, PAS and reticulin stains disclose complete investment of trapped glands and ducts in radial scars (Figure 11.13), whereas carcinomas lack that pattern with both methods.

Papillary Proliferations

Intraductal papillomas of the breast are usually biopsied before excision, and one may have trouble making a definitive diagnosis when only a small portion of the lesion is represented. A Masson trichrome stain can highlight the

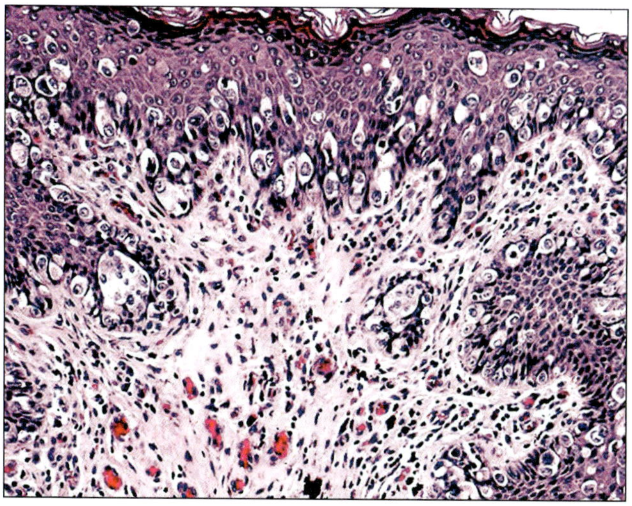

FIGURE 11.14: ***Paget's disease of the breast, in which the epidermis of the nipple is partially replaced by pale foreign epithelioid cells with a random intraepithelial disposition.***

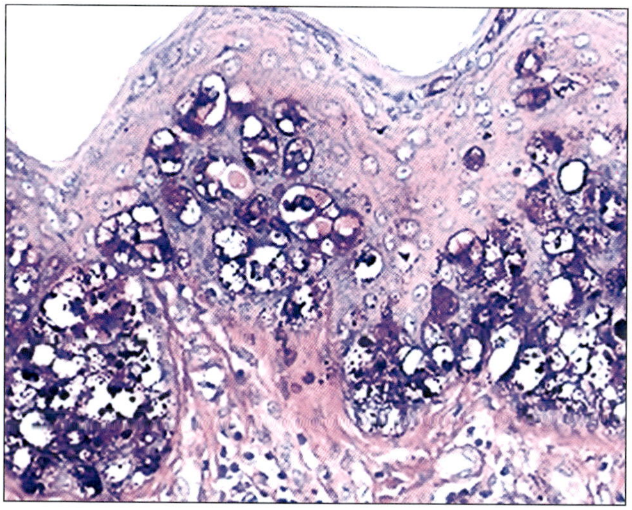

FIGURE 11.15: ***The tumor cells in mammary Paget's disease are reactive with the PAS method after diastase digestion.***

central fibrous stroma that is present in most papillomas. In contrast, papillary carcinomas have a sparse supporting matrix. The cells in a small percentage (~20%) of papillary carcinomas also show histochemical evidence of neuroendocrine differentiation by Grimelius staining, as well as mucin production with the MC, AB, and PAS procedures (31,41,42). Those findings are not expected in papillomas.

SELECTED MALIGNANT MAMMARY NEOPLASMS

Mammary Paget's Disease

The etiology of mammary Paget's disease (MPD) has been the focus of considerable debate since it was first described in 1874 (43). It has been commonly accepted that this

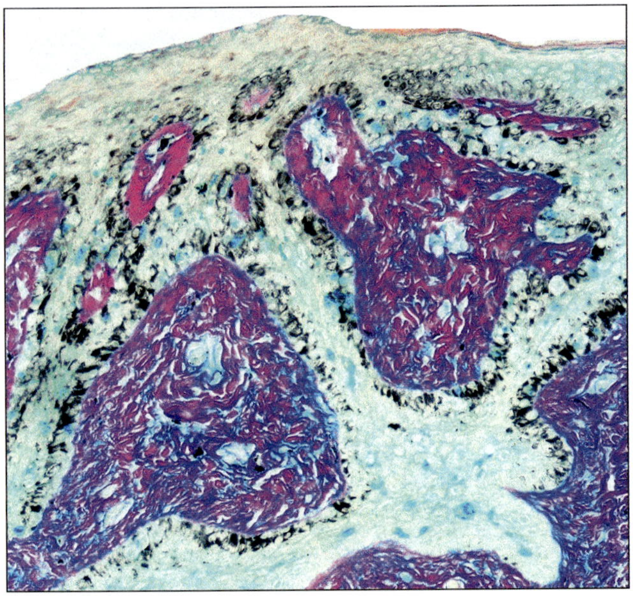

FIGURE 11.16: The CI method labels the neoplastic cells in mammary Paget's disease.

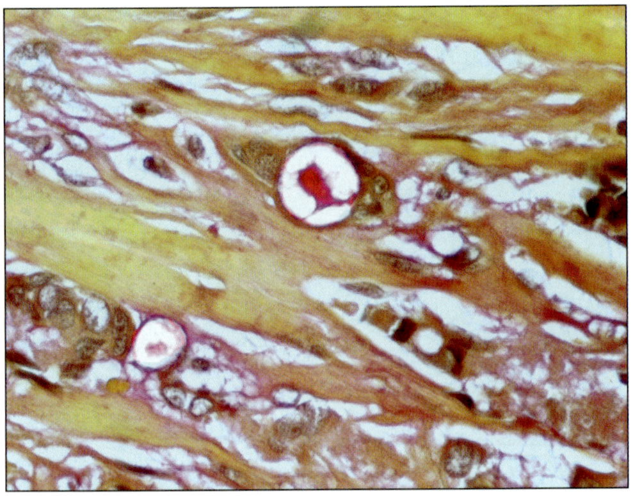

FIGURE 11.17: ILC of the breast with signet-ring cell differentiation, showing labeling of intracytoplasmic lumina with the mucicarmine method.

intraepidermal proliferation (Figure 11.14) is exclusively related to underlying DCIS, with an associated invasive component in a sizable percentage of cases. Histochemical preparations have figured prominently in the past into characterizations of the disorder.

In a 1985 study by Sitalkin and Ackerman, 10 cases of MPD were compared to 10 examples of extramammary Paget's disease (EPD) (44). The extramammary lesions demonstrated intense staining of the intraepidermal neoplastic cells with the PAS, PAS-diastase (PAS-D), mucicarmine, and CI methods (Figures 11.15 and 11.16) in 90% of cases, but only 50% of the cells in MPD were so labeled.

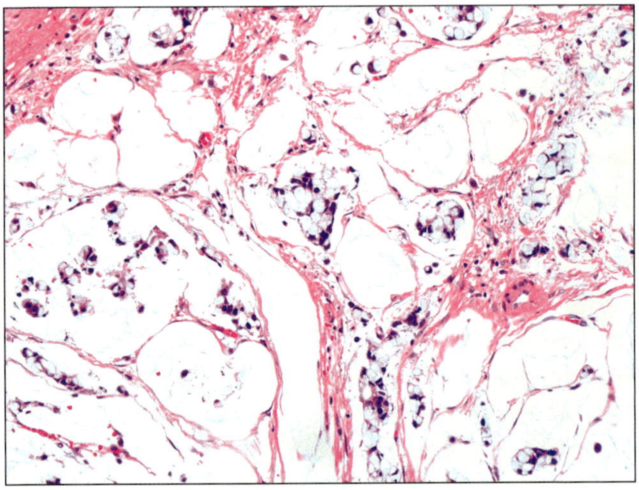

FIGURE 11.18: Mucinous (colloid) carcinoma of the breast, in which nests of tumor cells are suspended in pools of extracellular mucin.

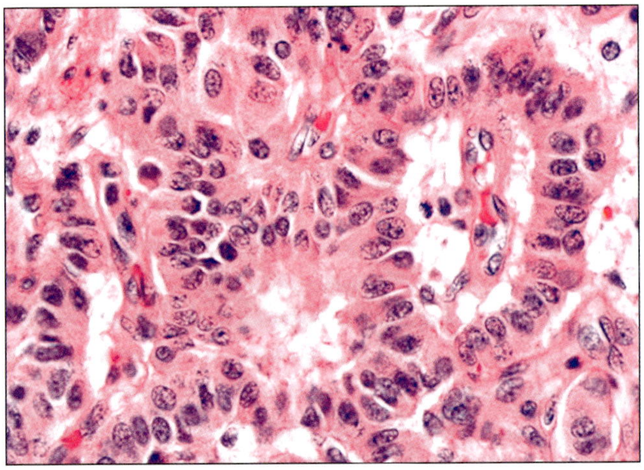

FIGURE 11.19: "Pure" low-grade neuroendocrine carcinoma (carcinoid) of the breast, in which nests and cords of monomorphic tumor cells assume an organoid growth pattern.

Another analysis produced similar results but exclusively considered MPD (45).

Superficial spreading malignant melanoma (SSMM) and primary intraepidermal squamous carcinoma (PISC) are the traditional differential diagnostic alternatives to MPD. Hitchcock et al. compared the usefulness of histochemistry to distinguish between these three entities (46). Mucin positivity, as evaluated with PAS-D staining, identified 6 of 13 MPD cases and an additional two stained positively with AB (62%). In five cases of SSMM and six of PISC, PAS-D, and AB stains were consistently negative. A potential drawback in this specific setting is the ability for tumor cells in MPD to passively phagocytize melanin from surrounding keratinocytes (47–51). This property may yield spurious reactivity with Fontana-Masson preparations and lead to possible diagnostic confusion with SSMM.

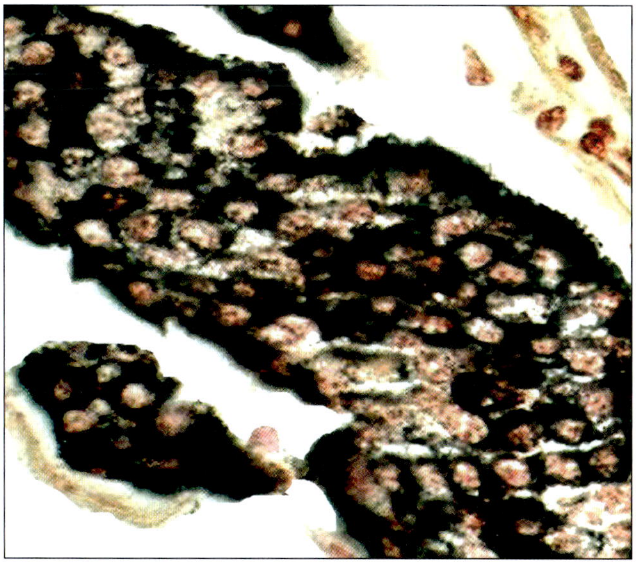

FIGURE 11.20: Positivity in neuroendocrine breast carcinoma with the Grimelius method.

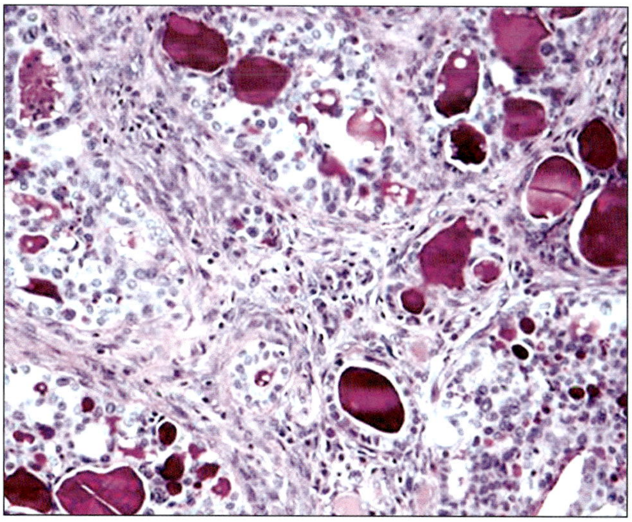

FIGURE 11.22: The digested PAS stain labels the extracellular material in secretory breast carcinoma.

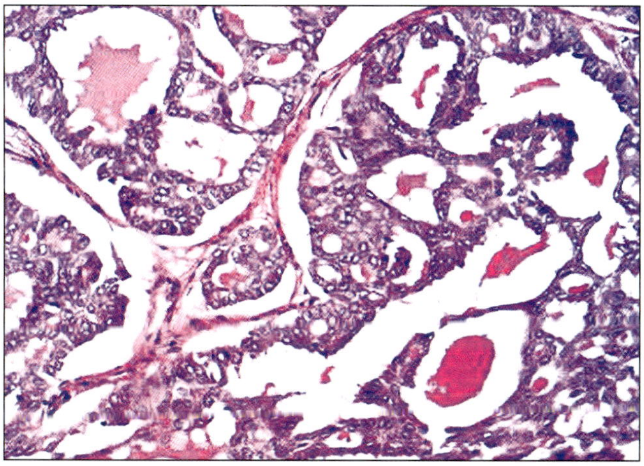

FIGURE 11.21: Secretory adenocarcinoma of the breast, in which intercellular accumulations of eosinophilic material are present.

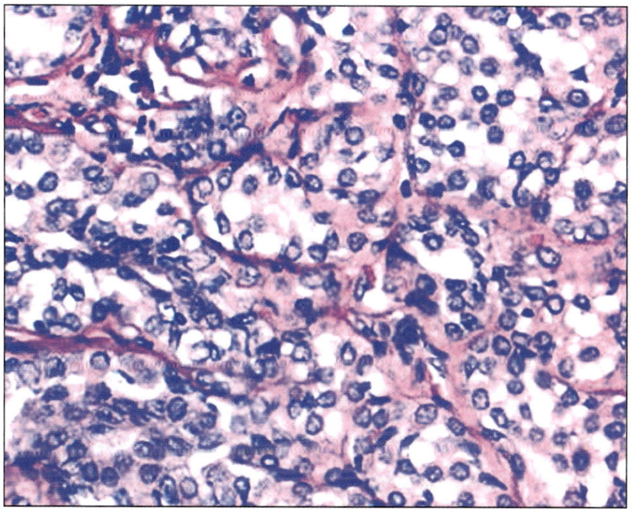

FIGURE 11.23: Glycogen-rich adenocarcinoma of the breast, showing cytoplasmic clarity.

Lobular Neoplasia

In small-needle biopsy specimens, it may be challenging to separate invasive breast carcinomas into ductal and lobular types. Mucicarmine, digested PAS, or CI stains can be used to highlight the presence of intracellular mucin vacuoles in lobular neoplasms. One study of invasive lobular mammary carcinomas demonstrated histochemical positivity for intracellular mucin in 90% of cases (52) (Figure 11.17).

Variants of Invasive Carcinoma

Angiolymphatic invasion is an important histologic finding in reference to many tumors, but this is particularly true in reference to breast carcinomas because of its prediction of lymph nodal metastasis. The Van Gieson or orcein stains for elastin can be used to outline the elastic fibers in veins. Reticulin stains also help to delineate lymphatic vessels by impregnating their basement membranes. If tumor cell groups are seen within spaces that are surrounded by elastin or reticulin, one can be certain of true vascular invasion and exclude the alternative possibility of artifactual tissue retraction spaces around neoplastic cellular aggregates (31).

As mentioned earlier, periductal elastosis has been associated with invasive carcinomas of the breast. Van Gieson stains can be used to elucidate this finding, allowing one to distinguish between adenosis or radial scar and infiltrating carcinoma.

Neuroendocrine differentiation has been recognized in a subset of invasive carcinomas with various cytomorphologic

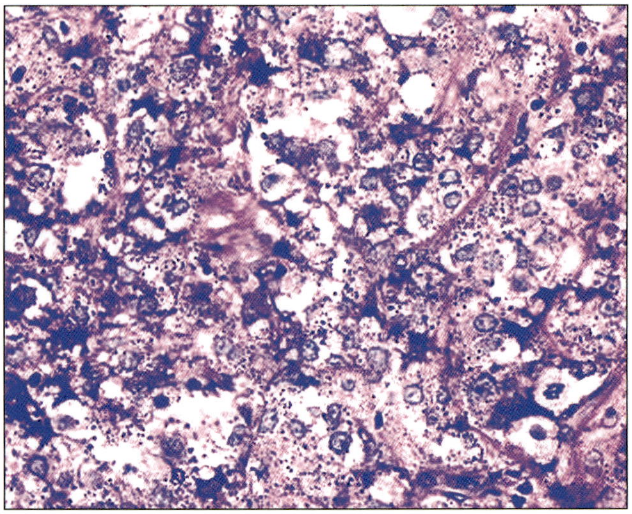

FIGURE 11.24: *An undigested PAS stain strongly labels the tumor cells of glycogen-rich mammary adenocarcinoma.*

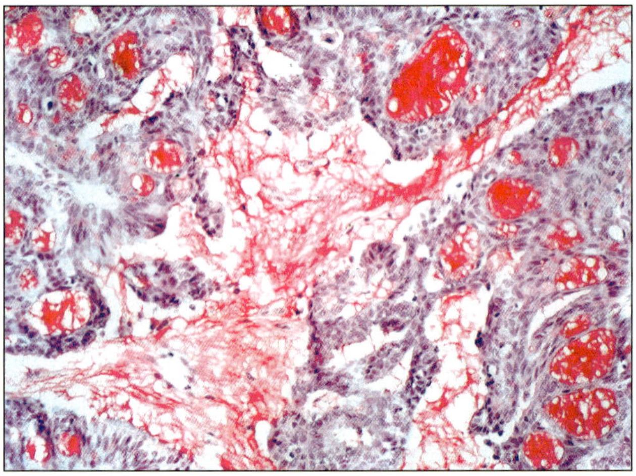

FIGURE 11.25: *Mucinous (colloid) carcinoma of the breast, manifesting positivity of extracellular stroma with the mucicarmine method.*

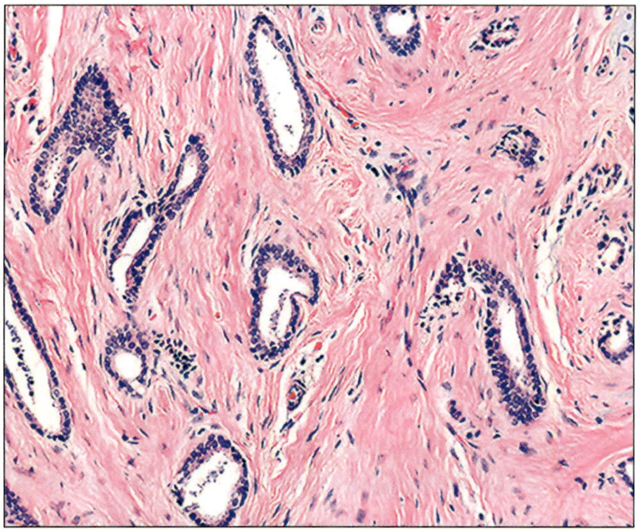

FIGURE 11.26: *Tubular carcinoma of the breast, in which angular ductal profiles of tumor cells show internal decapitation secretion and are separated by collagenized stroma.*

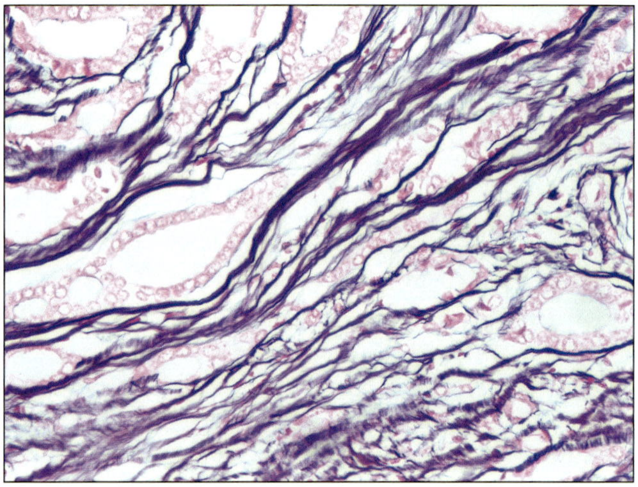

FIGURE 11.27: *Snook's method shows no reticulin investment of the ductal profiles of tumor cells in tubular carcinoma.*

images (53–65). Many of them are mucinous in nature ["CCs" (55)] as discussed previously (Figure 11.18); some manifest an overtly endocrine image with insular growth and dispersed nuclear chromatin (Figure 11.19), and others are "ordinary" ductal or lobular carcinomas. Any of the argyrophil stains (Sevier-Munger, Churukian-Schenk, Grimelius) are suitable for demonstrating neuroendocrine differentiation in breast tumors (6) (Figure 11.20). Argentaffin methods are consistently negative in such lesions. Aside from helping to identify "pure" neuroendocrine mammary carcinomas [e.g., "carcinoid" tumors, small-cell carcinomas (64–66)], which are uncommon, the prognostic and biological significance of neuroendocrine differentiation in other forms of breast cancer is still uncertain.

Secretory carcinoma of the breast is a rare clinicopathologic variant that is seen in children or young women (67,68); it is associated with a favorable prognosis in most instances. The prominent eosinophilic secretory material that is apparent by hematoxylin and eosin staining of this tumor (Figure 11.21) also labels intensely with the digested PAS (Figure 11.22) and CI or AB methods. In contrast, glycogen-rich carcinoma (which may be a subtype of invasive ductal adenocarcinoma of the usual type) (Figure 11.23) demonstrates PAS reactivity but with sensitivity to diastase pretreatment (31) (Figure 11.24).

Mucinous (colloid) and tubular carcinomas have been discussed earlier in this chapter. To reiterate, the stroma in the first of those neoplasms is consistently reactive with the

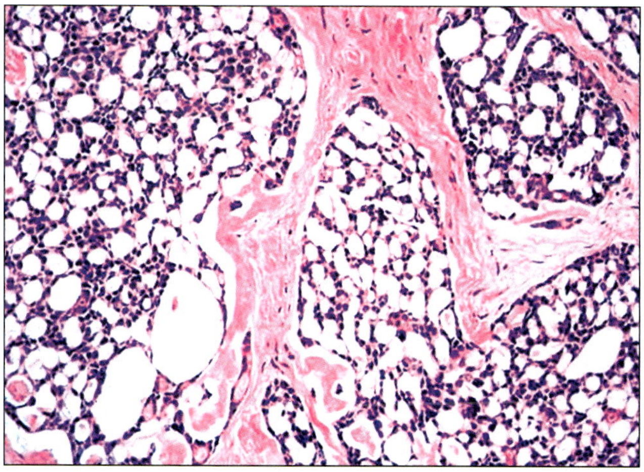

FIGURE 11.28: ACC of the breast, in which tubular arrays of basaloid tumor cells encompass mucoid or glycoproteinaceous stroma.

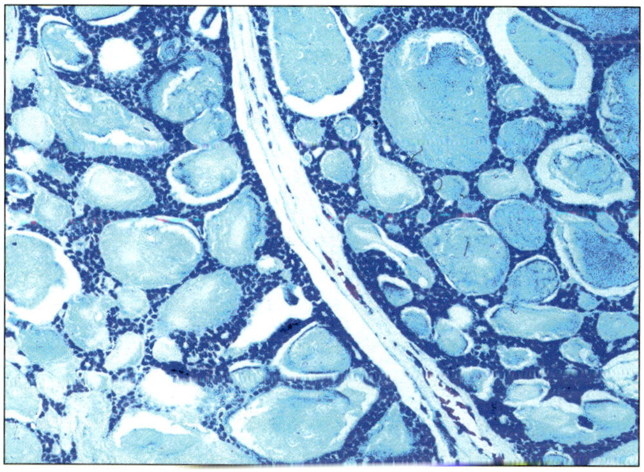

FIGURE 11.29: The CI stain labels the tubular contents of this ACC.

PAS, CI, AB, and mucicarmine techniques (Figure 11.25). In addition, the tumor cells in CCs often show "occult" neuroendocrine differentiation and argyrophilia. Neoplastic ducts in tubular carcinoma (Figure 11.26) are typically surrounded by a collagen- and reticulin-rich network of stroma, as shown by trichrome and Snook's stains, respectively. In contrast to its benign simulators (adenosis and radial scar), however, tubular carcinoma lacks PAS-positive or reticulin-reactive investments of its constituent epithelial tubules (31) (Figure 11.27).

ACC (Figure 11.28) represents another rare mammary carcinoma morphotype, with a distinctive histologic image that features cylinders of basaloid cells in which mucinous stroma is present. The latter labels with the PAS, AB, CI (Figure 11.29), and mucicarmine methods (69–73). Moreover, it often contains a central core of collagenized, trichrome-positive material.

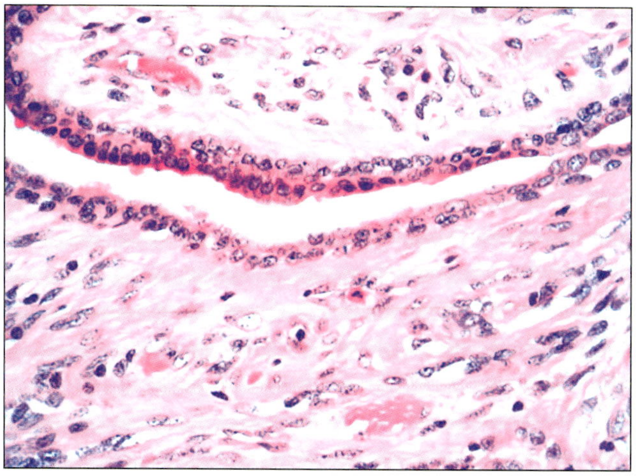

FIGURE 11.30: Low-grade phyllodes tumor of the breast, in which proliferating stromal cells are densely aggregated and show mitotic activity.

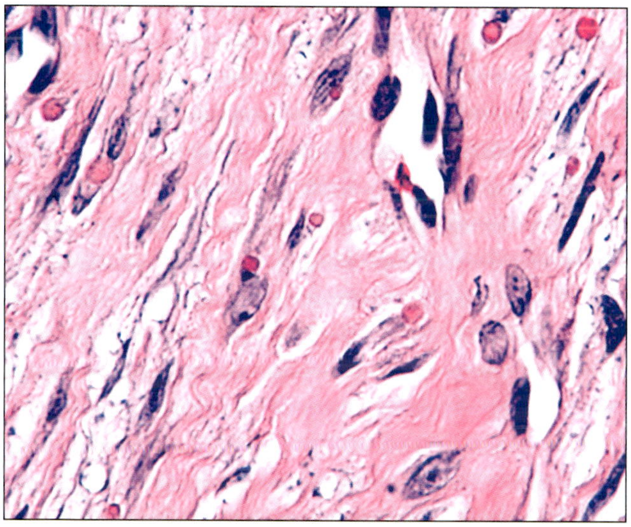

FIGURE 11.31: Perinuclear rounded eosinophilic inclusions in low-grade mammary phyllodes tumor, imitating the image of digital fibromatosis.

Phyllodes Tumor

A subtype of low-grade mammary phyllodes tumor (Figure 11.30) shows inclusions in proliferating stromal cells that simulate those of digital fibromatosis (74,75). They are rounded and paranuclear (Figure 11.31), staining red with hematoxylin and eosin and scarlet with the Masson trichrome procedure (Figure 11.32). Their significance is unknown.

Hematopoietic and Lymphoid Neoplasms

Although they are unusual in the breast, most if not all hematopoietic and lymphoreticular malignancies can occur there as either primary or secondary lesions. These disorders principally include lymphoid and myeloid leukemias, non-Hodgkin and Hodgkin lymphomas, and plasmacytic myeloma (76–82).

All proliferative lesions of the hematolymphoid system (Figure 11.33) are characterized by investment of individual tumor cells by reticulin, as demonstrated with Snook's method (Figure 11.34). Contrarily, carcinomas of the breast show reticulin fibers around *groups* of neoplastic cells (Figure 11.35). In addition, the methyl green–pyronine stain is usually intensely positive in hematopoietic neoplasms (Figure 11.36). Myeloid leukemia can be recognized selectively with the chloroacetate esterase procedure, and digested PAS stains may label some cases of myeloma.

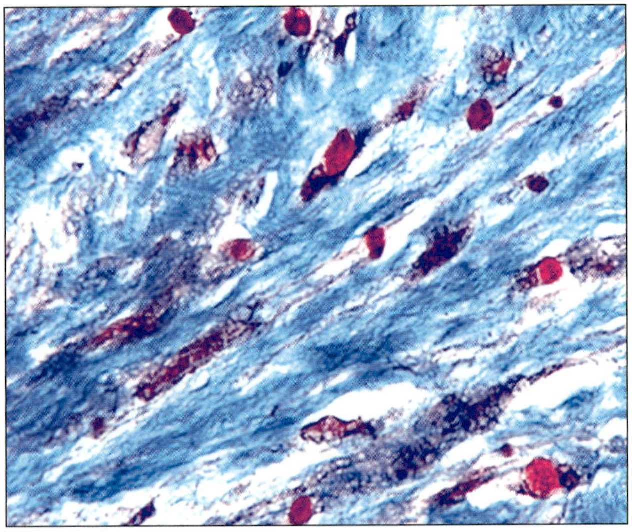

FIGURE 11.32: ***The inclusions seen in Figure 11.31 are eosinophilic with the trichrome stain.***

Metastatic Tumors in the Breast

Although a variety of malignant nonlymphoid neoplasms from other anatomic locations have been reported as metastases to the breast, only a few are reproducibly seen in surgical pathology in our experience. These include melanoma, rhabdomyosarcoma, and carcinoid tumors of the gut or lung (83–88). Respective histochemical stains for those lesions include the Fontana-Masson method, the PAS and PTAH procedures, and argyrophil and argentaffin silver impregnation techniques. Argentaffinity in a carcinoid tumor constitutes good evidence that it is a secondary lesion.

LECTIN HISTOCHEMISTRY IN BREAST PATHOLOGY

In general, lectin histochemistry has relatively limited use in the breast. However, four diagnostic areas lend themselves to productive application of that technique.

Mammary Paget's Disease

As mentioned earlier, the main differential diagnosis surrounding MPD also includes SSMM in situ and pagetoid squamous carcinoma in situ. In addition to the histochemical methods discussed above, one can also obtain stains with biotinylated wheat germ agglutinin and peanut agglutinin, as developed with the avidin-biotin-peroxidase complex method. Tamaki et al. (89) have shown that the tumor cells of MPD selectively bind to both of those lectins (Figure 11.37), whereas those of SSMM and PISC do not (90).

Angiosarcoma of the Breast

Angiosarcoma of the breast (Figure 11.38) is an often-fatal tumor that must be separated on one hand from

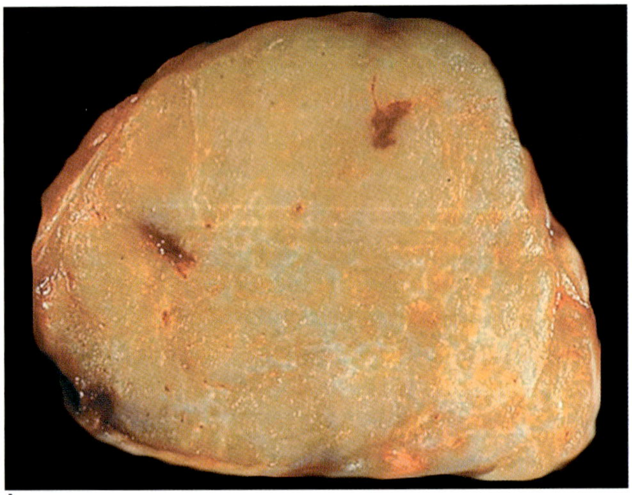

A

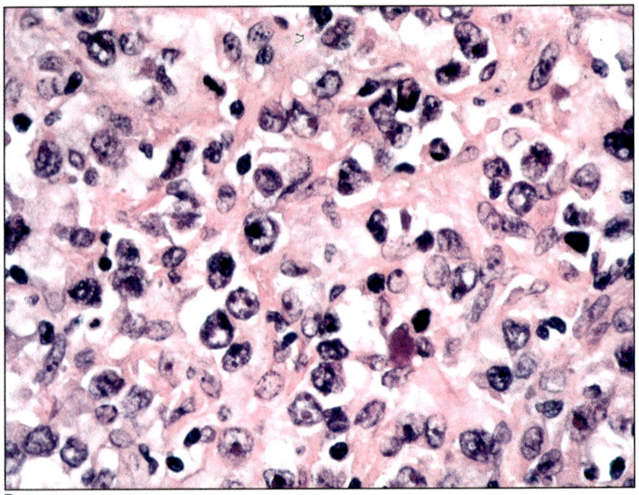

B

FIGURE 11.33: ***(A) Large-cell non-Hodgkin lymphoma (LCL) of the breast manifests a pale, "fish-flesh" cut surface in this excision specimen. (B) Microscopically, the tumor cells in LCL show irregular nuclear contours, variably prominent nucleoli, and amphophilic cytoplasm. The differential diagnosis also includes undifferentiated carcinoma, metastatic melanoma, sarcoma, and granulocytic sarcoma.***

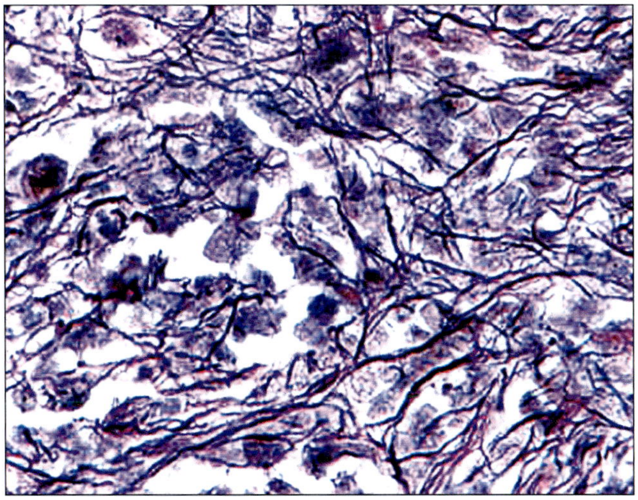

FIGURE 11.34: Snook's method in large-cell lymphoma of the breast demonstrates investment of individual tumor cells by reticulin fibers.

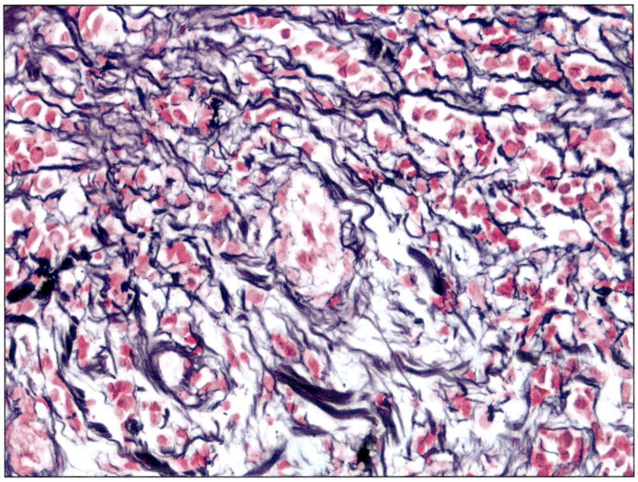

FIGURE 11.35: In contrast to Figure 11.34, Snook's procedure shows reticulin fibers around groups of tumor cells in poorly differentiated carcinoma of the breast.

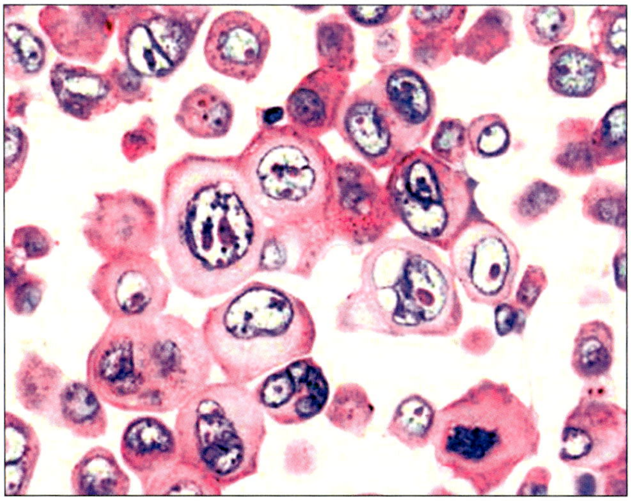

FIGURE 11.36: The methyl green–pyronine procedure shows intense labeling of tumor cells in large-cell non-Hodgkin lymphoma of the breast.

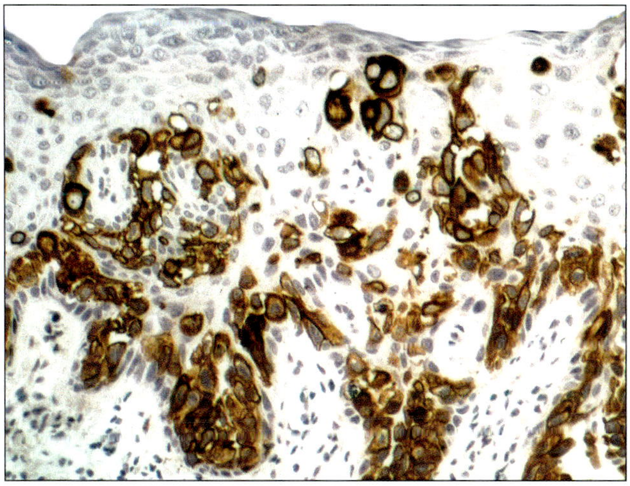

FIGURE 11.37: The tumor cells in mammary Paget's disease show uniform labeling by biotinylated wheat germ agglutinin.

reactive vascular proliferations or perilobular hemangiomas and, on the other, from pseudovascular mammary carcinomas. Lectin staining with *Ulex europaeus* I agglutinin (UEA), *Dolichos biflorus* agglutinin (DBA) (Figure 11.39), and *Bandeiraea simplicifolia* agglutinin (BSA) (Figure 11.40) is capable of labeling malignant vascular neoplasms of the breast (91–93). However, those three reagents likewise may bind to mammary carcinomas. Thus, the results favor endothelial differentiation only when two of the three lectins yield positive results. Also, epithelioid angiosarcomas and well-differentiated vascular sarcomas tend to be lectin reactive more often than tumors with ordinary morphotypes or high-grade endothelial lesions.

Hematopoietic Malignancies

The problems of recognizing malignant mammary hematopoietic neoplasms have been mentioned above (76). In particular, they may be difficult to separate diagnostically from undifferentiated carcinomas and metastatic melanomas. Lectin reactivity with *Ricinus communis* agglutinin (RCA) or peanut agglutinin (Figure 11.41), or both, is often seen in lymphomas and leukemias in the mammary parenchyma. However, other lectins, including UEA, concanavalin-A (CA), BSA, and DBA yield negative results (94). CA and RCA also may bind to melanomas (95) but UEA, BSA, and DBA do not.

Demonstration of Angiolymphatic Invasion by Carcinoma

Again, because they are potential markers for endothelial cells, lectin histochemical stains with UEA, DBA, and BSA

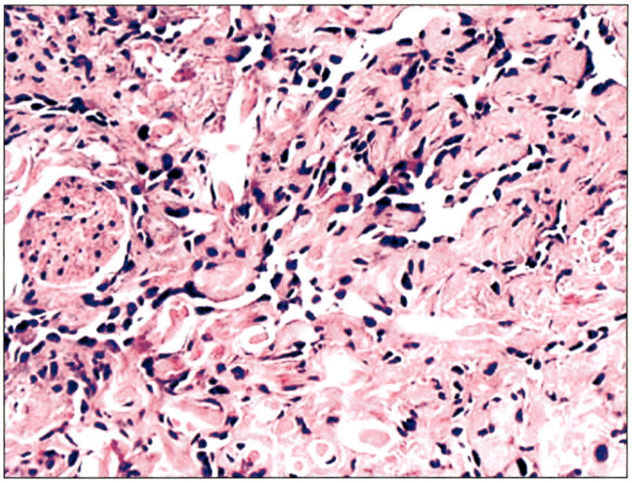

FIGURE 11.38: Angiosarcoma of the breast, in which interanastomosing vascular channels are lined by atypical hyperchromatic endothelial cells.

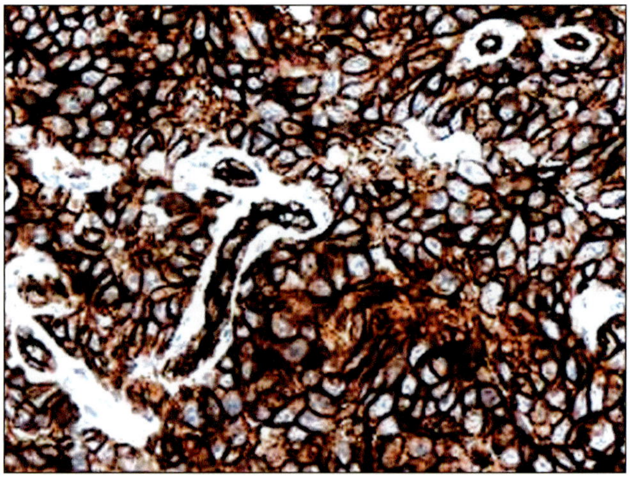

FIGURE 11.39: Reactivity with biotinylated UEA in mammary angiosarcoma.

can be used to outline blood vessels—both venous and lymphatic—in the breast parenchyma. This procedure is helpful in confirming the presence of tumor microemboli in cases of invasive carcinoma and excluding the alternative possibility of artifactual space formation around neoplastic cell groups (96). One looks for labeling of cells lining the spaces in question (i.e., endothelial elements) to confirm tumor invasion. The neoplastic cells may or may not stain as well (97).

ILLUSTRATIVE CASE STUDIES

CASE 1— *Collagenous Spherulosis:* A 44-year-old woman was found to have indeterminate calcifications in the right breast on routine screening mammography. Physical examination disclosed no palpable masses, and a stereotactic core needle biopsy was performed. Microscopic examination showed a monotonous population of cells expanding the lobular units and spreading in a pagetoid fashion into ductal structures (Figure 11.42). The ducts contained sharply demarcated spaces forming a cribriform arrangement. PAS staining highlighted a peripheral cuticle as well as internal positivity in the cribriform spaces (Figure 11.43). LCIS also was present.

The final diagnosis was that of CS, associated with LCIS. Because of the monotony of the cell population in the areas of spherulosis, however, this lesion could have been confused with cribriform ductal carcinoma in situ. ACC might also have been considered. Both of those lesions have a histochemical profile that differs from that of CS. ■

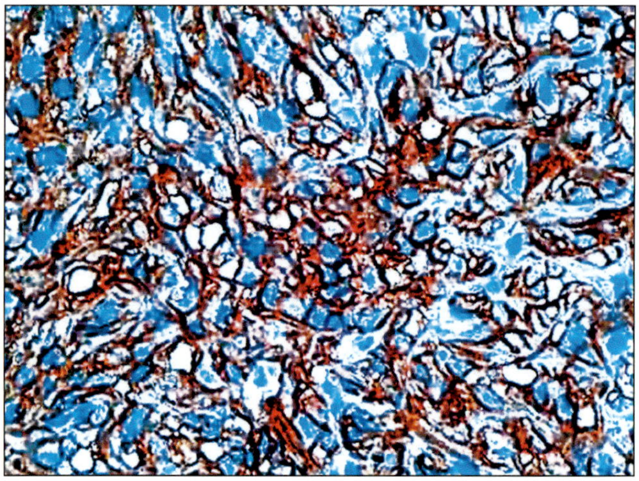

FIGURE 11.40: Positivity is seen with biotinylated DBA in angiosarcoma of the breast.

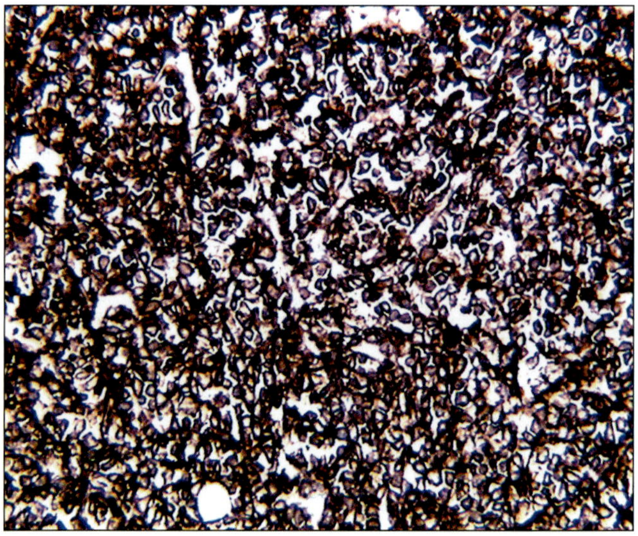

FIGURE 11.41: Large-cell lymphoma of the breast, demonstrating diffuse binding of biotinylated peanut agglutinin.

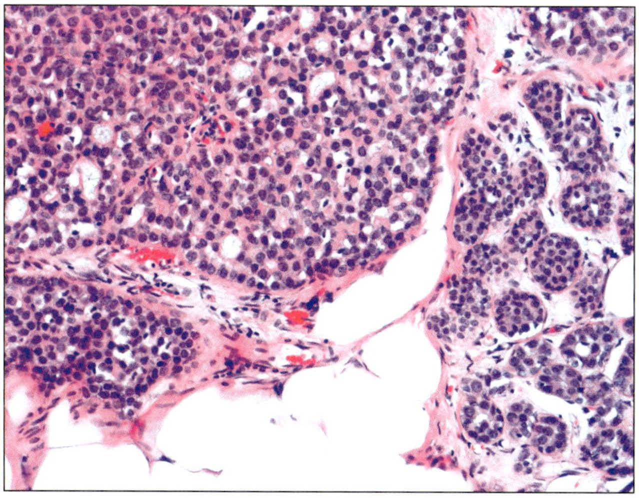

FIGURE 11.42: This breast biopsy shows a lobule on the left of the figure that contains small "cribriform" spaces among the proliferating epithelial cells.

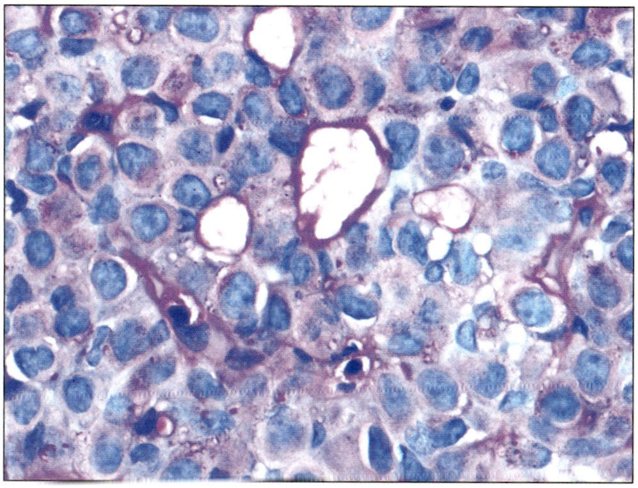

FIGURE 11.43: A PAS stain demonstrates peripheral "cuticular" labeling of the contents of cribriform spaces, together with internal staining as well. The final diagnosis was that of CC.

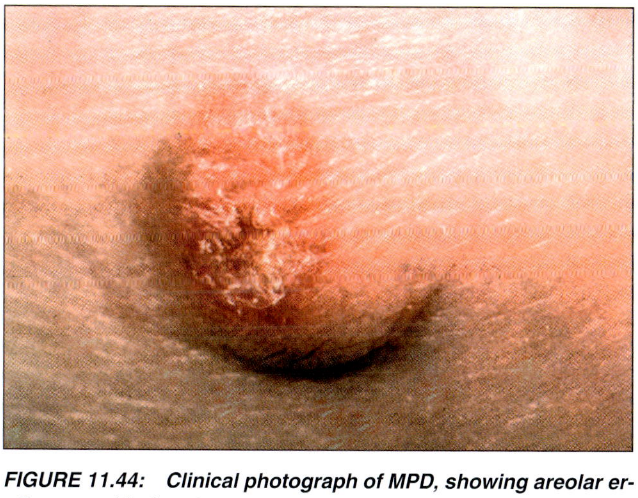

FIGURE 11.44: Clinical photograph of MPD, showing areolar erythema and induration.

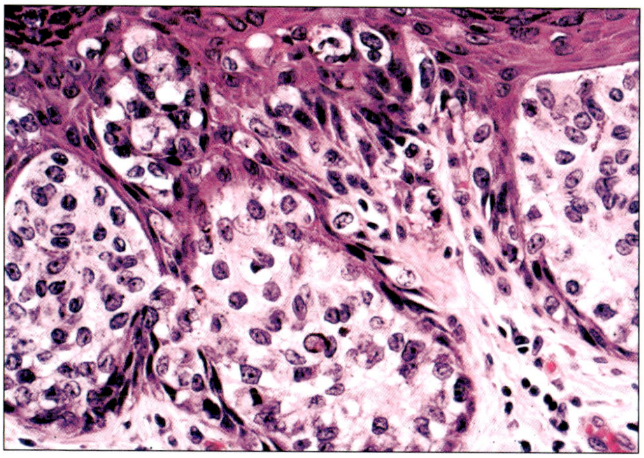

FIGURE 11.45: Mammary Paget's disease, demonstrating a "buckshot" pattern of atypical pale epithelioid tumor cells interspersed throughout the epidermis.

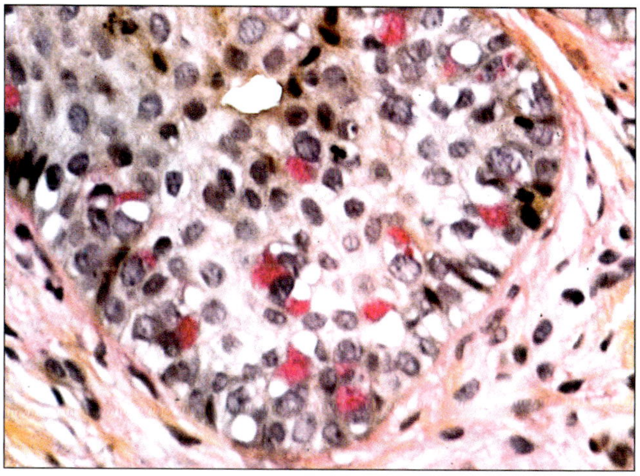

FIGURE 11.46: The mucicarmine method shows multifocal labeling of the tumor cells in mammary Paget's disease.

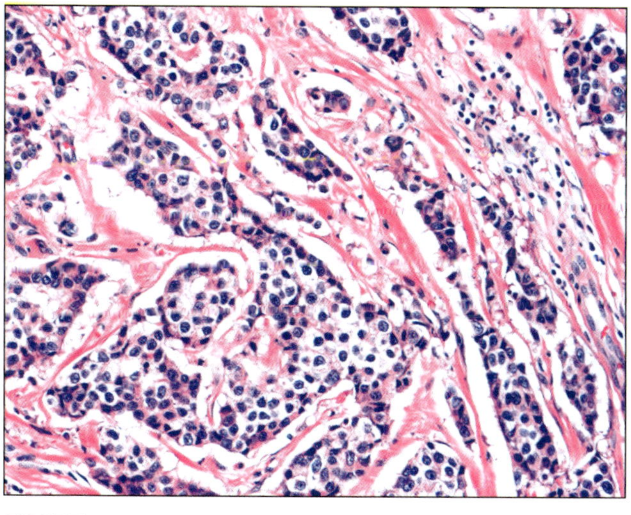

FIGURE 11.47: Invasive ductal carcinoma was present in the breast showing mammary Paget's disease.

CASE 2— *Mammary Paget's Disease:* A 55-year-old woman presented to her physician with erythema and excoriation of the left nipple-areolar complex (Figure 11.44), and a 3-cm palpable mass in the left breast. A biopsy of the skin was performed, showing infiltration by an atypical epithelioid cell population arranged in a "buckshot" pagetoid fashion (Figure 11.45). A mucicarmine stain labeled the cells in question (Figure 11.46), whereas the Fontana-Masson method yielded negative results. These findings confirmed the diagnosis of MPD. Excision of the underlying mass showed an invasive ductal adenocarcinoma (Figure 11.47).

The origin of MPD is still controversial. For years, the doctrinaire explanation of its pathogenesis has held that MPD represents upward "migration" of tumor cells into the skin from an underlying ductal breast carcinoma (98,99). However, we take conceptual issue with that paradigm. First, it is unclear how the tumor cells would permeate the epidermis in the observed pattern and "seal up" keratinocytic junctional complexes behind them. Second, examples do exist—in approximately 10% of all MPD cases—where *no* underlying breast carcinoma is found despite exhaustive sampling (100,101). As other authors have suggested (102), it would appear more likely to us that MPD and EPD are the same process but with greatly different levels of association with regional extracutaneous malignancies. ■

CASE 3— *Invasive Lobular Adenocarcinoma:* A 67-year-old woman found a 2-cm nonmobile palpable mass in the upper outer quadrant of her right breast. Mammography revealed an ill-defined mass in the same area, and an ultrasonographically directed core biopsy was obtained. Microscopically, it showed a population of atypical polygonal cells with a focal single-file arrangement (Figure 11.48); however, more solid nests were also present. The

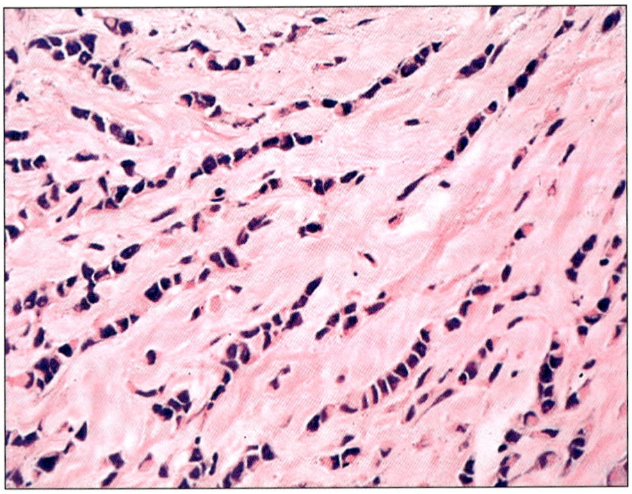

FIGURE 11.48: *Invasive mammary lobular carcinoma, demonstrating single-file growth of monotonous epithelioid cells.*

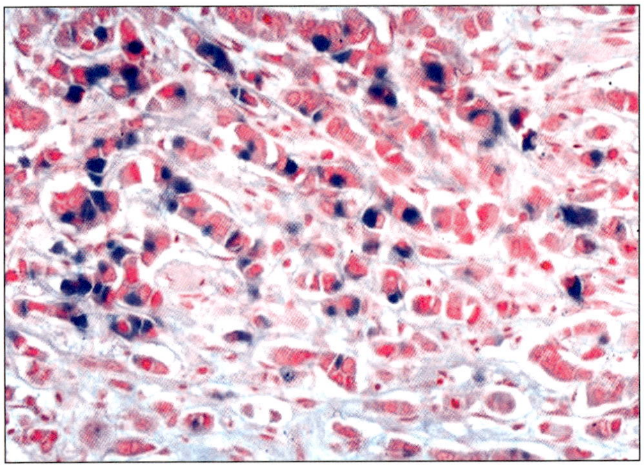

FIGURE 11.49: *The CI method highlights intracytoplasmic lumina with mucinous contents in ILC of the breast.*

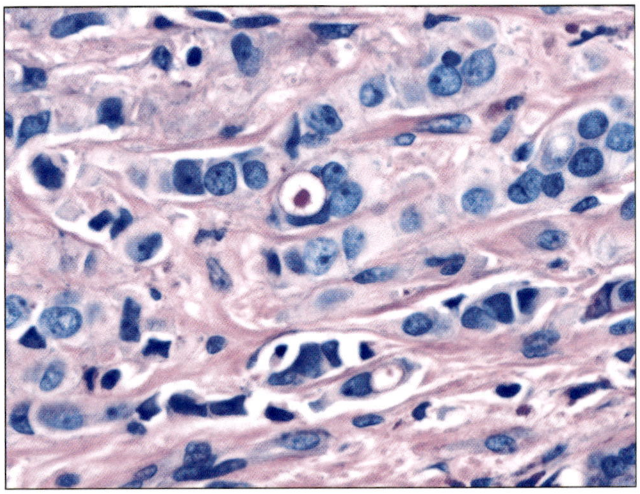

FIGURE 11.50: *The digested PAS technique also labels intracytoplasmic mucin in invasive mammary lobular carcinoma.*

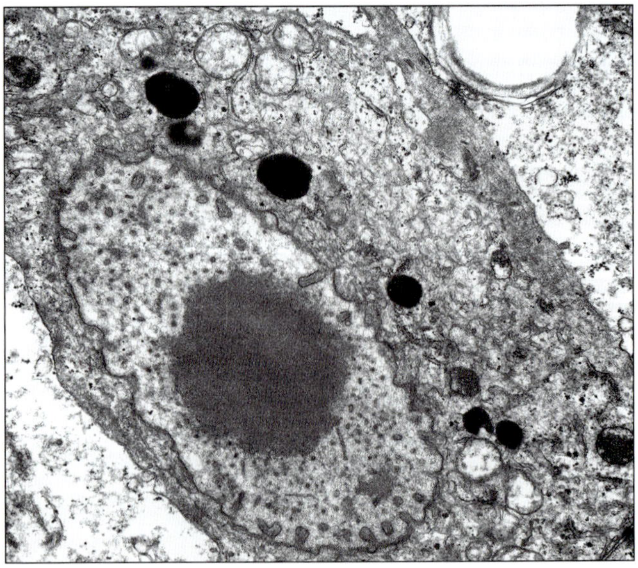

FIGURE 11.51: *Electron micrograph of ILC of the breast, demonstrating an intracytoplasmic lumen lined by microvilli.*

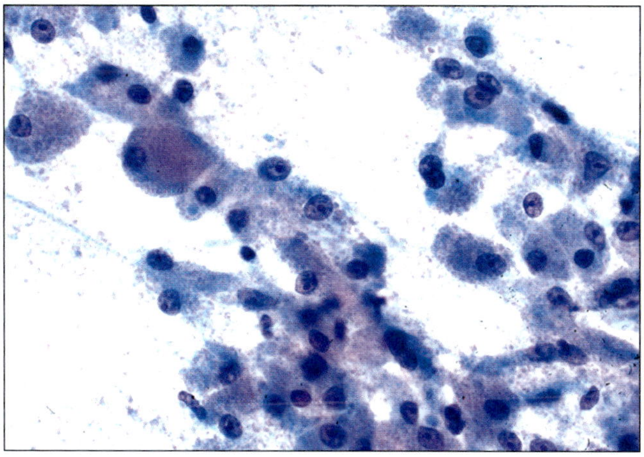

FIGURE 11.52: ***Fine-needle aspiration biopsy of mammary GCT, demonstrating dispersed tumor cells with eccentric nuclei and abundant granular cytoplasm.***

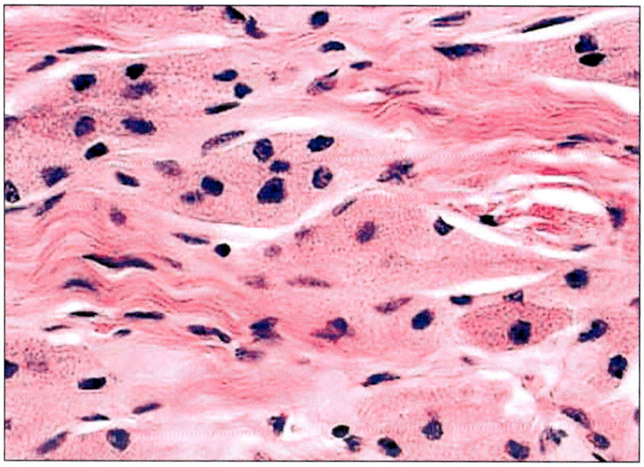

FIGURE 11.53: ***Mammary GCT, showing a uniform population of epithelioid cells with compact eccentric nuclei, and numerous cytoplasmic eosinophilic granules.***

differential diagnosis was that of pure invasive lobular carcinoma (ILC) versus a mixed lobular-ductal tumor. Digested PAS and CI stains showed discrete intracytoplasmic reactivity in the areas of single-file growth (Figures 11.49 and 11.50) as well as others with a nested arrangement. Hence, the final diagnosis was that of pure ILC.

Intracellular lumina are a peculiarity of ILC, among all invasive breast cancers. They are reflected at an ultrastructural level by intracytoplasmic spaces lined by microvillous projections, often containing central secretory droplets (103) (Figure 11.51). Although not absolutely pathognomonic, because some ductal mammary carcinomas also may show them focally (104,105), intracellular lumina help to define an invasive tumor as lobular if they are present uniformly throughout the mass. ■

<u>CASE 4</u>— ***Granular Cell Tumor:*** A 38-year-old woman found a mass in the left breast by self-examination.

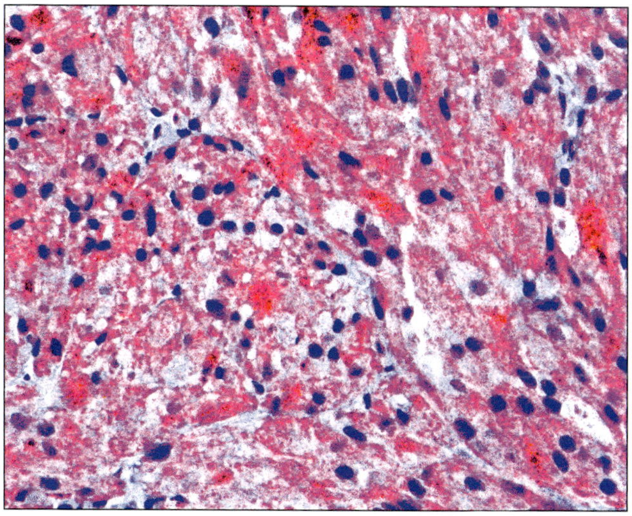

FIGURE 11.54: ***The cytoplasm of the neoplastic cells in mammary GCT manifests coarse granular labeling with the digested PAS method.***

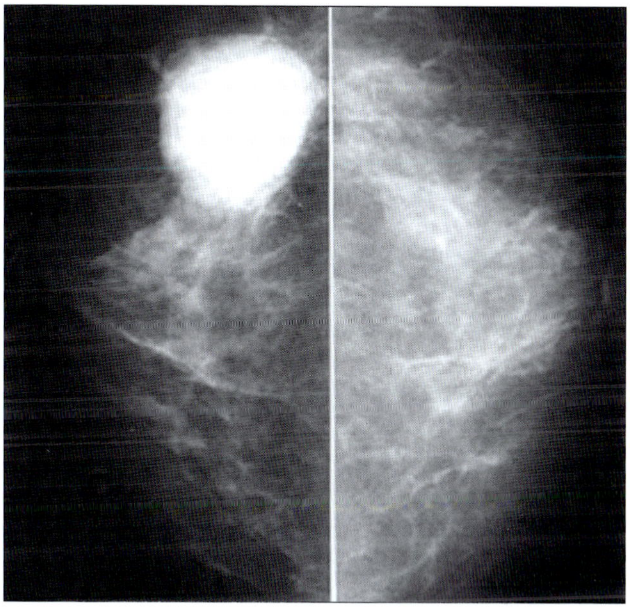

FIGURE 11.55: ***Mammograms showing a mass in the right breast in a 47-year-old woman.***

Mammography demonstrated a stellate lesion in the upper outer quadrant that was felt to be consistent with invasive carcinoma. A fine-needle aspiration biopsy was performed (Figure 11.52), following which a conservative lumpectomy was done. The latter specimen showed a uniform population of large epithelioid cells with eccentric small nuclei and small chromocenters, markedly granular cytoplasm, and a lack of lumen formation or mitotic activity (Figure 11.53). PAS staining demonstrated diastase-resistant positivity in the cytoplasmic granules (25) (Figure 11.54), and lectin histochemical stains were positive with

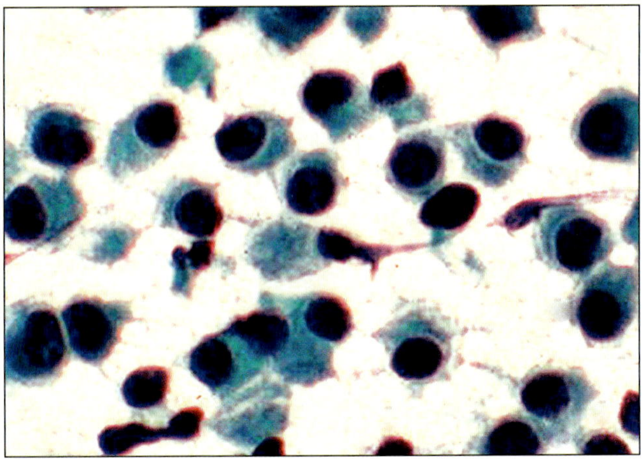

***FIGURE 11.56:* Fine-needle aspiration biopsy of the right breast mass shown in the previous figure, demonstrating a uniform population of dyshesive plasmacytoid cells.**

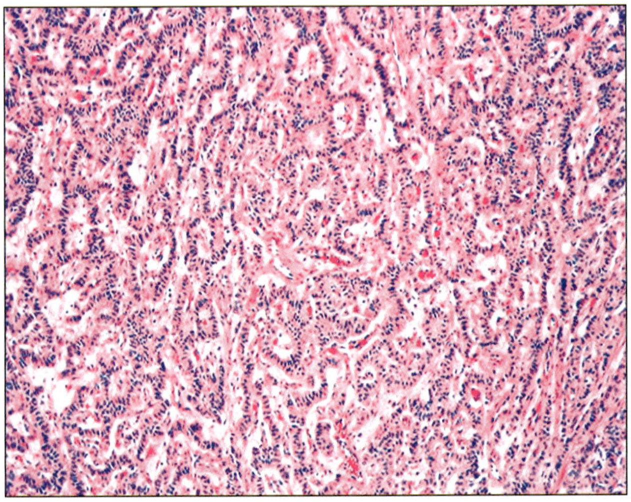

***FIGURE 11.57:* An excisional biopsy of the mass shown in Figure 11.55 revealed an epithelial neoplasm with an organoid growth pattern and dispersed nuclear chromatin, shown here.**

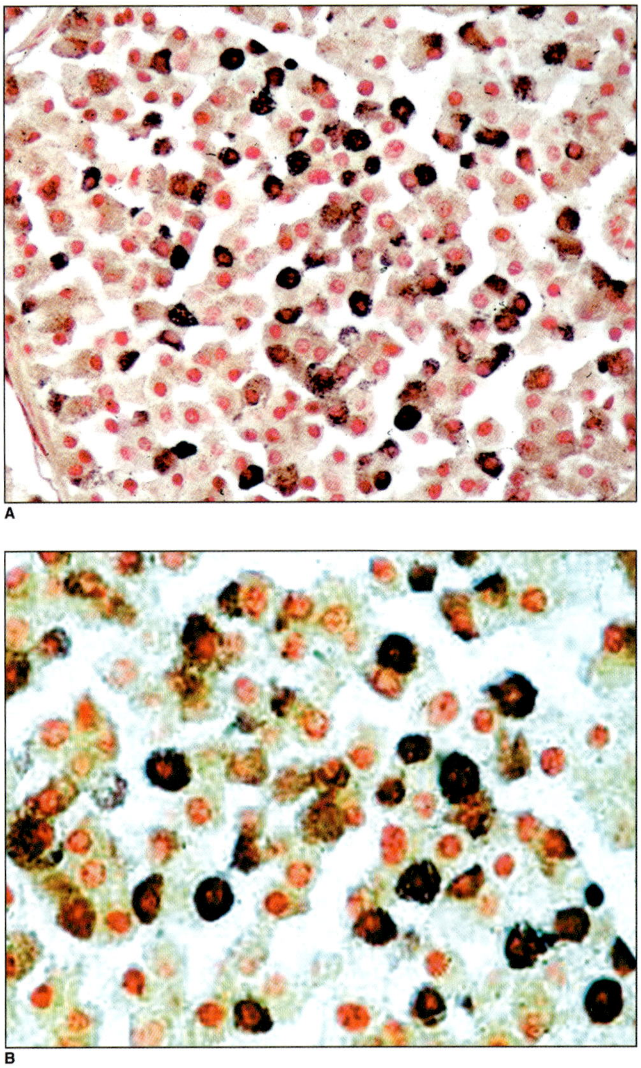

***FIGURE 11.58:* The Grimelius method (A) and the Fontana-Masson technique (B) both yielded positivity in the lesion shown in the previous figure, strongly supporting the diagnosis of metastatic well-differentiated neuroendocrine carcinoma.**

peanut agglutinin (106). UEA, DBA, and BSA stains were negative in the tumor cells. A diagnosis of GCT was made.

GCT is potentially common to several organ sites, including the tongue, breast, skin, esophagus, bronchus, neurohypophysis, and other locations (25). Because it grows in an irregularly permeative fashion, distorting the entrapped connective tissue, a macroscopic likeness to invasive carcinomas is often produced. Indeed, some carcinomas—in the kidneys, breasts, and skin, for example—do manifest secondary granular cell change. A positive digested PAS stain helps to exclude other lesions that may be microscopically similar but that are composed of oncocytes rather than granular cells (27). Negativity for UEA, DBA, and BSA would favor an interpretation of GCT over one of granular cell carcinoma; peanut lectin also labels most examples of the former of those tumor types. Mittal and True also have described reactivity with luxol fast blue in GCTs (107). ■

CASE 5— ***Metastatic Well-Differentiated Neuroendocrine Carcinoma (Carcinoid):*** A 47-year-old woman noted a rapidly enlarging mass in the upper outer portion of her right breast; this was confirmed by mammography (Figure 11.55). A fine-needle aspiration biopsy (Figure 11.56) was followed by a partial mastectomy. Based on the morphologic appearance of the tumor (Figure 11.57), Grimelius and Fontana-Masson stains were obtained, both of which were positive (Figure 11.58). A diagnosis of low-grade neuroendocrine carcinoma was made, and, based on the histochemical findings, questions were raised as to the primary nature of the lesion in the breast.

The patient volunteered the information that she had had an ileal resection 8 years previously at another hospital. Procurement and review of those slides showed a classical carcinoid tumor, with regional lymph node involvement. Computed tomography of the abdomen was done, showing multiple intrahepatic nodules that were compatible with metastases. In light of these data, the right breast mass was also felt to be a secondary tumor deposit.

To reiterate information presented earlier, neuroendocrine tumors of the breast may be primary or metastatic in nature. Reactivity with argyrophilic histochemical methods helps to confirm the general lineage of the tumor but is not especially helpful in localizing its anatomic origin. In contrast, Fontana-Masson positivity is not expected in primary mammary "carcinoids," or in foregut or hindgut tumors arising in endodermally derived organs (108–112). Hence, its presence in this case strongly suggested a metastasis. ■

REFERENCES

1. Fernandez BB, Hernandez FJ: Amyloid tumor of the breast. *Arch Pathol* 1973; 95: 102–105.
2. Lew W, Seymour AE: Primary amyloid tumor of the breast. Case report & literature review. *Acta Cytol* 1985; 29: 7–11.
3. Sahoo S, Reeves W, DeMay RM: Amyloid tumor: a clinical & cytomorphological study. *Diagn Cytopathol* 2003; 28: 325–328.
4. Ross MJ, Merino MJ: Sarcoidosis of the breast. *Hum Pathol* 1985; 16: 185–187.
5. Gansler TS, Wheeler JE: Mammary sarcoidosis: two cases and literature review. *Arch Pathol Lab Med* 1984; 108: 673–675.
6. Bodo M, Dobrossy L, Sugar J: Boeck's sarcoidosis of the breast: cytologic findings with aspiration biopsy cytology. A case clinically mimicking carcinoma. *Acta Cytol* 1978; 22: 1–2.
7. Farmer C, Stanley MW, Bardales RH, et al.: Mycoses of the breast: diagnosis by fine needle aspiration. *Diagn Cytopathol* 1995; 12: 51–55.
8. Opri F: Mammary mycoses. *Chemotherapy* 1982; 28 (Suppl 1): 61–65.
9. Truppman ES, Ellenby JD, Schwartz BM: Fungi in and around implants after augmentation mammoplasty. *Plast Reconstr Surg* 1979; 64: 804–806.
10. Salfelder K, Schwarz J: Mycotic "pseudotumors" of the breast: report of four cases. *Arch Surg* 1975; 110: 751–754.
11. Cohen C: Tuberculous mastitis: a review of 34 cases. *S Afr Med J* 1977; 52: 12–14.
12. Anim JT, van Herk EJ: Giant cell arteritis of the breast. *Med Princ Pract* 2004; 13: 234–236.
13. Kariv R, Sidi Y, Gur H: Systemic vasculitis presenting as a tumor-like lesion: four case reports and an analysis of 79 reported cases. *Medicine* 2000; 79: 349–359.
14. Pappo I, Beglaibter N, Amir G: Mammary arteritis mimicking cancer: case report. *Eur J Surg* 1992; 158: 191–193.
15. Potter BT, Housley E, Thomson D: Giant cell arteritis mimicking carcinoma of the breast. *Br Med J* 1981; 282: 1665–1666.
16. Pambakian H, Tighe JR: Breast involvement in Wegener's granulomatosis. *J Clin Pathol* 1971; 24: 343–347.
17. Veerysami M, Freeth M, Carmichael AR, Carmichael P: Wegener's granulomatosis of the breast. *Breast J* 2006; 12: 268–270.
18. Trueb RM, Pericin M, Kohler E, Barandun J, Burg G: Necrotizing granulomatosis of the breast. *Br J Dermatol* 1997; 137: 799–803.
19. Goulart RA, Mark EJ, Rosen S: Tumefactions as an extravascular manifestation of Wegener's granulomatosis. *Am J Surg Pathol* 1995; 19: 145–153.
20. Clement PB, Young RH, Azzopardi JG: Collagenous spherulosis of the breast. *Am J Surg Pathol* 1987; 11: 411–417.
21. Jan YJ, Li MC, Ho WL: Collagenous spherulosis presenting as a mass of the breast. *Zhonghua Yi Xue Za Zhi (Taipei)* 2002; 65: 494–497.
22. Resetkova E, Albarracin C, Sneige N: Collagenous spherulosis of the breast: morphologic study of 59 cases and review of the literature. *Am J Surg Pathol* 2005; 30: 20–27.
23. Grignon DJ, Jae RY, Mackay BN, et al.: Collagenous spherulosis of the breast: immunohistochemical and ultrastructural studies. *Am J Clin Pathol* 1989; 91: 386–392.
24. Mooney EE, Kayani N, Tavassoli FA: Spherulosis of the breast: a spectrum of mucinous and collagenous lesions. *Arch Pathol Lab Med* 1999; 123: 626–630.
25. Lack EE, Worsham GF, Callihan MD, et al.: Granular cell tumor: a clinicopathologic study of 110 patients. *J Surg Oncol* 1980; 13: 301–316.
26. DeMay RM, Kay S: Granular cell tumor of the breast. *Pathol Annu* 1984; 19(Part II): 121–148.
27. Damiani S, Dina R, Eusebi V: Eosinophilic and granular cell tumors of the breast. *Semin Diagn Pathol* 1999; 16: 117–125.
28. Adeniran A, Al-Ahmadie H, Mahoney MC, Robinson-Smith TM: Granular cell tumor of the breast: a series of 17 cases and review of the literature. *Breast J* 2004; 10: 528–531.
29. Leo C, Briest S, Pilch H, et al.: Granular cell tumor of the breast mimicking breast cancer. *Eur J Obstet Gynecol Reprod Biol* 2006; 127: 268–270.
30. Jayaram G, Hoon PB, Har YC: Granular cell tumor of breast: diagnosis by fine needle aspiration cytology. *Acta Cytol* 2006; 50: 593–595.
31. Rosen PP, Oberman HA: Tumors of the mammary gland. In: Atlas of Tumor Pathology, Series 3, Fascicle 7, Armed Forces Institute of Pathology, Washington, DC, 1993; pp. 162–244.
32. Rosen PP: Microglandular adenosis: a benign lesion simulating invasive mammary carcinoma. *Am J Surg Pathol* 1983; 7: 137–144.
33. James BA, Cranor ML, Rosen PP: Carcinoma of the breast arising in microglandular adenosis. *Am J Clin Pathol* 1993; 100: 507–513.
34. Resetkova E, Flanders DJ, Rosen PP: Ten-year follow-up of mammary carcinoma arising in microglandular adenosis treated with breast conservation. *Arch Pathol Lab Med* 2003; 127: 77–80.
35. James BA, Cranor ML, Rosen PP: Carcinoma of the breast arising in microglandular adenosis. *Am J Clin Pathol* 1993; 100: 507–513.

36. Michal M, Ludvikova M, Zamecnik M: Nodular mucinosis of the breast: report of three cases. *Pathol Int* 1998; 48: 542–544.
37. Rosen PP: Mucocele-like tumors of the breast. *Am J Surg Pathol* 1986; 10: 464–469.
38. Capella C, Eusebi V, Mann B, Azzopardi JG: Endocrine differentiation in mucoid carcinoma of the breast. *Histopathology* 1980; 4: 613–630.
39. Papotti M, Macri L, Finzi G, et al.: Neuroendocrine differentiation in carcinoma of the breast: a study of 51 cases. *Semin Diagn Pathol* 1989; 6: 174–188.
40. Del Vecchio M, Eusebi V: Tumors of the breast showing dual differentiation: a review. *Int J Surg Pathol* 2004; 12: 345–350.
41. Nassar H, Qureshi H, Volkanadsay N, Visscher D: Clinicopathologic analysis of solid papillary carcinoma of the breast and associated invasive carcinomas. *Am J Surg Pathol* 2006; 30: 501–507.
42. Nesland JM, Holm R, Johannessen JV, Gould VE: Neurone-specific enolase immunostaining in the diagnosis of breast carcinomas with neuroendocrine differentiation. Its usefulness and limitations. *J Pathol* 1986; 148: 35–43.
43. Paget J: On disease of the mammary areola preceding cancer of the mammary gland. *St. Bartholomew Hops Res Lond* 1874; 10: 87–89.
44. Sitalkin C, Ackerman AB: Mammary and extramammary Paget's disease. *Am J Dermatopathol* 1985; 7: 335–340.
45. Wood WS, Hegedus C: Mammary Paget's disease and intraductal carcinoma. *Am J Dermatopathol* 1988; 10: 183–188.
46. Hitchcock A, Topham S, Bell J, et al.: Routine diagnosis of mammary Paget's disease: a modern approach. *Am J Surg Pathol* 1992; 16: 58–61.
47. Culberson JD, Horn RC Jr.: Paget's disease of the nipple: review of twenty-five cases with special reference to melanin pigmentation of Paget cells. *Arch Surg* 1956; 72: 224–231.
48. Neubecker RD, Bradshaw RP: Mucin, melanin, and glycogen in Paget's disease of the breast. *Am J Clin Pathol* 1961; 36: 49–53.
49. Ho TC, St. Jacques M, Schopflocher P: Pigmented Paget's disease of the male breast. *J Am Acad Dermatol* 1990; 23: 338–341.
50. Menet E, Vabres P, Brecheteau P, et al.: Pigmented Paget's disease of the male nipple. *Ann Dermatol Venereol* 2001; 128: 649–652.
51. Requena L, Sangueza M, Sangueza OP, Kutzner H: Pigmented mammary Paget disease and pigmented epidermotropic metastases from breast carcinoma. *Am J Dermatopathol* 2002; 24: 189–198.
52. Breslow A, Brancaccio ME: Intracellular mucin production by lobular breast carcinoma cells. *Arch Pathol Lab Med* 1976; 100: 620–621.
53. Clayton F, Sibley RK, Ordonez NG, Hanssen G: Argyrophilic breast carcinomas: evidence of lactational differentiation. *Am J Surg Pathol* 1982; 6: 323–333.
54. Nesland JM, Memoli VA, Holm R, Gould VE, Johannessen JV: Breast carcinomas with neuroendocrine differentiation. *Ultrastruct Pathol* 1985; 8: 225–240.
55. Hull MT, Warfel KA: Mucinous breast carcinomas with abundant intracytoplasmic mucin and neuroendocrine features: light microscopic, immunohistochemical, and ultrastructural study. *Ultrastruct Pathol* 1987; 11: 29–38.
56. McCutcheon J, Walker RA: The significance of argyrophilia in human breast carcinomas. *Virchows Arch A Pathol Anat Histopathol* 1987; 410: 369–374.
57. Bussolati G, Papotti M, Sapino A, et al.: Endocrine markers in argyrophilic carcinomas of the breast. *Am J Surg Pathol* 1987; 11: 248–256.
58. Yao GY, Zhou JL, Zhao ZS, Ruan J: Biological characteristics of breast carcinomas with neuroendocrine cell differentiation. *Chin Med J* 2004; 117: 1536–1540.
59. Van Laarhoven HA, Gratama S, Wereldsma JC: Neuroendocrine carcinoid tumors of the breast: a variant of carcinoma with neuroendocrine differentiation. *J Surg Oncol* 1991; 46: 125–132.
60. Papotti M, Tanda F, Bussolati G, et al.: Argyrophilic neuroendocrine carcinoma of the male breast. *Ultrastruct Pathol* 1993; 17: 115–121.
61. Scopsi L, Andreola S, Pilotti S, et al.: Mucinous carcinoma of the breast: a clinicopathologic, histochemical, and immunocytochemical study with special reference to neuroendocrine differentiation. *Am J Surg Pathol* 1994; 18: 702–711.
62. Sapino A, Papotti M, Righi L, et al.: Clinical significance of neuroendocrine carcinoma of the breast. *Ann Oncol* 2001; 12 (Suppl 2): S115–S117.
63. Berruti A, Saini A, Leonardo E, et al.: Management of neuroendocrine differentiated breast carcinoma. *Breast* 2004; 13: 527–529.
64. Valdes EK, Feldman SM, Krassilnik N: Neuroendocrine tumor of the breast. *Am Surg* 2006; 72: 185–187.
65. Mirza IA, Shahab N: Small cell carcinoma of the breast. *Semin Oncol* 2007; 34: 64–66.
66. Taxy JB, Tischler AS, Insalaco SJ, Battifora H: "Carcinoid" tumor of the breast: a variant of conventional breast cancer? *Hum Pathol* 1981; 12: 170–179.
67. Colandrea JM, Shmookler BM, O'Dowd GJ, Cohen MH: Cystic hypersecretory duct carcinoma of the breast: report of a case with fine-needle aspiration. *Arch Pathol Lab Med* 1988; 112: 560–563.
68. Nguyen GK, Neifer R: Aspiration biopsy cytology of secretory carcinoma of the breast. *Diagn Cytopathol* 1987; 3: 234–237.
69. Cavanzo FJ, Taylor HB: Adenoid cystic carcinoma of the breast: an analysis of 21 cases. *Cancer* 1969; 24: 740–745.
70. Friedman BA, Oberman HA: Adenoid cystic carcinoma of the breast. *Am J Clin Pathol* 1970; 54: 1–14.
71. Ro JY, Silva EG, Gallagher HS: Adenoid cystic carcinoma of the breast. *Hum Pathol* 1987; 18: 1276–1281.
72. Rosen PP: Adenoid cystic carcinoma of the breast: a morphologically-heterogeneous neoplasm. *Pathol Annu* 1989; 24(Part II): 237–254.
73. Page DL: Adenoid cystic carcinoma of breast, a special histopathologic type with excellent prognosis. *Breast Cancer Res Treat* 2005; 93: 189–190.
74. Hiraoka N, Mukai M, Hosoda Y, Hata J: Phyllodes tumor of the breast containing intracytoplasmic inclusion bodies

identical with infantile digital fibromatosis. *Am J Surg Pathol* 1994; 18: 506–511.
75. Ortega E, Aranda FI, Chulia MT, et al.: Phyllodes tumor of the breast with actin inclusions in stromal cells: diagnosis by fine-needle aspiration cytology. *Diagn Cytopathol* 2001; 25: 115–117.
76. Lin Y, Govindan R, Hess JL: Malignant hematopoietic breast tumors. *Am J Clin Pathol* 1997; 107: 177–186.
77. Lawler MR Jr., Riddell DH: Hodgkin's disease of the breast. *Arch Surg* 1966; 93: 331–334.
78. Oberman HA: Primary lymphoreticular neoplasms of the breast. Surg *Gynecol Obstet* 1966; 123: 1047–1051.
79. Beaupre EM: Extraosseous manifestations of plasma cell myeloma. *J Maine Med Assoc* 1975; 66: 38–45.
80. Proctor NS, Rippey JJ, Shulman G, Cohen C: Extramedullary plasmacytoma of the breast. *J Pathol* 1975; 116: 97–100.
81. Sears HF, Reid J: Granulocytic sarcoma: local presentation of a systemic disease. *Cancer* 1976; 37: 1808–1813.
82. Chua ET: A case of granulocytic sarcoma of the breast and review of the literature. *Singapore Med J* 1989; 30: 311–312.
83. McFarlane ME: Metastasis to the breast: a rare site for secondary malignancy. *Int J Surg* 2006; 4: 204–205.
84. Domanski HA: Metastases to the breast from extramammary neoplasms: a report of six cases with diagnosis by fine needle aspiration cytology. *Acta Cytol* 1996; 40: 1293–1300.
85. Cangiarella J, Symmans WF, Cohen JM, et al.: Malignant melanoma metastatic to the breast: a report of seven cases diagnosed by fine-needle aspiration cytology. *Cancer* 1998; 84: 160–162.
86. Persic M, Roberts JT: Alveolar rhabdomyosarcoma metastatic to the breast: long-term survivor. *Clin Oncol* 1999; 11: 417–418.
87. Dehner LP, Hill DA, Deschryver K: Pathology of the breast in children, adolescents, and young adults. *Semin Diagn Pathol* 1999; 16: 235–247.
88. Vergier B, Trojani M, de Mascarel I, Coindre JM, Le Treut A: Metastases to the breast: differential diagnosis from primary breast carcinoma. *J Surg Oncol* 1991; 48: 112–116.
89. Tamaki K, Hino H, Ohara K, Furue M: Lectin binding sites in Paget's disease. *Br J Dermatol* 1985; 113: 17–24.
90. Venkataseshan VS, Budd DC, Un Kim D, Hutter RV: Intraepidermal squamous carcinoma (Bowen's disease) of the nipple. *Hum Pathol* 1994; 25: 1371–1374.
91. Miettinen M, Holthofer H, Lehto VP, Miettinen A, Virtanen I: *Ulex europaeus* I lectin as a marker for tumors derived from endothelial cells. *Am J Clin Pathol* 1983; 79: 32–36.
92. Walker RA. *Ulex europaeus* I-peroxidase as a marker of vascular endothelium: its application in routine histopathology. *J Pathol* 1985; 146: 123–127.
93. Leader M, Collins M, Patel J, Henry K. Staining for factor VIII-related antigen and *Ulex europaeus* agglutinin I (UEA-I) in 230 tumors: an assessment of their specificity for angiosarcoma and Kaposi's sarcoma. *Histopathology* 1986; 10: 1153–1162.
94. Ree HJ, Raine L, Crowley JP: Lectin binding patterns in diffuse large cell lymphoma. *Cancer* 1983; 52: 2089–2099.
95. Kohchiyama A, Oka D, Ueki H: Differing lectin-binding patterns of malignant melanoma and nevocellular and Spitz nevi. *Arch Dermatol Res* 1987; 279: 226–231.
96. Lee AK, DeLellis RA, Wolfe HJ: Intramammary lymphatic invasion in breast carcinomas: evaluation using ABH isoantigens as endothelial markers. *Am J Surg Pathol* 1986; 10: 589–594.
97. Baranyay F, Bethlen K: Diffuse splenic metastasis of occult breast cancer with incompatible blood group antigenic determinants. *Orvosi Hetilap* 2002; 143: 139–143.
98. Nicolau SG: Cancerization of the skin of the breast in Paget's disease: a further interpretation of its mechanism. *Ann Dermatol Syphil* 1966; 93: 147–158.
99. Kanitakis J: Mammary and extramammary Paget's disease. *J Eur Acad Dermatol Venereol* 2007; 21: 581–590.
100. Jones RE Jr.: Mammary Paget's disease without underlying carcinoma. *Am J Dermatopathol* 1985; 7: 361–365.
101. Yim HJ, Wick MR, Philpott GW, Norton JA, Doherty GM: Underlying pathology in mammary Paget's disease. *Ann Surg Oncol* 1997; 4: 287–292.
102. Nagle RB, Lucas DO, McDaniel KM, Clark VA, Schmalzel GM: Paget's cells: new evidence linking mammary and extramammary Paget cells to a common cell phenotype. *Am J Clin Pathol* 1985; 83: 431–438.
103. Battifora H: Intracytoplasmic lumina in breast carcinoma: a helpful histopathological feature. *Arch Pathol* 1975; 99: 614–617.
104. Quincey C, Raitt N, Bell J, Ellis IO: Intracytoplasmic lumina—a useful diagnostic feature of adenocarcinomas. *Histopathology* 1991; 19: 83–87.
105. Kamiya M, Mizuguchi K, Yoshimoto M, et al.: Cytologic diagnosis of signet-ring cell carcinoma of the breast. *Acta Cytol* 1998; 42: 650–656.
106. Nishioka H, Ii K, Liena JF, Hirano A: Immunohistochemical study of granular cell tumors of the neurohypophysis. *Virchows Arch B Cell Pathol* 1991; 60: 413–417.
107. Mittal KR, True LD: Origin of granules in granular cell tumor. Intracellular myelin formation with autodigestion. *Arch Pathol Lab Med* 1988; 112: 302–303.
108. Schurch W, Lamoureux E, Lefebvre R, Fauteux JP: Solitary breast metastasis: first manifestation of an occult carcinoid of the ileum. *Virchows Arch A Pathol Anat Histol* 1980; 386: 117–124.
109. van Bogaert LJ: The diffuse neuroendocrine system and derived tumors: histological and histochemical characteristics. *Acta Histochem* 1982; 70: 122–129.
110. Azzopardi JG, Muretto P, Goddeeris P, Eusebi V, Lauweryns JM: "Carcinoid" tumors of the breast: the morphological spectrum of argyrophil carcinomas. *Histopathology* 1982; 6: 549–569.
111. Ordonez NG, Manning JT Jr., Raymond AK: Argentaffin endocrine carcinoma (carcinoid) of the pancreas with concomitant breast metastasis: an immunohistochemical and electron microscopic study. *Hum Pathol* 1985; 16: 746–751.
112. Lozowski MS, Faegenburg D, Mishriki Y, Lundy J: Carcinoid tumor metastatic to breast diagnosed by fine needle aspiration: case report and literature review. *Acta Cytol* 1989; 33: 191–194.

CHAPTER 12

Applied Histochemistry in Disorders of the Endocrine System

Mauro Papotti, M.D.,
Marco Volante, M.D., and
Gianni Bussolati, M.D., F.R.C. Path

The pathologic diagnosis of endocrine disorders has greatly benefited from the introduction of immunohistochemical techniques that enable the pathologist to define the phenotypic profile of (neuro)endocrine cells and tumors with specific regard to the expression of hormones and other endocrine markers, using specific antibodies. Before the immunohistochemical era, the characterization of (neuro)endocrine cells and tumors was mainly based on the demonstration of chemical properties of cells and tumors using histochemical techniques. At present, some of those procedures have been replaced by more specific immunohistochemical methods, but others (e.g., Congo red staining for amyloid) still have a very relevant role in the diagnostic pathology of endocrine diseases. This short review describes the histochemical techniques that are useful in the evaluation of neoplastic and nonneoplastic disorders of the endocrine glands (Table 12.1). We have specially focused on practical applications in surgical pathology, rather than the biochemical bases of individual chemical reactions, for which the reader is referred to other technology-oriented references.

THE DIFFUSE NEUROENDOCRINE SYSTEM

Historical Background

The diffuse neuroendocrine system (DNS) encompasses several cells and tissues—including the pituitary gland, C-cells of the thyroid, parathyroid glands, the endocrine pancreas, gastrointestinal neuroendocrine cells, adrenal medullary tissue, and other scattered neuroendocrine cells of the skin and bronchi—which have common and specific features from embryologic, morphologic, and functional viewpoints. The DNS was originally identified as a system composed of clear cells (Hellezellen) by Feyrter (1) and Champaneria et al. (2), and it was subsequently reinterpreted by Pearse (3) as predicated on common histochemical properties summarized in the acronym "APUD." This abbreviation stands for "amine precursor uptake and decarboxylation." Indeed, the cells of the APUD system share the ability to process amine precursors using decarboxylases. Other properties shared by this cell group (including cells in the antral and fundic mucosa, C-cells of the thyroid, etc.) include the possession of cholinesterase and alpha-glycerophosphate dehydrogenase enzymatic activities (4,5). These cells also contain neurosecretory granules that can be demonstrated by electron microscopy, as well as specific antibody-mediated immunostains for endocrine markers, but they were initially detected by silver impregnation methods. Such procedures were developed by the pioneering studies of Ramon-y-Cajal in Spain and Golgi in Italy, which led to the discovery of the "black reaction." At first, these were used to investigate the nervous system. The application of similar silver impregnation methods to assessments of other tissues subsequently led to the identification of gastroenteropancreatic neuroendocrine cells. These elements were found to have different properties; some were argentaffinic, that is, with an ability to bind and reduce silver salts, whereas others were argyrophilic—that is, able to bind silver salts but not reduce them. Argentaffinity was linked to the production of serotonin, at least in the intestine (6). These same cells were also shown to be fluorescent after formaldehyde fixation, as a result of their biogenic amine content. They were termed "enterochromaffin cells," as typically seen in the gastrointestinal mucosa (7,8). Conversely, the property of argyrophilia was typical of neuroendocrine cells that produced exportable molecules other than serotonin. They are fluorescent after formaldehyde fixation only if predefined precursors (dihydroxyphenylalanine [DOPA], 5-hydroxytryptophan) are present (Figure 12.1). The latter designation was chosen to distinguish the cellular elements in question from others that contain serotonin. Enterochromaffin-like (ECL) cells classically lack serotonin and their argyrophilia has been linked to the presence of chromogranins, a group of acidic proteins in neurosecretory granules (9). Several

TABLE 12.1: Common Histochemical Stains Used in Studies of the Endocrine System

Stain	Target or Purpose
Chromaffin reaction (bichromate salts)	Paraganglionic elements
Silver impregnation methods	
Masson-Fontana	Argentaffinic NE cells
Grimelius	Argyrophilic NE cells
Sevier-Munger	Argyrophilic NE cells
CS	Argyrophilic NE cells
Thiosulfated mucin/aldehyde-fuchsin reaction	Pancreatic endocrine cells
PAS-orange-G	Pituitary cell types
Lipid stain (Sudan black)	Parathyroid gland fat cells
Iron (Perl's, Prussian blue)	Iron deposition diseases
Congo red, Lieb's, Pagoda red, sulfated alcian blue	Amyloid deposition
Reticulin (Snook's)	Glandular stromal infrastructure
Verhoeff–Van Gieson, Masson trichrome	General stromal architecture
Tissue Gram, silver methenamine, Ziehl-Neelsen	Infections
Endogenous biotin staining	Mitochondrion-rich (oncocytic) cells
Lectin histochemistry (UEA-I)	Endothelial cells in blood vessels
Sialic acid–binding lectins	Follicular thyroid differentiation
Receptor binding assay (ghrelin receptors)	Pituitary receptors

NE = neuroendocrine.

histochemical methods were developed to label these two cell populations, beginning in the early 1960s.

Diagnostic Applications

Among the available histochemical reactions for defining argentaffin cells, the Fontana-Masson stain has been the most commonly used. It is highly specific but not very sensitive. Other techniques include the Gomori-Burtner and Churukian-Schenk (CS) stains, with the latter procedure being claimed to sensitively label hindgut carcinoids (10). Several stains are capable of detecting argyrophilic cells, and they all require the addition of a reducing agent to reduce impregnated silver salts. These methods include the Bodian, Sevier-Munger, modified Davemport, and Grimelius techniques (5,8) (Figures 12.1 and 12.2). The last of these procedures has the advantage of being easily applicable to formalin-fixed and paraffin-embedded tissues, and it has been widely applied in diagnostic pathology laboratories (10,11). Both histologic and cytologic substrates are potentially labeled by the Grimelius procedure (12,13). It is based on the incubation of tissue sections in a silver nitrate solution for several hours (14) at a controlled temperature, followed by a 1-hour incubation in reducing solution and a final brief re-incubation in freshly prepared silver nitrate solution. Argyrophilic methods are generally more sensitive than argentaffin procedures, but they are less specific. This is because silver salts in argyrophilic techniques may also bind to exocrine cell granules (15). The type of fixative also affects the reactivity of argyrophilic methods. Bouin's fixative is optimal for silver impregnation, especially for the Sevier-Munger technique. This is manifested by stains with that procedure of thyroid C-cells (16), cells in the endocrine pancreas (17) (Figure 12.3), and those in the parathyroid glands (18).

The labeling of neuroendocrine tumors by silver impregnation methods is generally readily accomplished in well-differentiated lesions (e.g., classical "carcinoids," pituitary and parathyroid adenomas) (Figure 12.2). In contrast, reactivity is more limited in poorly differentiated neuroendocrine carcinomas (of both small and large cell types) because of the limited number of secretory granules that the cells in such tumors possess (19). Regardless of this point, all silver-based stains produce finely granular cytoplasmic deposits that have a brown or a black color. The intensity of labeling, however, cannot be used as a criterion for grading neuroendocrine tumors. In addition, specific hormonal products of these lesions, if any, cannot be determined by histochemistry alone.

Occasionally, mixed mucinous and neuroendocrine carcinomas are encountered (formerly termed "goblet cell carcinoid" or "mucinous carcinoid"), especially in the vermiform appendix (20). These tumors can be stained with silver methods, as well as the periodic acid–Schiff (PAS), alcian blue, colloidal iron, and mucicarmine reactions (Figure 12.4). The first of these techniques recognizes neuroendocrine elements, whereas the second preferentially labels exocrine and mucin-producing cells.

Specific Endocrine Organs and Their Lesions

Pituitary Gland: Histochemical stains can be used to outline the various endocrine cell populations in the pituitary gland. The combined PAS-orange-G stain separates basophilic cells (producing the glycoprotein hormones, follicle-stimulating hormone, luteinizing hormone, and thyroid-stimulating hormone) from acidophilic cells

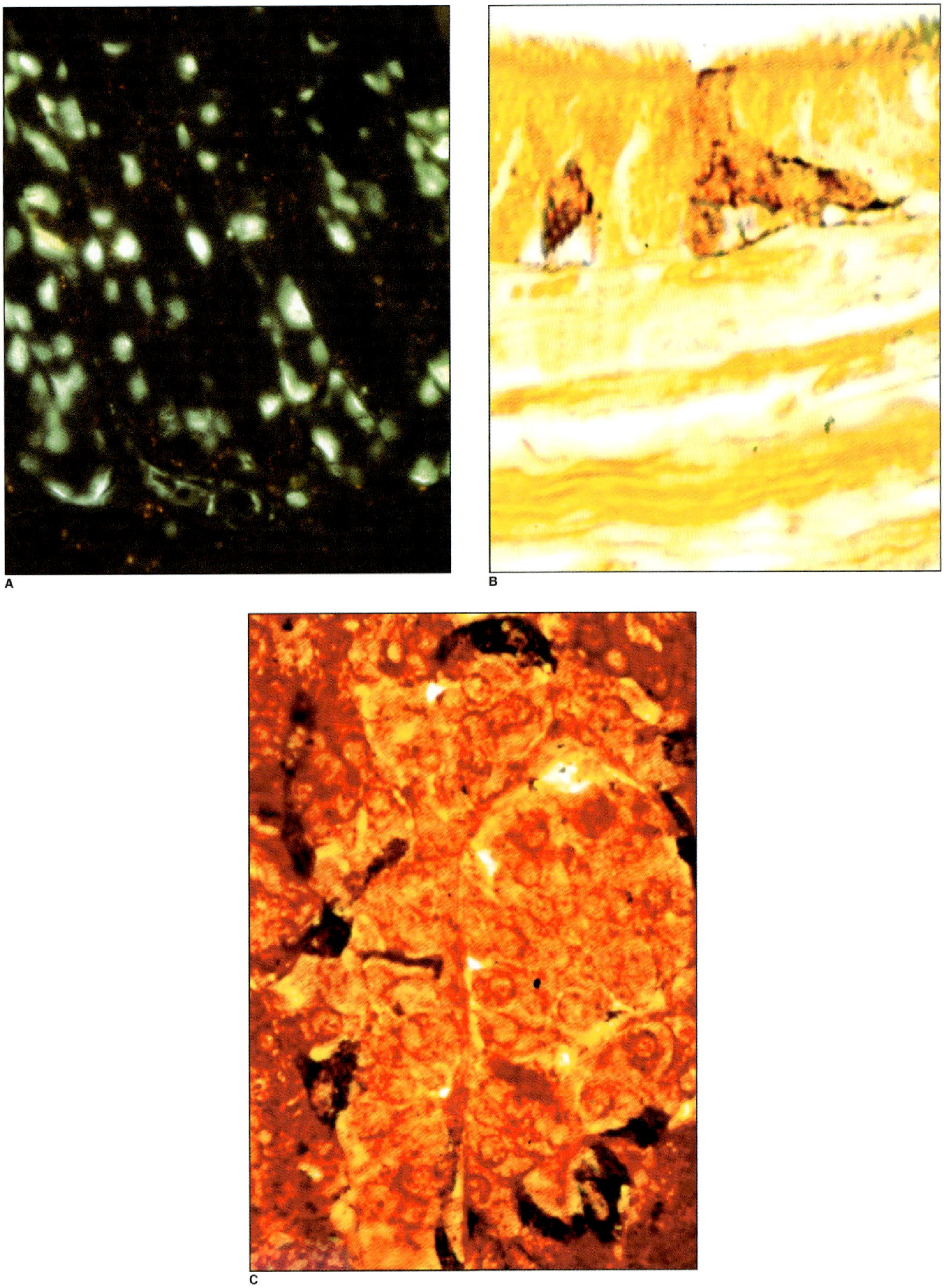

FIGURE 12.1: Histochemical demonstration of neuroendocrine cells in the gastric mucosa, bronchial tree, and pancreas. Neuroendocrine ECL cells of the gastric mucosa (A) are fluorescent after formaldehyde fixation and application of appropriate biochemical precursors (DOPA, 5-hydroxytryptophan). Neuroendocrine cells of the bronchial epithelium (B) and α-cells of a pancreatic islet (which are generally more strongly reactive with silver impregnation methods than are β-cells) (C) are visualized by techniques such as the Grimelius and Davemport procedures (B and C, respectively).

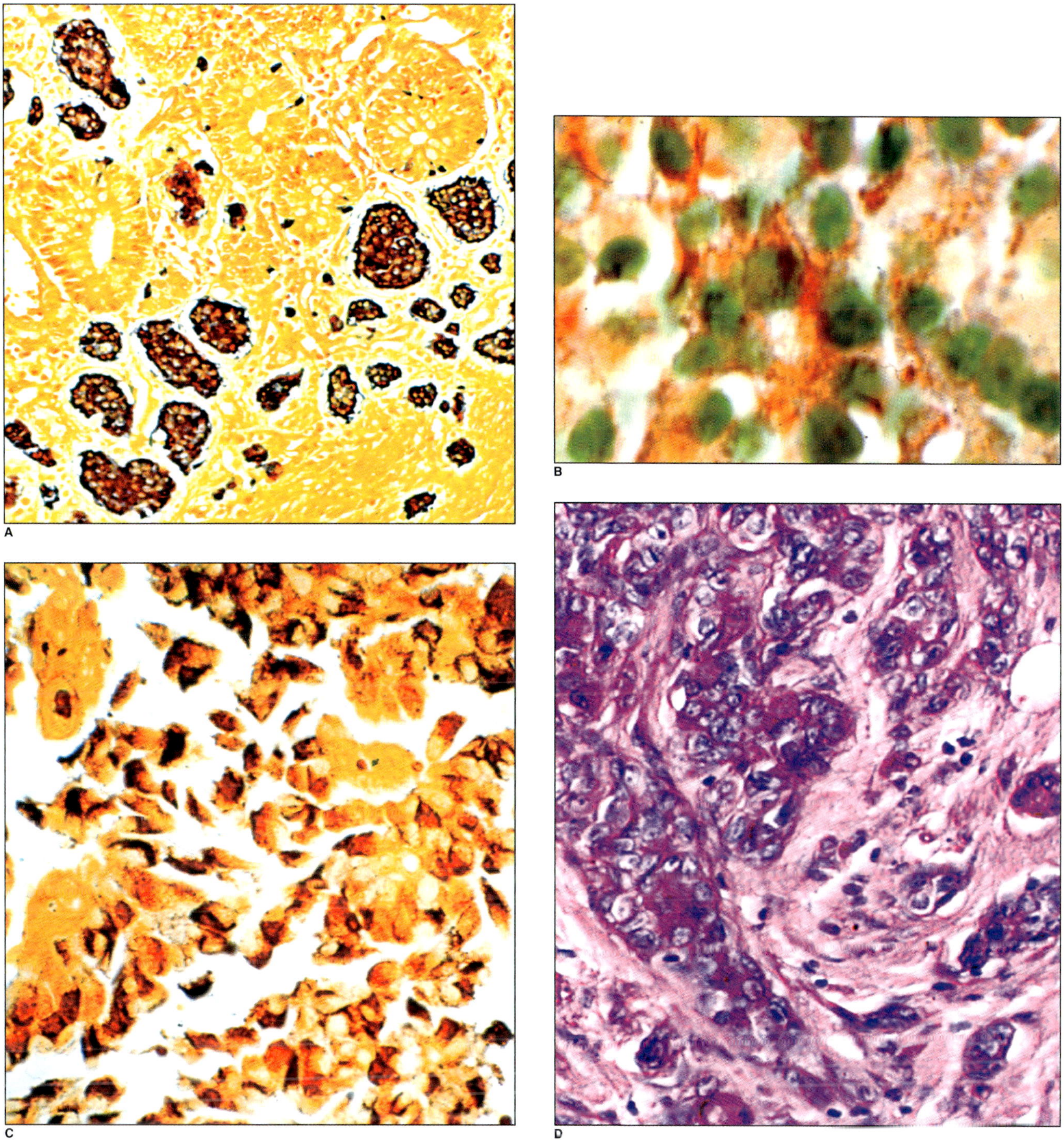

***FIGURE 12.2:** Demonstration of a neuroendocrine phenotype in gastroenteropancreatic tumors. Grimelius staining of a well-differentiated neuroendocrine carcinoma (carcinoid) of the ileum (A) and a hepatic metastasis of pancreatic insulinoma in a fine-needle aspiration biopsy (B). Neuroendocrine argentaffin cells in a well-differentiated ileal neuroendocrine carcinoma are revealed by the Masson-Fontana stain (C). PAS staining demonstrates mucinous material in a "carcinoid" of the appendix with divergent glandular differentiation (D).*

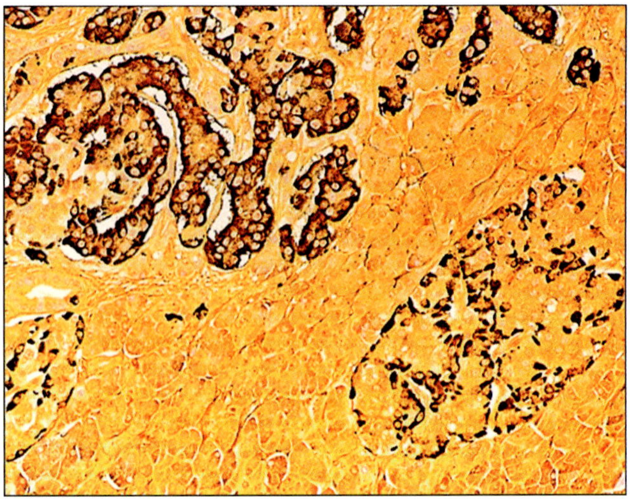

FIGURE 12.3: ***The Sevier-Munger argyrophil stain decorates the tumor cells of a pancreatic endocrine neoplasm (top left), as well as the cells in an adjacent normal islet of Langerhans (bottom right).***

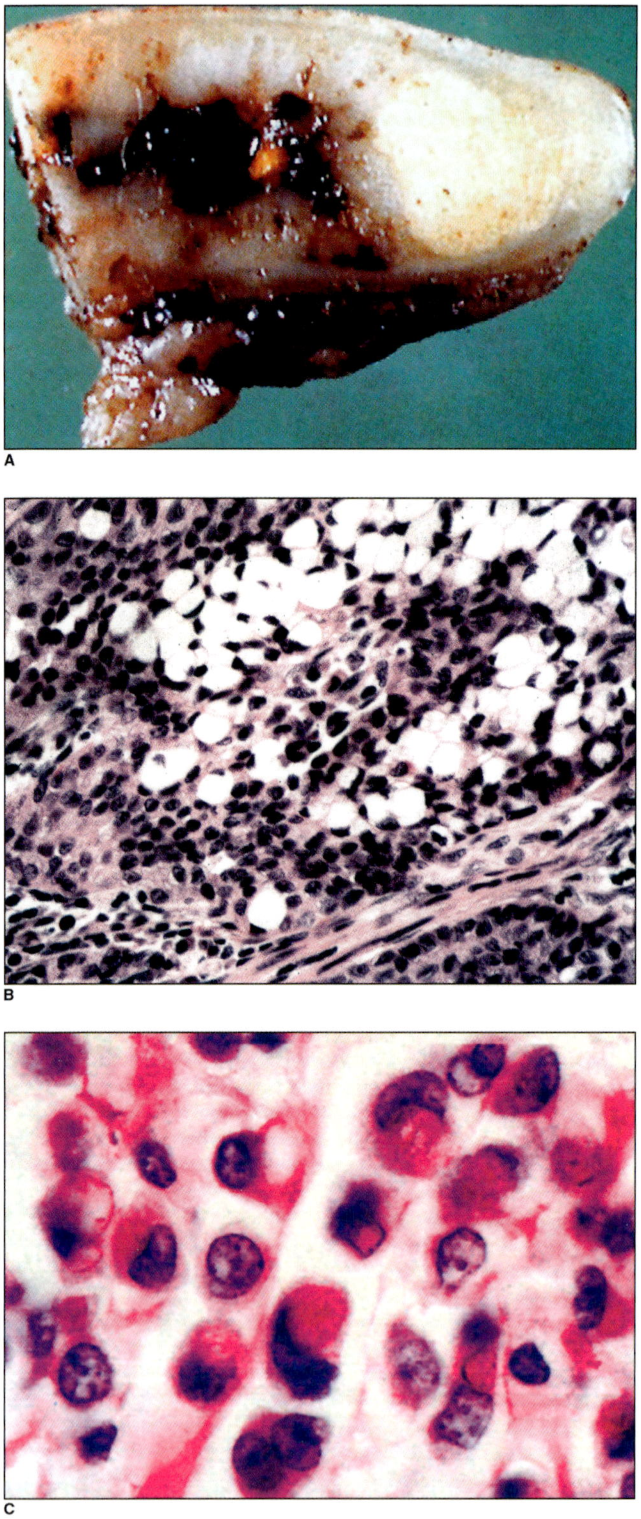

FIGURE 12.4: ***This neuroendocrine tumor of the appendix (A) demonstrates divergent mucinous differentiation (B). A diastase-digested PAS stain shows intense reactivity in the mucin-producing cellular component (C).***

[elaborating prolactin and growth hormone (GH)], as well as chromophobe cells (synthesizing adrenocorticotrophic hormone) (21). Fluorescent reagents have also been employed to identify specific cell types. For example, fluorescamine has been shown to induce fluorescence in GH-producing pituitary cells, as well endocrine cells in other organ systems (e.g., gastrin-producing cells in the stomach, insulin-producing cells in the endocrine pancreas, C-cells in the thyroid) (22).

Currently, one may identify such cell types using specific antibodies and immunohistology. However, histochemistry has maintained a role in the recognition of pituitary hyperplasia, using the reticulin stain. Because of various causes, the cells in the adenohypophysis may become hyperplastic with diffuse or nodular patterns of growth. These changes are well outlined by Snook's method, which highlights the delicate stromal tissue between cell groups. In contrast, neoplastic conditions (usually adenomas) in the pituitary are accompanied by a reticulin framework that is much more irregular (23). Amyloid and iron deposition may also be seen in the adenohypophysis, and other specific histochemical stains can reveal the presence of those substances (see below).

The pituitary also contains several neuropeptide receptors, which mediate the regulatory actions of several hormones and releasing factors, and antireceptor immunostains allow for their detection. However, hormone-binding sites can be also determined by functional binding assays, based on radioactive ligands and autoradiographic technique. We have developed a nonradioactive, nonimmunologic binding assay to detect receptors for ghrelin, a pituitary and gastroenteropancreatic hormone that affects GH release (24). After labeling ghrelin with biotin, the complex can be applied to frozen tissue sections or cell cultures,

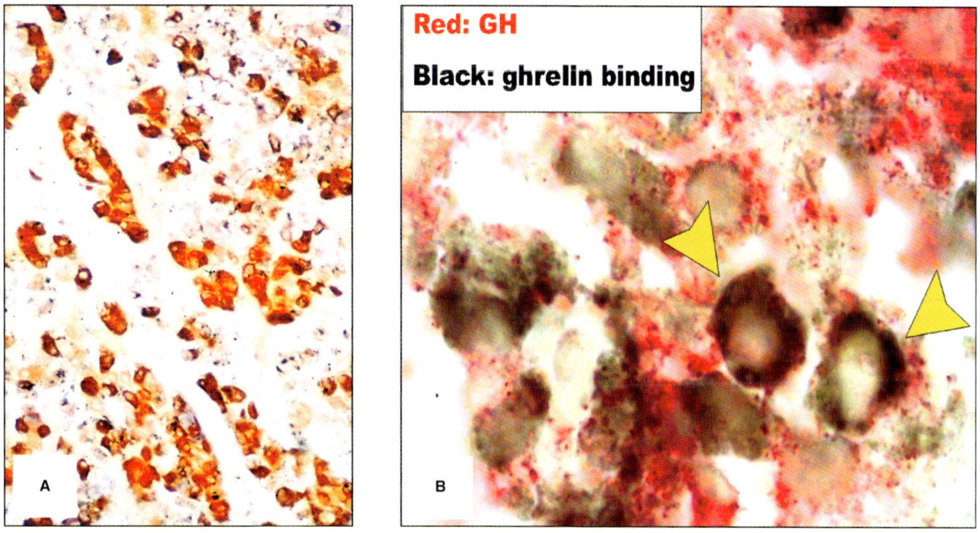

FIGURE 12.5: Grimelius staining showing neuroendocrine differentiation in anterior pituitary cells (A). Double immunohistochemical and ghrelin binding procedure (B) to demonstrate coexpression by the same cells (arrowheads) of GH (red) and ghrelin binding sites (brown).

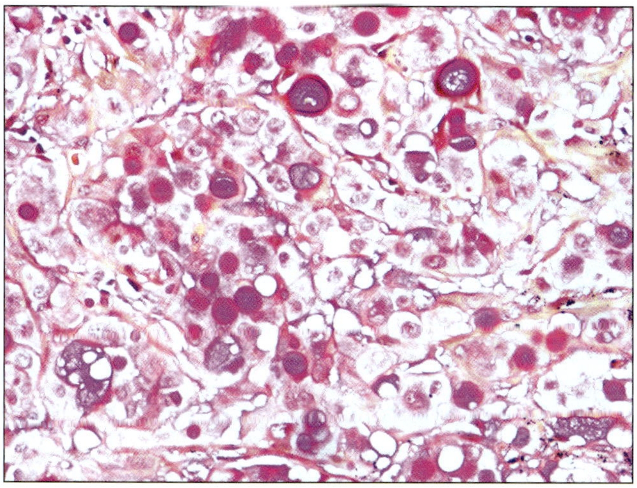

FIGURE 12.6: Metastatic adenocarcinoma of the pituitary region, showing intense positivity with the mucicarmine method.

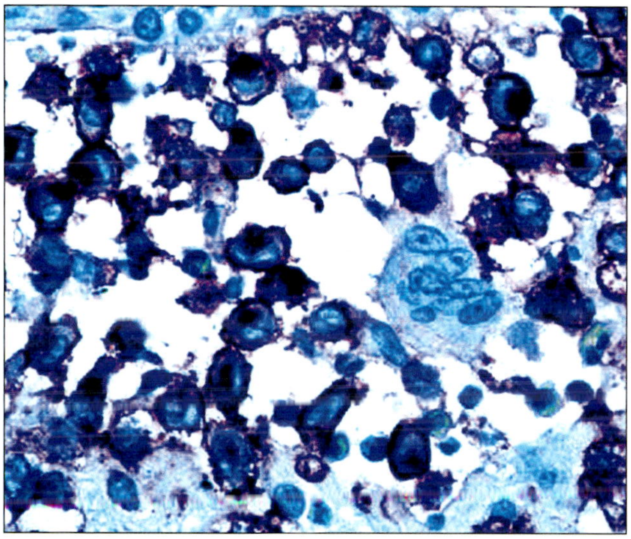

FIGURE 12.7: Langerhans cell histiocytosis of the pituitary region, showing diffuse binding of peanut agglutinin to the tumor cells.

and binding is detected with an avidin-tyramide-peroxidase method. The same procedure can also be combined with immunohistochemistry to define colocalization of the receptor with specific native hormones (Figure 12.5).

The pituitary and the parahypophyseal region may be involved by other pathologic processes as well. In particular, metastatic breast, renal cell, or lung carcinoma may be seen in that location (25–30), as well as infiltrates of Langerhans' cell histiocytosis (31,32) and extragonadal germ cell tumors (33). Secondary adenocarcinomas can be recognized by their cytologic atypia, together with histochemical reactivity for intracytoplasmic mucin, as shown with the mucicarmine, colloidal iron, and alcian blue techniques (Figure 12.6). The unique histiocytes of Langerhans' cell disease (histiocytosis-X) exhibit preferential binding of peanut agglutinin (34) (Figure 12.7), and PAS positivity is typical of extragonadal germ cell tumors in this location (35) (Figure 12.8).

Thyroid Gland: Thyroid colloid can be highlighted by staining with the PAS method, but this is generally not required in routine histopathologic specimens. In cytologic smears from fine-needle aspiration biopsies, colloid is readily visible in slides stained with Romanowsky dyes. It assumes an intense blue color, or sometimes is light pink. Iron accumulation in the thyroid is relatively common as a result of systemic disorders of iron metabolism or regressive changes in goitrous nodules, in which repeated hemorrhagic events lead to hemosiderosis in fibrotic or cystic areas. The classical dark-brown color of iron pigment is easily appreciated in standard hematoxylin and eosin (H&E)–stained preparations, but it can be specifically labeled with the Perl's or Prussian blue method (Figure 12.9). These techniques can be useful to eliminate consideration of other pigments that can be found in the normal thyroid or in thyroid tumor cells [e.g., melanin in variants of medullary thyroid carcinoma (MTC)] (36).

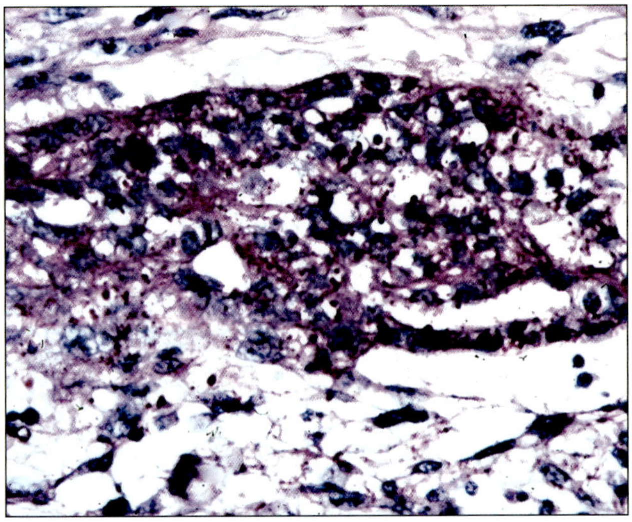

***FIGURE 12.8:** Yolk sac tumor of the pineal region, showing diastase-labile positivity with the PAS procedure.*

Amyloid deposits can be found in the thyroid as a consequence of four principal pathologic conditions. Rarely, primary (AL) or secondary (AA) amyloidosis can affect this gland, in association with chronic inflammatory disorders or plasma cell dyscrasias. Two primary conditions of the thyroid are also associated with intraglandular amyloid deposits—amyloid goiter and MTC (Figure 12.10). The former of these disorders is extremely rare; it is characterized by follicular nodules that are surrounded by irregular amyloid deposits with a classical amorphous appearance, often seen together with adipocytic infiltrates. MTC is a malignant C-cell neoplasm that produces calcitonin; in the vast majority of cases, it contains easily detected amyloid deposits in intratumoral stroma. In both amyloid goiter and MTC, amyloid can be visualized by specific stains, including the Lugol solution; metachromatic methods such as crystal violet; "cotton dyes" such as Pagoda red, and thioflavine-T (a fluorescent dye). The most sensitive and widely employed staining procedure for amyloid, however, is the Congo red method (Figure 12.11). It is associated with the demonstration of typical green birefringence with polarization microscopy (Figure 12.12). After an original description in 1922 by Bennhold, the Congo red stain has undergone several modifications.

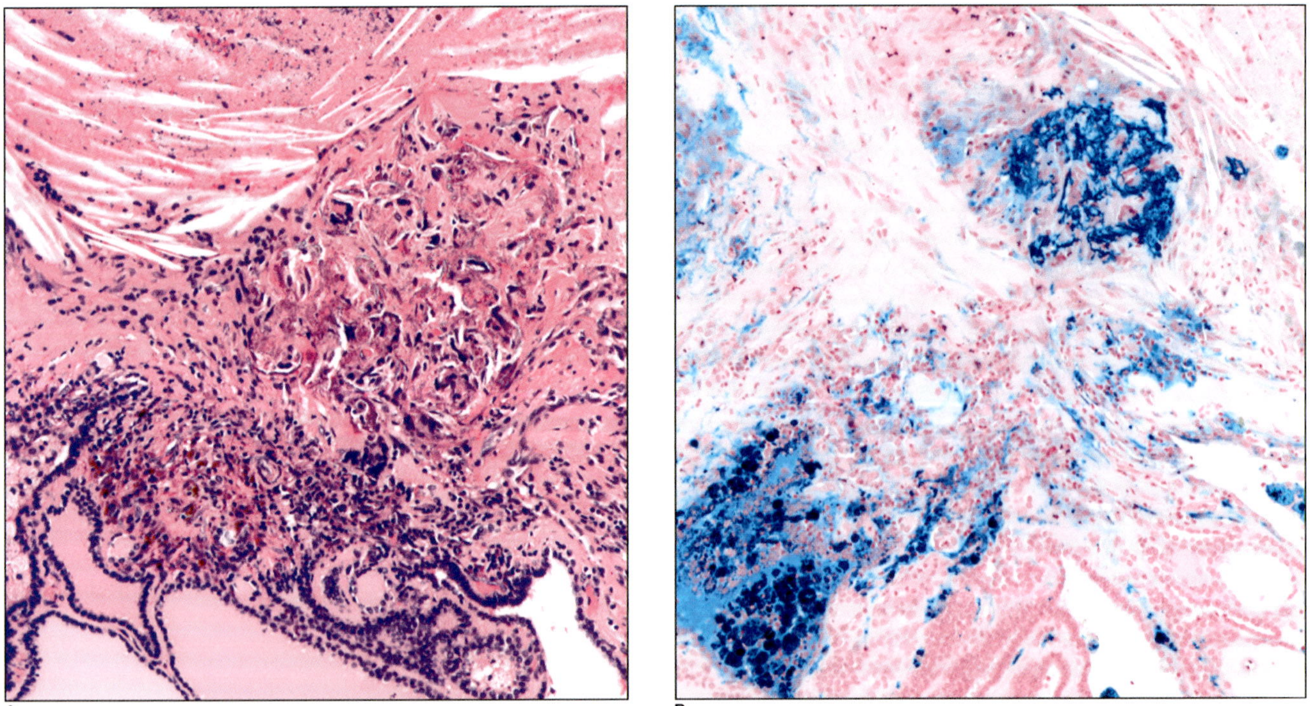

A B

***FIGURE 12.9:** Hemorrhage, fibrosis, and hemosiderin deposition are frequent in nodular goiter (H&E, A). The iron may be revealed with Perl's stain and distinguished from other pigments (B).*

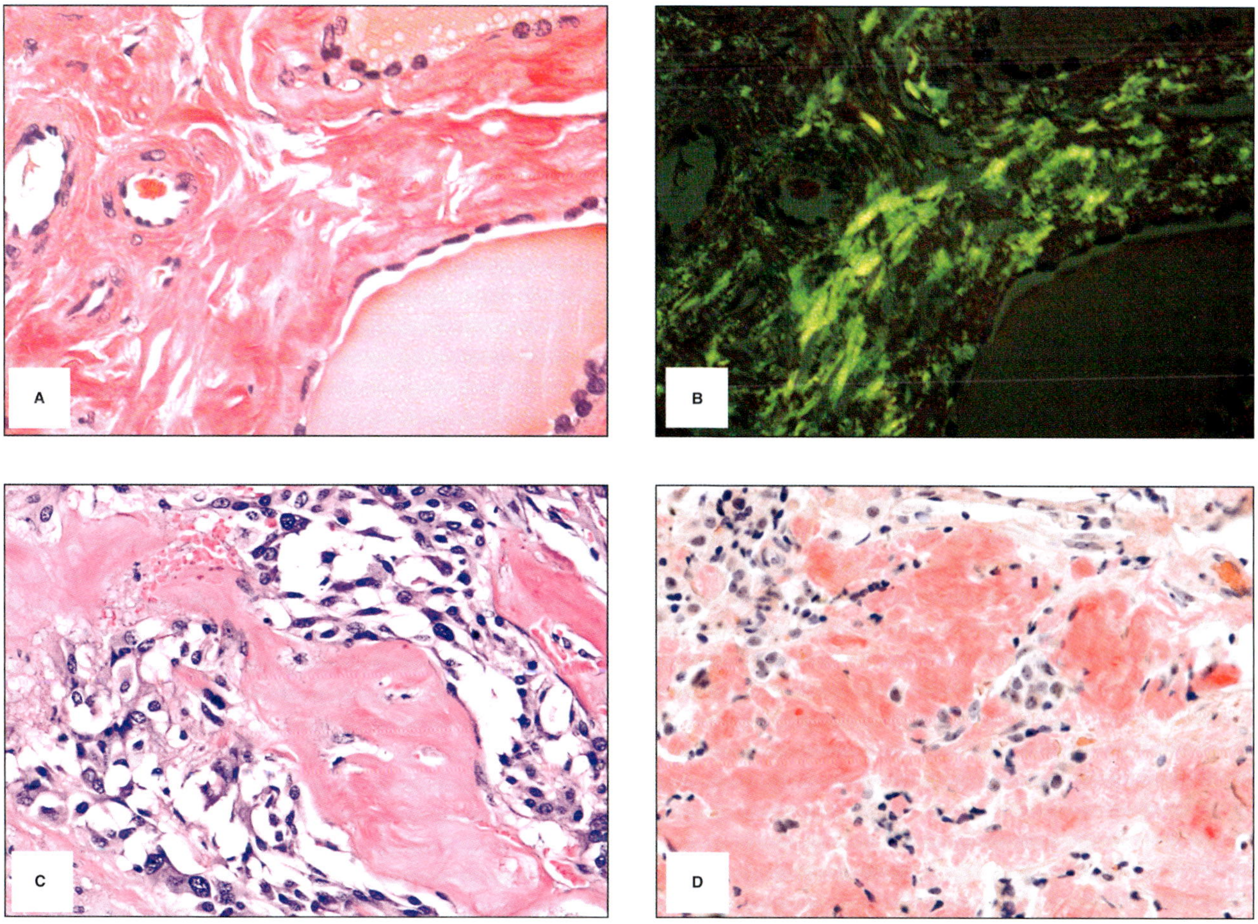

***FIGURE 12.10:* Amyloid deposits can be found in the thyroid gland as a consequence of different conditions, principally amyloid goiter and medullary carcinoma. They may be demonstrated by Congo red staining. Amyloid goiter is rare and is characterized by irregular amyloid deposits within follicular structures (A—Congo red stain; B—green birefringence of Congo red stain with polarized light). Medullary carcinoma is a C-cell tumor (H&E, C) that produces calcitonin. In the vast majority of cases, it contains amyloid deposits that are demonstrable by Congo red staining (D).**

Today, the method of Puchtler and Sweat (37), which uses alkaline Congo red solution and alcoholic differentiation, is the most commonly employed (38). In this procedure, alcoholic differentiation may decrease the intensity of birefringence owing to Congo red extraction. The sensitivity of the method can be improved by the use of a Congo red solution that is fully saturated with sodium chloride, to avoid nonspecific binding of the dye to tissues other than amyloid (39). In other amyloid-labeling methods, such as Romhanyi's technique, no dye molecules are removed in the absence of alcohol immersion, and a higher sensitivity is obtained (40). Unfortunately, in the latter procedure, a polar mounting medium is required, such as Arabic gum. The Congo red stain produces a red-orange color of amyloid; as viewed with polarized light, it is birefringent. The use of Congo red staining may be useful in the diagnostic evaluation of trabecular lesions of the thyroid that contain abundant amorphous or hyaline stroma. In fact, different thyroid conditions may simulate amyloid goiter and MTC. They include hyalinizing trabecular tumor, in which amorphous amyloid-like material contains proteins of basement membranes, and follicular adenomas or hyperplastic nodules with regressive changes. In these lesions, the amorphous stromal material is generally represented by dense and hypocellular collagen.

Lectin histochemistry also has been applied to the study of thyroid lesions, as well as other endocrine disorders. In this context, *Ulex europaeus* agglutinin I (UEA-I) was found to bind preferentially to endothelial cells and was proposed as a marker of vascular tissue. This premise has been confirmed in the thyroid gland, where some authors have found that UEA-I histochemistry was superior to factor VIII–related antigen immunostaining in delineating tumor invasion of blood vessels by follicular carcinomas (41,42). Several other lectins have likewise been studied in the thyroid (42,43). Follicular thyroid tumor cells are labeled by several of those markers, but none is completely effective in distinguishing follicular from papillary carcinoma or benign from malignant follicular lesions. This is because of an inconsistent and focal nature of the observed affinity

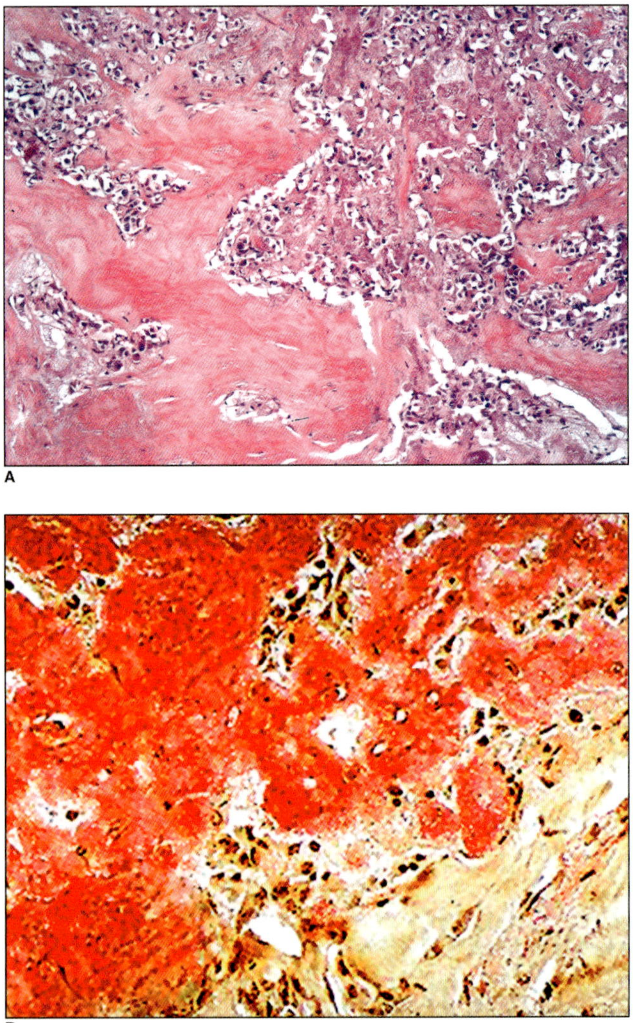

A

B

FIGURE 12.11: (A and B) Intense Congo red positivity in the stromal amyloid of another MTC (also see Figure 12.10).

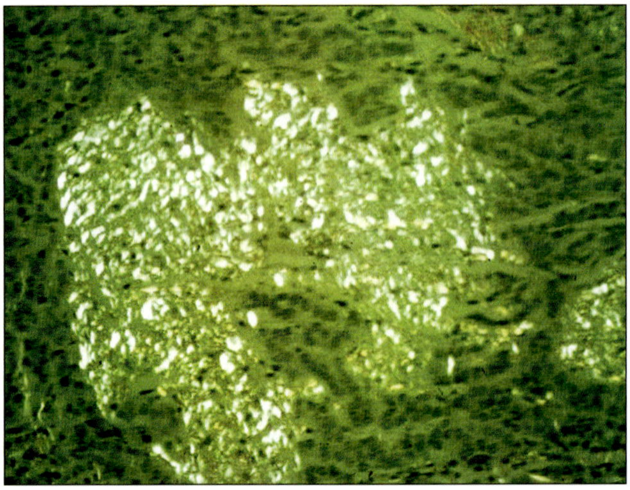

FIGURE 12.12: Green birefringence is present on polarization microscopy of a Congo red stain of another MTC (also see Figure 12.10).

profiles of lectin ligands (42). Different histochemical results have been obtained with lectins that are specific for glycidic chains and that reliably bind to predefined glycoproteins. Using sialic acid–binding lectins, changes in sialylation are detectable with special reference to terminal linkages. These alterations parallel neoplastic transformation in some thyroid cancers. Among them, follicular thyroid tumors display a differential pattern of luminal surface sialic acid reactivity, with *Tritrichomonas mobilensis*, *Maackia amurensis* leukoagglutinin, *Helix pomatia*, and *Sambucus nigra* agglutinin. Papillary thyroid carcinomas, and, to a lesser extent, follicular carcinomas, show a marked increase of sialic acid positivity at luminal cell surfaces, as compared with that seen in normal and adenomatous follicular cells (44) (Figure 12.13).

The lectins just listed can therefore be useful markers of potential malignant transformation in the diagnosis of follicular thyroid tumors. However, one must bear in mind

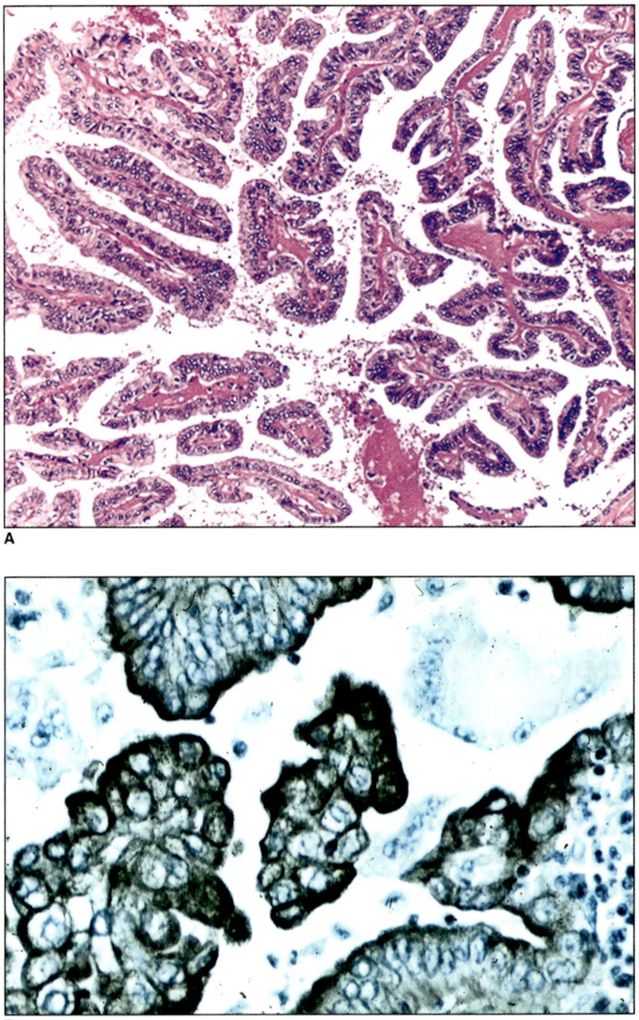

A

B

FIGURE 12.13: Papillary thyroid carcinoma (A) demonstrates binding of Helix pomatia lectin because of sialic acid residues on the surfaces of the tumor cells.

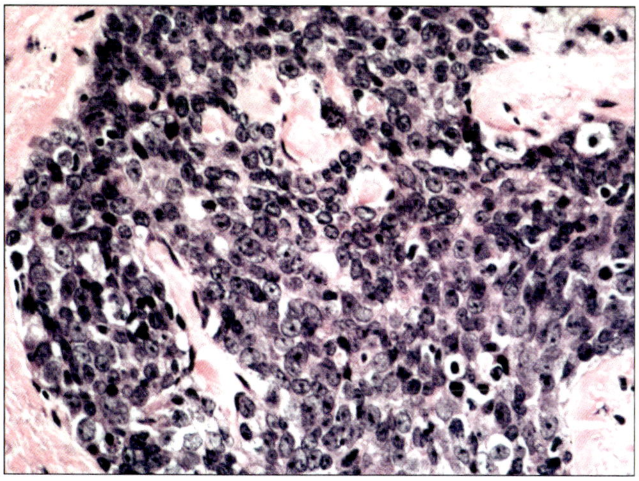

FIGURE 12.14: *Insular carcinoma is one of the potentially undifferentiated small cell tumors of the thyroid.*

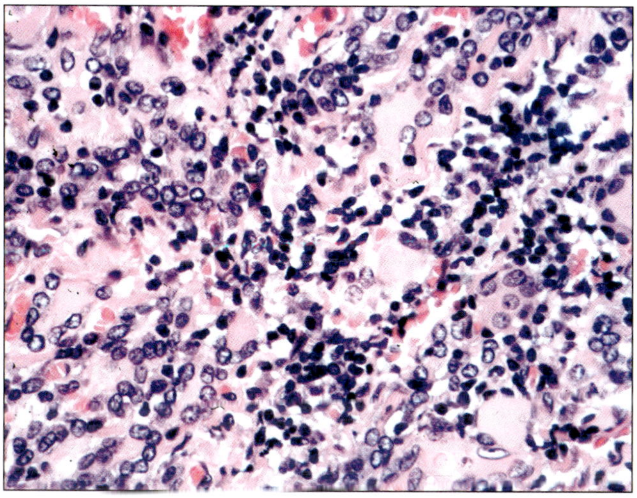

FIGURE 12.15: *Non-Hodgkin lymphoma may also present as a morphologically ambiguous small cell thyroid tumor.*

that vascular endothelia and epithelial follicular cells in autoimmune thyroiditis are also potentially reactive with the same plant-derived agglutinins (45).

Oncocytic large cell (Hurthle cell) tumors of the thyroid may be characterized by detecting endogenous cytoplasmic biotin. This can be accomplished by means of a histochemical nonimmunologic reaction using the affinity of biotin for avidin. Biotin is contained in mitochondria, and oncocytic cells are strongly biotin positive because they harbor an abundance of those organelles. This reaction is also useful for characterizing eosinophilic tumors comprising small cells, in which oncocytic features are not fully developed (46). This is true, for example, of poorly differentiated oxyphilic carcinomas in which oncocytic and granular cytoplasm may not be readily appreciated (47). It also applies to cases in which mitochondria-rich cells show extensive clear cell changes, masking the typically finely granular,

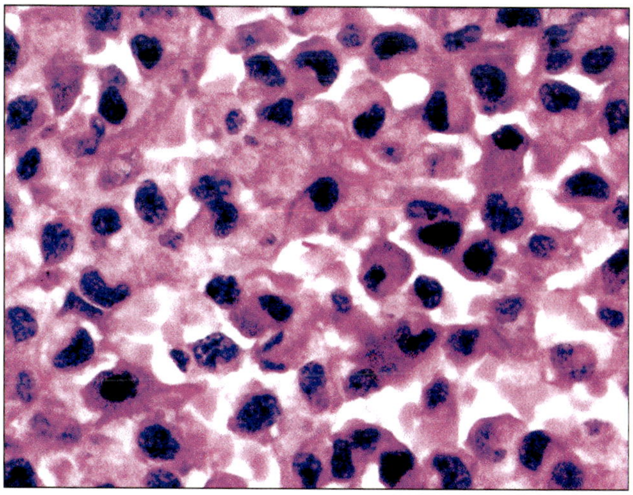

FIGURE 12.16: *The methyl green-pyronine technique yields nucleocytoplasmic positivity in non-Hodgkin lymphoma.*

deeply eosinophilic appearance of Hurthle cells (see methodologic details in the section on adrenal cortex below).

Another diagnostic problem in thyroid pathology concerns the distinction of small cell epithelial malignancies from hematolymphoid tumors. "Insular" carcinomas (48) (Figure 12.14) and small cell variants of MTC (49,50) comprise the former of those groups, whereas non-Hodgkin lymphomas—often arising in the context of Hashimoto's thyroiditis (51–54)—account for the latter (Figure 12.15). As shown by Tobler et al. (51) and others (52–55), thyroid lymphomas do not bind to *U. europaeus*, *H. pomatia*, peanut, lectin, or soybean agglutinin, whereas thyroid carcinomas tend to do so. Moreover, MTCs exhibit arygyophilia with the Sevier-Munger and Grimelius stains, as opposed to negative results in lymphomas. Occasionally, they may also be mucin positive with the PAS-diastase or mucicarmine method. Conversely, hematolymphoid proliferations are more likely to show positivity with the methyl green-pyronine technique (Figure 12.16).

Parathyroid Gland: The differential diagnosis of parathyroid lesions generally centers on the differences between hyperplastic and neoplastic conditions. In all such cases, normal glandular structure is altered, with the physiologic amount of intraglandular fat being absent or markedly reduced. Hyperplastic parathyroid lesions potentially exhibit a range of histologic changes, from a mild increase of chief cells to frankly adenomatous patterns. Adenomas and parathyroid carcinomas (which are rare) both usually show nodular growth that compresses the adjacent normal parenchyma, including fat cells, toward the periphery of the tumors. The recognition and quantification of adipocytes in enlarged parathyroid glands is important to correct pathologic interpretation. Conventional H&E stains generally allow for a confident diagnosis of nodular

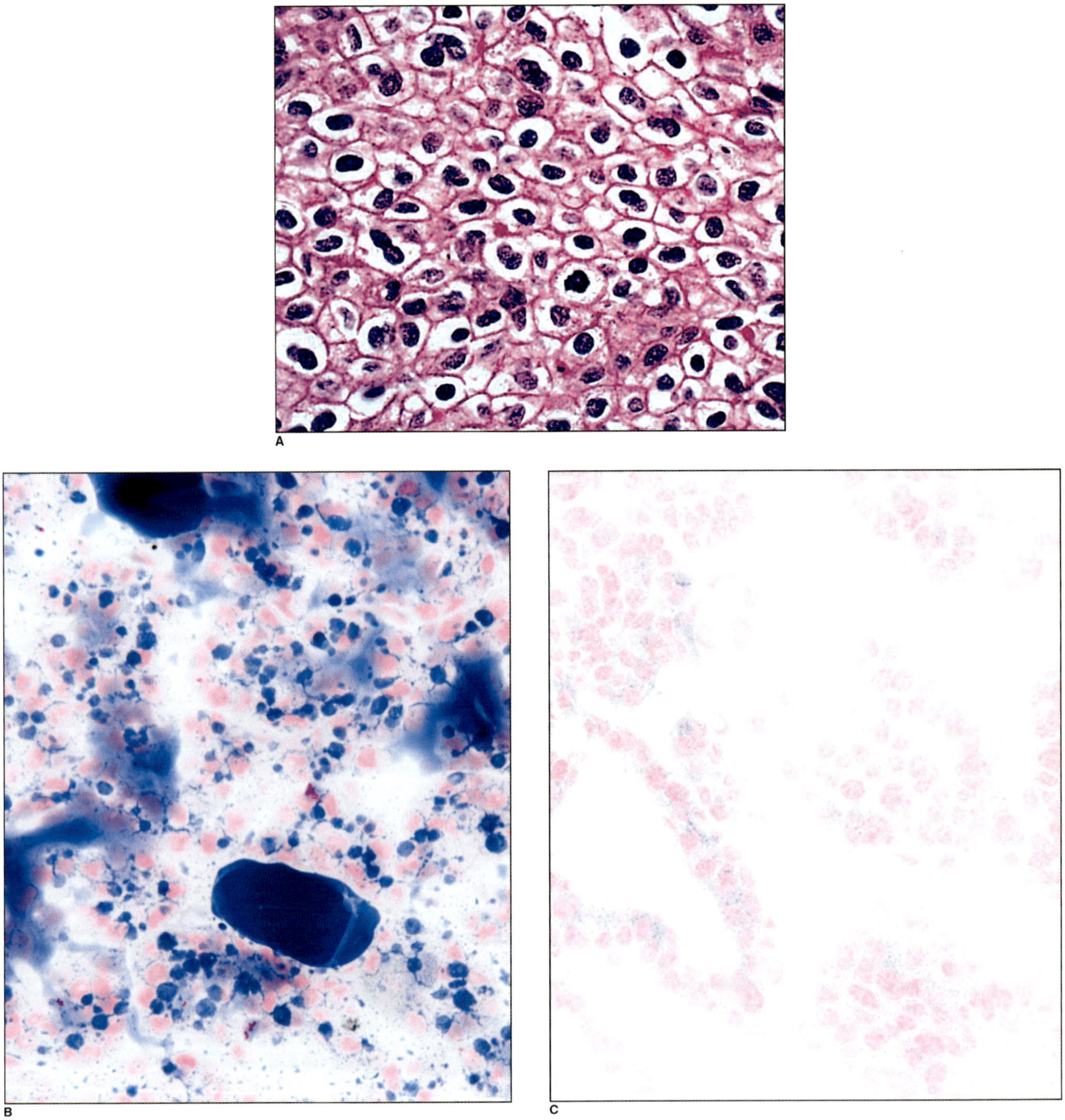

***FIGURE 12.17:** The recognition and quantification of adipocytes in enlarged parathyroid glands are important for obtaining a correct interpretation of proliferative disease (A), with special reference of intraoperative frozen section examination. Histochemical stains for fat such as Sudan black B and oil red O may help in the differential diagnosis between normal parathyroid tissue (B) containing black-stained adipocytes and hyperplastic or adenomatous tissue (C) that may contain "holes" or other artifacts but lacks fat cells.*

hyperplasia, based on a marked reduction of fat in a hypercellular gland weighing >40 mg.

This statement does not hold true in some cases submitted for intraoperative frozen section examination and, more rarely, in permanent tissue sections. Histochemical stains for fat may help to resolve problems, as an adjunct to H&E, but the need for fresh tissue is a major limitation to their application. In frozen sections, various methods for detecting neutral fats have been developed, including the oil red O and Sudan black B procedures (Figure 12.17). Both of these techniques are based on the greater solubility of the applicable dyes in fat, than in the low-concentration

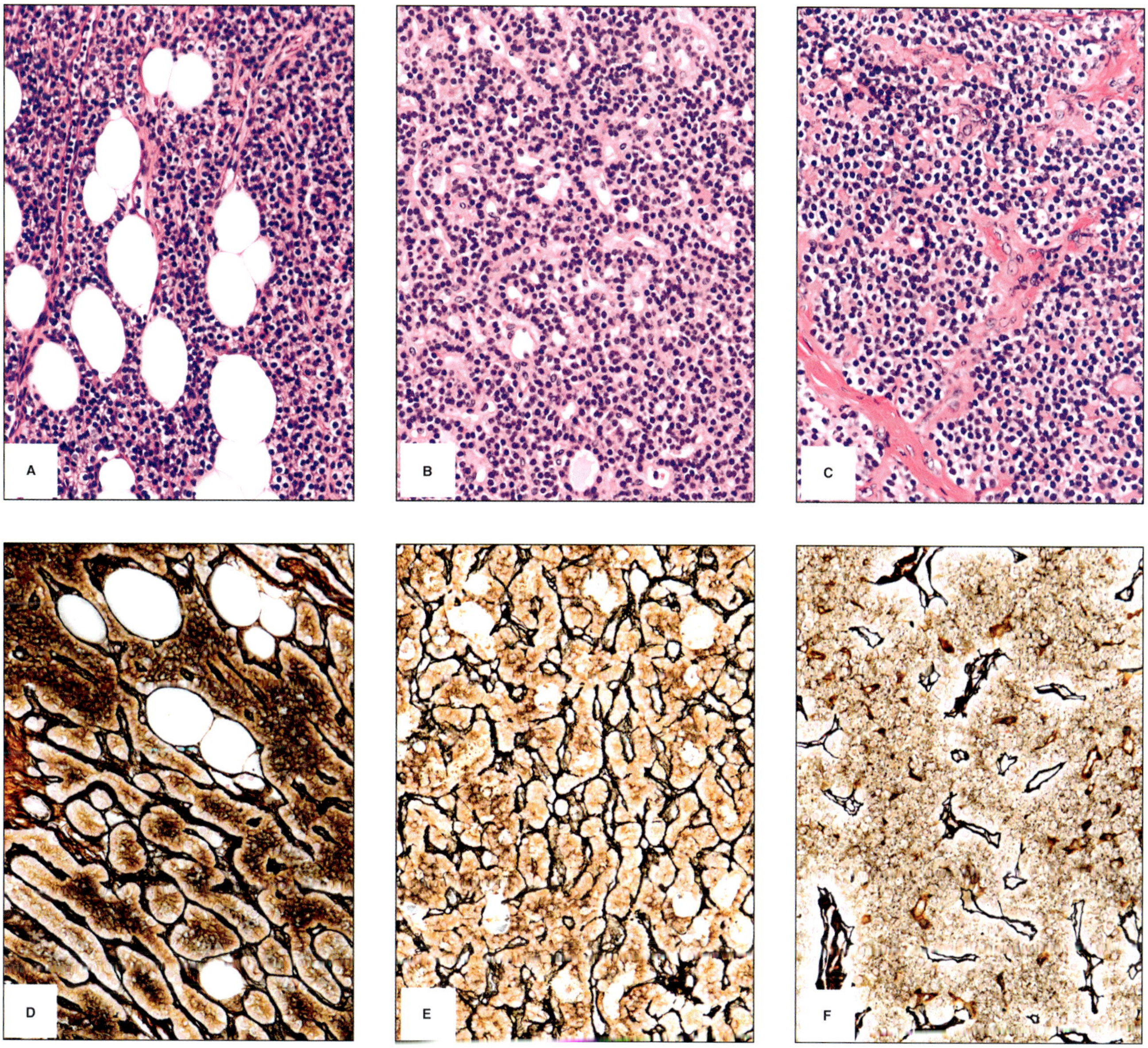

FIGURE 12.18: *Reticulin staining is a diagnostic adjunct in the differential diagnosis of neoplastic parathyroid lesions. A fine reticular network is maintained in parathyroid adenoma (B—H&E and E—Snook's stain) as compared with normal parathyroid tissue (A—H&E and D—reticulin stain). Parathyroid carcinoma usually manifests a much more irregular and disrupted reticulin framework (C—H&E and F—Snook's stain).*

alcohols in which they are dissolved. Positive reactions produce a red or blue-black color in adipocytes (with the oil red O and Sudan black B methods, respectively), reflecting uptake by fat molecules of the dyes from their original alcoholic solutions. The applicability of these techniques to paraffin-embedded tissues is very limited because of the extensive lipid extraction that occurs during tissue processing procedures (56).

Another useful histochemical stain in assessment of the parathyroids is the reticulin (Snook's) method. It may help to outline the stromal network of the gland, as it does in the pituitary (Figure 12.18). Hyperplastic parathyroid lesions have a preserved architecture with single chief cells or small clusters of them, surrounded by a fine reticulin network. In contrast, neoplastic conditions show a reticulin framework that is disrupted to a noticeable extent, especially in parathyroid carcinoma.

A relatively common differential diagnostic problem in cytopathology and surgical pathology concerns the distinction of cellular hyperplastic thyroid nodules from parathyroid proliferations. In this context, two additional histochemical methods may be helpful, namely, the PAS and CS procedures. Parathyroid proliferations commonly contain substantial amounts of cytoplasmic glycogen (57,58); therefore, they demonstrate particulate intracellular PAS positivity (Figure 12.19). In contrast, thyroid nodules

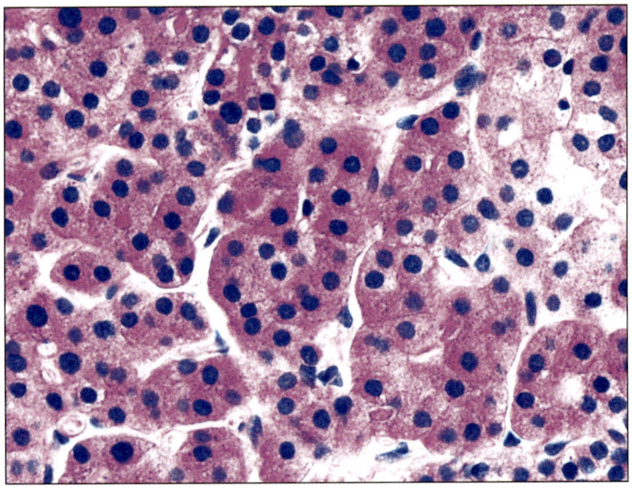

***FIGURE 12.19:** Parathyroid adenoma, showing diastase-labile reactivity with the PAS method.*

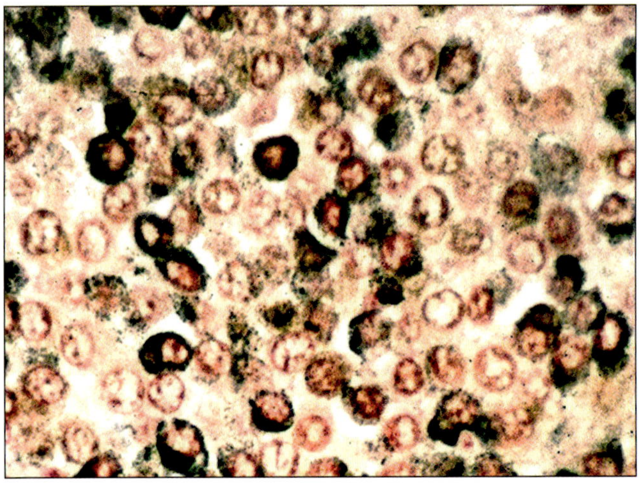

***FIGURE 12.21:** Multifocal positivity of a parathyroid adenoma with the CS method.*

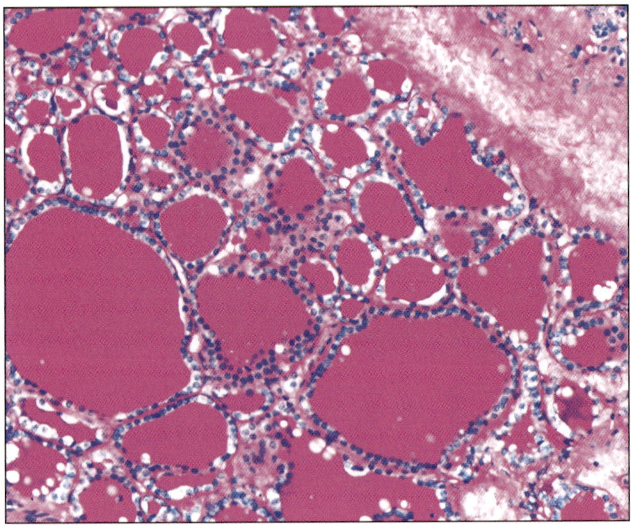

***FIGURE 12.20:** Diastase-resistant labeling of thyroid colloid is seen with the PAS procedure.*

do not show the same staining properties with PAS, although colloid will be labeled (Figure 12.20). Parallel comments apply to the CS method, which preferentially labels parathyroid cells but not follicular thyroid elements (59,60) (Figure 12.21).

Endocrine Pancreas: Pancreatic islet cells are also argyrophilic and therefore can be stained by silver impregnation methods such as the CS, Grimelius, Sevier-Munger, and Davemport procedures. Individual islet cell types (i.e., alpha, beta, delta, pancreatic polypeptide producing) do not have the same affinity for silver salts presumably because the content and structure of their neurosecretory granules are variable (61). Glucagon-producing (alpha) and somatostatin-producing (delta) cells are generally those that are most strongly reactive with argyrophil methods using alcoholic solutions. Other histochemical techniques, including—among others—the xanthydrol method, the lead-hematoxylin, and phosphotungstic acid-hematoxylin (PTAH) stains, have been reported to more or less specifically label alpha cells and related neuroendocrine tumors (glucagonomas) (62,63).

In contrast, insulin-producing beta cells can be recognized by detecting protein-bound disulfide and sulfhydryl groups, through thiosulfation and subsequent staining with the alcian blue or toluidine blue methods (64) (Figure 12.22). This technique is not specific for pancreatic beta cells because similar chemically reactive groups are also present in the skin and stomach. As an alternative, beta cells are detectable with a stain that is based on the aldehyde-fuchsin reaction (65,66) (Figure 12.23). It can be optimized by treating embedded formalin-fixed tissue sections with Bouin's fixative or glutaraldehyde before deparaffination to stabilize intracellular neurosecretory granules (65). Notably, aldehyde-fuchsin positivity can also be seen in exocrine pancreatic acinar cells, owing to reactivity of this reagent with zymogen granules. A double histochemical stain that combines aldehyde-fuchsin with the Grimelius silver impregnation method has been proposed to investigate the spatial distribution of the two main pancreatic endocrine cell populations (67). Islet cells do not possess any argentaffinity with the Fontana-Masson procedure.

Rarely, the pancreatic islets can be affected by stromal deposition of amyloid or iron. These substances can be visualized by Congo red and Perl's stains, respectively (see above). Amyloid is principally observed in the context of early-onset diabetes mellitus.

Adrenal Gland and Related Structures

Adrenal Medulla and Extra-Adrenal Paraganglia: Adrenomedullary cells produce adrenalin and noradrenalin,

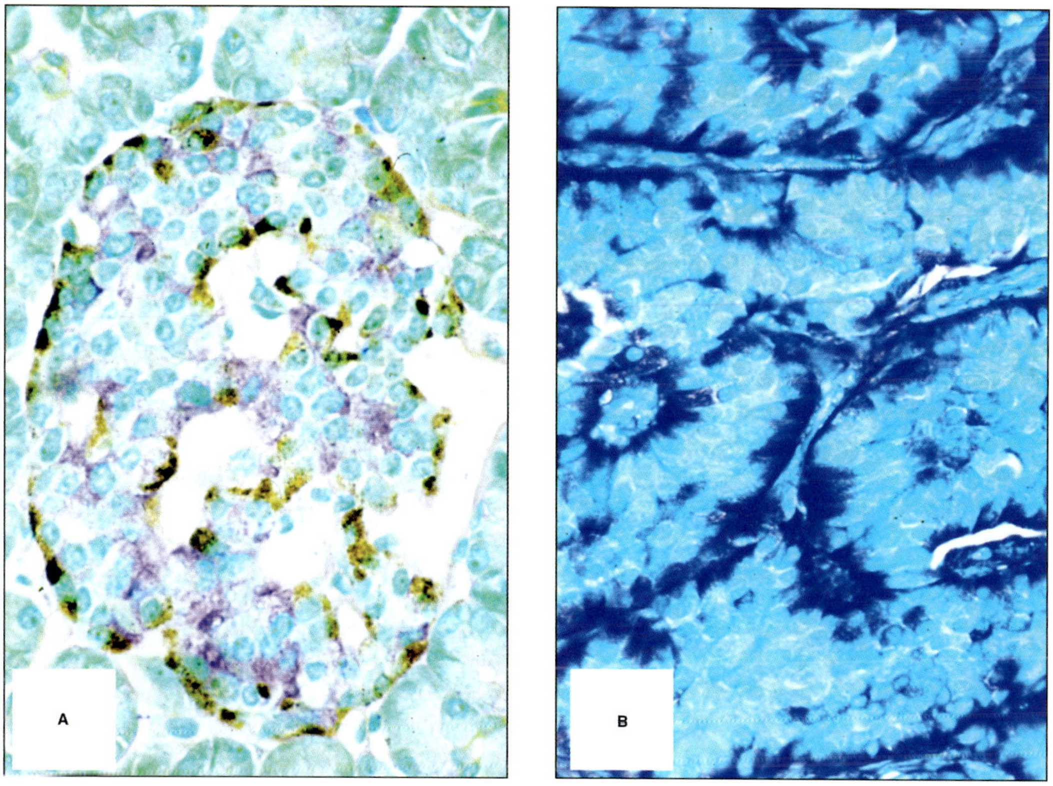

FIGURE 12.22: ***Insulin-producing β-cells are specifically labeled by the aldehyde-fuchsin reaction in a normal islet (A), or by detecting protein-bound disulfide and sulfhydryl groups by thiosulfation and subsequent staining with alcian blue, as seen in an insulinoma (B). In comparison, α-cells are sensitively labeled with the Grimelius stain (A—brown color).***

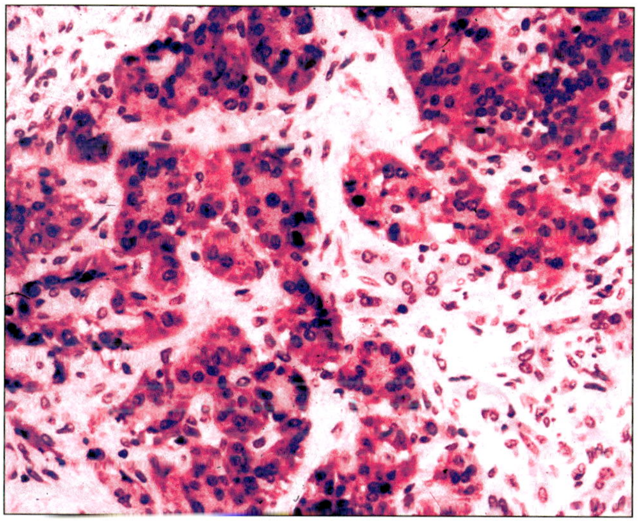

FIGURE 12.23: ***The aldehyde-fuchsin method stains this pancreatic endocrine tumor.***

which, together with serotonin and histamine, represent the most common biogenic amines. Although histamine is difficult to detect in tissue sections because of its high solubility, the other molecules just cited can be visualized as a result of their fluorescence after formalin fixation. Serotonin-containing cells develop a characteristic yellow fluorescence. As an alternative, the so-called chromaffin reaction is helpful in identifying biogenic amine–containing cells. Fresh tissue sections are treated with a chromium salt, such as potassium bichromate, after which the presence of biogenic amines (either catecholamines or serotonin) induces a brown-yellowish coloration in the tissue. Such color changes led to the terms "chromaffin reaction" and "chromaffin cells," to indicate the presence of cell populations that were able to react with chromium; these include elements in the adrenal medulla and in other paraganglia. Despite the name "chromaffin," the observed color change was subsequently shown to result from catecholamine oxidation rather than chromium affinity.

The term "pheochromocyte" was subsequently proposed to replace "chromaffin cell." Interestingly, the former of these designations survives only in regard to neoplastic counterparts of chromaffin cells in the adrenal medulla (i.e., pheochromocytoma). However, normal and neoplastic extra-adrenal catecholamine-producing cells likewise share the property of being able to reduce chromium salts (5,8).

Chromaffin cells also can bind and reduce silver salts, and therefore, they are argentaffinic and argyrophilic. These properties are responsible for positivity that is seen with the Masson-Fontana and Grimelius (or other argyrophilic) reactions, respectively, in the adrenal medulla, pheochromocytomas, and some extra-adrenal paragangliomas.

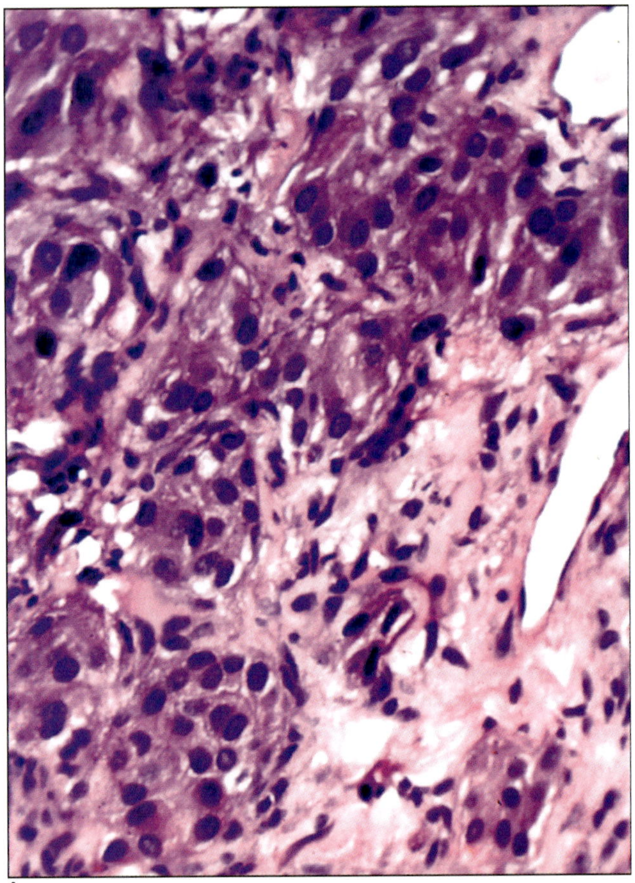

A

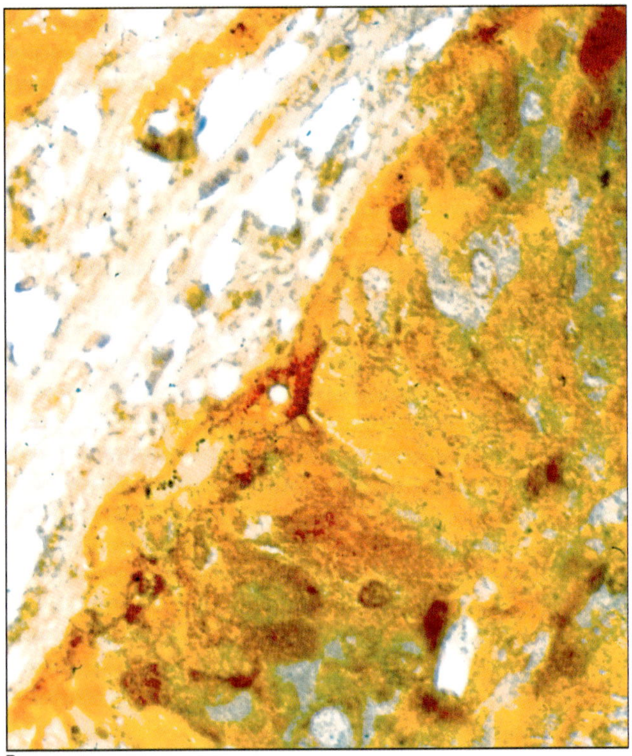

B

FIGURE 12.24: In adrenal pheochromocytoma (A–H&E), silver impregnation methods such as the Grimelius stain (B) are helpful in establishing the neuroendocrine nature of this tumor.

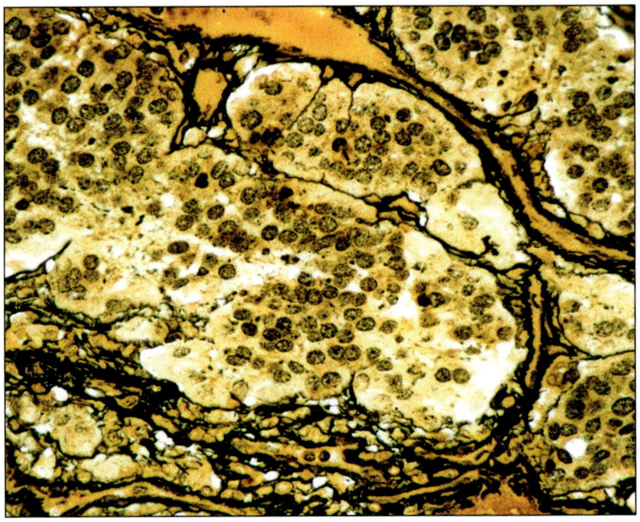

FIGURE 12.25: Snook's method reveals investment of cellular nests by reticulin fibers in neuroblastoma.

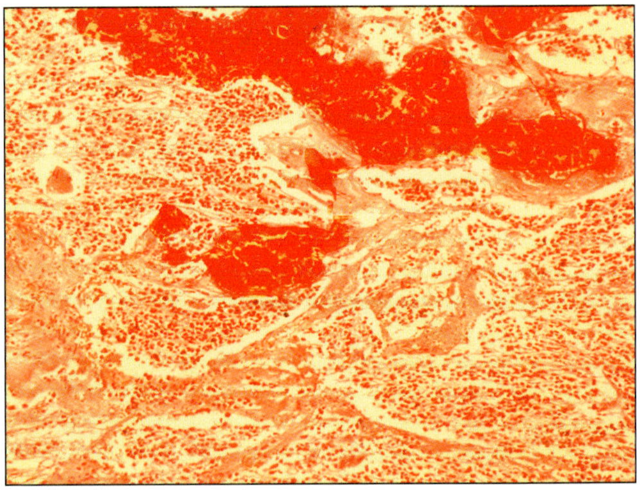

FIGURE 12.26: The alizarin red procedure decorates foci of dystrophic calcification in neuroblastoma.

Silver impregnation methods may be used in place of immunostains for chromogranin because the same intracellular molecules are targeted in all of those procedures. Hence, tumor nodules with a suspected neuroendocrine nature in the adrenals (Figure 12.24), the retroperitoneum, the posterior mediastinum, and other sites can be identified as paragangliomas with the CS, Grimelius, or Sevier-Munger techniques.

In young children (68), and occasionally in older patients as well (69), neuroblastoma may also arise in the adrenal medulla. This small cell "blastomatous" malignancy is *not* often arygyrophilic, and it demonstrates a reticulin pattern that differs from that of lymphoma—with the latter representing the principal differential diagnostic consideration in that age group. Reticulin fibers invest large groups of neuroblastoma cells (Figure 12.25) as

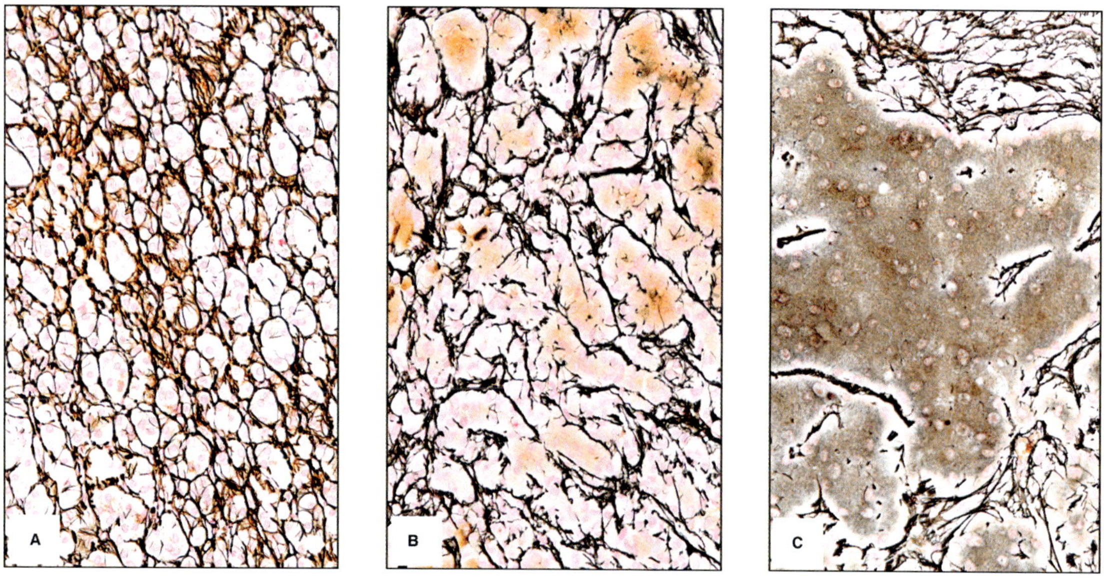

***FIGURE 12.27:* As described for parathyroid lesions, Snook staining of normal adrenal cortex (A) outlines a fine reticulin network. It is also maintained in adrenocortical adenoma (B), but is disrupted in ACC (C). Reticulin staining may be a useful means of recognizing a nodular growth pattern, which is a part of the Weiss scoring system for ACC.**

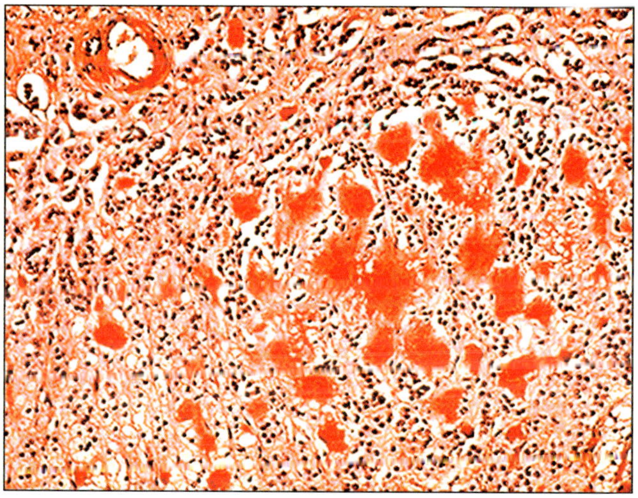

***FIGURE 12.28:* Adrenal amyloidosis shows Congo red positivity around cords of cortical epithelial cells and in stromal blood vessels.**

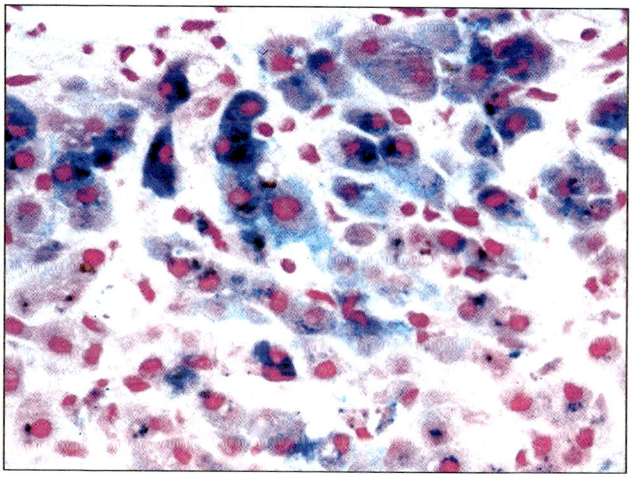

***FIGURE 12.29:* (A) Adrenal involvement by hemochromatosis is seen in this H&E stain, and the pigment is confirmed as iron with a Perl's stain (B).**

opposed to their presence around each tumor cell, or small groups of them, in hematolymphoid tumors. In addition, neuroblastoma often contains foci of dystrophic calcification, as highlighted with the von Kossa or alizarin red methods (Figure 12.26)—unlike lymphomas.

Adrenal Cortex: In diseases of the adrenal cortex, the evaluation of small nodules in the context of hyperplastic or neoplastic conditions may be facilitated by use of the reticulin stain, which can better highlight the cortical stromal framework. In contrast, the reticulin network of adrenocortical carcinomas (ACCs) is irregular and disrupted, as compared with that seen in benign conditions (M. Volante, unpublished observation, 2006) (Figure 12.27). Because the presence of a nodular or diffuse growth pattern is one of the parameters considered in the diagnostic evaluation of malignant lesions using the system of Weiss et al. (70), some cases in which nodularity is not obvious may benefit from reticulin staining.

In adrenocortical neoplasms, the vascular network of the cortex can be outlined by lectin histochemistry using ligands with a high affinity for endothelial cells. The principal agent used for this purpose is biotinylated *U. europaeus* agglutinin type I (see above).

In the course of systemic secondary amyloidosis, amyloid may accumulate in the adrenal gland. Both the zona fasciculata and zona reticularis are the primary sites for this process. They may expand, even assuming a nodular pattern, and eventually obliterate the inner portion of the gland. This phenomenon can lead to adrenal insufficiency if deposition is bilateral and extensive. Congo red staining generally helps to confirm the amyloidaceous nature of the amorphous hyaline material in this condition, which generally appears first around capillaries and cortical cells (35) (Figure 12.28).

Pigmentation of the adrenal gland can also be observed in rare cases. Iron overload can result from prior hemorrhage or as the result of systemic hemochromatosis or hemosiderosis. Melanin or neuromelanin pigment can be seen in rare cortical adenomas or in the rare familial condition (which is generally part of the "Carney complex") that is called "primary pigmented adrenocortical disease" (71). Specific stains for these pigments, that is, the Fontana-Masson and Perl procedures (Figure 12.29), allow for correct interpretation and differentiation from adrenal lipofuscinosis. In the latter condition, the pigment generally is reactive with the PAS method after diastase digestion, and it may label with luxol fast blue as well.

The adrenal cortex contains steroid-producing cells that are mitochondrion rich, as their neoplastic counterparts also may be, regardless of whether they are adenomas or carcinomas. The differential diagnosis of ACCs from tumors that are metastatic *to* the adrenals can be difficult. Thus, a demonstration of mitochondrion-rich cells may help to distinguish primary ACC from metastatic melanoma, metastatic large cell lung carcinoma, metastatic renal cell carcinoma (RCC), and other neoplasms (46). The aforementioned histochemical method for biotin (46) is useful in this context because it allows for detection of mitochondria and therefore will highlight oncocytic neoplastic cells such as those of ACC. Biotin positivity is also common in hepatocellular carcinoma, but the latter neoplasm only very rarely presents with isolated adrenal metastasis. A heat-based retrieval procedure can be used to unmask biotin, and it is suggested before applying avidin with biotinylated peroxidase reagents or peroxidase-conjugated streptavidin (46). Slides labeled with these reagents can then be incubated in diaminobenzidine as a chromogen and observed light microscopically. A brown-black signal in the cytoplasm indicates the presence

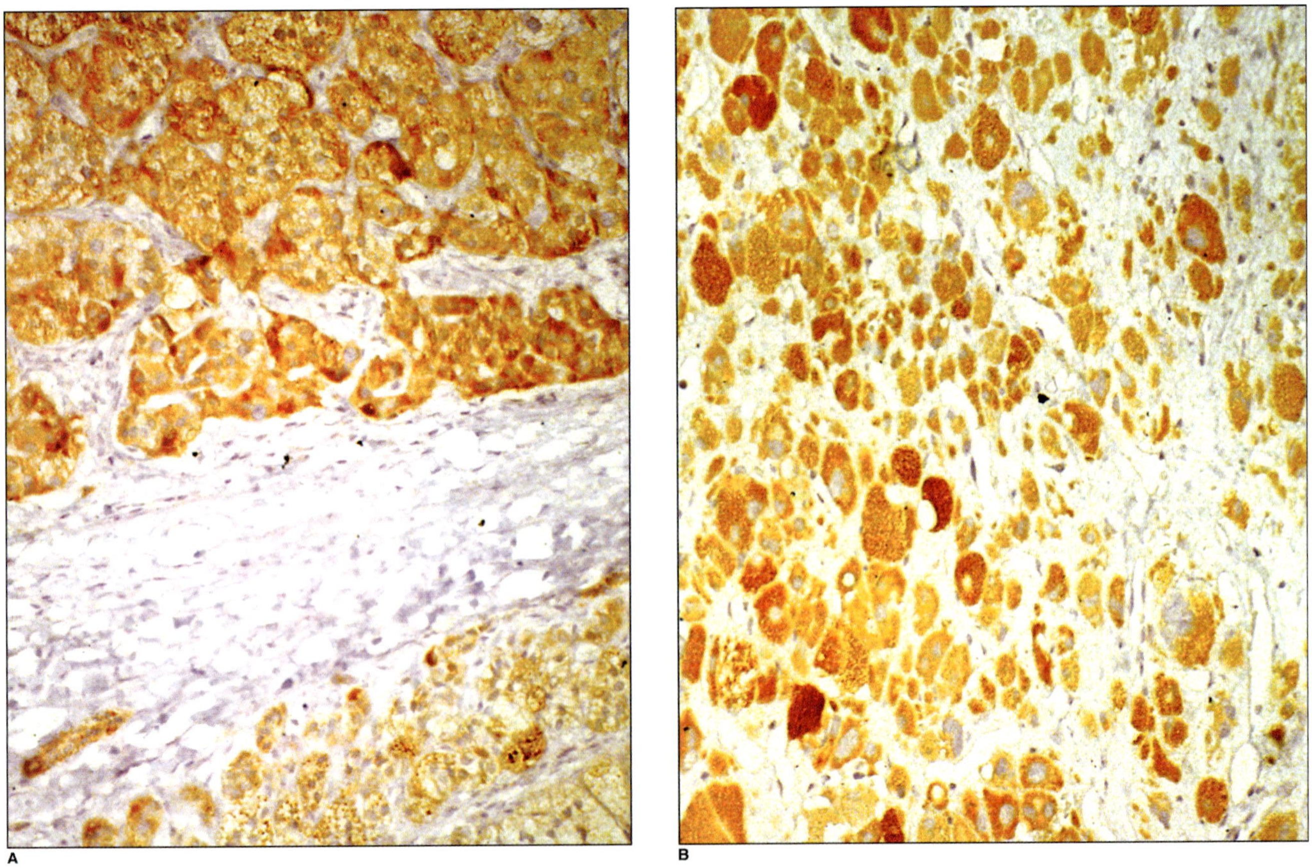

A B

FIGURE 12.30: Adrenocortical tissues and related neoplasms are rich in mitochondria. Therefore, they have a high cytoplasmic biotin content. Heat-based "antigen retrieval" procedures are able to unmask biotin, and a nonimmunologic histochemical method can then be employed to label biotin-rich cells and tumors such as those of the adrenal cortex [A—adrenal adenoma (top) and normal adrenal cortex (bottom); B—ACC]. This method may help to distinguish ACC from metastatic tumors involving the adrenal.

of mitochondrion-rich cells (Figure 12.30). The Fontana-Masson method and Schmorl's technique are used to identify metastatic melanoma, and intracellular mucin in metastatic adenocarcinomas can be recognized with the PAS stain after diastase digestion, as well as the mucicarmine procedure or the alcian blue and colloidal iron methods.

Occasionally, an epithelial neoplasm will be encountered that effaces both the adrenal gland and the upper pole of the adjacent kidney. In this circumstance, one may be hard pressed to determine whether the lesion is a primary RCC that invades the adrenal, or an ACC involving the kidney. RCC may also selectively metastasize to the contralateral adrenal in rare cases (72). Biotin labeling may also be helpful in these scenarios, but it is also worthwhile to obtain a PAS stain with and without diastase digestion. RCC typically contains abundant glycogen and is therefore intensely PAS positive (Figure 12.31), whereas ACC shows very scant cytoplasmic glycogen or none at all. Both tumor types often are reactive with the oil red O and Sudan black B methods

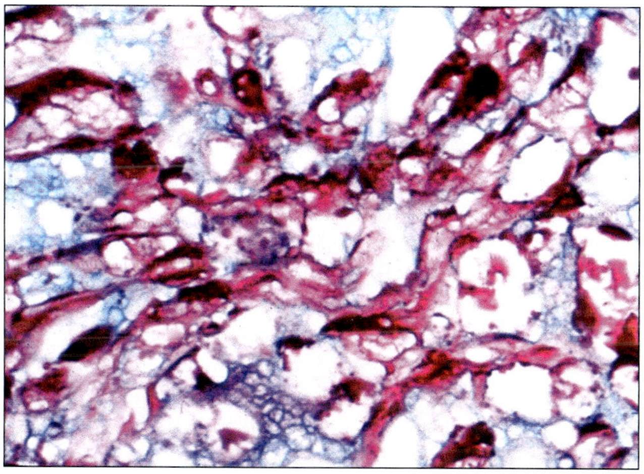

FIGURE 12.31. RCC is intensely positive with the PAS stain, but diastase predigestion abrogates the reaction.

NEUROENDOCRINE DIFFERENTIATION IN NONNEUROENDOCRINE TUMORS

Adenocarcinomas that have combined or mixed exocrine (e.g., mucin producing) and neuroendocrine features have been described in several organs, such as the breast, prostate, colon and rectum, stomach, lung, and others. These lesions include "hybrid" exocrine-endocrine carcinomas in which one or the other of the two components comprises at least one third of the tumor mass, according to a definition proposed in the World Health Organization classification of endocrine tumors (73). Another possibility is that conventional adenocarcinomas may demonstrate a variable degree of neuroendocrine differentiation that is seen in <30% of the tumor cells and is usually in the range of 2–5%. In other words, neuroendocrine cells are randomly scattered among the glandular components and typically are undetectable on conventional H&E microscopy (74). Such intratumoral endocrine cells are argyrophilic and can therefore be labeled with silver impregnation methods such as the Grimelius procedure. They appear as polygonal or elongated cells with a brown-black color, corresponding to the neurosecretory granules that are present in the cytoplasm (Figure 12.32). As already mentioned, argyrophil reactions are highly sensitive, but less than totally specific, because of their potential to label glandular tumor cells that contain cytoplasmic granules of mucin or other

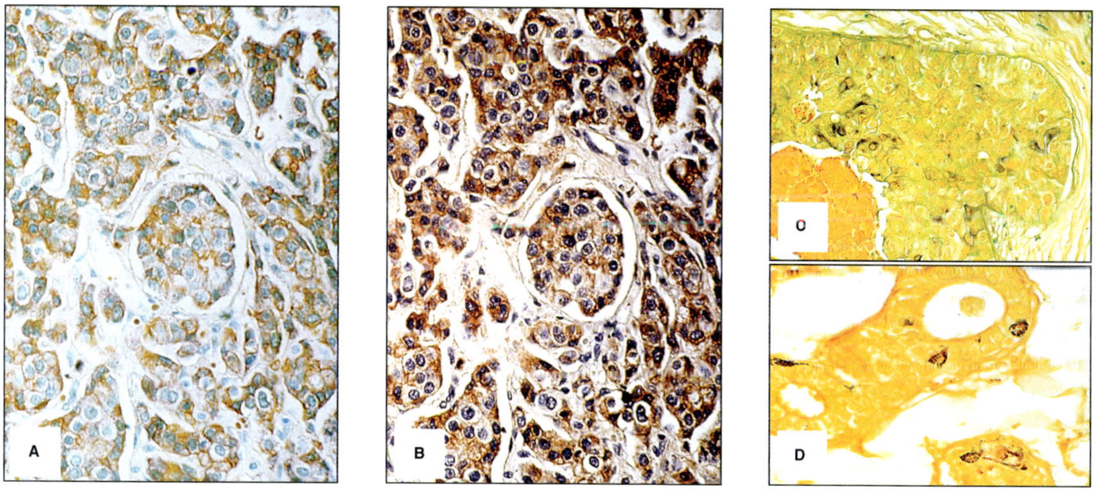

FIGURE 12.32: Adenocarcinomas with neuroendocrine features have been described in various organs, including the breast. The extent of neuroendocrine differentiation may be variable in such lesions, with argyrophilic cells being detectable with silver impregnation methods such as the Grimelius procedure. "Pure" neuroendocrine tumors of the breast (A), tumors with "occult" neuroendocrine differentiation (B), and cells in normal breast structures (C) may all be argyrophilic.

products such as lactalbumin (15). A double-staining method that combines silver impregnation and PAS or Best's mucicarmine can better delineate mixtures of the two cell populations (exocrine and neuroendocrine), including the possibility of detecting both characteristics in the same cells (so-called amphicrine cells) (15). The prognostic significance, if any, of "divergent" neuroendocrine differentiation in exocrine tumors is still being studied.

INFECTIONS OF THE ENDOCRINE GLANDS

Infectious diseases are generally rare in the endocrine system. However, some diseases have a longstanding historical tendency to affect the endocrine glands (e.g., adrenal tuberculosis) and others have become relevant in recent years because of the increased number of immunocompromised hosts. Their presence is a consequence of the human immunodeficiency virus pandemic, as well as prolonged therapy of neoplastic disorders and the advent of organ transplantation. The most common endocrine infections and the role of histochemical techniques in detecting them will be briefly summarized here.

Acute inflammation is potentially observed in the pituitary (usually as an extension of bacterial meningitis) as well as in the thyroid and the adrenal gland. Granulomatous inflammation may be secondary to tuberculosis or fungal infections, and the pituitary may also be affected by syphilis with granulomatous features. Although stains for bacteria, such as the tissue Gram stain, are generally not needed in these rare cases, stains for fungi are more useful. This is especially so in the presence of an acute necrotizing lesion with granulomatous features. Some fungi are visible on standard H&E slides (e.g., *Zygomyces* spp.), whereas others are better delineated with PAS or methenamine silver (Gomori's or Grocott's) stains. Likewise, some viral agents are recognizable in H&E slides, most notably cytomegalovirus and *Herpes simplex* (Figure 12.33). Viral inclusions can also be highlighted with the Machiavello, Feulgen, methenamine silver, or methyl green-pyronine stains (Figure 12.34).

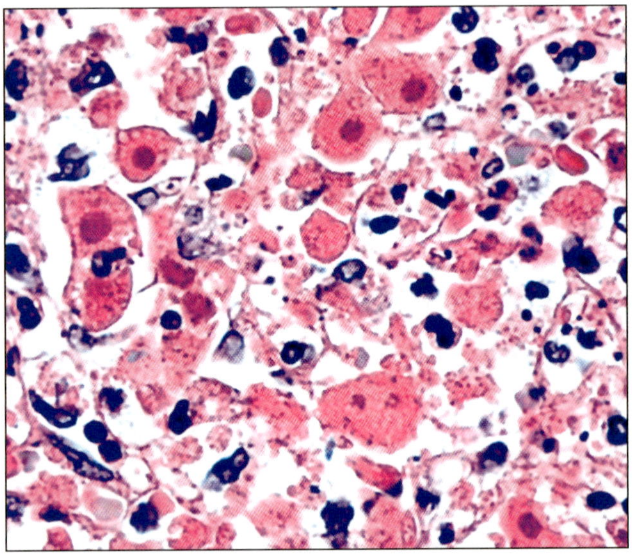

FIGURE 12.33: **Herpes simplex** ***adrenalitis shows homogenized nuclear chromatin with focal intranuclear viral inclusions, as seen in this H&E stain.***

Tuberculosis used to be common in the adrenal gland and less so in the pituitary and thyroid. This disease causes large caseating nodules that destroy the glandular parenchyma and may produce endocrine insufficiency. Several methods have been developed to detect mycobacteria, some being suitable to reveal "typical" species, whereas others (e.g., the Fite method) are more effective for so-called atypical organisms (Figure 12.35). Unfortunately, the sensitivity

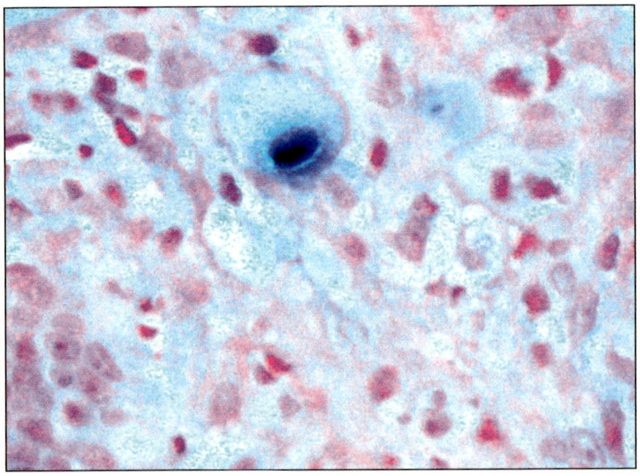

FIGURE 12.34: A viral inclusion in herpetic adrenalitis is positive with the Machiavello procedure.

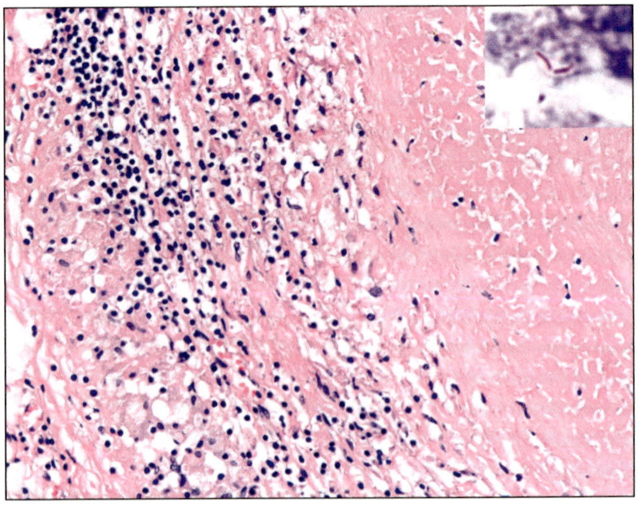

FIGURE 12.35: Histochemistry still represents a valuable tool for detection of microorganisms in tissue sections. Tuberculosis used to be common in the adrenal glands, causing caseating nodular lesions that are recognizable with H&E stains. The Ziehl-Neelsen method is applicable to formalin-fixed, paraffin-embedded sections to detect the causative mycobacteria (inset).

of these techniques is not high because it is largely dependent on the bacterial density in each case. "Productive" granulomatous lesions may not contain large amounts of bacteria, as opposed to "nonreactive" tissue reactions that are generally characterized by a high number of mycobacteria, in which the Ziehl-Neelsen method is generally positive for tubercle bacilli in nearly 50% of cases (56). In immunocompromised patients, the detection rate is much higher, although it is still below the threshold of more sensitive procedures such as cultures or polymerase chain reaction. This fact should be communicated to clinicians, who are prone to "exclude" infection from differential diagnosis if histochemical stains are reported as negative.

ILLUSTRATIVE CASE STUDIES

CASE 1— ***Amyloid Goiter:*** A 55-year-old woman had had a left thyroid nodule for several years, with slow growth. It was firm on palpation and measured approximately 3 cm in greatest diameter. By ultrasonography, a well-demarcated, hypoechoic 29-mm nodule was found. Fine-needle aspiration biopsy was performed, showing hyperplastic follicles in a background of minimal colloid. A left hemithyroidectomy was performed, showing a well-demarcated grayish nodule. Histologically, scattered follicular cell aggregates were seen in an amorphous eosinophilic background stroma that was interpreted as possible regressive changes in a long-standing nodular goiter. Microfollicles were also observed, and although the lesion had no well-defined capsule, a clear demarcation from the surrounding parenchyma was present. The stroma was markedly hypocellular and deeply eosinophilic with H&E staining (Figure 12.36). It surrounded small blood vessels as well as isolated follicles, and scattered adipocytes were also embedded in the matrix. The lesional thyrocytes were cuboidal, with no nuclear atypia and scant colloid production. Focally, they were arranged in trabeculae. The differential diagnosis for this lesion was thought to include regressive change in a nodular goiter, medullary carcinoma, hyalinizing trabecular tumor, and amyloid goiter. Results of CS and Sevier-Munger stains were negative, thus tending to exclude medullary carcinoma. Hyalinizing trabecular tumor was not favored because of a lack of intratrabecular hyalinization, which is regarded as the hallmark of this entity; the amorphous material in this case was

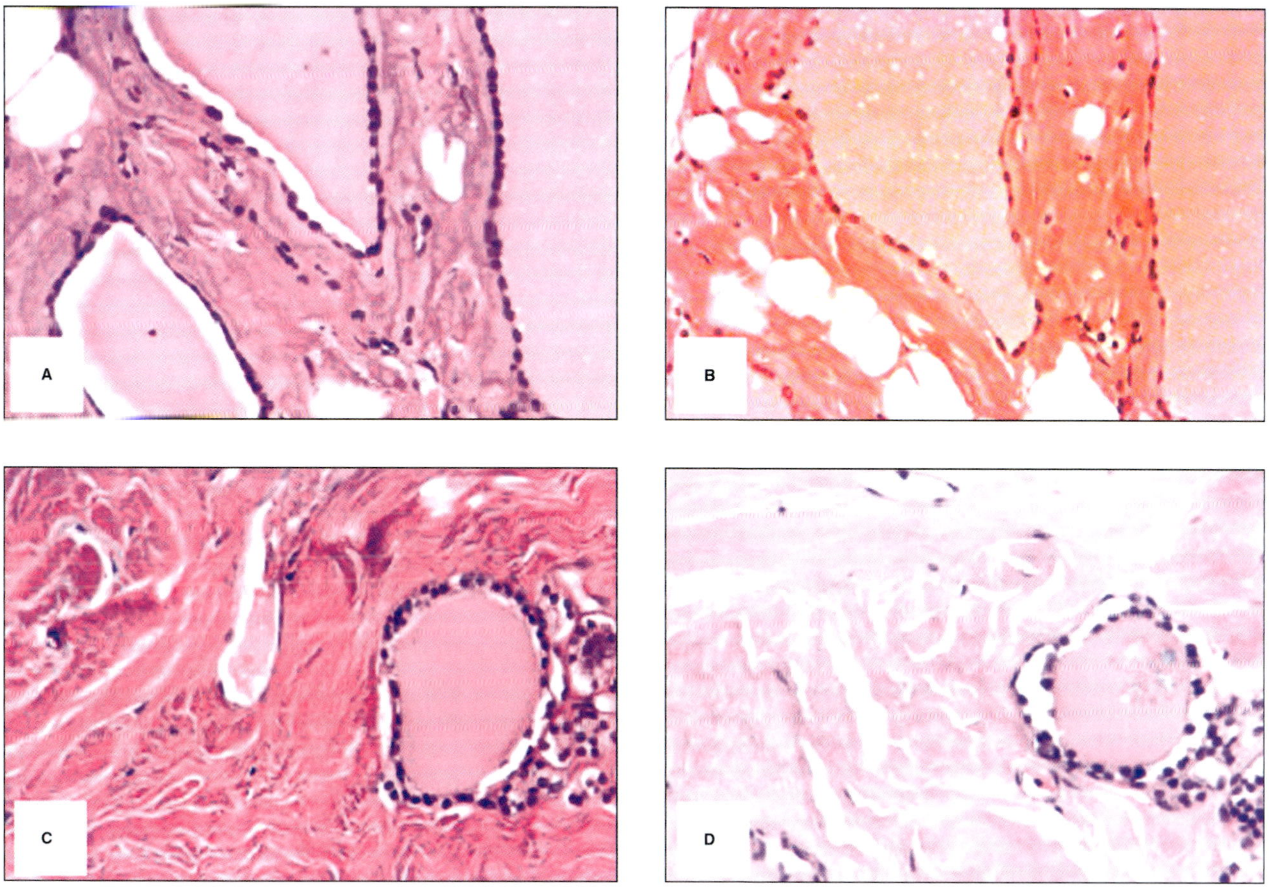

FIGURE 12.36: *Amyloid goiter, showing broad zones of amorphous material around hyperplastic follicular tissue (A), which is intensely positive with the Congo red stain (B). A control case of hyperplastic goiter with extensive regressive change shows collagen deposition instead of amyloid (C).*

extratrabecular. The favored interpretation was that of nodular thyroid hyperplasia with degenerative stromal change. These findings are common in areas of the world where goiters are endemic, such as northern Italy.

Nodular thyroid hyperplasia may undergo intralesional hemorrhage, followed by fibrous organization and partial loss of the follicular architecture. The diagnostic clues to this sequence are the presence of densely fibrotic hypocellular areas, in which fibrillar collagen is admixed with amorphous hyaline material. Rarely, the thyroid gland can be the seat of amyloid deposition, perhaps superimposed on a preexisting follicular proliferation, in the context of a systemic paraproteinemia. These cases of so-called amyloid goiter, in which intrathyroidal fat deposits may also be prominent, can be recognized by Congo red staining. In the present case, this technique rather surprisingly yielded strongly positive results, and the matrical material also demonstrated confirmatory green birefringence with polarization microscopy. Regressive zones in nodular goiters contain collagen—as demonstrable with the Masson trichrome technique—rather than amyloid. The absence of argyrophilia in the lesional cells again allowed one to exclude the possibility of medullary carcinoma with amyloid stroma. Subsequent investigation showed that the patient did have a circulating monoclonal paraprotein. ■

<u>CASE 2—</u> ***Parathyroid Carcinoma:*** A 76-year-old woman was found to have primary hyperparathyroidism by biochemical testing. A 3-cm nodule was subsequently found in the upper right parathyroid gland. It was resected, showing a reddish, well-demarcated 2.9-cm nodule weighing 4.4 g. The tumor was composed of chief cells arranged in cords and nests, separated by thick fibrous bands (Figure 12.37). A thin peripheral lesional capsule was present with focal penetration by the tumor; however, there was no transcapsular invasion or extraglandular involvement. The mitotic count was 1 per 10 high-power (400×) microscopic fields, and no significant nuclear anaplasia was appreciable. However, because of the size and weight of the lesion, as well as the presence of internal fibrous bands and questionable invasiveness, the possibility of parathyroid carcinoma was raised diagnostically.

In analogy with reported observations regarding anterior pituitary proliferations, the reticulin framework of normal and hyperplastic parathyroid glands shows regular investment of small groups of parathyroid cells, as shown with Snook's stain (Figure 12.37, inset). Conversely, malignant parathyroid tumors are characterized by disruption of the reticulin network, with Snook positivity that is generally restricted to perivascular areas. Large neoplastic islands are devoid of a reticulin network in parathyroid carcinoma. Thus, reticulin staining may be useful for supporting a diagnosis of malignancy in parathyroid neoplasms with "borderline" pathologic features.

Two years after excision of her tumor, the patient in this case is alive. However, she has increasing blood levels of parathyroid hormone, recurrent disease in the neck, and suspected pulmonary metastases. ■

<u>CASE 3—</u> ***Metastatic Carcinoma in the Adrenal:*** A 57-year-old man developed vague intermittent right-sided abdominal pain, which worsened over a 1-month period. There were no other symptoms, and physical examination was unremarkable. Computed tomography (CT) of the abdomen and pelvis demonstrated a mass in the right adrenal that measured 3 cm in greatest dimension. A laparotomy and adrenalectomy were performed, yielding the specimen shown in Figure 12.38. Microscopically, the lesion was obviously malignant on cytomorphologic grounds, with geographic zones of necrosis (Figures 12.39 and 12.40). An oil red O stain done on fresh tumor tissue was minimally positive, whereas a PAS stain showed

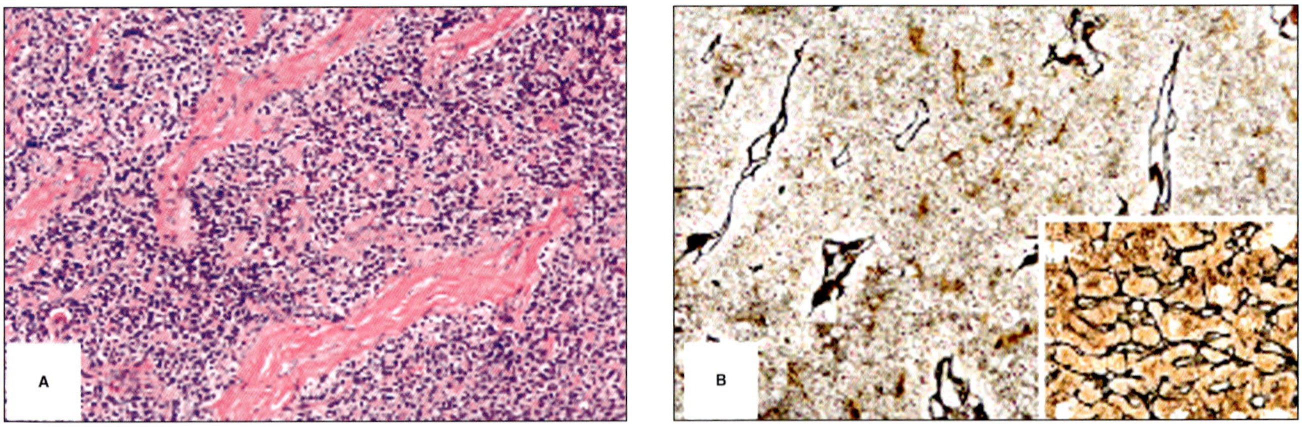

FIGURE 12.37: This parathyroid tumor comprises chief cells that are arranged in cords and nests, separated by thick fibrous bands (A). A parallel section stained with the Snook technique shows disruption of the reticulin network; large neoplastic islands are devoid of reticulin (B). This pattern contrasts with that seen in normal parathyroid glands, in which a fine reticulin network regularly surrounds groups of chief cells (B, inset).

the multifocal presence of intracytoplasmic reactivity that remained after diastase digestion. The biotin method of Bussolati et al. (46) was negative in the neoplastic cells. Based on these findings, a diagnosis of metastatic poorly differentiated adenocarcinoma was made. Additional history revealed that the patient had a 70 pack-year history of cigarette smoking; subsequent CT of the thorax showed

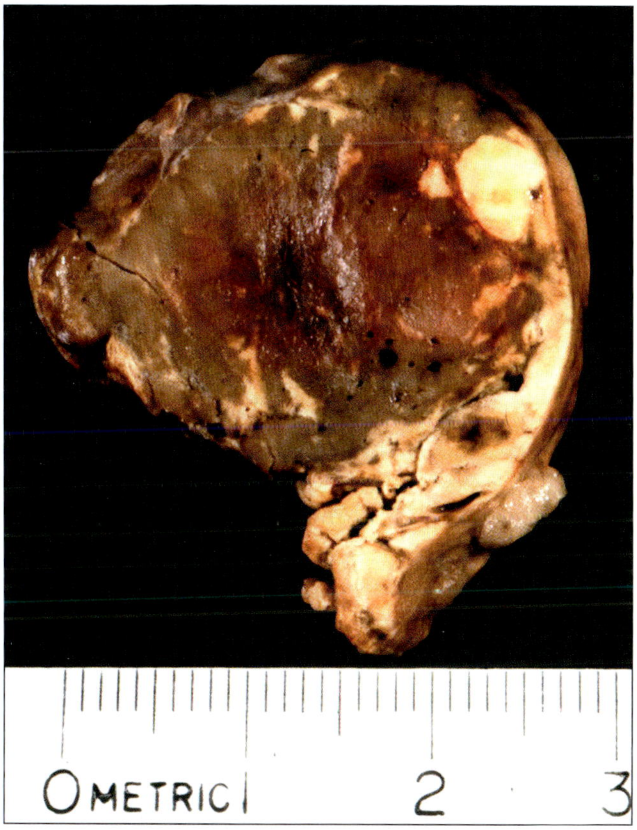

FIGURE 12.38: Gross photograph of a multinodular adrenal tumor, taken from a 57-year-old man.

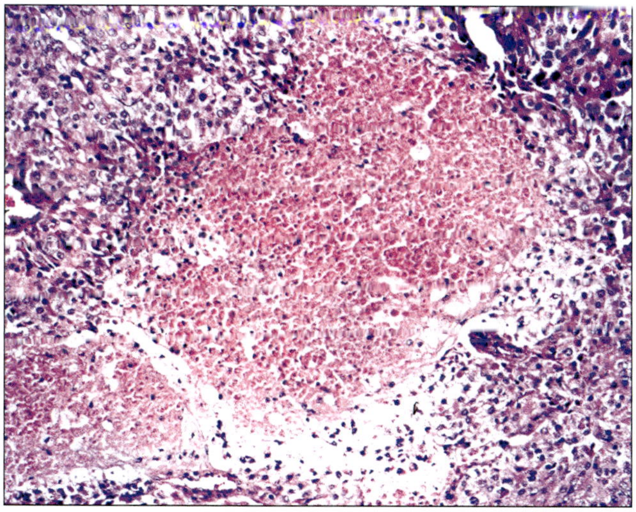

FIGURE 12.39: The lesion shown in the previous figure manifests broad zones of necrosis.

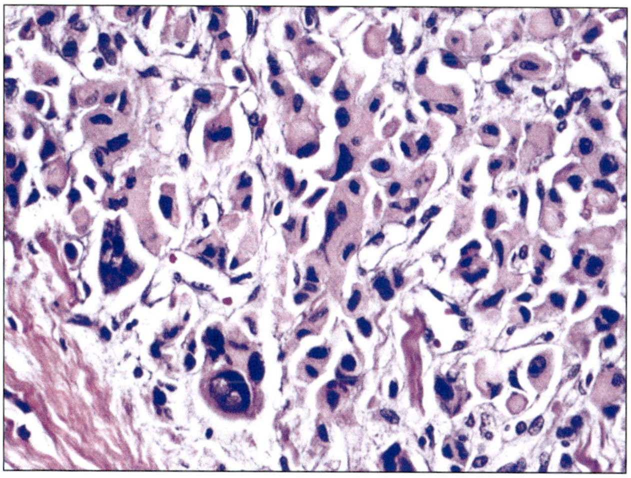

FIGURE 12.40: Overt cytologic atypia is also present in the adrenal mass. The cytoplasm of the tumor cells was multifocally reactive with the digested PAS procedure, but the lesion failed to label with the oil red O method. A final diagnosis of metastatic carcinoma was made.

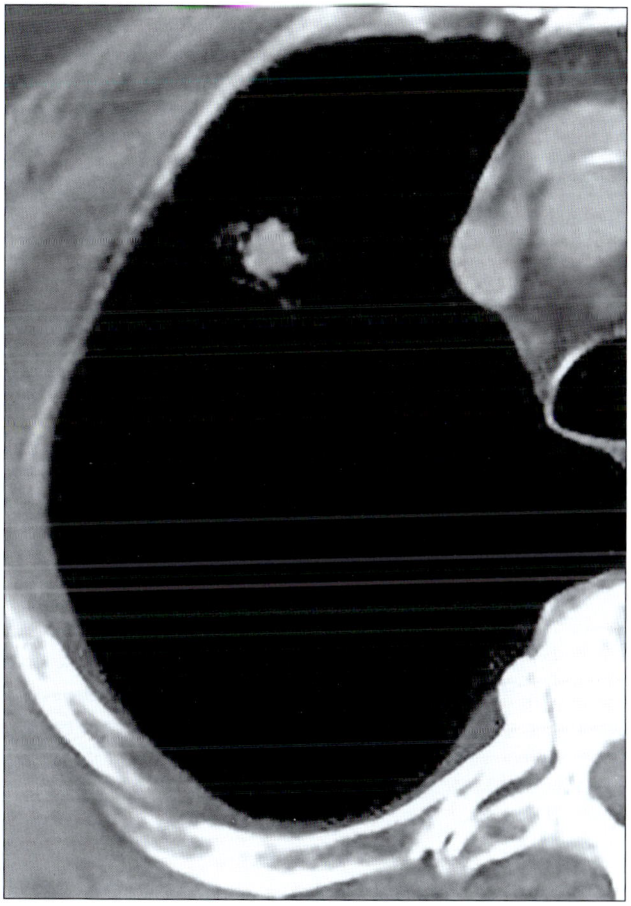

FIGURE 12.41: A computed tomogram of the chest revealed a 9-mm spiculated lesion in the right lung field, representing the primary tumor in this case.

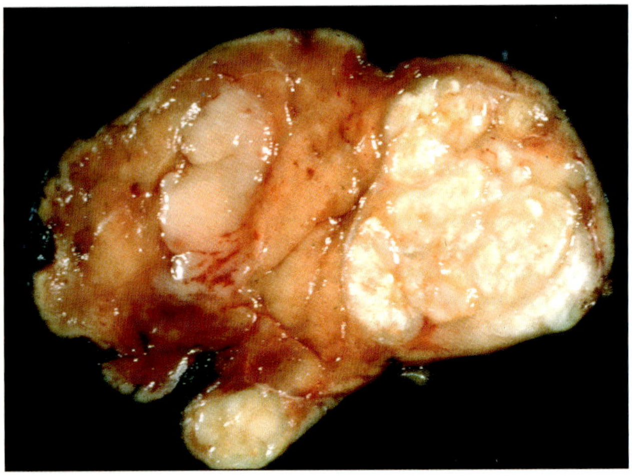

FIGURE 12.42: This gross photograph of a total thyroidectomy is from the case of a 72-year-old woman with a recently rapid growth in an enlarged thyroid of long standing.

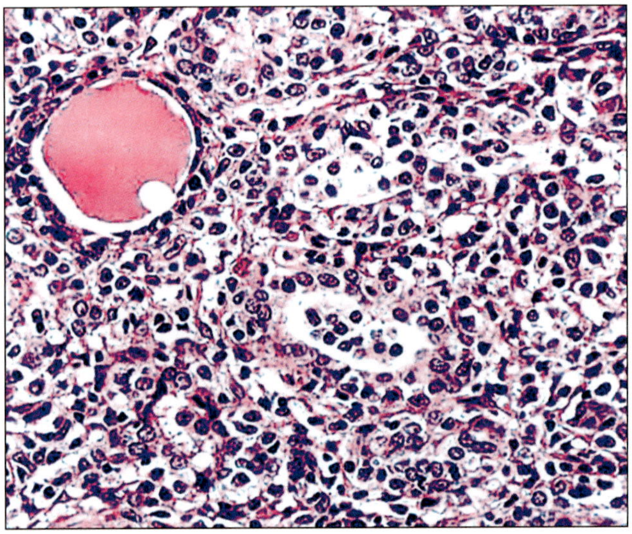

FIGURE 12.43: Microscopy of the lesion showed an undifferentiated polygonal cell tumor that effaced much of the glandular architecture.

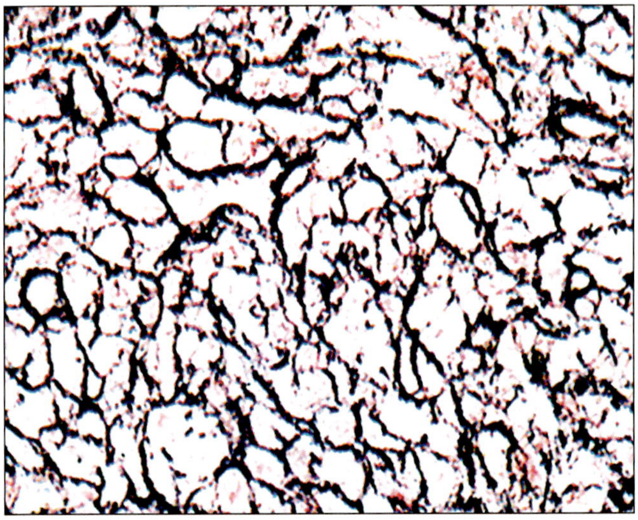

FIGURE 12.44: Snook's method demonstrates investment of individual tumor cells by reticulin.

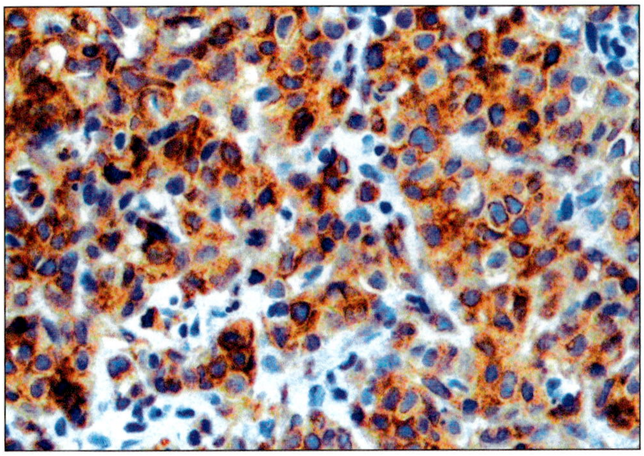

FIGURE 12.45: The neoplastic cells show binding to Ricinus communis *lectin, but they failed to label with other agglutinins including* Ulex europaeus I, Helix pomatia, Dolichos biflorus, Arachis hypogaea, *and* Glycine max. *The aggregate histochemical profile of the tumor supports the diagnosis of non-Hodgkin lymphoma.*

a peripheral spiculated 9-mm mass in the right lung field (Figure 12.41). ■

CASE 4— ***Malignant Lymphoma of the Thyroid, Ex Hashimoto's Thyroiditis:*** A 72-year-old woman had had a slowly enlarging goiter for several years, accompanied by gradually increasing hypothyroidism; she was given exogenous thyroxine. Serum antibodies to thyroglobulin were detected. Then, abruptly over a 2-month period, the patient's thyromegaly rapidly increased. A fine-needle aspiration biopsy of the thyroid was obtained, showing vaguely cohesive and atypical cells that were diagnostic of malignancy. A specific tumor lineage could not be determined. Because of worries over the possibility of anaplastic carcinoma, a total thyroidectomy was performed (Figure 12.42). Sections of the gland showed a large cell malignant tumor that effaced much of the thyroid parenchyma (Figure 12.43), with interspersed foci of Hashimoto's thyroiditis. The tumor cells were nonreactive with the PAS and mucicarmine methods. Snook's stain demonstrated investment of most individual tumor cells by reticulin (Figure 12.44). Lectin histochemical labeling showed tumor positivity with *Ricinus communis* (Figure 12.45) but not with *U. europaeus*, *H. pomatia*, *Dolichos biflorus*, peanut, lectin, or soybean agglutinin. A diagnosis of non-Hodgkin's lymphoma was made, arising in a background of lymphocytic thyroiditis. ■

CASE 5— ***Metastatic Renal Cell Carcinoma in the Thyroid:*** A 49-year-old man noticed a mass in his left neck while shaving; he was otherwise well. Physical examination subsequently showed that the lesion was in the left thyroid lobe, and, on ultrasonography, it measured 3.5 cm in greatest dimension. A fine-needle aspiration biopsy of the mass was obtained (Figure 12.46) and was interpreted as a "highly atypical epithelial lesion." Accordingly, a partial thyroidectomy was done, yielding the specimen shown in Figure 12.47. Histologic examination of the tumor demonstrated a composition by large epithelioid cells, arranged in tubules, with clear cytoplasm and foci of hemorrhage (Figures 12.48 and 12.49). Oil red O and PAS stains were intensely positive in the neoplastic cells (Figures 12.50 and 12.51), and diastase abrogated the PAS reaction. Lectin histochemistry

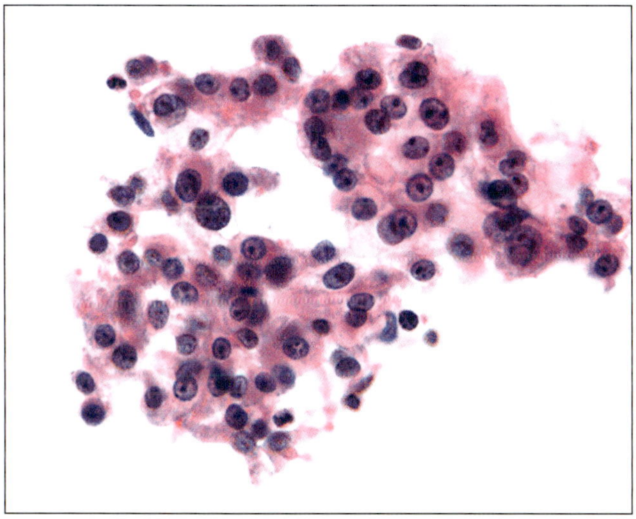

FIGURE 12.46: *Fine-needle aspiration biopsy of an intrathyroidal mass in a 49-year-old man. The specimen shows obviously atypical and clustered epithelioid cells with prominent nucleoli, consistent with a diagnosis of carcinoma.*

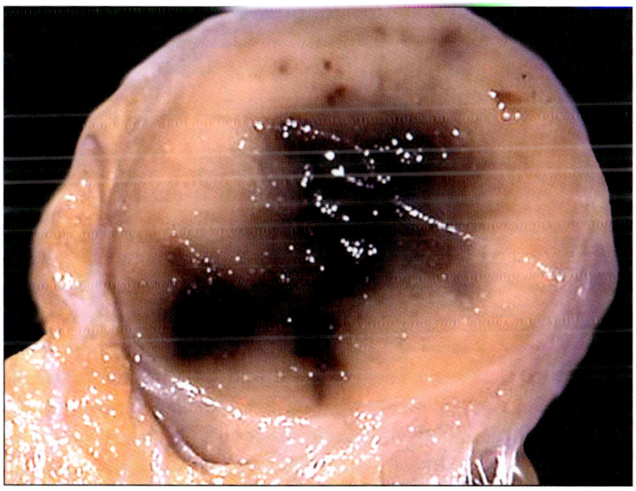

FIGURE 12.47: *This gross photograph is from an ensuing partial thyroidectomy. It shows a large nodular mass with central hemorrhage and necrosis.*

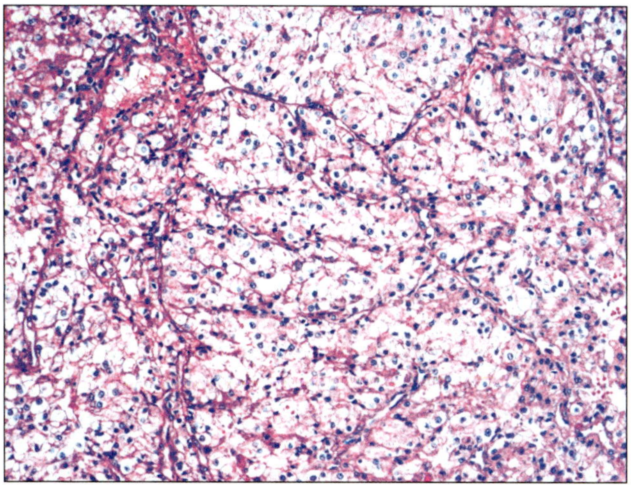

FIGURE 12.48: *Histologically, a clear cell neoplasm with obvious nuclear atypia was present, as shown in this H&E stain.*

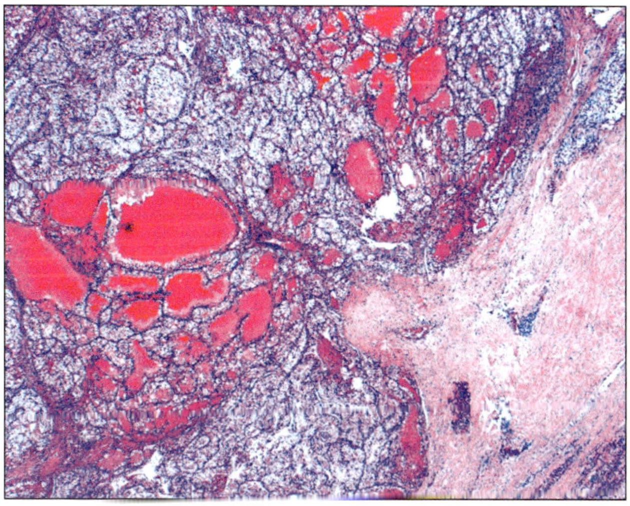

FIGURE 12.49: *The tumor manifested hemorrhage into constituent neoplastic tubules (H&E).*

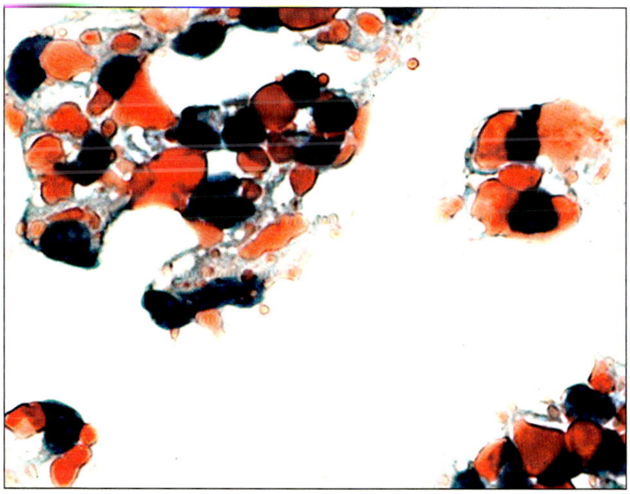

FIGURE 12.50: *Touch preparations of fresh tumor tissue were utilized to do an oil red O stain, as shown here. The neoplastic cells were intensely positive.*

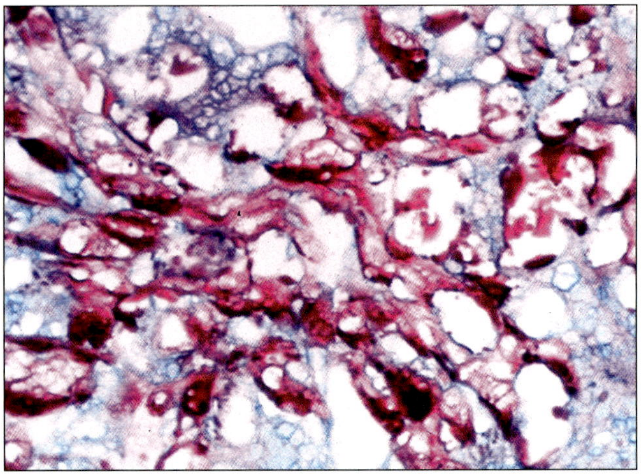

FIGURE 12.51:* *An undigested PAS stain was intensely reactive with the tumor cells, and positivity was abrogated by prior treatment with diastase.

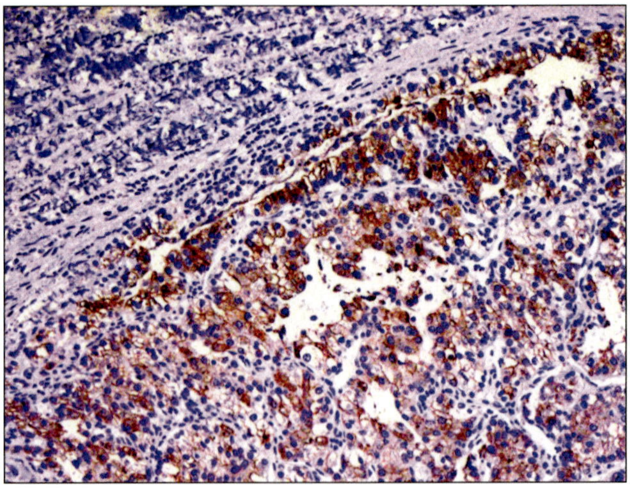

FIGURE 12.52:* *The intrathyroidal tumor bound to* Lotus tetragonolobus *lectin, but not to* Ulex europaeus I, Dolichos biflorus, *or* Glycine max. *The histochemical profile supported a diagnosis of metastatic RCC presenting as a thyroid mass, and a 5-cm mass was subsequently found radiographically in the right kidney.

showed diffuse reactivity with *Lotus tetragonolobus* (Figure 12.52), but there was no labeling by *U. europaeus*, *D. biflorus*, and soybean agglutinin (75). Given all the available information, a diagnosis was made of metastatic clear cell carcinoma of probable renal origin. CT of the abdomen showed a 5-cm mass in the right kidney. ▪

REFERENCES

1. Feyrter F: *Uber diffuse endocrine epitheliale organe*. Johan Ambrosius Barth, Leipzig, Germany, 1938.
2. Champaneria MC, Modlin IM, Eick GN: Friedrich Feyrter: a precise intellect in a diffuse system. *Neuroendocrinology* 2006; 83:394–404.
3. Pearse AGE: The cytochemistry and ultrastructure of polypeptide hormone-producing cells of the APUD series and the embryologic, physiologic and pathologic implication of the concept. *J Histochem Cytochem* 1969; 17:303–313.
4. Monga G, Bonfanti S, Bussolati G: Cholinesterase rich cells in the rat fundic mucosa. A light and electron microscopical investigation. *Histochemical J* 1974; 6:559–565.
5. Bussolati G, Volante M, Papotti M: Classic and recent special stains used in differential diagnosis of endocrine tumors. *Endocrine Pathol* 2001; 12:379–387.
6. Portela-Gomes GM, Grimelius L: Identification and characterisation of the entero-chromaffin cells with different staining techniques. *Acta Histochem* 1986; 79:161–174.
7. Hakanson R, Owman C, Sjoberg N, Sporrong B: Amine mechanisms in enterochromaffin and enterochromaffin-like cells of gastric mucosa in various mammals. *Histochemie* 1970; 21:189–220.
8. Bussolati G: Cellule endocrine producenti polipeptidi e amine biogene. *Riv Istoch Norm Pat* 1974; 18:95–118.
9. Lundqvist M, Arnberg H, Candell J, Malmgren M, Wilander E, Grimelius L, Oberg K: Silver stains for the identification of neuroendocrine cells. A study of the chemical background. *Histochem J* 1990; 22:615–623.
10. Smith DM Jr., Haggitt RC: A comparative study of generic stains for carcinoid secretory granules. *Am J Surg Pathol* 1983; 7:61–68.
11. Taxy JB, Mendelsohn G, Gupta PK: Carcinoid tumor of the rectum. Silver reactions, fluorescence and serotonin content of cytoplasmic granules. *Am J Clin Pathol* 1980; 74:791–795.
12. Wilander E, Norheim I, Oberg K: Application of silver stains to cytological specimens of neuroendocrine tumors metastatic to the liver. *Acta Cytol* 1985; 19:1053–1057.
13. Ascoli V, Newman GA, Kline TS: Grimelius stain for cytodiagnosis of carcinoid tumor. *Diagn Cytopathol* 1986; 2:157–159.
14. Grimelius L: A silver nitrate stain for alpha-2 cells in human pancreatic islets. *Acta Soc Med Ups* 1968; 73:243–70.
15. Bussolati G, Papotti M, Sapino A, Gugliotta P, Ghiringhello B, Azzopardi JG: Endocrine markers in argyrophilic carcinomas of the breast. *Am J Surg Pathol* 1987; 11:248–256.
16. Wilander E, Junnti-Berggren L, Lundqvist M, Grimelius L: Staining of rat thyroid parafollicular C-cells with the Sevier-Munger silver technique. *Acta Pathol Microbiol Scand* 1980; 88:339–340.
17. Wilander E, Lundqvist M, Westermark P, Grimelius L: Immunocytochemical study of argyrophil (Sevier-Munger) endocrine cells of adult human pancreas. *Histochemistry* 1980; 68:95–98.
18. Frigerio B, Capella C, Wilander E, Grimelius L: Argyrophil reaction in parathyroid glands: a light and electron microscopic study. *Acta Pathol Microbiol Immunol Scand* 1982;90: 323–326.
19. Loy TS, Darkow GV, Quesenberry JT: Immunostaining in the diagnosis of pulmonary neuroendocrine carcinomas: an immunohistochemical study with ultrastructural correlations. *Am J Surg Pathol* 1995; 19:173–182.
20. Burke AP, Sobin LH, Federspiel BH, Shekitka KM, Helwig EB: Goblet cell carcinoids and related tumors of the vermiform appendix. *Am J Clin Pathol* 1990; 94:27–35.

21. Lloyd RV: Morphologic methods. In: Kovacs K, Asa SL (eds.): *Functional Endocrine Pathology*, Blackwell Scientific publications, Boston, 1991, pp. 94–108.
22. Larsson LI, Sundler F, Hakanson R: Fluorescamine as a histochemical reagent: demonstration of polypeptide hormone secreting cells. *Histochemistry* 1975; 44:245–251.
23. Horvath E: Pituitary hyperplasia. *Pathol Res Pract* 1988; 183:623–625.
24. Volante M, Allia E, Fulcheri E, Cassoni P, Ghigo E, Muccioli G, Papotti M: Ghrelin in fetal thyroid and follicular tumors and cell lines: expression and effects on tumor growth. *Am J Pathol* 2003; 162:645–654.
25. Aaberg TM Jr., Kay M, Sternau L: Metastatic tumors to the pituitary. *Am J Ophthalmol* 1995; 119:779–785.
26. Komninos J, Vlassopoulou V, Protopapa D, Korflas S: Tumors metastatic to the pituitary gland: case report and literature review. *J Clin Endocrinol Metab* 2004; 89:574–580.
27. Van de Velde A, Wassenaar H, Strubbe A, Janssen P, Verhaert G: Metastatic breast cancer presenting with diabetes insipidus. *JBR-BTR* 2000; 83:68–70.
28. Rao SR, Rao RS: Pituitary metastases in carcinoma of the breast. *J Postgrad Med* 2001; 47:135–136.
29. Basaria S, Westra WH, Brem H, Salvatori R: Metastatic renal cell carcinoma to the pituitary presenting with hyperprolactinemia. *J Endocrinol Invest* 2004; 27:471–474.
30. Metivier AC, Mennecier B, Grunenberger F, et al.: Pituitary metastases from lung cancer: five cases. *Rev Pneumol Clin* 2006; 62:231–236.
31. Essais O, Jomaa R, Bougerra R, et al.: Hypothalamic-pituitary Langerhans cell histiocytosis in adults. *Presse Med* 2006; 35:1263–1266.
32. Ouyang DL, Roberts BK, Gibbs IC, Katznelson L: Isolated Langerhans cell histiocytosis in an adult with central diabetes insipidus: case report and review of literature. *Endocr Pract* 2006; 12:660–663.
33. Scheithauer BW: Pathology of the pituitary and sellar region, exclusive of pituitary adenoma. *Pathol Annu* 1985, 20(Part 1):67–155.
34. McLelland J: A comparison study of two methods of peanut agglutinin staining with S100 immunostaining in 29 cases of histiocytosis X. *Arch Pathol Lab Med* 1991; 115:107–108.
35. Koide O, Iwai S: An ultrastructural study of germinoma cells. *Acta Pathol Jpn* 1981; 31:755–766.
36. Khan A, Nosè V: Pathology of the thyroid. In: Lloyd RV (ed.): *Endocrine Pathology*, Humana Press, Totowa, New Jersey, 2004, pp. 153–190.
37. Puchtler H, Sweat F: Congo red as a stain for fluorescence microscopy of amyloid. *J Histochem Cytochem* 1965; 13:693–694.
38. Elghetany MT, Saleem A: Methods for staining amyloid in tissues: a review. *Stain Technol* 1988; 63:201–212.
39. Elghetany MT, Saleem A, Barr K: The Congo red stain revisited. *Ann Clin Lab Sci* 1989; 19:190–195.
40. Bely M, Makovitzky J: Sensitivity and specificity of Congo red staining according to Romhanyi. Comparison with Puchtler's or Bennhold's methods. *Acta Histochem* 2006; 108:175–180.
41. Stephenson TJ, Griffiths DW, Mills PM: Comparison of Ulex europaeus I lectin binding and factor VIII-related antigen as markers of vascular endothelium in follicular carcinoma of the thyroid. *Histopathology* 1986; 10:251–260.
42. Gonzales-Campora R, Sanchez Gallego F, Martin Lacave I, Mora Marin J, Montero Linares C, Galera-Davidson H: Lectin histochemistry of the thyroid gland. *Cancer* 1988; 62:2354–2362.
43. Wick MR, Sawyer MD: Antigenic alterations in autoimmune thyroid diseases. Observations and hypotheses. *Arch Pathol Lab Med* 1989; 113:77–81.
44. Babal P, Janega P, Cerna A, Kholova I, Brabencova E: Neoplastic transformation of the thyroid gland is accompanied by changes in cellular sialylation. *Acta Histochem* 2006; 108:133–140.
45. Janega P, Cerna A, Kholova I, Brabencova E, Babal P: Sialic acid expression in autoimmune thyroiditis. *Acta Histochem* 2002; 104:343–347.
46. Bussolati G, Gugliotta P, Volante M, Pace M, Papotti M: Retrieved endogenous biotin: a novel marker and a potential pitfall in diagnostic immunohistochemistry. *Histopathology* 1997; 31:400–407.
47. Papotti M, Torchio B, Grassi L, Favero A, Bussolati G: Poorly differentiated Hurthle cell carcinomas of the thyroid. *Am J Surg Pathol* 1996; 20:686–694.
48. Carcangiu ML, Zampi G, Rosai J: Poorly-differentiated ("insular") thyroid carcinoma: a reinterpretation of Langerhans "wucherdne Struma." *Am J Surg Pathol* 1984; 8:655–668.
49. Mendelsohn G, Baylin SB, Bigner SH, Wells SA Jr., Eggleston JC: Anaplastic variants of medullary thyroid carcinoma: a light-microscopic and immunohistochemical study. *Am J Surg Pathol* 1980; 4:333–341.
50. Mambo NC, Irwin SM: Anaplastic small-cell neoplasms of the thyroid: an immunoperoxidase study. *Hum Pathol* 1984; 15:55–60.
51. Tobler A, Maurer R, Hedinger CE: Undifferentiated thyroid tumors of diffuse small cell type: histological and immunohistochemical evidence for their lymphomatous nature. *Virchows Arch A Pathol Anat Histopathol* 1984; 404:117–126.
52. Anscombe AM, Wright DH: Primary malignant lymphoma of the thyroid—a tumor of mucosa-associated lymphoid tissue: review of seventy-six cases. *Histopathology* 1985; 9:81–97.
53. Sasano H, Rojas M, Silverberg SG: Analysis of lectin binding in benign and malignant thyroid nodules. *Arch Pathol Lab Med* 1989; 113:186–189.
54. Ree HJ, Hsu SM: Lectin histochemistry of malignant tumors. I. Peanut agglutinin (PNA) receptors in follicular lymphoma and follicular hyperplasia: an immunohistochemical study. *Cancer* 1983; 51:1631–1638.
55. Ree HJ, Raine L, Crowley JP: Lectin binding patterns in diffuse large cell lymphoma. *Cancer* 1983; 52:2089–2099.
56. Lloyd RV, Douglas BR, Young WF: *Atlas of Non-Tumor Pathology. Endocrine Diseases*, AFIP Publications, Washington, D.C., 2002, pp. 49, 108, 149, 191.
57. Weymouth RJ, Sheridan MN: Fine structure of human parathyroid glands: normal and pathological. *Acta Endocrinol* 1966; 53:529–546.
58. Szakacs JE, Bryant M: Ultrastructure of parathyroid adenomas. *Ann Clin Lab Sci* 1980; 10:13–25.
59. Halbauer M, Crepinko I, Tomc-Brzac H, Simonovic I: Fine needle aspiration cytology in the preoperative diagnosis of

ultrasonically-enlarged parathyroid glands. *Acta Cytol* 1991; 35:728–735.

60. Weymouth RJ, Baker BL: The presence of argyrophilic granules in the parenchymal cells of the parathyroid glands. *Anat Rec* 1954; 119:519–527.
61. Bussolati G, Capella C, Vassallo G, Solcia E: Histochemical and ultrastructural studies on pancreatic A cells. Evidence for glucagon and non-glucagon components of the alpha granule. *Diabetologia* 1971; 7:181–188.
62. Bordi C, Bussolati G: Immunofluorescence, histochemical and ultrastructural studies for the detection of multiple endocrine polypeptide tumors of the pancreas. *Virchows Arch B Cell Path* 1974; 17:13–27.
63. Heitz PU, Kasper M, Polak JM, Kloppel G: Pathology of the endocrine pancreas. *J Histochem Cytochem* 1979; 27:1401–1402.
64. Castino F, Bussolati G: Thiosulphation for the histochemical demonstration of protein-bound sulphydryl and disulphide groups. *Histochemistry* 1974; 39:93–96.
65. Lyon H, Prento P: Aldehyde fuchsin staining of pancreatic B cells. Reproducible high-contrast staining of formalin-fixed and paraffin-embedded material. *Histochem J* 1980; 12:97–105.
66. Cole LJ, Nettleton GS: A histochemical investigation of the mechanism of aldehyde fuchsin staining of pancreatic B-cell granules. *Histochem J* 1988; 20:635–641.
67. Giove S, Bussolati G: Double simultaneous staining of B and argyrophil cells of the pancreatic islets: a sequential thiosulphation-aldehyde fuchsin and Grimelius silver procedure. *Basic Appl Histochem* 1980; 24:75–78.
68. Varela-Duran J, Bohm N, Diaz-Flores L, et al.: Neuroblastoma: a study of the clinicopathological features influencing prognosis, based on the analysis of 54 cases. *Histol Histopathol* 1994; 9:583–590.
69. Genc H, Haciyanli M, Haciyanli SG, et al.: An adult adrenal neuroblastoma: a case report. *Acta Chir Belg* 2005; 105:673–676.
70. Weiss LM, Medeiros LJ, Vickery AL Jr.: Pathologic features of prognostic significance in adrenocortical carcinoma. *Am J Surg Pathol* 1989; 13:202–206.
71. Goncalves FT, Feibelmann TC, Mendes CM, et al.: Primary pigmented nodular adrenocortical disease associated with Carney complex: case report and literature review. *Sao Paulo Med J* 2006; 124:336–339.
72. Foucar E, Dehner LP: Renal cell carcinoma occurring with contralateral adrenal metastasis: a clinical and pathological trap. *Arch Surg* 1979; 114:959–963.
73. Solcia E, Kloppel G, Sobin LH: *World Health Organization International Histological Classification of Tumors. Histological Typing of Endocrine Tumors*, Springer, Berlin, 2000, pp. 56–68.
74. Volante M, Rindi G, Papotti M: The grey zone between pure (neuro)endocrine and non-(neuro)endocrine tumours: a comment on concepts and classification of mixed exocrine-endocrine neoplasms. *Virchows Arch* 2006; 449:499–506.
75. Ulrich W, Horvat R, Krisch K: Lectin histochemistry of kidney tumors and its pathomorphological relevance. *Histopathology* 1985; 9:1037–1050.

CHAPTER 13

Diagnostic Histochemistry in Dermatopathology

Mark R. Wick, M.D., and
James W. Patterson, M.D.

As true of the lungs, the skin represents a major interface with the ambient environment, and this fact likely explains the great number and scope of diseases one sees in dermatology. These include infections, noninfectious inflammatory conditions (with many potential causes), metabolic and genetic diseases, degenerative disorders, and neoplasms. With regard to the integration of histochemistry into dermatopathology, this branch of adjunctive investigation continues to be widely used and is consistently helpful in well-defined diagnostic settings. The following chapter summarizes the major applications of histochemical methods in studying diseases of the skin.

NONNEOPLASTIC DISEASES OF THE SKIN

Dermatitides and Dermatoses

For practical purposes, dermatitides can be divided nosologically into several discrete categories. These include vesiculopustular, spongiotic, lichenoid, psoriasiform, dermal lymphocyte–rich, granulomatous and necrobiotic, bullous acantholytic, vasculitic, adnexocentric inflammatory, and primarily panniculitic processes. Respective examples of disorders in each of those groups are listed in Table 13.1. Of course, the histologic details of these conditions will not be reviewed here because several textbooks of dermatopathology do so thoroughly (1–5). However, we will discuss reasonable histochemical studies that can be ordered in the evaluation of such diseases, touching upon the differential diagnostic reasons for them.

Vesiculopustular Dermatitis: Vesiculopustules in the epidermis may be the sign of bacterial or fungal infections with staphylococci, streptococci, or *Candida*. Alternatively, several noninfectious or parainfectious diseases (e.g., pemphigus variants, Ritter's disease, Sneddon-Wilkinson syndrome, erythema toxicum neonatorum) produce similar lesions in the skin. Tissue Gram stains, using the Brown-Hopps or Brown-Brenn methods, are often effective at demonstrating bacteria, and the silver methenamine (SM) procedure is helpful in labeling fungi. For a more complete list of histochemical methods used to recognize specific infectious organisms, see Table 13.2.

Spongiotic and Psoriasiform Dermatitides: Chronic spongiotic dermatitides, exemplified by eczema, contact dermatitis, pityriasis rosea, and others, merge morphologically with psoriasiform processes, most notably treated psoriasis. These disorders may be confused with chronic superficial dermatophytoses because of their shared microscopic features. Accordingly, and especially when neutrophils are present in the stratum corneum (a possible sign of infection), it is wise to obtain an SM or alcian blue–periodic acid-Schiff (ABPAS) stain (Figure 13.1). Both methods are sensitive techniques for labeling fungal organisms, and whether one chooses one or the other is a matter of personal preference (6).

Lichenoid Dermatitides: Lichenoid and interface dermatitides encompass several disease entities, including lupus erythematosus (LE), dermatomyositis, lichen sclerosus et atrophicus, lichen simplex chronicus, lymphoplasmacytic mucositis-dermatitis, syphilis, Lyme disease, lichen planus, and others. Depending upon their stages of development, these may all resemble one another substantially at a morphologic level.

LE, dermatomyositis, and lichen sclerosus often demonstrate thickening of the epidermal basement membrane (EBM), along with epidermal atrophy. This change in the EBM can be visualized by the use of the periodic acid-Schiff (PAS) stain, with prior diastase digestion (PAS-D) (7,8) (Figure 13.2). Because glycoproteinaceous material accumulates in it as the EBM is damaged over time, the membrane shows noticeable widening when compared with that of nonlesional skin using the PAS-D method. Other lichenoid processes fail to demonstrate such a change.

Another group of histochemical techniques that aid in the recognition of LE and dermatomyositis are those that detect deposits of acid mucopolysaccharide (stromal mucin) in the dermis (Figure 13.3). Both the ABPAS and colloidal

TABLE 13.1: Selected Dermatitides and Dermatoses

Type of Dermatitis	Examples

Vesiculopustular dermatitis	Bacterial or fungal infections
	  Impetigo
	  Dermatophytosis
	  Candidiasis
	Pemphigus
	Staphylococcal scalded skin syndrome
	Erythema toxicum neonatorum

Spongiotic and psoriasiform dermatitis	Eczematous dermatoses e.g. contact dermatitis Pityriasis rosea Psoriasis (including partly treated) Superficial dermatophytosis
Lichenoid dermatitis	Lichen planus Lupus erythematosus Dermatomyositis Lichen sclerosus et atrophicus Lymphoplasmacytic mucositis-dermatitis Syphilis
Dermal lymphocyte-rich infiltrates	Infection (syphilis, Lyme disease) Lupus erythematosus Cutaneous lymphoid hyperplasia Granuloma annulare (early) Pityriasis lichenoides Annular erythemas (e.g. erythema annulare centrifugum)
Granulomatous and histiocytic infiltrates	Sarcoidosis Actinic granuloma Granuloma annulare Necrobiosis lipoidica Rheumatoid nodule Interstitial granulomatous dermatitis Foreign body granulomas (including gout) Bacterial and fungal infections
Subepidermal bullous disorders	Bullous pemphigoid Epidermolysis bullosa acquisita Porphyria cutanea tarda Bullous lupus erythematosus Bullous lichen planus Bullous lichen sclerosus Bullous mastocytosis Bullous cutaneous amyloidosis
Vasculopathic disorders	Leukocytoclastic vasculitis Granuloma faciale Erythema elevatum diutinum "Temporal" arteritis Lymphomatoid granulomatosis Wegener's granulomatosis Churg-Strauss syndrome Polyarteritis nodosa Buerger's disease Atrophie blanche Degos' disease Calciphylaxis Superficial thrombophlebitis Pigmented purpuric eruptions Cryoglobulinemia Macroglobulinemia Infectious vasculitides

TABLE 13.1: *Continued*

Type of Dermatitis	Examples
Adnexocentric dermatitides (folliculitides)	Inflammatory- non-scarring Alopecia areata Alopecia mucinosa Some forms of infectious folliculitis Inflammatory-scarring Lupus erythematosus Lichen planopilaris Central-centrifugal alopecia Some forms of infectious folliculitis Non-inflammatory-scarring Chronic trichotillomania Traction alopecia Non-inflammatory-non-scarring Anagen and telogen effluvium
Panniculitides	Erythema nodosum Cold-induced panniculitis Infectious panniculitis Calciphylaxis Alpha-1 antitrypsin deficiency Pancreatic panniculitis Erythema induratum Cryoglobulinemia Morphea and scleroderma Lupus profundus
Deposition disorders	Calcinosis cutis and calciphylaxis Argyria Hemochromatosis Amiodarone or clofazimine-induced pigmentation (lipofuscin) Ochronosis (homogentisic acid) Minocycline-induced pigmentation (iron and melanin-like reducing substance) Melanin pigmentation (melasma, post-Inflammatory hyperpigmentation, diffuse or nodular melanosis Mucin deposition (myxedema, pretibial myxedema, focal mucinosis, reticular erythematous mucinosis, scleredema, scleromyxedema, nephrogenic fibrosing dermopathy, mucopolysaccharidoses Lipoid proteinosis (lipid and protein) Colloid milium, nodular colloid degeneration (protein, amyloid-like) Amyloidosis
Defects in dermal connective tissue	Lichen sclerosus, morphea Cutis laxa Anetoderma (macular atrophy) Erythema ab igne Pseudoxanthoma elasticum Elastofibroma dorsi Collagenous & elastotic plaques of the hands and ears Perforating disorders Elastosis perforans serpiginosa Reactive perforating collagenosis Perforating folliculitis Kyrle's disease Perforating dermatosis of chronic renal failure

TABLE 13.2: Stains for Specific Infectious Organisms

Pathogen Group	Genus or Category of Agent	Representative Disease or Species	Stain
Bacteria	Conventional organisms		Brown and Hopps Brown and Brenn
	Klebsiella rhinoscleromatis	Rhinoscleroma	PAS Warthin-Starry
	Spirochetes	Syphilis (*Treponema pallidum*) Lyme disease (*Borrelia burgdorferi*)	Warthin-Starry Dieterle Steiner and Steiner
	Acid-fast bacilli	Tuberculosis (*M. tuberculosis*) Atypical mycobacerioses	Fite (Wade-Fite, Fite-Faraco) Ziehl-Neelsen
	Mycobacterium leprae	Hansen's disease	Fite
	Nocardia sp.	Nocardia infections, including actinomycotic mycetoma	
	Bartonella sp.	Bacillary angiomatosis Cat scratch disease	Warthin-Starry
	Corynebacterium sp.	Erythrasma Trichomycosis	Gram (+) PAS-D GMS Giemsa
	Calymmatobacterium granulomatis	Granuloma inguinale, Donovanosis	Warthin-Starry, GMS, Dieterle Giemsa
	Rickettsia	Rocky Mountain spotted fever (*R. rickettsii*)	Giemsa Macchiavello Castaneda Pinkerton
Fungi**	Dermatophytes	Dermatophytoses	PAS
	Candida sp.	Candidiasis	PAS-Alcian blue
	Organisms of common systemic mycoses	Systemic mycoses	GMS
	Cryptococcus sp.	Cryptococcosis* (*C. neoformans*)	Stains listed above Fontana-Masson† Mucicarmine
	Pigment-forming fungi	Chromomycosis Phaeohyphomycosis	Fontana-Masson Schmorl Ziehl-Neelsen Wade-Fite
Protozoa	*Leishmania* sp.	Leishmaniasis	Giemsa
	Toxoplasma sp.	Toxoplasmosis	Weigert iron hematoxylin (visceral leishmaniasis)
	Acanthamoeba sp.	Cutaneous amebiasis Amebic keratitis	PAS-D GMS
	Entamoeba sp.	Cutaneous amebiasis	PAS-D Mallory trichrome
Viruses	Herpesvirus	*Herpes simplex*	Lendrum (phloxine-tartrazine)
	Poxvirus	Varicella	
	Human papillomavirus (HPV)	*Cytomegalovirus* *Molluscum contagiosum* Verrucae due to HPV	Feulgen Methyl green-pyronine

PAS = periodic acid-Schiff, PAS-D = periodic acid-Schiff with diastase digestion, GMS = Gomori methenamine silver.

† *Cryptococcus neoformans* produces a melanin-like pigment, derived from dihydroxyphenylalanine, that stains positively with the Fontana-Masson method. This stain is not dependent on capsular mucopolysaccharide. Positive results can also be seen with *Sporothrix Schenckii* and some forms of *Coccidioides immitis*.

* Kwon-Chung KJ, Hill WB, Bennett JE. New, special stain for histopathological diagnosis of cryptococcosis. *J Clin Microbiol* 1981; 13: 383–387.

** Wheeler MH, Bell AA. Melanins and their importance in pathogenic fungi. *Curr Topics Med Mycol* 1987; 2: 338–387.

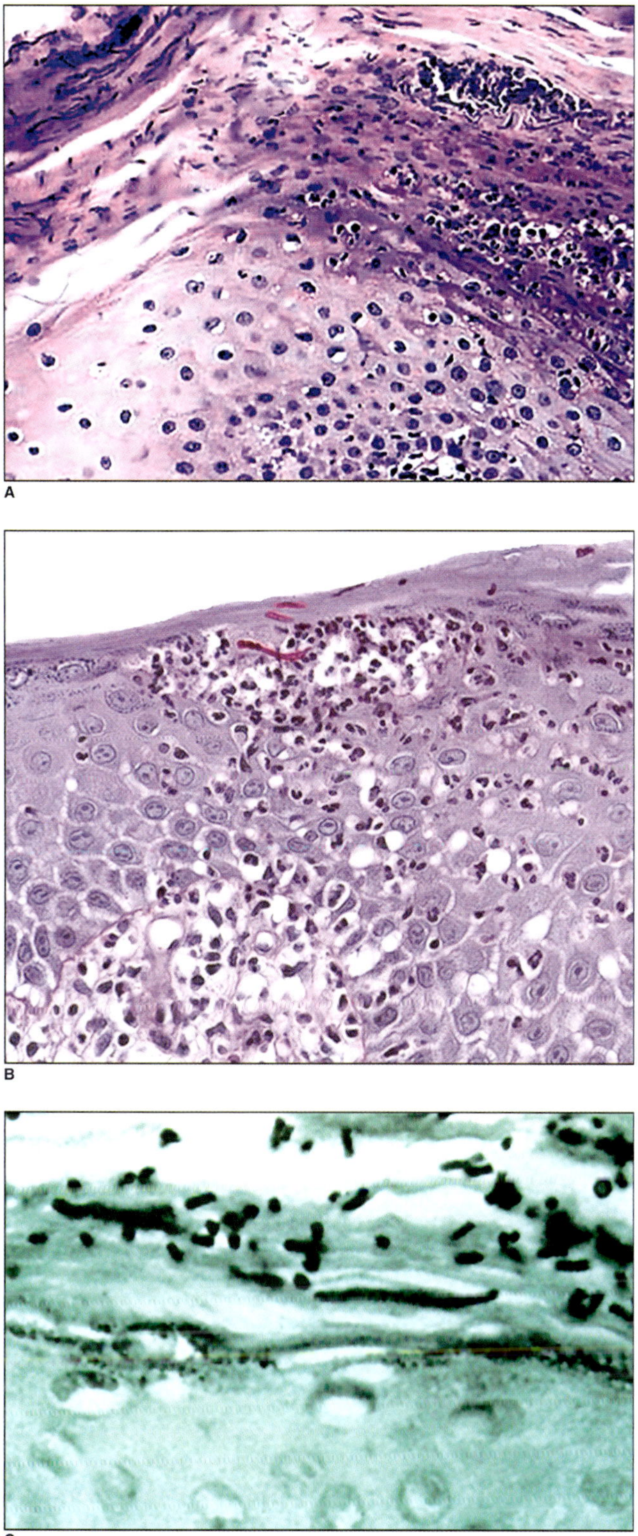

A

B

C

FIGURE 13.1: (A and B) Chronic spongiotic dermatitis with acanthosis and neutrophils in the stratum corneum. The differential diagnosis principally includes psoriasis and dermatophytosis. (C) The Gomori methenamine stain demonstrates fungal organisms in the superficial epidermis, establishing the diagnosis of dermatophytosis.

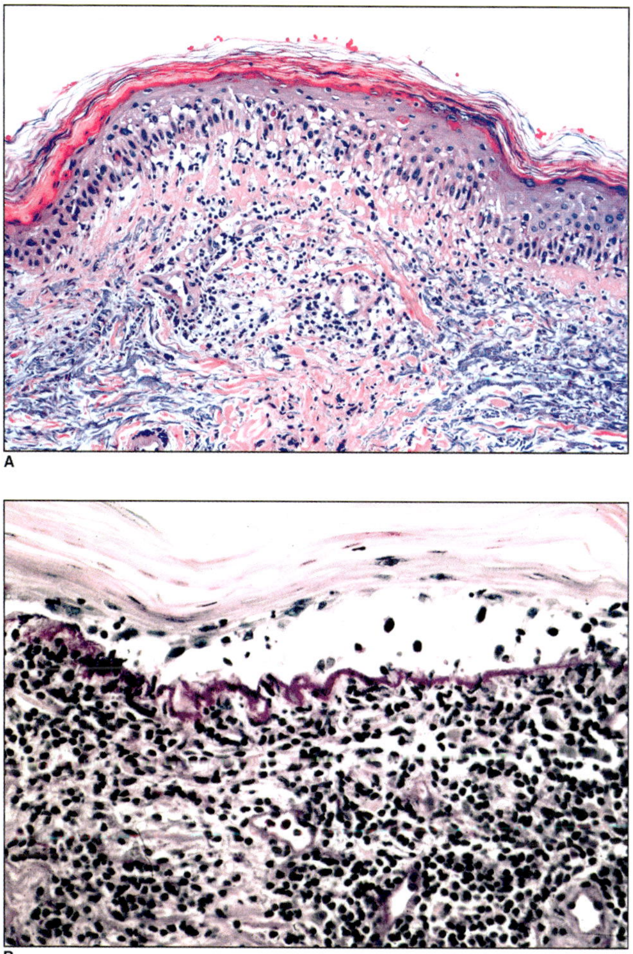

A

B

FIGURE 13.2: (A) LE, typified by effacement of rete ridges, basal epidermal vacuolization, and interface inflammation. (B) Chronic interface damage in LE is reflected by thickening of the EBM, as shown here with a PAS stain after diastase digestion.

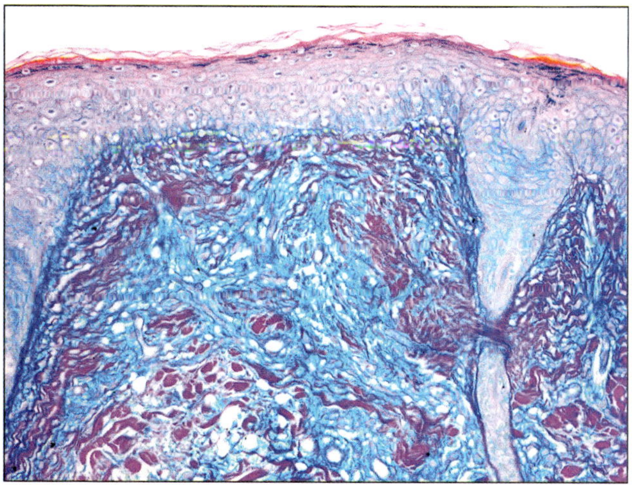

FIGURE 13.3: Dermal mucin deposition in LE can be shown with the CI technique (seen here) or the alcian blue method.

iron (CI) methods are effective in this regard, both demonstrating multifocal interstitial reactivity in the two disorders in question (9–11). In our opinion, there is no effective histochemical procedure to separate them one from another. Positivity for stromal mucin can be abrogated by pretreatment of tissue sections with hyaluronidase.

Lichen sclerosus manifests a peculiar homogenization of the upper dermal stroma, which is demonstrable with the Mallory trichrome (MT) technique (12) (Figure 13.4). In addition, this condition is associated with a diminution in elastic fibers in the same area of the corium, as shown with the Verhoeff–Van Gieson (VVG) procedure (13,14) (Figure 13.5). Lichen simplex chronicus—and a closely related disorder known as prurigo nodularis—are also characterized by increased deposition of collagen along the sides of epidermal rete ridges. This change can likewise be emphasized with the MT stain (15).

Lichen planus is typified by a dense zone of chronic inflammation that centers on the EBM. This infiltrate produces apoptotic change in basal keratinocytes, eventuating in degeneration of those cells and their encrustation by plasma proteins, producing so-called Civatte or cytoid bodies. These globular eosinophilic structures can be accentuated with the PAS-D stain (16).

Syphilis, Lyme disease, and other spirochetal/treponemal processes commonly produce a lichenoid pattern of chronic inflammation in the skin that may simulate most, if not all, of the other disorders considered here. Thus, procurement of histochemical stains for the causative organisms is prudent if the clinical features of a particular case raise the possibility of infection. Despite being classified as bacteria, treponemes do not label with the tissue Gram stain, and must be detected with silver impregnation methods. The Warthin-Starry, Dieterle, and Steiner techniques are all suitable for this purpose (17,18) (Figure 13.6). Spirochete-mediated diseases are not always lichenoid in character; they may alternatively assume psoriasiform or dermal lymphocyte–rich patterns as well. Hence, silver stains also comprise part of the evaluation of the latter lesions.

Dermal Lymphocyte–Rich Infiltrates: As just mentioned, dense benign lymphoid infiltrates in the corium may reflect the presence of selected infections. It should be understood that syphilis and Lyme disease do not often manifest stainable bacteria when they incite a brisk lymphocytic infiltration of the dermis; nonetheless, such organisms are still worthwhile seeking in this circumstance. More common

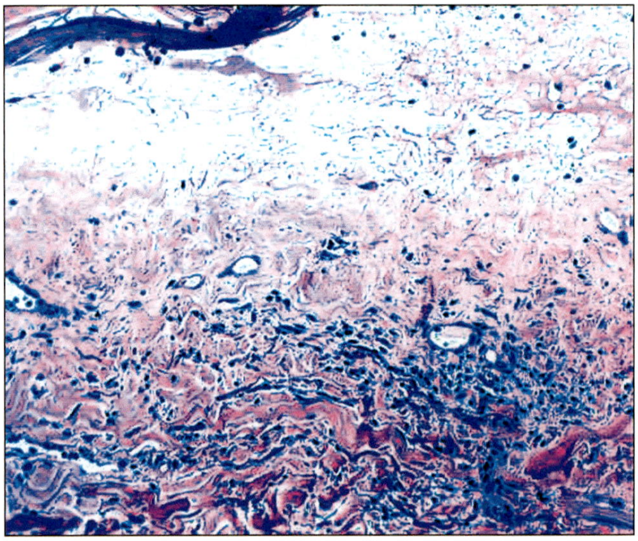

***FIGURE 13.5:** Verhoeff–Van Gieson staining shows a paucity of elastic fibers in the upper dermis in lichen sclerosus et atrophicus, as opposed to an important differential diagnostic alternative, that of morphea.*

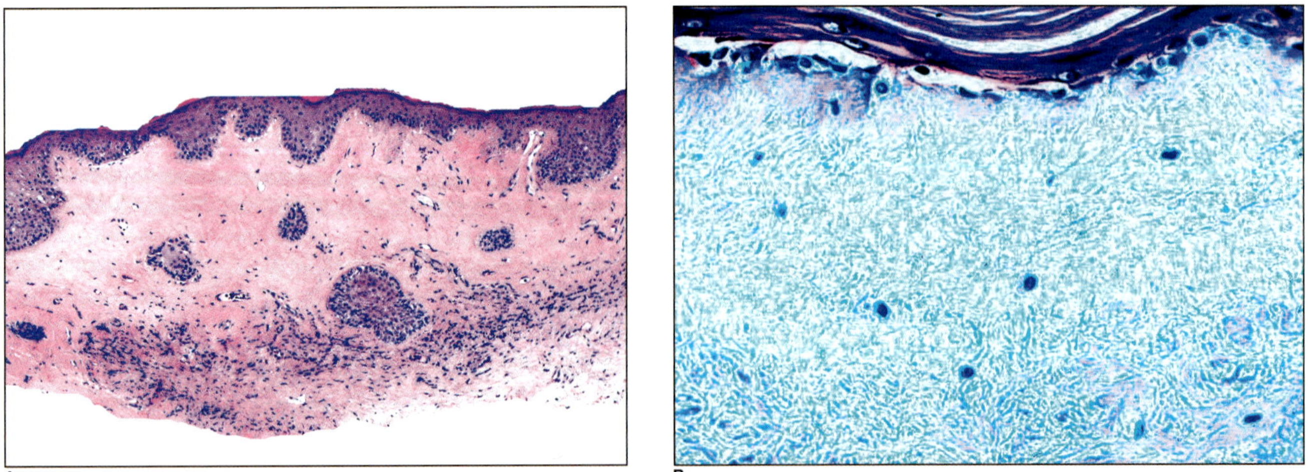

A B

***FIGURE 13.4:** (A) Lichen sclerosus et atrophicus (LS&A), showing homogenization of dermal collagen with chronic inflammation at the junction between the mid- and deep dermis. (B) The Masson trichrome method highlights the homogenized upper dermal collagen in LS&A.*

etiologies for dermal lymphoid infiltrates include LE and cutaneous lymphoid hyperplasia, the latter of which may represent a possible reaction to medications, cutaneous solar damage, or arthropod assaults. Other considerations include early granuloma annulare (GA), pityriasis lichenoides, erythema annulare centrifugum, erythema gyratum repens, and erythema chronicum migrans. Of these disorders, the only noninfectious ones that show specific histochemical abnormalities are LE and GA. As stated earlier, lupus manifests deposits of stromal mucin in the dermis and may show thickening of the EBM and intradermal vascular basement membranes, as seen with CI and PAS-D methods, respectively. GA also features stromal mucin deposition (19) (Figure 13.7) but differs from LE in showing an excessive number of interstitial histiocytes in the corium (20).

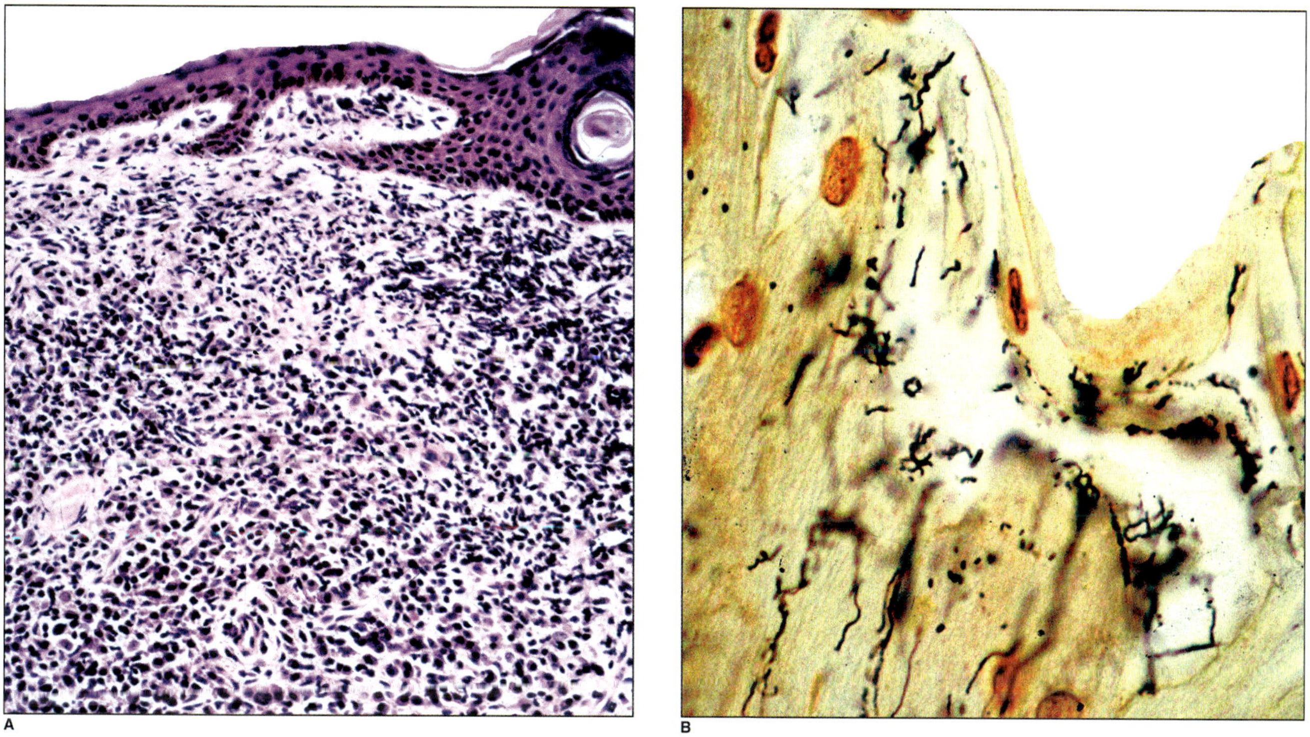

A B

FIGURE 13.6: (A) Either primary or secondary syphilis may show intense dermal lymphoplasmacytosis and variable alterations in the epidermis. (B) Numerous spirochetes are present in the epidermis in secondary syphilis, in this Warthin-Starry stain.

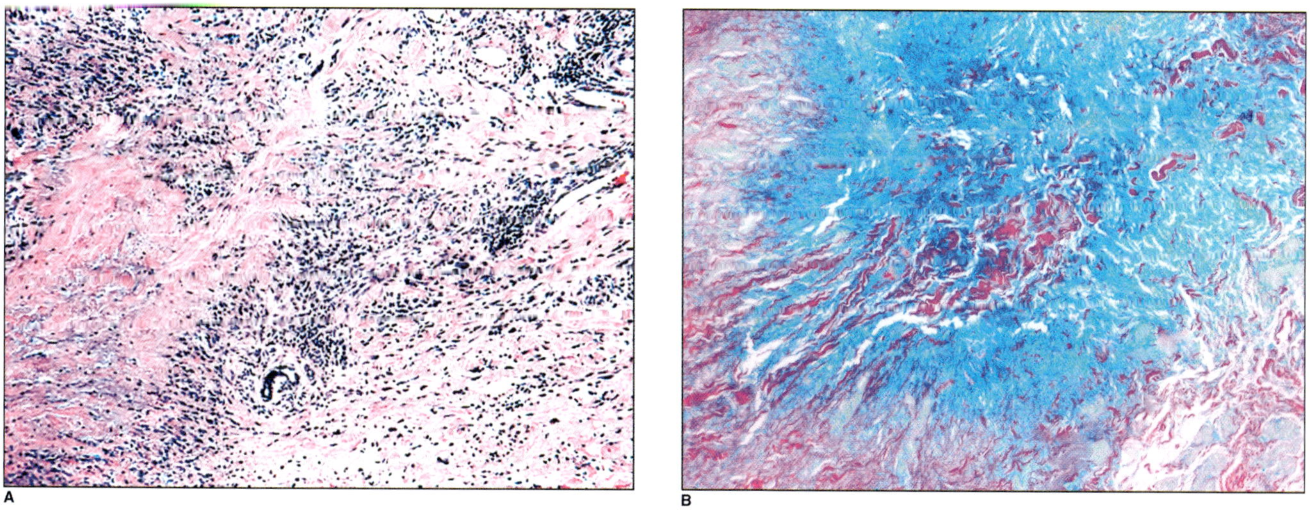

A B

FIGURE 13.7: (A) Arrays of dermal histiocytes surround zones of fragment or liquefied collagen in this example of necrobiotic granuloma. (B) The diagnosis of GA is secured in this case by CI staining, showing abundant mucin in the areas of necrobiosis.

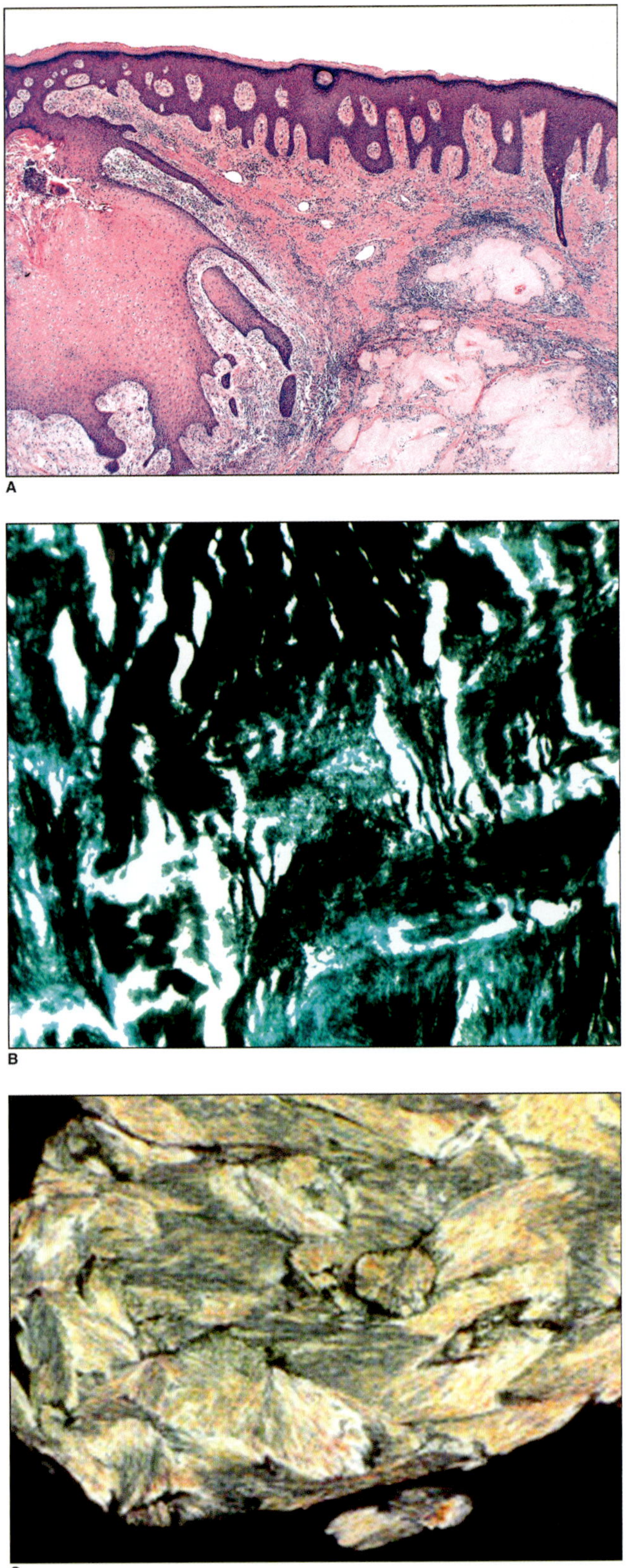

FIGURE 13.8: (A) Gout is typified by dermal granulomatous inflammation, associated with crystalline deposits in the corium. (B) In ethanol-fixed specimens, the urate crystals in gout can be labeled with the Gomori methenamine silver method. (C) Polarization microscopy demonstrates birefringence of the urate crystals.

Necrobiotic Dermatoses, Granulomatous Diseases, and Dermal Histiocytic Infiltrates: Histiocytic and granulomatous infiltrates in the dermis are pathogenetically diverse. They include idiopathic inflammatory or degenerative processes such as sarcoidosis, actinic (Miescher's) granuloma, GA, and necrobiosis lipoidica (NL); reflections of collagen-vascular disease [e.g., rheumatoid nodule (RN) and interstitial granulomatous dermatitis (IGD)]); foreign-body responses—including gout—and infections with various bacteria and fungi.

The necrobiotic diseases (GA, NL, and RN) are diagnostically characterized by the central contents of granulomas in the corium. As stated earlier, GA is associated with stromal mucin deposits; necrobiotic foci in NL contain extracellular fat (as seen in frozen sections with the oil red O or Sudan IV methods); and the central aspects of RNs demonstrate deposits of fibrin, which can be elucidated histochemically with the phosphotungstic acid-hematoxylin (PTAH) or Fraser-Lendrum procedures (19,21–24). Actinic granulomas and IGD also reveal their nature through their lesional contents. Respectively, these are represented by degenerated elastic fibers, as demonstrated with the VVG method (25), and fragmented collagen, as shown with the MT stain (26,27).

Wada et al. (28) have considered the specific histologic differential diagnosis of early Kaposi's sarcoma and interstitial (early) GA, both of which show diffuse dermal hypercellularity without necrobiosis. In that study, staining for iron with the Prussian blue method was helpful in distinguishing between the two conditions, yielding strong positivity in each case of Kaposi's sarcoma but in no examples of GA. Somewhat surprisingly, stains for stromal mucin were nondiscriminatory in this particular context.

Sarcoidosis is the prototypical noncaseating idiopathic granulomatous disease. Nevertheless, one routinely needs to stain the lesions in this condition for mycobacteria—with the Fite and Ziehl-Neelsen methods—and for fungi with the SM, PAS-D, or ABPAS techniques because infectious granulomas are often indistinguishable from those of sarcoid at a morphologic level (29). Schaumann bodies in sarcoid are small laminated calcifications in the lesional cells, and they stain with the von Kossa method (30). Asteroid bodies differ from the latter inclusions because they have a "star-burst" appearance, and they are best labeled with the PTAH procedure (31).

Foreign body granulomas are almost always self-evident after routine and polarization microscopy, but gout is a special exception to this statement. Deposits of urates in the dermis are often absent in formalin-fixed tissues because they dissolve during processing, whereas alcohol fixation will preserve them. Hence, the inciting factor for a gouty foreign body reaction is commonly no longer visible. If one does have an alcohol-fixed substrate in hand, the SM technique will label urate crystals particularly well (32) (Figure 13.8).

Granulomatous infections of the skin include those caused by mycobacteria (both "typical" and "atypical"), fungi in many genera, parasites, and selected bacteria (especially *Yersinia*, *Brucella*, and *Bartonella*). In addition, some infectious diseases are associated with *sheets* of histiocytes in the corium, without organization into granulomatous arrays. This subgroup of conditions includes selected cases of histoplasmosis, rhinoscleroma, leishmaniasis, lepromatous leprosy, and granuloma inguinale (33) (Figure 13.9). Again, the standard panel of histochemical stains that should be applied in this setting is as specified in the preceding paragraph; in addition, the Giemsa, Warthin-Starry, Dieterle, and Steiner methods are necessary to label the unusual microbial genera cited immediately above (34).

Another cutaneous disease that features sheets of histiocytes is malacoplakia. This disorder is caused by chronic infection, usually with a gram-negative bacterium, and the constituent histiocytes contain unusual concentric inclusions—Michaelis-Gutmann bodies—that comprise altered lysosomes, which are impregnated with calcium and iron (Figure 13.10). Accordingly, the von Kossa and Prussian blue methods are effective in recognizing such structures and supporting the diagnosis (35).

A peculiar group of fungal infections that is overrepresented in the skin are the chromomycoses. These are caused by pigment-forming fungi that synthesize melanin-like chromophores. Hence, such organisms can be labeled with the Fontana-Masson (FM) or Schmorl procedure (36) (Figure 13.11), which are usually utilized to

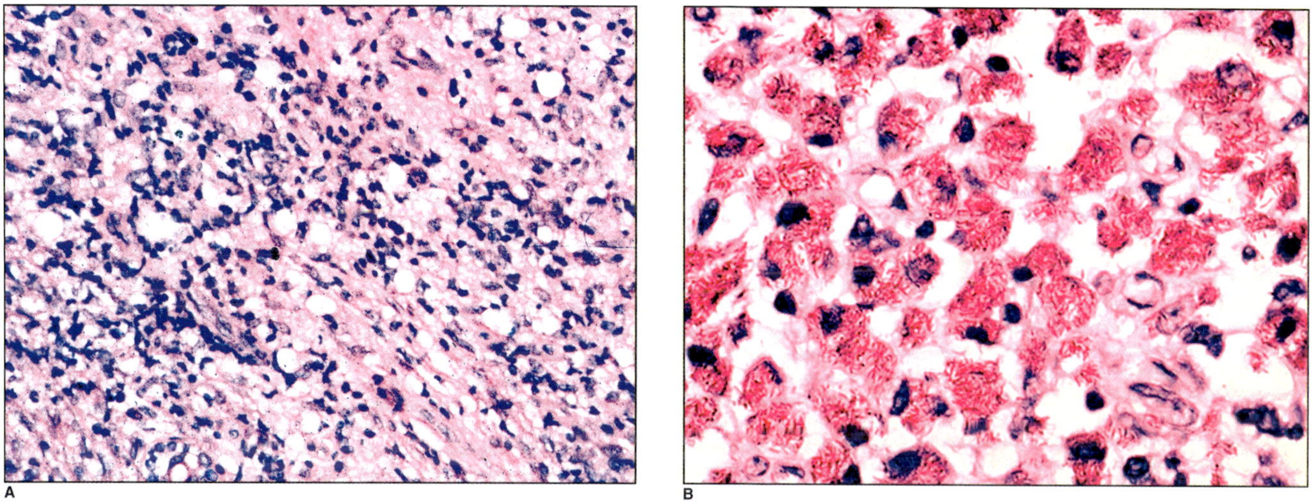

A B

FIGURE 13.9: (A) Lepromatous leprosy, showing groups of vacuolated (parasitized) histiocytes in the dermis. (B) Numerous Mycobacterium leprae *organisms are present in the histiocytes, as shown in this Fite stain.*

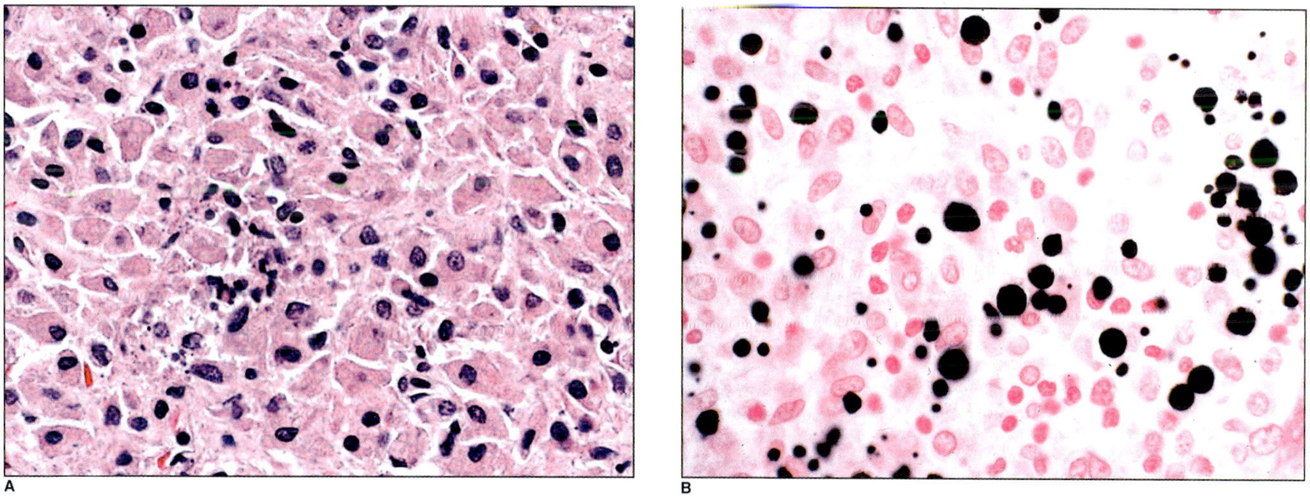

A B

FIGURE 13.10: (A) Sheets of pale eosinophilic histiocytes in the dermis, raising the diagnostic possibility of malacoplakia. (B) The von Kossa method demonstrates spherical inclusions of calcium (Michaelis-Gutmann bodies) in the histiocytes, supporting the interpretation of malacoplakia. The inclusions are also positive with iron stains and with the PAS technique.

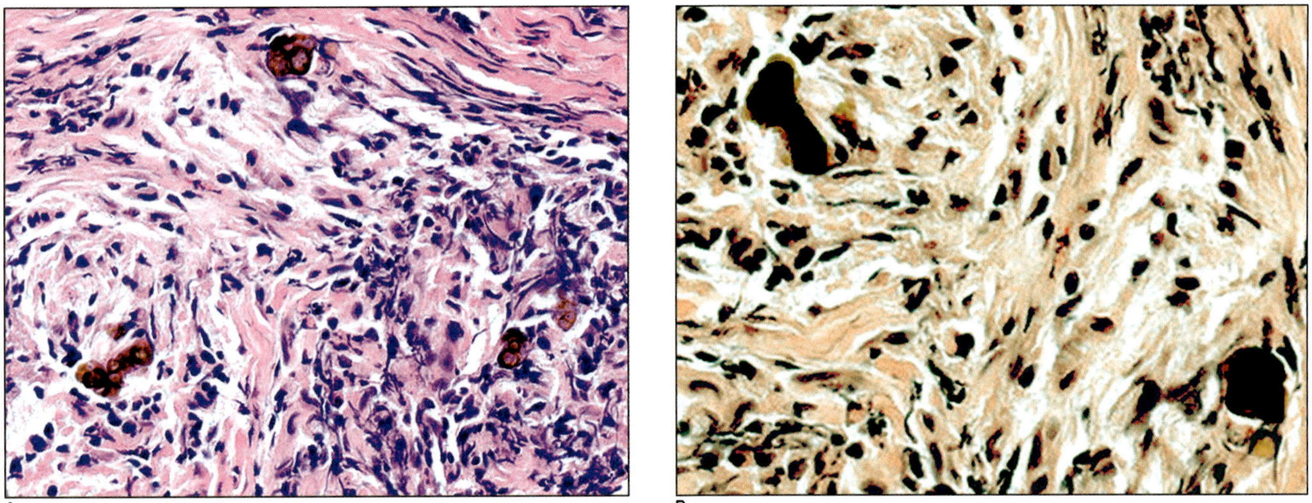

FIGURE 13.11: (A) This biopsy from a patient with chronically draining ulcers on the arms shows rounded pigmented organisms in the dermis. (B) The organisms are labeled with the FM method, indicating that they contain melanin and establishing the diagnosis of chromomycosis.

detect pigment in melanocytes. Lokuhetty et al. have also reported that the Ziehl-Neelsen and Wade-Fite stains are effective in this setting (37).

Subepidermal Bullous Disorders: Intraepidermal acantholytic blistering diseases have been addressed earlier in this discussion, in the section on vesiculopustular disorders. Another major group of bullous dermatoses is represented by those featuring a cleavage plane below the EBM. They include bullous pemphigoid, epidermolysis bullosa acquisita (EBA), porphyria cutanea tarda, bullous LE, bullous lichen planus, bullous lichen sclerosus, bullous mastocytosis, and bullous cutaneous amyloidosis. Among these conditions, concomitant inflammation typically allows for separation of bullous LE, bullous lichen planus, pemphigoid, and EBA from the other possibilities. In contrast, porphyria—which is noninflammatory—features a singular "festooning" pattern of the vascular cores of dermal papillae in the floor of lesional blisters, owing to reduplication and thickening of vascular basement membranes (Figure 13.12). These can be accentuated with the PAS-D method (38). Moreover, many examples of porphyria demonstrate the presence of so-called caterpillar bodies (39) in the blister cavity, near its roof (Figure 13.13). Raso et al. (40) have shown that such bodies are conjointly formed by fragments of epidermal cytoid bodies, collagen, and EBM. These inclusions are also reactive with the PAS-D stain and are felt to be virtually diagnostic of porphyria cutanea tarda.

Bullous lichen sclerosus is characterized by abnormalities in upper dermal collagen and elastin content, as described above. Mastocytic infiltrates are recognizable by their metachromatic labeling with toluidine blue or Giemsa procedure, as well as staining by the naphthol AS-D chloracetate esterase (Leder) method (41–45). Amyloid can be detected with the Congo red, sulfated alcian blue, or crystal violet (Lieb's) technique (46).

Vasculopathic Disorders: Cutaneous vasculopathies potentially affect blood vessels of all sizes and locations, from the superficial dermis through the subcutis. They include "hypersensitivity" (leukocytoclastic, Zeek's) vasculitis, granuloma faciale, erythema elevatum diutinum, idiopathic granulomatous (temporal) arteritis, lymphomatoid granulomatosis, Wegener's granulomatosis, Churg-Strauss syndrome, polyarteritis nodosa, Buerger's disease, atrophie blanche, Degos' disease, calciphylaxis, superficial thrombophlebitis, pigmented purpuric eruptions/lichen aureus, cryoglobulinemia, macroglobulinemia, and infectious vasculitides.

VVG stains are useful in this assemblage of diseases to see whether the internal and external elastic laminae of blood vessels are disrupted (47–50) (Figure 13.14). If so, such a finding would support the general diagnosis of vasculitis or vasculopathy. However, there are two exceptions to the latter rule, represented by Buerger's disease (51) and atrophie blanche (livedoid vasculopathy) (52). Stains that can be used to demonstrate fibrin deposition in and around vessel walls (fibrinoid change) in vasculopathies include the PTAH method and the Fraser-Lendrum (Martius-scarlet-blue trichrome, also called the Picro-Mallory) technique (53).

Aside from infectious vasculitides, which may or may not reveal positivity for microorganisms in vessels or perivascular tissue (Table 13.2), several of the above-cited conditions can be associated with additional histochemical findings. Deposits of calcium salts in dermal and subcuticular vessels are seen in calciphylaxis with the von Kossa and alizarin red methods (54–56). In pigmented purpuric eruptions (purpura pigmentosa chronica) and lichen aureus, long-term extravasation of erythrocytes from small vessels eventually yields deposits of dermal iron, as seen with the Prussian blue or Perl procedure (57,58) (Figure 13.15). Lastly, malignant atrophic papulosis (Degos' disease) is a rare vasculopathy that is characterized by micro-infarcts in the skin with surrounding small vessel thrombi and stromal mucin deposition (59). The MT method can be used to better outline zones of ischemic damage in the corium, and CI staining labels the adjacent foci of dermal mucinosis in this disorder.

Adnexocentric (Particularly Folliculocentric) Dermatitides: The alopecias can be divided into inflammatory-nonscarring disorders, inflammatory-scarring diseases, non-inflammatory-scarring diseases, and conditions that are neither inflammatory nor scarring. Alopecia areata and alopecia mucinosa belong to the first of those categories; lupus alopecia and lichen planopilaris are in the second. Infectious folliculitis with alopecia may be in either the first or the second group, and chronic trichotillomania, central centrifugal alopecia, and traction alopecia are in the third. Finally, anagen or telogen effluvia are examples of the fourth group.

Histochemical methods assist in recognizing several of these conditions. Stains for stromal mucin, such as CI or alcian blue, are useful in identifying alopecia mucinosa (which is associated with mycosis fungoides) (60–62) (Figure 13.16) and the alopecia of LE. Folliculitides caused by bacteria or fungi are obviously recognizable as such with Gram and SM stains (63,64). Scarring alopecias are typified by the loss of elastic fibers in the alopecic zone as shown in VVG stains (65).

Panniculitides: Inflammatory conditions that are centered in the subcutaneous fat are relatively rare. Several are secondary and are associated with disease elements in the dermis or even in other organs. Examples include infectious panniculitis, calciphylaxis, alpha-1-antitrypsin deficiency (AATD), panniculitis of pancreatic disease, and some examples of erythema induratum, cryoglobulinemia,

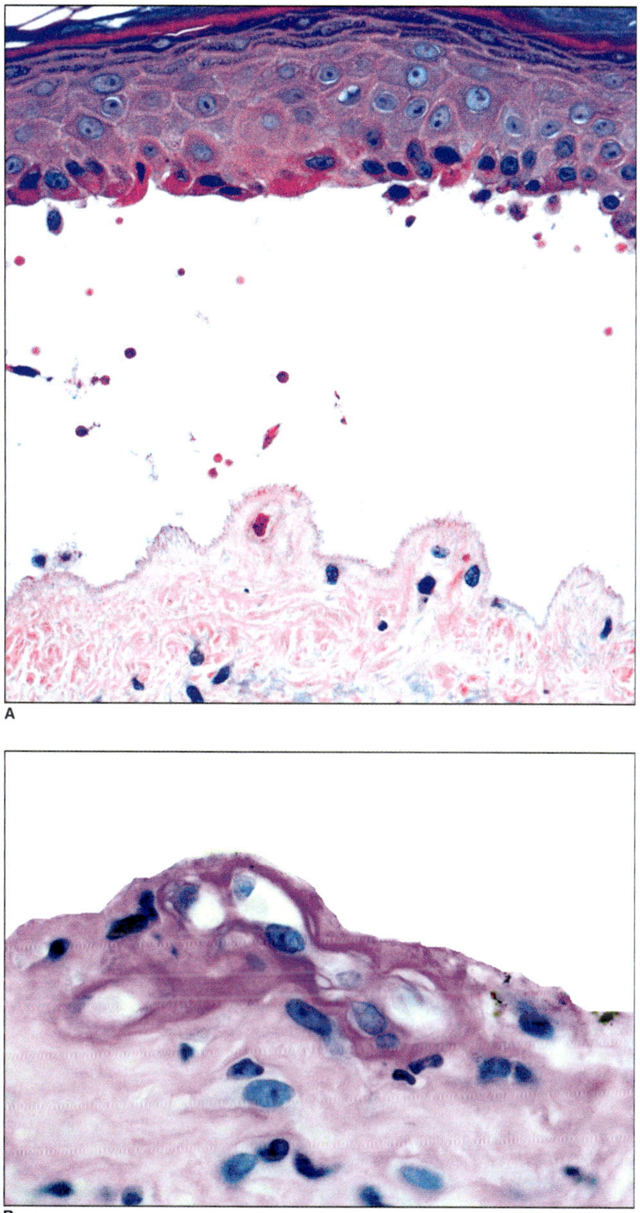

A

B

FIGURE 13.12: *(A) A "clean" subepidermal blister is present in this biopsy of a bullous lesion of the hand, and porphyria cutanea tarda (PCT) is a diagnostic possibility. (B) Staining with the PAS procedure after diastase digestion shows redundant basement membrane material around blood vessels at the base of the blister, supporting the diagnosis of PCT.*

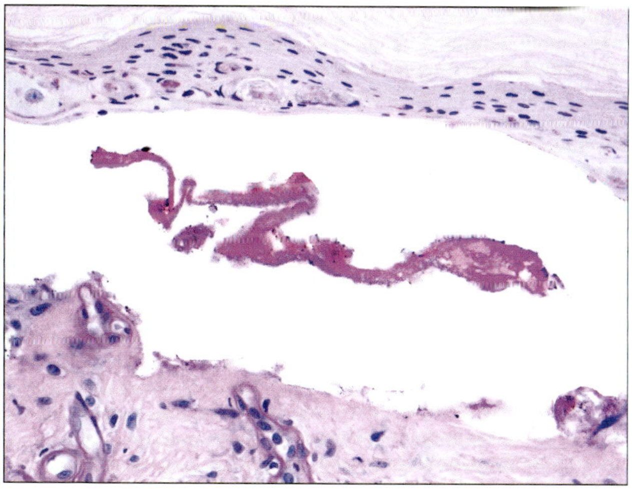

FIGURE 13.13: *A "caterpillar" body, probably derived from elements of the EBM, is stained with the PAS method after diastase digestion. This finding supports the diagnosis of porphyria cutanea tarda.*

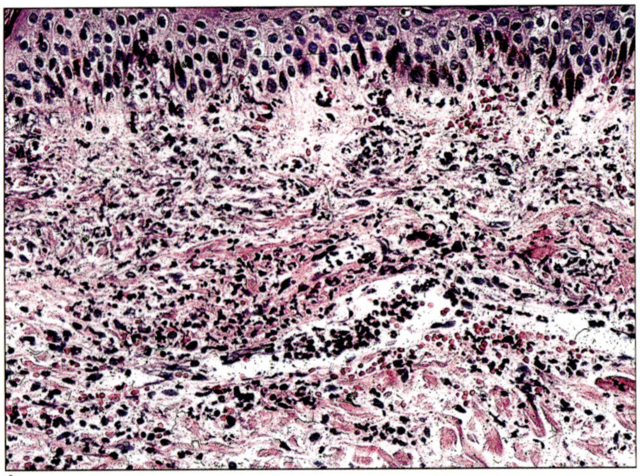
A

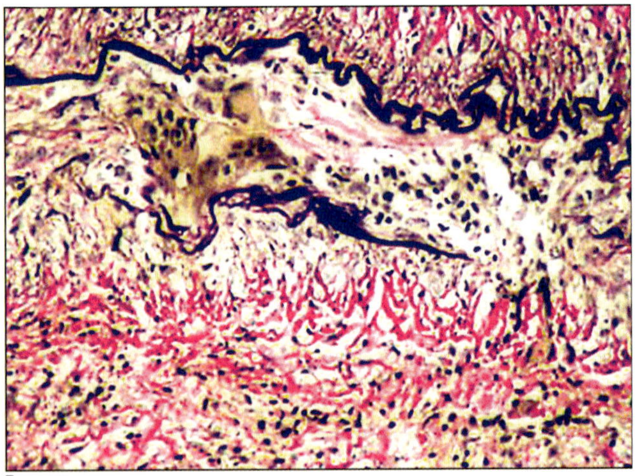
B

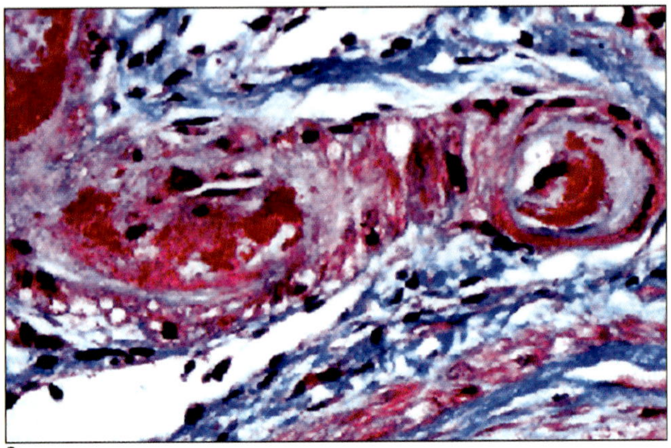
C

FIGURE 13.14:* *(A) Perivascular and intramural dermal blood vessel inflammation by neutrophils indicates the possible presence of vasculitis. (B) Verhoeff–Van Gieson staining of a cutaneous vessel shows disruption of its elastic laminae, supporting the diagnosis of vasculitis. (C) Masson trichrome staining demonstrates intravascular thrombi and intimal fibrosis in the affected vessels.

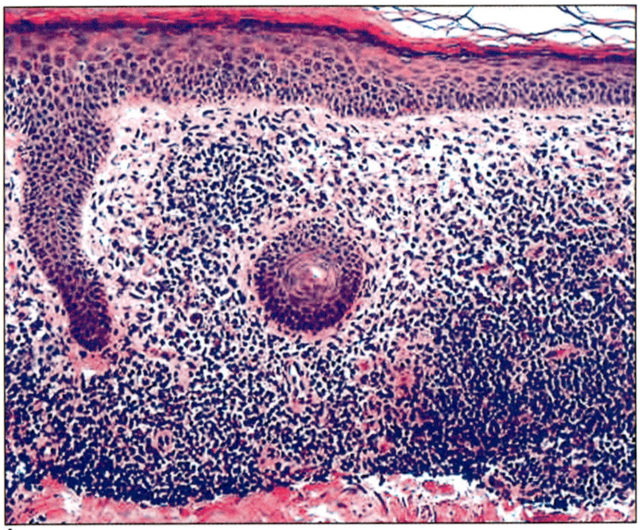
A

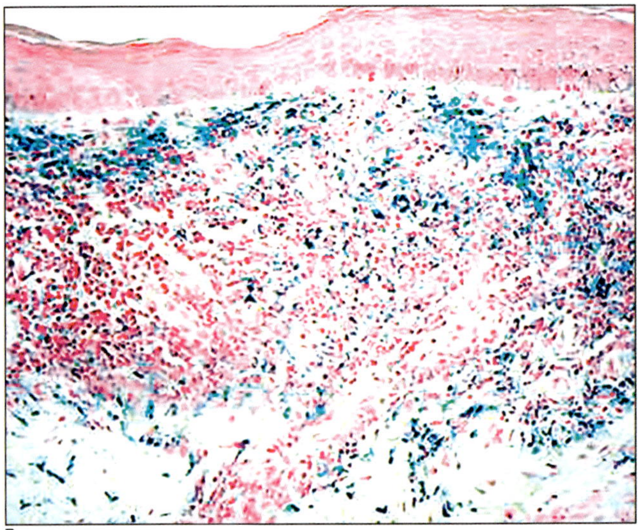
B

***FIGURE 13.15:* *(A) Intense band-like (lichenoid) chronic inflammation in the dermis, invoking diagnostic consideration of several diseases that have this histologic image. (B) Prussian blue staining of the specimen demonstrates diffuse deposition of iron throughout the corium, supporting the diagnosis of* lichen aureus.**

scleroderma, and lupus profundus (a form of LE). Others are primary panniculitides, such as erythema nodosum and cold-induced panniculitis.

Some of these disorders have been addressed earlier in regard to their attendant histochemical features, specifically, infections, calciphylaxis, LE, and cryoglobulinemia. In addition, morphea profunda—as well as its dermal counterpart—shows dense collagen deposition in MT stains, and normal or increased quantities of elastin with the VVG method (13,66). No specific histochemical observations have been made in erythema nodosum; AATD; most cases of erythema induratum; cold panniculitis; or pancreatic panniculitis.

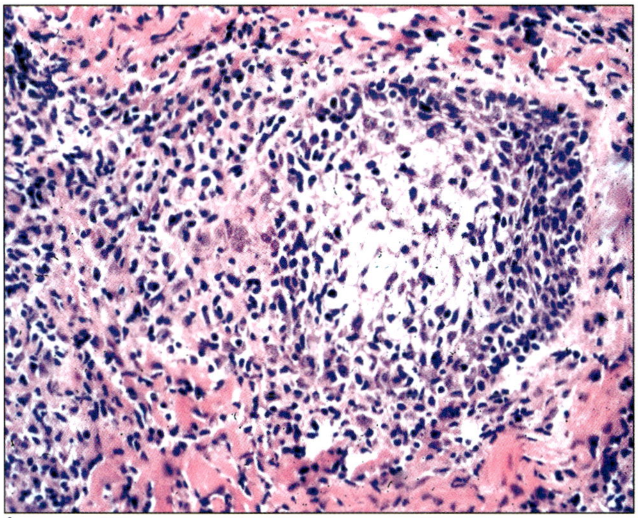

A

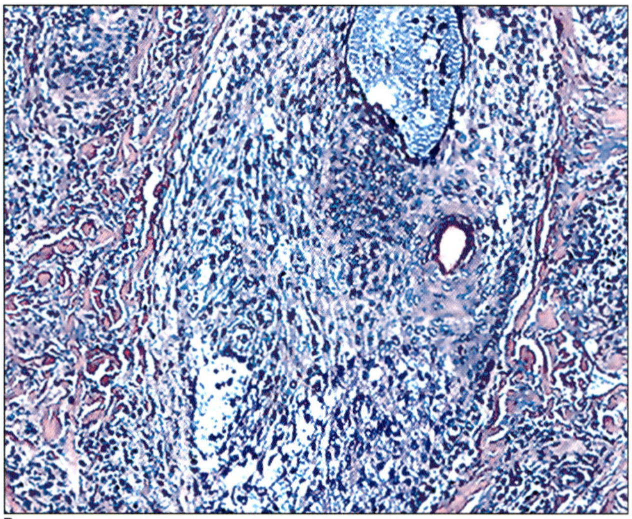

B

FIGURE 13.16: (A) This hair follicle is permeated by lymphocytes, and the constituent keratinocytes are separated by stromal material. (B) This alcian blue stain shows that the matrical material is stromal mucin, supporting the diagnosis of follicular mucinosis. The patient was later shown to have mycosis fungoides.

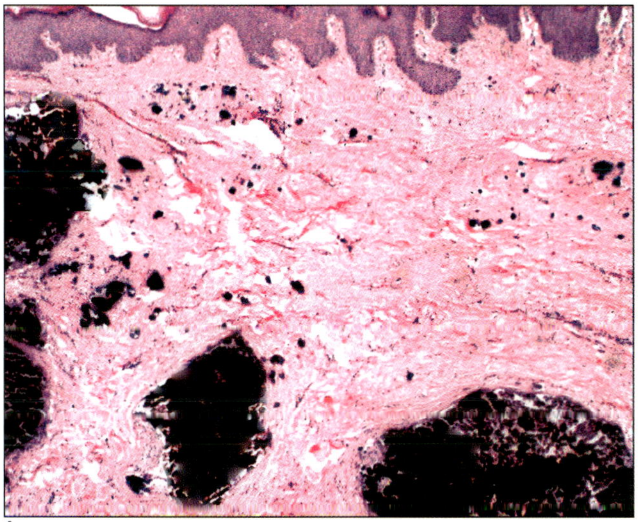

A

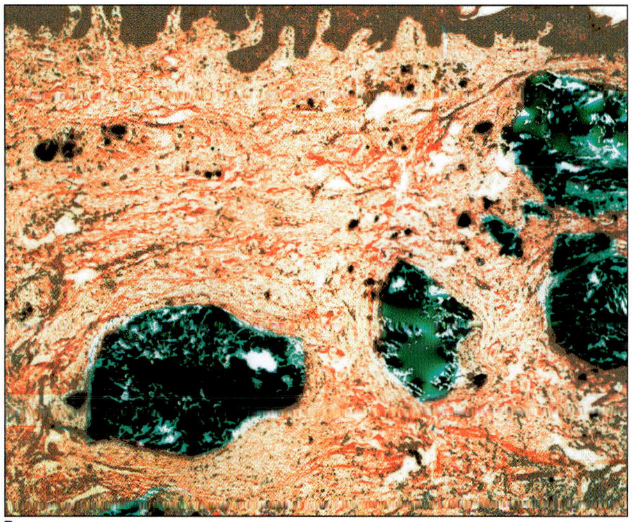

B

FIGURE 13.17: (A) Large basophilic deposits in the dermis, in a patient with nodular masses in the scrotal skin. (B) The von Kossa method labels the deposits, indicating a diagnosis of tumoral calcinosis.

Deposition Disorders

Several inorganic elements and chromophoric organic compounds may be deposited in the dermis, causing clinical hyperpigmentation or palpable lesions. These deposits can include calcium (in calcinosis cutis or calciphylaxis), silver (in argyria), or iron salts (in hemochromatosis); lipofuscin (in amiodarone- or clofazimine-induced drug-related hyperpigmentation); homogentisic acid (a product of tyrosine and phenylalanine metabolism in ochronosis); a hemosiderin-melanin complex (in minocycline-associated hyperpigmentation), or melanin alone (in melasma, post-inflammatory hyperpigmentation, and diffuse or nodular melanosis).

Calcium is typically recognizable in conventional hematoxylin and eosin stains because of its intense hematoxylin affinity. However, should one desire to confirm the chemical nature of such deposits, either the von Kossa (Figure 13.17) or alizarin red method is effective in labeling them (67,68). Iron can be stained with the Prussian blue or Perl technique, as mentioned repeatedly in the foregoing discussion. Silver is not typically stainable by histochemical methods that can be applied to formalin-fixed tissue, but it is refractile on polarization microscopy. Melanin is best detected with the FM and Schmorl procedures (69–73), whereas lipofuscin can be stained with either PAS-D or a modification of the Ziehl-Neelsen acid-fast technique

(74–77). Homogentisic acid is metachromatically stained black with the cresyl violet or methylene blue procedure in ochronosis (32).

Conversely, one may wish to determine what the cause of acquired *hypo*pigmentation might be, as seen in vitiligo (78,79). In this circumstance, procurement of lesional and nonlesional biopsies of skin, and performance of the FM or Schmorl method on each of them, will allow a semiquantitative comparison to be made of the number of melanocytes and the density of melanin pigment.

Another major group of cutaneous deposition disorders is represented by those in which interstitial mucin content is increased in the dermis or subcutis (generalized myxedema, pretibial myxedema, focal mucinosis, reticular erythematous mucinosis, scleredema, the mucopolysaccharidoses, scleromyxedema, nephrogenic fibrosing dermopathy) (80–87) (Figure 13.18), others where lipid and protein accumulate in the extracellular matrix (lipoid proteinosis) (88), conditions in which deposits of protein (amyloid-like or other forms) and elastic tissue occur (colloid milium, nodular colloid degeneration) (89–91) (Figure 13.19), and macular or nodular amyloidosis. We have previously considered the use of CI and alcian blue methods in recognizing acid mucopolysaccharide–rich mucins, and the PAS-D technique is effective at staining the lesions of lipoid proteinosis and colloid milium (88–90). The last of these conditions also contains altered elastic fibers, as true of nodular colloid degeneration as well, and therefore, one will observe VVG reactivity in such lesions (90,91). Regardless of the specific type of amyloid seen in the skin, it consistently labels with the Congo red, sulfated alcian blue, or crystal violet technique (92,93).

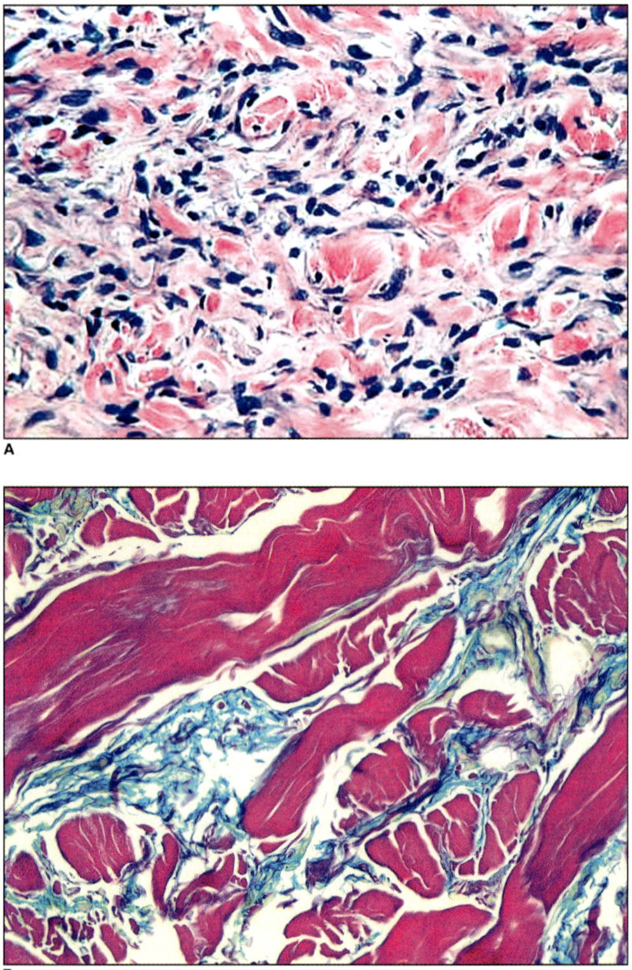
A
B

FIGURE 13.18: ***(A) The dermis in this biopsy is hypercellular, owing to a proliferation of spindle cells that are separated by a bluish matrix. (B) A CI stain shows reactivity with the stromal deposits, supporting a diagnosis of scleromyxedema.***

Defects in Formation or Maintenance of Dermal Connective Tissue

Several cutaneous disorders are typified by defective formation or preservation in integrity of the dermal connective tissue. Lichen sclerosus and morphea have been mentioned earlier, but for completeness, they will be summarized again here. The first of these conditions shows homogenization of upper dermal collagen with a concurrent diminution of elastic fibers in VVG stains; morphea, in contrast, manifests an increased density of collagen throughout the dermis in MT preparations (Figure 13.20), with little or no abnormality in elastic content (13). We also believe that atrophoderma is a form of morphea (94,95), with similar histochemical properties.

Cutis laxa also shows a decrement in dermal elastic fibers, regardless of whether it is congenital or acquired in nature. In addition, however, the remaining elastic fibers are fragmented and malformed in VVG stains (96,97). Findings in anetoderma (focal dermal atrophy) (98) and erythema ab igne (99) are comparable, except that no morphologic aberrations in elastic structure are seen in the lesions of these conditions. Erythema ab igne also is associated with melanin incontinence in the upper dermis, which may be accentuated with the FM method (99).

In contrast to the foregoing diseases, pseudoxanthoma elasticum (PXE) shows an increase in deformed, fragmented, and bulbous elastic fibers in the dermis (100–102). These structures also are abnormal in appearance in hematoxylin and eosin stains, wherein they assume a lavender-blue and tangled image. Secondary calcification of elastin in PXE is common, as revealed with the von Kossa method (Figure 13.21). Also, the lesions of elastofibroma dorsi (a pseudotumor caused by repetitive mechanical trauma) (103,104) and collagenous and elastotic plaques of the hands and ears (105,106) demonstrate closely similar features to those of PXE, despite the fact that their clinical presentations are quite different.

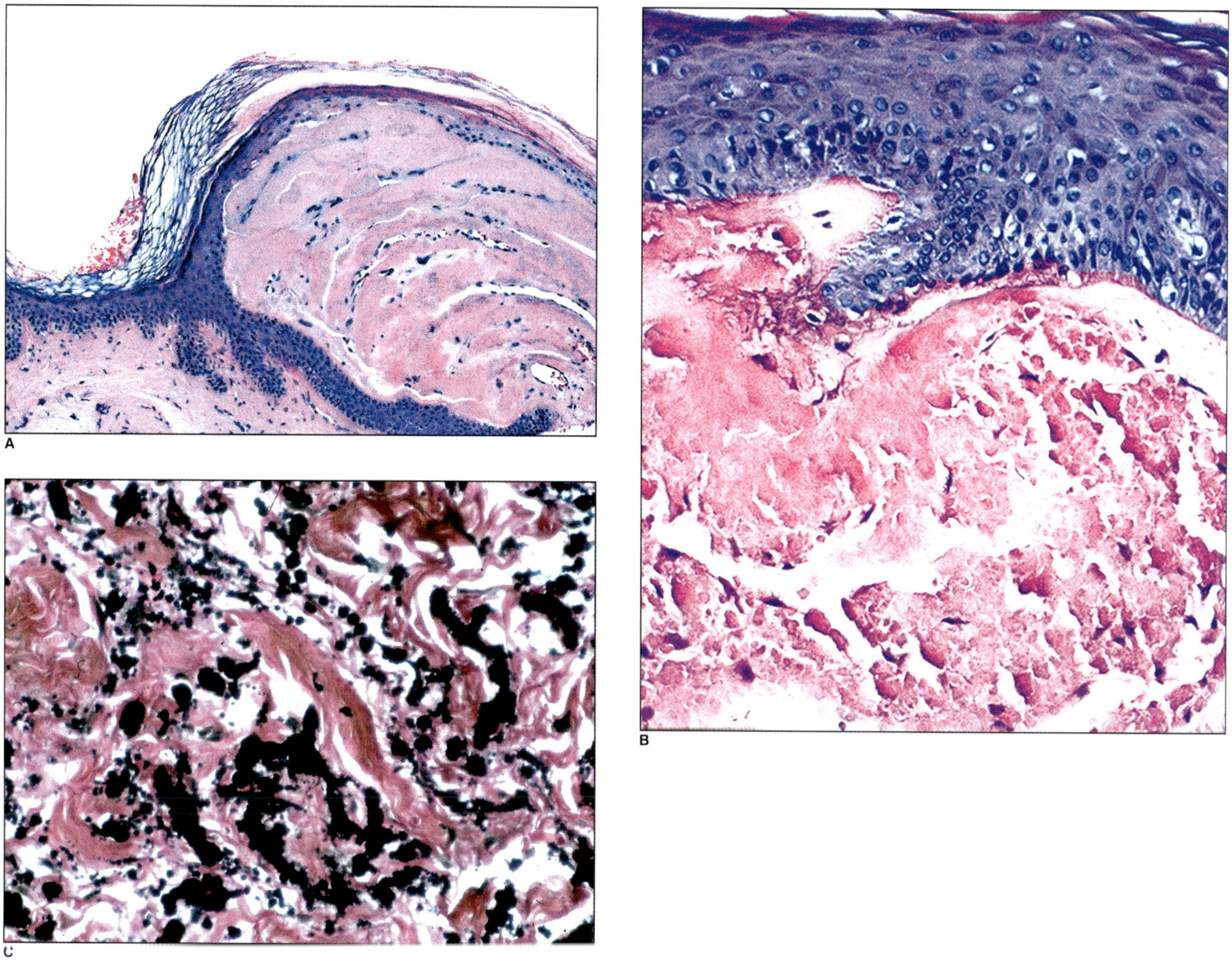

FIGURE 13.19: ***(A) A localized hypocellular nodular expansion is seen in the dermis in this biopsy from the paranasal skin. (B) The lesion is reactive with the PAS method after diastase digestion. (C) Verhoeff–Van Gieson staining demonstrates increased numbers of aberrantly shaped elastic fibers, suggesting the diagnosis of nodular colloid degeneration.***

The "perforating disorders" are also among those in which dermal connective tissue abnormalities are encountered, but in unique form (107). Here, components of the corium—potentially including adnexal contents—are extruded or eliminated through the skin surface, via an epidermal defect. Four diseases that have been considered essential perforating disorders include elastosis perforans serpiginosa (Figure 13.22), reactive perforating collagenosis, perforating folliculitis (PF), and Kyrle's disease (KD). A review of the literature, including reports of perforating diseases associated with chronic renal failure (108,109), suggests that there may be considerable clinical and histologic overlap among PF, KD, and the adult form of perforating collagenosis. With that having been said, the histochemical approach to diagnosis of the perforating dermatoses is similar regardless of the specific condition being considered. MT and VVG stains are useful in this context, with particular attention to the contents of the epidermal "crater." If fragmented collagen fibers or elastin, or both, are seen in that location, one has objective support for the interpretation of a perforating disorder in general (110).

Special Infections of the Skin

Infectious processes have been cited several times thus far and are summarized histochemically in Table 13.2. Nevertheless, some of them are unusual enough to merit special comments.

Erysipelas is principally a streptococcal infection, although changing epidemiologic features of this disease have altered the bacterial pathogens in some cases. Nonetheless, regardless of the etiological agent(s), one very rarely if ever sees microorganisms in Gram or silver impregnation stains of biopsies in erysipelas (111). Hence,

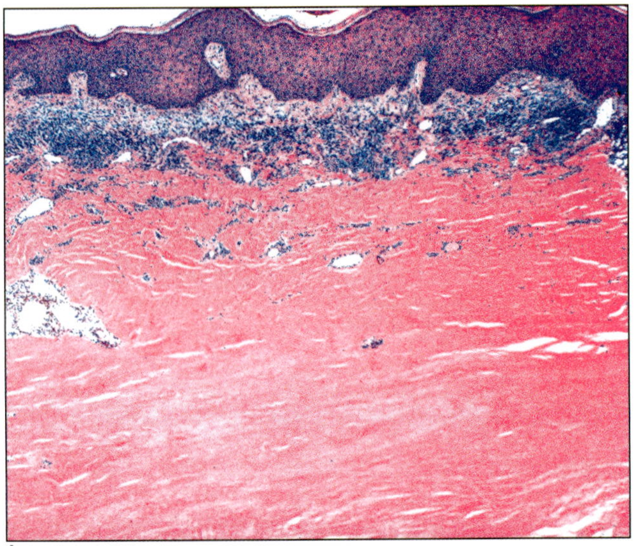

A

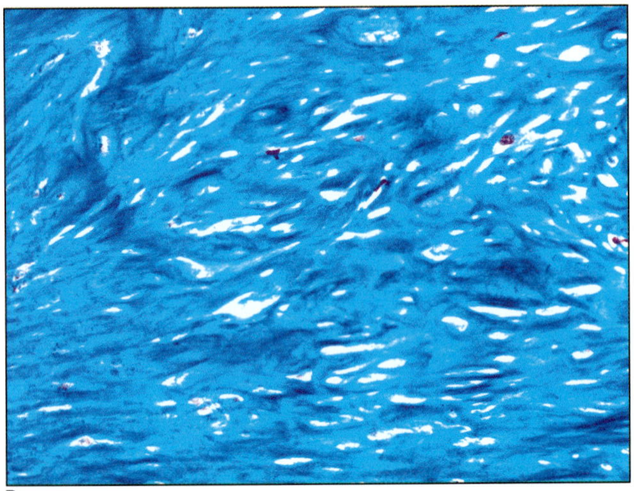

B

FIGURE 13.20: (A) Morphea (localized scleroderma) shows dense sclerosis of the dermis with a loss of dermal appendages. This example also demonstrates lichenoid chronic inflammation. (B) The Masson trichrome stain in this case demonstrates replacement of the dermis by mature collagen.

this diagnosis is not precluded by negative histochemical results.

Erythrasma manifests itself as an intertriginous eruption caused by infection with *Corynebacterium minutissimum*, a diphtheroid. In addition to its gram positivity, this organism may be labeled with the PAS-D, SM, or Giemsa method as well, in the form of coccobacilli in the stratum corneum (112). A similar infection is that of trichomycosis, again caused by *Corynebacteria*, but this time inhabiting hair follicles. Staining properties of the latter disease are identical to those of erythrasma.

Infection with *Calymmatobacterium granulomatis* is the cause of granuloma inguinale. This microbe is a pleomorphic gram-negative bacterium, and, as such, virtually never demonstrates labeling with the tissue Gram stain. It must be recognized with silver methods such as the Dieterle or Warthin-Starry procedures (Figure 13.23), or, alternatively, with the Giemsa technique (113,114). Comparable statements apply to *Klebsiella rhinoscleromatis*, the etiologic agent in rhinoscleroma (115) as well as, *Bartonella henselae* and *Bartonella quintana*, the causes of cat-scratch disease and bacillary angiomatosis (BA), respectively (116,117).

Rickettsiae represent still other gram-negative bacteria that are even more difficult to demonstrate in tissue sections, in the lesions of Rocky Mountain spotted fever or other tick-borne rickettsioses. Any one of three histochemical methods are suitable for this purpose; they are the Machiavello technique (Figure 13.24), the Giemsa stain, and the Castaneda procedure (118,119). One observes red-purple intracellular bacillary inclusions in each of the latter preparations.

Viral diseases are only generically identifiable by histochemical means, using stains that are selective for deoxyribonucleic acid (DNA)—such as the Lendrum (phloxine-tartrazine) and Feulgen methods—or which recognize both DNA and ribonucleic acid (RNA), as typified by the methyl green-pyronine (MGP) procedure. In the Feulgen and Lendrum methods, DNA stains bright pink (Figure 13.25), whereas it is blue green and RNA is pink red in the MGP technique (120–123). Hence, viral inclusions of *Herpes simplex*, *Varicella*, *Cytomegalovirus*, *Molluscum*, and human papillomavirus (HPV) can be labeled in skin lesions caused by these agents. Lucia et al. have described a modification of the Feulgen method that was done after digestion of paraffin sections of skin with 0.01% deoxyribonuclease for 16 hours; it was intended to discriminate viral DNA from native cellular nucleic acid (124). In this procedure, the inclusions of *Molluscum* and HPV were labeled, whereas others associated with *Cytomegalovirus* and *Herpes simplex* were not.

The pathogenic organisms in cutaneous parasitic infestations are usually obvious in hematoxylin and eosin-stained sections, unless they are protozoans such as *Leishmania*, *Toxoplasma*, *Acanthamoeba*, or *Entamoeba*. Giemsa staining is effective at demonstrating leishmanial and *Toxoplasma* microbes in the cytoplasm of infected cells (125) (Figure 13.26). *Acanthamoeba* is potentially stained by the PAS-D or SM procedure (126), and *Entamoeba* cysts or trophozoites can be labeled by the PAS-D or MT technique (127).

CUTANEOUS NEOPLASMS

It would be untenable to attempt a discussion here of histochemical findings in all neoplasms of the skin. Thus, we will confine our attention to particular differential diagnostic complexes of such lesions in which histochemistry is beneficial.

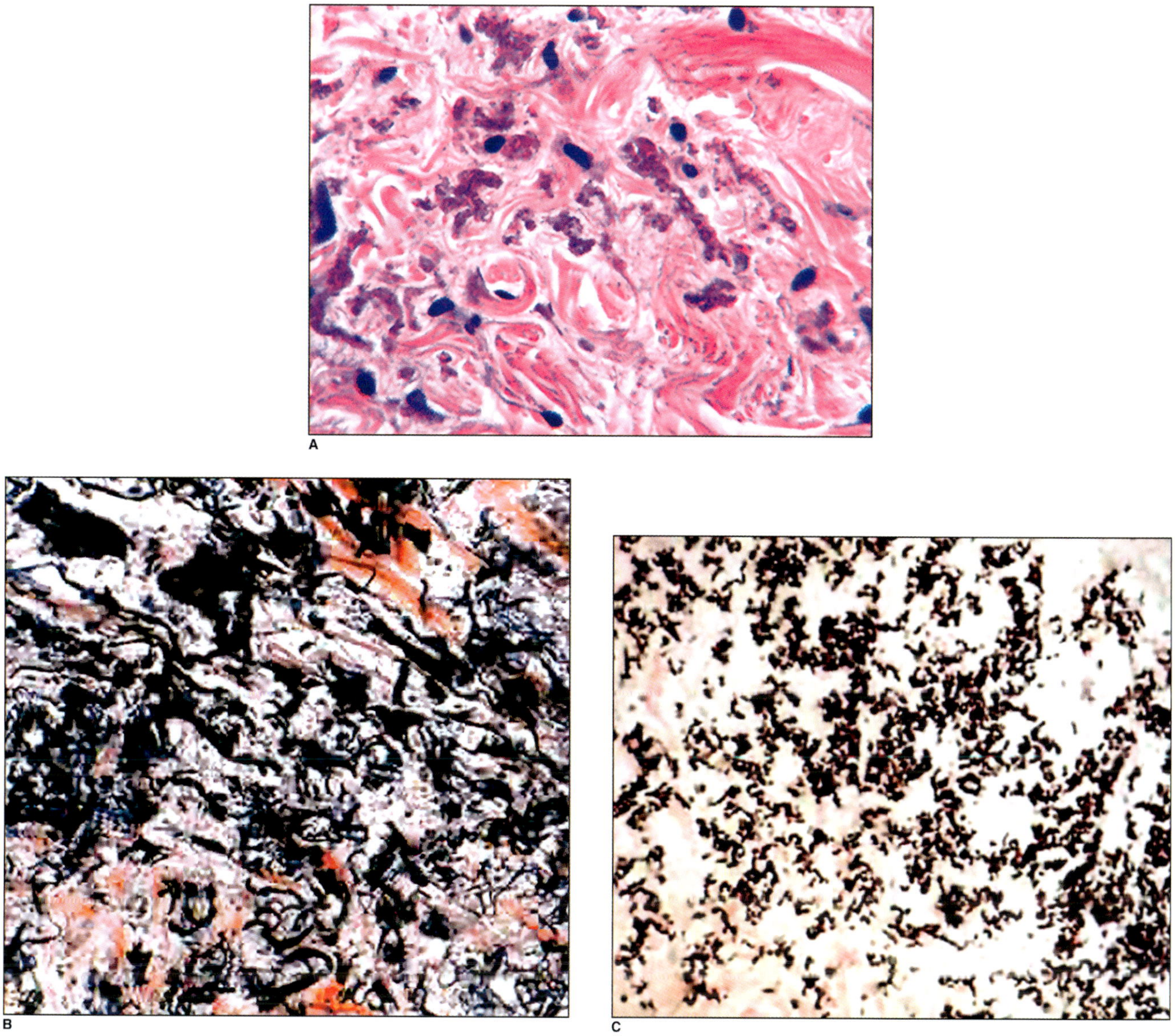

FIGURE 13.21: (A) Irregular, bluish, convoluted structures are seen in the dermis in this biopsy from a patient with PXE. (B) The Verhoeff–Van Gieson stain labels the structures seen in (A), indicating that they are abnormal elastic fibers. (C) A von Kossa stain further demonstrates dystrophic calcification of the elastic fibers in PXE.

Intraepidermal Tumors with Similar Morphologic Features

Cytologically Bland "Pale" Cell Proliferations: Several neoplasms cause a fusiform or irregular thickening of the epidermis and comprise populations of cytologically bland pale epithelioid cells. The three principal considerations in this group are "flat" seborrheic keratosis, intraepidermal poroma (hidroacanthoma simplex), and pale cell or clear cell acanthoma (128). Among these possibilities, only the last tumor entity demonstrates abundant intracellular glycogen in undigested PAS stains (129) (Figure 13.27). Parenthetically, the cells in both seborrheic keratosis and intraepidermal poroma may contain melanin, as revealed with FM preparations (130,131).

"Clonal" or "Pagetoid" Intraepidermal Malignancies: Some cytologically malignant proliferations in the epidermis show a distinctly clustered cellular growth pattern, which may, in addition, be coupled with scattered single epithelioid tumor cells in the surface integument (132). These "clonal" or "pagetoid" lesions include mammary and extramammary Paget's disease (PD), pagetoid melanoma, and clonal squamous cell carcinoma in situ (SCCIS). The

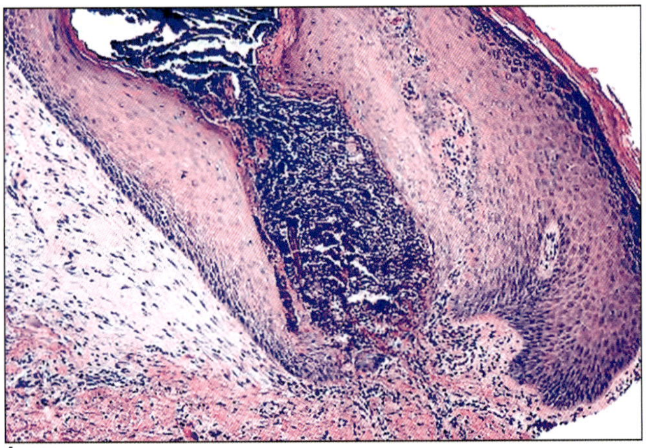

A

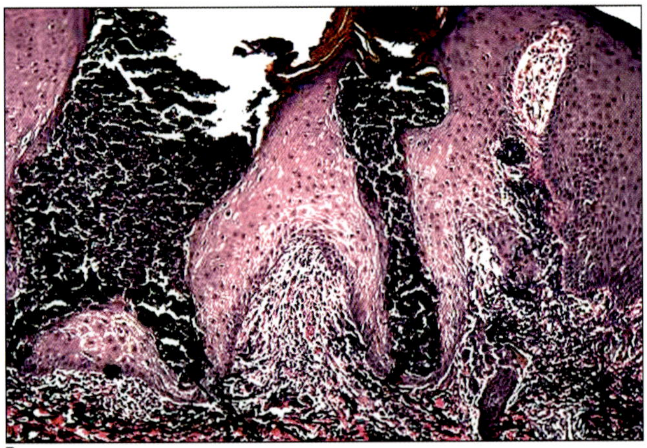

B

FIGURE 13.22: (A) An epidermal defect is filled with amorphous basophilic debris in this biopsy from a patient with renal failure. The suggested diagnosis is that of a perforating dermatosis. (B) A Verhoeff–Van Gieson stain labels the contents of the defect, supporting an interpretation of perforating elastosis.

neoplastic cells in PD are commonly reactive with the PAS-D or mucicarmine method, or both (Figure 13.28). They also may label with the alcian blue technique at pH 2.5 but not at pH 0.4, as well as the CI method, because of sialomucin content (133–138). All of these procedures are wise inclusions in the histochemical evaluation of PD. This is so because the constituent cells also may "parasitize" melanin from surrounding melanocytes in the epidermis and thereby acquire spurious positivity with the FM method (139). In contrast, neither intraepidermal melanoma nor SCCIS is reactive with PAS-D stain. Melanoma is typically labeled only by the FM and Schmorl procedures (140–142), and SCCIS is characteristically negative with all the techniques just mentioned. PD of the eyelids—and occasionally in extraocular sites as well—may show *sebaceous* differentiation in lieu of sudoriferous features, as true of other examples of PD (132,143). Therefore, if frozen tissue is available, oil–red O or Sudan IV stains may be helpful in identifying the cytoplasmic lipid content in sebaceous lesions (144,145).

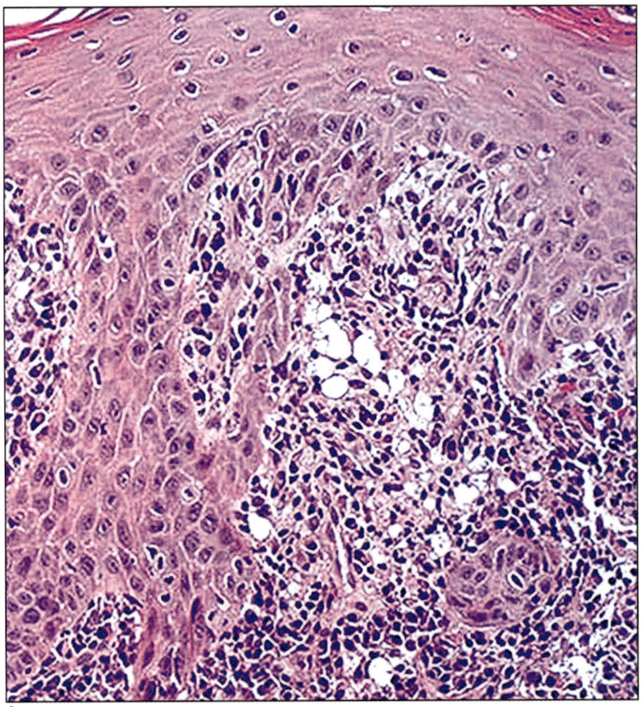

A

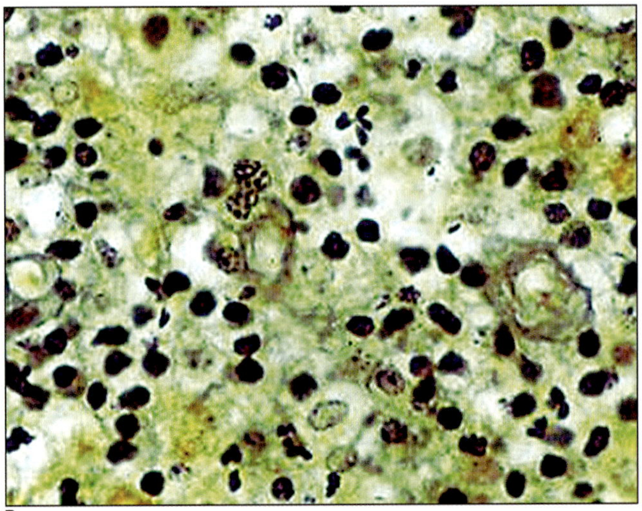

B

FIGURE 13.23: (A) Granuloma inguinale is another infectious disorder that features "parasitized" histiocytes in the dermis. (B) The Dieterle method labels short coccobacillary organisms in the lesional histiocytes.

Atypical Pigmented Lentiginous Proliferations: A rather common differential diagnostic problem in biopsies from the facial skin in adults concerns confluent proliferations of cytologically atypical polygonal cells at the epidermal base, which also demonstrate cytoplasmic pigment content. The distinction between lentiginous melanoma in situ (LMIS) and pigmented actinic keratosis (solar lentigo) is often quite difficult in this context (146). In this circumstance, staining with the FM method, or with azure-B (which metachromatically labels melanin green) (147) can be

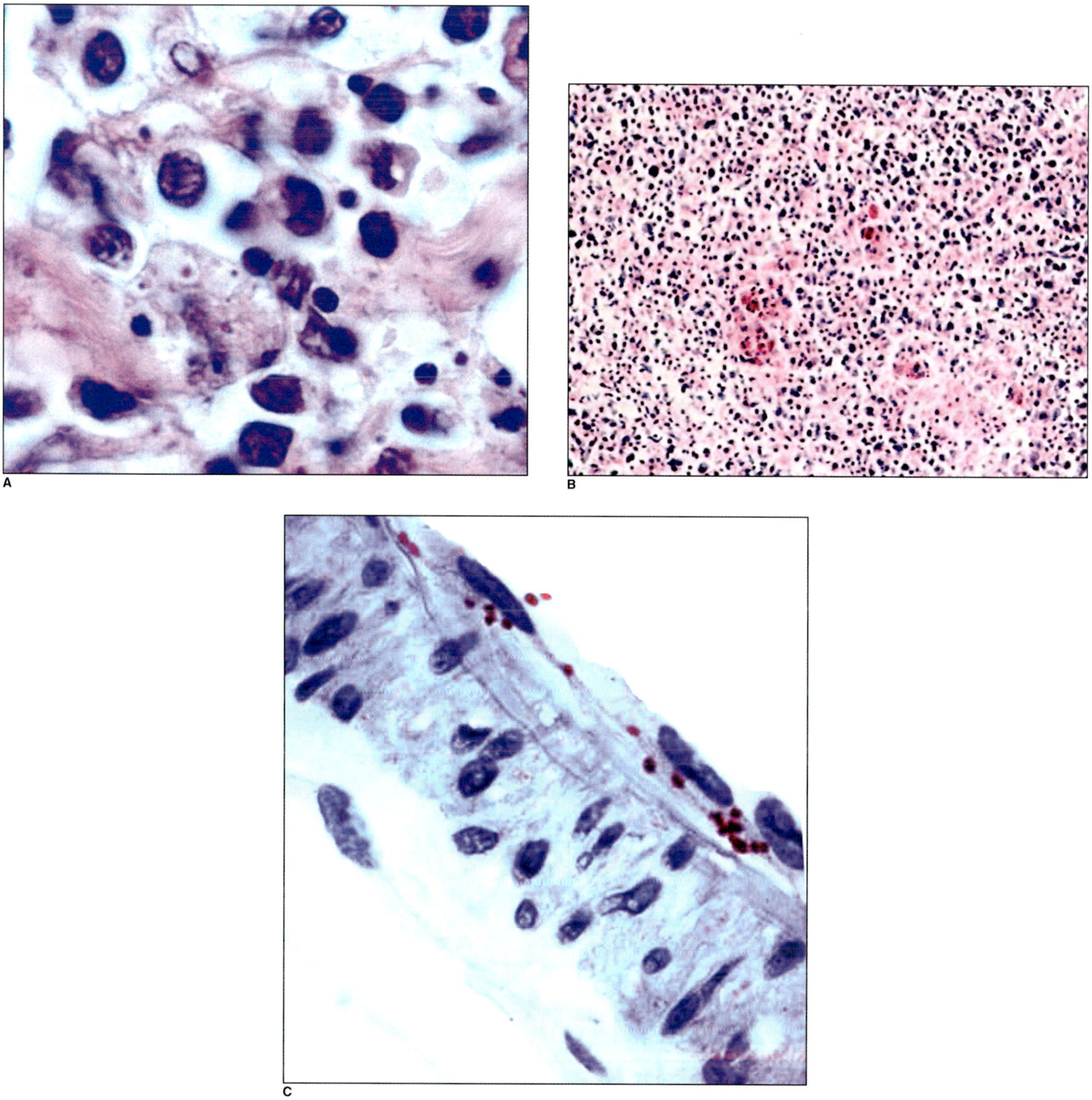

FIGURE 13.24: ***(A) This oil immersion photomicrograph of a skin lesion from a patient with diffuse papulohemorrhagic lesions shows granular inclusions in swollen endothelial cells. (B and C) The Castaneda method stains the endothelial cells magenta, supporting the interpretation of a rickettsiosis. The patient had Rocky Mountain spotted fever.***

helpful in further delineating cytologic details. If a solid confluence of melanin-containing cells is present at the epidermal base—especially if dendritic projections are inconspicuous in such cells—a diagnosis of LMIS would be favored .

"Spitzoid" Melanocytic Neoplasms

Melanocytic proliferations that involve the dermoepidermal junction may be composed of large epithelioid cells or fusiform cells, or both. If architectural and cytologic attributes of these lesions are indeterminate, one may have significant doubts about whether they represent Spitz nevi or Spitz-like melanomas. Although they are not pathognomonic (and can indeed be observed in some melanomas), intratumoral "Kamino bodies" in significant numbers are felt to give diagnostic support to a diagnosis of Spitz nevus (148). These structures—described by Kamino in 1979—are extracellular eosinophilic globules that are seen between lesional melanocytes in hematoxylin and eosin

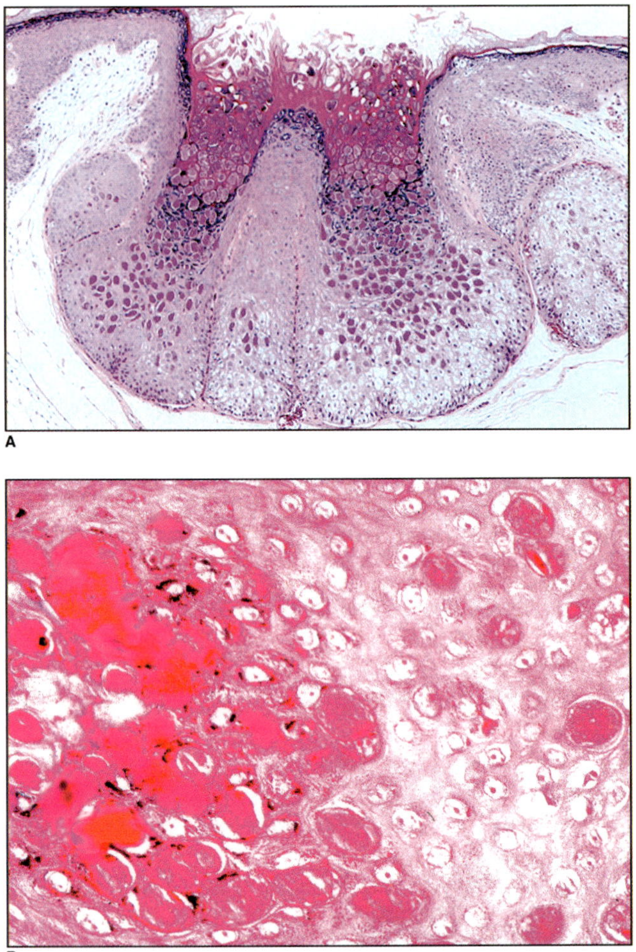

A

B

FIGURE 13.25: ***(A) Molluscum contagiosum is characterized by eosinophilic nuclear inclusions in follicular keratinocytes. (B) The Feulgen method labels the inclusions red, owing to their content of abundant viral deoxyribonucleic acid.***

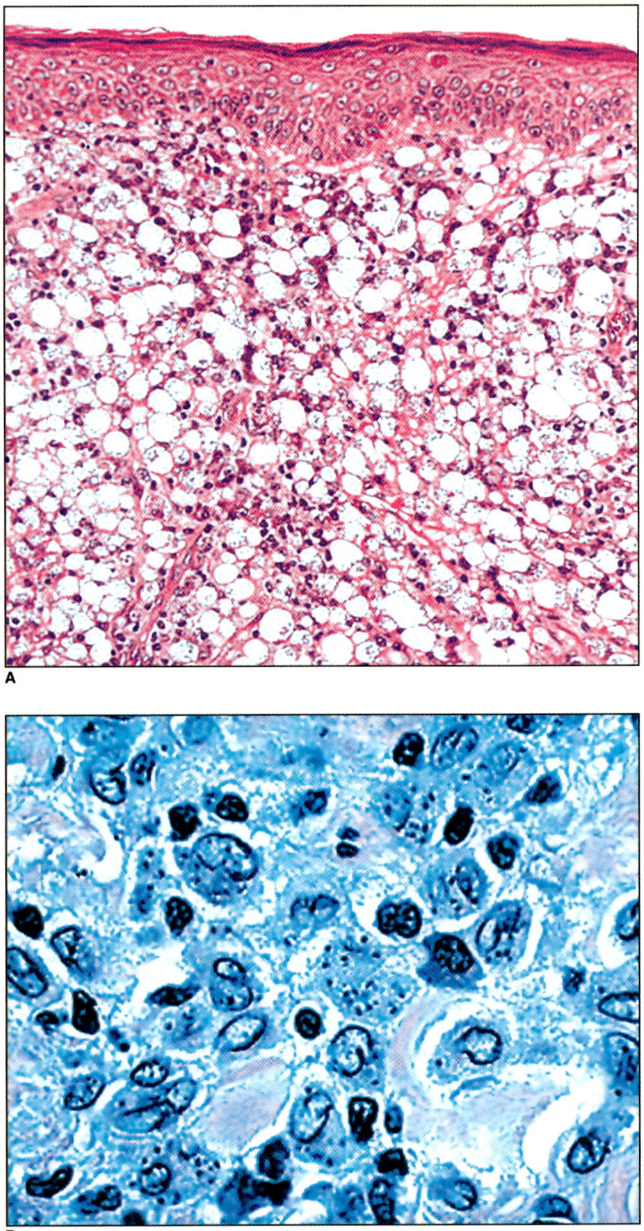

A

B

FIGURE 13.26: ***(A) Cutaneous leishmaniasis, yet another infectious disease with "parasitized" dermal histiocytes. (B) Protozoan intracellular organisms in leishmaniasis are labeled with the Giemsa technique.***

stains. They have been shown to contain basement membrane materials and, as such, are reactive with the PAS-D and MT methods (148) (Figure 13.29).

Glandular and Pseudoglandular Tumors of the Skin

Another relatively frequent problem in dermatopathology is the separation of carcinomas that contain internal spaces—possibly glandular lumina—from one another. Truly glandular tumors include primary sweat gland adenocarcinomas, adenosquamous carcinomas, and metastatic adenocarcinomas, whereas primary adenoid basal cell carcinomas and adenoid (acantholytic) squamous cell carcinomas are simulants of the former neoplasms.

Stains for epithelial and stromal mucins, such as the PAS-D and mucicarmine methods, and the alcian blue or CI techniques, respectively, are useful in this setting. Neoplasms containing true glands will demonstrate PAS-D positivity or mucicarmine reactivity (Figure 13.30), or they show labeling with alcian blue or CI that persists after hyaluronidase pretreatment (149–151). In contrast, adenoid basal cell carcinomas and acantholytic squamous cell carcinomas are expected to demonstrate the converse of the histochemical profile just described (152,153).

A related but separate conundrum concerns the diagnostic separation of cutaneous mixed tumors with abundant mucomyxoid stroma from mucinous sweat gland carcinomas of the skin. Here, both of the neoplastic entities in question undeniably contain relatively bland glandular components, but the nature of the intervening

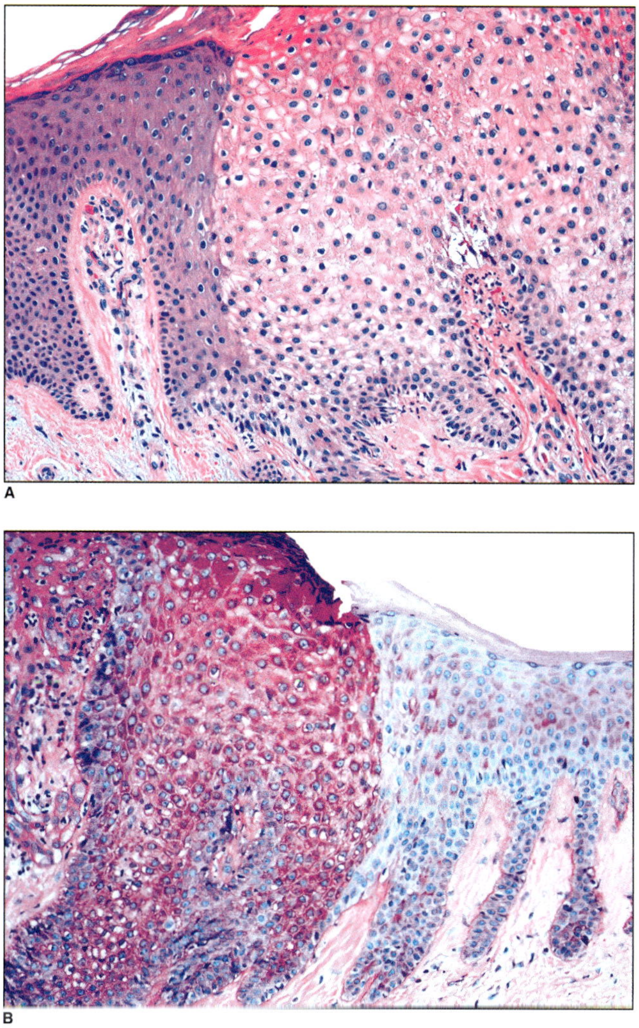

FIGURE 13.27: ***(A) A well-demarcated proliferation of epidermal keratinocytes with clear cytoplasm is seen in this clear cell acanthoma of Degos. (B) The PAS stain yields intense positivity; it was abrogated by predigestion with diastase, indicating the presence of glycogen in the tumor cells.***

matrical material is dissimilar. In mixed tumors, the stroma comprises a mixture of sulfated acid mucopolysaccharide and collagen; it reacts with alcian blue, CI, mucicarmine, picrosirius red, MT, and toluidine blue procedures, and staining is not appreciably decreased by pretreatment with hyaluronidase (154,155). In mucinous carcinoma, the intercellular matrix has mixed properties of epithelial and stromal mucins; it is labeled by PAS-D (Figure 13.31), mucicarmine, alcian blue, and CI, and is hyaluronidase sensitive (149,156,157). However, there is no staining with the MT and picrosirius red procedures because of an absence of collagen.

Granular Cell and Oxyphilic Intradermal Tumors

Granular cell tumors (GCTs) of the skin are relatively uncommon, and they have several potential lineages of differentiation (158–161). Most are neural, but others are variants of basal cell carcinoma, leiomyoma, angiosarcoma, or leiomyosarcoma (161). Finally, rare examples are apparently "null-cell" lesions (162).

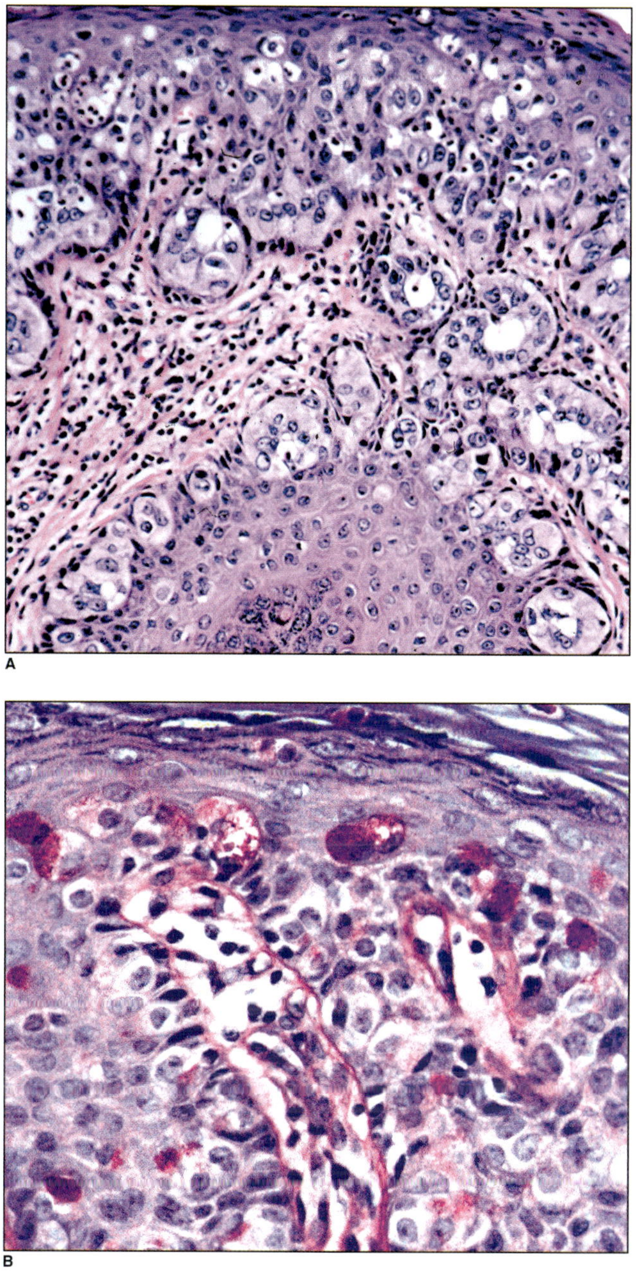

FIGURE 13.28: ***(A) "Foreign" pale cells are seen throughout the epidermis and in the follicular sheaths, with focal gland formation, in cutaneous PD. (B) The Paget cells label with the PAS method after diastase digestion, indicating the presence of epithelial mucin.***

Histochemistry is helpful in confirming the fact that GCTs are, in fact, "granular"; that is to say, they are typified by large numbers of cytoplasmic phagolysosomes that

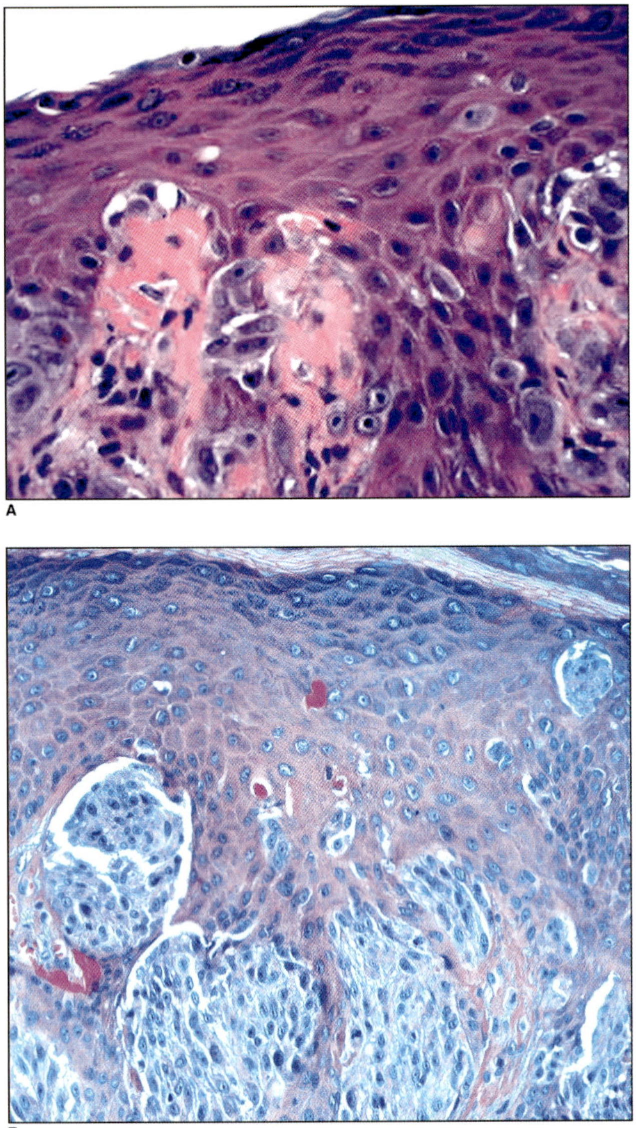

A

B

FIGURE 13.29: (A) Spitz nevi often show pale eosinophilic rounded inclusions at the base of the epidermis. (B) These structures, known as "Kamino bodies" are typically positive with the PAS method after diastase digestion.

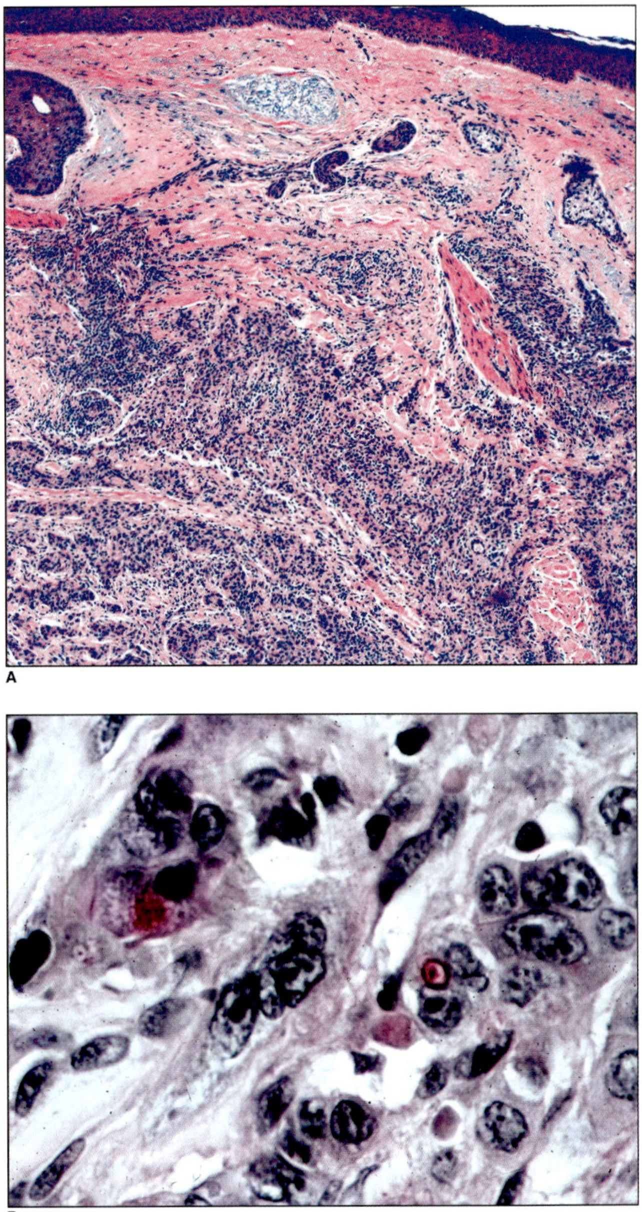

A

B

FIGURE 13.30: (A) This ductal eccrine adenocarcinoma is composed of narrow branching cords of atypical epithelial cells in the dermis. The differential diagnosis principally centers on basal cell carcinoma. (B) Mucicarmine staining shows intracellular reactivity in the tumor cells, establishing the glandular nature of the neoplasm.

contain lipofuscin and similar glycolipoproteins. The latter moieties are associated with PAS-D reactivity (158) (Figure 13.32) and sometimes are labeled by acid-fast (e.g., Ziehl-Neelsen) or luxol fast blue methods as well (77). The principal differential diagnostic alternatives to GCT, represented by oxyphilic neoplasms such as apocrine tumors, metastatic carcinomas from the kidney or breast, variants of melanoma, and cutaneous rhabdomyoma, do not show the same histochemical phenotype. In particular, rhabdomyoma demonstrates distinctive cytoplasmic labeling of sarcomeric proteins with the PTAH procedure (163).

Large Cell Undifferentiated Dermal Neoplasms

Overtly malignant tumors in the corium comprising large epithelioid cells, but lacking other distinguishing morphologic features, are challenging diagnostically. As true of other anatomic sites, the main interpretative possibilities are those of undifferentiated carcinoma, melanoma, and large cell lymphoma or leukemia. Uncommonly, mesenchymal tumors such as epithelioid sarcoma (ES) or clear cell sarcoma also enter consideration.

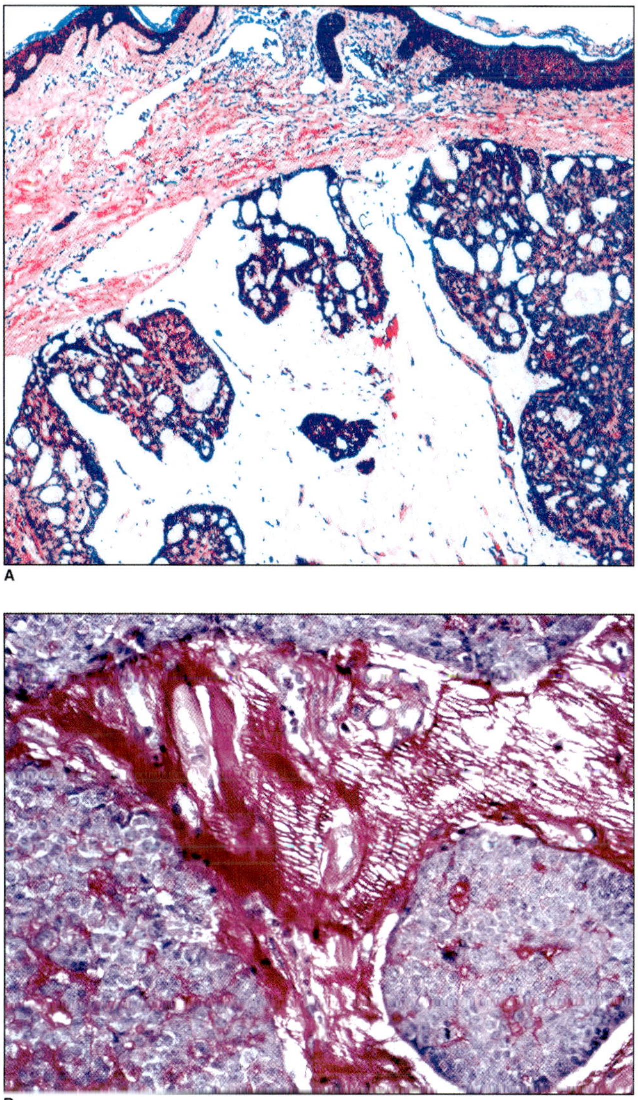

A

B

FIGURE 13.31: ***(A) Mucinous carcinoma of the skin shows cords and nests of tumor cells that are suspended in a mucomyxoid matrix. The differential diagnostic alternative to this tumor is that of myxoid-rich cutaneous mixed tumor. (B) The PAS stain labels the stroma of mucinous carcinoma after diastase digestion, establishing that it is epithelial mucin.***

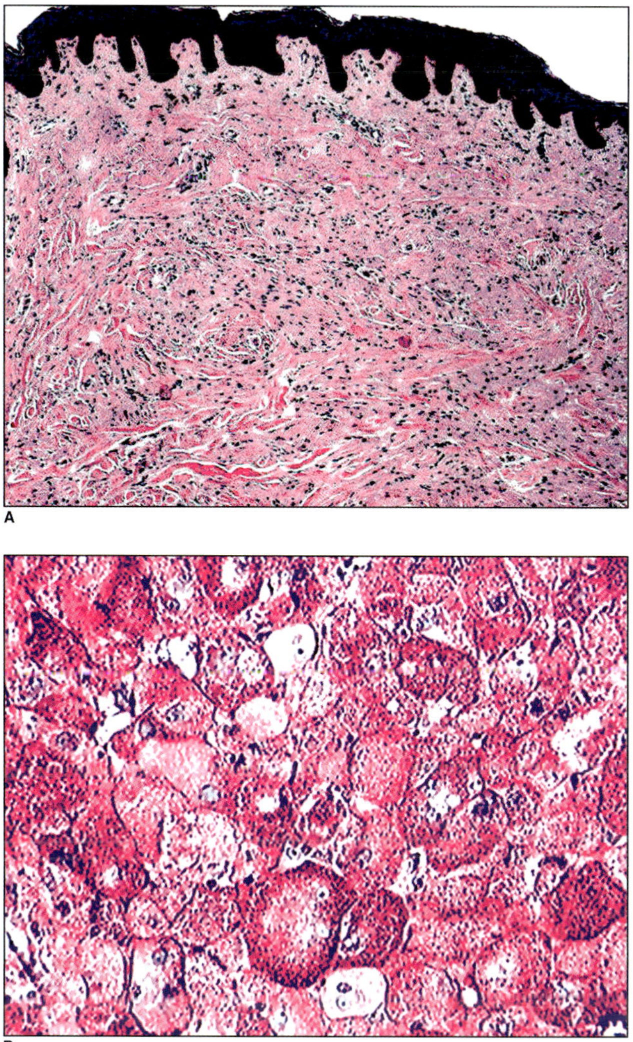

A

B

FIGURE 13.32: ***(A) This lesion shows nests and sheets of pale epithelioid cells in the dermis. The differential diagnosis principally centers on GCT versus a histiocytic proliferation. (B) The PAS stain, with diastase digestion, labels each of the intracellular granules in the lesional cells. This finding supports the diagnosis of GCT.***

Histochemical evaluation can be an important first step in addressing this problematic area. Glycogen content in the tumor cells, as reflected by PAS positivity with diastase lability, tends to argue strongly against the possibility of lymphoma (164,165) but can be seen in any of the other lesions in this categorical group. Intracellular PAS-D reactivity or mucicarmine labeling of a neoplasm militates for an interpretation of carcinoma (Figure 13.33). Melanin content, as judged by the results of the FM stain, the Schmorl technique (Figure 13.34), or metachromaticity with the azure-B method, obviously favors melanoma but is also a characteristic of clear cell sarcoma in >70% of cases (166,167).

A particular pitfall attending some examples of ES concerns their reactivity for hyaluronidase-sensitive stromal mucin with the CI or alcian blue procedure. Such staining is again observed in >70% of ES cases (168). Because ES can be closely simulated morphologically by cellular GA, which also contains stromal mucin (169), one must be particularly attentive to the level of cytologic atypia in the proliferating cells in order to avoid a mistake.

Large cell lymphoma and extramedullary leukemia (granulocytic sarcoma) tend to show the strongest labeling in this tumor group with the MGP method; although admittedly, that is a matter of degree. These two hematopoietic processes can often be separated from one another with the Leder, Giemsa, or toluidine blue stain. Granular

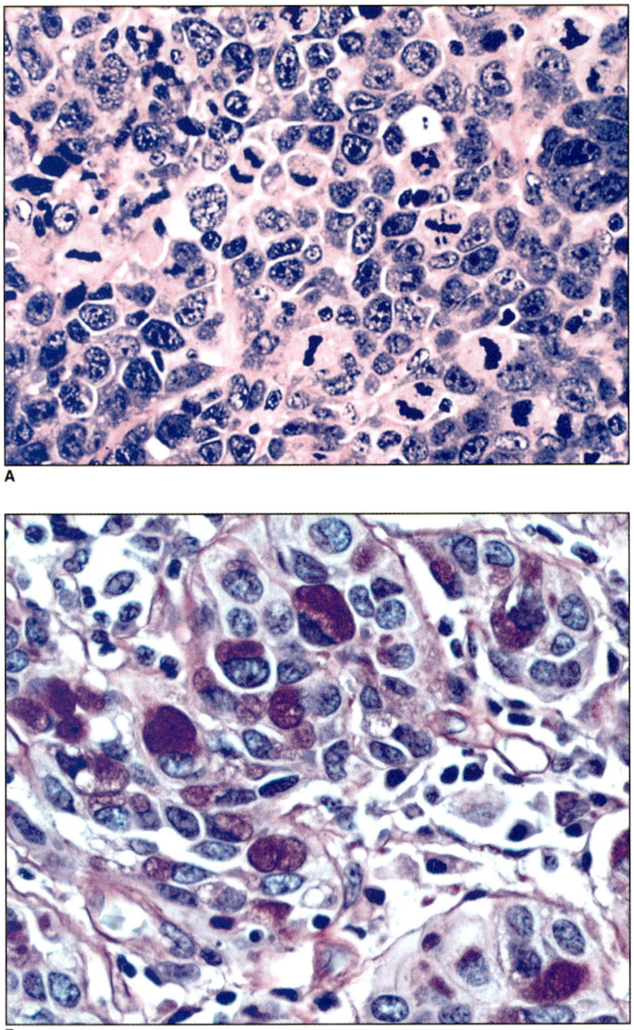

A

B

FIGURE 13.33: (A) This dermal neoplasm is undifferentiated on hematoxylin and eosin–stained sections. (B) PAS staining, after diastase digestion, shows intracellular mucin inclusions, indicating a diagnosis of poorly differentiated adenocarcinoma. The lesion was a metastasis from the lung.

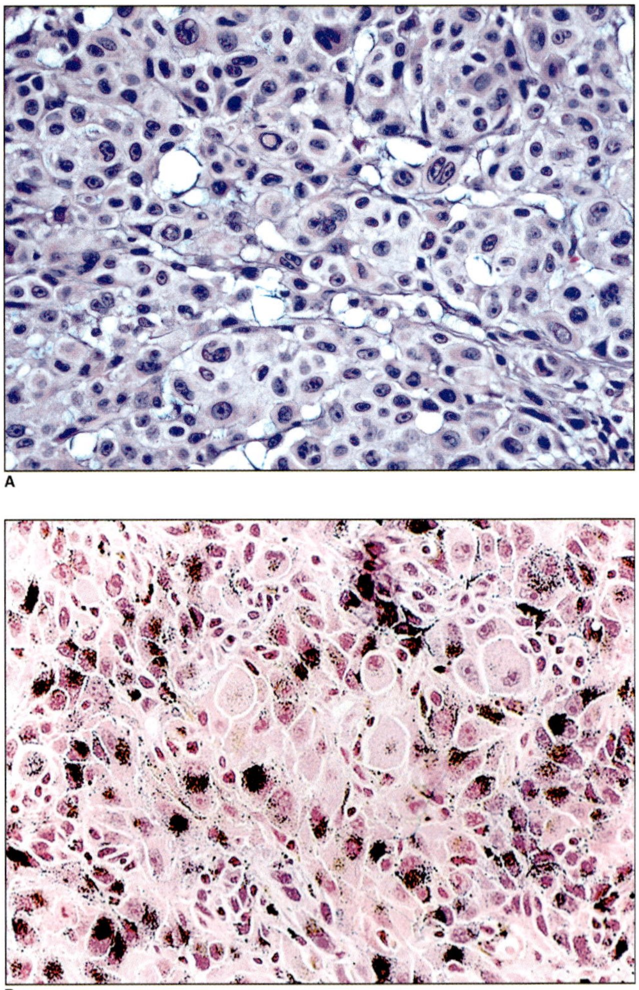

A

B

FIGURE 13.34: (A) This nodular deep dermal malignancy is composed of nests of epithelioid cells without other distinguishing morphologic characteristics. (B) An FM stain demonstrates multifocal positivity in the lesion, supporting the diagnosis of metastatic melanoma.

cytoplasmic positivity with any of the latter three techniques is limited to myeloid leukemia in this context (170).

Small Cell Undifferentiated Tumors

Similar comments can be made to those in the preceding section, in reference to small cell undifferentiated cutaneous tumors. In this morphologic category, differential diagnostic considerations include small cell (basaloid) squamous carcinoma, "anaplastic" basal cell carcinoma variants, small cell sweat gland carcinoma, Merkel cell (primary neuroendocrine) carcinoma (MCC), metastatic small cell neuroendocrine carcinoma (usually from the lung), small cell melanoma, lymphoblastic or Burkitt-type lymphoma, primitive neuroectodermal tumors (PNETs), and, rarely, primary cutaneous rhabdomyosarcoma (171).

Once again, reactivity for glycogen conflicts with a possible diagnosis of lymphoma but can be observed in essentially all the other lesions in this category. Nonetheless, a particular differential diagnosis in which strong PAS positivity is useful is that of MCC versus PNET. The latter of these tumors is often intensely reactive, whereas MCC contains sparse amounts of glycogen if any at all (171–173). Rhabdomyosarcomas are likewise commonly PAS positive, but they show a greater degree of nuclear heterogeneity than that seen in PNETs. Fibrillary cytoplasm may be demonstrable in some rhabdomyosarcomas with the PTAH method (Figure 13.35), whereas neither Merkel cell carcinoma nor PNET shows that feature (174).

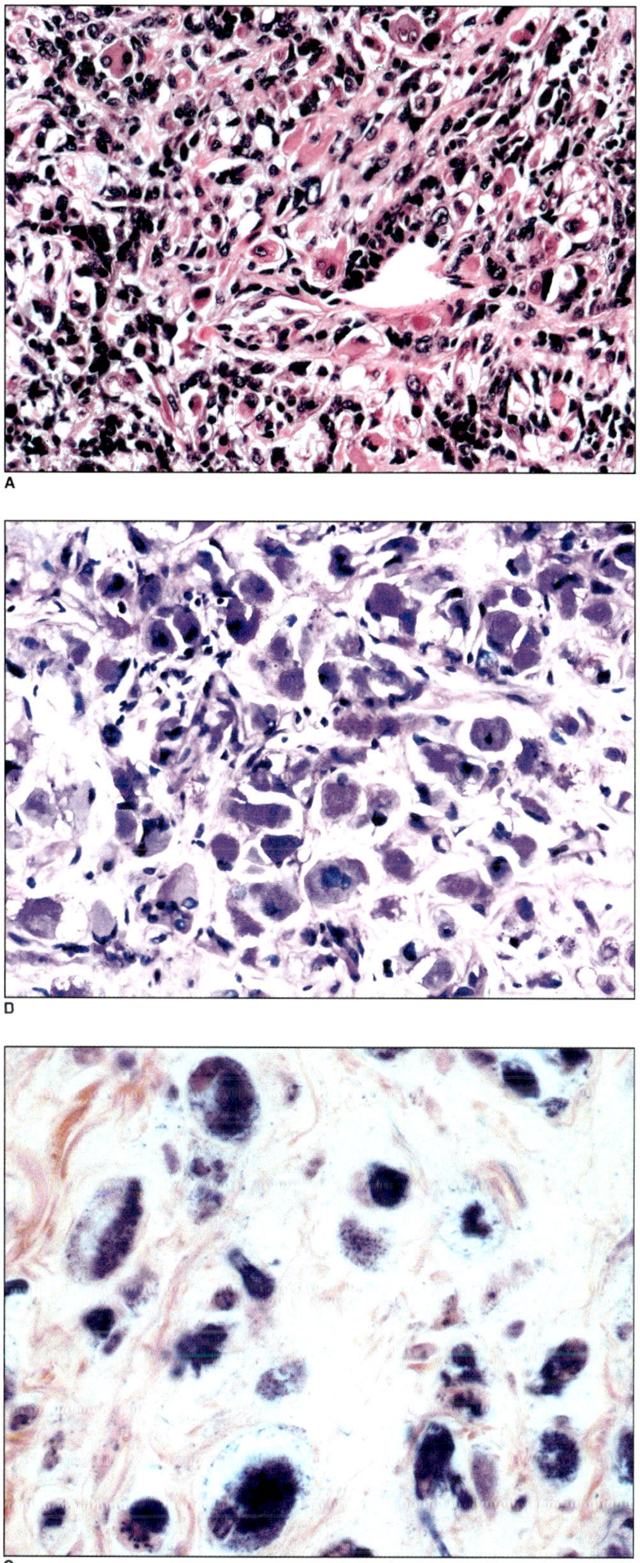

FIGURE 13.35: ***(A) This subcutaneous small round cell malignancy from an adolescent is morphologically ambiguous. It labels with the PAS method (B) and is diastase labile. The PTAH procedure (C) also shows positivity. The tumor is an alveolar rhabdomyosarcoma.***

Another caveat in this tumor group is that both basal cell carcinoma and melanoma may be reactive for melanin in the tumor cells (175). Although each of those neoplasms also may demonstrate the synthesis of stromal mucin, basal cell carcinoma much more consistently does so and in greater quantity than that seen in melanomas (176,177).

Differential silver impregnation methods are potentially helpful in separating MCC, PNET, metastatic small cell neuroendocrine carcinoma, and small cell melanoma, all of which have a generic neuroectodermal lineage. MCC and metastatic small cell carcinoma are positive with argyrophilic techniques such as the Grimelius, Sevier-Munger, and Churukian-Schenk procedures, in approximately 50–60% of cases (178–181) (Figure 13.36). Primary fixation of the tumor tissue in Bouin's solution enhances the argyrophilia of such lesions. In contrast, neither PNET nor melanoma has this property. Instead small cell melanoma is often argentaffinic, labeling with the FM technique (140,171).

Lymphomas demonstrate the most intense MGP reactivity among the lesions in this category. However, the differential gradient of labeling with this method is less than that seen among large cell undifferentiated tumors (182,183).

Spindle Cell Proliferations

Several benign and malignant neoplasms are potential dermal lesions. These may be approached in two groups, separated by their degrees of cytologic atypia. Only those tumors that lend themselves to histochemical characterization are considered here.

Cytologically Bland Lesions: Digital (congenital) fibroma (-tosis), calcifying aponeurotic fibroma, leiomyoma cutis, arteriovenous hemangioma, hemosiderotic dermatofibroma, neurofibroma, and neurilemmoma are benign spindle cell tumors that have distinctive special staining characteristics.

The cells of digital fibroma consistently contain peculiar paranuclear globular inclusions that are eosinophilic in hematoxylin and eosin stains. These structures are also labeled blue violet with the MT method, and violet with the PTAH technique (Figure 13.37). They are felt to represent altered actin filaments (184,185).

Calcifying aponeurotic fibroma has a predilection for the distal extremities in children and adolescents. It is typified histochemically by multifocal reactivity with the von Kossa procedure or the alizarin red stain, and by dense matrical labeling with the MT method (186,187).

Leiomyomas of the skin demonstrate delicate intracellular fibrillation. The fibrils are red with the MT stain and blue violet with the PTAH technique (188).

The walls of vascular channels in arteriovenous hemangiomas (acral arteriovenous tumors) have closely similar

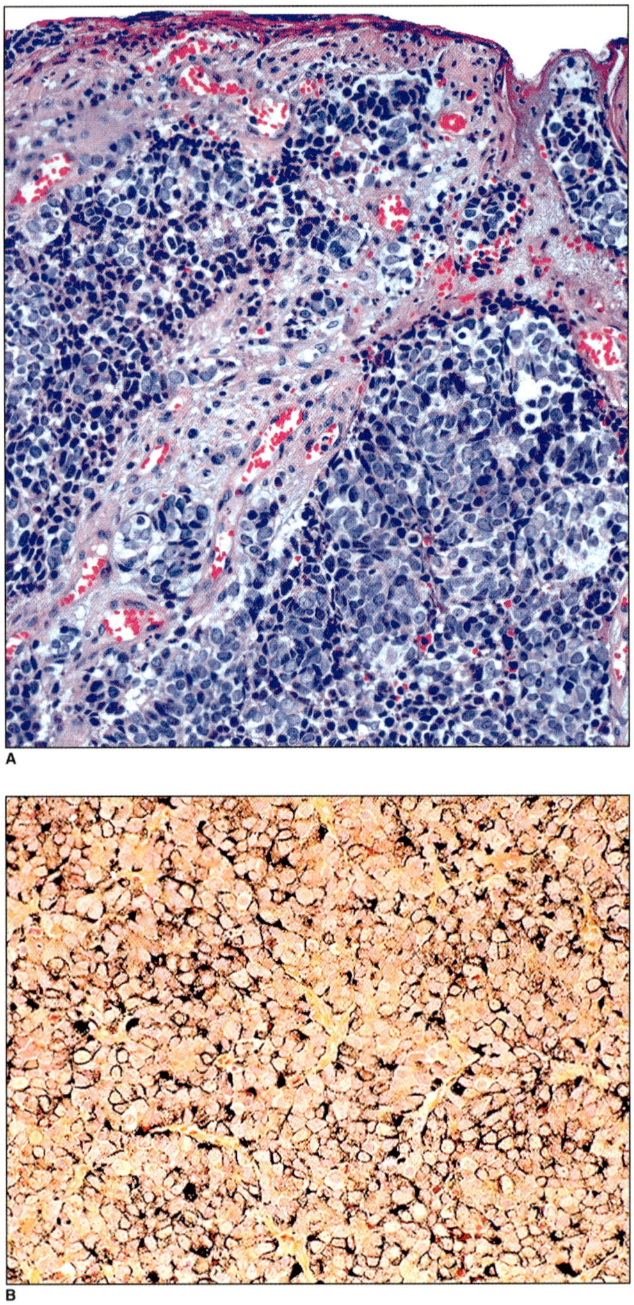

FIGURE 13.36: (A) This malignant small round cell tumor of the skin is from the face of an elderly man. (B) The Churukian-Schenk technique shows positivity in many of the neoplastic cells, after fixation in Bouin's solution. The diagnosis is MCC.

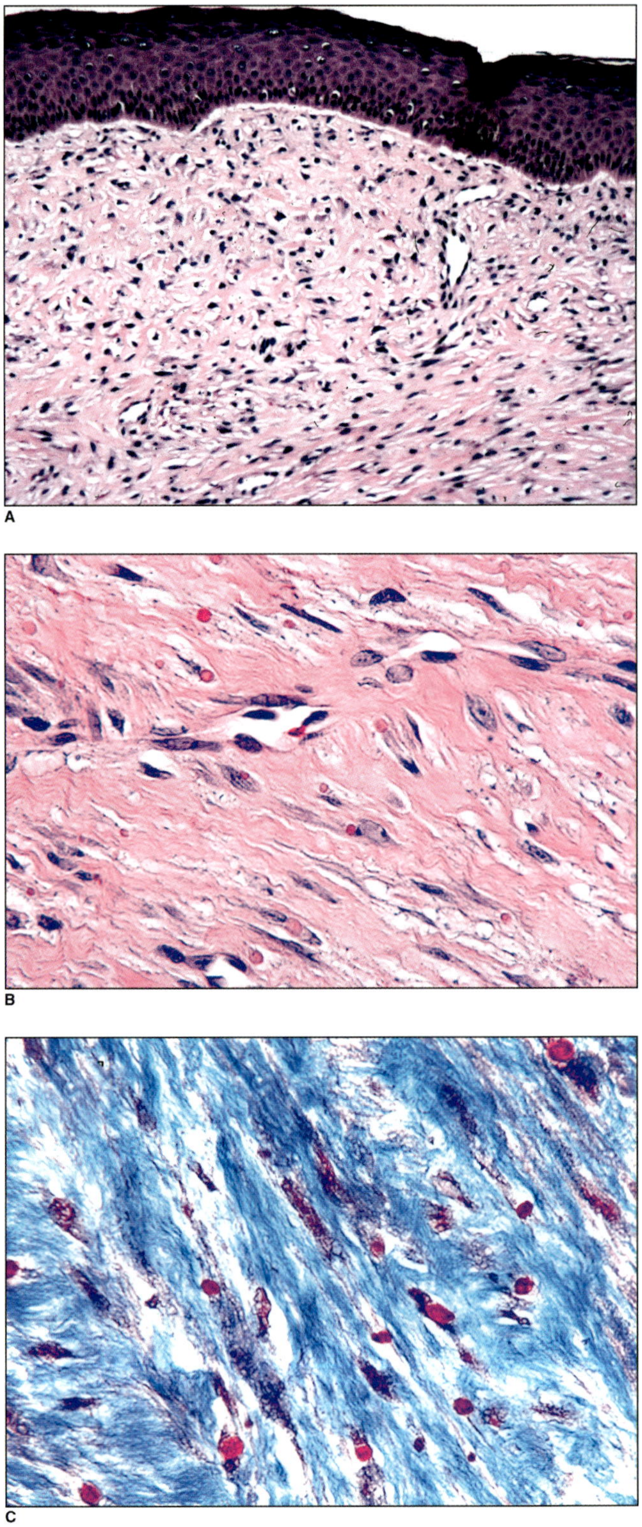

FIGURE 13.37: (A) This spindle cell tumor is from the periungual skin of a finger in a 5-year-old child. The neoplastic cells show rounded eosinophilic intracellular inclusions (B) that are labeled red with the Masson trichrome stain (C). The diagnosis is digital fibroma(-tosis).

properties to those of leiomyomas, owing to their content of smooth muscle. However, they lack elastin as seen in VVG stains (Figure 13.38) or nerve bundles as demonstrated with the Bodian technique. These features are unlike those seen in the principal differential diagnostic alternative, true arteriovenous malformations (189–191).

As its name suggests, hemosiderotic dermatofibroma contains abundant deposits of iron salts. These may be

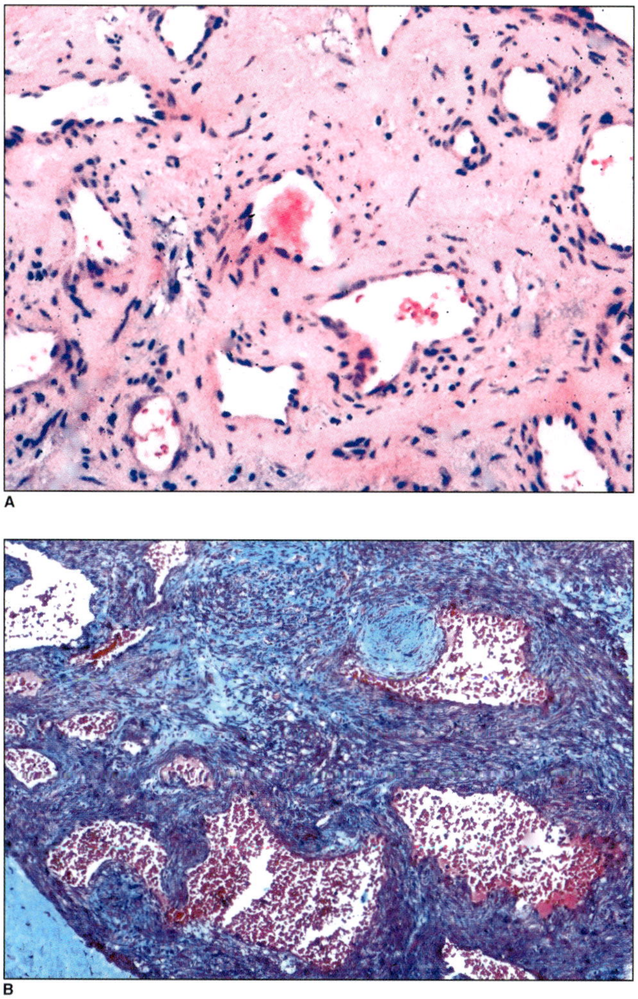

FIGURE 13.38: (A) This lesion from the face of an adult man features relatively large vascular channels that are surrounded by a spindle cell mantle. The Masson trichrome stain (B) shows no evidence of elastic tissue in the mantle, supporting the diagnosis of arteriovenous hemangioma (acral arteriovenous tumor).

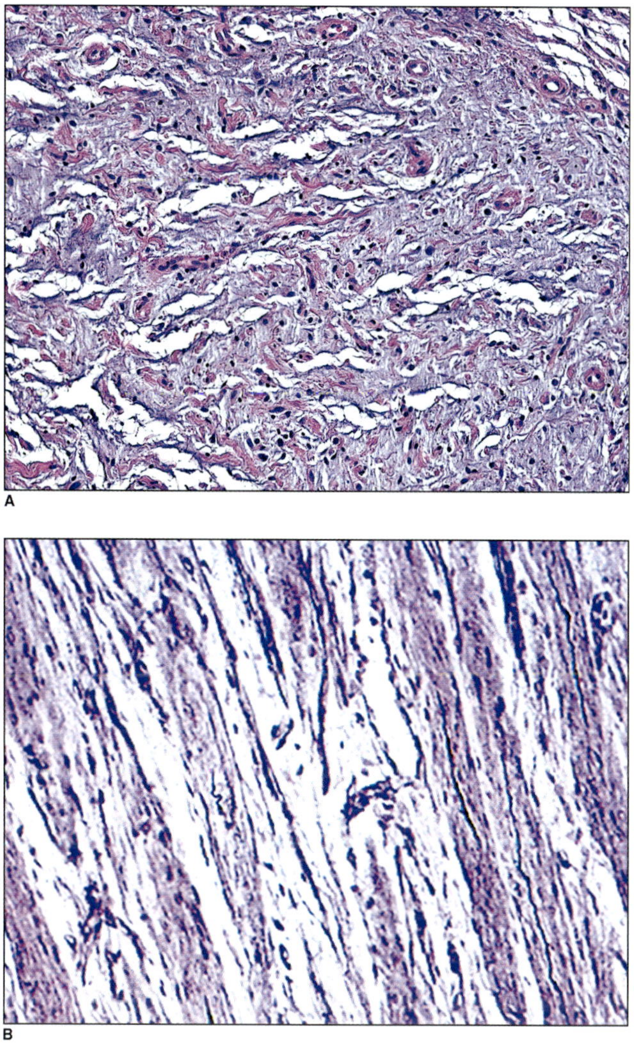

FIGURE 13.39: (A) A bland spindle cell proliferation in the dermis is seen here, with "serpiginous" nuclear contours. (B) Bodian staining demonstrates linear positivity throughout the lesion, supporting the diagnosis of neurofibroma.

accentuated with the Perl or Prussian blue method (192,193).

Neurofibroma can be separated from neurilemmoma with stains that impregnate neural processes, such as the Bielschowsky or Bodian technique. Neurites are present randomly throughout the substance of neurofibromas (Figure 13.39), whereas they are seen only at the peripheral aspects of neurilemmomas (schwannomas) (194,195). Encapsulated (solitary palisading) neuromas have a histochemical profile that is similar to that of neurofibroma (196).

Cytologically Atypical Proliferations: Three malignant spindle cell neoplasms have histochemical characteristics that may be helpful diagnostically. The features of leiomyosarcoma in MT and PTAH stains are essentially the same as those of leiomyoma, except that fibrillary differentiation is more limited in the former of those tumors (197). Second, pigmented dermatofibrosarcoma protuberans (also known as the "Bednar tumor") contains melanin-producing spindle cells that are scattered throughout the lesion (198). This fact may be biologically enigmatic, but it is paradoxically helpful in excluding a diagnosis of spindle cell melanoma, which is virtually always amelanotic (140,199,200). The FM and Schmorl methods are effective at labeling the melanin-bearing elements in the Bednar tumor (198,201) (Figure 13.40). Third, Kaposi's sarcoma demonstrates the formation of eosinophilic globules that are principally intercellular, but which may also be seen in the cytoplasm of the tumor cells. They are absent in differential diagnostic alternatives such as Kaposiform hemangioendothelioma, angiomatoid fibrous histiocytoma, hobnail hemangioma, acroangiodermatitis, or proliferating dermal scar (202). The globules are believed to be fragments of erythrocytes; they can be labeled

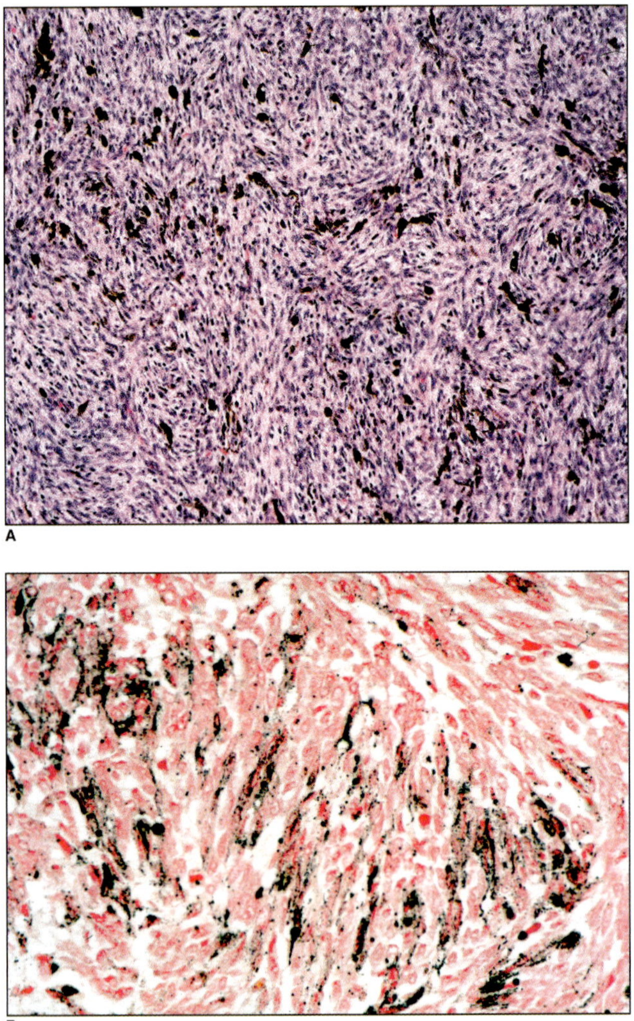

FIGURE 13.40: (A) This spindle cell tumor of the dermis extended widely into the subcutaneous tissue, and it contains numerous pigmented cells. They are decorated by the FM technique (B) indicating the presence of melanin and establishing the diagnosis of Bednar tumor (pigmented dermatofibrosarcoma protuberans).

with the benzidine reaction, the PTAH method, and the PAS-D stain (203,204) (Figure 13.41).

Metastatic Carcinomas in the Skin

Once a metastatic carcinoma of unknown anatomic origin has been identified in the skin, one may wish to undertake additional studies in an attempt to localize its primary site (205). In that context, the alcian blue method at pH 2.5; the PAS procedure, with and without diastase; the high-iron diamine–alcian blue method; the periodic acid–sodium borohydride–potassium hydroxide–PAS technique; and the galactose oxidase-Schiff stain may be helpful. The first two of these procedures are generic stains for glycogen or epithelial mucin. All the remaining techniques just listed are aimed at detecting sulfated mucins that predominate in enteric epithelial cells and their tumors (206–208). The oil red O or Sudan IV methods may be included as well, if frozen tissue is available (209). Examples of typical histochemical phenotypes in selected tumor types are presented in Table 13.3.

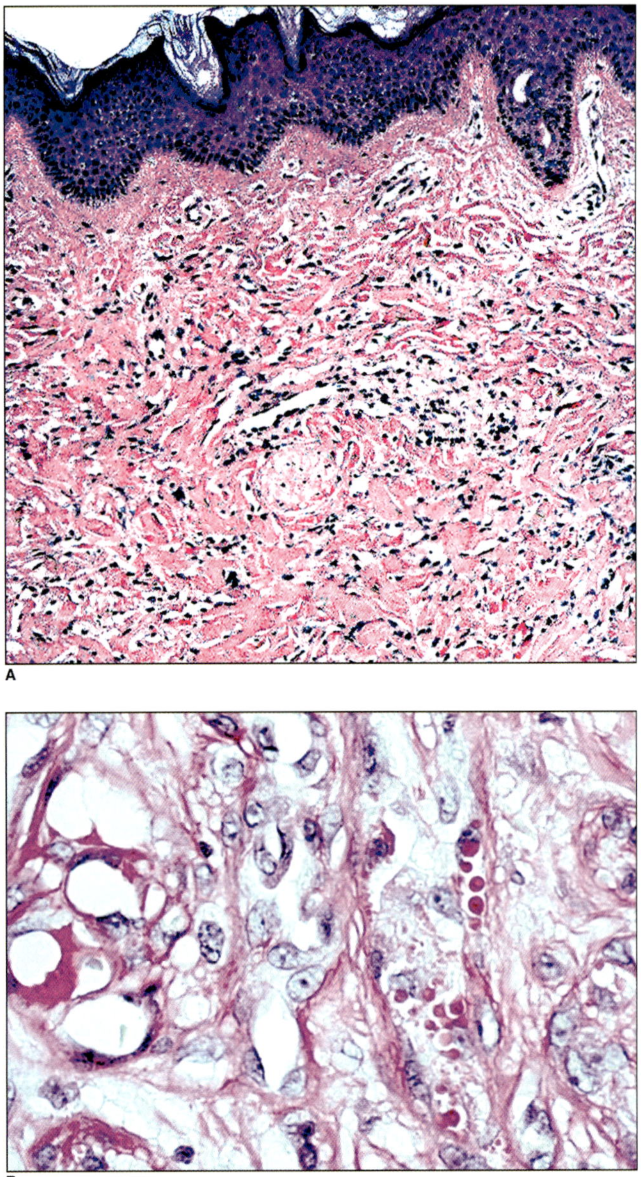

FIGURE 13.41: (A) Subtle hypercellularity is seen throughout the dermis in this ecchymosis-like lesion of the chest from a 40-year-old man. (B) PAS staining after diastase digestion reveals numerous globular inclusions in the lesional cells, supporting the diagnosis of Kaposi's sarcoma.

Lectin Histochemistry in Tumors of the Skin

Several studies have been done regarding lectin binding to cutaneous neoplasms, but, for the most part, their results have shown so much variation that few diagnostic applications have been realized. The principal plant agglutinins

TABLE 13.3: Histochemical Features of Metastatic Adenocarcinomas in the Skin & Differential Diagnostic Alternatives

Tumor Type	FM	PAS	PAS-D	MUCI	AB2.5	HID	SPPAS	GOS
Breast ACA	0	+/–	+/–*	+/–*	+/–*	0	0	0
Lung ACA	0	+/–	+/–	+/–	+/–	0**	0**	0**
GI Tract ACA	0	+	+	+	+	+	+	+
Kidney ACA	0	+	0	0	0	0	0	0
Melanoma	+	+/–	0	0	0	0	0	0
Lymphoma & Leukemia	0	0	0	0	0	0	0	0

FM = Fontana-Masson, PAS = Undigested periodic acid-Schiff, PAS = Digested PAS, MUCI = Mucicarmine, AB2.5 = Alcian blue at pH 2.5, HID = High iron-diamine alcian blue, SPPAS = Sodium borohydride-potassium hydroxide-PAS, GOS = Galactose oxidase-Schiff, ACA = Adenocarcinoma.

* Lobular carcinoma of breast often shows *intracellular* reactivity for mucin.

** Rare adenocarcinomas of the lung have an enteric phenotype and may show reactivity with these stains.

that appear to have some usefulness in dermatopathology are *Ulex europaeus* I agglutinin (UEA), *Arachis hypogaea* (peanut) agglutinin (PNA), *Helix pomatia* agglutinin (HPA), *Dolichos biflorus* agglutinin (DBA), concanavalin-A (CON-A), and *Ricinus communis* agglutinin (RCA). As stated in other chapters, all these lectins can be obtained commercially in biotinylated form, such that their binding to tissue sections can be assessed histologically with the avidin-biotin-peroxidase procedure.

Large Cell Undifferentiated Neoplasms: As considered earlier in this chapter, lymphomas, melanomas, carcinomas, and even sarcomas may assume morphologically indeterminate appearances and be confused with one another on hematoxylin and eosin stains. In addition to the techniques discussed above, lectin studies may prove beneficial in this context (Table 13.4). UEA and DBA commonly decorate carcinomas, whereas receptors for these lectins are lacking in melanomas and lymphomas (210–215). Similarly, sarcomas typically fail to bind the latter two markers, with the exception of epithelioid angiosarcoma (Figure 13.42) and ES (216–219). Thus, these two tumors are potential pitfalls, vis-a-vis their misinterpretation as carcinomas. In contrast, large cell lymphomas and melanomas are possibly reactive with RCA (in 100% and approximately 60% of cases, respectively), but this agglutinin does not recognize poorly differentiated carcinomas or the specified sarcomas (220,221). CON-A potentially labels melanomas and carcinomas, but not large cell lymphomas (220–224). PNA binding can be seen in any of the lesions considered above except for angiosarcoma.

Small Cell Undifferentiated Neoplasms: Exceptions to the information just presented pertain to selected lesions in the small cell undifferentiated tumor category. Lymphoblastic and Burkitt's lymphomas are potentially reactive with RCA, CON-A, and PNA, but so are small cell melanomas (221,225). In contrast to large cell carcinomas, MCC and metastatic small cell neuroendocrine carcinomas in the skin do not appear to label with UEA and DBA, but, like melanomas and lymphomas, both of these epithelial tumors can bind to PNA (226,227) (Figure 13.43). This fact, however, can be put to advantage in the differential diagnosis of MCC versus PNET, in which only the former tumor is PNA reactive. In addition, MCC appears to differ from visceral small cell carcinoma in regard to differential affinities for CON-A; MCC is labeled by this lectin consistently, whereas small cell lung cancer is not (228,229).

Vascular Tumors: As just stated in the preceding paragraph, UEA is effective in recognizing the great majority of endothelial neoplasms, including hemangiomas, hemangioendotheliomas, and angiosarcomas (230,231). Nevertheless, epithelial tumors commonly bind this moiety as well (215,232), and they cannot be reliably separated diagnostically from vascular proliferations only on the basis of lectin histochemistry.

Basal Cell Carcinoma Versus Basaloid Squamous Carcinoma: In superficial biopsies of the skin, a common problem involves the diagnostic separation of "metatypical" (squamoid) basal cell carcinoma from basaloid squamous cell carcinoma. Heng et al. found that only the latter tumor type bound to UEA (233). Results regarding PNA in the two neoplasms are not nearly as clear-cut (212,234–236), but they appear to demonstrate slightly more affinity for this marker by squamous carcinomas.

Prognosis of Melanoma: Cochran et al. (237) have examined the significance of cytoplasmic PNA labeling in cutaneous melanomas. They concluded that binding of that lectin was predominantly associated with vertical growth phase tumors in Clark level III or beyond. PNA reactivity was negatively correlated with tumor-related survival as well; however, it is not clear whether statistical independence from Breslow depth exists for binding of this lectin in melanomas. Thies et al. have reported similar findings in reference to HPA binding in melanomas (238).

TABLE 13.4: Lectin Binding Profiles of Cutaneous Tumors

Tumor Category	Specific Tumor Type	UEA	PNA	HPA	DBA	CON-A	RCA
Large cell undifferentiated neoplasms	Lymphoma	−	+		−	−	+
	Melanoma	−	+		−	+	+
	Carcinoma	+	+		+	+	−
	Sarcoma*	−	+		−		
Small cell undifferentiated neoplasms	Lymphoblastic & Burkitt's lymphomas		+			+	+
	Small cell melanoma		+			+	+
	Merkel cell carcinoma	−	+		−	+	
	Visceral small cell carcinoma	−	+		−	−	
	PNET		−				
Vascular tumors	Hemangiomas, hemangioendotheliomas, Angiosarcoma	+					
Basal cell carcinoma vs basaloid SCC**	Basal cell carcinoma	−					
	Basaloid SCC	+					
Melanoma growth phases	Radial growth phase melanoma		−	−			
	Vertical growth phase melanoma		+	+			
Histiocytic infiltrates	Langerhans cell histiocytosis		+				
	JXG, other xanthomatoses and histiocytoses		−				

UEA = *Ulex europaeus* I agglutinin, PNA = *Arachis hypogaea* (peanut) agglutinin, HPA = *Helix pomatia* agglutinin, DBA = *Dolichos biflorus* agglutinin, CON-A = Concanavalin-A, RCA = *Ricinus communis* agglutinin, PNET = primitive neuroectodermal tumor, SCC = squamous cell carcinoma, JXG = juvenile xanthogranuloma.

* Sarcomas typically fail to bind UEA and DBA, with the exception of epithelioid angiosarcoma and epithelioid sarcoma. RCA does not recognize the latter two sarcomas.

** Basaloid squamous cell carcinomas demonstrate slightly more affinity for PNA than is the case for basal cell carcinoma.

Langerhans Cell Histiocytosis Versus Other Histiocytic Infiltrates: Langerhans cell histiocytosis (histiocytosis-X) may be mistaken for other histiocytic infiltrates and neoplasms in the skin. In contrast to the lesional cells of juvenile xanthogranuloma, necrobiotic xanthogranuloma, and other xanthomatoses and histiocytoses, those of Langerhans cell disease consistently exhibit an affinity for PNA (238–241).

ILLUSTRATIVE CASE STUDIES

CASE 1— ***Macular Amyloidosis:*** A 45-year-old man complained of a progressively more pruritic eruption on his arms, of approximately 1 year's duration. Physical examination showed hyperpigmented 1- to 2-mm macular lesions over both forearms (Figure 13.44). Empirical treatment with topical corticosteroids produced no benefit, and a biopsy was performed.

This pathologic process was characterized by amorphous, hypocellular, hyaline eosinophilic deposits of material in the papillary and upper reticular dermis (Figure 13.45). There was no associated inflammation in the interstitium or around blood vessels, and the epidermis was unremarkable. Congo red stains yielded marked positivity in the deposits in question (Figure 13.46), and they were birefringent with an apple-green color on polarization microscopy. A diagnosis of macular amyloidosis was made. The patient was treated with dermabrasion of the affected skin areas, but the lesions recurred over the ensuing 12 months.

Regardless of whether amyloidosis in the skin is caused by a plasma cell dyscrasia (242), chronic systemic inflammatory processes (243), tumor-related deposition (as in basal cell carcinoma) (244), or genetic factors and chronic skin friction (245), the histochemical properties of the abnormal protein are comparable. The Congo red, crystal violet, sulfated alcian blue, Sirius red, or thioflavin-T

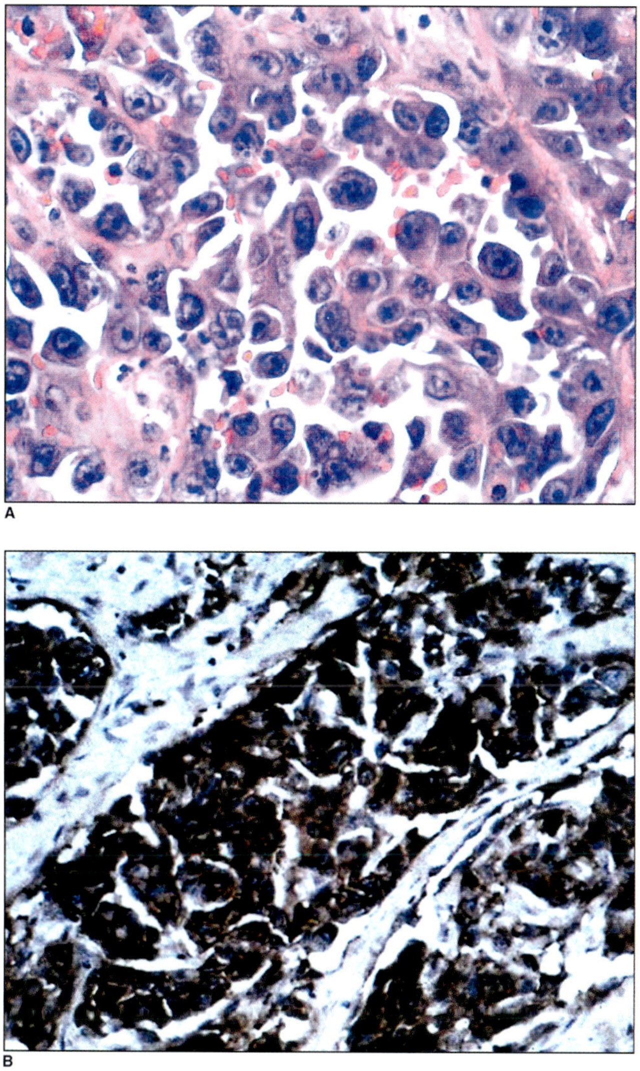

FIGURE 13.42: ***(A) Epithelioid angiosarcoma is difficult to separate diagnostically from other malignant polygonal cell tumors of the dermis. (B) Binding of UEA to the tumor cells aids in recognizing the endothelial nature of this neoplasm.***

procedures are all potentially effective in labeling amyloid. Labeling for amyloid AA, which is implicated when chronic inflammation is the etiology of amyloidosis, is greatly diminished by pretreatment of tissue sections with 0.5% potassium permanganate solution (246–248). Nevertheless, staining for other forms of amyloid protein—including AL and tumor-related types—may rarely be abrogated by permanganate as well. ■

<u>CASE 2—</u> ***Reaction to Monsel's Solution:*** A 73-year-old woman had had a crusted whitish 2-cm lesion in the skin of the face for 1 year. A biopsy of the lesion showed a squamous cell carcinoma, which was apparently in situ (Figure 13.47), but the base of the tumor had been transected in the sample and invasion could not be excluded with certainty.

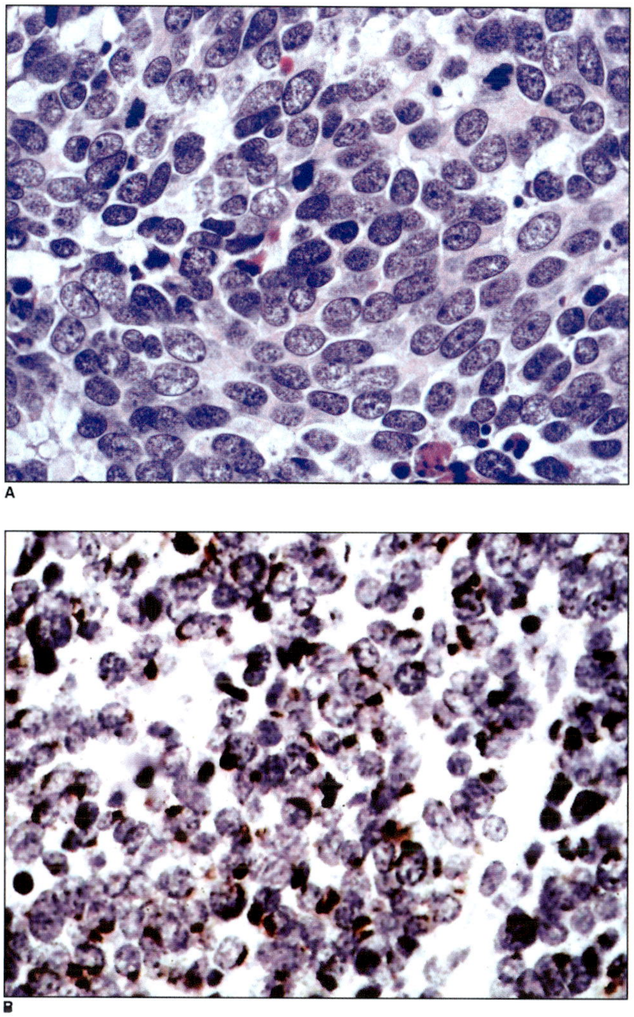

FIGURE 13.43: ***(A) This small cell dermal malignancy has cytologic features suggesting a neuroendocrine lineage. This conclusion is supported by punctate perinuclear binding of PNA to the tumor cells (B). The final diagnosis was that of metastatic small cell neuroendocrine carcinoma of the lung.***

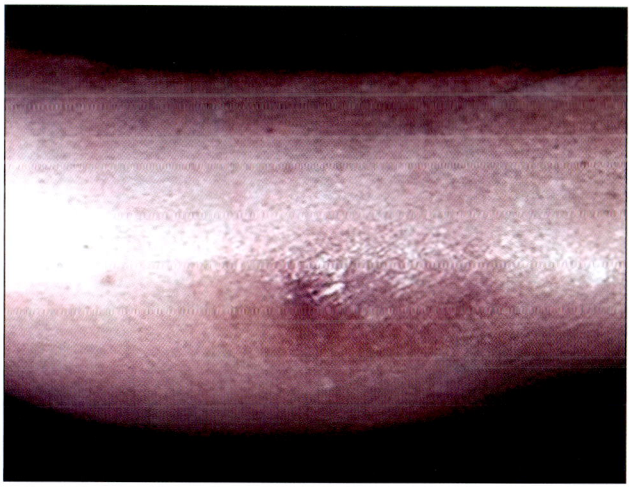

FIGURE 13.44: ***Hyperpigmented pruritic macules in the skin of the arms of a 45-year-old man.***

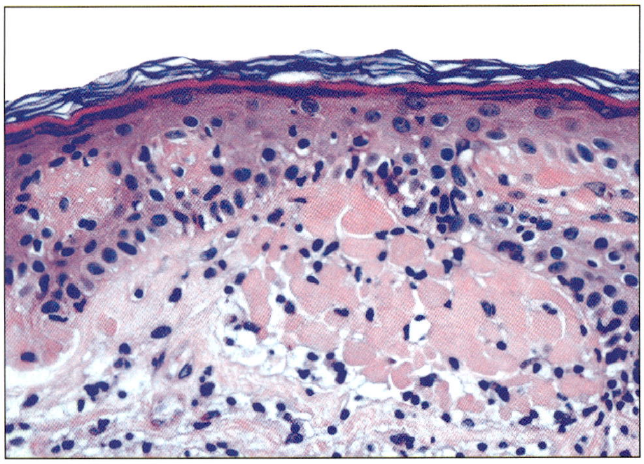

FIGURE 13.45: Hyalinized hypocellular eosinophilic deposits are seen in the upper dermis in this hematoxylin and eosin stain.

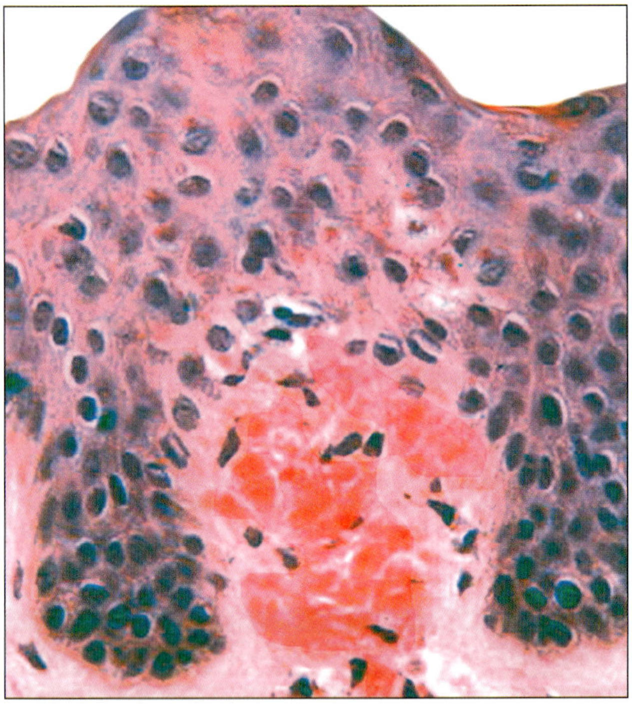

FIGURE 13.46: The deposits are strongly positive with the Congo red method. They also showed green birefringence on polarization microscopy. The final diagnosis was that of macular amyloidosis.

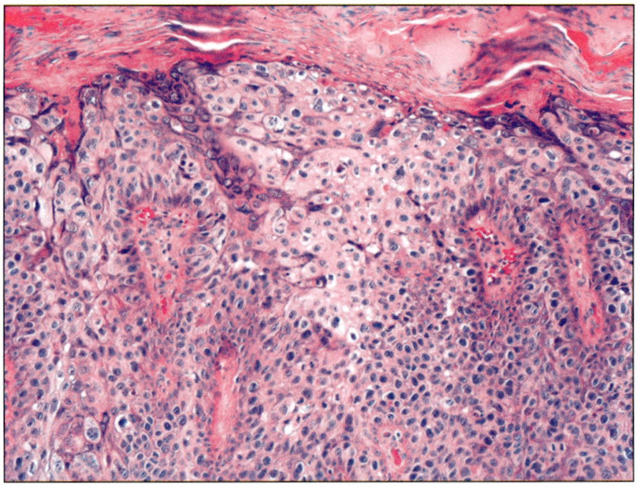

FIGURE 13.47: Squamous carcinoma in situ, as seen in this biopsy from a 73-year-old woman. It was treated by curettage.

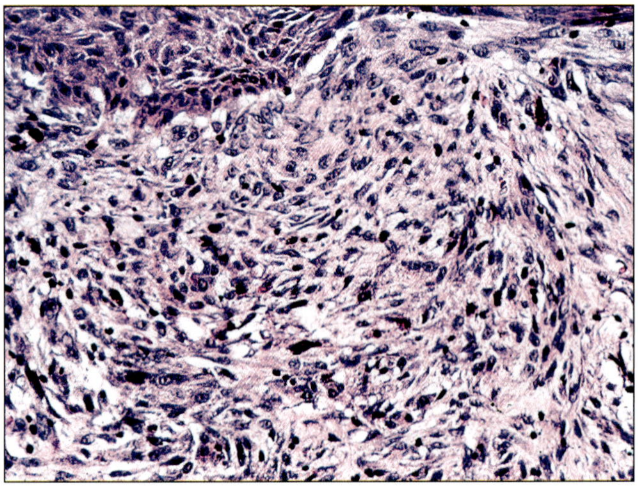

FIGURE 13.48: Re-biopsy of the lesional site, done 1 month later, showed a spindle cell proliferation in the dermis. Concern was raised over the possibility of sarcomatoid transformation of the original tumor.

The lesion was treated with curettage and left to heal by secondary intent.

One month later, the patient returned with a recurrent and enlarging plaque at the tumor site. Another biopsy was procured, which demonstrated a spindle cell proliferation, with variable but modest cytologic atypia, that undermined the adjacent epidermis (Figures 13.48 and 13.49). Differential diagnosis centered on the spindle cell transformation of a squamous cell carcinoma versus a floridly reactive fibroblastic process. Notice was taken that relatively large clumps of brownish pigment were interspersed throughout the lesional cell population histologically. These were found to react strongly with the Prussian blue method (Figure 13.50). The latter finding prompted a question to the treating dermatologist as to whether Monsel's solution had been used as a styptic after curettage, and it had. The final diagnosis was that of an idiosyncratic fibroproliferation, in reaction to the application of topical iron salt solution.

Monsel's solution is employed to staunch bleeding after dermatologic surgery or colposcopic biopsy of the uterine cervix. It is usually well tolerated with no untoward effects, but occasional patients have idiosyncratic reactions of the sort described here (249,250). ■

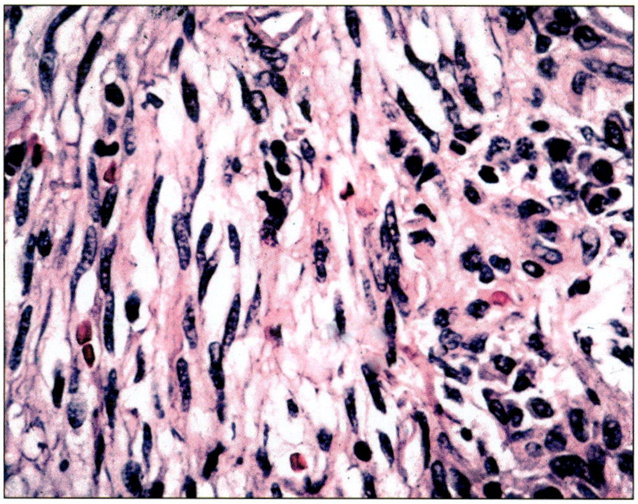

FIGURE 13.49: ***Mild nuclear atypia was present in the lesional spindle cells.***

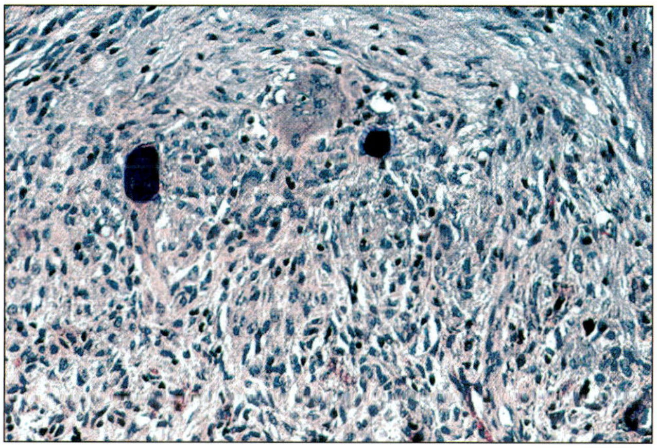

FIGURE 13.50: ***Perl's stain for iron shows multifocal globular positivity among the spindle cells, suggesting a diagnosis of reaction to Monsel's solution. It was subsequently learned that this styptic agent had indeed been used after the initial curettage.***

CASE 3— *Cutaneous Wegener's Granulomatosis:* A 51-year-old man had recurrent bouts of epistaxis over a 6-month period, along with "chronic sinusitis." He was treated for these problems supportively. Thereafter, tender 2- to 3-cm plaques began to develop on the lower legs and dorsal feet, which eventually ulcerated (Figure 13.51). Concurrently, the patient involuntarily lost 20 lbs and developed intermittent fevers to 101°F.

Physical examination confirmed the dermatologic findings just described. Chest radiographs showed reticular interstitial infiltrates throughout both lung fields, along with nodular densities. Examination of the urine sediment demonstrated the presence of erythrocytes and neutrophils, with cellular casts.

A skin biopsy was performed, showing lymphocytic vasculitis affecting medium-sized arteries in the deep dermis and subcutis, along with granulomatous infiltrates in the corium (Figures 13.52 and 13.53). MT and VVG stains revealed disruption of the elastic laminae in the affected blood vessels, together with mural fibrosis (Figures 13.54 and 13.55). A diagnosis of probable Wegener's granulomatosis was made; it was confirmed after serologic studies showed the presence of antineutrophil cytoplasmic autoantibodies.

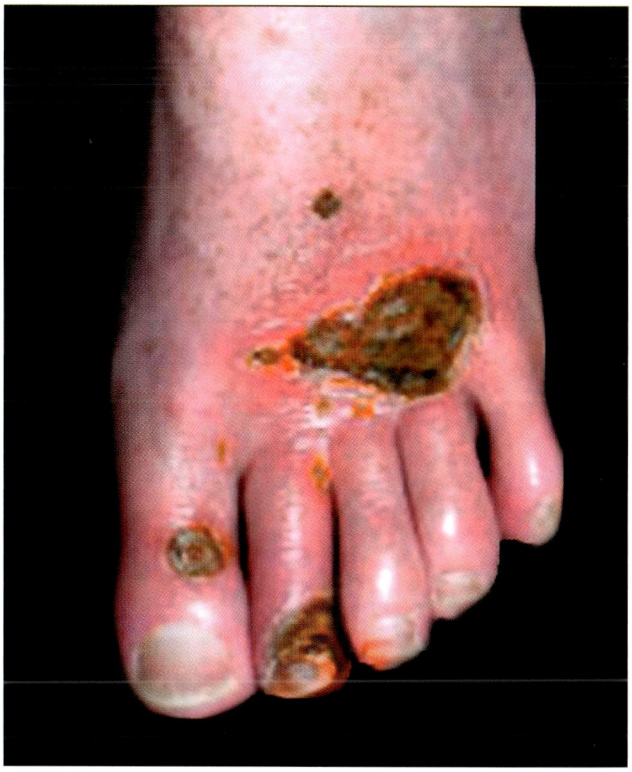

FIGURE 13.51: ***Ulcerated lesions of the foot are present in an elderly patient.***

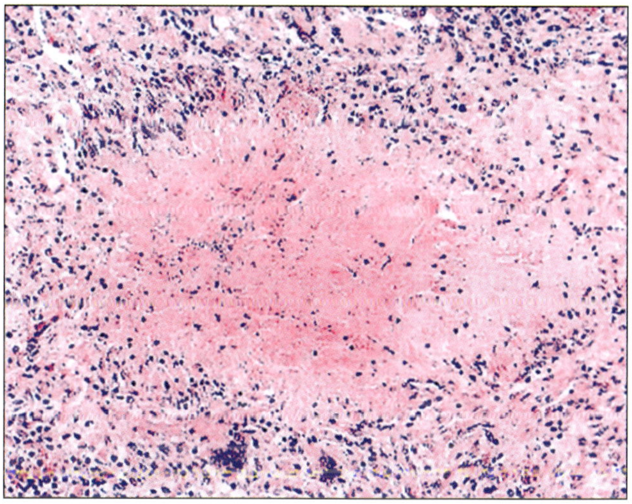

FIGURE 13.52: ***Necrotizing granulomatous inflammation effaces a large portion of the dermis.***

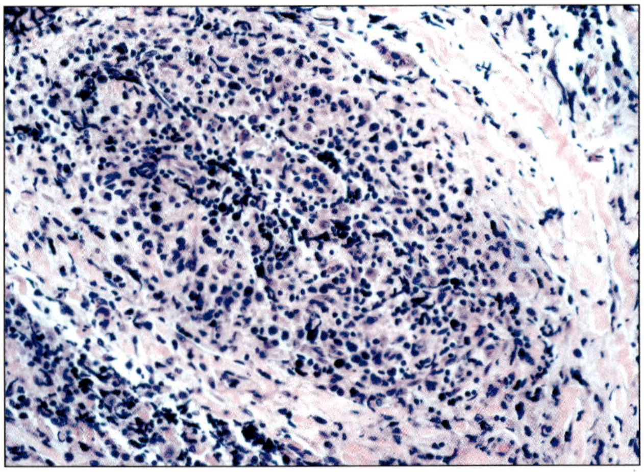

FIGURE 13.53: Lymphohistiocytic inflammation permeates the wall of this small dermal artery.

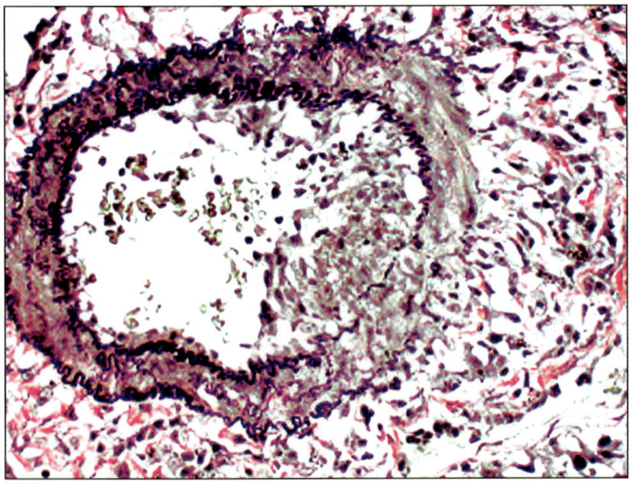

FIGURE 13.54: A Verhoeff–Van Gieson stain demonstrates breaks in the elastic laminae of dermal blood vessels, supporting the diagnosis of vasculitis.

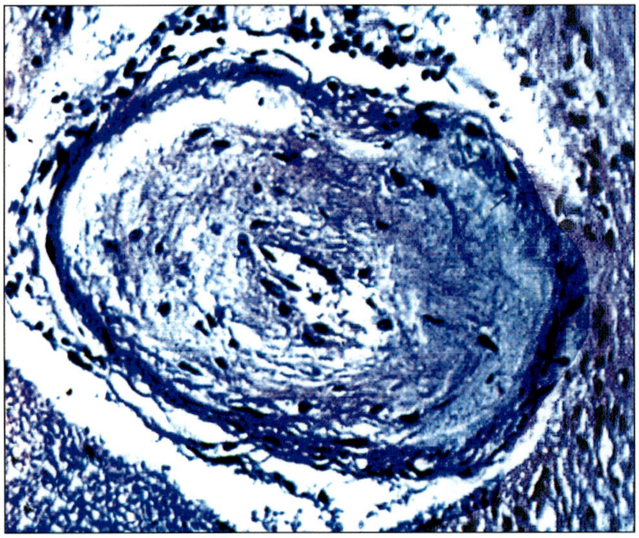

FIGURE 13.55: Trichrome staining shows luminal fibrosis in this blood vessel, reflecting prior inflammatory damage to it. The final diagnosis in this case was that of Wegener's granulomatosis. A test for circulating antineutrophil cytoplasmic antibody was positive.

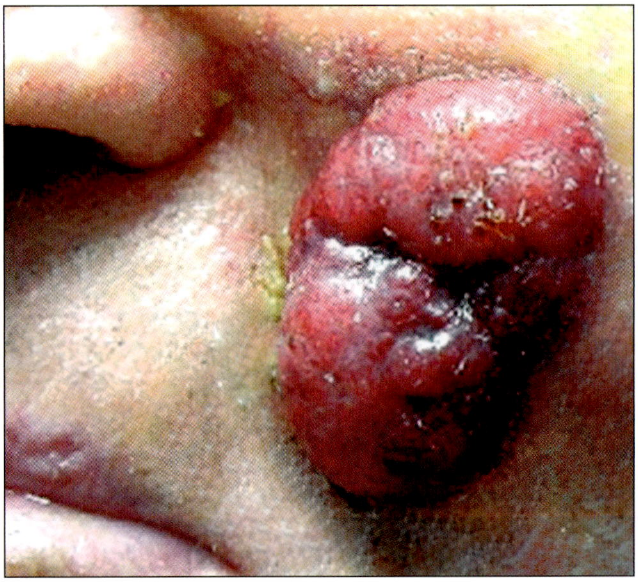

FIGURE 13.56: A violaceous nodular lesion is present on the face of an elderly woman.

Although the airway and kidneys are the usual sites of presentation for Wegener's granulomatosis, the skin is involved in roughly 30% of patients at some time during the course of the disease (251). Cutaneous biopsies may allow for relatively noninvasive but definitive diagnosis of the disorder, permitting effective treatment to be undertaken with corticosteroids and cyclophosphamide (251–253). ■

CASE 4— *Cutaneous Large Cell Lymphoma:* A 67-year-old woman developed a progressively enlarging 3.5-cm violaceous nodule in the skin of the left cheek (Figure 13.56), over a 6-month period She was otherwise asymptomatic, and only sought medical attention when the lesion became ulcerated.

No other abnormalities were found on physical examination; results of a chest radiograph, a hemogram, and screening biochemical studies were all normal. A biopsy of the cutaneous nodule was obtained.

It showed a dense infiltrate of large, markedly atypical, and lineage-indeterminate cells, arranged in sheets and effacing much of the dermis (Figures 13.57 and 13.58). The Snook method showed investment of individual cells and small cell groups by reticulin (Figure 13.59), and the tumor was strongly labeled by the MGP method (Figure 13.60). Lectin histochemical preparations with biotinylated RCA and PNA were both positive (Figure 13.61), but others done with UEA, CON-A, and DBA were negative.

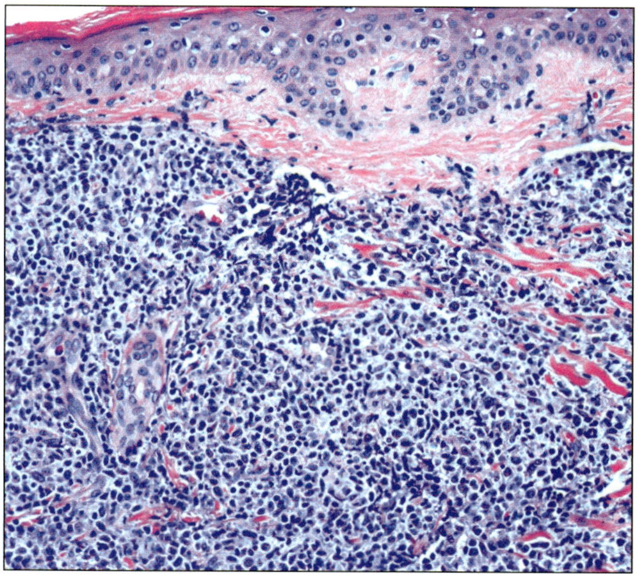

FIGURE 13.57: A morphologically ambiguous large polygonal cell malignancy effaces the dermis.

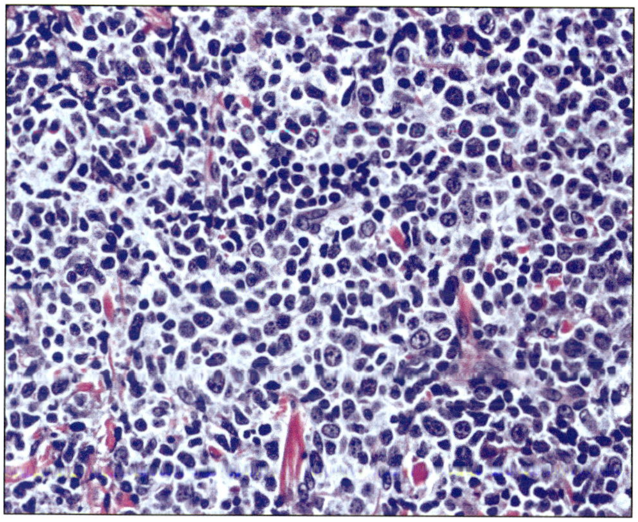

FIGURE 13.58: The lesional cells show discernible nuclear atypia and irregularity of the nuclear membranes.

A diagnosis of large cell lymphoma cutis was made. The patient underwent radiographic staging procedures that demonstrated no extracutaneous disease. She was given local irradiation as well as chemotherapy, which produced prompt involution of the skin lesion.

Primary large cell lymphoma that is limited to the skin, staged as IE, is potentially curable by prompt and aggressive treatment. The survival of patients with this disease varies depending on whether they have B-cell or T-cell lesions, but the therapeutic approach to such tumors is similar regardless of their specific lineages (254–256). ■

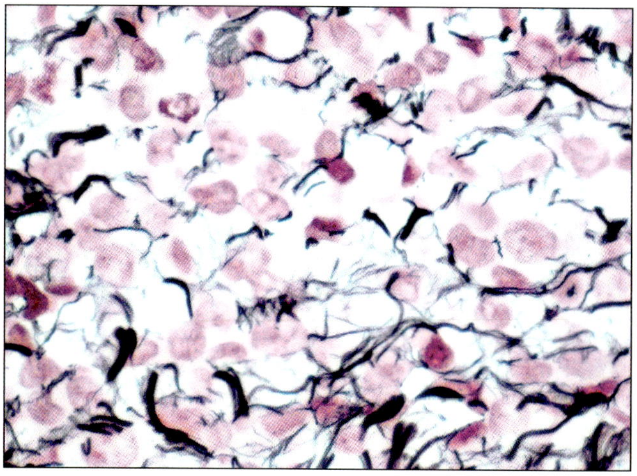

FIGURE 13.59: A Snook's stain demonstrates multifocal investment of individual tumor cells by reticulin.

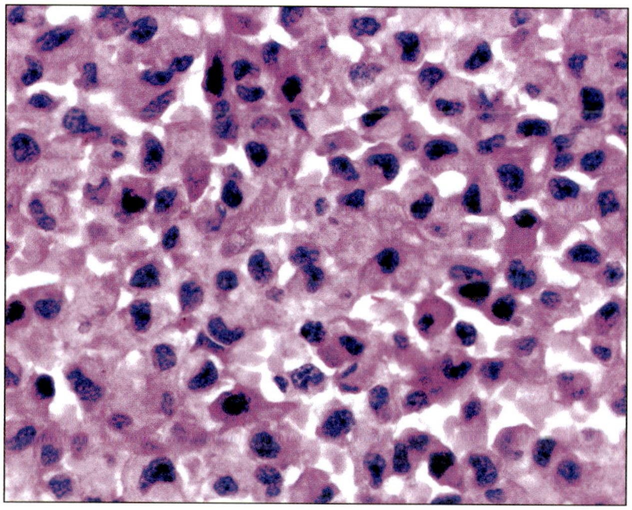

FIGURE 13.60: The MGP method shows nucleocytoplasmic positivity in the neoplastic cells.

FIGURE 13.61: The tumor cells uniformly bind Ricinus communis lectin. The final diagnosis was that of large cell lymphoma of the skin.

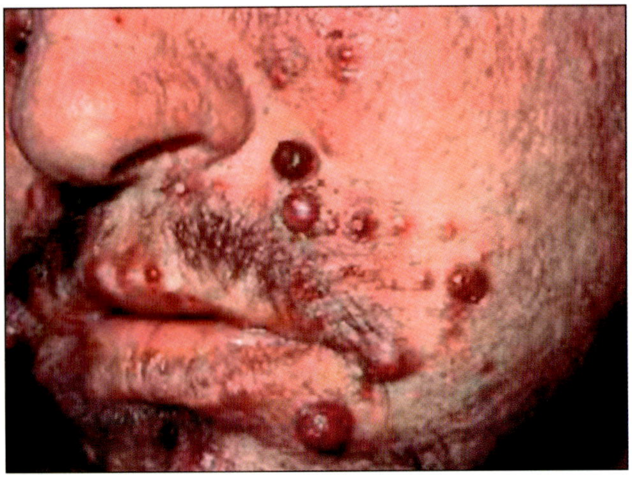

FIGURE 13.62: Multiple violaceous nodules in the facial skin of a 35-year-old man who is seropositive for antibody to human immunodeficiency virus.

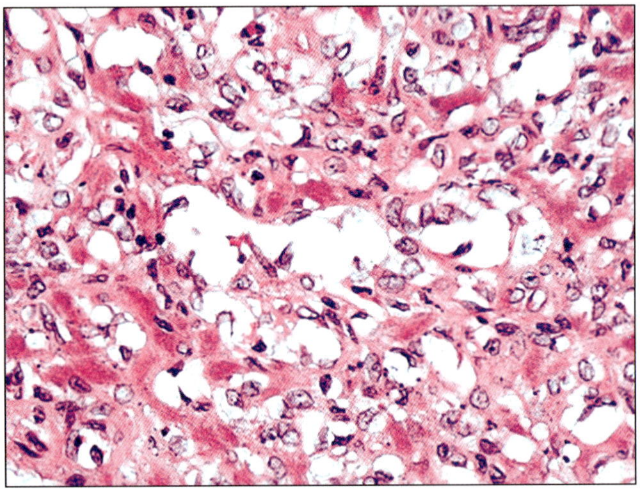

FIGURE 13.64: Vascular proliferation is well seen in this photomicrograph. The differential diagnosis centers on Kaposi's sarcoma versus bacillary angiomatosis.

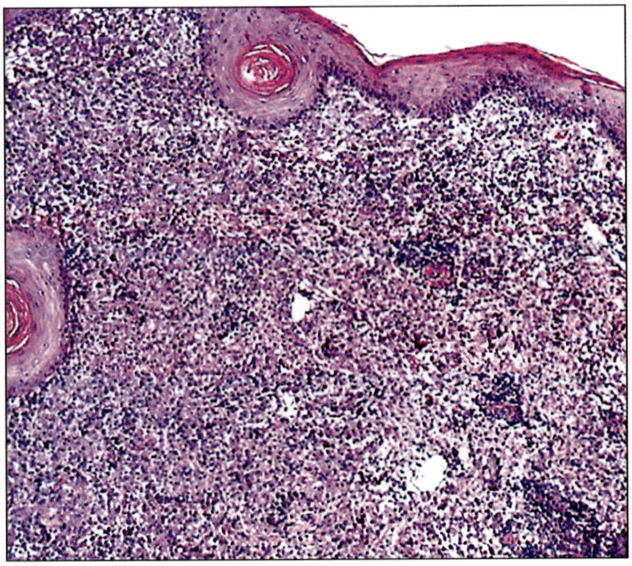

FIGURE 13.63: Microscopically, the lesions show marked dermal hypercellularity and vasogenesis.

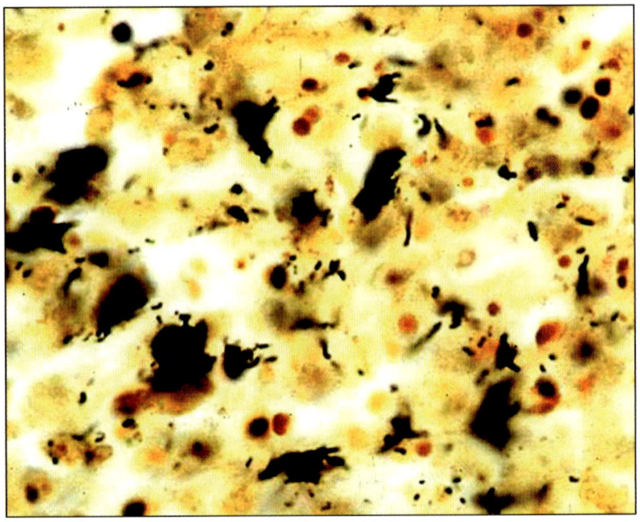

FIGURE 13.65: Numerous bacilliform organisms are seen in this Warthin-Starry stain of the lesion, establishing the diagnosis of bacillary angiomatosis.

CASE 5— ***Bacillary Angiomatosis:*** A 35-year-old man who was infected with the human immunodeficiency virus developed multiple 1- to 2-cm violaceous nodules in the skin of the face (Figure 13.62). Clinically, he was felt to have Kaposi's sarcoma, and chemotherapy was planned. However, because the patient's CD4 lymphocyte count was relatively normal, this diagnosis was questioned and a skin biopsy was obtained. It showed an organoid proliferation of dermal blood vessels that were lined by epithelioid endothelial cells and associated with an infiltrate of neutrophils (Figures 13.63 and 13.64). No interanastomosing vascular channels or extracellular hyaline globules were present. A Warthin-Starry stain demonstrated numerous intralesional bacilli that had the morphologic features of *Bartonella* species (Figure 13.65). A diagnosis of BA was made, followed by therapy with erythromycin. Regression of the skin lesions promptly followed.

BA is an infectious clinical simulant of Kaposi's sarcoma that was originally described by Stoler et al. in 1983 (257). Subsequently, it has been shown that the causative agents are members of the *Bartonella* genus and that the histologic lesions in this condition simulate those of epithelioid (histiocytoid) hemangioma rather than Kaposi's sarcoma (258–261). Stains for atypical gram-negative bacilli—principally including the Warthin-Starry or Dieterle method—are positive in cases of BA (262), and treatment with tetracyclines or erythromycin is usually effective in eradicating the skin lesions in this condition. ■

REFERENCES

1. Weedon D: *Skin Pathology*, 2nd ed., Churchill-Livingstone, New York, 2002.
2. McKee PH: *Pathology of the Skin*, Mosby, St. Louis, 1988.
3. Barnhill RL: *Textbook of Dermatopathology*, McGraw-Hill, Columbus, 2004.
4. Elder DE, et al. (Eds.): *Lever's Histopathology of the Skin*, 9th ed., Lippincott-Williams & Wilkins, Philadelphia, 2004.
5. Ackerman AB, et al. (Eds.): *Histologic Diagnosis of Inflammatory Skin Diseases:* An Algorithmic Method Based on Pattern Analysis, Ardor Scribendi, New York, 2005.
6. Berman EL, Laudate A, Carter HW: Comparison of selective staining of fungi in paraffin sections by light microscopy, scanning electron microscopy, and backscatter electron imaging. *Scan Electron Microsc* 1981; 2: 115–122.
7. LeBoit PE: A thickened basement membrane is a clue to lichen sclerosus. *Am J Dermatopathol* 2000; 22: 457–458.
8. Fung MA, LeBoit PE: Light microscopic criteria for the diagnosis of early vulvar lichen sclerosus: a comparison with lichen planus. *Am J Surg Pathol* 1998; 22: 473–478.
9. Gisslen H, Heyden G: Histological and histochemical investigations of lupus erythematosus in the skin and the oral mucosa. *Acta Dermatol Venereol* 1975; 55: 57–63.
10. Roper CA, English JC III: Tumid lupus erythematosus: an unusual scleredema-like presentation. *Skin Med* 2006; 5: 92–94.
11. Edward M, Fitzgerald L, Thind C, Leman J, Burden AD: Cutaneous mucinosis associated with dermatomyositis and nephrogenic fibrosing dermopathy: fibroblast hyaluronan synthesis and the effect of patient serum. *Br J Dermatol* 2007; 156: 473–479.
12. Regauer S, Liegl B, Reich O: Early vulvar lichen sclerosus: a histopathological challenge. *Histopathology* 2005; 47: 340–347.
13. Rahbari H: Histochemical differentiation of localized morphea-scleroderma and lichen sclerosus et atrophicus. *J Cutan Pathol* 1989; 16: 342–347.
14. Frances C, Wechsler J, Meimon G, et al.: Investigation of intercellular matrix macromolecules in lichen sclerosus. *Acta Dermatol Venereol* 1983; 63: 483–490.
15. Marks R, Wells GC: A histochemical profile of lichen simplex. *Br J Dermatol* 1973; 88: 557–562.
16. Gebhart W: Cytoid bodies in human skin. *Wien Klin Wochenschr* 1976; 60 (Suppl): 1–24.
17. DeKoning J, Bosma RD, Hoogkamp-Korstanje JA: Demonstration of spirochetes in patients with Lyme disease with a modified silver stain. *J Med Microbiol* 1987; 23: 261–267.
18. Hoang MP, High WA, Molberg KH: Secondary syphilis: a histologic and immunohistochemical evaluation. *J Cutan Pathol* 2004; 31: 595–599.
19. Patterson JW: Rheumatoid nodule and subcutaneous granuloma annulare: a comparative histologic study. *Am J Dermatopathol* 1988; 10: 1–8.
20. Dabski K, Winkelmann RK: Generalized granuloma annulare: histopathology and immunopathology. Systematic review of 100 cases and comparison with localized granuloma annulare. *J Am Acad Dermatol* 1989; 20: 28–39.
21. Ugo A: Necrobiosis lipoidica: clinical and histochemical observations. *G Ital Dermatol* 1963; 104: 337–355.
22. Wilson-Jones E: Necrobiosis lipoidica presenting on the face and scalp: an account of 29 patients and a detailed consideration of recent histochemical findings. *Trans St. Johns Hosp Dermatol Soc* 1971; 57: 202–220.
23. Gray HR, Graham JH, Johnson WC: Necrobiosis lipoidica: a histopathological and histochemical study. *J Invest Dermatol* 1965; 44: 369–380.
24. Maldyk E: Histopathological and histochemical comparative studies on fibrinoid necrosis. *Rheumatologia* 1969; 7: 101–109.
25. Al-Hogail IA, Al-Ghamdi AM, Martinka M, Crawford RI: Actinic granuloma is a unique and distinct entity: a comparative study with granuloma annulare. *Am J Dermatopathol* 2002; 24: 209–212.
26. Tomasini C, Pippione M: Interstitial granulomatous dermatitis with plaques. *J Am Acad Dermatol* 2002; 46: 892–899.
27. Magro CM, Crowson AN: The spectrum of cutaneous lesions in rheumatoid arthritis: a clinical and pathological study of 43 patients. *J Cutan Pathol* 2003; 30: 1–10.
28. Wada DA, Perkins SL, Tripp S, Coffin CM, Florell SR: Human herpesvirus-8 and iron staining are useful in differentiating Kaposi's sarcoma from interstitial granuloma annulare. *Am J Clin Pathol* 2007; 127: 263–270.
29. Ramasoota T, Johnson WC, Graham JH: Cutaneous sarcoidosis and tuberculoid leprosy: a comparative histopathologic and histochemical study. *Arch Dermatol* 1967; 96: 259–268.
30. Frizzera G, Mancini AM: Histochemical investigation on the nature and origin of Schaumann bodies. *Arch Ital Anat Istol Patol* 1966; 40: 59–92.
31. Azar HA, Lunardelli C: Collagen nature of asteroid bodies of giant cells in sarcoidosis. *Am J Pathol* 1969; 57: 81–92.
32. Touart DM, Sau P: Cutaneous deposition diseases. *Part II. J Am Acad Dermatol* 1998; 39: 527–544.
33. Ramdial P, Mosam A, Dlova NC, et al.: Disseminated cutaneous histoplasmosis in patients infected with human immunodeficiency virus. *J Cutan Pathol* 2002; 29: 215–225.
34. Kurban AK, Malak JA, Farah F, Chaglassian HT: Histopathology of cutaneous leishmaniasis. *Arch Dermatol* 1966; 93: 396–401.
35. Lowitt MH, Kariniemi AL, Niemi KM, Kao GF: Cutaneous malakoplakia: a report of two cases and review of the literature. *J Am Acad Dermatol* 1996; 34: 325–332.
36. Wood C, Russel-Bell B: Characterization of pigmented fungi by melanin staining. *Am J Dermatopathol* 1983; 5: 77–81.
37. Lokuhetty MD, Alahakoon VS, Kularatne BD, DeSilva MV: Ziehl Neelsen and Wade-Fite stains to demonstrate medlar bodies of chromoblastomycosis. *J Cutan Pathol* 2007; 34: 71–72.
38. Kemmer C, Reitman I, Kostler E, Riedel H: The morphology of skin damage in porphyria cutanea tarda. *Zentralbl Allg Pathol* 1988; 134: 227–234.
39. Fung MA, Murphy MJ, Hoss DM, Berke A, Grant-Kels JM: The sensitivity and specificity of "caterpillar bodies" in the differential diagnosis of subepidermal blistering disorders. *Am J Dermatopathol* 2003; 25: 287–290.
40. Raso DS, Greene WB, Maize JC, McGown ST, Metcalf JS: Caterpillar bodies of porphyria cutanea tarda

ultrastructurally represent a unique arrangement of colloid and basement membrane bodies. *Am J Dermatopathol* 1996; 18: 24–29.
41. Orkin M, Good RA, Clawson CC, Fisher I, Windhorst DB: Bullous mastocytosis. *Arch Dermatol* 1970; 101: 547–564.
42. Horny HP, Ruck P, Krober S, Kaiserling E: Systemic mast cell disease (mastocytosis): general aspects and histopathological diagnosis. *Histol Histopathol* 1997; 12: 1081–1089.
43. Chiu H, Lagunoff D: Histochemical comparison of vertebrate mast cells. *Histochem J* 1972; 4: 135–144.
44. Hennekenser HH: Histochemical demonstration of naphthol-AS-D-chloroacetate esterase activity in methacrylate embedded tissue sections. *Klin Wochenschr* 1972; 50: 213–214.
45. Leder LD: The chloroacetate esterase reaction: a useful means of histological diagnosis of hematological disorders from paraffin sections of skin. *Am J Dermatopathol* 1979; 1: 39–42.
46. Northover JM, Pickard JD, Murray-Lyon IM, et al.: Bullous lesions of the skin and mucous membranes in primary amyloidosis. *Postgrad Med J* 1972; 48: 351–353.
47. Lie JT: Wegener's granulomatosis: histological documentation of common and uncommon manifestations in 216 patients. *Vasa* 1997; 26: 261–270.
48. Bonsib SM: Polyarteritis nodosa. *Semin Diagn Pathol* 2001; 18: 14–23.
49. Magro CM, Crowson AN: The cutaneous neutrophilic vascular injury syndromes: a review. *Semin Diagn Pathol* 2001; 18: 47–58.
50. Burke AP, Virmani R: Localized vasculitis. *Semin Diagn Pathol* 2001; 18: 59–66.
51. Fujimoto J, Usui N, Kakinoki E, Uemichi A, Kinoshita H: Histologic characterization of the pathologic processes in thromboangiitis obliterans (Buerger's disease). *J Jpn Assoc Thorac Surg* 1990; 91: 1628–1635.
52. Hairston BR, Davis MD, Pittelkow MR, Ahmed I: Livedoid vasculopathy: further evidence for procoagulant pathogenesis. *Arch Dermatol* 2006; 142: 1413–1418.
53. Buk SJ: Simultaneous demonstration of connective tissue elastica and fibrin by a combined Verhoeff's elastic-Martius-scarlet blue trichrome stain. *Stain Technol* 1984; 59: 1–5.
54. Essary LR, Wick MR: Cutaneous calciphylaxis: an underrecognized clinicopathologic entity. *Am J Clin Pathol* 2000; 113: 280–287.
55. Nikko AP, Dunningham M, Cockerell CJ: Calciphylaxis with histologic changes of pseudoxanthoma elasticum. *Am J Dermatopathol* 1996; 18: 396–399.
56. Laurent R, Thiery F, Saint-Hillier Y, Blanc D, Agache P: Calcifying panniculitis associated with renal insufficiency: a tissue calciphylaxis syndrome. *Ann Dermatol Venereol* 1987; 11: 1073–1081.
57. Ratnam KV, Su WPD, Peters MS: Purpura simplex (inflammatory purpura without vasculitis): a clinicopathologic study of 174 cases. *J Am Acad Dermatol* 1991; 25: 642–647.
58. Aoki M, Kawana S: Lichen aureus. *Cutis* 2002; 69: 145–148.
59. Su WPD, Schroeter AL, Lee DA, Hsu T, Muller SA: Clinical and histologic findings in Degos' syndrome (malignant atrophic papulosis). *Cutis* 1985; 35: 131–138.
60. Boer A, Guo Y, Ackerman AB: Alopecia mucinosa is mycosis fungoides. *Am J Dermatopathol* 2004; 26: 33–52.
61. Sentis HJ, Willemze R, Scheffer E: Alopecia mucinosa progressing into mycosis fungoides: a long-term followup study of two patients. *Am J Dermatopathol* 1988; 10: 478–486.
62. Mehregan DA, Gibson LE, Muller SA: Follicular mucinosis: histopathologic review of 33 cases. *Mayo Clin Proc* 1991; 66: 387–390.
63. Marcon MJ, Powell DA: Human infection due to *Malassezia* species. *Clin Microbiol Rev* 1992; 5: 101–119.
64. Piamphonsant T: Pustular acne. *Int J Dermatol* 1985; 24: 441–443.
65. Sperling LC: Scarring alopecia and the dermatopathologist. *J Cutan Pathol* 2001; 28: 333–342.
66. Peters MS, Su WPD: Eosinophils in lupus panniculitis and morphea profunda. *J Cutan Pathol* 1991; 18: 189–192.
67. Cornelius CE III, Tenenhouse A, Weber JC: Calcinosis cutis: metabolic, sweat, histochemical, X-ray diffraction, and electron microscopic study. *Arch Dermatol* 1968; 98: 219–229.
68. Puchtler H, Meloan SN: Demonstration of phosphates in calcium deposits: a modification of von Kossa's reaction. *Histochemistry* 1978; 56: 177–185.
69. Sanchez NP, Pathak MA, Sato SS, et al.: Circumscribed dermal melaninoses: classification, light, histochemical, and electron microscopic studies on three patients with the erythema dyschromicum perstans type. *Int J Dermatol* 1982; 21: 25–31.
70. Sanchez NP, Pathak MA, Sato SS, et al.: Melasma: a clinical, light microscopic, ultrastructural, and immunofluorescence study. *J Am Acad Dermatol* 1981; 4: 698–710.
71. Grimes PE, Yamada N, Bhawan J: Light microscopic, immunohistochemical, and ultrastructural alterations in patients with melasma. *Am J Dermatopathol* 2005; 27: 96–101.
72. Zappi E, Lombardo W: Combined Fontana Masson/Perl's staining. *Am J Dermatopathol* 1984; 6(Suppl): 143–145.
73. McGrae JD Jr., Zelickson AS: Skin pigmentation secondary to minocycline therapy. *Arch Dermatol* 1980; 116: 1262–1265.
74. Brazzelli V, Borroni G, Dal Tio R, et al.: Amiodarone-induced pigmentation: a histological and ultrastructural study and review of the literature. *G Ital Dermatol Venereol* 1990; 125: 521–526.
75. Korting HC, Kolz R, Schmoeckel C, Bahda BR: Amiodarone pigmentation: a rare but typical drug side effect. *Hautarzt* 1981; 32: 301–305.
76. Delage C, Lagace R, Huard J: Pseudocyanotic pigmentation of the skin induced by amiodarone: a light and electron microscopic study. *Can Med Assoc J* 1975; 112: 1205–1208.
77. Barden H: Acid-fast staining of oxidized neuromelanin and lipofuscin in the human brain. *J Neuropathol Exp Neurol* 1979; 38: 453–462.
78. Arya SM: Histochemical, biochemical, and neurohistological observation on human vitiliginous skin. *Indian J Dermatol* 1982; 27: 139–142.
79. Garner J, Everett MA: A histochemical study of vitiligo. *J Okla State Med Assoc* 1962; 55: 489–491.
80. Kupchella CE, Matsuoka LY, Bryan B, Wortsman J, Dietrich JG: Histochemical evaluation of glycosaminoglycan deposition in the skin. *J Histochem Cytochem* 1984; 32: 1121–1124.

81. Matsuoka LY, Worsman J, Carlisle KS, Kupchella CK, Dietrich JG: The acquired cutaneous mucinoses. *Arch Intern Med* 1984; 144: 1974–1980.
82. Abd-Ed-Aal H, Salem SZ, Salem A: Lichen myxedematosus: histochemical study. *Dermatologica* 1981; 162: 273–276.
83. Mackay-Wiggan JM, Cohen DJ, Hardy MA, Knobler EH, Grossman ME: Nephrogenic fibrosing dermopathy (scleromyxedema-like illness of renal disease). *J Am Acad Dermatol* 2003; 48: 55–60.
84. Godby A, Bergstresser PR, Chaker B, Pandva AG: Fatal scleromyxedema: report of a case and review of the literature. *J Am Acad Dermatol* 1998; 38: 289–294.
85. Dinneen AM, Dicken CH: Scleromyxedema. *J Am Acad Dermatol* 1995; 33: 37–43.
86. Greenberg LM, Geppert C, Worthen HG, Good RA: Scleredema "adultorum" in children: report of three cases with histochemical study and review of the world literature. *Pediatrics* 1963; 32: 1044–1054.
87. Spicer SS, Garvin AJ, Simson JA, Wertelecki V: Cytochemistry of the skin of patients with mucopolysaccharidoses. *Histochem J* 1978; 10: 137–150.
88. Harper JI, Filipe MI, Staughton RC: Lipoid proteinosis: variations in the histochemical characteristics. *Clin Exp Dermatol* 1983; 8: 135–141.
89. Graham JH, Marques AS: Colloid milium: a histochemical study. *J Invest Dermatol* 1967; 49: 497–507.
90. Hashimoto K, Miller F, Bereston ES: Colloid milium: histochemical and electron microscopic studies. *Arch Dermatol* 1972; 105: 684–694.
91. Patterson JW, Wilkin JK, Schatski PF: Nodular colloid degeneration: distinctive histochemical and ultrastructural features. *Cutis* 1985; 36: 355–358.
92. Kurban AK, Malak JA, Afifi AK, Mire J: Primary localized macular cutaneous amyloidosis: histochemistry and electron microscopy. *Br J Dermatol* 1971; 85: 52–60.
93. Kakani RS, Goldstein AE, Meisher I, Hoffman C: Nodular amyloidosis: case report and literature review. *J Cutan Med Surg* 2001; 5: 101–104.
94. Tuffanelli DL: Localized scleroderma. *Semin Cutan Med Surg* 1998; 17: 27–33.
95. Kencka D, Blaszczyk M, Jablonska S: Atrophoderma of Pasini & Pierini is primary atrophic abortive morphea. *Dermatology* 1995; 190: 203–206.
96. Rybojad M, Baumann C, Godeau G, et al.: Congenital generalized cutis laxa: 5 cases. *Ann Dermatol Venereol* 1999; 126: 317–319.
97. Hashimoto K, Kanzaki T: Cutis laxa: ultrastructural and biochemical studies. *Arch Dermatol* 1975; 111: 861–873.
98. Venencie PY, Winkelmann RK: Histopathologic findings in anetoderma. *Arch Dermatol* 1984; 120: 1040–1044.
99. Finlayson GR, Sams WM Jr., Smith JG Jr.: Erythema ab igne: a histopathological study. *J Invest Dermatol* 1966; 46: 104–108.
100. Avogaro P, Castellari A: Pseudoxanthoma elasticum: clinical report of a case with histochemical and biochemical characterization of the affected substrate. *Clin Orthop Relat Res* 1966; 44: 231–239.
101. Adnet JJ, Kalis B, Hopfner C, et al.: Histochemical and ultrastructural-cytochemical study of pseudoxanthoma elasticum: apropos of 4 cases. *Ann Dermatol Syphiligr* 1973; 100: 530–535.
102. Lebwohl M, Phelps RG, Yannuzzi L, et al.: Diagnosis of pseudoxanthoma elasticum by scar biopsy in patients without characteristic skin lesions. *N Engl J Med* 1987; 317: 347–350.
103. Winkelmann RK, Sams WM Jr.: Elastofibroma: report of a case with special histochemical and electron microscopic studies. *Cancer* 1969; 23: 406–415.
104. Gun BD, Bahadir B, Behzatoglu K, Gun MO, Ozdamar SO: Elastofibroma: a clinicopathologic and immunohistochemical study of seven cases and literature review. *APMIS* 2007; 115: 115–119.
105. Rahbari H: Collagenous and elastotic marginal plaques of the hands (CEMPH). *J Cutan Pathol* 1991; 18: 353–354.
106. Jordaan HF, Rossouw DJ: Digital popular calcific elastosis: a histopathological, histochemical, and ultrastructural study of 20 patients. *J Cutan Pathol* 1990; 17: 358–370.
107. Patterson JW: The perforating disorders. *J Am Acad Dermatol* 1984; 10: 561–581.
108. Saray Y, Seckin D, Bilezikci B: Acquired perforating dermatosis: clinicopathological features in twenty-two cases. *J Eur Acad Dermatol Venereol* 2006; 20: 679–688.
109. Robinson-Bostrom L, DiGiovanni JJ: Cutaneous manifestations of end-stage renal disease. *J Am Acad Dermatol* 2000; 43: 975–986.
110. Fretzin DF, Beal DW, Jao W: Light and ultrastructural study of reactive perforating collagenosis. *Arch Dermatol* 1980; 116: 1054–1058.
111. Schneider I, Bertenyi C: Histochemical studies of the skin in recurrent erysipelas. *Int J Dermatol* 1974; 13: 76–83.
112. Marks R: Seeing through the stratum corneum. *Keio J Med* 2000; 49: 80–83.
113. Kampmeier RH: Granuloma inguinale. *Sex Transm Dis* 1984; 11: 318–321.
114. Schwarz RH: Chancroid and granuloma inguinale. *Clin Obstet Gynecol* 1983; 26: 138–142.
115. Meyer PR, Shum TK, Becker TS, Taylor CR: Scleroma (rhinoscleroma): a histologic and immunohistochemical study with bacteriologic correlates. *Arch Pathol Lab Med* 1983; 107: 377–383.
116. Margileth AM: Recent advances in diagnosis and treatment of cat scratch disease. *Curr Infect Dis Rep* 2000; 2: 141–146.
117. Schwartz RA, Nychay SG, Janniger CK, Lambert WC: Bacillary angiomatosis: presentation of six patients, some with unusual features. *Br J Dermatol* 1997; 136: 60–65.
118. Roger F, Roger A: Stain affinities of *Rickettsiae* and large visible viruses (lymphogranuloma-psittacosis group). I. Mechanisms of differentiation with Macchiavello stain and various potential applications. *Ann Inst Pasteur* 1958; 94: 126–128.
119. Burnet FM, Freeman M: Experimental studies on the virus of "Q" fever. *Med J Australia* 1937; 2: 299–305.
120. Choi DL, Lee HS, Choi HJ, et al.: Viral gametocytic hypertrophy caused by a papova-like virus infection in the Pacific oyster *Crassostrea gigas* in Korea. *Dis Aquat Organ* 2004; 59: 205–209.

121. Chacho MS, Eppich E, Wersto R, Koss LG: Influence of human papillomavirus on DNA ploidy determination in genital condylomas. *Cancer* 1990; 65: 2291–2294.
122. David-West TS: Cytological, cytochemical, and immunofluorescence studies with Dugbe virus—a new Nigerian tick-borne virus. *Acta Microbiol Acad Sci Hung* 1975; 22: 33–40.
123. Shariff S, Thomas JA: Herpetic cervicitis—a histomorphologic study. *Indian J Pathol Microbiol* 1989; 32: 81–91.
124. Lucia HL, LiVolsi VA, Lowell DM: A histochemical method for demonstrating papillomavirus infection in paraffin-embedded tissue. *Am J Clin Pathol* 1984; 82: 589–593.
125. Barlow D, Lonergan M, Tio FO, Meyerhoff WL: American mucocutaneous leishmaniasis. *South Med J* 1977; 70: 246–248.
126. Grossniklaus HE, Waring GO IV, Akor C, Castellano-Sanchez AA, Bennett K: Evaluation of hematoxylin and eosin and special stains for the detection of *Acanthamoeba* keratitis in penetrating keratoplasties. *Am J Ophthalmol* 2003; 136: 520–526.
127. Hulman G, Taylor LA: *Entamoeba histolytica*: demonstration by PAS/Martius yellow technique. *Med Lab Sci* 1987; 44: 396–397.
128. Steffen C, Ackerman AB: Intraepithelial epithelioma of Borst-Jadassohn. *Am J Dermatopathol* 1985; 7: 5–24.
129. Desmons F, Breuillard F, Thomas P, Leonardelli J, Hildebrand HF: Multiple clear-cell acanthomas (Degos): histochemical and ultrastructural study of two cases. *Int J Dermatol* 1977; 16: 203–213.
130. Pariser RJ: Benign neoplasms of the skin. *Med Clin North Am* 1998; 82: 1285–1307.
131. Mousawi A, Kibbi AG: Pigmented eccrine poroma: a stimulant of nodular melanoma. *Int J Dermatol* 1995; 34: 857–858.
132. Kohler S, Rouse RV, Smoller BR: The differential diagnosis of pagetoid cells in the epidermis. *Mod Pathol* 1998; 11: 79–92.
133. Inokuchi K, Sasai Y: Histochemical analysis of sialomucin in Paget cells of mammary and extramammary Paget's disease. *Acta Histochem* 1992; 92: 216–223.
134. Balducci L, Crawford ED, Smith GF, et al.: Extramammary Paget's disease: an annotated review. *Cancer Invest* 1988; 6: 293–303.
135. Sasai Y, Kaja H, Natsuaki M, Nakama T: The cytoplasmic mucin in Paget cells of extramammary Paget's disease. *Acta Histochem* 1981; 69: 50–56.
136. Wood WS, Culling CF: Perianal Paget's disease: histochemical differentiation utilizing the borohydride-KOH-PAS reaction. *Arch Pathol* 1975; 99: 442–445.
137. Sasaki M, Terrada T, Nakanuma Y, et al.: Anorectal mucinous adenocarcinoma associated with latent perianal Paget's disease. *Am J Gastroenterol* 1990; 85: 199–202.
138. Helm KF, Goellner JR, Peters MS: Immunohistochemical stains in extramammary Paget's disease. *Am J Dermatopathol* 1992; 14: 402–407.
139. Chiba H, Kazama T, Takenouchi T, et al.: Two cases of vulval pigmented extramammary Paget's disease: histochemical and immunohistochemical studies. *Br J Dermatol* 2000; 142: 1190–1194.
140. Nakhleh RE, Wick MR, Rocamora A, Swanson PE, Dehner LP: Morphologic diversity in malignant melanomas. *Am J Clin Pathol* 1990; 93: 731–740.
141. Warkel RL, Luna LG, Helwig EB: A modified Warthin-Starry procedure at low pH for melanin. *Am J Clin Pathol* 1980; 73: 812–815.
142. Kao GF, Helwig EB, Graham JH: Balloon cell malignant melanoma of the skin: a clinicopathologic study of 34 cases with histochemical, immunohistochemical, and ultrastructural observations. *Cancer* 1992; 69: 2942–2952.
143. Pereira PR, Odashiro AN, Rodrigues-Reyes AA, et al.: Histopathological review of sebaceous carcinoma of the eyelid. *J Cutan Pathol* 2005; 32: 496–501.
144. Rumelt S, Hogan NR, Rubin PA, Jakobiec FA: Four-eyelid sebaceous cell carcinoma following irradiation. *Arch Ophthalmol* 1998; 116: 1670–1672.
145. Wolfe JT III, Yeatts RP, Wick MR, Campbell RJ, Waller RR: Sebaceous carcinoma of the eyelid: errors in clinical and pathologic diagnosis. *Am J Surg Pathol* 1984; 8: 597–606.
146. Braun-Falco O, Schmoeckel C, Geyer C: Pigmented actinic keratosis. *Hautarzt* 1986; 37: 676–678.
147. Kamino H, Tam ST: Immunoperoxidase technique modified by counterstain with azure B as a diagnostic aid in evaluating heavily pigmented melanocytic neoplasms. *J Cutan Pathol* 1991; 18: 436–439.
148. Kamino H, Flotte TJ, Misheloff E, Greco MA, Ackerman AB: Eosinophilic globules in Spitz's nevi: new findings and a diagnostic sign. *Am J Dermatopathol* 1979; 1: 319–324.
149. Wick MR, Goellner JR, Wolfe JT III, Su WPD: Adnexal carcinomas of the skin. *I. Eccrine carcinomas. Cancer* 1985; 56: 1147–1162.
150. Urso C, Bondi R, Paglierani M, et al.: Carcinomas of sweat glands: report of 60 cases. *Arch Pathol Lab Med* 2001; 125: 498–505.
151. Berg JW, McDivitt RW: Pathology of sweat gland carcinomas. *Pathol Annu* 1968; 3: 123–144.
152. Herring SM: Giant adenoid basal cell carcinoma. *Ann Plast Surg* 1994; 32: 544–547.
153. Nappi O, Pettinato G, Wick MR: Adenoid (acantholytic) squamous cell carcinoma of the skin. *J Cutan Pathol* 1989; 16: 114–121.
154. Triantafyllou AG, Rapidis AD: Chondroid syringoma of the upper lip: report of a case. *J Oral Maxillofac Surg* 1986; 44: 744–748.
155. Kunikane H, Ishikura H, Yamaguchi J, et al.: Chondroid syringoma (mixed tumor of the skin): a clinicopathological study of 13 cases. *Acta Pathol Jpn* 1987; 37: 615–625.
156. Mendoza S, Helwig EB: Mucinous (adenocystic) carcinoma of the skin. *Arch Dermatol* 1971; 103: 68–78.
157. Santa Cruz DJ, Meyers JH, Gnepp DR, Perez BM: Primary mucinous carcinoma of the skin. *Br J Dermatol* 1978; 98: 645–653.
158. Lack EE, Worsham GF, Callihan BD, et al.: Granular cell tumor: a clinicopathologic study of 110 patients. *J Surg Oncol* 1980; 13: 301–316.
159. Apisarnthanarax P: Granular cell tumor: an analysis of 16 cases and review of the literature. *J Am Acad Dermatol* 1981; 5: 171–182.

160. Grekin JN, Schwartz OD, Elton RF: Granular cell myoblastoma. *Arch Dermatol* 1969; 100: 246–247.
161. LeBoit PE, Barr RJ, Burall S, et al.: Primitive polypoid granular cell tumor and other cutaneous granular cell neoplasms of apparent non-neural origin. *Am J Surg Pathol* 1991; 15: 48–58.
162. Lazar AJ, Fletcher CDM: Primitive non-neural granular cell tumors of skin: clinicopathologic analysis of 13 cases. *Am J Surg Pathol* 2005; 29: 927–934.
163. Bjondal-Sorensen K, Godballe C, Ostergaard B, Krogdahl A: Adult extracardiac rhabdomyoma: light and immunohistochemical studies of two cases in the parapharyngeal space. *Head Neck* 2006; 28: 275–279.
164. Osborne BM, Robboy SJ: Lymphomas and leukemias presenting as ovarian tumors: an analysis of 42 cases. *Cancer* 1983; 52: 1933–1943.
165. Krishnan J, Li CY, Su WPD: Cutaneous lymphomas: correlation of histochemical and immunohistochemical characteristics and clinicopathologic features. *Am J Clin Pathol* 1983; 79: 157–165.
166. Enzinger FM: Clear cell sarcoma of tendons and aponeuroses: an analysis of 21 cases. *Cancer* 1965; 18: 1163–1174.
167. Chung EB, Enzinger FM: Malignant melanoma of soft parts: a reassessment of clear cell sarcoma. *Am J Surg Pathol* 1983; 7: 405–413.
168. Enzinger FM: Epithelioid sarcoma: a sarcoma simulating a granuloma or a carcinoma. *Cancer* 1970; 26: 1029–1041.
169. Trotter MJ, Crawford RI, O'Connell JX, Tron VA: Mitotic granuloma annulare: a clinicopathologic study of 20 cases. *J Cutan Pathol* 1996; 23: 537–545.
170. Long JC, Mihm MC: Multiple granulocytic tumors of the skin: report of six cases of myelogenous leukemia with initial manifestations in the skin. *Cancer* 1977; 39: 2004–2016.
171. Wick MR, Scheithauer BW: Primary neuroendocrine carcinoma of the skin. In: *Pathology of Unusual Malignant Cutaneous Tumors* (Wick MR, Ed.), Marcel Dekker, New York, 1985; pp. 107–180.
172. Sexton CW, White WL: Primary cutaneous Ewing's family sarcoma: report of a case with immunostaining for glycoprotein p30/32- MIC2. *Am J Dermatopathol* 1996; 18: 601–605.
173. Patterson JW, Maygarden SJ: Extraskeletal Ewing's sarcoma with cutaneous involvement. *J Cutan Pathol* 1986; 13: 46–58.
174. Sarnat HB, DeMello DE, Siddiqui SY: Diagnostic value of histochemistry in embryonal rhabdomyosarcoma. *Am J Surg Pathol* 1979; 3: 177–183.
175. Goldberg LH, Friedman RH, Silapunt S: Pigmented speckling as a sign of basal cell carcinoma. *Dermatol Surg* 2004; 30: 1553–1555.
176. Hutcheson AC, Fisher AH, Lang PG Jr.: Basal cell carcinomas with unusual histologic features. *J Am Acad Dermatol* 2005; 53: 833–837.
177. Urso C: Cutaneous melanoma with pseudomyxoid features. *J Cutan Pathol* 2006; 33: 312–314.
178. Pilotti S, Rilke F, Lombardi L: Neuroendocrine (Merkel cell) carcinoma of the skin. *Am J Surg Pathol* 1982; 6: 243–254.
179. Balaton AJ, Capron F, Baviera EE, et al.: Neuroendocrine carcinoma (Merkel cell tumor) presenting as a subcutaneous tumor: an ultrastructural and immunohistochemical study of three cases. *Pathol Res Pract* 1989; 184: 211–216.
180. Partanen S, Syrjanen KJ, Hjelt LH: Light microscopical and fluorescence histochemical observations on the endocrinological activity of human pulmonary neoplasia. *Arch Geschwulstforsch* 1980; 50: 97–104.
181. Tateishi R, Horai T, Hattori S: Demonstration of argyrophil granules in small cell carcinoma of the lung. *Virchows Arch A Pathol Anat Histol* 1978; 377; 203–210.
182. Attah EB, Ani AN, Odor EI: Decay of methyl green pyroninophilia in Burkitt lymphoma. *Acta Cytol* 1977; 21: 151–153.
183. Grogan TM, Warnke RA, Kaplan HS: A comparative study of Burkitt and non-Burkitt "undifferentiated" malignant lymphoma: immunologic, cytochemical, ultrastructural, cytologic, histopathologic, clinical, and cell culture features. *Cancer* 1982; 49: 1817–1828.
184. Iwasaki H, Kikuchi M, Mori R, et al.: Infantile digital fibromatosis: ultrastructural, histochemical, and tissue culture observations. *Cancer* 1980; 46: 2238–2247.
185. Zardawi IM, Earley MJ: Inclusion body fibromatosis. *J Pathol* 1982; 137: 99–107.
186. Fetsch JF, Miettinen M: Calcifying aponeurotic fibroma: a clinicopathologic study of 22 cases arising in uncommon sites. *Hum Pathol* 1998; 29: 1504–1510.
187. Keasbey LE: Juvenile aponeurotic fibroma (calcifying fibroma): a distinctive tumor arising in the palms and soles of young children. *Cancer* 1953; 6: 338–346.
188. Mustakallio KK, Levonen E, Niemi M: Histochemical studies on cutaneous leiomyomatosis. *Br J Dermatol* 1963; 75: 60–70.
189. Carapeto FJ, Garcia-Perez A, Winkelmann RK: Acral arteriovenous tumor. *Acta Dermatol Venereol* 1977; 57: 155–158.
190. Connelly MG, Winkelmann RK: Acral arteriovenous tumor: a clinicopathologic review. *Am J Surg Pathol* 1985; 9: 15–21.
191. Adegbovega PA, Qui S: Hemangioma vs. vascular malformation: presence of nerve bundles is a diagnostic clue for vascular malformation. *Arch Pathol Lab Med* 2005; 129: 772–775.
192. Zelger B: Fibrohistiocytic skin lesions. *Verh Deutsch Ges Pathol* 1998; 82: 290–300.
193. Zaballos P, Llambrich A, Ara M, et al.: Dermoscopic findings of hemosiderotic and aneurysmal dermatofibroma: report of six patients. *Br J Dermatol* 2006; 154: 244–250.
194. Garcia CA: Combined Bodian-Masson trichrome stain for peripheral nerves. *Arch Pathol Lab Med* 1980; 104: 115.
195. Kao GF, Laskin WB, Olsen TG: Solitary cutaneous plexiform neurilemmoma (schwannoma): a clinicopathologic, immunohistochemical, and ultrastructural study of 11 cases. *Mod Pathol* 1989; 2: 20–26.
196. Argenyi ZB: Immunohistochemical characterization of palisaded encapsulated neuroma. *J Cutan Pathol* 1990; 17: 329–335.

197. Dahl I, Angervall L: Cutaneous and subcutaneous leiomyosarcomas: a clinicopathologic study of 47 patients. *Pathol Eur* 1974; 9: 307–315.
198. Dupree WB, Langloss JM, Weiss SW: Pigmented dermatofibrosarcoma protuberans (Bednar tumor): a pathologic, ultrastructural, and immunohistochemical study. *Am J Surg Pathol* 1985; 9: 630–639.
199. Gibson LE, Goellner JR: Amelanotic melanoma: cases studied by Fontana stain, S100 immunostain, and ultrastructural examination. *Mayo Clin Proc* 1988; 63: 777–782.
200. Conley J, Lattes R, Orr W: Desmoplastic malignant melanoma: a rare variant of spindle-cell melanoma. *Cancer* 1971; 28: 914–936.
201. Bednar B: Nature of the pigment in multilayer neurofibromas. *Cesk Patol* 1978; 14: 16–18.
202. Chor PJ, Santa Cruz DJ: Kaposi's sarcoma: a clinicopathologic review and differential diagnosis. *J Cutan Pathol* 1992; 19: 6–20.
203. Fukunaga M, Silverberg SG: Hyaline globules in Kaposi's sarcoma: a light microscopic and immunohistochemical study. *Mod Pathol* 1991; 4: 187–190.
204. Kao GF, Johnson FS, Sulica VI: The nature of hyaline (eosinophilic) globules and vascular slits of Kaposi's sarcoma. *Am J Dermatopathol* 1990; 12: 256–267.
205. Brownstein MH, Helwig EB: Metastatic tumors of the skin. *Cancer* 1972; 29: 1298–1307.
206. Culling CF, Reid PE, Burton JD, Dunn WL: A histochemical method of differentiating lower gastrointestinal tract mucin from other mucins in primary or metastatic tumors. *J Clin Pathol* 1975; 28: 656–658.
207. Cook HC: A histochemical characterization of malignant tumor mucins as a possible aid in the identification of metastatic deposits. *Med Lab Technol* 1973; 30: 217–224.
208. Bloom GD, Carlsoo B, Gustafsson A, Henriksson R: Distribution of mucosubstances in adenoid cystic carcinoma. *Virchows Arch A Pathol Anat Histol* 1977; 375: 1–12.
209. Heffess CS, Wenig BM, Thompson LD: Metastatic renal cell carcinoma of the thyroid gland: a clinicopathologic study of 36 cases. *Cancer* 2002; 95: 1869–1878.
210. Louis CJ, Wyllie RG, Chou ST, Sztynda T: Lectin-binding affinities of human epidermal tumors and related conditions. *Am J Clin Pathol* 1981; 75: 642–647.
211. Hyun KH, Nakai M, Kawamura K, Mori M: Histochemical studies of lectin binding patterns in keratinized lesions, including malignancy. *Virchows Arch A Pathol Anat Histol* 1984; 402: 337–351.
212. Matalanis G, Gardner ID, Whitehead RH: Lectin binding patterns and monoclonal antibodies to epidermal antigens in tumors of the skin. *Pathology* 1986; 18: 206–211.
213. Ramirez-Bosca A, Reano A, Valcuende-Cavero F, Thivolet J, Castells-Rodellas A: Lectin binding sites in squamous cell carcinomas and keratoacanthomas. *Acta Dermatol Venereol* 1988; 68: 480–485.
214. Tsubura A, Fujita Y, Sasaki M, Mori S: Lectin-binding profiles for normal skin appendages and their tumors. *J Cutan Pathol* 1992; 19: 483–489.
215. Ko T, Muramatsu T, Shirai T: Distribution of lectin *Ulex europaeus* agglutinin-I in trichilemmal carcinoma, squamous cell carcinoma, and other epithelial tumors of the skin. *J Dermatol* 1996; 23: 389–393.
216. Hitchcock MG, Hurt MA, Santa Cruz FJ: Cutaneous granular cell angiosarcoma. *J Cutan Pathol* 1994; 21: 256–262.
217. Holden CA: Histogenesis of Kaposi's sarcoma and angiosarcoma of the face and scalp. *J Invest Dermatol* 1989; 93(Suppl 2): 119S–124S.
218. Holden CA, Spaull J, Das AK, McKee PH, Wilson-Jones E: The histogenesis of angiosarcoma of the face and scalp: an immunohistochemical and ultrastructural study. *Histopathology* 1987; 11: 37–51.
219. Wick MR, Manivel JC: Epithelioid sarcoma and epithelioid hemangioendothelioma: an immunocytochemical and lectin-histochemical comparison. *Virchows Arch A Pathol Anat Histol* 1987; 410: 309–316.
220. Ree HJ, Raine L, Crowley JP: Lectin binding patterns in diffuse large cell lymphoma. *Cancer* 1983; 52: 2089–2099.
221. Kohchiyama A, Oka D, Ueki H: Differing lectin-binding patterns of malignant melanoma and nevocellular and Spitz nevi. *Arch Dermatol Res* 1987; 279: 226–231.
222. Romani N, Schuler G, Fritsch P: Identical lectin binding patterns of human melanocytes and melanoma cells in vitro. *J Invest Dermatol* 1983; 80: 272–277.
223. Slesak B, Boguslawska-Jaworska J, Pejcz J, Harlozinska A: Binding of lectins to human leukemic cells. *Folia Hematol Int Mag Klin Morphol Blutforsch* 1989; 116: 251–259.
224. Bramwell VH, Crowther D, Gallagher J, Stoddart RW: Studies of lectin binding to normal and neoplastic lymph nodes. II. Non-Hodgkin lymphoma. *Br J Cancer* 1982; 46: 582–592.
225. Jacobson W, Stoddart RW, Collins RD: Lectin staining of carbohydrates of hemic cells. II. The cells of normal lymphoid origin, of lymphatic leukemias, and related diseases. *Histopathology* 1980; 4: 491–500.
226. Hou WY: Immunohistochemical and immunoelectron microscopic observations of human lung carcinomas using biotinylated peanut lectin (PNA). *Chin J Tuberc Respir Dis* 1990; 13: 85–127.
227. Pajor L, Balazs M, Balogh J, et al.: Ultrastructural, lectin-histochemical, and immunohistological observations on Merkel cell tumors. *Pathol Res Pract* 1986; 181: 45–49.
228. Sames K, Schumacher U, Halata Z, et al.: Lectin and proteoglycan histochemistry of Merkel cell carcinomas. *Exp Dermatol* 2001; 10: 100–109.
229. Li PX: Histochemical localization of glycoconjugates and CEA-glycoprotein in human lung neoplasms. *Chin J Oncol* 1990; 12: 183–185.
230. Miettinen M, Holthofer H, Lehto VP, Miettinen A, Virtanen I: *Ulex europaeus* I lectin as a marker for tumors derived from endothelial cells. *Am J Clin Pathol* 1983; 79: 32–36.
231. Ordonez NG, Batsakis JG: Comparison of *Ulex europaeus* I lectin and factor VIII-related antigen in vascular lesions. *Arch Pathol Lab Med* 1984; 108: 129–132.
232. Nappi O, Wick MR, Pettinato G, Ghiselli RW, Swanson PE: Pseudovascular adenoid squamous cell carcinoma of the skin: a neoplasm that may be mistaken for angiosarcoma. *Am J Surg Pathol* 1992; 16: 429–438.
233. Heng MC, Fallon-Friedlander S, Bennett R: Expression of *Ulex europaeus* agglutinin I lectin-binding sites in squamous

cell carcinomas and their absence in basal cell carcinomas. Indicator of tumor type and differentiation. *Am J Dermatopathol* 1992; 14: 216–219.
234. Vigneswaran N, Peters KP, Hornstein OP, Diepgen TL: Alteration of cell surface carbohydrates associated with ordered and disordered proliferation of oral epithelia: a lectin-histochemical study in oral leukoplakias, papillomas, and carcinomas. *Cell Tissue Kinet* 1990; 23: 41–55.
235. Lalwani AK, Carey TE, Goldstein IJ, Peters BP: Lectin binding characteristics of squamous cell carcinomas of the head and neck. *Acta Otolaryngol* 1996; 116: 125–131.
236. Vigneswaran N, Whitaker SB, Budnick SD, Waldron CA: Expression patterns of epithelial differentiation antigens and lectin-binding sites in ameloblastomas: a comparison with basal cell carcinomas. *Hum Pathol* 1993; 24: 49–57.
237. Cochran AJ, Wen DR, Berthier-Vergnes O, et al.: Cytoplasmic accumulation of peanut agglutinin-binding glycoconjugates in the cells of primary melanoma correlates with clinical outcome. *Hum Pathol* 1999; 30: 556–561.
238. Thies A, Moll I, Berger J, Schumacher U: Lectin binding to cutaneous malignant melanoma: *Helix pomatia* agglutinin is associated with metastasis formation. *Br J Cancer* 2001; 84: 819–823.
239. Fartasch M, Goerdt S, Hornstein OP: Possibilities and limits of paraffin-embedded cell markers in diagnosis of primary cutaneous histiocytosis. *Hautarzt* 1995; 46: 144–153.
240. Fartasch M, Vigneswaran N, Diepgen TL, Hornstein OP: Immunohistochemical and ultrastructural study of histiocytosis-X and non-X histiocytoses. *J Am Acad Dermatol* 1990; 23: 885–892.
241. Zelger B, Cerio R, Orchard G, Fritsch P, Wilson-Jones E: Histologic and immunohistochemical study comparing xanthoma disseminatum and histiocytosis-X. *Arch Dermatol* 1992; 128: 1207–1212.
242. Wong CK, Wang WJ: Systemic amyloidosis: a report of 19 cases. *Dermatology* 1994; 189: 47–51.
243. Kyle RA, Bayrd ED: Amyloidosis: review of 236 cases. *Medicine* 1975; 54: 271–299.
244. Satti MB, Azzopardi JG: Amyloid deposits in basal cell carcinoma of the skin: a pathologic study of 199 cases. *J Am Acad Dermatol* 1990; 22: 1082–1087.
245. Eswaramoorthy V, Kaur I, Das A, Kumar B: Macular amyloidosis: etiological factors. *J Dermatol* 1999; 26: 305–320.
246. Wright JR, Calkins E, Humphrey RL: Potassium permanganate reaction in amyloidosis: a histologic method to assist in differentiating forms of this disease. *Lab Invest* 1977; 36: 274–281.
247. van Rijswijk MH, van Heusden CW: The potassium permanganate method: a reliable method for differentiating amyloid AA from other forms of amyloid in routine laboratory practice. *Am J Pathol* 1979; 97: 43–58.
248. Janssen S, Elema D, van Rijswijk MH, et al.: Classification of amyloidosis: immunohistochemistry vs. the potassium permanganate method in differentiating AA from AL amyloidosis. *Appl Pathol* 1985; 3: 29–38.
249. Amazon K, Robinson MJ, Rywlin AM: Ferrugination caused by Monsel's solution: clinical observations and experimentations. *Am J Dermatopathol* 1980; 2: 197–205.
250. Olmstead PM, Lund HZ, Leonard DD: Monsel's solution: a histologic nuisance. *J Am Acad Dermatol* 1980; 3: 492–498.
251. Patten SF, Tomecki KJ: Wegener's granulomatosis: cutaneous and oral mucosal disease. *J Am Acad Dermatol* 1993; 28: 710–718.
252. Barksdale SK, Hallahan CW, Kerr GS, et al.: Cutaneous pathology in Wegener's granulomatosis: a clinicopathologic study of 75 biopsies in 46 patients. *Am J Surg Pathol* 1995; 19: 161–172.
253. Gibson LE: Granulomatous vasculitides and the skin. *Dermatol Clin* 1990; 8: 335–345.
254. Willemze R: Primary cutaneous B-cell lymphoma: classification and treatment. *Curr Opin Oncol* 2006; 18: 425–431.
255. Dreno B: Standard and new treatments in cutaneous B-cell lymphomas. *J Cutan Pathol* 2006; 33(Suppl): 47–51.
256. Rosen ST, Querfeld C: Primary cutaneous T-cell lymphomas. *Hematology Am Soc Hematol Educ Progr* 2006: 1: 323–330.
257. Stoler MH, Bonfiglio TA, Steigbigel RT, Pereira M: An atypical subcutaneous infection associated with acquired immune deficiency syndrome. *Am J Clin Pathol* 1983; 80: 714–718.
258. Cotell SL, Noskin GA: Bacillary angiomatosis: clinical and histologic features, diagnosis, and treatment. *Arch Intern Med* 1994; 154: 524–528.
259. LeBoit PE: Bacillary angiomatosis: a systemic opportunistic infection with prominent cutaneous manifestations. *Semin Dermatol* 1991; 10: 194–198.
260. Cockerell CJ, LeBoit PE: Bacillary angiomatosis: a newly characterized, pseudoneoplastic, infectious, cutaneous vascular disorder. *J Am Acad Dermatol* 1990; 22: 501–512.
261. LeBoit PE, Berger TG, Egbert BM, et al.: Bacillary angiomatosis: the histopathology and differential diagnosis of a pseudoneoplastic infection in patients with human immunodeficiency virus disease. *Am J Surg Pathol* 1989; 13: 909–920.
262. Relman DA, Loutit JS, Schmidt TM, Falkow S, Tompkins LS: The agent of bacillary angiomatosis: an approach to the identification of uncultured pathogens. *N Engl J Med* 1990; 323: 1573–1580.

CHAPTER 14

Diagnostic Histochemistry in Myopathic Diseases and Soft Tissue Lesions

Richard A. Prayson, M.D.,
Xiaofei Qiu, M.D., Ph.D., and
Elizabeth Montgomery, M.D.

Although traditional histochemistry has been largely supplanted by immunohistochemical and molecular biology techniques in evaluation of soft tissue neoplasms, it retains tremendous value in evaluating muscle diseases. In this chapter, we begin with and stress the use of histochemical techniques in evaluating myopathies, but histochemistry is the foundation of many observations in mesenchymal neoplasms that will be outlined in the second section.

INTRODUCTION TO MUSCLE BIOPSY

Utilization of muscle biopsy in the evaluation of neuromuscular disorders is well established. Although much valuable information can be gleaned by careful examination of hematoxylin-eosin (H&E)–stained sections, the diagnosis of many neuromuscular disorders requires additional testing including histochemistry, ultrastructural examination, and genetic or molecular analysis. In contrast to a number of other areas of diagnostic surgical pathology in which histochemical stains may be ordered in selected circumstances, utilization of histochemical stains is a routine part of the evaluation of a neuromuscular biopsy. The selection of stains is somewhat dictated by personal preference and the potential diagnostic information that can be obtained. The focus of this discussion will be on those stains which are most commonly utilized in the evaluation of neuromuscular disorders.

Before discussing specific disease entities and histochemical stains that may be helpful in elucidating various pathologies, a brief mention of the utility of H&E staining in evaluation of muscle biopsies is in order. Valuable information can be obtained by careful examination of the H&E-stained sections of skeletal muscle. However, the extent to which a specific diagnosis can be made based on this staining alone is somewhat limited. H&E staining is useful in the identification of inflammatory myopathic conditions including myositis, vasculitis, and granulomatous myositis. Further distinction of certain inflammatory myopathic diagnoses (such as inclusion body myositis that requires the identification of rimmed vacuoles) is dependent on the evaluation of other stains, which will be discussed. One can also observe variability in muscle fiber size, noting features of individual cell atrophy, grouped atrophy, or fascicular atrophy. Vacuolar myopathic changes may also be observable, although evaluating the nature of the vacuoles requires ancillary studies. These patterns elicit differential diagnoses that may include a variety of pathologies, which require additional stains for further evaluation.

Many of the histochemical stains that are used in the evaluation of muscle biopsies are enzyme stains. Fresh frozen tissue is required in order to properly perform these stains: tissue that has been fixed in formalin does not lend itself to satisfactory evaluation by most of these stains (1,2). This underscores the importance of properly freezing muscle tissue in order to adequately evaluate these specimens. In cases that have been improperly frozen and marked by subsequent artifactual vacuolar changes, thawing and properly refreezing the tissue may eliminate much of this artifact (Figure 14.1). In doing this however, many of the enzyme stains do not perform well (Figure 14.2). Although there are advantages to processing specimens in a timely fashion, particularly for maintenance of ultrastructural findings, delays in processing may occur. For many of the routinely performed enzyme and special stains that will be discussed, staining does not seem to be significantly affected with delays in fixation up to 24 hours (3).

Inflammatory Myopathic Diseases

The inflammatory myopathies are a group of disorders of muscle marked by inflammation, frequently accompanied by some degree of muscle fiber necrosis or degeneration and muscle fiber regeneration. The identification of inflammation and the pattern or distribution of inflammation, which are important in the recognition of some of these disorders, can be readily identified on H&E staining.

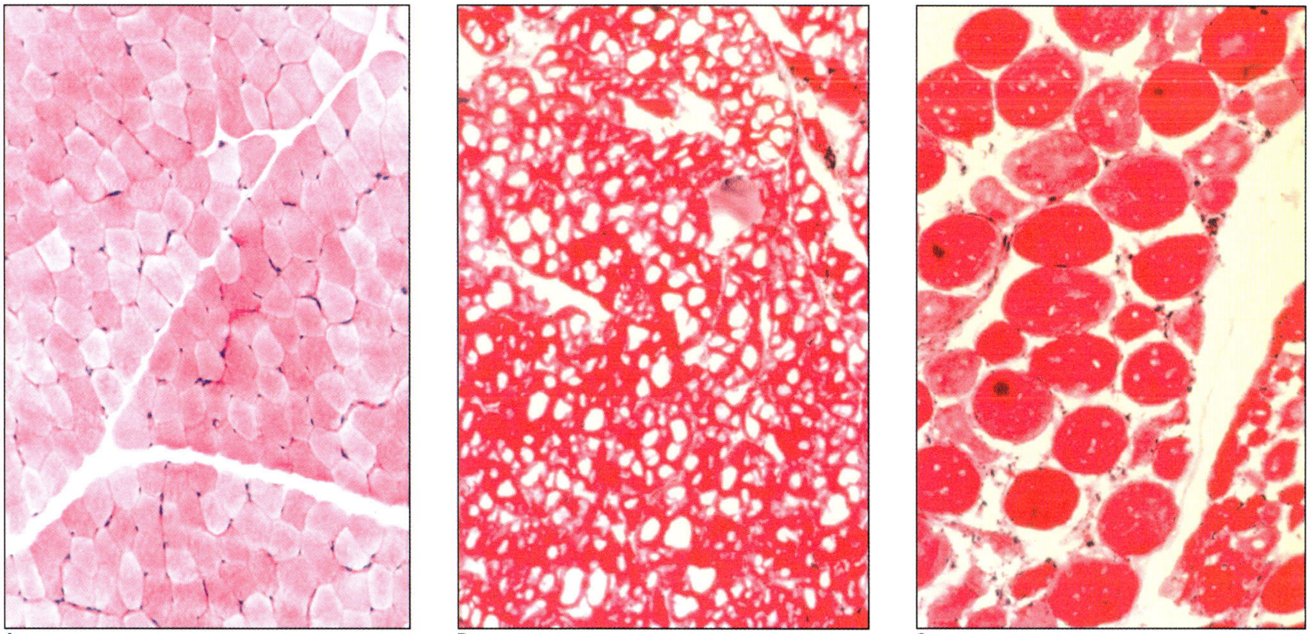

A B C

***FIGURE 14.1:* Normal muscle stained with H&E. (A) Normal staining. (B) Vacuolar change secondary to improper freezing of the biopsy specimen. (C) Much of the freeze artifact can be eliminated by thawing and refreezing the muscle (original magnification of all figures 400×).**

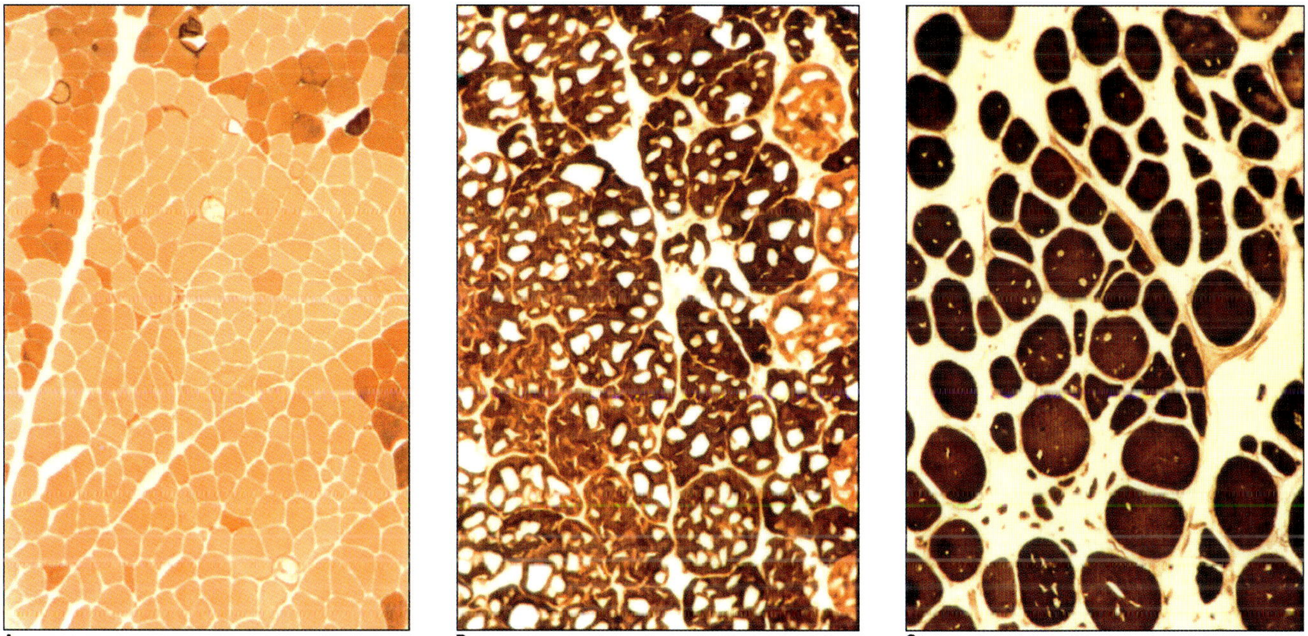

A B C

***FIGURE 14.2:* ATPase stains at pH 9.8. (A) Fiber type grouping. (B) Freeze artifact with thawing and refreezing. (C) The freeze artifact is mostly corrected but the enzyme staining is compromised (i.e., no longer get differential staining between type I and type II myofibers) (original magnification of all figures 100×).**

There is, however, a role for histochemical stains in the further evaluation of these cases.

Necrotic or degenerating muscle fibers are typically identified on H&E staining by their pale staining and infiltration by macrophages (Figure 14.3). An acid phosphatase stain may be useful in highlighting these fibers (Figure 14.4). Acid phosphatase is a lysosomal enzyme prominently noted in the inflammatory cells. Macrophages phagocytosing necrotic fibers will stain positively, as will infiltrating inflammatory cells (Figure 14.5).

Regenerating muscle fibers are also a common feature of many inflammatory myopathic conditions. On H&E staining, regenerating muscle fibers are marked by nuclear enlargement with prominent nucleolation. The cytoplasm acquires increased basophilia. Alkaline phosphatase is a useful stain in highlighting regenerating muscle fibers (Figure 14.6). The fibers stain positively (gray-black) when regenerating (Figure 14.7). In addition, blood vessel endothelial cells demonstrate positive staining. Nonspecific increased staining is also noted in association with connective tissue between muscle fascicles in certain inflammatory disorders.

The distinction between inflammatory myopathic conditions, based on acid and alkaline phosphatase stains, is not generally possible. Other features such as the distribution of the inflammation (more endomysial in polymyositis versus perivascular in dermatomyositis) and the presence of perifascicular atrophy (dermatomyositis) are more useful in this regard (4–6) (Figure 14.8). In dermatomyositis, the degenerating and regenerating muscle fibers tend to be most prevalent in the perifascicular region.

Inclusion body myositis warrants distinction from polymyositis and dermatomyositis because it is clinically refractory to immunosuppressive and steroid therapy (7–10). In addition to the chronic endomysial inflammation, muscle fiber degeneration, and muscle fiber regeneration, inclusion body myositis is also marked by the presence of autophagic or rimmed vacuoles. The vacuoles may be present in only a few or in a majority of muscle fibers and may be present as a single vacuole or multiple vacuoles in a given muscle fiber. Although the vacuoles can be visualized upon careful inspection on routine H&E staining, the Gomori trichrome stain can be helpful in highlighting their presence (Figures 14.9 and 14.10). Processing the muscle specimen with formalin fixation and paraffin embedding may result in dissolution of the granules within the vacuoles, making identification of the

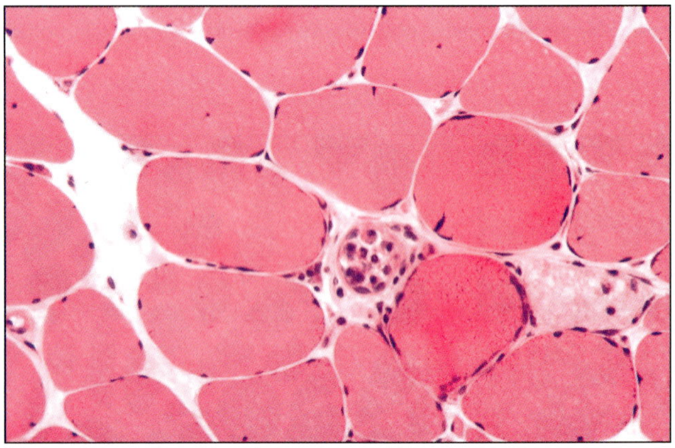

***FIGURE 14.3:** Scattered degenerating myofibers characterized by pale staining and intrafiber macrophages (H&E, original magnification 400×).*

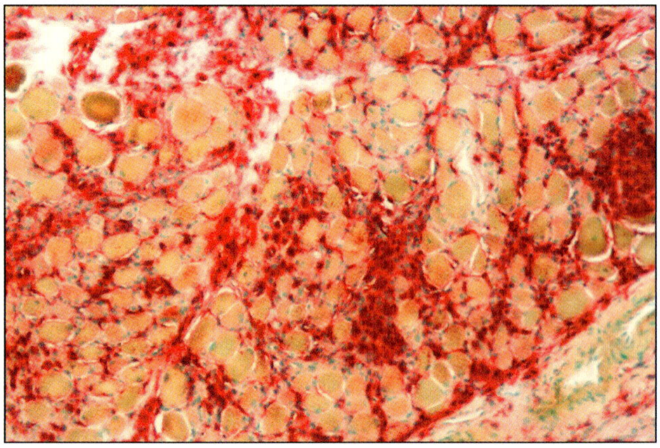

***FIGURE 14.5:** Acid phosphatase staining in an inflammatory myopathy. Inflammatory cells and intrafiber macrophages phagocytosing necrotic fibers are highlighted (original magnification 100×).*

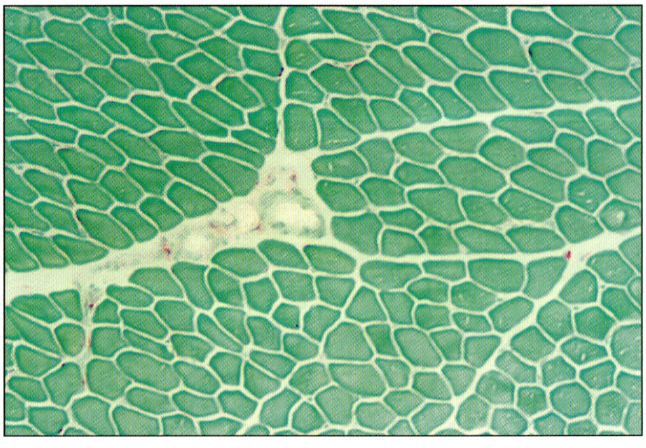

***FIGURE 14.4:** Normal acid phosphatase staining (original magnification 100×).*

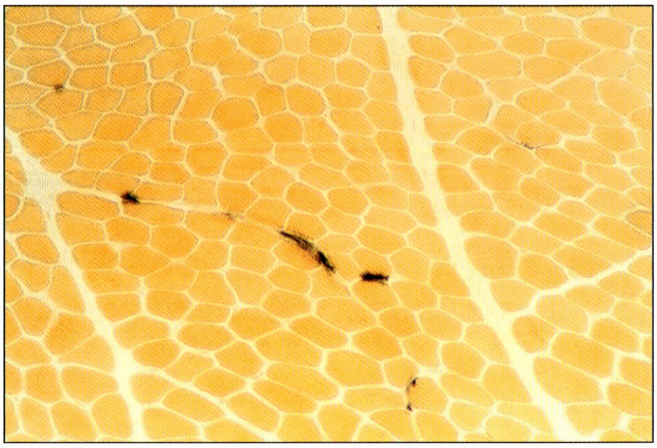

***FIGURE 14.6:** Normal alkaline phosphatase staining. Normal vascular endothelial cells stain positively (gray-black) (original magnification 100×).*

autophagic vacuoles more difficult. When the vacuoles are visualized in the setting of inflammatory myopathic condition, a diagnosis of inclusion body myositis should be seriously considered. The vacuoles, however, are somewhat nonspecific and may also be encountered in a variety of other neuromuscular disorders including muscular dystrophies and neurogenic atrophy (11,12). The vacuoles may also be highlighted by immunostaining with ubiquitin (13,14). Additionally, biopsies of inclusion body myositis show angular, atrophic, esterase-positive muscle fibers, indicative of possible denervation atrophy. Likewise, the expression of other proteins including beta amyloid, phosphorylated tau protein, and apolipoprotein E may be noted (15–18). In most cases, the amount of amyloid visualized may be small and not always readily discernible with routine histochemical stains for amyloid such as Congo red. More sensitive staining for amyloid including immunofluorescence methodologies or ultrastructural examination will reliably reveal the presence of amyloid. Additionally, a significant percentage of patients with inclusion body myositis have concomitant mitochondrial DNA alterations, which may result in abnormalities in a number of stains (19,20). The classic ragged red fibers, which are best visualized on a Gomori trichrome–stained section, may be evidenced in a minority of cases. These fibers are marked by increased eosinophilic granular material, most predominantly in the subsarcolemmal region, but are often accompanied by increased staining throughout the affected muscle fiber. In cases in which certain enzymes such as cytochrome oxidase or succinate dehydrogenase are affected by the DNA mutations, alterations in staining patterns may be evident on these stains as well. In the case of cytochrome oxidase, absence of staining in scattered muscle fibers may be noted. With succinate dehydrogenase, there is increased subsarcolemmal staining (so-called ragged blue fibers). The diagnosis of inclusion

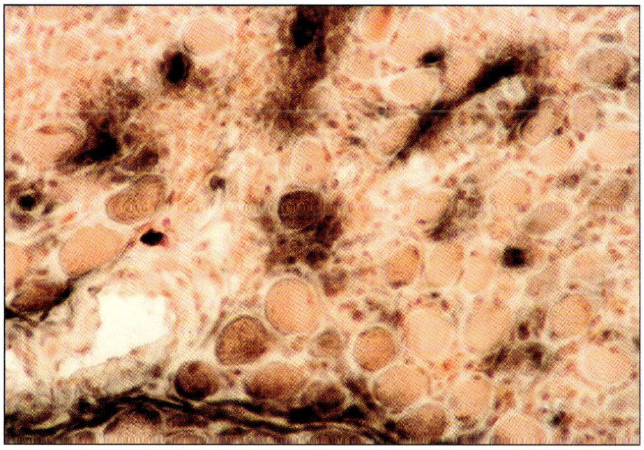

FIGURE 14.7: ***Alkaline phosphatase staining in an inflammatory myopathy highlighting regenerating muscle fibers (original magnification 200×).***

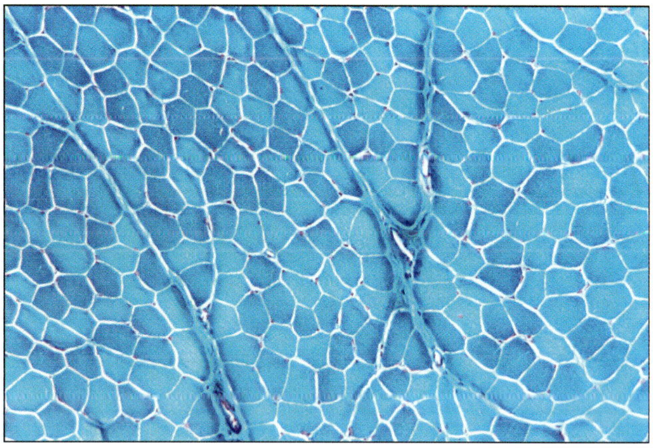

FIGURE 14.9: ***Normal Gomori trichrome staining (original magnification 100×).***

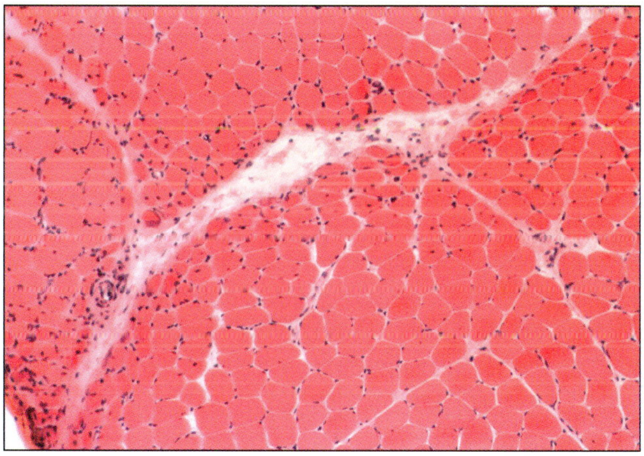

FIGURE 14.8: ***Perifascicular atrophy is a hallmark finding in dermatomyositis. Most of the muscle fiber degeneration and regeneration observed in dermatomyositis is situated in the areas of prominent atrophy (original magnification 200×).***

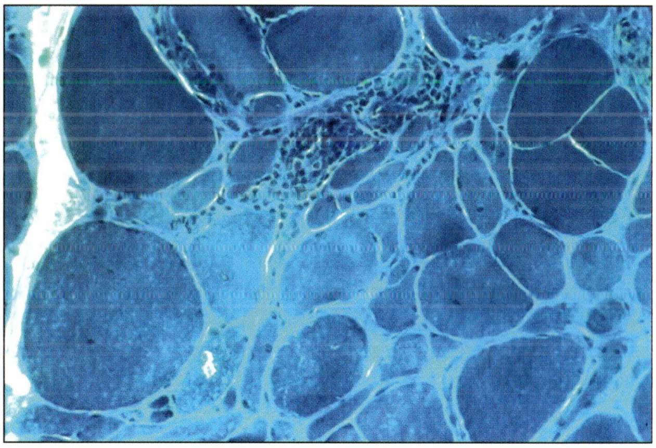

FIGURE 14.10: ***Inclusion body myositis with a rimmed (autophagic) vacuole highlighted on a trichrome-stained section (original magnification 400×).***

body myositis is confirmed by ultrastructural examination demonstrating the presence of either cytoplasmic or intranuclear tubulofilamentous structures measuring 12–18 nm.

A number of other collagen vascular disorders or autoimmune diseases may manifest themselves with inflammatory myopathic changes, usually consisting of some number of degenerating and regenerating muscle fibers. With few exceptions, the morphologic findings in these biopsies are not particularly specific for disease entity but may resemble polymyositis.

Infectious causes of inflammatory myopathy are well recognized. Bacterial infection typically results in an acute inflammatory cell infiltrate. Staining with gram stains may be useful in identifying bacterial organisms, although microbiologic cultures still remain the gold standard for identification of causative organisms. Viral myositis may morphologically overlap with polymyositis. The diagnosis is often one of exclusion.

Rhabdomyolysis with myoglobinuria may accompany a subset of viral myositis cases or may be encountered in other settings such as acute alcohol myopathy. The main morphologic feature of rhabdomyolysis is widespread myonecrosis, which can be highlighted with acid phosphatase staining (Figures 14.11 and 14.12). Over time, the process will resolve into widespread muscle fiber regeneration, which can be nicely highlighted with alkaline phosphatase staining. There typically is minimal chronic inflammation associated with rhabdomyolysis in contrast to most inflammatory myopathies. The process is often temporally synchronized in contrast to most other inflammatory myopathic conditions that generally show a mixture of both degenerating and regenerating muscle fibers, indicating an ongoing process.

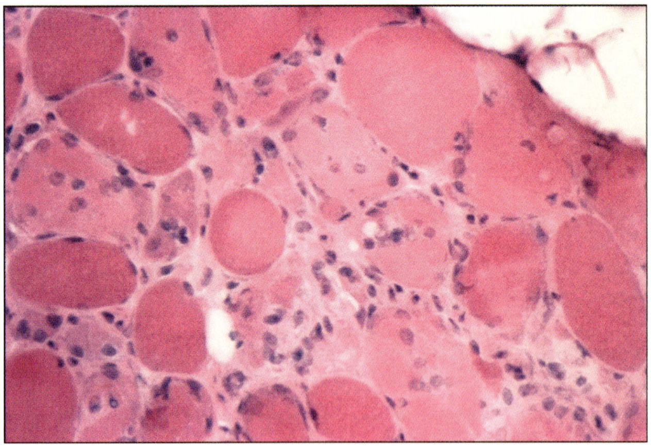

FIGURE 14.11: ***Widespread, temporarily synchronized myonecrosis in rhabdomyolysis (original magnification 400×).***

Of note is the association of human immunodeficiency virus type I (HIV-1) infection with the formation of nemaline rods (21,22). Nemaline rods are crystalloid structures that are comprised of Z-band material (alpha-actinin) (Figures 14.13 and 14.14). These rods are best visualized on a Gomori trichrome stain as small, dark purple crystalloid structures (Figures 14.13 and 14.14). These rods may also be present in large numbers in a rare congenital myopathy known as nemaline rod myopathy. Treatment of HIV infection with Zidovudine has been associated with muscle weakness and myalgia and is morphologically marked by the presence of ragged red fibers and accumulation of lipid material (23,24). Lipid may be readily highlighted with an oil red O stain or Sudan black staining.

Fungal infections of skeletal muscle are extraordinarily rare. In suspected cases, a variety of histochemical stains including periodic acid-Schiff (PAS) or methenamine silver stain may be helpful in highlighting fungal organisms. Parasitic infections of muscle generally are identified by careful inspection of H&E-stained sections.

Granulomatous inflammation, in the background of an inflammatory myopathy or myositis, is a relatively rare occurrence (25). Most commonly, the underlying etiology is sarcoidosis, where muscle involvement is relatively

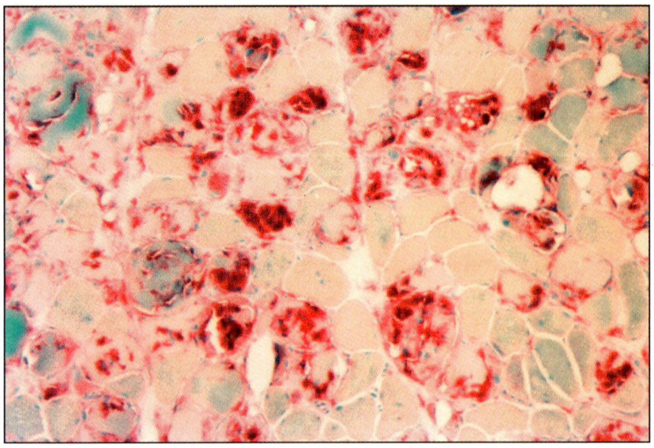

FIGURE 14.12: ***An acid phosphatase stain highlighting numerous degenerating muscle fibers in rhabdomyolysis (original magnification 200×).***

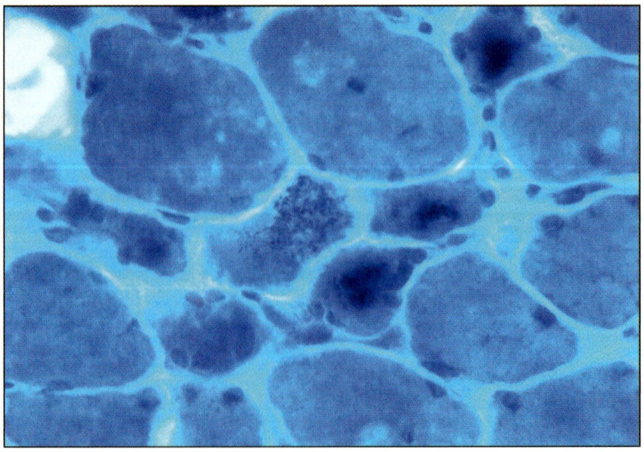

FIGURE 14.13: ***Increased dark staining nemaline rods in a patient with AIDS. The rods are best visualized on a trichrome-stained section (original magnification 400×).***

common and is present in a majority of patients with the disease. In addition to chronic endomysial inflammation, muscle fiber degeneration, and muscle fiber regeneration, nonnecrotizing granulomatous inflammation is also observed. Because the differential diagnosis includes infectious agents, staining of tissue for fungal and mycobacterial organisms utilizing methenamine silver and Ziehl-Neelsen stains is warranted (Figure 14.15). Polarization of tissue looking for foreign body giant cell reaction is also advised.

Focal myositis is a disorder marked by a localized inflammatory reaction and often presents as a tumorlike lesion (26,27). The process is usually self-limited, painful, and most commonly involves the lower extremities. The pathology is principally that of an inflammatory myopathy and is marked by chronic endomysial inflammation accompanied by muscle fiber degeneration and regeneration that can be highlighted on acid and alkaline phosphatase stains.

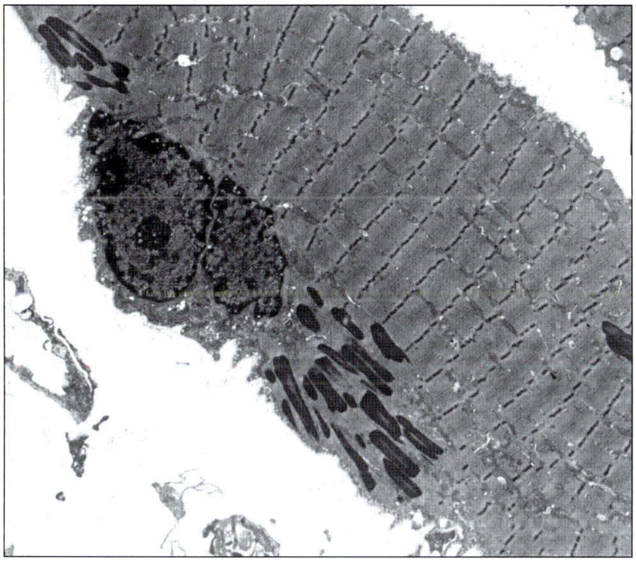

FIGURE 14.14: ***Ultrastructural appearance of nemaline rods as electron dense, crystalloid structures comprised of alpha-actinin (Z-band) material (original magnification 1800×).***

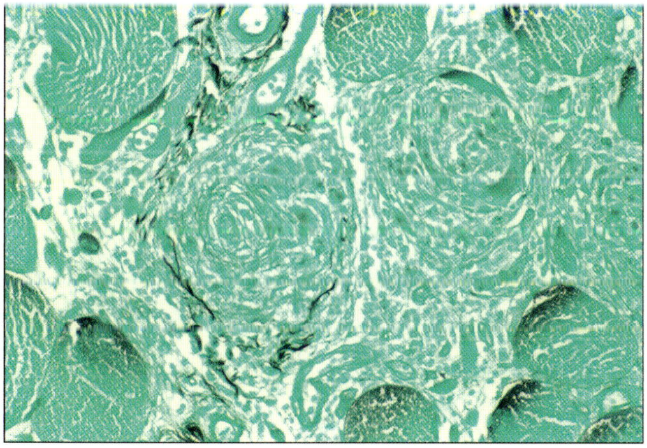

FIGURE 14.15: ***A methenamine silver–stained section of sarcoidal granuloma; fungal organisms are not identified. Granulomatous inflammation in most instances should prompt an evaluation for microorganisms (original magnification 200×).***

Neurogenic Muscle Disease

Neurogenic muscle disease represents a group of abnormalities in which the muscle pathology is secondary to primary disease involving the anterior horn cells of the spinal cord, nerve roots, and peripheral nerves. Most of the changes observed in the muscle, since secondary, are not specific for a particular disorder and provide few clues, if any, as to the nature of the nerve pathology. There are several enzyme histochemical stains that are useful in the evaluation of denervation atrophy. With acute denervation, muscle fibers undergo atrophy. In adults, the atrophic muscle fibers acquire an angulated contour. These muscles show increased staining with a nonspecific esterase stain (Figures 14.16 and 14.17). This is due to a concentration of the cell's esterase in a smaller surface

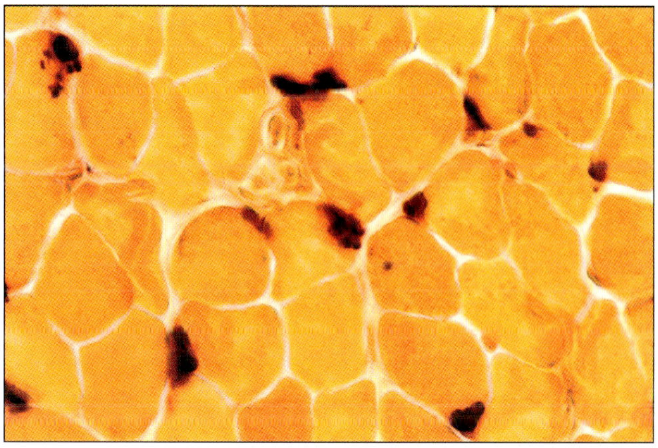

FIGURE 14.16: ***Normal nonspecific esterase staining highlighting neuromuscular junctions (original magnification 400×).***

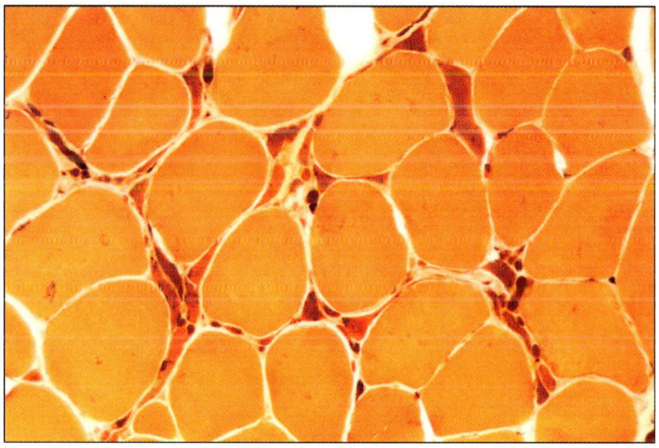

FIGURE 14.17: ***Acute denervation atrophy marked by angular, atrophic, esterase-positive staining myofibers (original magnification 200×).***

area of the atrophic muscle fibers. These fibers also show increased intensity of enzyme staining with NADH.

Additionally, with denervation atrophy, the NADH stain also highlights the presence of target fiber changes (28). Normal NADH stained fibers demonstrate a homogeneous, finely fibrillar pattern of staining, reflecting evenly distributed oxidative enzyme activity throughout the fiber cytoplasm (Figure 14.18). Target fibers are marked by alterations in oxidative enzyme activity, usually in a zonal pattern (Figure 14.19). The classic target fiber is characterized by a central pale zone surrounded by a darker, band-like intermediate zone and rimmed peripherally by normal staining. Target fiber changes are particularly prevalent in denervation atrophy cases. A less specific finding that may be observed is the presence of targetoid fibers. Targetoid fibers are characterized by a central pale zone or decreased oxidative activity surrounded by a normal outer zone (Figure 14.20). Moth-eaten fiber change, which is generally a nonspecific finding, is marked by uneven staining with NADH, giving the muscle fiber a mottled appearance (Figure 14.21).

In normal muscle, there is an alternating pattern arrangement (checkerboard or mosaic) of type I and type II muscle fibers in a given muscle. Some muscles show a predominance of one fiber type versus another, depending on the function of the muscle. Type I fibers or slow-twitch fibers are predominant in muscles that function in weight bearing or sustained activity. These fibers typically have a greater number of mitochondria and are higher in oxidative enzyme activity (stain more darkly on stains such as NADH). In contrast, the type II muscle fibers or fast-twitch fibers are more numerous in muscles that are involved with purposeful movement or sudden action. These muscle fibers typically have less oxidative enzyme activity, fewer mitochondria, and more glycogen.

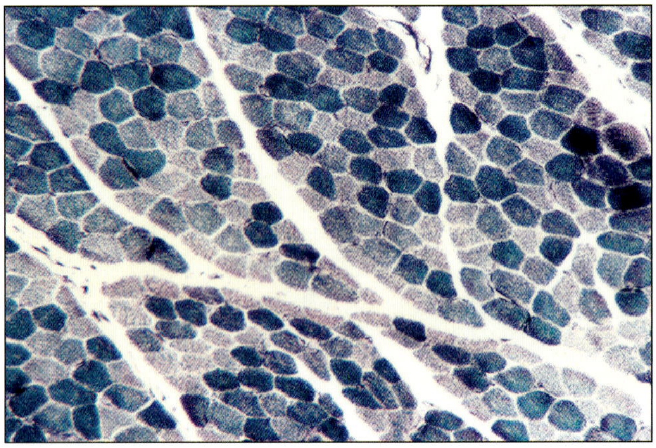

***FIGURE 14.18:* Normal NADH staining (original magnification 100×).**

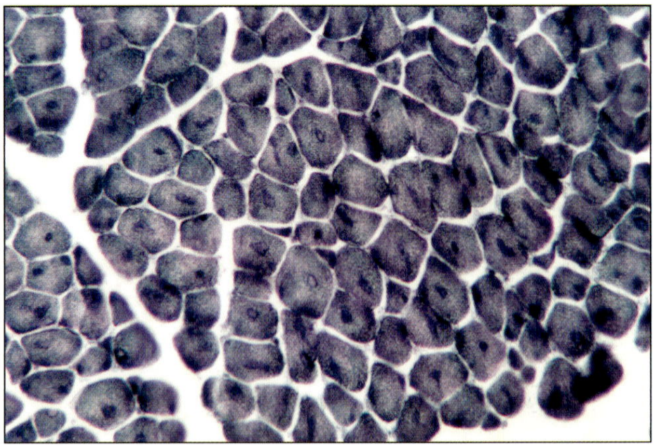

***FIGURE 14.19:* Target fibers involving most muscle fibers in this biopsy in the setting of neurogenic atrophy (NADH stain, original magnification 200×).**

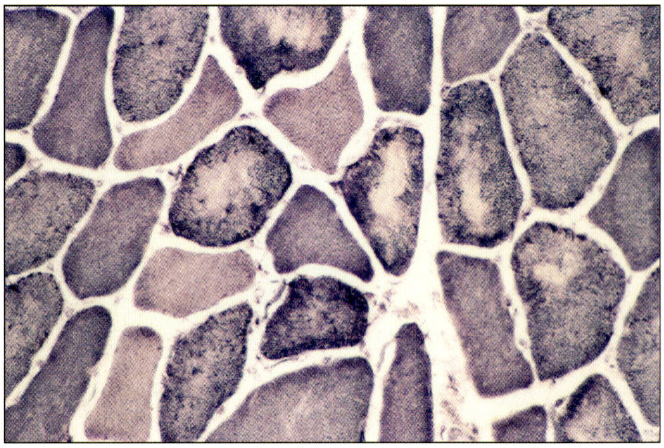

***FIGURE 14.20:* Central clearing on NADH staining characteristic of targetoid fiber change (original magnification 400×).**

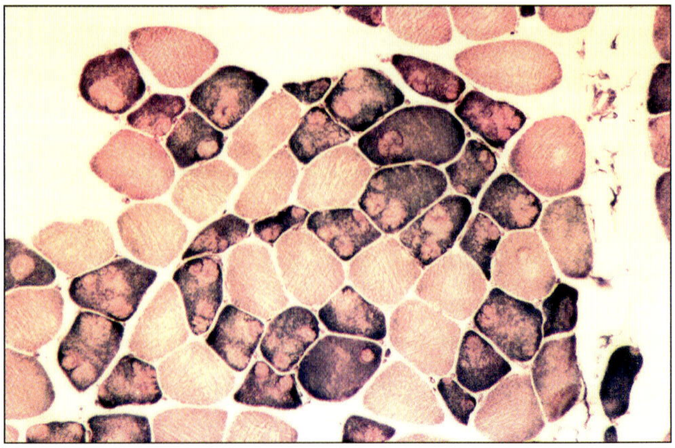

***FIGURE 14.21:* Mottled areas of decreased NADH staining mark moth-eaten fiber change (original magnification 200×).**

ATPase staining can be readily used in order to differentiate between fiber types. At acid pHs, typically pH 4.6 or 4.3, the ATPase staining (reverse ATPase) is characterized by dark staining type I muscle fibers and lighter staining type II muscle fibers (Figure 14.22). At the acidic pH, the type II muscle fibers can be further differentiated into type IIA fibers which are light staining, type IIB fibers, which stain an intermediate intensity between type IIA and type I fibers and type IIC fibers (fetal muscle) that stain lightly.

At an alkaline pH (standard ATPase), typically pH 9.8 or 10.4, the type II fibers stain darkly and the type I fibers stain lightly (Figure 14.23). The ATPase stain performed at the alkaline pH does not differentiate between type IIA and type IIB muscle fibers. Muscle fiber type is dictated by the nerve that innervates the fiber.

In the case of chronic denervation atrophy, the reinnervation of previously denervated muscle fibers is a frequent occurrence. The reinnervated muscle fiber will assume the fiber type dictated by the nerve that is reinnervating that fiber. Consequently, there is frequently the development of clustering of muscle fibers of the same type, referred to as fiber type grouping (Figure 14.24). A grouping of 13–15 muscle fibers of the same type is sufficient to make the diagnosis. This pathology manifests as a loss in the normal mosaic pattern of fiber type distribution.

There are other disease processes that can cause alteration in the distribution of fiber types in a muscle. For example, disuse atrophy preferentially involves type II muscle fibers (Figure 14.25). This is often best highlighted on utilizing the ATPase stains. Steroid myopathy classically results in type IIB muscle fiber atrophy (29); an ATPase stain performed at an acidic pH is useful in highlighting this alteration. Certain endocrine myopathies such as hyperthyroid associated myopathy and hyperparathyroid associated myopathy may show nonspecific changes including type II muscle fiber atrophy (30).

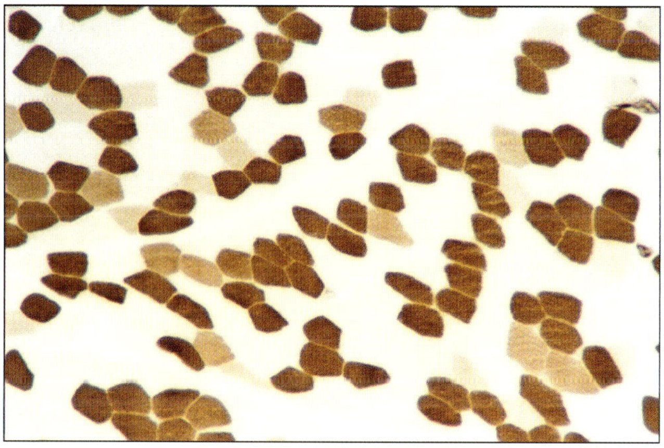

FIGURE 14.22: *Normal ATPase staining performed at pH 4.6; type I fibers stain darkly, type IIA fibers lightly, and type IIB fibers intermediately (original magnification 100×).*

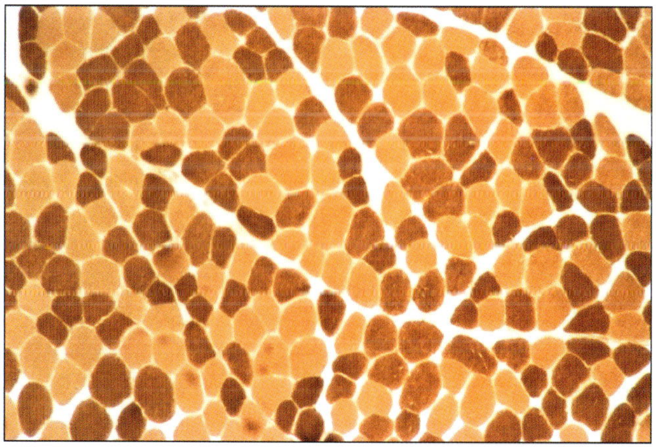

FIGURE 14.23: *Normal ATPase staining performed at pH 9.8; type II fibers stain darkly and type I fibers lightly (original magnification 100×).*

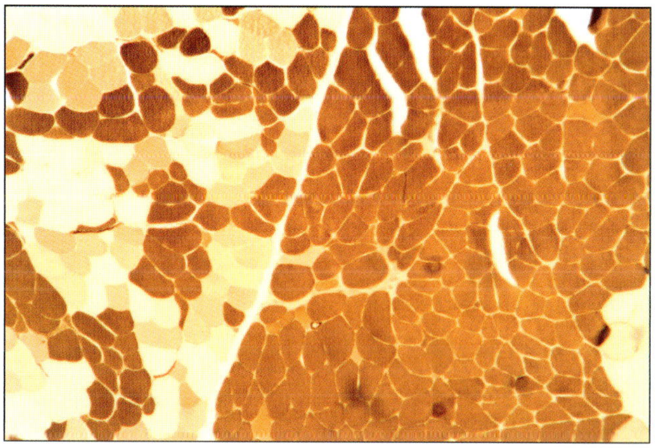

FIGURE 14.24: *Loss of normal mosaic pattern of fiber type distribution in fiber type grouping secondary to chronic neurogenic muscle disease (ATPase pH 4.6, original magnification 100×).*

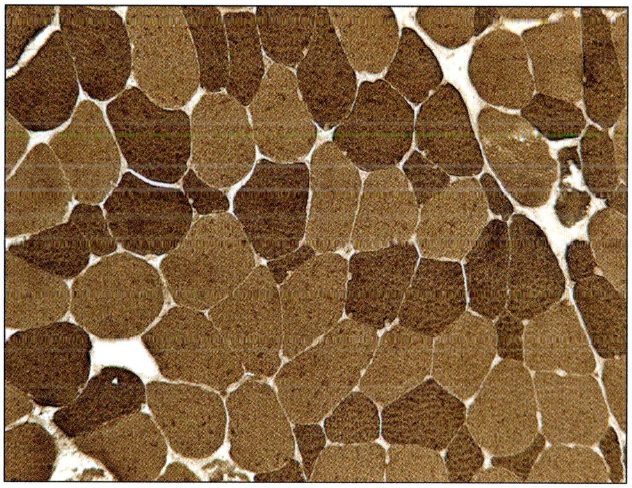

FIGURE 14.25: *Type II muscle fiber atrophy most commonly associated with disease (ATPase pH 9.8, original magnification 400×).*

Metabolic Myopathies

Ancillary staining of muscle biopsies in patients with suspected metabolic disorders can be very helpful. Often, the first clue that a patient may have a metabolic disorder is based on the light microscopic findings on various enzyme histochemical and special stains.

Mitochondrial myopathies are among the most common of the metabolic disorders. This rather diverse group of diseases is marked by mitochondrial DNA mutations that may affect a variety of mitochondrial enzymes (31–36). There are a number of clues one can look for which may indicate a mitochondrial abnormality. On routine H&E staining, many muscles look normal or demonstrate minimal, nonspecific abnormalities (Figure 14.26). Enzyme and special stains are particularly helpful in identifying an abnormality. The most classic finding is the so-called ragged red fiber, which is best visualized on the Gomori trichrome–stained section (Figure 14.27). This fiber is marked by increased granular eosinophilic staining, most pronounced in the subsarcolemmal region in a circumferential fashion and often accompanied by increased, more diffuse granular staining throughout the affected muscle fiber. Ultrastructural examination of ragged red fibers shows an abnormal number and frequently morphologically atypical mitochondria (Figure 14.28). Succinate dehydrogenase (SDH) staining, another mitochondrial enzyme, may demonstrate evidence of ragged blue fibers (35) (Figures 14.29 and 14.30). In a similar fashion, increased subsarcolemmal staining, manifested by dark blue staining, is predominantly noted in the subsarcolemmal region, frequently circumferentially, in affected muscle fibers. Muscle stained with the mitochondrial enzyme stain, cytochrome oxidase (Figures 14.31 and 14.32), if involved by a

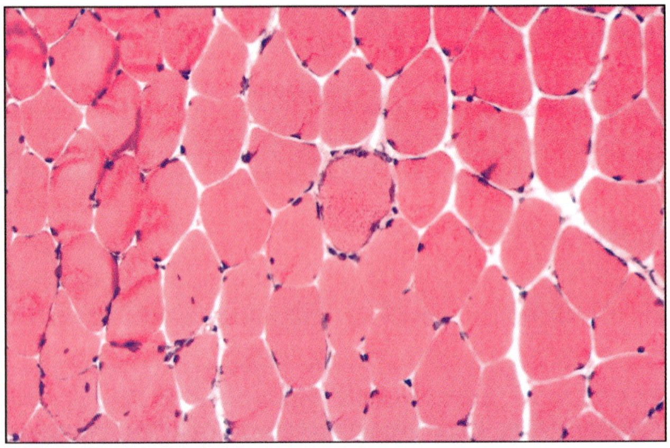

FIGURE 14.26: *Mitochondrial myopathy with normal appearing muscle on H&E staining (original magnification 200×).*

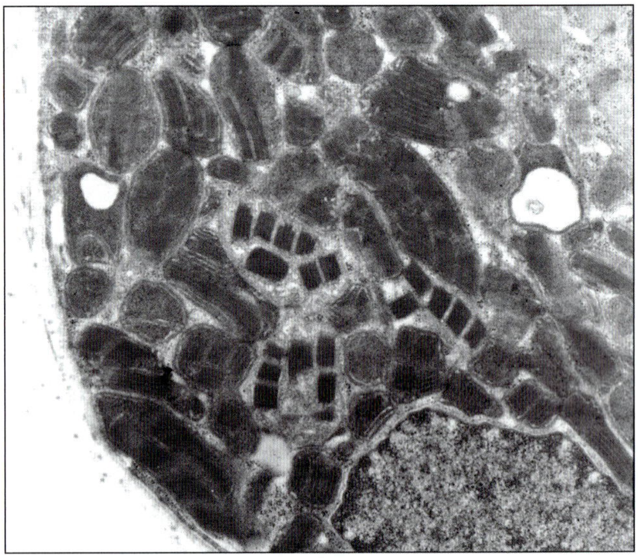

FIGURE 14.28: *Electron macrophage showing paracrystalline mitochondrial inclusions consistent with mitochondrial myopathy (original magnification 3000×).*

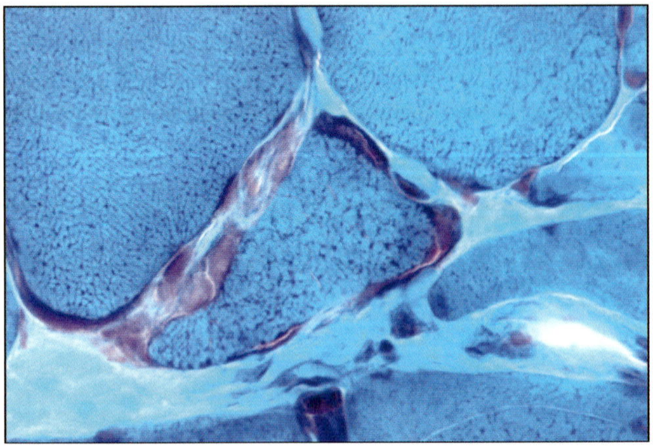

FIGURE 14.27: *Trichrome-stained section showing a ragged red fiber suggestive of a mitochondrial abnormality (original magnification 100×).*

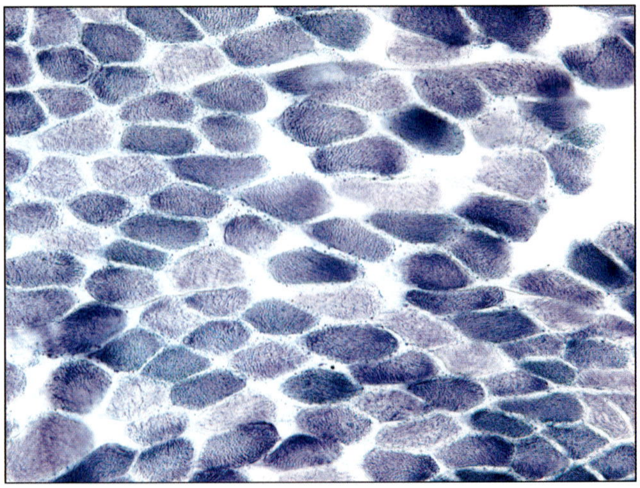

FIGURE 14.29: *Normal SDH staining (original magnification 200×).*

DNA mutation, may show scattered fibers that are devoid of staining. It is important to remember that each of these mitochondrial-associated abnormalities may be encountered (in small numbers) in normal elderly individuals and is associated with a variety of other neuromuscular disorders (such as inclusion body myositis as previously described) (37).

The glycogenoses represent a group of disorders that result from metabolic abnormalities in either the synthesis or metabolism of glycogen (38–41). Those disorders that can affect skeletal muscle include most commonly type II (acid maltase)—Pompe's disease, type III (debranching enzyme)—Cori-Forbes disease, type IV (branching enzyme)—Andersen's disease, type V (myophosphorylase)—McArdle's disease, and type VII (phosphofructokinase)—Tarui's disease. Specific enzyme histochemistry for phosphofructokinase and myophosphorylase exist and can be utilized in suspected cases, looking for an absence of staining in scattered muscle fibers. In many instances, the microscopic changes observed are those of a vacuolar myopathy (Figure 14.33); the vacuoles are filled with glycogen material. PAS staining with and without diastase may be useful in confirming the presence of increased glycogen in these vacuoles (Figures 14.34 and 14.35). Frequently, the glycogen is lost during tissue processing and the vacuoles appear empty or only sparsely filled with PAS-positive material. In such cases, confirmation of the diagnosis by ultrastructural examination and demonstration of increased glycogen accumulation can be useful (Figure 14.36). Of particular note is the fact that type II glycogenosis or acid maltase deficiency is a lysosomal process. Consequently, it is the only one of the glycogenoses in which the vacuoles, which are filled with glycogen, also demonstrate a peripheral increased acid phosphatase staining (Figure 14.37).

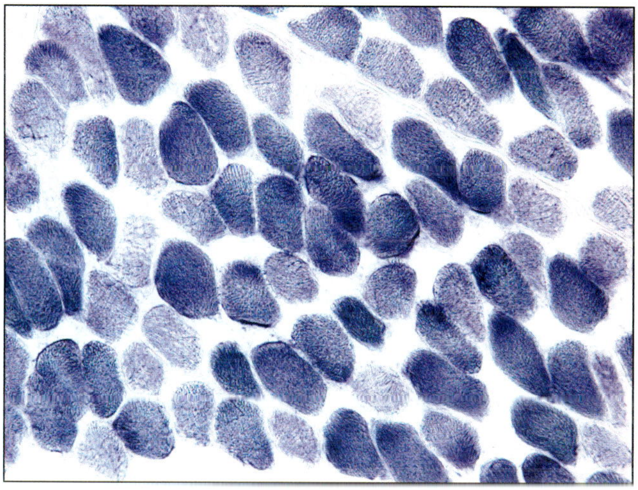

FIGURE 14.30: *Scattered fibers with increased subsarcolemmal staining on SDH corresponding to increased numbers of mitochondria in a mitochondrial myopathy (original magnification 200×).*

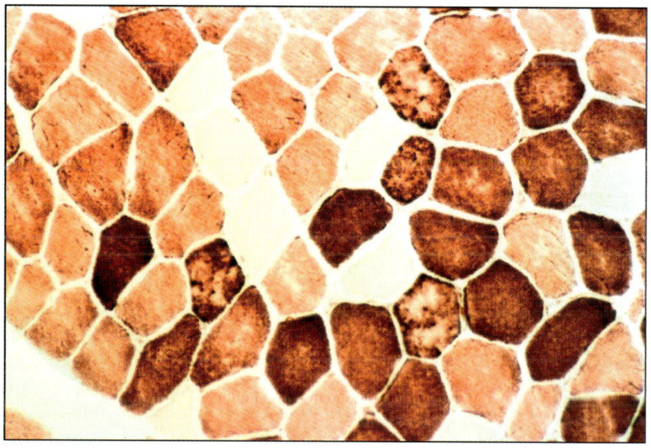

FIGURE 14.32: *Scattered cytochrome oxidase–negative fibers and nonspecific moth-eaten fiber changes in a case of mitochondrial myopathy (original magnification 200×).*

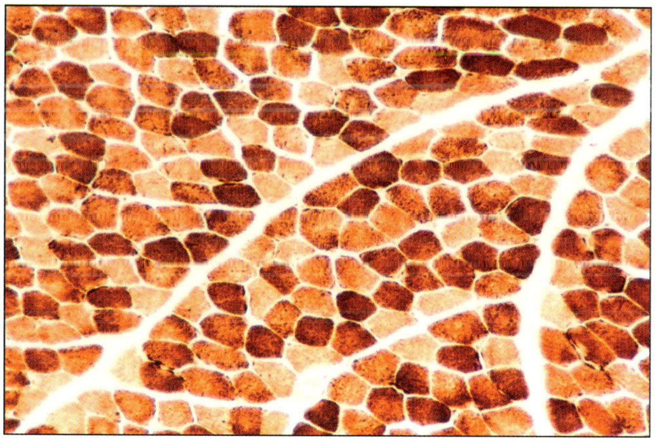

FIGURE 14.31: *Normal cytochrome oxidase staining (original magnification 100×).*

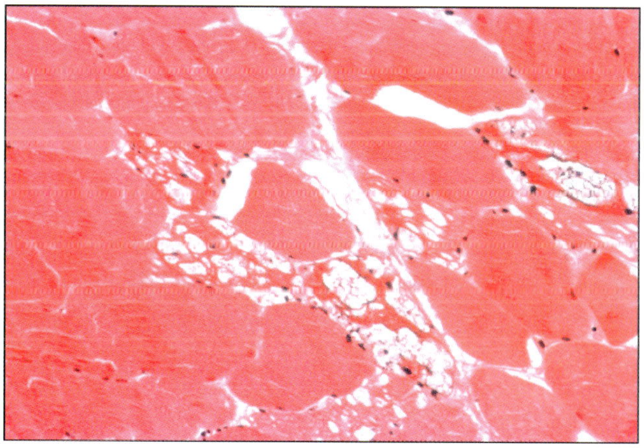

FIGURE 14.33: *H&E-stained section showing vacuolar myopathy change in a case of acid maltase deficiency (glycogenesis type II) (original magnification 400×).*

Lipidoses represent another group of metabolic disorders associated with abnormal lipid metabolism (42,43). Similar to the glycogenoses, these frequently present as vacuolar myopathies. In contrast to the glycogenoses, the vacuoles in these disorders are filled with lipid material. Oil red O– or Sudan black–positive staining material within the vacuoles confirms the lipid nature of the accumulated substance, which can also be confirmed by ultrastructural study (Figure 14.38).

Myoadenylate deaminase deficiency is a relatively poorly understood muscle disorder (44). Clinical manifestations range from fatigability, muscle cramps, and myalgias following exercise to being relatively asymptomatic. Enzyme histochemistry exists for this disorder and patients with a deficiency show reduced or absent staining in suspected muscle fibers (Figure 14.39). The vast majority of patients demonstrating deficiency are asymptomatic and this may be secondary to a variety of other disorders.

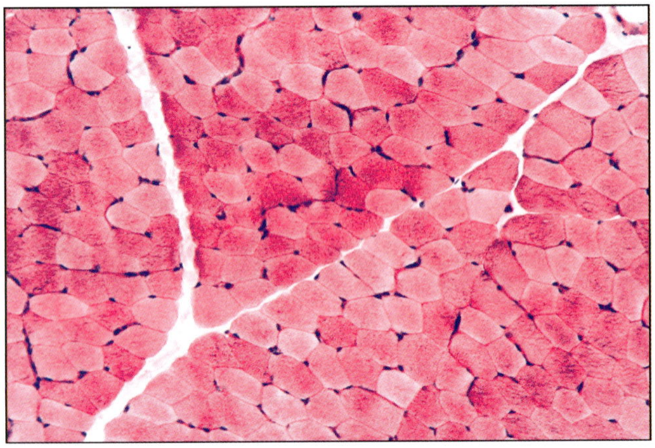

***FIGURE 14.34:* Normal PAS staining (original magnification 100×).**

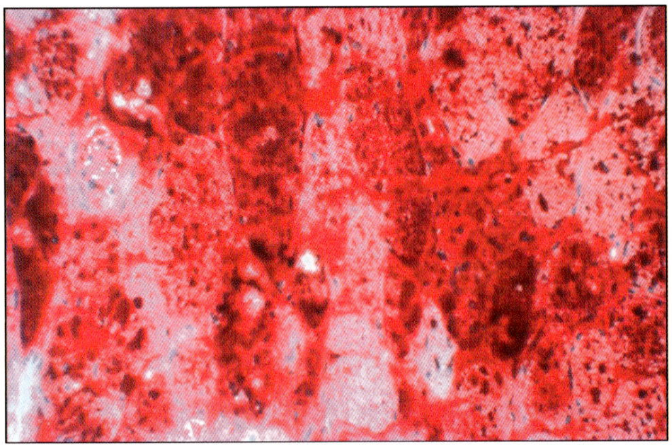

***FIGURE 14.35:* Increased PAS staining corresponding to glycogen accumulation in acid maltase deficiency (original magnification 400×).**

Muscular Dystrophy

The muscular dystrophies represent a group of genetic disorders marked by protein abnormalities. As a group, these disorders are progressively degenerative and may involve multiple organ systems. Historically, the dystrophies were categorized according to clinical phenotype and patterns of inheritance. More recently, genetic alterations and alterations in protein production and expression have been discovered in many of these disorders. Most of the staining is immunohistochemistry or immunofluorescence based. Routine enzyme histochemistry including acid and alkaline phosphatase stains may be helpful in highlighting the presence of degenerating and regenerating muscle fibers,

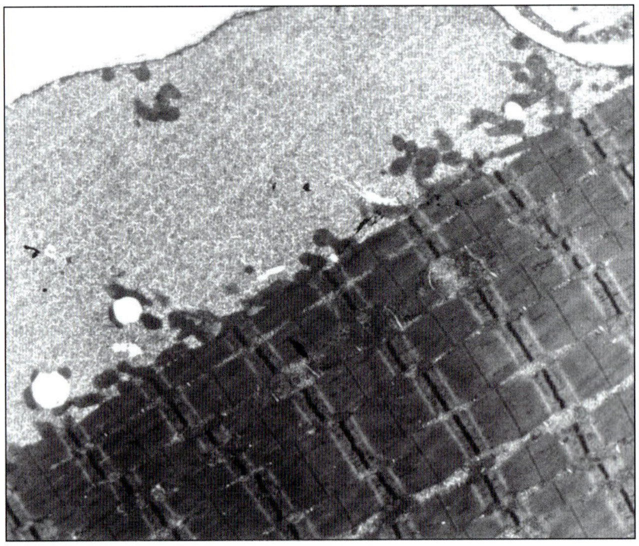

***FIGURE 14.36:* Ultrastructural examination showing increased subsarcolemmal glycogen accumulation in acid maltase deficiency (original magnification 2400×).**

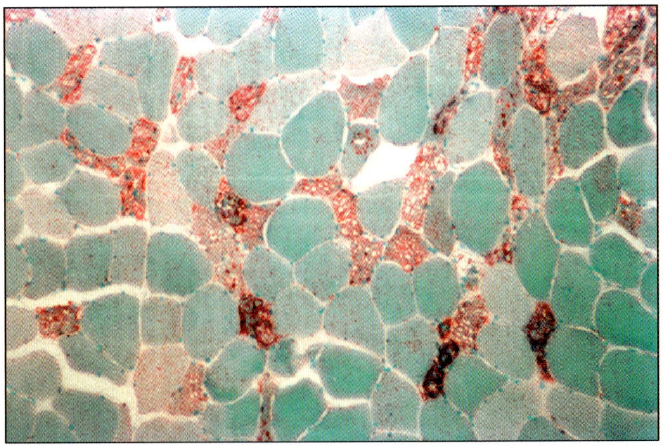

***FIGURE 14.37:* Acid maltase deficiency is marked by acid phosphatase positivity associated with vacuoles filled with glycogen. This is the only one of the glycogenoses that is lysosomal associated (original magnification 200×).**

respectively in certain dystrophic disorders (45,46) (Figures 14.40 and 14.41). Certain disorders, such as oculopharyngeal muscular dystrophy, tend to show a tendency toward the development of increased numbers of atrophic type I muscle fibers (47,48). The presence of prominent numbers of ring fibers, which are identifiable on NADH staining, is a characteristic of many cases of myotonic dystrophy (49). The ring fibers are marked by a zone in which the myofibrils in the cytoplasm are oriented perpendicularly with respect to the muscle fiber axis. Sarcoplasmic masses, which can be highlighted as red-staining irregularly shaped cytoplasmic masses on the Gomori trichrome stain, are also fairly frequently observed in myotonic muscular dystrophy.

Congenital Myopathies

The congenital myopathies represent a group of disorders characterized by morphologic abnormalities of skeletal muscle (50–53). Most of them present early in life with

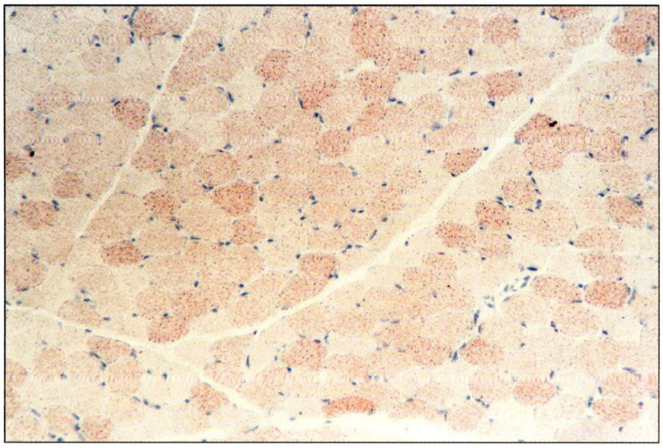

***FIGURE 14.38:* Normal lipid in muscle highlighted on an oil red O stain (original magnification 100×).**

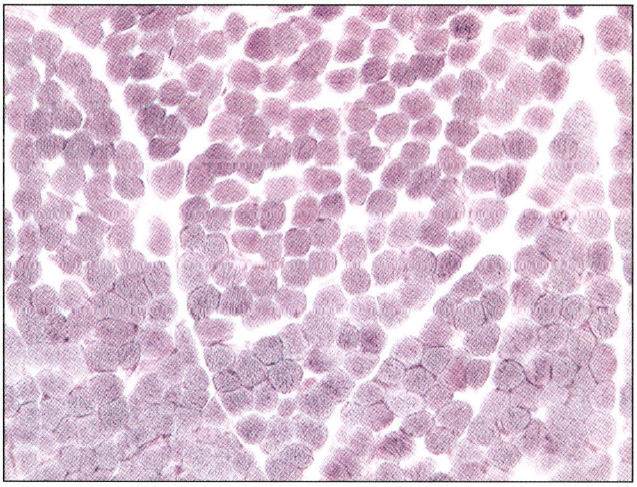

***FIGURE 14.39:* Normal myoadenylate deaminase staining (original magnification 200×).**

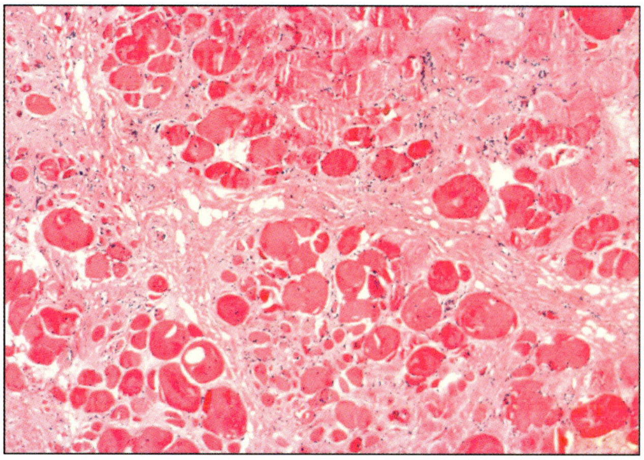

***FIGURE 14.40:* Chronic myopathic changes in a case of Becker's muscular dystrophy (dystrophinopathy) (original magnification 100×).**

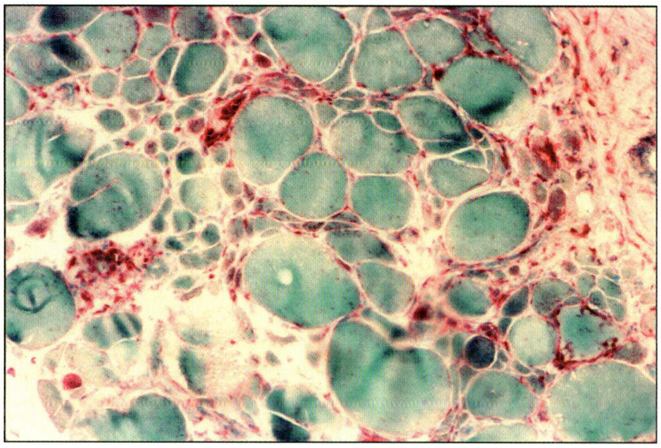

***FIGURE 14.41:* Acid phosphate staining highlighting occasional degenerating myofibers and inflammatory cells in Becker's muscular dystrophy (original magnification 200×).**

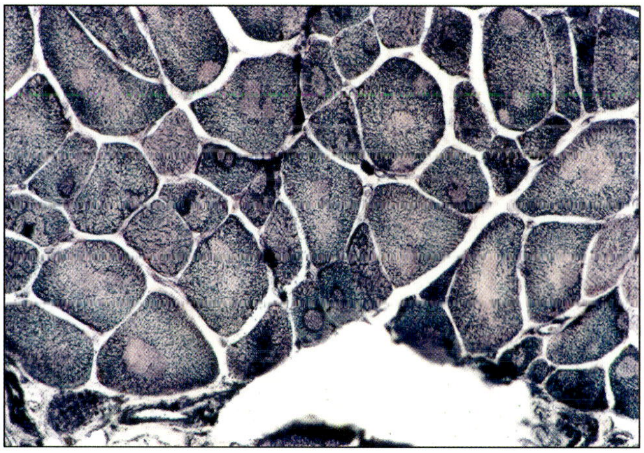

***FIGURE 14.42:* Central core congenital myopathy with core fiber changes noted in most of the muscle fibers (NADH stain, original magnification 400×).**

TABLE 14.1: Congenital Myopathies

Type	Morphologic Findings	Stains
Nemaline rod myopathy	Nemaline rods—crystalloid structure 1–7 μm in length, more common in type I fibers	Highlighted on Gomori trichrome or PTAH
Centered core myopathy and minicore myopathy	Core(s) marked by well-circumscribed circular areas predominantly in type I fibers	Decreased staining on NADH
Centronuclear (myotubular) myopathy	Increased number of fibers with central nuclei	ATPase stain highlights central nuclei as nonstaining, cytochrome oxidase with granular paranuclear staining
Tubular aggregate myopathy	Aggregation of sarcoplasmic reticulin material	Irregular areas of red staining on Gomori trichrome
Congenital fiber type disproportion	Bimodal distribution of muscle fiber size with type II fibers larger than type I fibers	ATPase stains
Spheroid body myopathy	Spherical masses comprised 12- to 15-nm filaments	Stains green—Gomori trichrome

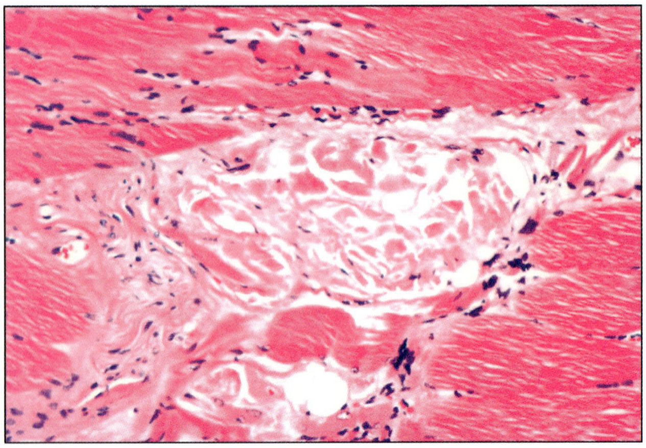

FIGURE 14.43: ***Amorphous, eosinophilic amyloid deposition in skeletal muscle (H&E, original magnification 400×).***

hypertonia and either nonprogressive or slowly progressive muscle weakness. These disorders have a variety of inheritance patterns and are frequently accompanied by dysmorphic skeletal abnormalities. As a group, many of the congenital myopathies are marked by the presence of a morphologic abnormality in the majority of muscle fibers in any given biopsy (Figure 14.42). Many of these abnormalities are readily highlighted with ancillary staining techniques.

Table 14.1 highlights some of the more common congenital myopathies with associated abnormalities and a list of stains that are useful in highlighting these findings.

Amyloid Myopathy

Deposition of amyloid material in skeletal muscle is a relatively rare occurrence. Amyloid refers to any protein that arranges itself in a beta-pleated configuration. This configuration contributes to the characteristic staining features that amyloid material has on a variety of stains. Amyloid myopathy is most commonly associated with immunoglobulin and light chain accumulation in the background setting of plasma cell dyscrasia or a lymphomatous process. It may be deposited around blood vessels, in endomysial connective tissue or rarely in a tumoral fashion (54,55) (Figures 14.43 and 14.44). There is a whole variety of well-known staining techniques that can be used to highlight the presence of amyloid material. The Congo red stain under polarization optics that demonstrates apple-green birefringence is classically employed (Figure 14.45). Metachromatic staining with methyl or crystal violet can be utilized. Immunofluorescent markers thioflavin S or T can also be helpful in highlighting amyloid material.

HISTOCHEMICAL STAINS IN SOFT TISSUE TUMORS AND TUMEFACTIONS

Although histochemical stains have lost their "edge" for the definitive evaluation of soft tissue neoplasms and proliferations, they remain interesting and useful in reference to a number of neoplasms. Indeed, histochemistry may be critical in some cases. For example, the Congo red stain can confirm the impression of an amyloid tumor or amyloid myopathy, as discussed above, and an acid-fast stain is essential in diagnosing mycobacterial spindle cell pseudotumors (56).

Adipose Tissue Lesions

Oil red O staining of frozen section materials had been touted as a method to demonstrate adipose tissue differentiation in the past but has largely fallen from favor. Finding lipid in the cytoplasm of neoplasms is hardly specific for liposarcomas because renal cell carcinomas can contain fat, and histiocytes and even neoplastic cells can engulf lipid as well. Lipoblasts show lipid droplets that crisply indent the nucleus and are generally readily identified on routine H&E stains.

The mucoid matrix of myxoid liposarcoma is composed of glycosaminoglycans that readily stain with alcian blue and colloidal iron preparations. These stains can be inhibited by hyaluronidase. In contrast, the sulfated glycosaminoglycans of myxoid chondrosarcoma and chordoma are resistant to hyaluronidase, so this technique can be exploited to separate such tumors (57).

Fibrous Lesions

Diagnosis of fibroblastic lesions is generally not aided by the use of histochemical stains, with notable exceptions. Trichrome stains are prepared using the Gomori or Masson method. With the Gomori technique, which uses iron hematoxylin, chromotrope 2R, light green stain, and phosphotungstic acid, smooth and skeletal muscle fibers are red, collagen is green, and nuclei are blue to black (58). The Masson method requires iron hematoxylin, a solution of Biebrich scarlet and acid fuscin, a phosphomolybdic-phosphotungstic acid solution, and either an aniline blue solution or light green solution. Using this version, nuclei are black, muscle, cytoplasm, and keratin are red; and collagen is either blue or green (59). In lesions that are myofibroblastic, such as the various fasciitis and fibromatoses, there are often delicate reddish fibrils on both types of trichrome preparations (Figures 14.46–14.50), whereas

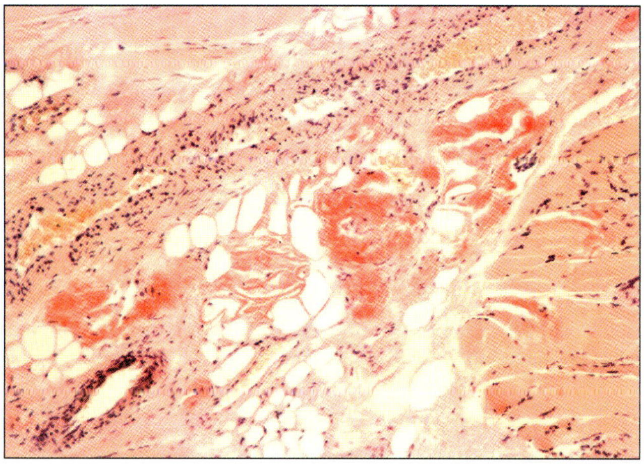

***FIGURE 14.44:** Congo red staining in an amyloid myopathy (original magnification 100×).*

***FIGURE 14.45:** Polarized Congo red staining with apple-green birefringence consistent with amyloid deposition in skeletal muscle (original magnification 200×).*

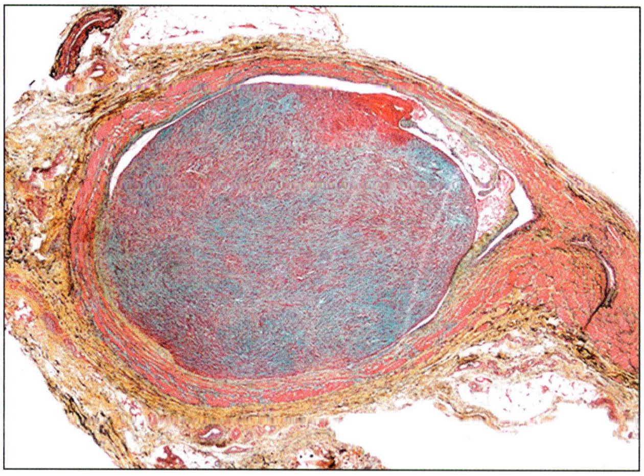

***FIGURE 14.46:** Trichrome stain of intravascular fasciitis, a myofibroblastic lesion. Compare the tinctorial quality of the intravascular lesion to that of the surrounding vessel.*

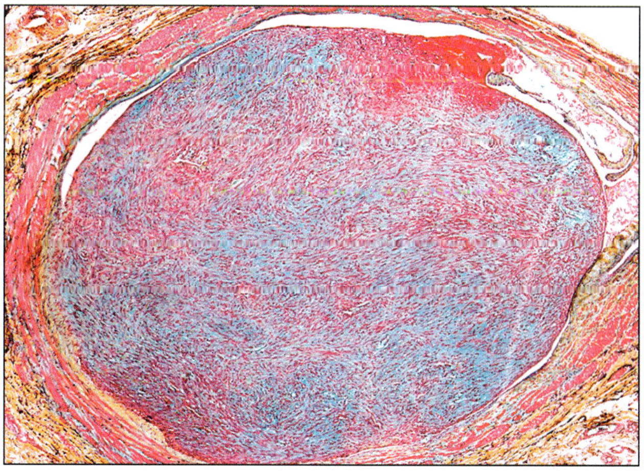

***FIGURE 14.47:** Higher magnification of the image depicted in Figure 14.46.*

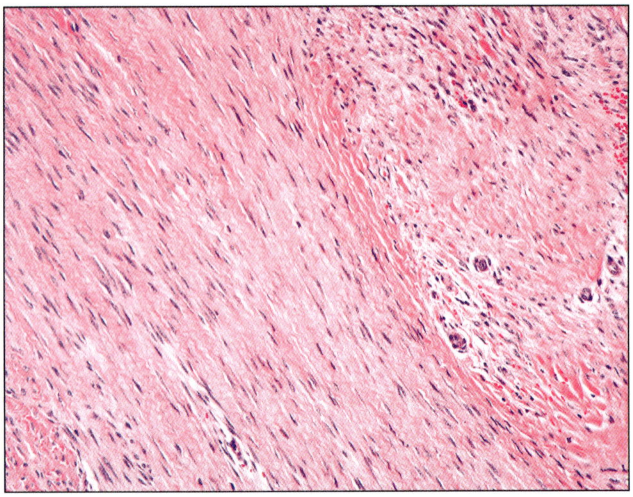

FIGURE 14.48: Low-magnification H&E stain of a fibromatosis (desmoid tumor). The lesion consists of sweeping fascicles of pale myofibroblasts.

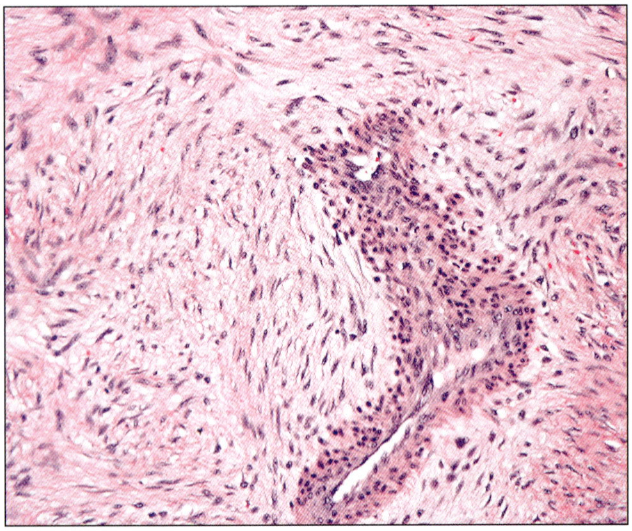

FIGURE 14.49: Higher magnification of an H&E stain of a fibromatosis. Note the brightly eosinophilic appearance of the vascular smooth muscle compared to the paler cytoplasm of the proliferating leisonal cells.

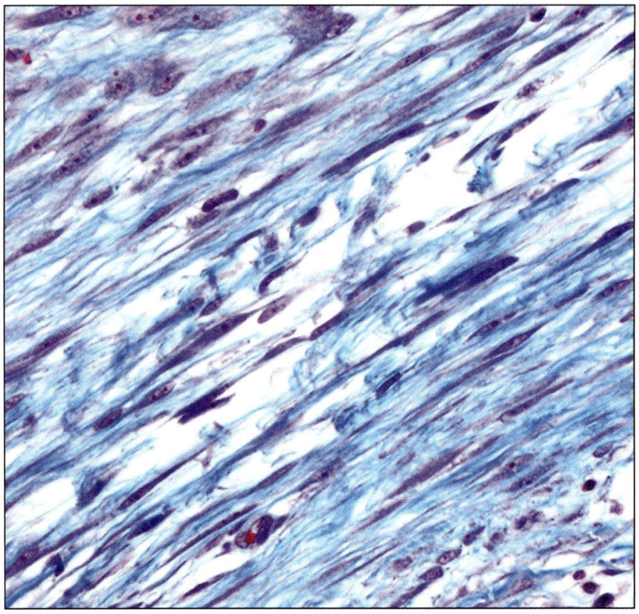

FIGURE 14.50: Trichrome stain of a fibromatosis. The cytoplasm is mostly blue with rare delicate red filaments. Compare this to the smooth muscle and leiomyosarcoma seen in Figures 14.63, 14.64, and 14.67.

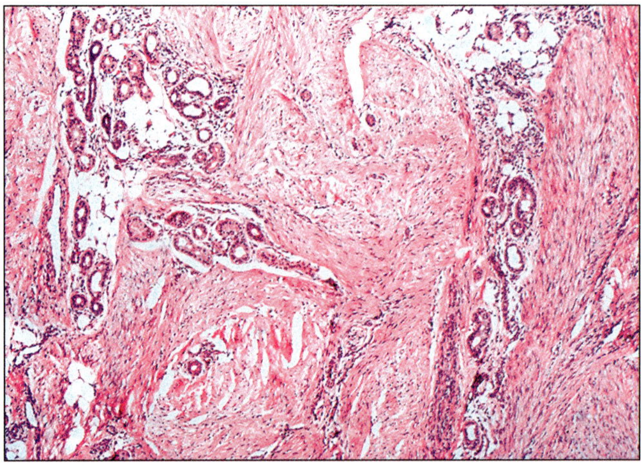

FIGURE 14.51: Inclusion body fibromatosis (infantile digital fibromatosis), H&E stain. At low magnification, this superficial lesion infiltrates skin appendages and appears similar to the deep fibromatosis depicted in Figure 14.48.

the fibrils are more organized in true smooth muscle see below).

Inclusion Body Fibromatosis (Infantile Digital Fibromatosis): Infantile digital fibromatosis (DF) is a distinct tumor of the dermis and subcutis of the fingers and toes of infants. Approximately 30% of cases are congenital. The salient feature of this tumor is the presence of cytoplasmic inclusion bodies. The lesion may be single or multiple, and has a tendency to recur. The ultimate prognosis is excellent and most tumors will eventually regress spontaneously after an initial growth period. However, these tumors are capable of causing contracture deformities.

Grossly, DF is poorly circumscribed and is attached to the overlying skin. Microscopically, there are fascicles of bland-appearing fibroblasts that appear similar to those of other superficial fibromatoses. Characteristic globular, eosinophilic cytoplasmic inclusion bodies can be seen on routine H&E staining but appear bright red with the trichrome stain (Figures 14.51–14.53). These inclusions correspond to actin-reactive globules immunohistochemically (60). The inclusions are variable in number and are seen less frequently in cases with copious fibrosis.

Elastofibroma: Elastofibroma dorsi (61–63) is a slowly growing fibroelastic proliferation believed to result from mechanical friction between the scapula and the chest wall. It typically occurs on the back in women with a history of mechanical labor, and ultrastructurally, the deposition of elastic tissue is believed to occur on a background of preexisting fibrils (62). Although these lesions have been presumed to be reactive in the past, recent studies suggest that it has a variety of chromosomal alterations and could instead be a neoplasm (64–66). Bilaterality is also known (67). Grossly, the excised mass is a firm ill-defined lesion with a glistening surface and foci of cystic degeneration and interspersed fatty areas. On microscopic examination, it consists of intertwining eosinophilic colonized collagen and thickened, serrated, deeply eosinophilic elastic fibers that have a degenerated beaded appearance or are fragmented into globules (chenille [French for caterpillar] bodies) or flower-like arrangements (Figures 14.54 and 14.55). Elastic stains, which contain an iodine solution and ferric chloride, characteristically highlight the degenerated appearance of the elastic fibers (68) (Figure 14.56).

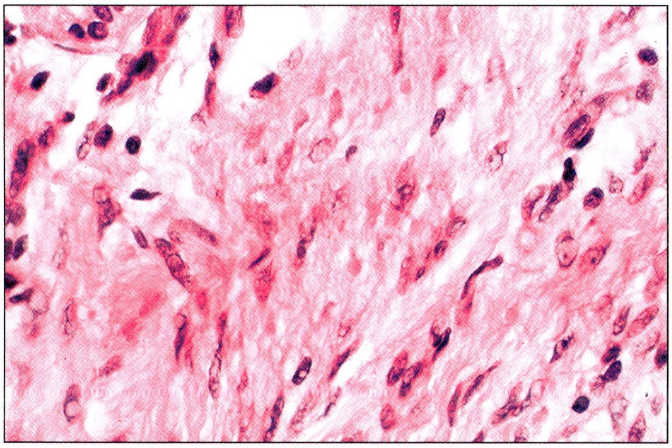

FIGURE 14.52: Higher magnification of the H&E stain seen in Figure 14.51. At this magnification, eosinophilic cytoplasmic inclusions are apparent.

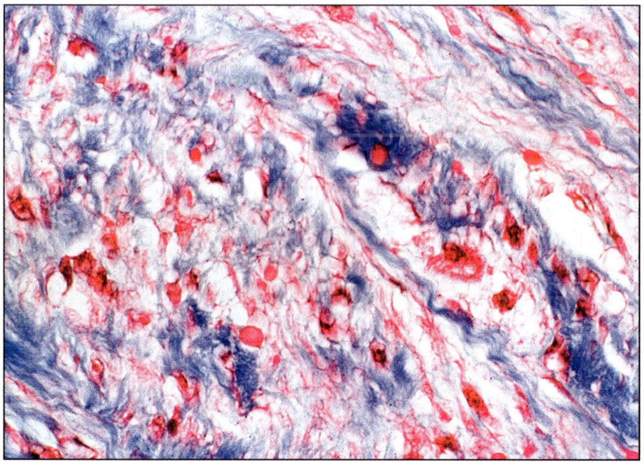

FIGURE 14.53: The trichrome stain highlights the cytoplasmic inclusions of inclusion body fibromatosis to advantage.

So-Called "Fibrohistiocytic Lesions"

Myxofibrosarcoma/Myxoid Malignant Fibrous Histiocytoma Versus Myxoid Liposarcoma: A common differential diagnostic problem is between myxoid malignant fibrous histiocytoma (MFH) (myxofibrosarcoma) and myxoid liposarcoma. Myxoid MFH, as described by Weiss and Enzinger (69), comprises 20% of all MFHs and is defined as having over 50% of myxoid change, with characteristic nodularity, vascular pattern, atypical spindle cells, and a tendency for superficial location. The myxoid

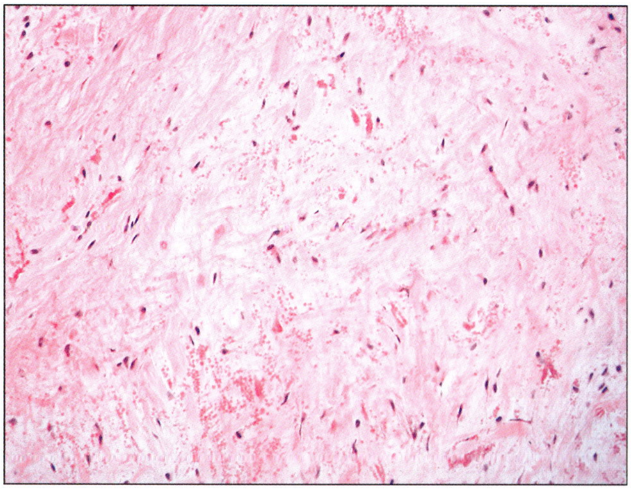

FIGURE 14.54: Low magnification of an H&E-stained section from an elastofibroma. The aberrant elastic fibers appear eosinophilic. The lesion is otherwise hypocellular.

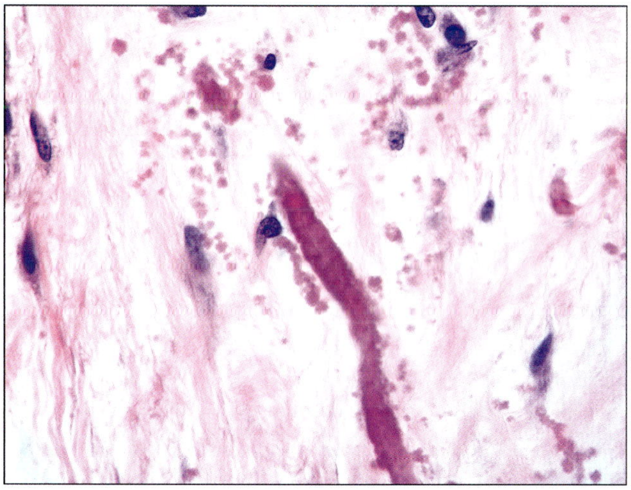

FIGURE 14.55: Higher magnification of the elastofibroma seen in Figure 14.54.

nodules may merge into areas of pleomorphic MFH. The likelihood of metastasis relates to depth from the skin surface and is inversely proportional to the amount of myxoid change. Myxofibrosarcoma was described at the same time (70,71) as a subcutaneous or deep multinodular tumor of older adults composed of thin tapered spindle cells in a myxoid pattern, with a prominent vascular pattern. The tumor cells are sometimes vacuolated, resembling lipoblasts, but they are irregularly shaped, with alcianophilic secretion. Four grades were initially recognized depending on degree of cellularity and pleomorphism; high-grade myofibrosarcoma is considered equivalent to myxoid MFH. Comparing grade with outcome in a large series, metastases were found in 0%, 21%, 47%, and 38% of patients with grades I–IV, respectively (71).

Up to half of all liposarcomas are of myxoid/round cell type. Although those two variants had historically been considered to be separate histologic entities, it is now apparent that the two diagnostic terms describe different histologic and clinical ends of a spectrum that is genetically homogenous (72). Tumors with pure myxoid liposarcoma morphology have a better prognosis than those with round cell features. Pure round cell morphology is extremely uncommon and most tumors classified as round cell liposarcoma are composed of mixed myxoid and round cell components. These tumors typically occur in adults and are most common in the fifth decade. Rare myxoid/round cell liposarcomas occur in children. Males are affected more commonly than females. The deep soft tissue of the thigh is the single most common location for these neoplasms. Primary tumors also arise in the soft tissue of the arms, trunk, and retroperitoneum. The majority are deep to the fascia. Patients typically complain of a painless soft tissue mass. Myxoid/round cell liposarcoma is treated by wide surgical excision with or without radiation therapy. Approximately 30% of patients develop distant metastases. Like most other sarcomas, these often involve the lungs; however, myxoid/round cell liposarcoma often metastasizes to nonpulmonary sites including the retroperitoneum, soft tissue, and skeleton (73). Histologic grading appears to have value in predicting those patients at risk for metastases. Specifically, as the cellularity of the tumor increases and the degree of tumor composed of round cell liposarcoma or so-called cellular myxoid liposarcoma increases, there is an increased risk for metastasis.

Both lesions contain myxoid matrix, but it is extracellular in myxoid liposarcoma and intracellular in myxoid MFH (myxofibrosarcoma). Myxoid liposarcomas also lack atypical nuclei and, like most translocation-associated sarcomas, have very uniform nuclear features (74). Lipid droplets in myxoid liposarcomas crisply indent the nuclei (Figure 14.57), whereas the mucopolysaccharide matrix of myxoid MFH (myxofibrosarcoma) compresses nuclei in an indistinct fashion (Figures 14.58 and 14.59). Intramuscular myxomas (IMs) are broadly similar histologically to

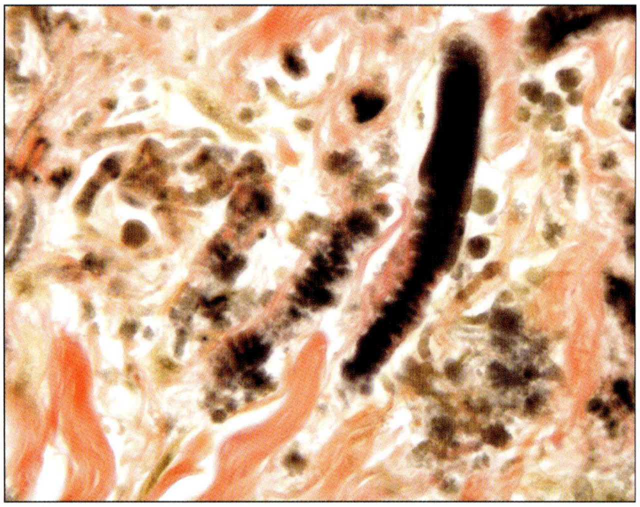

FIGURE 14.56: *Elastic stain of an elastofibroma. The appearance of the fibrils has been likened to that of a caterpillar (hence the term "Chenille" bodies; chenille is the French word for caterpillar).*

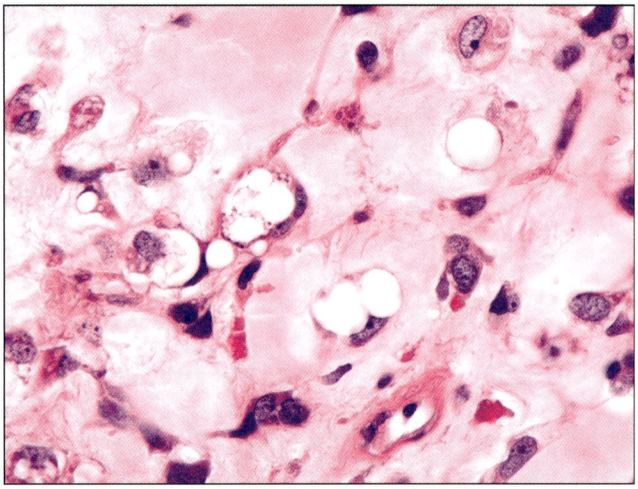

FIGURE 14.57: *In myxoid liposarcomas (and other liposarcomas), lipoblasts show crisp nuclear indentation by cytoplasmic lipid droplets (H&E stain).*

myxoid MFH (myxofibrosarcoma) but lack their rich vascularity.

Tenosynovial Giant Cell Tumor Versus Clear Cell Sarcoma: Iron stains can be reassuring when tenosynovial giant cell tumors are encountered. The differential diagnosis of densely cellular lesions in the hands includes clear cell sarcoma, a lesion with melanocytic differentiation both histochemically and immunohistochemically (75); acral melanoma; as well as epithelioid sarcoma (76). Usually, there is little morphologic overlap between epithelioid sarcoma and tenosynovial giant cell tumors, but clear cell sarcomas often feature abundant giant cells. Although most observers use immunohistochemistry to address this differential diagnosis, clear cell sarcomas generally lack hemosiderin and tenosynovial giant cell tumors contain it (Figures 14.60–14.62).

Smooth Muscle Tumors

Smooth muscle differentiation has the same histochemical features whether it is neoplastic or nonneoplastic. Both leiomyomas and leiomyosarcomas display histochemical features of smooth muscle differentiation. These include both delicate longitudinal striations in the trichrome stain (Figures 14.63–14.67) and glycogen with the PAS stain. The Schiff reagent in the PAS stain consists of basic fuchsin to which potassium metabisulfite and hydrochloric acid is added. The nuclei are counterstained with hematoxylin. With this stain, glycogen found in both smooth and skeletal muscle neoplasms (Figure 14.68), mucin, and some basement membranes are red to purple (usually a magenta shade) (68).

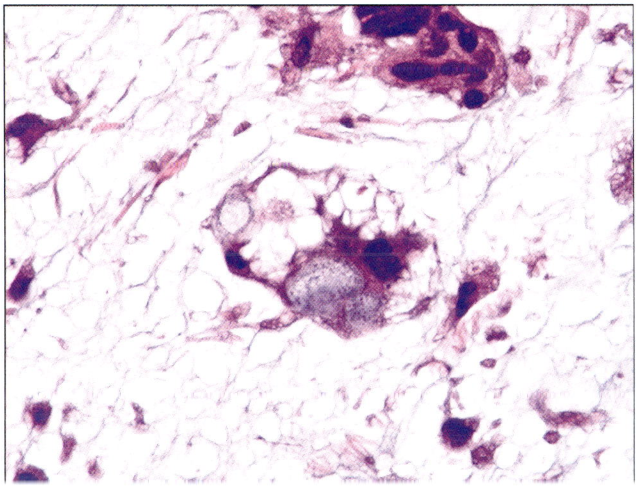

FIGURE 14.58: This myxofibrosarcoma/myxoid MFH has pleomorphic nuclei (in contrast to the uniform ones in myxoid liposarcoma) and the cytoplasm of the cell in the center of the field contains mucopolysaccharide matrix rather than lipid (H&E).

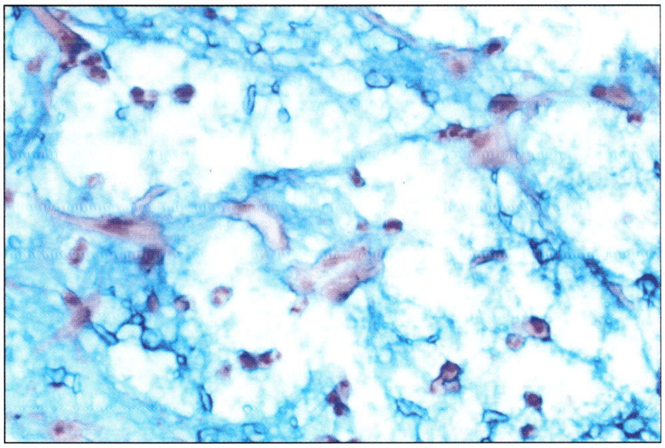

FIGURE 14.59: An alcian blue stain of a myxofibrosarcoma/myxoid MFH.

Skeletal Muscle Tumors

Skeletal muscle lesions share with smooth muscle lesions an accentuation of myofilaments with the trichrome stain and the presence of glycogen (Figures 14.68 and 14.69). In addition, cross striations, the hallmark of skeletal muscle differentiation, are more readily seen with the phosphotungstic acid hematoxylin (PTAH) stain (Figure 14.70) than routine staining but these can also be seen with trichrome stains. However, in the present era in which immunohistochemical and molecular techniques have simplified the diagnosis of skeletal muscle lesions, there is a lessened need to search for cross striations in order to diagnose rhabdomycosarcoma.

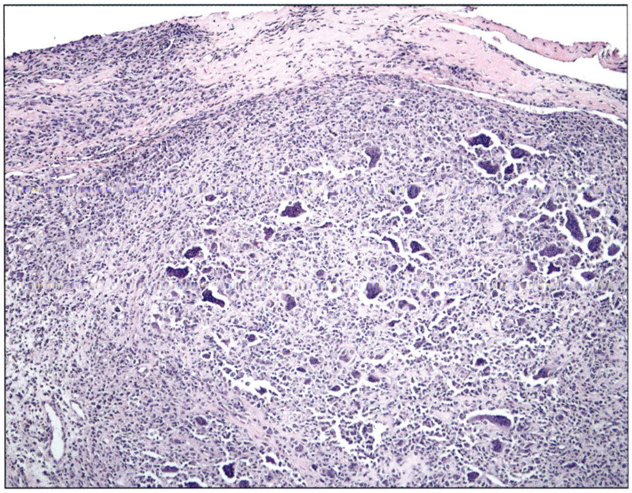

FIGURE 14.60: Giant cell of tendon sheath (tenosynovial giant cell tumor), H&E stain. At low magnification, there is a lobulated appearance.

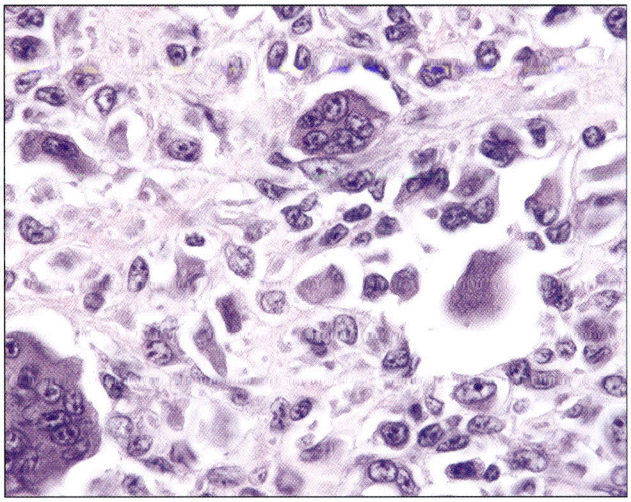

FIGURE 14.61: High magnification of the tenosynovial giant cell tumor seen in Figure 14.60. Compare the nuclei to those of the clear cell sarcoma in Figure 14.77, which feature more prominent nucleoli.

Tumors of Uncertain Histogenesis

Myxomas: IM affects individuals over an age range of 20–84 years but has a peak incidence in middle-aged adults, with a female majority (77,78). The tumor develops in deep skeletal muscle. The thigh and pelvic girdle, shoulder, and upper arm are the most common sites, in decreasing order of frequency. The lesion usually presents as a slow-growing, painless mass, with less than 25% of tumors associated with discomfort or pain. IM is moveable when the affected muscle relaxes but becomes fixed when the muscle contracts. The duration of signs and symptoms prior to presentation varies from months to years.

The presence of multiple lesions is a rare phenomenon and is usually associated with coexistent fibrous dysplasia of bone. In such cases, the manifestations of fibrous dysplasia typically appear first. In most reported cases, it involves more than one bone (polyostotic) and, in some instances, can be associated with Albright's syndrome (polyostotic fibrous dysplasia, cafe au lait spots on the skin, and endocrine abnormalities) (79). IM typically arises in the same general vicinity as the bone(s) that are involved by fibrous dysplasia. IM is a benign tumor that rarely recurs and is managed with local excision.

On microscopic examination, IM is characterized by scattered, bland spindled and stellate-shaped cells, sparse small vessels, and numerous thin collagen fibers suspended in a richly myxoid stromal matrix. The spindled and stellate-shaped cells have a small nucleus and a meager amount of pale, occasionally vacuolated, eosinophilic cytoplasm. Ill-defined cytoplasmic processes of the lesional cells are often continuous with delicate strands of collagen that run

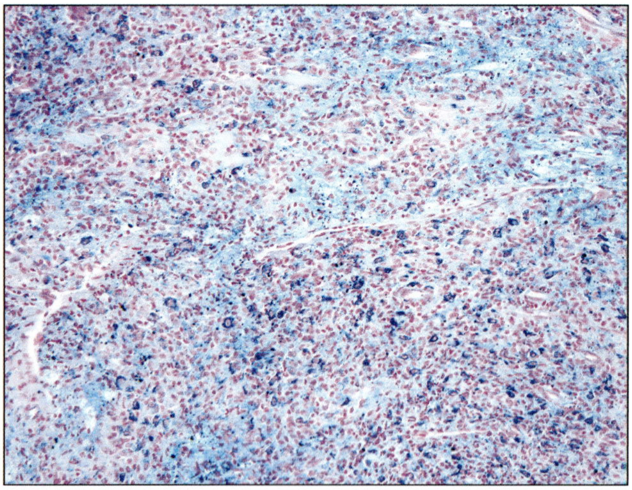

FIGURE 14.62: Often tenosynovial giant cell tumors contain abundant hemosiderin, as seen on this iron stain.

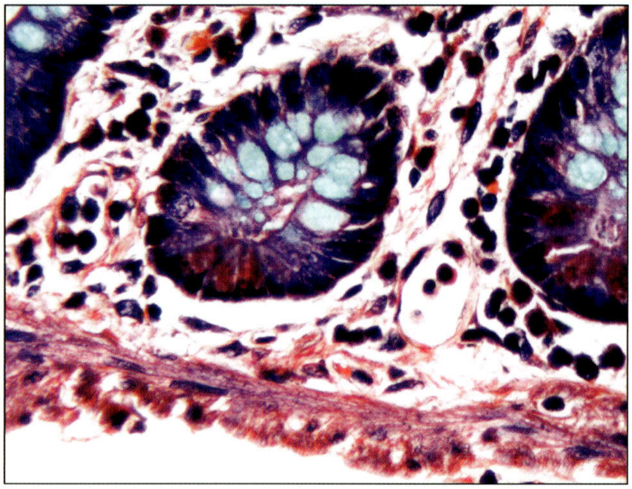

FIGURE 14.63: Smooth muscle appears bright red on the trichrome stain, as seen in this normal muscularis mucosae of the colon.

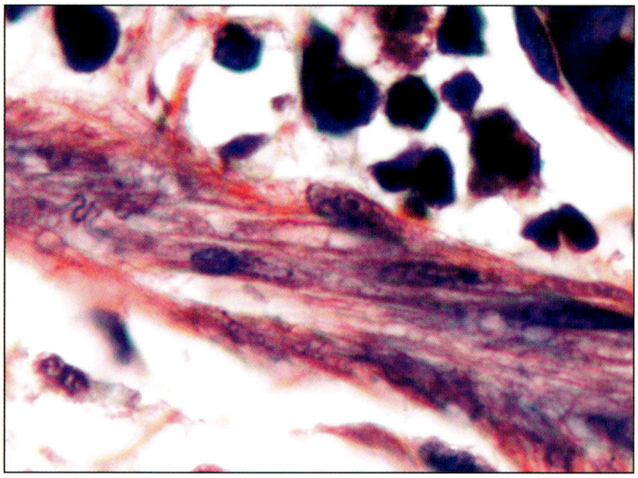

FIGURE 14.64: At higher magnification of the muscularis mucosae depicted in Figure 14.63, delicate longitudinal striations can be seen in the smooth myocytes.

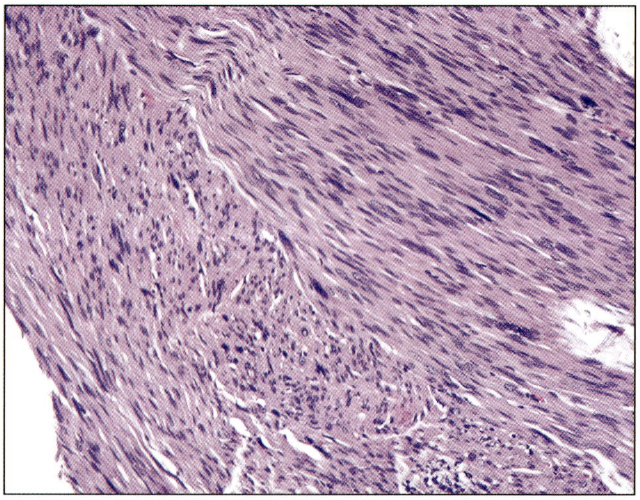

FIGURE 14.65: This leiomyosarcoma displays all the features of smooth muscle differentiation on H&E staining, namely, perpendicularly oriented brightly eosinophilic fascicles of cells with blunt-ended nuclei.

haphazardly throughout the tumor. Mitotic activity is virtually nonexistent. The hyaluronic acid–rich mucinous stroma often contains small cystic spaces and has the same features as in myxofibrosarcoma/myxoid MFHs (see Figure 14.59). Occasionally, residual atrophic skeletal muscle fibers, foamy histiocytes, and, rarely, mast cells are identified in the myxoid matrix. Simple, nonarborizing, capillary-sized vessels are scattered throughout. At the periphery of the lesion, skeletal muscle fibers adjacent to the tumor are atrophic and separated by edema fluid or infiltrating tumor. Fat cells are commonly interspersed in the skeletal muscle.

Ewing's Sarcoma/Primitive Neuroectodermal Tumor: It is only within the past two decades that we have realized that two histologically distinct neoplasms, Ewing sarcoma and peripheral primitive neuroectodermal tumor (PNET; also referred to as "peripheral neuroepithelioma"), are biologically related tumors. Ewing described a peculiar undifferentiated round cell tumor of bone in adolescents (80), and Stout published a case report of a primitive rosette-forming neoplasm of the ulnar nerve of a young adult (81). Ewing's sarcoma became a standard diagnosis by the

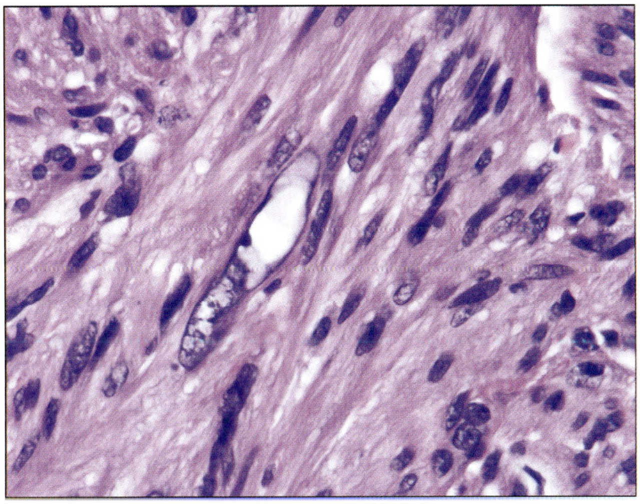

FIGURE 14.66: ***This image depicts another characteristic feature of smooth muscle differentiation in leiomyosarcomas. There is a paranuclear vacuole that indents the nucleus.***

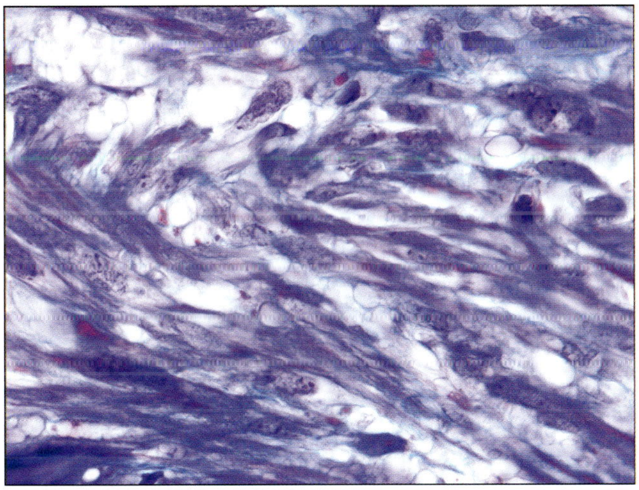

FIGURE 14.67: ***The trichrome stain of the leiomyosarcoma in Figures 14.65 and 14.66 displays delicate red longitudinal striations in the neoplastic cells that recapitulate those in the image of normal smooth muscle seen in Figure 14.64.***

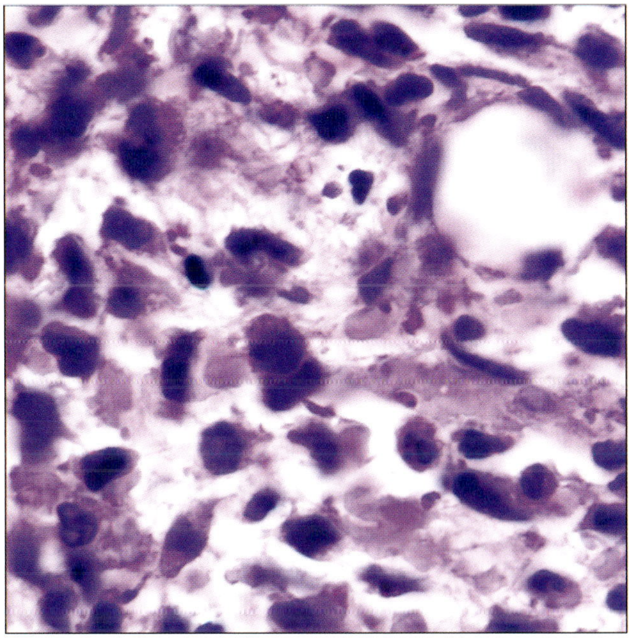

FIGURE 14.68: ***This embryonal rhabdomyosarcoma contains abundant glycogen on this PAS stain, which appears as pink dots.***

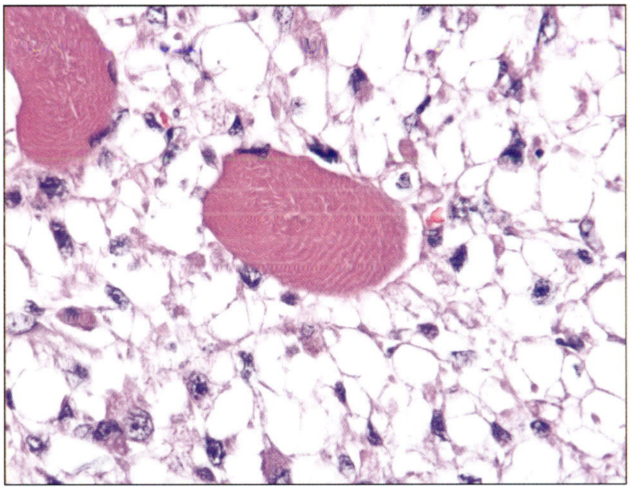

FIGURE 14.69: ***This highly differentiated rhabdomyosarcoma has invaded skeletal muscle of the neck of a child. Compare the features of the rhabdomyoblasts on this H&E-stained section to the entrapped skeletal muscle in the center of the field.***

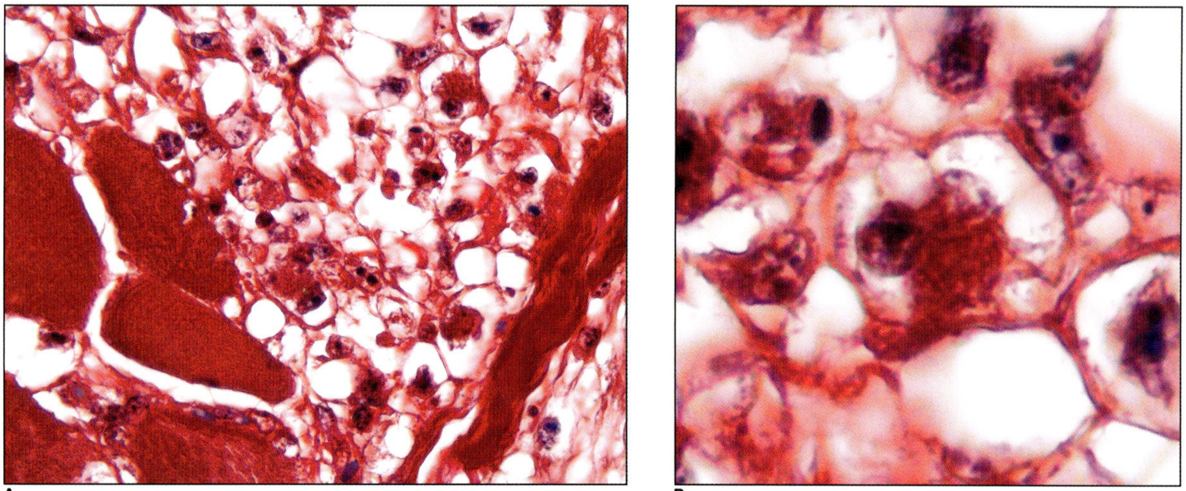

***FIGURE 14.70:* (A) Cross striations are readily identified on the PTAH stain of the highly differentiated rhabdomyosarcoma in Figure 14.69. The lower left of the field contains entrapped skeletal muscle with cross striations. Rhabdomyoblasts in the center and upper right contain a more primitive version of these striations. (B) An enlargement of the cell on the upper right.**

1950s, but PNET was largely unrecognized until Askin et al. described a peculiar small-cell tumor of the chest wall in adolescents (Askin tumor) (82). Subsequently, based on cytogenetic, ultrastructural, and biologic studies, the commonality of these three lesions, Ewing sarcoma, PNET, and Askin tumors, became apparent and they are presently referred to as the PNET or Ewing family of tumors (83,84). Soft tissue examples of these neoplasms are now recognized as the second most common pediatric soft tissue sarcoma, following rhabdomyosarcoma. Ewing family tumors are primarily bone lesions; only a relatively small percentage arises in soft tissue. These involve the chest wall, paravertebral region, pelvis, and proximal extremities of adolescents and young adults, although small children may be affected.

Ewing family tumors are aggressive lesions that are best considered systemic rather than localized. Systemic symptoms such as fever and elevated erythrocyte sedimentation rate may occur. As with all sarcomas, metastases at presentation (typically to the lungs) are indicators of poor prognosis.

Homer Wright rosettes are the hallmark of PNETs, whereas Ewing's sarcomas usually lack them. These microscopic structures comprise wreaths of dark, oval nuclei that circumscribe wispy, lightly pink, neurofibrillary cores. By electron microscopy, the fibrillary cores correspond with elongated, dendritic tumor cell processes that intertwine in a fashion similar to a skein of yarn. The presence of well-defined rosettes suffices to categorize a lesion histologically as PNET.

On the other end of the differentiation spectrum lies typical Ewing sarcoma, which, by its primitive nature, is the quintessential "small round blue cell" tumor. It contains dense cellular masses that usually grow in diffuse sheets. The tumor cell nuclei typically possess round, even contours and smooth chromatin with inconspicuous nucleoli, being larger than lymphoid cells. One prominent distinction is their cytoplasm, which is often clear or bubbly due to an abundance of glycogen that is demonstrable by PAS stains (Figures 14.71–14.74). At first glance, cytoplasmic vacuolization may resemble the neurofibrillary material found in rosettes, but neural similarities are otherwise lacking. Another prominent feature is the presence of interspersed amorphous clusters of cells with more condensed nuclei and lightly eosinophilic cytoplasm. This creates a pattern of alternating "light" cells and "dark" cells. Although Ewing's sarcomas are rapidly growing lesions, the mitotic index is often paradoxically low.

Clear Cell Sarcoma: Clear cell sarcoma (CCS) was described by Enzinger in 1965 (75) as a unique tumor

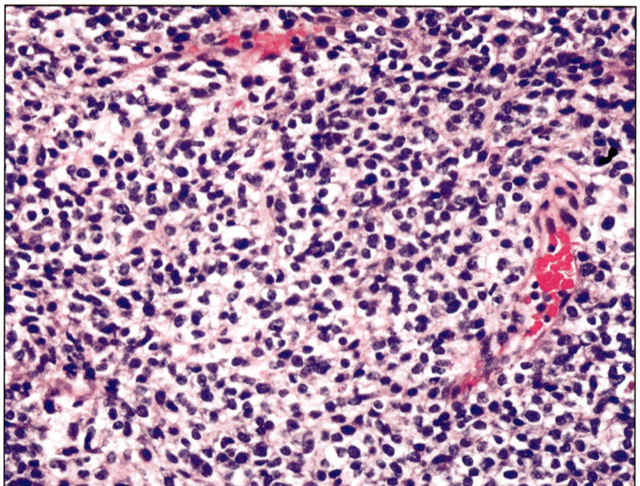

***FIGURE 14.71:* Low magnification of Ewing's sarcoma (H&E).**

associated with tendons and aponeuroses of the hands and feet. At the time of Enzinger's initial description, the nature of this lesion was unclear although he raised the possibility of a melanocytic proliferation. Subsequent studies reported the presence of melanosomes and the name "melanoma of soft parts" was adopted (85,86). CCS mainly affects the distal extremities, particularly the feet and ankles (75,85–87).

Histologically, CCS manifests a packeted and fascicular arrangement of uniform round to spindled cells with clear to eosinophilic cytoplasm (Figure 14.75). The individual cells have uniformly prominent nucleoli (Figure 14.76). Tumor giant cells with a peripheral wreath-like arrangement of nuclei are found in some cases (Figure 14.77). In Enzinger's original series, he described finding glycogen using PAS staining (Figure 14.78) in about 50% of cases and he detected melanin using the Fontana stain in half the cases as well (Figure 14.79). Melanin can also be demonstrated in CCS using the Warthin-Starry melanin preparation (Figure 14.80).

Some observers have speculated that CCS is simply melanoma of connective tissue. However, this is incorrect for several reasons. 1) Their histologic features are monotonous, whereas those of skin melanomas include more prominent cytologic pleomorphism. 2) Their clinical behavior is also markedly different. Whereas an excellent

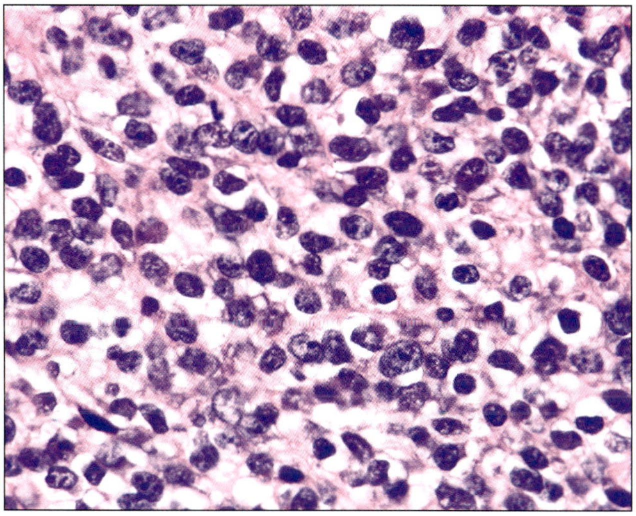

***FIGURE 14.72:** Higher magnification of an H&E image of Ewing's sarcoma. The lesional cells are monotonous (a characteristic of translocation-associated sarcomas) and have inconspicuous nucleoli.*

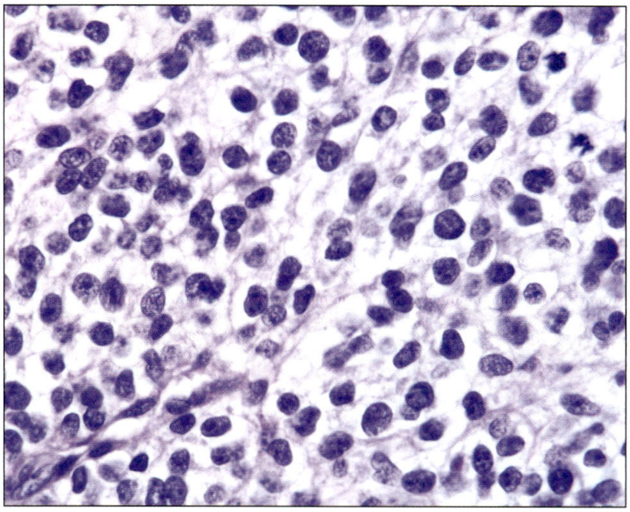

***FIGURE 14.74:** Pretreatment with diastase eradicated the PAS staining in the Ewing's sarcomas depicted in Figures 14.71–14.73 since the PAS staining was imparted by glycogen.*

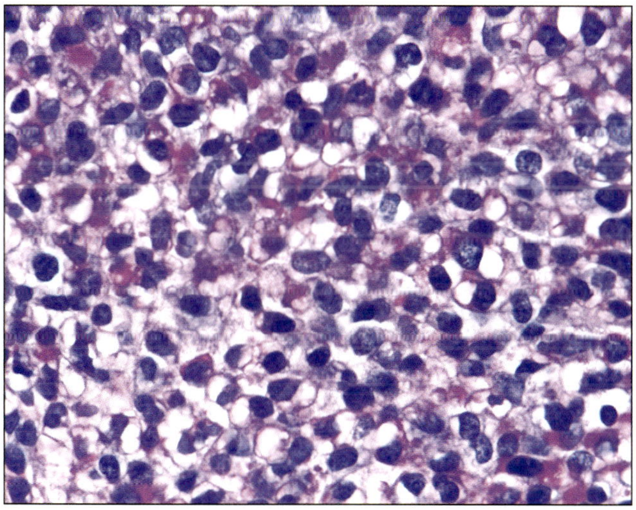

***FIGURE 14.73:** PAS stain of Ewing's sarcoma. This neoplasm is loaded with glycogen.*

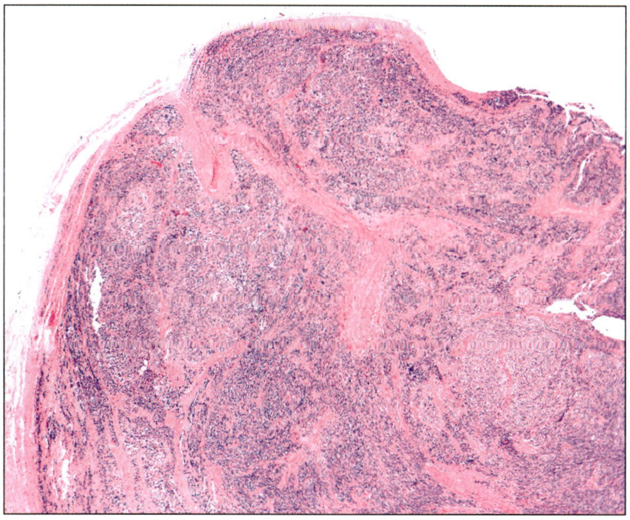

***FIGURE 14.75:** Low magnification of a clear cell sarcoma showing a packeted appearance (H&E).*

prognosis would be expected for a 1-cm CCS, a nodular cutaneous melanoma this size would be rapidly lethal. 3) Finally, CCS has a characteristic 12;22 translocation, whereas skin melanomas have complex and heterogeneous karyotypes.

Alveolar Soft Part Sarcoma: Despite the fact that alveolar soft part sarcoma (ASPS) accounts for fewer than 1% of all soft tissue sarcomas (88,89), it has been the grist for numerous studies attempting to establish histogenesis, which has remained obscure since the entity was recognized (90). Early reports claimed that ASPS was a variant of paraganglioma or granular cell tumor, a neoplasm exhibiting skeletal muscle differentiation, or a lesion capable of producing renin. Like other tumors with monotonous cytologic features, it has a characteristic translocation and gene fusion product. Ladanyi et al. identified the *ASPL-TFE3* fusion gene in ASPS characterized by der(17)t(X;17)(p11.2;q25). This unbalanced translocation results in fusion of the *TFE3* gene on Xp11.2 to a novel gene named *ASPL*. The protein product can be detected immunohistochemically (91). Therefore, ASPS can also be diagnosed by molecular techniques.

ASPS principally affects adolescents and young adults with a slight female predominance, but pediatric cases are known and older adults may be affected. The tumor arises primarily in the deep soft tissues of the lower extremity, particularly the anterior thigh and the buttock, followed by the chest and abdominal wall. In children, ASPS has a proclivity for the head and neck region, especially the periorbital soft tissue and tongue. ASPS clinically presents as a slowly growing, painless mass that may be apparent for months to years before the patient seeks medical attention.

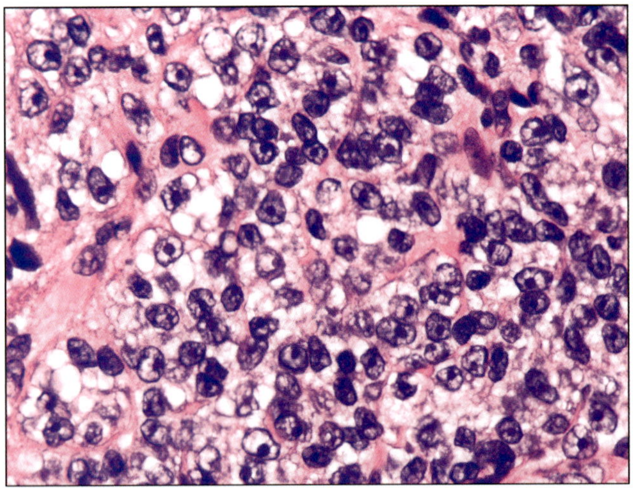

***FIGURE 14.76:** Clear cell sarcoma, a translocation-associated sarcoma, features monotonous cells, each with a prominent nucleolus (H&E).*

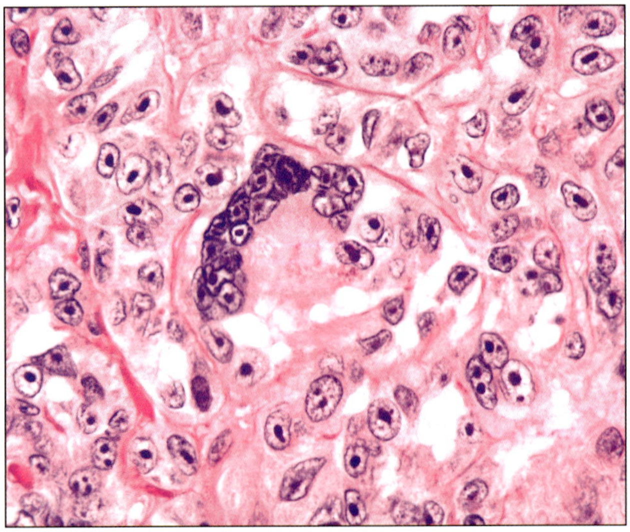

***FIGURE 14.77:** This clear cell sarcoma had many wreath-like giant cells. This feature results in confusion with tenosynovial giant cell tumors (compare this H&E image with Figure 14.61).*

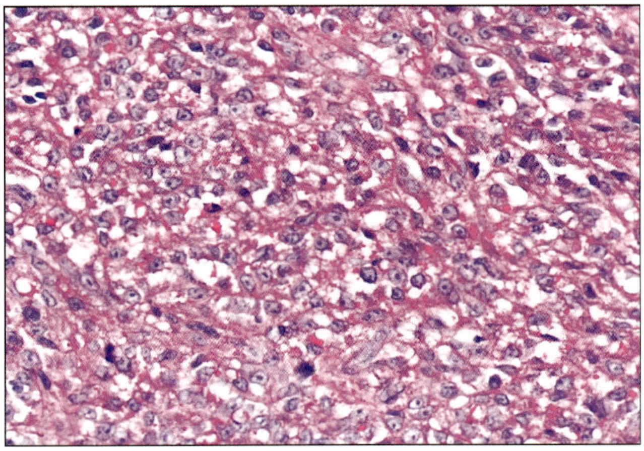

***FIGURE 14.78:** A PAS stain of a clear cell sarcoma highlights abundant glycogen.*

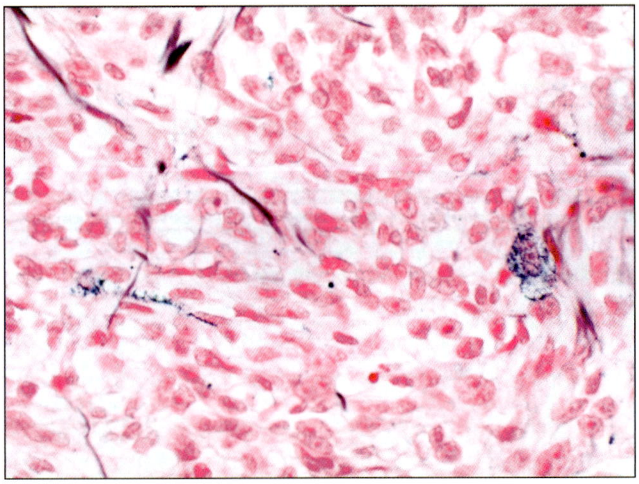

***FIGURE 14.79:** Fontana staining demonstrates scattered melanin in a subset of clear cell sarcomas.*

Unfortunately, metastatic disease to the lung or brain sometimes heralds the presence of an occult sarcoma.

ASPS has a poor long-term survival. In a clinicopathologic study covering >60 years, Lieberman et al. (88) documented a 5-year survival for patients presenting without metastatic disease of 60%, a 10-year survival of 38%, and a 20-year survival of only 15%. Death usually resulted from metastases to vital organs.

On low-magnification microscopy, ASPS displays a distinctive nested or organoid arrangement of large, polygonal cells with eosinophilic cytoplasm (Figure 14.81). The nests are separated from one another by thin-walled, sinusoidal vascular channels. Protrusion of the nests of tumor cells into vessels is a common finding in ASPS. Loss of cellular cohesion results in degenerated cells floating in the center of tumor cell nests and imparts a pseudoalveolar appearance to the otherwise ball-like arrangement. ASPS can also manifest small, compact nests of cells without prominent vascularity, an architectural pattern seen more commonly in children.

The neoplastic cells have a uniform, oval to polygonal shape with distinct cell borders, finely granular, eosinophilic cytoplasm (Figure 14.82), one or two eccentric round to reniform nuclei, vesicular chromatin, and prominent nucleoli. Mitotic activity is characteristically low. PAS staining detects intracytoplasmic glycogen and highlights elongated and rhomboid-shaped crystalline structures in sheaf-like aggregates within the cytoplasm (Figure 14.83). The crystals are diastase resistant in contrast to glycogen in renal cell carcinomas, with which ASPS can be confused.

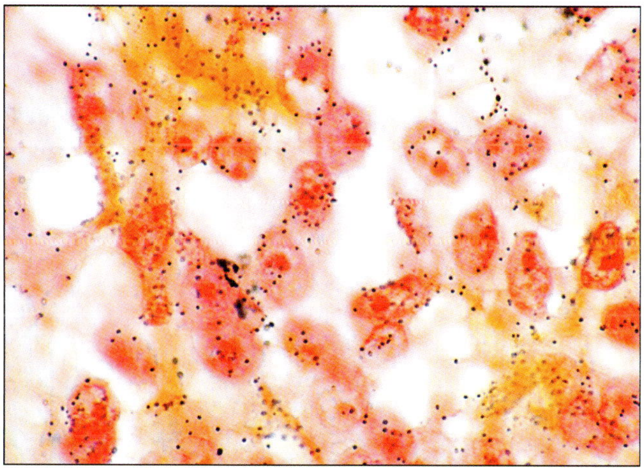

FIGURE 14.80: ***Warthin-Starry stain for melanin can also highlight melanin in clear cell sarcomas.***

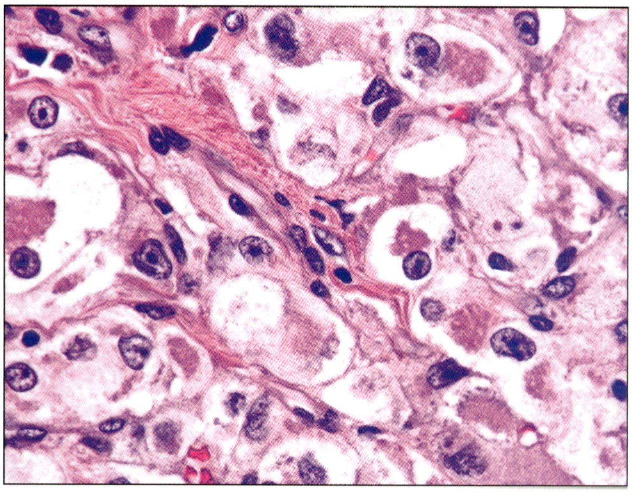

FIGURE 14.82: ***At higher magnification, alveolar soft part sarcomas are composed of uniform cells with abundant cytoplasm (H&E).***

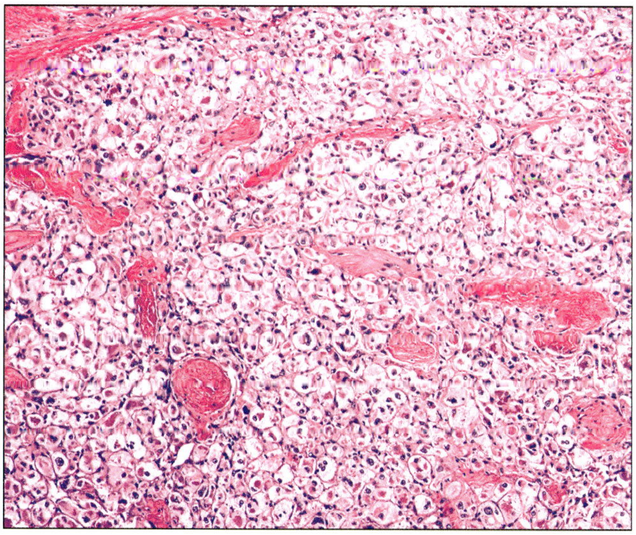

FIGURE 14.81: ***Low magnification of an H&E stain of an alveolar soft part sarcoma. Like the other translocation-associated sarcomas, again these neoplasms display homogeneous cells.***

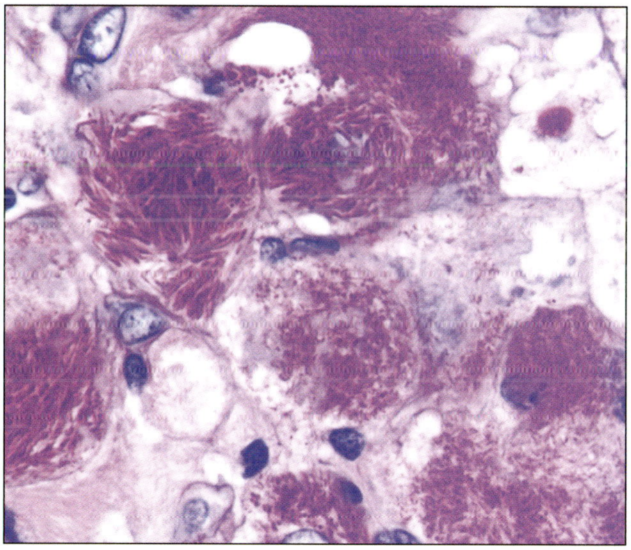

FIGURE 14.83: ***This PAS stain highlights sheaves of crystals in the alveolar soft part sarcoma depicted in Figures 14.81 and 14.82.***

Other Lesions with Characteristic Histochemical Findings: Several lesions other than those outlined above have characteristic histochemical findings and have been covered in the chapter on skin tumors. These include the presence of PAS reactive, diastase-resistant granules in granular cell tumors; melanin in pigmented dermatofibrosarcoma protuberans ("Bednar tumors," Figures 14.84 and 14.85); and PAS-reactive hyaline globules in Kaposi's sarcoma, representing erythrophagolysosomes in various stages of degeneration. Additionally, parachordomas also contain glycogen as demonstrated by the PAS stain (92,93). Reticulin stains have been described in a host of lesions but generally offer little specific information. Crystal storing histiocytosis, in which the crystals are composed of immunoglobulin (94) or even Charcot-Leyden crystals from eosinophils (95) can imitate fibrillar aggregates in rhabdomyoma. The crystals are PAS reactive and diastase resistant in the case of immunoglobulin storage and are deep blue with the PTAH stain (94), in contrast to the histochemical results for rhabdomyomas. They contain diastase-sensitive glycogen and red fibrils on PTAH stains.

Unusual Lesions Mimicking Soft Tissue Tumors

Mycobacterial Spindle Cell Tumor (MSCT): MSCTs perfectly mimic various fibrohistiocytic neoplasms and have been encountered in individuals with HIV/AIDS and other forms of immunocompromise (56,96–98). Histologic examination shows nodules of a spindle cell proliferation, admixed with scattered inflammatory cells (Figures 14.86 and 14.87). The main diagnostic clue is

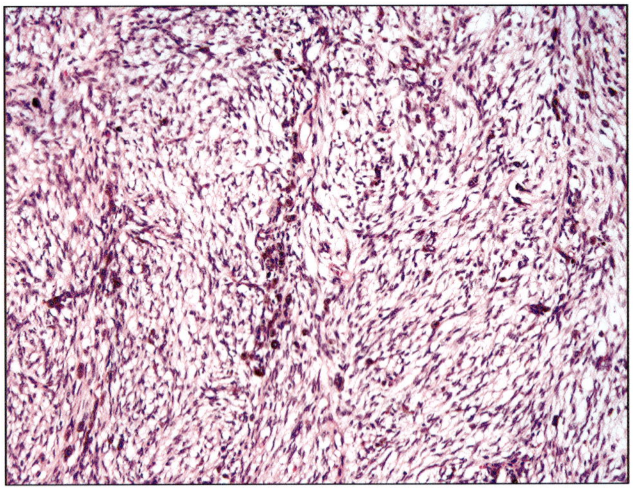

FIGURE 14.84: Pigmented dermatofibrosarcoma protuberans (Bednar tumor); this H&E stain shows melanin pigment. This phenomenon is more likely in individuals with pigmented skin than in those with pale skin.

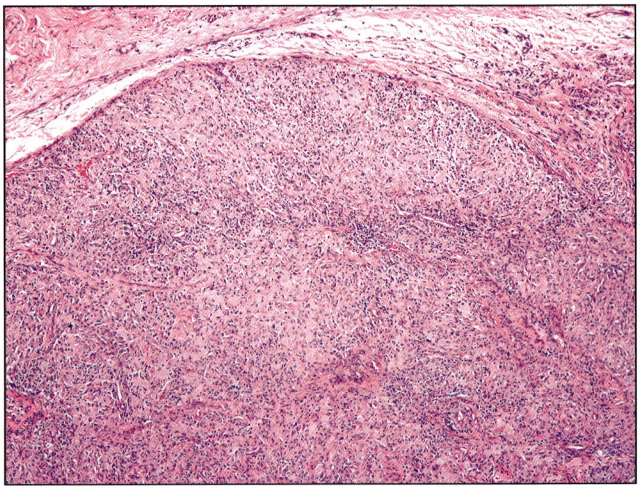

FIGURE 14.86: Mycobacterial spindle cell tumor. Such tumors are readily mistaken for a host of fibrohistiocytic neoplasms (H&E).

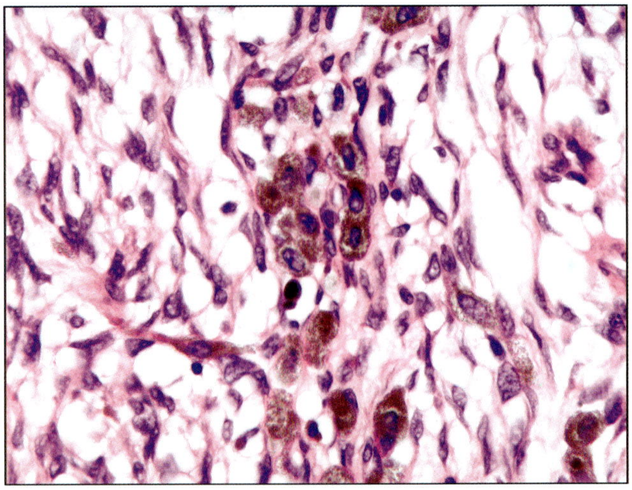

FIGURE 14.85: Higher magnification of the image depicted in Figure 14.84.

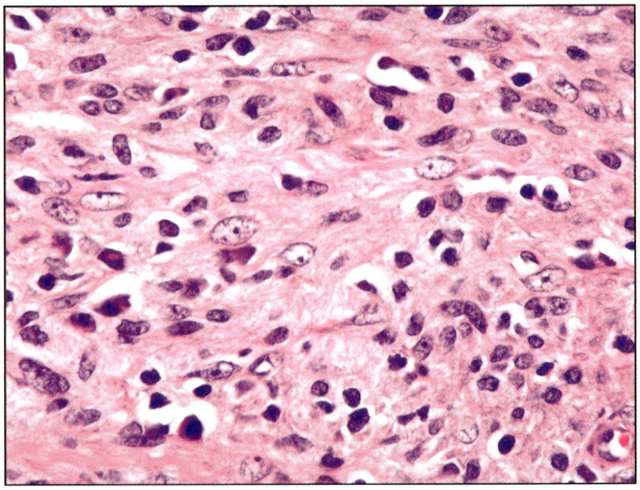

FIGURE 14.87: Higher magnification of the tumor seen in Figure 14.86.

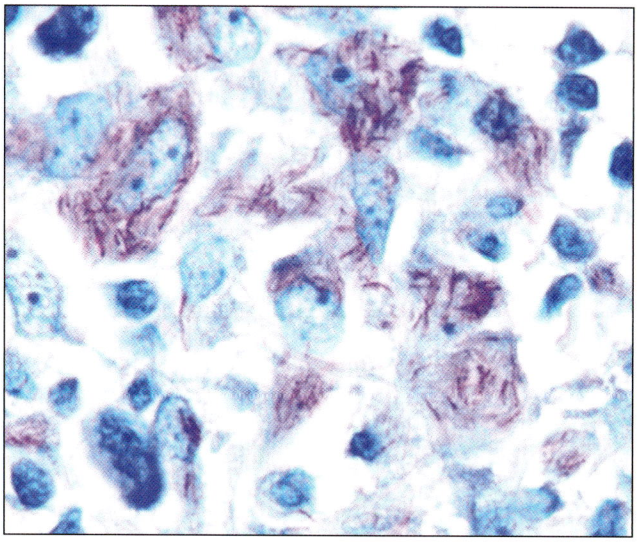

FIGURE 14.88:* *Acid-fast stain of the mycobacterial spindle cell tumor in Figures 14.86 and 14.87.

the history of immunosuppression, and mycobacterial stains are confirmatory (Figure 14.88).

REFERENCES

1. Anderson JR. Recommendations for the biopsy procedure and assessment of skeletal muscle biopsies. *Virchows Arch* 1997;431:227–233.
2. Skinner R. A working guideline for transport and processing of multiple portion muscle, skin, and nerve biopsies. *J Histotechnol* 1988;11:51–53.
3. Goss G, Prayson RA, Pavloski P. Delayed processing of muscle biopsy specimens: does it really compromise enzyme histochemistry? *J Histotechnol* 1998;21:305–308.
4. Bohan A, Peter JB. Polymyositis and dermatomyositis (first of two parts). *N Engl J Med* 1975;292:344–347.
5. Carpenter S, Karpati G. The pathological diagnosis of specific inflammatory myopathies. *Brain Pathol* 1992;2: 13–19.
6. Dalakas MC. Polymyositis, dermatomyositis and inclusion-body myositis. *N Engl J Med* 1991;325:1487–1498.
7. Askanas V, Engel WK, Mirabella M. Idiopathic inflammatory myopathies: inclusion body myositis, polymyositis, and dermatomyositis. *Curr Opin Neurol* 1994;7:448–456.
8. Carpenter S. Inclusion body myositis, a review. *J Neuropathol Exp Neurol* 1996;55:1105–1114.
9. Griggs RC, Askanas V, DiMauro S, Engel A, Karpati G, Mendell JR, et al. Inclusion body myositis and myopathies. *Ann Neurol* 1995;38:705–713.
10. Sekul EA, Dalakas MC. Inclusion body myositis: new concepts. *Semin Neurol* 1993;13:256–263.
11. Fukuhara N, Kumamoto T, Tsubaki T. Rimmed vacuoles. *Acta Neuropathol (Berl)* 1980;51:229–235.
12. Jongen PJ, Ter Laak HJ, Stadhouders AM. Rimmed basophilic vacuoles and filamentous inclusions in neuromuscular disorders. *Neuromuscul Disord* 1995;5:31–38.
13. Albrecht S, Bilbao JM. Ubiquitin expression in inclusion body myositis. An immunohistochemical study. *Arch Pathol Lab Med* 1993;117:789–793.
14. Prayson RA, Cohen ML. Ubiquitin immunostaining and inclusion body myositis: study of 30 patients with inclusion body myositis. *Hum Pathol* 1997;28:887–892.
15. Askanas V, Engel WK, Alvarez RB. Light and electron microscopic localization of beta-amyloid protein in muscle biopsies of patients with inclusion-body myositis. *Am J Pathol* 1992;141:31–36.
16. Askanas V, Engel WK, Alvarez RB. Enhanced detection of congo-red-positive amyloid deposits in muscle fibers of inclusion body myositis and brain of Alzheimer's disease using fluorescence technique. *Neurology* 1993;43:1265–1267.
17. Garlepp MJ, Tabarias H, van Bockxmeer FM, Zilko PJ, Laing B, Mastaglia FL. Apolipoprotein E epsilon 4 in inclusion body myositis. *Ann Neurol* 1995;38:957–959.
18. Mendell JR, Sahenk Z, Gales T, Paul L. Amyloid filaments in inclusion body myositis. Novel findings provide insight into nature of filaments. *Arch Neurol* 1991;48:1229–1234.
19. Nelson R, Prayson RA. Mitochondrial abnormalities and inclusion body myositis: a histopathological and untrastructural study of 36 patients. *J Surg Pathol* 1997;2:63–68.
20. Oldfors A, Moslemi AR, Fyhr IM, Holme E, Larsson NG, Lindberg C. Mitochondrial DNA deletions in muscle fibers in inclusion body myositis. *J Neuropathol Exp Neurol* 1995; 54:581–587.
21. Cabello A, Martinez-Martin P, Gutierrez-Rivas E, Madero S. Myopathy with nemaline structures associated with HIV infection. *J Neurol* 1990;237:64–65.
22. Maytal J, Horowitz S, Lipper S, Poiesz B, Wang CY, Siegal FP. Progressive nemaline rod myopathy in a woman coinfected with HIV-1 and HTLV-2. *Mt Sinai J Med* 1993;60: 242–246.
23. Dalakas MC, Illa I, Pezeshkpour GH, Laukaitis JP, Cohen B, Griffin JL. Mitochondrial myopathy caused by long-term zidovudine therapy. *N Engl J Med* 1990;322:1098–1105.
24. Grau JM, Masanes F, Pedrol E, Casademont J, Fernandez-Sola J, Urbano-Marquez A. Human immunodeficiency virus type 1 infection and myopathy: clinical relevance of zidovudine therapy. *Ann Neurol* 1993;34:206–211.
25. Prayson RA. Granulomatous myositis. Clinicopathologic study of 12 cases. *Am J Clin Pathol* 1999;112:63–68.
26. Heffner RR Jr., Armbrustmacher VW, Earle KM. Focal myositis. *Cancer* 1977;40:301–306.
27. Heffner RR Jr., Barron SA. Denervating changes in focal myositis, a benign inflammatory pseudotumor. *Arch Pathol Lab Med* 1980;104:261–264.
28. Kovarsky J, Schochet SS Jr., McCormick WF. The significance of target fibers: a clinicopathologic review of 100 patients with neurogenic atrophy. *Am J Clin Pathol* 1973;59:790–797.
29. Victor M, Sieb J. Myopathies due to drugs, toxins, and nutritional. In: Engel A, Franzini-Armstrong C, editors. *Myology: Basic and Clinical*, 2nd Edition. New York: McGraw Hill, 1994. pp. 21–30.
30. Kaminski J, Ruff R. Endocrine myopathies (hyper- and hypofunction of the adrenal, thyroid, pituitary and parathyroid glands and iatrogenic corticosteroid myopathy). In: Engel A, Franzini-Armstrong C, editors. Myology: Basic

and Clinical, 2nd Edition. New York: McGraw Hill,1994. pp. 111–123
31. Chinnery PF, Turnbull DM. Mitochondrial DNA and disease. *Lancet* 1999;354(Suppl 1):SI17–SI21.
32. Cohen BH, Gold DR. Mitochondrial cytopathy in adults: what we know so far. *Cleve Clin J Med* 2001;68:625–626, 629–642.
33. Collins S, Byrne E, Dennett X. Contrasting histochemical features of various mitochondrial syndromes. *Acta Neurol Scand* 1995;91:287–293.
34. DiMauro S, Bonilla E, Davidson M, Hirano M, Schon EA. Mitochondria in neuromuscular disorders. *Biochim Biophys Acta* 1998;1366:199–210.
35. Reichmann H, Vogler L, Seibel P. Ragged red or ragged blue fibers. *Eur Neurol* 1996;36:98–102.
36. Vogel H. Mitochondrial myopathies and the role of the pathologist in the molecular era. *J Neuropathol Exp Neurol* 2001;60:217–227.
37. Rifai Z, Welle S, Kamp C, Thornton CA. Ragged red fibers in normal aging and inflammatory myopathy. *Ann Neurol* 1995;37:24–29.
38. Fattah SM, Rubulis A, Faloon WW. McArdle's disease. Metabolic studies in a patient and review of the syndrome. *Am J Med* 1970;48:693–699.
39. McComas CF, Schochet SS Jr., Morris HH 3rd, Romansky SG, Gutmann L. The constellation of adult acid maltase deficiency: clinical, electrophysiologic, and morphologic features. *Clin Neuropathol* 1983;2:182–187.
40. Reuser AJ, Kroos MA, Hermans MM, Bijvoet AG, Verbeet MP, Van Diggelen OP, et al. Glycogenosis type II (acid maltase deficiency). *Muscle Nerve* 1995;3:S61–S69.
41. Tarui S, Okuno G, Ikura Y, Tanaka T, Suda M, Nishikawa M. Phosphofructokinase deficiency in skeletal muscle. A new type of glycogenosis. *Biochem Biophys Res Commun* 1965; 19:517–523.
42. DiMauro S, DiMauro PM. Muscle carnitine palmityltransferase deficiency and myoglobinuria. *Science* 1973;182:929–931.
43. Karpati G, Carpenter S, Engel AG, Watters G, Allen J, Rothman S, et al. The syndrome of systemic carnitine deficiency. Clinical, morphologic, biochemical, and pathophysiologic features. *Neurology* 1975;25:16–24.
44. Fishbein WN, Armbrustmacher VW, Griffin JL. Myoadenylate deaminase deficiency: a new disease of muscle. *Science* 1978;200:545–548.
45. Hoffman EP. Clinical and histopathological features of abnormalities of the dystrophin-based membrane cytoskeleton. *Brain Pathol* 1996;6:49–61.
46. Kaido M, Arahata K, Hoffman EP, Nonaka I, Sugita H. Muscle histology in Becker muscular dystrophy. *Muscle Nerve* 1991;14:1067–1073.
47. Tome FM, Chateau D, Helbling-Leclerc A, Fardeau M. Morphological changes in muscle fibers in oculopharyngeal muscular dystrophy. *Neuromuscul Disord* 1997;7(Suppl 1): S63–S69.
48. Tome FM, Fardeau M. Oculopharyngeal muscular dystrophy. In: Engel A, Franzini-Armstrong C, editors. *Myology: Basic and Clinical*, 2nd Edition. New York: McGraw-Hill, 1994. pp. 223–241
49. Tohgi H, Kawamorita A, Utsugisawa K, Yamagata M, Sano M. Muscle histopathology in myotonic dystrophy in relation to age and muscular weakness. *Muscle Nerve* 1994;17:1037–1043.
50. Bodensteiner JB. Congenital myopathies. *Muscle Nerve* 1994;17:131–144.
51. Bornemann A, Goebel HH. Congenital myopathies. *Brain Pathol* 2001;11:206–217.
52. Fardeau M, Tome FM. Congenital myopathies. In: Engel A, Franzini-Armstrong C, editors. *Myology: Basic and Clinical*, 2nd Edition. New York: McGraw-Hill, 1994. pp. 313–351.
53. Sarnat HB. New insights into the pathogenesis of congenital myopathies. *J Child Neurol* 1994;9:193–201.
54. Gertz MA, Kyle RA. Myopathy in primary systemic amyloidosis. *J Neurol Neurosurg Psychiatry* 1996;60:655–660.
55. Prayson RA. Amyloid myopathy: clinicopathologic study of 16 cases. *Hum Pathol* 1998;29:463–468.
56. Suster S, Moran CA, Blanco M. Mycobacterial spindle-cell pseudotumor of the spleen. *Am J Clin Pathol* 1994;101:539–542.
57. Winslow DJ, Enzinger FM. Hyaluronidase-sensitive acid mucopolysaccharides in liposarcomas. *Am J Pathol* 1960; 37:497–505.
58. Gomori G. A rapid one-step trichrome stain. *Am J Clin Pathol* 1950;20:661–664.
59. Masson P. Trichrome stainings and their preliminary techniques. *J Tech Met* 1929;12:75.
60. Iwasaki H, Kikuchi M, Ohtsuki I, Enjoji M, Suenaga N, Mori R. Infantile digital fibromatosis. Identification of actin filaments in cytoplasmic inclusions by heavy meromyosin binding. *Cancer* 1983;52:1653–1661.
61. Jarvi O, Saxen E. Elastofibroma dorse. *Acta Pathol Microbiol Scand Suppl* 1961;51(Suppl 144):83–84.
62. Kindblom LG, Spicer SS. Elastofibroma a correlated light and electron microscopic study. *Virchows Arch A Pathol Anat Histol* 1982;396:127–140.
63. Kransdorf MJ, Meis JM, Montgomery E. Elastofibroma: MR and CT appearance with radiologic-pathologic correlation. *AJR Am J Roentgenol* 1992;159:575–579.
64. Batstone P, Forsyth L, Goodlad J. Clonal chromosome aberrations secondary to chromosome instability in an elastofibroma. *Cancer Genet Cytogenet* 2001;128:46–47.
65. McComb EN, Feely MG, Neff JR, Johansson SL, Nelson M, Bridge JA. Cytogenetic instability, predominantly involving chromosome 1, is characteristic of elastofibroma. *Cancer Genet Cytogenet* 2001;126:68–72.
66. Nishio J, Iwasaki H, Ohjimi Y, Ishiguro M, Koga T, Isayama T, et al. Gain of Xq detected by comparative genomic hybridization in elastofibroma. *Int J Mol Med* 2002;10:277–280.
67. Turna A, Yilmaz MA, Urer N, Bedirhan MA, Gurses A. Bilateral elastofibroma dorsi. *Ann Thorac Surg* 2002;73:630–632.
68. Prophet E, Mills B, Arrington J, Sobin L. *Laboratory Methods in Histotechnology*. Washington, DC: American Registry of Pathology, 1994.
69. Weiss SW, Enzinger FM. Myxoid variant of malignant fibrous histiocytoma. *Cancer* 1977;39:1672–1685.
70. Angervall L, Kindblom LG, Merck C. Myxofibrosarcoma. A study of 30 cases. *Acta Pathol Microbiol Scand [A]* 1977;85A: 127–140.

71. Merck C, Angervall L, Kindblom LG, Oden A. Myxofibrosarcoma. A malignant soft tissue tumor of fibroblastic-histiocytic origin. A clinicopathologic and prognostic study of 110 cases using multivariate analysis. *Acta Pathol Microbiol Immunol Scand Suppl* 1983;282:1–40.
72. Tallini G, Akerman M, Dal Cin P, De Wever I, Fletcher CD, Mandahl N, et al. Combined morphologic and karyotypic study of 28 myxoid liposarcomas. Implications for a revised morphologic typing (a report from the CHAMP Group). *Am J Surg Pathol* 1996;20:1047–1055.
73. Spillane AJ, Fisher C, Thomas JM. Myxoid liposarcoma—the frequency and the natural history of nonpulmonary soft tissue metastases. *Ann Surg Oncol* 1999;6:389–394.
74. Montgomery E, Wilentz RE, Argani P, Fisher C, Hruban RH, Kern SE, et al. Analysis of anaphase figures in routine histologic sections distinguishes chromosomally unstable from chromosomally stable malignancies. *Cancer Biol Ther* 2003;2:248–252.
75. Enzinger F. Clear cell sarcoma of tendons and aponeuroses: an analysis of 21 cases. *Cancer* 1965;18:1163–1174.
76. Enzinger F. Epithelioid sarcoma. A sarcoma simulating a granuloma or a carcinoma. *Cancer* 1970;26:1029–1041.
77. Kindblom LG, Stener B, Angervall L. Intramuscular myxoma. *Cancer* 1974;34:1737–1744.
78. Nielsen GP, O'Connell JX, Rosenberg AE. Intramuscular myxoma: a clinicopathologic study of 51 cases with emphasis on hypercellular and hypervascular variants. *Am J Surg Pathol* 1998;22:1222–1227.
79. Wirth WA, Leavitt D, Enzinger FM. Multiple intramuscular myxomas. Another extraskeletal manifestation of fibrous dysplasia. *Cancer* 1971;27:1167–1173.
80. Ewing J. Diffuse endothelioma of bone. *Proc NY Pathol Soc* 1921;21:17–24.
81. Stout A. A tumor of the ulnar nerve. *Proc NY Pathol Soc* 1918;18:2–22.
82. Askin FB, Rosai J, Sibley RK, Dehner LP, McAlister WH. Malignant small cell tumor of the thoracopulmonary region in childhood: a distinctive clinicopathologic entity of uncertain histogenesis. *Cancer* 1979;43:2438–2451.
83. Dehner LP. Primitive neuroectodermal tumor and Ewing's sarcoma. *Am J Surg Pathol* 1993;17:1–13.
84. Yunis EJ. Ewing's sarcoma and related small round cell neoplasms in children. *Am J Surg Pathol* 1986;10 (Suppl 1):54–62.
85. Chung EB, Enzinger FM. Malignant melanoma of soft parts. A reassessment of clear cell sarcoma. *Am J Surg Pathol* 1983;7:405–413.
86. Sara AS, Evans HL, Benjamin RS. Malignant melanoma of soft parts (clear cell sarcoma). A study of 17 cases, with emphasis on prognostic factors. *Cancer* 1990;65:367–374.
87. Montgomery E, Meis J, Ramos A, Frisman D, Martz K. Clear cell sarcoma of tendons and aponeuroses. A clinicopathologic study of 58 cases with analysis of prognostic factors. *Int J Surg Pathol* 1993;1:89–100.
88. Lieberman PH, Brennan MF, Kimmel M, Erlandson RA, Garin-Chesa P, Flehinger BY. Alveolar soft-part sarcoma. A clinico-pathologic study of half a century. *Cancer* 1989;63:1–13.
89. Portera CA Jr., Ho V, Patel SR, Hunt KK, Feig BW, Respondek PM, et al. Alveolar soft part sarcoma: clinical course and patterns of metastasis in 70 patients treated at a single institution. *Cancer* 2001;91:585–591.
90. Christopherson W, Foote F, Stewart F. Alveolar soft-part sarcoma: structurally characteristic tumors of uncertain histogenesis. *Cancer* 1952;5:100–111.
91. Ladanyi M, Lui MY, Antonescu CR, Krause-Boehm A, Meindl A, Argani P, et al. The der(17)t(X;17)(p11;q25) of human alveolar soft part sarcoma fuses the TFE3 transcription factor gene to ASPL, a novel gene at 17q25. *Oncogene* 2001;20:48–57.
92. Fisher C. Parachordoma exists—but what is it? *Adv Anat Pathol* 2000;7:141–148.
93. Fisher C, Miettinen M. Parachordoma: a clinicopathologic and immunohistochemical study of four cases of an unusual soft tissue neoplasm. *Ann Diagn Pathol* 1997;1:3–10.
94. Kapadia SB, Enzinger FM, Heffner DK, Hyams VJ, Frizzera G. Crystal-storing histiocytosis associated with lymphoplasmacytic neoplasms. Report of three cases mimicking adult rhabdomyoma. *Am J Surg Pathol* 1993;17:461–467.
95. Lewis JT, Candelora JN, Hogan RB, Briggs FR, Abraham SC. Crystal-storing histiocytosis due to massive accumulation of charcot-leyden crystals: a unique association producing colonic polyposis in a 78-year-old woman with eosinophilic colitis. *Am J Surg Pathol* 2007;31:481–485.
96. Chen KT. Mycobacterial spindle cell pseudotumor of lymph nodes. *Am J Surg Pathol* 1992;16:276–281.
97. Morrison A, Gyure KA, Stone J, Wong K, McEvoy P, Koeller K, et al. Mycobacterial spindle cell pseudotumor of the brain: a case report and review of the literature. *Am J Surg Pathol* 1999;23:1294–1299.
98. Wood C, Nickoloff BJ, Todes-Taylor NR. Pseudotumor resulting from atypical mycobacterial infection: a "histoid" variety of *Mycobacterium avium-intracellulare* complex infection. *Am J Clin Pathol* 1985;83:524–527.

CHAPTER 15

Diagnostic Histochemistry of Hematolymphoid Diseases

Mark R. Wick, M.D.,
Nathan R. Shumaker, M.D., and
JoAnne M. Davis, MT (ASCP), SH

One of the earliest applications of diagnostic histochemistry concerned the discipline of hematopathology. In the late 1800s, Dr. Paul Ehrlich—subsequent Nobel laureate and pioneer in laboratory medicine and immunology—first applied aniline dyes to the morphologic characterization of leukocytes (1,2). This monumental undertaking set the stage for the development of hematology as we now know it, including the entire spectrum of cytochemical and immunohistochemical methods that are in current use. This chapter will consider those histochemical methods that continue to contribute to diagnostic interpretations in hematopathology.

HISTOCHEMICAL STAINS FOR HEMATOPOIETIC CELLS

The routine morphologic examination of peripheral blood smears and bone marrow aspirate specimens is predicated on the use of Romanowsky stains, particularly Wright's method and the Giemsa technique. These are typically applied to air-dried smears and can be viewed with or without slide coverslips (3).

A principal tenet for the application of other cytochemical procedures in hematology—indeed, in laboratory medicine in general—is that they are best used as dictated by morphologic abnormalities seen with standard Romanowsky preparations. In other words, one must be a skilled morphologist in order to utilize histochemistry optimally in hematopathology. In addition, correlation with clinical findings is another essential step in this process.

The following section summarizes those histochemical procedures that contribute to assessments of the hematopoietic system. Specific applications are mentioned briefly here, and are considered in more depth later in the discussion.

Nonenzymatic Methods

Wright-Giemsa Procedure: The "Wright-Giemsa procedure" is actually a family of Romanowsky dye methods that incorporate combinations of eosin or azure-B with methylene blue. As mentioned above, some of these techniques are used with air-dried smears (Figure 15.1), whereas others are applicable to paraffin sections as well. They are used for basic assessments of morphology, although some differential staining of cytoplasmic contents can be appreciated among cells of differing lineages.

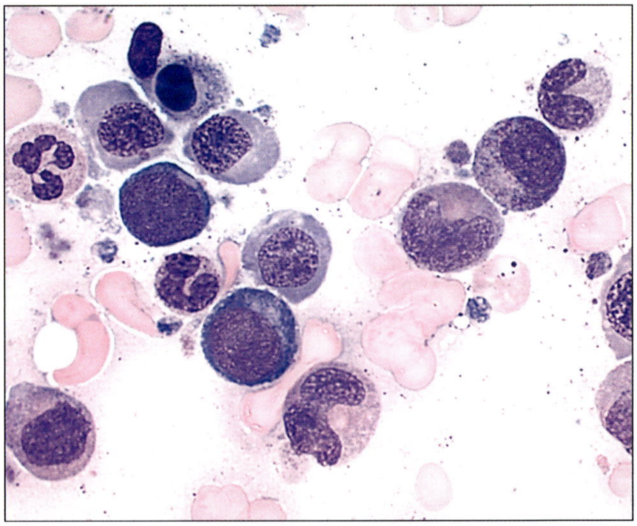

FIGURE 15.1: Wright-Giemsa stain of normal bone marrow aspirate, showing a mixture of normoblastic and myeloid cellular elements.

"Supravital" Dyes: Supravital dyes must be added to samples of blood before slide preparations are made, in order for the dyes to bind properly to cellular components. The most commonly used stains in this category are methylene blue, toluidine blue, and brilliant cresyl blue. They label ribonucleic acid (RNA), precipitates of hemoglobin, intracellular nuclear remnants, and cytoplasmic siderotic granules (4). The usual applications of such dyes are to demonstrate the increased amount of RNA in reticulocytes or to label Heinz bodies (hemoglobin precipitates) in certain enzymopathy-related hemolytic anemias. Some supravital methods also serve as metachromatic stains; for example, toluidine blue and methylene blue label the granules of mast cells and basophils a deep lavender or violet color (Figure 15.2).

Nitroblue Tetrazolium: Another supravital method is the nitroblue tetrazolium (NBT) technique. It employs a reducible "redox" dye that can be precipitated to yield formazan, a purple compound. Blood is incubated with the NBT reagent; a smear is subsequently made and stained with the Wright-Giemsa procedure, and the percentage of neutrophils containing formazan is determined (5). In the past, it was claimed that an NBT-positive granulocyte population of >50% indicated infection with bacterial organisms (particularly in the context of chronic granulomatous disease of childhood), whereas lesser labeling was associated with viral processes or noninfectious disorders. However, these assertions have been called into some question, and the NBT test is no longer frequently obtained as a predictor of bacterial infection.

Methyl Green-Pyronine: The methyl green-pyronine (MGP) procedure relies on the selective chemical affinity of methyl green (a cationic dye) for deoxyribonucleic acid and similar preferential binding of pyronine to RNA (6). Actually, according to Kiernan (7), the reagent currently marketed as "methyl" green is actually *ethyl* green, but the underlying principle of its action is the same. The MGP method is employed to recognize cells with dense concentrations of nucleic acid; it especially labels plasma cells intensely (Figure 15.3), as well as actively replicating hematopoietic elements (8).

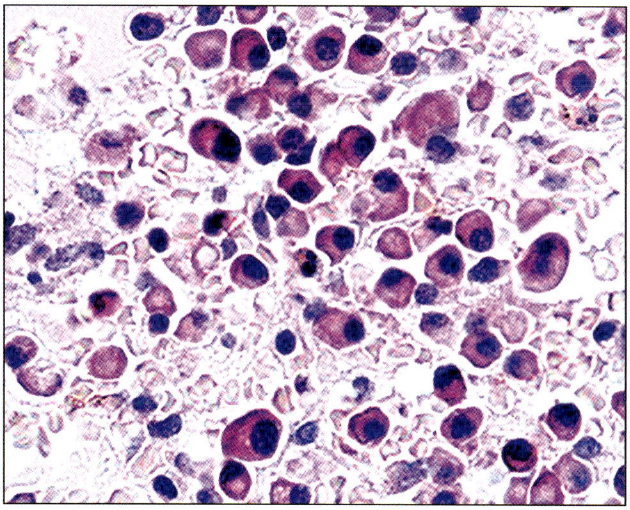

FIGURE 15.3: MGP stain of plasmacytic myeloma in a bone marrow aspirate.

Iron Stains (Perl's Method, Prussian Blue Procedure): Because of the centrality of iron storage and processing in the bone marrow to adequate red blood cell production, accurate identification of this element is crucial. Perl's procedure of the Prussian blue technique can be applied for that purpose to air-dried smears or tissue sections with equal efficacy (9) (Figure 15.4). Disturbances of iron trafficking—such as those observed in some chronic anemias and in myelodysplasias—are reflected by increased iron storage or abnormal intracellular iron distribution in the marrow.

Periodic Acid-Schiff Technique: The periodic acid-Schiff (PAS) method is used to demonstrate carbohydrates, glycolipids, and glycoproteins in cells and tissues. All these compounds contain sugar residues that can be oxidized with periodic acid to form aldehyde groups. The latter moieties have an affinity for Schiff's reagent, a red dye. Pretreatment of the samples being studied with diastase will remove free carbohydrates, thereby abrogating their

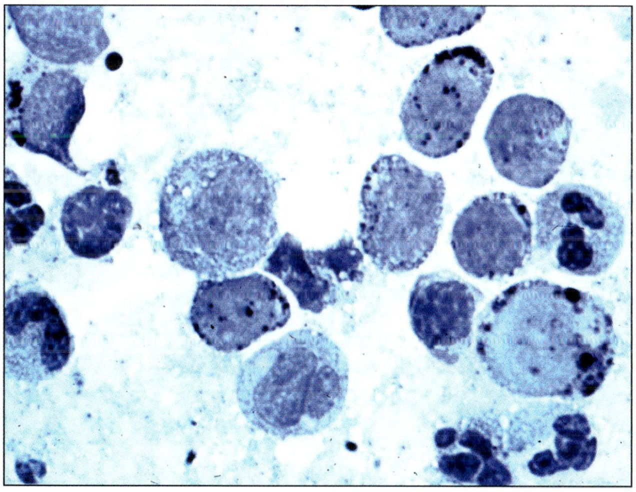

FIGURE 15.2: Toluidine blue stain of acute basophilic leukemia, demonstrating the presence of metachromatic cytoplasmic granules in the neoplastic myeloid cells.

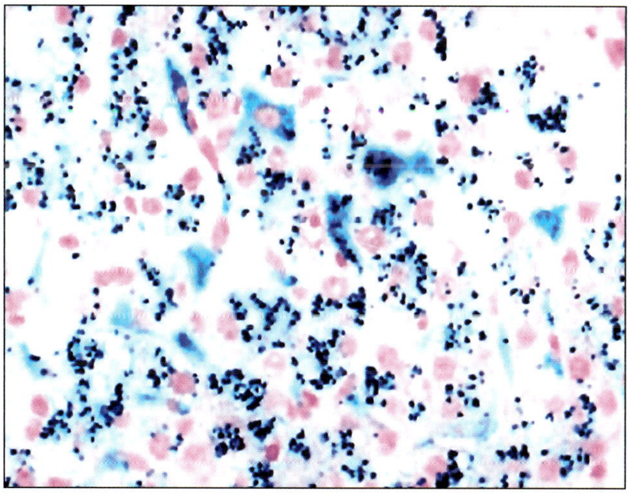

FIGURE 15.4: Perl's stain for iron in a bone marrow biopsy specimen, from a patient with hereditary hemochromatosis.

ability to react in the PAS sequence. In contrast, conjugated sugars—as in glycolipids and glycoproteins—retain their PAS positivity (10). The cells in several hematolymphoid malignancies, including acute lymphoblastic leukemia (ALL) (Figure 15.5), erythroleukemia, and acute megakaryoblastic leukemia, contain PAS-reactive and diastase-labile glycogen (11,12). This characteristic helps to distinguish them from nonneoplastic bone marrow elements and likewise assists in lineage classification. Plasma cell proliferations with a copious cytoplasmic immunoglobulin content also are PAS positive but diastase resistant. Finally, *normal* neutrophils and granulocytic precursors are labeled with the PAS method, but erythroid cells are not. Therefore, this stain can be used to help in estimating the myeloid to erythroid cell ratio in examinations of bone marrow aspirates and biopsies.

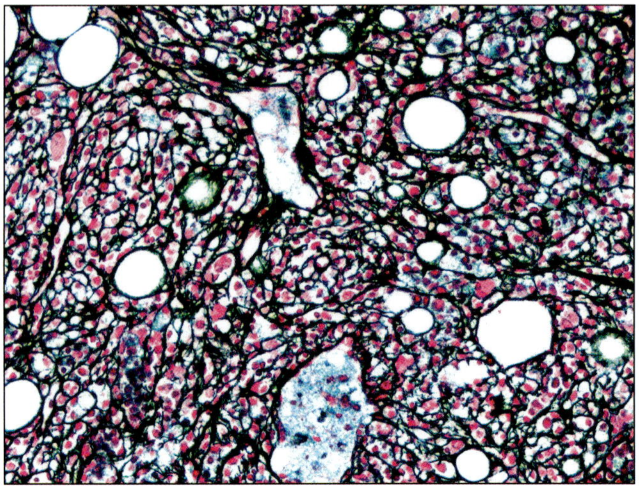

FIGURE 15.6: Snook's stain of a bone marrow biopsy, showing increased reticulin deposition in a case of myelofibrosis.

Reticulin Methods (Wilder's, Snook's): Reticulin comprises a group of connective tissue scleroproteins that are rich in collagen type III. It is increased in the stroma of the bone marrow and lymphoid organs in several hematopoietic proliferations of diverse types, and can be recognized with silver impregnation methods such as Snook's and Wilder's procedures (13). Augmented reticulin density is a clue to the diagnosis of such disorders as myelofibrosis (Figure 15.6), myelodysplasias, myeloid leukemias, and malignant lymphomas (14).

Masson's Trichrome Method: The trichrome stain is used to differentiate epithelium, muscle, elastic tissue, and collagen from one another. The procedure involves immersion of paraffin sections in Weigert's iron hematoxylin, and subsequent application of fuchsin-xylidine-Ponceau, phosphomolybdic acid, and methyl or aniline blue. The applications of this technique in hematopathology include elucidation of collagenous stroma in various lymphoproliferative diseases (Figure 15.7) and the labeling of cytoplasmic immunoglobulin crystalloids in neoplastic plasma cell lesions (15).

Lysochrome Stains (Oil Red O, Sudan Black-B): Lysochromes are dyes that demonstrate fat solubility and lipophilia. Oil red O (ORO) and Sudan black-B (SBB) are the principal representatives of this group that are used in hematopathology. Because lipid is dissolved from paraffin-embedded tissues during processing, lipophilic stains can only be applied to air-dried peripheral blood smears or marrow aspirate smears. The cells of myelogenous leukemias typically contain primary lysosomal granules, which are invested by membranes that contain glycolipid and glycoprotein. Consequently, these structures show an affinity for lipophilic dyes (16) (Figure 15.8). Similarly, macrophages in some storage disorders—for example, Niemann-Pick disease and Gaucher's disease—are filled

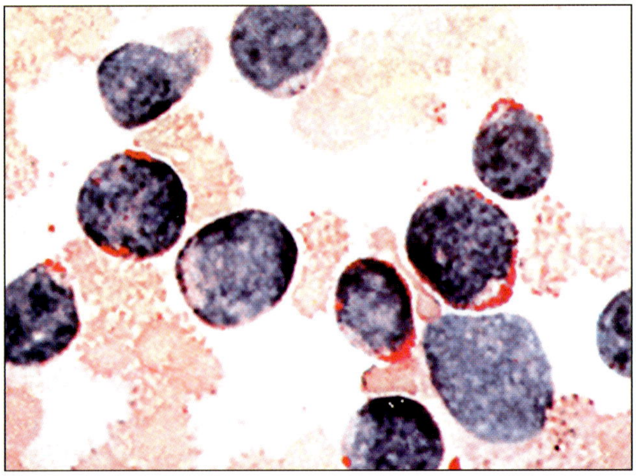

FIGURE 15.5: "Blocky" or "chunky" cytoplasmic labeling for glycogen is seen in the cytoplasm of these neoplastic blasts with the PAS stain, in a case of ALL.

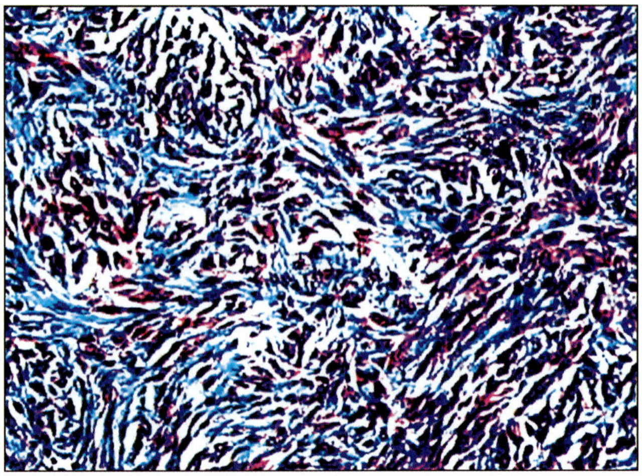

FIGURE 15.7: Mallory's trichrome stain shows stromal fibrosis among the tumor cells, in a case of sclerosing large cell NHL.

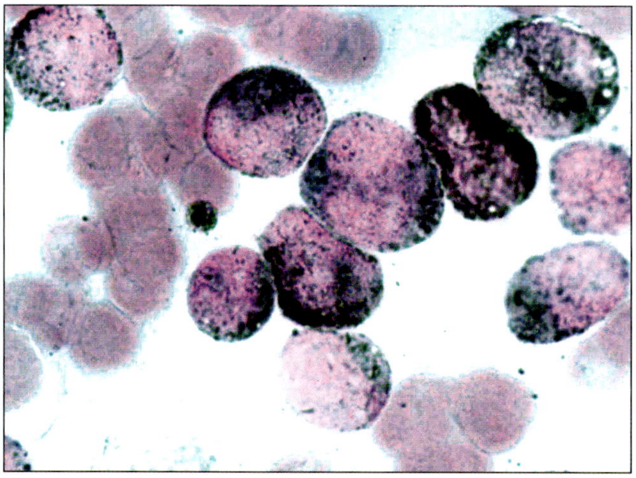

FIGURE 15.8: ***Cytoplasmic labeling with the SBB method, in a case of AML.***

with the same types of compounds. In addition, the cells of "L3" ALL possess cytoplasmic deposits of neutral lipid, with comparable positivity using the ORO or SBB technique (17).

Stains for Mucopolysaccharides (Colloidal Iron, Alcian Blue): At acidic pH levels, a colloidal suspension of ferric ions demonstrates an affinity for sulfated and carboxylated mucopolysaccharides. After application of this reagent in the *colloidal iron* procedure, the Prussian blue method (see above) is used to visualize the binding of ferric ions, if any. In contrast, *alcian blue* is a composite of polyvalent basic dyes, containing copper ions. They also form complexes with sulfated and carboxylated mucosubstances, probably through salt linkage to acidic groups in the latter molecules. These two methods are applied in hematopathology to recognize metastatic adenocarcinomas (Figure 15.9) in the bone marrow, lymph nodes, and other sites, and also to detect macrophage storage products in the mucopolysaccharidoses (e.g., Hunter's and Hurler's syndromes) (18,19).

Amyloid Stains (Congo Red, Crystal Violet): The bone marrow, spleen, liver, and lymph nodes may all be involved in cases of systemic amyloidosis, especially those that are related to plasma cell dyscrasias with paraprotein production (20,21). Congo red (CR) and other cotton dyes show selective binding to amyloid (Figure 15.10); pretreatment of amyloid-containing tissues with acidic potassium permanganate (0.25% sulfuric acid and 5% potassium permanganate) for 3 minutes will prevent CR binding if the amyloid is of the AA ("secondary"; nonimmunologic) type, but not if it is AL (immunoglobulin related) amyloid (20,22). Crystal violet (CV) similarly binds to amyloid but is not as selective as CR. After linking to amyloid, CV demonstrates metachromasia to a magenta hue; permanganate pretreatment has no effect.

Stains for Microorganisms (Silver Methenamine, Ziehl-Neelsen, Fite): The lymphoreticular system may harbor several infectious processes, *Leishmania*, fungi, and mycobacteria are important pathogens. Leishmanial microbes can be recognized satisfactorily with the Wright-Giemsa stain; that is fortunate because no other histochemical methods reproducibly label these organisms (23). In contrast, fungi such as *Cryptococcus* and *Histoplasma* are satisfactorily delineated with the SM procedure (Figure 15.11) (or the PAS stain) (24); mycobacteria can be detected with the Ziehl-Neelsen or Fite technique (Figure 15.12). All these stains can be done on either air-dried smears or tissue sections.

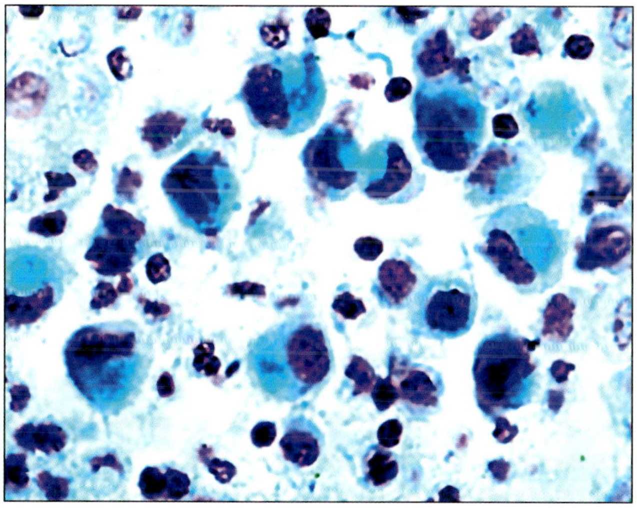

FIGURE 15.9: ***These tumor cells contain mucin that is shown with the colloidal iron method, in a case of metastatic adenocarcinoma involving the bone marrow.***

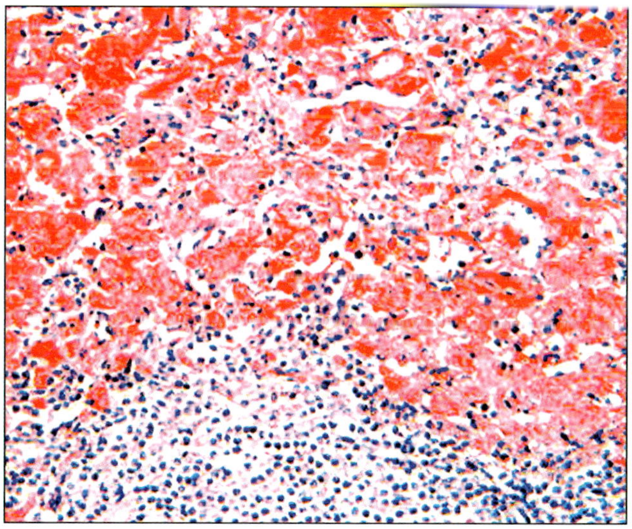

FIGURE 15.10: ***Splenic amyloidosis, as seen with the CR method.***

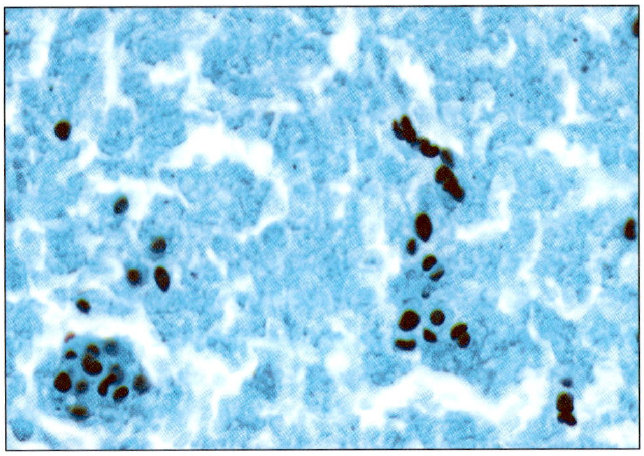

FIGURE 15.11:* *Splenic histoplasmosis, demonstrated with an SM stain.

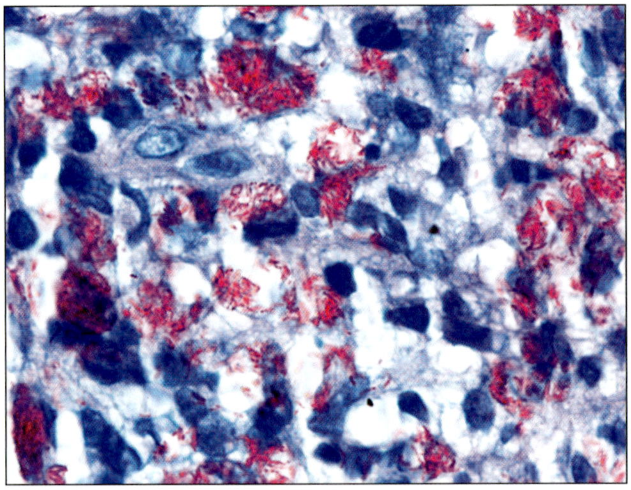

FIGURE 15.12:* *Lymph nodal mycobacterial infection with* Mycobacterium avium-intracellulare*, seen with a Fite stain.

Enzymatic Staining Procedures

Forty years ago, the seminal efforts of Yam, et al. elucidated a group of enzyme-based histochemical methods that allowed observers to separate leukocyte types from one another biochemically (25). This accomplishment may not seem overly impressive in the current climate of molecular technology, but it was a true landmark. Before this time, leukemias were pathologically classified solely on the basis of their morphologic features (26,27), and, predictably, this subjective process produced errors. Enzyme cytochemistry provided a level of objectivity and interpretative reproducibility that was clearly superior. Indeed, it eventually formed the underpinning of the French-American-British (FAB) classification of leukemias in 1976 (28), the basic elements of which are still in use today. Peripheral blood smears or bone marrow aspirate preparations can be air dried, fixed in cold (4 °C) calcium-formalin or chloroform-acetone for 5 minutes, and then used as substrates for enzyme histochemistry.

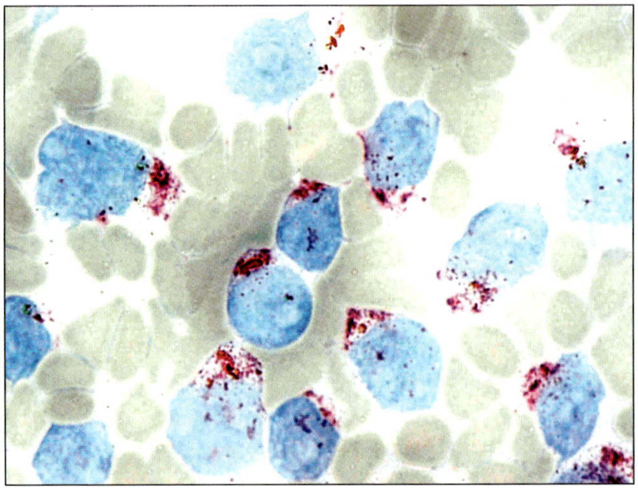

FIGURE 15.13:* *Granular red staining of the cytoplasm for ACP, in ALL.

Acid Phosphatases: Acid phosphatases (ACPs) are phosphohomoesterases, isozymes of which typify several organ sites. The basic chemical reaction that is catalyzed by such enzymes in an acid environment is orthophosphoric monoester + H_2O → alcohol + H_3PO_4. In hematopathology, naphthol AS-BI phosphate is used as a substrate, with hexazonium pararosaniline as a coupler chromogen (29). A red stain results (Figure 15.13). Hematopoietic cells that contain ACP include lymphocytes, macrophages, osteoclasts, and megakaryocytes (30). The particular localization of the enzyme in histochemical preparations is also important; for example, dot-like paranuclear (Golgi-zone) reactivity in lymphocytes is a selective pattern for T-cells (31,32). Tartrate is an effective inhibitor of ACP isoforms, except for those found in osteoclasts and the cells of hairy-cell leukemia (HCL). The latter cell types can easily be distinguished from one another morphologically. Hence, tartrate-resistant ACP (TRAP) emerges a useful marker of HCL among mononuclear cell proliferations (33).

"Nonspecific" (α-Naphthyl Acetate and α-Naphthyl Butyrate) Esterases: Esterases comprise a group of enzymes with variable specificity for a number of substrates containing ester linkages. α-Naphthyl acetate or α-naphthyl butyrate can be enzymatically hydrolyzed by "nonspecific" esterase (NSE) isozymes (designated as 3,4,5, and 6), liberating a free naphthol compound. This joins with a diazonium salt to yield brown-black deposits at the site of intracellular esterase activity (3) (Figure 15.14). NSEs are localized primarily in monocytes, megakaryocytes, and macrophages, but normally are not seen in granulocytes (11,34). Some lymphocytes may show NSE activity as well. Again, the pattern of reactivity is notable; diffuse cytoplasmic labeling is typical of monocytes, whereas globular paranuclear staining is characteristic of nonneoplastic T-lymphocytes (35,36). Stains for α-naphthyl acetate

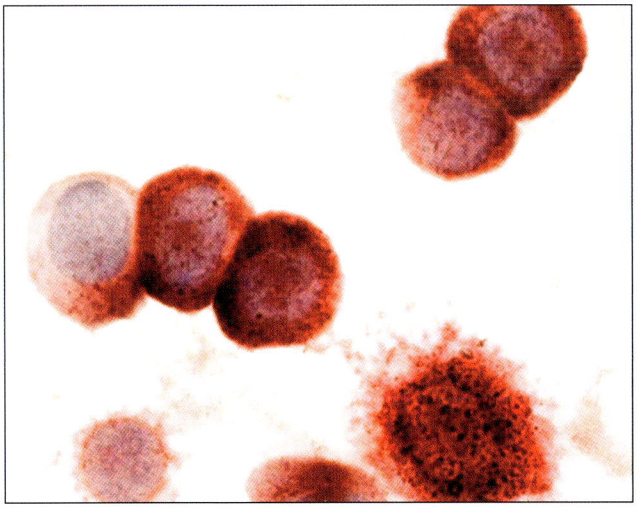

FIGURE 15.14: "Nonspecific" (α-naphthyl acetate) esterase staining of acute myeloid leukemia, type M4.

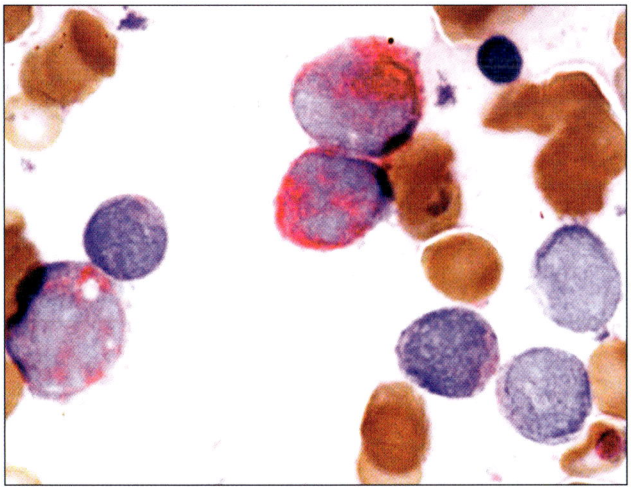

FIGURE 15.15: Labeling of leukemic blasts in type M2 acute myeloid leukemia, with Leder's stain (CAE).

esterase that are done for a prolonged time in an acidic medium appear to preferentially label the cells of some T-cell chronic lymphocytic leukemias (CLLs) and T-cell lymphomas.

Exposure of smears to sodium fluoride will inhibit NSE virtually completely in monocytes, but less so in other cell types (34). The NSE procedure may be applied to peripheral blood smears or bone marrow aspirate preparations.

"Specific" (Chloroacetate) Esterase (Leder's Stain): In contrast to NSEs, "specific" esterases are not observed in either monocytes or lymphocytes; rather, these isozymes (designated as 1,2,7,8, and 9) are restricted to granulated cells of the neutrophilic and mast cell/basophil lineages (34). Specific esterase hydrolyzes naphthol AS-D chloroacetate as a substrate, liberating naphthol. The latter moiety reacts with a diazonium salt, fast red violet LB, yielding red deposits (3) (Figure 15.15). Because of the limited cellular distribution of specific esterases, the chloroacetate esterase (CAE) method—also known as Leder's stain (37)—is an effective marker of granulocytic differentiation among hematopoietic cells with ambiguous morphologic features (11,12,30,32,38,39). Another attractive aspect of this procedure is that specific esterase activity survives routine tissue-processing procedures, making paraffin sections viable substrates for the Leder stain (38,39).

Myeloperoxidase: Myeloperoxidase (MPO)—also known as leukocyte peroxidase or simply "peroxidase"—is a hydrolytic enzyme that is largely restricted to granulocytic white cells (40). MPO produces hypochlorous acid from hydrogen peroxide and chloride anion in neutrophilic respiratory "bursts." Hypochlorous acid is cytocidal for bacteria and other pathogenic microorganisms. MPO requires heme as a cofactor; this complex confers tissues that possess high levels of this enzyme with a greenish hue in their unfixed state (41). When incubated with H_2O_2 in diaminobenzidine (DAB) solution, MPO cleaves peroxide and oxidizes DAB to form a brown-black precipitate in cells that possess the enzyme (Figure 15.16). Again, a procedural strength of the MPO method is that peroxidase activity is retained in fixed paraffin-embedded specimens.

Leukocyte Alkaline Phosphatase: Alkaline phosphatases (ALPs) are again a family of hydrolytic isozymes that are responsible for removing phosphate groups from nucleotides, proteins, and other molecules. In humans, ALPs are seen in virtually all tissues, but they are especially concentrated in the liver, bile duct, kidneys, bone, and placenta (42). Leukocytes also harbor the enzyme, which can be demonstrated by incubation of blood or marrow smears with naphthol AS-BI phosphate and hexazonium new fuchsin in an alkaline environment (3,43,44). Activity of ALP increases as neutrophils mature. A leukocyte alkaline

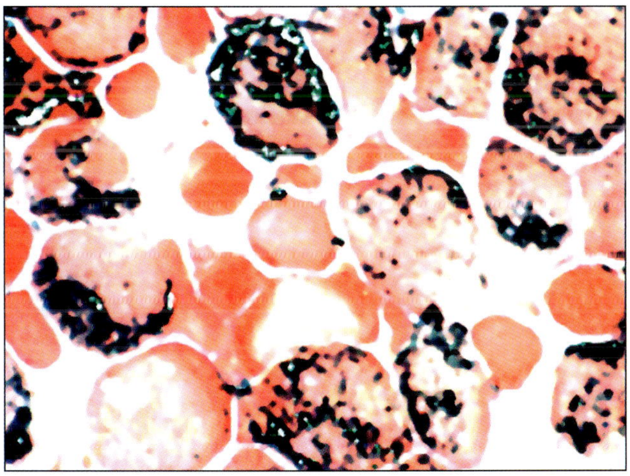

FIGURE 15.16: MPO positivity in the neoplastic cells of AML, type M2.

phosphatase (LAP) score is calculated by performing the cytochemical stain on a monolayer smear of peripheral blood, prepared quickly after venipuncture. One hundred segmented and band-type neutrophils are counted, and the level of LAP reactivity in each cell is graded from 0 to 4+. The number of cells with each of those five scores is multiplied by the grade of the reaction, and the total sum represents the final score. The lower limit of normal is generally approximately 15 (3). Patients with chronic myelogenous leukemia (CML) typically have a decreased LAP score (45). Elevated levels (>180) are seen in leukemoid reactions, stress-related leukocytosis, polycythemia vera, or myeloid metaplasia-myelofibrosis, and after glucocorticoid administration or during the third trimester of pregnancy.

DIAGNOSTIC APPLICATIONS OF CYTOCHEMISTRY IN HEMATOPATHOLOGY

The utility of cytohistochemistry in hematopathology continues to be great. The remainder of this chapter divides diseases in which this technology plays a role into two broad categories—those principally pertaining to the bone marrow and peripheral blood, and others that manifest primarily in lymph nodes, spleen, and liver.

BONE MARROW AND PERIPHERAL BLOOD

Anemias

Anemias in which cytochemical studies are critical include those that are caused by oxidation of hemoglobin, causing formation of Heinz bodies; non-Heinz body-related hemolytic processes; and others that reflect disturbances in iron metabolism.

Heinz-Body Hemolytic Anemias: Heinz bodies are precipitates of oxidized hemoglobin. The underlying conditions that cause them include glucose-6-phosphate dehydrogenase (G6PD) deficiency, hemoglobin-H disease, glutathione reductase deficiency, glutathione peroxidase deficiency, and triosephosphate isomerase deficiency. Of those disorders, G6PD deficiency is the most common (46). This X-linked condition likely results in protection against malaria, explaining its prevalence in certain geographic regions (47). G6PD mediates the oxidation of glucose-6-phosphate to 6-phosphogluconate, concurrently reducing nicotinamide adenine dinucleotide phosphate (NADP) to NADPH. The latter compound keeps glutathione in a reduced form, so that it can "scavenge" oxidative intracellular moieties that may be produced by various drugs such as antimalarials, sulfa-based medications, and certain anti-inflammatory agents (48).

Heinz bodies are not visible in Wright-Giemsa-stained slides. Instead, they must be visualized with supravital dyes such as brilliant cresyl blue, CV, or methylene blue, appearing as small dense ovoid inclusions that are typically attached to the erythrocyte membrane (Figure 15.17).

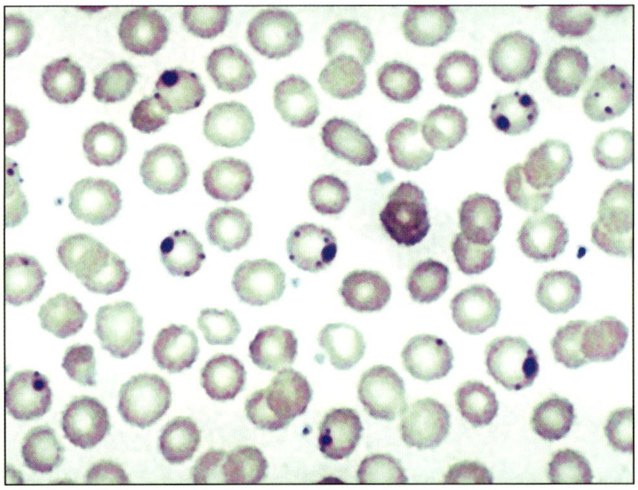

FIGURE 15.17: Brilliant cresyl blue staining of a peripheral blood smear shows the presence of peripheral intraerythrocytic Heinz bodies (precipitated hemoglobin) in an example of G6PD deficiency.

Other Hemolytic Anemias: Other hemolytic anemias do not demonstrate formation of Heinz bodies; they may be predicated on immune, autoimmune, or hemoglobinopathy-based mechanism, as well as structural defects in the erythrocyte membrane (49,50). The hallmark of hemolysis is the appearance of increased numbers of circulating reticulocytes, proving that production of red blood cells by the marrow is adequate. Reticulocytes are "early" forms of erythrocytes that contain abundant RNA (51). Accordingly, they can be seen on Wright-Giemsa stains as lavender-blue (Figure 15.18), as opposed to the pink-red color of

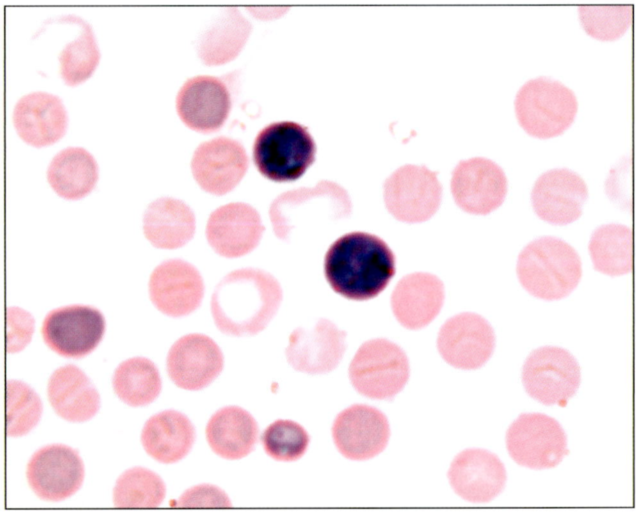

FIGURE 15.18: Blue-staining reticulocytic erythrocytes are present in this peripheral blood smear, stained with the Wright-Giemsa method.

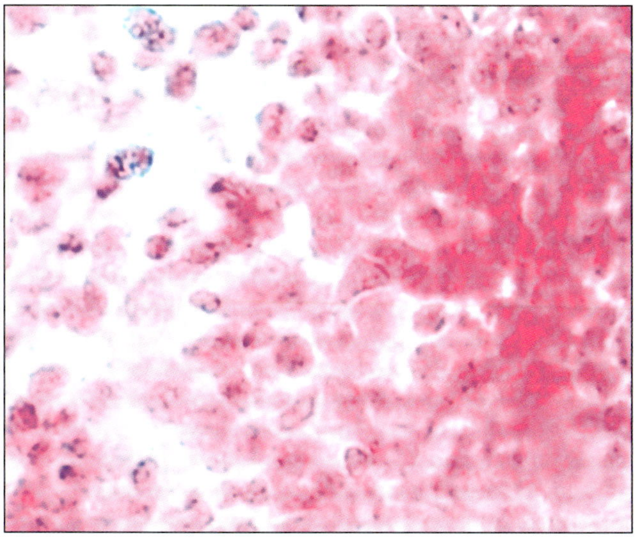

FIGURE 15.19: No stainable iron is present in this Perl's stain of a bone marrow aspirate from a patient with iron deficiency anemia.

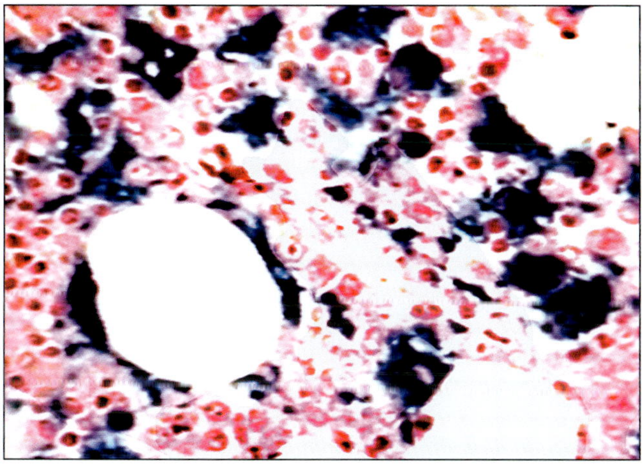

FIGURE 15.20: ACD features an increase in histiocyte-associated iron, as seen in this Perl's stain of a bone marrow aspirate.

mature red cells. In addition, the methylene blue or MGP method also are capable of labeling reticulocytes selectively (3).

Anemias Related to Disturbed Iron Metabolism: *Iron Deficiency Anemia:* Iron deficiency is the most common cause of anemia, owing either to primary dietary deficiency or chronic blood loss with inadequate compensatory intake of iron. In either of these circumstances, insufficient stores of iron are present in marrow macrophages and stromal cells. Hence, iron cannot be transferred to developing erythroid colonies properly (52). Perl's technique or the Prussian blue method will demonstrate a virtual absence of stainable iron in aspirates or trephine sections of the marrow (9,53) (Figure 15.19).

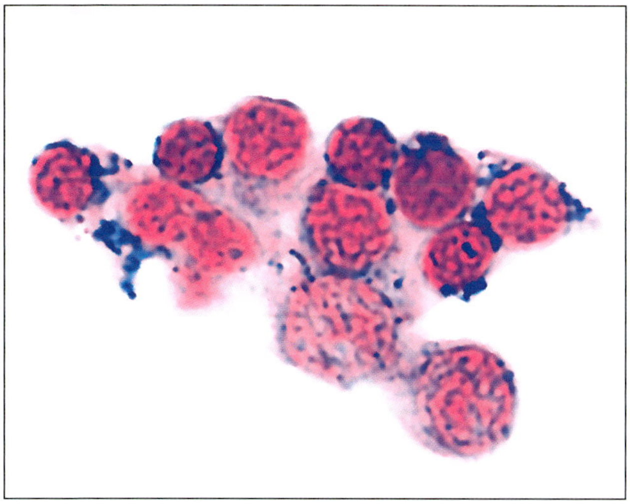

FIGURE 15.21: Perl's stain showing iron-containing mitochondria in circumferential distribution around normoblastic nuclei, in SBA.

Anemia of Chronic Disease: Anemia of chronic disease (ACD) may accompany illnesses of an inflammatory, infectious, degenerative, or neoplastic nature, any of which lasts for >2 months. The basic defect underlying this disorder appears to be the relative nonrelease of iron stores by stromal cells in the marrow, perhaps under the influence of increased cytokine (e.g., interleukin-1 or tumor necrosis factor-α) levels (54,55). Macrophages in the bone marrow are seen to contain increased iron with Perl's method or the Prussian blue technique (56) (Figure 15.20), but normoblasts are virtually devoid of this mineral in ACD.

Sideroblastic Anemias: Sideroblastic anemias (SBAs) are typified by inadequate marrow utilization of iron in the face of normal or increased marrow iron stores. They are often accompanied by polychromatophilic, stippled erythrocytes in circulation, together with dimorphic (microcytic and normocytic) red cell morphology (57). SBAs most often represent a part of myelodysplastic syndromes, but they also can be congenital or secondary to the effects of drugs or toxins. Bone marrow examination is required to make the diagnosis of SBA; it shows generalized erythroid hyperplasia, and the Prussian blue technique or silver impregnation (58,59) discloses iron-engorged paranuclear mitochondria in developing erythroid precursors (ringed sideroblasts) (Figure 15.21). General marrow iron stores are also increased.

Benign Leukocyte Disorders

Storage Diseases: Some inherited disorders of metabolism create leukocytic abnormalities that are associated with cytochemical abnormalities. In particular, these include Hunter's syndrome, Hurler's syndrome, Niemann-Pick disease, and Gaucher's disease.

Hunter's syndrome, or mucopolysaccharidosis II, is an X-linked recessive condition. Affected individuals are unable to metabolize selected glycosaminoglycan mucopolysaccharides (GMs). These substances accumulate in a systemic fashion because of a deficiency or absence of iduronate-2-sulfatase in cells of the reticuloendothelial system (60). Hurler syndrome is similar phenotypically to the Hunter complex, but it has a different mode of genetic transmission (autosomal recessive) and a different enzymopathy (α-L-iduronidase) (61,62). Two GMs in particular—dermatan sulfate and heparan sulfate—are the principal storage products in Hunter's and Hurler's syndromes (63,64). They are seen as inclusions in leukocytes of all types, but are usually most notable in lymphocytes (Figure 15.22). The colloidal iron and alcian blue methods are effective at labeling the intracellular mucopolysaccharide in both conditions (65).

Niemann-Pick syndrome (NPS) is another autosomal recessive condition potentially featuring an enzymopathy, namely, that involving sphingomyelinase (66,67). However, some forms of the disease are caused by defective proteins that influence intracellular lipid transport, rather than by an enzyme deficiency (68). With respect to hematologic findings in NPS, one sees foamy macrophages in the marrow that contain small cytoplasmic droplets. These inclusions stain minimally with the PAS method; however, the droplets may be birefringent on polarization microscopy, and they can be labeled with the ORO or SBB techniques (69,70).

Gaucher's disease is caused by a deficiency of functional lysosomal glucocerebrosidase (also known as β-glucosidase). It catalyzes the metabolism of glucocerebroside, which is a cell membrane constituent of erythrocytes and leukocytes. Macrophages phagocytize the cellular debris that derives from degeneration of such defective hematopoietic cells. They are unable to clear it, and the material subsequently accumulates in fibrillary form in macrophage cytoplasm in the marrow, spleen (Figure 15.23), lymph nodes, and liver. The glucocerebroside-rich storage product has been said to resemble "crumpled tissue paper" on Wright-Giemsa stains (71). It can be weakly but discernibly labeled with the PAS method (Figure 15.24), owing to its carbohydrate content (72,73).

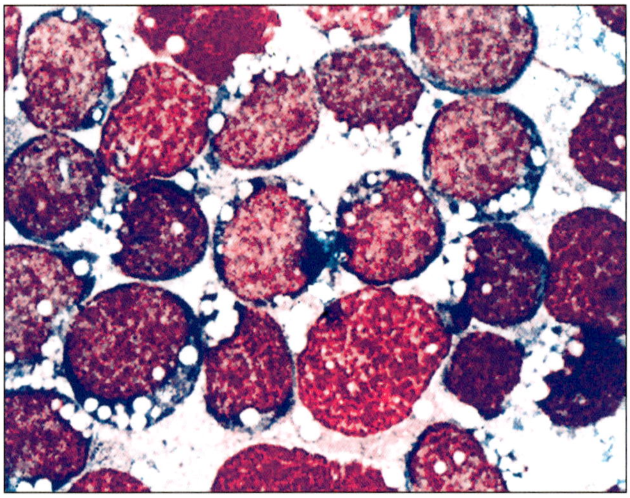

FIGURE 15.22: *Vacuolar inclusions are present in the cytoplasm of lymphocytes, in a touch preparation of lymph node stained with the Wright-Giemsa method from a case of Hunter's syndrome. These represent mucopolysaccharide deposits.*

Leukemoid and Leukoerythroblastic Reactions: Examples of benign but extreme leukocytosis are termed "leukemoid" (leukemia-like) reactions, usually defined by a white blood cell count of $\geq$50,000/mm^3. The peripheral blood smear typically reveals myelocytes, metamyelocytes, promyelocytes, and, potentially, some myeloblasts (74,75). Nevertheless, mature leukocytes are present as well, with

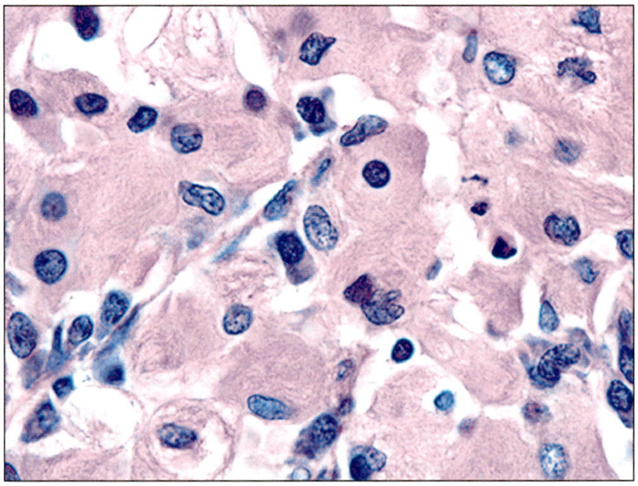

FIGURE 15.23: *Macrophages contain cytoplasm resembling "crinkled tissue paper" in Gaucher's disease, as seen in this hematoxylin and eosin–stained section of spleen.*

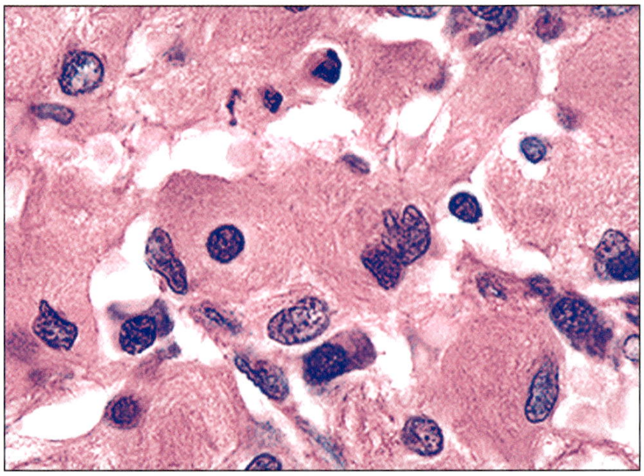

FIGURE 15.24: *Gaucher's cells are slightly positive with the PAS method, as shown here.*

a normal level of circulating basophils (74). If erythroblasts are present in the periphery as well, the term "leukoerythroblastic reaction" is used (76). The marrow in such reactive conditions is hypercellular but otherwise morphologically unremarkable. Leukemoid reactions are caused by infections, the administration of some medications, or underlying solid malignancies (as a paraneoplastic phenomenon) (75). In contrast, leukoerythroblastosis reflects the presence of myelophthisis, usually due to myelofibrosis, replacement of the marrow space by an infectious process, or metastatic tumor (77).

The most important differential diagnosis in this context is with CML. In this regard, the LAP score and the presence of basophilia are helpful. The LAP score is high (typically >100) in reactive states (Figure 15.25) but reduced (usually <15) in CML (3,44,78). Basophilia also favors the latter condition.

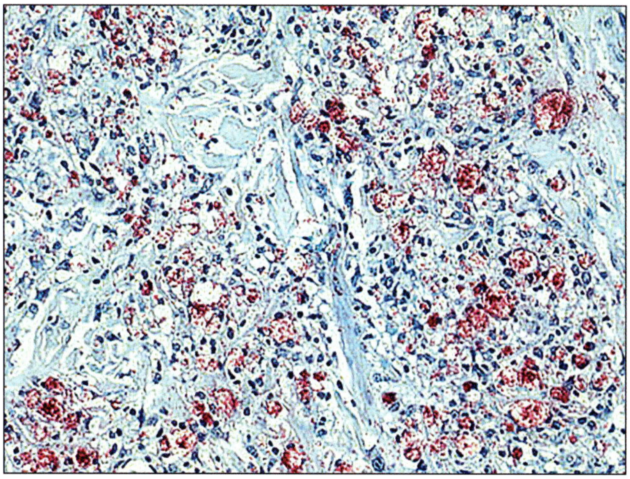

FIGURE 15.26: ***Many histiocytes contain mycobacteria, as shown with this Fite stain, in a case of lepromatous lymphadenitis.***

Granulomatous and Benign Histiocytic Infiltrates

Several granulomatous infectious processes have the ability to affect the marrow, yielding secondary deficiencies of blood cell production. Some of them are typified by well-formed granulomas including mononuclear histiocytes, multinucleated phagocytic cells, and lymphocytes, with or without central necrosis. Others are characterized by unorganized infiltrates of histiocytes, the cytoplasm of which often contains the causative organisms. Mycobacteria of various species, *Histoplasma capsulatum*, *Cryptococcus neoformans*, *Leishmania donovani*, *Brucella* species, and selected other organisms are capable of causing these lesions. The Ziehl Neelsen and Fite procedures (for mycobacteria) (Figure 15.26), the silver methenamine (SM) or PAS methods (for fungi), and the Giemsa stain (for protozoans) can be utilized effectively to visualize most of these pathogens. However, *Brucella* and other conventional bacteria are poorly seen by any histochemical procedure in this setting (79–84).

Sarcoidosis and Whipple's disease also may produce granulomatous and histiocytic bone marrow lesions that contain no histochemically demonstrable microorganisms (85,86). Wilder's stain done on sarcoidal granulomas demonstrates virtually complete investment of them by reticulin; this finding is typical but not diagnostic (15). On the other hand, the histiocytic infiltrates of Whipple's disease are intensely reactive with the PAS procedure (Figure 15.27), both before and after diastase pretreatment (86).

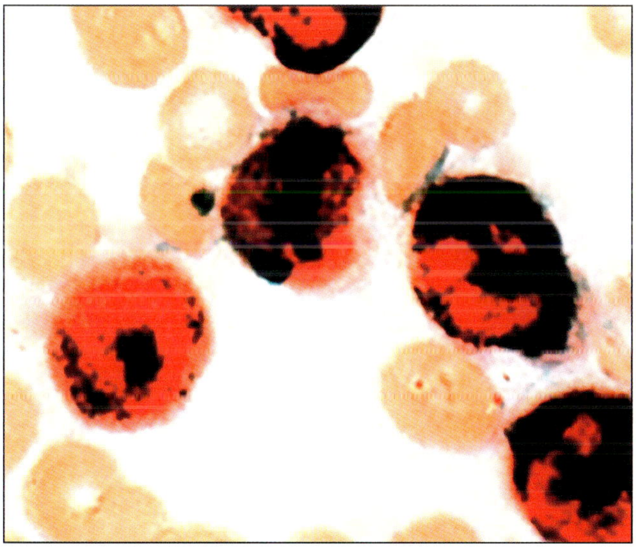

FIGURE 15.25: ***Intense labeling of granulocytic elements is seen for LAP in a peripheral blood smear, in an example of leukemoid reaction.***

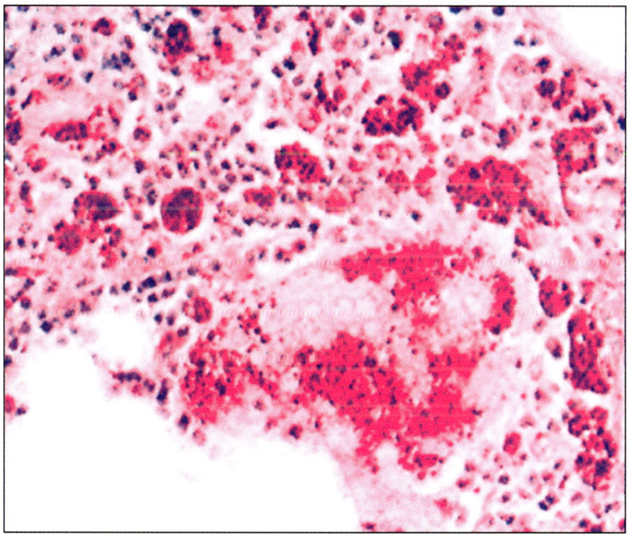

FIGURE 15.27: ***Intense labeling of bone marrow histiocytes with the PAS technique, in a case of Whipple's disease.***

Leukemias

As mentioned earlier, a major advance in the classification of acute leukemias was made with the systematic application of enzyme cytochemistry to that group of diseases. In fact, based on the latter technology, the FAB system for the categorization of acute leukemias was published by Bennett et al. (28), bringing a level of objectivity to the process that had theretofore been lacking. The histochemical features of these disorders are synopsized in Tables 15.1–15.5, as summarized by Foucar (87), Loffler (88), and Loffler and Gassmann (89). They are also discussed in the following sections.

Acute Lymphoblastic Leukemia: ALLs were categorized into three groups in the FAB scheme, numbered L1 through L3 (28). Their distinguishing attributes are both morphologic—based on nucleocytoplasmic ratios, overall cell sizes, and nucleolar prominence (87)—and cytochemical. In general, ALL demonstrates "blocky" cytoplasmic reactivity with the PAS method (Figure 15.28), which is removed by prior diastase treatment. More than 50% of cases of ALL also exhibit positivity for ACP, which is inhibitable by tartrate (92–92). Specifically, T-cell ALL typically manifests punctate paranuclear reactivity for this enzyme (93,94) (Figure 15.29). Conversely, ALL usually lacks CAE and MPO, and stains variably for NSEs (91,92,95). The L3 variant (Burkitt leukemia) is a special one histochemically because it shows prominent cytoplasmic vacuoles in Wright-Giemsa preparations; these exhibit consistent and unique positivity with

TABLE 15.1: From Foucar (87), with Permission

French-American-British Classification of Acute Myelogenous Leukemia*

Subtype	Bone Marrow Definition
AML-MO	⩾30% blasts (type I) <3% Sudan black B/myeloperoxidase positivity Myeloid antigen expression by immunophenotyping or myeloperoxidase expression by electron microscopy
AML-M1	⩾30% blasts (types I, II, rarely III) ⩾3% Sudan black B/myeloperoxidase positivity in blasts <10% cells exhibiting maturation beyond blast stage
AML-M2	⩾30% blasts (types I, II, III) ⩾3% Sudan black B/myeloperoxidase positivity in blasts >10% granulocytic cells exhibiting maturation beyond blast stage <20% monocytic cells
AML-M3	⩾30% blasts (types I, II, III) + hypergranular promyelocytes† Intense myeloperoxidase/Sudan black B reaction in virtually all cells
AML-M3m	Same criteria as AML-M3, except that granules within most promyelocytes are inconspicuous and nuclei are highly grooved and reniform
AML-M4	⩾30% myeloblasts + monoblasts + promonocytes >20% Sudan black B/myeloperoxidase positive cells >20% nonspecific esterase-positive cells
AML-M4 eos	Same criteria as AML-M4, with the addition of abnormal eosinophils in the bone marrow
AML-M5	⩾30% myeloblasts + monoblasts + promonocytes <20% Sudan black B/myeloperoxidase positive cells >80% nonspecific esterase-positive cells Monoblasts predominate in M5a Promonocytes predominate in M5b
AML-M6	⩾30% of nonerythroid cells are myeloblasts (types I, II, III) >50% erythroid elements
AML-M7	⩾30% blasts (myeloblasts + megakaryoblasts) >30% megakaryocytic elements defined by immunophenotyping or ultrastructural electron microscopy

* The National Cancer Institute recommends classification as acute leukemia when blood blasts exceed 30% regardless of bone marrow blast percentage. Blood definitions are variable except for AML-M4, which requires monocytosis (⩾5000/μm^3 [⩾5 × 10^9/L]) and increased lysozymes.

† Promyelocytes usually predominate in AML-M3.

the ORO procedure (Figure 15.30) and, in a proportion of cases, by the SBB method as well (16,17). The PAS stain is negative in L3 leukemia. "Granular cell" ALL is singular in its demonstration of azurophilic cytoplasmic granules in >5% of blasts in Wright-stained smears (96). Moreover, it differs from other ALL morphotypes cytochemically, showing positivity for ACP, α-naphthyl acetate esterase and/or α-naphthyl butyrate esterase and, occasionally, for SBB (97,98). Conjoint cytoplasmic granularity and SBB positivity represents a diagnostic pitfall because these findings are typical of acute *myelogenous* leukemia (AML) (see below).

Acute Myelogenous Leukemias: The AMLs are considerably more variable than ALLs, both clinically and morphologically. The FAB scheme partitioned AML into seven subtypes (28), which have subsequently been further subdivided (99). These can be summarized descriptively as acute minimally differentiated (M0), "ordinary" (M1), and "differentiating" (M2) myeloblastic leukemia; promyelocytic (M3) leukemia; myelomonocytic (M4) leukemia; monocytic (M5) leukemia; erythroleukemia (M6—also known as DiGuglielmo's leukemia); and megakaryoblastic (M7) leukemia (28).

Cytochemical reactions are effective in recognizing the M1 through M6 variants of AML, but M0 and M7 must be defined immunologically. This is because the latter two subtypes do not demonstrate consistent histochemical profiles (100).

In general terms, >3% of blasts in M1 through M6 AML can be labeled for SBB, CAE, and MPO, singly or in combination. These three markers are particularly intensely demonstrated in M3 (promyelocytic) AML (91,92,101–103) (Figure 15.31). Auer rods—representing linear aggregates of primary lysosomal granules that are pathognomonic of acute myeloid leukemias—also show reactivity for SBB, CAE, and MPO (104) (Figure 15.32).

Acute basophilic leukemia is a special variant of AML. It lacks MPO and SBB but demonstrates intense CAE reactivity with the Leder method. In addition, metachromatic granules are seen in the neoplastic blasts with the toluidine blue method (105).

The M4 and M5 varieties of AML exhibit concurrent myeloblastic and monoblastic characteristics, reflected by their positivity for MPO, SBB, CAE, and at least one of the NSEs in >20% of blasts (106,107) (Figure 15.33). The M4-E subtype also is accompanied by an increase in eosinophilic leukocytes in the marrow, and these cells are morphologically atypical (108). In the M6 form, >50% of marrow elements are erythroid, together with >30% myeloblasts, as defined above (109,111). Abnormal

TABLE 15.2: From Foucar (87), with Permission

French-American-British Classification of Acute Lymphoblastic Leukemia (ALL)

Subtype	Features
L1*	High nuclear/cytoplasmic ratio Nuclei have inconspicuous nucleoli Blasts variable in size but often small
L2*	Moderate to abundant amounts of cytoplasm Nuclei have one or more distinct, prominent nucleoli Blasts medium to large
L3	Scant to abundant amounts of deeply basophilic cytoplasm containing abundant distinct vacuoles Nuclear chromatin homogeneous with one or more, generally indistinct nucleoli Blasts medium to large

* Amount of cytoplasm and prominence of nucleoli are the key features useful in distinguishing cases of L1 and L2 ALL. If overlap occurs, large overall size and coarse nuclear lobulations favor L2 ALL.

TABLE 15.3: From Foucar (87), with Permission

Cytochemical Features of Acute Lymphoblastic Leukemia (ALL)*

Cytochemical Stain	Comments
PAS	Blocks and globules of PAS-positive glycoprotein/glycolipid often present in L1/L2; same pattern also seen in certain AMLs (M6, M5, rare M1)
Acid phosphatase	Focal, paranuclear positivity originally noted in L1/L2 ALL of T-cell origin, later found in other immunologic subtypes of ALL
Sudan black B	Detected in 2%–3% of L1/L2 ALL specimens; may result in misdiagnosis of AML
Naphthol ASD chloroacetate esterase	Myeloid enzyme detected in very rare ALL
Oil red O	Vacuoles in L3 blasts positive

PAS = periodic acid–Schiff, AML = acute myelogenous leukemia.

* No cytochemical stain is specific for ALL.

TABLE 15.4: From Foucar (87), with Permission

Morphologic, Cytochemical, Ultrastructural, and Clinical Features of Granular Acute Lymphoblastic Leukemia (ALL)

Definition/morphology
- >5% blasts contain distinct azurophilic granules
- L1 and L2 FAB subtypes

Incidence
- 2%–7% of cases of ALL in children*

Cytochemistry
- Acid phosphatase positive
- α-Naphthyl acetate esterase positive
- α-Naphthyl butyrate esterase sometimes positive
- Periodic acid-Schiff positive (weak, diffuse)
- Myeloperoxidase negative, chloroacetate esterase negative
- Usually Sudan black B negative (≤20% cases positive)

Immunophenotype
- Usually B-cell precursor phenotype
- Express CD10, CD19
- Lack surface immunoglobulin

Ultrastructure
- Membrane-bound organelles
- Could be mitochondria or lysosomes
- Single report of ultrastructural myeloperoxidase positive granules (negative by light microscopy)

Significance/comments
- Important not to confuse with acute myelogenous leukemia
- Recent report describes prevalence of FAB L2 subtype, poor response to induction chemotherapy, and association with poor prognosis

FAB = French-American-British.

* Analyses of granular ALL in large series of adults have not been performed.

normoblasts in the latter AML subtype are also PAS positive (111) (Figure 15.33).

M7 (megakaryoblastic) AML is, perhaps, the most difficult variant to characterize histochemically. This is because consistent enzymatic or cytochemical markers for immature megakaryocytes do not exist. It is true that many cases of M7 AML show PAS positivity, in likeness to M6, but this is not a universal finding (112,113).

Chronic Myelogenous Leukemia: As discussed earlier in this chapter, cytochemical evaluations are helpful in the distinction between CML and leukemoid reactions (114). CML is a myeloproliferative disorder (MPD) in which there is an increased proliferation of granulocytic cells, with a retained ability for differentiation. The peripheral blood shows a spectrum of myeloid cells, including occasional blasts, with a "bulge" at the myelocyte stage of development (115,116). The LAP score is decreased in the abnormal leukocytes of CML (Figure 15.34), as opposed to a normal or increased score in leukemoid reactions (44). After several years of disease activity, CML may undergo clonal evolution (blastic transformation), yielding a morphologic image that is indistinguishable from that seen in one of the AML or ALL subtypes (117,118). The cytochemical profiles of the CML blast phase parallel those of acute leukemia variants in their *de novo* forms.

Other Chronic "Myeloproliferative" Disorders and Myelodysplasias: Besides CML, the other chronic MPDs include essential thrombocythemia, polycythemia vera, and myelofibrosis-myeloid metaplasia (119,120). In addition, myelodysplastic syndromes also pursue a chronic course and are related conceptually to the MPDs. They comprise refractory anemia, with or without ringed sideroblasts or excess blasts, and chronic myelomonocytic leukemia (121,122). Cytochemical analysis does not play a significant role in the pathologic definition of most myelodysplasias, with the exception of SBA. As stated earlier, "ringed" sideroblasts in this condition are recognizable in marrow aspirates using Perl's method or silver impregnation (123–125). Nevertheless, histochemical staining of marrow biopsies with Wilder's or Snook's methods is an

TABLE 15.5: Classification of Acute Leukemias by Morphologic and Cytochemical Criteria [From Löffler and Gassmann (88,89)]

Stem cell leukemia		
Peroxidase	0	Undifferentiated type
PAS	0	
α-naphthyl acetate esterase	0	
Myeloblastic leukemia (AML)		
Peroxidase	(−)	Peroxidase types 1 and 2
PAS	0–(+)	
α-naphthyl acetate esterase	(+)	
Promyelocytic leukemia		
Peroxidase	+	Peroxidase type 3
PAS	(+)	
α-naphthyl acetate esterase	(+)	
Myelomonocytic leukemia		
Peroxidase	+	Peroxidase esterase type
PAS	(+)	
α-naphthyl acetate esterase	+	
Monocytic leukemia		
Peroxidase	(+)	Esterase type
PAS	0–(+)	
α-naphthyl acetate esterase	+	
Lymphoblastic leukemia (ALL)		
Peroxidase	0	PAS type
PAS	+	
α-naphthyl acetate esterase	0	

0 = negative, (+) = weakly positive <25%, + = strongly positive >50.

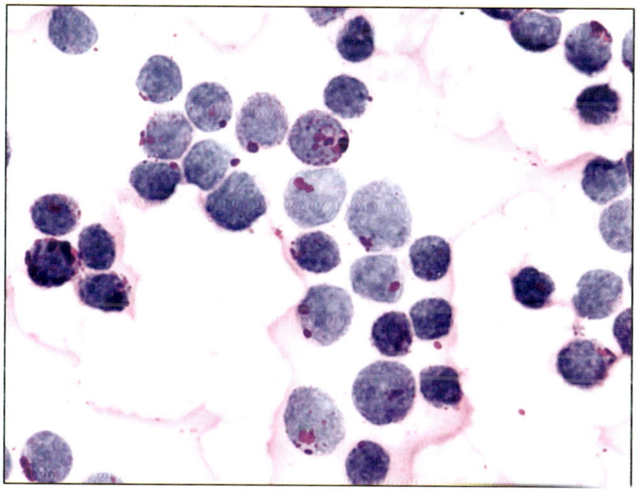

***FIGURE 15.28:** Glycogen is present in blasts in ALL, type L1, as seen with the PAS technique.*

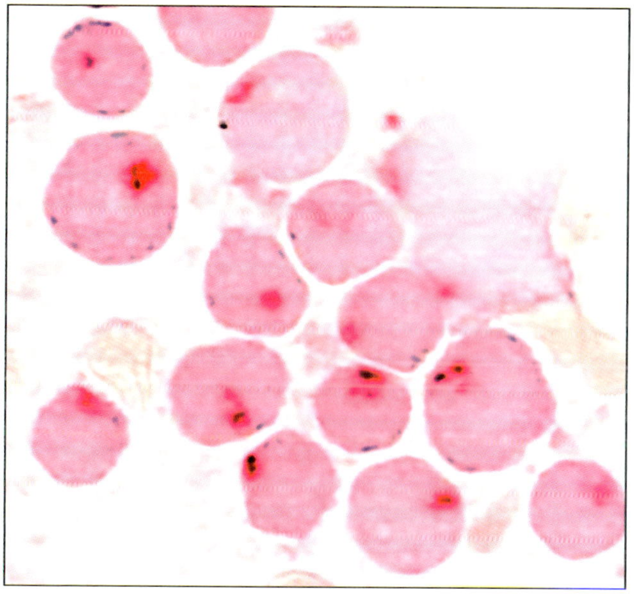

***FIGURE 15.29:** Discrete paranuclear reactivity is present for ACP, in an example of T-cell ALL, type L2.*

important step in separating MPDs or myelodysplasias from more innocuous conditions (Figure 15.35). Increased reticulin deposition in the marrow suggests a diagnosis of one of the former conditions, and it is particularly notable in myelofibrosis-myeloid metaplasia (126,127).

Chronic Lymphoid Leukemias: Most examples of CLL, regardless of whether they have a B-cell or T-cell lineage, usually do not require cytochemical characterization.

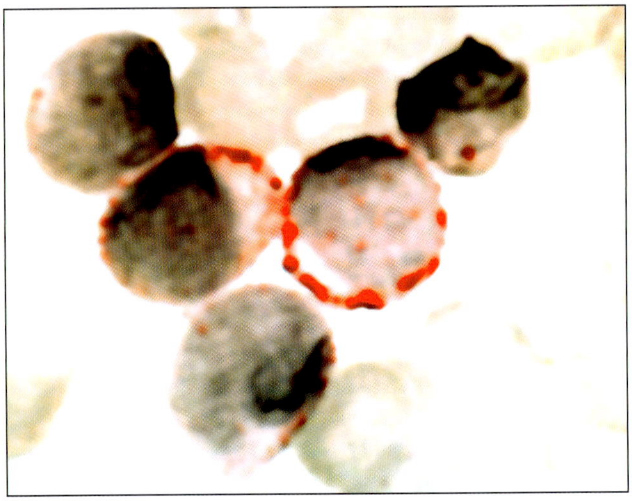

FIGURE 15.30: Globular intracytoplasmic reactivity with the ORO procedure is seen in this example of ALL, type L3 (Burkitt leukemia).

FIGURE 15.31: Intense reactivity is present for SBB, in a case of acute myeloid leukemia, type M2.

A provisional diagnosis of CLL can be assigned when the peripheral absolute lymphocyte count is ≥5000/μL and where bone marrow examination shows increased numbers of mature lymphocytes. HCL (another chronic lymphoid proliferation) has a different cytomorphologic appearance than that of CLL in smear preparations—with small projections (hairs) on the cellular surface in HCL—and a "fried-egg" appearance in bone marrow biopsy specimens. Reticulin fibers are also much more significantly increased in HCL than in CLL in marrow biopsies. As a caveat, however, the authors have seen examples of *acute* leukemia that were mistaken for CLL because the blasts were small and nucleoli were not particularly conspicuous. The latter mistake is avoidable only if a complete cytochemical panel is applied that would be appropriate for the evaluation of ALL and AML. With that having been said, one also may encounter extraordinary cases where CLL evolves *into* ALL or AML.

Other Lymphoid Infiltrates in the Bone Marrow: Other lymphoid aggregates in the bone marrow are diagnostically problematic, especially if the patient is known to have malignant lymphoma at another anatomic site. In general, the situation of intramedullary lymphoid infiltrates—with lymphomas preferring a paratrabecular localization—and their cytologic attributes are sufficient to make confident diagnoses (132,133). However, in ambiguous cases, Snook's or Wilder's procedure may provide additional useful information. This is because marrow lymphoma is associated with increased deposition of reticulin, whereas benign lymphoid aggregates do not show this finding. In fact, in many lymphomas—of both Hodgkin and non-Hodgkin types—individual tumor cells will be invested by circumferential reticulin networks (134,135) (Figure 15.36). Proliferative cells in these conditions are also potentially highlighted with the MGP method (136) (Figure 15.37).

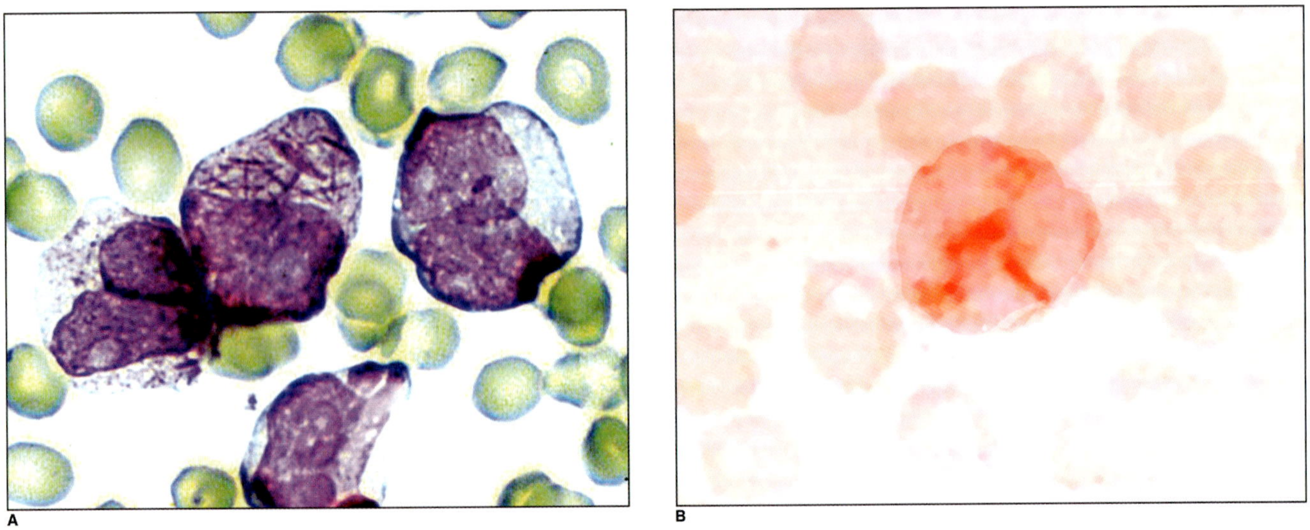

A B

FIGURE 15.32: Several Auer rods are present in the blast in the center of this photograph (A), in a case of acute myeloid leukemia (type M2). These structures are cytochemically positive for CAE (B).

Metastatic Malignant Tumors in the Bone Marrow: The bone marrow may be involved by metastatic carcinoma, melanoma, neuroblastoma, or sarcoma, with rhabdomyosarcoma principally representing the last of these categories. All these lesions are again associated with increased reticulin deposition around the secondary tumor implants (137–140). Moreover, adenocarcinomas manifest intracellular positivity with the PAS, alcian blue, colloidal iron, or mucicarmine technique (Figure 15.38); melanomas are reactive with the Fontana-Masson or Schmorl procedure; and rhabdomyosarcomas demonstrate PAS positivity as well as possible cytoplasmic labeling with the phosphotungstic acid–hematoxylin (PTAH) procedure.

Plasmacytic (Multiple) Myeloma: Plasmacytic myeloma is defined minimally by the presence of ⩾10% abnormal plasma cells in the bone marrow or histologic proof of an intraosseous plasmacytoma, along with the typical clinical features of myeloma and a monoclonal protein in the serum (usually >3 g/dL) or the urine (141).

Cytochemically, neoplastic plasma cells in myeloma are characterized by intense labeling with the MGP method,

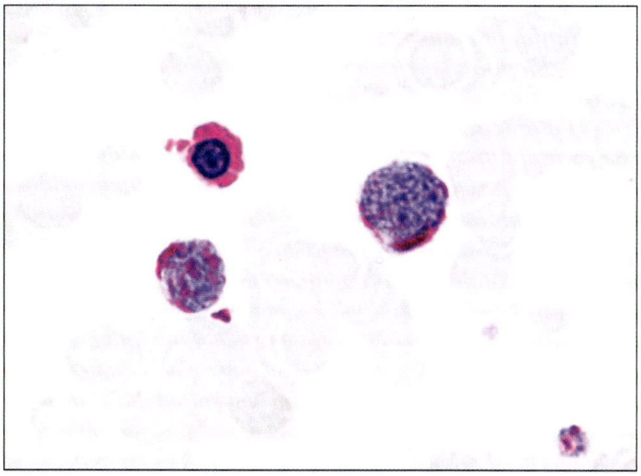

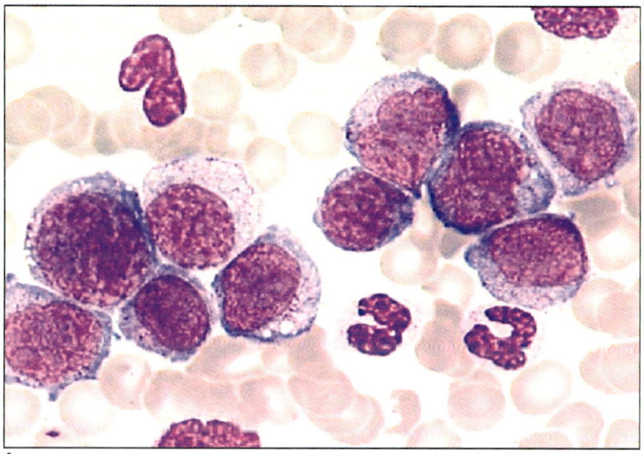

A

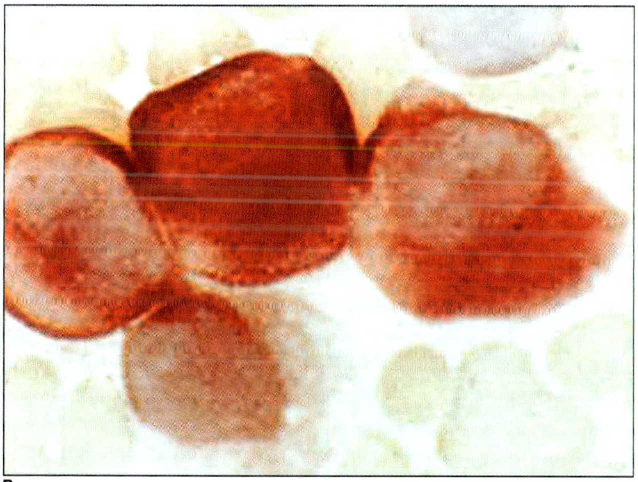

B

FIGURE 15.33: ***Acute monocytic leukemia, seen in a Wright-Giemsa stain (A). The neoplastic blasts are intensely positive for α-naphthyl butyrate esterase (B). Acute erythroleukemia, showing PAS–positivity in circulating normoblasts (C).***

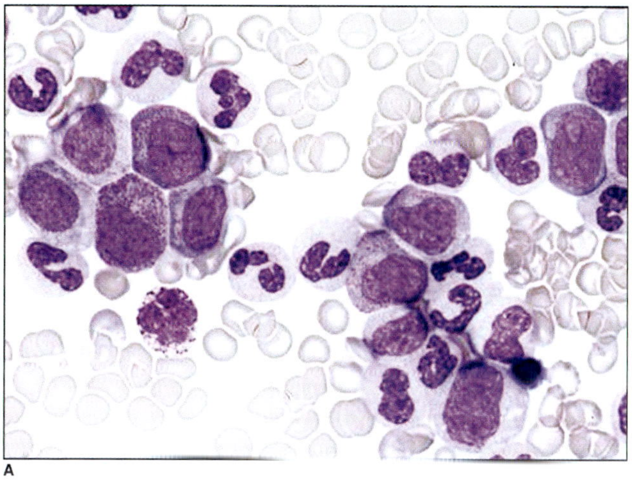

A

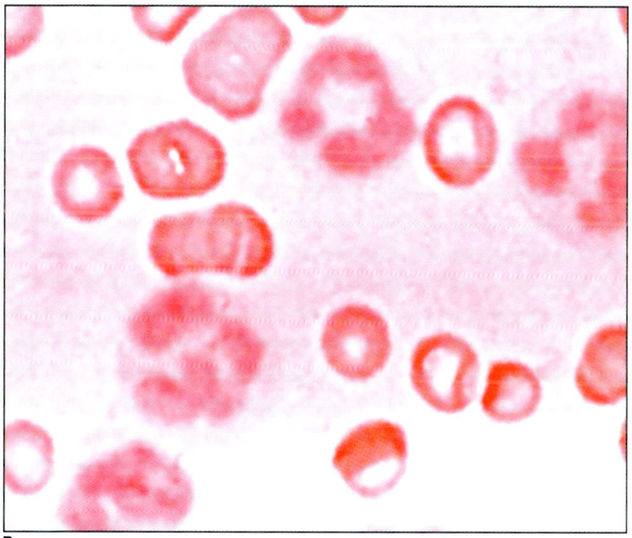

B

FIGURE 15.34: ***No staining for LAP is present in circulating granulocytic cells, in CML (A—Wright-Giemsa stain; B—LAP method).***

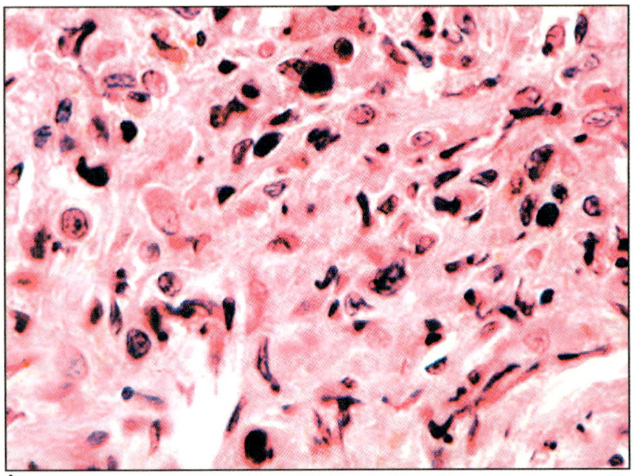

A

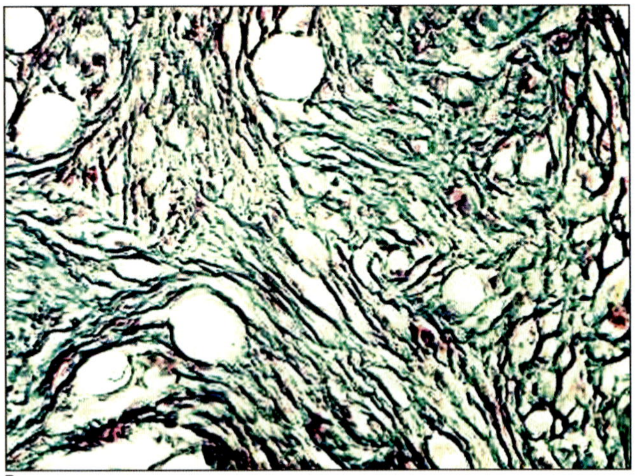

B

FIGURE 15.35: Late-stage polycythemia vera is seen in this bone marrow biopsy, typified by significant fibrosis (A). The latter finding is confirmed with a reticulin stain (B).

frequent PAS positivity, and the possible presence of cytoplasmic immunoglobulin crystalloids, as visualized with the Masson-trichrome procedure (142–145) (Figure 15.39).

Amyloid deposits occasionally accompany myelomatous infiltrates in the marrow as accumulations of amorphous pink-red interstitial material. They may be observed in the absence of an obvious plasmacellular proliferation as well (146–148). The CR and CV methods (among others) are effective at labeling amyloid in this setting.

Systemic Mast Cell Disease: Systemic mastocytosis is typified by mast cell infiltrates in several organs, including the skin, spleen, liver, bone marrow, and others. Eosinophilia is a common associated finding in the marrow, as is osteosclerosis and paratrabecular fibrosis (149,150). The most common peripheral blood profile is that of anemia; eosinophilia, leukocytosis, basophilia, thrombocytosis, and monocytosis may be seen in some instances (149).

Mast cells have a typical cytologic appearance, with a bluntly fusiform shape; uniform, ovoid nuclei; and granular amphophilic cytoplasm. Histochemically, they are strongly reactive with Leder's method (44) (Figure 15.40); methylene blue or Giemsa stain demonstrates numerous distinct cytoplasmic granules (151–153) (Figure 15.41). The MGP method stains the cytoplasm of mast cells a peculiar orange hue (15). No labeling is observed with MPO, SBB, PAS, or NSE. Snook's or Wilder's method demonstrates increased reticulin fibers in the foci of paratrabecular sclerosis in which aggregates of mast cells are typically enmeshed (154,155).

LYMPH NODES, SPLEEN, AND OTHER EXTRANODAL LOCATIONS

The remainder of the hematopoietic system is affected by many of the same disorders that were discussed above in

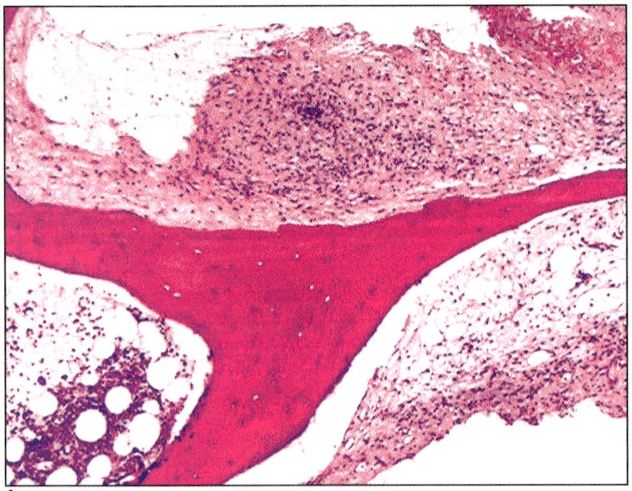

A

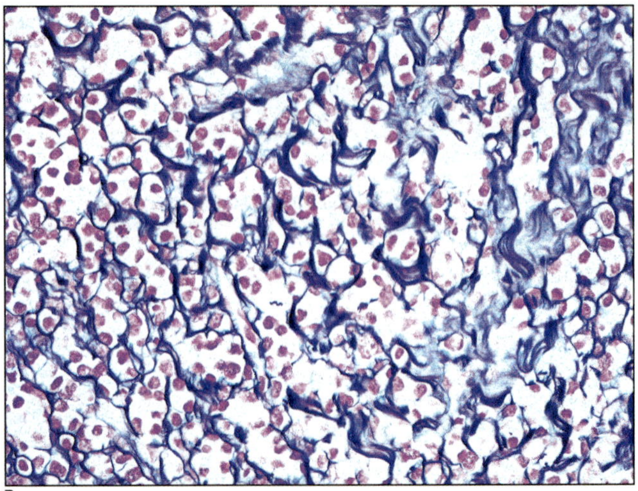

B

FIGURE 15.36: Involvement of the bone marrow by HL is seen in this biopsy specimen, in a paratrabecular location (A). The lymphoid cells in this focus demonstrate pericellular reticulin deposition (B).

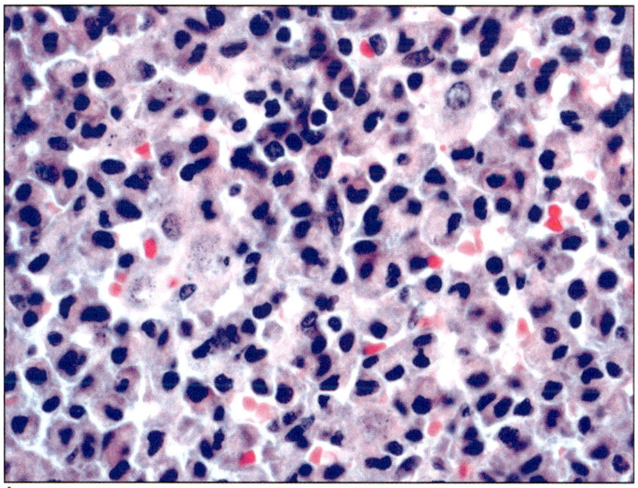

A

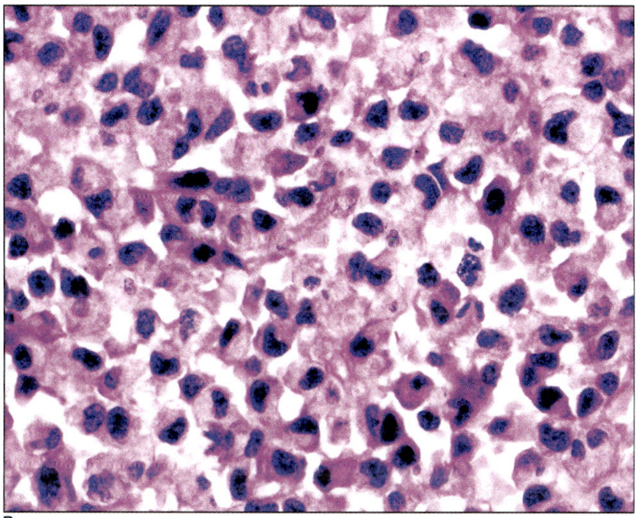

B

***FIGURE 15.37:** Large cell lymphoma in the bone marrow (A), labeling intensely with the MGP technique (B).*

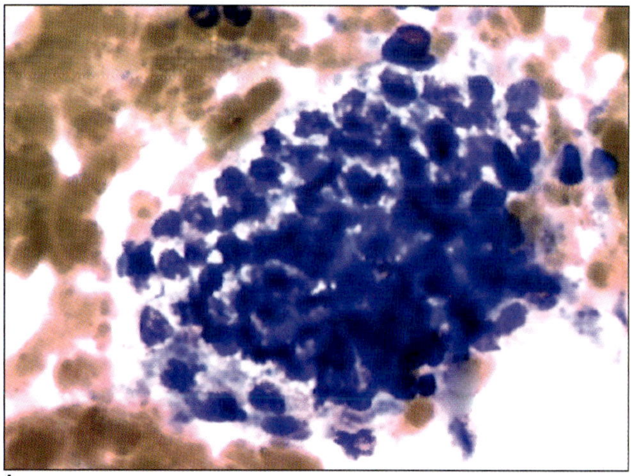

A

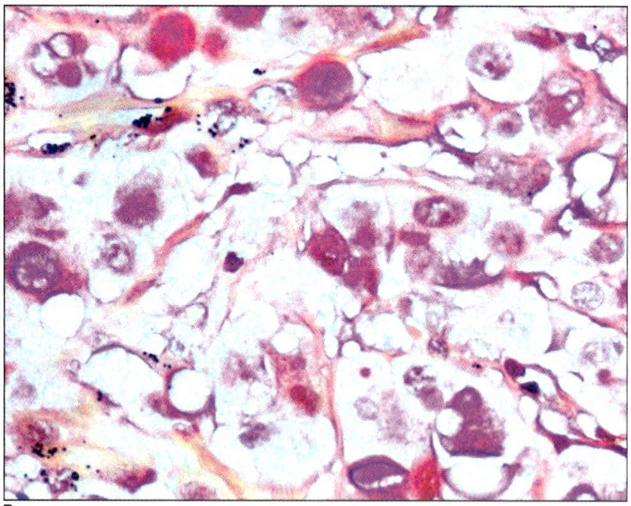

B

***FIGURE 15.38:** Metastatic adenocarcinoma is seen in this bone marrow aspirate, in the form of a cohesive group of epithelioid cells (A—Wright-Giemsa stain). The tumor cells in a corresponding biopsy specimen are labeled with the mucicarmine method (B).*

reference to the bone marrow. In particular, metabolic storage disorders, Whipple's disease, amyloidosis (Figure 15.42), mastocytosis, leukemic infiltrates (especially of HCL), and metastatic nonhematopoietic tumors can also be seen in the lymph nodes, spleen, liver, and other anatomic sites (156–158). Hence, they will not be considered again in the following sections.

Pathologic conditions in the lymph nodes and spleen occupy much of the attention of any hematopathologist. Other than the conditions just cited, the principal lesions that may be encountered in such specimens include infections, chronic autoimmune or idiopathic lymphadenitides, lymphoid hyperplasias, malignant lymphomas, ectopic cellular "rests," and metastatic neoplasms.

Infectious Lymphadenitis

In the past, infectious lymphadenopathy was a common process. Today, aside from involvement of lymph nodes in systemic viral diseases, it is relatively rare. *Tuberculous lymphadenitis* is mostly commonly encountered in the cervical, mediastinal, and axillary nodes, featuring the presence of granulomatous inflammation with or without necrosis. Ziehl-Neelsen or Fite stain reveals the presence of mycobacteria in approximately 50% of cases, with the latter of these methods being selective for "atypical" organisms (e.g., *Mycobacterium kansasii, M. genavense, M. marinum, M. simiae, M. scrofulaceum, M. szulgai, M. haemophilum, M. avium-intracellulare, M. malmoense, M. ulcerans, M. xenopi, M. abscessus, M. chelonae, M. fortuitum*) (159–161). The lesions in leprosy-related lymphadenitis also stain with the Fite procedure, and the constituent histiocytes often label with colloidal iron or alcian blue as well (162,163). Syphilitic lymphadenitis accompanies either the primary or secondary form of lues, but treponemal organisms are principally visible (usually in intranodal postcapillary venular endothelium) in primary syphilis with the Warthin-Starry or Steiner

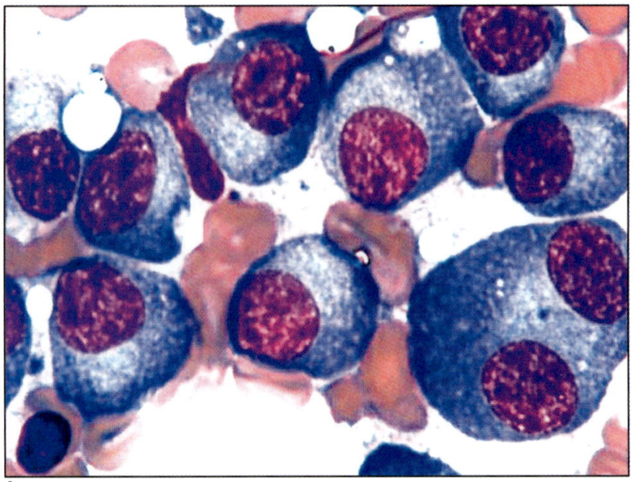

A

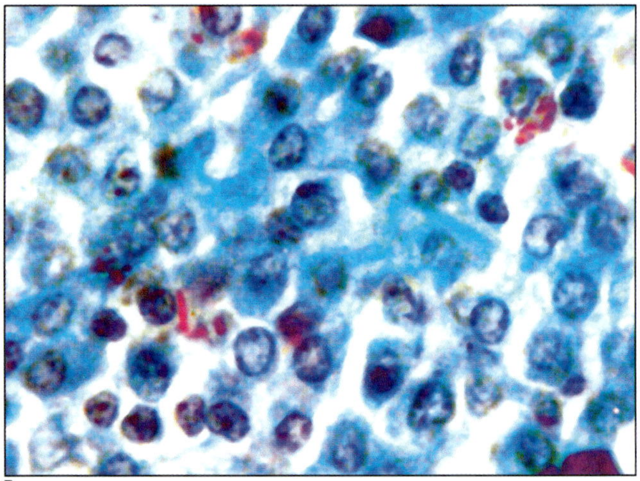

B

FIGURE 15.39: Plasma cell myeloma, as seen in a bone marrow aspirate (A—Wright-Giemsa stain). The tumor cells contain intracytoplasmic crystalloids, demonstrated with the Masson trichrome technique (B).

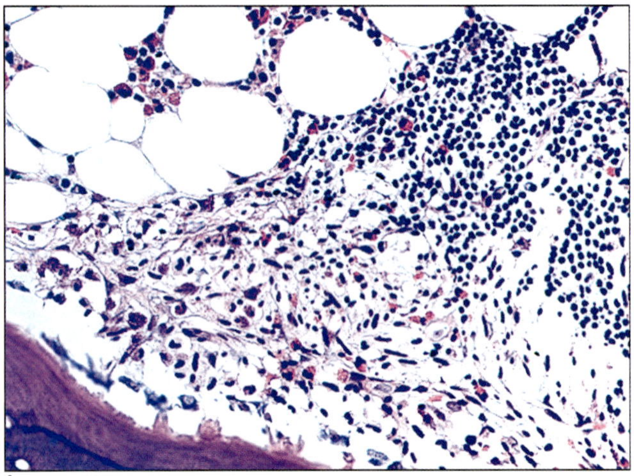

A

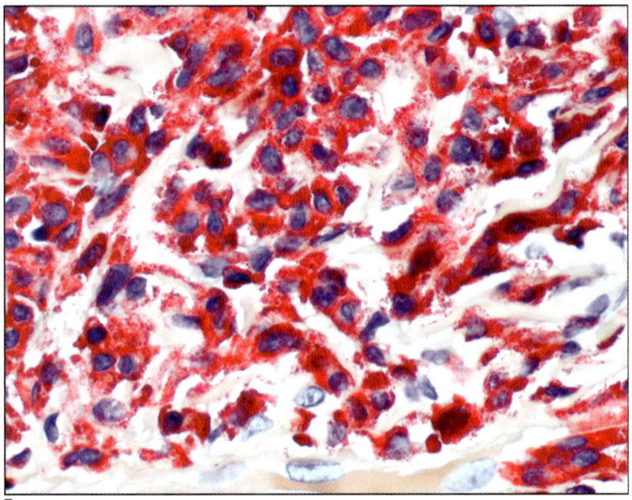

B

FIGURE 15.40: A paratrabecular lesion of mast cell disease is present in this bone marrow biopsy (A—hematoxylin and eosin). The lesional cells are strongly reactive for CAE (B).

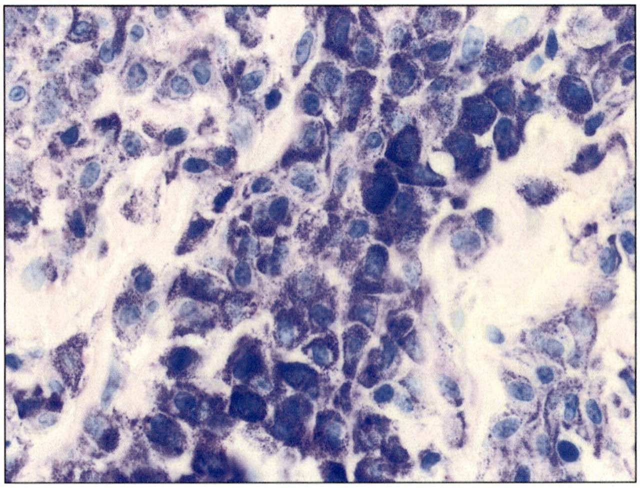

FIGURE 15.41: Numerous metachromatic cytoplasmic granules are seen in mastocytosis, with the toluidine blue procedure.

procedure (164–166). The latter two stains are also potentially positive in cat-scratch lymphadenitis, caused by *Bartonella* species (167,168) (Figure 15.43). Fungal infections of the nodes are demonstrable using the SM or PAS technique, both of which can label the organisms in histoplasmosis, cryptococcosis, and coccidiomycosis (169). Leishmanial lymphadenitis shows positivity of the intracellular protozoans with the Giemsa stain (170). In contrast, no microbes are recognizable histochemically in nodal infections by *Yersinia*, *Brucella*, *Chlamydia trachomatis* (in lymphogranuloma venereum), or *Toxoplasma gondii*. Intranodal viral inclusions caused by infection with cytomegalovirus, *Herpes simplex* virus, or Varicella-zoster virus can usually be seen satisfactorily with hematoxylin and eosin staining (171,172), but the MGP procedure can also be used to highlight such structures.

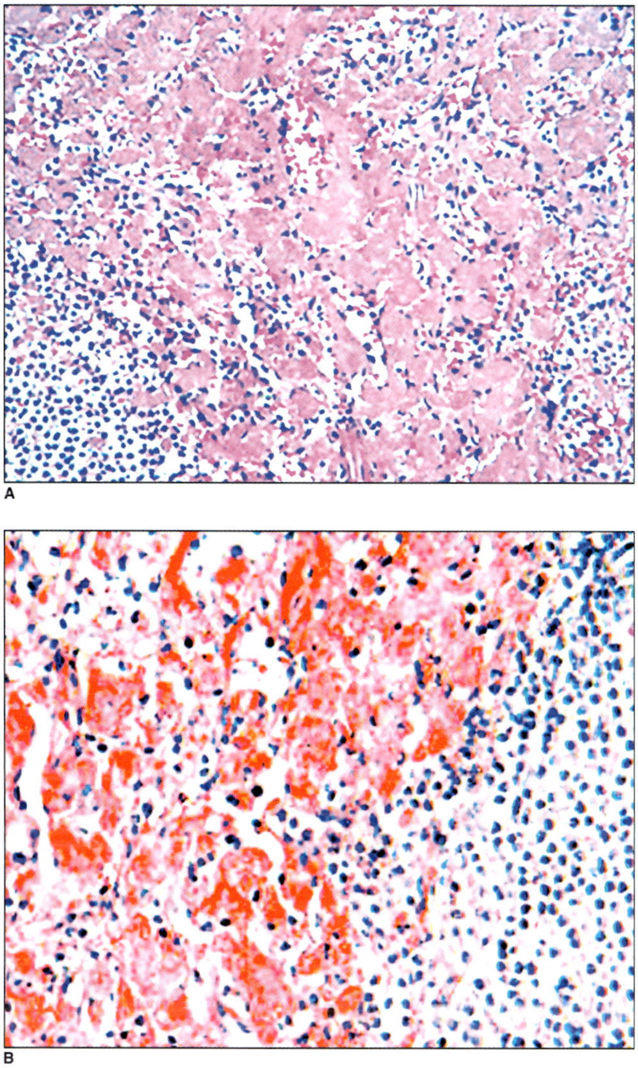

FIGURE 15.42. *Splenic amyloidosis, showing replacement of parenchyma by hypocellular eosinophilic material (A—hematoxylin and eosin). It labels with the CR technique (B).*

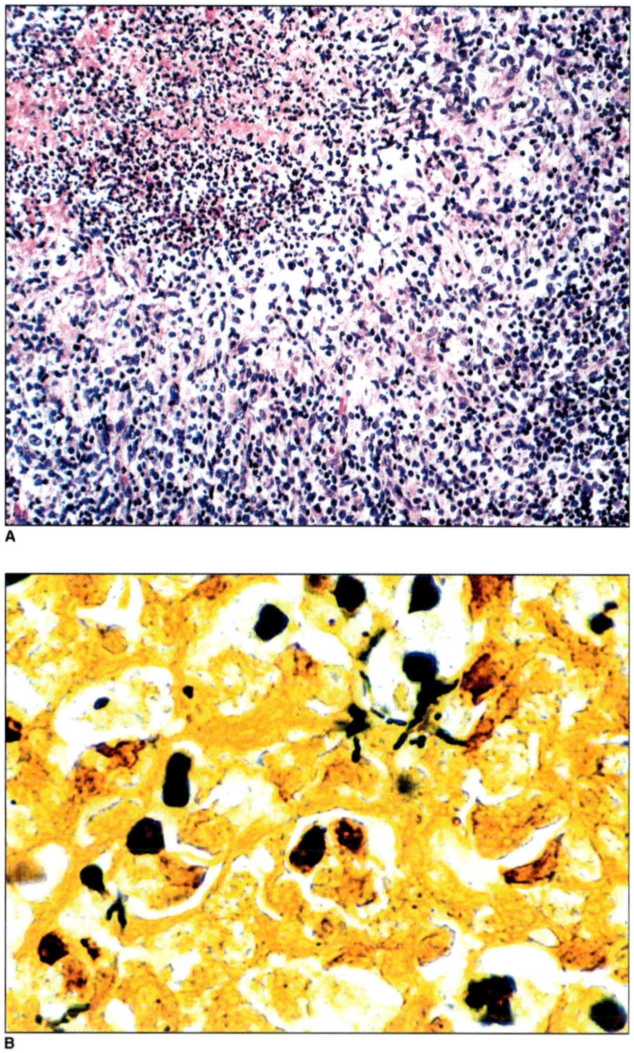

FIGURE 15.43: *Cat-scratch lymphadenitis, showing "stellate" microabscess formation in a lymph node (A—hematoxylin and eosin). A Steiner stain (B) demonstrates the presence of bacilliform* Bartonella *organisms.*

A potential diagnostic trap exists with regard to misclassification of some noninfectious reactive lymphadenopathies, based on findings in Ziehl-Neelsen or PAS stain. If sinus histiocytosis is prominent, regardless of its cause, constituent cells in the nodal sinuses may contain ceroid-rich structures, also known as Hamazaki-Wesenberg (HW) bodies (173,174) (Figure 15.44). These nonspecific inclusions of glycolipid are degenerative in nature, but they label with the two histochemical stains just mentioned. Hence, HW bodies can be mistaken for microorganisms in this setting. In contrast, another condition in which HW bodies are indirect indicators of infection is granulomatous disease of childhood. This disorder features a congenital inability of neutrophils to kill bacteria, leading to accumulations of foamy, faintly pigmented histiocytes in lymph nodes and other lymphoreticular organs (175). These cells may contain HW bodies, and they are also autofluorescent (176,177).

Autoimmune Lymphadenitides and Splenitides

The lymph nodes and spleen in patients with autoimmune diseases such as rheumatoid arthritis, lupus erythematosus, and thrombotic thrombocytopenic purpura usually demonstrate nonspecific lymphoid hyperplasia. Occasionally, however, they show foci of vasculitis, with or without parenchymal necrosis and "hematoxylinophilic bodies" (nuclear debris) (178–180). The latter structures can be labeled with the MGP method. In addition, plasma cell infiltrates may be prominent in such conditions, and they likewise will stain brightly with the MGP procedure. Foci of vasculitis can be highlighted with the PTAH procedure or Martin's scarlet blue stain, provided that the intramural "fibrinoid

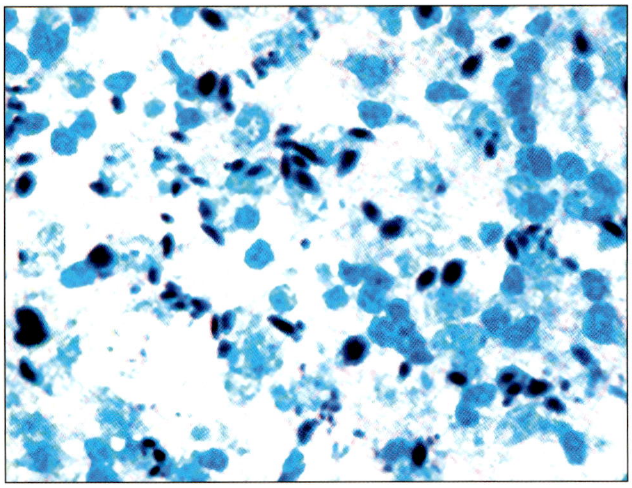

FIGURE 15.44: Numerous ceroid (HW) bodies are present in a lymph node showing reactive sinus histiocytosis, as demonstrated with a Ziehl-Neelsen stain. These may be mistaken for microorganisms.

necrosis" is apparent morphologically. In the spleen, the PTAH and reticulin methods may also show "onion-skin" lamellation of connective tissue around central arterioles in the white pulp (Figure 15.45), especially in lupus erythematosus (156–158,181).

Sinus histiocytosis with massive lymphadenopathy (SHML; Rosai-Dorfman disease) is also currently considered to be an autoimmune condition, although its causative mechanism is still obscure (182,183). Reticulin fibers are particularly prominent in the walls of intranodal sinusoids in SHML, and the nodal capsule demonstrates laminated collagen deposition with the trichrome method (182).

Dermatopathic Lymphadenopathy

Patients with chronic pruritic dermatitides or mycosis fungoides may develop regional lymphadenopathy in nodes that drain the affected skin areas (184,185). Excision of these structures again usually demonstrates only nonspecific lymphoid hyperplasia, but sometimes pigmented histiocytes are scattered throughout the interfollicular zones. These "dermatopathic" cells contain glycolipid, glycoprotein, and melanin, and therefore can be labeled to varying degrees with the SBB, PAS, and Fontana-Masson methods (15).

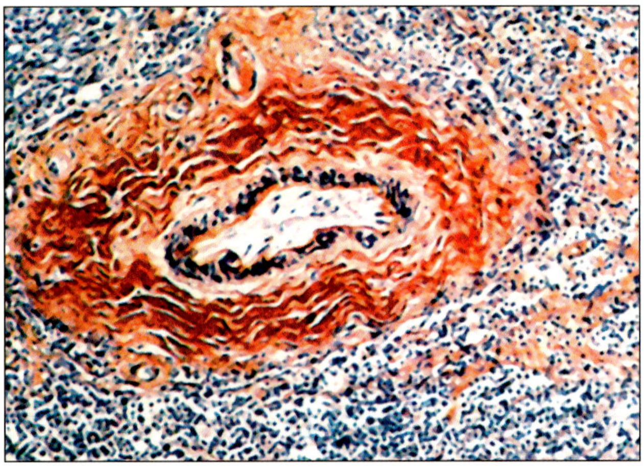

FIGURE 15.45: "Onion-skin" fibrosis is present around an arteriole in the splenic white pulp, as seen with the PTAH method in a case of lupus erythematosus.

Nodal and Splenic Sarcoidosis

Lymph nodes in any anatomic group may potentially be involved by sarcoidosis, but those in the supraclavicular

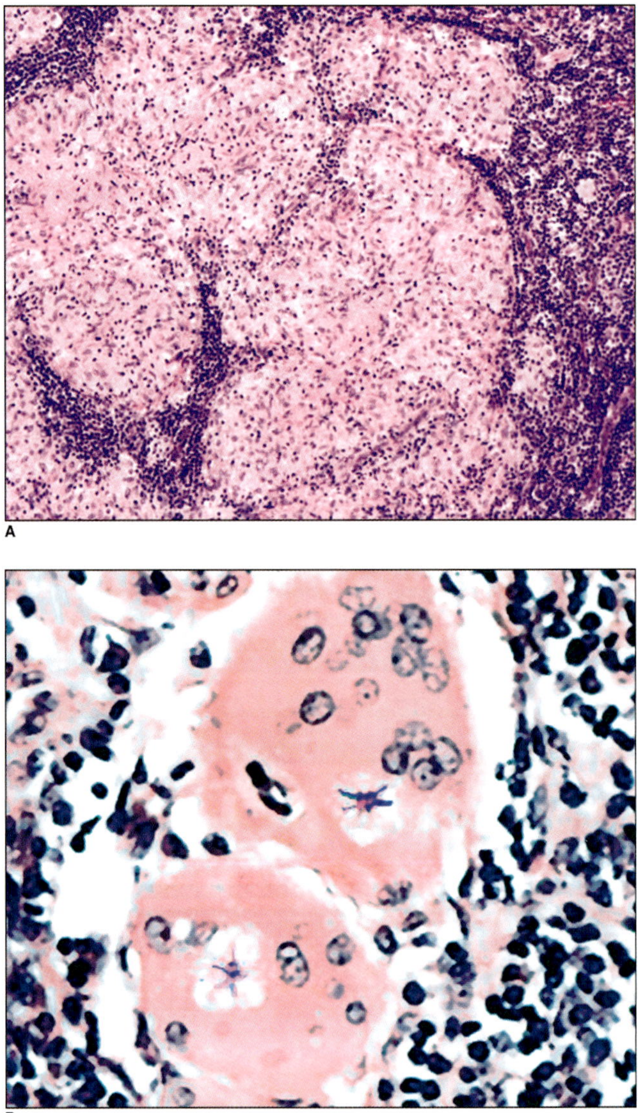

A

B

FIGURE 15.46: Noncaseating granulomatous inflammation in a lymph node, in an example of sarcoidosis (A—hematoxylin and eosin). Giant cells in the lesion contain "asteroid" bodies that are labeled with the PTAH stain (B).

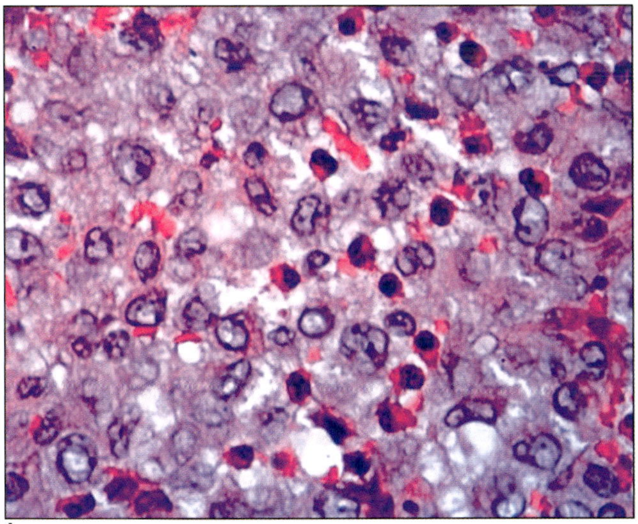

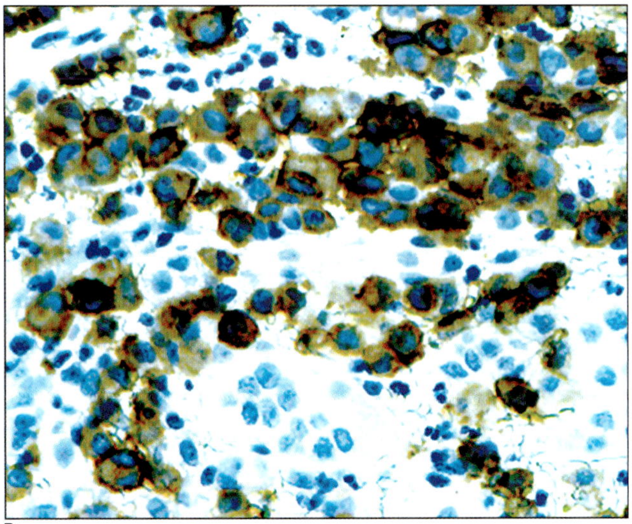

FIGURE 15.47: LCH in a lymph node (A—hematoxylin and eosin), demonstrating affinity of the lesional cells for peanut agglutinin (B).

region and the mediastinum are most often affected. The spleen is less commonly affected (186). As described earlier with reference to the bone marrow, one sees potentially confluent noncaseating granulomas in this condition, featuring numerous epithelioid histiocytes and multinucleated giant cells. The latter elements may contain Schaumann bodies, which are concentric microcalcifications that can be labeled with the von Kossa procedure (187,188). In addition, sarcoidal histiocytes may harbor asteroid bodies, represented by floret-like structures with delicate extensions that radiate from a central point. These are stainable with the PTAH method (15) (Figure 15.46) and are reasonably diagnostically specific. Of course, negative stains for mycobacteria and fungi are *de rigeur* before making such an interpretation.

Langerhans' Cell Histiocytosis

Langerhans' cell histiocytosis (LCH; "histiocytosis-X," "eosinophilic granuloma") may affect lymph nodes primarily, or apparently represent a secondary or inducible proliferation therein, in association with other lesions such as non-Hodgkin lymphoma (NHL) (189–191). The specialized dendritic-type histiocytes of LCH are recognizable by lectin histochemistry with biotin labeled peanut (*Arachis hypogaea*) and concanavalin-A agglutinins (see section on lectin histochemistry below) (192,193) (Figure 15.47). Affinity for the latter markers is also shared by the histiocytes of SHML (194), as discussed above, but they differ markedly from the cells in LCH at a morphologic level and therefore do not represent a diagnostic conundrum.

Intranodal Rests

Lymph nodes in a variety of anatomic sites may contain rests composed of grouped melanocytes (Figure 15.48). These are typically—but not always—associated with the nodal capsule or parasinusoidal fibrous septa (195,196). If their identity is in doubt, vis-a-vis confusion with hematopoietic elements, one can apply the Fontana-Masson method for melanin to label nevocytic rests (197) (Figure 15.49).

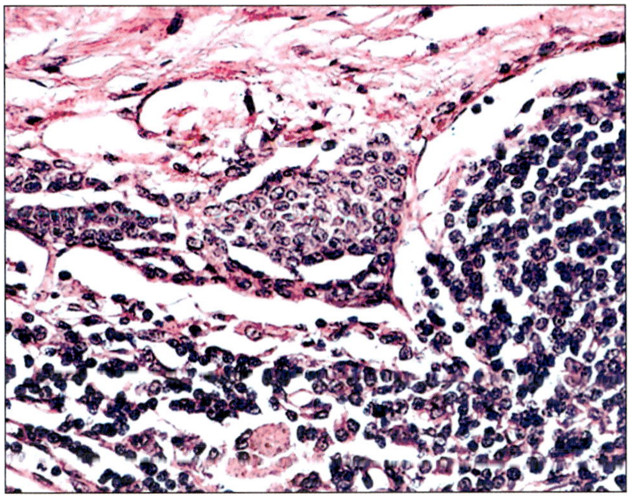

FIGURE 15.48: A capsular nevocyte "rest" is present in this lymph node (hematoxylin and eosin).

Even more problematically, lymph nodes in the mediastinum—and also in other node groups—can harbor gland-like inclusions comprising mesothelial cells or epithelial profiles (198–200). These are easily confused with metastatic adenocarcinoma. Unlike true glandular elements, mesothelial rests fail to label with the PAS or mucicarmine techniques. Instead, contents of the gland-like structures often demonstrate positivity with the colloidal iron or alcian blue method, which is removed by prior hyaluronidase treatment of the tissue sections.

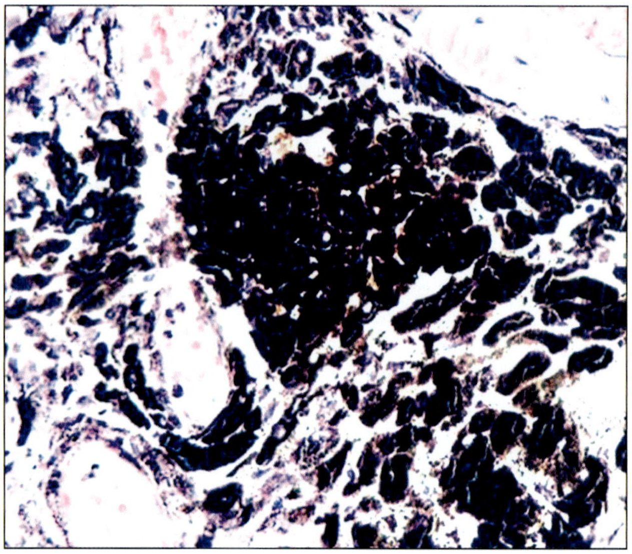

FIGURE 15.49: Another lymph nodal nevocyte rest demonstrates reactivity with the Fontana-Masson method.

Extramedullary Hematopoiesis

Extramedullary hematopoiesis (EMH) can be observed in the lymph nodes, spleen, liver, soft tissue, or various viscera. It is most marked in patients with primary bone marrow failure owing to myelophthesis and occasionally may be so florid as to produce a grossly tumefactive mass (201). All three hematopoietic cell lines are typically represented in lesions of EMH; therefore, positivity is expected with the Leder stain (Figure 15.50) and the PAS method, with the latter labeling both myeloid and megakaryocytic elements.

Nodal and Extranodal Hodgkin Lymphomas

Histochemistry in Hodgkin lymphomas (HLs) may provide useful ancillary information, which supplements that gained thorough morphologic evaluation. For example, Snook's and Wilder's methods can be employed to help delineate the nodal capsule in fragmented node biopsies or to accentuate the stromal architectural pattern that one may see in nodular lymphocyte–predominant HL. Likewise, trichrome stains may allow one to better visualize the internal fibrous septations that separate lymphoid cell groups in nodular sclerosis HL. Moreover, such septa are birefringent with polarization microscopy (156) (Figure 15.51). MGP staining also is capable of drawing attention to Reed-Sternberg cells and their variants, inasmuch as they contain prominent pyroninophilic nucleoli that can easily be seen on screening microscopy (15) (Figure 15.52).

Nodal and Extranodal Non-Hodgkin Lymphomas

The same benefits can be derived from reticulin, trichrome, and MGP procedures in the analysis of NHLs. In particular, the internal architecture of "nodular" (follicular) lymphomas is well seen with Snook's or Wilder's stain (Figure 15.53), and the MGP procedure can assist in separating immunoblastic lymphoid cells from activated macrophages. Indeed, in some high-grade NHLs such as "immunoblastic lymphoma" (diffuse large cell B-cell NHL with immunoblastic features), anaplastic large cell lymphoma (ALCL), and Burkitt's lymphoma, virtually all tumor cells exhibit florid MGP positivity (202). In accord with the basic identity of Burkitt's NHL to L3-type ALL, one also can appreciate cytoplasmic vacuolar positivity with ORO or SBB stain using touch-imprint preparations of such tumors. Granulocytic sarcoma (extramedullary myeloid tumor) is always a differential diagnostic consideration in cases of diffuse large cell NHL, but only the former of these lesions is expected to show Leder stain

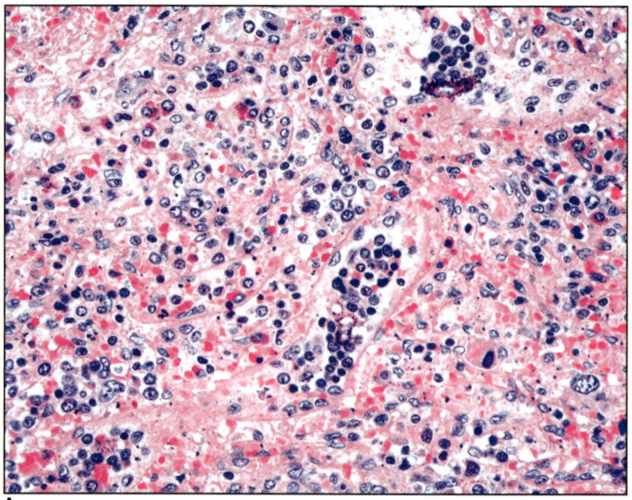

A

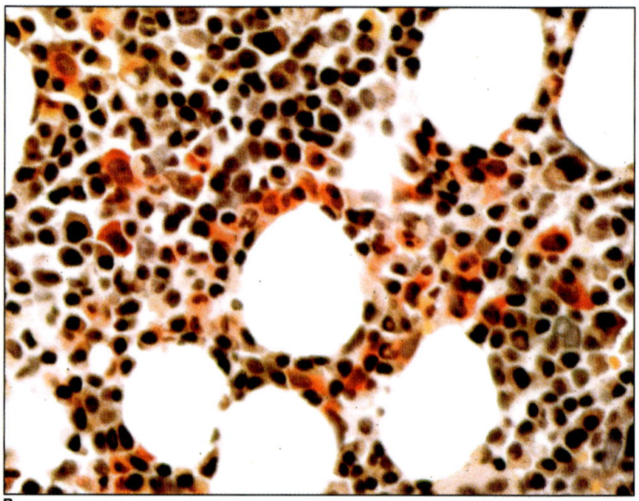

B

FIGURE 15.50: EMH in the paraspinal soft tissue, in a patient with myelofibrosis (A—hematoxylin and eosin). The lesional myeloid cells show positivity with Leder's stain (B).

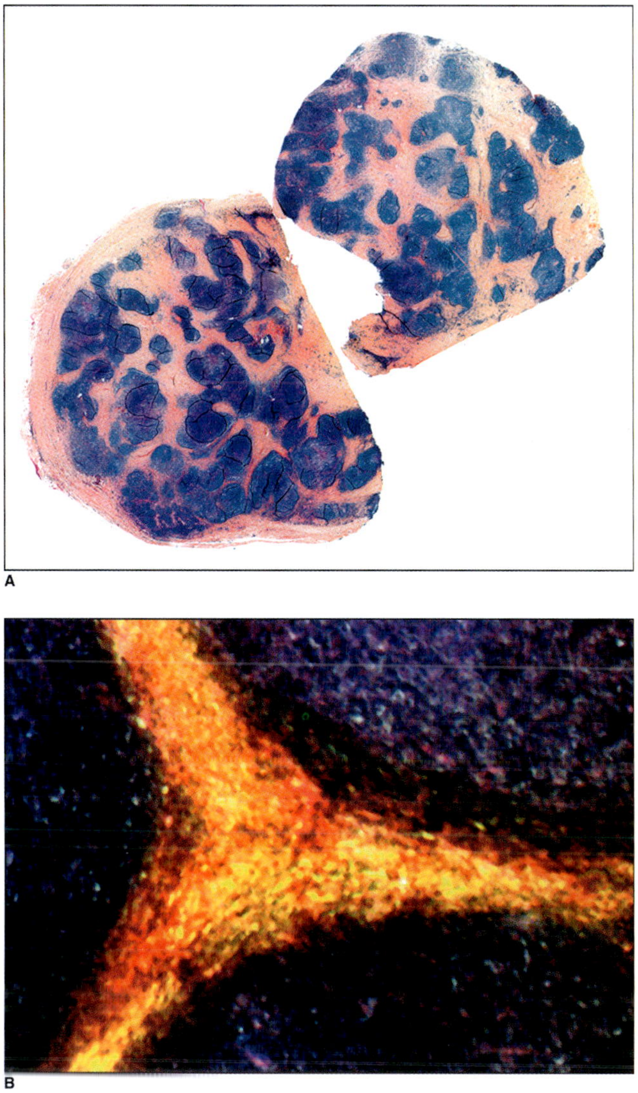

A

B

FIGURE 15.51: ***Nodular sclerosing HL, demonstrating numerous intranodal collagenous bands (A—hematoxylin and eosin). These are birefringent on polarization microscopy (B).***

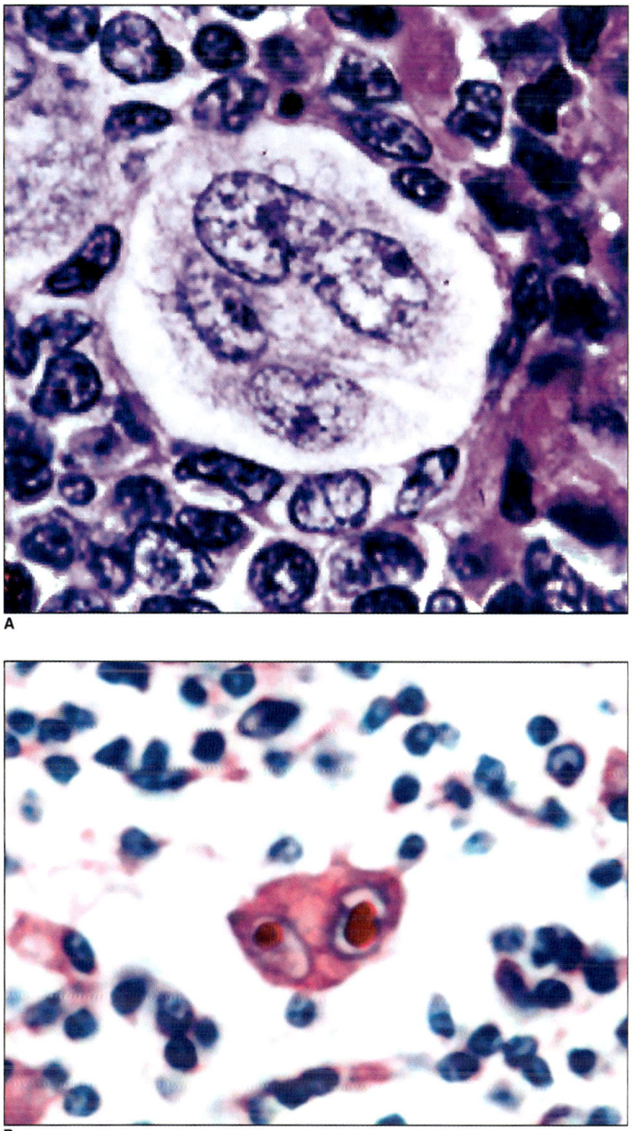

A

B

FIGURE 15.52: ***Reed-Sternberg cells in HL (A—hematoxylin and eosin) can be labeled with the MGP technique (B).***

positivity and potential reactivity with the DAB (tissue peroxidase) method (203,204). Similarly, touch preparations of lesional tissue can be used to assess the presence of TRAP in atypical lymphoid cells, as seen in HCL but not in NHLs (15,156). Otherwise, enzyme cytochemistry of the NHLs has been less productive than in the acute leukemias, in regard to distinguishing between diagnostic categories of lymphoma or use in lineage typing (205).

LECTIN HISTOCHEMISTRY IN HEMATOPATHOLOGY

Plant lectins have been studied rather extensively in regard to their affinity for hematopoietic proliferations. Although none of these moieties is specific for any given disease entity, combinations of them can be employed to good advantage in a differential diagnostic context, especially with respect to malignant large polygonal cell neoplasms (Table 15.6). Large cell lymphomas show reasonably reproducible binding to peanut agglutinin and *Ricinus communis* lectin but not to *Dolichos biflorus*, concanavalin-A, or *Ulex europaeus* I agglutinin (206–208). This profile differs from those attending large cell melanoma and large cell carcinoma (209,210). However, selected epithelioid cell sarcomas may indeed replicate the lectin phenotype of lymphoma.

In the small cell tumor group, this approach is not as helpful. This is true because lymphoblastic lymphoma/leukemia and Burkitt lymphoma/leukemia exhibit lectin binding profiles that essentially recapitulate that of small cell melanoma (210,211). In addition, there is substantial overlap with the pattern seen in small cell carcinomas and rhabdomyosarcoma (212,213).

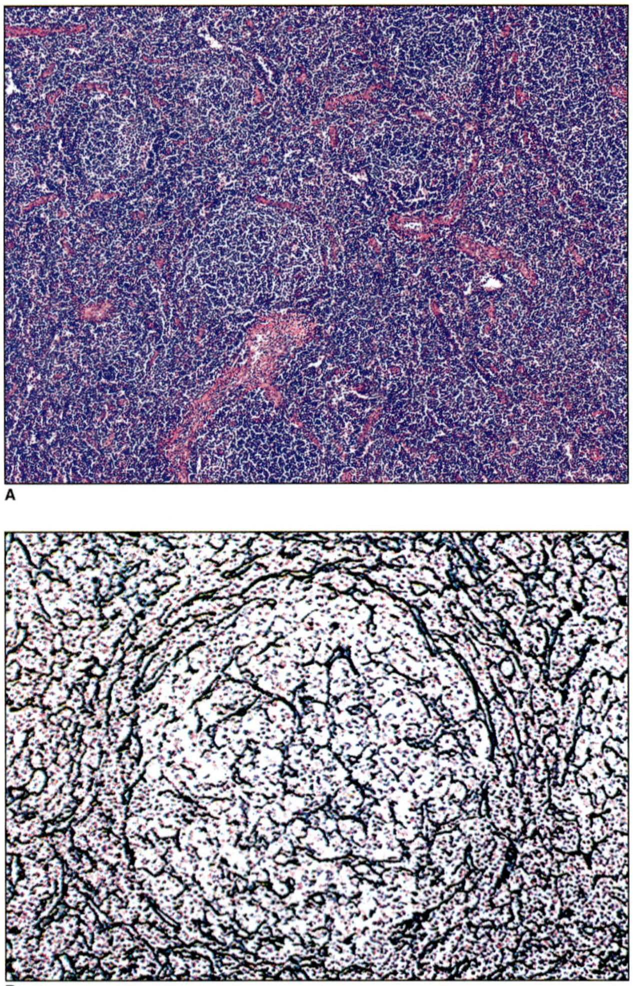

FIGURE 15.53: Grade I follicular NHL (A—hematoxylin and eosin), with demonstration of the nodular architecture with a reticulin stain (B).

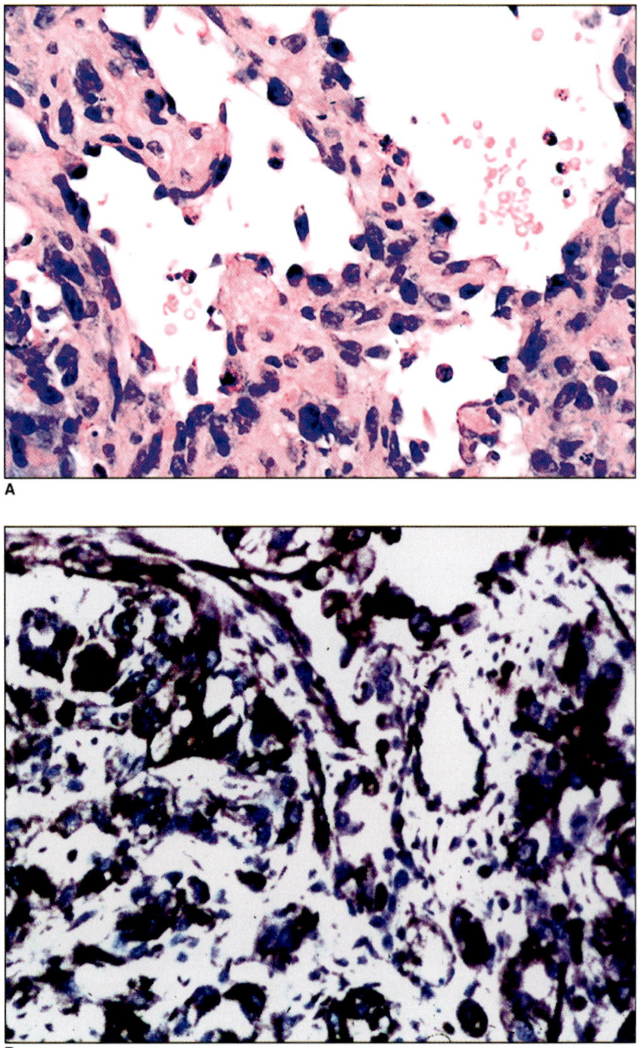

FIGURE 15.54: Splenic angiosarcoma (A—hematoxylin and eosin), showing binding of Ulex europaeus *I lectin (B).*

TABLE 15.6: Lectin Binding Profiles of Hematopoietic Proliferations

Tumor Category	Specific Tumor Type	UEA	PNA	HPA	DBA	CON-A	RCA
Large cell undifferentiated neoplasms	Lymphoma	–	+	+/–	–	–	+
	Melanoma	–	+	–	–	+	+
	Carcinoma	+	+	+/–	+	+	–
	Sarcoma*	–	+	–	–	+/–	+/–
Small cell undifferentiated neoplasms	Lymphoblastic and Burkitt's lymphomas	–	+	–	–	+	+
	Small cell melanoma	–	+	–	–	+	+
	Merkel cell carcinoma	–	+	–	–	+	–
	Visceral small cell carcinoma	–	+	–	–	–	–
	PNET	–	–	–	–	–	+/–
Histiocytic infiltrates	LCH	–	+	–	–	+	+
	Other histiocytoses	–	+/–	–	–	+	+

CON-A = concanavalin-A, DBA = *Dolichos biflorus* agglutinin, HPA = *Helix pomatia* agglutinin, PNA = *Arachis hypogaea* (peanut) agglutinin, PNET = primitive neuroectodermal tumor, RCA = *Ricinus communis* agglutinin, UEA = *Ulex europaeus* I agglutinin.

* Sarcomas typically fail to bind UEA and DBA, with the exception of epithelioid angiosarcoma and epithelioid sarcoma.

Similarly, LCH shows consistent binding to peanut agglutinin (192,193), but the same may be said of other histiocytic proliferations (214,215). Thus, morphologic differences are still paramount in the diagnosis of such lesions.

Angiosarcoma may be observed as a primary splenic malignancy (216). This tumor is reactive with *Ulex europaeus* I agglutinin (Figure 15.54), distinguishing it from other sarcoma morphotypes (217).

ILLUSTRATIVE CASE STUDIES

CASE 1— *Splenic Hairy Cell Leukemia:* A 61-year-old man developed slowly progressive fatigue and left-sided abdominal discomfort. Examination by his physician revealed significant splenomegaly, with the spleen tip being palpable 8 cm beneath the left costal margin. A hemogram was obtained, demonstrating pancytopenia with abnormal circulating lymphoid cells (Figure 15.55); these were positive cytochemically for TRAP (Figure 15.56). Based on this finding and results of a bone marrow biopsy (Figure 15.57), the diagnosis of HCL was made. The patient was treated with 2-chlorodeoxyadenosine, but it failed to eliminate all circulating HCL cells; moreover, organomegaly persisted. Therefore, a splenectomy was done. Histologic examination of this specimen showed HCL cells in the splenic cords and sinusoids, using reticulin and PAS-diastase (PAS-D) methods (Figures 15.58 and 15.59). The patient had symptomatic benefit from the splenectomy; rituximab was subsequently administered, with improvement in his peripheral blood counts. ■

CASE 2— *Splenic Extramedullary Myeloid Tumor:* A 72-year-old woman began to notice early satiety and a feeling of abdominal fullness. Physical examination by her physician detected the presence of splenomegaly, there were no other abnormalities. A hemogram and biochemical survey of the blood yielded normal results. Reexamination after 1 month revealed an increase in the splenic size, and it was removed surgically. Pathologic assessment showed a splenic weight of 550 g, and several parenchymal greenish nodules were apparent in fresh-cut sections (Figure 15.60). Touch-imprint preparations of the lesion demonstrated sheets of blastic cells that were reactive for MPO and SBB (Figure 15.61). The Leder stain was also positive in the neoplastic cells in paraffin sections (Figure 15.62). The diagnosis of granulocytic sarcoma (extramedullary myeloid tumor) was made. Subsequent bone marrow biopsy showed no abnormalities; after 6 months the patient was still well and being monitored carefully. ■

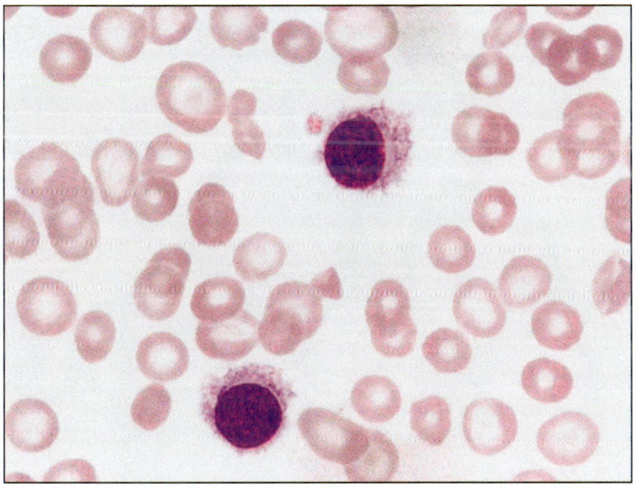

FIGURE 15.55: ***Atypical circulating mononuclear cells, in a blood smear from a patient with HCL.***

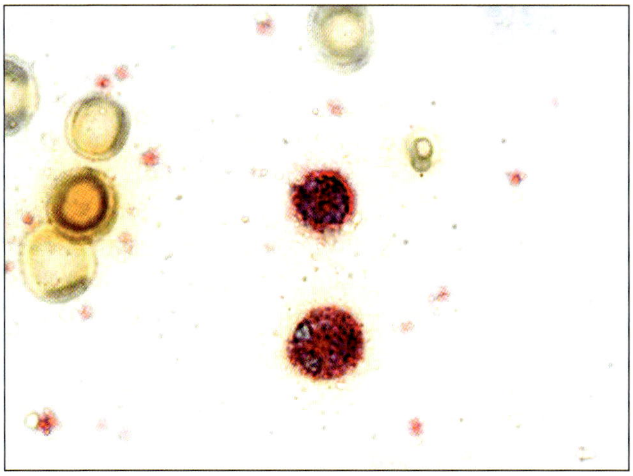

FIGURE 15.56: ***Positivity for TRAP, in circulating cells of HCL.***

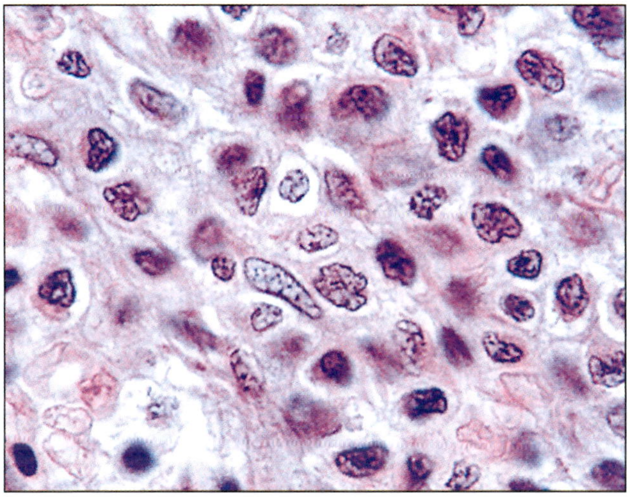

FIGURE 15.57: ***HCL, as seen in a bone marrow biopsy.***

CASE 3— *Metastatic Nodal Pleomorphic Lobular Breast Carcinoma:* A 55-year-old woman noted a mass under

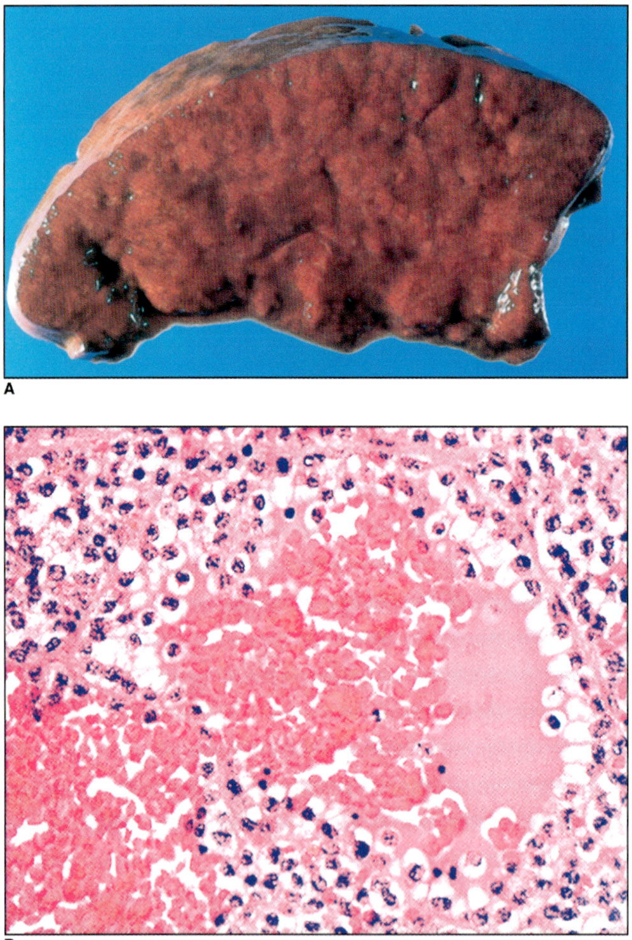

A

B

FIGURE 15.58: (A) Splenectomy specimen from a case of HCL. (B) The tumor cells are present in the splenic cords as well as sinuses.

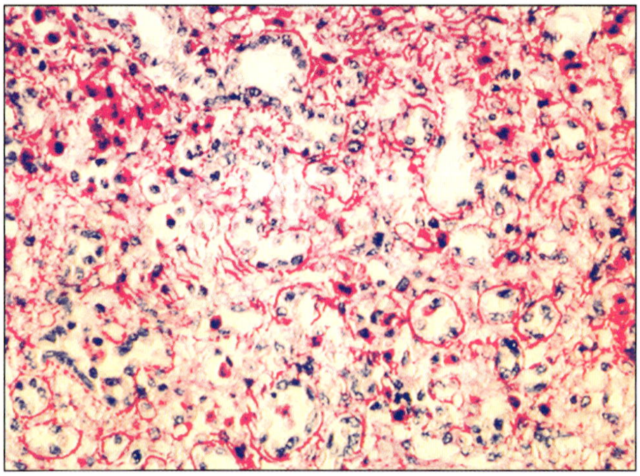

FIGURE 15.59: The splenic cords are expanded by infiltrates of HCL, as shown by this PAS stain. It labels basement membranes of the intervening sinuses.

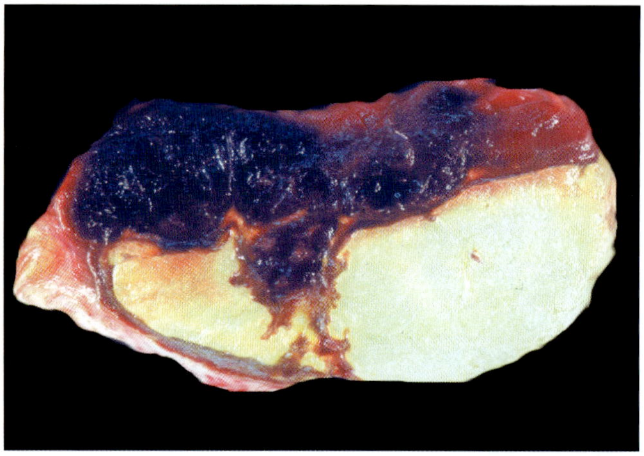

FIGURE 15.60: This fresh-cut section of a splenectomy specimen demonstrates large greenish nodules of tumor.

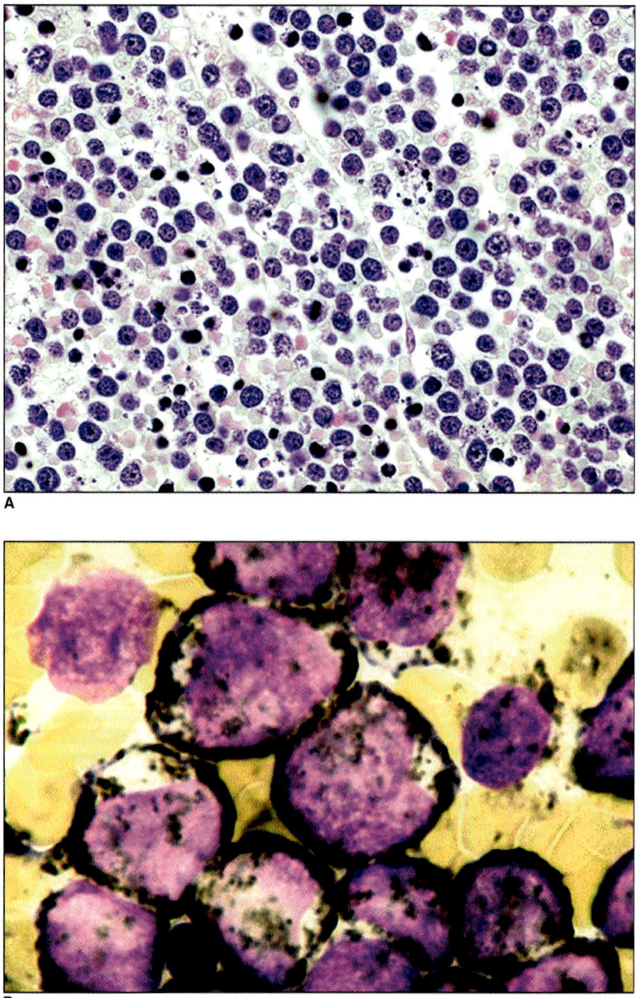

A

B

FIGURE 15.61: A touch preparation of the splenic lesions (see prior figure) shows a sheet of blast-like cells (A—hematoxylin and eosin). They are intensely reactive for MPO (B).

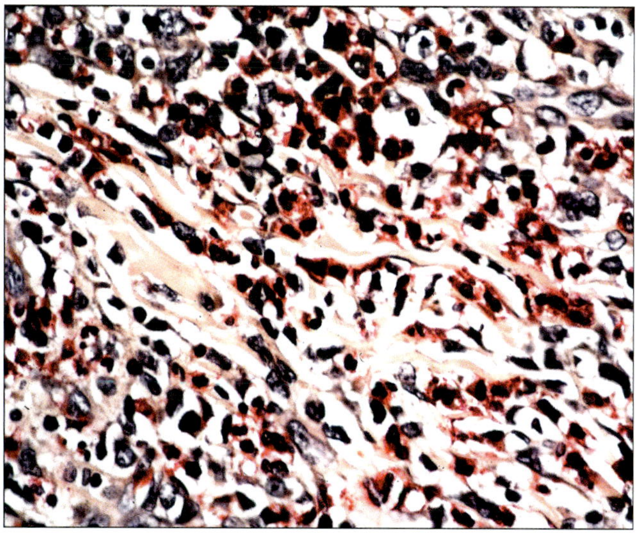

FIGURE 15.62: A Leder stain on a paraffin section of a splenic lesion (see Figure 15.60) demonstrates reactivity in many of the lesional cells, supporting the diagnosis of extramedullary myeloid tumor (granulocytic sarcoma).

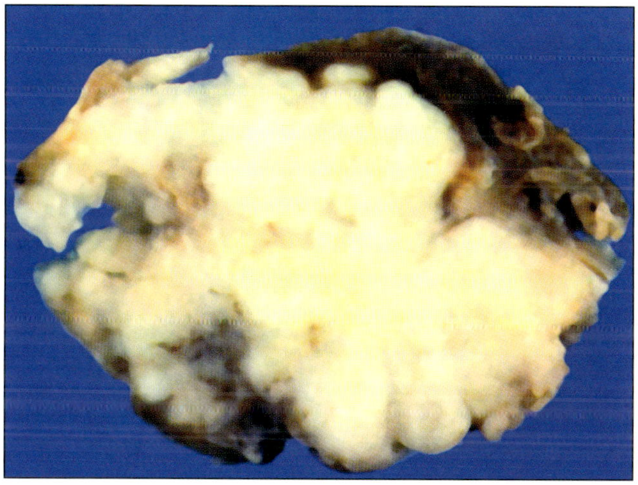

FIGURE 15.63: This axillary lymph node biopsy specimen shows effacement of the normal gross architecture.

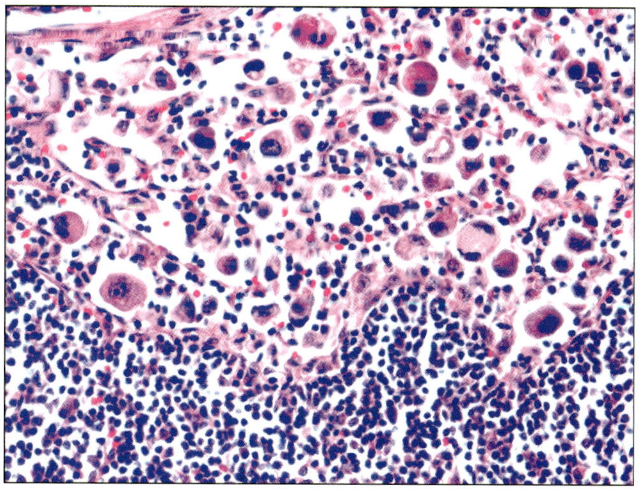

FIGURE 15.64: The nodal sinuses contain dyshesive groups of large atypical cells with pleomorphic nuclei.

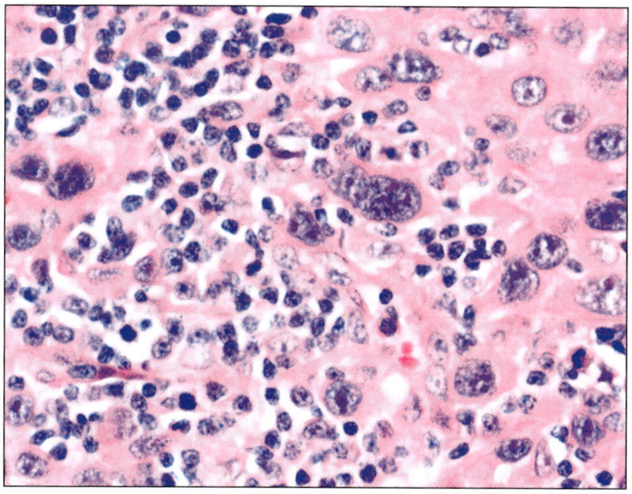

FIGURE 15.65: Multinucleation is apparent in some of the tumor cells. No squamous features or luminal profiles are evident.

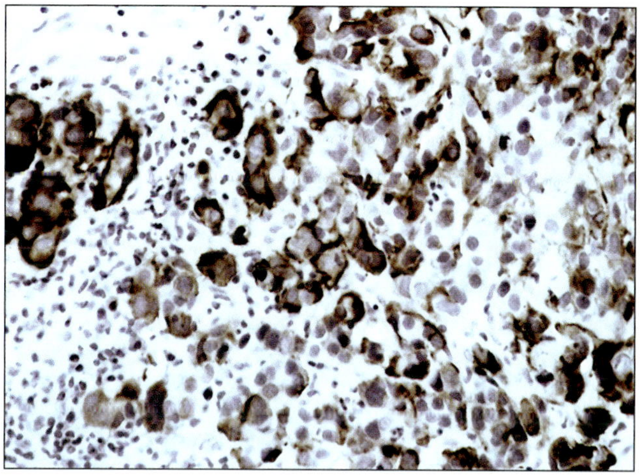

FIGURE 15.66: The neoplastic cells seen in the prior two figures bind to Ulex europaeus *I lectin, supporting the diagnosis of metastatic adenocarcinoma. A tumor was subsequently found in the ipsilateral breast.*

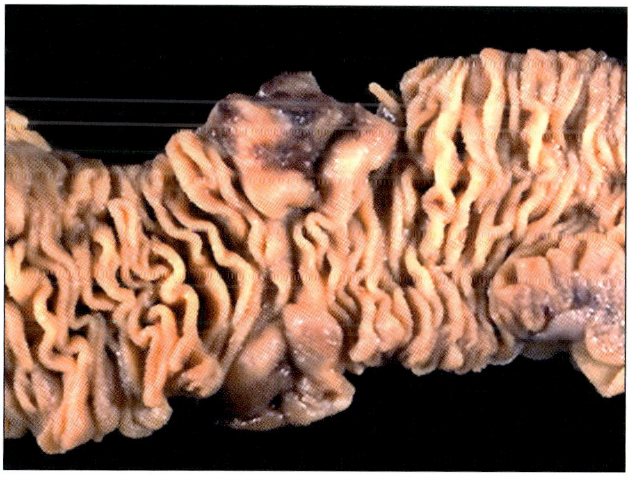

FIGURE 15.67: This partial small bowel resection specimen demonstrates two discrete mucosally based masses, which had caused clinical symptoms of obstruction.

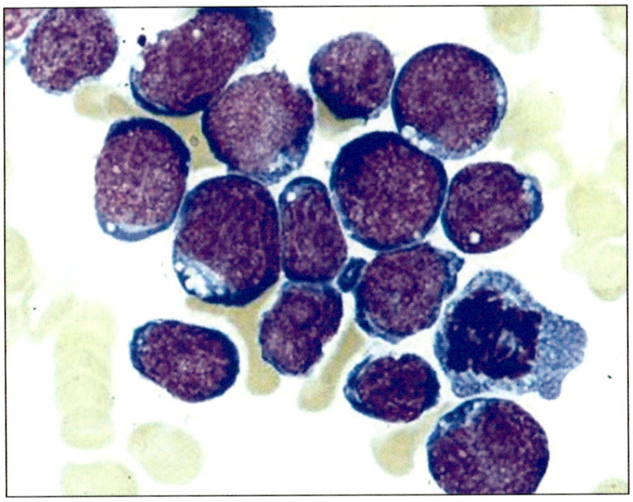

FIGURE 15.68: A touch-imprint preparation of the masses shown in the previous figure demonstrates a sheet of blast-like cells with cytoplasmic vacuoles, as seen in this Wright-Giemsa stain.

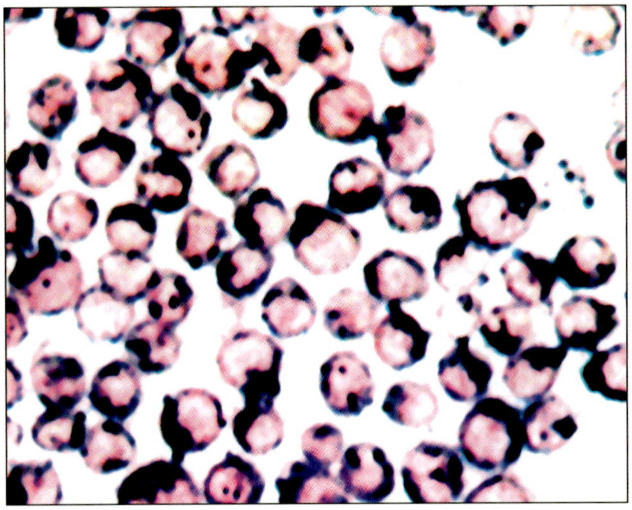

FIGURE 15.69: The tumor cells seen in the previous figure were strongly reactive with SBB.

her left arm while bathing and consulted her physician. Physical examination confirmed the presence of axillary lymphadenopathy; the node was firm to palpation and painless. The patient was referred to a surgeon for an excisional node biopsy. Pathologic examination of the latter (Figure 15.63) disclosed the presence of malignant, pleomorphic, dyshesive polygonal cells, confined to, and expanding the nodal sinuses (Figures 15.64 and 15.65). Mucicarmine and PAS-D stains yielded no reactivity with the tumor cells, but the neoplasm was strongly MGP positive. A preliminary diagnosis of sinusoidal ALCL was made. However, lectin histochemistry subsequently showed uniform intense labeling of the tumor with *U. europaeus* I agglutinin (Figure 15.66). The morphologic interpretation was revised to that of metastatic adenocarcinoma,

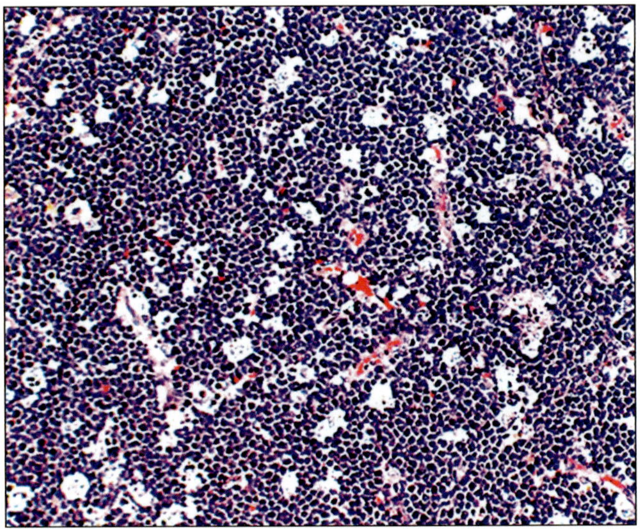

FIGURES 15.70: The neoplasm seen in Figures 15.67–15.69 comprised a uniform population of lymphoid cells with a "starry sky" growth pattern, as shown here in a paraffin section stained with hematoxylin and eosin.

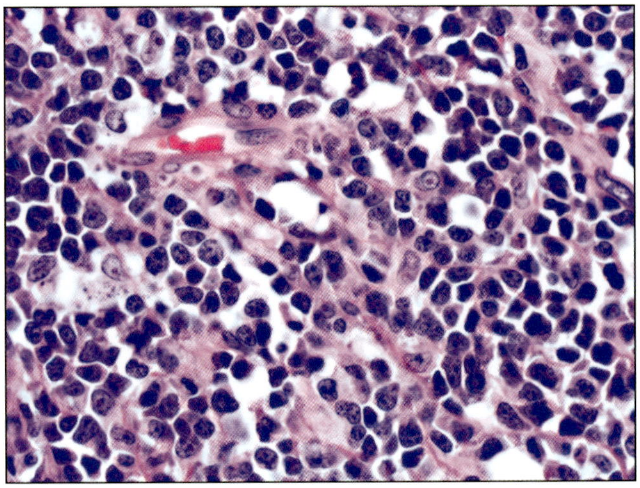

FIGURE 15.71: The tumor cells seen in Figure 15.70 have high nucleocytoplasmic ratios, brisk mitotic activity, and discernible nucleoli. Based on the clinicopathologic presentation, the morphologic features, and the cytochemical attributes of the lesion, a diagnosis of Burkitt lymphoma was made.

simulating ALCL. Mammography subsequently revealed a mass in the upper outer quadrant of the left breast, and a fine-needle aspiration biopsy showed pleomorphic invasive lobular adenocarcinoma. ■

CASE 4— *Extranodal Burkitt's Lymphoma:* A 41-year-old man who had been infected with the human immunodeficiency virus for 15 years became refractory to antiretroviral therapy. His circulating levels of CD4+ lymphocytes were low, and the patient was treated for three opportunistic infections within a 6-month period. He then began to complain of intermittent abdominal pain that was worsened by eating. Radiographic barium-contrast studies of the bowels showed an area of luminal narrowing in the distal ileum, with an associated mass effect. An exploratory laparotomy confirmed this finding, and a partial ileal resection was performed. The resulting specimen (Figure 15.67) showed two mucosal-based masses that thickened the intestinal wall and compromised the lumen. Touch-imprint preparations of the lesions contained blastic cells with cytoplasmic vacuoles (Figure 15.68); they were reactive for SBB (Figure 15.69) and ORO, but the cells were negative for MPO, ACP, NSEs, and CAE. Histologic sections of the masses showed a "starry sky" image (Figure 15.70) and a lesional composition by intermediately sized lymphoid cells with uniformly large nucleoli, high nucleocytoplasmic ratios, and many mitotic figures (Figure 15.71). The diagnosis of extranodal Burkitt's lymphoma was made and the patient was treated with multiagent chemotherapy. However, he subsequently developed bacterial sepsis and died. ■

REFERENCES

1. Wick MR: Paul Ehrlich: the prototypic clinical pathologist. *Am J Clin Pathol* 1988; 90: 329–332.
2. Jay V: Paul Ehrlich. *Arch Pathol Lab Med* 2001; 125: 724–725.
3. Turgeon ML: Clinical Hematology: *Theory & Procedures* (2nd ed.), Little, Brown & Co., Boston, 1993; pp. 15–263.
4. Hyuan BH: *Practical Hematology*, WB Saunders, Philadelphia, 1975; pp. 300–333.
5. Ayatollah M, Tabei Z, Ramzi M, Keshef S, Haghshenas M: A fast and easy nitroblue tetrazolium method for carrier screening and prenatal detection of chronic granulomatous disease. *Arch Iran Med* 2006; 9: 335–338.
6. Johansen J, Frederiksen K: A simple and rapid whole-mount staining method with methyl green-pyronine-G, applied to a molluscan brain preparation. *Stain Technol* 1979; 54: 125–128.
7. Kiernan JA: Methyl green pyronine staining: methyl green source. http://www.histosearch.com/histonet/Apr05A/Re:Histonetmethylgreenpyr.html (accessed 7/6/2007).
8. Gondos B, Miller TR, King EB: Cytologic diagnosis of multiple myeloma and macroglobulinemia in effusions. *Ann Clin Lab Sci* 1978; 8: 11–16.
9. Stuart-Smith SE, Hughes DA, Bain BJ. Are routine iron stains on bone marrow trephine biopsy specimens necessary? *J Clin Pathol* 2005; 58: 269–272.
10. Meloan SN, Puchtler H: A combined periodic acid-Schiff-resorcin-fuchsin stain. *Stain Technol* 1978; 53: 182–183.
11. Crook L, Lin PI, Cannon A, Walker EM: Histochemistry of bone marrow aspirations. *Ann Clin Lab Sci* 1980; 10: 290–304.
12. Elghetany MT, MacCallum JM, Davey FR: The use of cytochemical procedures in the diagnosis and management of acute and chronic myeloid leukemia. *Clin Lab Med* 1990; 10: 707–720.
13. Dekmezian R, Kantarjian HM, Keating MJ, et al.: The relevance of reticulin stain-measured fibrosis at diagnosis in chronic myelogenous leukemia. *Cancer* 1987; 59: 1739–1743.
14. Krzyzaniak RL, Buss DH, Cooper MR, Wells HB: Marrow fibrosis and multiple myeloma. *Am J Clin Pathol* 1988; 89: 63–68.
15. Henry K, Farrer-Brown G: *A Colour Atlas of Thymus & Lymph Node Histopathology*, Wolfe Medical Publishers, Netherlands, 1981; pp. 75–305.
16. Minowada J, Mathe G, Barcos M, et al.: Cytological and immunologic study of 139 patients with acute leukemia. *Med Oncol Tumor Pharmacother* 1984; 1: 3–7.
17. Hann IM, Evans DI, Palmer MK, Morris-Jones PJ, Haworth C: The prognostic significance of morphological features in childhood acute lymphoblastic leukemia. *Clin Lab Haematol* 1979; 1: 215–226.
18. Yagihashi N, Koyama M, Yagihashi S: Rectal adenocarcinoid with lymph node metastasis. *Pathol Int* 1999; 49: 563–565.
19. Resnick JM, Whitley CB, Leonard AS, Krivit W, Snover DC: Light and electron microscopic features of the liver in mucopolysaccharidosis. *Hum Pathol* 1994; 25: 276–286.
20. Buck FS, Koss MN: Hepatic amyloidosis: morphologic differences between systemic AL and AA types. *Hum Pathol* 1991; 22: 904–907.
21. Reed SB, Morris GT: Amyloidosis: current approaches for diagnosis and treatment. *J Ky Med Assoc* 1992; 90: 68–72.
22. Wright JR, Calkins E, Humphrey RL: Potassium permanganate reaction in amyloidosis: a histologic method to assist in differentiating forms of this disease. *Lab Invest* 1977; 36: 274–281.
23. Mbati PA, Abok K, Anjili CO, et al.: Comparison of Giemsa and acridine orange stains for the diagnosis of *Leishmania donovani* in biopsies of infected hamsters. *Afr J Health Sci* 1995; 2: 228–231.
24. Lopez JF, Grocott RG: Demonstration of *Histoplasma capsulatum* in peripheral blood by the use of methenamine silver nitrate stain (Grocott's). *Am J Clin Pathol* 1968; 50: 692–694.
25. Yam LT, Li CY, Crosby WH: Cytochemical identification of monocytes and granulocytes. *Am J Clin Pathol* 1971; 55: 283–290.
26. Engelbreth-Holm J: Pathological aspects of leukemia. *Acta Haematol* 1960; 24: 36–41.
27. Duhamel G: Histological lesions of the bone marrow during the initial period of acute leukemia in adults: study by marrow biopsy. *Presse Med* 1965; 73: 2211–2214.
28. Bennett JM, Catovsky D, Daniel MT, et al.: Proposals for the classification of the acute leukemias: French-American-British (FAB) cooperative group. *Br J Haematol* 1976; 33: 451–458.
29. Kaplow LS, Burstone MS: Cytochemical demonstration of acid phosphatase in hematopoietic cells in health and in various hematological disorders using azo dye techniques. *J Histochem Cytochem* 1964: 12: 805–811.
30. Leder LD: Lymph nodes and bone marrow: recommended diagnostic enzyme-histochemical methods for hematopoietic neoplasms. *Beitr Pathol* 1970; 141: 286–287.

31. Huhn D, Rodt H, Thiel E: Acid phosphatase in acute lymphatic leukemia (ALL). *Haematol Blood Transfus* 1977; 20: 169–170.
32. Catovsky D, Galton DA: Cell markers and the classification of acute leukemia. *Haematol Blood Transfus* 1977; 20: 25–31.
33. Katayama I, Li CY, Yam LT: Ultrastructural cytochemical demonstration of tartrate-resistant acid phosphatase isoenzyme activity in "hairy cells" of leukemic reticuloendotheliosis. *Am J Pathol* 1972; 69: 471–482.
34. Kass L: Cytochemistry of esterases. *CRC Crit Rev Clin Lab Sci* 1979; 10: 205–223.
35. Pinkus GS, Hargreaves HK, McLeod JA, et al.: Alpha-naphthyl acetate esterase activity—a cytochemical marker for T-lymphocytes. Correlation with immunologic studies of normal tissues, lymphocytic leukemias, non-Hodgkin's lymphomas, Hodgkin's disease, and other lymphoproliferative disorders. *Am J Pathol* 1979; 97: 17–41.
36. Pinkus GS, Said JW: Characterization of non-Hodgkin's lymphomas using multiple cell markers: immunologic, morphologic, and cytochemical studies of 72 cases. *Am J Pathol* 1979; 94: 349–380.
37. Leder LD: On the selective enzyme-cytochemical demonstration of neutrophilic myeloid cells and tissue mast cells in paraffin sections. *Klin Wochenschr* 1964; 42: 553.
38. Bourgeois N, Buyssens N: Non-lymphoid hematopoietic malignant disease of the lymph nodes. *Hum Pathol* 1981; 12: 879–886.
39. Beckstead JH, Halverson PS, Ries CA, Bainton DF: Enzyme histochemistry and immunohistochemistry on biopsy specimens of pathologic human bone marrow. *Blood* 1981; 57: 1088–1098.
40. Koeffler HP, Ranyard J, Pertscheck M: Myeloperoxidase: its structure and expression during myeloid differentiation. *Blood* 1985; 65: 484–491.
41. McCarty KS Jr., Wortman J, Daly J, Rundles RW, Hanker JS: Chloroma (granulocytic sarcoma) without evidence of leukemia: facilitated light microscopic diagnosis. *Blood* 1980; 56: 104–108.
42. Plaut D: Timely topics in clinical chemistry: enzymes and isoenzymes. *Am J Med Technol* 1969; 35: 320–326.
43. Warnock ML: Characterization of tissue and serum alkaline phosphatases. *Clin Chim Acta* 1966; 14: 156–165.
44. Meisliin AG, Lee SL, Wasserman LR: Leukocyte alkaline phosphatase activity in hematopoietic disorders: a correlative study by biochemical and cytochemical techniques. *Cancer* 1959; 12: 760–766.
45. Rubin H: Chronic neutrophilic leukemia. *Ann Intern Med* 1966; 65: 93–100.
46. Sackey K: Hemolytic anemia: Part 1. *Pediatr Rev* 1999; 20: 152–158.
47. Verma M, Singla D, Crowell SB: Glucose-6-phosphate dehydrogenase in neonates: a prospective study. *Indian J Pediatr* 1990; 57: 385–388.
48. Bloom KE, Brewer GJ, Magon AM, Wetterstroem N: Microsomal incubation test of potentially hemolytic drugs for glucose-6-phosphate dehydrogenase deficiency. *Clin Pharmacol Ther* 1983; 33: 403–409.
49. Hobbs J, Wright CS: The hemolytic anemias. *Am Fam Physician* 1979; 20: 83–93.
50. Tabbara IA: Hemolytic anemias: diagnosis and management. *Med Clin North Am* 1992; 76: 649–668.
51. Marik T, Kodicek M, Brabec V: Crude spectrin extraction from reticulocyte-rich blood samples. *Clin Chim Acta* 1985; 153: 15–21.
52. Killip S, Bennett JM, Chambers MD: Iron deficiency anemia. *Am Fam Physician* 2007; 75: 671–678.
53. Ganti AK, Moazzam N, Laroia S, et al.: Predictive value of absent bone marrow iron stores in the clinical diagnosis of iron deficiency anemia. *In Vivo* 2003; 17: 389–392.
54. Thomas L: Anemia of chronic disease—pathophysiology and laboratory diagnosis. *Lab Hematol* 2004; 10: 163–165.
55. Krantz SB: Pathogenesis and treatment of the anemia of chronic disease. *Am J Med Sci* 1994; 307: 353–359.
56. Junca J, Fernandez-Aviles F, Oriol A, et al.: The usefulness of the serum transferrin receptor in detecting iron deficiency in the anemia of chronic disorders. *Haematologica* 1998; 83: 676–680.
57. Kushner JP, Cartwright GE: Sideroblastic anemia. *Adv Intern Med* 1977; 22: 229–249.
58. Tham KT, Cousar JB, Macon WR: Silver stain for ringed sideroblasts: a sensitive method that differs from Perl's reaction in mechanism and clinical application. *Am J Clin Pathol* 1990; 94: 73–76.
59. Tham KT, Cousar JB: Combined silver-Perl's stain for differential staining of ringed sideroblasts and marrow iron. *J Clin Pathol* 1993; 46: 766–768.
60. Timms KM, Bondeson ML, Ansari-Lari MA, et al.: Molecular phenotypic variation in patients with severe Hunter syndrome. *Haem Molec Genet* 1997; 6: 479–486.
61. Gocer C, Linthicum FH Jr.: Hurler disease. *Otol Neurotol* 2004; 25: 81–82.
62. Bassyouni HT, Afifi HH, El-Awadi MK, Meguid NA: Mucopolysaccharidosis type I: clinical and biochemical study. *East Mediterr Health J* 2000; 6: 359–366.
63. Nagashima K, Endo H, Sakakibara K, et al.: Morphological and biochemical studies of a case of mucopolysaccharidosis II (Hunter's syndrome). *Acta Pathol Jpn* 1976; 26: 115–132.
64. Clarke JT, Willard HF, Teshima I, Chang PL, Skomorowski MA: Hunter disease (mucopolysaccharidosis type II) in a karyotypically-normal girl. *Clin Genet* 1990; 37: 355–362.
65. Peterson L, Parkin J, Nelson A: Mucopolysaccharidosis type VII: a morphological, cytochemical, and ultrastructural study of the blood and bone marrow. *Am J Clin Pathol* 1982; 78: 5454–548.
66. Koranyi G, Rajk A: Sphingomyelin lipoidosis. *Monatsschr Kinderheilkd* 1976; 124: 712–716.
67. Elleder M, Hrodek J, Cihula J: Niemann-Pick disease: lipid storage in bone marrow macrophages. *Histochem J* 1983; 15: 1065–1077.
68. Chikh K, Vey S, Simonot C, Vanier MT, Millat G: Niemann-Pick type C disease: importance of N-glycosylation sites for function and cellular location of the NPC2 protein. *Mol Genet Metab* 2004; 83: 220–230.
69. Elleder M, Lojda Z: Studies in lipid histochemistry. XII: Histochemical detection of sphingomyelin. *Histochemie* 1973; 37: 371–373.

70. Lake BD: A histochemical study of Gaucher's disease and Niemann-Pick's disease. *J R Microsc Soc* 1967; 86: 417–425.
71. Parkin JL, Brunning RD: Pathology of the Gaucher cell. *Prog Clin Biol Res* 1982; 95: 151–175.
72. Elleder M, Lojda Z: Histochemical diagnosis of lipidoses. *Acta Univ Carol Med Monogr* 1977; 79: 117–123.
73. Tabak L, Yilmazbaydan D, Killicaslan Z, Tascioglu C, Agan M: Value of bronchoalveolar lavage in lipidoses with pulmonary involvement. *Eur Respir J* 1994; 7: 409–411.
74. Ward PC: The myeloid leukocytoses. *Postgrad Med* 1980; 67: 219–229.
75. Ramos FJ, Zamora F, Perez-Sicilia M, Sang MA, del Villar R: Chronic granulocytic leukemia versus neutrophilic leukemoid reaction. *Am J Med* 1990; 88: 83–84.
76. Lazar GS, Forman SJ: Leukoerythroblastic reaction in Still's disease in an adult. *West J Med* 1979; 131: 152–155.
77. Dingli D, Mesa RA, Tefferi A: Myelofibrosis with myeloid metaplasia: new developments in pathogenesis and treatment. *Intern Med* 2004; 43: 540–547.
78. Okun DB, Tanaka KR: Leukocyte alkaline phosphatase. *Am J Hematol* 1978; 4: 293–299.
79. Rice L, Shenkenberg T, Lynch EC, Wheeler TM: Granulomatous infections complicating hairy cell leukemia. *Cancer* 1982; 49: 1924–1928.
80. Al-Eissa Y, Al-Nasser M: Haematological manifestations of childhood brucellosis. *Infection* 1993; 21: 23–26.
81. Roelandt PR, Dendooven M, DeGroef K, Van Couter A: Hidden tuberculosis revealed by bone marrow biopsy as a cause of fever of unknown origin. *Acta Clin Belg* 2006; 61: 363–366.
82. Kumar PV, Vasei M, Sadeghipour A, et al.: Visceral leishmaniasis: bone marrow biopsy findings. *J Pediatr Hematol Oncol* 2007; 29: 77–80.
83. Basu D, Saravana R, Purushotham B, Ghotakar LH: Granulomas in bone marrow—a study of fourteen cases. *Indian J Pathol Microbiol* 2005; 48: 13–16.
84. Chang KL, Gaal KK, Huang Q, Weiss LM: Histiocytic lesions involving the bone marrow. *Semin Diagn Pathol* 2003; 20: 226–236.
85. Browne PM, Sharma OP, Salkin D: Bone marrow sarcoidosis. *JAMA* 1978; 240: 2654–2655.
86. Krobar SM, Kaiserling E, Horny HP, Weber A: Primary diagnosis of Whipple's disease in bone marrow. *Hum Pathol* 2004; 35: 522–525.
87. Foucar K: *Bone Marrow Pathology* (2nd ed.), ASCP Press, Chicago, 2001; pp. 25–350.
88. Loffler H: Morphology, immunology, cytochemistry, and cytogenetics and the classification of subtypes of acute myelogenous leukemia. *Haematol Blood Transfus* 1990; 33: 239–242.
89. Loffler H, Gassmann W: Morphology and cytochemistry of acute lymphoblastic leukemia. *Baillieres Clin Haematol* 1994; 7: 263–272.
90. McKenna RW, Brynes RK, Nesbit ME, et al.: Cytochemical profiles in acute lymphoblastic leukemia. *Am J Pediatr Hematol Oncol* 1979; 1: 263–275.
91. McKenna RW: Multifaceted approach to the diagnosis and classification of acute leukemias. *Clin Chem* 2000; 46: 1252–1259.
92. Rosenthal NS, Farhi DC: Special stains in the diagnosis of acute leukemia. *Clin Lab Med* 2000; 20: 29–38.
93. Gaedicke G, Drexler HG: The use of enzyme marker analysis for subclassification of acute lymphocytic leukemia in childhood. *Leuk Res* 1982; 6: 437–448.
94. Savage RA, Valenzuela R, Hoffman GC: Acid phosphatase staining pattern as an indicator of T-cell acute leukemia. *Am J Clin Pathol* 1981; 76: 760–764.
95. Knowles DM II, Halper JP, Machin GA, Sherman W: Acid alpha-naphthyl acetate esterase activity in human neoplastic lymphoid cells: usefulness as a T-cell marker. *Am J Pathol* 1979; 96: 257–277.
96. Stein P, Peiper S, Butler D, et al.: Granular acute lymphoblastic leukemia. *Am J Clin Pathol* 1983; 79: 426–430.
97. Grogan TM, Insalaco SJ, Savage RA, Vail ML: Acute lymphocytic leukemia with prominent azurophilic granulation and punctuate acidic nonspecific esterase and phosphatase activity. *Am J Clin Pathol* 1981; 75: 716–722.
98. Tricot G, Broeckaert-Van Orshoven A, Van Hoof A, Verwilghen RL: Sudan Black B positivity in acute lymphoblastic leukemia. *Br J Haematol* 1982; 51: 615–621.
99. Bennett JM: World Health Organization classification of the acute leukemias and myelodysplastic syndrome. *Int J Hematol* 2000; 72: 131–133.
100. Catovsky D, Matutes E: The classification of acute leukemia. *Leukemia* 1992; 6(Suppl 2): 1–6.
101. Sharma JS, Mohindroo S: FAB classification of leukemia: a cytochemical study. *Indian J Pathol Microbiol* 2004; 47: 336–339.
102. Catovsky D: Classification of leukemia. I. The classification of acute leukemia. *Pathology* 1982; 14: 277–281.
103. Castoldi GL, Liso V, Specchia G, Tomasi P: Acute promyelocytic leukemia: morphological aspects. *Leukemia* 1994; 8: 1441–1446.
104. Hayhoe FG: Cytochemistry of the acute leukaemias. *Histochem J* 1984; 16: 1051–1059.
105. Wick MR, Li CY, Pierre RV: Acute nonlymphocytic leukemia with basophilic differentiation. *Blood* 1982; 60: 38–45.
106. Dunphy CH: Comprehensive review of adult acute myelogenous leukemia: cytomorphological, enzyme-cytochemical, flow-cytometric, immunophenotypic, and cytogenetic findings. *J Clin Lab Anal* 1999; 13: 19–26.
107. Li CY, Phyliky RL, Yam LT: Acute myelomonocytic leukemia: an unusual variant with both granulocytic and monocytic esterases in the leukemic cells. *Mayo Clin Proc* 1986; 61: 104–109.
108. Stark EW, Garvin DF, Surana RB, et al.: Acute myelomonocytic leukemia (M4 subtype) with abnormal marrow eosinophilia and a normal chromosome 16. *Am J Clin Pathol* 1985; 84: 551–554.
109. Domingo-Claros A, Larriba I, Rozman M, et al.: Acute erythroid neoplastic proliferations: a biological study based on 62 patients. *Haematologica* 2002; 87: 148–153.
110. Breton-Gorious J, Villeval JL, Mitjavila MT, et al.: Ultrastructural and cytochemical characterization of blasts from early erythroblastic leukemias. *Leukemia* 1987; 1: 173–181.

111. Davey FR, Abraham N, Brunetto VL, et al.: Morphologic characteristics of erythroleukemia (acute myeloid leukemia, FAB-M6): a CALGB study. *Am J Hematol* 1995; 49: 29–38.
112. Gassmann W, Loffler H: Acute megakaryoblastic leukemia. *Leuk Lymphoma* 1995; 18(Suppl 1): 69–73.
113. Shukla J, Rai S, Singh VP: Acute megakaryoblastic leukemia: a clinicohematological profile of five cases. *Indian J Pathol Microbiol* 2004; 47: 266–268.
114. Li CY: The role of morphology, cytochemistry, and immunohistochemistry in the diagnosis of chronic myeloproliferative diseases. *Int J Hematol* 2002; 76(Suppl 2): 6–8.
115. Quintas-Cardama A, Cortes JE: Chronic myeloid leukemia: diagnosis and treatment. *Mayo Clin Proc* 2006; 81: 973–988.
116. Shepherd PC, Ganesan TS, Galton DA: Haematological classification of the chronic myeloid leukaemias. *Baillieres Clin Haematol* 1987; 1: 887–906.
117. Nemoto K, Honma K, Fukuda T, Ohnishi Y: Blast crisis of chronic myelogenous leukemia: morphological and immunological features. *Acta Pathol Jpn* 1986; 36: 1441–1454.
118. Melo JV, Barnes DJ: Chronic myeloid leukemia as a model of disease evolution in human cancer. *Nat Rev Cancer* 2007; 7: 441–453.
119. Michiels JJ, De Raeve H, Hebeda K, et al.: World Health Organization bone marrow features and European clinical, molecular, and pathological (ECMP) criteria for the diagnosis of myeloproliferative disorders. *Leuk Res* 2007; 31: 1031–1038.
120. Michiels JJ, Bernema Z, Van Bockstaele D, De Raeve H, Schroyens W: Current diagnostic criteria for the chronic myeloproliferative disorders (MPD): essential thrombocythemia (ET), polycythemia vera (PV), and chronic idiopathic myelofibrosis (CIMF). *Pathol Biol* 2007; 55: 92–104.
121. Orazi A: Histopathology in the diagnosis and classification of acute myeloid leukemia, myelodysplastic syndromes, and myelodysplastic-myeloproliferative diseases. *Pathobiology* 2007; 74: 97–114.
122. Nair R, Iyer RS, Nair CN, et al.: Myelodysplastic syndromes: a clinical and pathological analysis of 88 patients. *Indian J Cancer* 1993; 30: 169–175.
123. D'Angelo G: Refractory anemia with ringed sideroblasts and chronic myelomonocytic leukemia: myelodysplastic/myeloproliferative disease. *Lab Hematol* 2005; 11: 171–173.
124. Lau LG, Chng WJ, Liu TC, et al.: Clinicopathological analysis of myelodysplastic syndromes according to French-American-British classification and international prognostic scoring system. *Ann Acad Med Singapore* 2004; 33: 589–595.
125. Steensma DP, Bennett JM: The myelodysplastic syndromes: diagnosis and treatment. *Mayo Clin Proc* 2006; 81: 104–130.
126. Thiele J, Kvasnicka HM: Myelofibrosis in chronic myeloproliferative disorders—dynamics and clinical impact. *Histol Histopathol* 2006; 21: 1367–1378.
127. Michiels JJ, Kutti J, Stark P, et al.: Diagnosis, pathogenesis, and treatment of the myeloproliferative disorders: essential thrombocythemia, polycythemia vera, and essential megakaryocytic granulocytic metaplasias and myelofibrosis. *Neth J Med* 1999; 54: 46–62.
128. Yee KWL, O'Brien SM: Chronic lymphocytic leukemia: diagnosis and treatment. *Mayo Clin Proc* 2006; 81: 1105–1129.
129. Ravandi F, O'Brien SM: Chronic lymphoid leukemias other than chronic lymphocytic leukemia: diagnosis and treatment. *Mayo Clin Proc* 2005; 80: 1660–1674.
130. Janckila AJ, Li CY, Lam KW, Yam LT: The cytochemistry of tartrate-resistant acid phosphatase: technical considerations. *Am J Clin Pathol* 1978; 70: 45–55.
131. Nanba K, Jaffe ES, Soban EJ, Braylan RC, Berard CW: Hairy cell leukemia: enzyme-histochemical characterization, with special reference to splenic stromal changes. *Cancer* 1977; 39: 2323–2336.
132. Thiele J, Zirbes TK, Kvasnicka HM, Fischer R: Focal lymphoid aggregates (nodules) in bone marrow biopsies: differentiation between benign hyperplasia and malignant lymphoma—a practical guideline. *J Clin Pathol* 1999; 52: 294–300.
133. Thiele J, Kvasnicka HM: Reactive lymphoid nodule or malignant lymphoma: a brief guideline for the differential diagnosis of bone marrow infiltrates in non-Hodgkin's lymphomas. *Pathologe* 2000; 21: 16–23.
134. Thiele J, Langohr J, Skorupka M, Fischer R: Reticulin fibre content of bone marrow infiltrates of malignant non-Hodgkin's lymphomas (B-cell type, of low malignancy)—a morphometric evaluation before and after therapy. *Virchows Arch A Pathol Anat Histopathol* 1990; 417: 485–492.
135. Burston J, Pinniger JL: The reticulin content of bone marrow in haematological disorders. *Br J Haematol* 1963; 9: 172–184.
136. Dimmock VM: A rapid method for methyl green-pyronine staining of blood and bone marrow smears. *Med Lab Sci* 1977; 34: 173–174.
137. Lugassy G, Vorst EJ, Varon D, et al.: Carcinocythemia: report of two cases, one simulating a Burkitt lymphoma. *Acta Cytol* 1990; 34: 265–268.
138. Carey RW, Taft PD, Bennett JM, Kaufman S: Carcinocythemia (carcinoma cell leukemia): an acute leukemia-like picture due to metastatic carcinoma cells. *Am J Med* 1976; 60: 273–278.
139. Kumar L, Majhi U, Shanta V: Frequency of bone marrow involvement in non-haematological malignancies. *J Assoc Physicians India* 1990; 38: 553–555.
140. Anner RM, Drewinko B: Frequency and significance of bone marrow involvement by metastatic solid tumors. *Cancer* 1977; 39: 1337–1344.
141. Rajkumar SV, Kyle RA: Multiple myeloma: diagnosis and treatment. *Mayo Clin Proc* 2005; 80: 1371–1382.
142. Hayhoe FG, Neuman Z: Cytological, cytometric, and cytochemical studies in myeloma. *Br J Haematol* 1968; 15: 320–321.
143. Sundstrom C, Nilsson K: Cytochemical profile of human hematopoietic cells and derived cell lines. *Br J Haematol* 1977; 19: 5–13.
144. Grozdea J, De Boissezon JF, Guiraud M, Martin J: Periodic acid-Schiff staining and immunoglobulin classes in myeloma. *Lancet* 1980; 2: 1036.
145. Nomura S, Kanoh T, Uchino H: Intracellular formation of amyloid fibrils in myeloma: cytochemical, immunochemical, and electron microscopic observations. *Cancer* 1984; 54: 303–307.

146. Limas C, Wright JR, Matsuzaki M, Calkins E: Amyloidosis and multiple myeloma: a reevaluation using a control population. *Am J Med* 1973; 54: 166–173.
147. Krause JR: Value of bone marrow biopsy in the diagnosis of amyloidosis. *South Med J* 1977; 70: 1072–1079.
148. Kyle RA: Clinical aspects of multiple myeloma and related disorders, including amyloidosis. *Pathol Biol* 1999; 47: 148–157.
149. Horny HP, Sotlar K, Valent P: Mastocytosis: state of the art. *Pathobiology* 2007; 74: 121–132.
150. Valent P, Akin C, Sperr WR, et al.: Mastocytosis: pathology, genetics, and current options for therapy. *Leuk Lymphoma* 2005; 46: 35–48.
151. Li CY: Diagnosis of mastocytosis: value of cytochemistry and immunohistochemistry. *Leuk Res* 2001; 25: 537–541.
152. Browarsky IL, Lotz MJ: Systemic mast cell disease. *South Med J* 1983; 76: 508–511.
153. Travis WD, Li CY, Bergstralh EJ, Yam LT, Swee RG: Systemic mast-cell disease: analysis of 58 cases and literature review. *Medicine* 1988; 67: 345–368.
154. Li CY, Baek JY: Mastocytosis and fibrosis: role of cytokines. *Int Arch Allergy Immunol* 2002; 127: 123–126.
155. Horny HP, Parwaresch MR, Lennert K: Bone marrow findings in systemic mastocytosis. *Hum Pathol* 1985; 16: 808–814.
156. Jaffe ES (Ed.): *Surgical Pathology of the Lymph Nodes & Related Organs* (2nd ed.), WB Saunders, Philadelphia, 1995; pp. 98–533.
157. Wilkins BS, Wright DH: *Illustrated Pathology of the Spleen*, Cambridge University Press, Cambridge, UK, 2000; pp. 77–176.
158. Neiman RS, Orazi A: *Disorders of the Spleen* (2nd ed.), WB Saunders, Philadelphia, 1999; pp. 65–286.
159. Lazarus AA, Thilagar B: Tuberculous lymphadenitis. *Dis Mon* 2007; 53: 10–15.
160. Patel NC, Minifee PK, Dishop MK, Munoz FM: *Mycobacterium simiae* cervical lymphadenitis. *Pediatr Infect Dis J* 2007; 26: 362–363.
161. Polesky A, Grove W, Bhatia G: Peripheral tuberculous lymphadenitis: epidemiology, diagnosis, treatment, and outcome. *Medicine* 2005; 84: 350–362.
162. Bagla N, Patel MM, Patel RD, Jarag M: Lepromatous lymphadenitis masquerading as lymphoma. *Lepr Rev* 2005; 76: 87–90.
163. Chakrabarty AK, Maire MA, Lambert PH: SDS-PAGE analysis of *Mycobacterium leprae* protein antigens reacting with antibodies from sera from lepromatous patients and infected armadillos. *Clin Exp Immunol* 1982; 49: 523–531.
164. Kraus Z, Vortel V, Vanasek J: Diagnostic errors in syphilitic lymphadenitis—importance of the histological examination of the node for a correct diagnosis. *Cesk Dermatol* 1974; 49: 201–205.
165. Vortel V, Krauz Z, Vanasek J: Syphilitic lymphadenitis. *Cesk Patol* 1975; 11: 72–77.
166. Choi YJ, Reiner L: Syphilitic lymphadenitis: immunofluorescent identification of spirochetes from imprints. *Am J Surg Pathol* 1979; 3: 553–555.
167. Margileth AM: Recent advances in diagnosis and treatment of cat scratch disease. *Curr Infect Dis Rep* 2000; 2: 141–146.
168. Miller-Catchpole R, Variakojis D, Vardiman JL, Loew JM, Carter J: Cat-scratch disease: identification of bacteria in seven cases of lymphadenitis. *Am J Surg Pathol* 1986; 10: 276–281.
169. Barbyal RS, Basu D, Roy S, Kumar P: Cryptococcal lymphadenitis: report of a case with fine needle aspiration cytology. *Acta Cytol* 2005; 49: 58–60.
170. Azadeh B: "Localized" leishmania lymphadenitis: a light and electron microscopic study. *Am J Trop Med Hyg* 1985; 34: 447–455.
171. Miliauskas JR, Leong ASY: Localized *Herpes simplex* lymphadenitis: report of three cases and review of the literature. *Histopathology* 1991; 19: 355–360.
172. Sinha AK, Covett M, Pillay G: Cytomegalovirus infection with lymphadenopathy. *Br Med J* 1970; 3: 163.
173. Ro JY, Luna MA, Mackay B, Ramos O: Yellow-brown (Hamazaki-Wesenberg) bodies mimicking fungal yeasts. *Arch Pathol Lab Med* 1987; 111: 555–559.
174. Seiracki JC, Fisher ER: The ceroid nature of the so-called "Hamazaki-Wesenberg bodies." *Am J Clin Pathol* 1973; 59: 248–253.
175. Fakan F: Chronic granulomatosis in childhood. *Cesk Patol* 1977; 13: 101–107.
176. Tanaka T, Takahashi K, Morita H, et al.: Chronic granulomatous disease of childhood and sea-blue histiocytosis: a pathologic study of an autopsy case. *Acta Pathol Jpn* 1984; 34: 1385–1401.
177. Bianco P: A histochemical study of lipoid pigment storage in chronic granulomatous disease. *Basic Appl Histochem* 1983; 27: 35–43.
178. Burke JS: Reactive lymphadenopathies. *Semin Diagn Pathol* 1988; 5: 312–316.
179. Nosanchuk JS, Schnitzer B: Follicular hyperplasia in lymph nodes from patients with rheumatoid arthritis. *Cancer* 1969; 24: 343–354.
180. Butler JJ: Non-neoplastic lesions of lymph nodes of man to differentiated from lymphomas. *Natl Cancer Inst Monogr* 1969; 32: 233–255.
181. Kitamura H, Kitamura H, Ito T, Kanisawa M, Kato K: Systemic lupus erythematosus with multiple calcified fibrous nodules of the spleen. *Acta Pathol Jpn* 1985; 35: 213–226.
182. Foucar E, Rosai J, Dorfman RF: Sinus histiocytosis with massive lymphadenopathy (Rosai-Dorfman disease): review of the entity. *Semin Diagn Pathol* 1990; 7: 19–73.
183. Foucar E, Rosai J, Dorfman RF, Eyman JM: Immunologic abnormalities and their significance in sinus histiocytosis with massive lymphadenopathy. *Am J Clin Pathol* 1984; 82: 515–525.
184. Gould E, Porto R, Albores-Saavedra J, Ibe MJ: Dermatopathic lymphadenitis: the spectrum and significance of its morphologic features. *Arch Pathol Lab Med* 1988; 112: 1145–1150.
185. Burke JS, Colby TV: Dermatopathic lymphadenopathy: comparison of cases associated and unassociated with mycosis fungoides. *Am J Surg Pathol* 1981; 5: 343–352.
186. Bunn DT, Johnston RN: A ten-year study of sarcoidosis. *Br J Dis Chest* 1972; 66: 45–52.
187. Eckert H, Zaumseil I, Sehrt I: Enzyme-histochemical and immunopathological findings in intrathoracic lymph node sarcoidosis. *Acta Histochem* 1978; 61: 98–117.

188. Kirkpatrick CJ, Curry A, Bisset DL: Light- and electron microscopic studies on multinucleated giant cells in sarcoid granuloma: new aspects of asteroid and Schaumann bodies. *Ultrastruct Pathol* 1988; 12: 581–597.
189. Favara BE, Jaffe R: Pathology of Langerhans cell histiocytosis. *Hematol Oncol Clin North Am* 1987; 1: 75–97.
190. Chhabra UO, Desai SS, Jambhekar NA: Langerhans cell histiocytosis: a clinicopathological study of 50 cases. *Indian J Pathol Microbiol* 2004; 47: 370–376.
191. Neumann MP, Frizzera G: The coexistence of Langerhans cell granulomatosis and malignant lymphoma may take different forms: report of seven cases and a review of the literature. *Hum Pathol* 1986; 17: 1060–1065.
192. Barbey S, Jaubert F, Nogues C, Nezelof C, Gane P: Histiocytes-X and X-body reactivity with concanavalin-A, peanut agglutinin, and BSPT. *Pathol Res Pract* 1987; 182: 805–809.
193. Hajdu I, Zhang W, Gordon GB: Peanut agglutinin binding as a histochemical tool for diagnosis of eosinophilic granuloma. *Arch Pathol Lab Med* 1986; 110: 719–721.
194. Paulli M, Rosso R, Kindl S, et al.: Immunophenotypic characterization of the cell infiltrate in five cases of sinus histiocytosis with massive lymphadenopathy (Rosai-Dorfman disease). *Hum Pathol* 1992; 23: 647–654.
195. Ridolfi RL, Rosen PP, Thaler H: Nevus cell aggregates associated with lymph nodes: estimated frequency and clinical significance. *Cancer* 1977; 39: 164–171.
196. Johnson WT, Helwig EB: Benign nevus cells in the capsule of lymph nodes. *Cancer* 1969; 23: 747–753.
197. Busam KJ, Lohmann CM: Congenital paucimelanotic cellular blue nevus. *J Cutan Pathol* 2004; 31: 312–317.
198. Parkash V, Vidwans M, Carter D: Benign mesothelial cells in mediastinal lymph nodes. *Am J Surg Pathol* 1999; 23: 1264–1269.
199. Horn LC, Bilek K: Atypical (carcinoma-like) and simple glandular inclusions in retroperitoneal lymph nodes. *Gen Diagn Pathol* 1996; 141: 319–325.
200. Maassen V, Hiller K: Glandular inclusions in lymph nodes: pattern of distribution and metaplastic transformation. *Arch Gynecol Obstet* 1994; 255: 1–8.
201. Nappi O, Boscaino A, Wick MR: Extramedullary hematopoietic proliferations, extraosseous plasmacytomas, and ectopic splenic implants (splenosis). *Semin Diagn Pathol* 2003; 20: 338–356.
202. Osborne BM, Robboy SJ: Lymphomas or leukemia presenting as ovarian tumors: an analysis of 42 cases. *Cancer* 1983; 52: 1933–1943.
203. Menasce LP, Banerjee SS, Beckett E, Harris M: Extramedullary myeloid tumor (granulocytic sarcoma) is often misdiagnosed: a study of 26 cases. *Histopathology* 1999; 34: 391–398.
204. Ballon SC, Donaldson RC, Berman ML, Swanson GA, Byron RL: Myeloblastoma (granulocytic sarcoma) of the ovary. *Arch Pathol Lab Med* 1978; 102: 474–476.
205. Chilosi M, Pizzolo G, Janossy G, Bofill M, Fiore-Donati L: Enzyme-histochemical analysis on cryostat sections of human bone marrow. *J Clin Pathol* 1982; 35: 1220–1226.
206. Ree HJ, Raine L, Crowley JP: Lectin binding patterns in diffuse large-cell lymphoma. *Cancer* 1983; 52: 2089–2099.
207. Wick MR, Mills SE, Scheithauer BW, et al.: Reassessment of malignant "angioendotheliomatosis:" evidence in favor of its reclassification as "intravascular lymphomatosis." *Am J Surg Pathol* 1986; 10: 112–123.
208. Tsutsumi Y: Receptors for *Helix pomatia* agglutinin on normal and neoplastic lymphocytes—a histochemical study using paraffin-embedded tissue sections. *Acta Pathol Jpn* 1981; 31: 641–661.
209. Lalwani AK, Carey TE, Goldstein IJ, Peters BP: Lectin binding characteristics of squamous cell carcinomas of the head and neck. *Acta Otolaryngol* 1996; 116: 125–131.
210. Kohchiyama A, Oka D, Ueki H: Differing lectin-binding patterns of malignant melanoma and nevocellular and Spitz nevi. *Arch Dermatol Res* 1987; 279: 226–231.
211. Slesak B, Boguslawska-Jaworska J, Pejcz J, Harlozinska A: Binding of lectins to human leukemic cells. *Folia Haematol Int Mag Klin Morphol Blutforsch* 1989; 116: 251–259.
212. Pajor L, Balazs M, Balogh J, et al.: Ultrastructural lectin-histochemical and immunohistological observations on Merkel cell tumors. *Pathol Res Pract* 1986; 181: 45–49.
213. Swanson PE, Dehner LP, Wick MR: Small round-cell tumors in children: a lectin-histochemical and immunohistochemical study of 30 cases. *Surg Pathol* 1993; 5: 5–16.
214. Ree HJ: Lectin-histochemistry of malignant tumors. II. Concanavalin-A: a new histochemical marker for macrophage-histiocytes in follicular lymphoma. *Cancer* 1983; 51: 1639–1646.
215. Howard DR, Batsakis JG: Peanut agglutinin: a new marker for tissue histiocytes. *Am J Clin Pathol* 1982; 77: 401–408.
216. Wick MR, Scheithauer BW, Smith RL, Beart RW: Primary nonlymphoreticular malignant neoplasms of the spleen. *Am J Surg Pathol* 1982; 6: 229–242.
217. Falk S, Krishnan J, Meis JM: Primary angiosarcoma of the spleen: a clinicopathologic study of 40 cases. *Am J Surg Pathol* 1993; 17: 959–970.

CHAPTER 16

Diagnostic Histochemistry in Diseases of the Nervous System

M. Beatriz S. Lopes, M.D.

Special stains are very frequently required in neuropathology, probably more than in most other areas of diagnostic pathology. This statement applies equally to "medical" (nonneoplastic) and tumor-related neuropathology, as well as to neuromuscular pathology. Routine hematoxylin and eosin (H&E) staining is generally adequate in the study of neural tumors. However, when the pathologic process represents an inflammatory, demyelinating, infectious, neurodegenerative, or neuromuscular condition, histochemical analysis is an integral part of laboratory evaluation. The special stains that are most frequently employed in surgical and medical neuropathology can be grouped into four categories: 1) methods for demonstration of neurons, 2) techniques to delineate neuroglial elements, 3) procedures used to label connective tissues and reticulin, and 4) methods used especially in rapid intraoperative diagnosis.

The objectives of this chapter are to briefly discuss each of those categories, and to illustrate them with case examples that emphasize the diagnostic usefulness of histochemical methods. An important subset of histochemical stains that are used in neuropathology is represented by specific enzyme histochemical procedures as applied to myopathic disorders. That particular topic is considered in a separate chapter in this monograph, by Prayson et al. Similarly, stains used to identify infectious processes are comparable to those employed in other organ systems, as are the histochemical attributes of primary neural neoplasms that can also take origin outside the nervous system (e.g., germ cell tumors, lymphomas, Langerhans cell histiocytosis, primitive neuroectodermal tumors, metastatic carcinomas melanomas, and sarcomas). Hence, they are likewise not included in this discussion.

CAVEATS REGARDING HISTOCHEMISTRY OF THE NERVOUS SYSTEM

In most other anatomic sites, thin tissue sections (cut at 2–4 μm) are preferred for histopathologic analysis. In the nervous system, however, 6- to 8-μm sections are routinely used; indeed, some pathologic processes can be best investigated with preparations as thick as 10–14 μm. These precepts must be considered when sampling materials from both surgical and autopsy brain tissue. For example, in surgical specimens in which a demyelinating disorder is suspected, serial sectioning of the tissue for H&E (at 4–6 μm), myelin stains (e.g., luxol fast blue [LFB], at 10 μm), and axonal stains (e.g., Bielschowsky silver impregnation, at 6 μm) is commonly performed. Careful selection of these staining methods must be done when small surgical specimens from needle biopsies are evaluated, so that the tissue is not exhausted and a definitive diagnosis can still be achieved. Submission of tissue blocks from autopsy brains should be at least as thick as a U.S. 25-cent coin, for the diagnosis of dementia and other neurodegenerative disorders. In such cases, H&E, Bielschowsky silver, and Congo red stains (at 8 μm) are routinely obtained for the detection of neurofibrillary tangles (NFTs), neuritic plaques (NPs), neuronal/glial inclusions, and cerebrovascular amyloid deposits (Figures 16.1 and 16.2). Another important stipulation for the preparation of optimal histochemical stains in neuropathology is that they require the knowledge and skill of experienced histotechnologists. Most of the applicable methods are demanding, with several steps of differentiation; consequently, they are "manual" stains that are not as well prepared in automated platforms.

METHODS FOR THE DEMONSTRATION OF NEURONS

Procedures used to label neurons include various techniques for the demonstration of the neuronal cell body, axons and neurites, and myelin sheaths.

Neuronal Cell Body and Nissl Substance—Nissl Stains

Nissl bodies, which are primarily composed of ribonucleic acid–rich rough endoplasmic reticulum, are basophilic structures in neurons. They are stained with basic aniline dyes including cresyl violet, cresyl fast violet, cresyl etch violet, cresyl acetate, toluidine blue, thionin, azure-B, or carbol-azure. The amount of Nissl substance varies in different types of neurons, and it may disappear altogether in the face of neuronal damage (chromatolysis).

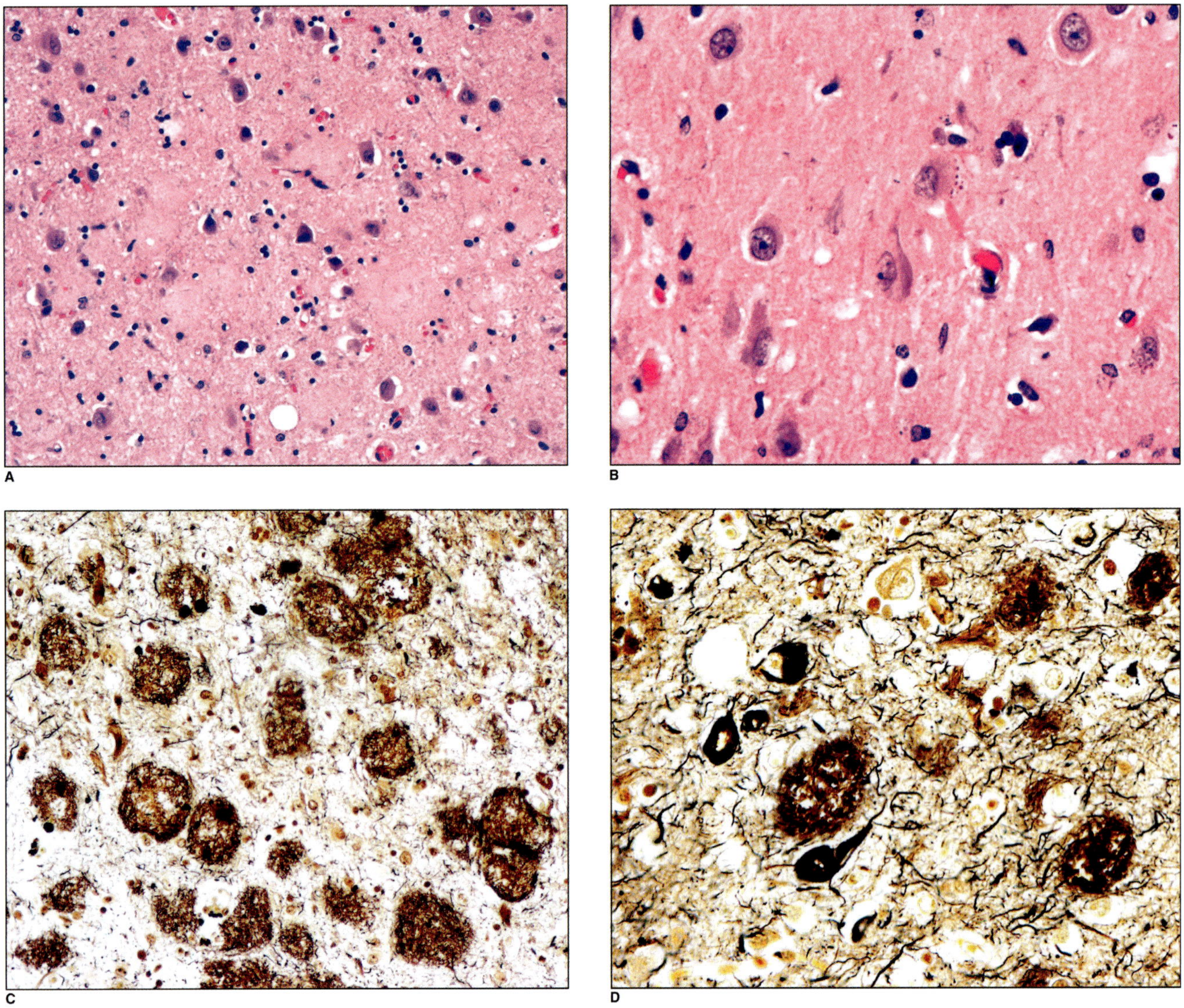

FIGURE 16.1: (A) This section of frontal cortex shows several irregular areas of disrupted neuropil, suggestive of NPs (H&E). (B) Hippocampal neurons demonstrate NFTs and granulovacuolar degeneration (H&E). Several NPs (C) and NFTs (D) are seen with the Bielschowsky silver stain in sections of the isocortex and hippocampus. There are also disseminated irregular neuritic processes (D) in the neuropil, also known as neurofibrillary threads.

The classical Nissl stain (cresyl etch violet) identifies neurons in formalin-fixed, paraffin-embedded (FFPE) tissue sections (1,2). By varying the pH and the degree of differentiation of that stain, both Nissl bodies and nuclei, or Nissl bodies alone, may be demonstrated. Nissl staining is helpful in the identification of neurons or ganglion cells in neoplastic neuropathology. However, it has its greatest value for the assessment of cortical abnormalities including dyslaminations, gray matter heterotopias, and other cortical dysplasias. This is because Nissl stains provide a general overview of the cortical layering that is seen in 12-μm sections of the brain, with strong staining of the neuronal cell bodies. Nevertheless, nowadays, Nissl stains are principally used in experimental neuropathology because pathologists have eschewed them in favor of more "modern" technologies. However, we continue to believe that Nissl staining does still have a role in diagnostic neuropathology (Figure 16.3).

Stains for Axons and Neuritic Processes—Silver Impregnation Methods (Bielschowsky, Bodian, Holmes)

Argyrophilic impregnation techniques continue to have an important role in the examination of neuronal processes in diagnostic neuropathology. Axonal and neuritic processes in most portions of the central nervous system (CNS) are easily demonstrated in paraffin sections by methods in which tissue is impregnated with solutions containing a low concentration of free silver ions (3). Several techniques have been used for this purpose, including Bodian's

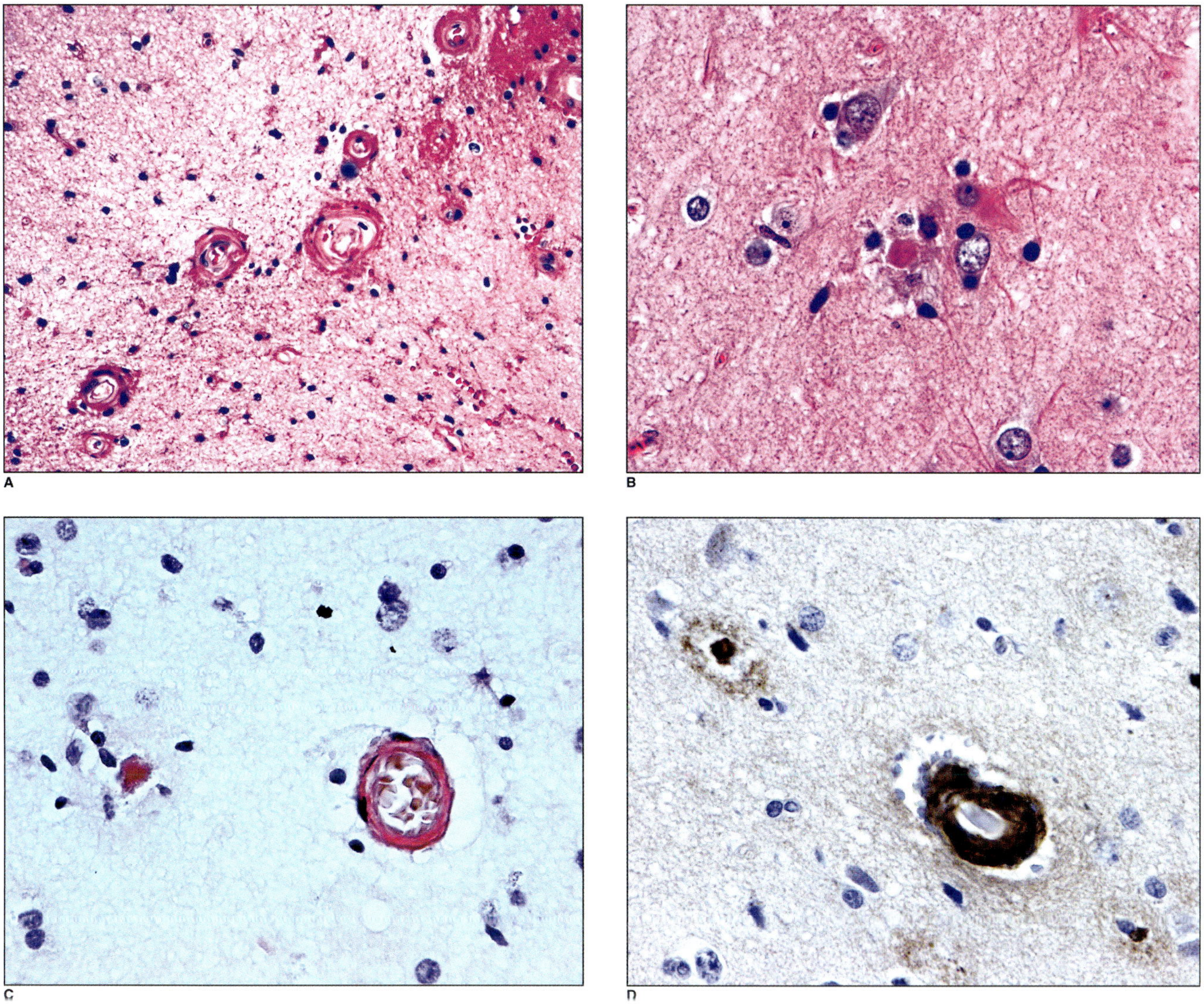

FIGURE 16.2: ***(A) Small cortical vessels have thick, hyalinized walls in this section. (B) NPs with hyaline cores are also seen. (C) Congo red staining demonstrates amyloid deposition in vascular walls and in the cores of NPs. (D) Immunostains for Aβ-amyloid confirm the diagnosis of the cerebral amyloid angiopathy.***

and Holmes's methods (4); nevertheless, most laboratories currently employ modifications of the Bielschowsky procedure as applicable to FFPE sections (4,5). The latter technique visualizes axons and dendrites, and also shows details of intracellular neurofibrillary networks in mature neuronal cells. Bielschowsky staining involves a silver ion complex reaction formed by adding ammonium hydroxide to silver nitrate, toning with a developing solution containing formalin and nitric or citric acid, and development with sodium thiosulfate. Intracellular neurofibrils, axons, and dendrites stain black, whereas neuronal cell bodies and the neuropil are labeled golden brown.

As mentioned earlier, silver impregnation techniques are technically challenging and require close attention to detail. Their successful application is also significantly affected by tissue fixation and embedding procedures. Bielschowsky stains can generally be done successfully using FFPE tissues. Nonetheless, optimal delineation of delicate cell processes in immature neurons seems to be best achieved with sections of formalin-fixed tissue embedded in low-melting point (37 °C) polyester wax or by using frozen sections (FSs) of formalin fixed tissue (6). Other fixatives such as Bouin's, Zenker's, and ethanol-based solutions are unsuitable for silver impregnation. Interestingly, treatment of deparaffinized tissue sections with buffered 10% formalin markedly enhances neuritic argyrophilia, regardless of how the specimen was primarily fixed (Figures 16.1, 16.4, and 16.5).

Staining of Myelin Sheaths—Luxol Fast Blue

The histochemical methods used to label myelin sheaths depend upon the presence of complex lipids—including phospholipids—in those structures. LFB, a stain

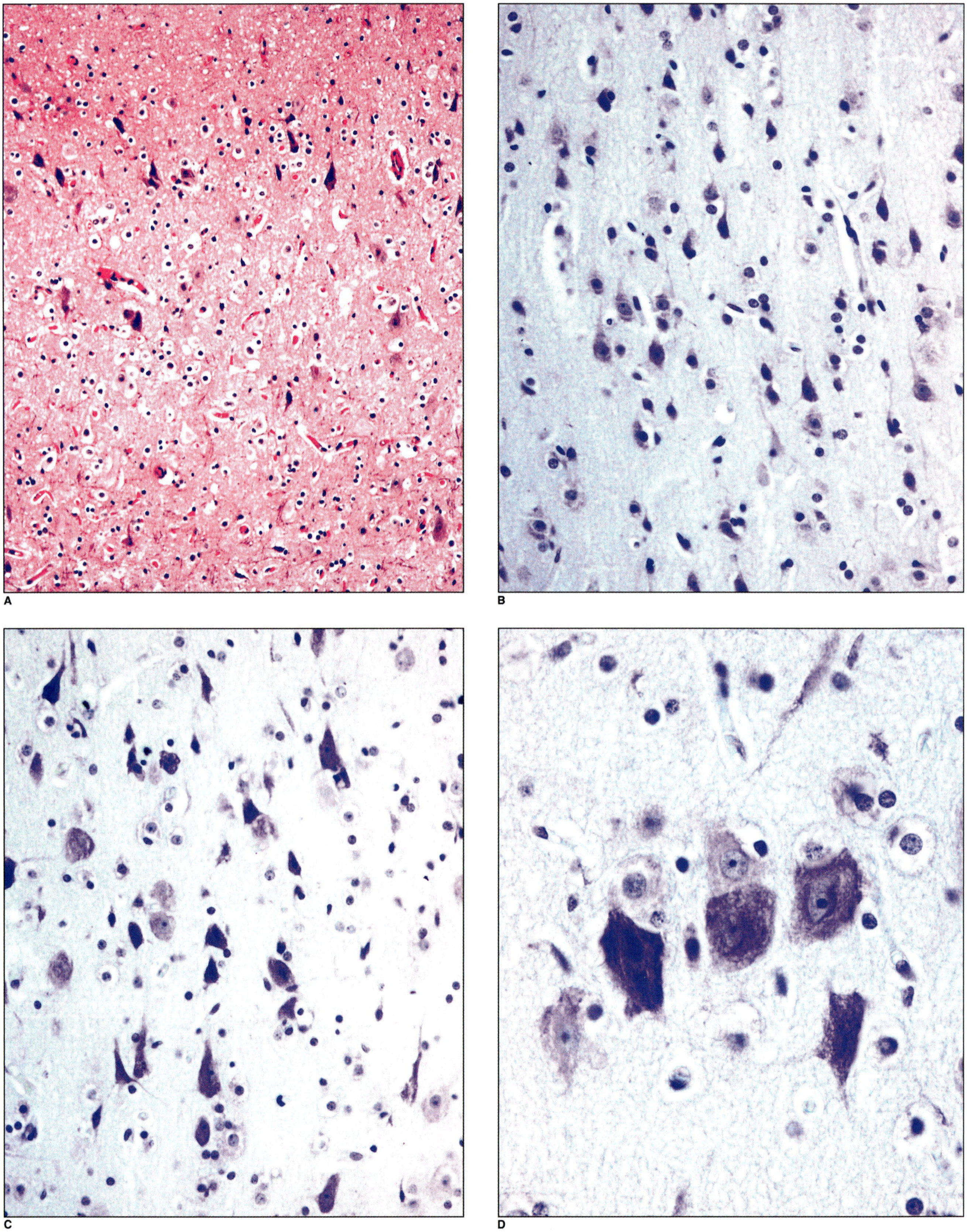

FIGURE 16.3: *(A) This section of cortical gray and white matter, stained with H&E, shows normal layering in the majority of the specimen. (B) and (C) Nissl staining, however, demonstrates several areas of irregular lamination (B) and large dysplastic neurons (C). Nissl staining also highlights the presence of large neurons with prominent cytoplasmic labeling (D, F) that was only faintly seen by H&E staining (E).*

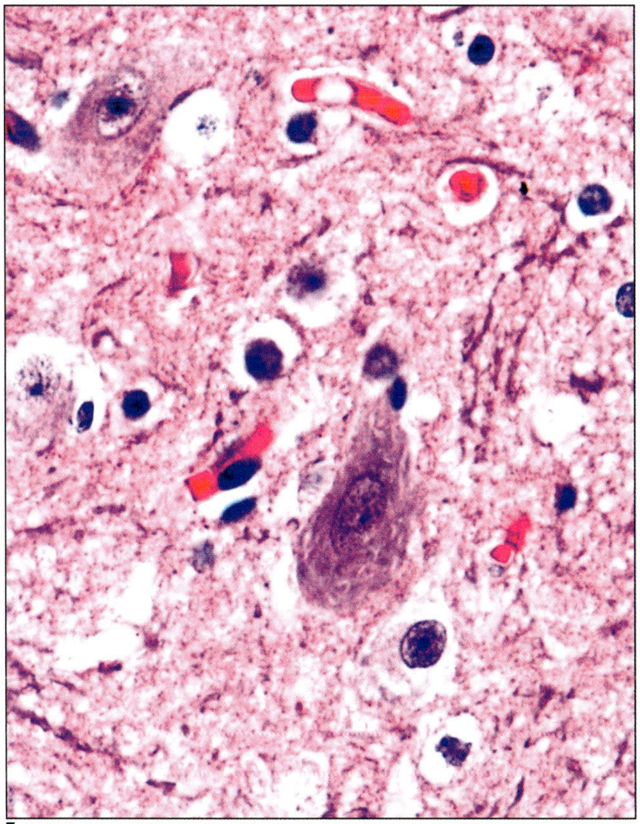
E

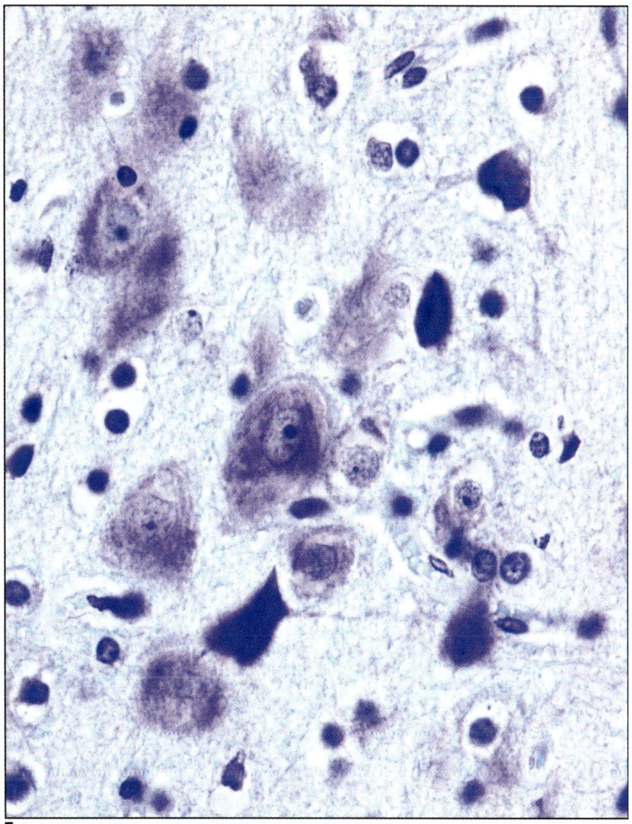
F

FIGURE 16.3: (Continued).

introduced by Klüver and Barrera (7), is the most commonly employed histochemical procedure for the demonstration of myelin. Several other classical stains for myelin, including Weil's and Speilmeyer's methods, are rarely used in current clinical practice. One advantage of LFB is its flexibility; it can be done in combination with several other stains, including those used for neurons (LFB-cresyl etch violet, also know as the Klüver-Barrera procedure), and neuritic processes (LFB-Holmes) (2,7,8). When combined with the periodic acid–Schiff technique, H&E, or simply hematoxylin, LFB provides excellent discrimination between cellular structures of normal and abnormal nervous tissue. Again, LFB staining involves a differentiation step and can be technically difficult (Figures 16.1, 16.4, and 16.5).

METHODS FOR DEMONSTRATION OF NEUROGLIA

In the past two decades, histochemical stains for the demonstration of neuroglial elements (astrocytes, oligodendrocytes, and ependymal cells) are only uncommonly used in clinical neuropathology. However, from a historical perspective, mention should be made of important procedures that were developed by Ramon y Cajal and his student Del Rio Hortega at the beginning of the 20th century. Using metal-based staining methods, those investigators demonstrated several types of nonneuronal cells in the CNS. Such traditional stains for neuroglial cells depended upon their impregnation by silver or gold solutions. The gold chlo ride sublimate method of Cajal for astrocytes is a classic example of those techniques (2,9). As explained by Kiernan (9), colloidal gold is deposited on or near glial fibrillary acidic protein, the major intermediate filament of astrocytes. The Cajal technique is not specific for astrocytic elements, but they are more conspicuously labeled than are other glial cells. From a practical perspective, metallic impregnation techniques are of limited use in hospital practice because they can be applied only to frozen tissue.

Rubinstein (10) discussed the possible uses of variants of Hortega's silver carbonate method for neuroglial cells in FFPE tissues. He furthermore provided several illustrations of their application to the diagnosis of glial tumors. The methods in question include the double silver carbonate impregnation method described by Penfield, the double or triple silver carbonate impregnation procedure of Scharenberg, and the triple silver impregnation technique of Polak. Those stains demonstrate astroglial and oligodendroglial cells as well as glial fibrils. Nevertheless, Rubinstein emphasized that tissue must be either fixed in formol-ammonium bromide, or, if originally fixed in formalin, it must undergo bromidation in order to be reactive with Penfield's and Scharenberg's methods (10).

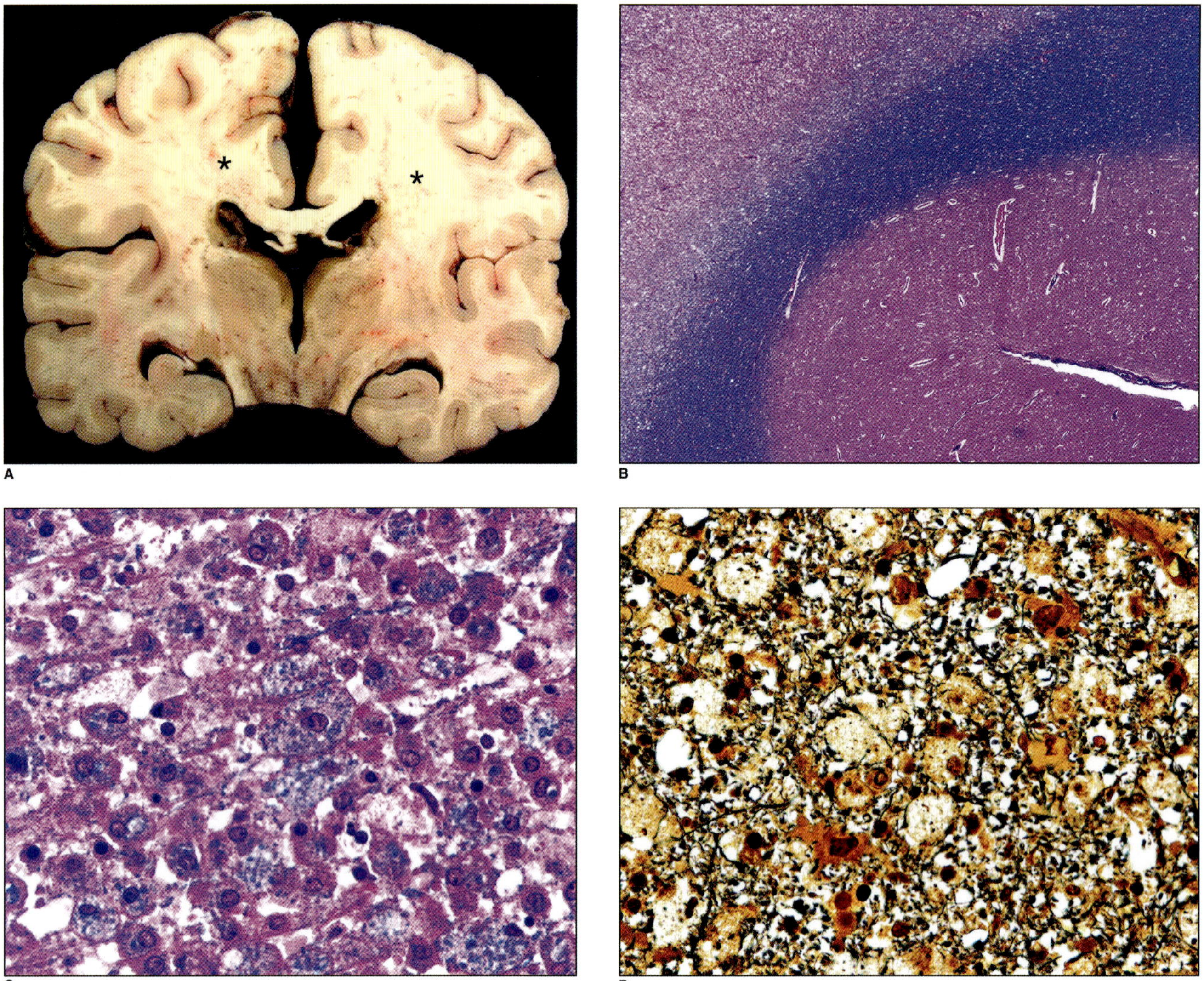

FIGURE 16.4: ***(A) A formalin-fixed slice of the brain shows bilateral softening of the white matter (*) in periventricular areas. (B) Intense demyelination of the cortical white matter can be seen with the LFB method. (C) Higher magnification of the demyelinated area demonstrates active phagocytosis of myelin debris by macrophages in this LFB stain. (D) In the same foci, partial axonal preservation is observed with the Bielschowsky silver stain. (E and F) In the spinal cord, zones of acute demyelination were seen in the anterior, posterior, and lateral columns by LFB staining.***

Other histochemical methods that may be used to identify neuroglial cells or glial processes include the Holzer procedure and the Mallory phosphotungstic acid–hematoxylin (PTAH) stain (Figure 16.6) (2,4,5). Both Holzer's technique and the PTAH stain label all glial fibers irrespective of whether they are astroglial or ependymal in nature. Moreover, PTAH decorates myelin sheaths. At the present time, these stains are of limited value in daily practice but may still contribute to diagnosis in selected cases.

METHODS FOR THE DEMONSTRATION OF CONNECTIVE TISSUES AND RETICULIN FIBERS

In similarity to other organ systems, several histochemical stains emphasize mesenchymal tissues and permit a facile distinction to be made between them and neuroectodermal elements. Common methods for staining collagen fibers, reticulin, and elastic laminae are numerous. The selection of preferred techniques among such stains is often a matter of personal preference, based on the experience of histotechnologists or available options of automated staining devices. In our laboratory, the connective tissue stains that are typically used are Masson's trichrome method for collagen fibers, the Verhoef–Van Gieson method for elastic fibers, and Wilder's stain for reticulin (Figures 16.7–16.10).

Reticulin Fiber Stains—Wilder, Gordon-Sweet's

Histochemical stains for reticulin fibers are essential for the recognition of stroma that is produced by cells of mesodermal origin (e.g., meningothelial cells or a fibrous

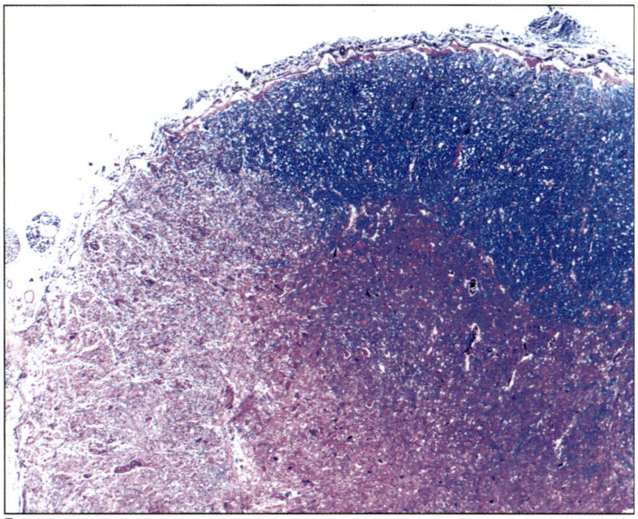

E

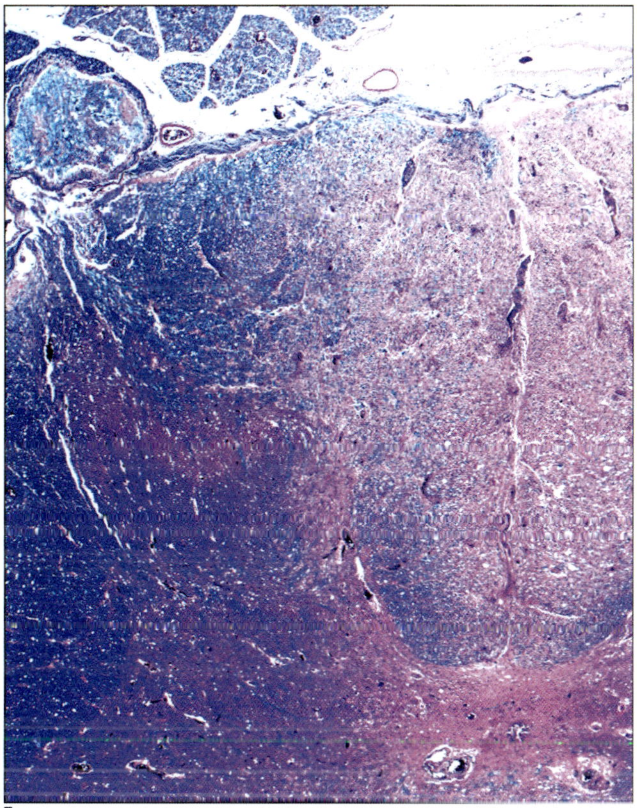

F

***FIGURE 16.4:* (Continued).**

component of a glial tumor) and also to determine whether neoplastic tissue is present in the parenchyma of the nervous system, the leptomeninges, or perivascular spaces of the CNS. Several methods for the labeling of reticulin fibers have been described; the most popular of which are Wilder's and Gordon-Sweet's procedures.

Mention should be made of the usefulness of reticulin staining in the diagnosis of special categories of CNS tumors. These include glial and mixed glioneuronal tumors with prominent reticulin deposition or desmoplasia. Some examples are pleomorphic xanthoastrocytoma (PXA), desmoplastic ganglioglioma of infancy (DIG) and desmoplastic infantile astrocytoma, gliofibroma, and gliosarcoma 16.11–16.14. Mesenchymal tumors of the dura that are delineated in part by reticulin labeling include meningioma and hemangiopericytoma-solitary fibrous tumor. Lastly, the reticulin pattern of pituitary adenomas is also distinctive, vis-a-vis that of normal adenohypophysis or pituitary hyperplasia (Figures 16.8–16.11).

RAPID DIAGNOSTIC METHODS

Smear Preparations

Opinions differ on the value and reliability of cytologic preparations for rapid diagnosis of brain tumors during intraoperative consultations. Smear techniques have a long history and have been used in the majority of neuropathology laboratories as an effective method for intraoperative evaluation of CNS biopsies (6,15). These techniques allow for selective examination of several areas of small biopsy specimens—now a routine practice in diagnostic neuropathology—eliminating the need to perform FSs on tiny samples. Moreover, smears also prevent the risk of distorting or losing valuable diagnostic material. As is true of any other cytologic procedure, familiarity with the various cytologic appearances and tintorial qualities of normal and neoplastic nervous tissues is required for accurate diagnosis by the smear technique.

Any of several well-known histochemical stains for cytologic preparations may be used in neuropathology, including H&E, the Papanicolaou method, and the DiffQuik^{R} procedure. In our laboratory, we favor the use of the modified Morris stain, which combines a water-based eosin with methylene blue (16). Unlike H&E staining, Morris's technique highlights delicate neuroglial fibers in CNS neoplasms and simultaneously provides good nuclear detail. The final preparations show brilliantly pink glial fibers, slight gray to pink cytoplasm, and bluish nuclei (Figures 16.12–16.14). The stain also accentuates endothelial cells in proliferating blood vessels of malignant gliomas; they are stained pale blue, differing clearly from pinkish glial cells (Figure 16.14). In contrast, although Morris's stain is certainly suitable for the cytologic study of mesenchymal and metastatic tumors, some neuropathologists may prefer to use H&E-stained or Papanicolaou-stained slides to study such lesions (Figures 16.15 and 16.16).

Frozen Sections

The FS technique, using the cryostat after rapid freezing of tissue specimens, has long been a staple of surgical neuropathology. It is usually most effective in regard to tumors that are hard or dense, such as meningiomas. Soft, friable, or gelatinous neuroglial neoplasms are much more difficult

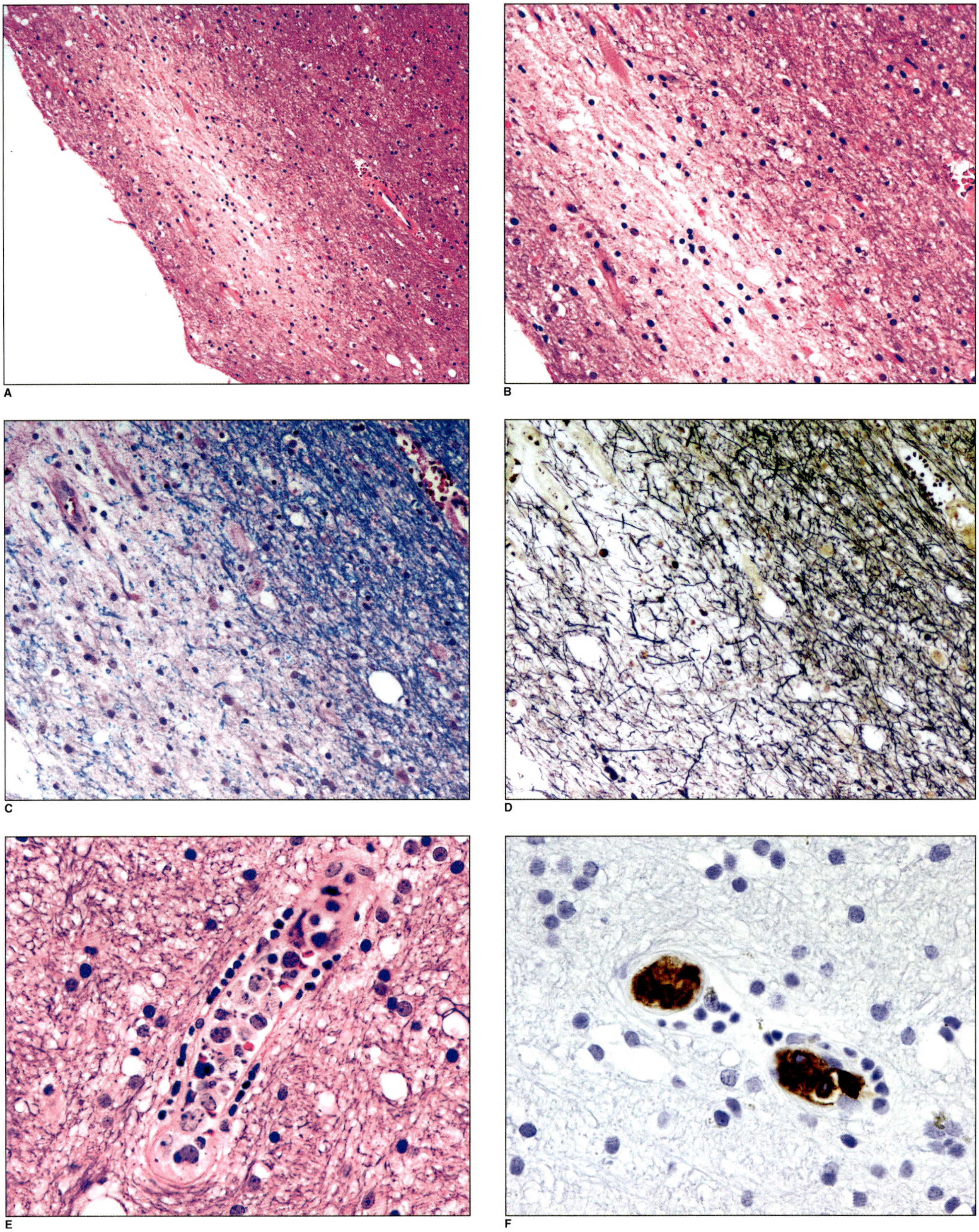

FIGURE 16.5: (A and B): H&E staining shows small areas of parenchymal loss in the white matter, with mild reactive gliosis. Myelin loss (C) and a moderate degree of axonal dropout (D) are clearly seen in an LFB stain and a Bielschowsky silver stain. (E) Intravascular accumulations of atypical lymphoid cells are present in other areas of the biopsy. (F) These cells were immunoreactive for CD79a.

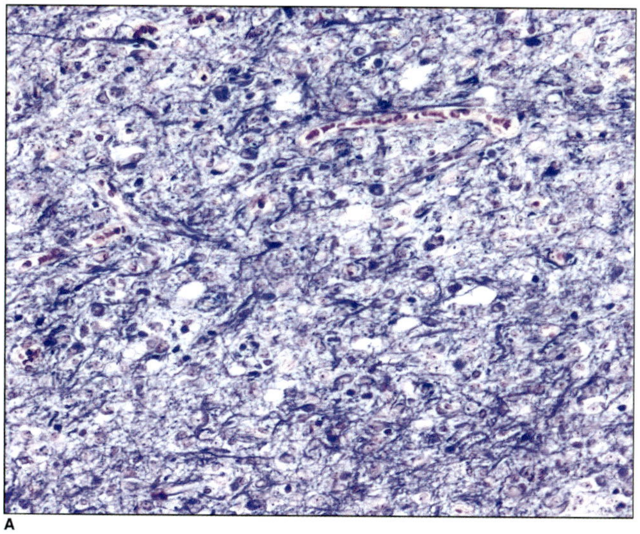

A

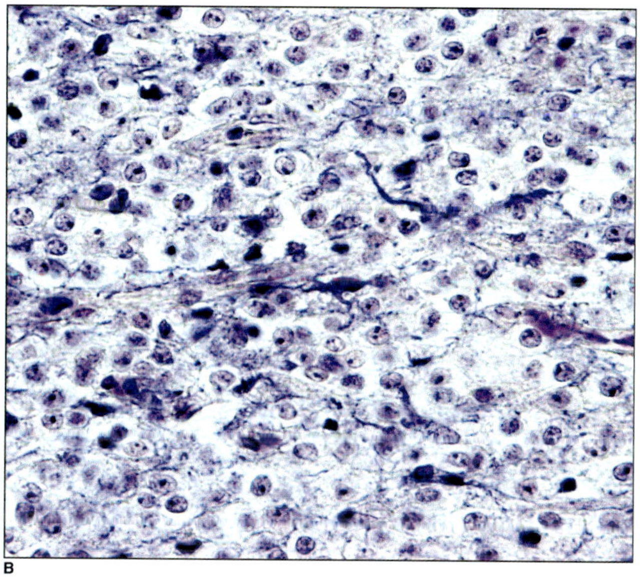

B

FIGURE 16.6: ***(A) A fibrillary low-grade astrocytoma (WHO grade II) shows the presence of neuroglial fibers, as intensely stained by the PTAH method. (B) In an oligodendroglioma (WHO grade II), reactive stromal astrocytes are again highlighted by PTAH staining (arrows).***

to handle by FS, and, in our opinion, those lesions are better diagnosed cytologically during intraoperative consultation. In addition, needle biopsies are often small, and conventional FS of them may produce unwanted freezing artifact that interferes with subsequent paraffin section analysis. H&E is the most commonly employed stain for FS examination of neural tissues and tumors.

HISTOCHEMISTRY OF PERIPHERAL NERVE BIOPSY SPECIMENS

Peripheral nerve biopsy is one of several methods for the diagnosis of possible neuropathies. Indications for nerve biopsy are beyond the scope of this chapter and have been reviewed in several other specialized publications. However, there are very limited clinical scenarios that warrant a peripheral nerve biopsy in current practice. This is true in part because of advances in molecular testing for inherited neuropathies. Nonetheless, nerve biopsy is particularly helpful for the recognition of interstitial intraneural processes including vasculitis or vasculopathy, specific diseases such as amyloidosis and leprosy, and neuropathies that cannot be identified from the results of clinical or electrophysiologic studies.

A variety of methods exists for preparing and examining specimens of peripheral nerves, with each procedure potentially providing different information. In particular, a knowledge of detailed clinical information is important before processing a nerve biopsy.

Routinely, such biopsies are divided into at least three parts for different processing methods: 1) formalin fixation and paraffin embedding for general morphologic and histochemical analysis of the nerve components, 2) glutaraldehyde fixation and resin embedding for possible ultrastructural analysis, and 3) cold isopentane immersion and cryostat sectioning for detection of neural enzymopathies and storage diseases. In cases where a distinction between axonal degeneration and demyelinating diseases is necessary, biopsy material may also be reserved for "teased fiber" processing [for details of this procedure, see references (17) and (18)].

Routine histochemical stains that can be done using FFPE nerve biopsies comprise the H&E, trichrome, LFB, Bielschowsky, and Congo red procedures. Resin-embedded semithin sections are commonly stained with toluidine blue—with or without an aldehyde fuchsin counterstain—for light microscopic and ultrastructural analysis. Together, those assessments provide a great deal of morphologic information in regard to demyelinating and remyelinating processes (Figure 16.17).

LECTIN HISTOCHEMISTRY OF NEURAL DISEASES

A relatively limited body of data exists on the use of plant lectins in histochemistry of the nervous system. However, such reagents have indeed been assessed as possible markers of lineage or functionality.

Schill and Cervos-Navarro (19) suggested that concanavalin-A (CON-A) preferentially labeled astrocytes and that peanut agglutinin (PNA) was selective for oligodendroglial cells, based on studies of mixed gliomas comprising both of those cell types. However, other publications by Yang et al. (20) and Cruz-Sanchez et al. (21) have failed to support this contention. Nevertheless, the latter analyses do confirm that CON-A and PNA bind to the majority of glial neoplasms in a generic sense (Figure 16.18).

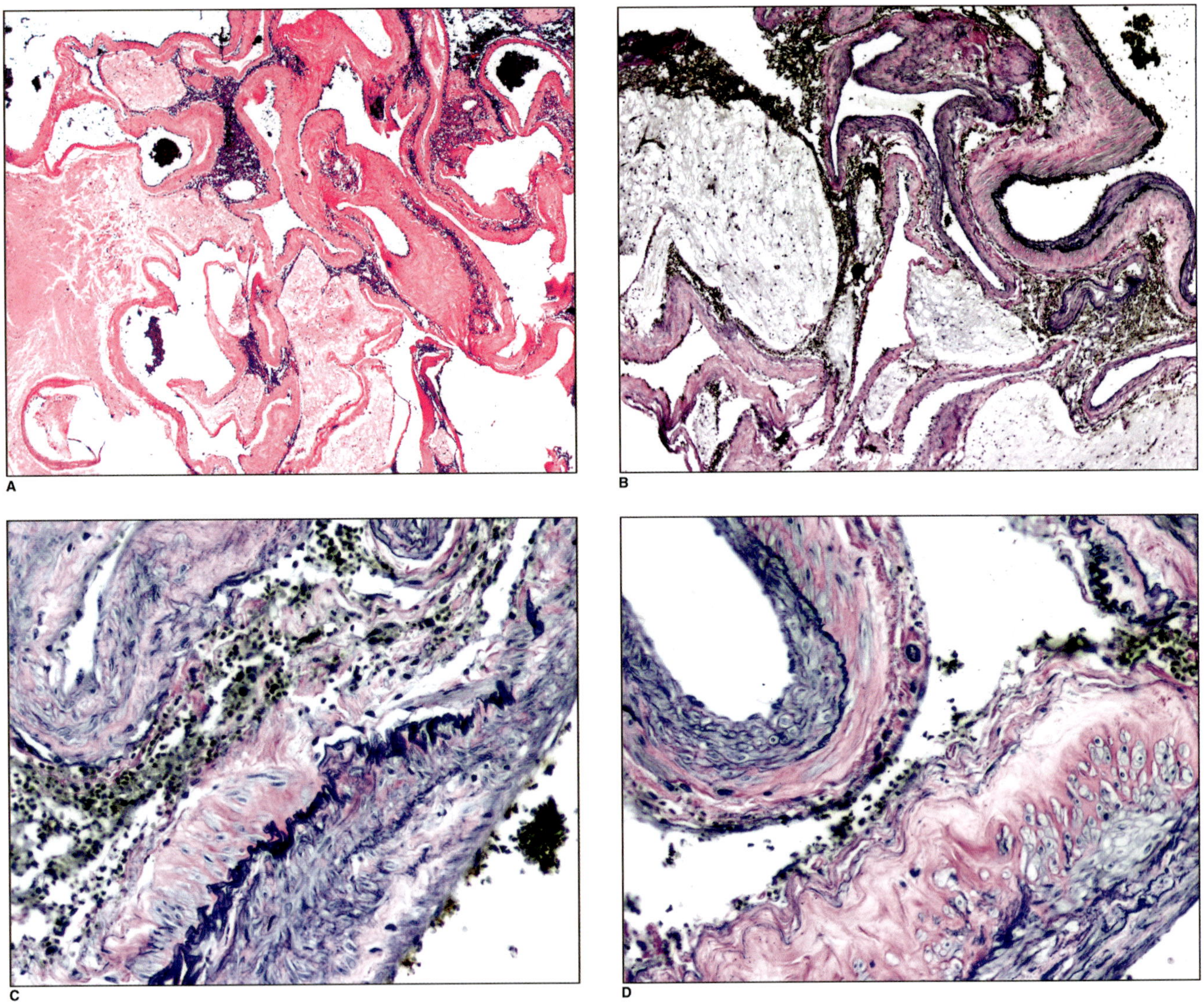

FIGURE 16.7: (A) An arteriovenous malformation of the brain shows tortuous and disorganized vascular structures with variable calibers and mural thicknesses, admixed with gliotic brain parenchyma. (B) Verhoeff–Van Gieson staining delineates arteries, veins, and abnormal vascular structures. (C and D) Aberrant vessels, with disrupted internal elastic lamina, and others with thick walls lacking elastic lamina are appreciated by elastic staining. The abnormal vessels have been described as "arterialized veins."

Another area of investigation has centered on lectin affinities for pituicytes and pituitary adenomas. Kurosaki et al. observed a selective relationship between PNA-binding (Figure 16.19) and prolactin-producing adenomas (22). IN contrast, Behncken and Saeger (23) were unable to reproduce that finding in an analysis of several lectins and various functional types of anterior pituitary tumors.

Finally, cerebellar hemangioblastoma (CHB) is well known as a potential simulant of metastatic clear-cell renal cell carcinoma, particularly in patients with von Hippel–Lindau syndrome (Figure 16.20) (24). Studies of low-grade clear-cell renal carcinomas have shown frequent labeling of those neoplasms by *Lotus tetragonolobus* agglutinin or PNA (25,26). However, to date, no comparative analyses of CHB have been done using those markers.

CASE HISTORIES ASSOCIATED WITH CHAPTER FIGURES

The following cases are examples showing the value of specific histochemical stains in neural diseases. They do not encompass the entire spectrum of disorders that may affect the CNS or the peripheral nervous system, but they provide an overview of the most common histochemical methods that are used in the practice of diagnostic neuropathology.

CASE 1— *Focal Cortical Dysplasia*: A 3-year-old boy had epilepsy since he was 3 months of age; it was refractory to pharmacologic suppression and was accompanied by progressive left-sided hemiparesis. Neurological imaging studies did not show any specific lesions, but intracranial

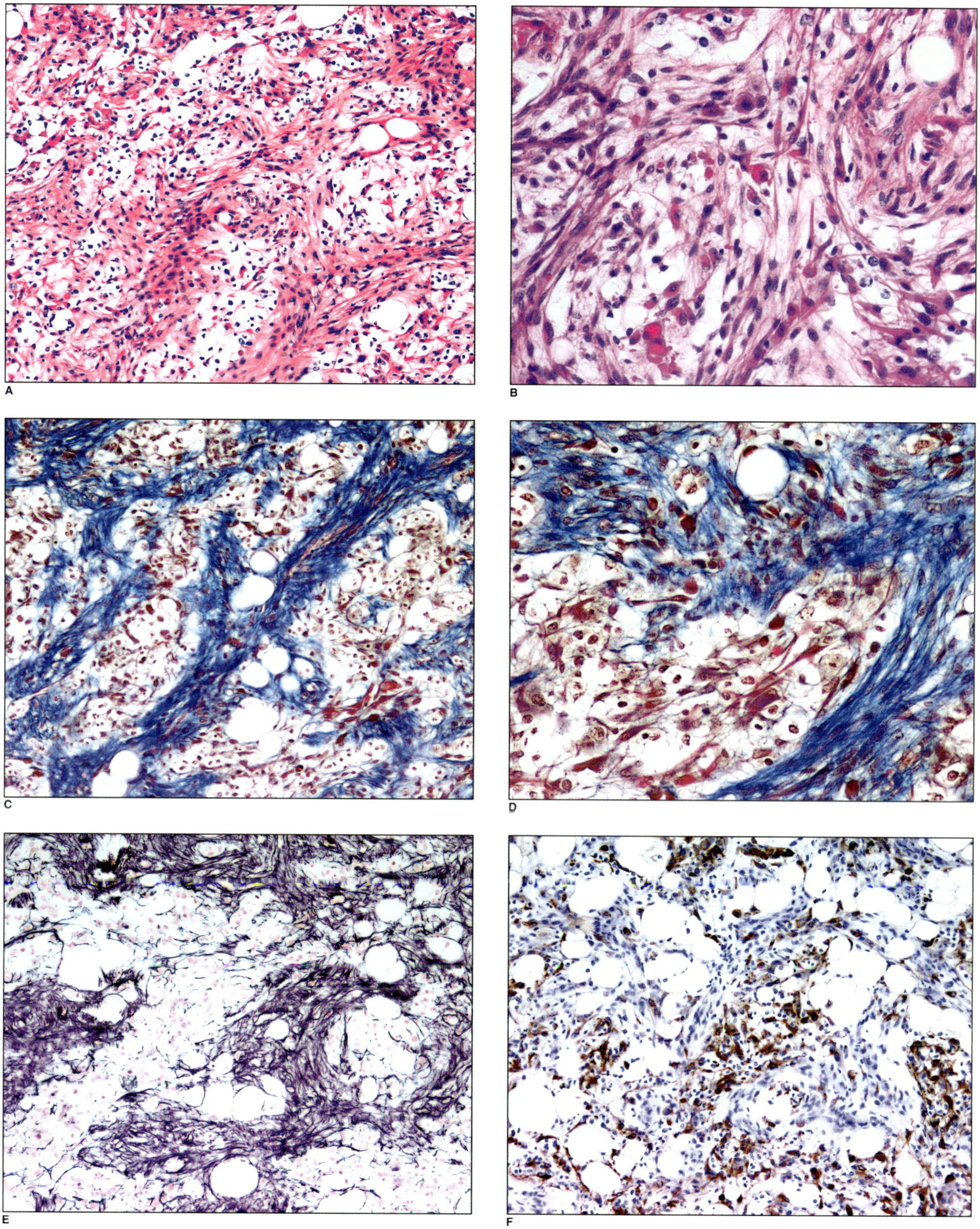

FIGURE 16.8: *(A) Sections of this tumor clearly demonstrate two distinct cell populations. (B) A dense spindle-cell component is clearly separate from a gemistocytic glial element. (C and D) Trichrome staining better delineates the spindle-cell component. (E) Wilder's procedure highlights the deposition of reticulin fibers in the spindle-cell zones. (F) An immunostain for glial fibrillary acidic protein (GFAP) demonstrates a neoplastic astroglial component, interdigitating with GFAP-negative sarcomatous elements.*

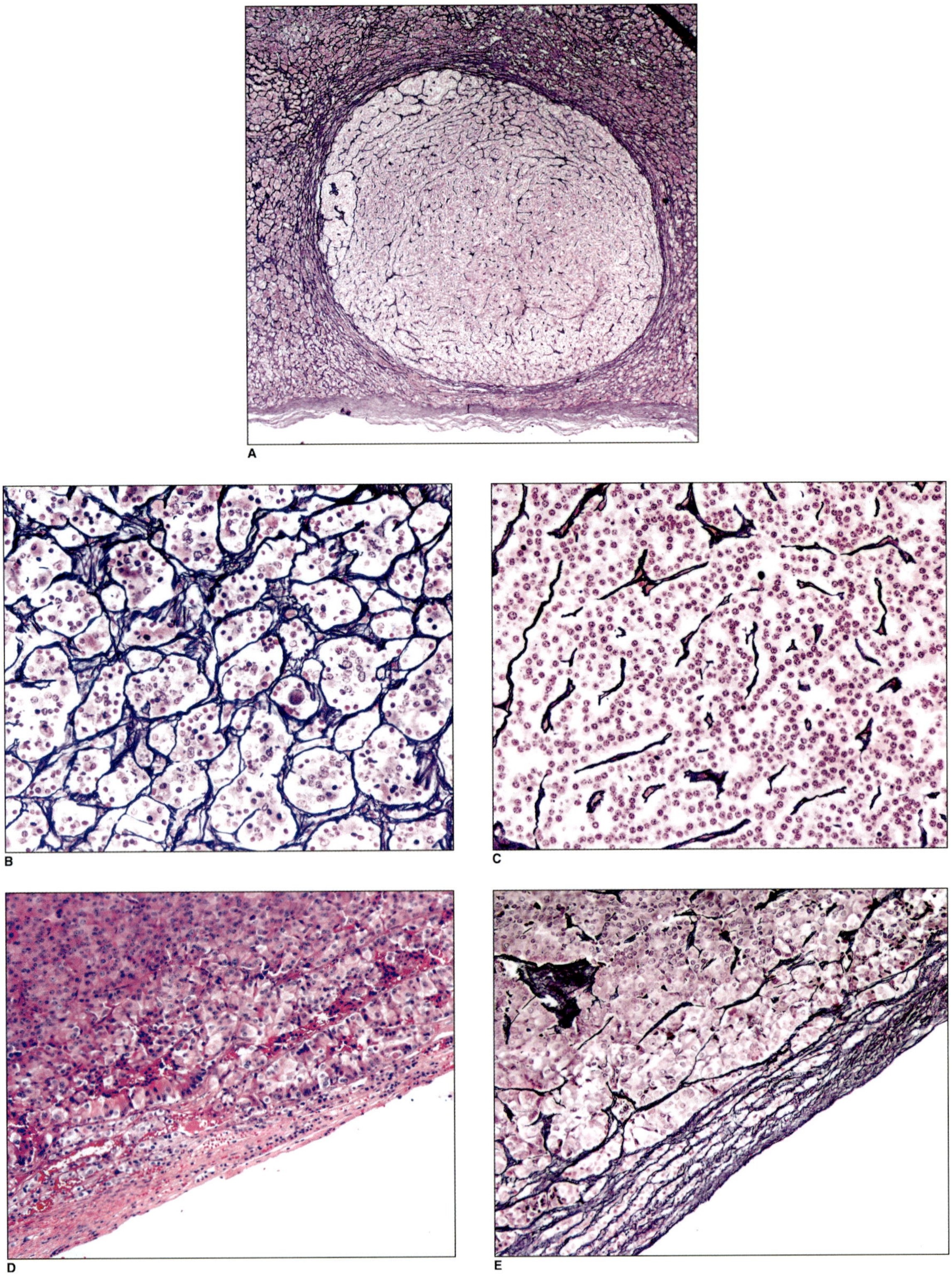

FIGURE 16.9: (A) Reticulin staining of the pituitary gland shows a well-demarcated microadenoma. (B and C) The normal acinar pattern of the adenohypophysis (B) is easily appreciated in a Wilder's stain. In contrast, the microadenoma (C) manifests a reticulin network that is effaced. (D and E) This macroadenoma compresses the adjacent anterior pituitary, forming a pseudocapsule around the tumor as seen with Wilder's procedure.

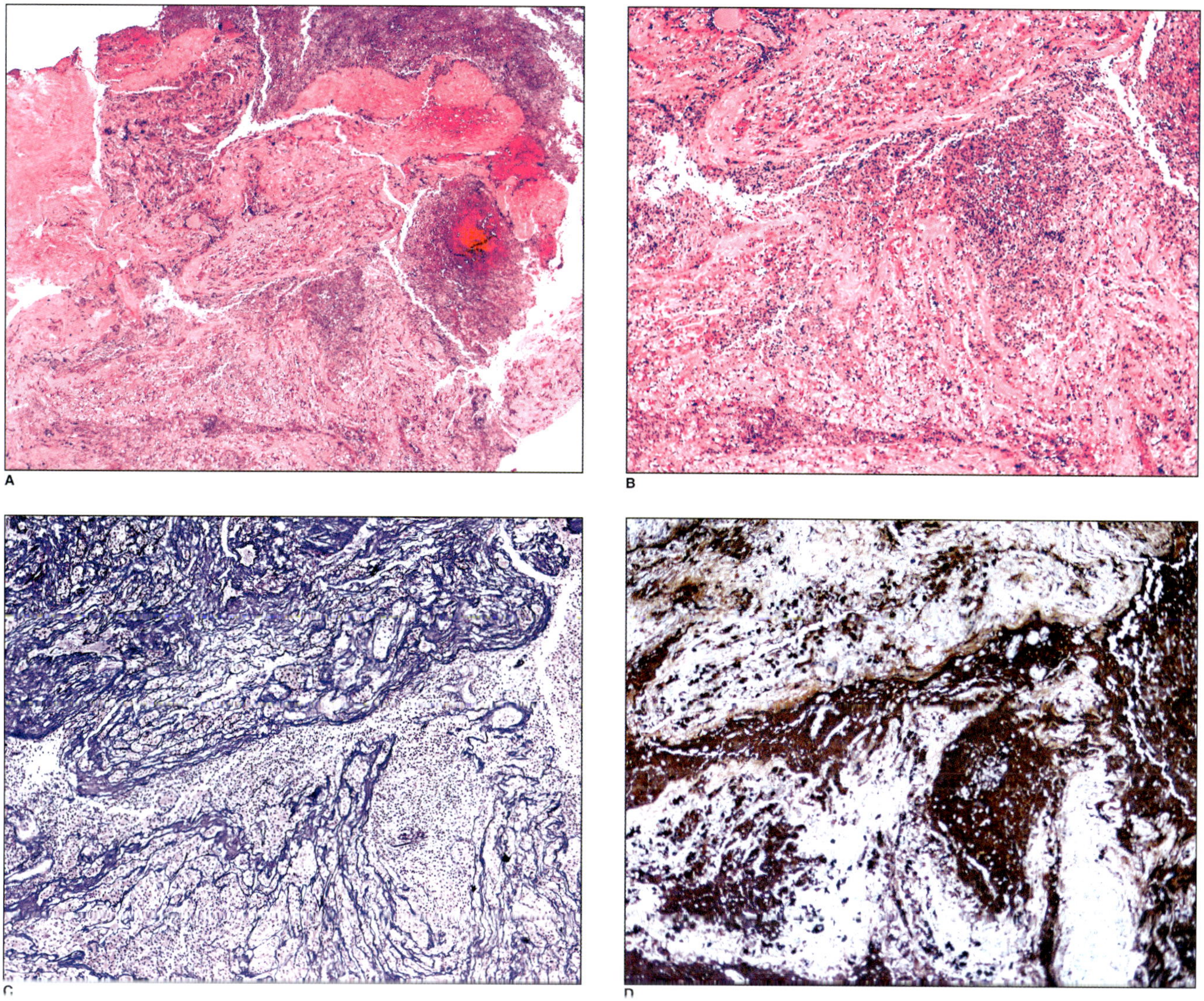

FIGURE 16.10: ***(A and B) Nonneoplastic pituitary gland, connective tissue, and possibly adenomatous areas are intermixed in this specimen and are not easily separated from one another morphologically. (C) Reticulin staining clearly delineates the areas where there is effacement of the reticulin network, supporting a diagnosis of pituitary adenoma. (D) Corresponding areas that show a lack of reticulin are strongly immunoreacitve for ACTH, supporting the diagnosis of corticotrophic adenoma. (E–G) Higher magnification views of the transition area between the adenoma and the normal pituitary gland in a reticulin stain clearly delineate the adenomatous areas.***

electrophysiologic monitoring was successful in capturing evidence of multiple seizures with focal onset in the right frontoparietal area. A resection of that area was performed, using intraoperative electrophysiologic monitoring.

The surgical specimen comprised a fragment of cerebral cortex measuring 4.0 × 2.0 × 1.0 cm. Initial H&E analysis demonstrated that the majority of the resected brain manifested normal lamination of the cortical ribbon (Figure 16.3A). However, there were several areas of laminar disorganization (Figure 16.3B,C) and numerous large dysplastic neurons. The dyslamination was best seen with a Nissl stain (Figure 16.3C), which also highlighted the dysplastic neurons (Figure 16.3D–F). A diagnosis of focal cortical dysplasia (FCD type IIA) (27,28) was made. ■

CASE 2— *Acute Multiple Sclerosis:* A 50-year-old man had a 1-week history of weakness in both lower extremities with difficulty in coordinating them, blurred vision, and urinary incontinence. Computed tomography of the brain demonstrated numerous ill-defined, low-attenuation intracerebral lesions of the right internal capsule, right occipital lobe, and right temporal lobe. Magnetic resonance imaging showed multiple periventricular high-FLAIR signals and enhancement of the lesions. In addition, similar abnormalities were seen in the cervical and thoracic regions of the spinal cord. The patient's condition worsened rapidly and he suffered complete paralysis of the left upper extremity. Several serologic investigations failed to reveal evidence of an autoimmune process, and death occurred after only 3 weeks of disease activity.

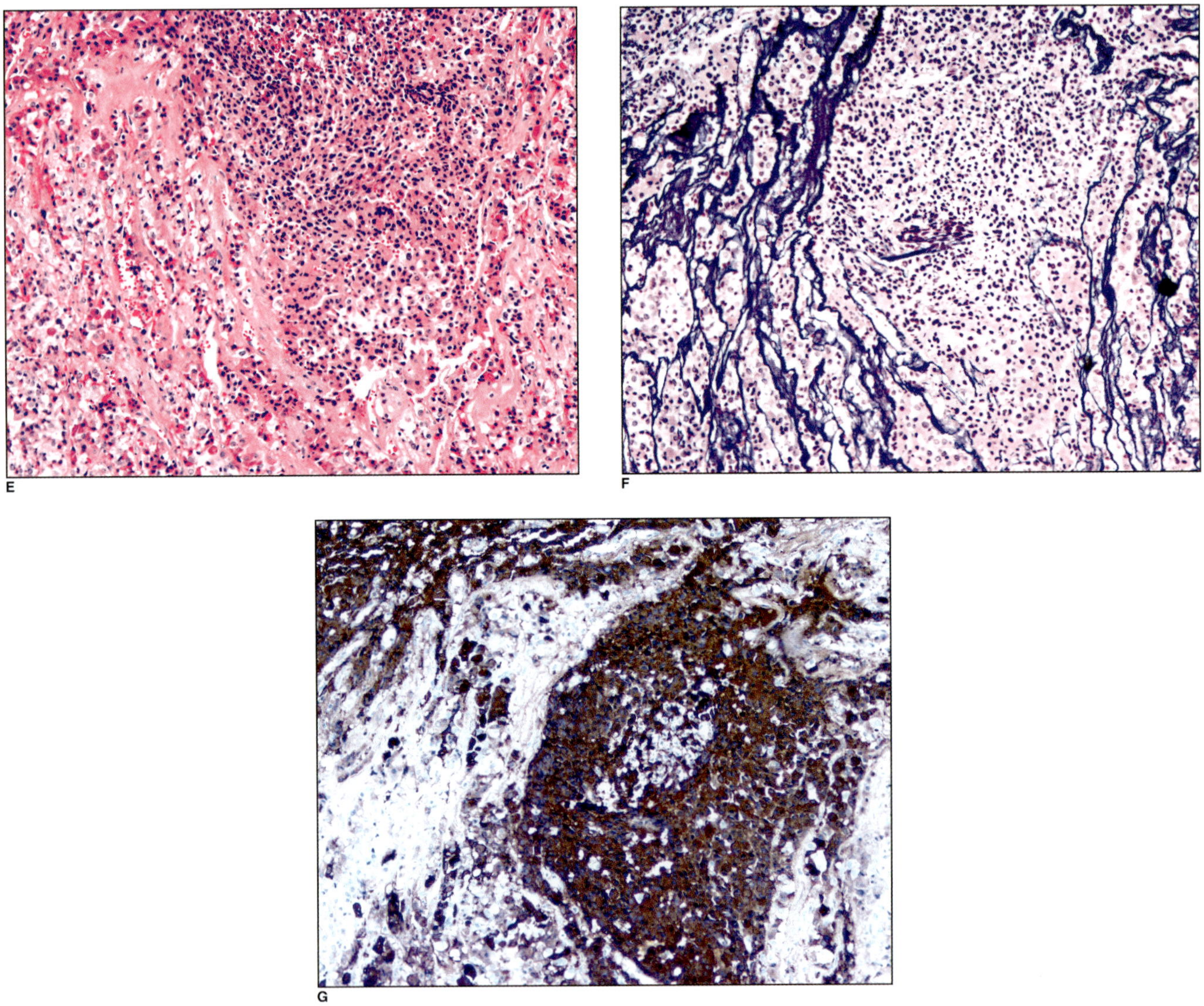

FIGURE 16.10: **(*Continued*).**

Examination of the brain and spinal cord showed large irregular areas of softening in the cerebral white matter (Figure 16.4A). Microscopically, these areas showed marked demyelination as demonstrated by an LFB stain (Figure 16.4B), with partial preservation of the axons seen with the Bielschowsky method (Figure 16.4C). Dense histiocytic infiltrates were seen in the demyelinated areas, with the presence of myelin debris in the cytoplasm of the macrophages (Figure 16.4D). The spinal cord showed similar findings; namely, acute demyelination in the anterior, posterior, and lateral columns (Figure 16.2E–F). The diagnosis was that of an acute, multifocal, and demyelinating leukoencephalopathy, most consistent with acute multiple sclerosis. ■

CASE 3— ***Alzheimer's Disease:*** A 78-year-old woman had a history of progressive dementia over the past 5 years. She had been in a nursing facility for the last 18 months, where she ultimately developed pneumonia with bilateral pleural effusions. Appropriate medical management failed to produce any improvement, and the patient died. Because the nature of the dementia was uncertain, a brain-only autopsy was requested.

A neurodegenerative evaluation was performed that included representative sections of the neocortex, hippocampus, deep gray matter, brainstem, and cerebellum, according to guidelines from the National Institute of Aging-Reagan Institute (29). In addition to H&E staining (Figure 16.1A,B), sections of the neocortex and hippocampal region were screened for the presence of NPs and NFTs with Bielschowsky silver staining (Figure 16.1C,D). The latter procedure revealed frequent NP, corresponding to an age-related plaque score of "C" (30). Frequent NFTs were also observed, corresponding to

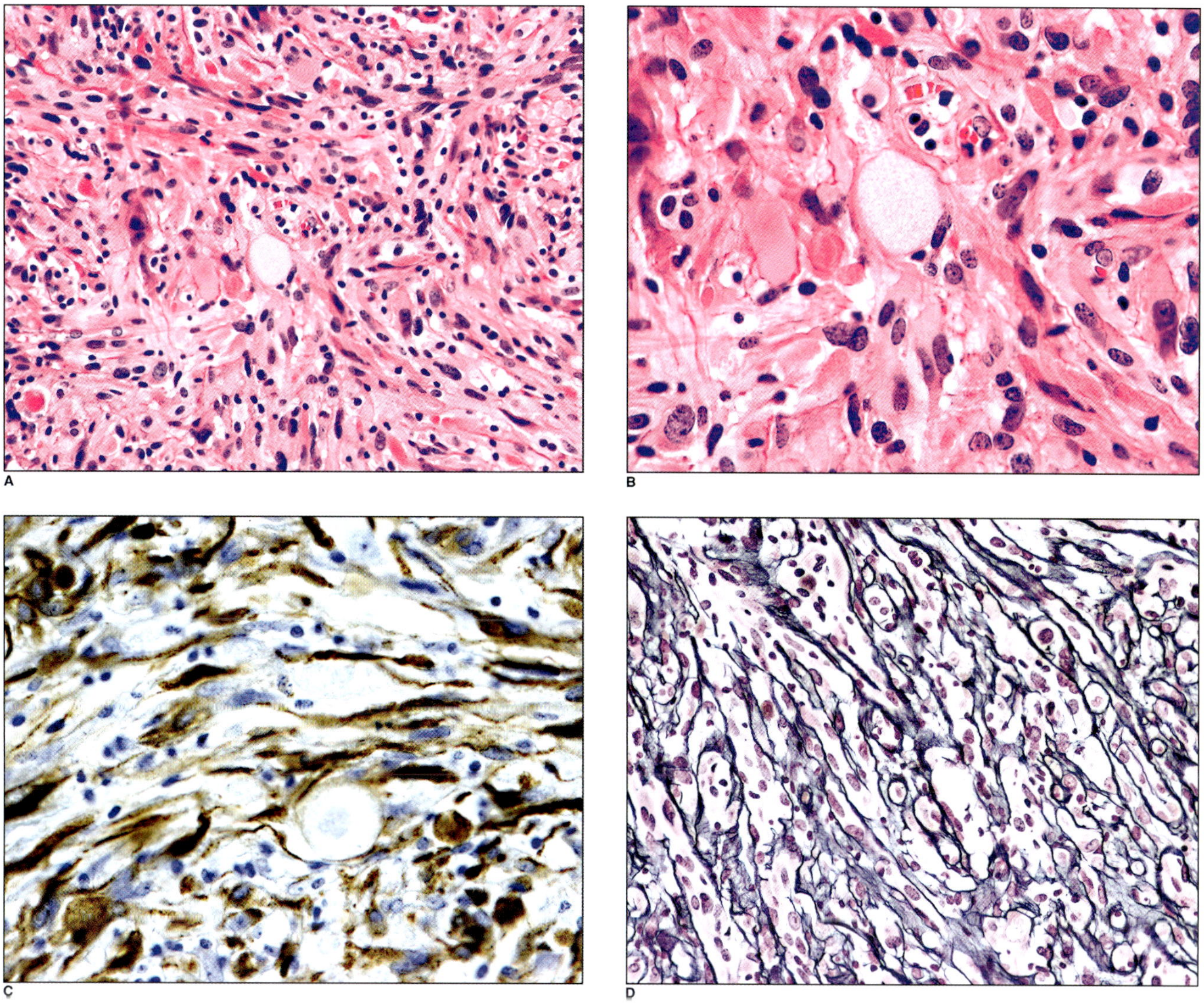

FIGURE 16.11: (A) Tumor cells in this lesion range from polygonal to elongated and fibrillary in nature. (B) Xanthomatous changes were prominent, and hyaline droplets were also present throughout the lesion. (C) Glial fibrillary acidic protein immunoreactivity in the mass confirms its glial nature. (D) Reticulin deposition is seen around individual tumor cells or cellular clusters with Wilder's stain.

a Braak and Braak stage of "V" (31). In addition, immunostains for phosphorylated Tau and Aβ-amyloid confirmed the presence of NFT and NP. No other specific morphologic abnormalities were seen. The combination of the neuropathologic findings and the clinical course of the dementia supported a definitive diagnosis of Alzheimer's disease. ■

CASE 4— ***Intravascular B-Cell Lymphoma and White Matter Infarct-Like Lesions:*** A 72-year-old woman had a complex medical history, including symptomatic coronary arterial atherosclerosis, hypothyroidism, and essential hypertension. She developed acute numbness and tingling of her lower extremities from the waist down, with progressive weakness in her legs over a 7-month period, finally becoming wheelchair bound. Magnetic resonance imaging of the brain showed several subcortical lesions and a presumptive diagnosis of a demyelinating process was made. Corticosteroids were administered without benefit, and a stereotactic brain biopsy was finally done in an attempt to secure a tissue diagnosis.

Microscopic examination demonstrated foci of parenchymal pallor in the cerebral white matter, with mild reactive astrocytosis (Figure 16.5A,B). Myelin loss was confirmed by an LFB stain (Figure 16.5C) and a moderate degree of axonal loss was apparent on Bielschowsky silver staining (Figure 16.5D). However, in other areas of the biopsy, blood vessels were filled with atypical, pleomorphic, dyshesive cells (Figure 16.5E) that were proven to be abnormal B-lymphocytes by immunohistologic studies (Figure 16.5F). The final diagnosis was that of intravascular large-cell B-cell lymphoma, with secondary infarct–like

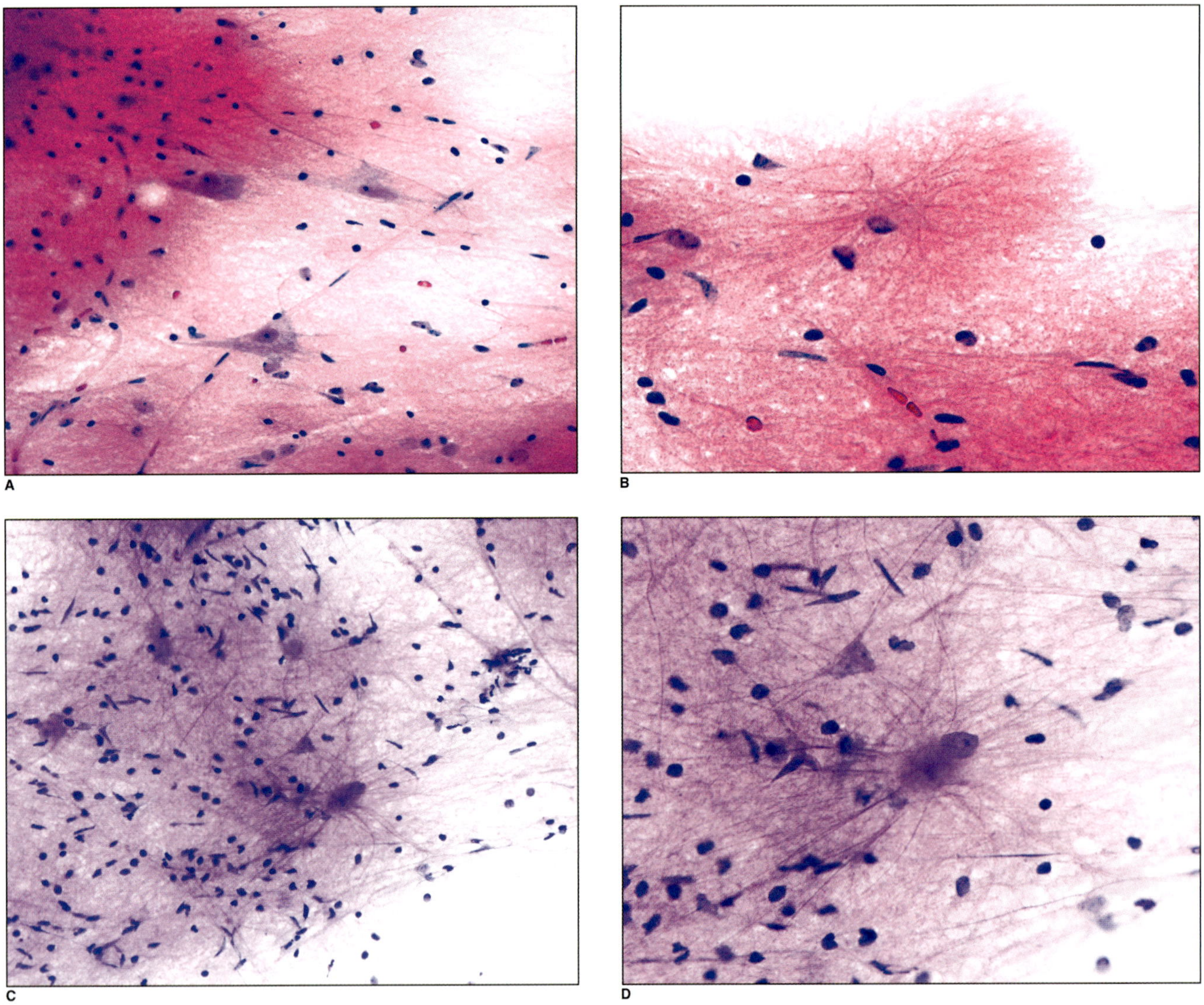

FIGURE 16.12: (A and B) This smear preparation of normal cerebral cortex demonstrates a delicate fibrillary background (neuropil) with pyramidal neurons (A) and reactive astrocytes (B). The pinkish delicate processes of the glial cells contrast with the blue neuronal cytoplasm and nuclei. Thin, delicate capillary vessels are also present. (C and D) This smear of a brain infarct shows large, reactive astrocytes with enlarged nuclei, prominent cytoplasm, and elongated fibrillary processes, in a Morris-stained preparation.

areas in the white matter with demyelination and axonal loss (32,33). Extensive clinical and radiologic evaluation failed to reveal skin lesions or visceral involvement by the lymphoma. The patient was given appropriate chemotherapy, and she demonstrated a good clinical response. ■

CASE 5— *Pleomorphic Xanthoastrocytoma*: A 33-year-old woman had a long history of headaches. She was felt to have chronic sinusitis and presumed migraines. Over the past year, the headaches worsened in frequency and were accompanied by right facial numbness. Magnetic resonance imaging of the brain showed an apparent intracranial tumor, represented by a 2.0 × 4.2 × 2.5–cm extraaxial right parietooccipital lesion. It was intensely enhancing with solid and cystic components, significant peritumoral edema, and a mass effect on the adjacent brain. These findings were felt to suggest a diagnosis of meningioma. At surgery, the tumor was located superficially in the cortex but with a definitive intraaxial component. The lesion was constituted by spindle-shaped cells arranged in fascicles, as well as large, globoid, eosinophilic cells with atypical nuclei (Figure 16.11A). Other cells with lipidized or xanthomatized cytoplasm were also prominent (Figure 16.11B). The tumor cells were shown to be glial in nature by immunohistochemistry (Figure 16.11C). Reticulin staining (Figure 16.11D) demonstrated atypical single cells and cellular clusters that were surrounded by reticulin, a feature that is typically seen in pleomorphic xanthoastrocytoma. The WHO grade of the neoplasm was II. ■

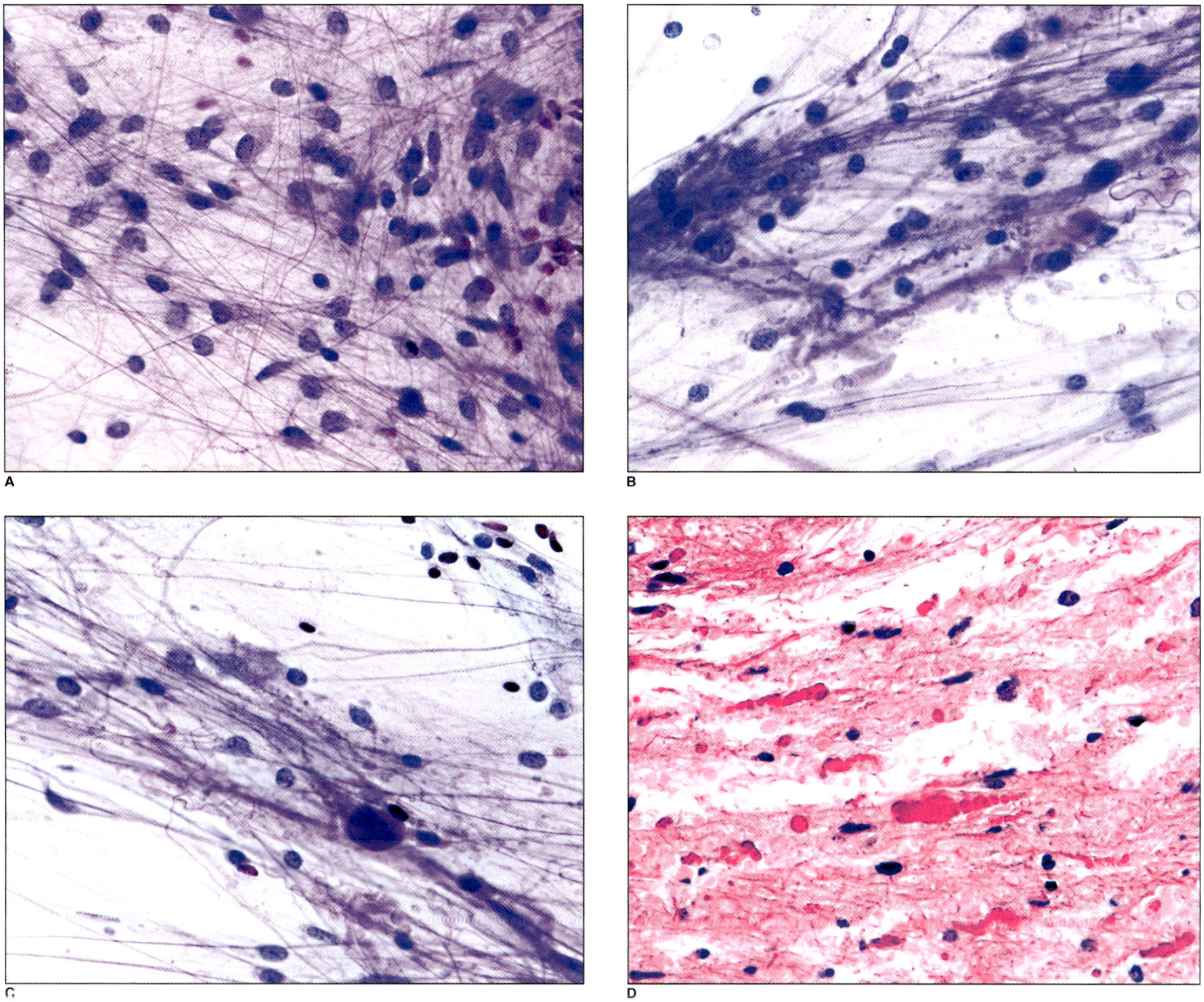

FIGURE 16.13: (A–C): This smear preparation of a pilocytic astrocytoma (WHO grade I) shows typical cytologic features of the constituent astroctytes. The dense, eosinophilic cellular processes impart a "wire-like" appearance to the background of the smear, and thick processes with features of Rosenthal fibers can also be seen throughout. (D): A corresponding paraffin section shows the fibrillary nature of the tumor cells, intermixed with prominent Rosenthal fibers.

CASE 6— *Gliosarcoma:* A 72-year-old woman was brought to an emergency room after having had a seizure at home; there was no history of epilepsy. The patient was disoriented, with signs of left hemiparesis. Computed tomography and magnetic resonance imaging of the brain showed a large ring-enhancing lesion in the right frontoparietal area. Imaging of the thorax and abdomen was unrevealing. The patient underwent a craniotomy with subtotal resection of the tumor. Histologic sections demonstrated an infiltrative high-grade astrocytic neoplasm with a biphasic growth pattern. Tumor cells with gemistocytic-like features alternated with spindle cells that were arranged in fascicles (Figure 16.8A,B). Trichrome staining clearly demarcated the two cellular components (Figure 16.8C,D) and the Gordon-Sweet procedure revealed an accumulation of reticulin fibers around tumor cells in the spindle-cell areas (Figure 16.8E). Immunohistology showed that the gemistocytic cells were glial in nature (Figure 16.8F), but that was not true of the spindle cells. A diagnosis of gliosarcoma (WHO grade IV) was made. ■

CASE 7— *Subclinical Pituitary Adenoma*: A 76-year-old man had a 3-week history of a gait disturbance. Imaging studies of the brain showed a hemorrhagic 2.5-cm cerebellar lesion, with a smaller lesion in the temporal lobe. A presumptive diagnosis of metastatic intracranial malignancy was made. Radiographic studies of the chest, abdomen, and pelvis revealed a right renal mass that was felt to be suspicious for renal cell carcinoma. Surgical decompression of

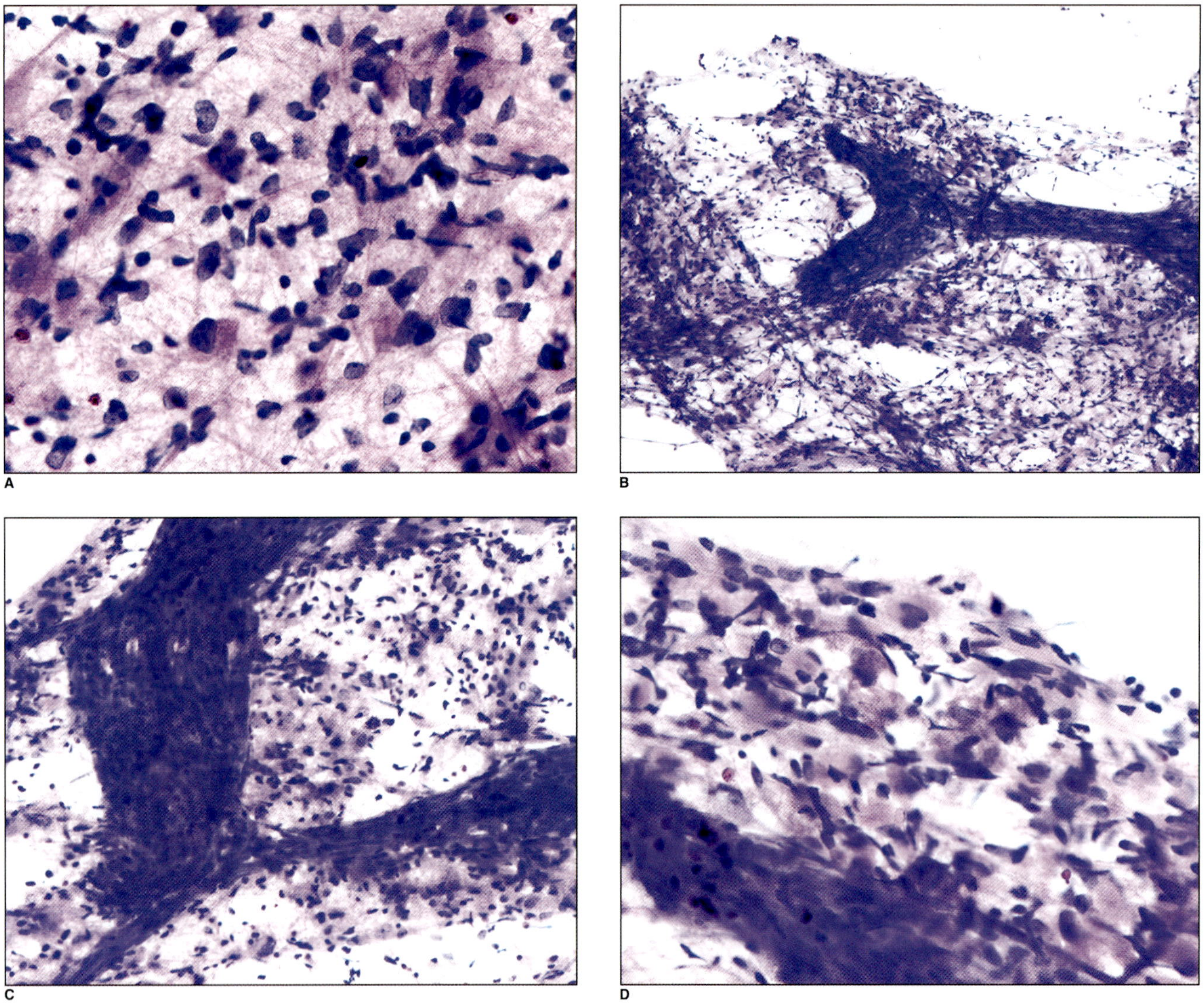

FIGURE 16.14: (A): This smear preparation of a glioblastoma (astrocytoma WHO grade IV) demonstrates tumor cells with extensive cytoplasmic and nuclear pleomorphism, as well as eosinophilic cytoplasm. (B–D): Prominent intratumoral blood vessels show a glomeruloid-like appearance. The endothelial and pericytic cells in the lesion, stained blue in a Morris preparation, are easy to separate from the pink neuroglial elements (D).

the posterior fossa was scheduled, but the patient had a cardiopulmonary arrest and died before it could be done.

Autopsy examination confirmed the presence of primary renal cell carcinoma in the right kidney with several metastases to the cerebrum, cerebellum, and lungs, as well as tonsillar herniation of the brain. In addition, an incidental pituitary adenoma was found (Figure 16.9A). The latter lesion was immunonegative for all pituitary hormones and had a "null-cell" phenotype.

The normal anterior pituitary gland had a distinctive acinar structure composed of an admixture of secretory cell types. Its architecture was easily demonstrated by reticulin staining (Figure 16.9B). In the pituitary adenoma, the regular acinar arrangement of the cells was disrupted; indeed, reticulin stains revealed a complete loss of a nested substructure (Figure 16.9C).

Irregular bundles of fibers may also separate large groups of adenomatous cells, and areas of desmoplasia can likewise be seen. Most adenomas compress the adjacent normal pituitary, and, as a result, reticulin fibers form a "pseudocapsule" at the periphery of the tumors (Figure 16.9D,E). Reticulin stains are also useful in distinguishing adenohypophyseal hyperplasia from normal glandular tissue because the former lesion shows an expanded but retained reticulin network; it is not effaced as seen in adenomas (34). ■

CASE 8— ***Pituitary Adenoma:*** A 32 year-old woman with insulin-resistant diabetes mellitus and hypertension presented with symptoms and signs of Cushing's syndrome. Magnetic resonance imaging of the brain revealed

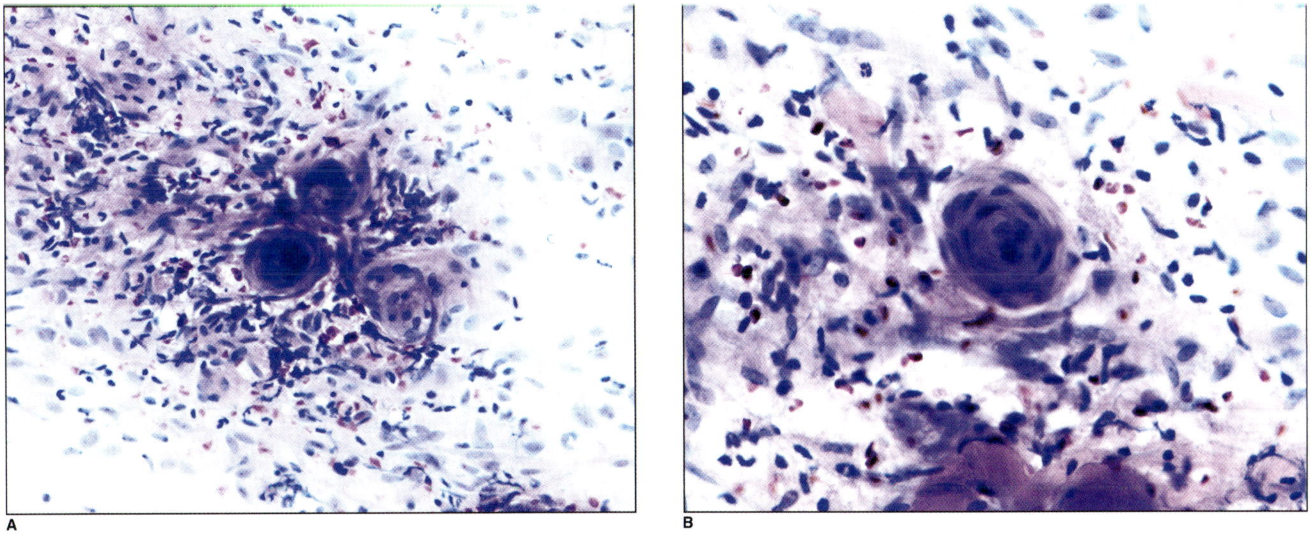

FIGURE 16.15: (A and B) A smear of meningioma shows typical meningothelial cells with ill-defined cytoplasmic borders and prominent cellular whorl formation. Evenly distributed chromatin and small nucleoli are readily appreciated in this cytologic preparation stained with Morris's procedure.

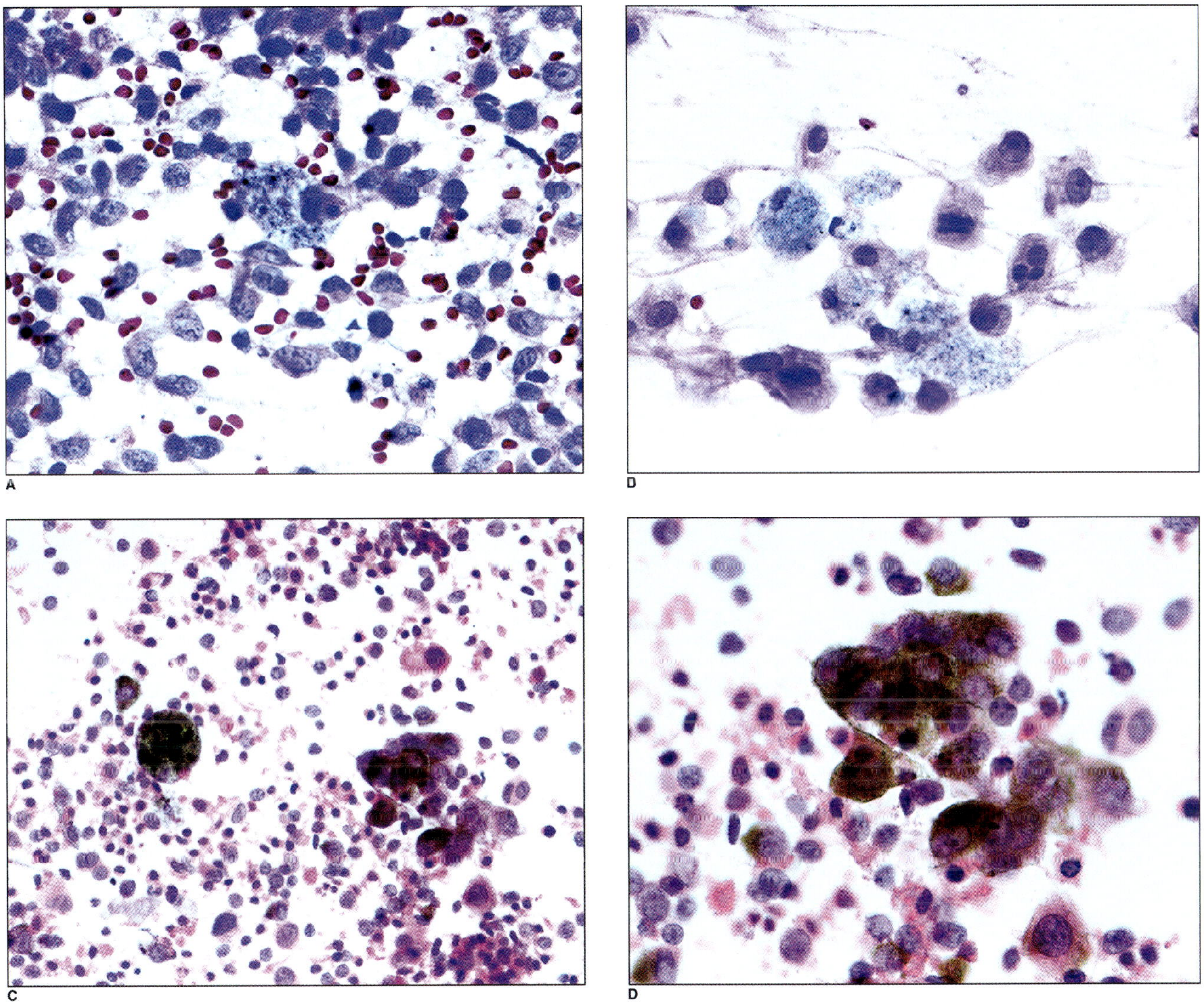

FIGURE 16.16: (A–D) Smears of a metastatic melanoma, stained with Morris's and H&E methods. In addition to the different tintorial quality of the cytoplasm and nuclei that are produced in these two stains, notice should be taken of the distinctive labeling of melanin that is seen in the H&E (golden brown) and Morris-stained (gray black) preparations.

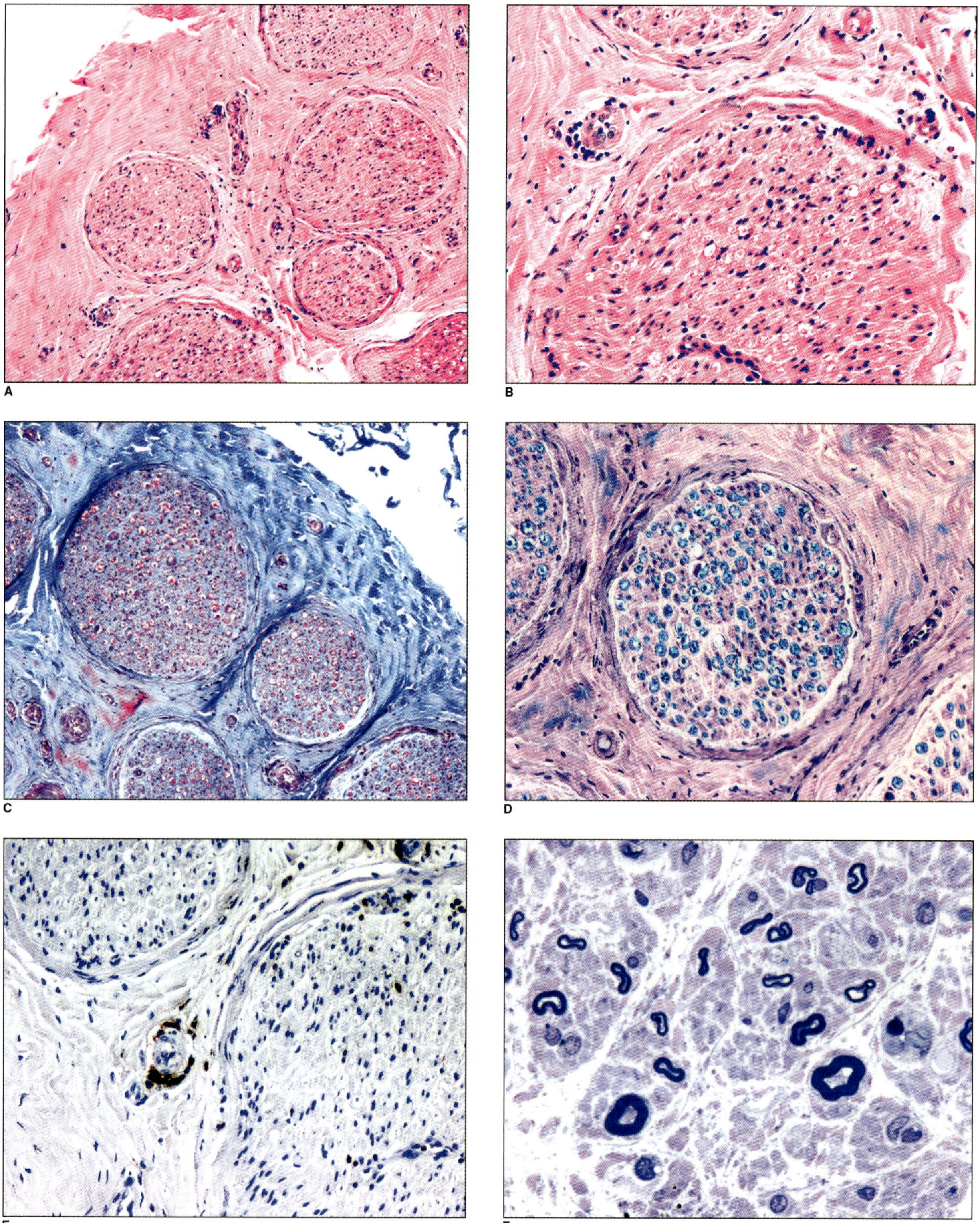

FIGURE 16.17: (A and B) This transverse section of a paraffin-embedded sural nerve biopsy shows perineurial fibrosis (A) and an epineurial perivascular mononuclear infiltrate without fibrinoid necrosis (B). (C) Trichrome staining better outlines the perineurial fibrosis. (D) The LFB method demonstrates demyelination of the nerve. (E) Immunohistochemistry for CD45 accentuates the lymphocytic inflammatory infiltrate around epineurial and endoneurial vessels. (F) An epoxy-embedded and toluidine blue–stained transverse section of the nerve biopsy shows a decreased density of myelinated fibers with a particular decrement in large myelinated fibers. The myelin is thin and some fibers are completely demyelinated. "Onion bulb" formations are not appreciated.

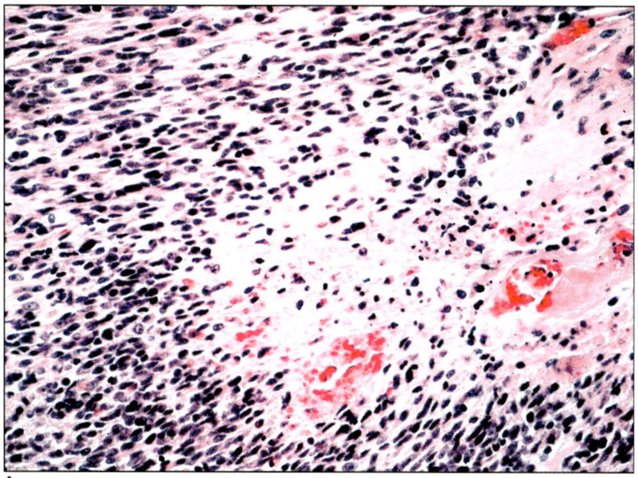

A

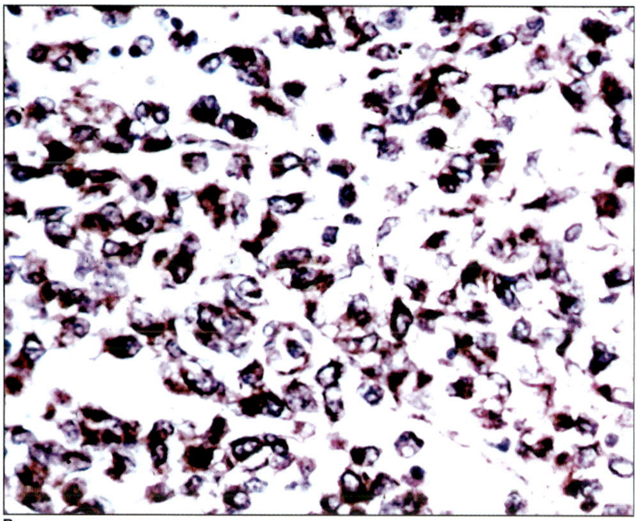

B

FIGURE 16.18: ***A glioblastoma multiforme (high-grade astrocytoma) is shown here (A), which demonstrates lectin affinity for CON-A (B).***

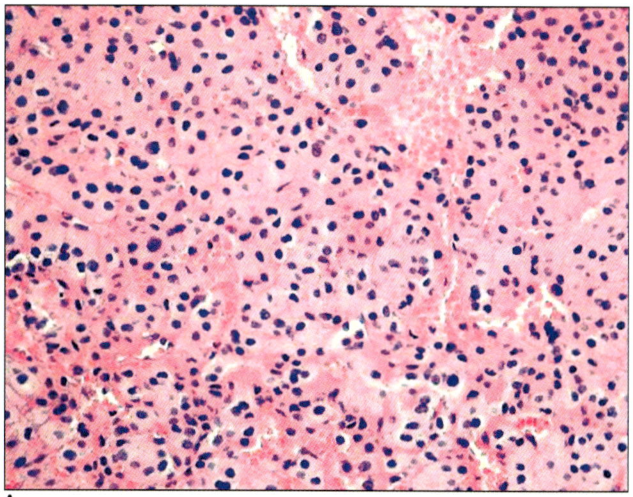

A

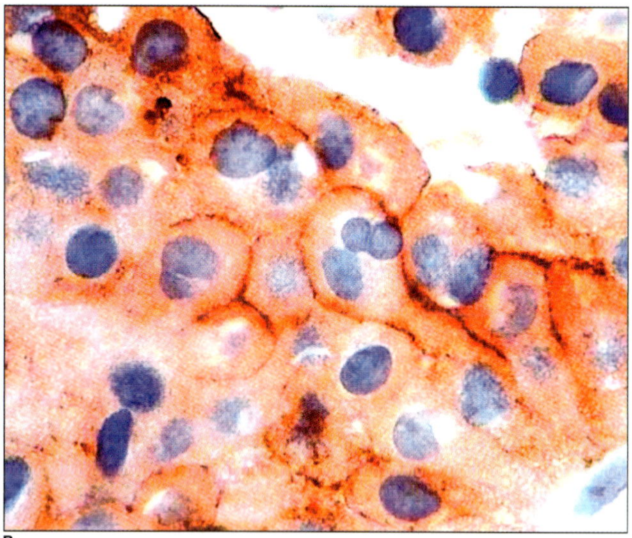

B

FIGURE 16.19: ***This pituitary adenoma (A) demonstrates binding of the tumor cells to PNA (B). This finding has been correlated in some studies with prolactin production.***

a sellar lesion consistent with a pituitary adenoma. The patient underwent transsphenoidal surgery and a pituitary adenoma was indeed found. The tumor was immunoreactive for adrenocorticotrophic hormone (ACTH) and a diagnosis of corticotroph cell adenoma was made. After surgery, the patient lost 20 lb and regained a normal menstrual pattern. However, 2 years later, she complained of worsening control of her diabetes, with weight gain and amenorrhea. Plasma cortisol and ACTH levels were again elevated. Although imaging studies of the sella showed no abnormalities, transsphenoidal exploration of the sellar region was again undertaken.

Histopathologic examination of specimens from the second surgery showed pituitary tissue with extensive fibrosis (Figure 16.10A). Intermixed with areas of normal anterior hypophysis, there were foci of a monomorphic cellular proliferation with no discernible acinar architecture (Figure 16.10B). Those areas were well seen on reticulin staining, which demonstrated the lack of an acinar pattern and confirmed the diagnosis of recurrent pituitary adenoma (Figure 16.10C). The tumor cells were again immunoreactive for ACTH (Figure 16.10D). Reticulin stains are particularly helpful in evaluating recurrent pituitary adenomas. Neurosurgeons customarily pack the sellar cavity with fascia or fat after transsphenoidal surgery, and this practice may distort the histologic appearance of tissues that are sampled in subsequent procedures. Because of potentially extensive fibrosis of the pituitary gland and surrounding tissues, reticulin staining helps to distinguish normal adenohypophysis from adenomatous tissue. ■

CASE 9— ***Peripheral Neuropathy***: A 68-year-old woman had a long cigarette smoking history; she developed

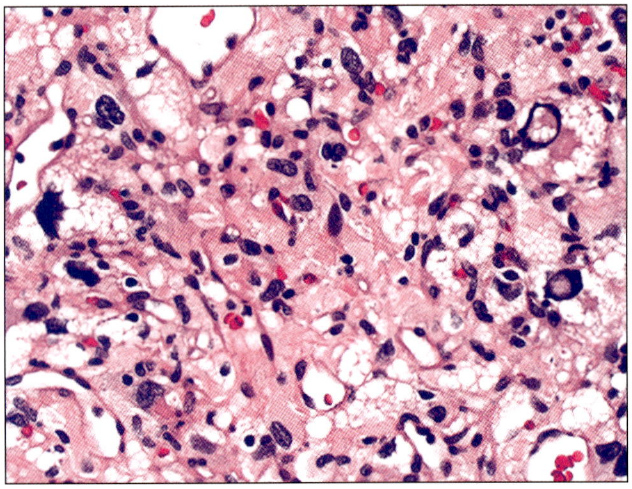

FIGURE 16.20: CHB, a clear-cell neoplasm that may be confused histologically with metastatic renal cell carcinoma. It remains to be seen whether lectin histochemistry can separate those two lesions.

dysesthesia of the right foot with a sural distribution, followed thereafter by pain in the entire foot and right leg weakness. Her history and sequential electrodiagnostic studies suggested a diagnosis of mononeuritis multiplex, preferentially affecting the right lower extremity. An extensive evaluation for possible infectious or metabolic etiologies was unrevealing, leading to a preferred clinical diagnosis of vasculitis or paraneoplastic neuropathy. A sural nerve biopsy was performed.

Examination of H&E-stained sections of the sampled nerve showed chronic inflammatory infiltrates surrounding blood vessels in the endoneurium and perineurium (Figure 16.16A,B). In several foci, the vascular inflammation was transmural, but no evidence of fibrinoid necrosis was apparent. The Masson trichrome stain showed a moderate degree of perineurial fibrosis, with no appreciable increase in endoneurial collagen content (Figure 16.16C). There was a mild loss of large myelinated axons, with some areas showing myelin degeneration on LFB staining (Figure 16.16D,E). Examination of Epon resin–embedded thick sections confirmed the loss of myelinated fibers, with thinning of myelin sheaths and regenerating Schwannian clusters (Figure 16.16F). A diagnosis of vasculitic neuropathy was made, with axonal loss and demyelination. The patient subsequently underwent immunomodulatory therapy with beneficial results. ■

<u>CASE 10</u>— ***Cerebral Amyloid Angiopathy:*** A 67-year-old woman had a 1-year history of slowly progressive dementia but with an abrupt recent change in her mentation. Neurologic imaging studies revealed significant T2-hyperintense signal abnormalities involving the frontal white matter bilaterally and the right temporooccipital region. A mild mass effect was noted, with no enhancement of the lesional foci. Differential diagnostic considerations included cerebral vasculitis, the adult form of leukodystrophy, an opportunistic infection, and gliomatosis cerebri. A biopsy was obtained.

The specimen showed thickened walls of small- to intermediate-sized arteries in the cerebrum. They contained amorphous, eosinophilic, intramural material (Figure 16.2A). In addition, numerous NPs with central cores were seen in H&E-stained sections (Figure 16.2B). Congo red staining was strongly positive in both the vascular walls and the cores of NPs (Figure 16.2C). Immunostaining for Aβ-amyloid confirmed the presence of that protein in the abnormal vessels and plaques (Figure 16.2D). A diagnosis of cerebral amyloid angiopathy was made (35). ■

REFERENCES

1. Vacca L. *Laboratory Manual of Histochemistry*. 1985, Raven Press, New York.
2. Carson FL. *Histotechnology: A Self-Instructional Text*. 2nd Edition. 1997, American Society for Clinical Pathology ASCP Press, Chicago.
3. Kiernan JA. *Histological and Histochemical Methods: Theory and Practice*. 3rd Edition. 1999, Butterworth-Heinemann, Oxford, pp. 363–376.
4. Luna LG. *Manual of Histologic Staining Methods of the AFIP*. 3rd Edition. 1968.
5. Luna LG. *Histopathologic Methods and Color Atlas of Special Stains and Tissue Artifacts*. 1992, American Histolabs, Inc., Gaithersburg.
6. Lopes MBS, Vandenberg SR. Tumors of the Central Nervous System. In: *Diagnostic Histopathology of Tumors*. 3rd Edition (Fletcher CDM, Ed.). 2007, Churchill Livingstone, London, pp. 1653–1732.
7. Klüver H, Barrera E. A method for the combined staining of cells and fibers in the nervous system. *J Neuropathol Exp Neurol* 12: 400–403, 1953.
8. Margolis G, Pickett JP. New applications of the luxol fast blue myelin stain. *Lab Invest* 5: 459–474, 1956.
9. Kiernan JA. *Histological and Histochemical Methods: Theory and Practice*. 3rd Edition. 1999, Butterworth-Heinemann, Oxford, pp. 355–389.
10. Rubinstein LJ. Diagnostic aids for the histologic and cytologic examination of tumors of the central nervous system. In: *Tumors of the Central Nervous System, AFIP Atlas of Tumor Pathology*. Armed Forces Institute of Pathology, Washington, D.C. 1972, pp. 361–378.
11. Vandenberg SR. Desmoplastic infantile ganglioglioma and desmoplastic cerebral astrocytoma of infancy. *Brain Pathol* 3: 275–281, 1993.
12. Kepes JJ, Rubinstein LJ, Eng LF. Pleomorphic xanthoastrocytoma: a distinctive meningocerebral glioma of young subjects with relatively favorable prognosis. A study of 12 cases. *Cancer* 44: 1839–1852, 1979.
13. Kepes JJ. Pleomorphic xanthoastrocytoma: the birth of a diagnosis and a concept. *Brain Pathol* 3: 269–274, 1993.
14. Cerda-Nicolas M, Kepes JJ. Gliofibromas (including malignant forms), and gliosarcomas: a comparative study and review of the literature. *Acta Neuropathol* (Berl) 85: 349–361, 1993.

15. Meyermann R, Kletter G. Comparative evaluation of various histological techniques for the rapid diagnosis of brain tumors. *Acta Neurochir* 35: 171–180, 1976.
16. Morris AA. The use of smear technique in the rapid histological diagnosis of tumors of the central nervous system. *J Neurosurg* 4: 497–504, 1947.
17. Midroni G, Bilbao JM. *Biopsy Diagnosis of Peripheral Neuropathy*. 1995, Butterworth-Heineman, Newton, pp. 1–105.
18. Dyck PJ, Dyck PJB, Englestad J. Pathologic alterations of nerves. In: *Peripheral Neuropathy*. 4th Edition (Dyck PJ, Thomas PK, Eds.). 2005, Elsevier-Saunders, Philadelphia, pp. 733–829.
19. Schill J, Cervos-Navarro J. Lectin histochemistry of mixed gliomas demonstrating an intermediate cell type. *Histo Histopathol* 13: 73–79, 1998.
20. Yang S, Liu Z, Zhu Y, Chen X. Labeling and quantitative analysis of six lectin receptors of intracranial gliomas. *Chin Med J* 108: 37–43, 1995.
21. Cruz-Sanchez FF, Rossi ML, Buller JR, et al. Oligodendrogliomas: a clinical, histological, immunocytochemical, and lectin-binding study. *Histopathology* 19: 361–367, 1991.
22. Kurosaki M, Hori T, Takata K, Kawakami H, Hirano H. Peanut agglutinin binding to human prolactin-producing pituitary adenomas. *Acta Neuropathol* 89: 475–482, 1995.
23. Behncken A, Saeger W. Lectin binding in pituitary adenomas and normal pituitaries. *Pathol Res Pract* 187: 629–631, 1991.
24. Ironside JW, Stephenson TJ, Royds JA, et al. Stromal cells in cerebellar hemangioblastomas: an immunocytochemical study. *Histopathology* 12: 29–40, 1988.
25. Ulrich W, Horvat R, Krisch K. Lectin histochemistry of kidney tumours and its pathomorphological relevance. *Histopathology* 9: 1037–1050, 1985.
26. Hiasa Y, Kitamura M, Nakaoka S, et al. Antigen immunohistochemistry of renal cell adenomas in autopsy cases: relevance to histogenesis. *Oncology* 52: 97–105, 1995.
27. Palmini A, Avanzini G, Babb T, Guerrini R, Foldvary-Schaefer N, Jackson G, Luders HO, Prayson R, Spreadico R, Vinters HV. Terminology and classification of the cortical dysplasia. *Neurology* 62 (6 Suppl 3): S2–S8, 2004.
28. Englund C, Folkert RD, Born D, Lacy JM, Hevner RF. Aberrant neuronal-glial differentiation in Taylor-type focal cortical dysplasia (type IIA/B). *Acta Neuropathol* 109: 519–533, 2005.
29. Newell KL, Hyman BT, Growdon JH, Hedley-Whyte ET. Application of the National Institute on Aging (NIA)-Reagan Institute criteria for the neuropathological diagnosis of Alzheimer disease. *J Neuropathol Exp Neurol* 58: 1147–1155, 1999.
30. Mirra SS, Heyman A, McKeel D, Sumi SM, Crain BJ, Brownlee LM, Vogel FS, Hughes JP, van Belle G, Berg L. The Consortium to Establish a Registry for Alzheimer's Disease (CERAD). Part II: standardization of the neuropathologic assessment of Alzheimer's disease. *Neurology* 41: 479–486, 1991.
31. Braak B. Neuropathologic staging of Alzheimer related changes. *Acta Neuropathol* 82: 239–259, 1991.
32. Baehring JM, Henchcliffe C, Ledezma J, Fulbright R, Hochberg FH. Intravascular lymphoma: magnetic resonance imaging correlates of disease dynamics within the central nervous system. *J Neurol Neurosurg Psych* 76: 540–544, 2005.
33. Greenberg SM, Parisi JE, Keegan BM. A 63-year-old man with headaches and behavioral deterioration. *Neurology* 68: 782–787, 2007.
34. Kovacs K, Horvath E. *Tumors of the Pituitary Gland. AFIP Atlas of Tumor Pathology*. Armed Forces Institute of Pathology, Washington, D.C. 2nd Series. 1986, pp. 1–5.
35. Salat DH, Smith EE, Tuch DS, Benner T, Pappu V, Schwab KM, Gurol ME, Rosas HD, Rosand J, Greenberg SM. White matter alterations in cerebral amyloid angiopathy measured by diffusion tensor imaging. *Stroke* 37: 1759–1764, 2006.

CHAPTER 17

Diagnostic Histochemistry in Ophthalmic Pathology

Amy Lin, M.D., and
Robert Folberg, M.D.

Although "routine" histochemistry has fallen out of favor in many subspecialties in surgical pathology, the use of the periodic acid-Schiff (PAS) stain remains an essential component of the practice of surgical ophthalmic pathology. It is fair to say that the full extent of pathologic processes within the eye cannot be appreciated without the pathologist reviewing a PAS-stained slide. In these, it is unnecessary to stain control tissue, such as liver, because many internal controls (such as basement membranes) are present.

The first portion of this chapter will focus on the use of PAS in ophthalmic pathology. The remainder of the chapter illustrates the application of a variety of histochemical stains to the diagnosis of ocular diseases.

PERIODIC ACID-SCHIFF STAIN

Cornea

Two basement membrane structures—the epithelial basement membrane and Descemet's membrane—are present in the cornea. The epithelial basement membrane is thin and not readily seen in routine hematoxylin and eosin (H&E)–stained sections but is identified easily with PAS-stained slides. In cases of corneal edema, resolution of bullous separation of the epithelium may result in entrapment of epithelial tags beneath newly formed cells and duplication of basement membrane (Figure 17.1). Similar findings may be seen in epithelial basement membrane dystrophy (map-dot-fingerprint dystrophy).

Descemet's membrane—the basement membrane of the corneal endothelium—thickens with age. Discontinuities of Descemet's membrane may occur in a variety of conditions including Peter's anomaly (Figure 17.2), obstetrical forceps injury, congenital glaucoma, keratoconus with hydrops (Figure 17.3), and perforating corneal wounds, either accidental or surgical (Figure 17.4).

Fuchs' endothelial dystrophy and bullous keratopathy are common indications for corneal transplantation. Both are the result of corneal endothelial loss with subsequent

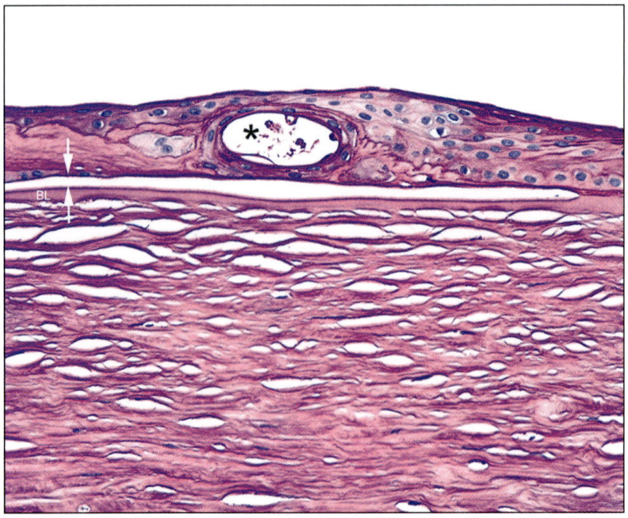

***FIGURE 17.1:** Intraepithelial basement membrane deposition in a patient with corneal edema. There is bullous separation of the epithelium (between arrows) and an intraepithelial "cyst" (asterisk). Notice that Bowman's layer (BL) does not stain with PAS (PAS, 200×).*

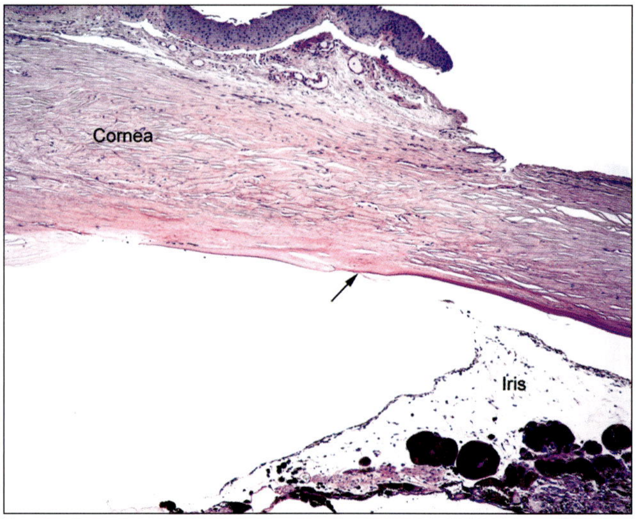

***FIGURE 17.2:** Central absence of Descemet's membrane in Peter's anomaly, a congenital malformation of the anterior segment. A strand of iris tissue is adherent to the cornea at the termination of Descemet's membrane (arrow) (PAS, 40×).*

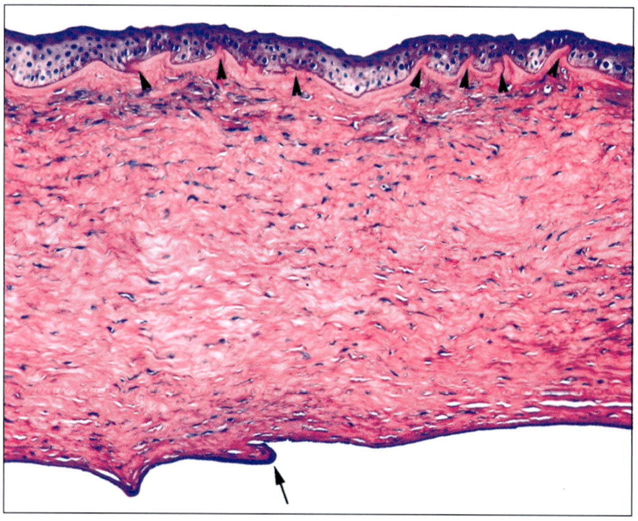

FIGURE 17.3: Keratoconus with hydrops. Multiple breaks in Bowman's layer are present (arrowheads), consistent with keratoconus. Progressive thinning of the cornea led to rupture of Descemet's membrane and endothelium. Descemet's membrane retracted and coiled on itself (arrow). Endothelial cells resurfaced the exposed stroma and generated new basement membrane. The overlying corneal stroma is edematous and scarred (PAS, 200×).

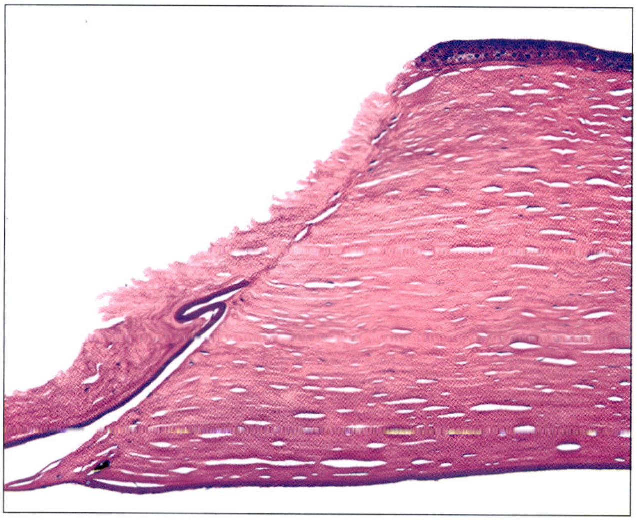

FIGURE 17.4: Corneal graft failure. A perforating corneal wound associated with a break in and entrapment of Descemet's membrane at the edge of a corneal button, consistent with previous penetrating keratoplasty (corneal transplantation) (PAS, 100×).

stromal edema and bullous epithelial changes. In bullous keratopathy, the thickness of Descemet's membrane is appropriate for the patient's age and is uniform in thickness, whereas in Fuchs' endothelial dystrophy, Descemet's membrane is diffusely thickened and drop-like excrescences (guttata) are typically identified (Figure 17.5).

Ciliary Body

Thickening of the basement membrane of the ciliary epithelium in the pars plicata of the ciliary body is characteristic

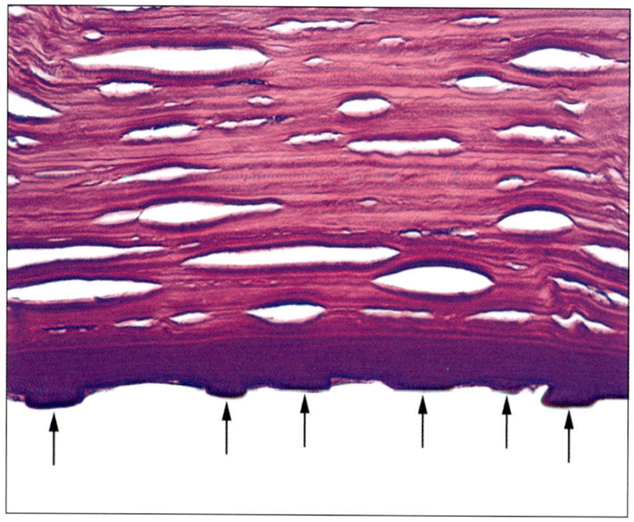

FIGURE 17.5: Fuchs' endothelial dystrophy. Descemet's membrane is thick, and multiple guttata are identified (arrows) (PAS 400×).

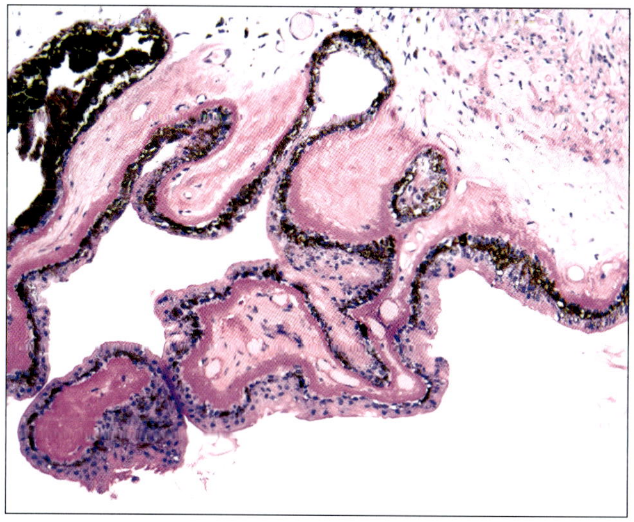

FIGURE 17.6: Thickening of the basement membrane of the ciliary epithelium of the pars plicata in the ciliary body of a diabetic patient (PAS, 100×).

of diabetes mellitus (Figure 17.6). The finding is analogous to mesangial thickening in the renal glomerulus in chronic diabetes.

Lens

The lens capsule is the basement membrane of the lens epithelium. Pseudoexfoliation syndrome [sometimes known as "exfoliation" syndrome (1)] is characterized by the accumulation of abnormal acellular material on the surface of the lens, zonules, ciliary processes, and posterior surface of the iris. Pseudoexfoliation material is a glycoprotein-proteoglycan complex that incorporates many varieties of extracellular matrix material, including laminin (2), heparan sulfate (3), chondroitin sulfate, and fibronectin (4). On

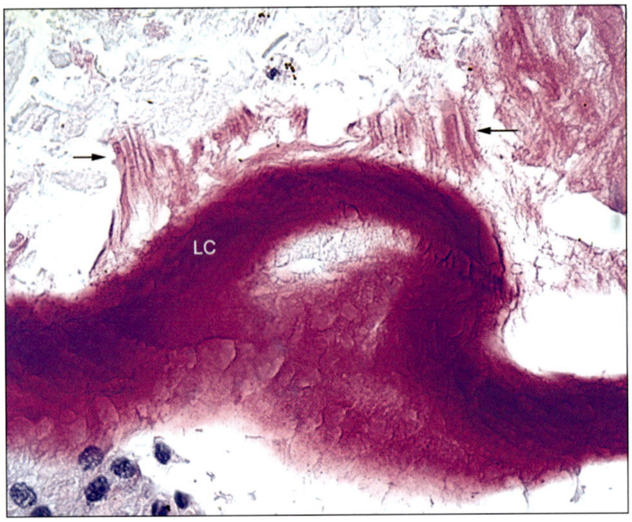

FIGURE 17.7: Pseudoexfoliation material (arrows) on the surface of the lens capsule (LC) (PAS, 400×).

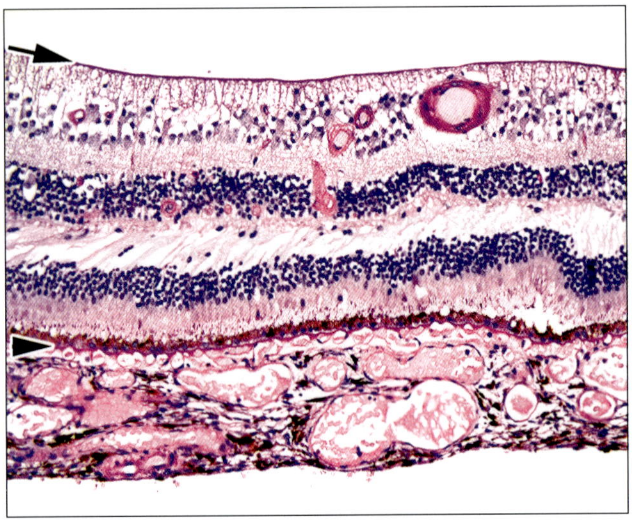

FIGURE 17.8: Normal retina and choroid in the macula. PAS highlights the ILM (arrow) and Bruch's membrane (arrowhead) (PAS, 100×).

H&E sections, the material appears as eosinophilic, brush-like excrescences (similar to iron filings on the surface of a magnet). PAS stain highlights both the lens capsule and the exfoliation material (Figure 17.7).

Retina

The internal limiting membrane (ILM) (Figure 17.8) is the basal lamina of Muller cells. The ILM is attenuated in the foveola, over major retinal vessels, and at the optic nerve head. Normally, cells are absent on the internal (vitreous) surface of the ILM. Epiretinal membranes (ERMs) are composed of retinal glial cells, retinal pigment epithelial cells, or combinations of these two cell types along the surface of the ILM (5,6). Contraction of these glial cells wrinkles the ILM and underlying retina (Figure 17.9), with associated visual disturbance. ERMs may be idiopathic, or associated with retinal vascular disease, retinal breaks and detachments, ocular trauma, uveitis, and intraocular surgery.

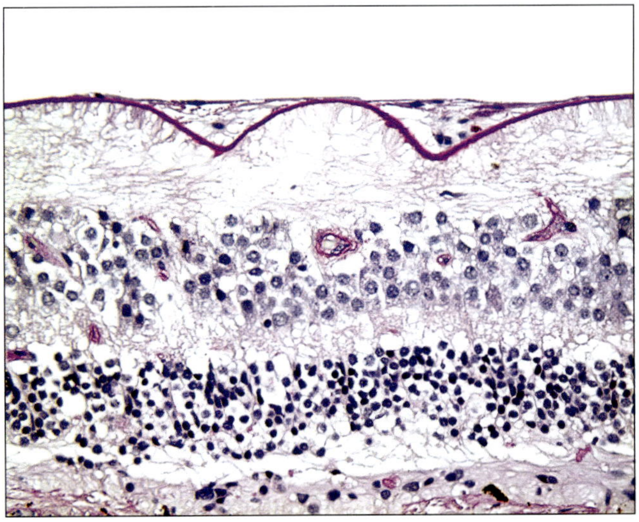

FIGURE 17.9: Retina with ERM. Cells lining the ILM have contracted and have distorted (wrinkled) the retinal surface (PAS, 200×).

The ILM is also an important landmark for evaluating retinal vascular diseases such as diabetic retinopathy. Abnormal angiogenic vessels may develop because of retinal ischemia. When these new vessels are confined to the retina, they are referred to as intraretinal microvascular abnormalities or non-(pre)-proliferative diabetic retinopathy. Proliferative diabetic retinopathy (PDR) refers to angiogenic vessels that have broken through the ILM, either over the retina (neovascularization elsewhere) or optic disk (neovascularization of the disk). Patients with PDR, in particular, or retinal neovascularization, in general, are at increased risk for visual loss due to vitreous hemorrhage, retinal detachment, and neovascular glaucoma.

Choroid

Bruch's membrane separates the retinal pigment epithelium from the choroid, serving as the basement membranes of both the retinal pigment epithelium (RPE) and the choriocapillaris. Composed of collagen and elastic fibers, Bruch's membrane normally thickens with increasing age, may calcify in advanced age, and be the site of basement membrane deposition on the surface beneath the retina lined by the retinal pigment epithelium: these excrescences are known as retinal pigment epithelial drusen. Identification of diffuse drusen (Figure 17.10) is associated with age-related macular degeneration (AMD). Somewhat analogous to diabetic retinopathy, AMD is classified as nonneovascular or "dry," or as neovascular or "wet." In wet AMD, angiogenic vessels from the choroid occupy the sub-RPE space

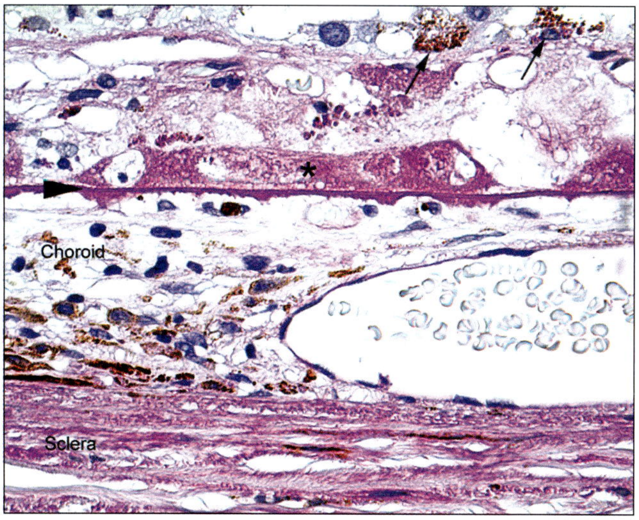

FIGURE 17.10: Diffuse drusen, also known as "basal laminar drusen" (*) in a patient with AMD. The overlying retinal pigment epithelium (arrows) and photoreceptor outer segments are replaced by a neovascular membrane (CNV). Compare with Figure 17.7 (PAS, 400×).

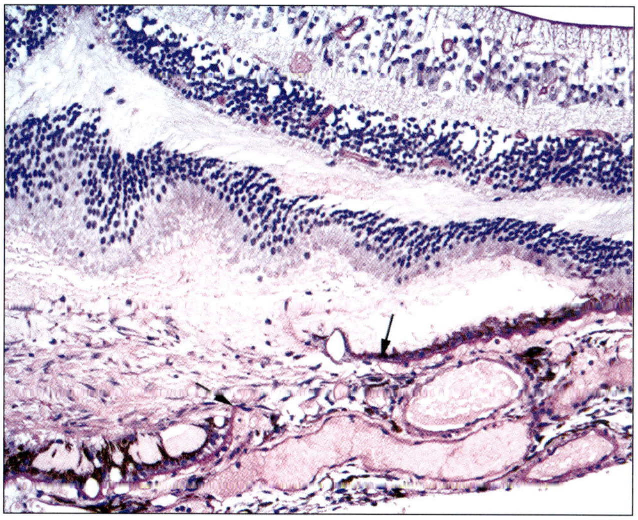

FIGURE 17.11: Traumatic choroidal rupture in the macula. Notice the break in the RPE and Bruch's membrane (between arrows) (PAS, 100×).

(Figure 17.10). Patients with choroidal neovascular membranes are at increased risk for visual loss from subretinal hemorrhage and scarring. Other causes of choroidal neovascular membranes, such as traumatic choroidal rupture (Figure 17.11), high myopia, angioid streaks, and presumed ocular histoplasmosis syndrome, are associated with breaks in Bruch's membrane.

IRON STAIN

In the eye, iron deposits usually in epithelial structures or structures derived from the epithelia (such as the neurosensory

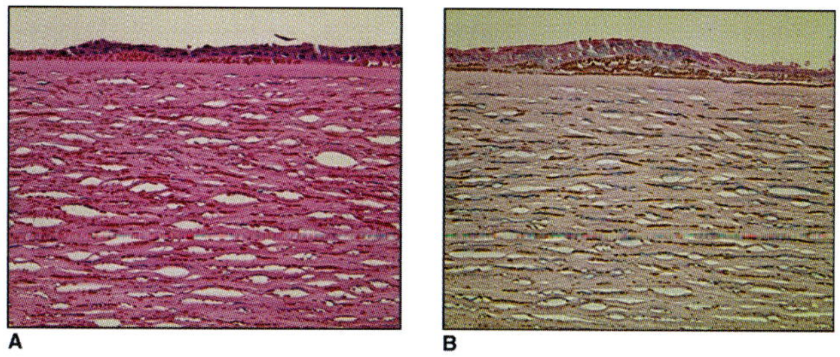

FIGURE 17.12: (A) Corneal blood staining. The red stippling in the corneal stroma represents hemoglobin. (B) Iron is detected within stromal fibroblasts (keratocytes) (A—H&E, 200×; B—iron, 200×).

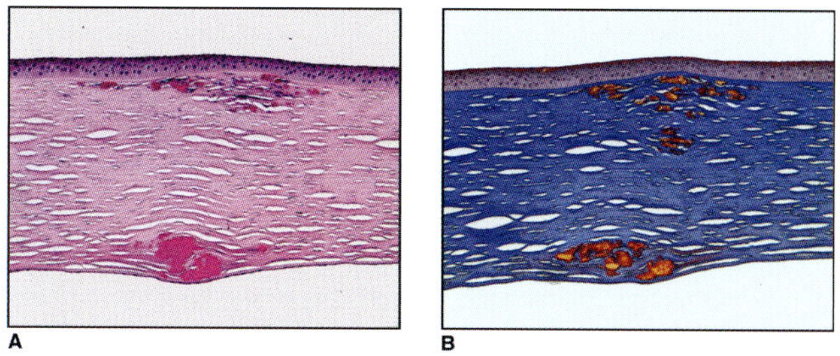

FIGURE 17.13: Granular dystrophy. (A) Superficial and deep stromal eosinophilic deposits of granular dystrophy. (B) The deposits stain red with MT stain (A—H&E, 100×; B—MT, 100×).

retina). Iron in the corneal epithelium originates from a stagnant tear film, seen clinically as a Fleischer ring in keratoconus, the Ferry line in advance of a glaucoma filtering bleb, the Stocker line at the advancing margin of a pterygium, or the Hudson Stähli line seen in the aging cornea.

Iron may deposit in ocular tissues from resolving hemorrhages or as a result of retained intraocular foreign bodies. In the context of elevated intraocular pressure or defect in Descemet's membrane, hemorrhage in the anterior chamber (hyphema) may diffuse into the corneal stroma. Hemoglobin particles may be detected in the corneal stroma (Figure 17.12) on H&E-stained sections, and iron is detected within stromal fibroblasts known as "keratocytes" (Figure 17.12). As a vitreous hemorrhage resolves, iron deposits in epithelia, including the iris and ciliary body epithelia, lens epithelium, RPE, and retina.

HISTOCHEMISTRY IN NONNEOPLASTIC OPHTHALMIC CONDITIONS

Corneal Dystrophies

Corneal stromal dystrophies are bilateral hereditary conditions. The deposition of stromal proteins may compromise vision. Granular, lattice, and Avellino dystrophies are related autosomal dominant disorders associated with defects of the β-keratoepithelin gene (7–13). In granular dystrophy (Figure 17.13), the deposits are morphologically

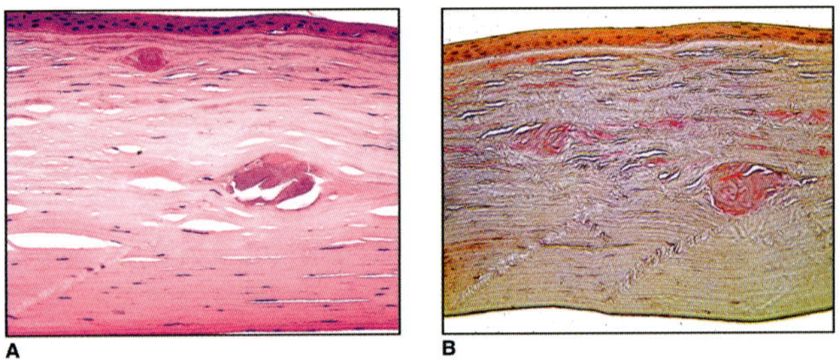

FIGURE 17.14: ***LD. (A) Fusiform eosinophilic deposits are present in the stroma. (B) The deposits are composed of amyloid and stain positive with CR (A—H&E, 200×; B—CR, 200×).***

"chunky" and are described as "hyaline" in H&E-stained sections. The deposits of granular dystrophy appear bright red with the Masson trichrome (MT) stain (Figure 17.13). In lattice dystrophy (LD), the deposits are needle shaped and composed of amyloid (Figure 17.14). Avellino dystrophy is a combination of the lattice and granular dystrophies (Figure 17.15).

Macular corneal dystrophy is an autosomal recessive disorder with defects of keratan sulfate synthesis (14–16). Focal deposits of keratan sulfate accumulate in the stroma early in the disease with diffuse deposition and involvement of the corneal endothelium later in the course. Deposits are demonstrated by alcian blue or colloidal iron stains (Figure 17.16).

Infectious Diseases

Various bacterial, viral, fungal (Figure 17.17), and parasitic organisms may infect the eye and periocular tissues. Many infections are diagnosed on the basis of clinical findings and laboratory testing, such as culture or serology, and are treated aggressively because of potential vision- and sometimes life-threatening sequelae. Thus, organisms are frequently not seen in histologic preparations despite histochemical staining. The histochemical stains used are the same as those discussed elsewhere.

NEOPLASMS

Melanocytic Lesions

Melanocytic lesions may occur in various ocular tissues including the eyelids, conjunctiva, and uveal tract, and range from benign nevi to precursor lesions, such as primary acquired melanosis with atypia of the conjunctiva, to malignant melanoma. Melanin bleaching of pigmented lesions aids in the evaluation of cytologic features. Bleaching can be accomplished by the potassium permanganate/oxalate or dilute hydrogen peroxide methods. The permanganate/oxalate technique results in faster bleaching times but may interfere with subsequent immunohistochemistry

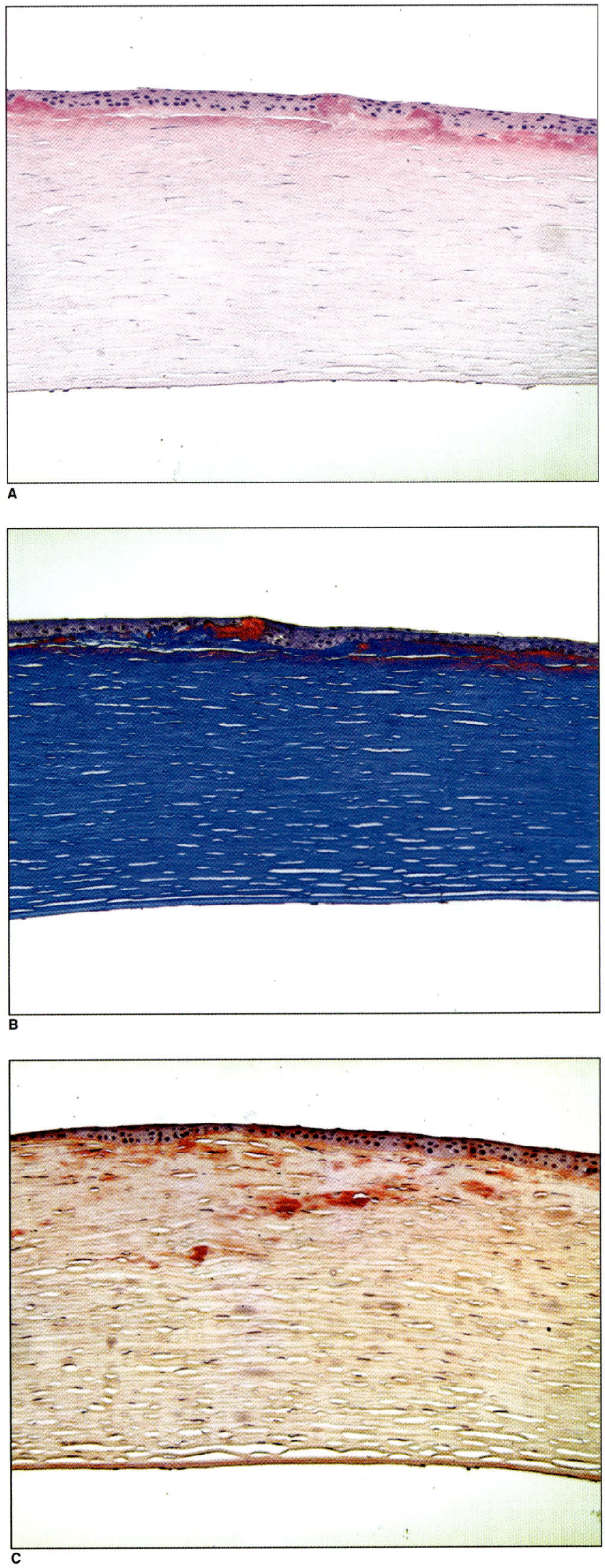

FIGURE 17.15: ***Avellino dystrophy. (A) Subepithelial and superficial stromal eosinophilic deposits are present. (B) Some of the deposits stain red with MT stain. (C) Other deposits are positive for CR (A—H&E, 100×; B—MT, 100×; C—CR, 100×).***

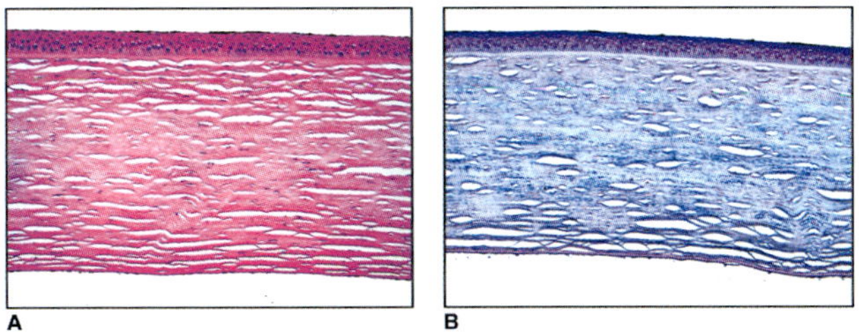

***FIGURE 17.16:** Macular corneal dystrophy. (A) Cornea of a patient with macular dystrophy, which appears unremarkable on H&E. (B) Colloidal iron stain of the same cornea showing diffuse staining of the stroma (A—H&E, 100×; B—colloidal iron, 100×).*

on the bleached section, whereas the dilute hydrogen peroxide method increases the bleaching time but allows the use of many antibodies (17). Kivela favors the peroxide technique in ophthalmic pathology (18).

Cellular morphology is particularly important in distinguishing ciliochoroidal melanocytoma from melanoma, and in providing prognostic information in melanoma. Melanocytoma (magnocellular nevus) is composed of heavily pigmented, plump polygonal cells (Figure 17.18A,B). Although benign, these tumors may demonstrate characteristics of malignant melanoma, such as necrosis and extraocular extension (19). Uveal malignant melanoma is composed of varying proportions of spindle (Callender types A and B) and epithelioid tumor cells (Figure 17.18C,D). An increased fraction of epithelioid cells confers a worse prognosis (20–22).

The histologic identification of looping vasculogenic mimicry patterns is associated with adverse outcome (23–28). Although these patterns may be identified by immunohistochemical staining for laminin, most diagnostic immunohistochemistry laboratories do not include laminin in the menu of available stains. These patterns were originally described by staining with a modification of the PAS stain in which hematoxylin counterstaining is omitted (Figure 17.19), thus eliminating visual confusion from nuclei. It is helpful to view the sections with a green filter in the light path, thus rendering the patterns black (24).

Nonmelanocytic Neoplasms

Histochemical stains have limited value in evaluating nonmelanocytic neoplasms of the eye and periocular tissues. Sebaceous carcinoma is the second most common eyelid malignancy. The diagnosis can be made from H&E-stained sections that show atypical cells with vacuolated cytoplasm (Figure 17.20). The tumor may be misdiagnosed histologically as squamous cell carcinoma or basal cell carcinoma. Oil red O staining on frozen sections may be helpful in these cases but is rarely necessary.

Mucoepidermoid carcinoma is a rare neoplasm of the conjunctiva, lacrimal gland, and lacrimal sac. Mucin stains, such as mucicarmine, may be useful in distinguishing mucoepidermoid carcinoma from squamous cell carcinoma or sebaceous carcinoma (Figure 17.21).

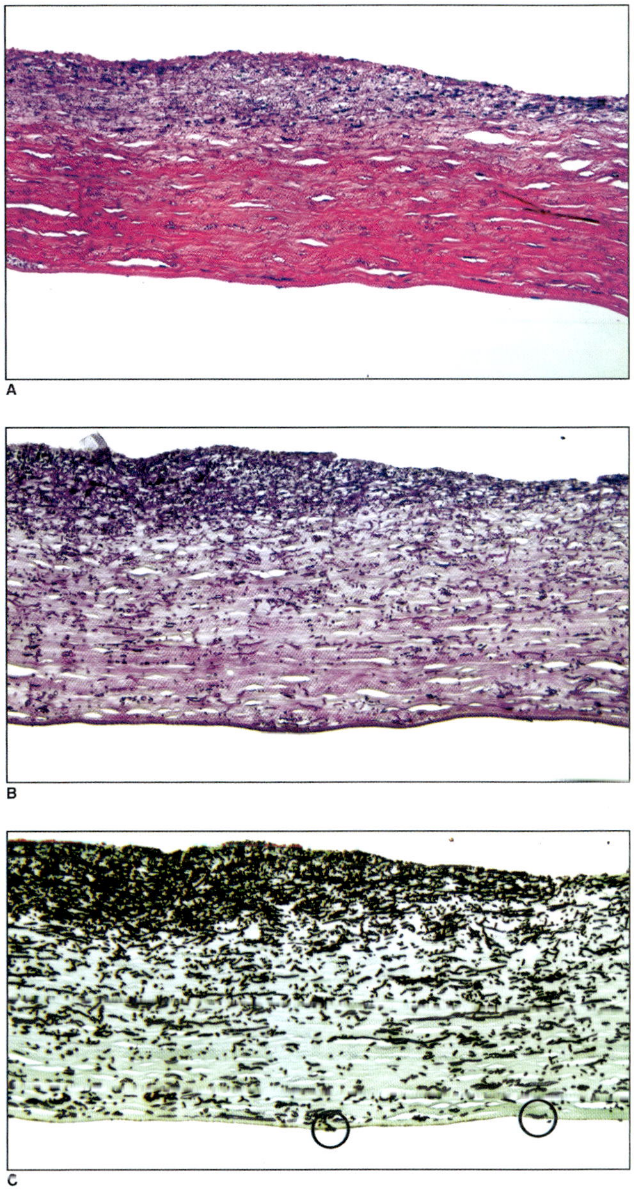

***FIGURE 17.17:** Fungal keratitis. (A) In this necrotizing corneal ulcer, some fungal hyphae may be appreciated. (B) PAS stain of the specimen shows fungal hyphae throughout the entire thickness of the stroma. (C) Unlike most organisms that cause corneal ulcers, fungi can penetrate an intact Descemet's membrane (circle) (A—H&E; B—PAS; C—GMS).*

ILLUSTRATIVE CASE STUDIES

<u>CASE 1—</u> ***Corneal Lattice Dystrophy:*** A 35-year-old man complained of increasing photophobia and decreasing visual acuity, over the past year. Physical examination showed multiple opacities in both corneas (Figure 17.22), confirmed by slit-lamp examination, with refractile lines by

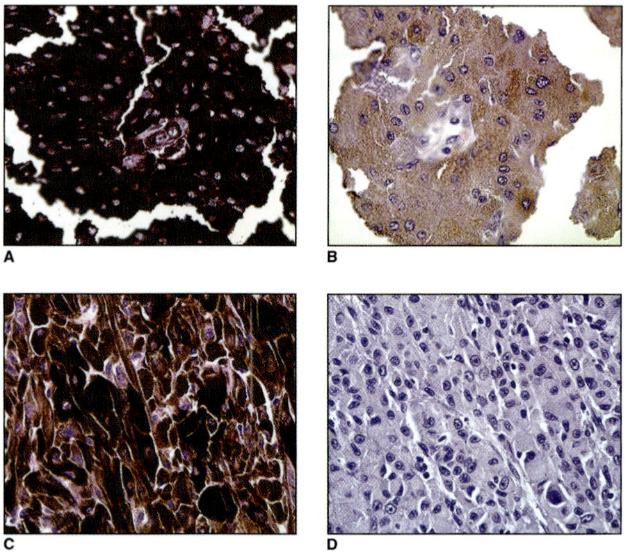

FIGURE 17.18: *(A) High-power photomicrograph of a heavily pigmented choroidal lesion. (B) Bleached section of tumor in (A) showing polygonal cells with uniform oval nuclei and abundant cytoplasm, consistent with melanocytoma. (C) High-power photomicrograph of a different heavily pigmented choroidal lesion. (D) Bleached section of tumor in (C) showing epithelioid melanoma cells with prominent nucleoli (A and C—H&E, 400×; B and D—bleached H&E, 400×).*

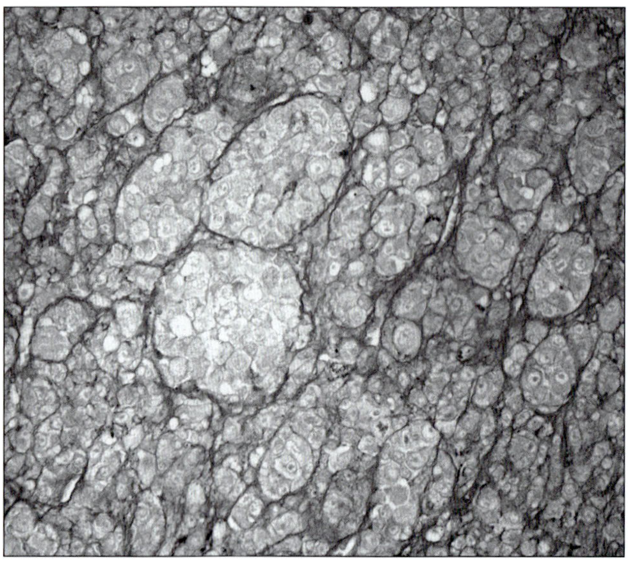

FIGURE 17.19: *Looping vasculogenic mimicry patterns in choroidal melanoma (PAS without hematoxylin, 400×).*

retroillumination. The patient's father had also had the same problem and had required corneal transplantation. The decision was made to undertake that procedure in this case as well.

Histologic examination of the excised corneas showed evidence of endothelial decompensation, with accentuation of the normal "basketweave" pattern of the corneal stroma. In addition, however, numerous deposits of

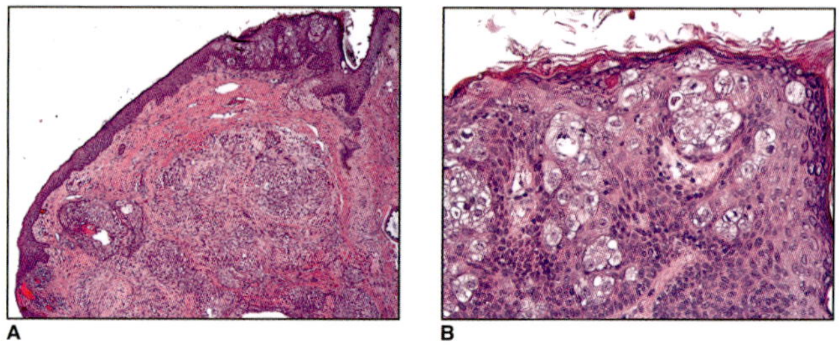

FIGURE 17.20: *(A) Eyelid margin showing sebaceous carcinoma. Focal intraepithelial (pagetoid) involvement is present. (B) Higher power view of intraepithelial component demonstrating small groups and individual atypical cells with enlarged nuclei, prominent nucleoli, and foamy cytoplasm (A—H&E, 40×; B—H&E, 200×).*

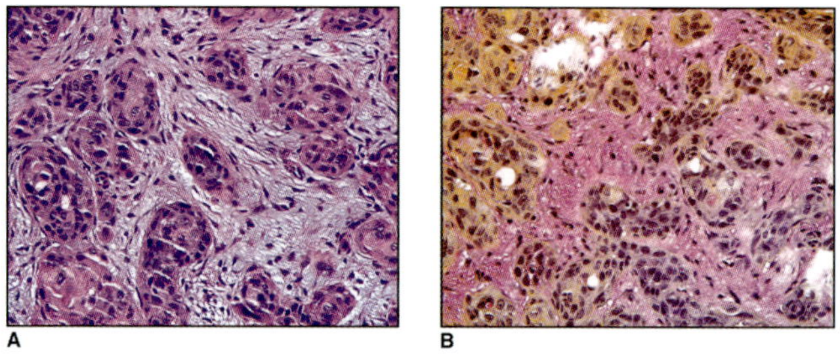

FIGURE 17.21: *Mucoepidermoid carcinoma of the conjunctiva. (A) Nests of tumor cells invade a myxoid stroma. (B) Intracytoplasmic mucin is identified using mucicarmine stain (A—H&E, 200×; B—mucicarmine, 200×).*

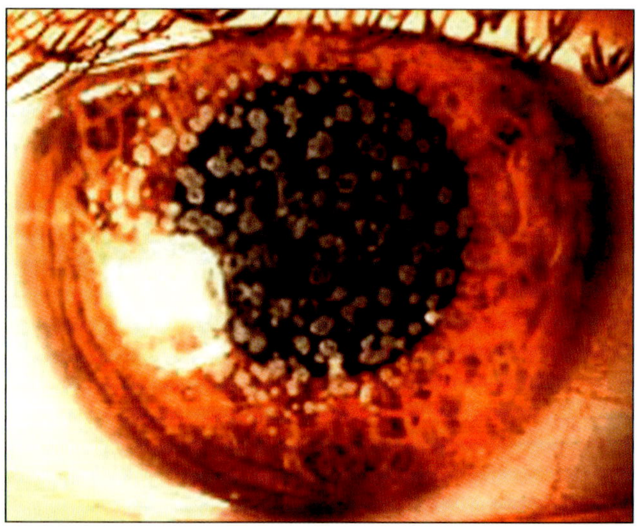

FIGURE 17.22: *Multiple corneal opacities, in lattice corneal dystrophy.*

amorphous eosinophilic material were present, randomly disposed between Bowman's and Descemet's membranes (Figure 17.23). It stained with the Congo red (CR) (Figure 17.24), MT (Figure 17.25), and PAS techniques; polarization microscopy of the CR-labeled sections demonstrated

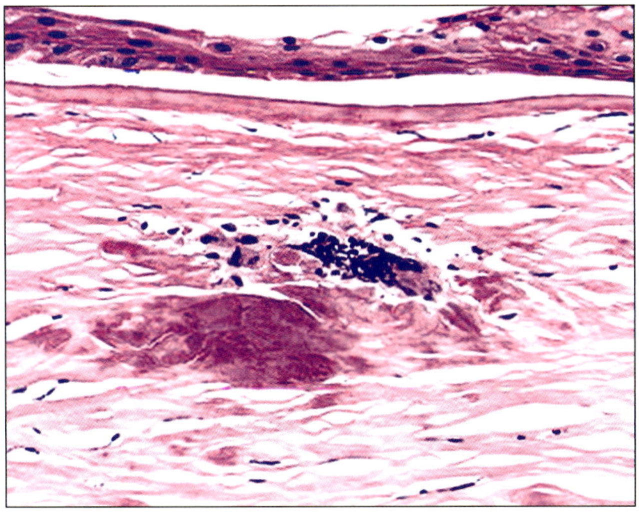

FIGURE 17.23: Deposition of amorphous, eosinophilic, hyaline material in the corneal stroma, in corneal LD.

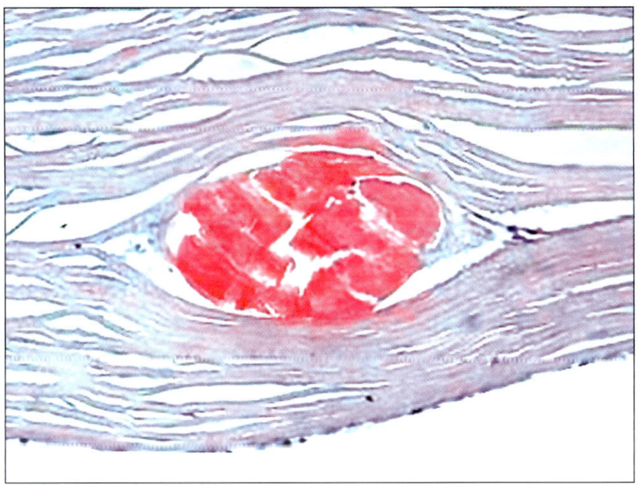

FIGURE 17.24: The stromal deposits in LD stain with CR.

green-yellow dichroism (Figure 17.26). A final diagnosis of LD was made.

Like most of the corneal dystrophies, LD is an autosomal dominant condition, the genetic defect for which has been localized to the *BIG H3* gene on chromosome 5q (29). Patients become symptomatic with this condition during adolescence or early adulthood, but not all of them will ultimately need corneal transplantation. Various other dystrophies of the cornea exist; they are generally classified as surface epithelial centered (Meesman's, "dot," "fingerprint," and "geographic" dystrophies; Fabry's disease); Bowman's membrane centered (Reis-Buckler dystrophy), stromal (granular, macular, lattice, congenital hereditary, hereditary fleck, and central crystalline dystrophies; mucopolysaccharidoses; mucolipidoses; sphingolipidoses; ochronosis; cystinosis; keratoconus, keratoglobus, and

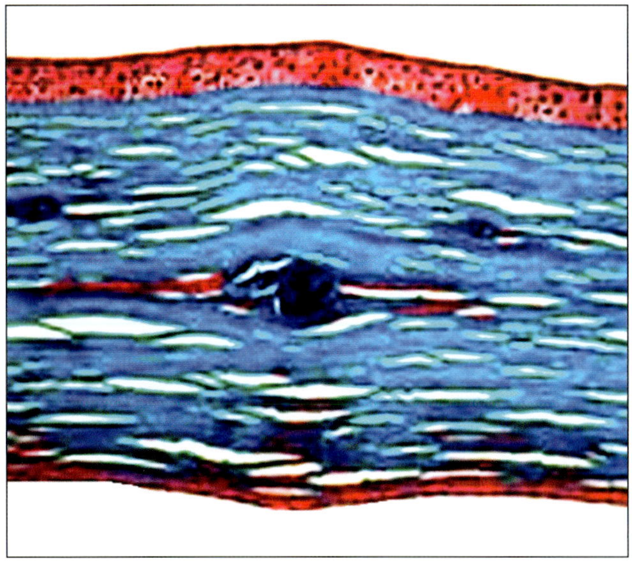

FIGURE 17.25: MT labels the stromal deposits blue.

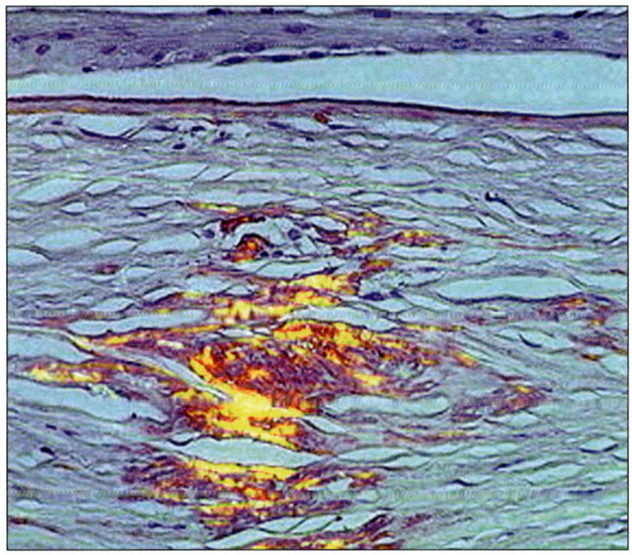

FIGURE 17.26: With polarization microscopy, CR-stained stromal deposits in corneal LD demonstrate yellow-green dichroism.

pellucid marginal degeneration), and endothelial centered (Fuchs', posterior polymorphous, hereditary endothelial, and nonguttate endothelial dystrophies). Many of these show histochemical abnormalities (30).

The histologic corneal deposits seen in LD have been called "amyloid," and they do have many of the histochemical and physicochemical properties of that substance. However, unlike amyloid lesions in most other anatomic sites, the corneal deposits in LD stain not only with CR but also with MT and PAS methods (30,31). Thus, there is a likelihood that additional matrical materials are coprecipitated with true amyloid in this condition. ■

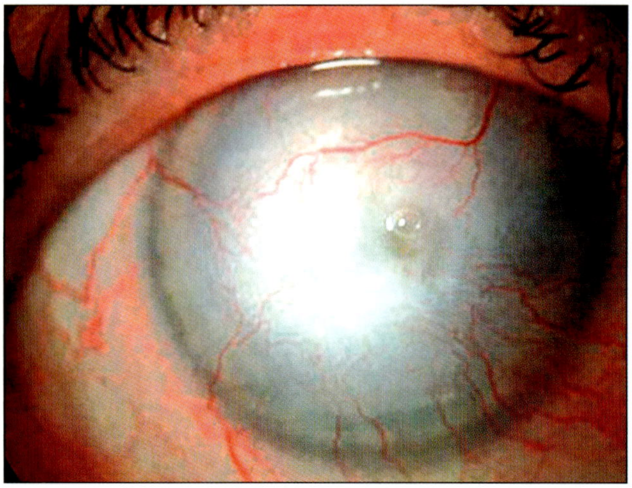

FIGURE 17.27: Corneal erosions, corneal clouding, and conjunctivitis in AK in a soft contact lens wearer.

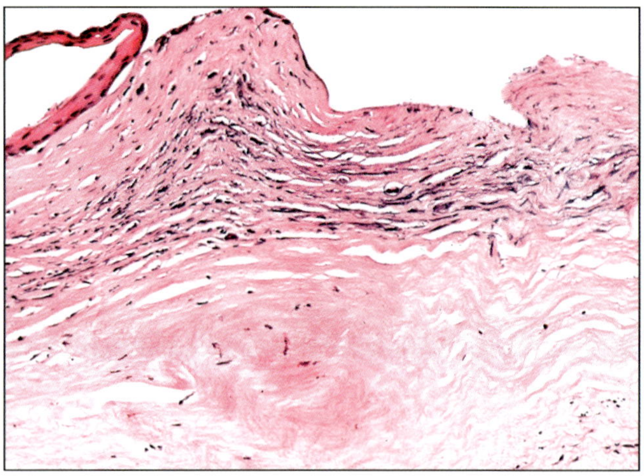

FIGURE 17.28: Erosion of the corneal epithelium is accompanied by acute and chronic stromal inflammation in AK.

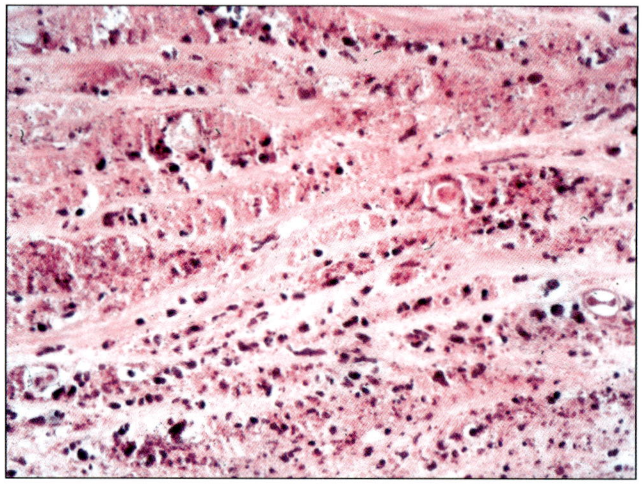

FIGURE 17.29: Inflammation, necrosis, and interstitial organisms in AK.

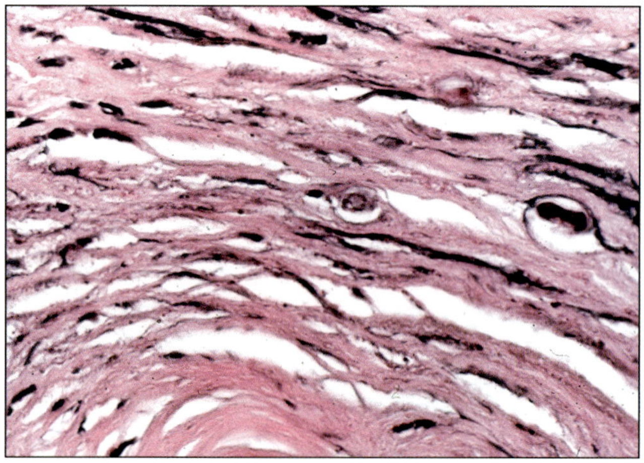

FIGURE 17.30: The amebic organisms are visible on H&E stains.

CASE 2— ***Acanthamoeba polyphaga Keratitis:*** A 33-year-old man had worn soft contact lenses for only a short time. He went swimming at a resort lake in northern Wisconsin during the summer months and did not remove his contacts while in the water; he also did not wear a mask or goggles. One week later, the patient developed rapidly progressive bilateral ocular pain and blurred vision. Physical examination showed bilateral corneal erosions and marked vascular injection of the conjunctivae (Figure 17.27).

Corneal scrapings were taken and sent for pathologic examination. Staining with calcofluor white (a chemoluminescent dye) (32) was positive, showing the presence of numerous rounded to oval inclusions in the corneal stroma, and a diagnosis of *Acanthamoeba* keratitis (AK) was made. Aggressive medical treatment was pursued with polyhexamethylene biguanide, pentamidine, and propamidine (33), but it was ineffective. Bilateral corneal transplantation was performed.

Histologic examination of the excised corneas showed widespread epithelial erosion and stromal acute and chronic inflammation with focal necrosis (Figures 17.28 and 17.29). Multiple irregularly ovoid organisms were identified in the stromal interstices (Figure 17.30). They stained with the PAS, MT, and Gomori methenamine silver methods (Figures 17.31 and 17.32), confirming the diagnosis of AK. Additional immunohistologic studies indicated that the pathogenic species of *Acanthamoeba* in this case was *A. polyphaga*.

Several microbes are capable of infecting the cornea under selected conditions. These include viruses—*Herpes simplex and Varicella-zoster*; bacteria—*Escherichia coli*, *Haemophilus influenzae*, *Klebsiella pneumoniae*, *Nocardia* sp., *Proteus* sp., *Pseudomonas* sp., *Staphylococcus aureus*, *Streptococcus*

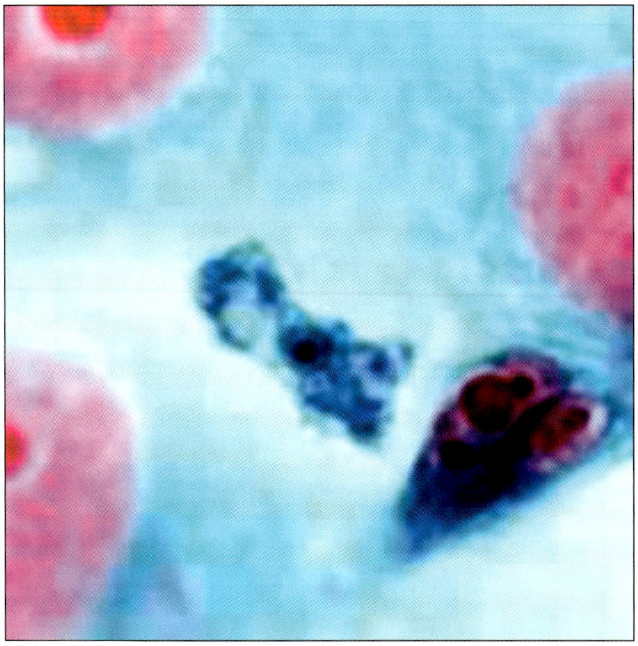

FIGURE 17.31: ***Acanthamoeba* labels with the MT method, as shown here.**

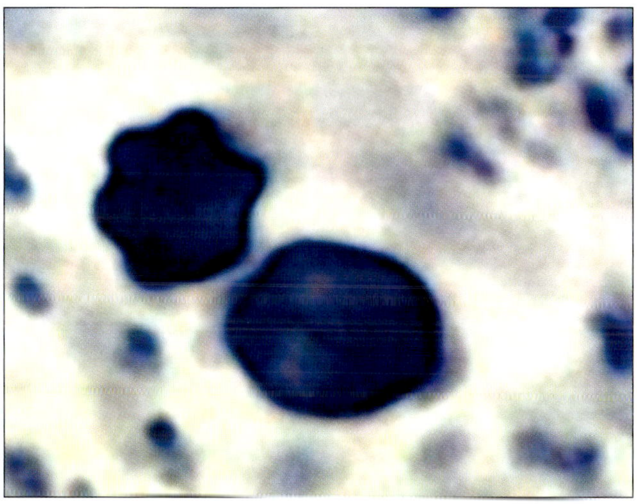

FIGURE 17.32: ***The Gomori methenamine silver technique will also stain* Acanthamoeba.**

sp., and *Treponema pallidum*; fungi—*Fusarium* sp., *Aspergillus* sp., *Candida* sp., *Curvularia* sp., *Myrothecium* sp., *Paecilomyces* sp., *Phialophora* sp., and *Petriellidium boydii*; and protozoan parasites such as several species of *Acanthamoeba* and Nosema (30).

AK is seen principally in the wearers of "soft" contact lenses (34), especially if they improperly clean them, use homemade cleaning solutions, or wear the lenses while swimming without eye protection in natural bodies of water. Under the microscope, the amebic trophozoites may resemble macrophages and are often difficult to recognize with certainty in the absence of histochemical studies. Any of the staining procedures cited above is effective at demonstrating the organisms; the Giemsa method has been used in this context as well (30). ■

<u>CASE 3—</u> ***Retinoblastoma:*** A 14-month-old boy had immigrated to the United States with his parents, from Afghanistan. The family was indigent, and, although the child developed progressive proptosis of the right eye, they did not seek medical care for several months. Physical examination demonstrated a large violaceous mass that essentially effaced both eyelids on the right side and involved the periocular soft tissues (Figure 17.33). Computed tomography showed that the lesion had an intraocular base (Figure 17.34). An evaluation for possible systemic metastasis showed none, and the boy was given neoadjuvant chemoirradiation before undergoing a right orbital exenteration.

Examination of the bisected globe showed a huge intraocular mass that was broadly attached to the retina, but

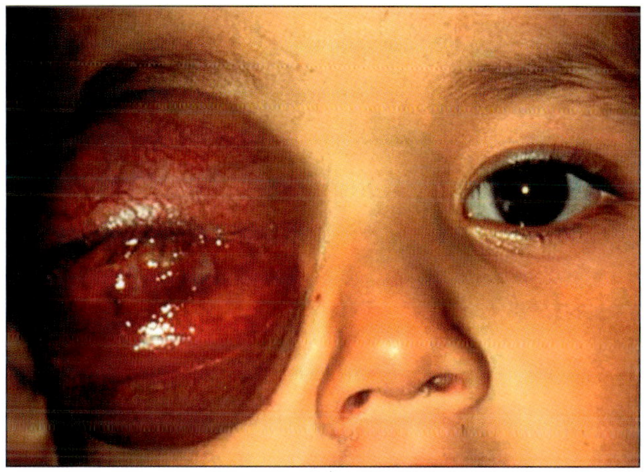

FIGURE 17.33: Massive proptosis of the right eye, with involvement of the periocular soft tissue, is present in this young child with RB.

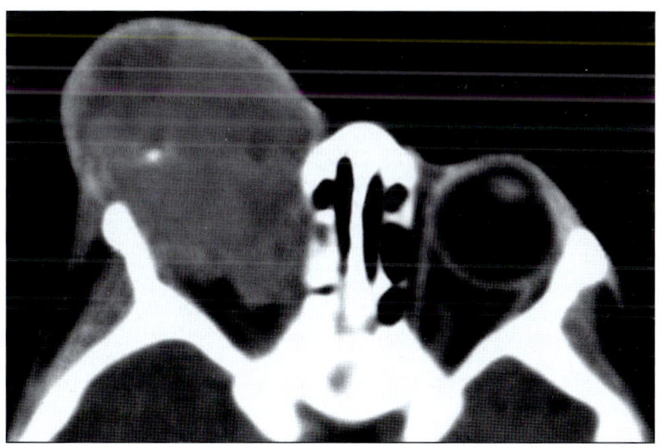

FIGURE 17.34: Computed tomography shows the extent of tumor growth, essentially effacing the right orbit.

with minimal involvement of the optic nerve. Microscopically, it was composed of small round cells, arranged in sheets and broad clusters, with interspersed foci of necrosis (Figures 17.35 and 17.36). The cytologic appearance of the lesion was monotonous, but focally, the tumor cells demonstrated formation of rosettes (Figure 17.37). A Fontana-Masson stain was negative, but PAS and colloidal iron stains were positive in the centers of the rosettes (Figure 17.38). A lectin histochemical stain with *Lens culinaris* also demonstrated diffuse reactivity (Figure 17.39). The diagnosis of retinoblastoma (RB) was made.

RB is the third most common intraocular malignancy—behind melanoma and metastatic tumors—but, in young children, it is, by far, the leading consideration (35,36). Patients with this tumor are usually less than 3 years of age, with no particular racial predisposition. When the neoplasms are bilateral, it is likely that the autosomal dominant RB syndrome is operative (37). The *Rb* gene, which is important in pathogenesis, is located on chromosome 13 (13q14), and is a tumor suppressor gene. Accordingly, patients are at risk for RB when a homozygous deletion occurs, particularly in concert with amplification of N-*myc* (38–40).

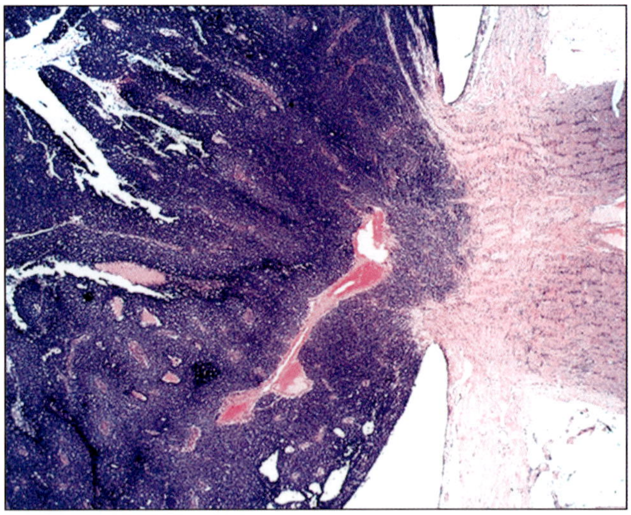

FIGURE 17.35: ***After neoadjuvant therapy and orbital exenteration, RB is still present but only minimally invading the optic nerve.***

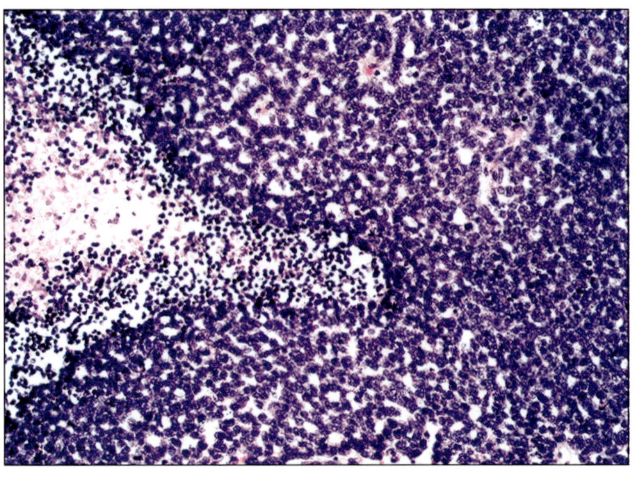

FIGURE 17.36: ***The tumor is characterized by a monotonous small round cell composition, with interspersed foci of necrosis.***

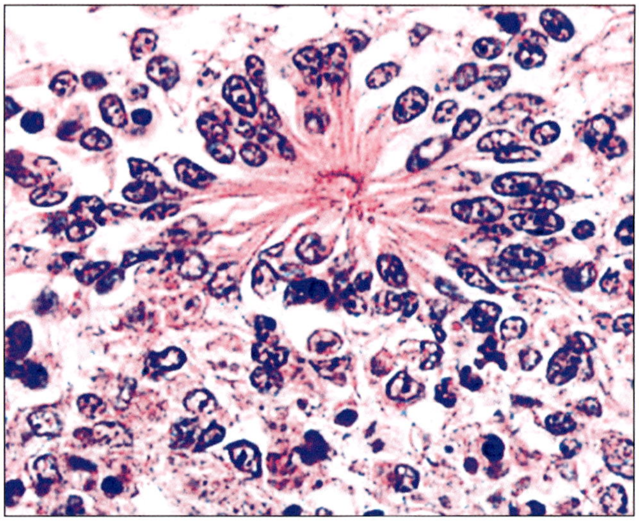

FIGURE 17.37: ***Focal formation of Flexner-Wintersteiner rosettes is seen in RB.***

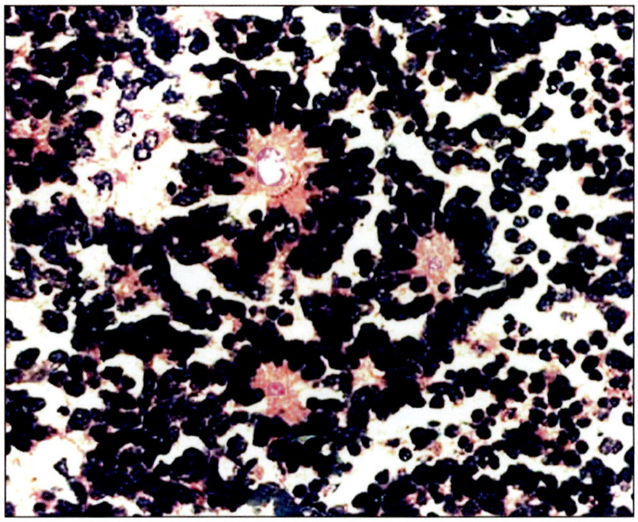

FIGURE 17.38: ***The internal spaces of the rosettes are labeled with the PAS stain (shown here), as well as colloidal iron or alcian blue.***

Pathologically, RB is regarded as a form of primitive neuroectodermal tumor (PNET). Hence, it is not surprising that the histochemical features of RB are virtually identical to those of PNETs in other areas of the neuraxis and in the soft tissues (41,42). RB often manifests Flexner-Wintersteiner rosettes—with small virtual spaces in their centers—and they have the histochemical properties

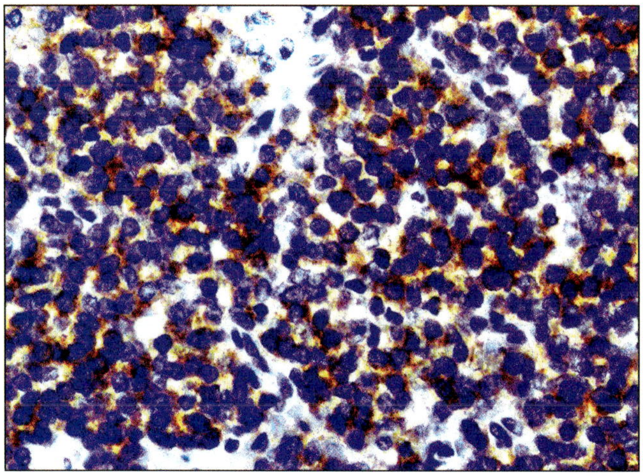

FIGURE 17.39: ***Diffuse lectin histochemical binding of*** **Lens culinaris** ***in RB mirrors the expected properties of PNETs in general.***

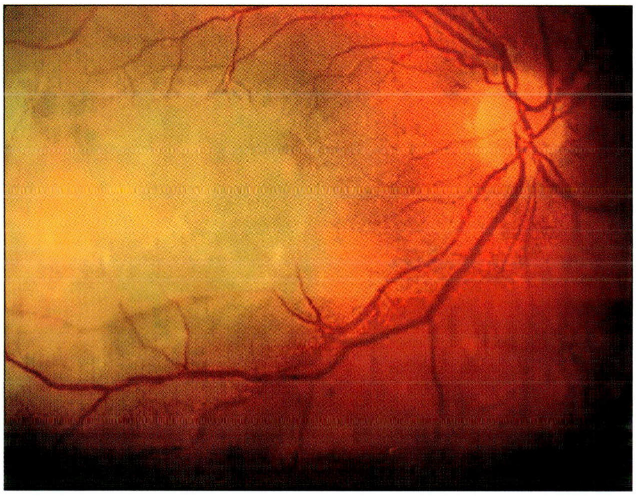

FIGURE 17.40: ***Funduscopic examination of the right eye shows an ill-defined uveal mass.***

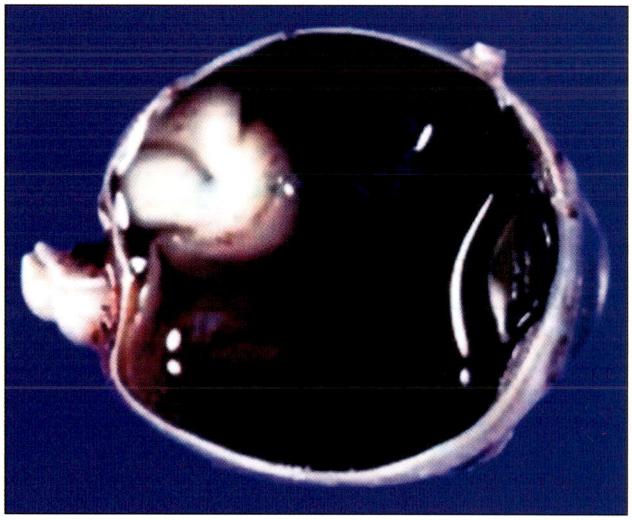

FIGURE 17.41: ***Enucleation specimen showing an eccentric tumor mass in the posterior uveal tract.***

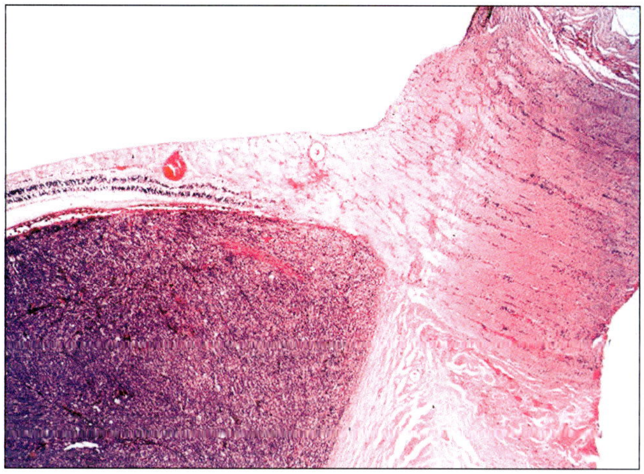

FIGURE 17.42: ***The tumor is confined to the sclera and has not produced retinal detachment.***

demonstrated in this case (43). Necrosis, the Azzopardi effect (encrustation of intratumoral blood vessels by DNA), dystrophic calcification, and brisk apoptosis are commonly seen as well (44).

Distant metastasis and tumor "seeding" of the central nervous system may occur in RB cases. However, this is true principally of tumors that have extended outside the globe at presentation. Thus, enucleation, with or without preoperative radiotherapy or chemotherapy, radiotherapy alone, or a combination of those treatments, is used, depending upon the specific clinical findings (45–47). ■

CASE 4— *Uveal Melanoma:* A 47-year-old man noticed blurring of vision in his right eye over a several-month period. Funduscopic examination showed an ill-defined mass in the posterior choroid (Figure 17.40). After a systematic evaluation failed to show any evidence of distant metastasis, an enucleation was performed.

Gross examination demonstrated an eccentric ovoid mass in the posterior chamber (Figure 17.41). Microscopically, it demonstrated a predominant composition by bluntly fusiform cells, with moderate nuclear atypia (Figures 17.42 and 17.43), but a small component of the tumor showed obvious pleomorphism, a brisk mitotic rate, and epithelioid cytology (Figure 17.44). Fontana-Masson and lectin histochemical stains with concanavalin-A were positive (Figures 17.45 and 17.46), particularly in the epithelioid focus, whereas another lectin histochemical procedure with *Ulex europaeus* I yielded negative results (48). The diagnosis of mixed spindle "B" and epithelioid cell uveal melanoma was made.

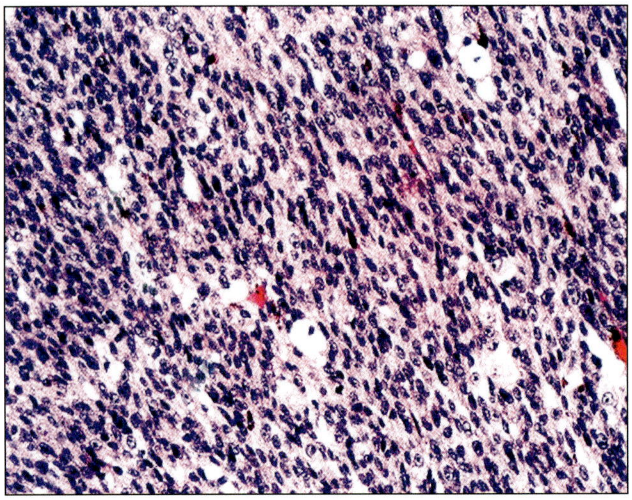

FIGURE 17.43: ***The bulk of the tumor was constituted by bluntly fusiform cells with noticeable nuclear atypia but limited mitotic activity.***

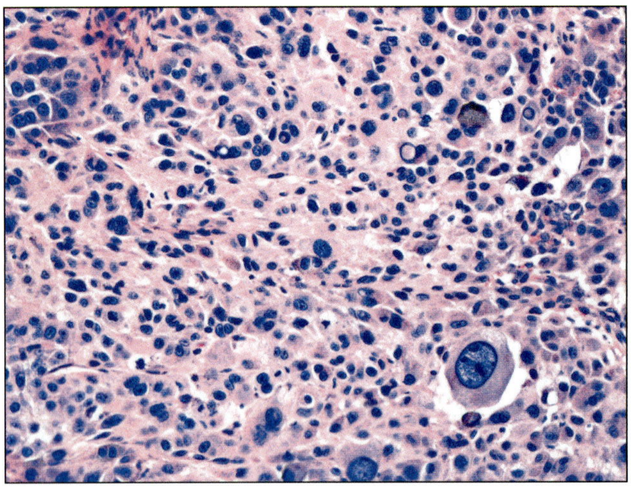

FIGURE 17.44: ***Focally epithelioid and pleomorphic tumor cells were present.***

Melanoma of the eye (ME) is the most common intraocular malignancy in adults. It is typically solitary and unilateral, and is classified according to the site of origin (i.e., choroidal, iris, or ciliary body). Caucasian, blue-eyed persons are preferentially affected; there may be a clinical association between ME and the atypical cutaneous mole syndrome, neurofibromatosis, and multiple uveal melanocytic nevi (49).

Pathologically, ME is usually obviously cytologically malignant (50–52). However, in largely amelanotic tumors, one may have difficulty in eliminating the differential diagnostic possibilities of metastatic carcinoma in the uveal tract or anaplastic carcinoma of the ciliary apparatus (53,54). In these settings, the Fontana-Masson and Schmorl stains are helpful, inasmuch as positivity with either one secures the diagnosis of ME.

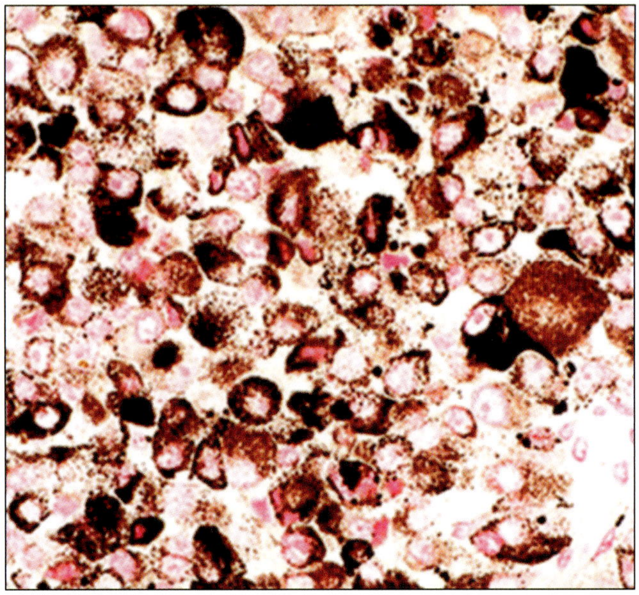

FIGURE 17.45: ***The epithelioid focus stained with the Fontana-Masson procedure.***

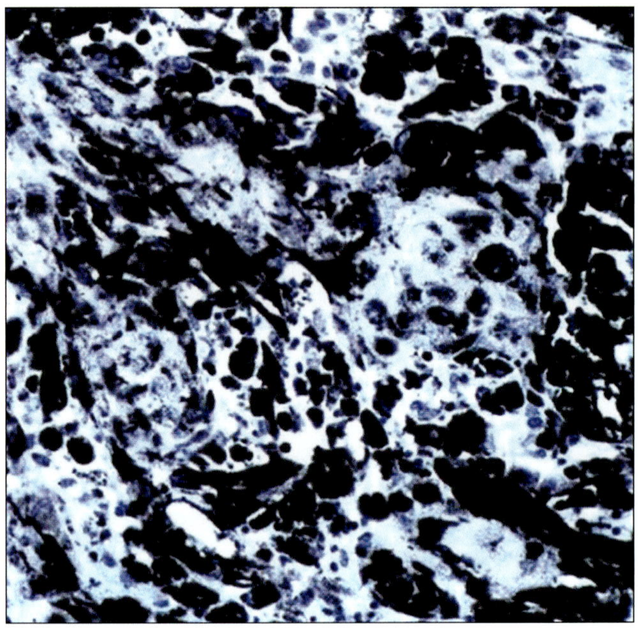

FIGURE 17.46: ***Lectin histochemical positivity for concanavalin-A is present in the epithelioid area.***

Tumor size, the proportion of epithelioid tumor cells, the presence of extraocular extension, mitotic rate, necrosis, and specific intraocular location are all prognostic factors for ME (55). The usual treatment is that of enucleation, although "plaque" radiotherapy or external irradiation are also employed (56,57). The 10-year survival varies between 50% and 70% (57). ■

REFERENCES

1. Sunde OA. On the so-called senile exfoliation of the anterior lens capsule; a clinical and anatomical study. *Acta Ophthalmol Suppl* 1956:1–85.
2. Konstas AG, Marshall GE, Lee WR. Immunogold localisation of laminin in normal and exfoliative iris. *Br J Ophthalmol* 1990;74:450–7.
3. Harnisch JP, Barrach HJ, Hassell JR, Sinha PK. Identification of a basement membrane proteoglycan in exfoliation material. *Albrecht Von Graefes Arch Klin Exp Ophthalmol* 1981;215:273–8.
4. Schlotzer-Schrehardt U, Dorfler S, Naumann GO. Immunohistochemical localization of basement membrane components in pseudoexfoliation material of the lens capsule. *Curr Eye Res* 1992;11:343–55.
5. Clarkson JG, Green WR, Massof D. A histopathologic review of 168 cases of preretinal membrane. *Am J Ophthalmol* 1977;84:1–17.
6. Kampik A, Kenyon KR, Michels RG, Green WR, De La Cruz ZC. Epiretinal and vitreous membranes. Comparative study of 56 cases. *Arch Ophthalmol* 1981;99:1445–54.
7. Munier FL, Korvatska E, Djemai A, et al. Kerato-epithelin mutations in four 5q31-linked corneal dystrophies. *Nat Genet* 1997;15:247–51.
8. Gupta SK, Hodge WG, Damji KF, Guernsey DL, Neumann PE. Lattice corneal dystrophy type 1 in a Canadian kindred is associated with the Arg124 ->Cys mutation in the kerato-epithelin gene. *Am J Ophthalmol* 1998;125:547–9.
9. Okada M, Yamamoto S, Watanabe H, et al. Granular corneal dystrophy with homozygous mutations in the keratoepithelin gene. *Am J Ophthalmol* 1998;126:169–76.
10. Fujiki K, Hotta Y, Nakayasu K, et al. A new L527R mutation of the betaig-h3 gene in patients with lattice corneal dystrophy with deep stromal opacities. *Hum Genet* 1998;103:286–9.
11. Fujiki K, Hotta Y, Nakayasu K, Kanai A. Homozygotic patient with betaig-h3 gene mutation in granular dystrophy. *Cornea* 1998;17:288–92.
12. Okada M, Yamamoto S, Inoue Y, et al. Severe corneal dystrophy phenotype caused by homozygous R124H keratoepithelin mutations. *Invest Ophthalmol Vis Sci* 1998;39: 1947–53.
13. Mashima Y, Nakamura Y, Noda K, et al. A novel mutation at codon 124 (R124l) in the bigh3 gene is associated with a superficial variant of granular corneal dystrophy. *Arch Ophthalmol* 1999;117:90–3.
14. Klintworth GK, Smith CF. Macular corneal dystrophy. Studies of sulfated glycosaminoglycans in corneal explant and confluent stromal cell cultures. *Am J Pathol* 1977; 89:167–82.
15. Hassell JR, Newsome DA, Krachmer JH, Rodrigues MM. Macular corneal dystrophy: failure to synthesize a mature keratan sulfate proteoglycan. *Proc Natl Acad Sci USA* 1980;77:3705–9.
16. Klintworth GK, Smith CF. Abnormalities of proteoglycans and glycoproteins synthesized by corneal organ cultures derived from patients with macular corneal dystrophy. *Lab Invest* 1983;48:603–12.
17. Orchard GE. Heavily pigmented melanocytic neoplasms: comparison of two melanin-bleaching techniques and subsequent immunohistochemical staining. *Br J Biomed Sci* 1999;56:188–93.
18. Kivela T. Immunohistochemical staining followed by bleaching of melanin: a practical method for ophthalmic pathology *Br J Biomed Sci* 1995;52:325–6.
19. Rummelt V, Nauman Goh, Folberg R, Weingeist TA. Surgical management of melanocytoma of the ciliary body with extrascleral extension—reply. *Am J Ophthalmol* 1994;118:404–5.
20. Mclean IW, Foster WD, Zimmerman LE. Prognostic factors in small malignant melanomas of choroid and ciliary body. *Arch Ophthalmol* 1977;95:48–58.
21. Mclean IW, Foster WD, Zimmerman LE.Uveal melanoma: location, size, cell type, and enucleation as risk factors in metastasis. *Hum Pathol* 1982;13:123–32.
22. Seddon JM, Albert DM, Lavin PT, Robinson N. A prognostic factor study of disease-free interval and survival following enucleation for uveal melanoma. *Arch Ophthalmol* 1983; 101:1894–9.
23. Seregard S, Spangberg B, Juul C, Oskarsson M. Prognostic accuracy of the mean of the largest nucleoli, vascular patterns, and pc-10 in posterior uveal melanoma. *Ophthalmology* 1998;105:485–91.
24. Folberg R, Rummelt V, Parys-Van Ginderdeuren R, et al. The prognostic value of tumor blood vessel morphology in primary uveal melanoma. *Ophthalmology* 1993;100:1389–98.
25. Sakamoto T, Sakamoto M, Yoshikawa H, et al. Histologic findings and prognosis of uveal malignant melanoma in Japanese patients. *Am J Ophthalmol* 1996;121:276–83.
26. Schaling DF, Van Der Pol JP, Schlingemann RO, Parys-Van Ginderdeuren R, Jager MJ, Schaling DF. Vascular density and vascular patterns in the prognosis of choroidal melanoma. In: *Radionucleotides dnd Radiolabelled Antibodies in Choroidal Melanoma (Diagnosis and Therapy)*. Rijksuniversiteit Te Leiden, Leiden, 1996; pp. 43–54.
27. Makitie T, Summanen P, Tarkannen A, Kivela T. Microvascular loops and networks as prognostic indicators in choroidal and ciliary body melanomas. *J Nat Cancer Inst* 1999;91:359–67.
28. Mclean IW, Keefe KS, Burnier MN. Uveal melanoma: comparison of the prognostic value of fibrovascular loops, mean of the ten largest nucleoli, cell type and tumor size. *Ophthalmology* 1997;104:777–80.
29. Schmitt-Bernard CF, Pouliquen Y, Argiles A. BIG-H3 protein: mutation of codon 124 and corneal amyloidosis. *J Fr Ophthalmol* 2004;27:510–22.
30. Lee WR. *Ophthalmic Histopathology* (2nd ed.), Springer, New York, 2002; pp. 401–442.
31. Loeffler KU. An immunohistochemical study of gelsolin immunoreactivity in corneal amyloidosis. *Am J Ophthalmol* 1992;113:546–54.
32. Wilhelmus KR, Osato MS, Font RL, Robinson NM, Jones DB. Rapid diagnosis of *Acanthamoeba* keratitis using calcofluor white. *Arch Ophthalmol* 1986;104:1309–12.
33. Singhal T, Bajpal A, Kaira V. Successful treatment of *Acanthamoeba* meningitis with combination oral antibiotics. *Pediatr Infect Dis* 2001;20:623–7.
34. Kaji Y, Hu B, Kawana K, Oshika T. Swimming with soft contact lenses: danger of *Acanthamoeba* keratitis. *Lancet Infect Dis* 2005;5:392.

35. Abramson DH, Frank CM, Susman M, et al. Presenting signs of retinoblastoma. *J Pediatr* 1998;132:505–8.
36. Albert DM. Historic review of retinoblastoma. *Ophthalmology* 1987;94:654–62.
37. Cowell JK. The genetics of retinoblastoma. *Br J Cancer* 1991;63:333–6.
38. Harbour JW. Overview of RB gene mutations in patients with retinoblastoma: implications for clinical genetic screening. *Ophthalmology* 1998;105:1442–7.
39. Kaelin WG. Functions of the retinoblastoma protein. *Bioessays* 1999;21:950–8.
40. Lipinski MM, Jacks T. The retinoblastoma gene family in differentiation and development. *Oncogene* 1999;18:7873–82.
41. Swanson PE, Dehner LP, Wick MR. Small round-cell tumors in children: lectin-histochemical and immunohistochemical study of 30 cases. *Surg Pathol* 1993;5:5–16.
42. Kivela T. Glycoconjugates in retinoblastoma: a lectin-histochemical study of ten formalin-fixed and paraffin-embedded tumors. *Virchows Arch A Pathol Anat Histopathol* 1987;410: 471–9.
43. Yi YZ, Yang WZ, Zhen HL. Retinoblastoma: cell origin and differentiation. *Chin J Ophthalmol* 1994;30:214–7.
44. Sang DN, Albert DM: Retinoblastoma: clinical & histopathologic features. *Hum Pathol* 1982;13:133–47.
45. Ferris FL III, Chew EY. A new era for the treatment of retinoblastoma. *Arch Ophthalmol* 1996;114:1412.
46. Finger PT, Czechonska G, Demirci H, Rausen A. Chemotherapy for retinoblastoma: a current topic. *Drugs* 1999;58: 983–96.
47. Honavar SG, Singh AD. Management of advanced retinoblastoma. *Ophthalmol Clin North Am* 2005;18:65–73.
48. Kohchiyama A, Oka D, Ueki H. Differing lectin-binding patterns of malignant melanoma and nevocellular and Spitz nevi. *Arch Dermatol Res* 1987;279:226–31.
49. Grin-Jorgensen C, Berke A, Grin M. Ocular melanoma. *Dermatol Clin* 1992;10:663–8.
50. Zimmerman LE. Changing concepts concerning the malignancy of ocular tumors. *Arch Ophthalmol* 1967;78:166–73.
51. McCartney AC. Pathology of ocular melanomas. *Br Med Bull* 1995;51:678–93.
52. Collaborative Ocular Melanoma Study Group. Histopathologic characteristics of uveal melanomas in eyes enucleated from the Collaborative Ocular Melanoma Study (report number 6). *Am J Ophthalmol* 1998;125:745–66.
53. Grossniklaus HE, Zimmerman LE, Kachmer ML. Pleomorphic adenocarcinoma of the ciliary body: immunohistochemical and electron microscopic features. *Ophthalmology* 1990;97:763–8.
54. DePotter P. Ocular manifestations of cancer. *Curr Opin Ophthalmol* 1998;9:100–4.
55. McLean IW, Ainbinder DJ, Gamel JW, McCurdy JB. Choroidal-ciliary body melanoma: a multivariate survival analysis of tumor location. *Ophthalmology* 1995;102:1060–4.
56. Shields JA, Shields CL. Current management of posterior uveal melanoma. *Mayo Clin Proc* 1993;68:1196–200.
57. Shelds JA, Shields CL, DePotter P, Singh AD. Diagnosis and treatment of uveal melanoma. *Semin Oncol* 1996;23: 763–7.

Index